AF616058

Mesozoic Resource Potential in the Southern Permian Basin

Geological Society books refereeing procedures

The Society makes every effort to ensure that the scientific and production quality of its books matches that of its journals. Since 1997, all book proposals have been refereed by specialist reviewers as well as by the Society's Books Editorial Committee. If the referees identify weaknesses in the proposal, these must be addressed before the proposal is accepted.

Once the book is accepted, the Society Book Editors ensure that the volume editors follow strict guidelines on refereeing and quality control. We insist that individual papers can only be accepted after satisfactory review by two independent referees. The questions on the review forms are similar to those for *Journal of the Geological Society*. The referees' forms and comments must be available to the Society's Book Editors on request.

Although many of the books result from meetings, the editors are expected to commission papers that were not presented at the meeting to ensure that the book provides a balanced coverage of the subject. Being accepted for presentation at the meeting does not guarantee inclusion in the book.

More information about submitting a proposal and producing a book for the Society can be found on its website: www.geolsoc.org.uk.

It is recommended that reference to all or part of this book should be made in one of the following ways:

Kilhams, B., Kukla, P.A., Mazur, S., McKie, T., Mijnlieff, H.F. & van Ojik, K. (eds) 2018. *Mesozoic Resource Potential in the Southern Permian Basin*. Geological Society, London, Special Publications, **469**.

Seidel, E., Meschede, M. & Obst, K. 2018. The Wiek Fault System east of Rügen Island: origin, tectonic phases and its relationship to the Trans-European Suture Zone. *In*: Kilhams, B., Kukla, P.A., Mazur, S., McKie, T., Mijnlieff, H.F. & van Ojik, K. (eds) *Mesozoic Resource Potential in the Southern Permian Basin*. Geological Society, London, Special Publications, **469**, 59–82. First published online January 24, 2018, https://doi.org/10.1144/SP469.10

GEOLOGICAL SOCIETY SPECIAL PUBLICATION NO. 469

Mesozoic Resource Potential in the Southern Permian Basin

EDITED BY

B. KILHAMS
Nederlandse Aardolie Maatschappij, The Netherlands

P. A. KUKLA
RWTH Aachen University, Germany

S. MAZUR
Getech, UK

T. MCKIE
Shell, UK

H. F. MIJNLIEFF
TNO, The Netherlands

and

K. VAN OJIK
Argo Geological Consultants, The Netherlands

2018
Published by
The Geological Society
London

THE GEOLOGICAL SOCIETY

The Geological Society of London (GSL) was founded in 1807. It is the oldest national geological society in the world and the largest in Europe. It was incorporated under Royal Charter in 1825 and is Registered Charity 210161.

The Society is the UK national learned and professional society for geology with a worldwide Fellowship (FGS) of over 10 000. The Society has the power to confer Chartered status on suitably qualified Fellows, and about 2000 of the Fellowship carry the title (CGeol). Chartered Geologists may also obtain the equivalent European title, European Geologist (EurGeol). One fifth of the Society's fellowship resides outside the UK. To find out more about the Society, log on to www.geolsoc.org.uk.

The Geological Society Publishing House (Bath, UK) produces the Society's international journals and books, and acts as European distributor for selected publications of the American Association of Petroleum Geologists (AAPG), the Indonesian Petroleum Association (IPA), the Geological Society of America (GSA), the Society for Sedimentary Geology (SEPM) and the Geologists' Association (GA). Joint marketing agreements ensure that GSL Fellows may purchase these societies' publications at a discount. The Society's online bookshop (accessible from www.geolsoc.org.uk) offers secure book purchasing with your credit or debit card.

To find out about joining the Society and benefiting from substantial discounts on publications of GSL and other societies worldwide, consult www.geolsoc.org.uk, or contact the Fellowship Department at: The Geological Society, Burlington House, Piccadilly, London W1J 0BG: Tel. +44 (0)20 7434 9944; Fax +44 (0)20 7439 8975; E-mail: enquiries@geolsoc.org.uk.

For information about the Society's meetings, consult *Events* on www.geolsoc.org.uk. To find out more about the Society's Corporate Affiliates Scheme, write to enquiries@geolsoc.org.uk.

Published by The Geological Society from:
The Geological Society Publishing House, Unit 7, Brassmill Enterprise Centre, Brassmill Lane, Bath BA1 3JN, UK

The Lyell Collection: www.lyellcollection.org
Online bookshop: www.geolsoc.org.uk/bookshop
Orders: Tel. +44 (0)1225 445046, Fax +44 (0)1225 442836

British Library Cataloguing in Publication Data

A catalogue record for this book is available from the British Library.
ISBN 978-1-78620-384-7
ISSN 0305-8719

Distributors
For details of international agents and distributors see:
www.geolsoc.org.uk/agentsdistributors

Typeset by Nova Techset Private Limited, Bengaluru & Chennai, India
Printed and bound by CPI Group (UK) Ltd, Croydon CR0 4YY

Contents

Jurassic resources

Cretaceous resources

Acknowledgements

The editorial team extends its appreciation to all the authors and collaborators that contributed to this book. The time and effort taken represents considerable determination and perseverance. Everyone that contributed to the original Geological Society conference is also appreciated, the open and collaborative atmosphere formed the basis of this book, with special thanks going to: Laura Griffiths, Gary Hampson, Howard Johnson, James Maynard, Robert Schöner, Martin Wells and Sarah Woodcock. We thank various managers who allowed the editors and authors to spend time making a significant contribution. Ben Kilhams particularly thanks Max Brouwers, Ramon Loosveld, Carlo Nicolai and Edwin Verdonk, who have supported the venture through their patience and sponsorship. Sincere thanks go to all the reviewers who took time and care to give excellent, constructive feedback: Oscar Abbink, Kresten Anderskouv, Katrine Andresen, Stuart Archer, Renaud Bouroullec, Dan Carruthers, Teresa Sabato Ceraldi, Joel Corcoran, Daan den Hartog Jager, Ramues Gallois, Mark Geluk, Thomas Gerling, Graham Goffey, Andrew Green, Cor Hofstee, Christian Hubscher, Chris Jackson, Marek Jarosinski, Fabian Jahne-Klingberg, Marek Jarosinski, Jason Jeremiah, Matthias Keym, Marloes Kortekaas, Piotr Krzywiec, Pepijn Kole, Juliette Lamarche, Herald Ligtenberg, James Maynard, Yuriy Maystrenko, Adam McArthur, Inga Moeck, Carlo Nicolai, Lars Nielsen, Mette Olivarius, Richard Porter, Horst Reuter, Ian Saikia, Frank Strozyk, Leo van Borren, Frans van Buchem, Eva van der Voet, Matthijs van Winden, Reinhoud Veenhof, Bruno Vendeville, Hanneke Verweij, Thomas Voigt, Jonathan Wonham and others who wished to remain anonymous. Finally, the editors wish to thank the Geological Society Publishing House, especially Tamzin Anderson, Lucy Bell and Angharad Hills for their support.

This book is dedicated to the memory of Mark Geluk (1958–2018)

Mark Geluk was a passionate geologist who spent most of his professional career studying the regional geology of the Paleozoic and Mesozoic of the Southern Permian Basin. During his career Mark moved several times between jobs in both academic, non-profit and commercial organisations, always adding significant value based on his ability to apply his huge knowledge and understanding of the subsurface across Northern Europe. Most of all, however, Mark was a very kind person who was fond of sharing that deep-rooted passion for geology, which is reflected in his extensive legacy of publications. As a regional geologist, Mark understood the important details of a geological interpretation of a well, a piece of core, or a seismic section and to put those details into the context of the dynamic evolution of the Southern Permian Basin on a much larger scale. His long publishing career spanned from 1988 to 2018, and included papers in high impact journals and special publications (ranging from early contributions on geothermal energy to Triassic paleogeography). His most cited work includes his PhD thesis (Permo–Triassic tectonostratigraphy) and contributions to the Dutch lithostratigraphic nomenclature, 'Geology of the Netherlands' and 'Permian Rotliegend of the Netherlands', forming cornerstones to our current knowledge of the subsurface across the Netherlands, Germany, Poland and the United Kingdom. His passion for geoscience will be sorely missed by the community.

Photograph courtesy of Ilse Urlings.

Mesozoic resource potential in the Southern Permian Basin area: the geological key to exploiting remaining hydrocarbons whilst unlocking geothermal potential

BEN KILHAMS[1]*, PETER A. KUKLA[2], STAN MAZUR[3], TOM McKIE[4], HARMEN F. MIJNLIEFF[5], KEES VAN OJIK[6] & EVELINE ROSENDAAL[7]

[1]*Nederlandse Aardolie Maatschappij (NAM), PO Box 28000, 9400 HH Assen, The Netherlands*

[2]*Geological Institute, Energy & Mineral Resources Group (EMR), RWTH Aachen University, Wuellnerstraße 2, 5206 Aachen, Germany*

[3]*Getech UK, Kitson House, Elmete Hall, Elmete Lane, Leeds LS8 2LJ, UK*

[4]*Shell UK Exploration & Production, 1 Altens Farm Road, Nigg, Aberdeen AB12 3FY, UK*

[5]*TNO (Geological Survey of the Netherlands), PO Box 80015, NL-3508 TA Utrecht, The Netherlands*

[6]*Argo Geological Consultants, Bachlaan 46, 3706 BD Zeist, The Netherlands*

[7]*EBN B.V., Daalsesingel 1, 3511 SV Utrecht, The Netherlands*

**Correspondence: b.kilhams@shell.com*

It is generally accepted that hydrocarbon exploration in northern Europe has reached a mature stage. A basin's maturity is defined by the underlying number of new discoveries and the declining production rate of mature fields (SPE 2015). For geoscientists, a mature basin has well-defined characteristics in terms of, for example, reservoir presence or trap formation (e.g. Byrne 2012). It is interesting, therefore, to note how much is still unknown about certain stratigraphic intervals in northern Europe. The Mesozoic overburden of the Southern Permian Basin (*sensu* Maystrenko *et al.* 2008; Doornenbal & Stevenson 2010) continues to provide fresh insights into the geological history of an area where, as the name suggests, historical hydrocarbon exploration has focused on the Paleozoic. The aim of this Special Publication is to increase knowledge of the Mesozoic overburden as a driver for further hydrocarbon exploration/production and the development of new geothermal energy sources.

The succeeding chapters are introduced by tectonic framework overviews that give a context for the papers that follow. The remaining articles are organized in approximate stratigraphic order, from old to young, and include a variety of examples from semi-regional to localized field or sub-basin studies. An overview of the study area for all the following chapters is given in Figure 1.

It is recognized that the overprinting Mesozoic systems have different naming conventions across the study area (e.g. the Central European Basin System *sensu* Littke *et al.* 2008; Maystrenko *et al.* 2008). Here we use the title 'Southern Permian Basin area' to emphasize the resource opportunities associated with a region that has, at least in large part, been associated with hydrocarbon exploration, and very large gas fields, in the Permian Rotliegend. For our purposes, this includes part of the Polish Trough, the highs surrounding the basin (e.g. the Ringkøbing-Fyn High), and analogous elements of the Danish and Norwegian offshore.

In this introduction, an overview of the geological history and resource base of the area is presented as a framework for the chapters that follow. A description of further prospective resources is then offered (oil and gas plus geothermal potential).

Geological history

The geological evolution of northern Europe, specifically the Southern Permian Basin area, has been described at length by various authors (e.g. Ziegler 1990*a*; Geluk 2005; Littke *et al.* 2008; McCann 2008*a*, *b*; Doornenbal & Stevenson 2010). Here a

From: KILHAMS, B., KUKLA, P. A., MAZUR, S., MCKIE, T., MIJNLIEFF, H. F. & VAN OJIK, K. (eds) 2018. *Mesozoic Resource Potential in the Southern Permian Basin*. Geological Society, London, Special Publications, **469**, 1–18.
First published online May 3, 2018, https://doi.org/10.1144/SP469.26

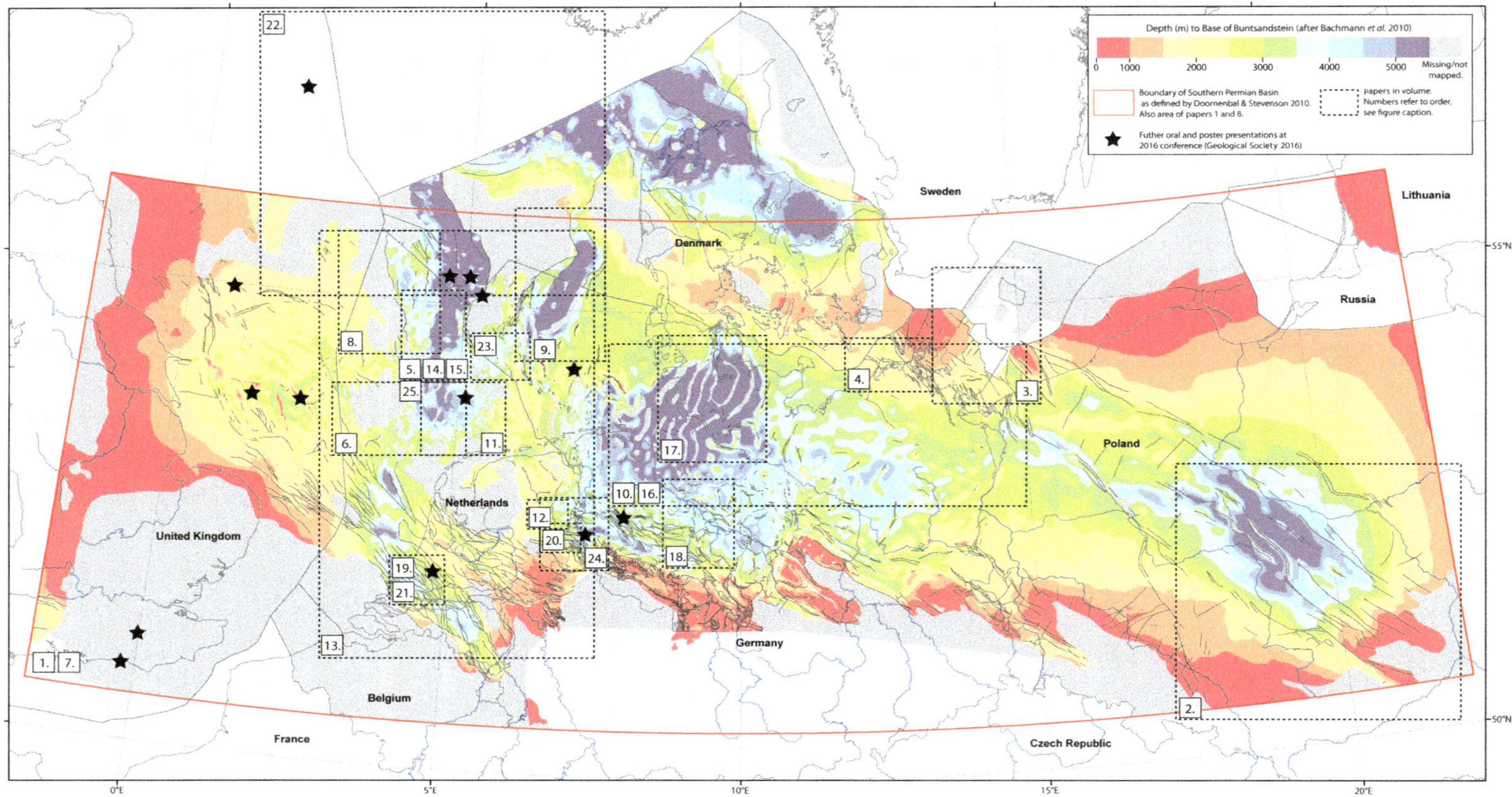

Fig. 1. Location of the Southern Permian Basin in NW Europe (bounding red box, *sensu* Doornenbal & Stevenson 2010) and its associated Mesozoic sub-basins (illustrated by coloured map of depth (in metres) to base Buntsandstein interval: Bachmann *et al.* 2010). The approximate areas of studies included in this volume are shown with a black dashed outline and associated number related to their book order. Note that numbers 1 and 8 cover the entire basinal area. Black stars relate to further presentations detailed in Geological Society (2016). Numbers relate to: 1, **Kley (2018)**; 2, **Krzywiec *et al.* (2018)**; 3, **Seidel *et al.* (2018)**; 4, **Deutschmann *et al.* (2018)**; 5, **van Winden *et al.* (2018)**; 6, **Hernandez *et al.* (2018)**; 7, **Geluk *et al.* (2018)**; 8, **Kortekaas *et al.* (2018)**; 9, **Kilhams *et al.* (2018)**; 10, **Franz *et al.* (2018)**; 11, **Peeters *et al.* (2018)**; 12, **Goswami *et al.* (2018)**; 13, **van Kempen *et al.* (2018)**; 14, **Bouroullec *et al.* (2018)**; 15, **Verreussel *et al.* (2018)**; 16, **Barth *et al.* (2018)**; 17, **Sachse & Littke (2018)**; 18, **Stock & Littke (2018)**; 19, **Vondrak *et al.* (2018)**; 20, **Vis *et al.* (2018)**; 21, **Porter *et al.* (2018)**; 22, **Zwaan (2018)**; 23, **Wolf *et al.* (2018)**; 24, **Strozyk *et al.* (2018)**; 25, **van Lochem (2018)**.

generalized and simplified overview is presented as a framework for the following papers.

The time period from 500 Ma (Cambro-Ordovician boundary) to 300 Ma (latest Carboniferous) saw a period of large-scale plate reorganization which was fundamental to the development of the NW European basins (e.g. Ziegler & Dèzes 2006; Krawczyk *et al.* 2008; McCann 2008*c*; Pharaoh *et al.* 2010). In Ordovician–Devonian times (*c.* 460–380 Ma), the Caledonian collision led, through a series of phased events, to the amalgamation of the Avalonian, Laurentian and Baltica plates which formed Laurussia (e.g. Cocks *et al.* 1997; Cocks 2000; Holdsworth *et al.* 2002; Krawczyk *et al.* 2008; Smit *et al.* 2016). To make a simplistic statement, for what was in reality a complex structural situation, the Caledonian Front between Avalonia and Baltica is expressed as the Trans-European Fault Zone in northern Germany/southern Denmark and as the Tornquist–Teisseyre Zone through NE Germany and Poland, with the relict crustal domains defined by gravity/magnetic data and long-offset seismic studies (e.g. EUGENO-S 1988; Zielhuis & Nolet 1994; Thybo 1997; Abramovitz & Thybo 2000; Thybo 2001; Banka *et al.* 2002; Grad *et al.* 2002, 2009; Yegerova *et al.* 2007; Krawczyk *et al.* 2008; Kaban *et al.* 2010; Maystrenko & Scheck-Wenderoth 2013). This sutured margin and its related structures continued to have a strong influence on the subsequent Mesozoic sequence, as discussed in the Pomerian area by **Deutschmann *et al.* (2018)** and **Seidel *et al.* (2018)**. In Late Devonian–Permian times (*c.* 370–270 Ma), the final stages of collision between Gondwana to the south and Laurussia to the north occurred, leading to the amalgamation of Pangaea and the generation of the Variscan mountain belt, and associated structures, in Central Europe (e.g. Gast 1988; Ziegler 1990*b*; Kroner *et al.* 2008; Maystrenko *et al.* 2008). The foreland of the Variscides became the main Carboniferous basin, allowing the deposition of large-scale deltaic and marine systems associated with the coal-rich Westphalian intervals and their related gas-prone source rocks (e.g. Kombrink *et al.* 2010). The gradual fill of that basin, combined with changes in palaeogeographical setting, access to marine gateways and global climate shifts, led to the deposition of continental sequences of the Rotliegend in the Mid–Late Permian followed by thick evaporites in the late Permian–Early Triassic (e.g. Ziegler 1990*a*; Verdier 1996; Geluk 2005; Peryt *et al.* 2010; Fryberger *et al.* 2011; McKie 2017). These evaporite sequences later (typically, from the Early–Mid Triassic onwards) mobilized into significant salt pillows, walls and diapirs which impacted Mesozoic depocentres and allowed hydrocarbon trap formation (as discussed by **Bouroullec *et al.* 2018**; **Hernandez *et al.* 2018**; **van Winden *et al.* 2018**).

The Mesozoic structural evolution of the Southern Permian Basin area is defined by the relict foreland basin and subsequent tectonic movement along the terrane boundaries (Thybo 2001; Pharaoh *et al.* 2010). For example, the Triassic is notable for the generation, or accelerated evolution, of various overprinting rift systems (Thybo 2001). These led to thick Triassic sequences in areas such as the Glückstadt Graben (Maystrenko *et al.* 2005) and the Horn Graben (e.g. **Kilhams *et al.* 2018**) with subsequent thick Jurassic sequences in the Central Graben and adjacent areas (e.g. **Verreussel *et al.* 2018**). These fast rifting phases, and variable filling of the associated basins, have an impact on aspects such as overpressure distribution related to depth, reservoir quality and fluid fills (e.g. **Peeters *et al.* 2018**). Subsequent inversion and erosion phases led to the exposure of some Triassic intervals. **Wolf *et al.* (2018)** discuss the associated dissolution of discrete evaporite bands and the possible impact on hydrocarbon prospectivity.

Dry continental conditions were prevalent in the Early Triassic, forming various fluvial–aeolian reservoir units (e.g. Geluk 2005; **Kortekaas *et al.* 2018**), with a gradual shift towards marine conditions through time (more details provided by **Geluk *et al.* 2018**). In generalized terms, the Jurassic sequence continues a marine-dominated trend across the basin with some fluvio-deltaic intervals (Pieńkowski & Schudak 2008; **Barth *et al.* 2018**), shoreface deposits (e.g. Munsterman *et al.* 2012) and high-TOC (total organic carbon) marine bands forming oil-prone source rocks (e.g. Posidonia and Kimmeridge shales: e.g. Oschmann 1991; McArthur *et al.* 2008; Pierce *et al.* 2008; **Sachse & Littke 2018**; **Stock & Littke 2018**). The Jurassic sequence is, however, only preserved in areas of the basin that were protected from a late Jurassic–Early Cretaceous inversion episode (related to the Alpine collisional phase: e.g. Rosenbaum *et al.* 2002; Pharaoh *et al.* 2010) or areas that only suffered from subsequent Late Cretaceous–Cenozoic inversion (related to further southern European plate convergence: Dèzes *et al.* 2004; Kley & Voigt 2008; Pharaoh *et al.* 2010; **Kley 2018**). In The Netherlands, this includes discrete rifts such as the Roer Valley Graben/West Netherlands Basin and the Central Graben (e.g. Winstanley 1993; de Jager *et al.* 1996; de Jager 2003; **Verreussel *et al.* 2018**). In Germany, Jurassic preservation is often associated with halokinetic rim synclines and other localized areas such as the Lower Saxony Basin (Schwarzkopf 1990; Baldschuhn *et al.* 1996; Doehler 2005; Lott *et al.* 2010; Bruns *et al.* 2014). Although the Cretaceous interval suffered various discrete inversion phases (e.g. Vejbaek *et al.* 2010; **Kley 2018**; **Krzywiec *et al.* 2018**; **Wolf *et al.* 2018**), which allowed hydrocarbon traps to form, there is a general basin sag during this time.

This, coupled with rising eustatic sea level (e.g. Miller *et al.* 2005; Ramkumar 2016) and generally warm global temperatures (e.g. Steuber *et al.* 2005), led to an evolution from a clastic-dominated succession (marine shorefaces/pelagic mudstones/deep-water sandstones: e.g. Milton-Worssell *et al.* 2006; Jeremiah *et al.* 2010; **Vis *et al.* 2018**; **Zwaan 2018**) to chalk-rich seas (including important reservoir units across the region: e.g. **van Lochem 2018**). The Cenozoic then saw a switch back to clastic conditions. Shallow-marine sands were deposited around the basin edges (e.g. Knox *et al.* 2010) and deep-water deposits are recorded in the Central/Northern North Sea in Paleocene–Eocene times (e.g. den Hartog Jager *et al.* 1993; Ahmadi *et al.* 2003). Subsequently, large-scale fluvio-deltaic systems prograded into the basin from the SE/east towards the NW/west during the Oligo-Miocene (e.g. Huuse 2002; Rasmussen & Dybkjaer 2014; Gibbard & Lewin 2016). In the Plio-Pleistocene, glacial phases developed (e.g. Moreau *et al.* 2012) which are postulated by **Sachse & Littke (2018)** to have impacted Jurassic hydrocarbon systems via pressure variation.

Historical resource focus: conventional hydrocarbons

Boigk (1981), van Hulten (2009) and Breunese *et al.* (2010) provide detailed overviews of the hydrocarbon exploration and production history of this basin. Here a short, simplified summary is presented to illustrate the relative historical importance of the Paleozoic and Mesozoic intervals to hydrocarbon exploration. It is recognized that there are further Mesozoic geological resources that are not covered here or in subsequent chapters. These include, but are not limited to, quarrying (e.g. limestone at Winterswijk in the eastern Netherlands; Faber 1959) and salt extraction (e.g. Wassmann & Brouwer 1987).

The discovery and use of hydrocarbons in NW Europe can be traced back to at least 1628 when oil products were used as cart lubrication and medical remedies in the region of Wietze, Lower Saxony (Boigk 1981; Arndt 2017). Although oil remained a useful product throughout the subsequent centuries, it was not until the late 1930s that hydrocarbon exploration started to boom in the region with the discovery of the Bentheim gas field (Zechstein), and subsequently (in 1943) the Schoonebeek oil and gas accumulations (van Hulten 2009). At that time exploration was focused on Mesozoic and Permian Zechstein targets with, for example, discoveries at the Goldenstedt (in 1959: Zechstein), Adorf (in 1959: Triassic) and Hengstlage (in 1963: Triassic) fields (Breunese *et al.* 2010). However, the discovery of a series of giant Rotliegend gas fields was to change the focus of explorers across the basin. In 1959, the Slochteren-1 well proved up 2900 Bcm (billion cubic metres) (*c.* 102 Tcf (trillion cubic ft)) at Groningen, at that time the largest gas field in the world (Grötsch *et al.* 2011). This was followed up by the 1965 Groothusen and 1969 Salzwedel discoveries, the latter containing 200 Bcm (*c.* 7.1 Tcf) of gas, the largest field in Germany (Breunese *et al.* 2010). The very large hydrocarbon volumes associated with the Paleozoic have subsequently made it the key economic driver for operators. However, the Mesozoic has also had an important role to play in the Southern Permian Basin area. Figure 2a illustrates that around 10% of gas reserves are hosted within Mesozoic reservoirs. These are typically, but not always, associated with the Triassic interval, being charged by Westphalian Type III coals. Field examples include Apeldorn (Kus *et al.* 2005), Caister (Ritchie & Pratsides 1993), De Wijk (Bruijn 1996; **Goswami *et al.* 2018**) and F15-A (Fontaine *et al.* 1993). Perhaps of more consequence, Figure 2b illustrates that around 90% of the basin's oil reserves are found in the Mesozoic interval. These are typically, but not always, associated with the Jurassic and Cretaceous successions, being charged by Posidonia and/or Kimmeridgian Type II marine shales. Field examples include Bramstedt (Jurassic: Boigk 1981), M07-B (Jurassic gas: ONE 2014), Mittelplate (Jurassic–Cretaceous: Doehler 2005), Rotterdam (Lower Cretaceous clastics: **Porter *et al.* 2018**) and Dan (Cretaceous chalk: Jorgensen 1992).

Figure 2c, d illustrates that these Mesozoic conventional gas and oil reserves, within the Southern Permian Basin, are predominantly found in The Netherlands, Germany and Denmark (97% for gas and 99% for oil), with only minor amounts associated with the UK and other countries such as Poland. This geographical distribution is a result of the geological history described above, including the spatial extent of rift systems (e.g. the Central Graben and associated Jurassic source rocks in a discrete area running through the Dutch, German and Danish offshore), areas of inversion, source-rock presence and charge timing (cf. Pletsch *et al.* 2010). However, this does not rule out further conventional hydrocarbon discoveries in the Mesozoic of, for example, Poland or other historically less attractive areas (e.g. **Kilhams *et al.* 2018**; **Kortekaas *et al.* 2018**).

An example, from The Netherlands, of the number of historical exploration wells and their stratigraphic targets (post-1940) is shown in Figure 3a. Early exploration, as described above, focused on the Triassic and Zechstein intervals, with the former considered the most attractive. However, after the 1959 Groningen discovery, there was a distinct upswing in both the annual total number of exploration wells and the proportion of Paleozoic targets which continued throughout the 1960s and 1970s.

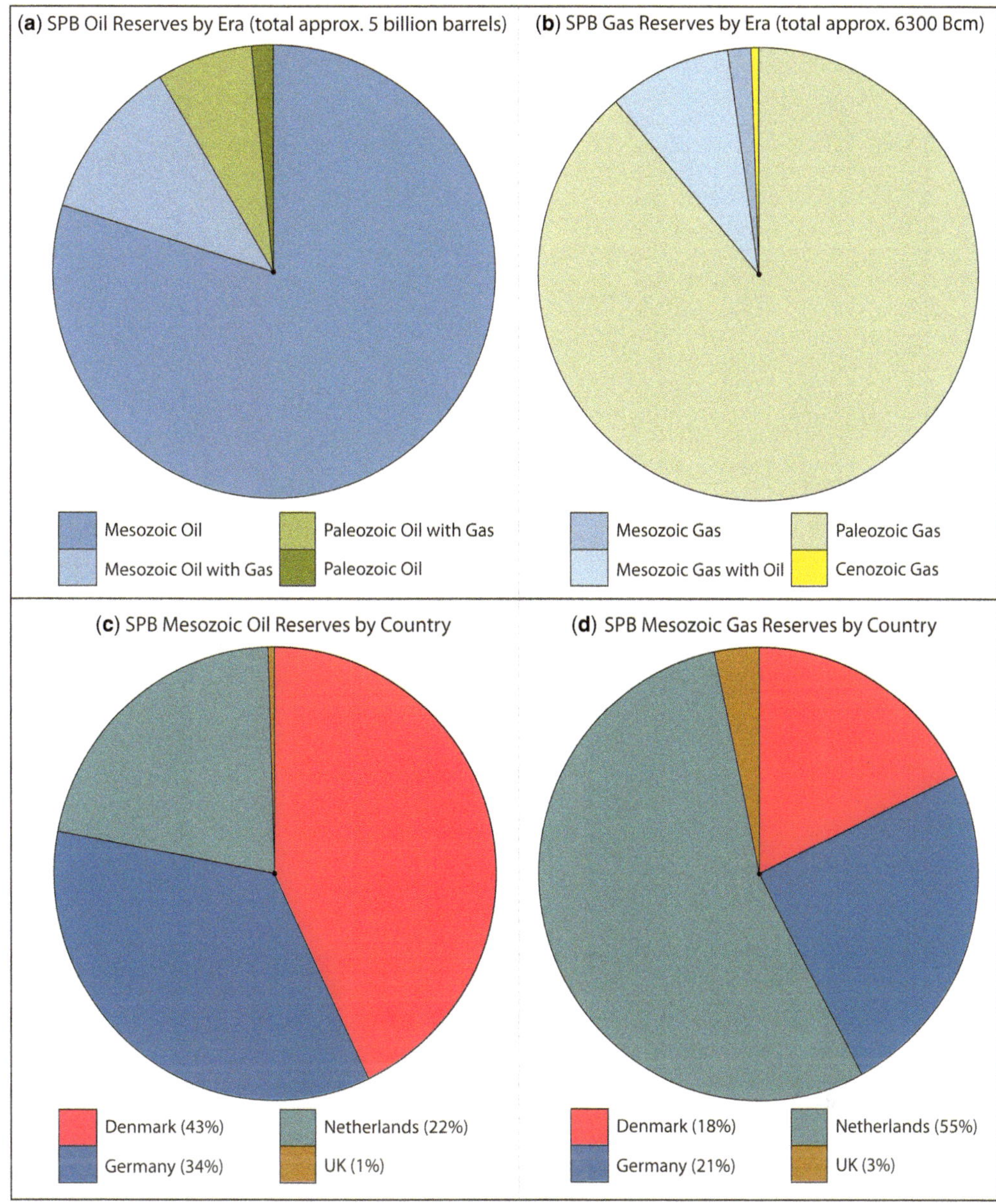

Fig. 2. Overview of hydrocarbon reserves across the Southern Permian Basin (SPB) area, based on estimates of Doornenbal & Stevenson 2010. (**a**) Oil reserves by era across the basin illustrating the proportion of discovered Mesozoic oil (blue shades) in comparison to Paleozoic oil (green shades) (total *c.* 5 Bbbl). (**b**) Gas reserves by era across the basin illustrating the proportion of discovered Mesozoic gas (blue shades) in comparison to Paleozoic gas (green shades) and Cenozoic gas (yellow) (total *c.* 6300 Bcm or 222 Tcf). (**c**) Distribution of discovered Mesozoic oil by country across the SPB. (**d**) Distribution of discovered Mesozoic gas by country across the SPB.

In the period from 1980 to 1990, a combination of high to moderate oil prices, new 3D seismic technology and a desire to explore new plays saw an increase in the annual total number of exploration wells to record numbers. In that period, the proportion of Mesozoic targets also increased, and the success rate of all exploration wells jumped from around 35 to 65% (Breunese & Rispens 1996). Figure 3b illustrates which Mesozoic interval these Dutch wells targeted (post-1940). Historically,

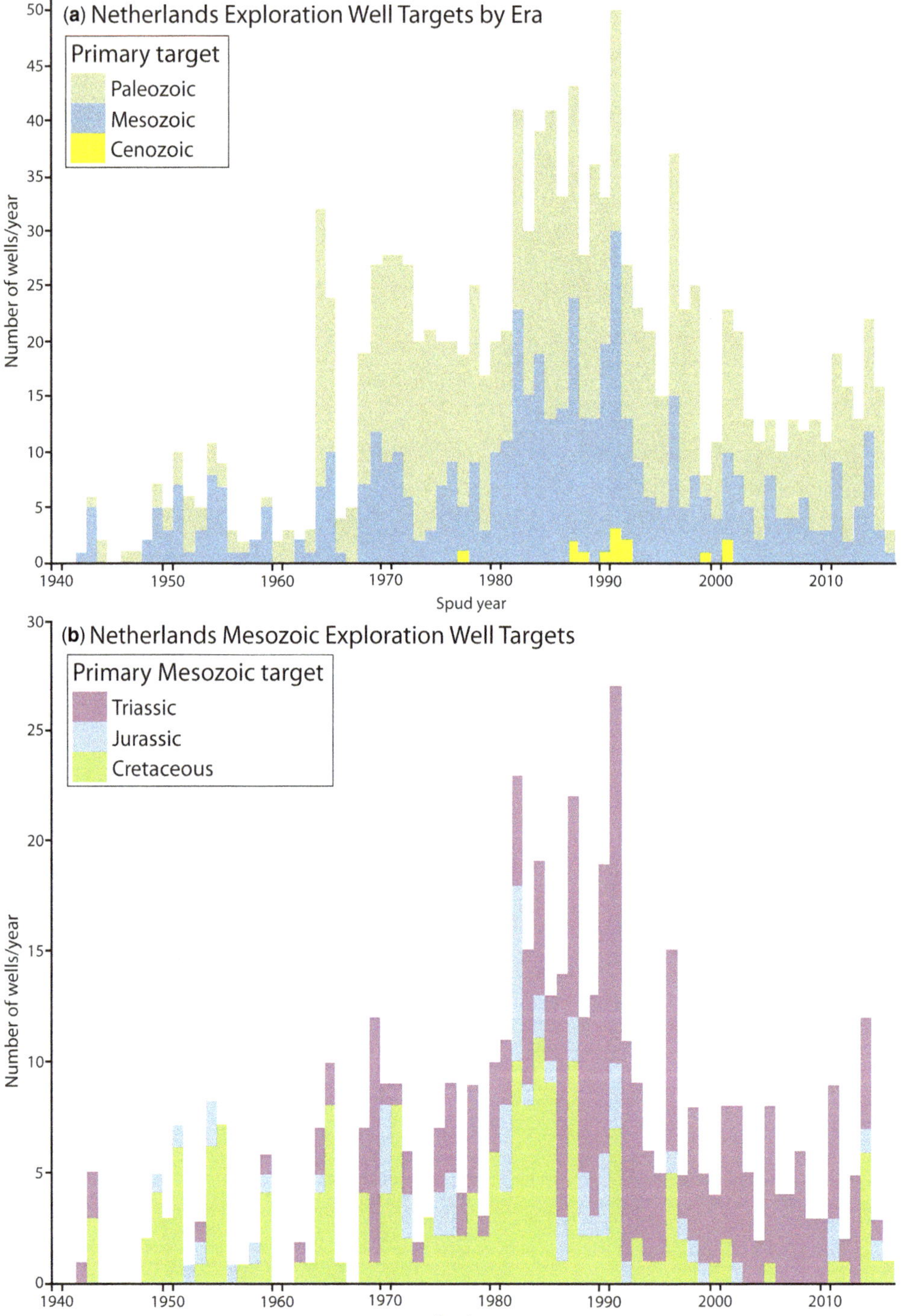

Fig. 3. Overview of The Netherlands exploration well targets between 1940 and 2016, based on EBN database. (**a**) The Netherlands exploration well (primary) targets by geological era including Paleozoic (green), Mesozoic (blue) and Cenozoic (yellow). (**b**) For the subset of wells which targeted Mesozoic intervals, primary targets by geological period including Triassic (purple), Jurassic (blue) and Cretaceous (green).

the early (pre-1986) discoveries in The Netherlands were associated with the relatively shallow Cretaceous structures (and, to a lesser extent, the Jurassic) which could be defined and drilled using 2D seismic data (Breunese & Rispens 1996). With the advent of 3D seismic technology, there was a switch to deeper Triassic targets and those prospects that could be understood and de-risked by new techniques such as amplitude analysis (e.g. Breunese & Rispens 1996; Bruijn 1996). Note that in the period 2011–16 there was a resurgence in Cretaceous exploration drilling. This is associated with the chalk interval in the Dutch Central Graben where the application of advanced seismic inversion techniques, developed in the adjacent Danish offshore, gave confidence to chalk reservoir quality and fluid fill predictions (e.g. Megson 1992; Megson & Tygesen 2005; Abramovitz *et al.* 2010; **van Lochem 2018**).

For Germany, the historical discovered recoverable volumes per era is shown in Figure 4a and per Mesozoic period in Figure 4b. This illustrates a similar early (pre-1964) focus on Mesozoic exploration before a switch to Paleozoic targets and subsequent discoveries. With the exception of the 1980 Mittelplate discovery (cf. Doehler 2005), only very small volumes (<10 MMboe UR (million barrels of oil equivalent ultimate recovery)) have been discovered in the Mesozoic since 1968. This is likely to be due to a combination of geological factors (e.g. limited Jurassic source-rock presence: cf. Pletsch *et al.* 2010) and the more limited acquisition of 3D seismic data in comparison to The Netherlands. The historical perception of relatively small-volume promise in Germany has also led to high oil-price sensitivity, with an increase in discoveries related to peaks in global oil prices (e.g. between 1978 and 1980 the Brent crude price (inflation adjusted) rose from $50/bbl (barrel) to $105/bbl, by 1985 the inflation adjusted price was around $31/bbl: Macrotrends 2017). Prior to the latest 2014–16 oil price fall (from *c.* $105/bbl to $30/bbl) there had been significant interest from smaller operators to redevelop (mainly Jurassic) historical oil fields and to chase exploration upside (including unconventional plays). The oil-price fall and general social movement against unconventionals caused companies to downscale activities or exit the country. Examples include PRD Energy at Volkensen (Arndt 2015; MarketWired 2015), Kimmeridge GmbH in Lower Saxony (iDeals 2016) and Central Anglia A/S in the Sterup licence, Schleswig-Holstein (Central Anglia 2016).

Future resource focus: hydrocarbons and geothermal energy

It is clear after at least 70 years of intense exploration across the Southern Permian Basin area that there is increased pressure to apply new geological ideas or new techniques to the exploration and exploitation of conventional hydrocarbons. The development of unconventional oil plays also gives an opportunity to extend hydrocarbon production within the basin. However, environmental concerns and societal pressure has, in recent years, seen an upsurge in government support for geothermal projects. In all these resource scenarios, a sound geological understanding is key to successful exploitation. Here, each of these existing and future resources is considered within the framework of the chapters that follow.

Maximizing production efficiency from existing hydrocarbon fields

It has become standard practice in conventional oil and gas fields across the Southern Permian Basin area and elsewhere, if considered economic, to drill extra infill wells and apply enhanced recovery techniques to achieve the highest possible hydrocarbon yield (e.g. Lake 2010). Such techniques require a good geological understanding of how, for example, sedimentary layers cause differential fluid flow. **Porter *et al.* (2018)** give an example from the Cretaceous shoreface reservoirs of the Rotterdam oil field, including a comparison to analogous outcrop examples in southern England. Additionally, **Vis *et al.* (2018)** consider how local tectonic phases influenced reservoir distribution (which, therefore, has a possible impact on hydrocarbon extraction strategies) in the Dutch Schoonebeek Field.

Typically, enhanced recovery techniques have been focused on oil reservoirs via water injection. However, gas injection is increasingly being used. **Goswami *et al.* (2018)** give an example of enhanced recovery via nitrogen injection at the De Wijk Field (Triassic reservoir interval), illustrating how new technologies can be utilized to increase gas yield. It is possible that existing or future technologies, such as cheaper drilling or enhanced recovery techniques, could unlock further resources.

Exploring for further conventional hydrocarbon resources

An estimate of the remaining prospective conventional hydrocarbon resources for The Netherlands is shown in Figure 5, which shows considerable potential left in all the Mesozoic intervals, including Triassic gas (Fig. 5a) and both Jurassic (Fig. 5b) and Cretaceous oil (Fig. 5c). Chasing further exploration opportunities in a mature basin demands a high level of geological understanding of all the play elements. This starts with consideration of the potential of large existing datasets. An example is presented by **van Kempen *et al.* (2018)**, utilizing public well

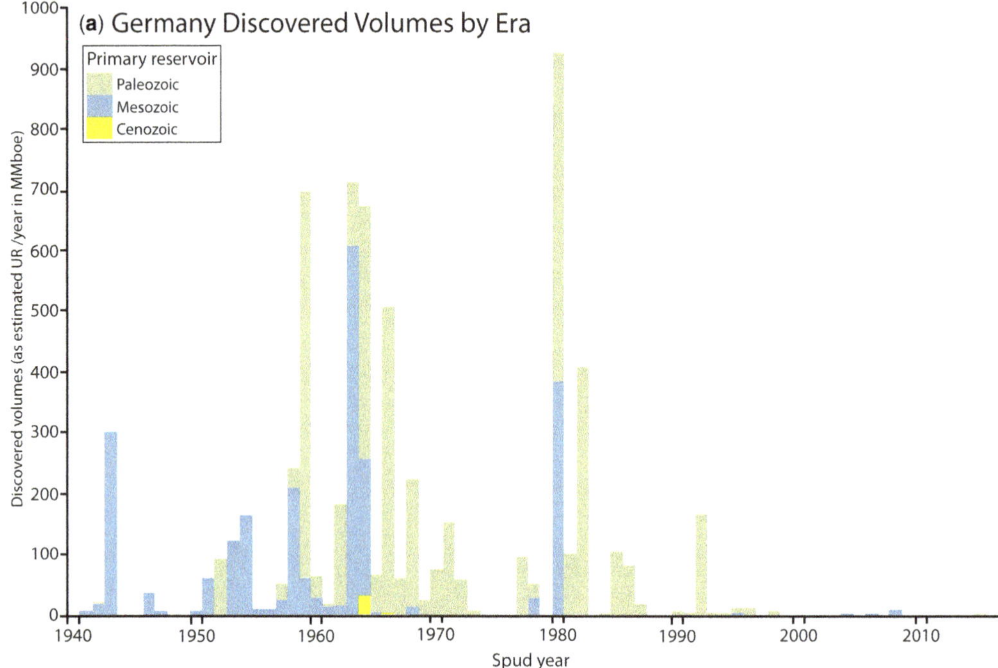

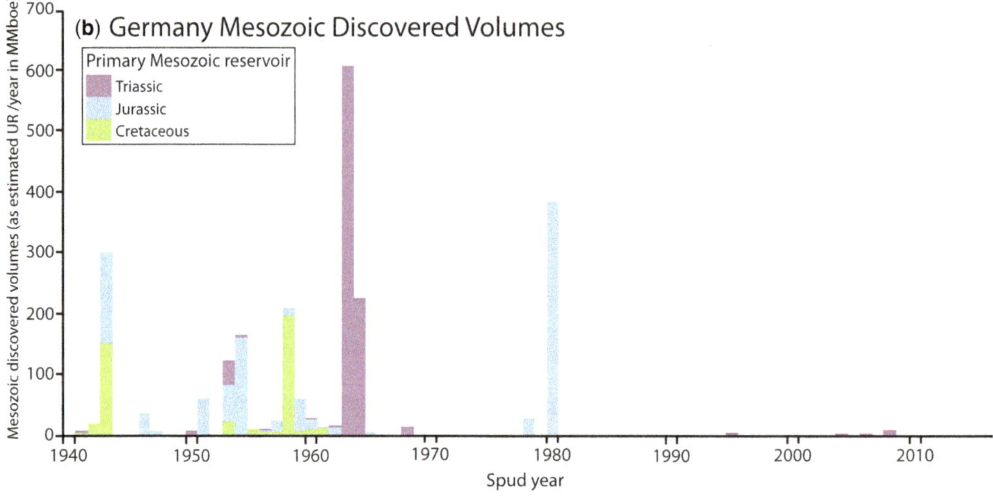

Fig. 4. Overview of German discovered volumes between 1940 and 2016, based on IHS and LBEG data compilation. (**a**) Discovered volumes (all hydrocarbons, as estimated ultimate recoverable (UR)/year in MMboe) by geological era including Paleozoic (green), Mesozoic (blue) and Cenozoic (yellow). (**b**) For the subset of wells which targeted Mesozoic intervals, discovered volumes (all hydrocarbons, as estimated ultimately recoverable (UR)/year in MMboe) by geological period including Triassic (purple), Jurassic (blue) and Cretaceous (green).

log data to consider Triassic Bunter interval reservoir properties and trends across The Netherlands. Detailed consideration of existing seismic data can also reveal interesting features. For example, **Strozyk *et al.* (2018)** identify a series of pockmarks which could refine the gas-charge timing story in the eastern Netherlands. It is also possible to question the fundamental tectonic framework of an area and the impact this might have on exploration models. An example is presented by **Krzywiec *et al.* (2018)** for the Polish Trough. Dogmas around reservoir (e.g. **Kortekaas *et al.* 2018** for the Bunter of the

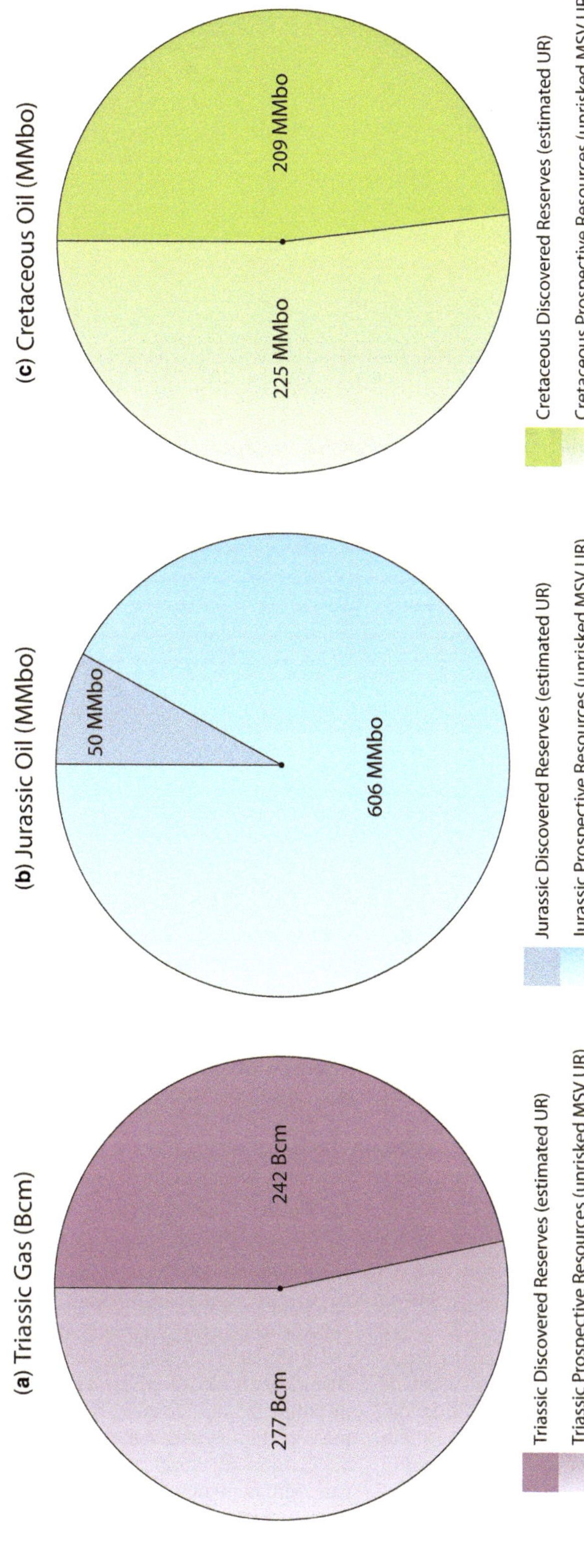

Fig. 5. Mesozoic gas and oil reserves and prospective resources in the EBN (Netherlands) portfolio, designed to be representative of a subset of further Mesozoic hydrocarbon potential across the basin. Discovered reserves as estimated ultimately recoverable (UR), prospective resources as unrisked mean success volume ultimately recoverable (MSV UR). (**a**) Triassic gas in Bcm: dark purple discovered volumes (242 Bcm; *c.* 8.5 Tcf) and light purple prospective resources (277 Bcm; *c.* 9.8 Tcf). (**b**) Jurassic oil in million barrels of oil (MMbo): dark blue discovered volumes (50 MMbo) and light blue prospective resources (606 MMbo). (**c**) Cretaceous oil: dark green discovered volumes (209 MMbo) and light green prospective resources (225 MMbo).

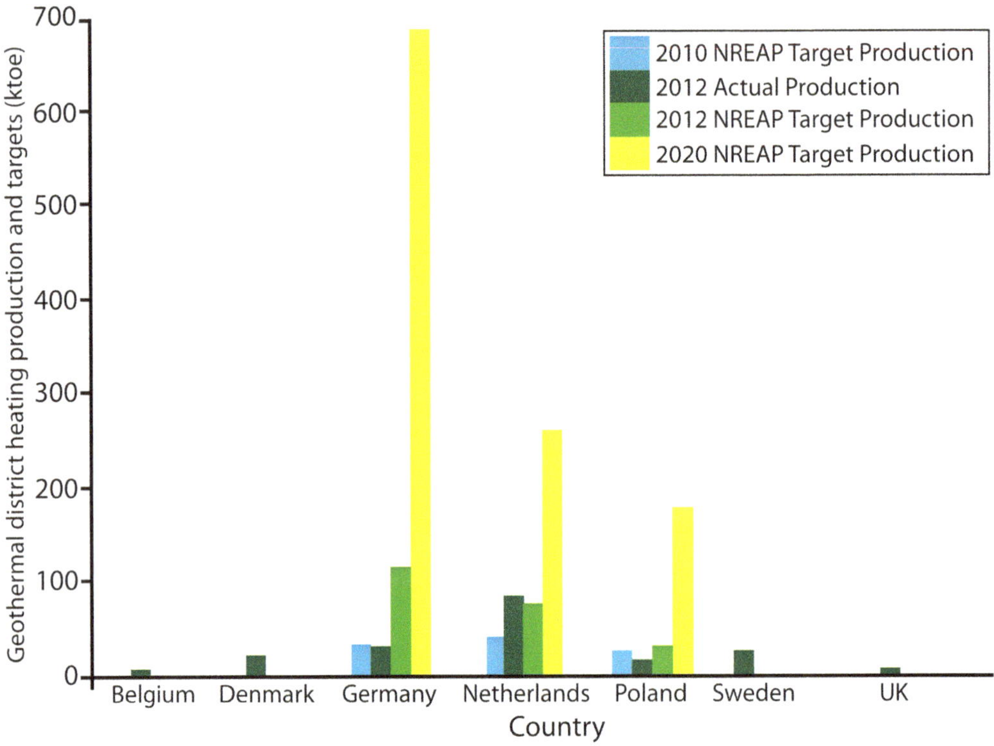

Fig. 6. Exploitation of geothermal reserves v. published targets for countries in the Southern Permian Basin (SPB) area (after Dumas & Bartosik 2014), designed to be representative of the large prospective energy resources available. Values are district heating production/targets expressed as thousand tonnes of oil equivalent (ktoe). Targets based on the National Renewable Energy Action Plans (NREAP) for 2010 (blue), 2012 (light green) and 2020 (yellow). Production estimates are for 2012 (dark green).

Dutch Northern Offshore; **Zwaan 2018** for the Cretaceous of the adjacent Norwegian and Danish offshore) or source-rock development (e.g. ***Kilhams et al.* 2018** for the Triassic play of the German Horn Graben) can be challenged in areas perceived to be fallow through the integrated evaluation of well results and conceptual geological models. Conventional hydrocarbon opportunities remain. For example, for the aforementioned German Jurassic oil play opportunities, the geological elements remain the same even if the economic boundaries have shifted.

Exploring for unconventional hydrocarbon resources

The presence of various Mesozoic source rock intervals and associated high TOC shale units (in addition to similar Paleozoic intervals) suggested that the recent unconventional oil and gas boom in the USA could be transferred to NW Europe (e.g. Schulz *et al.* 2010). A combination of societal pressure, economics and geological factors mean that this has not yet happened (e.g. Selley 2012; Johnson & Boersma 2013; Weijermars 2013). However, geoscience is central to the identification of resources and can contribute to the continuing debate over its extraction. **Stock & Littke (2018)** give an example of how the Posidonia shale may be associated with unconventional resources in the Lower Saxony Basin.

Efficient exploration for, and exploitation of, geothermal energy resources

Northern Europe is often considered to be at the forefront of the energy transition from hydrocarbons to renewable sources. Here, we consider the role that geoscience can play in unlocking further geothermal energy reserves. In the area of the Southern Permian Basin, there are a number of ongoing initiatives to promote geothermal energy. Figure 6 shows an estimate of the geothermal district heating production and targets for various countries in the study area and, although not specific to the Mesozoic, gives an example of the gap to potential for this energy

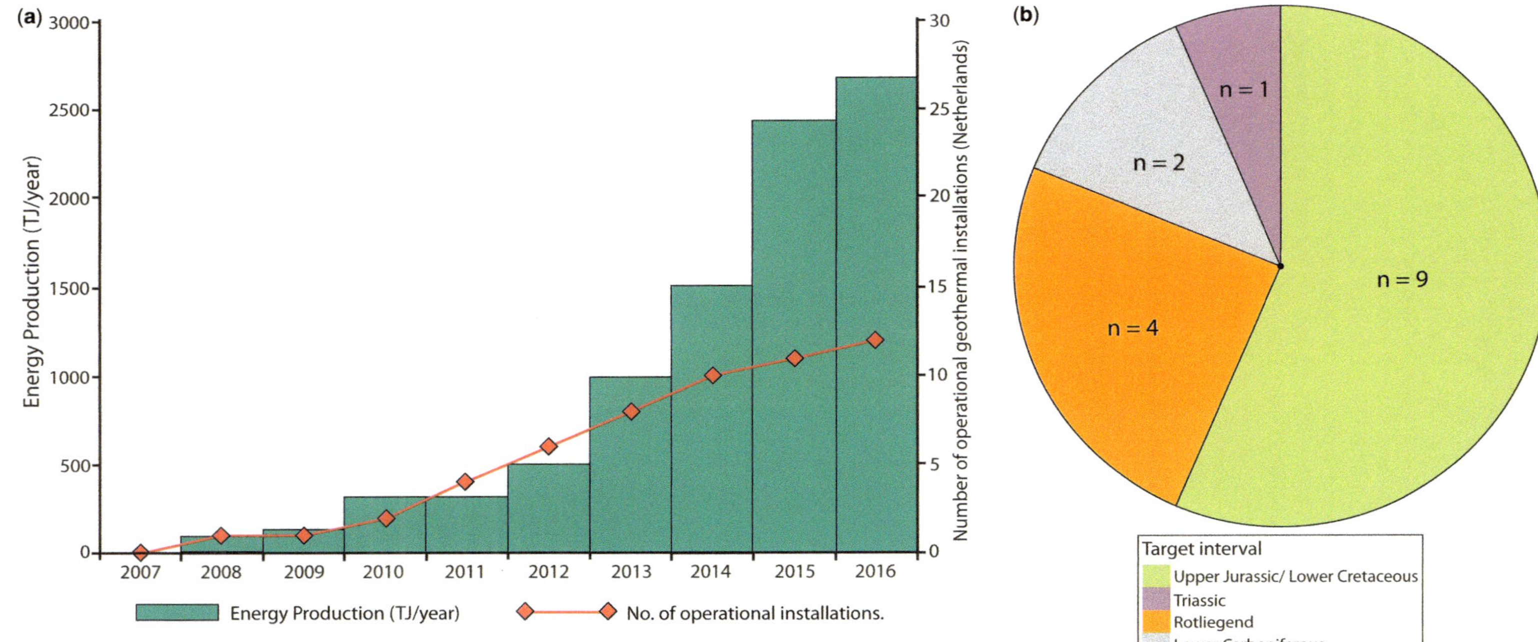

Fig. 7. Geothermal heat production and number of projects in The Netherlands (after Ministrie v. Economische Zaken 2016). (**a**) Energy (heat) production (from geothermal projects) profile 2007–16 expressed as Terajoules (TJ) per year (blue histogram) and discrete number of operational geothermal installations for each year (red line and diamonds). (**b**) Total number of drilled geothermal projects in The Netherlands by main geological target age, the majority are aimed at the Mesozoic interval. Note that this plot includes projects not yet producing, hence the larger number than in (a).

source. As suggested by **Franz *et al.* (2018)**, there is considerable potential for the use of geothermal energy in Germany, a statement which can also be extended to The Netherlands and Poland. The resources of, for example, Denmark and the UK are not yet fully defined. In some areas of the basin, there is social and economic demand for clean energy (mainly for urban greenhouse heating). Pilot projects, often with a combination of Triassic and Cretaceous targets, have been undertaken which have highlighted the importance of sound geological understanding, through facies and porosity prediction in achieving economic water production rates (e.g. Pluymaekers *et al.* 2012; De Vaal 2017; **Franz *et al.* 2018**; **Vondrak *et al.* 2018**). For example, Figure 7a illustrates the growth of geothermal systems in The Netherlands (both in operational projects (2016, $n = 12$) and associated energy production). This effort began with a 2005–07 pilot project by A+G van den Bosch horticultural company at Bleiswijk (West Netherlands Basin), with the aim of utilizing heat for growing via water production from the Upper Jurassic–Lower Cretaceous interval (Platform Geothermie 2017; VleesTomat 2017) The success of this project has led to an acceleration of geothermal investment (focused on the Mesozoic, as demonstrated by all the drilled projects (2016, $n = 16$) in Fig 7b) aided by a wealth of subsurface data in the public domain (see also **Vondrak *et al.* 2018**). Various projects have also been undertaken in NE Germany. For example, in Neubrandenburg, production has been achieved from Triassic Rhaetian sandstones (Wolfgramm *et al.* 2009; **Franz *et al.* 2018**), with further potential in the Middle Jurassic fluvial sandstone sequence (**Barth *et al.* 2018**; **Franz *et al.* 2018**). Various regional and local governmental organizations are supporting both further research into geothermal energy (with a number of technical and mapping tools now available: e.g. TNO 2013; Peta 2015) and subsidizing projects (e.g. EBN 2017). It is clear that a large range of techniques originally developed for the hydrocarbon industry, such as basin modelling (e.g. Nelskamp & Verweij 2012) and seismic attribute analysis (e.g. Dierkhising 2015), can also be applied to geothermal projects to improve pre-drill prediction of, for example, reservoir temperature and fluid fill.

Conclusions

The Mesozoic of the Southern Permian Basin area continues to provide fresh insights which can be applied to energy resource exploitation and identification. The general tectonic history of this area is considered well known, but the papers that follow illustrate that new observations can still be made. It is demonstrated throughout this publication that the key to unlocking remaining resources is built on a foundation of solid geological understanding. This applies to the efficient production of discovered hydrocarbons, especially when attempting to apply new engineering techniques, as well as defining and exploring for new conventional reserves. If the exploitation of unconventional hydrocarbons becomes socially acceptable in northern Europe, an underlying geological understanding of various Mesozoic age units will become increasingly important. However, this mantra also applies to the development of geothermal energy reserves, where accurate predictions of porosity and geothermal gradients are one of the many keys to a successful project. It is hoped that this Special Publication will spur further exploration for, and efficient extraction of, the remaining resources in the basin.

The editorial team extends its appreciation to all the authors and collaborators that contributed to this book. The time and effort taken represents considerable determination and perseverance. Everyone that contributed to the original Geological Society conference is also appreciated, the open and collaborative atmosphere formed the basis of this book, with special thanks going to: Laura Griffiths, Gary Hampson, Howard Johnson, James Maynard, Robert Schöner, Martin Wells and Sarah Woodcock. We thank various managers who allowed the editors and authors to spend time making a significant contribution. Ben Kilhams particularly thanks Max Brouwers, Ramon Loosveld, Carlo Nicolai and Edwin Verdonk, who have supported the venture through their patience and sponsorship. Sincere thanks go to all the reviewers who took time and care to give excellent, constructive feedback: Oscar Abbink, Kresten Anderskouv, Katrine Andresen, Stuart Archer, Renaud Bouroullec, Dan Carruthers, Teresa Sabato Ceraldi, Joel Corcoran, Daan den Hartog Jager, Ramues Gallois, Mark Geluk, Thomas Gerling, Graham Goffey, Andrew Green, Cor Hofstee, Christian Hubscher, Chris Jackson, Marek Jarosinski, Fabian Jahne-Klingberg, Marek Jarosinski, Jason Jeremiah, Matthias Keym, Marloes Kortekaas, Piotr Krzywiec, Pepijn Kole, Juliette Lamarche, Herald Ligtenberg, James Maynard, Yuriy Maystrenko, Adam McArthur, Inga Moeck, Carlo Nicolai, Lars Nielsen, Mette Olivarius, Richard Porter, Horst Reuter, Ian Saikia, Frank Strozyk, Leo van Borren, Frans van Buchem, Eva van der Voet, Matthijs van Winden, Reinhoud Veenhof, Bruno Vendeville, Hanneke Verweij, Thomas Voigt, Jonathan Wonham and others who wished to remain anonymous. Finally, the editors wish to thank the Geological Society Publishing House, especially Tamzin Anderson, Lucy Bell and Angharad Hills for their support.

References

Abramovitz, T. & Thybo, H. 2000. Seismic images of Caledonian, lithosphere-scale collision structures in the southeastern North Sea along MONA LISA profile 2. *Tectonophysics*, **317**, 27–54.

Abramovitz, T., Andersen, C., Jakobsen, F., Kristensen, L. & Sheldon, E. 2010. 3D seismic mapping and porosity variation of intra-chalk units in the southern Danish North Sea. *In*: VINING, B. & PICKERING, S. (eds) *Petroleum Geology: From Mature Basins to New Frontiers – Proceedings of the 7th Petroleum Geology Conference*. Geological Society, London, **7**, 537–548, https://doi.org/10.1144/0070537

AHMADI, Z., SAWYERS, M., KENYON-ROBERTS, S., STANWORTH, C., KUGLER, K., KRISTENSEN, J. & FUGELLI, E. 2003. Paleocene. *In*: EVANS, D., GRAHAM, C., ARMOUR, A. & BATHURST, P. (eds) *The Millenium Atlas: Petroleum Geology of the Central and Northern North Sea*. Geological Society, London, 235–259.

ARNDT, S. 2015. Schwierige ökonomische, politische und gesellschaftliche Rahmenbedingungen: PRD Energy zieht sich aus Deutschland zurück. *Erdöl und Erdgas in Deutschland*, http://www.erdoel-erdgas-deutschland.de/2015/08/02/schwierige-oekonomische-politische-und-gesellschaftliche-rahmenbedingungen-prd-energy-zieht-sich-aus-deutschland-zurueck/

ARNDT, S. 2017. Chronik. *Erdöl und Erdgas in Deutschland*, http://www.erdoel-erdgas-deutschland.de/chronik/

BACHMANN, G.H., GELUK, M.C. *ET AL*. 2010. Triassic. *In*: DOORNENBAL, H. & STEVENSON, A. (eds) *Petroleum Geological Atlas of the Southern Permian Basin Area*. European Association of Geoscientists and Engineers (EAGE), Houten, The Netherlands, 148–173.

BALDSCHUHN, R., KOCKEL, F. & FRISCH, U. 1996. *Geotektonischer Atlas von NW-Deutschland [Geotectonic Atlas of NW Germany]*. BGR, Hannover, Germany.

BANKA, D., PHARAOH, T.C., WILLIAMSON, J.P. & TESZ PROJECT POTENTIAL FIELD CORE GROUP 2002. Potential field imaging of Palaeozoic orogenic structure in northern and central Europe. *Tectonophysics*, **360**, 23–45.

BARTH, G., PIEŃKOWSKI, G., ZIMMERMANN, J., FRANZ, M. & KUHLMANN, G. 2018. Palaeogeographical evolution of the Lower Jurassic: high-resolution biostratigraphy and sequence stratigraphy in the Central European Basin. *In*: KILHAMS, B., KUKLA, P.A., MAZUR, S., MCKIE, T., MIJNLIEFF, H.F. & VAN OJIK, K. (eds) *Mesozoic Resource Potential in the Southern Permian Basin*. Geological Society, London, Special Publications, **469**. First published online January 4, 2018, https://doi.org/10.1144/SP469.8

BOIGK 1981. *Erdöl und Erdölgas in der Bundesrepublik Deutschland*. Ferdinand Enke, Stuttgart, Germany.

BOUROULLEC, R., VERREUSSEL, R.M.C.H. *ET AL*. 2018. Tectonostratigraphy of a rift basin affected by salt tectonics: syn-rift Middle Jurassic–Lower Cretaceous Dutch Central Graben, Terschelling Basin and neighbouring platforms, Dutch offshore. *In*: KILHAMS, B., KUKLA, P.A., MAZUR, S., MCKIE, T., MIJNLIEFF, H.F. & VAN OJIK, K. (eds) *Mesozoic Resource Potential in the Southern Permian Basin*. Geological Society, London, Special Publications, **469**. First published online March 15, 2018, https://doi.org/10.1144/SP469.22

BREUNESE, J. & RISPENS, F. 1996. Natural gas reserves in the Netherlands: exploration and development in historic and future perspective. *In*: RONDEEL, H., BATJES, D. & NIEUWENHUIJS, W. (eds) *Oil and Gas under the Netherlands*. Kluwer Academic, Dordrecht, The Netherlands, 19–30.

BREUNESE, J., ANDERSEN, J. *ET AL*. 2010. Reserves and production history. *In*: DOORNENBAL, J.C. & STEVENSON, A.G. (eds) *Petroleum Geological Atlas of the Southern Permian Basin Area*. European Association of Geoscientists and Engineers (EAGE), Houten, The Netherlands, 271–281.

BRUIJN, A. 1996. De Wijk gas field (Netherlands): reservoir mapping with amplitude anomalies. *In*: RONDEEL, H., BATJES, D. & NIEUWENHUIJS, W. (eds) *Oil and Gas under the Netherlands*. Kluwer Academic, Dordrecht, The Netherlands, 243–253.

BRUNS, B., LITTKE, R., GASPARIK, M., VAN WEES, J.-D. & NELSKAMP, S. 2014. Thermal evolution and shale gas potential estimation of the Wealden and Posidonia Shale in NW-Germany and the Netherlands: a 3D basin modelling study. *Basin Research*, **28**, 2–33.

BYRNE, P. 2012. Now is the time to invest in mature oil basins: an interview with Josh Young. The Energy Report, http://oilprice.com/Finance/investing-and-trading-reports/Now-is-the-Time-to-Invest-in-Mature-Oil-Basins-An-Interview-with-Josh-Young.html

CENTRAL ANGLIA 2016. Erlabnisfeld Sterup. Central Anglia AS, Oslo, http://www.central-anglia.com/6.html

COCKS, L.R.M. 2000. The Early Palaeozoic geography of Europe. *Journal of the Geological Society, London*, **157**, 1–10, https://doi.org/10.1144/jgs.157.1.1

COCKS, L.R.M., MCKERROW, W.S. & VAN STAAL, C.R. 1997. The margins of Avalonia. *Geological Magazine*, **130**, 711–724.

DE JAGER, J. 2003. Inverted basins in the Netherlands, similarities and differences. *Netherlands Journal of Geosciences*, **82**, 339–349.

DE JAGER, J., DOYLE, M., GRANTHAM, P. & MABILLARD, J. 1996. Hydrocarbon habitat of the West Netherlands Basin. *In*: RONDEEL, H., BATJES, D. & NIEUWENHUIJS, W. (eds) *Oil and Gas under the Netherlands*. Kluwer Academic, Dordrecht, The Netherlands, 191–209.

DE VAAL, E. 2017. Oil exploration and the geothermal dual play in the west Netherlands. Presented at the Geothermal Cross Over Technology Workshop, 25–26 April 2017, Durham, UK, https://www.eiseverywhere.com/file_uploads/15ab7cf9793f64917635ebdbabc568cf_programmedurhamworkshop.pdf?eb=399828

DEN HARTOG JAGER, D., GILES, M. & GRIFFITHS, G. 1993. Evolution of Paleogene submarine fans of the North Sea in space and time. *In*: PARKER, J. (ed.) *Petroleum Geology of Northwest Europe: Proceedings of the 4th Conference*. Geological Society, London, 59–71, https://doi.org/10.1144/0040059

DEUTSCHMANN, A., MESCHEDE, M. & OBST, K. 2018. Fault system evolution in the Baltic Sea area west of Rügen, NE Germany. *In*: KILHAMS, B., KUKLA, P.A., MAZUR, S., MCKIE, T., MIJNLIEFF, H.F. & VAN OJIK, K. (eds) *Mesozoic Resource Potential in the Southern Permian Basin*. Geological Society, London, Special Publications, **469**. First published online March 9, 2018, https://doi.org/10.1144/SP469.24

DÈZES, P., SCHMID, S.M. & ZIEGLER, P.A. 2004. Evolution of the Cenozoic Rift System: interaction of the Alpine and Pyrenean orogens with their foreland lithosphere. *Tectonophysics*, **389**, 1–33.

DIERKHISING, J. 2015. *Development and application of AVO methods and seismic attribute analysis for characterisation of the San Emidio Geothermal reservoir*. MSc thesis, University of Nevada, Reno, tNV, USA.

Doehler, M. 2005. The Mittelplate oil field. *In*: Doré, A. & Vining, B. (eds) *Petroleum Geology: North-West Europe and Global Perspectives – Proceedings of the 6th Petroleum Geology Conference*. Geological Society, London, 461–468, https://doi.org/10.1144/0060461

Doornenbal, J.C. & Stevenson, A.G. 2010. *Petroleum Geological Atlas of the Southern Permian Basin Area*. European Association of Geoscientists and Engineers (EAGE), Houten, The Netherlands.

Dumas, P. & Bartosik, A. 2014. Geothermal District Heating potential in Europe. *Geothermal District Heating (GeoDH)*, European Geothermal Energy Council (EGEC), Brussels, http://www.geodh.eu

EBN 2017. *EBN treedt toe tot de Green Deal Geothermie Brabant*, EBN, Utrecht, The Netherlands, https://www.ebn.nl/ebn-treedt-toe-tot-green-deal-geothermie-brabant/

EUGENO-S Working Group 1988. Crustal structure and tectonic evolution of the transition between the Baltic Shield and the North German Caledonides (the Eugeno-S Project). *Tectonophysics*, **150**, 253–348.

Faber, F. 1959. De Winterswijkse Muschelkalk. *Netherlands Journal of Geoscience*, **21**, 25–31.

Fontaine, J., Guastella, G., Jouault, P. & de la Vega, P. 1993. F15-A: a Triassic gas field on the eastern limit of the Dutch Central Graben. *In*: Parker, J. (ed.) *Petroleum Geology of Northwest Europe: Proceedings of the 4th Conference*. Geological Society, London, 583–593, https://doi.org/10.1144/0040583

Franz, M., Barth, G., Zimmermann, J., Budach, I., Nowak, K. & Wolfgramm, M. 2018. Geothermal resources of the North German Basin: exploration strategy, development examples and remaining opportunities in Mesozoic hydrothermal reservoirs. *In*: Kilhams, B., Kukla, P.A., Mazur, S., McKie, T., Mijnlieff, H.F. & van Ojik, K. (eds) *Mesozoic Resource Potential in the Southern Permian Basin*. Geological Society, London, Special Publications, **469**. First published online April 13, 2018, https://doi.org/10.1144/SP469.11

Fryberger, S., Knight, R., Hern, C., Moscariello, A. & Kabel, S. 2011. Rotliegend facies, sedimentary provinces, and stratigraphy, Southern Permian Basin UK and the Netherlands: A review with new observations. *In*: Grötsch, J. & Gaupp, R. (eds) *The Permian Rotliegend of the Netherlands*. SEPM, Special Publications, **98**, 51–88.

Gast, R. 1988. Rifting im Rotliegenden Niedersachsens. *Die Geowissenschaften*, **4**, 115–122.

Geluk, M. 2005. *Stratigraphy and tectonics of Permo-Triassic basins in the Netherlands and surrounding areas*. PhD thesis, Utrecht University, Utrecht, The Netherlands.

Geluk, M., McKie, T. & Kilhams, B. 2018. An introduction to the Triassic: current insights into the regional setting and energy resource potential of NW Europe. *In*: Kilhams, B., Kukla, P.A., Mazur, S., McKie, T., Mijnlieff, H.F. & van Ojik, K. (eds) *Mesozoic Resource Potential in the Southern Permian Basin*. Geological Society, London, Special Publications, **469**. First published online January 26, 2018, https://doi.org/10.1144/SP469.1

Geological Society 2016. *Mesozoic Resource Potential in the Southern Permian Basin, Conference Abstract Book*. Geological Society, London, https://www.geolsoc.org.uk/~/media/shared/documents/specialist%20and%20regional%20groups/petroleum/Mesozoic%20Resource%20Potential%20in%20the%20Southern%20Permian%20Basin%20Abstract%20book.pdf?la=en

Gibbard, P. & Lewin, J. 2016. Filling the North Sea Basin: cenozoic sediment sources and river styles. *Geologica Belgica*, **19**, 201–217.

Goswami, R., Seeberger, F.C. & Bosman, G. 2018. Enhanced gas recovery of an ageing field utilising N_2 displacement: De Wijk Field, The Netherlands. *In*: Kilhams, B., Kukla, P.A., Mazur, S., McKie, T., Mijnlieff, H.F. & van Ojik, K. (eds) *Mesozoic Resource Potential in the Southern Permian Basin*. Geological Society, London, Special Publications, **469**. First published online January 16, 2018, https://doi.org/10.1144/SP469.2

Grad, M., Guterch, A. & Mazur, S. 2002. Seismic refraction evidence for crustal structure in the central part of the Trans-European Suture Zone in Poland. *In*: Winchester, J.A., Pharaoh, T.C. & Verniers, J. (eds) *Palaeozoic Amalgamation of Central Europe*. Geological Society, London, Special Publications, **201**, 295–309, https://doi.org/10.1144/GSL.SP.2002.201.01.14

Grad, M., Tiira, T. & ESC Working Group 2009. The Moho depth map of the European Plate. *Geophysical Journal International*, **176**, 279–292.

Grötsch, J., Sluijk, A., Van Ojik, K., De Keijzer, M., Graaf, J. & Steenbrink, J. 2011. The Groningen Gas Field: Fifty years of exploration and gas production from a Permian dryland reservoir. *In*: Grötsch, J. & Gaupp, R. (eds) *The Permian Rotliegend of the Netherlands*. SEPM, Special Publications, **98**, 11–33.

Hernandez, K., Mitchell, N.C. & Huuse, M. 2018. Deriving relationships between diapir spacing and salt-layer thickness in the Southern North Sea. *In*: Kilhams, B., Kukla, P.A., Mazur, S., McKie, T., Mijnlieff, H.F. & van Ojik, K. (eds) *Mesozoic Resource Potential in the Southern Permian Basin*. Geological Society, London, Special Publications, **469**. First published online February 26, 2018, https://doi.org/10.1144/SP469.16

Holdsworth, R.E., Woodcock, N.H. & Strachan, R.A. 2002. Geological framework of Britain and Ireland. *In*: Woodcock, N.H. & Strachan, R. (eds) *Geological History of Britain and Ireland*. Blackwell, Oxford, 19–37.

Huuse, M. 2002. Cenozoic uplift and denudation of southern Norway: insights from the North Sea Basin. *In*: Doré, A., Cartwright, J., Stoker, M., Turner, J. & White, N. (eds) *Exhumation of the North Atlantic Margin: Timing, Mechanism and Implications for Petroleum Exploration*. Geological Society, London, Special Publications, **196**, 209–233, https://doi.org/10.1144/GSL.SP.2002.196.01.13

iDeals 2016. Kimmeridge Energy sells German E&P arm to Petroleum Equity, https://energy.ideals.net/feed/20b4dba7-ce35-407d-b048-cbe040543c58

Jeremiah, J., Duxbury, S. & Rawson, P. 2010. Lower Cretaceous of the southern North Sea Basins: reservoir distribution within a sequence stratigraphic framework. *Netherlands Journal of Geosciences*, **89**, 203–237.

JOHNSON, C. & BOERSMA, T. 2013. Energy (in)security in Poland the case of shale gas. *Energy Policy*, **53**, 389–399.

JORGENSEN, L. 1992. Dan field – Denmark, Central Graben, Danish North Sea. *In*: BEAUMONT, E. (ed.) *Structural Traps VI*. AAPG Treatise of Petroleum Geology: Atlas of Oil and Gas Fields. American Association of Petroleum Geologists (AAPG), Tulsa, OK, USA, 199–218.

KABAN, M.K., TESAURO, M. & CLOETINGH, S. 2010. An integrated gravity model for Europe's crust and upper mantle. *Earth and Planetary Science Letters*, **296**, 195–209.

KILHAMS, B., STEVANOVIC, S. & NICOLAI, C. 2018. The 'Buntsandstein' gas play of the Horn Graben (German, Danish offshore): dry well analysis, remaining hydrocarbon potential. *In*: KILHAMS, B., KUKLA, P.A., MAZUR, S., MCKIE, T., MIJNLIEFF, H.F. & VAN OJIK, K. (eds) *Mesozoic Resource Potential in the Southern Permian Basin*. Geological Society, London, Special Publications, **469**. First published online January 11, 2018, https://doi.org/10.1144/SP469.5

KLEY, J. 2018. Timing and spatial patterns of Cretaceous and Cenozoic inversion in the Southern Permian Basin. *In*: KILHAMS, B., KUKLA, P.A., MAZUR, S., MCKIE, T., MIJNLIEFF, H.F. & VAN OJIK, K. (eds) *Mesozoic Resource Potential in the Southern Permian Basin*. Geological Society, London, Special Publications, **469**. First published online March 9, 2018, https://doi.org/10.1144/SP469.12

KLEY, J. & VOIGT, T. 2008. Late Cretaceous intraplate thrusting in Central Europe: effect of Africa–Europe–Iberia convergence, not Alpine collision. *Geology*, **36**, 839–842.

KNOX, R., BOSCH, J. *ET AL*. 2010. Cenozoic. *In*: DOORNENBAL, J.C. & STEVENSON, A.G. (eds) *Petroleum Geological Atlas of the Southern Permian Basin Area*. European Association of Geoscientists and Engineers (EAGE), Houten, The Netherlands, 211–233.

KOMBRINK, H., BESLY, B. *ET AL*. 2010. Carboniferous. *In*: DOORNENBAL, J.C. & STEVENSON, A.G. (eds) *Petroleum Geological Atlas of the Southern Permian Basin Area*. European Association of Geoscientists and Engineers (EAGE), Houten, The Netherlands, 81–99.

KORTEKAAS, M., BÖKER, U., VAN DER KOOIJ, C. & JAARSMA, B. 2018. Lower Triassic reservoir development in the Dutch northern offshore. *In*: KILHAMS, B., KUKLA, P.A., MAZUR, S., MCKIE, T., MIJNLIEFF, H.F. & VAN OJIK, K. (eds) *Mesozoic Resource Potential in the Southern Permian Basin*. Geological Society, London, Special Publications, **469**. First published online April 13, 2018, https://doi.org/10.1144/SP469.19

KRAWCZYK, C.M., MCCANN, T., COCKS, L.R.M., ENGLAND, R.W., MCBRIDE, J.H. & WYBRANIEC, S. 2008. Caledonian tectonics. *In*: MCCANN, T. (ed.) *The Geology of Central Europe. Volume 1: Precambrian and Palaeozoic*. Geological Society, London, 303–381.

KRONER, U., MANSY, J.-L. *ET AL*. 2008. Variscan tectonics. *In*: MCCANN, T. (ed.) *The Geology of Central Europe: Precambrian and Palaeozoic*. Geological Society, London, 559–664.

KRZYWIEC, P., STACHOWSKA, A. & STYPA, A. 2018. The only way is up – on Mesozoic uplifts and basin inversion events in SE Poland. *In*: KILHAMS, B., KUKLA, P.A., MAZUR, S., MCKIE, T., MIJNLIEFF, H.F. & VAN OJIK, K. (eds) *Mesozoic Resource Potential in the Southern Permian Basin*. Geological Society, London, Special Publications, **469**. First published online March 20, 2018, https://doi.org/10.1144/SP469.14

KUS, J., CRAMER, B. & KOCKEL, F. 2005. Effects of a Cretaceous structural inversion and a postulated high heat flow event on petroleum system of the western Lower Saxony Basin and the charge history of the Apeldorn gas field. *Netherlands Journal of Geosciences*, **84**, 3–24.

LAKE, L. 2010. *Enhanced Oil Recovery*. Society of Petroleum Engineers, Richardson, TX, USA.

LITTKE, R., BAYER, U., GAJEWSKI, D. & NELSKAMP, S. 2008. *Dynamics of Complex Intracontinental Basins: The Central European Basin System*. Springer, Berlin.

LOTT, G., WONG, T., DUSAR, M., ANDSBJERG, J., MÖNNIG, E., FELDMAN-OLSZEWSKA, A. & VERREUSSEL, R. 2010. Jurassic. *In*: DOORNENBAL, J.C. & STEVENSON, A.G. (eds) *Petroleum Geological Atlas of the Southern Permian Basin Area*. European Association of Geoscientists and Engineers (EAGE), Houten, The Netherlands, 175–193.

MACROTRENDS 2017. Historical Brent crude price, http://www.macrotrends.net/2503/brent-crude-oil-prices-historical-chart

MARKETWIRED 2015. PRD Energy announces operational update, http://www.marketwired.com/press-release/prd-energy-announces-operational-update-2012610.htm

MAYSTRENKO, Y., BAYER, U. & SCHECK-WENDEROTH, M. 2005. The Glueckstadt Graben, a sedimentary record between the North and Baltic Sea in north Central Europe. *Tectonophysics*, **397**, 113–126.

MAYSTRENKO, Y., BAYER, U., BRINK, H.-J. & LITTKE, R. 2008. The Central European Basin System – an overview. *In*: LITTKE, R., BAYER, U., GAJEWSKI, D. & NELSKAMP, S. (eds) *Dynamics of Complex Intracontinental Basins: The Central European Basin System*. Springer, Berlin, 15–34.

MAYSTRENKO, Y. & SCHECK-WENDEROTH, M. 2013. 3D lithosphere-scale density model of the Central European Basin System and adjacent areas. *Tectonophysics*, **601**, 53–77.

MCARTHUR, J.M., ALGEO, T.J., VAN DE SCHOOTBRUGGE, B., LI, Q. & HOWARTH, R.J. 2008. Basinal restriction, black shales, Re–Os dating, and the Early Toarcian (Jurassic) oceanic anoxic event. *Paleoceanography*, **23**, PA4217.

MCCANN, T. 2008*a*. *The Geology of Central Europe. Volume 1: Precambrian and Palaeozoic*. Geological Society, London.

MCCANN, T. 2008*b*. *The Geology of Central Europe. Volume 2: Mesozoic and Cenozoic*. Geological Society, London.

MCCANN, T. 2008*c*. Introduction and overview. *In*: MCCANN, T. (ed.) *The Geology of Central Europe. Volume 1: Precambrian and Palaeozoic*. Geological Society, London, 1–20.

MCKIE, T. 2017. Palaeogeographic evolution of latest Permian and Triassic salt basins in North West Europe. *In*: SOTO, J., FLINCH, J. & TARI, G. (eds) *Permo-Triassic salt provinces of Europe, North Africa and the Atlantic Margins: Tectonics and Hydrocarbon Potential*. Elsevier, Amsterdam.

MEGSON, J. 1992. The North Sea Chalk Play: examples from the Danish Central Graben. *In*: HARDMAN, R. (ed.)

Exploration Britain. Geological Insights for the Next Decade. Geological Society, London, Special Publications, **67**, 247–282, https://doi.org/10.1144/GSL.SP.1992.067.01.10

Megson, J. & Tygesen, T. 2005. The North Sea Chalk: an underexplored and underdeveloped play. *In*: Doré, A. & Vining, B. (eds) *Petroleum Geology: North-West Europe and Global Perspectives – Proceedings of the 6th Petroleum Geology Conference*. Geological Society, London, 159–168, https://doi.org/10.1144/0060159

Miller, K., Kominz, M. *et al.* 2005. The phanerozoic record of global sea-level change. *Science*, **310**, 1293–1298.

Milton-Worssell, R.J., Stoker, S.J. & Cavill, J.E. 2006. Lower Cretaceous deep-water sandstone plays in the UK Central Graben. *In*: Allen, M., Goffey, G., Morgan, R. & Walker, I. (eds) *The Deliberate Search for the Stratigraphic Trap*. Geological Society, London, Special Publications, **254**, 169–189, https://doi.org/10.1144/GSL.SP.2006.254.01.09

Ministrie v. Economische Zaken 2016. Jaarverslag 2016, http://www.nlog.nl/jaarverslagen

Moreau, J., Huuse, M., Janszen, A., van der Vegt, P., Gibbard, P.L. & Moscariello, A. 2012. The glaciogenic unconformity of the southern North Sea. *In*: Huuse, M., Redfern, J., Le Heron, D., Dixon, R., Moscariello, A. & Craig, J. (eds) *Glaciogenic Reservoirs and Hydrocarbon Systems*. Geological Society, London, Special Publications, **368**, 99–110, https://doi.org/10.1144/SP368.5

Munsterman, D., Verreussel, R., Mijnlieff, H., Witmans, N., Kerstholt-Boegehold, S. & Abbink, O. 2012. Revision and update of the Callovian–Rhyazanian stratigraphic nomenclature in the northern Dutch offshore, i.e. Central Graben Subgroup and Scruff Group. *Netherlands Journal of Geosciences*, **91**, 555–590.

Nelskamp, S. & Verweij, J. 2012. *Using Basin Modelling for Geothermal Energy Exploration in the Netherlands – An Example from the West Netherlands Basin and Roer Valley Graben*. TNO Report, TNO-060-UT-2012-00245, http://nlog.nl/papers

ONE 2014. Oranje-Nassau Energie B.V. has successfully appraised their M07-B Jurassic field, http://www.onebv.com/node/216

Oschmann, W. 1991. Distribution, dynamics and palaeoecology of Kimmeridgian (Upper Jurassic) shelf anoxia in western Europe. *In*: Tyson, R. & Pearson, T. (eds) *Modern and Ancient Continental Shelf Anoxia*. Geological Society, London, Special Publications, **58**, 381–395, https://doi.org/10.1144/GSL.SP.1991.058.01.24

Peeters, S., Asschert, A. & Verweij, H. 2018. Towards a better understanding of the highly overpressured Lower Triassic Bunter reservoir rocks in the Terschelling Basin. *In*: Kilhams, B., Kukla, P.A., Mazur, S., McKie, T., Mijnlieff, H.F. & van Ojik, K. (eds) *Mesozoic Resource Potential in the Southern Permian Basin*. Geological Society, London, Special Publications, **469**. First published online March 22, 2018, https://doi.org/10.1144/SP469.13

Peryt, T.M., Geluk, M.C., Mathiesen, A., Paul, J. & Smith, K. 2010. Zechstein. *In*: Doornenbal, J.C. & Stevenson, A.G. (eds) *Petroleum Geological Atlas of the Southern Permian Basin Area*. European Association of Geoscientists and Engineers (EAGE), Houten, The Netherlands, 123–147.

Peta 2015. Pan European Thermal Atlas, http://maps.heatroadmap.eu/berndmoller/maps/31157/Renewable-Res ources-Map-for-EU28?preview=true#;

Pharaoh, T.C., Dusar, M. *et al.* 2010. Tectonic evolution. *In*: Doornenbal, J.C. & Stevenson, A.G. (eds) *Petroleum Geological Atlas of the Southern Permian Basin Area*. European Association of Geoscientists and Engineers (EAGE), Houten, The Netherlands, 25–57.

Pieńkowski, G. & Schudak, M. 2008. The Jurassic of Central Europe. *In*: McCann, T. (ed.) *The Geology of Central Europe. Volume 2: Mesozoic and Cenozoic*. Geological Society, London, 823–922.

Pierce, C.R., Cohen, A.S., Coe, A.L. & Burton, K.W. 2008. Molybdenum isotope evidence for global ocean anoxia coupled with perturbations to the carbon cycle during the Early Jurassic. *Geology*, **36**, 231–234.

Platform Geothermie 2017. Platform Geothermie. *Projecten in Nederland, Van den Bosch*. Platform Geothermie, Delft, The Netherlands, https://www.geothermie.nl/index.php/nl/geothermie-aardwarmte/projecten-in-nederland/38-van-den-bosch-1-2

Pletsch, T., Appel, J. *et al.* 2010. Petroleum generation and migration. *In*: Doornenbal, J.C. & Stevenson, A.G. (eds) *Petroleum Geological Atlas of the Southern Permian Basin Area*. European Association of Geoscientists and Engineers (EAGE), Houten, The Netherlands, 225–253.

Pluymaekers, M., Kramers, L., van Wees, J.-D., Kronimus, A., Nelskamp, S., Boxem, T. & Bonté, D. 2012. Reservoir characterisation of aquifers for direct heat production: methodology and screening of the potential reservoirs for the Netherlands. *Netherlands Journal of Geosciences*, **91**, 621–636.

Porter, R.J., Munoz Rojas, A.M. & Schlüter, M. 2018. The impact of heterogeneity on waterflood developments in clastic inner shelf reservoirs: an example from the Holland Greensand Member, Rotterdam Field, The Netherlands. *In*: Kilhams, B., Kukla, P.A., Mazur, S., McKie, T., Mijnlieff, H.F. & van Ojik, K. (eds) *Mesozoic Resource Potential in the Southern Permian Basin*. Geological Society, London, Special Publications, **469**. First published online April 16, 2018, https://doi.org/10.1144/SP469.20

Ramkumar, M. 2016. *Cretaceous Sea Level Rise*. Elsevier, Amsterdam.

Rasmussen, E. & Dybkjaer, K. 2014. Patterns of Cenozoic sediment flux from Western Scandinavia: discussion. *Basin Research*, **26**, 338–346.

Ritchie, J. & Pratsides, P. 1993. The Caister Fields, Block 44/23a, UK North Sea. *In*: Parker, J. (ed.) *Petroleum Geology of Northwest Europe: Proceedings of the 4th Conference*. Geological Society, London, 759–769, https://doi.org/10.1144/0040759

Rosenbaum, G., Lister, G.S. & Duboz, C. 2002. Relative motion of Africa, Iberia and Europe during the Alpine orogeny. *Tectonophysics*, **359**, 117–129.

Sachse, V.F. & Littke, R. 2018. The impact of Quarternary glaciation on temperature and pore pressure in Jurassic troughs in the Southern Permian Basin, northern Germany. *In*: Kilhams, B., Kukla, P.A., Mazur, S., McKie, T., Mijnlieff, H.F. & van Ojik, K. (eds) *Mesozoic Resource Potential in the Southern Permian Basin.* Geological Society, London, Special Publications, **469**. First published online January 16, 2018, https://doi.org/10.1144/SP469.7

Schulz, H., Horsfield, B. & Sachsenhofer, R. 2010. Shale gas in Europe: a regional overview and current research activities. *In*: Vining, B. & Pickering, S. (eds) *Petroleum Geology: From Mature Basins to New Frontiers – Proceedings of the 7th Petroleum Geology Conference.* Geological Society, London, 1079–1085, https://doi.org/10.1144/0071079

Schwarzkopf, T. 1990. Relationship between petroleum generation, migration and sandstone diagenesis, Middle Jurassic, Gifhorn Trough, N. Germany. *Marine and Petroleum Geology*, **7**, 153–170.

Seidel, E., Meschede, M. & Obst, K. 2018. The Wiek Fault System east of Rügen Island: origin, tectonic phases and its relationship to the Trans-European Suture Zone. *In*: Kilhams, B., Kukla, P.A., Mazur, S., McKie, T., Mijnlieff, H.F. & van Ojik, K. (eds) *Mesozoic Resource Potential in the Southern Permian Basin.* Geological Society, London, Special Publications, **469**. First published online January 24, 2018, https://doi.org/10.1144/SP469.10

Selley, R. 2012. UK Shale gas: the story so far. *Marine and Petroleum Geology*, **31**, 100–109.

Smit, J., van Wees, J.D. & Cloetingh, S. 2016. The Thor suture zone: from subduction to intraplate basin setting. *Geology*, **44**, 707–710.

SPE 2015. Mature fields. *PetroWiki*, http://petrowiki.org/Mature_fields

Steuber, T., Rauch, M., Masse, J.-P., Graaf, J. & Malkoč, M. 2005. Low-latitude seasonaility of Cretaceous temperatures in warm and cold episodes. *Nature*, **437**, 1341–1344.

Stock, A.T. & Littke, R. 2018. The Posidonia Shale of northern Germany: unconventional oil and gas potential from high-resolution 3D numerical basin modelling of the cross-junction between the eastern Lower Saxony Basin, Pompeckj Block and Gifhorn Trough. *In*: Kilhams, B., Kukla, P.A., Mazur, S., McKie, T., Mijnlieff, H.F. & van Ojik, K. (eds) *Mesozoic Resource Potential in the Southern Permian Basin.* Geological Society, London, Special Publications, **469**. First published online April 5, 2018, https://doi.org/10.1144/SP469.21

Strozyk, F., Reuning, L, Back, S. & Kukla, P. 2018. Giant pockmark formation from Cretaceous hydrocarbon expulsion in the western Lower Saxony Basin, The Netherlands. *In*: Kilhams, B., Kukla, P.A., Mazur, S., McKie, T., Mijnlieff, H.F. & van Ojik, K. (eds) *Mesozoic Resource Potential in the Southern Permian Basin.* Geological Society, London, Special Publications, **469**. First published online January 11, 2018, https://doi.org/10.1144/SP469.6

Thybo, H. 1997. Geophysical characteristics of the Tornquist Fan area, northwest Trans-European Suture Zone: indication of late Carboniferous to early Permian dextral transtension. *Geological Magazine*, **134**, 597–606.

Thybo, H. 2001. Crustal structure along the EGT profile across the Tornquist Fan interpreted from seismic, gravity and magnetic data. *Tectonophysics*, **334**, 155–190.

TNO 2013. *ThermoGIS*. TNO, Utrecht, The Netherlands, http://www.thermogis.nl/

van Hulten, F.F.N. 2009. Brief history of petroleum exploration in the Netherlands. Conference note from the symposium Fifty Years of Petroleum Exploration in the Netherlands after the Groningen Discovery, 15–16 January 2009, Utrecht, The Netherlands, http://www.f-van-hulten.com/Geology/van_Hulten_2009.pdf

van Kempen, B.M.M., Mijnlieff, H.F. & van der Molen, J. 2018. Data mining in the Dutch Oil and Gas Portal: a case study on the reservoir properties of the Volpriehausen Sandstone. *In*: Kilhams, B., Kukla, P.A., Mazur, S., McKie, T., Mijnlieff, H.F. & van Ojik, K. (eds) *Mesozoic Resource Potential in the Southern Permian Basin.* Geological Society, London, Special Publications, **469**. First published online March 15, 2018, https://doi.org/10.1144/SP469.15

van Lochem, H. 2018. F17-Chalk: new insights in the tectonic history of the Dutch Central Graben. *In*: Kilhams, B., Kukla, P.A., Mazur, S., McKie, T., Mijnlieff, H.F. & van Ojik, K. (eds) *Mesozoic Resource Potential in the Southern Permian Basin.* Geological Society, London, Special Publications, **469**. First published online February 19, 2018, https://doi.org/10.1144/SP469.17

van Winden, M., de Jager, J., Jaarsma, B., & Bouroullec, R. 2018. New insights into salt tectonics in the northern Dutch offshore: a framework for hydrocarbon exploration. *In*: Kilhams, B., Kukla, P.A., Mazur, S., McKie, T., Mijnlieff, H.F. & van Ojik, K. (eds) *Mesozoic Resource Potential in the Southern Permian Basin.* Geological Society, London, Special Publications, **469**. First published online January 29, 2018, https://doi.org/10.1144/SP469.9

Vejbaek, O., Andersen, C. et al. 2010. Cretaceous. *In*: Doornenbal, J.C. & Stevenson, A.G. (eds) *Petroleum Geological Atlas of the Southern Permian Basin Area.* European Association of Geoscientists and Engineers (EAGE), Houten, The Netherlands, 195–209.

Verreussel, R.M.C.H., Bouroullec, R. et al. 2018. Stepwise basin evolution of the Middle Jurassic–Early Cretaceous rift phase in the Central Graben area of Denmark, Germany and The Netherlands. *In*: Kilhams, B., Kukla, P.A., Mazur, S., McKie, T., Mijnlieff, H.F. & van Ojik, K. (eds) *Mesozoic Resource Potential in the Southern Permian Basin.* Geological Society, London, Special Publications, **469**. First published online March 15, 2018, https://doi.org/10.1144/SP469.23

Verdier 1996. The Rotliegend sedimentation history of the southern North Sea and adjacent countries. *In*: Rondeel, H., Batjes, D. & Nieuwenhuijs, W. (eds) *Oil and Gas under the Netherlands.* Kluwer Academic, Dordrecht, The Netherlands, 45–56.

Vis, G.-J., Smoor, W.D., Rutten, K.W., de Jager, J. & Mijnlieff, H.F. 2018. Tectonic control on the Early Cretaceous Bentheim Sandstone sediments in the Schoonebeek oil field, The Netherlands. *In*: Kilhams, B., Kukla, P.A., Mazur, S., McKie, T., Mijnlieff, H.F. & van Ojik, K. (eds) *Mesozoic Resource Potential in the Southern Permian Basin.* Geological Society,

London, Special Publications, **469**. First published online March 15, 2018, https://doi.org/10.1144/SP469.25

VLEESTOMAT 2017. Tomatoes grown with renewable energy. http://vleestomaat.nl/index.php/onze-kwekerij/4-aardwarmte

VONDRAK, A.G., DONSELAAR, M.E. & MUNSTERMAN, D.K. 2018. Reservoir architecture model of the Nieuwerkerk Formation (Early Cretaceous, West Netherlands Basin): diachronous development of sand-prone fluvial deposits. *In*: KILHAMS, B., KUKLA, P.A., MAZUR, S., MCKIE, T., MIJNLIEFF, H.F. & VAN OJIK, K. (eds) *Mesozoic Resource Potential in the Southern Permian Basin*. Geological Society, London, Special Publications, **469**. First published online March 15, 2018, https://doi.org/10.1144/SP469.18

WASSMANN, T. & BROUWER, M. 1987. The mining of rock salt. *In*: VISSER, W., ZONNEVELD, J. & VAN LOON, A. (eds) *Seventy-Five Years of Geology and Mining in the Netherlands*. Royal Geological and Mining Society of the Netherlands, Den Haag, 137–146.

WEIJERMARS, R. 2013. Economic appraisal of shale gas plays in Continental Europe. *Applied Energy*, **106**, 100–115.

WINSTANLEY, A. 1993. A review of the Triassic play in the Roer Valley Graben, SE onshore Netherlands. *In*: PARKER, J. (ed.) *Petroleum Geology of Northwest Europe: Proceedings of the 4th Conference*. Geological Society, London, 595–607, https://doi.org/10.1144/0040595

WOLF, M., VIS, A. & ASSCHERT, A. 2018. Erosional valleys at a major Late Jurassic–Early Cretaceous unconformity offshore Germany and The Netherlands: potential reservoirs or deteriorated seals? *In*: KILHAMS, B., KUKLA, P.A., MAZUR, S., MCKIE, T., MIJNLIEFF, H.F. & VAN OJIK, K. (eds) *Mesozoic Resource Potential in the Southern Permian Basin*. Geological Society, London, Special Publications, **469**. First published online January 25, 2018, https://doi.org/10.1144/SP469.4

WOLFGRAMM, M., OBST, K., BEICHEL, K., BRANDES, J., KOCH, R., RAUPPACH, K. & THORWART, K. 2009. Produktivitätsprognosen Geothermischer Aquifere in Deutschland. Presented at Der Geothermiekongress 2009, 17–19 November 2009, Bochum, Germany.

YEGEROVA, T., BAYER, U., THYBO, H., MAYSTRENKO, Y., SCHECK-WENDEROTH, M. & LYNGSIE, S.B. 2007. Gravity signals from the lithosphere in the Central European Basin System. *Tectonophysics*, **429**, 133–163.

ZIEGLER, P.A. 1990*a*. *Geological Atlas of Western and Central Europe*. 2nd edn. Shell Internationale Petroleum Maatschappij, The Hague. Geological Society, London.

ZIEGLER, P.A. 1990*b*. Tectonic and palaeogeographic development of the North Sea rift system. *In*: BLUNDELL, D.J. & GIBBS, A.D. (eds) *Tectonic Evolution of the North Sea Rifts*. Clarendon Press, Oxford, 1–36.

ZIEGLER, P.A. & DÈZES, P. 2006. Crustal evolution of Western and Central Europe. *In*: GEE, D.G. & STEPHENSON, R.A. (eds) *European Lithosphere Dynamics*. Geological Society, London, Memoirs, **32**, 43–56, https://doi.org/10.1144/GSL.MEM.2006.032.01.03

ZIELHUIS, A. & NOLET, G. 1994. Deep seismic expression of an ancient plate boundary in Europe. *Science*, **265**, 79–81.

ZWAAN, F. 2018. Lower Cretaceous reservoir development in the North Sea Central Graben, and potential analogue settings in the Southern Permian Basin and South Viking Graben. *In*: KILHAMS, B., KUKLA, P.A., MAZUR, S., MCKIE, T., MIJNLIEFF, H.F. & VAN OJIK, K. (eds) *Mesozoic Resource Potential in the Southern Permian Basin*. Geological Society, London, Special Publications, **469**. First published online January 4, 2018, https://doi.org/10.1144/SP469.3

Timing and spatial patterns of Cretaceous and Cenozoic inversion in the Southern Permian Basin

JONAS KLEY

Georg-August-Universität Göttingen, Geowissenschaftliches Zentrum, Goldschmidtstrasse 3, 37077 Göttingen, Germany

jonas.kley@geo.uni-goettingen.de

Abstract: Mesozoic extensional basins of the Southern Permian Basin (SPB) System became inverted from Late Cretaceous time onwards. Following a first Cretaceous 'Subhercynian' pulse of contractional deformation and basin uplift, several distinct inversion events of Cenozoic age were often described. The oldest of these is the 'Laramide' event of Paleocene age which coincides with the termination of chalk deposition and widespread regression around the North Sea Basin, whose axial part continued to subside. The spatial extent of these effects is too wide to be compatible with inversion by folding and reverse faulting. The width of the uplifting and subsiding regions was also too large to be consistent with folding of the entire lithosphere under tangential compression. There appears to be no unequivocal evidence of discrete structures formed or reactivated in the Laramide event. By contrast, well-documented younger inversion of approximately Late Eocene to Late Oligocene–(Miocene?) age affected the region from the Celtic Sea to the western Netherlands. The associated deformation is weaker than that of the Late Cretaceous event and spatially overlaps with it only in the Southern North Sea. Structural inversion of the SPB thus comprised only two events separated in time and mostly also in space.

The Southern Permian Basin (SPB) is a large intracontinental basin within the Central European Basin System. Its tectonic evolution is characterized by contrasting tectonic regimes: a long period of slow subsidence and intermittent extension of varying intensity and direction, lasting from the Permian to the Early Cretaceous, was followed by a second period where basin subsidence was punctuated by contraction and basin inversion beginning in middle Late Cretaceous time. A characteristic feature of the SPB is the distributed nature of both extensional and contractional deformation. A complex array of numerous faults covers much of the basin but, with a few exceptions, their offsets are less than 1 km and often much smaller. The areas affected by extension and later inversion largely coincide.

There is a consensus that the first inversion phase is bracketed between about 90 and 70 Ma, and was caused by increased far-field horizontal stress resulting from the convergence between Africa, Europe and intervening microplates. Many authors have argued that the Late Cretaceous pulse either continued into the Paleogene or was followed by several younger inversion events in Cenozoic time, as documented by unconformities. In this paper, characteristics of these events, such as wavelengths, amplitudes and associated structures, are first described. From a comparison of these characteristics with those predicted for different uplift mechanisms, it is proposed that there was only one Cenozoic event caused by horizontal contraction: thus, matching the definition of basin inversion. In particular, there seems to be no clear evidence of Paleocene 'Laramide' shortening. This tentative conclusion is based on a non-exhaustive literature review and the discussion of a published key seismic line. The main purpose of this short contribution is to present a working hypothesis and, hopefully, trigger a discussion. The emphasis is on structural and stratigraphic observations rather than (plate) tectonic interpretation.

Basin inversion v. basin uplift

It is often considered that four major, temporally distinct inversion events affected the SPB in Late Cretaceous through to Cenozoic time. These are the Subhercynian (Late Cretaceous), Laramide (mid-Paleocene), Pyrenean (end Eocene) and Savian (end Oligocene) phases (e.g. de Jager 2007). Since the pioneering work of Ziegler (e.g. 1987, 1989), they were commonly interpreted to be caused by temporally changing tangential forces arising from plate tectonic processes: in particular, the opening of the Atlantic Ocean and convergence/collision in the Mediterranean realm (e.g. Dèzes *et al.* 2004; Sissingh 2006).

Following the most widely accepted definition (Cooper & Williams 1989), the term 'basin inversion' is used herein to indicate a change from subsidence to uplift that is due to a change in the tectonic regime from extension to shortening. (The term

From: Kilhams, B., Kukla, P. A., Mazur, S., McKie, T., Mijnlieff, H. F. & van Ojik, K. (eds) 2018. *Mesozoic Resource Potential in the Southern Permian Basin*. Geological Society, London, Special Publications, **469**, 19–31.
First published online March 9, 2018, https://doi.org/10.1144/SP469.12

'negative inversion' for a change from shortening to extension never became widely used.) It makes sense to insist on the second, genetic, part of this definition because there are several configurations where a transition from subsidence to uplift is not the effect of a switch from extension to contraction (Fig. 1): if parts of a rift basin become elevated due to shoulder uplift, this is not inversion. If a segment of a foreland basin first subsides due to thrust loading and then is uplifted on top of another thrust sheet, this is not inversion either. The same would be true of uplift caused by delamination of the mantle lithosphere. Folding on the scale of the entire crust or lithosphere, although caused by tangential shortening, enhances the subsidence of existing depocentres and induces uplift of highs (Cloetingh 1986): thus, contrasting with the typical pattern of inversion.

Uplift events in the SPB

The Subhercynian event

The Subhercynian event of Late Cretaceous age caused uplift of extensional basins in a wide area of central Europe. Basin uplift was typically connected to folding and faulting, often involving the reverse reactivation of normal faults. New syntectonic depocentres and clastic fringes formed adjacent to uplifts (Voigt *et al.* 2008). Uplift and subsidence locally attained magnitudes of many kilometres. The shortening deformation affecting a faulted stratigraphic succession that contains multiple evaporitic levels in Permian and Triassic strata has often created structures of bewildering complexity when studied in detail. However, thin-skinned deformation on a large scale has not been documented, and neither the bulk extension nor the contraction accommodated across the SPB is thought to exceed a few tens of kilometres or *c.* 5% strain. Shortening during inversion was rather uniformly directed SSW–NNE (Kley & Voigt 2008).

The effects of the Subhercynian event are centred on a NW–SE-trending, 600 km-wide swath that fringes the Baltic Shield in the SW (Fig. 2). No major structures of Late Cretaceous age appear to be present west of the Southern/Central North Sea. The timing of the Subhercynian event, well constrained through stratigraphic ages in the basins (Badley *et al.* 1989; Baldschuhn *et al.* 2001; Kockel 2003; Voigt *et al.* 2004; Jackson *et al.* 2013) and thermochronological data from uplifting areas, is mostly bracketed between 90 and 70 Ma (Hejl *et al.* 1997; Thomson & Zeh 2000; Senglaub *et al.* 2005; von Eynatten *et al.* 2008; Danisik *et al.* 2010). Its onset varies between the Turonian and the Santonian. The termination of strong shortening is dated in the Lower Saxony and Sole Pit basins to within the Campanian by strata of Late Campanian age that unconformably overlie inversion structures (Badley *et al.* 1989; Kockel 2003). Very gentle folding of these strata, which are themselves overstepped by a Late Paleocene transgression, documents weak deformation in the Maastrichtian–Middle Paleocene interval (Baldschuhn *et al.* 2001). Information gleaned from an unpublished map of the most recent movements on faults in northern Germany (Brückner-Röhling *et al.* 2002) indicates that weak Paleocene–Eocene activity was focused on a swath just north of the

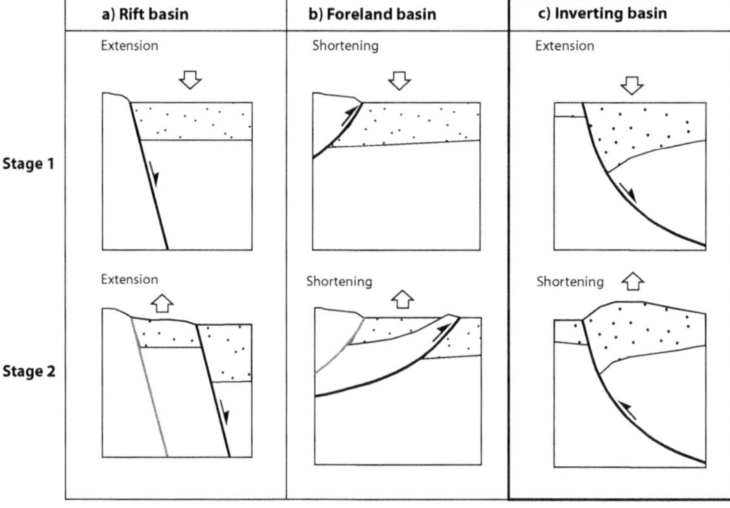

Fig. 1. Three different scenarios for a tectonically induced change from basin subsidence to basin uplift. Following common usage, the term 'inversion' is only used for (**c**) in this paper.

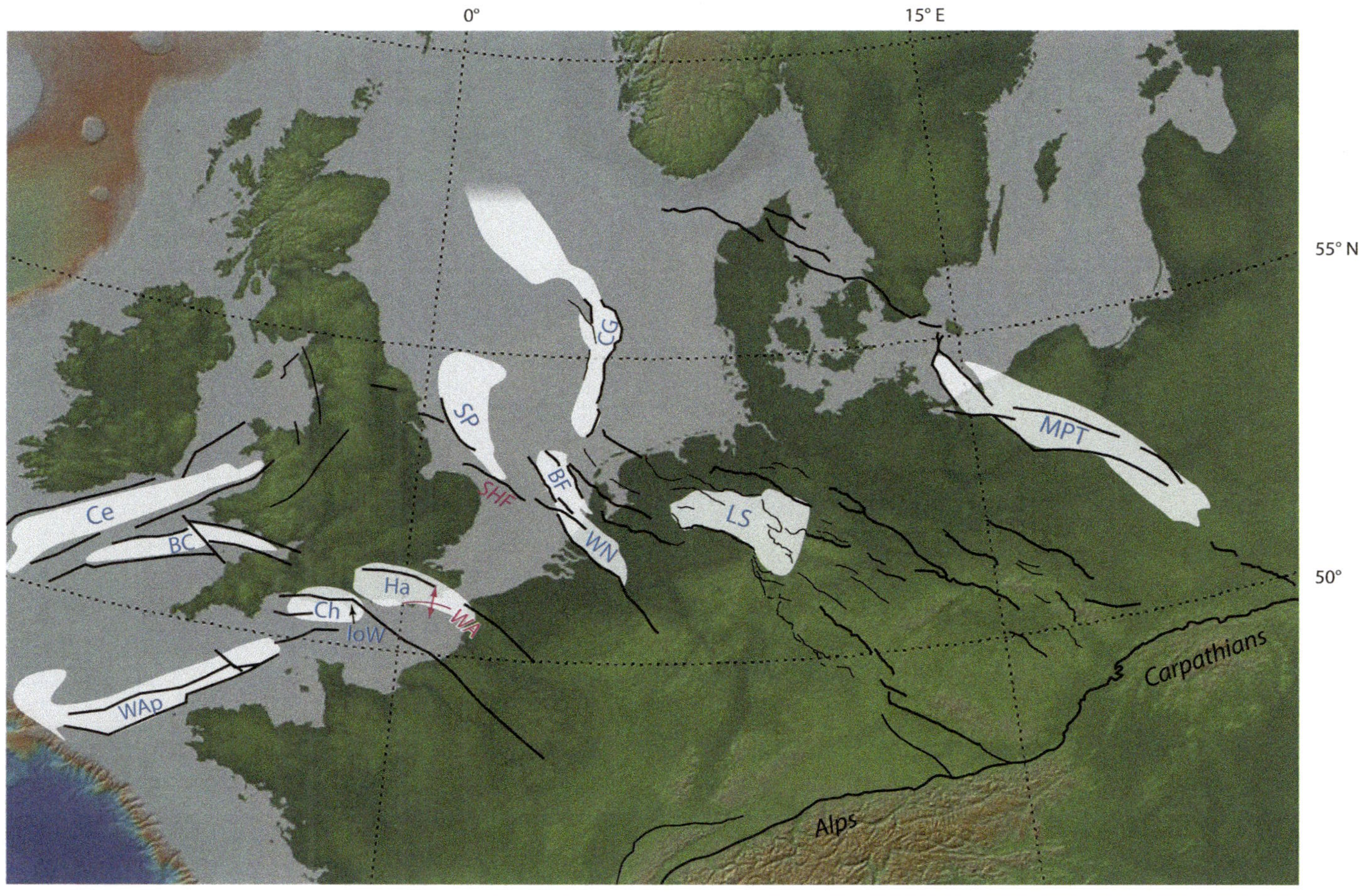

Fig. 2. Map of inversion structures in central and western Europe, based on Ziegler (1989), de Jager (2007), Kley & Voigt (2008) and Doornenbal & Stevenson (2010). Sub-basins, locations and structures mentioned in the text are highlighted. Basins: Ce, Celtic Sea; BC, Bristol Channel; WAp, Western Approaches; Ch, Channel; Ha, Hampshire; SP, Sole Pit; CG, North Sea Central Graben; BF, Broad Fourteens; WN, West Netherlands; LS, Lower Saxony; MPT, Mid-Polish Trough. Structures and locations: SHF, South Hewett Fault; WA, Weald Anticline; IoW, Isle of Wight. The topographical base is from Ryan *et al.* (2009) as implemented in GeoMappApp (http://www.geomapapp.org).

Lower Saxony Basin. No distinction is made on this map between reverse and normal faulting, but the oldest movements probably reflect contraction. Paleocene thickness patterns in the German North Sea sector still resemble those of the Late Cretaceous, although thickness variations are of much smaller amplitude (Arfai *et al.* 2014). Inversion may also continue into the early Paleogene in Poland (Krzywiec 2006).

While the Subhercynian event was long interpreted to be a far-field effect of collisional orogeny in the Alps (Ziegler 1987), Kley & Voigt (2008) used timing, shortening directions and plate tectonic reconstructions to argue that it resulted from the beginning of convergence between Africa, Iberia and Europe, and thus a global plate tectonic reorganization.

The Laramide event

The Laramide event of Middle Paleocene age (*c.* 60 Ma) is considered the strongest inversion phase in a number of Dutch basins by some authors (de Jager 2007; Doornenbal & Stevenson 2010) (Fig. 3). Between the early and late Paleocene, almost 2×10^6 km^2 of previously flooded western and central Europe became emergent and subject to erosion, and a distribution of land and sea similar to the present day was established. Thermochronological and thermal maturity data indicate an onset of regional Cenozoic exhumation in southern England and the adjacent North Sea around 63–59 Ma. The regression coincided with a transient, *c.* 30–50 m fall in global sea level (Haq *et al.* 1987; Miller *et al.* 2005), but the following transgression did not re-establish the extent of the Late Cretaceous sea. These effects are best observed by comparing palaeogeographical reconstructions before and after the Laramide event (Ziegler 1990) (Fig. 4). The uplift pattern is very different from the Subhercynian inversion event where strongly deforming areas became elevated but no wholesale regression took place. In the remaining North Sea Basin, chalk deposition was replaced by shale and sandy marginal facies. The area affected in total, including the subsiding central area and surrounding highs, was about 2000 × 1700 km in size. Apparently, there is no unambiguous evidence of structures linked to the Laramide uplift event. The classical maps by Ziegler (1989) and later interpretations by Dèzes *et al.* (2004) show Laramide deformation extending from the Celtic Sea to the Mid-Polish Trough. The area affected comprises upland terrains of central Europe where stratigraphic evidence is unavailable and the majority of thermochronological data indicate Late Cretaceous cooling, as well as regions where stratigraphic evidence suggests Laramide inversion is insignificant or absent (the Lower Saxony and West Netherlands basins, and the Southern North Sea; see above), and areas further west where inversion did not begin before the Eocene (the Hampshire, Channel, Western Approaches and Celtic Sea basins; see below). A more recent map (Doornenbal & Stevenson 2010, p. 216) shows Laramide inversion from the Southern North Sea to the Mid-Polish Trough. Even this smaller region includes large areas where the inversion is predominantly of Subhercynian age and proof of a distinct Laramide event is lacking (Mid-Polish Trough: Krzywiec 2006).

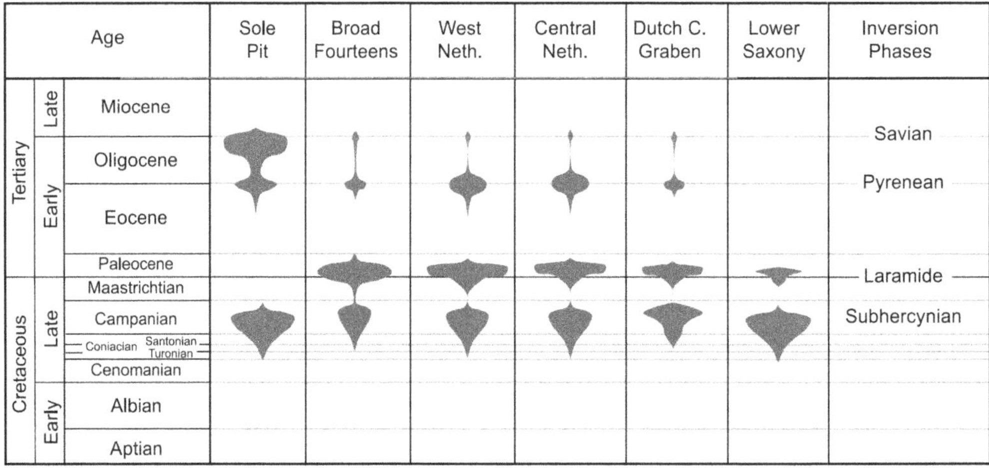

Fig. 3. The relative intensity and duration of inversion events proposed for several sub-basins of the SPB (after de Jager 2007). See Figure 2 for the locations.

(a)

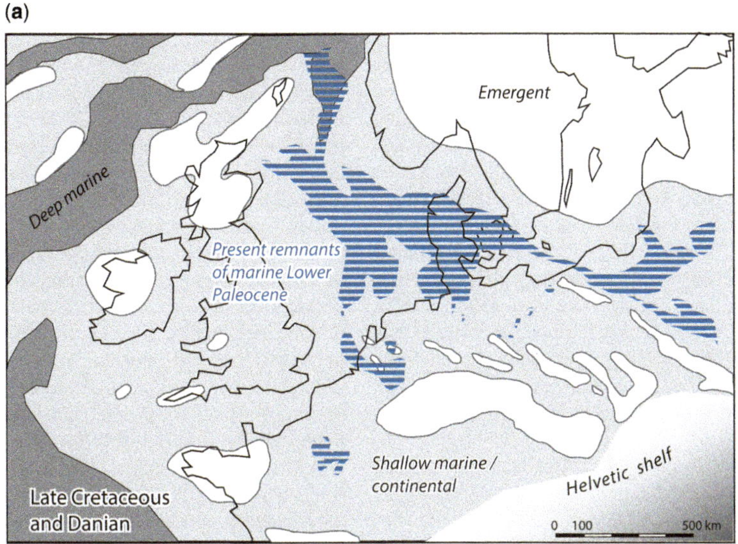

(b)

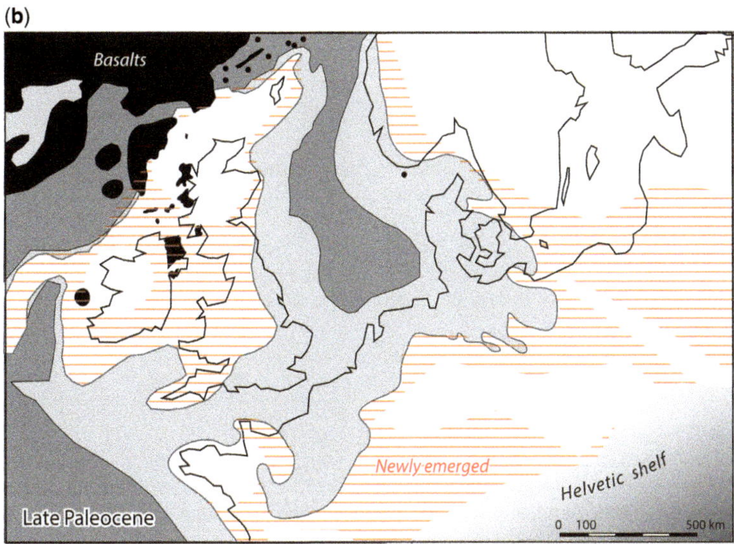

Fig. 4. Palaeogeographical maps showing the situation (**a**) before and (**b**) after Laramide uplift (maps modified from Ziegler 1990). The actual inner limit of uplift and erosion is constrained by preserved marine Early Paleocene strata (Danian and Montian: Gramann & Kockel 1988). The Late Paleocene map includes the effects of renewed flooding during the Landen transgression, which was substantially less extensive than in Late Cretaceous time.

The Pyrenean and Savian events

There is clear evidence for inversion structures formed in Eocene time and later. These occur in the southern British Isles and Channel region, the Southern North Sea as far north as the Sole Pit Basin, and in the western Dutch basins. The largest structure produced is the Weald Anticline, while the most spectacular outcrops are probably those of steeply dipping to vertical chalk and Eocene strata on the Isle of Wight. It has been demonstrated on seismic lines and from exposures on land that the anticlines formed through the inversion of Jurassic and Lower Cretaceous depocentres that were covered with a relatively uniform thickness of Upper Cretaceous chalk and Paleogene strata (Chadwick 1993). The main structures form a northwards convex arc, with strikes gradually changing from ENE in the west to ESE in the east. Fold-axis orientations and fault-slip data (Vandycke 2002) suggest approximately north–south-directed shortening, rotated counterclockwise by 20°–30° with respect to the shortening direction of Late Cretaceous inversion structures farther east. The structural relief attains around 1200 m in the largest anticlines (Chadwick 1993). The timing of Cenozoic inversion is well constrained in a few locations. Over the South Hewett Fault, the master fault of the Sole Pit Basin, a fold began to grow in late Eocene–Oligocene time (Badley *et al.* 1989). The growth strata are sealed by an intra-Miocene unconformity, indicating that inversion took place between late Eocene and middle Miocene time. Circumstantial evidence suggests that inversion probably did not reach far into the Miocene: around early Miocene time, the regional major horizontal stress rotated from approximately north–south to a NW–SE direction, which persists to the present (Dèzes *et al.* 2004; Bourgeois *et al.* 2007). The east–west-trending inversion structures of southern Britain are more consistent with the pre-Miocene stress field, with the possible exception of the WSW-striking Celtic Sea, Western Approaches and western Bristol Channel basins in the far west. The well-documented timing constraints extractable from the literature do not warrant the separation of distinct Pyrenean and Savian deformation events.

Case study: the Broad Fourteens and West Netherlands basins

The Broad Fourteens Basin in the offshore Netherlands is a classical inverted sub-basin of the SPB (van Wijhe 1987; Brun & Nalpas 1996) and is one of the areas considered to exhibit the effects of all four inversion events (Fig. 3) (de Jager 2007). Figure 5 shows a stratigraphic scheme and an interpreted seismic line across the Broad Fourteens Basin and the adjacent West Netherlands Basin reproduced from Doornenbal & Stevenson (2010). There, the line is presented as an example of 'Pyrenean' inversion without providing further explanations. Adopting the stratigraphic assignments from the original paper, a new interpretation of structures, unconformities and the timing of deformation events is provided in the following paragraph. The seismic line exhibits two clear angular unconformities and three groups of structures: the older unconformity separates folded and faulted Upper Cretaceous and older strata in the centre from a much less deformed succession above. Growth strata occur in the Upper Cretaceous chalk. The unconformity surface is located *c.* 100 ms below the Landen clay and grades into a non-angular contact on both ends of the line. The younger unconformity has Rupel clay as the youngest formation below the contact and Neogene strata above. The contact is only markedly angular in the SW portion of the seismic line. SW-prograding Neogene strata downlap onto it from above.

The first group of structures is a train of relatively tight folds in the centre. These folds are associated with many faults, some of which have substantial offsets. The second group of structures comprises folds of larger wavelength bordering the first set in the SW. No large faults are present in this area. The third group of structures consists of two fault zones offsetting Neogene and older strata with a down-to-the-SW sense of motion.

As regards the timing of deformation, the position of the older unconformity at some distance below the Landen clay and growth strata in the chalk suggests that the deformation predates the supposed Laramide event and is probably of Late Cretaceous (Subhercynian) age. A similar situation with Maastrichtian unconformably overlying older chalk was recently demonstrated in boreholes from the Southern North Sea (van Lochem 2018). There is no evidence in this seismic line of an angular unconformity between the Landen Clay Member and the directly underlying strata, and thus no evidence of Laramide inversion.

The second angular unconformity must be younger than the Oligocene Rupel Formation which underlies it in places. Below the unconformity, there is some evidence of growth folding in the Eocene Brussels and Asse members of the Dongen Formation; and the Rupel Formation overlaps both of them, as well as the substantially eroded Ieper Member, in the central part of the line. These observations indicate that the deformation spanned the Middle Eocene–Oligocene or even Miocene, depending on the basal age of the overlying Neogene strata. Evidence for long-lasting, slow inversion during this period has already been presented by Worum & Michon (2005). Comparison with growth strata in the chalk indicates that the folds of the second set resulted from the tightening of pre-existing folds of the first set. In contrast, the Broad Fourteens Basin, which had been strongly deformed in Late Cretaceous time, was uplifted with little internal deformation.

The structures of the third set are the only ones affecting Neogene strata. There they produce down-to-the-SW fault offsets or drape folding of upwards-diminishing amplitude. Large offsets in Cretaceous and older strata below indicate that these faults were active as SW-dipping reverse faults in Late Cretaceous time. The SW dip suggests that the Cenozoic offsets were caused by extension and normal faulting, not shortening. The overall configuration is one of two broad half-graben superimposed on a regional NE dip of the strata. Similar Neogene extensional reactivation of reverse faults has been reported from westernmost Germany (Doornenbal & Stevenson 2010, p. 44).

The structural evolution of the Broad Fourteens and West Netherlands basins can thus be summarized as follows: the first and strongest inversion of the Mesozoic extensional depocentres took place

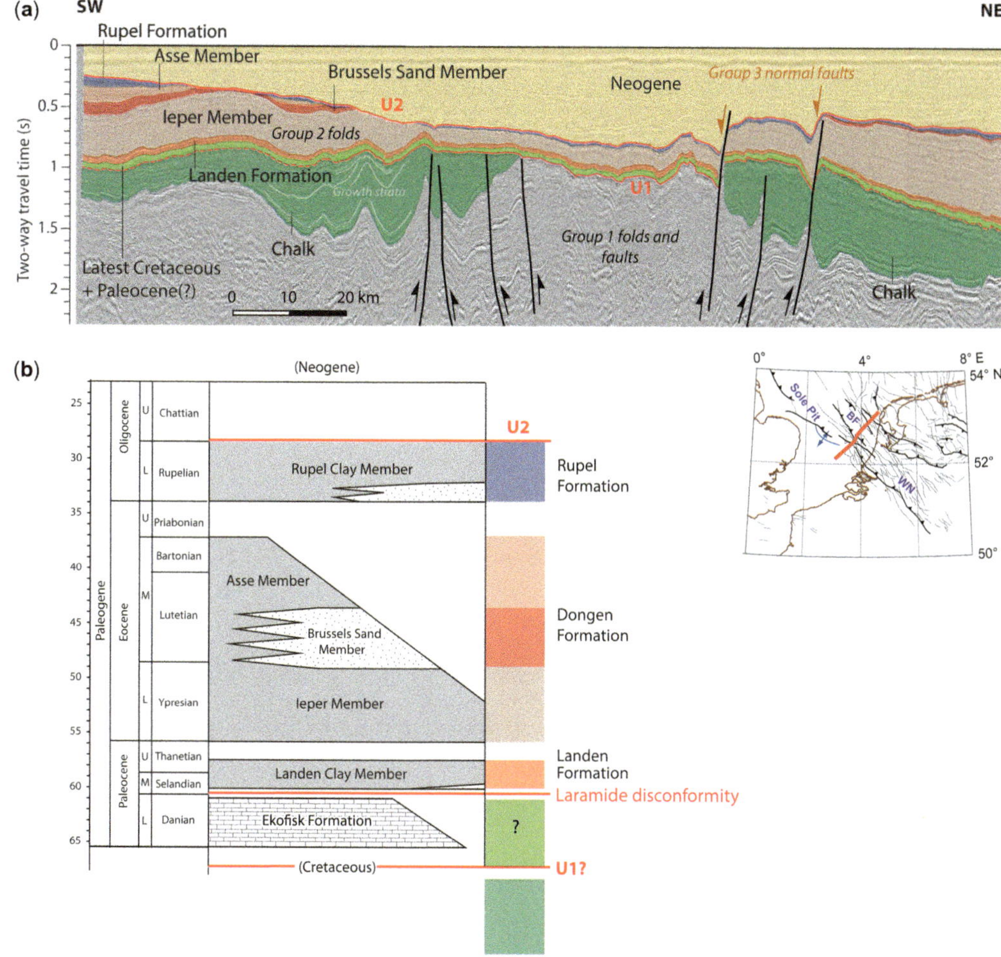

Fig. 5. (**a**) Interpreted seismic section across the West Netherlands and Broad Fourteens basins (modified from Doornenbal & Stevenson 2010 after de Lugt *et al.* 2003). Two unconformities U1 (shown dashed where conformable), U2 and growth strata (all marked by the author of this paper) suggest two phases of basin inversion in Late Cretaceous and Middle Eocene–Late Oligocene or Miocene time. Inset shows structures in the Southern North Sea and the location of the seismic section (thick red line). BF, Broad Fourteens Basin; WN, West Netherlands Basin. (**b**) Stratigraphic scheme of the Paleogene in the Southern North Sea (simplified from Doornenbal & Stevenson 2010). The 'Laramide disconformity' also marks the level used in seismic mapping as 'near base Cenozoic'.

in Late Cretaceous time. The inverted and partly eroded basins were then covered by probably Latest Cretaceous–Danian chalk and Paleogene clastics. A second, much weaker, pulse of inversion led to wholesale uplift of the area most deformed during the Cretaceous phase, and to tightening of folds from that time in the West Netherlands Basin. This second phase of inversion lasted from the Middle Eocene to the Oligocene or possibly early Miocene, and encompasses the periods assigned to both the Pyrenean and Savian events. The timing of the inversion phases is remarkably similar to the Sole Pit Basin (van Hoorn 1987; Badley *et al.* 1989). The Cenozoic inversion phase was followed by renewed extension of small magnitude in Neogene time.

Discussion

If we restrict the term inversion to comprise effects of horizontal shortening, we can identify a few processes that are capable of translating shortening to uplift. These processes, summarized in Table 1, are: (1) localized uplift over folds or thrust ramps;

Table 1. *Mechanisms of basin uplift*

Mechanism	Description	Amplitude (vertical)	Wavelength	Compression-induced?	Remarks
Folding and thrusting	Localized uplift associated with discrete structures	Up to many kilometres	Kilometres to a few tens of kilometres	Yes	Narrow uplifts interspersed with subsiding basins
Crustal thickening	Isostatic compensation of thickened continental crust	Up to many kilometres	Tens to hundreds of kilometres	Yes	Regional uplift only possible where crustal flow redistributes material. Expected to be associated with localized folding and thrusting in the upper crust
Lithospheric folding	Folding of the entire lithosphere or its mechanically strong layers (upper crust, upper mantle) under tangential compression	1.5 km	60–400 km (270 km)	Yes (but basin uplift occurs under tension; cf. interpretation of Laramide subsidence by Nielsen *et al.* (2005)	Alternating linear zones of uplift and subsidence. Wavelength depends on the age of the lithosphere: the indicated maximum value is for folding of the entire, Variscan-age lithosphere, the minimum value is for the folding of the upper crust alone (Cloetingh *et al.* 2005); the value in brackets is for Neogene central Europe (Bourgeois *et al.* 2007)
Thinning of mantle lithosphere	Isostatic compensation of thinned mantle lithosphere. Thinning may be thermally induced or involve instability/delamination	<2 km	?	No	Thermal thinning is slow; convective instability is difficult to achieve in continental interiors
Dynamic topography	Uplift dynamically supported by rising mantle (not necessarily a plume)	<2 km	>1000 km for plumes, several thousand kilometres for mantle convection (Flament *et al.* 2013)	No	Does not create permanent uplift

(2) isostatic uplift of thickened crust; and (3) uplift over lithospheric- or (upper) crustal-scale anticlines.

Of these, localized structural uplift is the most relevant in inverting basins. Reverse faulting and folding can create structural relief of the order of many kilometres. The Laramide event was interpreted by Ziegler (1987, 1989) to reflect a second pulse of such compression-induced inversion after the Subhercynian event. Nielsen *et al.* (2005) argued the opposite by proposing that Paleocene deposition reflects the lowering of flexural forebulges and, hence, relaxation of the far-field tectonic stress. This analysis was recently shown to be based on the erroneous assumption that the present-day erosional remnants of Cretaceous basins represent original basin geometries (Krzywiec & Stachowska 2016). At any rate, both models predict uplift and subsidence signals on maximum wavelengths of several tens of kilometres. However, the crests of emergent areas surrounding the North Sea Basin after the Laramide regression, whether measured in a NW–SE or SW–NE direction, are at least 1200 km apart.

Isostatic uplift seems to play a minor role in most cases of inversion, because the prevalence of steeply dipping faults results in limited and localized crustal shortening and thickening. Jarosinski *et al.* (2011) used numerical modelling of the Pannonian Basin to suggest that thin, hot continental lithosphere can pervasively thicken at a few per cent of shortening, causing hundreds of metres of regional isostatic uplift. However, the basin flanks in these models, with characteristics more akin to the central European lithosphere in early Cenozoic time, recorded no thickening or uplift. In natural examples, one would expect some faulting in the brittle upper crust to result from shortening of the modelled magnitude (maximum of 6% or 40 km over a distance of 600 km).

Lithospheric or crustal folding (buckling) has been modelled to induce uplift and subsidence at wavelengths ranging from a few tens to several hundred kilometres, depending on lithospheric age and whether the crust and lithospheric mantle are mechanically coupled (Cloetingh *et al.* 2005; Bourgeois *et al.* 2007; Jarosinski *et al.* 2011; Jarosinski 2012). Previously downwarped areas (basins) buckle downwards and subside under compression, whereas pre-existing upwarps are raised. This pattern matches that of the Laramide uplift, but even wholesale lithospheric folding with a predicted wavelength of 400 km (Table 1) would fail to produce the observed wavelength of 1200 km. Since the wavelength of lithospheric folding saturates at around 600 km for lithosphere older than about 1000 Ma (Cloetingh *et al.* 2005), this conclusion does not depend on the age assumed for the central European lithosphere.

Uplift mechanisms unrelated to horizontal compression (Table 1) include the isostatic effect of thinning the mantle lithosphere and dynamic topography reflecting mantle convection (Flament *et al.* 2013). Thermal thinning of the lithospheric mantle as a cause of Laramide uplift is compatible with coeval emplacement of the Iceland plume and basalt volcanism (Dèzes *et al.* 2004), but too slow to account for the rapid Laramide regression. Even so, the slowly increasing isostatic effect of a thinning mantle lithosphere may help to explain why most of the elevated regions did not subside again later. The Laramide uplift appears similar to enigmatic uplift events observed on many passive margins long after the termination of rifting (Japsen & Chalmers 2000; Japsen *et al.* 2012; Green *et al.* 2017). Mantle convection has been interpreted as causing dynamic topography (Fig. 6) with deviations from isostatic equilibrium

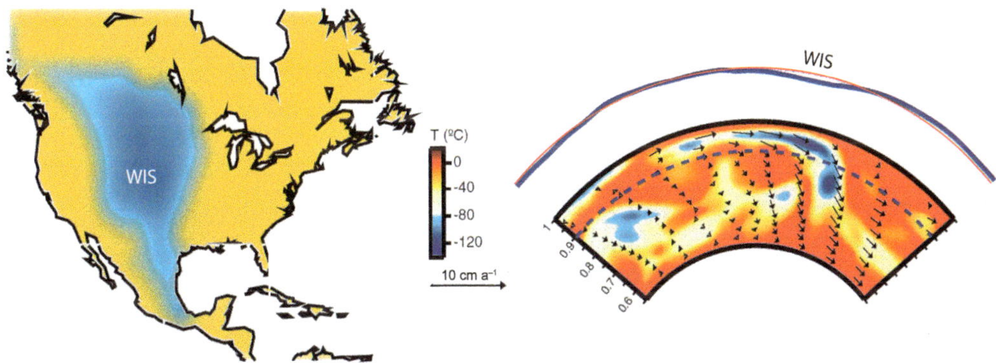

Fig. 6. Example of numerically modelled dynamic topography. The Cretaceous Western Interior Seaway (WIS) of North America (left) can be explained as the result of flat subduction of the Farallon Plate and the associated convection pattern in a vertical section of the Earth's mantle (right; scale is the normalized Earth radius). The maximum basin depth (difference between the thick blue and thin red lines) is about 700 m. Modified from Spasojevic *et al.* (2009).

down to a wavelength of 1300 km (Hoggard *et al.* 2017). This mechanism may be the best explanation presently available for regional uplift events, even though reliable estimates of past dynamic topography are not (yet) available.

Conclusions

The four 'inversion' phases commonly recognized in the Southern Permian Basin (SPB) were very probably caused by different mechanisms. There is little doubt that the Late Cretaceous (Subhercynian) event represents basin inversion resulting from horizontal shortening. Thrusting and folding account for much of the uplift that was areally widespread but localized and created strong structural relief. Inversion essentially ceased in latest Cretaceous time, although minor shortening persisted into the early Paleogene.

The Paleocene Laramide uplift is of much smaller magnitude but affects a vast area. It coincides with the termination of chalk deposition and the emergence of large stretches of west-central Europe. These changes occur indiscriminately across Mesozoic graben and older inversion structures. It is not clear if any discrete structure can be confidently ascribed to the Laramide event. These observations suggest that the Laramide event was not caused by horizontal shortening and inversion.

Younger inversion of late Eocene–Oligocene or early Miocene age has created structures from the western Netherlands to southern Britain. Timing constraints on the growth of these structures are insufficient to separate 'Pyrenean' and 'Savian' inversions. The north–south shortening direction is not greatly different to the NNE–SSW shortening direction of the Late Cretaceous inversion event. Cretaceous and Cenozoic inversion are separated in time (Fig. 7) and largely also in space, with spatial overlap occurring only in and around the Southern North Sea, in the southwesternmost part of the SPB (Fig. 8).

Inversion events associated with discrete structures differ from regional unconformity-creating events whose length scales are much larger. The causes of these regionally uniform events remain partly elusive. The Laramide uplift is likely to reflect a combination of different processes. The rapid drop in relative sea level over a few million years suggests a contribution from dynamic topography associated with mantle flow. The effect of uplift was amplified by a coeval, *c.* 30–50 m fall in global sea level. Thermal erosion of the mantle lithosphere by rising hot material may partly explain central Europe's thin lithosphere and the persistence of the established highs over tens of millions of years to the present. Despite advances in the understanding of dynamic topography, these hypotheses cannot be rigorously tested at present.

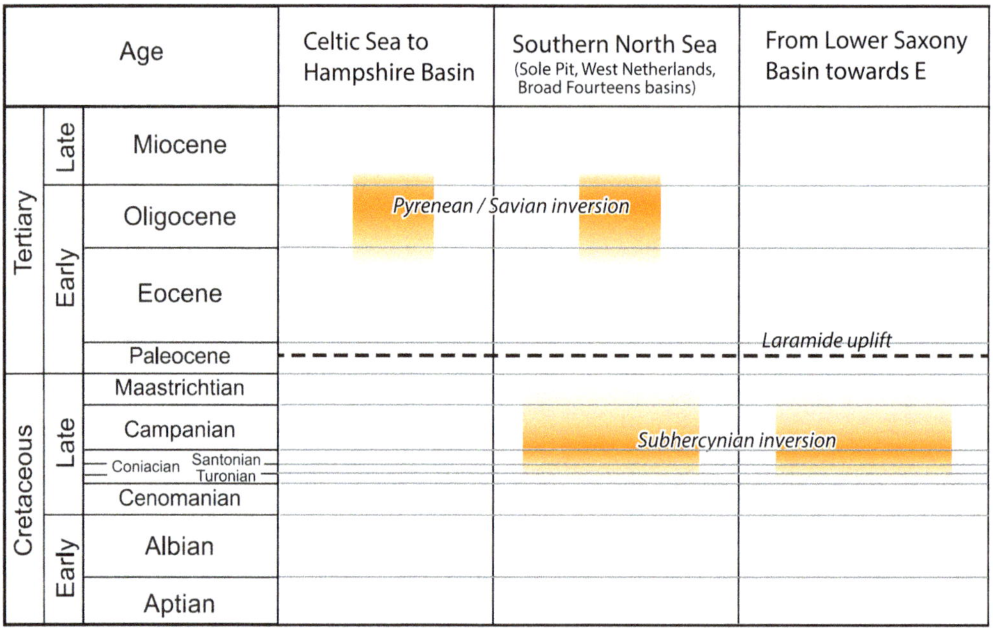

Age			Celtic Sea to Hampshire Basin	Southern North Sea (Sole Pit, West Netherlands, Broad Fourteens basins)	From Lower Saxony Basin towards E
Tertiary	Late	Miocene			
	Early	Oligocene	Pyrenean / Savian inversion	Pyrenean / Savian inversion	
		Eocene			
		Paleocene	Laramide uplift	Laramide uplift	Laramide uplift
Cretaceous	Late	Maastrichtian			
		Campanian		Subhercynian inversion	Subhercynian inversion
		Santonian			
		Coniacian			
		Turonian			
		Cenomanian			
	Early	Albian			
		Aptian			

Fig. 7. Modified scheme for tectonic events in the SPB. Compare with Figure 3.

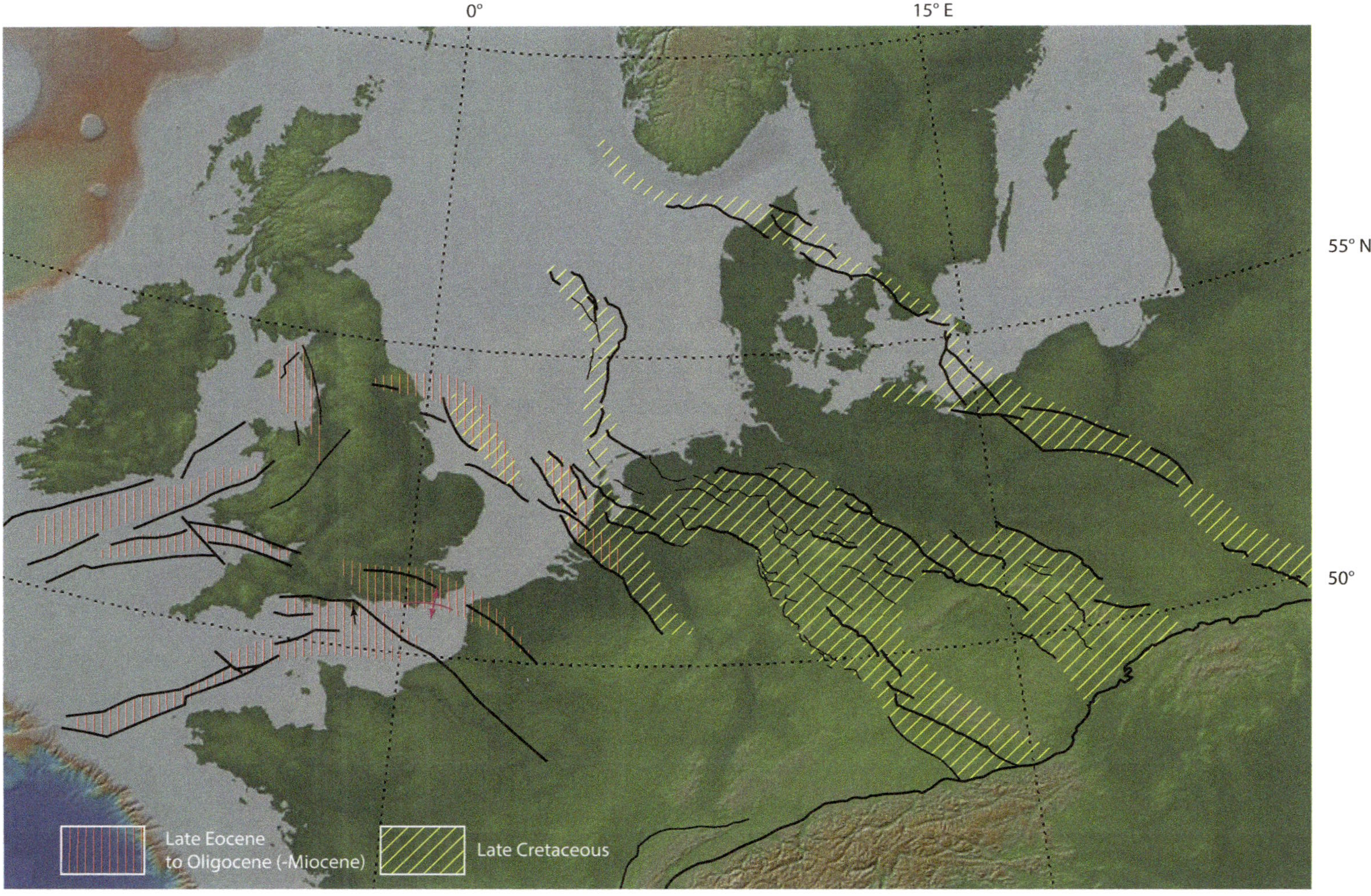

Fig. 8. Areal extents of Late Cretaceous and Late Eocene–Late Oligocene/Early Miocene inversion as proposed in this paper. The topographic base is from Ryan *et al.* (2009) as implemented in GeoMappApp (http://www.geomapapp.org).

I would like to thank Ben Kilhams for getting me started, Marek Jarosiński and Thomas Voigt for their reviews and helpful comments, and editor Stanislaw Mazur for additional advice.

References

Arfai, J., Jähne, F., Lutz, R., Franke, D., Gaedicke, C. & Kley, J. 2014. Late Palaeozoic to Early Cenozoic geological evolution of the northwestern German North Sea (Entenschnabel): new results and insights. *Netherlands Journal of Geosciences – Geologie En Mijnbouw*, **93**, 147–174, https://doi.org/10.1017/njg.2014.22

Badley, M.E., Price, J.D. & Backshall, L.C. 1989. Inversion, reactivated faults and related structures: seismic examples from the southern North Sea. *In*: Cooper, M.A. & Williams, G.D. (eds) *Inversion Tectonics*. Geological Society, London, Special Publications, **44**, 201–217, https://doi.org/10.1144/GSL.SP.1989.044.01.12

Baldschuhn, R., Binot, F., Fleig, S. & Kockel, F. 2001. *Geotektonischer Atlas von Nordwest-Deutschland und dem deutschen Nordsee-Sektor [Tectonic Atlas of NW-Germany and the German sector of the North Sea]*. Geologisches Jahrbuche Reihe A, **153**, 1–88.

Bourgeois, O., Ford, M. *et al.* 2007. Separation of rifting and lithospheric folding signatures in the NW-Alpine foreland. *International Journal of Earth Sciences*, **96**, 1003–1031.

Brückner-Röhling, S., Espig, M. *et al.* 2002. *Standsicherheitsnachweise Nachbetriebsphase: Seismische Gefährdung – Teil I: Strukturgeologie [Proofs of safe status in the post-operational period - Part I: Structural geology]*. Unpublished Report 1-92. Bundesanstalt für Geowissenschaften und Rohstoffe (BGR), Hannover, Germany.

Brun, J.-P. & Nalpas, T. 1996. Graben inversion in nature and experiments. *Tectonics*, **15**, 677–687.

Chadwick, R.A. 1993. Aspects of basin inversion in southern Britain. *Journal of the Geological Society, London*, **150**, 311–322, https://doi.org/10.1144/gsjgs.150.2.0311

Cloetingh, S. 1986. Intraplate stresses: a new tectonic mechanism for fluctuations of relative sea level. *Geology*, **14**, 617–620.

Cloetingh, S., Ziegler, P.A. *et al.* 2005. Lithospheric memory, state of stress and rheology: neotectonic controls on Europe's intraplate continental topography. *Quaternary Science Reviews*, **24**, 241–304.

Cooper, M.A. & Williams, G.D. (eds) 1989. *Inversion Tectonics*. Geological Society, London, Special Publications, **44**, https://doi.org/10.1144/GSL.SP.1989.044.01.25

Danisik, M., Migon, P., Kuhlemann, J., Evans, N.J., Dunkl, I. & Frisch, W. 2010. Thermochronological constraints on the long-term erosional history of the Karkonosze Mts., Central Europe. *Geomorphology*, **117**, 78–89.

de Jager, J. 2007. Geological development. *In*: Wong, T.E., Batjes, D.A.J. & de Jager, J. (eds) *Geology of the Netherlands*. Royal Netherlands Academy of Arts and Sciences, Amsterdam, 5–26.

de Lugt, I.R., van Wees, J.D. & Wong, T.E. 2003. The tectonic evolution of the southern Dutch North Sea during the Palaeogene: basin inversion in distinct pulses. *Tectonophysics*, **373**, 141–159.

Dèzes, P., Schmid, S.M. & Ziegler, P.A. 2004. Evolution of the European Cenozoic rift system: interaction of the Alpine and Pyrenean orogens with their foreland lithosphere. *Tectonophysics*, **389**, 1–33.

Doornenbal, H. & Stevenson, A. (eds) 2010. *Petroleum Geological Atlas of the Southern Permian Basin Area*. European Association of Geoscientists and Engineers (EAGE), Houten, The Netherlands.

Flament, N., Gurnis, M. & Müller, R.D. 2013. A review of observations and models of dynamic topography. *Lithosphere*, **5**, 189–210.

Gramann, F. & Kockel, F. 1988. Palaeogeographical, lithological, palaeoecological and palaeoclimatic development of the North West European Tertiary Basin. *In*: Vinken, R. (ed.) *The Northwest European Tertiary Basin: Results of the International Geological Correlation Programme, Project No. 124*. Geologisches Jahrbuche Reihe A, **100**, 428–442.

Green, P.F., Duddy, I.R. & Japsen, P. 2017. Multiple episodes of regional exhumation and inversion identified in the UK Southern North Sea based on integration of palaeothermal and palaeoburial indicators. *In*: Bowman, M. & Levell, B. (eds) *Petroleum Geology of NW Europe: 50 Years of Learning – Proceedings of the 8th Petroleum Geology Conference*. Geological Society, London. 47–65, https://doi.org/10.1144/PGC8.21

Haq, B.U., Hardenbol, J. & Vail, P.R. 1987. Chronology of fluctuating sea levels since the Triassic. *Science*, **235**, 1156–1167.

Hejl, E., Coyle, D., Lal, N., VandenHaute, P. & Wagner, G.A. 1997. Fission track dating of the western border of the Bohemian massif: thermochronology and tectonic implications. *Geologische Rundschau*, **86**, 210–219.

Hoggard, M.J., Winterbourne, J., Czarnota, K. & White, N. 2017. Oceanic residual depth measurements, the plate cooling model, and global dynamic topography. *Journal of Geophysical Research: Solid Earth*, **122**, 2328–2372, https://doi.org/10.1002/2016JB013457

Jackson, C.A.-L., Chua, S.T., Bell, R.E. & Magee, C. 2013. Structural style and early stage growth of inversion structures: 3D seismic insights from the Egersund Basin, offshore Norway. *Journal of Structural Geology*, **46**, 167–185.

Japsen, P. & Chalmers, J.A. 2000. Neogene uplift and tectonics around the North Atlantic: overview. *Global and Planetary Change*, **24**, 165–173.

Japsen, P., Chalmers, J.A., Green, P.F. & Bonow, J.M. 2012. Elevated, passive continental margins: not rift shoulders, but expressions of episodic, post-rift burial and exhumation. *Global and Planetary Change*, **90–91**, 73–86.

Jarosinski, M. 2012. Compressive deformations and stress propagation in intracontinental lithosphere: finite element modeling along the Dinarides–East European Craton profile. *Tectonophysics*, **526–529**, 24–41.

Jarosinski, M., Beekman, F., Matenco, L. & Cloetingh, S. 2011. Mechanics of basin inversion: finite element modelling of the Pannonian Basin System. *Tectonophysics*, **502**, 121–145.

Kley, J. & Voigt, T. 2008. Late Cretaceous intraplate thrusting in central Europe: effect of Africa–Iberia–Europe convergence, not Alpine collision. *Geology*, **36**, 839–842.

Kockel, F. 2003. Inversion structures in Central Europe – expressions and reasons, an open discussion. *Netherlands Journal of Geosciences – Geologie en Mijnbouw*, **82**, 367–382.

Krzywiec, P. 2006. Structural inversion of the Pomeranian and Kuiavian segments of the Mid-Polish Trough – lateral variations in timing and structural style. *Geological Quarterly*, **50**, 151–168.

Krzywiec, P. & Stachowska, A. 2016. Late Cretaceous inversion of the NW segment of the Mid-Polish Trough – how marginal troughs were formed, and does it matter at all? *Zeitschrift der Deutschen Geologischen Gesellschaft*, **167**, 107–119.

Miller, K.G., Kominz, M.A. *et al.* 2005. The Phanerozoic record of global sea-level change. *Science*, **310**, 1293–1298.

Nielsen, S., Thomsen, E., Hansen, D.L. & Clausen, O.R. 2005. Plate-wide stress relaxation explains European Palaeocene basin inversions. *Nature*, **435**, 195–198.

Ryan, W.B.F., Carbotte, S.M. *et al.* 2009. Global Multi-Resolution Topography synthesis. *Geochemistry, Geophysics, Geosystems*, **10**, Q03014, https://doi.org/10.1029/2008GC002332

Senglaub, Y., Brix, M.R., Adriasola, A.C. & Littke, R. 2005. New information on the thermal history of the southwestern Lower Saxony Basin, northern Germany, based on fission track analysis. *International Journal of Earth Sciences*, **94**, 876–896.

Sissingh, W. 2006. Syn-kinematic palaeogeographic evolution of the West European Platform: correlation with Alpine plate collision and foreland deformation. *Netherlands Journal of Geosciences – Geologie en Mijnbouw*, **85**, 131–180.

Spasojevic, S., Liu, L. & Gurnis, M. 2009. Adjoint models of mantle convection with seismic, plate motion, and-stratigraphic constraints: North America since the Late Cretaceous. *Geochemistry, Geophysics, Geosystems*, **10**, Q05W02, https://doi.org/10.1029/2008GC002345

Thomson, S.N. & Zeh, A. 2000. Fission-track thermochronology of the Ruhla Crystalline Complex: new constraints on the post-Variscan thermal evolution of the NW Saxo-Bohemian Massif. *Tectonophysics*, **324**, 17–35.

Vandycke, S. 2002. Palaeostress records in Cretaceous formations in NW Europe: extensional and strike-slip events in relationships with Cretaceous–Tertiary inversion tectonics. *Tectonophysics*, **357**, 119–136.

van Hoorn, B. 1987. Structural evolution, timing and tectonic style of the Sole Pit inversion. *Tectonophysics*, **137**, 239–284.

van Lochem, H. 2018. F17-Chalk: new insights in the tectonic history of the Dutch Central Graben. *In*: Kilhams, B., Kukla, P.A., Mazur, S., Mckie, T., Mijnlieff, H.F. & Van Ojik, K. (eds) *Mesozoic Resource Potential in the Southern Permian Basin*. Geological Society, London, Special Publications, **469**. First published February 19, 2018, https://doi.org/10.1144/SP469.17

van Wijhe, D.H. 1987. Structural evolution of inverted basins in the Dutch offshore. *Tectonophysics*, **137**, 171–219.

Voigt, T., von Eynatten, H. & Franzke, H.J. 2004. Late Cretaceous unconformities in the Subhercynian Cretaceous Basin (Germany). *Acta Geologica Polonica*, **54**, 673–694.

Voigt, T., Reicherter, K., von Eynatten, H., Littke, R., Voigt, S. & Kley, J. 2008. Sedimentation during inversion. *In*: Littke, R., Bayer, U., Gajewski, D. & Nelskamp, S. (eds) *Dynamics of Complex Intracontinental Basins*. Springer, Berlin, 211–232.

von Eynatten, H., Voigt, T., Meier, A., Franzke, H.J. & Gaupp, R. 2008. Provenance of Cretaceous clastics in the Subhercynian Basin: constraints to exhumation of the Harz Mountains and timing of inversion tectonics in Central Europe. *International Journal of Earth Sciences*, **97**, 1315–1330.

Worum, G. & Michon, L. 2005. Implications of continuous structural inversion in the West Netherlands Basin for understanding controls on Palaeogene deformation in NW Europe. *Journal of the Geological Society, London*, **162**, 73–85, https://doi.org/10.1144/0016-764904-011

Ziegler, P.A. 1987. Late Cretaceous and Cenozoic intraplate compressional deformations in the Alpine foreland. *Tectonophysics*, **137**, 389–420.

Ziegler, P.A. 1989. Geodynamic model for Alpine intraplate compressional deformation in Western and Central Europe. *In*: Cooper, M.A. & Williams, G.D. (eds) *Inversion Tectonics*. Geological Society, London, **44**, 63–85, https://doi.org/10.1144/GSL.SP.1989.044.01.05

Ziegler, P.A. 1990. *Geological Atlas of Western and Central Europe*. 2nd edn. Shell Internationale Petroleum Maatschappij, The Hague. Geological Society, London.

The only way is up – on Mesozoic uplifts and basin inversion events in SE Poland

PIOTR KRZYWIEC*, ALEKSANDRA STACHOWSKA & AGATA STYPA

Institute of Geological Sciences, Polish Academy of Sciences, 51/55 Twarda Street, 00-818 Warsaw, Poland

**Correspondence: piotr.krzywiec@twarda.pan.pl*

Abstract: The inversion of a sedimentary basin could be associated with compressional reactivation of basin-forming normal faults, upwards movement of the basement blocks and partial or complete erosion of its sedimentary infill. Basin inversion might be also related to whole-basin uplift that is not linked to the reactivation of basement faults, and results in the development of regional stratigraphic gaps and unconformities. Both types of basin inversion have been documented in SE Poland using seismic data. Regional NW–SE seismic profiles illustrate earliest Late Jurassic (earliest Oxfordian) and earliest Late Cretaceous (Cenomanian) regional unconformities related to regional basin-scale uplifts in the SE segment of the Polish Basin. Late Cretaceous (Turonian?–Maastrichtian) progressive uplift of the Mid-Polish Swell has been documented along the NE border zone of this regional anticlinal structure. The Upper Cretaceous inversion-related sedimentary succession is characterized by an overall progradational character directed from the SW towards the NE. Buried contourite drifts that were detected within the Upper Cretaceous succession using seismic data indicate the existence of contour currents encircling inversion-related intrabasinal morphological barriers. A new tectonic scenario of the Mesozoic evolution of SE Poland would have a significant impact on the modelling of tectonic subsidence and the history of petroleum systems.

Evolution of sedimentary basins, including their inversion phase, has been the subject of intense research since the early 1980s (cf. Cooper & Williams 1989; Buchanan & Buchanan 1995). The inversion of extensional sedimentary basins involves compressional–transpressional reactivation of their fault systems as reverse fault zones (Brun & Nalpas 1996) and uplift of the basin floor; it might also involve folding of the basin fill. Reverse faulting leads to gradual upwards movement of the basement blocks, which ultimately can result in complete destruction of the sedimentary basin, and associated erosion of its uplifted and exposed infill (Hayward & Graham 1989).

In his classic paper, Bally (1984) proposed a model for inverted half-graben and defined key geometrical characteristics of inversion tectonics, a term first used by Glennie & Boegner (1981). Inversion is usually associated with compressional reactivation of a previous extensional fault(s) and uplift – full or partial – of the extensional basin infill above the regional elevation (e.g. Williams *et al.* 1989; Cooper & Warren 2010). During the inversion of sedimentary basins, faults that did not have any previous extensional history during basin subsidence (i.e. were not associated with localized thickening of the synrift/synextensional infill) might also become active. These could be much older, deeply rooted basement faults activated by regional inversion-related compression. An example of such a scenario is provided by several crustal blocks from the Rhenohercynian Zone in southern Germany (Harz Mountains) which were uplifted in the Late Cretaceous along deeply rooted reverse faults such as the North Harz Fault, the Haldensleben Fault and the Gardelengen Fault (Hecht *et al.* 2003).

Another example of thick-skinned Late Cretaceous–early Cenozoic inversion of intracratonic basin, well documented by deep seismic data, is provided by the Dnipro–Donetsk Basin and the Donbas Fold Belt (Maystrenko *et al.* 2003). In this case, basin inversion is related to compressional reactivation of a major detachment fault transecting the entire crust and the formation of an associated large-scale backthrust. As a result, an internally deformed crustal-scale pop-up structure was formed.

Inversion of sedimentary basins may also be, to some extent, framed differently. It might be seen as a general tectonic process that halts subsidence, slows down and terminates creation of accommodation space and associated sedimentation, and eventually leads to variable regional uplift of an area previously occupied by a sedimentary basin. Such a process might be related to the compressional reactivation of extensional basement fault(s) but other scenarios might also be possible. One of them might be, for example, the inversion of a sedimentary basin developed above a thick salt layer. In such a

From: Kilhams, B., Kukla, P. A., Mazur, S., McKie, T., Mijnlieff, H. F. & van Ojik, K. (eds) 2018. *Mesozoic Resource Potential in the Southern Permian Basin*. Geological Society, London, Special Publications, **469**, 33–57.
First published online March 20, 2018, https://doi.org/10.1144/SP469.14

case, subsidence and an increase in accommodation space is caused by the lateral removal of salt, while progressive uplift and basin inversion might be related to the inflow of salt beneath the basin centre that could be triggered by some external tectonic forces (Jackson & Talbot 1994). As a result, the sedimentary basin becomes inverted but there are no basement faults involved in this inversion process. Another form of basin inversion, not fulfilling the requirement of compressional reactivation of basement faults, might be, as has already been pointed out by Dewey (1989), regional, whole-basin uplift. Reasons for such regional basement uplifts that lead to inversion of the overlying sedimentary basins include, but are not limited to, regional variations in stress patterns within plates and associated lithospheric-scale arching, or isostatic response to crustal stretching (Dewey 1989). This scenario of basin inversion invokes much more regional uplift that is not focused on discrete basement faults.

Reverse faulting leading to localized uplift of the basement blocks and erosion is, during early phases of basin inversion, often associated with the deposition of generally wedge-shaped sedimentary units characterized by local dispersal patterns directed away from the uplifted areas (growth strata), and by the presence of local progressive unconformities (Voigt *et al.* 2008), as can be seen either on seismic data (e.g. Cartwright 1989; Khriachtchevskaia *et al.* 2010) or in the field (e.g. Voigt *et al.* 2004; von Eynatten *et al.* 2008). Dating of the onset and main phases of basin inversion might be challenging as inversion-related uplift often leads to substantial erosion of the sedimentary cover formed during syn-/post-rift stages, as well as during early stages of basin inversion. On the other hand, regional basin-scale uplift might result in the formation of regional stratigraphic gaps that might not be related to any angular unconformities, etc. In such a case, there is no sedimentary record of basin inversion which could be stratigraphically bracketed only by the youngest strata located below such a regional unconformity and by the oldest strata located above it.

Two general scenarios described above are shown in Figure 1. Scenario 1 illustrates deformation of the crystalline basement and the pre-extensional sedimentary cover by a discrete deeply rooted fault. This basement fault controls lateral thickness changes of the synkinematic deposits: the synrift (pink) sedimentary succession is characterized by thickening above the hanging wall of this fault, and the syn-inversion succession (green) formed during reverse reactivation of a previously normal fault is characterized by thinning towards the uplifted block.

An alternative tectonosedimentary evolution is illustrated by scenario 2 (Fig. 1). In this case, subsidence and basin uplift (inversion) are both related to regional vertical movements not related to activity of any specific basement fault(s). Regional, whole-basin uplift results in tectonically driven cessation of sedimentation, erosion and the formation of a regional stratigraphic gap at the top of a partly eroded pre-uplift/pre-inversion sedimentary sequence (pink). In the case of largely vertical movements of crustal blocks, as shown in Figure 1, such a stratigraphic gap might not be related to any angular unconformity developed at the base of the sedimentary succession that was formed after the uplift phase (shown in green).

The Mesozoic geological evolution of Europe is punctuated by several episodes of regional uplifts that resulted in the formation of various regional unconformities and stratigraphic gaps, with the most recent regional inversion event taking place in the Late Cretaceous–Paleocene which reshaped much of its geological structure (Fig. 2) (e.g. see Ziegler 1990; Scheck-Wenderoth *et al.* 2008; Pharaoh *et al.* 2010 for synthetic overviews and further references). In this paper, results of the interpretation of seismic data are used to illustrate and analyse different modes of regional Jurassic and Cretaceous uplifts in SE Poland.

Geological setting

Evolution of the Polish Basin: an overview

The Polish Basin, together with its axial most subsiding part, the Mid-Polish Trough, formed the eastern part of the Permian–Mesozoic system of epicontinental basins of western and central Europe, the so-called Central European Basin System (Ziegler 1990; Scheck-Wenderoth *et al.* 2008; Pharaoh *et al.* 2010). During the Permian stage of its evolution, the Polish Basin formed the eastern part of the Southern Permian Basin (Kiersnowski *et al.* 1995; van Wees *et al.* 2000; Krzywiec *et al.* 2017*c*).

Following its Permian rifting, the Polish Basin experienced long-term thermal subsidence, commencing in the Permian and lasting until the Late Cretaceous, which was punctuated by Zechstein–Scythian, Oxfordian–Kimmeridgian and early Cenomanian major pulses of accelerated tectonic subsidence (Dadlez *et al.* 1995; Stephenson *et al.* 2003). It was filled with up to 8 km of Permian–Mesozoic sediments (Dadlez *et al.* 1998). Throughout the Mesozoic, regional subsidence patterns in the Polish Basin were dominated by the development of the NW–SE-trending Mid-Polish Trough, with local modifications imposed mainly by lateral salt movements in its central and NW segments.

After long-term Mesozoic subsidence, the Polish Basin was inverted during the Late Cretaceous–Paleogene. Inversion of the Polish Basin commenced in the Late Turonian and lasted until Maastrichtian–post-Maastrichtian times (e.g. Krzywiec

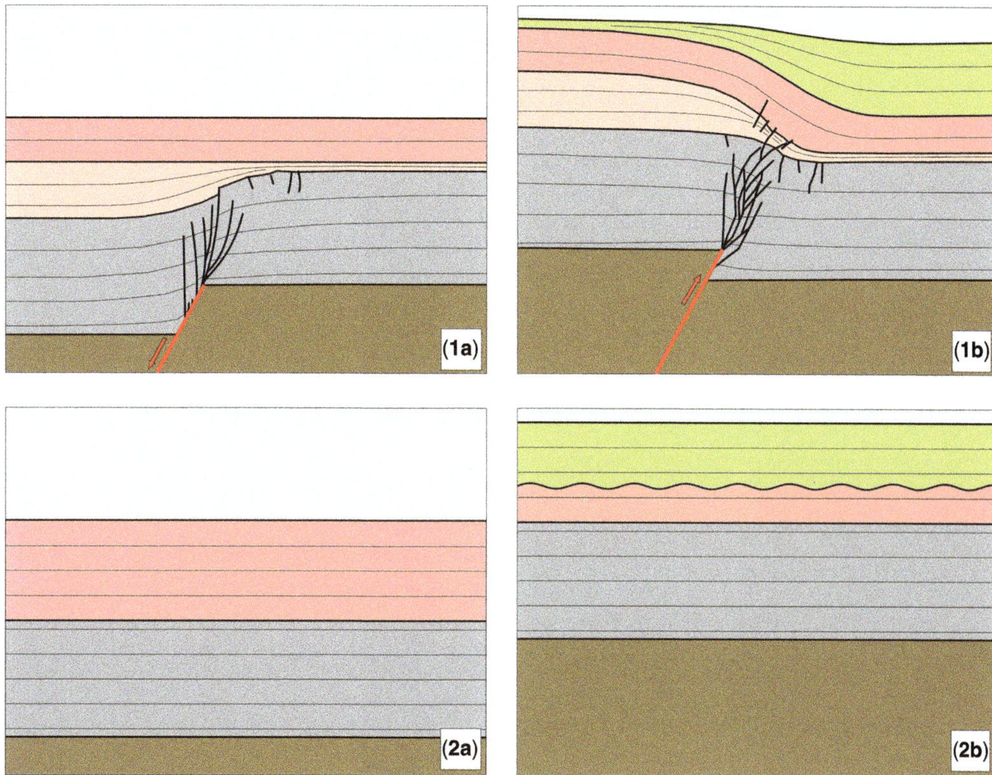

Fig. 1. Two models of basin inversion. Model 1: subsidence (1a) and consecutive basin inversion (1b) is realized along discrete faults rooted within the basement (brown). Grey. pre-extensional sedimentary cover; pink, synrift sedimentary successions; dark pink, post-rift/pre-inversion sedimentary cover; green, syn-inversion sedimentary successions (after Krzywiec *et al.* 2009, modified; based on results of sandbox modelling by Mitra & Islam 1994). Model 2: subsidence (2a) and basin uplift (inversion: 2b) are both related to regional vertical movements not associated with the activity of basement faults. After uplift and associated erosion, the post-uplift/post-inversion sedimentary succession is deposited above a regional stratigraphic gap.

2006; Resak *et al.* 2008; Krzywiec *et al.* 2009). It was associated with substantial uplift and erosion of the axial part of the Polish Basin (i.e. the Mid-Polish Trough), which presently forms a regional anticlinal structure referred to as the Mid-Polish Swell (Mid-Polish Anticlinorium), outlined by the Cenozoic subcrop of the Lower Cretaceous and older rocks (Fig. 3). As a combined effect of a regional inversion-driven uplift, erosion and deposition, the Upper Cretaceous succession is presently preserved only along both flanks of the Mid-Polish Swell (Fig. 3). Paleocene deposits occur almost exclusively along the NE flank of the Mid-Polish Swell, while Eocene deposits are present on both of its flanks (Piwocki 2004). This suggests that by Paleocene times, inversion movements had ended on the SW flank of the Mid-Polish Swell which formed an uplifted area with sedimentation continuing on its NE flank. By early Eocene times, both flanks had been inverted with Eocene sediments sealing the eroded inversion structures, including its axial part (Piwocki 2004).

A more detailed discussion of various aspects of subsidence and inversion in different segments of the Polish Basin, with an extensive reference list, is given by Marek & Pajchlowa (1997), Mazur *et al.* (2005), Krzywiec (2006), Resak *et al.* (2008), Scheck-Wenderoth *et al.* (2008), Krzywiec *et al.* (2009), Narkiewicz *et al.* (2010), Pharaoh *et al.* (2010) and Krzywiec & Stachowska (2016).

Permian–Cretaceous evolution of SE Poland

Late Permian–Early Cretaceous basin evolution. The SE segment of the Polish Basin (Fig. 3) was located, at least partly, within the broadly defined transition zone between the European epicontinental basin system and the Tethyan domain (Kutek & Głazek 1972; Ziegler 1990; Gutowski *et al.* 2005; Świdrowska *et al.* 2008). The preserved Mesozoic

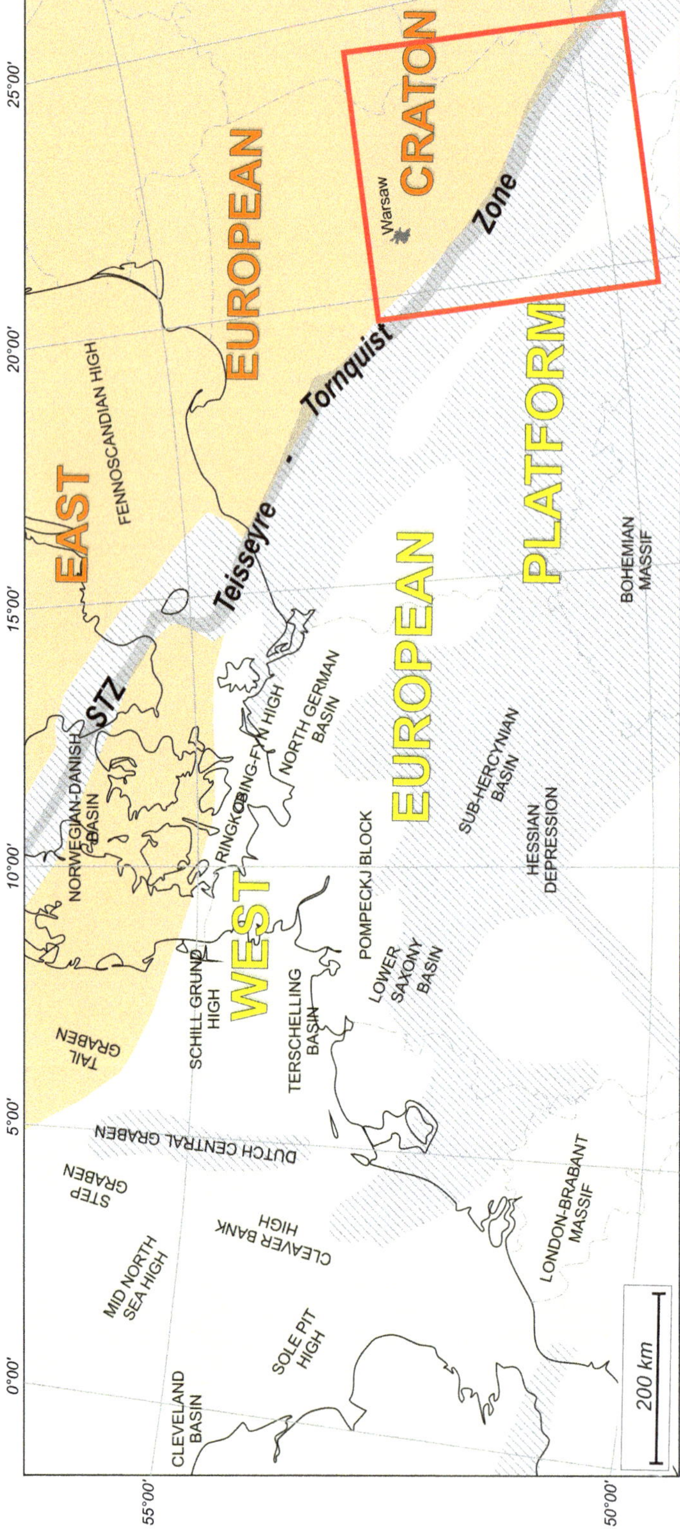

Fig. 2. Regional schematic map of western and central Europe (after Pharaoh *et al.* 2010; Mazur *et al.* 2015, simplified) with the schematic extent of the Late Cretaceous–Paleocene inversion (patterned area: after Ziegler 1990; Ziegler *et al.* 2002). Red rectangle, area shown in Figure 3.

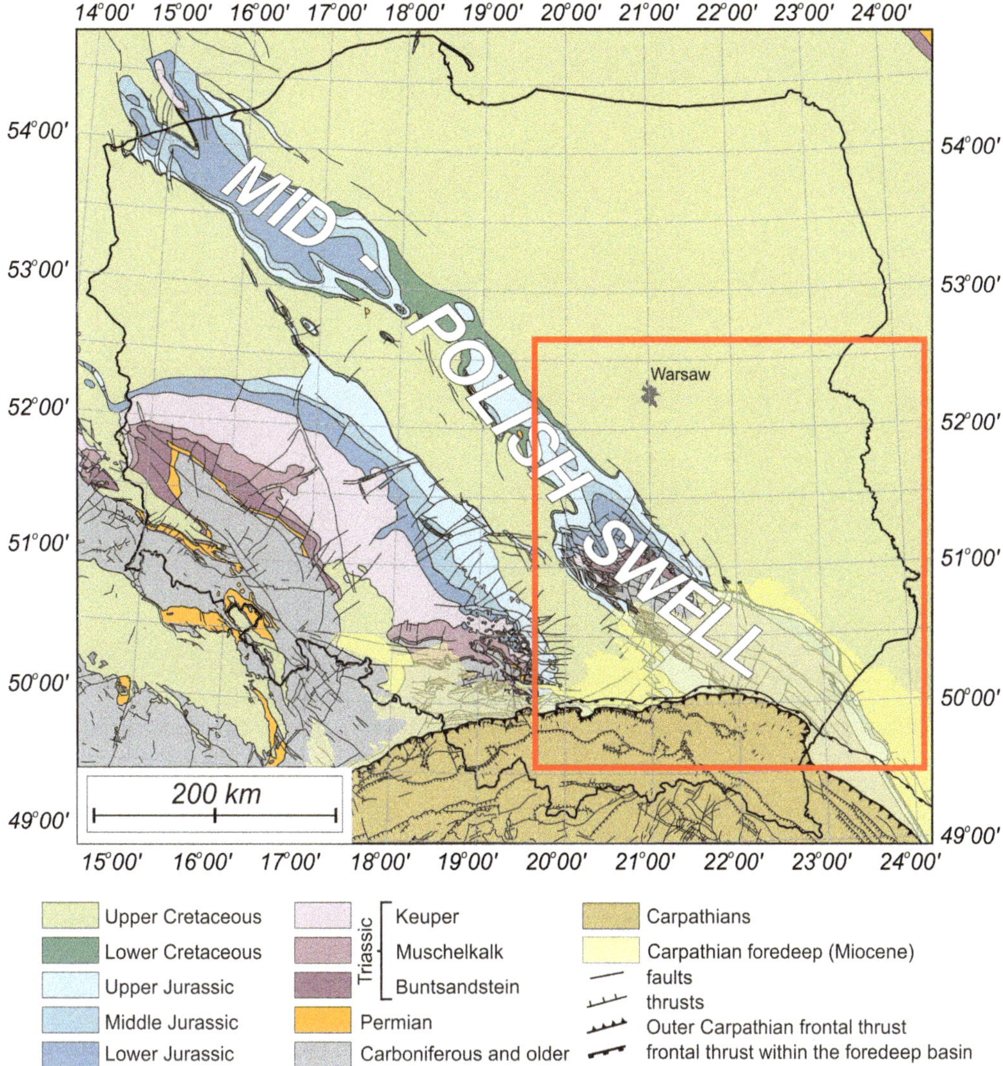

Fig. 3. Geological map of Poland (simplified after Pożaryski 1978; Dadlez *et al.* 2000). Red rectangle, study area shown in Figures 4 & 5.

sedimentary cover of the Lublin Basin consists almost exclusively of Upper Jurassic and Upper Cretaceous deposits. However, various previous studies, based mostly on well data, suggest that other parts of the Mesozoic cover could have been also deposited – and then eroded – in this part of the basin (e.g. Marek & Pajchlowa 1997; Dadlez *et al.* 1998; Świdrowska *et al.* 2008), as illustrated by paleothickness maps in Figure 4.

In the Late Permian (deposition of the Zechstein evaporites), the study area was located in the southeasternmost part of the evaporitic basin, characterized by incomplete evaporitic cyclothems and a relatively thin (0–400 m) Permian sedimentary cover (Fig. 4a) (see Krzywiec *et al.* 2017*c* for recent overview and further references).

The evaporitic units of the uppermost Zechstein Group were replaced in the Early Triassic by predominantly continental (fluvial, lacustrine, playa) deposits of the Buntsandstein Group (Szyperko-Teller 1997). According to Iwanow (1998*a*), sedimentation of the Lower Triassic deposits took place in most of SE Poland, with thicknesses varying from 100 m in the NW to 50 m in the SE, with only a very small area exposed during that time in the southeasternmost part of the study area (Fig. 4b).

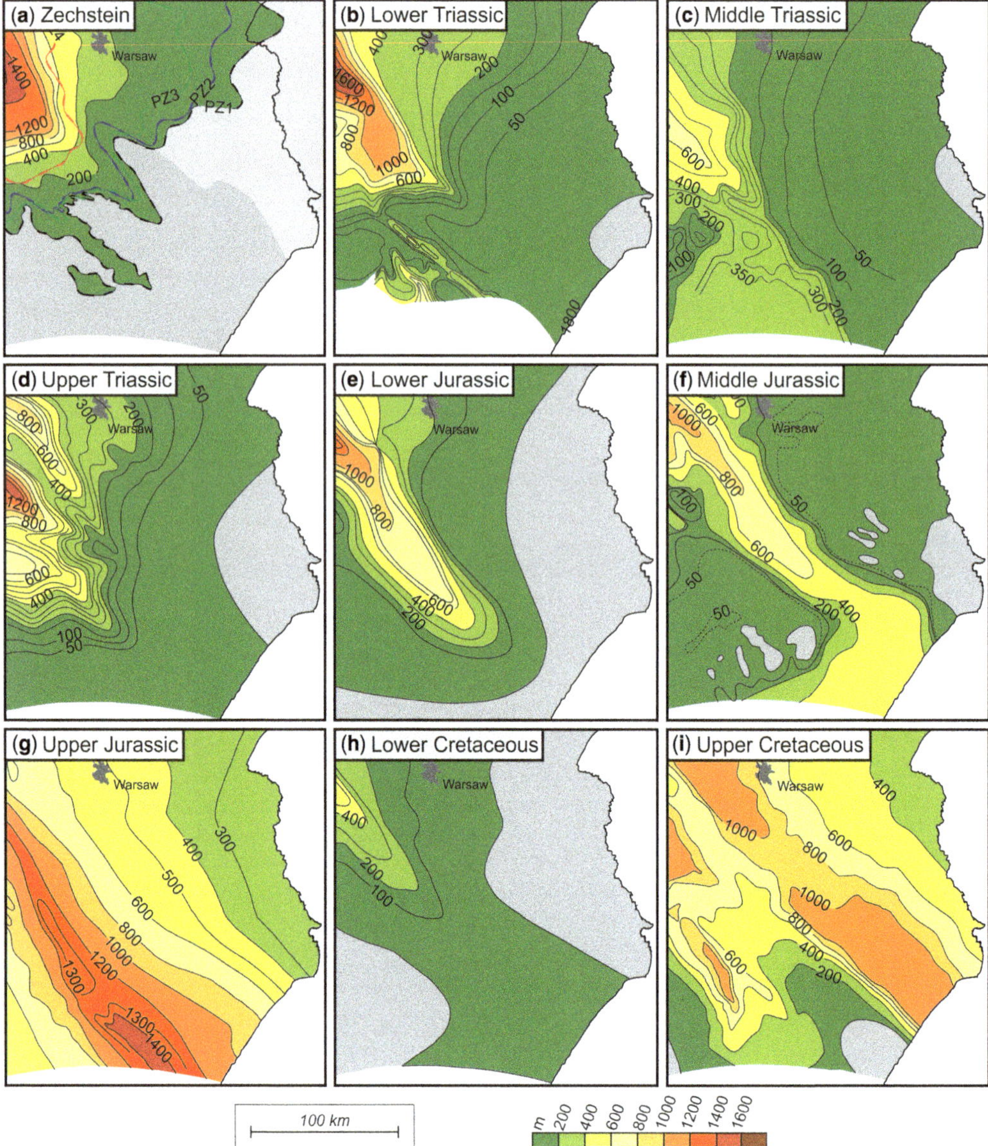

Fig. 4. Paleothickness maps for SE Poland: (**a**) Zechstein (based on Wagner 1998); (**b**) lower and middle Buntsandstein (based on Iwanow 1998*a*); (**c**) upper Buntsandstein, Muschelkalk and lower Keuper (based on Iwanow 1998*b*); (**d**) middle–upper Keuper and Rhaetian (based on Iwanow 1998c); (**e**) Lower Jurassic (based on Feldman-Olszewska 1998*a*); (**f**) Middle Jurassic (based on Feldman-Olszewska 1998*b*); (**g**) Upper Jurassic (including Lower Berriasian) (based on Gaździcka 1998); (**h**) Lower Cretaceous (excluding Lower Berriasian and Upper Albian) (based on Leszczyński 1998*a*); and (**i**) Upper Cretaceous (including Upper Albian) (based on Leszczyński 1998*a*, *b*).

Some other, more conservative authors, such as Szyperko-Teller (1997) and Świdrowska *et al.* (2008), assumed that almost the entire area of SE Poland was devoid of Lower Triassic sedimentation and that widespread denudation took place during that time period.

Middle Triassic (Anisian and early Ladinian) sedimentation of the carbonate-dominated Muschelkalk Group occurred across the entire Polish Basin (Gajewska 1971; Ziegler & Haag 1982; Gajewska 1997*a*). It was controlled by eustatic sea-level changes, and by episodes of opening and closing of the sea gateways in the SE and SW, resulting in connections between the Tethys Ocean and the Polish Basin (Gajewska 1997*a*; Szulc 2000). Iwanow (1998*b*) assumes that most of SE Poland was covered by Middle Triassic deposits, with thickness trends similar to those observed for the Lower Triassic (Fig. 4c). According to Gajewska (1997*a*), a large part of SE Poland was still uplifted, with the resulting land area surrounded by shaly marls to the NW, and by limestones to the west and SW. Świdrowska *et al.* (2008) also suggested a lack of Middle Triassic sedimentation in this part of Poland, with denudation dominant.

A Late Ladinian regressive phase led to a change in marine conditions of the Muschelkalk Group to a deltaic, hypersaline and non-marine depositional environment of the Keuper Group. Deposition of the Keuper Group lasted until the end of the Rhaetian and covered only the NW, and partly the central area, of the Polish Basin. Moreover, a phase of strong tectonic movements occurred in the Polish Basin area at the end of Carnian, which resulted in the uplift and erosion of a major part of the basin (Ziegler & Haag 1982; Gajewska 1997*b*; Feist-Burkhardt *et al.* 2008). Iwanow (1998*c*) suggested that Late Triassic sedimentation occurred within most of SE Poland, with the elevated land area being only slightly larger than in Early and Middle Triassic times (Fig. 4d). According to Iwanow (1998*c*), the thickness of the Upper Triassic succession was between 50 and 100 m. On the other hand, however, Gajewska (1997*b*), Deczkowski (1997*a*) and Świdrowska *et al.* (2008) suggested that in the Late Triassic most of SE Poland formed an uplifted and exposed land area with no sedimentation taking place.

The evolution of the Polish Basin in the Early Jurassic can be divided into three stages: Hettangian–early Sinemurian, late Sinemurian–Pliensbachian and Toarcian (Deczkowski 1997*b*). During the first stage, deposition of sandstone lithofacies occurred only in the NW and central parts of the Polish Basin. During the second stage, the basin widened towards the NE and SW, and, due to transgression from the west, marine shales, mudstone and sandstones were deposited. During the third stage, the lateral extent of the Polish Basin was reduced. It is thought that SE Poland was mostly not covered by Lower Jurassic sediments (Deczkowski 1997*b*; Feldman-Olszewska 1998*a*; Świdrowska *et al.* 2008; Fig. 4e).

The Middle Jurassic stratigraphic sequence, for the entire Polish Basin, is dominated by clastic deposits (Dayczak-Calikowska 1997*a*, *b*). The Middle Jurassic basin had a mainly transgressive character, with only two short regressive episodes in the Early–Middle Bajocian and earliest Callovian. According to numerous authors (Dayczak-Calikowska 1997*a*, *b*; Feldman-Olszewska 1998*b*; Świdrowska *et al.* 2008), Middle Jurassic uplifted and eroded areas in SE Poland were surrounded by local zones of sedimentation of Aalenian sandstones and conglomerates followed by Bajocian mudstones and sandy shales. A widespread Bathonian transgression resulted in the establishment of carbonate sedimentation. The maximum paleothickness of the Middle Jurassic complex in SE Poland reached, on average, 50–100 m (Feldman-Olszewska 1998*b*) (Fig. 4f).

In the Oxfordian, the northern part of the Polish Basin was dominated by clastic deposition that was replaced towards the SE by a carbonate-dominated shelf sedimentation (Dembowska 1979; Niemczycka & Brochwicz-Lewiński 1988; Niemczycka 1997*b*). This lateral, basin-scale change of depositional environments is considered to be due to the northwards expansion of the Tethys Ocean (Pieńkowski *et al.* 2008). The Kimmeridgian–early Tithonian interval was dominated by deep-water sedimentation. Ensuing eustatic sea-level fall created a restricted brackish basin in which carbonates and evaporites were deposited. The study area was covered in Oxfordian times by a shallow open sea in which carbonates up to 100–300 m thick accumulated (Niemczycka & Brochwicz-Lewiński 1988; Niemczycka 1997*a*, *b*). In the Kimmeridgian, this part of the Polish Basin was covered by a coastal lagoon characterized by the deposition of dolomites and evaporites in the range of 100–300 m in thickness (Niemczycka 1997*a*, *b*). In SE Poland, a Tithonian regression created a shallow open sea characterized by oolitic limestones (Niemczycka 1997*a*, *b*; Świdrowska *et al.* 2008) According to Gaździcka (1998), the total paleothickness of the Upper Jurassic succession was in the range of 300–600 m (Fig. 4g).

The Early Cretaceous Polish Basin was rather restricted, and was characterized by alternating marine and paralic sedimentation. Transgressive fine-grained shales and regressive sandstones dominated in the northern part of the basin; while in its SE part, sedimentation of clastic carbonates occurred, probably due to the influence of Tethyan warm oceanic currents (Marek & Pajchlowa 1997). In this part of the basin, the Lower Cretaceous sedimentary cover developed only locally and, when present, is characterized by relatively thin packages, up to a maximum of 100 m (Marek 1997*a*, *b*; Świdrowska *et al.* 2008; Leszczyński 1998*a*) (Fig. 4h).

Late Cretaceous inversion. Outcrop and well data have enabled the recognition of facies development and stratigraphy (mainly based on ammonites, inoceramids, belemnites and foraminifera) of the Upper Cretaceous succession in SE Poland (e.g. Pożaryski 1938; Cieśliński 1959; Błaszkiewicz 1980; Walaszczyk 1992; Remin 2015; Walaszczyk *et al.* 2016). The Late Cretaceous evolution of this part of the basin, including the timing of inversion tectonics, has been discussed in numerous publications (e.g. Pożaryski 1948; Walaszczyk 1992; Świdrowska & Hakenberg 1999; Świdrowska 2007; Świdrowska *et al.* 2008; Krzywiec 2009; Krzywiec *et al.* 2009; Leszczyński 2012; Walaszczyk & Remin 2015).

The Upper Cretaceous succession in SE Poland forms up to *c.* 1000 m of sedimentary cover (Fig. 4i) formed in a shallow epicontinental basin occupying the NE marginal part of the basin located within the flank of the Mid-Polish Swell. A Mid-Albian marine transgression was related to the development of a shallow siliciclastic shelf system (Cieśliński 1959; Marcinowski 1980; Leszczyński 1997, 2010, 2012). Following an early Cenomanian sea-level rise (cf. Hardenbol *et al.* 1998), marine sediments covered the entire area of SE Poland. As a result, the Cenomanian–Maastrichtian sedimentary cover is represented by a siliciclastic-carbonate shelf system and pelagic carbonates (cf. Leszczyński 1997, 2010, 2012). This sedimentary succession is rather monotonous and consists mostly of pelagic carbonates. An influx of terrigenous quartz has been observed locally that, together with the presence of phosphorite, hardgrounds, condensation surfaces and/or a stratigraphic gap, may suggest local episodes of uplift (Samsonowicz 1925; Walaszczyk 1987).

Some authors claim that in the Late Cretaceous, the axial part of the Polish Basin (i.e. the Mid-Polish Trough) continued to represent the deepest, most subsiding section of the basin (Kutek & Głazek 1972; Świdrowska 2007; Świdrowska & Hakenberg 1999; Świdrowska *et al.* 2008). Those classic views have been challenged by other authors that proposed the existence of morphological barriers (i.e. elevated areas) within the axial part of the basin since the Turonian (Remin *et al.* 2015, 2016; Walaszczyk & Remin 2015). Those observations are in line with the results of seismic data interpretation which proved that the Upper Cretaceous sedimentary cover is characterized by an overall thinning towards the SW: that is, towards the already partly uplifted Mid-Polish Swell (Krzywiec 2009; Krzywiec *et al.* 2009). Seismic data have also demonstrated the presence of Turonian(?)–Maastrichtian local unconformities and progadational sedimentary patterns directed from the SW towards the NE: that is from the axial part of the basin towards its NE flank (Krzywiec 2009; Krzywiec *et al.* 2009).

Within the Turonian succession, dominated by limestones and opokas (i.e. porous, flinty and calcareous sedimentary rock consisting of fine-grained silica and hardened by the presence of silica of organic origin (e.g. radiolaria, sponge spicula and diatoms)), locally developed Janików limestones are characterized by progradational patterns and transport of the bioclastic material orientated towards the NE (Pożaryski 1948; Walaszczyk 1992).

The Coniacian–Santonian deposits consist of gaizes (porous carbonate–siliceous rock containing detrital quartz and sponge spicules) with a significant admixture of quartz and glauconite (Pożaryski 1948; Radwański 1960; Walaszczyk 1992); Lower Santonian glauconitic opokas often contain fragments of non-glauconitic opokas that have been interpreted as a submarine slump breccia. They might have been formed due to inversion-related tectonic movements.

The Campanian–Maastrichtian carbonate–siliceous and carbonate facies are characterized by frequent occurrences of quartz and flora remains (Cieśliński & Milaković 1962). The fine-grained chalk facies were deposited only in the NE part of the study area, at a greater distance from the Mid-Polish Swell; while in the SW part of the basin, in proximity to the Mid-Polish Swell, Campanian deltaic deposits have been recognized (Remin *et al.* 2015; Walaszczyk & Remin 2015). They are characterized by a paleotransport direction from the SW towards the NE, determined using anisotropy of remnant magnetization (Remin *et al.* 2015). This suggests that during the Campanian–Maastrichtian, the sediment-source area was located within the axial part of the basin (Jaskowiak-Schoeneichowa & Krassowska 1988).

Data and methods

The seismic database used in this study included two different types of seismic profiles. A regional NW–SE profile of the PolandSPAN® seismic survey was used to study regional Mesozoic uplifts in SE Poland; particular focus was on the transition zone between the central and SE parts of the Polish Basin (cf. Figs 3 & 5): that is, on the area previously only scarcely studied using seismic reflection data. However, in this part of the basin, several deep wells are located that provided stratigraphic calibration. The PolandSPAN® regional high-end seismic survey was acquired by ION Geophysical above the SW edge of the East European Craton with a long recording time (12 s), high nominal fold (480), long offsets (12 km), tight receiver and shot spacing (25 m), and broadband sweep (2–150 Hz). Data processing included pre-stack time migration (PreSTM) and pre-stack depth migration (PSDM). Profile 1100 extends from the NW to SE Poland (cf. Krzywiec *et al.* 2014). In its central part, it is located in the area characterized by thick and almost complete Mesozoic sedimentary cover; while in its SE segment, it imaged a significantly reduced Mesozoic succession, both in terms of total thickness as

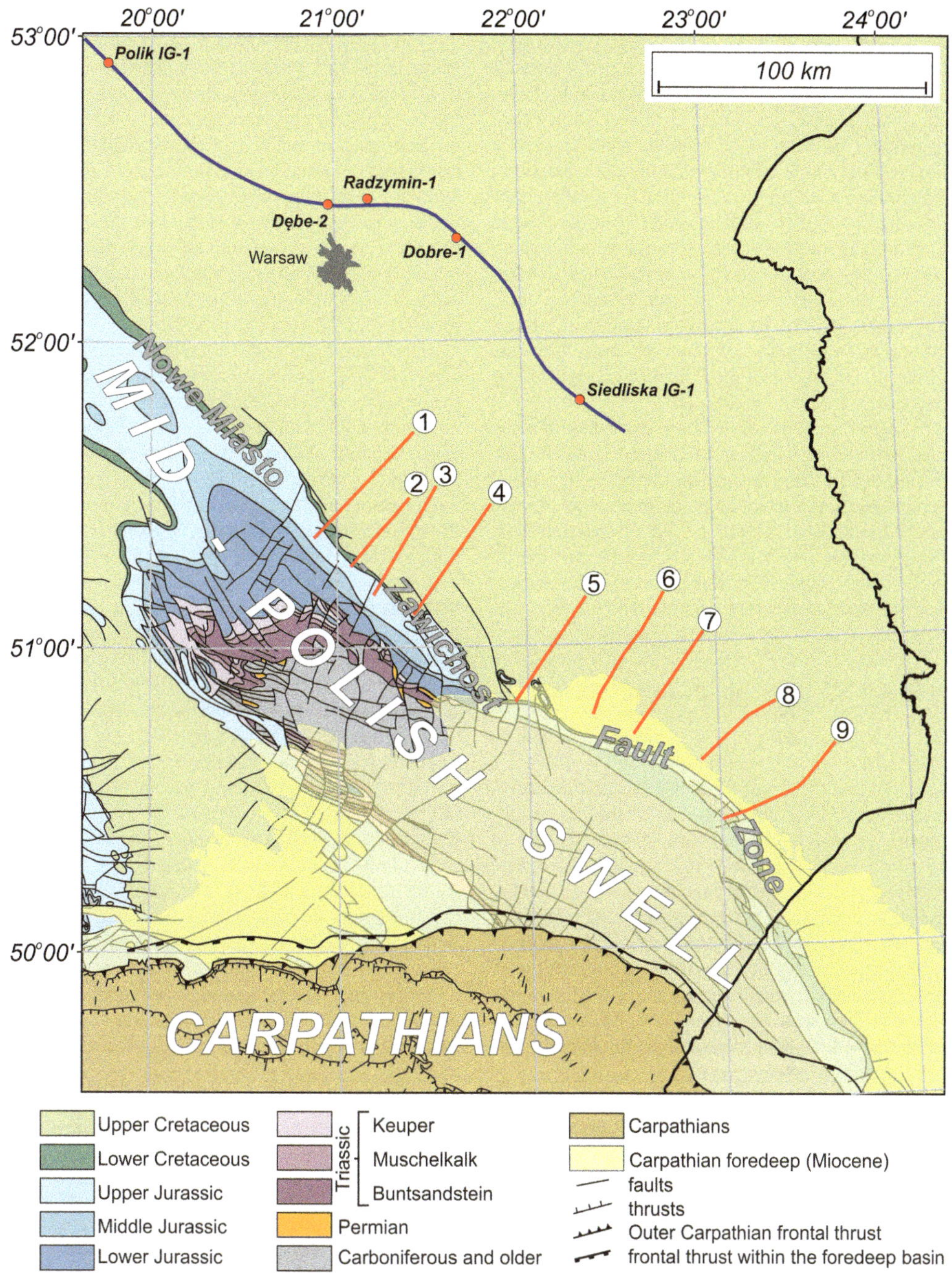

Fig. 5. Geological map of SE Poland (simplified after Pożaryski 1978; Dadlez *et al.* 2000). Dark blue line, regional seismic profile shown in Figure 7; red lines, seismic profiles shown in Figures 8 and 9; red dots, the locations of calibration wells.

well as stratigraphic profile (cf. Krzywiec *et al.* 2014). The quality of seismic imaging for the entire Permo-Mesozoic sedimentary cover is excellent, and provides detailed insight regarding, for example, long-distance thickness variations and the presence of regional unconformities.

In order to study Late Cretaceous inversion tectonics along the NE edge of the Mid-Polish Swell, nine seismic profiles were used (Fig. 5). Two of them (6 and 9) also belong to the PolandSPAN® survey; other profiles have been acquired during conventional exploration activities and processed through PreSTM. In order to guarantee compatibility of the obtained results, time versions of all these profiles, including two PolandSPAN® profiles, were used. For all these profiles, the quality of seismic imaging was considered high, with vertical resolution for the Permo-Mesozoic section of order of 10–30 m, which enabled the interpretation of numerous seismostratigraphic features related to the Late Cretaceous inversion of the Polish Basin and the formation of the Mid-Polish Swell.

A relatively rich well database consisting of more than 50 deep research wells drilled by the Polish Geological Institute and exploratory wells drilled by oil industry was used to calibrate the seismic data. Standardized formation tops were retrieved from the Central Geological Database operated by the Polish Geological Institute, loaded into the database and used to calibrate seismic data (cf. Fig. 6).

Regional Mesozoic uplifts in SE Poland

The central part of the Polish Basin underwent relatively steady Mesozoic subsidence and is characterized by a thick (3.5 km or more) Permo-Cretaceous sedimentary cover devoid of any significant stratigraphic gaps (Fig. 6). In contrast, the SE part of the basin is characterized by a rather incomplete and thinner (1 km or less) Mesozoic cover. Such lateral stratigraphic and thickness changes reflect consecutive stages in the evolution of various parts of the Polish Basin, as described above. Various previously published paleothickness maps (Fig. 4) have been based on well penetrations but with minimum input from the seismic data – some parts of the basin are sparsely covered by seismic lines (e.g. in the vicinity of, and SE of, Warsaw). This part of the Polish Basin is located between two domains characterized either by thick and complete Mesozoic sedimentary cover (NW of Warsaw) or by thin and incomplete Mesozoic sedimentary cover (SE of Warsaw). Therefore, it was reasonable to expect that, in this transition zone at least, a partial explanation regarding viable mechanisms of such lateral thickness and stratigraphic variations might be found. This area was covered by the PolandSPAN® seismic survey (cf. Krzywiec *et al.* 2014) and profile 1100 was used to study the detailed internal structure of the Mesozoic sedimentary cover (Fig. 7).

Polik IG-1, a deep research well drilled by the Polish Geological Institute in 1987, provided detailed stratigraphic calibration for profile 1100 (Fig. 6). Additional stratigraphic calibration for this profile was provided by exploratory wells Dębe-2 (1971), Radzymin-1 (1968) and Dobre-1 (1968), and research well Siedliska IG-1 (1970), all also located on this profile (Fig. 5). All key formation tops – base Permian, top Permian, top Lower Triassic, top Middle Triassic, top Upper Triassic, top Lower Jurassic, top Middle Jurassic, top Upper Jurassic and top Lower Cretaceous – were tied to the 1100 PolandSPAN® profile. The well-to-seismic tie was straightforward as depth (PSDM) seismic data were available. Stratigraphic information for all these wells was retrieved from the Central Geological Database.

The interpreted section of the regional seismic profile 1100 is shown in Figure 7. The most striking feature immediately visible on this profile is very a different depth of the top of the pre-Permian substratum in the two analysed segments of the Polish Basin (cf. also Krzywiec *et al.* 2014). In the central part, where Polik IG-1 well was drilled, the base of the Permo-Mesozoic succession is located at almost 4 km; while in the SE part, within the Lublin Basin, this stratigraphic boundary is at around 800–500 m. This trend continues towards the SE, towards the Polish–Ukrainian state border, where the base of the thin Mesozoic (mostly, or entirely, of Cretaceous age) succession is located at 200–300 m depth (cf. Krzywiec *et al.* 2014).

Another first-order observation, based on the internal depositional architecture of the Mesozoic cover, is the presence of two regional unconformities, numbered 1 and 2, visible in the transition zone between both studied segments of the Polish Basin, particularly in the area located SE from the Radzymin-1 well (Fig. 7). The lower unconformity (1), in the vicinity of the Polik IG-1 well, is located within the lowermost part of the Lower Jurassic (lowermost Oxfordian); in this part of the basin, it is not related to any angular relationship. Towards the SE, this unconformity variably truncates the Middle Jurassic–Lower Triassic section, in places also eroding the uppermost Carboniferous. The upper unconformity (2), in the vicinity of the Polik IG-1 well, is located within the lowermost part of the Upper Cretaceous (approximately top Cenomanian). Towards the SE, it truncates the Lower Cretaceous and Upper Jurassic. Angular contacts along this unconformity, within the hinge area between both analysed segments of the Polish Basin, are less pronounced than in the case of the Jurassic unconformity.

It is also important to note regional long-distance thickness variations within the Mesozoic cover. The Permian (mostly Zechstein) interval is characterized by a significant thinning from the central part of the basin towards its SE part that appears to be syndepositional. The Triassic succession exhibits less pronounced thinning, from *c.* 600 m in the central part of the basin to 400 m in the area located

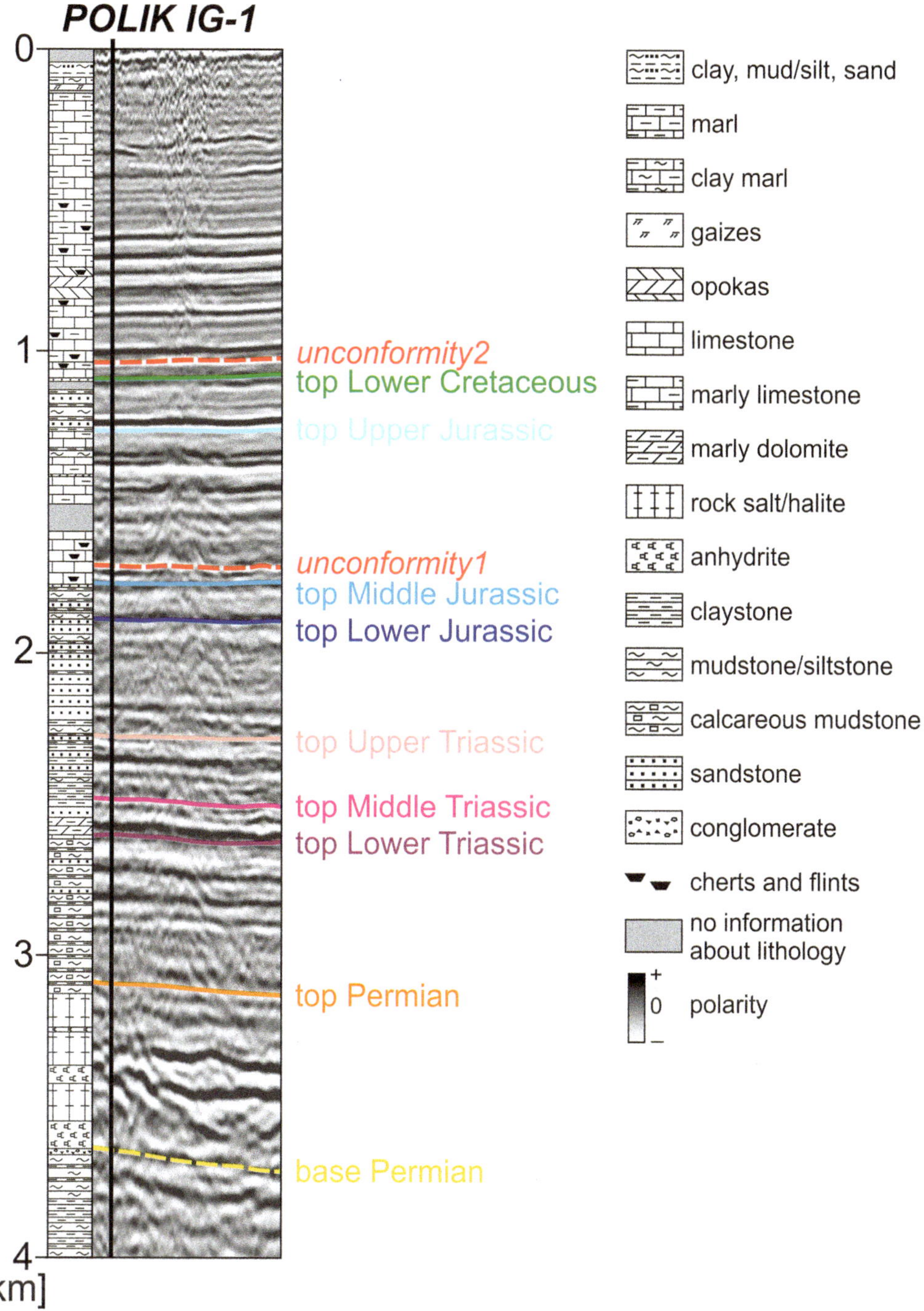

Fig. 6. Correlation of the Polik IG-1 well with PSDM seismic data.

immediately SE from Dobre-1 well, where a complete Triassic succession is still preserved (Fig. 7). Further towards the SE, the thickness eventually decreases to 0 m but this is due to erosion caused by unconformity 1. The Lower Jurassic interval exhibits significant depositional thinning towards

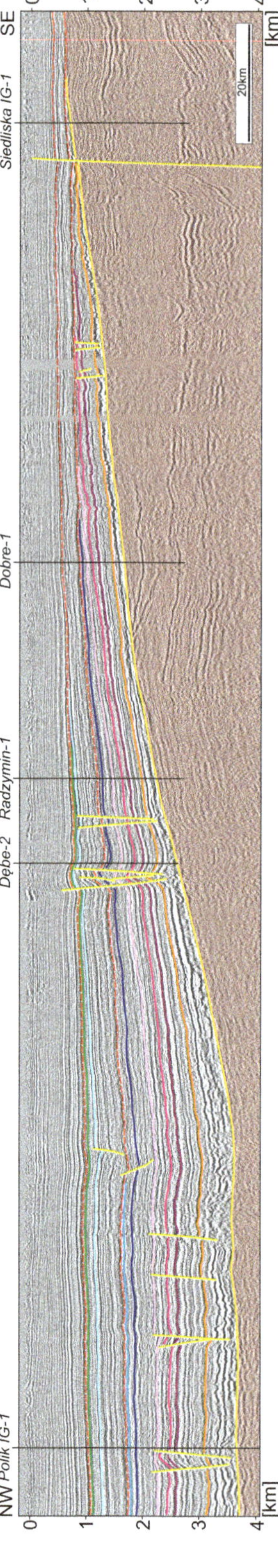

Fig. 7. Part of the regional PolandSPAN® seismic profile 1100. Vertical exaggeration ×10; the line is reflectivity in depth; see Figure 6 for the data polarity and stratigraphic key for the interpreted seismic horizons (cf. Krzywiec *et al.* 2014). The location is shown in Figure 5.

the SE, whereas the top of the Middle Jurassic is mostly erosional and the Upper Jurassic is characterized by only minor depositional thinning towards the SE. The Lower Cretaceous interval, fully truncated by unconformity 2 in the vicinity of the Radzymin-1 well, also – where preserved – exhibits depositional thinning towards the SE. A thick Upper Cretaceous succession is characterized by some thinning towards the SE. The observed thickness variations and angular unconformities within the Mesozoic succession have been used to recognize main uplift events related to the formation of both regional unconformities.

Seismostratigraphic record of Late Cretaceous inversion

Nine seismic profiles (Figs 8 & 9), perpendicular to the Mid-Polish Swell and located within its NE flank (Fig. 5), were selected in order to analyse Late Cretaceous inversion of the Polish Basin and the formation mode of the SE segment of the Mid-Polish Swell. Profiles 1–4 were presented by Krzywiec *et al.* (2009); in this study, an updated and modified model of structural control on Late Cretaceous inversion tectonics is presented.

All of these profiles are located within the border zone of two main Paleozoic units developed above the SW margin of the East European Craton: the Radom-Kraśnik Block and the Lublin Basin (e.g. Krzywiec 2009; Krzywiec *et al.* 2017*a*, *b*). The Radom-Kraśnik Block, built of the deformed Ediacaran–Devonian sedimentary succession, has been recently reinterpreted as the easternmost segment of the Variscan (Late Carboniferous) fold-and-thrust belt, with the most frontal compressional deformations developed within the Lublin Basin (Krzywiec *et al.* 2017*a*, *b*; see also Antonowicz *et al.* 2003).

Paleozoic units are overlain by the Permo-Mesozoic sedimentary cover formed within the SE segment of the Polish Basin. The NE edge of the Mid-Polish Swell, formed during inversion of this part of the basin and uplift of its axial part, is defined by the Nowe Miasto–Zawichost Fault Zone and its prolongation towards the SE (Fig. 5) (see also Krzywiec 2009; Krzywiec *et al.* 2009). Profile 9, in its SW end, imaged an erosional boundary between the deformed Meso-Paleozoic complex and the Miocene infill of the Carpathian foredeep basin (Krzywiec 2001).

Structural interpretation of seismic profiles from Figures 8 & 9 was focused on the tectonics of the Permo-Mesozoic complex. Faults within the Radom-Kraśnik Block and the Lublin Basin (shown by red hatched lines in Figs 8 & 9) have only been interpreted if they were related to Late Cretaceous inversion tectonics.

The top of the pre-Permian basement (substratum) is characterized by significant morphology. Local elevation changes could be related either to erosion (post-Variscan and enhanced in the Jurassic, see below) or localized uplift induced by Late Cretaceous inversion. This duality is best illustrated by profile 9 (Fig. 9). In its central part, local uplift of the top of the Paleozoic is related to mild reactivation of a deeper thrust fault located at the front of the Radom-Kraśnik Block (cf. Krzywiec *et al.* 2017*b*). This uplift was inversion-related as the entire Jurassic cover is also mildly folded and characterized by a constant thickness (*c.* 200 ms two-way time (TWT)) across this structure, which attests to post-depositional folding that took place in the latest Cretaceous (late Maastrichtian) as the entire Upper Cretaceous is also folded. Further to the NE, similar top Paleozoic surface morphology (top of Carboniferous of the Lublin Basin) can be observed. In this case, however, the origin is proposed to be different – here, the Jurassic is unconformably covering Paleozoic substratum and is characterized by ateral thickness variations: that is, reduced thickness (of the order of 20–30 ms TWT) above the highs and increased thickness (up to 100 ms TWT) above the lows. This attests to the erosional origin of these palaeomorphological features.

Significant gross thickness changes of the Permo-Mesozoic sedimentary cover are observed across the NW and central parts of the Nowe Miasto–Zawichost Fault Zone (profiles 1–6; see also Pożaryski 1997). In this part of the study area, the Permian–Jurassic succession is characterized by a prominently divergent pattern and associated thickness increase towards the SW (i.e. towards the Mid-Polish Swell). This thickness increase reflects increased subsidence within the basin's axial part related to the deposition of a thick sedimentary cover above the hanging wall of an extensional fault system that belongs to the border zone of the axial part of the Polish Basin i.e. of the Mid-Polish Trough (cf. Fig. 1, model 1). Profiles 7 and 8 (Fig. 9) do not cross the Mid-Polish Trough–Mid-Polish Swell border fault zone (cf. Fig. 3) and, hence, did not image thickness variations of the Jurassic succession. Profile 9 (Fig. 9) crosses this fault zone (cf. Fig. 3) but there post-Cretaceous erosion removed most of the Jurassic cover, so thickness variations are also not visible.

The Upper Cretaceous (Turonian?–Maastrichtian: cf. Krzywiec *et al.* 2009) succession is characterized by a different thickness pattern – within the hinge zone located immediately NE of the Nowe Miasto–Zawichost Fault Zone, seismic packages thin towards the SW and towards the NE (Figs 8 & 9). This low-angle progradational complex, previously recognized in the NW part of this area (profiles 1–4: Krzywiec *et al.* 2009), can be traced along the entire NE edge of the Mid-Polish Swell towards

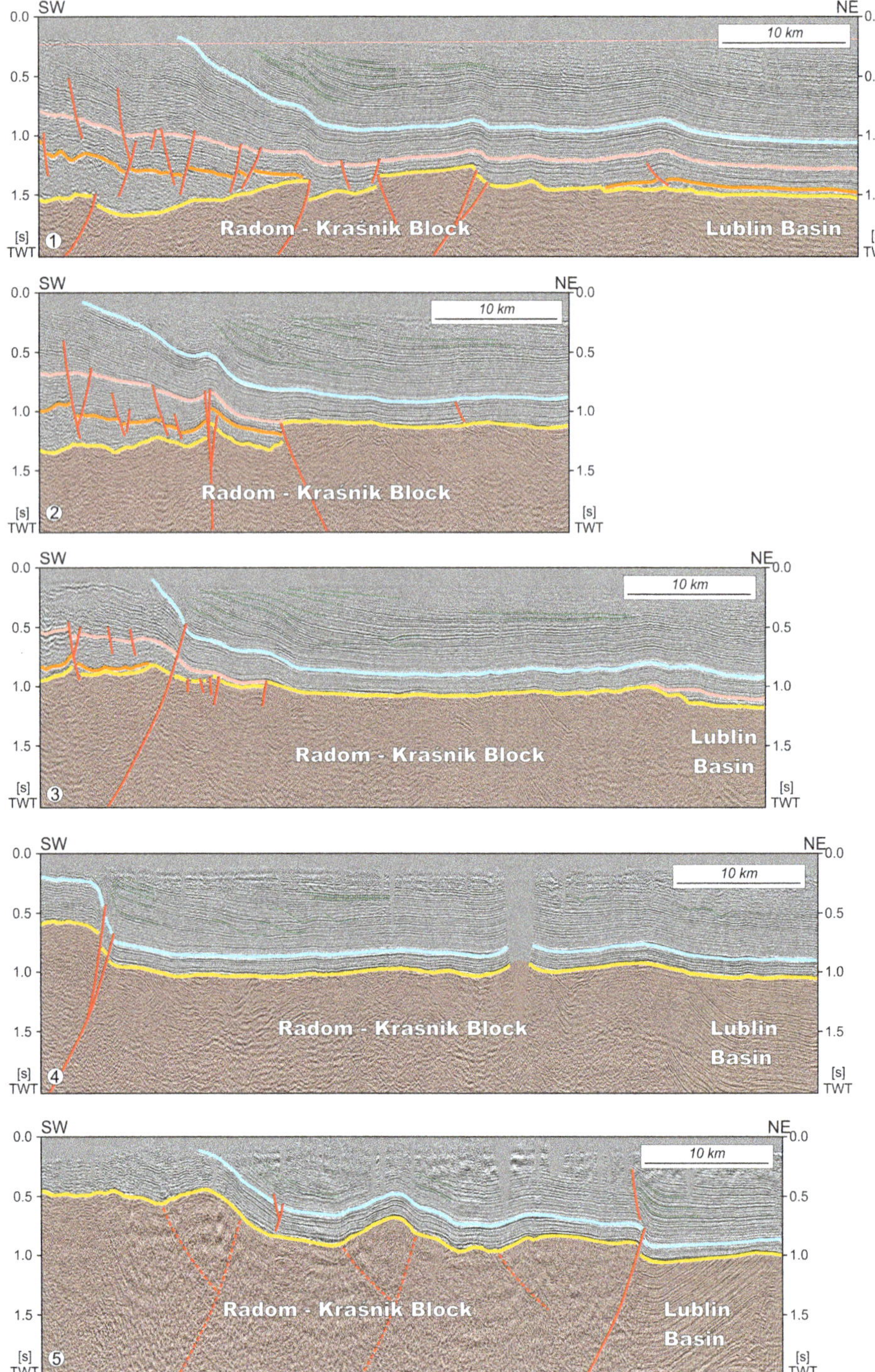

Fig. 8. Seismic profiles 1–5 documenting Late Cretaceous inversion along the NE edge of the SE segment of the Mid-Polish Trough (after Krzywiec 2015). Sections are reflectivity in two-way time (TWT). Dark yellow horizon, top of the pre-Permian basement (substratum); orange horizon, top of the Permian; pink horizon, top of the Triassic; blue horizon, top of the Jurassic; green dotted lines, seismostratigraphic features within the Upper Cretaceous (primarily Santonian–Maastrichtian: cf. Krzywiec *et al.* 2009). The location is shown in Figure 5.

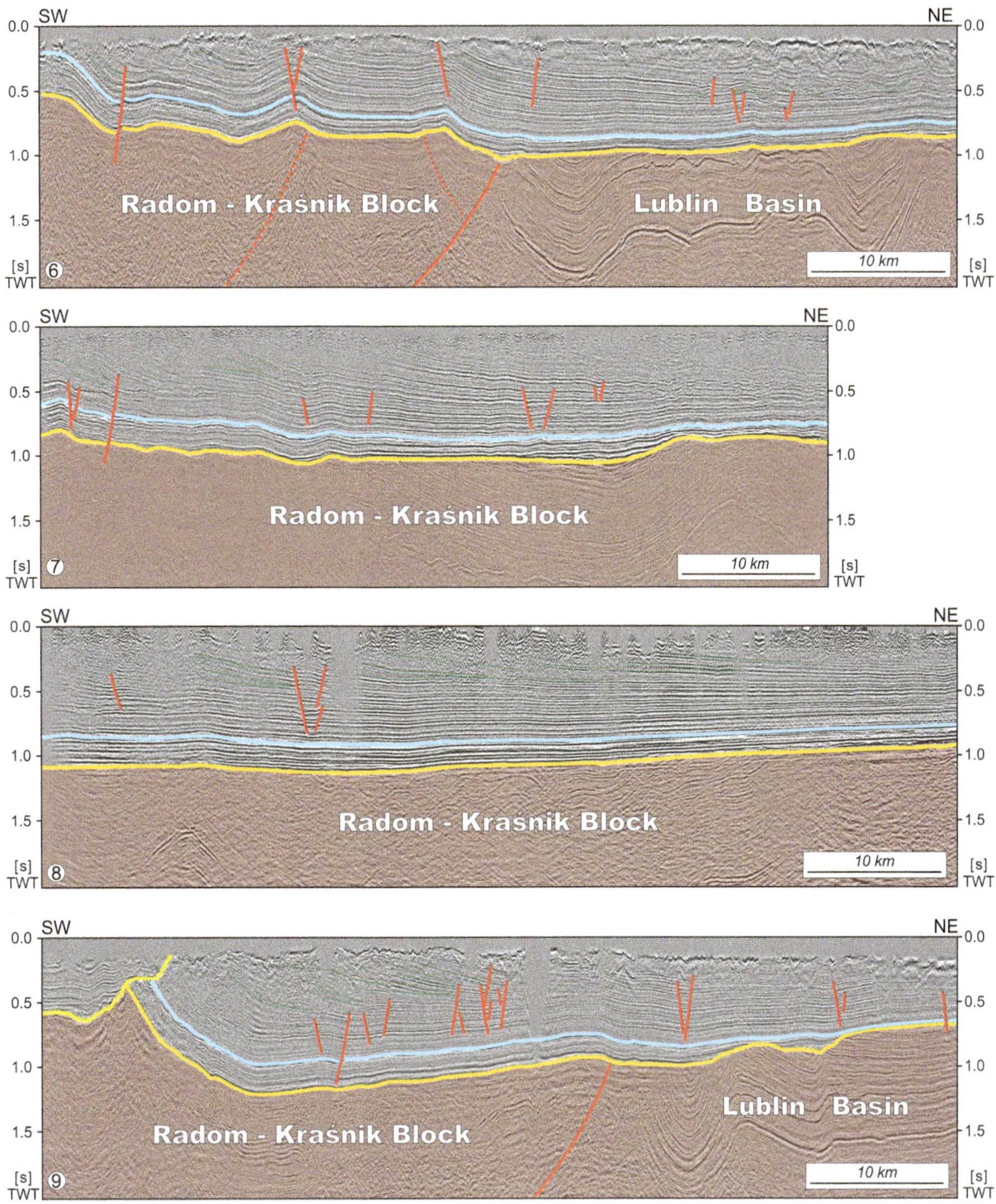

Fig. 9. Seismic profiles 6–9 documenting Late Cretaceous inversion along the NE edge of the SE segment of the Mid-Polish Trough (after Krzywiec 2015). Sections are reflectivity in two-way time (TWT). Dark yellow horizon, top of the pre-Permian basement (substratum); blue horizon, top of the Jurassic; yellow horizon, base of the Miocene (Carpathian foredeep); green dotted lines, seismostratigraphic features within the Upper Cretaceous succession (primarily Santonian–Maastrichtian: cf. Krzywiec *et al.* 2009). The location is shown in Figure 5.

the Polish–Ukrainian state border, and is likely to continue beyond this. Such depositional architecture attests to a general direction of sediment supply from the SW (i.e. from the Mid-Polish Swell) being progressively uplifted during the Late Cretaceous. Such a seismostratigraphic pattern could be compared to model 1b in Figure 1, with the Upper Cretaceous synkinematic strata formed due to uplift and erosion of the hanging wall of an inverted basement fault. The progradational and aggradational character of the Turonian?–Maastrichtian succession implies high sediment supply and high (rising) sea level. High sediment supply could have been related to the substantial inversion-related uplift of the axial part of the basin and subsequent erosion; a high sea level characterizes the entire Late Cretaceous (Hardenbol *et al.* 1998).

Characteristic sedimentary features have been recognized within the Upper Cretaceous (Campanian–Maastrichtian) succession which have been interpreted as buried contourite drifts (cf. Faugères *et al.* 1999; Krzywiec *et al.* 2009). They are visible on all seismic profiles located along the NE edge of the Mid-Polish Swell, between the city of Radom and the Polish–Ukrainian state border (cf. Fig. 5). These features may represent deposits of contour currents which were flowing along the slope that was formed by the Mid-Polish Swell, being progressively uplifted in the Late Cretaceous. This interpretation is compatible with a regional palaeocirculation model currently proposed by Remin *et al.* (2016).

Discussion

Seismic interpretation has shed new light regarding the Mesozoic history of SE Poland, in particular for its two aspects:

(1) gross depositional patterns of the Permian–Cretaceous sedimentary succession v. regional uplifts within the transition zone between the central and SE segments of the Polish Basin;

(2) Late Cretaceous inversion tectonics within the axial part of the Polish Basin and the formation of the Mid-Polish Swell

In order to analyse the first problem, a conceptual model illustrating consecutive stages of Mesozoic evolution of the Polish Basin along the regional transect located between its central and SE segments was constructed (Fig. 10). This model was based on the depth seismic profile from Figure 7 flattened on the top Triassic (Fig. 10a), unconformity 1 (Fig. 10b) and unconformity 2 (Fig. 10c; cf. Fig. 6); Figure 10d shows a simplified recent configuration. Note that the effect of compaction has not been taken into account in this model. The recent configuration is schematic; in particular, it does not include any localized inversion-related deformation visible on regional seismic profile (cf. Fig. 7).

The end-Triassic configuration (Fig. 10a) illustrates that along this regional transect, the entire Triassic sequence might have continued considerably farther to the SE, beyond the area of its present-day extent which is largely controlled by erosion. An estimate, based on seismic profile flattened on top Triassic, suggests that the Triassic thickness within the SE segment of the Polish Basin was considerably larger (of the order of 500–800 m) than previously assumed (100–200 m maximum: cf. Fig. 4b–d).

Figure 10b illustrates how unconformity 1 might have been formed during considerable uplift of the SE segment of the Polish Basin that led to extensive erosion. Stratigraphic correlation with the Polik IG-1 well shows that this uplift took place during the earliest Late Jurassic (earliest Oxfordian). It should be stressed, however, that several wells from the SE segment of the Polish Basin (including the Siedliska IG-1 well) will also have a thin veneer of Middle Jurassic (Bathonian–Callovian) deposits preserved beneath the unconformity (Pieńkowski *et al.* 2008). This may suggest that either uplift was diachronous, or there might have been another, earlier episode of regional uplift within this part of the basin. In order to fully assess this problem, extensive mapping of the Jurassic succession within the entire SE segment of the Polish Basin would be necessary which should be coupled with detailed re-evaluation of the Jurassic well stratigraphy. The magnitude of this uplift might be tentatively estimated in range of several hundreds of metres to 1 km.

The second phase of uplift along the analysed regional transect, although of lesser magnitude, took place in the earliest Late Cretaceous (Fig. 10c). It also produced an extensive erosional surface and associated unconformity encroaching into the SE segment of the Polish Basin, where it separates Upper Jurassic and Upper Cretaceous deposits. In a regional context, this uplift might have been related to an early phase of Late Cretaceous inversion of the Polish Basin.

The formation of both regional unconformities can be compared to model 2 in Figure 1, where regional uplift and basin inversion are not related to the reactivation of a discrete basement fault but to regional geodynamic processes, resulting in basin-wide uplift and ensuing widespread erosion. An example of such a process is the uplift of the North Sea Dome, which strongly influenced the geological evolution of large parts of western and central Europe in the Middle Jurassic (Underhill & Partington 1993). The North Sea Dome was a broad arc comprising the Central North Sea, the Ringkobing-Fyn High, the Danish Basin and the Bornholm areas (Fig. 2) (Pieńkowski *et al.* 2008; Lott *et al.* 2010). Thermal uplift of this extensive dome resulted in erosion of the Lower Jurassic and Triassic sedimentary cover; in the Central North Sea area, it created the so-called Mid-Cimmerian Unconformity. The Mid-Jurassic uplift, along with its conjoined major tectonic activity at the triple junction of the Viking, Central and Moray Firth graben, caused separation between the Tethys, the Central Atlantic oceans and the Arctic seas (Ziegler 1990; Pharaoh *et al.* 2010). The regional high formed by this uplift was subjected to extensive erosion. The Mid-Jurassic uplift events also caused erosion in other parts of Europe, sometimes even down to the Paleozoic

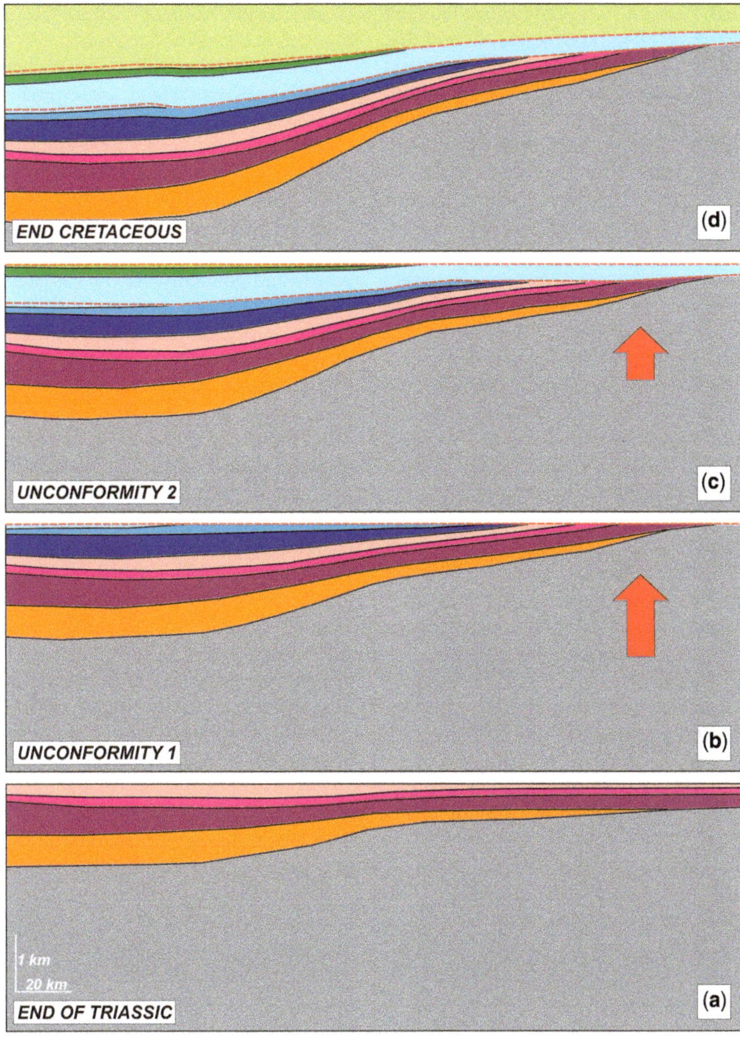

Fig. 10. A conceptual model illustrating the Permo-Mesozoic evolution of the transition zone between the central and SE segments of the Polish Basin based on seismic profile from Figure 7. Vertical exaggeration is approximately ×20, in order to better visualize internal geometries of the Triassic–Cretaceous sedimentary cover related to consecutive stages of regional subsidence and uplift. (**a**) Tentative and simplified end-Triassic configuration; (**b**) formation of unconformity 1; (**c**) formation of unconformity 2; (**d**) tentative and simplified end-Cretaceous configuration.

interval; for example, the Lower Carboniferous core of the London-Brabant Massif was exposed and, as indicated by thermochronological studies, up to 3000 m of sediments might have been eroded in this area during this pervasive erosional event (Van den Haute & Vercoutere 1990; Pharaoh *et al.* 2010). The first stage of the Central North Sea uplift took place towards the end of the Aalenian. Its inferred cause was possibly the transient mantle plume impinging on the lithosphere, which persisted through the Bajocian and Bathonian (Ziegler 1990; Japsen *et al.* 2007; Pieńkowski *et al.* 2008; Pharaoh *et al.* 2010). During this stage, the eastern part of the Central North Sea Dome was uplifted, and this resulted in a major supply of clastic sediments from the Baltic and the North Sea areas into the North German Basin and further south into the Lower Saxony Basin (Pharaoh *et al.* 2010). By the early Bajocian, the Ringkobing-Fyn High was re-established. As a result of this doming episode, up to 2000 m of older sediments were removed in the Central North Sea area, which is suggested to have been uplifted slightly earlier than the Pompeckj Block in the northern part of the South Permian

Basin (Baldschuhn *et al.* 1996; Jaritz 1987; Pharaoh *et al.* 2010). However, it should be mentioned that the eastern part of the North Sea Basin area was not subjected to a strong Jurassic uplift and erosion, as can be seen in Horn and Glueckstadt graben where erosion probably did not exceed 500 m (Mogensen 1995; Beha *et al.* 2008; Maystrenko *et al.* 2006, 2008). The North East German Basin was also unevenly affected by the Jurassic uplift and erosion; in its central part, *c.* 830 m of strata were eroded; while towards both basin margins, erosion decreased to 350 m. However, it was also suggested that towards the southernmost part of the basin, the amount of erosion rises to 600 m (Kossow & Krawczyk 2002). The uplift in the NE part of the North German Basin seems to be of Late Jurassic age, and probably lasted until the Early Cretaceous (Ziegler 1990; Pharaoh *et al.* 2010). Moreover, even though in the NW part of the North German Basin mid-Late Jurassic uplift took place, its SE segment reached its maximum extent during this time (Meyer & Schmidt-Kaler 1996; Pharaoh *et al.* 2010). Furthermore, while the North German Basin was subjected to Late Jurassic uplift and erosion, the North Sea area underwent the most intense phase of subsidence during this time (Badley *et al.* 1988; Mazur & Scheck-Wenderoth 2005). This phenomenon was explained by Mazur & Scheck-Wenderoth (2005) as having been caused by an initial, Middle Jurassic thermal doming in the axial zone of the North Sea Basin and subsequent outwards migration of the rift shoulder. The mid-Late Jurassic uplift led also to a reactivation of the basement faults that caused uplift and erosion of the Hessian Depression. The NW-trending Saxonian Straight transected the Bohemian Massif during the Callovian times, which led to the formation of a new marine connection between the North German Basin and the Tethyan shelves of the Carpathian domain (Pieńkowski *et al.* 2008). During the mid-Cimmerian event, the majority of the Dutch offshore area was also uplifted. In the north, adjacent to the Dutch Central Graben, the Cleaver Bank High underwent the most severe erosion episode which removed up to 2000 m of Triassic and Lower Jurassic sediments (Glennie 1986; Pharaoh *et al.* 2010). Also, the Terschelling Basin and Step Graben were subjected to uplift and erosion (Pharaoh *et al.* 2010). The mid-Cimmerian event also led to the uplift of the Norwegian–Danish Basin (Nielsen 2003; Pharaoh *et al.* 2010). As this short overview shows, the mid-Jurassic was a time of formation of regional uplifts associated with pervasive erosion and the formation of regional unconformities. Large-scale uplift documented in the SE segment of the Polish Basin might have been at last partly related to the same regional processes that were operating in western and central parts of the European epicontinental basins. Further studies would, however, be needed to fully document such a potential causal relationship.

Finally, it should also be mentioned that the proposed uplift events and the estimated thickness of eroded Mesozoic sedimentary cover would have a significant impact on the modelling of tectonic subsidence and the history of Paleozoic petroleum systems in the Lublin Basin. Previously completed modelling studies for SE Poland (Poprawa 2007*a*, *b*, 2008, 2012; Narkiewicz *et al.* 2010) relied on tectonic scenarios that generally adopted a palaeogeographical approach, which is summarized on maps shown in Figure 4. These studies, however, were mainly focused on examining the Paleozoic history of the Lublin Basin, and the Mesozoic evolution was regarded as less important due to its presumably small influence on the development of the petroleum systems. According to Poprawa (2007*a*, *b*, 2008, 2012), during Mesozoic times the area was subjected only to short pulses of increased subsidence, but no major uplift events occurred. A recent study by Schito *et al.* (2017), carried out in the adjacent Holy Cross Mountains area, revealed that, during the Late Cretaceous, a strong uplift event might have taken place and, as a result, about 3500–4300 m of sediment were eroded. Different assumptions regarding the pre-erosional thickness of Triassic and Jurassic based on results presented in this paper would then modify Mesozoic burial v. uplift scenarios, and, as a consequence, the burial of the Paleozoic source rock and other elements of petroleum system.

Late Cretaceous inversion led to the formation of a system of inverted Mesozoic basins and upthrust blocks within the pre-Permian basement in western and central Europe (Fig. 2) (e.g. Dadlez 1980; Liboriussen *et al.* 1987; Norling & Bergström 1987; Schröder 1987; Vejbæk & Andersen 1987; Stackenbrandt & Franzke 1989; Dronkers & Mrozek 1991; Michelsen & Nielsen 1993; Mogensen & Jensen 1994; Erlström *et al.* 1997; Michelsen 1997; Gras & Geluk 1999; Kossow & Krawczyk 2002; de Lugt *et al.* 2003; Kockel 2003; Otto 2003; Mazur *et al.* 2005; Krzywiec 2006; Scheck-Wenderoth *et al.* 2008; see also Krzywiec *et al.* 2009 and Pharaoh *et al.* 2010 for detailed overviews and further references). Most frequently, Late Cretaceous inversion within the European Plate is associated with compressional/transpressional stresses related to successive tectonic phases within the Alpine–Carpathian collisional zone transmitted into its foreland (Ziegler 1987; Ziegler *et al.* 2002). Within the Carpathian domain, numerous collisional events could be identified that might be linked with inversion tectonics in the Carpathian foreland, as during the Late Cretaceous both the Inner Carpathians and the Pieniny Klippen Belt were subjected to several shortening and subduction events (see Krzywiec 2002 for a detailed overview). These collisional events might have generated pulses of regional compressional stresses that could have been transferred into the foreland plate and might have led to inversion of the foreland intracontinental basins, such as the Polish Basin or the Dnipro–Donetsk Basin (Maystrenko *et al.* 2003). Such a causal link would, however, indicate that the Carpathian orogenic system was mechanically coupled with its foreland. In this context, it must be stressed that in the Late Cretaceous large sedimentary basins of the Outer Carpathians existed between the Inner Carpathian domain and its European foreland. These basins were underlain by either oceanic or thinned continental crust and characterized, especially during Albian–Cenomanian/Turonian, by rather uniform pelagic sedimentation in the abyssal depths (see Ślączka *et al.* 2006 for a detailed overview and further references). It remains questionable, therefore, as to what extent compressional stresses generated by collision within the Inner Carpathian and the Pieniny Klippen Belt domains could have been transferred across vast sedimentary basins without a major effect on their sedimentation and tectonic pattern, but which would have led at the same time to the significant inversion of much more distant basins within the European foreland. An alternative model for the Late Cretaceous inversion in Europe has been proposed by Kley & Voigt (2008). It links the onset of intraplate contraction and basin inversion with regional changes in relative motion between the European and African plates. Regardless of its cause, the Late Cretaceous inversion within the European Plate resulted in the uplift of previously subsiding areas and a major rearrangement of gross depositional patterns.

As the results of the interpretation of seismic profiles from the NE border zone of the Mid-Polish Swell demonstrate, such a process operated on large scale within the SE segment of the Polish Basin. Progressive uplift of the axial part of the basin resulted in a reversal of the depositional systems, with gross sediment supply directed from the SW towards the NE: that is, from the inverted axial part of the basin towards its flanking area. Results presented in this paper also provide information on the timing of inversion tectonics in this part of the basin – an issue widely discussed in the past (cf. Dadlez *et al.* 1995; Leszczyński & Dadlez 1999; Świdrowska & Hakenberg 1999, 2000; Hakenberg & Świdrowska 2001; Świdrowska *et al.* 2008). Świdrowska & Hakenberg (1999) suggested that the earliest evidence of inversion of the SE segment of the Polish Basin could be observed in the Maastrichtian. Here, it is suggested that inversion-related uplift of the Mid-Polish Swell commenced in the Turonian, terminating in latest Maastrichtian–Paleogene times. Studies based on subsidence and thermal modelling also point to the Late Turonian/Coniacian–Maastrichtian/Paleogene inversion of the Polish Basin (Resak *et al.* 2008), in general accordance with the results based on seismic data interpretation.

Due to a lack of any deep seismic data, previously proposed models of Late Cretaceous uplift along the Nowe Miasto–Zawichost Fault Zone tentatively associated the process with deeply rooted reverse faults (Krzywiec *et al.* 2009: cf. model 1 shown in Fig. 1). Recently acquired deep seismic reflection data, such as PolandSPAN® survey and Polcrust-01 profile (Krzywiec *et al.* 2017*a*, *b*), have shown that the SW part of the study area is underlain by a thin-skinned Variscan thrust belt which has been overthrust above the SW edge of the East European Craton. This observation has allowed for the proposal that some of the faults that controlled Late Cretaceous inversion might have been reactivated Variscan thrust faults of the Radom–Kraśnik Block. This is visible, for example, on profiles 5, 6 and 9 shown in Figures 8 & 9.

Conclusions

Interpretation of seismic data from the SE segment of the Polish Basin allowed an improved constraint of the major Jurassic and Cretaceous uplift phases that strongly affected this part of the Polish Basin. Earliest Late Jurassic (earliest Oxfordian) and earliest Late Cretaceous (Cenomanian) regional uplifts have been documented by the interpretation of regional NW–SE seismic profiles that led to the formation of two regional unconformities. Unconformity 1 may be either partly diachronous or may have been preceded by another phase of erosion, as suggested by the presence of locally preserved thin upper Middle Jurassic (Bathonian–Callovian) deposits beneath the Upper Jurassic and above the deformed Carboniferous or Devonian of the Lublin Basin and the Radom-Kraśnik Block. Large-scale Jurassic uplift of the SE segment of the Polish Basin might have been, at least partly, related to the mid-Jurassic uplift of the North Sea Dome. It strongly influenced the geological evolution of large parts of western and central Europe, and was associated with the pervasive erosion and formation of regional unconformities. Despite the general temporal coincidence, the causal relationship between these two tectonic events still needs to be proved.

Late Cretaceous inversion tectonics within the Polish Basin were focused within its axial part and led to the formation of the Mid-Polish Swell; during its early stages, it might also have influenced the NE flank of the basin and might have led to the formation of unconformity 2 documented using regional seismic data. Within the axial part of the basin, reverse faulting of the pre-Permian basement (substratum) was, at least partly, related to reactivation of the Variscan thrust faults present within the Radom–

Kraśnik Block. Uplift of the border zone of the Mid-Polish Swell was recorded by the Upper Cretaceous depositional cover formed along the Nowe Miasto–Zawichost Fault Zone. It is characterized by an overall progradational character directed from the SW towards the NE (i.e. away from the uplifted and eroded Mid-Polish Swell). Such a gross depositional pattern is fully compatible with recently published results of palaeocurrent studies based on the anisotropy of remnant magnetization. Seismostratigraphic features, possibly representing buried contourite drifts, have also been recognized within the Upper Cretaceous succession along the entire NE edge of the SE segment of the Mid-Polish Swell. They might have been formed by contour currents encircling intrabasinal morphological barriers formed due to inversion-related uplift of the basement blocks of the Mid-Polish Swell.

A new tectonic scenario of the Mesozoic evolution of SE Poland would have a significant impact on the modelling of tectonic subsidence and the history of petroleum systems. Inclusion of documented regional Jurassic and Cretaceous uplifts into the basin-modelling studies would modify Mesozoic burial v. uplift scenarios. This, in turn, would modify the burial history of the source rock and other elements of the petroleum system.

ION Geophysical and Chevron Polska Energy Resources are thanked for providing seismic data for research studies carried out at the Institute of Geological Sciences, Polish Academy of Sciences. IHS Kingdom kindly provided seismic interpretation software. Very helpful comments by Yuriy Maystrenko, an anonymous reviewer and the volume editor Stanisław Mazur are acknowledged with many thanks.

References

ANTONOWICZ, L., HOOPER, R. & IWANOWSKA, E. 2003. Lublin Syncline as a result of thin-skinned Variscan deformation (SE Poland). *Przegląd Geologiczny*, **51**, 344–350 [in Polish with English summary].

BADLEY, M.E., PRICE, J.D., RAMBECH DAHL, C. & AGDESTEIN, T. 1988. The structural evolution of the northern Viking Graben and its bearing upon extensional modes of basin formation. *Journal of the Geological Society, London*, **145**, 455–472, https://doi.org/10.1144/gsjgs.145.3.0455

BALDSCHUHN, R., FRISCH, U. & KOCKEL, F. (eds) 1996. *Geotektonischer Atlas von NW-Deutschland – Tectonic Atlas of NW-Germany*. Bundesanstalt für Geowissenschaften und Rohstoffe (BGR), Hannover, Germany.

BALLY, A.W. 1984. Tectogenèse et sismique réflexion. *Bulletin de la Société géologique de France*, **S7-XXVI**, 279–285, https://doi.org/10.2113/gssgfbull.S7-XXVI.2.279

BEHA, A., THOMSEN, R.O. & LITTKE, R. 2008. Thermal history, hydrocarbon generation and migration in the Horn Graben in the Danish North Sea: a 2D basin modelling study. *International Journal of Earth Sciences*, **97**, 1087–1100.

BŁASZKIEWICZ, A. 1980. *Campanian and Maastrichtian ammonites of the Middle Vistula River Valley, Poland: A Stratigraphic–Paleontological Study*. Prace Instytutu Geologicznego, **92** [in Polish with English summary].

BRUN, J.-P. & NALPAS, T. 1996. Graben inversion in nature and experiments. *Tectonics*, **15**, 677–687, https://doi.org/10.1029/95TC03853

BUCHANAN, J.G. & BUCHANAN, P.G. (eds). 1995. *Basin Inversion*. Geological Society, London, Special Publications, **88**, https://doi.org/10.1144/GSL.SP.1995.088.01.30

CARTWRIGHT, J.A. 1989. The kinematics of inversion in the Danish Central Graben. *In*: COOPER, M.A. & WILLIAMS, G.D. (eds) *Inversion Tectonics*. Geological Society, London, Special Publications, **44**, 153–175, https://doi.org/10.1144/GSL.SP.1989.044.01.10

CIEŚLIŃSKI, S. 1959. *The Albian and Cenomanian in the Northern Periphery of the Holy Cross Mountains (Stratigraphy Based on Cephalopods)*. Prace Instytutu Geologicznego, **28** [in Polish with English summary].

CIEŚLIŃSKI, S. & MILAKOVIĆ, B. 1962. Cretaceous vertebrates and Cretaceous flora from the Mesozoic rocks surrounding the Święty Krzyż Mts. *Biuletyn Instytutu Geologicznego*, **174**, 245–266 [in Polish with English summary].

COOPER, M. & WARREN, M.J. 2010. The geometric characteristics, genesis and petroleum significance of inversion structures. *In*: LAW, R.D., BUTLER, R.W.H., HOLDSWORTH, R.E., KRABBENDAM, M. & STRACHAN, R.A. (eds) *Continental Tectonics and Mountain Building: The Legacy of Peach and Horne*. Geological Society, London, Special Publications, **335**, 827–846, https://doi.org/10.1144/SP335.33

COOPER, M.A. & WILLIAMS, G.D. (eds). 1989. *Inversion Tectonics*. Geological Society, London, Special Publications, **44**, https://doi.org/10.1144/GSL.SP.1989.044.01.25

DADLEZ, R. 1980. Tectonics of the Pomeranian Swell (NW Poland). *Geological Quarterly*, **24**, 741–767.

DADLEZ, R., NARKIEWICZ, M., STEPHENSON, R.A., VISSER, M.T.M. & VAN WESS, J.-D. 1995. Tectonic evolution of the Mid-Polish Trough: modelling implications and significance for central European geology. *Tectonophysics*, **252**, 179–195, https://doi.org/10.1016/0040-1951(95)00104-2

DADLEZ, R., MAREK, S. & POKORSKI, J. (eds) 1998. *Palaeographical Atlas of the Epicontinental Permian and Mesozoic in Poland (1:2 500 000)*. Polish Geological Institute, Warszawa, Poland [in Polish with English summary].

DADLEZ, R., MAREK, S. & POKORSKI, J. (eds). 2000. *Geological Map of Poland without Cainozoic Deposits, 1:1 000 000*. Państwowy Instytut Geoleologiczny, Warszawa, Poland.

DAYCZAK-CALIKOWSKA, K. 1997*a*. Middle Jurassic. Formal and informal lithostratigraphic units. Polish Lowlands. *In*: MAREK, S. & PAJCHLOWA, M. (eds) *The Epicontinental Permian and Mesozoic in Poland. Prace Państwowego Instytutu Geologicznego*, **153**, 263–264 [in Polish with English summary].

DAYCZAK-CALIKOWSKA, K. 1997*b*. Middle Jurassic. Sedimentation, paleogeography and paleotectonics.

In: MAREK, S. & PAJCHLOWA, M. (eds) *The Epicontinental Permian and Mesozoic in Poland. Prace Państwowego Instytutu Geologicznego*, **153**, 269–282 [in Polish with English summary].

DECZKOWSKI, Z. 1997*a*. Upper Triassic. Norian-Rhaetian. Sedimentation, paleogeography and paleotectonics. *In*: MAREK, S. & PAJCHLOWA, M. (eds) *The Epicontinental Permian and Mesozoic in Poland. Prace Państwowego Instytutu Geologicznego*, **153**, 187–194 [in Polish with English summary].

DECZKOWSKI, Z. 1997*b*. Lower Jurassic. Sedimentation, paleogeography and paleotectonics. *In*: MAREK, S. & PAJCHLOWA, M. (eds) *The Epicontinental Permian and Mesozoic in Poland. Prace Państwowego Instytutu Geologicznego*, **153**, 208–217 [in Polish with English summary].

DEMBOWSKA, J. 1979. Systematization of lithostratigraphy of the Upper Jurassic in northern and central Poland. *Geological Quarterly*, **23**, 617–630 [in Polish with English summary].

DE LUGT, I.R., VAN WEES, J.D. & WONG, T.E. 2003. The tectonic evolution of the southern Dutch North Sea during the Palaeogene: basin inversion in distinct pulses. *Tectonophysics*, **373**, 141–159, https://doi.org/10.1016/S0040-1951(03)00284-1

DEWEY, J.F. 1989. Kinematics and dynamics of basin inversion. *In*: COOPER, M.A. & WILLIAMS, G.D. (eds) *Inversion Tectonics*. Geological Society, London, Special Publications, **44**, 352, https://doi.org/10.1144/GSL.SP.1989.044.01.20

DRONKERS, A.J. & MROZEK, F.J. 1991. Inverted basins of the Netherlands. *First Break*, **9**, 409–425, https://doi.org/10.3997/1365-2397.1991019

ERLSTRÖM, M., THOMAS, S.A., DEEKS, N. & SIVHED, U. 1997. Structure and tectonic evolution of the Tornquist Zone and adjacent sedimentary basins in Scania and the southern Baltic Sea area. *Tectonophysics*, **271**, 191–215, https://doi.org/10.1016/S0040-1951(96)00247-8

FAUGÈRES, J.-C., DORRIK, A.V., IMBERT, P. & VIANA, A. 1999. Seismic features diagnostic of counturite drifts. *Marine Geology*, **162**, 1–38, https://doi.org/10.1016/S0025-3227(99)00068-7.

FEIST-BURKHARDT, S., GÖTZ, A. ET AL. 2008. Triassic. *In*: MCCANN, T. (ed.) *The Geology of Central Europe. Volume 2: Mesozoic and Cenozoic*. Geological Society, London, 749–821, https://dx.doi.org/10.1144/CEV2P.1.

FELDMAN-OLSZEWSKA, A. 1998*a*. Lower Jurassic – thickness. *In*: DADLEZ, R., MAREK, S. & POKORSKI, J. (eds) *Palaeographical Atlas of the Epicontinental Permian and Mesozoic in Poland (1:2 500 000)*. Polish Geological Institute, Warszawa, Poland, plate 36 [in Polish with English summary].

FELDMAN-OLSZEWSKA, A. 1998*b*. Middle Jurassic – thickness. *In*: DADLEZ, R., MAREK, S. & POKORSKI, J. (eds) *Palaeographical Atlas of the Epicontinental Permian and Mesozoic in Poland (1:2 500 000)*. Polish Geological Institute, Warszawa, Poland, plate 49 [in Polish with English summary].

GAJEWSKA, I. 1971. Muschelkalk of the western Poland. *Geological Quarterly*, **15**, 77–86 [in Polish with English summary].

GAJEWSKA, I. 1997*a*. Middle Triassic (Muschelkalk-Lower Keuper). Sedimentation, paleogeography and paleotectonics. *In*: MAREK, S. & PAJCHLOWA, M. (eds) *The Epicontinental Permian and Mesozoic in Poland. Prace Państwowego Instytutu Geologicznego*, **153**, 144–151 [in Polish with English summary].

GAJEWSKA, I. 1997*b*. Upper Triassic. Keuper. Sedimentation, paleogeography and paleotectonics. *In*: MAREK, S. & PAJCHLOWA, M. (eds) *The Epicontinental Permian and Mesozoic in Poland. Prace Państwowego Instytutu Geologicznego*, **153**, 166–172 [in Polish with English summary].

GAŹDZICKA, E. 1998. Upper Jurassic (including Lower Berriasian) – thickness. *In*: DADLEZ, R., MAREK, S. & POKORSKI, J. (eds) *Palaeographical Atlas of the Epicontinental Permian and Mesozoic in Poland (1:2 500 000)*. Polish Geological Institute, Warszawa, plate 56 [in Polish with English summary].

GLENNIE, K.W. 1986. The structural framework and the pre-Permian history of the North Sea area. *In*: GLENNIE, K.W. (ed.) *Introduction to the Petroleum Geology of the North Sea*. Blackwell Scientific, Oxford, 25–62.

GLENNIE, K.W. & BOEGNER, P.L.E. 1981. Sole Pit inversion tectonics. *In*: ILLING, L.V. & HOBSON, G.D. (eds) *Petroleum Geology of the Continental Shelf of Northwest Europe*. Institute of Petroleum, London, 110–120.

GRAS, R. & GELUK, M. 1999. Late Cretaceous–Early Tertiary sedimentation and tectonic inversion in the southern Netherlands. *Geologie en Mijnbouw*, **78**, 1–19, https://doi.org/10.1023/A:1003709821901

GUTOWSKI, J., POPADYUK, I.V. & OLSZEWSKA, B. 2005. Late Jurassic–earliest Cretaceous evolution of the epicontinental sedimentary basin of southeastern Poland and Western Ukraine. *Geological Quarterly*, **49**, 31–44.

HAKENBERG, M. & ŚWIDROWSKA, J. 2001. Cretaceous basin evolution in the Lublin area along the Teisseyre-Tornquist Zone (SE Poland). *Annales Societatis Geologorum Poloniae*, **71**, 1–20.

HARDENBOL, J., THIERRY, J., FARLEY, M.B., JACQUIN, T., GRACIANSKY, P.C. & VAIL, P.R. 1998. Mesozoic and Cenozoic sequence chronostratigraphic framework of European basins. *In*: GRACIANSKY, P.C., HARDENBOL, J., JAQUIN, T. & VAIL, P.R. (eds) *Mesozoic and Cenozoic Sequence Stratigraphy of European Basins*. SEPM, Special Publications, **60**, 3–13, charts 1–8.

HAYWARD, R.H. & GRAHAM, A.B. 1989. Some geometrical characteristic of inversion. *In*: COOPER, M.A. & WILLIAMS, G.D. (eds) *Inversion Tectonics*. Geological Society, London, Special Publications, **44**, 17–39, https://doi.org/10.1144/GSL.SP.1989.044.01.03

HECHT, C.A., LEMPP, C. & SCHECK, M. 2003. Geomechanical model for the post-Variscan evolution of the Permocarboniferous–Mesozoic basins in Northeast Germany. *Tectonophysics*, **373**, 125–139, https://doi.org/10.1016/S0040-1951(03)00288-9

IWANOW, A. 1998*a*. Lower and Middle Bunter – thickness. *In*: DADLEZ, R., MAREK, S. & POKORSKI, J. (eds) *Palaeographical Atlas of the Epicontinental Permian and Mesozoic in Poland (1:2 500 000)*. Polish Geological Institute, Warszawa, Poland, plate 14 [in Polish with English summary].

IWANOW, A. 1998*b*. Upper Bunter, Muschelkalk and Lower Keuper – thickness. *In*: DADLEZ, R., MAREK, S. & POKORSKI, J. (eds) *Palaeographical Atlas of the Epicontinental Permian and Mesozoic in Poland*

(1:2 500 000). Polish Geological Institute, Warszawa, Poland, plate 20 [in Polish with English summary].

Iwanow, A. 1998*c*. Middle Keuper, Upper Keuper and Rhaetian – thickness. *In*: Dadlez, R., Marek, S. & Pokorski, J. (eds) *Palaeographical Atlas of the Epicontinental Permian and Mesozoic in Poland (1:2 500 000)*. Polish Geological Institute, Warszawa, Poland, plate 27 [in Polish with English summary].

Jackson, M.P.A. & Talbot, C.J. 1994. Advances in salt tectonics. *In*: Hancock, P.L. (eds) *Continental Deformation*. Pergamon Press, Oxford, 159–179.

Japsen, P., Green, P.F., Nielsen, L.H., Rasmussen, E.S. & Bidstrup, T. 2007. Mesozoic–Cenozoic exhumation events in the eastern North Sea Basin: a multidisciplinary study based on palaeothermal, palaeoburial, stratigraphic and seismic data. *Basin Research*, **19**, 451–490, https://doi.org/10.1111/j.1365-2117.2007.00329.x

Jaritz, W. 1987. The origin and development of salt structures in Northwest Germany. *In*: Lerche, I. & O'Brian, J. (eds) *Dynamical Geology of Salt and Related Structures*. Academic Press, Orlando, FL, 479–493.

Jaskowiak-Schoeneichowa, M. & Krassowska, A. 1988. Palaeothickness, lithofacies and palaeotectonic of the epicontinental Upper Cretaceous in Poland. *Geological Quarterly*, **32**, 177–198 [in Polish with English summary].

Khriachtchevskaia, O., Stovba, S. & Stephenson, R. 2010. Cretaceous–Neogene tectonic evolution of the northern margin of the Black Sea from seismic reflection data and tectonic subsidence analysis. *In*: Sosson, M., Kaymakci, N., Stephenson, R., Bergerat, F. & Starostenko, V. (eds) *Sedimentary Basin Tectonics from the Black Sea and Caucasus to the Arabian Platform*. Geological Society, London, Special Publications, **340**, 137–157, https://doi.org/10.1144/SP340.8

Kiersnowski, H., Paul, J., Peryt, T.M. & Smith, D.B. 1995. Facies, paleogeography, and sedimentary history of the Southern Permian Basin in Europe. *In*: Scholle, P., Peryt, T.M. & Ulmer-Scholle, D. (eds) *The Permian of Northern Pangea. Volume 2: Sedimentary Basins and Economic Resources*. Springer, Berlin, 119–136.

Kley, J. & Voigt, T. 2008. Late Cretaceous intraplate thrusting in central Europe: effect of Africa–Iberia–Europe convergence, not Alpine collision. *Geology*, **36**, 839–842, https://doi.org/10.1130/G24930A.1

Kockel, F. 2003. Inversion structures in Central Europe – expressions and reasons, an open discussion. *Netherlands Journal of Geosciences*, **82**, 367–382, https://doi.org/10.1017/S0016774600020187

Kossow, D. & Krawczyk, C.M. 2002. Structure and quantification of processes controlling the evolution of the inverted NE-German Basin. *Marine and Petroleum Geology*, **19**, 601–618, https://doi.org/10.1016/S0264-8172(02)00032-6

Krzywiec, P. 2001. Contrasting tectonic and sedimentary history of the central and eastern parts of the Polish Carpathian Foredeep Basin – results of seismic data interpretation. *Marine and Petroleum Geology*, **18**, 13–38, https://doi.org/10.1016/S0264-8172(00)00037-4

Krzywiec, P. 2002. Mid-Polish Trough inversion – seismic examples, main mechanisms and its relationship to the Alpine–Carpathian collision. *In*: Bertotti, G., Schulmann, K., Cloetingh, S. (eds) *Continental Collision and the Tectonosedimentary Evolution of Forelands*. European Geosciences Union, Stephan Mueller Special Publication Series, **1**, 151–165.

Krzywiec, P. 2006. Structural inversion of the Pomeranian and Kuiavian segments of the Mid-Polish Trough – lateral variations in timing and structural style. *Geological Quarterly*, **51**, 151–168.

Krzywiec, P. 2009. Devonian–Cretaceous repeated subsidence and uplift along the Tornquist-Teisseyre Zone in SE Poland – insight from seismic data interpretation. *Tectonophysics*, **475**, 142–159, https://doi.org/10.1016/j.tecto.2008.11.020

Krzywiec, P. 2015. Late Cretaceous inversion within the NE Permian-Mesozoic cover of the Holy Cross Mountains – results of seismic data interpretation. *In*: Skompski, S. & Mizerski, W. (eds) *Ekstensja i inwersja powaryscyjskich basenów sedymentacyjnych, LXXIV Zjazd Naukowy Polskiego Towarzystwa Geologicznego, Chęciny, September 9–11*. Polish Geological Society, Warsaw, 51–58.

Krzywiec, P. & Stachowska, A. 2016. Late Cretaceous inversion of the NW segment of the Mid-Polish Trough – how marginal troughs were formed, and does it matter at all? *Zeitschrift der Deutschen Gesellschaft für Geowissenschaften*, **167**, 107–119, https://doi.org/10.1127/zdgg/2016/0068

Krzywiec, P., Gutowski, J., Walaszczyk, I., Wróbel, G. & Wybraniec, S. 2009. Tectonostratigraphic model of the Late Cretaceous inversion along the Nowe Miasto–Zawichost Fault Zone, SE Mid-Polish Trough. *Geological Quarterly*, **53**, 27–48.

Krzywiec, P., Malinowski, M., Lis, P., Buffenmyer, V. & Lewandowski, M. 2014. Lower Paleozoic basins developed above the East European Craton in Poland: new insight from regional high-effort seismic reflection data. *Abstract presented at the SPE/EAGE European Unconventional Resources Conference & Exhibition*, 25–27 February 2014, Vienna, Austria.

Krzywiec, P., Gągała, Ł. *et al.* 2017*a*. Variscan deformation along the Teisseyre-Tornquist Zone in SE Poland: thick-skinned structural inheritance or thin-skinned thrusting? *Tectonophysics*, https://doi.org/10.1016/j.tecto.2017.06.008

Krzywiec, P., Mazur, S., Gągała, Ł., Kufrasa, M., Lewandowski, M., Malinowski, M. & Buffenmyer, V. 2017*b*. Late Carboniferous thin-skinned compressional deformation above the SW edge of the East European Craton as revealed by reflection seismic and potential fields data – correlations with the Variscides and the Appalachians. *In*: Law, R., Thigpen, H., Thigpen, J.R., Merschat, A.J. & Stowell, H. (eds) *Linkages and Feedbacks in Orogenic Processes*, Geological Society of America, Memoirs, **213**, 353–372, https://doi.org/10.1130/2017.1213(14)

Krzywiec, P., Peryt, T.M., Kiersnowski, H., Pomianowski, P., Czapowski, G. & Kwolek, K. 2017*c*. Permo-Triassic evaporites of the Polish Basin and their bearing on the tectonic evolution and hydrocarbon system, an overview. *In*: Soto, J.I., Flinch, J.F. & Tari, G. (eds) *Permo-Triassic Salt Provinces of Europe, North Africa and the Central Atlantic: Tectonics and Hydrocarbon Potential*. Elsevier, Amsterdam, 243–261, https://doi.org/10.1016/B978-0-12-809417-4.00012-4

Kutek, J. & Głazek, J. 1972. The Holy Cross area, Central Poland, in the Alpine cycle. *Acta Geologica Polonica*, **22**, 603–653.

Leszczyński, K. 1997. The Upper Cretaceous carbonate-dominated sequences of the Polish Lowland. *Geological Quarterly*, **41**, 521–532.

Leszczyński, K. 1998*a*. Lower Cretaceous (excluding Lower Berriasian and Upper Albian) – thickness. *In*: Dadlez, R., Marek, S. & Pokorski, J. (eds) *Palaeographical Atlas of the Epicontinental Permian and Mesozoic in Poland (1:2 500 000)*. Polish Geological Institute, Warszawa, Poland, plate 64 [in Polish with English summary].

Leszczyński, K. 1998*b*. Upper Cretaceous (including Upper Albian) – thickness. *In*: Dadlez, R., Marek, S. & Pokorski, J. (eds) *Palaeographical Atlas of the Epicontinental Permian and Mesozoic in Poland (1:2 500 000)*. Polish Geological Institute, Warszawa, Poland, plate 71 [in Polish with English summary].

Leszczyński, K. 2010. Lithofacies evolution of the late Cretaceous basin in the Polish Lowlands. *Biuletyn Państwowego Instytutu Geologicznego*, **443**, 33–54 [in Polish with English summary].

Leszczyński, K. 2012. The internal geometry and lithofacies pattern of the Upper Cretaceous–Danian sequence in the Polish Lowlands. *Geological Quarterly*, **56**, 363–386.

Leszczyński, K. & Dadlez, R. 1999. Subsidence and the problem of incipient inversion in the Mid-Polish Trough based on thickness maps and Cretaceous lithofacies analysis – discussion. *Przegląd Geologiczny*, **47**, 625–628 [in Polish with English summary].

Liboriussen, J., Ashton, P. & Tygesen, T. 1987. The tectonic evolution of the Fennoscandian Border Zone in Denmark. *Tectonophysics*, **137**, 21–29, https://doi.org/10.1016/0040-1951(87)90310-6

Lott, G.K., Wong, T.E., Dusar, M., Andsbjerg, J., Mönnig, E., Feldman-Olszewska, A. & Verreussel, R.M.C.H. 2010. Jurassic. *In*: Doornenbal, J.C. & Stevenson, A.G. (eds) *Petroleum Geological Atlas of the Southern Permian Basin Area*. European Association of Geoscientists and Engineers (EAGE), Houten, The Netherlands, 175–193.

Marcinowski, R. 1980. Cenomanian ammonites from German Democratic Republic, Poland, and the Soviet Union. *Acta Geologica Polonica*, **30**, 215–325.

Marek, S. 1997*a*. Lower Cretaceous. Formal and informal lithostratigraphic units. *In*: Marek, S. & Pajchlowa, M. (eds) *The Epicontinental Permian and Mesozoic in Poland. Prace Państwowego Instytutu Geologicznego*, **153**, 351–360 [in Polish with English summary].

Marek, S. 1997*b*. Lower Cretaceous. Sedimentation, paleogeography and paleotectonics. *In*: Marek, S. & Pajchlowa, M. (eds) *The Epicontinental Permian and Mesozoic in Poland. Prace Państwowego Instytutu Geologicznego*, **153**, 362–367 [in Polish with English summary].

Marek, S. & Pajchlowa, M. (eds). 1997. *The Epicontinental Permian and Mesozoic in Poland. Prace Państwowego Instytutu Geologicznego*, **153** [in Polish with English summary].

Maystrenko, Y., Stovba, S. et al. 2003. Crustal-scale pop-up structure in cratonic lithosphere: DOBRE deep seismic reflection study of the Donbas fold belt, Ukraine. *Geology*, **31**, 733–736, https://doi.org/10.1130/G19329.1

Maystrenko, Y., Bayer, U. & Scheck-Wenderoth, M. 2006. 3D reconstruction of salt movements within the deepest post-Permian structure of the Central European Basin System – the Glueckstadt Graben. *Netherlands Journal of Geosciences – Geologie en Mijnbouw*, **85**, 181–196.

Maystrenko, Y., Bayer, U., Brink, H.-J. & Littke, R. 2008. The Central European Basin System - an Overview. *In*: Littke, R., Bayer, U., Gajewski, D. & Nelskamp, S. (eds) *Dynamics of Complex Sedimentary Basins. The Example of the Central European Basin System*. Springer, Berlin, 15–34.

Mazur, S. & Scheck-Wenderoth, M. 2005. Constraints on the tectonic evolution of the Central European Basin System revealed by seismic reflection profiles from Northern Germany. *Netherlands Journal of Geosciences*, **84**, 389–401.

Mazur, S., Scheck-Wenderoth, M. & Krzywiec, P. 2005. Different modes of the Late Cretaceous–Early Tertiary inversion in the North German and Polish basins. *International Journal of Earth Sciences*, **94**, 782–798, https://doi.org/10.1007/s00531-005-0016-z

Mazur, S., Mikołajczyk, M., Krzywiec, P., Malinowski, M., Buffenmyer, V. & Lewandowski, M. 2015. Is the Teisseyre-Tornquist Zone an ancient plate boundary of Baltica? *Tectonics*, **34**, 2465–2477, https://doi.org/10.1002/2015TC003934

Meyer, R.K. & Schmidt-Kaler, H. 1996. Gesteinsabfolge des Deckgebirges nördlich der Donau und im Molasseuntergrund: Jura. *In*: Freudenberger, W. & Schwerd, K. (eds) *Erläterungen zur Geologischen Karte von Bayern 1:500 000*. Bayrisches Geologisches Landesamt, München, Germany, 90–111.

Michelsen, O. 1997. Mesozoic and Cenozoic stratigraphy and structural development of the Sorgenfrei-Tornquist Zone. *Zeitschrift der Deutschen Gesellschaft für Geowissenschaften*, **148**, 33–50.

Michelsen, O. & Nielsen, L.H. 1993. Structural development of the Fennoscandian Border Zone, offshore Denmark. *Marine and Petroleum Geology*, **10**, 124–134, https://doi.org/10.1016/0264-8172(93)90017-M

Mitra, S. & Islam, Q.T. 1994. Experimental (clay) models of inversion structures. *Tectonophysics*, **230**, 211–222, https://doi.org/10.1016/0040-1951(94)90136-8

Mogensen, T.E. 1995. Triassic and Jurassic structural development along the Tornquist Zone, Denmark. *Tectonophysics*, **252**, 197–220.

Mogensen, T.E. & Jensen, L.N. 1994. Cretaceous subsidence and inversion along the Tornquist Zone from Kattegat to the Egersund Basin. *First Break*, **12**, 211–222, https://doi.org/10.3997/1365-2397.1994016

Narkiewicz, M., Resak, M., Littke, R. & Marynowski, L. 2010. New constraints on the Middle Palaeozoic to Cenozoic burial and thermal history of the Holy Cross Mts. (Central Poland): results from numerical modelling. *Geologica Acta*, **8**, 189–205, https://doi.org/10.1344/105.000001529

Nielsen, L.H. 2003. Late Triassic–Jurassic development of the Danish Basin and Fennoscandian Border Zone, southern Scandnavia. *In*: Ineson, J.R. & Surlyk, F. (eds) *The Jurassic of Denmark and Greenland, Volume 1*. Geological Survey of Denmark and Greenland

Ministry of the Environment (GEUS), Copenhagen, 459–526.

Niemczycka, T. 1997*a*. Upper Jurassic. Formal and informal lithostratigraphic units. *In*: Marek, S. & Pajchlowa, M. (eds) *The Epicontinental Permian and Mesozoic in Poland. Prace Państwowego Instytutu Geologicznego*, **153**, 309–322 [in Polish with English summary].

Niemczycka, T. 1997*b*. Upper Jurassic. Sedimentation, paleogeography and paleotectonics. *In*: Marek, S. & Pajchlowa, M. (eds) *The Epicontinental Permian and Mesozoic in Poland. Prace Państwowego Instytutu Geologicznego*, **153**, 327–331 [in Polish with English summary].

Niemczycka, T. & Brochwicz-Lewiński, W. 1988. Evolution of the Upper Jurassic sedimentary basin in the Polish Lowland. *Geological Quarterly*, **32**, 137–156 [in Polish with English summary].

Norling, E. & Bergström, J. 1987. Mesozoic and Cenozoic tectonic evolution of Scania, southern Sweden. *Tectonophysics*, **137**, 7–19, https://doi.org/10.1016/0040-1951(87)90309-X

Otto, V. 2003. Inversion-related features along the southeastern margin of the North German Basin (Elbe Fault System). *Tectonophysics*, **373**, 107–123, https://doi.org/10.1016/S0040-1951(03)00287-7

Pharaoh, T.C., Dusar, M. *et al.* 2010. Tectonic evolution. *In*: Doornenbal, J.C. & Stevenson, A.G. (eds) *Petroleum Geological Atlas of the Southern Permian Basin Area*. European Association of Geoscientists and Engineers (EAGE), Houten, The Netherlands, 25–57.

Pieńkowski, G., Schudack, M.E. *et al.* 2008. Jurassic. *In*: McCann, T. (ed.) *The Geology of Central Europe. Volume 2: Mesozoic and Cenozoic*. Geological Society, London, 823–922.

Piwocki, M. 2004. Paleogene. *In*: Peryt, T. & Piwocki, M. (eds) *Budowa Geologiczna Polski, Stratygrafia, Kenozoik, I (3a)*. Polish Geological Institute, Warszawa, Poland, 22–71 [in Polish].

Poprawa, P. 2007*a*. Tectonic subsidence and sedimentation rate analysis. *In*: Waksmundzka, M.I. (ed.) *Lublin IG-1*. Profile Głębokich Otworów Wiertniczych PIG, **119**, 183–185 [in Polish with English summary].

Poprawa, P. 2007*b*. Analysis of tectonic subsidence and sedimentation rate. *In*: Pacześna, J. (ed.) *Busówno IG-1*. Profile Głębokich Otworów Wiertniczych PIG, **118**, 183–185 [in Polish with English summary].

Poprawa, P. 2008. Analysis of tectonic subsidence and sedimentation rate. *In*: Pacześna, J. (ed.) *Łopiennik IG-1*. Profile Głębokich Otworów Wiertniczych PIG, **123**, 194–196 [in Polish with English summary].

Poprawa, P. 2012. Analysis of tectonic subsidence and sedimentation rate. *In*: Pacześna, J. (ed.) *Białopole IG-1*. Profile Głębokich Otworów Wiertniczych PIG, **134**, 138–139 [in Polish with English summary].

Pożaryski, W. 1938. Senonstratigraphie im Durchbruch der Weichsel zwischen Rachów und Puławy in Mittelpolen. *Biuletyn Państwowego Instytutu Geologicznego*, **6**, 1–94 [in Polish with German summary].

Pożaryski, W. 1948. Jurassic and Cretaceous between Radom, Zawichost and Kraśnik (Central Poland). *Biuletyn Państwowego Instytutu Geologicznego*, **46**, 1–141 [in Polish with English summary].

Pożaryski, W. (ed.) 1978. *Geological Map of Poland and Surrounding Countries without Cenozoic (Carpathians without Quaternary), 1:1 000 000*. Wydawnictwa Geologiczne, Warszawa.

Pożaryski, W. 1997. Post-Variscan tectonics of the Holy Cross Mountains-Lublin area in the light of basement structure. *Przegląd Geologiczny*, **45**, 1265–1270.

Radwański, A. 1960. Submarine slides of epicontinental Upper Jurassic and Upper Cretaceous margins of the Holy Cross Mts (Central Poland). *Acta Geologica Polonica*, **10**, 221–252 [in Polish with English summary].

Remin, Z. 2015. The Belemnitella-based stratigraphy across Campanian–Maastrichtian boundary of the Middle Vistula section, central Poland. *Acta Geologica Polonica*, **59**, 783–813, https://doi.org/10.7306/gq.1257

Remin, Z., Cyglicki, M., Cybula, M. & Roszkowska-Remin, J. 2015. Deep versus shallow? Deltaically influenced sedimentation and new transport directions – case study from the Upper Campanian of the Roztocze Hills, SE Poland. *In*: *Abstract Book of the 31st IAS Meeting of Sedimentology*, 22–25 June 2015, Polish Geological Society, Kraków, 438.

Remin, Z., Gruszczyński, M. & Marshall, J.D. 2016. Changes in paleo-circulation and the distribution of ammonite faunas at the Coniacian–Santonian transition in central Poland and western Ukraine. *Acta Geologica Polonica*, **66**, 107–124, https://doi.org/10.1515/agp-2016-0006

Resak, M., Narkiewicz, M. & Littke, R. 2008. New basin modelling results from the Polish part of the Central European Basin system: implications for the Late Cretaceous–Early Paleogene structural inversion. *International Journal of Earth Sciences*, **97**, 955–972, https://doi.org/10.1007/s00531-007-0246-3

Samsonowicz, J. 1925. Esquisse géologique des environs de Rachów sur la Vistule et les transgressions de l' Albien et du Cénomanien dans le silon nord-européen. *Sprawozdania Państwowego Instytutu Geologicznego*, **3**, 45–118 [in Polish with French summary].

Scheck-Wenderoth, M., Krzywiec, P., Zülke, R., Maystrenko, Y. & Frizheim, N. 2008. Permian to Cretaceous tectonics. *In*: McCann, T. (ed.) *The Geology of Central Europe. Volume 2: Mesozoic and Cenozoic*. Geological Society, London, 999–1030.

Schito, A., Corrado, S. *et al.* 2017. Assessment of thermal evolution of Paleozoic successions of the Holy Cross Mountains (Poland). *Marine and Petroleum Geology*, **80**, 112–132, https://doi.org/10.1016/j.marpetgeo.2016.11.016

Schröder, B. 1987. Inversion tectonics along the western margin of the Bohemian Massif. *Tectonophysics*, **137**, 93–100, https://doi.org/10.1016/0040-1951(87)90316-7

Ślączka, A., Krugłov, S., Golonka, J., Oszczypko, N. & Popadyuk, I. 2006. Geology and hydrocarbon resources of the outer Carpathians, Poland, Slovakia, and Ukraine: general geology. *In*: Golonka, J., Picha, F.J. (eds) *The Carpathians and their Foreland: Geology and Hydrocarbon Resources*. American Association of Petroleum Geologists, Memoirs, **84**, 221–258.

Stackenbrandt, W. & Franzke, H.J. 1989. Alpidic reactivation of the Variscan consolidated lithosphere – the activity of some fracture zones in central Europe. *Zeitschrift fur Geologische Wissenschaften*, **17**, 699–712.

Stephenson, R.A., Narkiewicz, M., Dadlez, R., van Wees, J.-D. & Andriessen, P. 2003. Tectonic subsidence modelling of the Polish Basin in the light of new data on crustal structure and magnitude of inversion. *Sedimentary Geology*, **156**, 59–70, https://doi.org/10.1016/S0037-0738(02)00282-8

Świdrowska, J. 2007. Cretaceous in Lublin area – sedimentation and tectonic conditions. *Biuletyn Państwowego Instytutu Geologicznego*, **422**, 63–77 [in Polish with English summary].

Świdrowska, J. & Hakenberg, M. 1999. Subsidence and the problem of incipient inversion in the Mid-Polish Trough based on thickness maps and Cretaceous lithofacies analysis. *Przegląd Geologiczny*, **47**, 61–68 [in Polish with English summary].

Świdrowska, J. & Hakenberg, M. 2000. Paleotectonic conditions of Cretaceous basin development in the southeastern segment of the Mid-Polish Trough. *In:* Crasquin-Soleau, S. & Barrier, E. (eds) *Peri-Tethys Memoir 5: New Data on Peri-Tethyan Sedimentary Basins*. Mémoire du Muséum national d'Histoire naturelle, **182**, 239–256.

Świdrowska, J., Hakenberg, M., Poluhtovič, B., Seghedi, A. & Višnâkov, I. 2008. Evolution of the Mesozoic basins on the southwestern edge of the East European Craton (Poland, Ukraine, Moldova, Romania). *Studia Geologica Polonica*, **130**, 3–130.

Szulc, J. 2000. Middle Triassic evolution of the northern Peri-Tethys area as influenced by early opening of the Tethys ocean. *Annales Societatis Geologorum Poloniae*, **70**, 1–48.

Szyperko-Teller, A. 1997. Lower Triassic (*Buntsandstein*). Sedimentation, paleogeography and paleotectonics. *In*: Marek, S. & Pajchlowa, M. (eds) *The Epicontinental Permian and Mesozoic in Poland. Prace Państwowego Instytutu Geologicznego*, **153**, 121–133 [in Polish with English summary].

Underhill, J.R. & Partington, M.A. 1993. Jurassic thermal doming and deflation in the North Sea: implications of the sequence stratigraphic evidence. *In*: Parker, J.R. (ed.) *Petroleum Geology of Northwest Europe: Proceedings of the 4th Conference*. Geological Society of London, London, 337–346, https://doi.org/10.1144/0040337

Van den Haute, P. & Vercoutere, C. 1990. Apatite fission track evidence for a Mesozoic uplift of the Brabant Massif – preliminary results. *Annales Société Géologique de Belgique*, **112**, 443–452.

van Wees, J.-D., Stephenson, R.A. *et al.* 2000. On the origin of the Southern Permian Basin, Central Europe. *Marine and Petroleum Geology*, **17**, 43–59, https://doi.org/10.1016/S0264-8172(99)00052-5

Vejbæk, O.V. & Andersen, C. 1987. Cretaceous–Early Tertiary inversion tectonism in the Danish Central Trough. *Tectonophysics*, **137**, 221–238, https://doi.org/10.1016/0040-1951(87)90321-0

Voigt, T., von Eynatten, H. & Franzke, H.-J. 2004. Late Cretaceous unconformities in the Subhercynian Cretaceous Basin (Germany). *Acta Geologica Polonica*, **54**, 675–696.

Voigt, T., Reicherter, K., von Eynatten, H., Littke, R., Voigt, S. & Kley, J. 2008. Sedimentation during basin inversion. *In*: Littke, R., Bayer, U., Gajewski, D. & Nelskamp, S. (eds) *Dynamics of Complex Intracontinental Basins. The Central European Basin System*. Springer, Berlin, 211–232.

von Eynatten, H., Voigt, T., Meier, A., Franzke, H.-J. & Gaupp, R. 2008. Provenance of Cretaceous clastics in the Subhercynian Basin: constrains to exhumation of the Harz Mountains and timing of inversion tectonics in Central Europe. *International Journal of Earth Sciences*, **97**, 1315–1330.

Wagner, R. 1998. Zechstein – thickness. *In*: Dadlez, R., Marek, S. & Pokorski, J. (eds) *Palaeographical Atlas of the Epicontinental Permian and Mesozoic in Poland (1:2 500 000)*. Polish Geological Institute, Warszawa, Poland, plate 9 [in Polish with English summary].

Walaszczyk, I. 1987. Mid-Cretaceous events at the marginal part of the Central European Basin (Annopol-on-Vistula section, Central Poland). *Acta Geologica Polonica*, **37**, 61–74.

Walaszczyk, I. 1992. Turonian through Santonian deposits of the Central Polish Uplands; their facies development, inoceramid paleontology and stratigraphy. *Acta Geologica Polonica*, **42**, 1–122.

Walaszczyk, I. & Remin, Z. 2015. The Cretaceous of the Holy Cross Mountains surroundings. *In*: Skompski, S. & Mizerski, W. (eds) *Extension and inversion of the post-Variscan Sedimentary Basins, LXXIV Zjazd Naukowy Polskiego Towarzystwa Geologicznego, Chęciny, September 9–11*. Polish Geological Society, Warsaw, 41–50.

Walaszczyk, I., Dubicka, Z., Olszewska-Nejbert, D. & Remin, Z. 2016. Integrated biostratigraphy of the Santonian through Maastrichtian (Upper Cretaceous) of extra-Carpathian Poland. *Acta Geologica Polonica*, **66**, 313–350, https://doi.org/10.1515/agp-2016-0016

Williams, G.D., Powell, C.M. & Cooper, M.A. 1989. Geometry and kinematics of inversion tectonics. *In*: Cooper, M.A. & Williams, G.D. (eds) *Inversion Tectonics*. Geological Society, London, Special Publications, **44**, 3–15, https://doi.org/10.1144/GSL.SP.1989.044.01.02

Ziegler, P.A. (ed.) 1987. *Compressional Intra-Plate Deformations in the Alpine Foreland. Tectonophysics*, **137**, 1–420.

Ziegler, P.A. 1990. *Geological Atlas of Western and Central Europe*, 2nd edn. Shell Internationale Petroleum Maatschappij, The Hague. Geological Society, London.

Ziegler, P.A. & Haag, D. 1982. Triassic Rifts and Facies Patterns in Western and Central Europe. *Geologische Rundschau*, **71**, 747–774.

Ziegler, P.A., Bertotti, G. & Cloetingh, S. 2002. Dynamic processes controlling foreland development – the role of mechanical (de)coupling of orogenic wedges and forelands. *In*: Bertotti, G., Schulmann, K. & Cloetingh, S. (eds) *Continental Collision and the Tectono-Sedimentary Evolution of Forelands*. European Geosciences Union, Stephan Mueller Special Publication Series, **1**, 17–56.

The Wiek Fault System east of Rügen Island: origin, tectonic phases and its relationship to the Trans-European Suture Zone

ELISABETH SEIDEL[1]*, MARTIN MESCHEDE[1] & KARSTEN OBST[2]

[1]*Institute for Geography and Geology, Ernst-Moritz-Arndt University of Greifswald, Friedrich-Ludwig-Jahn-Straße 16, D-17489 Greifswald, Germany*

[2]*Geological Survey of Mecklenburg-Western Pomerania, LUNG M-V, Goldberger Straße 12, D-18273 Güstrow, Germany*

**Correspondence: elisabeth.seidel@uni-greifswald.de*

Abstract: The Tornquist Fan, reflecting the northern part of the Trans-European Suture Zone, comprises a series of fault zones and major single faults, striking mainly subparallel to the SW margin of the Fennoscandian Shield. The deep-seated faults of Wiek, Nord Jasmund and Schaabe, which cross the northern part of Rügen Island and areas of the adjacent Baltic Sea from NW to SE, originated in the late Paleozoic. They are accompanied by younger faults, especially in the Pomeranian Bay, that were formed by Mesozoic tectonic processes. Based on reprocessed offshore seismic lines east of Rügen, a polyphase evolution for the Wiek Fault System is proposed. It implies changes in the stress field since the Caledonian Orogeny. Crustal extension in the Middle Devonian led to the formation of basins along the SW margin of Laurussia. Subsequent compressional movements, induced by the distant Variscan Orogeny, resulted in segmentation and block faulting of the Rügen Basin prior to the late Carboniferous. These Paleozoic faults were reactivated by Mesozoic extensional stress regimes. In addition, new en echelon faults were generated, contemporaneously with the formation of the Western Pomeranian Fault System. Since the Late Cretaceous (Africa–Iberia–Europe convergence), selected major normal faults have been reactivated as reverse faults.

Faults crossing Rügen Island and its vicinity have been a focus of scientific investigation since the early twentieth century (Deecke 1906; von Zwerger 1948; Wegner 1966; Kurrat 1974; Franke & Hoffmann 1988 and others). During the 1960s, the area between the Darss Peninsula and Usedom Island was of particular interest due to hydrocarbon exploration in NE Germany (Müller *et al.* 1993). Deep wells and numerous seismic data (e.g. Rempel 1992; Hoth *et al.* 1993) suggest that the Rügen area is divided into several blocks, each showing a different structural evolution. In particular, offshore seismic surveys and drilling activities, performed by the former consortium Petrobaltic, in the southern Baltic Sea between 1975 and 1990 resulted in an extensive database (Rempel 1992, 2011). The SASO working group (1993–1998) reprocessed selected lines and published the first results of their structural analysis (Schlüter *et al.* 1997*b*). Subsequently, the block model was enhanced and a more detailed description provided for the structural framework around Rügen (Mayer *et al.* 1994; Schlüter *et al.* 1997*a*, *b*, 1998; Krauss & Mayer 2004). The results influenced the discussion regarding the geological evolution and character of the Tornquist Fan; a northwestward widening zone of dominantly Paleozoic faults in the transition between the East European Craton and the Western European Platform (Berthelsen 1992*a*, *b*; Franke 1993; Makris & Wang 1994; Erlström *et al.* 1997; Schlüter *et al.* 1997*a*, *b*, 1998; DEKORP-BASIN Research Group 1999; Thybo 2000; Krawczyk *et al.* 2002). The major NW- to WNW-trending faults of Rügen and the adjacent Baltic Sea area form part of this structural trend and strike almost parallel to the Tornquist Zone (Fig. 1). Due to a right lateral offset west of Bornholm, this major fault zone at the SW margin of the Fennoscandian Shield and the East European Craton is divided into a NW branch, the Sorgenfrei–Tornquist Zone (STZ), and a SE branch, the Tornquist–Teisseyre Zone (TTZ; Berthelsen 1992*a*; Erlström *et al.* 1997; Scheck-Wenderoth & Lamarche 2005).

Whereas the orientation and character of the major faults of the Tornquist Zone have been described in detail (Thomas & Deeks 1994; Makris & Wang 1994; Schlüter *et al.* 1997*a*; Mazur *et al.* 2015, 2016), the genesis of the complex fault pattern in the area between Bornholm and Rügen is not well understood. Published maps show a simplified fault pattern, and precise descriptions of fault origin and possible reactivation phases are missing. The deep-seated Wiek Fault, for example, has been illustrated

From: Kilhams, B., Kukla, P. A., Mazur, S., McKie, T., Mijnlieff, H. F. & van Ojik, K. (eds) 2018. *Mesozoic Resource Potential in the Southern Permian Basin*. Geological Society, London, Special Publications, **469**, 59–82.
First published online January 24, 2018, https://doi.org/10.1144/SP469.10

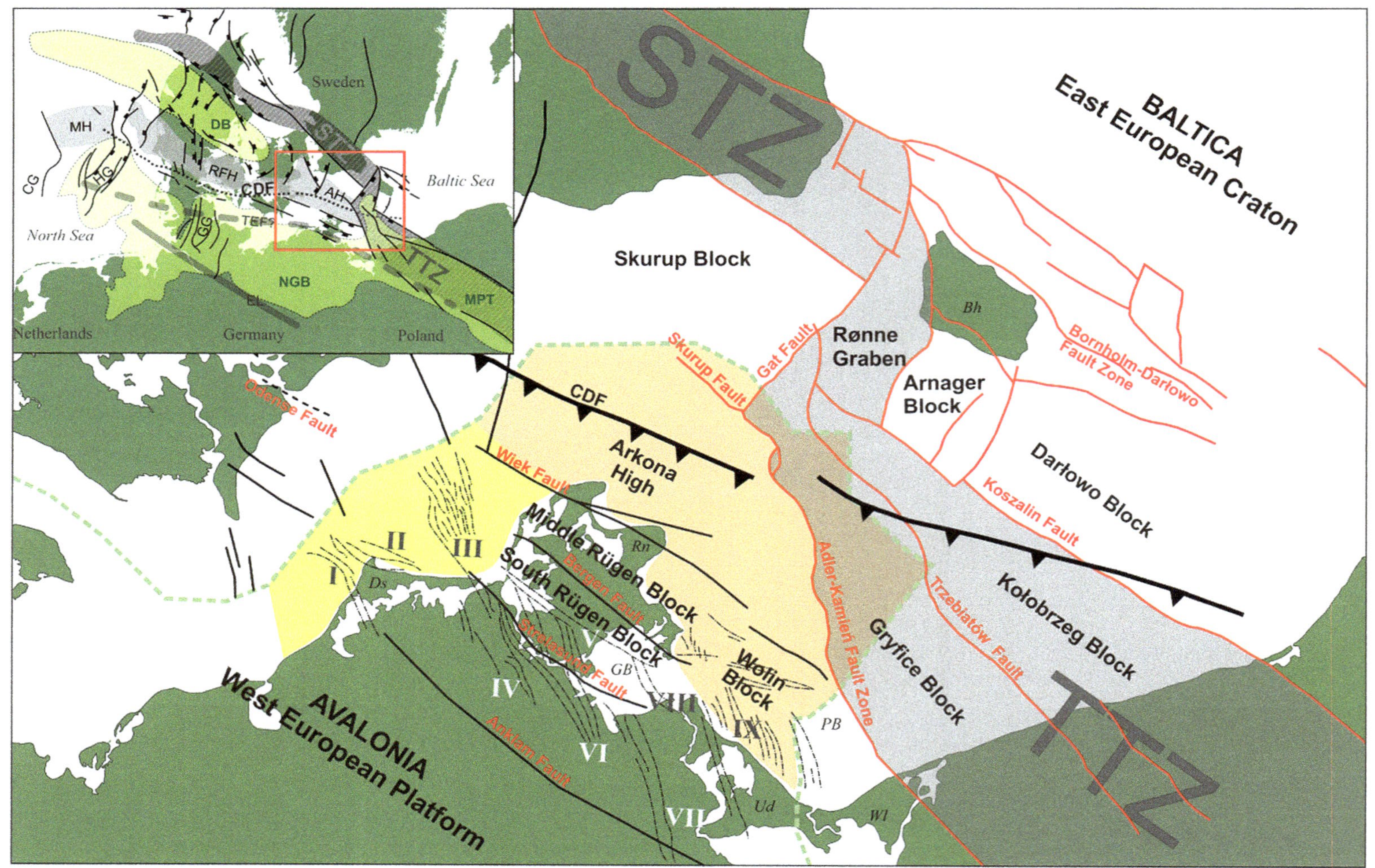
BALTICA
East European Craton
STZ
Skurup Block
Bh
Bornholm-Darłowo Fault Zone
Skurup Fault
Gat Fault
Rønne Graben
Arnager Block
Darłowo Block
Koszalin Fault
CDF
Arkona High
Wiek Fault
Odense Fault
Middle Rügen Block
Rn
Bergen Fault
South Rügen Block
Stralsund Fault
GB
Ds
I
II
III
IV
V
VI
VII
VIII
IX
Wolin Block
PB
Ud
Wl
Adler-Kamień Fault Zone
Gryfice Block
Trzebiatów Fault
Kołobrzeg Block
TTZ
Anklam Fault
AVALONIA
West European Platform
Sweden
Baltic Sea
North Sea
MH
DB
RFH
CDF
AH
HG
CG
GG
TEF?
NGB
EL
MPT
Netherlands
Germany
Poland

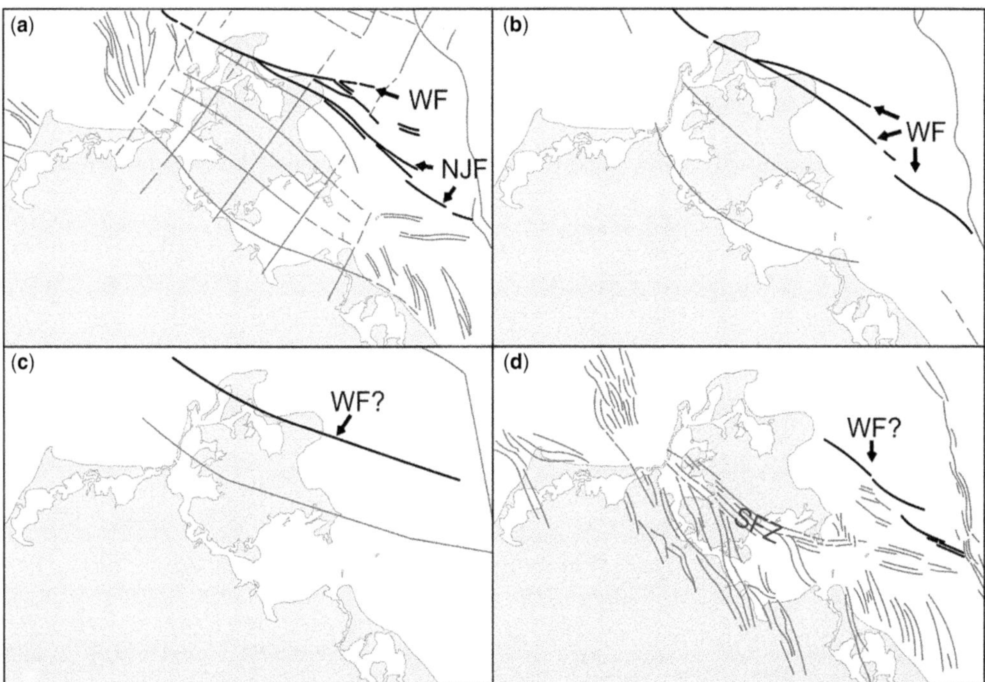

Fig. 2. Different maps illustrating the major Paleozoic and Mesozoic faults in the vicinity of Rügen, modified after: (**a**) Mayer *et al.* (1994); (**b**) Schlüter *et al.* (1997*a*); (**c**) Erlström *et al.* (1997); and (**d**) Krauss & Mayer (2004). The diversely interpreted locations of the Wiek Fault (WF) and the Nord Jasmund Fault (NJF) are indicated by arrows. SFZ, Samtens Fault Zone.

in a number of different ways. Mayer *et al.* (1994) (Fig. 2a) showed the Wiek Fault to be splitting into a splay of faults on northern Rügen. Whereas the northern branch is named the Wiek Fault, the southern branch is called the Nord Jasmund Fault, with both branches continuing offshore into the Pomeranian Bay. Schlüter *et al.* (1997*a*) (Fig. 2b) also showed a similar branching of the Wiek Fault without giving them separate names. Erlström *et al.* (1997) (Fig. 2c), in contrast, illustrates two single faults crossing Rügen, without further annotations. Krauss & Mayer (2004) investigated the Mesozoic faults of the Western Pomeranian Fault System (WPFS). They mapped about 20 additional faults north of the Samtens Fault Zone, which were not further discussed (Fig. 2d).

Due to the varying interpretations of the position and character of the Wiek Fault, the geometry and fault history is being addressed by the USO project ('Untergrundmodell Südliche Ostsee': Obst *et al.*

Fig. 1. Major fault systems of the southern Baltic Sea. The working area of USO-East and USO-West are marked in orange and yellow. The Tornquist Zone, marked in grey, is subdivided into the Sorgenfrei–Tornquist Zone (STZ) and the Tornquist–Teisseyre Zone (TTZ). Thin dashed black lines indicate faults of the Mesozoic Western Pomeranian Fault System (WPFS): I, Werre Fault Zone; II, Prerow Fault Zone; III, Agricola Fault Zone; IV, Reinberg Fault Zone; V, Samtens Fault Zone; VI, Greifswald–Poseritz Fault Zone; VII, Moeckow–Dargibell Fault Zone; VIII, Freest Fault Zone; IX, Usedom Fault Zone. Bh, Bornholm; Ds, Darss Peninsula; GB, Greifswalder Bodden; PB, Pomeranian Bay; Rn, Rügen Island; Ud, Usedom Island; Wl, Wolin Island. Insert: the research area is marked by a red rectangle with additional sub-basins of the Central European Basin System (CEBS): Danish Basin (DB), North German Basin (NGB) and Mid-Polish Trough (MPT). The central high can be divided into the Ringkøbing-Fyn High (RFH), the Møn High (MH) and the Arkona High (AH). Major grabens are the Central Graben (CG), the Horn Graben (HG) and the Glückstadt Graben (GG). The locations of the Caledonian Deformation Front (CDF), the Trans-European Fault (TEF) and the Elbe Line (EL) are also shown (Vejbaek & Britze 1994; Schlüter *et al.* 1997*a*; Bayer *et al.* 1999; Thybo 2000; Krawczyk *et al.* 2002; Krzywiec *et al.* 2003; Krauss & Mayer 2004; Scheck-Wenderoth & Lamarche 2005; Bachmann *et al.* 2010; Franke 2017 and citations therein).

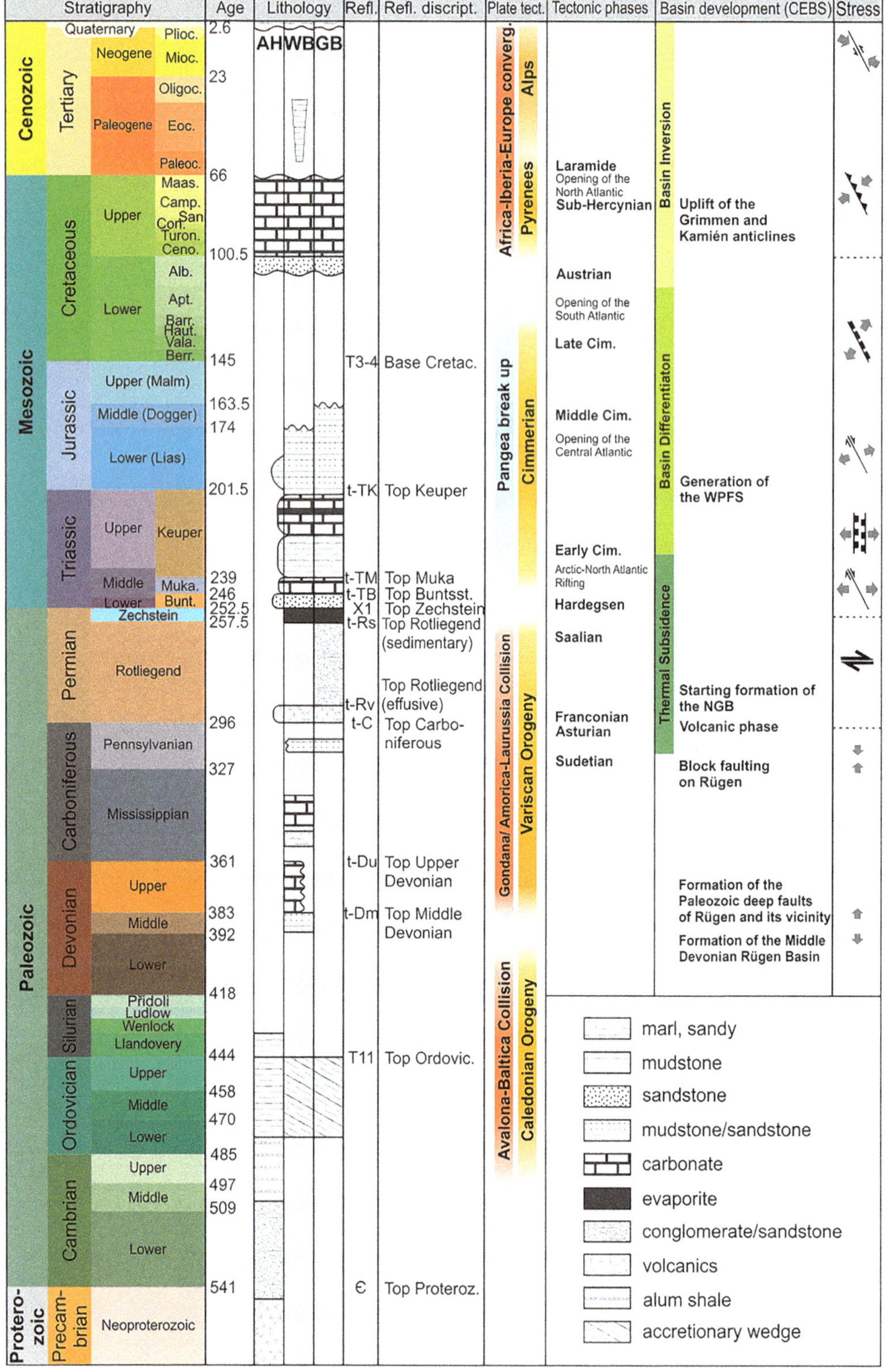
Stratigraphy
Age
Lithology
Refl.
Refl. discript.
Plate tect.
Tectonic phases
Basin development (CEBS)
Stress
Cenozoic
Mesozoic
Paleozoic
Protero-zoic
Quaternary
Tertiary
Neogene
Paleogene
Plioc.
Mioc.
Oligoc.
Eoc.
Paleoc.
Cretaceous
Upper
Lower
Maas.
Camp.
San
Con.
Turon.
Ceno.
Alb.
Apt.
Barr.
Haut.
Vala.
Berr.
Jurassic
Upper (Malm)
Middle (Dogger)
Lower (Lias)
Triassic
Upper
Middle
Lower
Keuper
Muka.
Bunt.
Zechstein
Permian
Rotliegend
Carboniferous
Pennsylvanian
Mississippian
Devonian
Upper
Middle
Lower
Silurian
Pridoli
Ludlow
Wenlock
Llandovery
Ordovician
Upper
Middle
Lower
Cambrian
Upper
Middle
Lower
Precam-brian
Neoproterozoic
2.6
23
66
100.5
145
163.5
174
201.5
239
246
252.5
257.5
296
327
361
383
392
418
444
458
470
485
497
509
541
AH WB GB
T3-4
t-TK
t-TM
t-TB
X1
t-Rs
t-Rv
t-C
t-Du
t-Dm
T11
Є
Base Cretac.
Top Keuper
Top Muka
Top Buntsst.
Top Zechstein
Top Rotliegend (sedimentary)
Top Rotliegend (effusive)
Top Carbo-niferous
Top Upper Devonian
Top Middle Devonian
Top Ordovic.
Top Proteroz.
Africa-Iberia-Europe converg.
Alps
Pyrenees
Pangea break up
Cimmerian
Gondana/ Amorica-Laurussia Collision
Variscan Orogeny
Avalona-Baltica Collision
Caledonian Orogeny
Laramide
Opening of the North Atlantic
Sub-Hercynian
Austrian
Opening of the South Atlantic
Late Cim.
Middle Cim.
Opening of the Central Atlantic
Early Cim.
Arctic-North Atlantic Rifting
Hardegsen
Saalian
Franconian
Asturian
Sudetian
Basin Inversion
Basin Differentiaton
Thermal Subsidence
Uplift of the Grimmen and Kamién anticlines
Generation of the WPFS
Starting formation of the NGB
Volcanic phase
Block faulting on Rügen
Formation of the Paleozoic deep faults of Rügen and its vicinity
Formation of the Middle Devonian Rügen Basin
marl, sandy
mudstone
sandstone
mudstone/sandstone
carbonate
evaporite
conglomerate/sandstone
volcanics
alum shale
accretionary wedge

2015) – a cooperation between the University of Greifswald and the Geological Survey of the German Federal State Mecklenburg-Western Pomerania (LUNG M-V). This project aims to resolve the following questions: Is the Wiek Fault a single fault, does it branches off, or is the deep fault part of a complex fault system? When was this fault or fault system first formed, and has it been reactivated? Is there a relationship with the WPFS?

Numerous Petrobaltic seismic lines, which were partly reprocessed during the SASO project but more recently by the Central European Petroleum Ltd (CEP), form the basis for reinterpretation and detailed structural analyses using modern software techniques. The working area of the USO project is located in the German Baltic Sea, near Rügen, and is divided into a western and an eastern area (cf. Fig. 1). Whilst USO-West is investigated in Deutschmann *et al.* (this volume, in press), the investigation results of USO-East are presented here.

Our structural analyses allow the characterization of a portion of a complex fault system which resulted from several phases of fault generation and reactivation that have occurred during a changing paleostress field at the border of the Trans-European Suture Zone. The faults in the vicinity of Rügen and Usedom provide detailed insights into the geological evolution of the southern Baltic Sea and its relationship to major plate tectonic events in the time span between the Caledonian and Alpine orogenies.

Tectonic evolution and nomenclature

The working area of the USO is located in the transition zone between Baltica (the East European Craton) in the north and the younger terrane of Avalonia, which forms part of the Western European Platform, to the south. The crustal border between Baltica and Avalonia, or between the Precambrian East European Craton and the younger Paleozoic Western European Platform, is termed the Trans-European Suture Zone (TESZ: Pharaoh 1999; Krawczyk *et al.* 2002). The USO research area covers, furthermore, the NE margin of the North German Basin (NGB) as part of the Central European Basin System (CEBS: Bayer *et al.* 2002; Scheck-Wenderoth & Lamarche 2005).

The southern Baltic Sea area between Sweden, northern Germany and Poland is segmented into various blocks with different Neoproterozoic–Cretaceous successions. These blocks were downfaulted and uplifted along NW- to WNW-trending faults that belong to the Tornquist Fan (Fig. 1). Thybo (2000) described the Tornquist Fan as a splay of faults that opens in a westerly direction. These faults occur in the NW part of the TESZ, SW of the STZ and TTZ – both parts of the Tornquist Zone (Berthelsen 1992*a*; Guterch *et al.* 2010).

The pre-Quaternary development of this area is characterized by several orogenic and tectonic phases (Fig. 3), resulting in a complex subsurface structure (Franke *et al.* 1989; Thomas *et al.* 1993; Krauss 1994; Berthelsen 1998; Thybo 2000; Katzung 2001; Bayer *et al.* 2002; Krawczyk *et al.* 2002; Torsvik & Rehnström 2003; Mazur *et al.* 2015, 2016). Due to the collision between the ancient continent of Baltica and the Avalonian terrane during the Ordovician–Silurian, an accretionary wedge was formed between Bornholm and Rügen. In northern Germany, especially in the area of Rügen, up to 3000 m of Ordovician marine sedimentary rocks were thrust onto the southern edge of Baltica (Beier & Katzung 2001). The northernmost rim of this accretionary wedge is marked by the Caledonian Deformation Front (CDF). It can be traced from the North Sea, across the Arkona High until the TTZ in Polish Pomerania (Pharaoh 1999; Mazur *et al.* 2016). Other authors suggest that it continues to the Holy Cross Mountains (Berthelsen 1998). Furthermore, the precise location and character of the plate boundary between Baltica and Avalonia, also called the Thor Suture or Trans-European Fault, is still a point of discussion (Krauss 1994; Pharaoh 1999; Torsvik & Rehnström 2003). For example, Mazur *et al.* (2016) recently proposed that the suture lies to the west of the TTZ.

Subsequently, the Variscan Orogeny led to the formation of Middle Devonian–lower Carboniferous foreland basins forming a sedimentary cover to the TESZ. Compressional movements at the early–late Carboniferous boundary led to block faulting along the NW- to WNW-trending faults of the Tornquist Fan. In the late Paleozoic, the CEBS underwent widespread thermal subsidence. Changing stress fields during the Mesozoic subsequently affected the area of the Tornquist Fan. Besides reactivation of existing faults, new faults were generated. The Pyrenean and Alpine orogenies were accompanied by compressional forces from Late Cretaceous times,

Fig. 3. Proterozoic–Cenozoic timescale with an overview of the lithology at the Arkona High (AH), the Wolin Block (WB), and the Gryfice Block (GB), as well as corresponding major reflectors, investigated and characterized in the USO-East project. Important orogenic cycles and tectonic phases of Europe are shown, which influenced the fault formation and reactivation in the southern Baltic Sea. Muka, Muschelkalk; Bunt., Buntsandstein; CEBS, Central European Basin System; NGB, North German Basin (based on Kley *et al.* 2008; Kley & Voigt 2008; Pharaoh *et al.* 2010; STG 2016; Al Hseinat & Hübscher 2017 and citations therein).

and triggered inversion movements along faults and graben systems (Thomas & Deeks 1994; Kley & Voigt 2008).

The southern Baltic Sea area between Bornholm and Rügen is characterized by several fault zones and systems (Fig. 1). The SE branch of the Tornquist Zone, the TTZ, strikes NNW and is located close to the German–Polish border. It incorporates the Gryfice Block with a set of NNW- to NW-trending faults, here called the Gryfice Fault Zone. The TTZ and the Gryfice Block are bordered to the west by the Adler–Kamień Fault Zone (AKFZ). The AKFZ comprises a couple of mainly NNW-striking faults. After a horizontal offset in the north, it continues into the Skurup Fault. The complex evolution of the TTZ can be traced back until the late Paleozoic (Schlüter *et al.* 1997*a*; Berthelsen 1998; Graversen 2009) or even until the Precambrian (Mazur *et al.* 2015, 2016).

The area SW of the TTZ, in the vicinity of Rügen, is intensely block faulted. According to Kurrat (1974) or Franke & Hoffmann (1988), Paleozoic normal faults occur that can be divided, on the basis of different strike directions, into NW–SE-, NNE–SSW- and NNW–SSE-orientated faults. The Wiek Fault, as well as the Bergen and Strelasund faults, belong to a set of dominate NW-trending faults. They cross Rügen and separate the Arkona High (North Rügen Block), the Middle Rügen Block (Wiek-Trent Block) and the South Rügen Block (Gingst-Garz Block: Schlüter *et al.* 1997*a*; Franke 2017). The Wolin Block is known as the SE extension of the Middle Rügen and South Rügen blocks. It is located in the German and Polish offshore area, between the islands of Rügen and Wolin. Thus, it is bordered by the Arkona High in the north and the Gryfice Block in the SE. The faults that cross Rügen and its vicinity are considered to be predominantly of middle–late Paleozoic age and related to the Variscan Orogeny (Kurrat 1974; Franke & Hoffmann 1988; Mayer *et al.* 1994; Piske *et al.* 1994; Schlüter *et al.* 1998).

The Wiek Fault can be traced from NW of Rügen towards the AKFZ in the SE. This fault is also known as the Odense–Wiek Fault (e.g. Mayer *et al.* 1994) or deep Wiek Fault (in German: Wieker Tiefenbruch; e.g. Franke 2017). It separates the deformed Ordovician succession of the Arkona High from the down-faulted Devonian and Carboniferous deposits of the Middle Rügen Block. The research wells in the north of Rügen (e.g. Rn 5/66, Dke 1/68 and Loh 2/70) imply a vertical offset of the Ordovician by about 3000 m (Franke & Hoffmann 1988; Piske *et al.* 1994).

Another system of NW–SE- to NNW–SSE-striking faults occurs within the Mesozoic succession between the Wiek and Anklam faults. This was originally termed 'the NE Mecklenburg fault system' (Wegner 1966) and was mapped onshore plus in the Greifswalder Bodden, during the projects VPSS I and II (Krauss & Mayer 1999, 2004; Mayer *et al.* 2000, 2001*a*, *b*). Latterly, it has been known as the 'Western Pomeranian Fault System' (WPFS; in German: Vorpommern-Störungssystem). These convergent faults often border small Y-shaped graben structures and are characterized by branching/sigmoidal strike patterns in plan view. Up to nine fault zones are distinguished on the German side: I, Prerow; II, Werre; III, Agricola; IV, Reinberg; V, Samtens; VI, Greifswald-Poseritz; VII, Moeckow-Dargibell; VIII, Freest; and IX, Usedom (Fig. 1) (Franke & Hoffmann 1988; Mayer *et al.* 2000, 2001*a*, *b*; Krauss & Mayer 2004).

Geological succession

The crystalline basement of the SW margin of Baltica, made up of Mesoproterozoic granites as exposed in the well G14 1/86 (Obst *et al.* 2004), is overlain by nearly undeformed Cambro-Silurian sediments. It can be traced until the Anklam Fault (e.g. Beier & Katzung 2001) but may also continue towards the Elbe Line (Bayer *et al.* 2002; Guterch *et al.* 2010) (Fig. 1). Ordovician clay and shale deposits were thrust onto Baltica during the Mid-European Caledonian Orogeny (Ordovician–Early Devonian: Beier *et al.* 2000; Torsvik & Rehnström 2003). The CDF outlines today the folded and slightly metamorphosed marine sediments of the Tornquist Sea (Beier & Katzung 2001; Katzung 2004). Very weak deformation of foreland basin sediments in the area of Bornholm suggests that the Caledonian thrust-and-fold belt may have actually been felt further to the north (Beier & Katzung 1999).

During the Early Devonian, the southern passive margin of Laurussia was tectonically stable and the former Caledonides were eroded. Extension in the north and thinning of the crust triggered the formation of basins, such as the Rhenohercynian Basin (Ziegler 1990). On Rügen, alluvial to limnic-fluviatile clastic deposits intercalated with shallow-marine sediments of the Middle Devonian ('Old Red facies') discordantly overlaid the deformed Ordovician succession and pass continuously into marine, calcareous sediments of the Upper Devonian (Aehnelt & Katzung 2009). Zagora & Zagora (2004) illustrated that the Devonian originally covered a larger area but the deposits, especially at the Arkona High, were subsequently eroded by late Carboniferous (Sudetian) movements. Nevertheless, Devonian deposition was restricted to a depression between the Strelasund and Wiek faults, with the depocentre in the northern Middle Rügen Block (Zagora & Zagora 2004), also called the Rügen Basin (Aehnelt

& Katzung 2009). In onshore (Rügen) and offshore (H9 1/87 and H2 1/90) wells, the Middle Devonian is characterized by alternating beds of sandstones and claystones, whereas the Upper Devonian comprises marl, dolomite and limestone.

In the early Carboniferous, while the Rheic Ocean was completely subducted, the increasing compressive stress field induced by the Variscan Orogeny further impacted the northern foreland. For example, the Middle Rügen swell was uplifted (Lindert & Hoffmann 2004). The Erzgebirgian and Asturian movements associated with the compression, and uplift, of single blocks, resulted in an angular discordance at the base of the Upper Carboniferous. Therefore, the lower Carboniferous of Rügen is characterized by dominantly marine carbonates and claystones, whereas the Upper Carboniferous represents the Variscan foreland molasse with limnic and fluviatile sediments (Lindert & Hoffmann 2004).

The effects of the Variscan Orogeny in this area were largely dissipated by the end of the late Carboniferous (Fig. 3), after the Variscan Fold Belt became wedged between Laurussia and Gondwana (Franke 2000). The subducted oceanic and subcontinental crust segments melted in the asthenosphere, resulting in high magmatic productivity (Benek *et al.* 1996). Thus, the beginning of Rotliegend was characterized by thermal relaxation, and the formation of thick volcanic sequences in central and NW Europe (Neumann *et al.* 2004; Wilson *et al.* 2004; Geissler *et al.* 2008). A dextral transtensional stress field subsequently affected this area (Kley *et al.* 2008; Pharaoh *et al.* 2010). This phase was followed by stabilization and cooling of the crust, resulting in an increase in rock density and the formation of intracontinental basins by thermal subsidence (van Wees *et al.* 2000). In central and northern Europe, the formation of the CEBS was initiated, which can be subdivided into the Anglo-Dutch Basin, the NGB, the Mid-Polish Trough and the Danish Basin (Fig. 1) (Maystrenko *et al.* 2008). At this time, the Gryfice Block subsided as part of the TTZ (Schlüter *et al.* 1997*a*). The wells east of Rügen reveal that the Rotliegend deposits of the Wolin Block are relatively thin (some tens of metres) and are represented only by volcanic successions of early Rotliegend age. This is in contrast to the hundreds of metres-thick sequence of volcanics and upper Rotliegend sediments deposited on the Gryfice Block.

During the late Permian (Zechstein), the subsiding NGB and Gryfice Block were submerged by a marine transgression from the NW via the Barents Sea and the North Sea (Katzung 2004). Sedimentation at this time was largely controlled by sea-level fluctuations, with seven recognizable evaporitic cycles occurring in the central part of the NGB (Pharaoh *et al.* 2010). Contemporaneously, the Ringkøbing-Fyn High, which was probably connected to the Arkona High, formed a topographical high or 'swell' bordering the marine Zechstein sedimentation to the north of the NGB, as indicated by thin claystone, dolomite and anhydrite deposits on Rügen (Katzung 2004).

The rate of thermal subsidence increased at the beginning of the Triassic (van Wees *et al.* 2000). At this time the supercontinent of Pangaea became unstable and began to break apart. Rifting of the Central Atlantic, and between Greenland and the North Sea, resulted in an east–west-orientated extensional stress regime that continued throughout early Mesozoic time (Pharaoh *et al.* 2010). Cimmerian movements led to the formation of complex graben systems (such as the Central Graben, Horn Graben and Glückstadt Graben). Since the Late Triassic (Keuper), a rifting in the Arctic–North Atlantic was accelerated and joined with the Central Atlantic Rift. During the Jurassic, the Penninic Ocean developed as part of the Atlantic Rift System, SE of the Western European Platform. Thus, the regional stress system of the latter one changed continuously toward a NE–SW orientation until the Early Cretaceous (Ziegler 1990; Kley *et al.* 2008; Pharaoh *et al.* 2010) (Fig. 3). The associated east–west to ENE–WSW extension induced strike-slip movements along pre-existing NW–SE-striking faults. The area of the TESZ, especially between the deep-seated Samtens Fault in the NE and the Anklam Fault in the SW, was partly reactivated and acted as a shear zone (Mayer *et al.* 2001*a*; Krauss & Mayer 2004). This led to the formation of en echelon faults within the WPFS. In offshore wells (e.g. well H2 1/90), the Triassic section is represented by alternating terrestrial sandstones, siltstones and claystones (Buntsandstein), subsequent marine marls, limestones and claystones (Muschelkalk), and by alternating terrestrial claystones, marls and sandstones (Keuper). The subsequent Lower Jurassic sequence consists of partly coal-bearing sandstones, siltstones and claystones, and the overlying Middle Jurassic of sandstones with some fossil plant remains, coal and ore minerals (offshore wells H9 1/87, H2 1/90 and K5 1/88).

A new stress regime developed contemporaneously with the opening of the North Atlantic and the northwards drift of Gondwana (Katzung 2004; Pharaoh *et al.* 2010). Since the Late Cretaceous, the Pyrenean phase of deformation induced SW–NE compression in central and northern Europe (Kley & Voigt 2008) (Fig. 3), whereby many NW–SE-striking faults were reactivated as reverse faults. This led to an erosion of Lower Cretaceous–Middle Jurassic deposits (Pharaoh *et al.* 2010). After this hiatus, a transgression during the Albian–Cenomanian led to sedimentation of thick limestone and chalk deposits of the Upper Cretaceous. During the Cenozoic (especially in the Oligocene–Miocene), the main stress direction changed from NE to NW, caused by the progressive development of the Alpine Orogeny (Kley & Voigt 2008). Offshore wells show a Cenozoic cover of dominantly Quaternary till and meltwater deposits, with some Tertiary clay locally preserved in small graben, such as that drilled in well H2 1/90.

Methods

Database

Altogether, 144 reflection seismic lines, with a total length of about 3200 km, were used for this study of the USO East area. These lines belong to the Petrobaltic database from 1975–1990 (Rempel 2011) which was primarily acquired for oil and gas exploration. During different phases of investigation, profile grids of 2 × 4, 2 × 1 and 1 × 1 km were measured using a shot-point distance of 25 m, a CDP fold of 48, an offset of 160–240 m, and a registration length of up to 5 s using an airgun or vaporshoc source (Scheidt *et al.* 1995; Arndt *et al.* 1996; Rempel 1992). A maximum penetration of 5 s (TWT) was reached. The Federal Institute for Geosciences and Natural Resources (BGR) provided 69 stacked or time-migrated seismic lines for USO-East. These lines had already been utilized and reprocessed during the SASO project in the 1990s (Schlüter *et al.* 1997*a*, *b*). More recently, Central European Petroleum Ltd (CEP) reprocessed further lines of the Petrobaltic database, and provided 75 migrated and stacked sections for the eastern part of the USO working area. Figure 4 shows the location of the reflection seismic lines. Additional lithological and geophysical information of four offshore exploration wells (G14 1/86, H9 1/87, H2 1/90 and K5 1/88) and five onshore research wells (E Rn 5/66, E Dke 1/68, E Loh2/70, E Sagd 1/70 and E Binz 1/73) were used for correlation with the seismic data (cf. Fig. 4). Simplified well lithologies are documented by Hoth *et al.* (1993) and Schlüter *et al.* (1997*b*). Detailed well reports and log data were provided by the Geological Survey of Mecklenburg-Western Pomerania.

Mapping and interpretation

The software SeisWare™ was used for the interpretation of the stacked and time-migrated seismic sections. Sonic and density logs of the four offshore wells were used for the generation of synthetic seismograms, thus allowing for well ties. Twelve of 18 main reflectors, which were recognized and mapped across the USO-East working area, are shown in the presented sections (Fig. 5). They represent key lithological boundaries lying between the top of the Baltica basement and the base of the Cretaceous (Fig. 3). Due to the complex fault pattern, the subdivision into different blocks and the location of the USO working area at the NE border of the NGB, the lithological composition and thickness changes rapidly within short distances. Within the USO working area, there are only four offshore wells, which all were drilled east of Rügen. Unfortunately, they all have different successions exposed, as they are located in different blocks: G14 1/86, Arkona High; K5 1/88, Gryfice Block; H9 1/87, Middle Rügen Block; and H2 1/90, Wolin Block. Therefore, a precise velocity model, which is necessary for a time-to-depth conversion, has not yet been realized. Nevertheless, to give an idea of the depth or vertical exaggeration, and to avoid misinterpretations of tectonic structures, the seismic sections are shown in TWT but are additionally labelled with depth. The latter one was calculated using a constant interval velocity of 3000 m s^{-1} (which is, according to Schlüter *et al.* 1997*b*, almost the average velocity of the illustrated succession). Based on our data, we can define the dip direction and the fault character (normal or reverse), and, thus, make statements regarding the stress system (compressional or extensional). Furthermore, it is also possible to analyse whether a succession is thickening in a certain direction, or if there are differences in thickness or geometry across a fault (e.g. growth faults). For more precise depth and thickness analyses, a velocity model has to be calculated with regard to not only offshore well data but also the numerous onshore well data of the oil and gas industry.

Restoration procedure

The restoration of two seismic lines was carried out with the software suite MOVE™ by Midland Valley (2D Kinematic Modelling). The restoration of 2D sections to their pre-deformational state is a useful tool used for structural balancing. A typical workflow is comprised of the following steps. First polygons are assigned for each stratigraphic section and faults selected as lines. The subsequent restoration is divided into three major steps: (1) The surface of the uppermost stratigraphic unit is unfolded; thus restoring its post-sedimentation condition. Midland Valley provides an ‘Unfold’ workflow within the ‘2D Kinematic Modelling’ module of MOVE™. (2) The uppermost unit can then be removed using the ‘Decompaction’ workflow. (3) The hanging block is then restored by moving it along a defined fault. Finally, the tops of one stratigraphic unit separated by the fault should be set to the same datum. These three steps were repeated for each stratigraphic unit down to the Middle Devonian. It should be noted that the applied decompaction method did not consider any variations in rock properties or volume changes due to the absence of information on

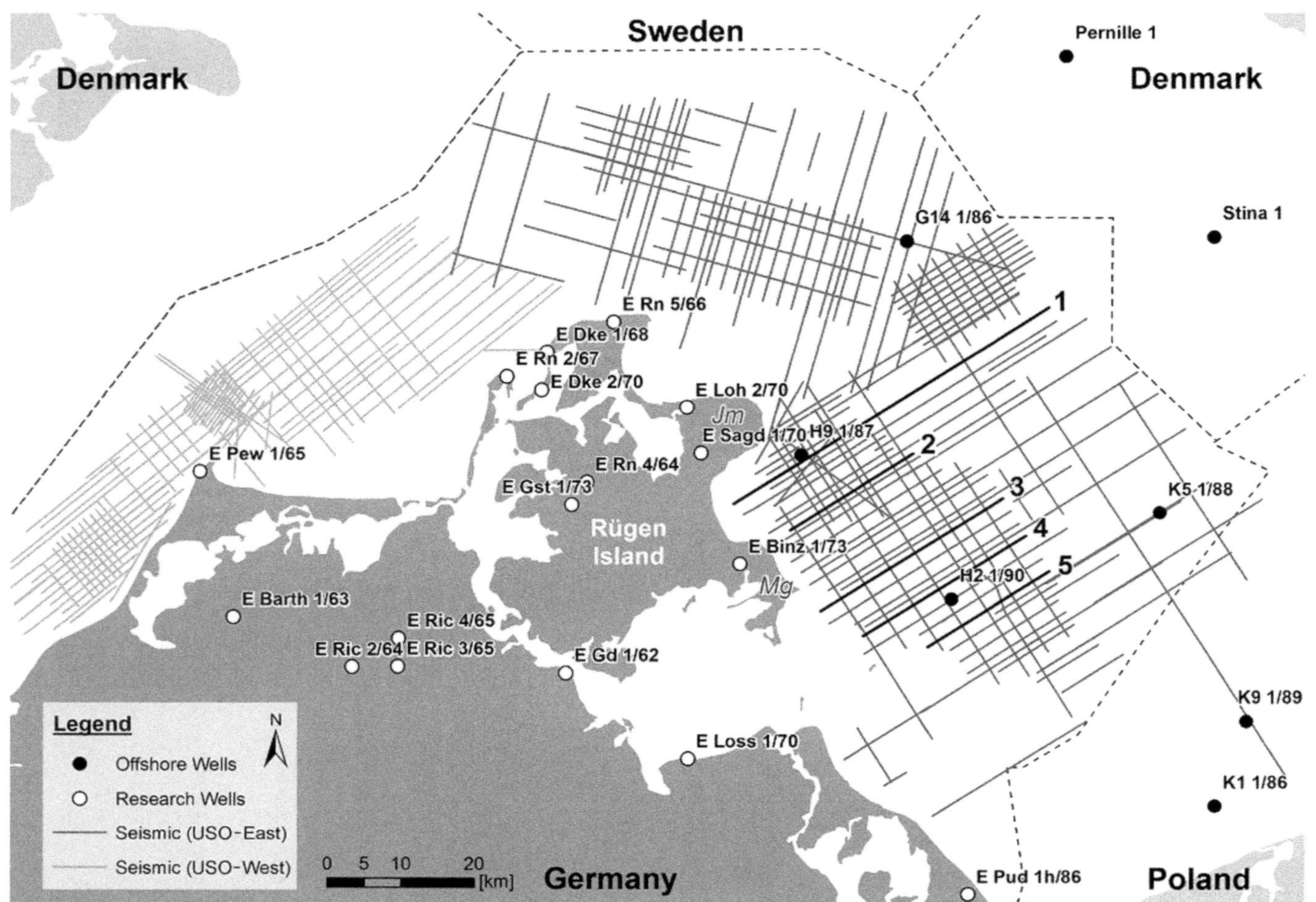

Fig. 4. Database (wells and reflection seismic lines) of the USO project. Seismic sections marked by thick black lines are shown and discussed in detail in this paper. Jm, Jasmund Peninsula; Mg, Mönchgut Peninsula.

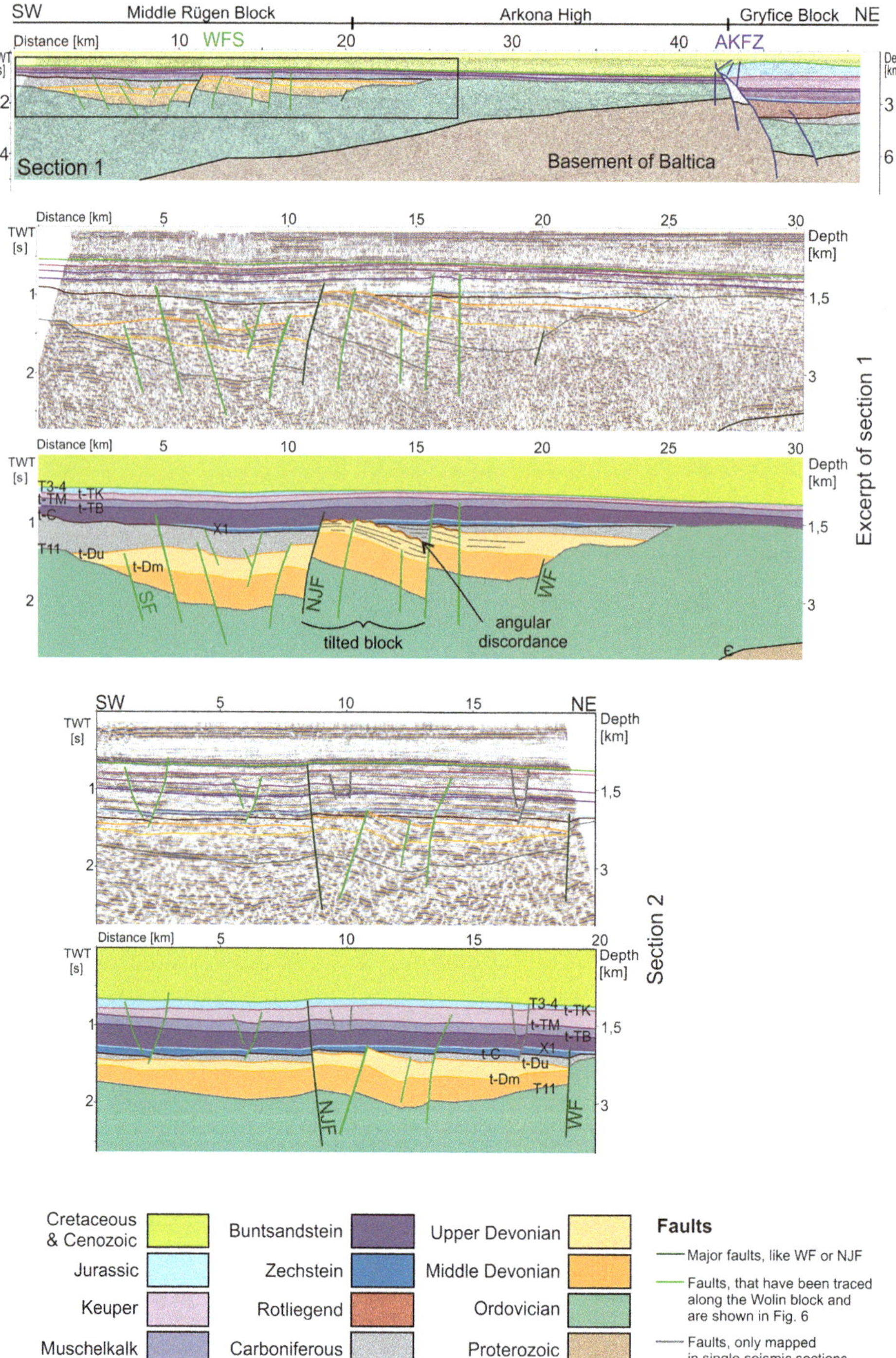

Fig. 5. Selected seismic sections east of Rügen with the main reflectors, as listed in Figure 3, and important faults (NJF, Nord Jasmund Fault; SF, Schaabe Fault; WF, Wiek Fault) and their geological interpretation. For the location of profiles, see Figure 4. The depth was calculated with a constant interval velocity of 3000 m s^{-1} (which is almost the average velocity of the illustrated successions). The depth calculation was made to give an idea of the true depth conditions. The detailed seismic sections are about two times vertically exaggerated.

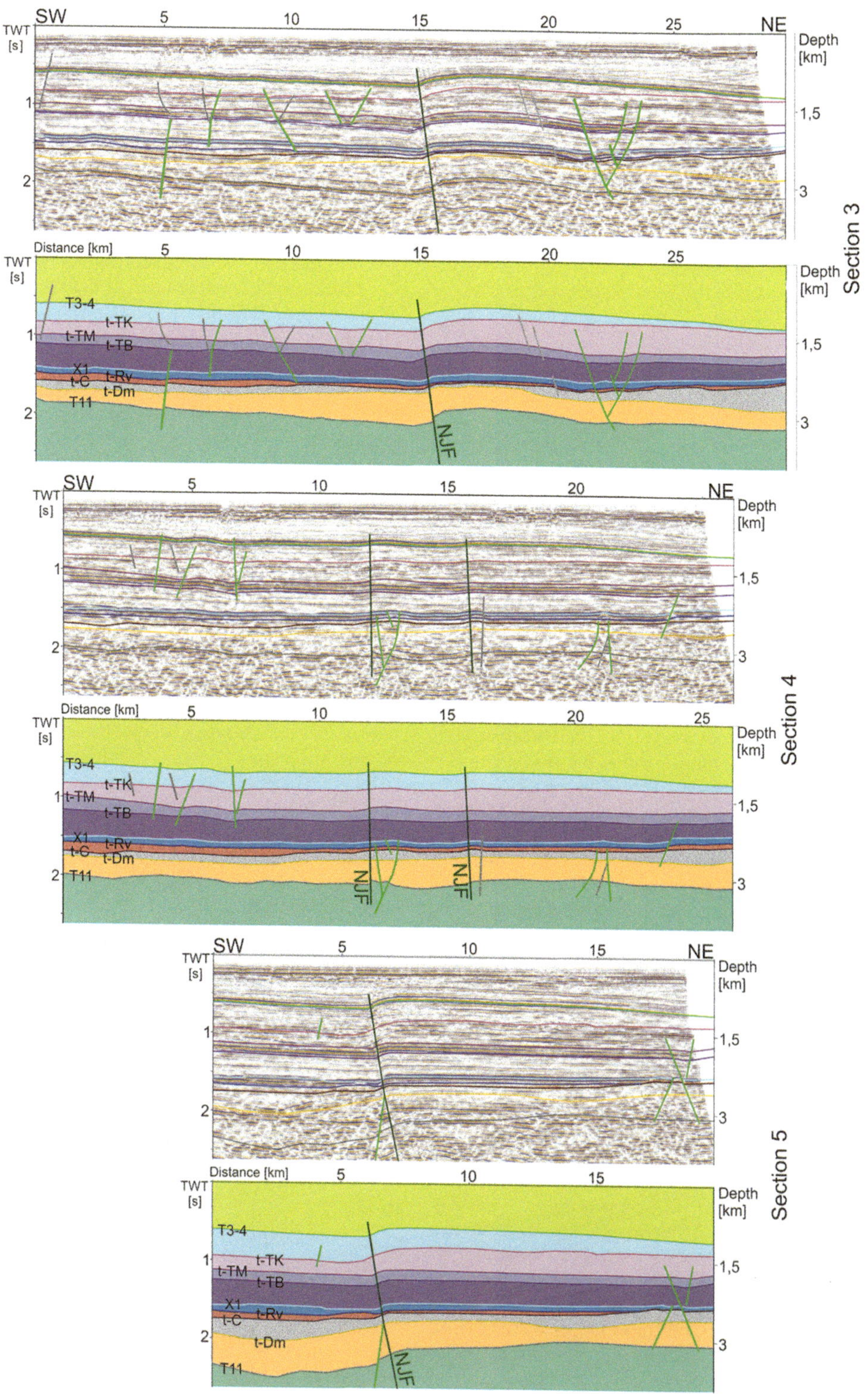

Fig. 5. *Continued.*

burial history or isostatic movements. During the 'Move-on-fault' workflow, the simple shear method was chosen, mainly with a shear angle of about 90° to the fault inclination.

The results of the restoration cannot be treated as truly quantitative with the dip angles of faults and the relative thicknesses of horizons are changing when converting the sections from time to depth. However, based on the restorations, extensional and compressional stress regimes are identified, and the restored sections provide valuable indications of the polyphase evolution of the Wiek Fault System.

Seismic stratigraphy NE of Rügen

Based on the reprocessed seismic data in the area north and east of Rügen, 12 seismic reflectors were used to recognize different lithostratigraphic units, as summarized in Figure 3. The geological interpretations of five seismic sections illustrate the major stratigraphic units, which were distinguished on the basis of well markers and, via well ties, associated to seismic reflectors (Fig. 5). Those seismic sections presented in detail were measured east of Rügen, in the western part of the Pomeranian Bay. The lines run SW–NE, and across the Wolin Block, the Arkona High and, in the case of section 1, also across the Gryfice Block, which is situated within the TTZ. These blocks are separated from each other by NW- to NNW-trending faults, such as the AKFZ between the Gryfice Block and the Wolin Block/Arkona High. Thus, the seismic sections run almost perpendicular to these bordering faults.

The Proterozoic basement forms the lowermost seismically mappable unit. The top Proterozoic reflector ε (cf. Fig. 3) has its highest position at the SE rim of the Arkona High (at about 2 s TWT or 3 km, assuming an average velocity of 3000 m s^{-1}) from where it gently dips towards the SW, probably down to the Moho (Guterch *et al.* 2010). As a result, this reflector is only detectable in seismic section 1 (Fig. 5). The basement of Baltica is overlain by folded Ordovician sediments representing the Caledonian accretionary wedge. The basement and the overlying sedimentary rocks are downfaulted along the AKFZ, which separates the Arkona High from the Gryfice Block.

Towards Rügen (SW part of sections 1 and 2 in Fig. 5), a structural depression above these two older units contains Devonian and Carboniferous strata, as recorded in wells H9 1/87, Sagd 1/70 and Loh 2/70. This depression is filled with Middle Devonian (Old Red) and Upper Devonian sediments, which are, in part, discordantly overlain by lower Carboniferous and upper Carboniferous (Variscan molasse) deposits. These different stratigraphic units cannot be clearly distinguished from each other in the profiles. In Figure 5 (section 1), however, an angular discordance is visible above the Devonian of the tilted blocks. Comparing all seismic sections in Figure 5, it is evident that the Upper Devonian thins towards the south and the Middle Devonian is covered by upper Carboniferous rocks, as shown in well H2 1/90. The Ordovician–Carboniferous units are block faulted along NW–SE-trending faults, as documented on Rügen (cf. Franke 2017).

Wells H2 1/90 and K5 1/88 illustrate that the lower Rotliegend volcanics are restricted to the Gryfice Block and the southern part of the Wolin Block, whereas upper Rotliegend sedimentary rocks occur only along the Gryfice Block. A thin cover of predominantly Zechstein deposits caps the Paleozoic succession along the Gryfice and Wolin blocks.

Triassic and Jurassic successions, which were detected in all four offshore wells (with the exception of well G 14 1/86 were Jurassic strata is missing), show large differences in thickness. Relatively thin (about 500 m) sediments are found on the Wolin and Arkona blocks west of the AKFZ, thickening (up to 2000 m) on the Gryfice Block in the east (section 1 in Fig. 5). Furthermore, the general thickness of the Mesozoic increases at the Wolin Block towards the SW (cf. all sections in Fig. 5: from *c.* 500 to 1500 m). In detail, Liassic (Lower Jurassic) deposits are widely distributed, whereas sediments of the Dogger (Middle Jurassic) only occur on the Gryfice Block. The Lower–Middle Jurassic succession is overlain by Lower Cretaceous (Albian: cf. Fig. 3) strata, suggesting an erosional hiatus in-between. All blocks are covered by Upper Cretaceous carbonate deposits. Due to an insufficient resolution within the uppermost 300 ms (TWT), the Cretaceous and Cenozoic successions could not be differentiated.

Structural elements north and east of Rügen

Based on seismic mapping, numerous faults can be traced north and east of Rügen. Based on their location, strike direction and age of formation, they are grouped into different fault zones and systems (Fig. 5). In addition to those structures (described above) that run parallel to the SW margin of the Fennoscandian Shield (along the Arkona High), two major fault zones can be distinguished:

- the NNW-striking AKFZ bordering the western edge of the Gryfice Block;
- the NNW- to NW-orientated Gryfice Fault Zone within the Gryfice Block (Fig. 6).

These zones are characterized by narrow sets of parallel-running faults. Between Rügen and the AKFZ, numerous structures accompany the deep-seated Wiek and Nord Jasmund faults. These

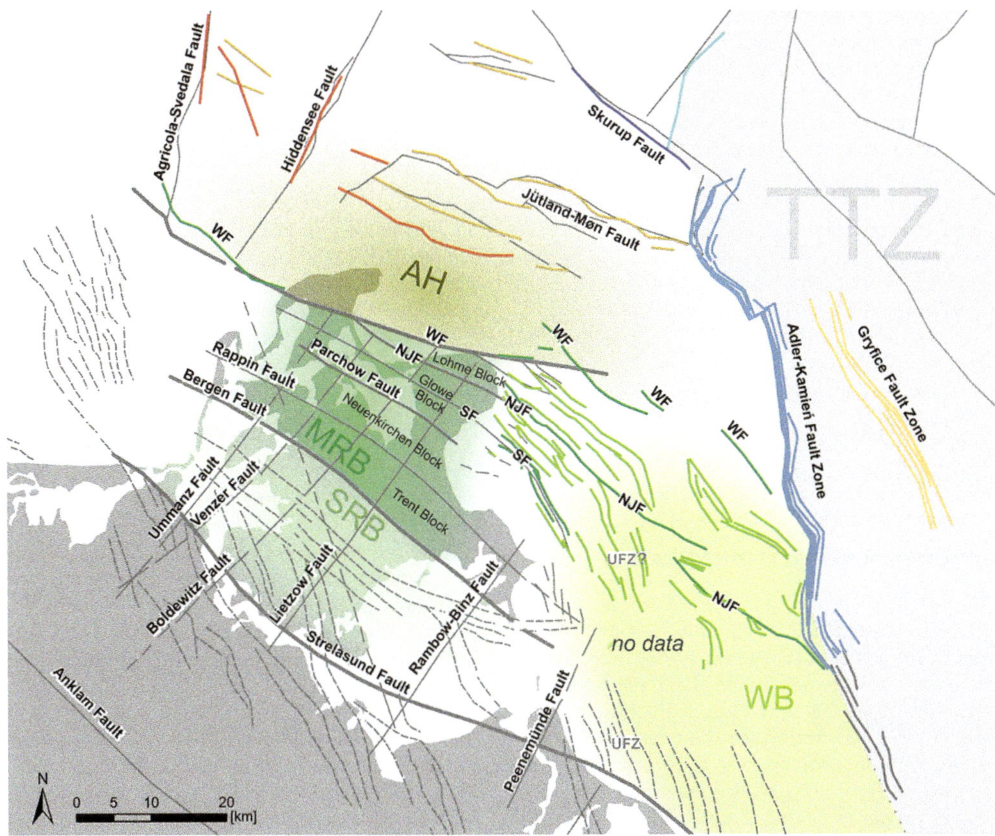

Fig. 6. The faults investigated in the USO-East area are coloured. Those of the Wiek Fault System (green), the Adler–Kamień Fault Zone (blue) and the Gryfice Fault Zone (orange) are compared with previously published patterns of Paleozoic faults (grey lines: Thomas *et al.* 1993; Mayer *et al.* 1994) and with the Mesozoic WPFS (dashed lines: Krauss & Mayer 2004). AH, Arkona High; MRB, Middle Rügen Block; SRB, South Rügen Block; important faults: NJF, Nord Jasmund Fault; SF, Schaabe Fault; UFZ, Usedom Fault Zone; WF, Wiek Fault.

differently orientated lineaments are grouped together as the Wiek Fault System (WFS), which crosses the Middle Rügen and Wolin blocks.

The WFS comprises about 60 single faults, most of which are only detected in the more recent offshore seismic lines (Figs 5 & 6). On Rügen, these structures strike NW–SE; whereas further to the SE, in the Pomeranian Bay, they are orientated NNW–SSE (Fig. 6). In addition, general differences in the age and character of individual faults can be observed. In the NW part of the WFS, pre-Permian units are predominantly affected and dislocated by normal faults, showing large offsets especially in the Upper Devonian successions. Towards the SE, the number of faults and flexures within the Mesozoic strata increases. Most of them have a normal character, but single faults also show reverse indicators. These structures, dominating in the Mesozoic successions, are observed to be conjugating, concave planes, with a partial listric character. The master and antithetic faults/flexures often border Y-shaped small graben (sections 2–4 in Fig. 5). These faults dominate in the Lower Jurassic and Keuper successions, and commonly terminate at depth in the Buntsandstein or uppermost Zechstein. Thus, only a few structures indicate formation by the reactivation of older faults that also displace Devonian or Ordovician strata.

Faults of the WFS, crossing the Wolin Block, show different lateral extensions, but only one dominant fault can be traced from Rügen over a long distance. Following a lateral, about 3 km-wide, offset east of the Mönchgut Peninsula (Rügen: section 4 in Fig. 5 & Fig. 6), it merges with faults of the AKFZ. This fault represents the offshore continuation of the Nord Jasmund Fault (Figs 5 & 6). It laterally changes its normal character towards the SE into a reverse offset. This is accompanied by a change in dip direction from SW to NE and an increasing impact on Mesozoic strata.

Further to the NE, the Wiek Fault is observed (sections 1 and 2 in Fig. 5 & Fig. 6). It separates the folded Ordovician of the Caledonian accretionary wedge in the NE against the Devonian and Carboniferous deposits of the Rügen depression. The overlying Mesozoic strata are not affected, suggesting a Paleozoic age for the fault displacement. As shown in Figure 6, the Wiek Fault is segmented into several individual faults, which can be traced from Jasmund (Rügen) to the AKFZ.

A set of several faults occurs SW of the Nord Jasmund Fault (Figs 5 & 6), which strike parallel to each other and dip towards the NE. The southernmost is considered to represent the offshore continuation of the Schaabe Fault on Rügen (section 1 in Fig. 5 & Fig. 6).

Observations from restored sections

Two profiles from the area east of Rügen were restored to illustrate the structural development of this part of the southern Baltic Sea. Profile 1 is SW–NE orientated, and crosses the WFS and the AKFS (for the location see Fig. 4). From this long profile, only the SW part was restored to show the development of the WFS in detail (Fig. 7). It shows a structural depression lying above Caledonian deformed sedimentary rocks of Ordovician age. The block-faulted Devonian strata are discordantly overlain by Carboniferous sedimentary rocks. Towards the surface, deposits of Triassic–Cretaceous age follow. The faults mainly intersect the Paleozoic successions, only a few were subsequently reactivated resulting in flexures in the basal layers of the Mesozoic.

The synsedimentary tectonic evolution could be reconstructed back to the Middle Devonian. Deposits of Devonian age (Fig. 7a, b) are restricted to a depression above the folded Ordovician (accretionary wedge). This Rügen Basin was formed as a result of regional extension in the foreland of the Variscan orogenic belt and filled with Devonian Old Red sediments (Aehnelt & Katzung 2009). During the Middle Devonian, the region was block faulted along en echelon structures and formed a basin that was partly restricted in its lateral extent. In the NE area of the Middle Rügen Block, synsedimentary displacements occurred along the Wiek Fault, which resulted in separation from the Arkona High. Furthermore, the Schaabe Fault was generated close to the SW border of this local depression (as part of the Rügen Basin). The Nord Jasmund Fault has also been active since the Middle Devonian as it separates the sub-blocks of Lohme and Glowe within the depression (Fig. 6). Then continental clastic sediments were accumulated during the Middle Devonian (Fig. 7a), and overlain by marine deposits after a transgression in the Late Devonian (Fig. 7b). All faults, with exception of the Wiek Fault, remained active until early Carboniferous times, as suggested by varying sedimentary thicknesses. During the transition from early to late Carboniferous times, the regional stress system changed into a compressional one (Fig. 7c), causing the reactivation of faults. The Devonian and the underlying Ordovician strata were block faulted. NE of the Nord Jasmund Fault, the southern part of the Lohme Block was tilted and uplifted as part of the Middle Rügen Block, with its top dipping towards NE. The largest offset is about 500 ms, which corresponds to 750 m, assuming an average velocity of 3000 m s^{-1}. Contemporaneously, the upper Carboniferous was deposited on a variety of older units, as documented by an angular discordance. Due to the rotation component, some faults SE of the uplifted block maintained their normal character but most of them were reactivated as reverse faults. The Schaabe Fault, in the SW part of the profile, dips towards the NE and shows reactivation in the form of a reverse flexure. The upper Carboniferous sediments accumulated post-tectonically above an angular unconformity at the top of the underlying Devonian succession. Sudetian and Asturian movements (Fig. 3) probably caused erosion and reduced the thickness of the Carboniferous strata. The erosion also reached the Upper Devonian strata in the tilted block. Due to a lack of data, only the present-day thickness and distribution was considered in the restoration workflow.

During the Permian, the stress regime changed and the area was affected by extension in the west–east direction. A large intracontinental basin formed, of which the NE margin extended towards the northern Rügen and was flooded by the Zechstein Sea (Fig. 7d). A few normal faults, especially those intersecting the Lohme Block, remained active during the latest Permian and Early Triassic, as suggested by thickness differences. Therefore, the Zechstein is missing above the tilted block in the centre, whereas it is 300 m thick nearby.

An extensional stress regime dominated almost the entire Mesozoic. Four normal faults were active during the Triassic (Fig. 7e). The Nord Jasmund Fault shows an offset, whereas the others form flexures primarily in the Buntsandstein and also partly in the Muschelkalk. During the Early Jurassic (Fig. 7f), no faults were active but the differences in thickness of the Jurassic strata indicate an increasing subsidence rate towards the SW.

After a hiatus of about 75 Ma, sedimentation started again after the Albian–Cenomanian transgression. The thickness of mainly Upper Cretaceous deposits (Fig. 7g) increases slightly towards the NE. Major faults were not observed in this profile, although the NE part of the Lohme Block may

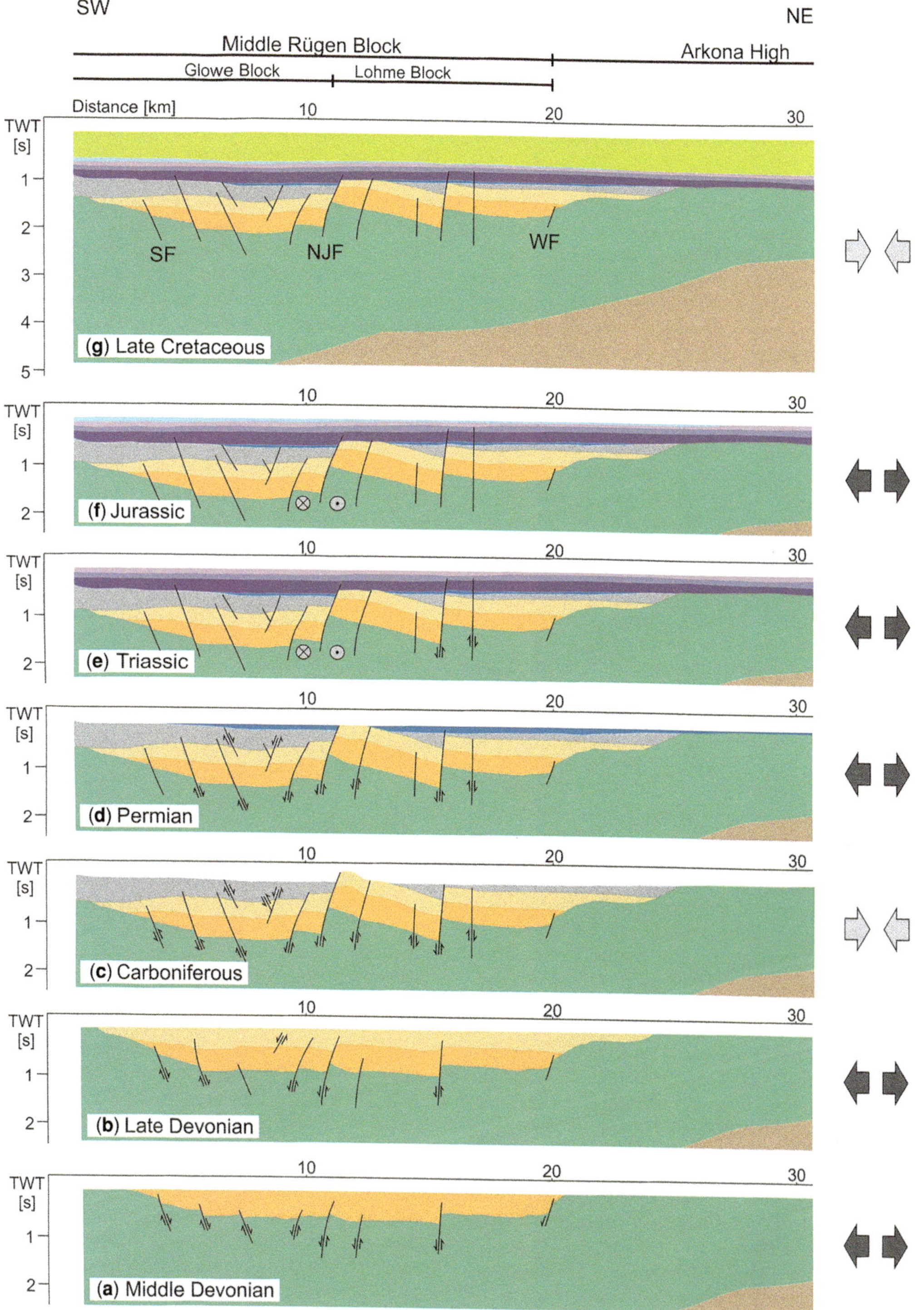

Fig. 7. Results of the restoration of profile 1. For the location, see Figure 4; for the legend, see Figure 8.

have been slightly uplifted. This is interpreted, based on observations from other profiles (Fig. 8), as indicative of a compressional stress regime in Late Cretaceous times.

Profile 3 (Fig. 8) was also used for restoration purposes. This section is located further to the SE of the study area and crosses the Wolin Block in a NE–SW direction. Comparing this line with seismic section 1 (Figs 5 & 7), the tectonic development appears to change in this part of the Pomeranian Bay. The Paleozoic–Mesozoic successions lie horizontally and show only minor tectonic displacements. The Upper Devonian is missing, either not deposited or eroded. In contrast, a succession of lower Rotliegend volcanic rocks is preserved (well H2 1/90). Furthermore, the thickness of the upper Permian (Zechstein), Triassic and Jurassic deposits increases towards the south. The faults mainly occur within the Mesozoic successions, bordering small depressions and graben structures with listric, partly Y-shaped flexures. These can be traced from the Keuper down to the Muschelkalk or even to the top of the Zechstein. Some of them also show displacement in Paleozoic units.

As such, Profile 3 illustrates the following evolution. Continental clastic sediments were deposited during the Middle Devonian (Fig. 8a) on top of the accretionary wedge of deformed Ordovician strata. The Devonian thickness increases towards the central part of the profile, indicating a basin-like structure. Two flexures with a normal displacement formed and show dip towards the depocentre. The fault in the centre of the profile represents the SE continuation of the Nord Jasmund Fault. The formation of an extensional basin and the generation of normal faults suggest a tensional regime during Middle and Late Devonian times.

The geometry of the upper Carboniferous strata (Fig. 8b) suggests that a compressional stress field developed subsequently. The former centre of the Devonian basin was uplifted and the relief inverted, as suggested by the thin (some tens of metres) Carboniferous deposits, which contrasts with that developed at the former basin margins in the NE and SW where the thickness of the Carboniferous sediments increased to some hundreds of metres. At this time, the Nord Jasmund Fault and the another fault were reactivated, and became reverse faults with different dip directions. The uplift might also be accompanied by local erosion of the Upper Devonian and lower Carboniferous successions (not considered in the restoration process).

The Permian (Fig. 8c) was again affected by an extensional stress regime, with the average thickness of the Rotliegend and Zechstein deposits increasing towards the SW, were a new depocentre was formed. The active faults, such as the Nord Jasmund Fault, became normal faults, whereas new faults were generated in the NE part of the profile. The Triassic and Jurassic are characterized by the formation of normal en echelon and listric faults (Fig. 8d–e). They occur mostly in the Keuper, and extend down to the Muschelkalk and Buntsandstein strata. Furthermore, a few faults are still active in the Jurassic. The conjugating flexures and faults, forming a Y-shaped graben, are similar in orientation to the faults of the WPFS (cf. Fig. 1), which are characterized by strike-slip motions. These Mesozoic faults are located in the northern prolongation of the Usedom Fault Zone (part IX of the WPFS in Fig. 1). The Nord Jasmund Fault shows vertical displacements, but a strike-slip component cannot be excluded. The Triassic and Jurassic therefore appear to have a character dominated by a transtensional regime. After a hiatus of 75 myr, sedimentation started again in the middle Cretaceous (Fig. 8f). The reactivation along the Nord Jasmund Fault in the centre of the profile and the Mesozoic fault in the SW part of the seismic section is indicated by reverse flexures that mark the upwards continuation of the deep-seated fault plane.

Development of the Wiek Fault System (WFS)

Based on the known tectonic evolution of the region and observations from 2D seismic line restoration, it is proposed that the formation and evolution of the WFS can be divided into five stages:

- The formation of the Rügen Basin began in the Middle Devonian following the Caledonian Orogeny (Aehnelt & Katzung 2009). Terrigenous, clastic sediments of Old Red facies were overlain by marine, calcareous sediments of the Upper Devonian. These sediments accumulated in synsedimentary basins, formed as the epi-Caledonian cap rock (Rügen Caledonides) on the Western European Platform (Katzung 2004). The NW–SE-elongated Rügen Basin between Hiddensee and Usedom was bordered and intersected by parallel, NW-trending en echelon faults (Figs 7a, b & 8a). The SW-dipping Wiek Fault bordered this basin to the NE. The Nord Jasmund Fault was formed dipping to the SW and the Schaabe Fault dipping to the NE.
- A compressive stress field, induced by the Variscan Orogeny (late Carboniferous), led to formation and reactivation of faults in the northern foreland. Thus, the part of the Lohme Block NE of the Nord Jasmund Fault was uplifted and tilted (Fig. 7c). This uplift can be traced further to the SE on the Wolin Block, as indicated by a local high NE of the Nord Jasmund Fault (Fig. 8b). Comparing Figures 7b, c and 8a, b, a change in the

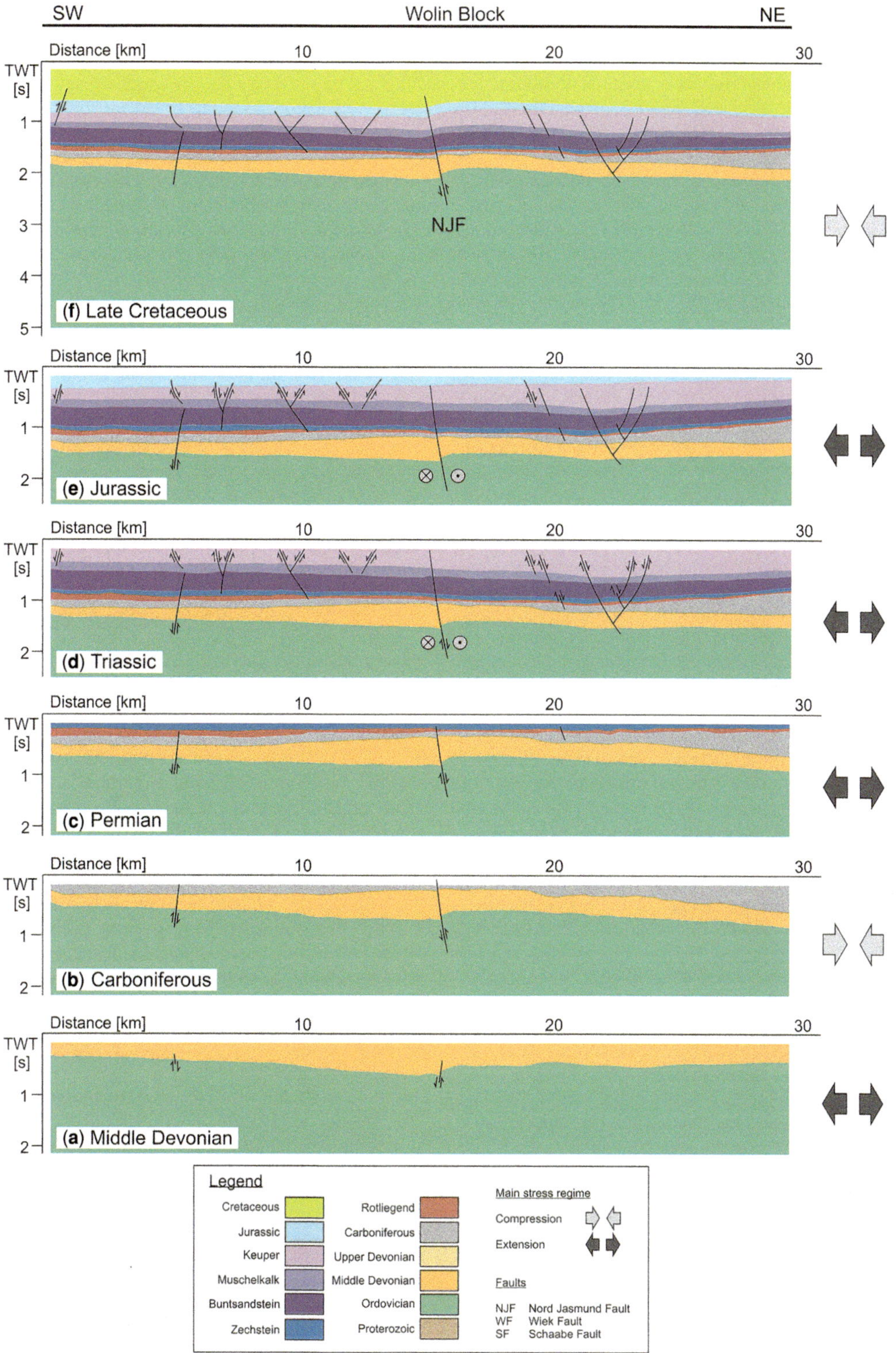

Fig. 8. Results of the restoration of profile 3. For the location, see Figure 4.

dip direction of the Nord Jasmund Fault is visible. During the Carboniferous, reactivation of the fault occurred and its dip direction changed in the SE part of the working area, from a SW-dipping normal fault to a NE-dipping reverse fault. The change in the stress regime also resulted in the formation of an angular discordance at the base of the upper Carboniferous (section 1 in Fig. 5).

- Thermal destabilization of the continental crust in Rotliegend times induced the formation of the Central European Basin System. Rügen was located at the NE margin of the subsiding North German Basin (cf. Fig. 1). The marine transgression of the Zechstein Sea was controlled by locally prominent blocks, forming shallow sedimentary facies or islands (Fig. 7d). The Arkona Block remained as a local high (as eastern extension of the Ringkøbing-Fyn High). The seismic sections show an increasing thickness of the Permian succession towards the SSW. During the formation of the NGB, the faults crossing the Wolin Block were mostly inactive, with only minimal movement along a few existing faults (Figs 7d & 8c). In the Early Triassic, the rate of subsidence was enhanced within an east–west-orientated extensional stress regime.
- From Keuper times, the extensional stress system rotated counter-clockwise into a NE–SW orientation (Kley *et al.* 2008; Pharaoh *et al.* 2010). This induced dextral strike-slip movements along existing NW–SE-trending faults (Figs 7e, f & 8d, e). Besides the reactivation of older Paleozoic faults, this shear stress induced the formation of new faults in the SE part of the WFS, striking NNW–SSE (Fig. 8e).
- Since the Late Cretaceous, central and northern Europe was affected by SW–NE compression, induced by the convergence in the Pyrenees (Kley & Voigt 2008). This caused NW–SE-striking faults to be reactivated as reverse faults (e.g. the Nord Jasmund Fault: Fig. 8f).

Discussion

The southern Baltic Sea area in the vicinity of Rügen is characterized by major NW–SE-striking faults, generated by extensional movements in the Middle Devonian. Whilst they are well described onshore Rügen, their SE offshore continuation has not been well documented (Kurrat 1974; Franke & Hoffmann 1988; Thomas *et al.* 1993; Mayer *et al.* 1994; Piske *et al.* 1994; Schlüter *et al.* 1998) (cf. Fig. 3a). Other NNW–SSE-orientated faults and fault zones were activated during the Mesozoic. Krauss & Mayer (2004) (Fig. 3d) showed that they occur in the middle and southern part of Rügen and adjacent offshore areas forming the WPFS. This paper provides a more detailed description of both the Paleozoic and Mesozoic faults and their interactions.

A comparison between the detailed fault pattern obtained by our new analysis and the results of former investigations is summarized in Figure 6. The area east of Rügen is characterized by a very complex fault pattern in which we have recognized a large number of new faults, as well as provided modifications to the position and orientation of many of the major faults, in particular that of the Wiek Fault.

A range of newly discovered faults occur in the Middle Rügen and Wolin blocks, and are grouped together as the WFS. Most of them strike NW–SE between the coast of Jasmund and the AKFZ. Towards the SE, faults also occur with a NNW–SSE trend. Among these faults, one dominant fault can be traced across the entire Wolin Block. This major fault is characterized in the north as a normal fault dipping to the SW, which borders an exposed and tilted Devonian block. Further to the SE, it is embodied by a reverse flexure, reactivated during Late Cretaceous inversion tectonics. East of Mönchgut (Rügen), it has a lateral offset of about 3 km to the south. In previously published maps (cf. Fig. 2), this fault was either shown as the Wiek Fault or the Nord Jasmund Fault. According to Franke (2017), the Wiek Fault borders the Arkona High to the south and is defined as a major displacement, separating the faulted Ordovician strata of the accretionary wedge in the north from the upper Paleozoic deposits in the south. Thus, the Wiek Fault east of Rügen is located further to the north, as shown in Figures 5 and 6. In contrast, the major fault, situated in the centre of the WFS, clearly represents the Nord Jasmund Fault, which separates the Lohme and Glowe blocks (Fig. 6).

It is proposed that the Wiek Fault can be traced via segmented fault planes from the north of Jasmund to the AKFZ. Relay ramps seem to be formed between the partly overlapping segments. Wells on Rügen suggest a vertical displacement of up to 3 km along the Wiek Fault. East of Rügen, however, offsets of between 100 and 1000 m (assuming an average velocity of 3000 m s^{-1}) occur. Accompanying the transition from the Middle Rügen to the Wolin Block, the vertical displacement between the Arkona High and the Middle Rügen Block decreases. The NW continuation of the Wiek Fault could also be detected west of Rügen (cf. Deutschmann *et al.* this volume, in press), which supports the idea that the Wiek Fault and Nord Jasmund Fault form part of the Odense–Wiek Fault Zone as described by Franke (2017).

A set of subparallel faults occurs SW of the Nord Jasmund Fault and dip towards the NW. The southernmost is considered to represent the offshore continuation of the Schaabe Fault that intersects the Glowe Block (Fig. 6). Due to its changing strike direction from NW towards NNW, it can be traced from Jasmund to the east of Mönchgut.

Several faults that change from a NW–SE into a NNW–SSE strike direction dominate the fault pattern in the SE part of the Wolin Block. These faults form typical listric flexures with a normal sense of movement. They often border small Y-shaped graben structures, and are therefore similar to faults analysed during the VPSS I and II projects and attributed to the WPFS (Mayer *et al.* 2000, 2001*a*, *b*; Krauss & Mayer 2004). These newly detected faults, especially those within the Keuper and Lower Jurassic successions, may represent the NW continuation of the Usedom Fault Zone (part IX of the WPFS: cf. Fig. 1). Due to a lack of data between the two areas, this assumption could not be validated. Considered altogether, the nine distinct fault zones of the WPFS (Krauss & Mayer 2004), and the newly mapped faults east of Jasmund, reflect a NE–SW-orientated tensional stress system. These structures have been discussed previously with some debate on the sense of lateral displacement. For example, Mayer *et al.* (2000) suggested a sinistral reactivation of the Trans-European Suture Zone due to early Cimmerian stress impulses of the Tornquist Zone, and Mayer *et al.* (2001*a*) proposed the formation of the WPFS by dextral late Cimmerian movements (Fig. 3) along the Trans-European Fault. In contrast, Krauss & Mayer (2004) described the WPFS as a system of NNW–SSE-trending pull-apart graben, originating as a result of early Cimmerian reactivation of pre-Permian WNW–ESE-orientated basement faults that remained active until the Early Cretaceous. The mega-shear zone of the TESZ was reactivated and acted as compensation zone. Thus, during a dextral displacement, the formation of the WPFS was influenced by dominant pre-existing faults (Krauss & Mayer 2004).

Although our observations are in agreement with Krauss & Mayer (2004) by illustrating a dextral sense of lateral displacement during the Middle Mesozoic (Keuper–Early Cretaceous), additional NNE-trending graben were detected north of the Samtens Fault Zone, at least in the offshore area (Figs 6 & 9). Thus, all these faults that border NNW–SSE-striking graben structures occur within a NW–SE-orientated zone between the Wiek and Anklam faults. This represents a typical transtensional shear zone, where dextral extensional forces are directed at an angle of 60° to it (Fig. 9a), and triggered block faulting. Similar to the geometry of small-scale S–C structures, several normal faults were generated with a strike direction of about 45° to the orientation of the shear zone (Fig. 9b). As a result, small graben structures developed perpendicular to the main extension direction (σ_3).

Finally, the term WFS summarizes all Paleozoic and Mesozoic faults in the offshore area east of Rügen, especially on the Middle Rügen and Wolin blocks. However, this is a regional attribution. The WFS comprises Paleozoic faults of the Tornquist Fan, striking NW and crossing Rügen Island (Wiek Fault, Nord Jasmund Fault and Schaabe Fault), as well as Mesozoic faults that were generated during a dextral transtensional stress regime forming the WPFS. The major Paleozoic faults show a reactivation during the Mesozoic transtensional stress regime, but also during the phase of Late Cretaceous inversion tectonics.

Deutschmann *et al.* (this volume, in press) analysed and discussed the westerly offshore continuation of the WFS and other fault zones west of Rügen, which belong to the WPFS. The deep Wiek Fault can be traced in several seismic profiles orientated subparallel to the western coast of Rügen, suggesting a late Paleozoic formation and Mesozoic reactivation of this NW–SE-striking structure. In contrast to the east of Rügen, almost no accompanying younger faults were detected. Other deep faults that strike oblique to the Wiek Fault are named the Plantagenet and Agricola faults. They were also generated during the late Paleozoic. Both belong to the Agricola Fault Zone which, together with the Werre Fault Zone, crosses the northern Darss Peninsula – part of the en echelon structures of the WPFS developed during the Mesozoic extensional tectonics.

As further work, the calculation of a precise velocity model is planned with regard to German onshore wells and other offshore wells in the vicinity of the USO research area. After this, a time-to-depth conversion will be possible, and therefore also an exact investigation of the different tectonic phases (containing assumptions for the lateral extension or shortening in metres, dip angles of faults or vertical displacements). Furthermore, the land–sea transition and the progression of the WFS towards Jasmund, and also towards Usedom, should be analysed.

Conclusion

The southern Baltic Sea area is characterized by a complex fault pattern above the Trans-European Suture Zone (TESZ), marking the Caledonian amalgamation of Avalonia and Baltica. This led to the formation of faults of different orientations that were generated during several tectonic phases throughout the late Paleozoic and the Mesozoic. Based on reprocessed offshore seismic, 60 different faults could be detected east of Rügen, which are grouped together as the Wiek Fault System (WFS).

Five stages can be recognized in the polyphase evolution of the area: (1) post-Caledonian extension during the Middle and Late Devonian; (2) Variscan compression and block faulting at the boundary

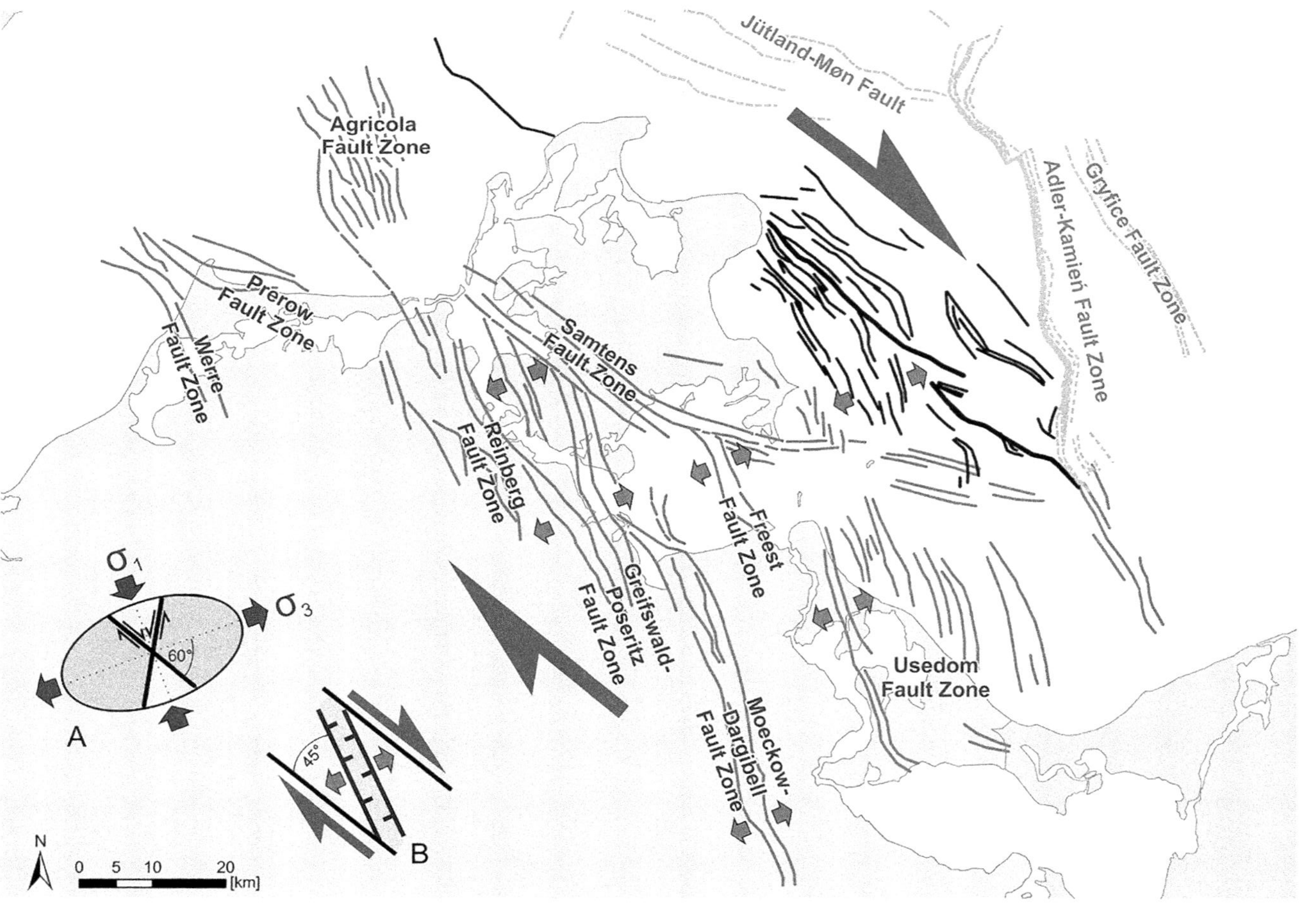

Fig. 9. Middle Mesozoic reactivation of the Trans-European Suture Zone due to transtensional stress resulted in the subsequent reactivation of Paleozoic faults and the generation of new faults, forming together the WFS (black lines). Other faults mapped during the USO-East project are dotted. Mesozoic faults, as shown by Krauss & Mayer (2004), are in grey. A: illustration of the deformation ellipsoid for the NE–SW-orientated main extension. B: the resulting formation of NW–SE-trending shear zones.

between early and late Carboniferous; (3) Permian thermal subsidence, which led to the formation of the North German Basin (NGB); (4) Triassic–Early Cretaceous transtension with reactivation of faults and the formation of the Western Pomeranian Fault System (WPFS); and (5) inversion tectonics from the Late Cretaceous onwards, especially along NW- to NNW-striking faults with a NE dip direction.

The WFS contains three major faults. The Wiek Fault crosses northern Rügen and continues NE of Jasmund in the Pomeranian Bay. Several segments of this fault could be traced to the Adler–Kamień Fault Zone (AKFZ). The Wiek Fault separates the Arkona High from the Middle Rügen and the Wolin Block. Compared to the results onshore, where a vertical displacement of about 3 km displaced Ordovician against Devonian and Carboniferous deposits, the vertical offsets along the offshore continuation of the Wiek Fault are rather small, with Upper Devonian–Carboniferous strata extending even further to the NE. This fault appears to have been inactive during the late Paleozoic. The deep Nord Jasmund Fault crosses Jasmund and can be traced nearly continuously to the AKFZ. Only one lateral offset about 3 km to the south is detected on the Wolin Block. The fault divides the Middle Rügen Block into the Lohme and the Glowe blocks. It originated during the Devonian as a SW-dipping normal fault. Due to Carboniferous compression, the dip direction changed towards the NE in the southern part of the Wolin Block. During Cretaceous inversion, this southern branch was reactivated as a reverse fault. The Schaabe Fault continues offshore east of Rügen, where it intersects the Rügen Basin above the Glowe Block. Due to the transition from the Rügen Basins towards the Wolin Block, the Schaabe Fault tapers off east of Mönchgut.

Numerous minor faults of the WFS represent conjugating Mesozoic faults, which often border Y-shaped small graben. Their formation is attributed to the reactivation of the TESZ. NE–SW extension, with an increasing dextral shear stress along existing NW–SE-orientated Paleozoic faults, induced the development of sigmoidal NNW–SSE-striking faults and an intervening graben during the Late Triassic–Early Cretaceous. These are genetically related to other faults of the WPFS, suggesting a northwards continuation of the Usedom Fault Zone. Therefore, the WPFS also covers the area between the Samtens Fault Zone and the Wiek Fault.

This project received financial support from Central European Petroleum Ltd (CEP). The authors thank CEP, the Federal Institute for Geosciences and Natural Resources (BGR), and the Geological Survey of Mecklenburg-Western Pomerania (LUNG M-V) for granting access to seismic and well data. The software SeisWare™ was provided by SeisWare International Inc.™ with an educational software licence. The authors acknowledge the use of the Move Software Suite granted by Midland Valley's Academic Software Initiative. Special thanks go to editors, reviewers and Laurence N. Warr for their constructive comments on the manuscript. Furthermore, we thank Arezki Ioughlissen, Manfred K. Hauptmann (both CEP), Christian Hübscher (University of Hamburg) and Jonas Kley (University of Göttingen) for useful discussions. Special thanks for further technical support are dedicated to Juliane Brandes (LUNG M-V) and Anna Gehrmann (Ernst-Moritz-Arndt University of Greifswald).

References

AEHNELT, M. & KATZUNG, G. 2009. Middle Devonian Old Red Rügen Basin in Western Pomerania, NE Germany: implications for post-Caledonian evolution and palaeogeography at southern margin of the Old Red Continent. *Zeitschrift der Deutschen Gesellschaft für Geowissenschaften*, **160**, 1–11.

AL HSEINAT, M. & HÜBSCHER, C. 2017. Late Cretaceous to recent tectonic evolution of the North German Basin and the transition zone to the Baltic Shield/southwest Baltic Sea. *Tectonophysics*, **708**, 28–55.

ARNDT, G., STOLLENWERK, M. & ZENKER, F. 1996. *Ergebnisbericht, Reprocessing von digitalseismischen Daten aus der südlichen Ostsee) [Final report, Reprocessing of digital seismic data of the southern Baltic Sea]*. Petrobaltic unpublished report

BACHMANN, G.H., GELUK, M.C. *ET AL.* 2010. Triassic. *In*: DOORNENBAL, J.C. & STEVENSON, A.G. (eds) *Petroleum Geological Atlas of the Southern Permian Basin Area*. European Association of Geoscientists and Engineers (EAGE), Houten, The Netherlands, 149–173.

BAYER, U., SCHECK, M. *ET AL.* 1999. An integrated study of the NE German Basin. *Tectonophysics*, **314**, 285–307.

BAYER, U., GRAD, M. *ET AL.* 2002. The southern margin of the East European Craton: new results from seismic sounding and potential fields between the North Sea and Poland. *Tectonophysics*, **360**, 301–314.

BEIER, H. & KATZUNG, G. 1999. Lithologie und Strukturgeologie des Altpaläozoikums in der Offshore-Bohrung G 14-1/86 (südliche Ostsee) [Lithology and structural geology of the early Paleozoic in the offshore well G14-1/86]. *Greifswalder Geowissenschaftliche Beiträge*, **6**, 327–345.

BEIER, H. & KATZUNG, G. 2001. The deformation history of the Rügen Caledonides (NE Germany) – implications from the structural inventory of the Rügen 5 borehole. *Neues Jahrbuch für Geologie und Paläontologie – Abhandlungen*, **222**, 269–300.

BEIER, H., MALETZ, J. & BÖHNKE, A. 2000. Development of an Early Palaeozoic foreland basin at the SW margin of Baltica. *Neues Jahrbuch für Geologie und Paläontologie – Abhandlungen*, **218**, 129–152.

BENEK, R., KRAMER, W. *ET AL.* 1996. Permo-Carboniferous magmatism of the Northeast German Basin. *Tectonophysics*, **266**, 379–404.

BERTHELSEN, A. 1992*a*. Mobile Europe. *In*: BLUNDELL, D., FREEMAN, R. & MUELLER, S. (eds) *A Continent Revealed. The European Geotraverse*. Cambridge University Press, Cambridge, 11–32.

BERTHELSEN, A. 1992*b*. From Precambrian to Variscan Europe. *In*: BLUNDELL, D., FREEMAN, R. & MUELLER, S.

(eds) *A Continent Revealed. The European Geotraverse*. Cambridge University Press, Cambridge, 153–164.

Berthelsen, A. 1998. The Tornquist Zone northwest of the Carpathians: An intraplate pseudosuture. *Geologiska Foreningens i Stockholm Forhandlingar*, **120**, 223–230.

Deecke, W. 1906. *Der Strelasund und Rügen. Eine tektonische Studie. Sitzungsbericht der Königlich Preussischen Akademie der Wissenschaften zu Berlin [The Strelasund and Rügen Island. A tectonic study. Minutes of Proceedings of the Royal Prussian Academy of Sciences in Berlin]*. Preussische Akademie der Wissenschaften, Berlin.

DEKORP-BASIN Research Group 1999. Deep crustal structure of the Northeast German basin: New DEKORP-BASIN '96 deep-profiling results. *Geology*, **27**, 55–58.

Deutschmann, A., Meschede, M. & Obst, K. In press. Fault system evolution in the Baltic Sea area west of Rügen, NE Germany. *In*: Kilhams, B., Kukla, P.A., Mazur, S., McKie, T., Mijnlieff, H.F. & van Ojik, K. (eds) *Mesozoic Resource Potential of the Southern Permian Basin*. Geological Society, London, Special Publications, **469**, https://doi.org/10.1144/SP469.24

Erlström, M., Thomas, S.A., Deeks, N. & Sivhed, U. 1997. Structure and tectonic evolution of the Tornquist Zone and adjacent sedimentary basins in Scania and the southern Baltic Sea area. *Tectonophysics*, **271**, 191–215.

Franke, D. 1993. The southern border of Baltica – a review of the present state of knowledge. *Precambrian Research*, **64**, 419–430.

Franke, D. 2017. *Regionale Geologie von Ostdeutschland – Ein Wörterbuch [Regional Geology of eastern Germany – A dictionary]*, http://www.regionalgeologie-ost.de [last accessed 29 March 2017].

Franke, D. & Hoffmann, N. 1988. Der Bruchschollenbau Rügens – ein Beispiel tafelrandparalleler Strukturentwicklung [Block faulting of Rügen Island – An example for the structural development at the platform margin]. *WTI (Wissenschaftlich Technischer Informationsdienst des Zentralen Geologischen Instituts)*, **29**, 50–59.

Franke, D., Kölbel, B. & Schwab, G. 1989. Zur Interpretation der Tornquist-Teisseyre-Zone nach plattentektonischen Aspekten [The interpretation of the Tornquist-Teisseyre-Zone considering plate-tectonic aspects]. *Zeitschrift für angewandte Geologie*, **35**, 193–198.

Franke, W. 2000. The mid-European segment of the Variscides: tectonostratigraphic units, terrane boundaries and plate tectonic evolution. *In*: Franke, W., Haak, V., Oncken, O. & Tanner, D. (eds) *Orogenic Processes: Quantification and Modelling in the Variscan Belt*. Geological Society, London, Special Publications, **179**, 35–61, https://doi.org/10.1144/GSL.SP.2000.179.01.05

Geissler, M., Breitkreuz, C. & Kiersnowski, H. 2008. Late Paleozoic volcanism in the central part of the Southern Permian Basin (NE Germany, W Poland): facies distribution and volcano-topographic hiati. *International Journal of Earth Sciences*, **97**, 973–989.

Graversen, O. 2009. Structural analysis of superposed fault systems of the Bornholm horst block, Tornquist Zone, Denmark. *Bulletin of the Geological Society of Denmark*, **57**, 25–49.

Guterch, A., Wybraniec, S. *et al.* 2010. Crustal structure and structural framework. *In*: Doornenbal, J.C. & Stevenson, A.G. (eds) *Petroleum Geological Atlas of the Southern Permian Basin Area*. European Association of Geoscientists and Engineers (EAGE), Houten, The Netherlands, 11–23.

Hoth, K., Rusbült, J., Zagora, K., Beer, H. & Hartmann, O. 1993. *Die tiefen Bohrungen im Zentralabschnitt der Mitteleuropäischen Senke – Dokumentation für den Zeitabschnitt 1962–1990 [The deep wells in the centre of the Central European Basin – documentation for the timeframe between 1962–1990]*. Schriftenreihe für Geowissenschaften, **2**.

Katzung, G. 2001. The Caledonides at the southern margin of the East European Craton. *Neues Jahrbuch für Geologie und Paläontologie – Abhandlungen*, **222**, 3–53.

Katzung, G. 2004. Regionalgeologische Entwicklung [Regional geological development]. *In*: Katzung, G. (ed.) *Geologie von Mecklenburg-Vorpommern [Geology of Mecklenburg-Western Pomerania]*. E. Schweizerbart'sche Verlagsbuchhandlung, Stuttgart, Germany, 15–37.

Kley, J. & Voigt, T. 2008. Late Cretaceous intraplate thrusting in central Europe: Effect of Africa–Iberia–Europe convergence, not Alpine collision. *Geology*, **36**, 839–842.

Kley, J., Franzke, H.-J. *et al.* 2008. Strain and stress. *In*: Littke, R., Bayer, U., Gajewski, D. & Nelskamp, S. (eds) *Dynamics of Complex Intracontinental Basins. The Central European Basin System*. Springer, Berlin, 97–124.

Krauss, M. 1994. The Tectonic Structure below the Southern Batic Sea and its Evolution. *Zeitschrift für Geologische Wissenschaften*, **22**, 19–32.

Krauss, M. & Mayer, P. 1999. Der präquartäre Strukturbau im Bereich des Greifswalder Boddens und sein Einfluß auf den quartären Sedimentkomplex (BMBF-Projekt SASO II) [The pre-quaternary structure in the area of the Greifswalder Bodden and its influence on the Quaternary sedimentary complex (BMBF project SASO II)]. *Zeitschrift für Geologische Wissenschaften*, **27**, 153–160.

Krauss, M. & Mayer, P. 2004. Das Vorpommern-Störungssystem und seine regionale Einordnung zur Transeuropäischen Störung [The Western Pomeranian Fault System and its relationship to the Trans-European Fault]. *Zeitschrift für Geologische Wissenschaften*, **32**, 227–246.

Krawczyk, C.M., Eilts, F., Lassen, A. & Thybo, H. 2002. Seismic evidence of Caledonian deformed crust and uppermost mantle structures in the northern part of the Trans-European Suture Zone, SW Baltic Sea. *Tectonophysics*, **360**, 215–244.

Krzywiec, P., Kramarska, R. & Zientara, P. 2003. Strike-slip tectonics within the SW Baltic Sea and its relationship to the inversion of the Mid-Polish Trough – evidence from high-resolution seismic data. *Tectonophysics*, **373**, 93–105.

Kurrat, W. 1974. *Komplexgeophysikalische Beiträge zur Erfassung der tektonischen, strukturellen und faziellen Gliederung des Präzechsteins in einem Teilgebiet am Südwestrand der Osteuropäischen Tafel: dargestellt am Beispiel der geophysikalisch-geologischen Erkundungsarbeiten im Gebiet Rügen [Complex geophysical characterisation of the tectonic, structural and facial development of the pre-Zechstein at the southwestern margin of the East-European Platform: illustrated for the example of the geophysical-geological prospection in the area of Rügen Island]*. Dissertation, Ernst-Moritz-Arndt Universität Greifswald, Greifswald, Germany.

LINDERT, W. & HOFFMANN, N. 2004. Karbon [Carboniferous]. *In*: KATZUNG, G. (ed.) *Geologie von Mecklenburg-Vorpommern [Geology of Mecklenburg-Western Pomerania]*. E. Schweizerbart'sche Verlagsbuchhandlung, Stuttgart, Germany, 79–95.

MAKRIS, J. & WANG, S.-R. 1994. Crustal Structure at the Tornquist–Teisseyre zone in the Southern Baltic Sea. *Zeitschrift für Geologische Wissenschaften*, **22**, 47–54.

MAYER, P., SEIFERT, M. & SCHEIBE, R. 1994. Geologisch-geophysikalische Ergebnisse im Schelfbereich der Insel Rügen [Geological-geophysical investigation results in the shelf area of Rügen Island]. *Zeitschrift für Geologische Wissenschaften*, **22**, 55–66.

MAYER, P., KRAUSS, M. & VORMBAUM, M. 2000. Der Strukturbau des Vorpommern-Störungssystems im Bereich der NE-Fortsetzung des DEKORP-Profils BASIN 9601 (DGF-Projekt VPSS I) [The structure of the Western Pomeranian Fault System in the NE continuation of the DEKORP profile BASIN 9601]. *Zeitschrift für Geologische Wissenschaften*, **28**, 397–404.

MAYER, P., KRAUSS, M., ZENKER, F. & ZÖLLNER, H. 2001*a*. Ergebnisse eines Reprocessings seismischer Industriedaten aus dem Bereich des Vorpommern-Störungssystems [Results of the reprocessing of seismic industrial data from the area of the Western Pomeranian Fault System]. *Zeitschrift für Geologische Wissenschaften*, **29**, 383–400.

MAYER, P., ZÖLLNER, H., SCHMIDT, V., MEYER, H.G. & SCHICKOWSKY, P. 2001*b*. Beiträge zum Strukturbau des Präzechsteins im Bereich des NE-Abschnittes des DEKORP-Profiles BASIN '96 auf der Grundlage reprozessierter industrieseismischer Felddaten (DFG-Projekt VPSS II) [Contributions to the structural setting of the pre-Zechstein in the northeastern part of the DEKORP profile BASIN'96 based on reprocessed industrial-seismic field data (DFG project VPSS II)]. *Zeitschrift für Geologische Wissenschaften*, **29**, 267–273.

MAYSTRENKO, Y., BAYER, U., BRINK, H.-J. & LITTKE, R. 2008. The Central European Basin System – an overview. *In*: LITTKE, R., BAYER, U., GAJEWSKI, D. & NELSKAMP, S. (eds) *Dynamics of Complex Intracontinental Basins. The Central European Basin System*. Springer, Berlin, 15–34.

MAZUR, S., MIKOLAJCZAK, M., KRZYWIEC, P., MALINOWSKI, M., BUFFENMYER, V. & LEWANDOWSKI, M. 2015. Is the Teisseyre-Tornquist Zone an ancient plate boundary of Baltica? *Tectonics*, **34**, 2465–2477.

MAZUR, S., MIKOLAJCZAK, M., KRZYWIEC, P., MALINOWSKI, M., LEWANDOWSKI, M. & BUFFENMYER, V. 2016. Pomeranian Caledonides, NW Poland: a collisional suture or thin-skinned fold-and-thrust belt? *Tectonophysics*, **692**, 29–43.

MÜLLER, E.P., DUBSLAFF, H., EISERBECK, P. & SALLUM, R. 1993. Zur Entwicklung der Erdöl- und Erdgasexploration zwischen Ostsee und Thüringer Wald [The development of the oil and gas exploration between Baltic Sea and Thuringian Forest]. *Geologisches Jahrbuch, A*, **131**, 5–30.

NEUMANN, E.-R., WILSON, M., HEEREMANS, M., SPENCER, E.A., OBST, K., TIMMERMAN, M.J. & KIRSTEIN, L. 2004. Carboniferous–Permian rifting and magmatism in southern Scandinavia, the North Sea and northern Germany: a review. *In*: WILSON, M., NEUMANN, E.-R., DAVIES, G.R., TIMMERMAN, M.J., HEEREMANS, M. & LARSEN, B.T. (eds) *Permo-Carboniferous Magmatism and Rifting in Europe*. Geological Society, London, Special Publications, **223**, 11–40, https://doi.org/10.1144/GSL.SP.2004.223.01.02

OBST, K., HAMMER, J., KATZUNG, G. & KORICH, D. 2004. The Mesoproterozoic basement in the southern Baltic Sea: insights from the G 14-1 offshore borehole. *International Journal of Earth Sciences*, **93**, 1–12.

OBST, K., DEUTSCHMANN, A., SEIDEL, E. & MESCHEDE, M. 2015. Steps towards a 3D model of the German Baltic Sea area – collaboration with academic research in the USO project. *In*: *Proceedings of the 8th European Congress on Regional Geoscientific Cartography and Information Systems – Geological 3D Modelling and Soils: Functions and Threats*. Institut Cartogràfic i Geològic de Catalunya, Barcelona, Spain, 44–45.

PHARAOH, T.C. 1999. Palaeozoic terranes and their lithospheric boundaries within the Trans-European Suture Zone (TESZ): a review. *Tectonophysics*, **314**, 17–41.

PHARAOH, T.C., DUSAR, M. *ET AL*. 2010. Tectonic evolution. *In*: DOORNENBAL, J.C. & STEVENSON, A.G. (eds) *Petroleum Geological Atlas of the Southern Permian Basin Area*. European Association of Geoscientists and Engineers (EAGE), Houten, The Netherlands, 25–57.

PISKE, J., RASCH, H.-J., NEUMANN, E. & ZAGORA, K. 1994. Geologischer Bau und Entwicklung des Präperms der Insel Rügen und des angrenzenden Seegebietes [Geological setting and development of the pre-Permian of Rügen Island and the adjacent offshore area]. *Zeitschrift für Geologische Wissenschaften*, **22**, 211–226.

REMPEL, H. 1992. Erdölgeologische Bewertung der Arbeiten der Gemeinsamen Organisation 'Petrobaltic' im deutschen Schelfbereich [Evaluation of the petroleum geological work of the consortium Petrobaltic in the German shelf area]. *Geologisches Jahrbuch*, **D99**, 3–32.

REMPEL, H. 2011. Petrobaltic: Erdölsuche in der Ostsee [Petrobaltic: Oil research in the Baltic Sea]. *Erdöl Erdgas Kohle*, **127**, 307–310.

SCHECK-WENDEROTH, M. & LAMARCHE, J. 2005. Crustal memory and basin evolution in the Central European Basin System: new insights from a 3D structural model. *Tectonophysics*, **397**, 143–165.

SCHEIDT, W., SCHMIDT, A., ZENKER, F. & ARNDT, G. 1995. *Bericht über das Reprocessing 2D-seismischer Daten aus der südlichen Ostsee [Reprocessing Report of the 2D seismic data from the southern Baltic Sea]*. Geophysik Seismik Leipzig GmbH Report

SCHLÜTER, H.U., BEST, G., JÜRGENS, U. & BINOT, F. 1997*a*. Interpretation reflexionsseismischer Profile zwischen baltischer Kontinentalplatte und kaledonischem Becken in der südlichen Ostsee – erste Ergebnisse [Interpretation of reflection seismic profiles between the continental Baltica plate and the Caledonian Basin in the southern Baltic Sea – first results]. *Zeitschrift der Deutschen Geologischen Gesellschaft*, **148**, 1–32.

SCHLÜTER, H.U., JÜRGENS, U., BEST, G., BINOT, F. & STAMME, H. 1997*b*. *Endbericht zum Teilprojekt 'Analyse geologischer und geophysikalischer Daten aus der südlichen Ostsee' [Final report of the subproject 'Analysis of geological and geophysical data from the southern Baltic Sea']*. Strukturatlas südliche Ostsee (SASO). Unpublished report Bundesanstalt für Geowissenschaften und Rohstoffe, Hannover, Germany.

SCHLÜTER, H.U., JÜRGENS, U., BINOT, F. & BEST, G. 1998. Die Bedeutung geologisch-tektonischer Strukturen in der südlichen Ostsee als Quellen natürlichen Stoffeintrags [The Importance of Geological Structures as Natural Sources of Potentially Hazardous Substances in the Southern Part of the Baltic Sea]. *Zeitschrift für angewandte Geologie*, **44**, 26–32.

STG 2016. *Stratigraphic Table of Germany 2016*. German Stratigraphic Commission (editing, coordination and layout: Menning, M. & Hendrich, A.). German Research Centre for Geosciences, Potsdam. Table plain 100 × 141 cm [in German].

THOMAS, S.A. & DEEKS, N.R. 1994. Seismic evidence for inversion tectonics in the strike-slip regime of the Tornquist zone, Southern Baltic Sea. *Zeitschrift für Geologische Wissenschaften*, **22**, 33–45.

THOMAS, S.A., SIVHED, U., ERLSTRÖM, M. & SEIFERT, M. 1993. Seismostratigraphy and structural framework of the SW Baltic Sea. *Terra Nova*, **5**, 364–374.

THYBO, H. 2000. Crustal structure and tectonic evolution of the Tornquist Fan region as revealed by geophysical methods. *Bulletin of the Geological Society of Denmark*, **46**, 145–160.

TORSVIK, T.H. & REHNSTRÖM, E.F. 2003. The Tornquist Sea and Baltica–Avalonia docking. *Tectonophysics*, **362**, 67–82.

VAN WEES, J.-D., STEPHENSON, R.A. *ET AL*. 2000. On the origin of the Southern Permian Basin, Central Europe. *Marine and Petroleum Geology*, **17**, 43–59.

VEJBAEK, O.V. & BRITZE, P. 1994. *Geological Map of Denmark 1:750 000. Top Pre-Zechstein (Two-Way Travel Time and Depth)*. Ministry of Environment and Energy, Map Series, **45**. Geological Survey of Denmark, Copenhagen, Denmark.

VON ZWERGER, R. 1948. Der tiefe Untergrund des westlichen Peribaltikums (Beitrag zur Deutung der regionalen Störgebiete der Schwere und des Erdmagnetismus) [The deeper sub surface of the western Peri-Baltic area (contribution to the interpretation of regional anomalies of gravity and earth magnetism)]. *Abhandlungen der Geologischen Landesanstalt Berlin, Akademie-Verlag*, **210**, 1–74.

WEGNER, J. 1966. Strukturbau und Tektonik im Nordosten der DDR [Structure and tectonics of the northeastern GDR]. *Geophysik und Geologie*, **9**, 44–56.

WILSON, M., NEUMANN, E.-R., DAVIES, G.R., TIMMERMAN, M.J., HEEREMANS, M. & LARSEN, B.T. 2004. Permo-Carboniferous magmatism and rifting in Europe: introduction. *In*: WILSON, M., NEUMANN, E.-R., DAVIES, G.R., TIMMERMAN, M.J., HEEREMANS, M. & LARSEN, B.T. (eds) *Permo-Carboniferous Magmatism and Rifting in Europe*. Geological Society, London, Special Publications, **223**, 1–10, https://doi.org/10.1144/GSL.SP.2004.223.01.01

ZAGORA, K. & ZAGORA, I. 2004. Devon [Devonian]. *In*: KATZUNG, G. (ed.) *Geologie von Mecklenburg-Vorpommern [Geology of Mecklenburg-Western Pomerania]*. E. Schweizerbart'sche Verlagsbuchhandlung, Stuttgart, Germany, 70–79.

ZIEGLER, P.A. (ed.) 1990. *Geological Atlas of Western and Central Europe*. 2nd edn, Shell Internationale Petroleum Maatschappij, The Hague. Geological Society, London.

Fault system evolution in the Baltic Sea area west of Rügen, NE Germany

ANDRÉ DEUTSCHMANN[1,2]*, MARTIN MESCHEDE[1] & KARSTEN OBST[2]

[1]*Ernst-Moritz-Arndt University of Greifswald, Institute for Geography and Geology, Friedrich-Ludwig-Jahn-Str. 17a, D-17487 Greifswald, Germany*

[2]*Geological Survey Mecklenburg-Western Pomerania, LUNG M-V, Goldberger Str. 12, D-18273 Güstrow, Germany*

**Correspondence: andre.deutschmann@lung.mv-regierung.de*

Abstract: Based on reprocessed offshore seismic lines acquired during oil and gas exploration in the 1980s, we reconstruct the formation and reactivation of major fault systems in the southern Baltic Sea area since the late Paleozoic. The geological evolution of different crustal blocks from the Caledonian Avalonia–Baltica collision until the Late Cretaceous–Paleogene inversion tectonics is also examined. The detected fault systems occur in the northern part of the Trans-European Suture Zone (TESZ) and belong either to the late Paleozoic Tornquist Fan or to the complex Western Pomeranian Fault System (WPFS) generated during Mesozoic extensional movements. While the NW–SE-trending deep Wiek Fault separates the Arkona High from the Middle Rügen Block, the NNW–SSE-trending Agricola Fault demarcates the Middle Rügen Block to the Falster Block in the west. Together with the Plantagenet Fault and numerous younger faults in the Mesozoic cover, it forms the Agricola Fault System. Furthermore, structural analyses of the Prerow Fault Zone above the Prerow salt pillow and the Werre Fault Zone crossing the Grimmen High indicate a complex fault history.

The subsurface geology of the southern Baltic Sea area is characterized by major faults and fault zones that separate different structural units with varying geological development. The faults were mainly generated during the late Paleozoic and reactivated during the Mesozoic. They have influenced the distribution and varying thicknesses of the Caledonian deformed Ordovician sediments, the Devonian and Carboniferous foreland basin fill, and the Permian–Cenozoic sediments of the Southern Permian Basin. Recent models of the tectonic evolution of the Baltic Sea area west and east of Rügen Island are based on seismic profiles obtained during offshore oil and gas exploration in the 1970s and 1980s of the former consortium Petrobaltic. The northern part of the Trans-European Suture Zone is the major focus of the USO ('Untergrundmodell Südliche Ostsee'; Obst *et al.* 2015) research project conducted by the Geological Survey of Mecklenburg–Western Pomerania (LUNG M-V) and the University of Greifswald. This zone was formed during the Caledonian collision between Baltica and Avalonia (Fig. 1). Results of the USO East area are presented in Seidel *et al.* (2018), while the current paper deals with the USO West area.

The focus of this study is the deep faults generated during the Middle Devonian–early Carboniferous extension followed by late Carboniferous compression, and the fault zones formed during Mesozoic extension processes. Their genetic relationship to other faults is examined, as well as their possible continuation into the Danish and Swedish sectors of the Baltic Sea. Finally, the differentiated sediment accumulation at the northeastern margin of the North German Basin (NGB) as part of the Southern Permian Basin is analysed in the context of the main tectonic phases.

Regional geological setting

The southern Baltic Sea area is located in the transition zone between the Fennoscandian Shield as part of the East European Craton (EEC, Baltica) and the West European Platform (e.g. Avalonia). This area is characterized by a mosaic of various geological blocks separated by several faults or fault zones formed since the late Paleozoic (Fig. 1; Liboriussen *et al.* 1987; Berthelsen 1992; Vejbæk *et al.* 1994; Pharaoh 1999; Thybo 2000; van Wees *et al.* 2000). The most prominent tectonic feature is the NW–SE-trending Tornquist Zone, crossing the southern Baltic Sea area between Scania in Sweden and Pomerania in Poland (Fig. 1; Pharaoh 1999). This zone is characterized by major faults bordering horst and

From: Kilhams, B., Kukla, P. A., Mazur, S., McKie, T., Mijnlieff, H. F. & van Ojik, K. (eds) 2018. *Mesozoic Resource Potential in the Southern Permian Basin*. Geological Society, London, Special Publications, **469**, 83–98.
First published online March 9, 2018, https://doi.org/10.1144/SP469.24

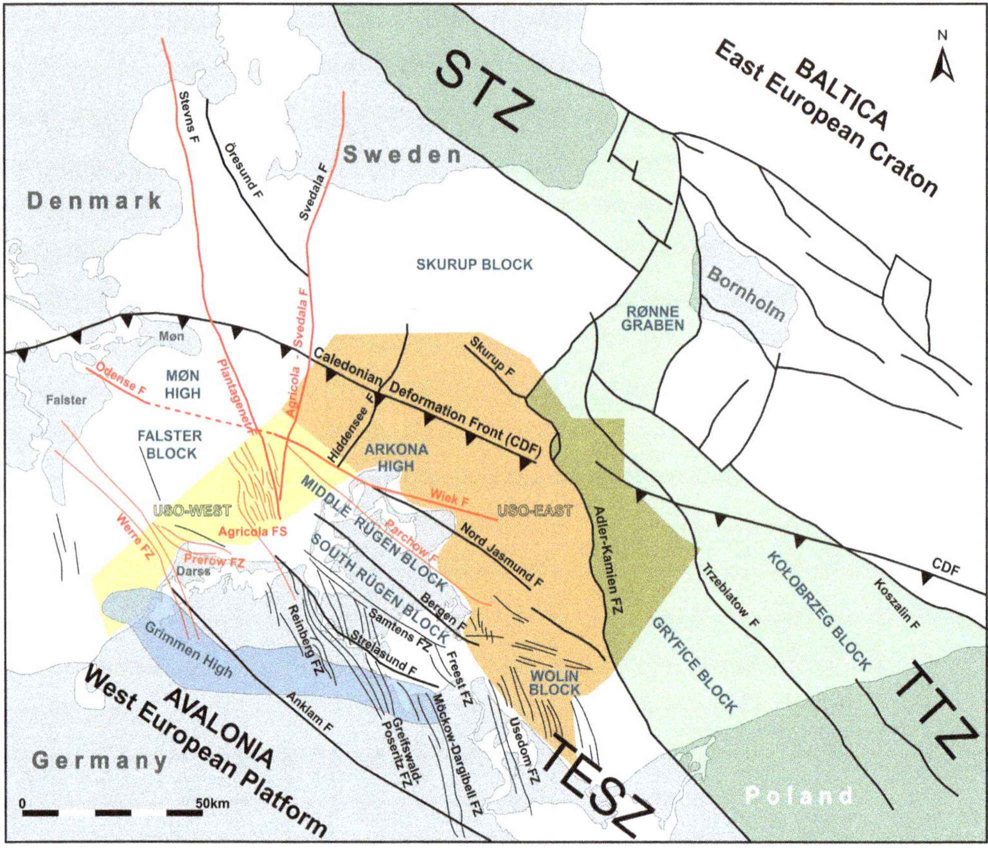

Fig. 1. Fault pattern (black and red lines) and tectonic blocks (named in blue) of the southern Baltic Sea (based on Seidel *et al.* 2018 and references therein). Overview of the study areas USO East and USO West in the vicinity of Rügen. Faults in red are discussed in this paper. STZ, Sorgenfrei-Tornquist Zone; TTZ, Tornquist-Teisseyre Zone; TESZ, Trans-European Suture Zone; F, fault; FZ, fault zone; FS, fault system.

graben structures that were generated during several tectonic phases in the late Paleozoic and Mesozoic–Cenozoic. It is subdivided into a northwestern branch, the Sorgenfrei–Tornquist Zone (STZ), and a southeastern branch, the Tornquist–Teisseyre Zone (TTZ). According to Berthelsen (1992), the TTZ marks the SW margin of the East European Craton, whereas the STZ separates the Fennoscandian Shield from the buried Danish massif (BABEL Working Group 1991) that also consists of crystalline rocks of the basement of Baltica. However, Berthelsen (1998) suggested that the TTZ may represent a pseudosuture as inferred by a significant step of the MOHO depth. New geophysical investigations by Mazur *et al.* (2015) support this view. The TTZ can therefore also be considered an intraplate feature within the EEC.

At the southwestern margin of Baltica, Mesoproterozoic granites and gneisses are overlain by Cambro-Silurian sediments as penetrated by the offshore well G14-1/1986 (Fig. 2; Franke *et al.* 1994; Beier & Katzung 1999). The closure of the Tornquist Ocean and the subsequent collision of Baltica with the micro-continent Avalonia during Late Ordovician–middle Silurian times led to the formation of the Caledonian fold-and-thrust belt (Katzung *et al.* 1993). In this context, the Cambro-Silurian successions were overthrusted by deformed marine Ordovician sediments (Katzung *et al.* 1993; Torsvik *et al.* 1993). Seismic offshore surveys have demonstrated the formation of an accretionary wedge north of Rügen Island (BABEL Working Group 1993; Thomas *et al.* 1993; Piske *et al.* 1994; Schlüter *et al.* 1997*b*; DEKORP-BASIN Research Group 1999). In addition, evidence of deformed Ordovician sediments of marine origin were found in several deep wells on Rügen Island (e.g. Katzung 2001). The northernmost extension of the Caledonian fold-and-thrust belt is marked by the Caledonian Deformation Front (CDF), which can be traced

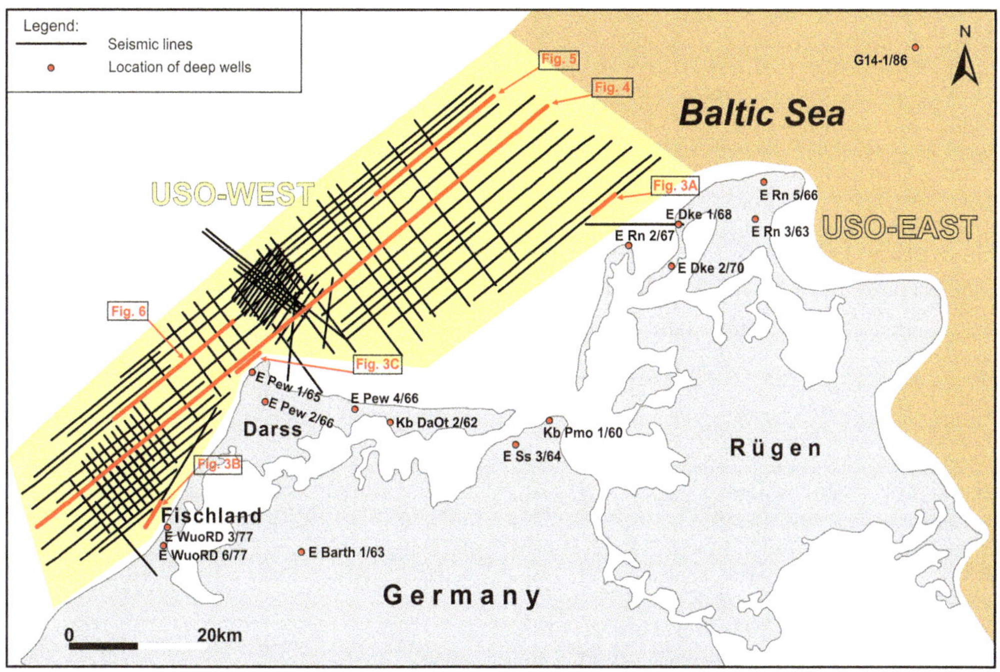

Fig. 2. Research area of USO West (yellow) and distribution of seismic lines within. Wells in red dots (WouRD, Wustrow; Pew, Prerow; DaOt, Darßer Ort; Pmo, Pramort; Ss, Stralsund; Rn, Rügen; Dke, Dranske; well prefix E, hydrocarbon exploration well; Kb, geological mapping well). USO East refers to the investigations of Seidel *et al.* (2018).

from Lolland and Møn (Denmark) to north of Rügen Island and further towards northwestern Poland (Fig. 1; Liboriussen *et al.* 1987; Erlström *et al.* 1997). The collision zone between the thick continental crust of Baltica and the thinner crust of Avalonia, termed the Trans-European Suture Zone (TESZ) by Pharaoh (1999) and Krawczyk *et al.* (2002), is assumed to occur in the area between Rügen Island and the NE German mainland. Towards the south, it can be traced at least until the NW–SE-trending Anklam fault (Fig. 1; Piske *et al.* 1994; Hoffmann *et al.* 2008). Other authors suggest that the Elbe Line marks the southern border of the EEC west of the TTZ (Bayer *et al.* 2002; Guterch *et al.* 2010; Mazur *et al.* 2015).

Following the Caledonian Orogeny and amalgamation of Baltica, Avalonia and Laurentia to the new continent Laurussia, the extension and thinning of the crust resulted in the formation of depocentres such as the Rhenohercynian Basin (Ziegler 1990). In the area of Rügen and Usedom, Middle Devonian clastic sediments in terrestrial 'Old Red' facies and Late Devonian–early Carboniferous platform carbonates were deposited in a shallow-marine environment (Aehnelt & Katzung 2009). They filled the so-called Rügen Basin that formed due to downfaulting of blocks along NW–SE-trending faults, mainly between the Strelasund and Wiek faults (Kurrat 1974; Franke & Hoffmann 1988). The basin depocentre was probably situated in the northern Middle Rügen Block as evidenced by the highest preserved thickness of Devonian sediments. The thickness of the Middle Devonian reaches 1857 m in the well E Binz 1/1973 (Aehnelt & Katzung 2009). The highest thickness of the Upper Devonian is preserved in the well E Rügen 2/1967 (Zagora & Zagora 2004). Further north, on the Arkona High, Middle Devonian–early Carboniferous sediments might have been eroded due to late Carboniferous inversion tectonics, triggered by compressional movements during the Variscan Orogeny. Closure of the Rheic Ocean and crustal shortening affected the northern foreland of the Variscan Orogen and led to wrench faulting between the NE margin of the London–Brabant Platform and the TTZ (Glennie & Underhill 2009). This was also accompanied by uplift of discrete blocks in the area of Rügen and Usedom that were separated by NW–SE-trending deep faults (see Seidel *et al.* 2018). These faults belong to the Tornquist Fan that is considered as a northwestwards widening splay of faults formed during the late Paleozoic (Thybo 2000). It comprises deep-seated

NW–WNW-oriented faults in the northern part of the TESZ between the Tornquist Zone and the Anklam Fault (Fig. 1). After the Variscan Orogeny, melting of the subducted oceanic crust and asthenospheric upwelling, especially along terrane borders, suture zones and major faults, led to high magmatic productivity. Remnants of this widespread Permo-Carboniferous rift magmatism occur in Central and NW Europe (Neumann *et al.* 2004; Wilson *et al.* 2004; Geissler *et al.* 2008). Different volcanic stages, mainly characterized by extrusion of rhyolitic to andesitic melts, have been documented in deep wells in NE Germany (Benek *et al.* 1996). In contrast, basaltic lava flows and subvolcanic intrusions are typical for the southern and central part of Rügen Island (Korich & Kramer 1994; Kramer 1995; Benek *et al.* 1996; Geissler *et al.* 2008).

Controlled by thermal subsidence (van Wees *et al.* 2000; Brink 2005) and long-lasting extensional movements, the North German Basin (NGB) was formed as part of the Central European Basin System (CEBS; Scheck-Wenderoth & Lamarche 2005; Maystrenko *et al.* 2008), one area of the Southern Permian Basin (van Wees *et al.* 2000; Guterch *et al.* 2010). The subsiding NGB was filled with terrestrial and marine sediments during the middle Permian–Late Cretaceous with a maximum thickness of over 10 km in the depocentre located between Hamburg, Lübeck and Schwerin (Ziegler 1990; van Wees *et al.* 2000; Katzung 2004). The northeastern basin margin is delimited in the Baltic Sea by the Ringköbing-Fyn-High that continues towards the SE into the Møn and Arkona highs (Fig. 1). Nearly all lithostratigraphic units decrease in thickness in the northern Rügen area.

Lithospheric stretching created a transtensional stress field during the Triassic and Jurassic periods (Gregersen *et al.* 2010), resulting in numerous minor faults between the Wiek Fault and the Anklam Fault, constituting the Western Pomeranian Fault System (WPFS; Krauss & Mayer 2004; Seidel *et al.* 2018). The mainly NW–NNW-trending faults often border grabens and half-grabens (e.g. the Werre and Prerow fault zones and the Agricola Fault System) and adjacent offshore areas (Krauss & Mayer 2004). In this context, active tectonic movements during the Late Triassic–Early Jurassic, termed the Mid-Cimmerian in Beutler *et al.* (2012), are documented by increased thicknesses of Keuper successions that were observed in seismic records and wells (Krauss & Mayer 1999). Beutler (2004) stated that the Keuper thickness in the graben system reaches more than 1000 m. This is in contrast to less than 100 m on Rügen and 600–650 m in the centre of the NGB. Their formation is related to transtensional movements above the TESZ, at least in the northern part (see Seidel *et al.* 2018). The distribution of sediments in northeastern Germany is also influenced by salt tectonic processes as described for several locations by Kossow *et al.* (2000) and Kossow & Krawczyk (2002). In the case of some fault zones of the WPFS, a relationship of increased sediment accumulation and halokinetic movements is proposed by Krull (2004). Compressional tectonics, generated by the Pyrenean orogenic event in the Late Cretaceous, led to the subsequent inversion of former basins in Central Europe (Mazur *et al.* 2005; Kley & Voigt 2008). As a consequence, anticline structures such as the Grimmen High (Fig. 1; Wegner 1966; Jubitz *et al.* 1981; Krull 2004) and the Mid-Polish Anticlinorium (or Mid-Polish Swell; Krzywiec *et al.* 2003; Mazur *et al.* 2005) were formed.

The polyphase tectonic evolution of the northeastern margin of the NGB and underlying units can be analysed and characterized in detail on basis of reprocessed seismic data measured in the southern Baltic Sea. This will improve the definition and timing of tectonic events in this area and support the correlation of major faults and fault zones west of Rügen with faults in the Danish and Swedish sectors.

Data and methods

The evaluation of seismostratigraphic units, separated from each other by main reflectors and major faults in the Baltic Sea area west of Rügen, was based on the interpretation of reprocessed 2D offshore seismic and onshore deep wells, tied by two onshore to offshore lines. A total of 86 seismic lines spanning a length of approximately 1517 km were used to detect seismostratigraphic units north of the German Baltic Sea coast between Rostock and Rügen to characterize faults or fault zones (investigation area of USO West; Fig. 2). The seismic lines were obtained from several 2D seismic surveys carried out by the organization ‘Petrobaltic’, which was a joint venture of the former German Democratic Republic, Poland and Soviet Union founded in the 1980s (Rempel 1992). Data were acquired down to 5 s two-way travel time with a sample rate of 4–2 ms and a sweep of 8–62.5 Hz provided by a 48-channel VAPORCHOC system (Teumer *et al.* 1990; Zöllner *et al.* 2008). These seismic lines were reprocessed by Central European Petroleum Ltd. (CEP) in 2009, resulting in a reduction of noise and multiples. In particular, the range of 1–3 s two-way travel time showed a significant improvement of the representation of marker horizons compared to the original data. The seismic lines were selected to ensure a regular grid with a lattice distance of 1 × 2 km across the study area. In addition, two sub-areas within the study area provided a profile lattice of 1 × 1 km. These additional lines represented areas in which hydrocarbon prospects were registered in preliminary investigations (Rempel 1992). As such, this regular seismic grid allowed uniform interpolation with low interpolation errors.

Well ties were based on data from 15 onshore wells (Fig. 2) located along the coast in the vicinity of the offshore seismic profiles: E Wustrow 6/1977; E Wustrow 3/1977; E Barth 1/1963; E Prerow 1/1965; E Prerow 2/1966; E Prerow 4/1966; Kb Darßer Ort 2/1962; Kb Pramort 1/1960; E Stralsund 3/1964; E Rügen 2/1967; E Dranske 1/1968; E Dranske 2/1970; E Rügen 3/1963; E Rügen 5/1966 and E G14-1/1986 (Hoth *et al.* 1993; and unpublished reports hosted by LUNG M-V). Of particular importance are the wells E Wustrow 3/1977, E Wustrow 6/1977 and E Prerow 1/1965 which are only 1–2 km away from the nearest seismic lines. The E Dranske 1/1968 research well is directly connected to the study area by onshore–offshore profile 88701.

The horizon and fault interpretations for this study were carried out using the SeisWare™ interpretation system. The well-seismic tie allows the comparison of well data, measured in depth units, with seismic data, measured in time. The sonic and density logs were used, if measured, to generate a synthetic seismic trace (Fig. 3a, c). This enabled a direct correlation between lithological boundaries identified in a well with distinct reflectors in the seismic profile. These boundaries were the top of the deformed Ordovician, top Devonian, top Carboniferous, top Rotliegend, top Zechstein, top Buntsandstein, top Muschelkalk, top Keuper and base Cretaceous.

It is recognized that an accurate time to depth conversion model is important to precisely define fault dips and offsets. This is especially true in this area because of changes in facies and sediment thicknesses (e.g. Schlüter *et al.* 1997*a*; Hansen *et al.* 2007). However, due to a lack of offshore wells and other reliable calibration points, this work is currently presented in TWT (two way time).

Results

Seismostratigraphic units

The deepest seismostratigraphic unit within the study area is considered to be the crystalline basement of the Baltica craton underneath folded and slightly metamorphosed Ordovician units (Katzung 2001). Middle Proterozoic granitic rocks are exposed NE of Rügen at 2 km depth in the German offshore well G14-1/1986 (Fig. 2; Franke *et al.* 1994; Obst *et al.* 2004). Towards the SW, these Precambrian rocks are downthrown several thousands of metres in a step-wise fashion (Piske *et al.* 1994; Beier 2001). Due to their considerable depth of more than 7 km (Hoffmann *et al.* 2008), seismic reflection events that may represent the top of the basement rocks were difficult to detect and interpret. Several reflections below 2.5 s TWT occurred north of the Wiek Fault on the Arkona High (Fig. 4). Due to their high reflection coefficient they are interpreted as Middle Cambrian Alum Shale, which overlays the basement rocks and Lower Cambrian sandstones of Baltica (based on Schlüter *et al.* 1997*b*).

South of the Caledonian Deformation Front (CDF), an accretionary wedge occurs that is thought to have developed due to the docking of Avalonia and Baltica during the Caledonian orogenic phase (Beier 2001; Beier & Katzung 2001). Folded layers and disturbed stratification caused characteristic dipping reflectors within the seismic sections (Fig. 4). This typical reflection pattern occurred on the Arkona High and is traceable within the seismic sections until the Wiek Fault in the south. This dipping was interpreted as folded Ordovician rocks which occur in the wells E Rügen 3/1963 and E Rügen 5/1966 (Fig. 2). Note that at the deepest part of the latter borehole, Neoproterozoic sediments of the Schwarbe–Buntschiefer Formation are claimed to belong to the overthrust southwestern margin of Baltica (Beier & Katzung 1999; Beier 2001), but they are not visible in the reprocessed seismic profiles.

The Devonian and Carboniferous strata of the Middle Rügen Block, which is subdivided by the Bergen Fault into the Wiek–Trent and Gingst–Garz sub-blocks, are visible in several onshore well penetrations (Lindert & Hoffmann 2004; Aehnelt & Katzung 2009). Seismic to well ties allowed correlations of several reflection events between 1.5 and 4 s TWT in the northern part of the seismic profiles with Middle Devonian (E Rügen 2/1967), Upper Devonian (E Dranske 1/1968; Fig. 3a), lower Carboniferous (E Barth 1/1963) and upper Carboniferous (E Prerow 1/1965) deposits. While the seismostratigraphic units of the upper Carboniferous were traceable from the southwestern part of the study area until the Parchow fault (Fig. 4), Devonian and lower Carboniferous units continued until the Wiek Fault. These differences in distribution are probably due to reverse faulting and partial erosion induced by the Variscan Orogeny further south.

The seismostratigraphic unit immediately above the upper Carboniferous represents Permian strata which showed a strong reflector at the base, especially in the southwestern part of the study area. Within the E Prerow 1/1965 well, this reflector correlates with the base of lower Rotliegend volcanic rocks overlying upper Carboniferous clastic sediments. Towards the NE, the Permian succession pinches out marking the northern margin of the North German Basin. The Rotliegend is therefore relatively thin on the Middle Rügen Block, where it is composed of conglomerates and sandstones only. As such, seismically traceable reflectors are not present.

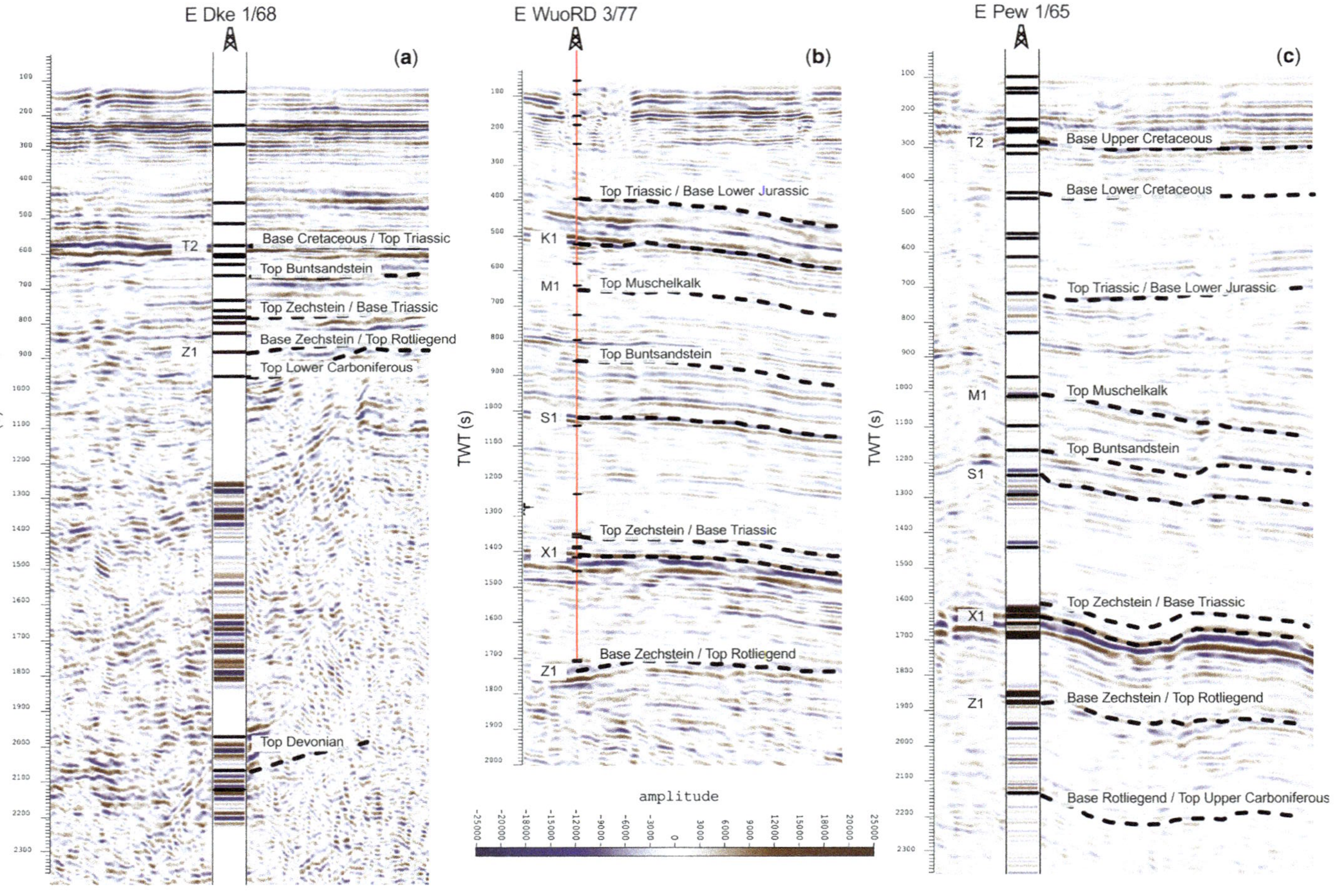

Fig. 3. Well-to-seismic tie (for location see Fig. 2) of upper Paleozoic to Mesozoic strata: (**a**) synthetic seismogram of the E Dranske 1/1968 well (*c.* 10 km off line; true vertical depth or TVD: 3006 m) is correlated with representative stratigraphic tops as typical for the northeastern research area; (**b**) stratigraphic tops characterizing the southwestern study area; major seismostratigraphic units are correlated with stratigraphic tops of the E Wustrow 3/1977 well (*c.* 2 km off line; TVD: 2759 m); and (**c**) synthetic seismogram of the E Prerow 1/1965 well (*c.* 2 km off line; TVD: 5250 m) with additional drill log marker from upper Carboniferous to Cretaceous. Reflection markers based on Mayer *et al.* (2000) and Zöllner *et al.* (2008). The seismic is zero-phase SEG normal and shown in TWT (two-way time).

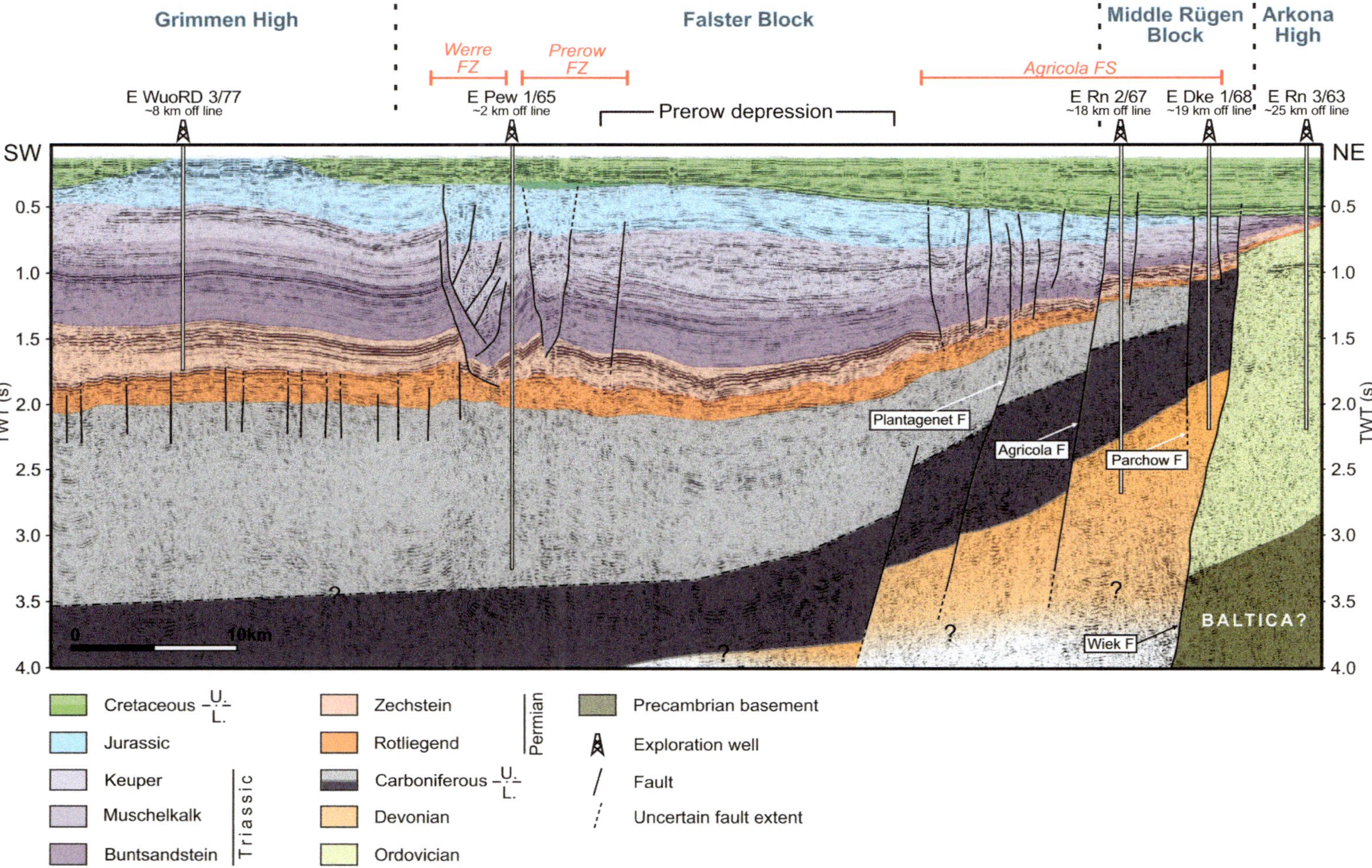

Fig. 4. SW–NE time-migrated Profile 86714 (for location see Fig. 2) showing seismostratigraphic units in the southern Baltic Sea west of Rügen Island. The major faults and fault zones discussed in the text are emphasized. Seismic is shown in TWT. The seismic section is about four times vertically exaggerated.

Other strong reflectors within the Permian interval are detectable at the base and top of the Zechstein succession, which comprise alternating beds of limestone, anhydrite and rock salt deposited during several evaporitic cycles according to Zagora & Zagora (2004). Two sharp maxima (Z3 and Z1 reflectors) are visible at the base of the Werra anhydrite and the top of the Stassfurt base anhydrite (Fig. 3b). The top of the uppermost Zechstein evaporite is marked by another maximum (X1 reflector). The basal and top planes of the Zechstein gently rise from the SW towards the NE, showing a decreasing thickness. They closely approach each other (less than 0.1 s TWT) north of the Agricola Fault System, marking the margin of the NGB (Fig. 4).

The Triassic succession is subdivided into Buntsandstein, Muschelkalk and Keuper intervals. A stratigraphic correlation with reflectors was possible using lines that intersect a position 1–2 km from the E Wustrow 3/1977 and E Prerow 1/1965 wells in the southern area and directly on the E Dranske 1/1968 well in the north (Fig. 2). The base of the Buntsandstein is marked by a rather weak reflector (X1) which may represent either the uppermost Zechstein succession or the base of the Calvörde Formation, which is the lowermost Buntsandstein unit. A more distinct seismic reflector (S1) occurs close to the top of the Upper Buntsandstein evaporites, reflecting the top of the uppermost anhydrite layer. This reflector is traceable across most of the study area.

The Muschelkalk interval is generally less reflective than the underlying units. A slight, broadly hard phase (M3 reflector) directly above the main Buntsandstein reflector was interpreted as representing the base of the Muschelkalk. Contrary to the observation in the eastern part of the NGB that evaporites of the Middle Muschelkalk should give a distinct reflector (Göthel 2016), there is a reflection horizon of poor quality which was difficult to follow. Near to the top of the Muschelkalk, a limestone bed could be correlated with the top Muschelkalk reflector (M1) as demonstrated by the E Wustrow 3/1977 well (Fig. 3b). The subsequent Keuper succession is characterized by alternating sand and silt deposits with thin evaporite intercalations. Two soft phases separated by a broad hard phase represent a reflection horizon (K1) that was correlatable with the base of the Upper Keuper based on the E Wustrow 3/1977 well. Due to a low-reflection coefficient, clastic sediments of the uppermost Keuper cannot be distinguished in the seismic profiles from the overlying sand–silt–clay successions of the Lower Jurassic. The top of the Keuper was estimated on the basis of a marker in the E Wustrow 3/1977 (Fig. 3b) and E Prerow 1/1965 wells (Fig. 3c).

With the exception of a thin Buntsandstein sequence, all Triassic units pinch out within the Middle Rügen Block. Increased thicknesses of the Keuper sediments between the Prerow Fault Zone and the Agricola Fault System indicate the formation of a local basin named Prerow Depression (Fig. 4).

The seismostratigraphic unit above the Triassic represents a Lower Jurassic succession, of which the upper part is eroded and missing as well as the entire Middle–Upper Jurassic. This is indicated by overlying Lower–Upper Cretaceous sediments. In most parts of the research area, the top of the Lower Jurassic is marked by a strong reflector (T2). This transgressive surface appears as a sharp double reflector with two sharp hard maxima separated by a rather broad soft phase within the seismic profile (Fig. 4). This was interpreted to represent the middle Albian–Cenomanian transgressive event (Fig. 3a). Within the Prerow Fault Zone another seismostratigraphic unit can be interpreted: a Lower Cretaceous unit of several hundred metres thickness, drilled by the E Prerow 1/1965 well. Its base is mainly transparent in seismic section. Another exception occurs in the southwestern part of the investigation area, where the uplift of the Grimmen High (Fig. 1) resulted in a complete erosion of the Cretaceous succession. The Lower Jurassic is therefore directly overlain by near-surface sediments of Tertiary age. A well tie of the base Cretaceous reflector is evidenced in the E Dranske 1/1968 with only 7 m thin Lower Cretaceous in the NE and in the E Prerow 1/1965 with 224 m thick Lower Cretaceous in the SW wells.

Major faults and fault systems

The study area in the southern Baltic Sea, west of Rügen, covers various crustal blocks separated or intersected by distinct faults. These faults, which have different orientation and deformation styles, reflect different tectonic phases of formation or reactivation. Most of them can be grouped into fault zones or fault systems, namely the NW–WNW-oriented Werre and Prerow fault zones in the SW and the mainly NNW–NW-striking Agricola Fault System in the NE. In addition, two major, deep-seated single faults occur in the northeasternmost part of the study area, which can be correlated with major NW–SE-trending faults crossing Rügen.

The prominent NW–SE-trending Wiek Fault can be traced from 0.55 s TWT to the base of seismic profile 84714 shown in Figure 4 (at least 4 s TWT). This major fault marks a sharp vertical offset of several thousand metres between the northern Arkona High and the southern Middle Rügen Block, as demonstrated by boreholes drilled on northern Rügen (Franke & Hoffmann 1988; Piske *et al.* 1994). The southern block is deeply downfaulted so that the lower Carboniferous is juxtaposed to the deformed Ordovician successions of the Caledonian accretionary wedge. The downfaulting probably occurred in

Middle Devonian–early Carboniferous times, and led to the formation of the Rügen Basin.

A reactivation of the Wiek Fault in post-Carboniferous times can also be observed in the investigation area. Uplifted and eroded Jurassic strata on the Arkona High and a slight flexure in the lowermost part of the Cretaceous (Fig. 4) suggest that the Wiek Fault was active during Late Cretaceous time, when inversion tectonics started in the NGB. Similar reverse faulting due to the SW–NE compression induced by the convergence of the Pyrenees (Kley & Voigt 2008) was observed at the Nord Jasmund Fault that branches off from the Wiek Fault on northeastern Rügen (Seidel *et al.* 2018).

Towards the SW, the Wiek Fault is accompanied by another NW–SE-oriented fault that separates two sub-blocks within the Middle Rügen Block (Figs 1 & 4). Correlations with the fault pattern from Rügen (Franke & Hoffmann 1988) suggest that this fault could be the offshore continuation of the Parchow Fault that separates the sub-blocks of Glowe and Neuenkirchen. South of the Wiek Fault, the lower Carboniferous is downfaulted and overlain by upper Carboniferous deposits; on the northern block the upper Carboniferous is missing, probably due to later uplift and erosion.

The central part of the investigation area is dominated by a complex fault pattern termed the Agricola Fault System, considered to be part of the WPFS (Krauss & Mayer 2004). Looking in detail at the reprocessed seismic sections, numerous NNW-oriented faults can be detected that border several thin graben structures within the Mesozoic succession. Two deep-rooted faults were also observed. One of these strikes NNE–SSW and is traceable from 0.4 s TWT to more than 3.2 s TWT. This fault is termed the Agricola–Svedala Fault by Thomas *et al.* (1993), which borders the Skurup Block to the west (Fig. 1), or Agricola Fault by Krauss (1994). East of this fault, the thickness of the Triassic succession decreases from more than 300 m as documented by the offshore well Smygehuk 1 south of Sweden, towards less than 30 m in the well G14-1/1986 NE of Rügen (Thomas *et al.* 1993).

Vertical offsets along the Agricola Fault and the varying thickness of different seismostratigraphic units allow two phases of extension or transtension to be distinguished. First, the Devonian–lower Carboniferous succession of the western block, here named Falster Block (Fig. 1), were downfaulted during extensional movements that resulted in the Variscan foreland basin formation. The second phase of downfaulting was triggered by Cimmerian transtensional tectonics during the Late Triassic–Early Jurassic. Furthermore, reactivation during the Late Cretaceous inversion is indicated by an off-set at the base of the Cretaceous interval (Fig. 5). The second major fault is termed the Plantagenet Fault with similar characteristics. This NNW–SSE-striking fault can be traced from 0.6 s TWT to 3.4 s TWT. An analysis of the vertical offset along this fault suggests downfaulting of the western sub-block during the early Carboniferous (Fig. 4) and Late Triassic transtension (Fig. 5).

All other mapped faults between these two deep major lineaments reflect only Mesozoic dextral transtensional movements. These movements resulted in the formation of en echelon fault zones between the Wiek Fault and the Anklam Fault, bordering small depressions and graben structures with Y-shaped fault geometries (Krauss & Mayer 2004; Seidel *et al.* 2018). Considering that the Paleozoic faults of the Tornquist Fan, *sensu* Thybo (2000), were generated during late Paleozoic extension in the northern Variscan foreland and reactivated during Mesozoic transtensional processes, we propose a new term for this set of faults: the Agricola Fault System.

The southwestern part of the research area is characterized by two faults that can be traced north of the Darss peninsula at time intervals from 0.3 s TWT to 1.6 s TWT. They mark a local graben above a shallow salt pillow named Prerow (Fig. 6). The graben was initially formed during the Triassic as indicated by increasing thickness of the Keuper. Local subsidence lasted until the Early Cretaceous as indicated by a more than 200 m thick Lower Cretaceous succession drilled in the E Prerow 1/1965 well. This Prerow Fault Zone shows a change in strike from WNW–ESE onshore to NW–SE offshore.

The Prerow Fault Zone is accompanied in the SW by the Werre Fault Zone (Figs 3 & 6), which is characterized by two bounding faults with a complex internal pattern. In general, the dominant southwestern fault is the main branch which is formed as a listric fault. An extensional regime during the Late Triassic–Early Jurassic led to Cimmerian movements and the formation of a roll-over structure. The sediments of the Keuper and Jurassic intervals show significantly increased thicknesses within the graben. From the SE towards the NW, the character of both faults of the Werre Fault Zone switches. The main fault dips towards the NE in the southeastern lines (Figs 4 & 6), whereas the main fault dips towards the SW in the northwestern lines. Furthermore, slight reverse faulting is also indicated by a flexure in the basal Cretaceous interval.

Geological development and fault evolution west of Rügen

The southern Baltic Sea area between Møn (Denmark) and Rügen (Germany) is segmented by several Paleozoic faults that border different blocks (Fig. 1). These faults, such as the Wiek and Parchow faults (Fig. 4), are part of the Tornquist Fan described by

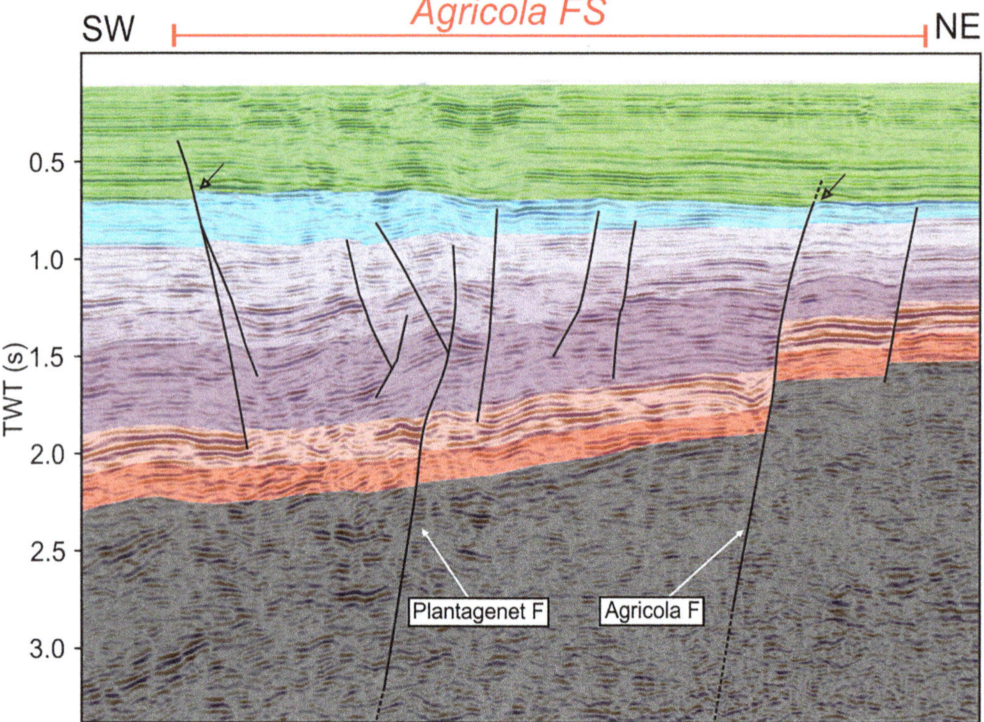

Fig. 5. SW–NE time-migrated and interpreted seismic section (for location, see Fig. 2) crosses the Agricola Fault System. Major faults are the Plantagenet Fault and the Agricola Fault that indicate downfaulting during Middle Devonian to lower Carboniferous. Several minor faults are observed in the Mesozoic deposits. Younger compressional tectonics is slightly indicated by flexures at the base of the Cretaceous (black arrows). For legend see Figure 4. Seismic is shown in TWT. The seismic section is about two times vertically exaggerated.

Thybo (2000). Downfaulting during the Middle Devonian–early Carboniferous led to the formation of the Rügen Basin in the northern foreland of the Variscan Orogeny. Sedimentation history in this basin was completed in late Visean time, accompanied by compressional tectonics, which led to the formation of different blocks in the area of Rügen and Usedom separated by NW–SE- and NE–SW-trending deep faults (Franke & Hoffmann 1988; Seidel *et al.* 2018).

The northernmost block in the study area is the Arkona High (Fig. 1). This uplifted block is bordered by the Skurup Fault in the north towards the Skurup Block. The southern border is marked by the prominent Wiek Fault which separates the Arkona High, with its deformed Ordovician sediments of the Caledonian accretionary wedge, from the downfaulted Middle Rügen Block. This block contains thick Middle Devonian–lower Carboniferous successions deposited in the former Rügen Basin (Aehnelt & Katzung 2009) that suggest a vertical offset of 5000–6000 m (Franke & Hoffmann 1988). The Middle Rügen Block is subdivided onshore into several sub-blocks by minor faults. One of these faults is the Parchow Fault that is also visible offshore (Fig. 4). The Wiek Fault is mapped to continue westwards until the Odense Fault. To the west, the Arkona High is separated from the Møn High either by the major Agricola Fault that continues into the Svedala Fault, as suggested by Thomas *et al.* (1993), or by the Plantagenet Fault, which appears to continue towards the Stevns Fault (Erlström & Sivhed 2012).

After the Variscan Orogeny, Permo-Carboniferous rift magmatism in central Europe was followed by thermal subsidence and formation of the CEBS and its sub-basins. The northern margin of the North German Basin reached as far as northern Rügen as indicated by the decreasing depth of Permian and Triassic lithostratigraphic horizons shown in time–structure maps (Fig. 7). While the depth of Zechstein base generally decreases towards the NE, the local Prerow Depression developed north of the Darss (Fig. 7a). A similar picture is visible for the base of the Triassic (Fig. 7b). The Prerow Depression is filled with Triassic sediments, mainly with Keuper

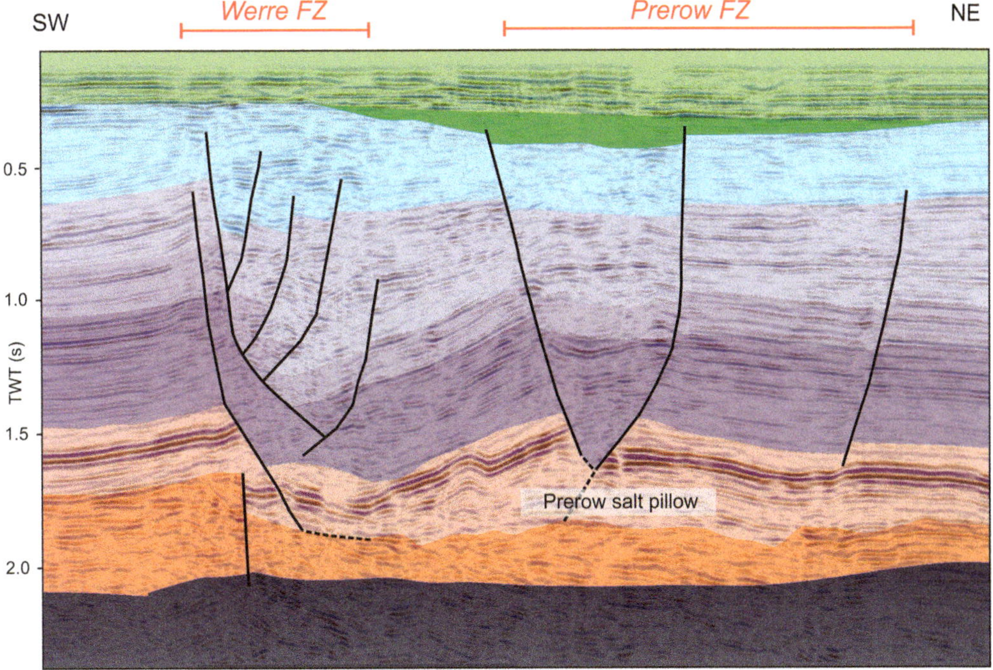

Fig. 6. SW–NE time-migrated interpreted seismic section (for location see Fig. 2) showing the Werre Fault Zone and the Prerow Fault Zone. The major fault of the Werre Fault Zone forms a roll-over structure, dipping towards the NW. Increasing thicknesses of the Keuper and Lower Jurassic indicates activation during the Late Triassic–Early Jurassic. For legend see Figure 4. Seismic is shown in TWT. The seismic section is about two times vertically exaggerated.

sediments up to 0.55 s (TWT) (Fig. 4), and is weakly indicated in the time–structure map for the top of Triassic (Fig. 7c).

Numerous minor faults of the Agricola Fault System as well as the Werre Fault Zone were formed during Mesozoic transtensional movements, which were probably induced by rifting processes in the Arctic–North Atlantic area (Pharaoh *et al.* 2010). They belong to the WPFS that borders NW–NNW-trending graben and half-grabens on- and offshore between the Wiek Fault and the Anklam Fault. This characterizes the en echelon fault system that developed above the northern TESZ due to strike-slip movements induced by the closure of the Tethys (Krauss & Mayer 2004; Seidel *et al.* 2018).

Late Cretaceous inversion tectonics induced by the onset of the Pyrenean Orogeny (Kley & Voigt 2008) resulted in the formation of anticline structures at the northern margin of the NGB (e.g. the Grimmen High). This is visible in the time–structure map for the base of the Cretaceous (Fig. 7d), where the Cretaceous is partly missing in the southwestern part of the investigation area. However, this WNW–ESE-trending anticline seems to wedge out in the west, about 2 km off the coast. Its offshore continuation can therefore only be inferred from the more reduced Upper Cretaceous succession in the offshore seismic profiles close to Fischland peninsula.

Comparing our investigation results with the known fault and structure pattern in the southern Baltic Sea area, it is recognized that many of the faults and blocks have already been described or shown in simplified maps (e.g. Surlyk 1980; Liboriussen *et al.* 1987; Thomas *et al.* 1993; Vejbæk *et al.* 1994; Surlyk *et al.* 1995; Thybo 2000; Krauss & Mayer 2004) but these previous descriptions often lack precise genetic analyses. Major deep faults generated during a Middle Devonian–lower Carboniferous extension, which outline the Rügen Basin, as part of the Rhenohercynian foreland basins were reactivated as reverse faults during the Variscan compressional phase. They were reactivated again during Mesozoic transtensional movements and are accompanied by the formation of minor faults, for example in the Agricola Fault System west of Rügen as presented here, and in the Wiek Fault System east of Rügen as discussed by Seidel *et al.* (2018). Based on variable structural evolution, the crust in the southern Baltic Sea area can be subdivided into various blocks. As well as the known uplifted blocks of Møn and Arkona and the downfaulted blocks of Middle and Southern Rügen, it has been possible to demonstrate the existence of a downfaulted block west of Rügen. This block is named Falster Block and is demarcated by the Agricola Fault in the east and Wiek–Odense Fault in the north (Fig. 1). Towards the SW, it might be bordered by the Werre Fault but can only be determined by new seismic investigations between Fischland (Germany) and Falster (Denmark). Furthermore, the western extension of the Grimmen High is now precisely shown, which contradicts a former interpretation presented by Hübscher *et al.* (2010).

Future work should be focused on the creation of a consistent velocity model for this offshore area in order to make a realistic time-to-depth conversion. A useful additional study would be the creation of 3D subsurface models of the southern Baltic Sea area, harmonized between Denmark, Germany, Poland and Sweden.

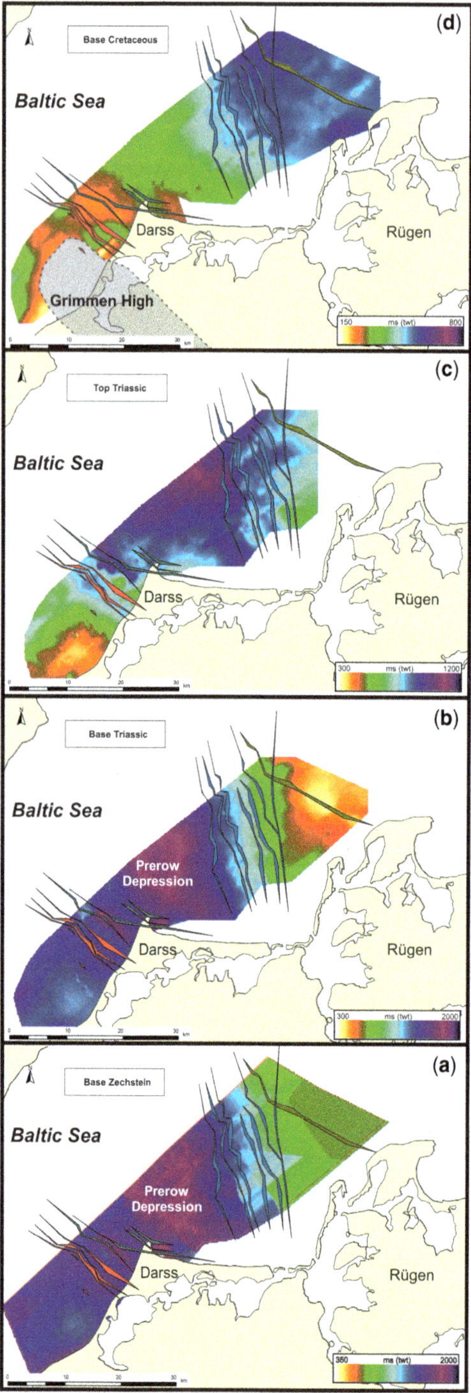

Fig. 7. Gridded time–structure maps with relevant faults within the investigation area: Werre Fault Zone (red), Prerow Fault Zone (green), Agricola Fault System incl. Plantagenet Fault and Agricola Fault (blue), Wiek Fault (brown). (**a**) The base Zechstein reflector is located at a depth of 1.7 s in the SW part (TWT, Two Way Time), but raised in the NE part up to 0.9 s (TWT). Looking carefully, this general trend seems to be disturbed locally in the Prerow Depression, between the Prerow Fault Zone and the Agricola Fault Systems. (**b**) The Base Triassic is characterized by decreasing depth towards the NE margin of the NGB. In the SW area a plateau exists at 1.4 s (TWT). In the north of Darss peninsula, the Prerow Depression with a maximum depth of 1.9 s (TWT) is visible. (**c**) The Top Triassic is mostly at the same level of 0.6–0.8 s (TWT). In the area of Wustrow the depth reaches only 0.3 s (TWT). A local depression is indicated north of Prerow by increasing depth up to 1 s (TWT). (**d**) The base Cretaceous is missing in the area of Wustrow, indicating non-deposition or later erosion of these deposits due to the uplift of Grimmen High. Increased depth of the Cretaceous base towards the NE up to 0.55 s (TWT) marks the formation of the Danish–North German–Polish Cretaceous Basin with its deposition between Rügen and Bornholm.

Conclusions

An analysis of the initiation and reactivation of faults in the offshore areas west of Rügen (presented in this paper) and east of Rügen (Seidel *et al.* 2018) reveal a geological history back to the Devonian. The fault evolution is coupled to several extensional and compressional phases that are either related to rift, subduction or collision processes in central and southern Europe. Major deep faults were initiated during Middle Devonian–lower Carboniferous extension and firstly reversely reactivated due to the long-distance effect of the Variscan Orogeny. The Permian–Mesozoic evolution of the Central European Basin and associated transtensional tectonics led to reactivation of these faults and the generation of further accompanying structures. Furthermore, Late Cretaceous inversion tectonics led to the youngest observable reactivation of these faults.

The geological development of the southern Baltic Sea area is influenced by major tectonic processes such as the Mid-European Caledonian continent collision, Variscan foreland basin formation, the generation of the North German Basin, the rifting and opening of the Atlantic Ocean and several Mesozoic–Cenozoic orogenic phases in the vicinity of the Mediterranean Sea. This complex history is visible in offshore seismic lines west of Rügen that suggest a polyphase crustal evolution between Baltica and Avalonia.

(1) Extension after the Caledonian Orogeny led to the formation of the Rügen Basin, filled with shallow-marine sediments in the Middle Devonian–lower Carboniferous. This tectonic phase is mainly documented by downfaulting of the Middle Rügen Block along the NW–SE-oriented faults of the Tornquist Fan.
(2) A compressional tectonic phase in the late Carboniferous interval, induced by the Variscan Orogeny, caused reverse faulting along these NW-trending faults, especially at the Wiek Fault that continues towards the west into the Odense Fault striking between Møn and Falster. This fault demarcates the uplifted blocks of the Møn and the Arkona High from the Middle Rügen Block and the Falster Block. These downfaulted blocks are separated from each other by the Agricola Fault that continues towards the north as the Svedala Fault.
(3) Transtension during the Mesozoic, mainly initiated by the rifting in the Arctic–North Atlantic region, resulted in the formation of the Western Pomeranian Fault System (WPFS), which also comprises the Agricola Fault System and the Werre Fault Zone. This fault system is considered to represent en echelon faults that developed above the northern part of the TESZ thanks to Cimmerian transtensional movements (Late Triassic–Early Jurassic). The Prerow Fault Zone that was generated above the Prerow salt pillow might also belong to this complex fault system.
(4) A new phase of faulting occurred along the Wiek Fault as suggested by uplifted and eroded Jurassic strata on the Arkona High and a slight flexure in the overlying lowermost part of the Cretaceous. Reactivation of the Agricola Fault is also indicated by a similar flexure in the same seismostratigraphic unit. These are indications for transpressional movements during the Late Cretaceous that also led to the uplift of distinct structures, such as the Grimmen High.

In order to verify these different tectonic phases and to restore the related deformation processes in the southern Baltic Sea area, especially west of Rügen, high-resolution velocity models are necessary. Together with new seismic investigations between Denmark, Sweden and Germany, they will allow the establishment of a harmonized 3D subsurface model that can be extended towards the area east of Rügen based on the structural analyses given by Seidel *et al.* (2018).

The authors would like to thank Central European Petroleum GmbH Berlin for providing the offshore seismic dataset. We thank Cornelius Rott and Arezki Ioghlissen for fruitful discussions. Seisware Inc. is thanked for providing the SeisWare Software package under the Academic User License Agreement. Stanislaw Mazur and Ben Kilhams are thanked for their useful comments.

References

Aehnelt, M. & Katzung, G. 2009. Middle Devonian Old Red Rügen Basin in Western Pomerania, NE Germany: implications for post-Caledonian evolution and palaeogeography at southern margin of the Old Red Continent. *Zeitschrift der deutschen Gesellschaft für Geowissenschaften*, **160**, 1–11.

BABEL Working Group 1991. Deep seismic survey images the structure of the Tornquist Zone beneath the Southern Baltic Sea. *Geophysical Research Letter*, **18**, 1091–1094.

BABEL Working Group 1993. Deep Seismic Reflection/Refraction Interpretation of Crustal Structure along Babel Profiles A and B in the Southern Baltic Sea. *Geophysical Journal International*, **112**, 325–343.

Bayer, U., Grad, M. *et al.* 2002. The southern margin of the East European Craton: new results from seismic sounding and potential fields between the North Sea and Poland. *Tectonophysics*, **360**, 301–314.

Beier, H. 2001. *Die strukturelle Entwicklung der Rügen-Kaledonieden und ihres nördlichen Vorlandes (Nordost-Deutschland und südliche Ostsee)*. PhD thesis, Ernst-Moritz-Arndt University Greifswald.

Beier, H. & Katzung, G. 1999. Lithologie und Strukturgeologie des Altpaläozoikums in der Offshore-Bohrung G 14-1/86 (südliche Ostsee). *Greifswalder Geowissenschaftliche Beiträge*, **6**, 327–345.

Beier, H. & Katzung, G. 2001. The deformation history of the Rügen Caledonides (NE Germany) – implications from the structural inventory of the Rügen 5 borehole. *Neues Jahrbuch für Geologie und Paläontologie - Abhandlungen*, **222**, 269–300.

Benek, R., Kramer, W. *et al.* 1996. Permo-Carboniferous magmatism of the Northeast German Basin. *Tectonophysics*, **266**, 379–404.

Berthelsen, A. 1992. From Precambrian to Variscan Europe. *In*: Blundell, D., Freeman, R. & Mueller, S. (eds) *A Continent Revealed. The European Geotraverse*. University Press, Cambridge, 153–164.

Berthelsen, A. 1998. The Tornquist Zone NW of the Carpathians – an intraplate pseudosuture. *Geologiska Föreningens I Stockholm Förhandlinga*, **120**, 223–230.

Beutler, G. 2004. Trias. *In*: Katzung, G. (ed.) *Geologie von Mecklenburg-Vorpommern*. E. Schweizerbart'sche Verlagsbuchhandlung, Stuttgart, 140–151.

Beutler, G., Junker, R., Niediek, S. & Rößler, D. 2012. Tektonische Diskordanzen und tektonische Zyklen im Mesozoikum Nordostdeutschlands. *Zeitschrift der deutschen Gesellschaft für Geowissenschaften*, **163**, 447–468.

Brink, H.J. 2005. The evolution of the North German Basin and the metamorphism of the lower crust. *International Journal of Earth Science*, **94**, 1103–1116.

DEKORP-BASIN Research Group 1999. Deep crustal structure of the Northeast German basin: new DEKORP-Basin '96 deep profiling results. *Geology*, **27**, 55–58.

Erlström, M. & Sivhed, U. 2012. *Pre-Rhaetian Triassic strata in Scania and adjacent offshore areas – stratigraphy, petrology and subsurface characteristics*. Sveriges geologiska untersökning. Rapporter och meddelanden 132.

Erlström, M., Thomas, S.A., Deeks, N. & Sivhed, U. 1997. Structure and tectonic evolution of the Tornquist Zone and adjacent sedimentary basins in Scania and the southern Baltic Sea area. *Tectonophysics*, **271**, 191- 215.

Franke, D. & Hoffmann, N. 1988. Der Bruchschollenbau Rügens – ein Beispiel tafelrandparalleler Strukturentwicklung. *WTI (Wissenschaftlich Technischer Informationsdienst des Zentralen Geologischen Instituts)*, **29**, 50–59.

Franke, D., Gründel, J., Lindert, W., Meissner, B., Schulz, E., Zagora, I. & Zagora, K. 1994. Die Ostseebohrung G 14 – eine Profilübersicht. *Journal for the Geological Science*, **22**, 235–240.

Geissler, M., Breitkreuz, C. & Kiersnowski, H. 2008. Late Paleozoic volcanism in the central part of the Southern Permian Basin (NE Germany, W Poland): facies distribution and volcano-topographic hiati. *International Journal of Earth Sciences*, **97**, 973–989.

Glennie, M.K. & Underhill, R. 2009. Origin, development and evolution of structural styles. *In*: Glennie, M.K. (ed.) *Petroleum Geology of the North Sea: Basic Concepts and Recent Advances*. Blackwell Science, Oxford, 42–84.

Göthel, M. 2016. Lithologische Interpretation und stratigraphisches Niveau der reflexionsseismischen Horizonte im Untergrund Brandenburgs einschließlich Berlins. Evaluation, lithological interpretation and stratigraphic correlation of the seismic reflection interfaces into the deep underground of the federal states Brandenburg and Berlin, Germany. *Brandenburger Geowissenschaftliche Beiträge*, **23**, 85–90.

Gregersen, S., Voss, P., Nielsen, L.V., Achauer, U., Busche, H., Rabbel, W. & Shomali, Z.H. 2010. Uniqueness of modeling results from teleseismic P-Wave tomography in Project Tor. *Tectonophysics*, **481**, 99–107.

Guterch, A., Wybraniec, S. *et al.* 2010. Crustal structure and structural framework. *In*: Doornenbal, J.C. & Stevenson, A.G. (eds) *Petroleum Geological Atlas of the Southern Permian Basin Area*. EAGE Publications b.v., Houten, 11–23.

Hansen, M.B., Scheck-Wenderoth, M., Hübscher, C., Lykke Andersen, H., Dehghani, A., Hell, B. & Gajewski, D. 2007. Basin evolution of the northern part of the Northeast German Basin – insights from a 3D structural model. *Tectonophysics*, **437**, 1–16.

Hoffmann, N., Hengesbach, L., Friedrichs, B. & Brink, H.-J. 2008. The contribution of magnetotellurics to an improved understanding of the geological evolution of the North German Basin – review and new results. *Zeitschrift der deutschen Gesellschaft für Geowissenschaften*, **159**, 591–606.

Hoth, K., Rusbült, J., Zagora, K., Beer, H. & Hartmann, O. 1993. Die tiefen Bohrungen im Zentralabschnitt der Mitteleuropäischen Senke – Dokumentation für den Zeitabschnitt 1962–1990. *Schriftenreihe Geowissenschaften*, **2**, 7–145.

Hübscher, C., Hansen, M.B., Trinanes, S.P., Lykke-Andersen, H. & Gajewski, D. 2010. Structure and evolution of the Northeastern German Basin and its transition onto the Baltic Shield. *Marine and Petroleum Geology*, **27**, 923–938.

Jubitz, K.B., Schwab, G. & Teschke, H.J. 1981. Geologische Entwicklungstrends am Südwestrand der Osteuropäischen Tafel – Ein Überblick. *Journal for the Geological Science*, **9**, 1113–1137.

Katzung, G. 2001. The Caledonides at the southern margin of the East European Craton. *Neues Jahrbuch für Geologie und Paläontologie – Abhandlungen*, **222**, 3–53.

Katzung, G. 2004. Regionalgeologische Entwicklung. *In*: Katzung, G. (ed.) *Geologie von Mecklenburg-Vorpommern*. E. Schweizerbart'sche Verlagsbuchhandlung, Stuttgart, 15–37.

Katzung, G., Giese, U., Walter, R. & Von Winterfeld, C. 1993. The Rügen Caledonides, northeast Germany. *Geological Magazine*, **130**, 725–730.

Kley, J. & Voigt, T. 2008. Late Cretaceous intraplate thrusting in central Europe: effect of Africa-Iberia-Europe convergence, not Alpine collision. *Geology*, **36**, 839–842.

Korich, D. & Kramer, W. 1994. Permosilesische Magmatite im Untergrund von Rügen und der östlich angrenzenden Ostsee. *Zeitschrift für Geologische Wissenschaften*, **20**, 249–256.

Kossow, D. & Krawczyk, Ch. 2002. Structure and quantification of processes controlling the evolution of the inverted NE-German Basin. *Marine and Petroleum Geology*, **19**, 601–618.

Kossow, D., Krawczyk, Ch., McCann, T., Strecker, M. & Negendank, J. 2000. Style and evolution of salt pillows and related structures in the northern part of the Northeast German Basin. *International Journal for Earth Science*, **89**, 652–664.

Kramer, W. 1995. Phanerozoic magmatic activity in the northwestern part of the Trans-European Suture Zone. *Studia Geophysica et Geodaetica*, **39**, 320–329.

Krauss, M. 1994. The tectonic structure below the southern Baltic Sea and its evolution. *Journal for the Geological Science*, **22**, 19–32.

Krauss, M. & Mayer, P. 1999. Der präquartäre Strukturbau im Bereich des Greifswalder Boddens und sein Einfluß auf den quartären Sedimentkomplex (BMBF-Projekt SASO II). *Zeitschrift für Geologische Wissenschaften*, **27**, 153–160.

Krauss, M. & Mayer, P. 2004. Das Vorpommern-Störungssystem und seine regionale Einordnung zur

Transeuropäischen Störung. *Journal for the Geological Science*, **32**, 227–246.

Krawczyk, C.M., Eilts, F., Lassen, A. & Thybo, H. 2002. Seismic evidence of Caledonian deformed crust and uppermost mantle structures in the northern part of the Trans-European Suture Zone, SW Baltic Sea. *Tectonophysics*, **360**, 215–244.

Krull, P. 2004. Epivariszisches Tafeldeckgebirge. *In*: Katzung, G. (ed.) *Geologie von Mecklenburg-Vorpommern*. E. Schweizerbart'sche Verlagsbuchhandlung, Stuttgart, 388–397.

Krzywiec, P., Kramarska, R. & Zientara, P. 2003. Strike-slip tectonics within the SW Baltic Sea and its relationship to the inversion of the Mid-Polish Trough – evidence from high-resolution seismic data. *Tectonophysics*, **373**, 93–105.

Kurrat, W. 1974. *Komplexgeophysikalische Beiträge zur Erfassung der tektonischen, strukturellen und faziellen Gliederung des Präzechsteins in einem Teilgebiet am Südwestrand der Osteuropäischen Tafel: dargestellt am Beispiel der geophysikalisch-geologischen Erkundungsarbeiten im Gebiet Rügen*. PhD thesis, Ernst-Moritz-Arndt Universität Greifswald, Greifswald.

Liboriussen, J., Ashton, P. & Tygesen, T. 1987. The tectonic evolution of the Fennoscandian Border Zone in Denmark. *Tectonophysics*, **137**, 21–29.

Lindert, W. & Hoffmann, N. 2004. Karbon. *In*: Katzung, G. (ed.) *Geologie von Mecklenburg- Vorpommern*. E. Schweizerbart'sche Verlagsbuchhandlung, Stuttgart, 79–95.

Mayer, P., Krauss, M. & Vorbaum, M. 2000. Der Strukturbau des Vorpommern-Störungssystems im Bereich der NE-Fortsetzung des DEKORP-Profils Basin 9601 (DFG-Projekt VPSS I). *Journal for the Geological Science*, **28**, 397–404.

Maystrenko, Y., Bayer, U., Brink, H.-J. & Littke, R. 2008. The Central European Basin system – an overview. *In*: Littke, R., Bayer, U., Gajewski, D. & Nelskamp, S. (eds) *Dynamics of Complex Intracontinental Basins – The Central European Basin System*. Springer-Verlag, Berlin-Heidelberg, 18–34.

Mazur, M., Scheck-Wenderoth, M. & Krzywiec, P. 2005. Different modes of the Late Cretaceous–Early Tertiary inversion in the North German and Polish basins. *International Journal for Earth Science*, **94**, 782–798.

Mazur, S., Mikolajczak, M., Krzywiec, P., Malinowski, M., Buffenmyer, V. & Lewandowski, M. 2015. Is the Teisseyre-Tornquist Zone an ancient plate boundary of Baltica? *Tectonics*, **34**, 2465–2477.

Neumann, E.-R., Wilson, M., Heeremans, M., Spencer, E.A., Obst, K., Timmerman, M.J. & Kirstein, L. 2004. Carboniferous-Permian rifting and magmatism in southern Scandinavia, the North Sea and northern Germany: a review. *In*: Wilson, M., Neumann, E.-R., Davies, G.R., Timmerman, M.J., Heeremans, M.J. & Larsen, B.T. (eds) *Permo-Carboniferous Magmatism and Rifting in Europe*. Geological Society, London, Special Publications, **223**, 11–40, https://doi.org/10.1144/GSL.SP.2004.223.01.02

Obst, K., Hammer, J., Katzung, G. & Korich, D. 2004. The Mesoproterozoic basement in the southern Baltic Sea: insights from the G14-1 off-shore borehole. *International Journal for Earth Science*, **93**, 1–12.

Obst, K., Deutschmann, A., Seidel, E. & Meschede, M. 2015. Steps towards a 3D model of the German Baltic Sea area – collaboration with academic research in the USO project. *8th European Congress on Regional Geoscientific Cartography and Information Systems – Geological 3D Modelling and Soils: Functions and Threats*, 44–45.

Pharaoh, T.C. 1999. Palaeozoic terranes and their lithospheric boundaries within the Trans-European Suture Zone (TESZ): a review. *Tectonophysics*, **314**, 17–41.

Pharaoh, T.C., Dusar, M. *et al.* 2010. Tectonic evolution. *In*: Doornenbal, J.C. & Stevenson, A.G. (eds) *Petroleum Geological Atlas of the Southern Permian Basin Area*. EAGE Publications b.v., Houten, 25–57.

Piske, J., Rasch, H.-J., Neumann, E. & Zagora, K. 1994. Geologischer Bau und Entwicklung des Präperms der Insel Rügen und des angrenzenden Seegebietes. *Zeitschrift für Geologische Wissenschaften*, **22**, 211– 226.

Rempel, H. 1992. Erdölgeologische Bewertung der Arbeiten der GO 'Petrobaltik' im deutschen Schelfbereich. *Geologisches Jahrbuch D*, **99**, 3–32.

Scheck-Wenderoth, M. & Lamarche, J. 2005. Crustal memory and basin evolution in the Central European Basin System – new insights from a 3D structural model. *Tectonophysics*, **397**, 143–165.

Schlüter, H.U., Jürgens, U., Best, G., Binot, F. & Stamme, H. 1997*a*. *Endbericht zum Teilprojekt 'Analyse geologischer und geophysikalischer Daten aus der südlichen Ostsee'*. Strukturatlas südliche Ostsee (SASO) Bundesanstalt für Geowissenschaften und Rohstoffe, Hannover, Report [unpublished].

Schlüter, H., Best, G., Jürgens, U. & Binot, F. 1997*b*. Interpretation reflexionsseismischer Profile zwischen baltischer Kontinentalplatte und kaledonischem Becken in der südlichen Ostsee – erste Ergebnisse. *Zeitschrift der Deutschen Geologischen Gesellschaft*, **148**, 1–32.

Seidel, E., Meschede, M. & Obst, K. 2018. The Wiek Fault System east of Rügen Island: origin, tectonic phases and its relationship to the Trans-European Suture Zone. *In*: Kilhams, B., Kukla, P.A., Mazur, S., McKie, T., Mijnlieff, H.F. & van Ojik, K. (eds) *Mesozoic Resource Potential in the Southern Permian Basin*. Geological Society, London, Special Publications, **469**. First published online January 24, 2018, https://doi.org/10.1144/SP469.10

Surlyk, F. 1980. Denmark. *In*: *The Geology of the European Countries, Denmark, Finland, Norway, Sweden*. Dunod, published in cooperation with the Comité National Francais de Géologie (C.N.F.G.) on the occasion of the 26th International Geological Congress, 1–50.

Surlyk, F., Arndorff, L. *et al.* 1995. High-resolution sequence stratigraphy of a Hettangian-Sinemurian paralic succession, Bornholm, Denmark. *Sedimentology*, **42**, 323–354.

Teumer, P., Müller, E.P. & Anclam, P. 1990. Aufsuchung und Gewinnung von Kohlenwasserstoffen in der DDR. *Erdöl Erdgas Kohle*, **106**, 285–292.

Thomas, S., Sivhed, U., Erlström, M. & Seifert, M. 1993. Seismostratigraphy and structural framework of the SW Baltic Sea. *Terra Nova*, **5**, 364–374.

Thybo, H. 2000. Crustal structure and tectonic evolution of the Tornquist Fan region as revealed by geophysical methods. *Bulletin of the Geological Society of Denmark*, **46**, 145–160.

Torsvik, T.H., Trench, A., Svensson, I. & Walderhaug, H.J. 1993. Palaeogeographic significance of mid-Silurian palaeomagnetic results from southern Britain – major revision of the apparent polar wander path for eastern Avalonia. *Geophysical Journal International*, **113**, 651–688.

van Wees, J.D., Stephenson, R.A. *et al.* 2000. On the origin of the Southern Permian Basin, Central Europe. *Marine and Petroleum Geology*, **17**, 43–59.

Vejbæk, O.V., Stouge, S. & Poulsen, K.D. 1994. Palaeozoic tectonic and sedimentary evolution and hydrocarbon prospectivity in the Bornholm area. *Danmarks Geologiske Undersøgelse*, **A34**, 1–23.

Wegner, J. 1966. Strukturbau und Tektonik im Nordosten der DDR. *Geophysik und Geologie*, **9**, 44–56.

Wilson, M., Neumann, E.-R., Davies, G.R., Timmerman, M.J., Heeremans, M. & Larsen, B.T. 2004. Permo-Carboniferous magmatism and rifting in Europe: introduction. *In*: Wilson, M., Neumann, E.-R., Davies, G.R., Timmerman, M.J., Heeremans, M. & Larsen, B.T. (eds) *Permo-Carboniferous Magmatism and Rifting in Europe.* Geological Society, London, Special Publications, **223**, 1–10, https://doi.org/10.1144/GSL.SP.2004.223.01.01

Zagora, K. & Zagora, I. 2004. Devon. *In*: Katzung, G. (ed.) *Geologie von Mecklenburg-Vorpommern*. E. Schweizerbart'sche Verlagsbuchhandlung, Stuttgart, 70–79.

Ziegler, P.A. 1990. *Geological Atlas of Western and Central Europe*. 2nd edn. Shell Internationale Petroleum Maatschappij B.V., Geological Society Publishing House, Bath, 56 enclos.

Zöllner, H., Reicherter, K. & Schikowsky, P. 2008. High-resolution seismic analysis of the coastal Mecklenburg Bay (North German Basin): the pre-Alpine evolution. *International Journal for Earth Science*, **97**, 1013–1027.

New insights into salt tectonics in the northern Dutch offshore: a framework for hydrocarbon exploration

MATTHIJS VAN WINDEN[1]*, JAN DE JAGER[1], BASTIAAN JAARSMA[2] & RENAUD BOUROULLEC[3]

[1]*Department of Earth Sciences, Utrecht University, Heidelberglaan 8, 3584 CS Utrecht, The Netherlands*

[2]*Energie Beheer Nederland, Daalsesingel 1, 3511 SV Utrecht, The Netherlands*

[3]*TNO, Princetonlaan 6, 3584 CB Utrecht, The Netherlands*

**Correspondence: Matthijsvanwinden@gmail.com*

Abstract: The northern Dutch offshore is an area that has seen less hydrocarbon exploration activity than other areas of The Netherlands. Acquisition of a new regional 3D seismic dataset allowed further testing and re-evaluation of established geological concepts in this area. It is recognized that the presence and movement of Upper Permian Zechstein evaporites had a major impact on depositional patterns in Mesozoic sediments, structural development and hydrocarbon migration. As such, this study looks specifically at the role of salt tectonics in tectonosedimentary development. To assess this salt tectonic evolution within its structural context, a restoration of the Step Graben and Dutch Central Graben was performed. It follows that depositional patterns are closely linked to the nature of salt structure movement and the timing of regional tectonism. For example, during Late Triassic rifting, salt pillows developed and sedimentation focused away from salt structures into depocentres along regional fault trends. Restoration results show that this interplay between salt movement and tectonism is needed to accommodate the sedimentation patterns associated with the formation of the Step Graben and Central Graben during the Triassic and Jurassic, and later during Late Cretaceous and Cenozoic inversion tectonics.

During the Late Permian, hundreds of metres of Zechstein evaporites, including salt, were deposited in the northern Dutch offshore (Ziegler 1990; Geluk 2005). Subsequent halokinesis played an important role in the geological development of the area (e.g. Van Wijhe 1987; Ziegler 1990; Scheck-Wenderoth *et al.* 2008; Ten Veen *et al.* 2012). The formation of salt diapirs, salt walls and salt pillows led to the development of a range of hydrocarbon trap types. These include four-way dip closures in strata above diapirs and pillows, as well as three-way dip closures on the flanks of salt structures (Hodgson *et al.* 1992; Wride 1995; Davison *et al.* 2000*a*, *b*; Stewart 2007; de Jager 2012). Halokinesis also had a potential impact on Jurassic source rock distribution, burial depths and maturity (Grassmann *et al.* 2005; Verweij *et al.* 2009; Abdul Fattah *et al.* 2012), as well as on hydrocarbon migration paths and intra-reservoir facies distributions (e.g. Farmer & Barkved 1999; Van der Molen *et al.* 2005; Magri *et al.* 2008; Hampton *et al.* 2010; Back *et al.* 2011). As such, in order to conduct an effective play evaluation in the Dutch Central Graben and Step Graben, it is crucial to understand the timing of salt movement episodes.

The primary aim of this study is to better constrain when the main periods of salt movement occurred and how the geometries of salt structures developed through geological time. A thorough understanding of the development of the Dutch Central Graben and Step Graben systems and their later restructuration through inversion tectonics is also required to accurately describe the relationship between regional structuration and salt tectonics. In this study, a transect through the Dutch Central Graben and Step Graben is structurally restored. This restoration integrates a regional inventory of salt structure interpretations in the northern Dutch offshore, in which salt structures are systematically listed and characterized according to a range of salt structure characteristics (see Table 1). Using this method, it is possible to test the role of Zechstein halokinesis in influencing adjacent stratigraphic geometries and to visualize how the different deformation stages evolved. This approach gives a framework for the analysis of the role of salt tectonics in hydrocarbon play definition and the tectonostratigraphic development within the Dutch Central Graben and Step Graben.

From: Kilhams, B., Kukla, P. A., Mazur, S., McKie, T., Mijnlieff, H. F. & van Ojik, K. (eds) 2018. *Mesozoic Resource Potential in the Southern Permian Basin*. Geological Society, London, Special Publications, **469**, 99–117.
First published online January 29, 2018, https://doi.org/10.1144/SP469.9

Table 1. *A selection of interpreted stratigraphic relationships around salt walls and salt diapirs in the study area, based on observations from regional 3D seismic data (Spectrum DEFAB 2010 survey)*

	Salt structure	Location (Schill Grund and Elbow Spit Platform, Step Graben, Central Graben)	Salt structure type	Youngest affected horizon	Oldest affected interval	Intervals thinning towards the salt structure	Intervals thickening towards the salt structure	Pre-salt fault orientation
1	A18-NORTH1	SG	Diapir	Base North Sea Supergroup	Upper Germanic Trias Group (RN)	RB, CK	RN, SL	North–south
2	B16F01-WEST1	SG	Wall	Middle Miocene	Upper Germanic Trias Group (RN)	RN, KN, CK, NS	AT	North–south
3	B16F01-EAST1	SG	Wall	Middle Miocene	Lower Germanic Trias Group (RB)	RN, CK, NS	RB	North–south
4	B17-SOUTH1	CG	Diapir	Base Quarternary	Upper Germanic Trias Group (RN)	AT, SL	RN, CK, NS	North–south
5	F02-NORTH1	SG	Wall	Base Quarternary	Upper Germanic Trias Group (RN)	CK, NS	RN, KN	North–south; ESE–WNW
6	F02-NORTH2	CG	Diapir	Base Quarternary	Upper Germanic Trias Group (RN)	AT, SL, CK, NS	RN	?
7	F03-EAST1	CG	Diapir	Base Quarternary	Upper Germanic Trias Group (RN)	RN, SL, KN, CK, NS	SL	North–south
8	F03-EAST2	SGP/CG	Diapir	Base Quarternary	Upper Germanic Trias Group (RN)	RN, AT, SL, KN	CK, NS	?
9	F05-EAST1	CG	Diapir	Base Quarternary	Upper Germanic Trias Group (RN)	RN, CK, NS	AT, SL	NNE–SSW
10	F05-WEST1	SG	Diapir	Base Quarternary	Upper Germanic Trias Group (RN)	RN, AT, SL, CK	None	North–south
11	F05F08-WEST1	SG	Diapir	Base Quarternary	Upper Germanic Trias Group (RN)	AT, SL, CK, NS	RN	North–south
12	F06b-EAST1	CG	Diapir	Middle Miocene	Upper Germanic Trias Group (RN)	RB, SL	RN, AT, NS	North–south
13	F06b-EAST2	SGP/CG	Wall	Middle Miocene	Lower Germanic Trias Group (RB)	RB, AT, SL, CK	RN, NS	North–south
14	F07F08-NORTH1	SG	Wall	Middle Miocene	Upper Germanic Trias Group (RN)	CK	RN, NS	NNE–SSW
15	F07F08-SOUTH1	SG	Diapir	Base Quarternary	Lower Germanic Trias Group (RB)	RN, AT, SL, CK, NS	RB,	North–south; NE–SW
16	F09-WEST1	CG	Diapir	Base Quarternary	Upper Germanic Trias Group (RN)	RN, CK, NS	AT, SL, KN	North–south
17	F09-EAST1	SGP/CG	Wall	Middle Miocene	Upper Germanic Trias Group (RN)	AT, SL, CK	RN, NS, KN	NNE–SSW
18	F10-EAST1	SG	Wall	Middle Miocene	Upper Germanic Trias Group (RN)	RN, KN, CK, NS	None	North–south; east–west; NW–SE

See Figure 2 for an overview of the interpreted diapirs and walls, and Figure 3 for a definition of the stratigraphic intervals.

Regional geology

Tectonic development

The study area, within the A, B, D, E and F blocks of the northern Dutch Central Graben and Step Graben, is located in the Southern North Sea, on the northern fringe of the Southern Permian Basin (Figs 1 & 2). The development of the wider Southern Permian Basin area was affected by three major periods of plate tectonics: (1) assembly of the Pangea supercontinent; (2) break-up of the Pangea supercontinent; and (3) distal inversion effects of the Alpine Orogeny (Nøttvedt *et al.* 1995; de Jager 2007; Breitkreuz *et al.* 2008; Krawczyk *et al.* 2008; Voigt *et al.* 2008; Pharaoh *et al.* 2010).

The assembly of Pangea is characterized by two major collisional events, resulting in the Caledonian and Variscan fold and thrust belts. The Caledonian collision occurred in the Early–Middle Paleozic between the continents of Laurentia and Baltica (Krawczyk *et al.* 2008). This resulted in the closure of the Tornquist Sea (Pharaoh *et al.* 2010) along the NW–SE-running Tornquist Suture Zone (Berthelsen 1998; Cocks & Torsvik 2006; Pharaoh *et al.* 2010). Subsequently, the microcontinent of Avalonia collided from the south, which closed the Iapetus Ocean along the NW–SE- to north–south-running Iapetus suture (Krawczyk *et al.* 2008; Pharaoh *et al.* 2010). This created a triple junction of plate boundaries just to the NW of the study area (Ziegler 1982, 1990; de Jager 2007). Generally, this junction is linked to the location of Mesozoic basins such as the Central Graben in the UK and The Netherlands (Ziegler 1990). A subsequent collision of the resulting continent (Laurussia) with the Gondwanan continent resulted in the Variscan Orogeny (Kroner *et al.* 2008). The Variscan thrust front associated with this collision moved northwards throughout the Carboniferous, with its final position trending east–west through present-day Belgium and to the NE into Germany (Fig. 1) (Ziegler 1990; Kroner *et al.* 2008; Pharaoh *et al.* 2010). Late Variscan tectonic activity (Wilson *et al.* 2004; Timmerman *et al.* 2009; Breitkreuz *et al.* 2008) induced widespread erosion in the Late Carboniferous (Geluk 2005). A phase of dextral translation of northern Africa relative to Europe then led to the onset of orogenic collapse and basin formation in NW Europe during the Late Carboniferous–Early Permian (Ziegler 1990; Geluk 2005; Pharaoh *et al.* 2010). In this period, an arid, desert-like area developed north of the Variscan front, where Lower Permian Rotliegend sediments accumulated (Stollhofen *et al.* 2008; Gast *et al.* 2010; Mijnlieff & Geluk 2011). Rotliegend isopach maps show subtle early structuration in the underlying basement, where several separate depocentres can be distinguished, including the large east–west-running Northern and Southern Permian basins (Geluk 2005; Gast *et al.* 2010). Continued isostatic subsidence (van den Belt & de Boer 2007) induced periodic flooding, resulting in the cyclic deposition of the Zechstein Group (Geluk 2005; Peryt *et al.* 2010). Rifting commenced during the Early Triassic in the Northern Atlantic domain (Ziegler 1990; Zanella & Coward 2003; Feist-Burkhardt *et al.* 2008; Stollhofen *et al.* 2008), which induced several phases of extension from the Triassic to the Early Cretaceous across the study area, although no continental break-up subsequently occurred (Ziegler 1990; Zanella & Coward 2003; Feist-Burkhardt *et al.* 2008; Stollhofen *et al.* 2008). In The Netherlands, a regional east–west orientation of extension is typically assumed, although basins like the Broad Fourteens Basin and the West Netherlands Basin follow a persistent NW–SE fault trend (Van Wijhe 1987), possibly due to older, reactivated basement fault trends (Ziegler 1990). Due to these extensional events, Triassic thermal subsidence was interrupted by periods of active faulting during the deposition of the Buntsandstein and Keuper formations (Geluk 2005). Subsequent development of a Mid North Sea thermal dome (Ziegler 1992; Underhill & Partington 1993), during the Middle Jurassic, induced deep erosion of Triassic–Jurassic sediments on platform and marginal areas, while deposition was limited to fault-bounded rift basins, such as the Central Graben (Husmo *et al.* 2002; Pieńkowski *et al.* 2008; Lott *et al.* 2010). Here, subsidence continued and a complete Jurassic sequence is typically present (Ziegler 1992; Bouroullec *et al.*, this volume, in press; Verreussel *et al.*, this volume, in press). Active rifting ceased in the Early Cretaceous and rift basins were filled with Lower Cretaceous sediments (Bouroullec *et al.*, this volume, in press; Verreussel *et al.*, this volume, in press). In the Mid-Cretaceous, continental break-up occurred in the Mid-Atlantic domain (Ziegler 1988, 1990). This caused extensional stresses to focus towards the Arctic and away from the Southern North Sea domain (e.g. Ziegler 1990; Scheck-Wenderoth *et al.* 2008). As such, background regional subsidence persisted in most of the Southern North Sea area throughout the Middle Cretaceous (Van Wijhe 1987; Littke *et al.* 2008). In southern Europe, the closure of the Tethys Ocean initiated in the Late Cretaceous with the collision of the African, Indian and Cimmerian plates from the south with the Eurasian continent in the north, resulting in the Alpine Orogeny (e.g. Reicherter *et al.* 2008; Pharaoh *et al.* 2010). Distal effects of this event significantly affected the Southern North Sea area, inducing several pulses of inversion within the Dutch Central Graben and the Broad Fourteens Basin (Van Wijhe 1987; Surlyk *et al.* 2003; de Jager 2003, 2007; Worum & Michon 2005). For the former, the most significant of these

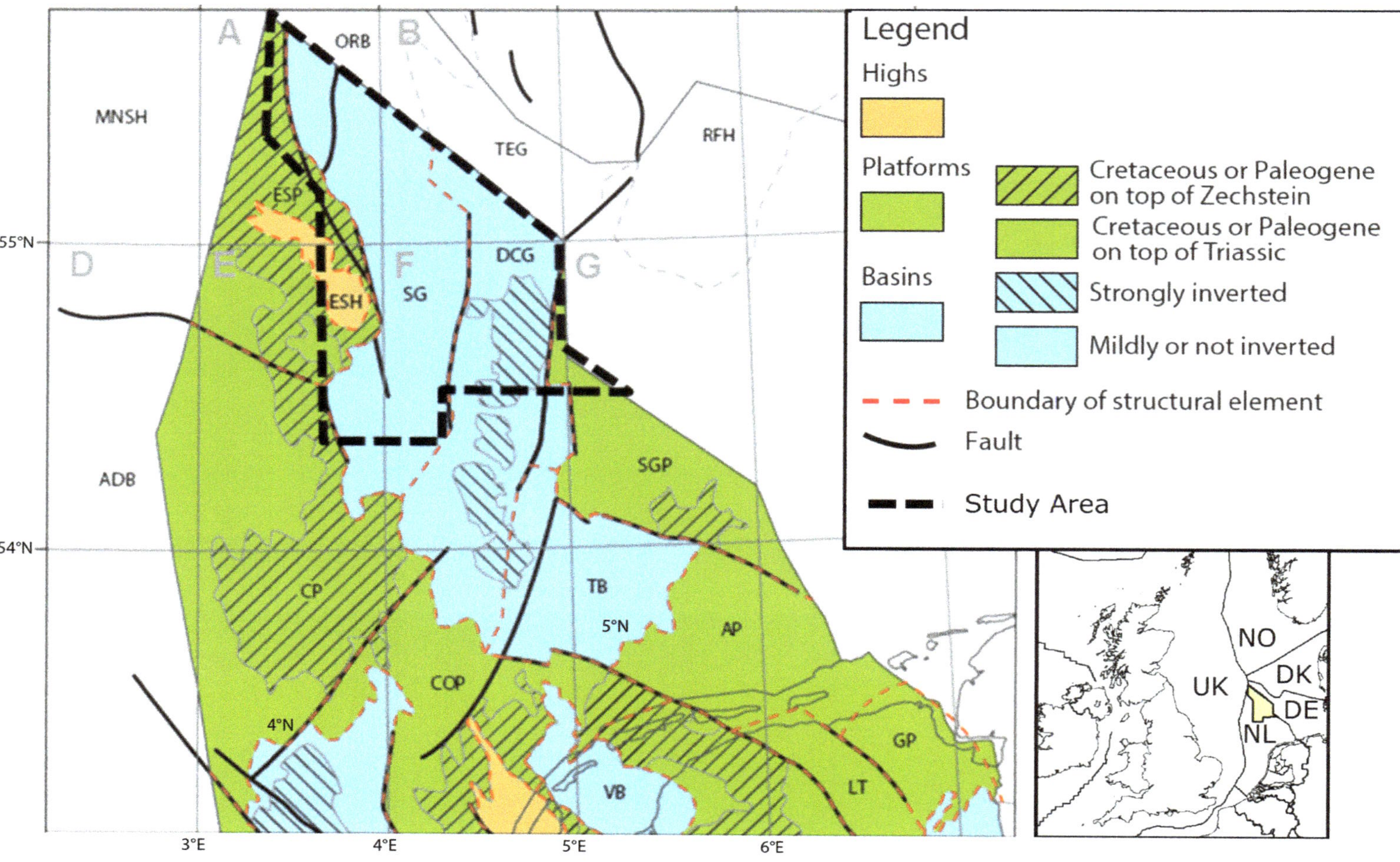

Fig. 1. Location of the study area (black outline) illustrating the main structural elements in the northern Dutch offshore (after Kombrink *et al.* 2012). ESP, Elbow Spit Platform; ESH, Elbow Spit High; SGP, Schill Grund Platform; DCG, Dutch Central Graben; SG, Step Graben; CP, Cleaverbank Platform.

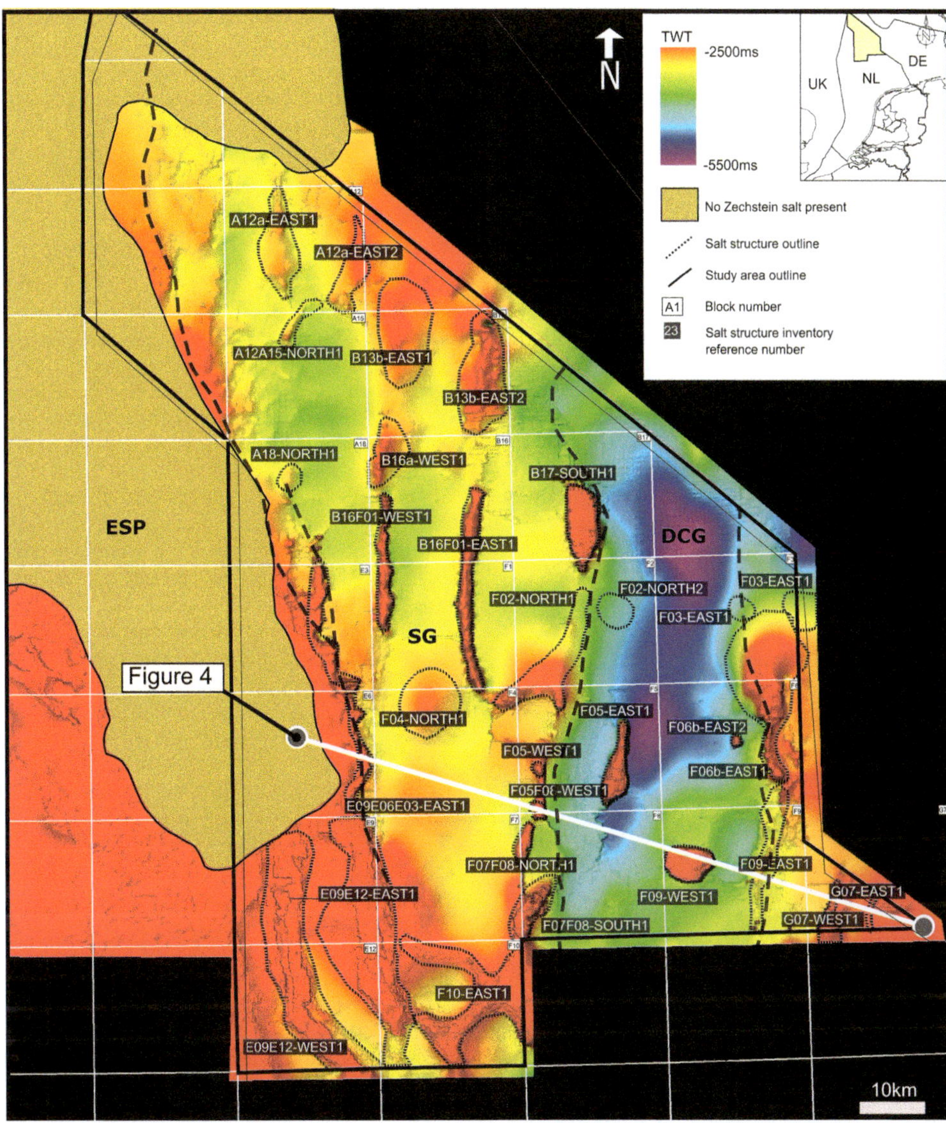

Fig. 2. Two-way time (TWT) map of the top Zechstein Group, based on the interpretation of regional 3D seismic data (Spectrum DEFAB 2010 survey). All 30 interpreted salt structures are indicated (dotted black lines); dashed black lines define the structural domain edges. All salt structures are listed in Table 1.

pulses occurred during the Campanian, Paleocene and Eocene (de Jager 2003, 2007; Surlyk *et al.* 2003; Van der Molen *et al.* 2005; Esmerode *et al.* 2008). While these inversions typically induced uplift of basin centres, most platform areas subsided (Littke *et al.* 2008; Pharaoh *et al.* 2010).

Structural and stratigraphic framework

The main structural elements in the study area are the Dutch Central Graben (DCG), the Step Graben (SG), and the margins of the Schill Grund Platform (SGP) and Elbow Spit Platform (ESP) (Duin *et al.* 2006). The Dutch Central Graben is the deepest section of the study area, and continues as the German Central Graben and the Tail End Graben to the north (Wride 1995). Most of the halite in this area is associated with the basinal Zechstein Group, which consists of evaporites, carbonates and clastics (Van Adrichem Boogaert & Kouwe 1994; Taylor 1998; Geluk 2005; Peryt *et al.* 2010). The Zechstein depositional cycles in this area are defined as Z1 (oldest)–Z5 (youngest), where the Z2 and Z3 members typically contain the thickest halite intervals (Ten Veen *et al.* 2012). Carbonate members are subdivided into shelf, slope and basinal facies (Van der Baan 1990; Tolsma 2014). Occurrences of slope-facies carbonates in the Zechstein Group are an indication of the location of the margins of the Southern Permian Basin and give an approximation of the extent of Z2 and Z3 halite deposition (Van der Baan 1990; Geluk 2005; Jenyon & Taylor 2005; Słowakiewicz *et al.* 2013). In general, halite thickness increases from west to east towards the basinal part of the Southern Permian Basin (Ten Veen *et al.* 2012), where the Zechstein Group sediments reaches an estimated initial thickness of 1500–2000 m (Ziegler 1990; Ten Veen *et al.* 2012; Hernandez *et al.*, this volume, in press).

The post-Permian stratigraphy in the study area is described using the nomenclature of Kombrink *et al.* (2012) with the key stratigraphic intervals shown in Figure 3.

Concepts of salt tectonics

In basins with a sufficiently thick, and therefore potentially mobile, salt sequence, depositional patterns and geometries are often controlled by halokinesis. This can typically be directly related to four distinct stages of salt movement (cf. Vendeville 2002):

- Layered salt stage: the salt layer is in its initial, depositional configuration.
- Pillowing stage: lateral salt movement within the salt layer leads to the development of a primary rim-syncline basin, away from the salt structure. Stratigraphic thinning occurs on top of, and adjacent to, the pillow.
- Piercing stage: salt moves vertically, piercing through younger stratigraphic layers. This stage is typically accompanied by withdrawal of the surrounding salt towards the piercing structure. This leads to the development of a secondary rim syncline where additional accommodation space is created. This results in a thickening of sediments towards the salt structure and into the rim syncline.
- Diapir rejuvenation stage: reactivation phase of the salt structure growth, typically associated with a third stage of rim-syncline development, where strata thickens away from the salt structure (Trusheim 1960; Vendeville 2002).

All four stages can be observed in the study area, as illustrated in Figure 3 for salt diapir F09-WEST1. Note that the model of Trusheim (1960), with only buoyancy being responsible for the creation of salt diapirs, is no longer considered appropriate with the generally accepted concept involving active faulting (Vendeville 2002). Furthermore, the depositional thickness of the salt is considered to have a major control on the style of subsequent faulting (the varying thickness of the salt results in variable degrees of decoupling between the basement and cover structures) and the distribution of synrift sediments (Stewart 2007; Ten Veen *et al.* 2012; Duffy *et al.* 2013).

Regional salt tectonic development

Based on the assessment of stratigraphic relationships around salt structures, described in the aforementioned salt structure inventory and supported by published material (see Table 1; Fig. 2), a first-order interpretation of salt tectonic development is presented with the aim of giving the reader a framework for understanding subsequent reconstructions.

The Early Triassic appears to have been a period of relative tectonic quiescence, although some syndepositional tectonics is suggested by the observation of variable thicknesses in the Lower Triassic of the northern Step Graben and the Dutch Central Graben (see also Dronkert *et al.* 1989; Ziegler 1990; Bachmann *et al.* 2010; Peryt *et al.* 2010; van Winden 2015). Middle and Late Triassic deposits show large thickness variations throughout the study area (Ziegler 1988, 1990; Remmelts 1995; Geluk 2005; Bachmann *et al.* 2010; van Winden 2015). Frequently observed thinning of the Upper Germanic Trias Group interval towards salt structures indicates a widespread salt-pillowing stage, interpreted as the first regional onset of halokinesis (Dronkert *et al.* 1989; Remmelts 1995; Geluk 2005; Bachmann *et al.* 2010; van Winden 2015). These salt movements are likely to be linked to the onset of the Early Cimmerian rifting phase, which allowed the formation of elongated pillows and locally detached faults above the salt layer (de Jager 2003, 2007, 2012; Ten Veen *et al.* 2012; Kombrink *et al.* 2012). In turn, this led to the development of depocentres in areas of salt withdrawal (Davison *et al.* 2000*a*, *b*; Ten Veen *et al.* 2012; Matthews *et al.* 2007).

The climax of the salt tectonics is observed to have occurred in the Jurassic period (see also Remmelts 1995; van Winden 2015). The presence of prominent rim synclines suggests the dominance of a piercing stage for many salt structures (Remmelts 1995; Lott *et al.* 2010; van Winden 2015; Vendeville 2002). In the Late Jurassic, rifting-related subsidence of the Dutch Central Graben was accompanied by widespread salt withdrawal, focusing deposition of sediments into the subsequent mini-basins (Lott *et al.* 2010). Because of the presence of thick Zechstein salt, Late Jurassic east–west rifting did not induce visible faults in the Upper Jurassic succession (Fig. 4) (Wijker 2014; van Winden 2015), although it is recognized that the main bounding

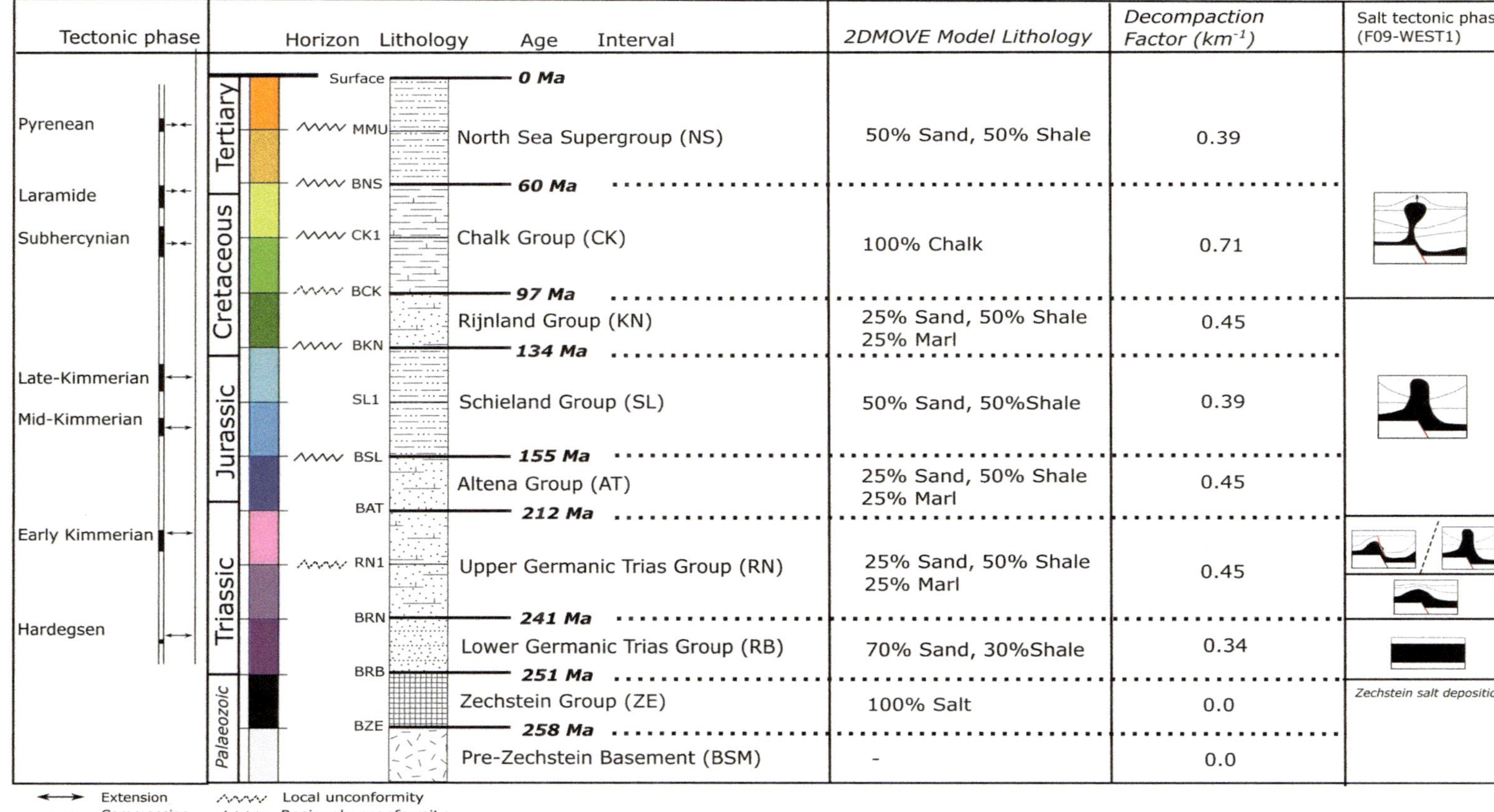

Fig. 3. Main tectonic events and restoration model stratigraphy, including lithological compositions and decompaction factors (after Verweij *et al.* 2009). Additionally, the interpreted salt tectonic phase of salt structure F09-WEST1 is plotted through time.

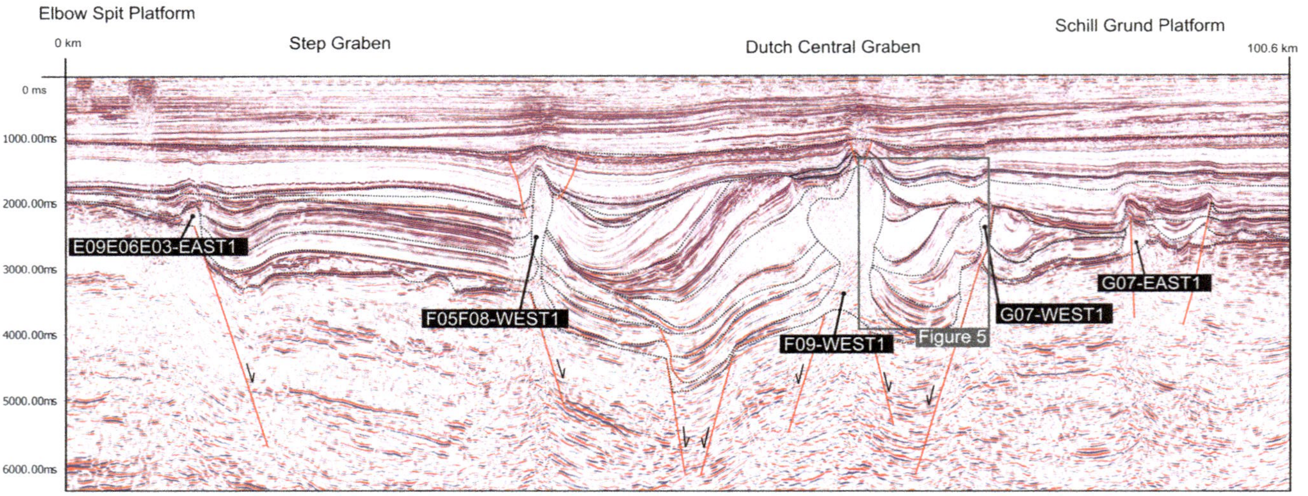

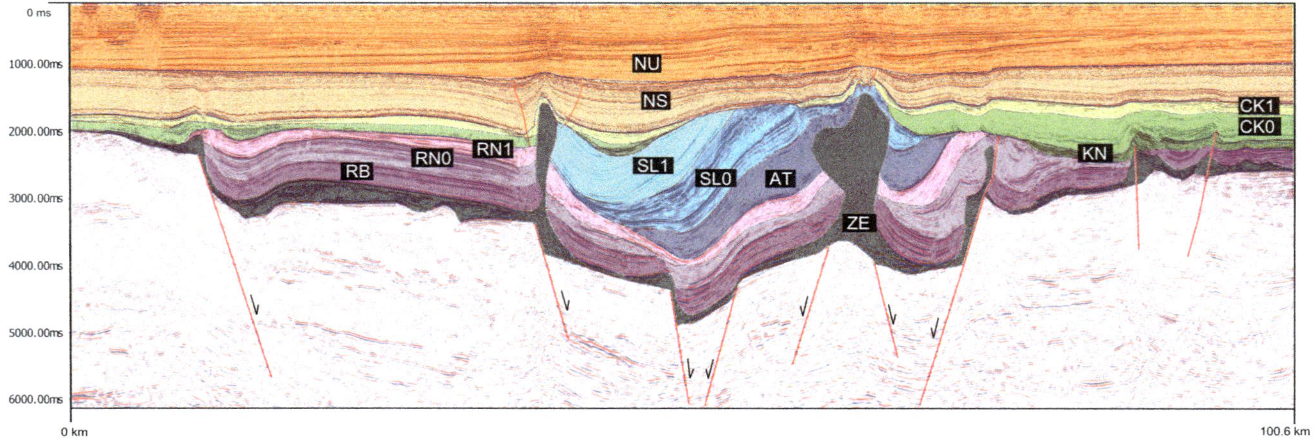

Fig. 4. Interpreted regional seismic section used for restoration (Fig. 6), from regional 3D seismic data (Spectrum DEFAB 2010 survey). For the location of the section see Figure 2. The abbreviations of stratigraphic intervals refer to Figure 3. The part of the section shown in Figure 5 is indicated on the seismic section (grey box).

faults of the Dutch Central Graben are overlain by salt walls, concealing much of the seismically definable subsalt structural detail (Fig. 4). However, subsalt basement faulting, through increased subsidence in the Dutch Central Graben, may have been responsible for a westwards shift of the Dutch Central Graben depocentre in Late Jurassic times (Wijker 2014; van Winden 2015). Thinning and local absence of Upper Cretaceous and Cenozoic strata over salt structures provides evidence for renewed salt movement during the Late Cretaceous and Cenozoic (Remmelts 1995; Van der Molen *et al.* 2005, Goffey *et al.* 2016).

The timing and development of salt structures, their interaction with faults and the effects on depositional patterns are not consistent throughout the entire study area, but vary depending on their location within the graben system. In order to better understand the implications of all post-Permian major structural deformation events on salt movement, a structural (palinspastic) restoration was performed.

Methods

A regional east–west seismic cross-section from the Spectrum DEFAB 2010 survey, through the Dutch Central Graben and Step Graben, was selected for a structural restoration with the aim of illustrating and testing hypotheses for the main stages of halokinesis and associated deformation. This particular section was chosen as it transects five key salt structures and all main structural domains (see Fig. 2). Additionally, this line trends perpendicular to the regional structural grain (NNE–SSW). Figure 4 illustrates this section with key stratigraphic intervals (listed in Fig. 3) and faults indicated (detailed section is shown in Fig. 5). Interpretations of horizons and

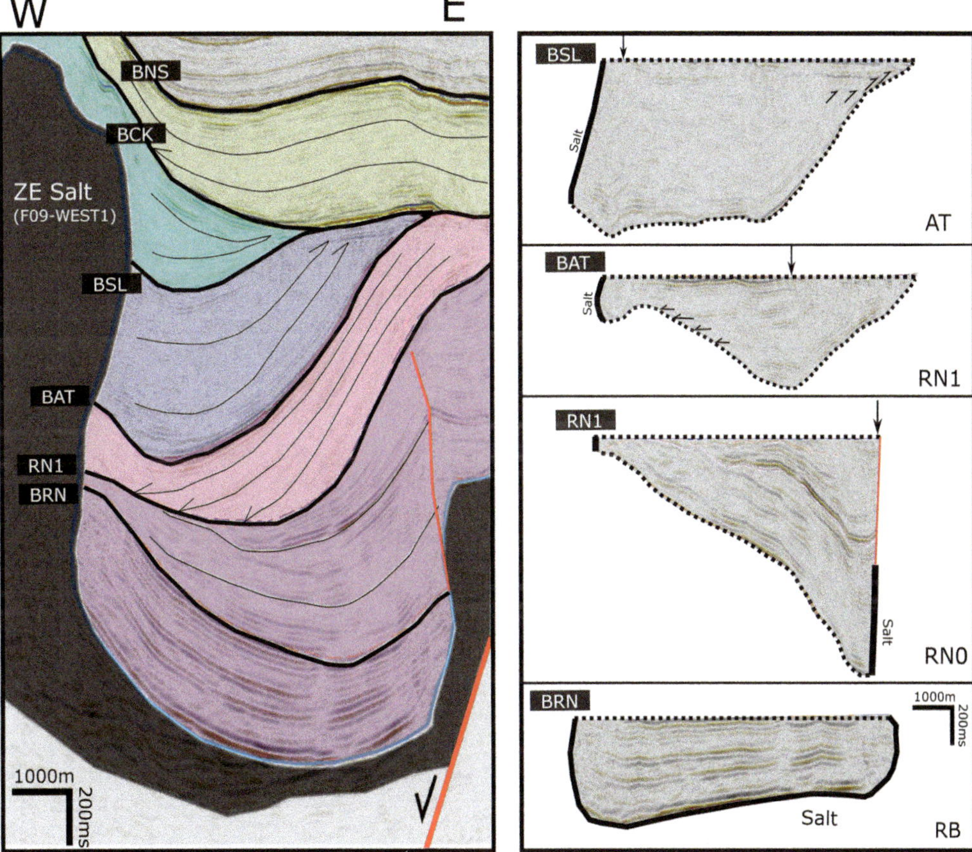

Fig. 5. Detailed interpretation on 3D seismic data (Spectrum DEFAB 2010 survey) of the Figure 4 seismic section, showing salt structure F09-WEST1 and adjacent intervals to the east. Four intervals are shown flattened on the corresponding top horizon: Lower Germanic Trias Group (RB), Upper Germanic Trias Group (RN0 and RN1) and the Altena Group (AT). The location of the palaeo-depocentre is interpreted (indicated by the black arrows).

Table 2. *The assumptions made for initial porosity, decompaction factors and densities of lithologies in the northern Dutch offshore (after Verweij* et al. *2009)*

Lithology	Initial porosity	Decompaction factor (km^{-1})	Density ($kg\ m^{-3}$)
Sandstone	0.49	0.27	2650
Shale	0.63	0.51	2720
Chalk	0.70	0.71	2200
Salt (halite)	0.00	0.00	2200
Marl	0.50	0.50	2700

faults were performed in time and subsequently depth-converted. The time–depth conversion was performed using a regional time–depth conversion model (Velmod v2.0: van Dalfsen *et al.* 2006). For all post-Permian intervals, a $V = V_0k$ function, where V = instantaneous velocity, V_0 = reference velocity and k = velocity gradient, was applied and Zechstein velocities were assumed to be constant. In order to appropriately model deformation and decompaction, rock properties were assigned to all 11 intervals of the model. For these intervals, the following properties were defined: (1) initial porosity; (2) decompaction factor; (3) compaction curve; and (4) bulk rock density (Table 2). These parameters were based on lithological information from the Terschelling Basin to the south of the study area (Verweij *et al.* 2009) and the Cleaverbank Platform to the SW of the study area (Abdul Fattah *et al.* 2012) (see Fig. 3). The Zechstein salt layer is treated as an incompressible layer. Vertical simple shear was applied in the restoration of faults and in unfolding. It is accepted that this may be unrealistic for restoration around salt structures, where strata dip steeply and a bed-length restoration algorithm may be more appropriate (Rowan & Ratliff 2012).

Several uncertainties are inherent to the structural restoration of salt sections. First, salt may have flowed in and out of the 2D section plane which would change the total area of the Zechstein salt layer, as represented in the 2D section. Secondly, dissolution and/or erosion of salt may have occurred, especially since models from this study suggest that salt was near, or even at, the surface in several locations and different moments in time (see also Hernandez *et al.*, this volume, in press). In this study, restoration of salt volumes and estimates of the total original thickness are further constrained by stratigraphic and structural geometries and by comparison to previous studies (e.g. Hossack 1995; Ten Veen *et al.* 2012).

Results

Figure 6a–h presents the results of the structural restoration of the regional transect shown in Figure 4. Eight restoration steps were performed in order to present a possible Early Triassic structural configuration. A chronological description of the resulting models is given below.

Early Triassic

The Lower Germanic Trias Group (RB) was deposited on top of the Zechstein Group (ZE) evaporites. Although the original thickness distribution of the ZE interval in the restored section is uncertain, it is likely that thinning of the interval occurred towards the edges of the Zechstein salt basin to the west of salt structure E09E06E03-EAST1 (Fig. 4), as is modelled in the restored section shown in Figure 6a. The location of this basin edge in this area is based on the occurrence of slope facies carbonates (Tolsma 2014). Towards the east of salt structure F05F08-WEST1 (Fig. 4), the original salt thickness may have reached up to 1500–2000 m (based on 50% salt dissolution and restored current-day salt thicknesses of up to 900 m: Ten Veen *et al.* 2012). The relatively constant thickness of the Lower Triassic sediments observed in this section suggests that the ZE salt was mostly unstructured at this time and is modelled as an interval with layered geometry. The RB interval is offset by younger faults, including the Dutch Central Graben boundary faults, at several locations. However, it is modelled that in this section the RB interval itself was deposited with a mostly homogeneous thickness (Fig. 6a). There are no indications that, during the deposition of the RB interval, active tectonism occurred in this part of the basin, even though in the north of the Step Graben (A and B blocks) there are local indications of active Early Triassic rifting (van Winden 2015). As such, it is expected RB was deposited without much palaeorelief present in the modelled part of the basin. The model includes some structuration in the pre-Zechstein basement, which may have controlled the location of the Southern Permian Basin margins and the Central Graben boundary faults, which show similar trends (north–south). This interpretation is speculative and based on seismic observations of a pre-Zechstein graben structure following the trend of the present-day Dutch Central Graben.

Late Triassic

Clear thickness variations near salt structures (mainly thinning: e.g. salt structure F09-WEST1) and thickening towards the eastern boundary fault of the Dutch Central Graben is observed in the lower part of the Upper Germanic Triassic Group (RN0) in this section. As the ZE salt had a significant thickness and was less deformed at this time, suprasalt faults were likely to be detached from pre-salt faults. This is consistent with the observed

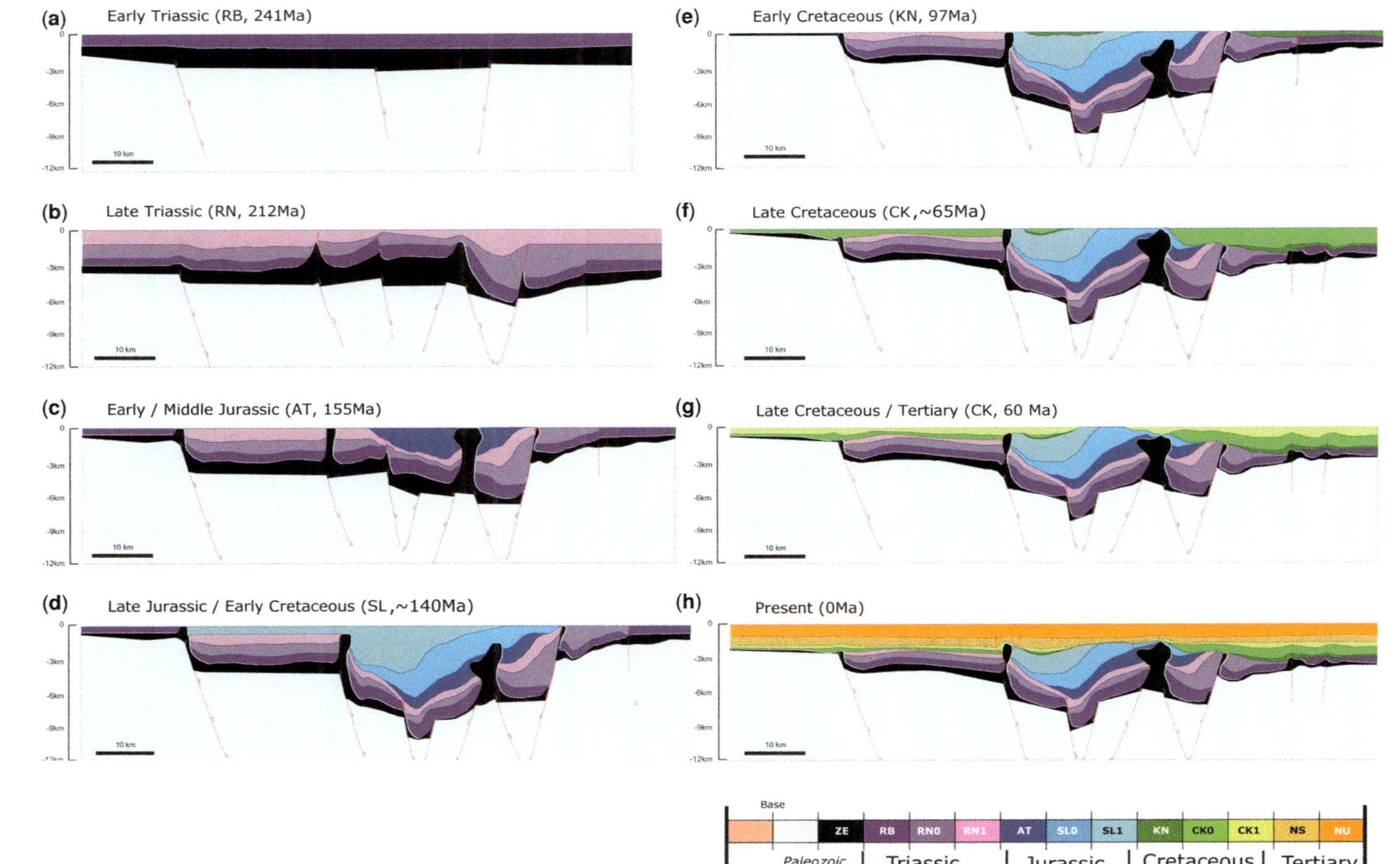

Fig. 6. Resulting cross-sections at various times in the structural restoration model (×2 vertical exaggeration) showing: **(a)** the estimated depositional salt thickness in the Early Triassic; **(b)** the initiation of salt tectonics in the Late Triassic; **(c)** & **(d)** the salt tectonic climax in the Jurassic; **(e)** & **(g)** Cretaceous erosion; and renewed salt movement in **(f)** the Late Cretaceous and **(h)** the present day.

stratigraphic geometries in the RN0 interval (e.g. Fig. 5). It is interpreted that most faulting throughout the Late Triassic was detached from basement faulting and the depositional pattern was controlled by non-linked or soft-linked faults above salt pillows. This is based both on the stratigraphic geometries throughout the RN interval (Fig. 5) and on the idea that the Zechstein salt layer probably had a greater, more homogenous thickness at the time of initiation of faulting. The RN0 interval shows clear stratigraphic thinning towards salt structure F09-WEST1. This geometry can be explained by the development of a depocentre away from a Zechstein salt pillow during deposition. A similar geometry, where strata thicken into a fault plane on one side of a salt structure and thins on the other side, can be observed to the west of salt structure F09-WEST1.

In this section, sediments of the RN0 interval within the Step Graben appear to have been deposited relatively undisturbed, as reflectors are parallel and continuous (Fig. 4). Some internal truncation can be observed near the western Step Graben boundary fault, which suggests minor syndepositional fault activity. On the Elbow Spit Platform, no Triassic strata can be observed today, but it is assumed the interval was deposited homogenously across the platform area and eroded at a later stage. This is based on the continuous nature of Triassic stratigraphy near the western Step Graben boundary fault (see also the palaeogeographical reconstructions of Doornenbal *et al.* 2010; after Ziegler 1990). From the Dutch Central Graben onto the Schill Grund Platform, the RN0 interval in this section shows stratigraphic thinning and is absent further to the east (Fig. 4). The restoration assumes relatively thin deposits, although it is uncertain how much Triassic was originally deposited on the stable Schill Grund Platform. The interval thickens rapidly into the Dutch Central Graben towards the west, resulting from subsidence of this area with respect to the adjacent platform areas and the Step Graben (Geluk 2005; de Jager 2007).

Fault activity gradually slowed during the later part of the Late Triassic. Stratigraphic thickness variations within the RN1 interval in the restored section shows a shift of depocentres at several locations within the Dutch Central Graben, in particular to the east of salt structure F09-WEST1 near the eastern graben boundary fault (Figs 5 & 6b). The depocentre of the RN1 interval is observed to shift from close to this boundary westwards towards salt structure F09-WEST1. This indicates an increased control of salt movement on the deposition of sediments. Reflectors of the RN1 interval downlap onto the top of the RN0 interval (Figs 4 & 5), illustrating the development of an eastwards-dipping slope during deposition of RN1. The flattened sections shown in Figure 5 confirm this westwards shift. This shift and associated downlapping sediments indicate the withdrawal of salt towards the west and an increase in salt thickness at the location of salt structure F09-WEST1, as modelled in Figure 6b. Away from the graben, RN1 is mostly absent at the present day, but is interpreted to have been deposited relatively undisturbed and homogenously (Ziegler 1992).

Jurassic

The Lower Jurassic Altena Group (AT) thickens towards the centre of the Dutch Central Graben in this section (Fig. 6c), where a secondary rim-syncline geometry is observed adjacent to salt structure F09-WEST1 (see Fig. 5). This implies that the salt structure pierced the overburden at this time. Towards the Dutch Central Graben boundary faults, the AT interval is truncated and absent on the platform areas and in the Step Graben. The AT interval is assumed to have been deposited regionally (in an open-marine setting: Doornenbal *et al.* 2010; after Ziegler 1990) and eroded from platform highs only at a later stage.

This erosion is believed to have been caused by uplift in the area of the restored section related to thermal doming in the Middle Jurassic (*c.* 155 Ma: Kombrink *et al.* 2012) which induced deep erosion (including of Middle–Upper Triassic deposits) on the platform areas and minor erosion within the Step Graben. During the same period, the Dutch Central Graben continued to subside and is, therefore, mostly unaffected by erosion, forming a restricted marine embayment (Ziegler 1990; Cope *et al.* 1992; Ineson & Surlyk 2003; Feldman-Olszewska 2006). Salt structures F05F08-WEST1 and E09E06E03-EAST1 were at, or near, the surface during this period (Fig. 6c), exposing Zechstein salt to erosion and dissolution. This is incorporated in the model as significant reduction in salt volume (up to 40%) throughout the Jurassic (Fig. 6c, d).

During deposition of the Upper Jurassic SL0 interval, a secondary rim-syncline geometry adjacent to salt structure F09-WEST1 persists (Fig. 6d), although a gradual shift of the depocentre occurs towards the west (Fig. 4). The shift of depocentre is accommodated in the restored model by fault-controlled subsidence (related to active accelerated rifting), combined with an increased withdrawal of salt in the west of the Dutch Central Graben. However, the true relative influence of these factors is uncertain.

A regional unconformity between SL1 and SL0 (Fig. 4) represents a major westwards shift of the depocentre, which starts during the deposition of the SL1 interval (Fig. 6d). As fault activity continued, deposition of both SL0 and SL1 remained restricted to the subsiding graben structures. To the north, Upper Jurassic sediments locally occur in rim synclines adjacent to salt structures within the Step Graben. This indicates that some Upper Jurassic deposition occurred in the Step Graben. The SL1 interval is interpreted to have been widely deposited and subsequently eroded in this area (Fig. 6d).

Cretaceous and Cenozoic

In the Early Cretaceous, erosion of almost all Jurassic strata outside of the Dutch Central Graben occurred, represented by the Base Cretaceous Unconformity (Fig. 6d) (Copestake *et al.* 2003; Pharaoh *et al.* 2010). On the Elbow Spit Platform, erosion removed all sediments overlying the Zechstein salt and is likely to have significantly eroded Zechstein sediments themselves, leading to a reduction in salt volume (Fig. 6e). Some erosion of Upper Triassic sediments occurred on the Schill Grund Platform and possibly within the Step Graben (Fig. 6e). The Rijnland Group (KN) was deposited during the Early Cretaceous, and was eroded from most of the Dutch Central Graben and Step Graben areas during the Late Cretaceous (Fig. 6f).

Within the Upper Cretaceous Chalk Group (CK), the Late Campanian Unconformity is interpreted to separate the Chalk Group into two intervals (CK0 and CK1: Fig. 3) (Van der Molen *et al.* 2005; Scheck-Wenderoth *et al.* 2008). This boundary is chosen as it is represents a significant Late Campanian inversion event, which can be identified as an unconformity in seismic (Fig. 3), and in CK0 and CK1 depositional trends. The CK0 interval onlaps towards the west on the eastern side of the Dutch Central Graben. On top of the Dutch Central Graben almost no CK0 is present, as the Late Campanian Subhercynian phase of uplift caused erosion down to the Base Cretaceous Unconformity (Figs 4 & 5) (Farmer & Barkved 1999; Van der Molen *et al.* 2005; Hampton *et al.* 2010). Subsequently, the post-Campanian sediments of the CK1 interval were deposited during the later part of the Late Cretaceous (Fig. 6g), which represents the preserved Chalk sediments of the Dutch Central Graben. After deposition of CK1, the central part of the Dutch Central Graben was inverted once more during the Laramide inversion pulse (Ziegler 1990; Ziegler & Dèzes 2007), causing local erosion of CK1. These inversions were not instantaneous and the deposition of sediments will have continued, as is suggested by the observed thinning of the CK0 and CK1 intervals towards the axis of main inversion.

The phases of inversion also induced renewed halokinesis, causing deformation of CK strata above salt structures. This resulted, for example, in an anticlinal geometry at the CK1 level above salt structures F05F08-WEST1 and F09-WEST1 (Fig. 4), and to a lesser extent above salt structures E09E06E03-EAST1, G07-WEST and G07-EAST (van Winden 2015). Faults on the Schill Grund Platform were also reactivated in a reverse sense, evidence of which is shown by locally thrusted CK0 strata to the east of salt structure G07-EAST1 (Fig. 4) and mass-flow deposits (e.g. near the eastern Central Graben boundary fault in Fig. 4 and above salt structure G07-EAST1: see also Arfai *et al.* 2016). Upper Cretaceous strata on top of the Late Campanian Unconformity appear less deformed, suggesting that no further significant inversion events occurred during the Cretaceous in this particular area of the Dutch Central Graben.

In the Cenozoic North Sea Group (NS), further vertical growth of salt structures F05F08-WEST1 and F09-WEST1 (Fig. 6h) is noted, indicated by a general thinning of the NS interval towards the axis of the Dutch Central Graben and above these salt structures (Fig. 4). Thinning of the pre-Miocene Lower North Sea Group sediments in this section also indicates another phase of inversion. Late Miocene–Pleistocene sediments exhibit stratigraphic thinning in anticline salt structures F05F08-WEST1 and F09-WEST1, suggesting a more recent reactivation of vertical salt structure growth.

Discussion

Considerations of the presented restoration models are discussed in this section, including inherent assumptions related to this restoration method and the limitations of the input data. Additionally, the main uncertainties and shortcomings of this model, related to the local geology and modelling approach, are highlighted.

Restoration model considerations

Whilst there are some indications of Early Triassic active tectonism towards the northern Step Graben (e.g. near salt structure A12-EAST2) and along the eastern Dutch Central Graben boundary fault, there are no clear indications of active rifting and associated salt movement in this period across the majority the study area. Faults, although mostly detached, can typically be correlated with active deeper basement faults. For example, major fault movements are suggested at the eastern Dutch Central Graben boundary fault (Figs 4 & 5). At the present day, this fault is observed to link from basement into the overburden through the salt cover, but it is unclear if this was the case in Late Triassic times. The restoration model assumes a consistent structural style across the Triassic basin and, therefore, models the fault as detached. It is likely that, at this stage, the location of salt structures was already determined and these Late Triassic salt pillows have a generic relationship to the salt structures present within the Dutch Central Graben today.

Zechstein salt thickness through time is interpolated between two end members: estimated thickness during the Early Triassic (1500–2000 m: Ten Veen *et al.* 2012) and present-day salt thickness. A reduction in salt volume is assumed to have been most intense in periods during which Zechstein salt was exposed to the surface (Early Jurassic and Cretaceous: Fig. 6c, e, f) and in periods during which most salt movement out of the plane of section is expected (Early Jurassic–Early Cretaceous: Fig. 6c, d, e).

The development of a secondary rim syncline during deposition of the Altena interval is interpreted as the piercing stage of the F09-WEST1 diapir (Fig. 6d). However, this creation of accommodation space can also be attributed to tectonic subsidence. The restoration model associates AT sediment accommodation with salt withdrawal which is supported by active salt movement during this period (transition from Triassic pillowing to a fully pierced salt structure in the Late Jurassic). Other, tectonic subsidence models could be investigated to understand how these processes interact. Note that this study interpreted a gradual thickening of the Altena Group into the Dutch Central Graben in contrast to published concepts of homogeneous deposition (Doornenbal *et al.* 2010; after Ziegler 1990), based on its interpreted thickness in 3D seismic data (Figs 4 & 5) (Wijker 2014; van Winden 2015).

The observation that the Altena Group is only preserved within the Dutch Central Graben is explained in the restoration model by continued subsidence subsequent to deposition. The accommodation space for the overlying SL0 and SL1 intervals can, again, be attributed to either salt withdrawal or tectonic subsidence. In this case, tectonic subsidence is used in the model because of known regional Late Jurassic rift acceleration (Ziegler 1990; amongst others), and a continuous thick (>6 km: Fig. 4) pile of SL0 and SL1 sediments. Some salt withdrawal is, nevertheless, likely and is incorporated locally into the restoration model. As such, the model assumes that accommodation was generated by salt withdrawal in the Altena interval, with tectonic subsidence more dominant in the SL0 and SL1 intervals.

In the restoration model, platform areas are deeply eroded below the Base Cretaceous Unconformity. The precise depositional area of KN is uncertain, here it is assumed that the interval had an approximately constant thickness, but with localized thickening onto the Schill Grund Platform and within the Dutch Central Graben. This variability is based on the palaeogeographical maps of Ziegler (1982), which suggest a transition from paralic to deep-marine deposition from platform to graben (see also Zwaan 2018).

The restoration of CK0 involves an estimate of the amount of eroded Jurassic and Cretaceous strata associated with the Late Campanian inversion. The restored geometry and thickness of these intervals is important, since this affects every older restoration step. The decision on how much strata to restore is based on the most realistic geometry in the context of local geology and extrapolation of truncated reflectors above the Late Campanian Unconformity (by flattening seismic data at this level). This interpretation suggests a maximum erosion of 900 m in the centre of the Dutch Central Graben. Basin modelling, constrained by vitrinite reflectance data, indicates this amount of uplift and erosion is in a realistic range (700 m: de Jager 2003). Alternative models with a thicker missing section would increase the compaction of underlying strata in the preceding restoration steps, which would, in turn, increase depositional thickness estimates of Jurassic and Triassic strata.

The role of Triassic evaporite intervals (e.g. Röt evaporites) and their interaction (e.g. co-mobilization or intermingling) with Zechstein evaporites (Niebuhr *et al.* 1999; Maystrenko *et al.* 2006; Pharaoh *et al.* 2010) was not within the scope of this study. Dedicated modelling work is suggested to investigate a possible relationship between the timing of Triassic evaporite deposition and the presence of Zechstein evaporites at or near the surface. This could include investigation of a potential correlation between the formation of lateral Zechstein salt intrusions and the presence of Röt evaporites (see also Pharaoh *et al.* 2010).

Uncertainties and assumptions of the restoration model

It is acknowledged that all restoration models have inherent uncertainties and assumptions. The main limitation for this study is the use of 2D sections that do not take into account salt and tectonic movements outside or at an angle to the section of the plane. Additionally, strike-slip components have not been taken into account, even though oblique components are documented in the tectonic phases (most notably Late Cretaceous and Cenozoic inversion: e.g. Van Wijhe 1987; Dronkers & Mrozek 1991; Nalpas *et al.* 1995; de Jager 2007; Pharaoh *et al.* 2010).

The authors acknowledge that it is likely that salt moved in and out of the plane of the restored 2D section (e.g. F05F08-WEST1 and F09-WEST1). At the same time, the main depocentres in the restored section (e.g. on both sides of the F09-WEST1 structure) were probably affected by very significant salt withdrawal to areas out of the section plane. Further work could be considered to build 3D restorations of such structures.

The presented model is based on a depth-converted seismic interpretation but, nevertheless, shows simplified geometries. Due to limited deep well control, the time-to-depth conversion is poorly constrained for the oldest intervals and for the top of the pre-salt. As such, the depth of the base Zechstein in the deepest part of the graben (up to 9 km in the restoration model; see also Geluk 2005) is associated with considerable uncertainty (at least 500 m). Lithologies, and associated decompaction factors, assigned to the deeper intervals also carry uncertainty due to limited or no well control.

Impact of models on hydrocarbon play definition

The model presented in this study, for the timing and nature of salt movement, can give insight into various aspects of the petroleum system of the northern Dutch offshore. For example, this model could be applied to define the depositional and preserved extent of Jurassic (Posidonia/Kimmeridge) source rocks, to investigate the timing of subsequent hydrocarbon charge though salt windows or to understand the distribution of shallow gas occurrences above salt diapirs. Understanding rim-syncline development can also help with the definition of reservoir distribution in Upper Triassic, Jurassic and Cretaceous strata (see also Bouroullec, this volume, in press; Verreussel, this volume, in press; Zwaan 2018). Zechstein salt movement also impacts other plays in this area, including the Triassic Volpriehausen play (charge through salt windows from Carboniferous source rocks, salt-controlled traps and lateral seals, and the risk of halite reservoir cementation: e.g. Fontaine *et al.* 1993; Purvis & Okkerman 1996), the Chalk play (Jurassic source rock maturation, reservoir facies distribution and fracturing, and diaper-controlled four-way dip closure traps: e.g. Van der Molen *et al.* 2005; Grassmann *et al.* 2005; Verweij *et al.* 2009; van Lochem, this volume, in press) and Upper Jurassic plays (local source rock and reservoir burial in salt-withdrawal basins, reservoir sand distribution and trap formation: Abbink *et al.* 2006; Grassmann *et al.* 2005). To further understand the impact of salt tectonics on Mesozoic hydrocarbon systems and reservoirs, a similar restoration study on a local scale is suggested. This would, for example, allow a more detailed analysis of the local effects of salt tectonics on Jurassic source rocks (e.g. around salt structure F09-WEST1 and F05F08-WEST1). Additionally, it could give more insight into the impact of late stages of salt structure growth on Chalk Group sediment facies and fracturing. The integration of further well data and the development of a more detailed regional velocity model will reduce the uncertainties of structural restoration in future studies, which would allow for better constrained assumptions on lithological properties and time-to-depth conversion.

Conclusions

This paper describes a detailed structural restoration of a regional seismic section through the Step Graben and the Dutch Central Graben. Combined with a regional analysis of salt structures and depositional patterns, this allowed the development of a new model, based on recent 3D seismic data, for the interaction between salt movement, tectonic activity and sedimentation at different stages in the post-Permian history of the northern Dutch offshore.

It is observed that periods of salt tectonic development are closely linked to regional tectonic events. It follows that depositional patterns are controlled by an interplay between salt movement and active faulting. This is exhibited during widespread Late Triassic rifting, which induced detached faulting above a layer of Zechstein evaporites. Deposition of Upper Triassic sediments occurred in fault-bounded depocentres along these soft-linked faults, while salt pillows formed simultaneously along the regional structural grain (NNE–SSW). This resulted in elongated depocentres away from salt structures and the thinning of sediments over the salt pillow crests. Early Jurassic deposition of thick Altena Group sediments resulted in continued halokinesis and the subsequent localized formation of diapirs. Middle Jurassic thermal doming eroded much of the sediment on the higher platform areas, which is likely to include a significant volume of Zechstein salt. Tectonic subsidence related to a Late Jurassic rifting phase provided the primary mechanism for the creation of accommodation space in the Step Graben and Dutch Central Graben. Acceleration of salt withdrawal into isolated salt diapirs and salt walls provided a secondary mechanism to accommodate the thick Upper Jurassic Schieland Group sediments filling the graben during Late Jurassic. Distal effects of Late Campanian (Subhercynian) and Early Cenozoic (Laramide) inversion phases led to regional uplift in the Dutch Central Graben, and local vertical salt movement in salt diapirs and pillows. This had a direct effect on the depositional characteristics and structural geometry of the Upper Cretaceous Chalk Group sediments above salt diapirs.

This study demonstrates that an analysis with a strong focus on halokinetic development can provide new insights into the local effects of structuration, which has implications for the position of local depocentres and for hydrocarbon play elements such as source rock and reservoir deposition or preservation.

I would like to thank Energie Beheer Nederland (EBN) and TNO for facilitating this research. Fugro are also thanked for permission to use the 3D seismic data (DEFAB-survey, 2010) and Midland Valley Exploration are thanked for providing a MOVE restoration software licence. In addition, reviewers Harald de Haan and Herald Ligtenberg are thanked.

References

Abbink, O.A., Mijnlieff, H.F., Munsterman, D.K. & Verreussel, R.M.C.H. 2006. New stratigraphic insights in the 'Late Jurassic' of the Southern Central North Sea Graben and Terschelling Basin (Dutch Offshore) and related exploration potential Netherlands. *Journal of Geosciences*, **85**, 221–238.

Abdul Fattah, R., Verweij, J.M. & Witmans, N. 2012. *Reconstruction of Burial History, Temperature, Source Rock Maturity and Hydrocarbon Generation for the NCP-2D Area*. Dutch Offshore TNO Report TNO-034-UT-2010-0223. TNO (Geological Survey of The Netherlands), Utrecht, The Netherlands.

Arfai, J., Lutz, R., Franke, D., Gaedicke, C. & Kley, J. 2016. Mass-transport deposits and reservoir quality of Upper Cretaceous Chalk within the German Central Graben, North Sea International. *Journal of Earth Sciences*, **105**, 797–818.

Bachmann, G.H., Geluk, M.C. *et al.* 2010. Triassic. *In*: Doornenbal, J.C. & Stevenson, A.G. (eds) *Petroleum Geological Atlas of the Southern Permian Basin Area*. European Association of Geoscientists and Engineers (EAGE), Houten, The Netherlands, 149–173.

Back, S., van Gent, H., Reuning, L., Grötsch, J., Niederau, J. & Kukla, P. 2011. 3D seismic geomorphology and sedimentology of the Chalk Group, southern Danish North Sea. *Journal of the Geological Society, London*, **168**, 393–406, https://doi.org/10.1144/0016-76492010-047

Berthelsen, A. 1998. The Tornquist Zone northwest of the Carpathians: an intraplate pseudosuture. *Geologiska Föreningens i Stockholm Förhandlingar*, **120**, 223–230.

Breitkreuz, C., Geißler, M., Schneider, J. & Kiersnowski, H. 2008. Basin initiation: volcanism and sedimentation. *In*: Littke, R., Bayer, U., Gajewski, D. & Nelskamp, S. (eds) *Dynamics of Complex Intracontinental Basins: The Central European Basin System*. Springer, 173–180.

Bouroullec, R., Verreussel, R. *et al.* In press. Tectonostratigraphy of rift basins affected by halokinetics: the synrift Middle Jurassic–Lower Cretaceous in the Dutch Central Graben, Terschelling Basin and neighbouring platforms, Dutch offshore. *In*: Kilhams, B., Kukla, P.A., Mazur, S., McKie, T., Mijnlieff, H.F. & van Ojik, K. (eds) *Mesozoic Resource Potential of the Southern Permian Basin*. Geological Society, London, Special Publications, **469**, https://doi.org/10.1144/SP469.22

Cocks, L.R.M. & Torsvik, T.H. 2006. European geography in a global context from the Vendian to the end of the Palaeozoic. *In*: Gee, D. & Stephenson, R. (eds) *European Lithosphere Dynamics*. Geological Society, London, Memoirs, **32**, 83–95, https://doi.org/10.1144/GSL.MEM.2006.032.01.05

Cope, J.C.W., Ingham, J.K. & Rawson, P.F. (eds). 1992. *Atlas of Palaeogeography and Lithofacies*. Geological Society, London, Memoirs, **13**, https://doi.org/10.1144/GSL.MEM.1992.012.01.16

Copestake, P., Sims, A.P., Crittenden, S., Hamar, G.P., Ineson, J.R., Rose, P.T. & Tringham, M.E. 2003. Lower Cretaceous. *In*: Evans, D., Graham, C., Armour, A. & Bathurst, P. (eds) *The Millennium Atlas: Petroleum Geology of the Central and Northern North Sea*. Geological Society, London, 191–211.

Davison, I., Alsop, I. *et al.* 2000*a*. Geometry and late-stage structural evolution of Central Graben salt diapirs. *North Sea Marine and Petroleum Geology*, **17**, 499–522.

Davison, I., Alsop, I., Evans, N.G. & Safaricza, M. 2000*b*. Overburden deformation patterns and mechanisms of salt diaper penetration in the Central Graben. *North Sea Marine and Petroleum Geology*, **17**, 601–618.

de Jager, J. 2003. Inverted basins in the Netherlands, similarities and differences. *Netherlands Journal of Geosciences*, **92**, 355–366.

de Jager, J. 2007. Geological development. *In*: Wong, T., Batjes, D.A. & de Jager, J. (eds) *Geology of the Netherlands*. Royal Netherlands Academy of Arts and Sciences, Amsterdam, 5–26.

de Jager, J. 2012. The discovery of the Fat Sand Play (Solling Formation, Triassic), Northern Dutch offshore: a case of serendipity Netherlands. *Journal of Geosciences*, **91**, 609–619.

Doornenbal, J.C., Abbink, O.A. *et al.* 2010. Introduction, stratigraphic framework and mapping. *In*: Doornenbal, J.C. & Stevenson, A.G. (eds) *Petroleum Geological Atlas of the Southern Permian Basin Area*. EAGE, Houten, 1–9.

Dronkers, A.J. & Mrozek, F.J. 1991. Inverted basins of The Netherlands. *First Break*, **9**, 409–425.

Dronkert, H., Nio, S.D., Kouwe, W., Van der Poel, N. & Baumfalk, Y. 1989. *Buntsandstein of the Netherlands Offshore. Non-Exclusive Study*. Intergeos BV, Leiderdorp.

Duffy, O.B., Gawthorpe, R.L., Docherty, M. & Broncklehorst, S.H. 2013. Mobile evaporite controls on the structural style and evolution of rift basins: Danish Central Graben, North Sea. *Basin Research*, **25**, 310–330.

Duin, E.J.T., Doornenbal, J.C., Rijkers, R.H.B., Verbeek, J.W. & Wong, T.E. 2006. Subsurface structure of the Netherlands; results of recent onshore and offshore mapping. *Netherlands Journal of Geosciences*, **85**, 245–276.

Esmerode, E.V., Lykke-Andersen, H. & Surlyk, F. 2008. Interaction between bottom currents and slope failure in the Late Cretaceous of the southern Danish Central Graben, North Sea. *Journal of the Geological Society, London*, **165**, 55–72, https://doi.org/10.1144/0016-76492006-138

Farmer, C.L. & Barkved, O.I. 1999. Influence of syndepositional faulting on thickness variations in chalk reservoirs – Valhall and Hod fields. *In*: Fleet, A.J. & Boldy, S.A.R. (eds) *Petroleum Geology of Northwest Europe: Proceedings of the 5th Conference*. Geological

Society, London, 949–957, https://doi.org/10.1144/0050949

Feldman-Olszewska, A. 2006. Sedimentary environment of the Middle Jurassic black shales from the central part of Polish Basin. *In*: Matyja, B., Pieńkowski, G. & Wierzbowski, A. (eds) *Sediment 2006. 21th Meeting of Sedimentologists and 4th Meeting of SEPM Central European Section*, June 6–11, 2006, Göttingen, Germany; Abstracts and Field Trips. Deutsche Gesellschaft für Geowissenschaften, Hannover, Germany, 1–67.

Feist-Burkhardt, S., Götz, A.E., Szulc, J., Borkhataria, R., Geluk, M., Haas, J. & Nawrocki, J. 2008. Triassic. *In*: McCann, T. (ed.) *The Geology of Central Europe. Volume 2: Mesozoic and Cenozoic*. Geological Society, London, 749–822.

Fontaine, J.M., Guastella, G., Jouault, P. & de la Vega, P. 1993. F15-A: a Triassic gas field on the eastern limit of the Dutch Central Graben. *In*: Parker, J.R. (ed.) *Petroleum Geology of Northwest Europe: Proceedings of the 4th Conference*. Geological Society, London, 583–593, https://doi.org/10.1144/0040583

Gast, R.E., Dusar, M. *et al.* 2010. Rotliegend. *In*: Doornenbal, J.C. & Stevenson, A.G. (eds) *Petroleum Geological Atlas of the Southern Permian Basin Area*. European Association of Geoscientists and Engineers (EAGE), Houten, The Netherlands, 101–121.

Geluk, M.C. 2005. *Stratigraphy and tectonics of Permo-Triassic basins in the Netherlands and surrounding areas*. PhD thesis, Utrecht University, Faculty of Geosciences, Utrecht, The Netherlands.

Goffey, G., Attree, M. *et al.* 2016. New exploration discoveries in a mature basin: offshore Denmark Geological Society. *In*: Bowman, M. & Levell, B. (eds) *Petroleum Geology of NW Europe: 50 Years of Learning – Proceedings of the 8th Petroleum Geology Conference*. Geological Society, London. First published online September 26, 2016, https://doi.org/10.1144/PGC8.1

Grassmann, S., Cramer, B., Delisle, G., Messner, J. & Winsemann, J. 2005. Geological history and petroleum system of the Mittelplate oil field, Northern Germany. *International Journal of Earth Sciences*, **94**, 979–989.

Hampton, M.J., Bailey, H.W. & Jones, A.D. 2010. A holostratigraphic approach to the chalk of the North Sea Eldfisk Field, Norway. *In*: Vining, B.A. & Pickering, S.C. (eds) *Petroleum Geology: from Mature Basins to New Frontiers – Proceedings of the 7th Petroleum Geology Conference*. Geological Society, London, 473–492, https://doi.org/10.1144/0070473

Hernandez, K., Mitchell, N.C. & Huuse, M. In press. Deriving relationships between diapir spacing and salt layer thickness in the Southern North Sea. *In*: Kilhams, B., Kukla, P.A., Mazur, S., McKie, T., Mijnlieff, H.F. & van Ojik, K. (eds) *Mesozoic Resource Potential of the Southern Permian Basin*. Geological Society, London, Special Publications, **469**, https://doi.org/10.1144/SP469.15

Hodgson, N.A., Farnsworth, J. & Fraser, A.J. 1992. Salt related tectonics, sedimentation and hydrocarbon plays in the Central Graben North Sea UKCS. *In*: Hardman, R.F.P. (ed.) *Exploration Britain: Geological Insights for the Next Decade*. Geological Society, London, Special Publications, **67**, 31–63, https://doi.org/10.1144/GSL.SP.1992.067.01.03

Hossack, J.R. 1995. Geometrical rules of section balancing of salt structures. *In*: Jackson, M.P.A., Roberts, D.G. & Snelson, S. (eds) *Salt Tectonics: A Global Perspective*. American Association of Petroleum Geologists, Memoirs, **65**, 29–40.

Husmo, T., Hamar, G.P., Hoiland, O., Johannessen, E.P., Romuld, A., Spencer, A.M. & Titterton, R. 2002. Lower and middle Jurassic. *In*: Evans, D., Graham, C., Armour, A. & Bathurst, P. (eds) *The Millenium Atlas: Petroleum Geology of the Central and Northern North Sea*. Geological Society, London, 129–155.

Ineson, J.R. & Surlyk, F. (eds). 2003. *The Jurassic of Denmark and Greenland. Geological Survey of Denmark and Greenland Bulletin*, **1**.

Jenyon, M.K. & Taylor, J.C.M. 2005. Dissolution effects and reef-like features in the Zechstein across the Mid North Sea High. *In*: Peryt, T.M. (ed.) *The Zechstein Facies in Europe*. Lecture Notes in Earth Sciences, **10**. Springer, Berlin, 51–75.

Kombrink, H., Doornenbal, J.C., Duin, E.J.T., den Dulk, M., ten Veen, J.H. & Witmans, N. 2012. New insights into the geological structure of the Netherlands; results of a detailed mapping project Netherlands. *Journal of Geosciences*, **91**, 419–446.

Krawczyk, C.M., McCann, T., Cocks, L.R.M., England, R., McBride, J. & Wybraniec, S. 2008. Caledonian tectonics. *In*: McCann, T. (ed.) *The Geology of Central Europe. Volume 1: Precambrian and Palaeozoic*. Geological Society, London, 301–381.

Kroner, U., Mansy, J.-L. *et al.* 2008. Variscan tectonics. *In*: McCann, T. (ed.) *The Geology of Central Europe. Volume 1: Precambrian and Palaeozoic*. Geological Society. London, 559–664.

Littke, R., Bayer, U., Gajewski, D. & Nelskamp, S. (eds). 2008. *Dynamics of Complex Intracontinental Basins. The Central European Basin System*. Springer, Berlin.

Lott, G.K., Wong, T.E., Dusar, M., Andsbjerg, J., Mönnig, E., Feldman-Olszewska, A. & Verreussel, R.M.C.H. 2010. Jurassic. *In*: Doornenbal, J.C. & Stevenson, A.G. (eds) *Petroleum Geological Atlas of the Southern Permian Basin Area*. European Association of Geoscientists and Engineers (EAGE), Houten, The Netherlands, 175–193.

Magri, F., Littke, R., Rodon, S., Bayer, U. & Urai, J.L. 2008. Temperature fields, petroleum maturation and fluid flow in the vicinity of salt domes. *In*: Littke, R., Bayer, U., Gajewski, D. & Nelskamp, S. (eds) *Dynamics of Complex Intracontinental Basins: The Central European Basin System*. Springer, 323–330.

Matthews, W.J., Hampson, G.J., Trudgill, B.D. & Underhill, J.R. 2007. Controls on fluviolacustrine reservoir distribution and architecture in passive salt-diapir provinces: Insights from outcrop analogs. *AAPG Bulletin*, **91**, 1367–1403.

Maystrenko, Y., Bayer, U. & Scheck-Wenderoth, M. 2006. 3D reconstruction of salt movements within the deepest post-Permian structure of the Central European Basin System – the Glueckstadt Graben Netherlands. *Journal of Geosciences*, **85**, 181–196.

Mijnlieff, H.F. & Geluk, M. 2011. Palaeotopography-governed sediment distribution – a new predictive model for the Permian Upper Rotliegend in the

Dutch sector of the Southern Permian Basin. *In*: Grötsch, J. & Gaupp, R. (eds) *The Permian Rotliegend of the Netherlands*. SEPM, Special Publications, **98**, 147–159.

Nalpas, T., Le Douaran, S., Brun, J.P., Unternehr, P. & Richert, J.P. 1995. Inversion of the Broad Fourteens Basin, a small scale model investigation. *Sedimentary Geology*, **95**, 237–250.

Niebuhr, B., Baldschuhn, R., Ernst, G., Walaszczyk, I., Weiss, W. & Wood, C.J. 1999. The Upper Cretaceous succession (Cenomanian–Santonian) of the Staffhorst Shaft, Lower Saxony, northern Germany: integrated biostratigraphic, lithostratigraphic and downhole geophysical log data. *Acta Geologica Polonica*, **49**, 175–213.

Nøttvedt, A., Gabrielsen, R.H. & Steel, R.J. 1995. Tectonostratigraphy and sedimentary architecture of rift basins, with reference to the northern North Sea. *Marine and Petroleum Geology*, **12**, 881–901.

Peryt, T.M., Geluk, M.C., Mathiesen, A., Paul, J. & Smith, K. 2010. Zechstein. *In*: Doornenbal, J.C. & Stevenson, A.G. (eds) *Petroleum Geological Atlas of the Southern Permian Basin Area*. European Association of Geoscientists and Engineers (EAGE), Houten, The Netherlands, 123–147.

Pharaoh, T., Dusar, M. *et al.* 2010. Tectonic evolution. *In*: Doornenbal, H. & Stevenson, A. (eds) *Petroleum Geological Atlas of the Southern Permian Basin Area*. European Association of Geoscientists and Engineers (EAGE), Houten, The Netherlands, 24–57.

Pieńkowski, G., Schudack, M.E., Bosak, P., Enay, R., Feldman-Olszewska, A., Golonka, J. & Lathuiliere, B. 2008. Jurassic. *In*: McCann, T. (ed.) *The Geology of Central Europe. Volume 2: Mesozoic and Cenozoic*. Geological Society, London, 823–922.

Purvis, K. & Okkerman, J.A. 1996. Inversion of reservoir quality by early diagenesis: an example from the Triassic Buntsandstein, offshore the Netherlands. *In*: Rondeel, H.E., Batjes, D.A.J. & Nieuwenhuijs, W.H. (eds) *Geology of Gas and Oil Under the Netherlands*. Kluwer Academic, Dordrecht, The Netherlands, 179–190.

Reicherter, K., Froitzheim, N. *et al.* 2008. Alpine tectonics north of the Alps. *In*: McCann, T. (ed.) *The Geology of Central Europe*. Geological Society, London, 1233–1295.

Remmelts, G. 1995. Fault-related salt tectonics in the southern North Sea, the Netherlands. *In*: Jackson, M.P.A., Roberts, D.G. & Snelson, S. (eds) *Salt Tectonics: A Global Perspective*. American Association of Petroleum Geologists, Memoirs, **65**, 261–272.

Rowan, M.G. & Ratliff, R.A. 2012. Cross-section restoration of salt-related deformation: Best practices and potential pitfalls. *Journal of Structural Geology*, **41**, 24–37.

Scheck-Wenderoth, M., Maystrenko, Y.P., Hansen, M. & Mazur, S. 2008. Dynamics of salt basins. *In*: Littke, R., Bayer, U., Gajewski, D. & Nelskamp, S. (eds) *Dynamics of Complex Intracontinental Basins. The Central European Basin System*. Springer, Berlin, 307–322.

Słowakiewicz, M., Tucker, M.E., Pancost, R.D., Perri, E. & Mawson, M. 2013. Upper Permian (Zechstein) microbialites: Supratidal through deep subtidal deposition, source rock, and reservoir potential. *AAPG Bulletin*, **97**, 1921–1936.

Stewart, S.A. 2007. Salt tectonics in the North Sea Basin: a structural template for seismic interpreters. *In*: Ries, A. C., Butler, R.W.H. & Graham, R.H. (eds) *Deformation of the Continental Crust: The Legacy of Mike Coward*. Geological Society, London, Special Publications, **272**, 361–396, https://doi.org/10.1144/GSL.SP.2007.272.01.19

Stollhofen, H., Bachmann, G.H., Barnasch, J., Bayer, U., Beutler, G., Franz, M. & Radies, D. 2008. Upper Rotliegend to Early Cretaceous basin development. *In*: Littke, R., Bayer, U., Gajewski, D. & Nelskamp, S. (eds) *Dynamics of Complex Intracontinental Basins. The Central European Basin System*. Springer, Berlin, 181–210.

Surlyk, F., Dons, T., Clausen, C.K. & Higham, J. 2003. Upper Cretaceous. *In*: Evans, D., Graham, C., Armour, A. & Bathurst, P. (eds) *The Millennium Atlas: Petroleum Geology of the Central and Northern North Sea*. Geological Society, London, 213–233.

Taylor, J.C.M. 1998. *Upper Permian–Zechstein*. *In*: Glennie, K.W. (ed.) Petroleum Geology of the North Sea, 4th edn. Blackwell Science, Oxford, 174–212.

Ten Veen, J.H., van Gessel, S.F. & den Dulk, M. 2012. Thin- and thick-skinned salt tectonics in the Netherlands; a quantitative approach. *Netherlands Journal of Geosciences*, **91**, 447–464.

Timmerman, M.J., Heeremans, M., Kirstein, L.C., Larsen, B.L., Spencer-Dunworth, E.A. & Sundvoll, B. 2009. Linking changes in tectonic style with magmatism in northern Europe during the late Carboniferous to latest Permian. *Tectonophysics*, **473**, 375–390.

Tolsma, S. 2014. *Seismic characterization of the Zechstein carbonates in the Dutch northern offshore*. MSc thesis, Utrecht University, Faculty of Geosciences, Utrecht, The Netherlands.

Trusheim, F. 1960. Mechanisms of salt migration in northern Germany. *AAPG Bulletin*, **44**, 1519–1540.

Underhill, J.T. & Partington, M.A. 1993. Jurassic thermal doming and deflation in the North Sea: implications of the sequence stratigraphic evidence. *In*: Parker, J.R. (ed.) *Petroleum Geology of Northwest Europe – Proceedings of the 4th Conference*. Geological Society, London, 337–345, https://doi.org/10.1144/0040337

Van Adrichem Boogaert, H.A. & Kouwe, W.F.P. 1994. Stratigraphic nomenclature of the Netherlands; revision and update by RGD and NOGEPA, Section D Permian. *Mededelingen Rijks Geologische Dienst*, **50**, 42.

van Dalfsen, W., Doornenbal, J.C., Dortland, S. & Gunnink, J.L. 2006. A comprehensive seismic velocity model for the Netherlands based on lithostratigraphic layers. *Netherlands Journal of Geosciences*, **85**, 277–292.

van den Belt, F.J.G. & de Boer, P.L. 2007. A shallow-basin model for 'saline giants' based on isostasy driven subsidence. *In*: Nichols, G., Williams, E. & Paola, C. (eds) *Sedimentary Processes, Environments and Basins – A Tribute to Peter Friend*. International Association of Sedimentologists, Special Publications, **38**, 241–252.

Van Der Baan, D. 1990. Zechstein reservoirs in The Netherlands. *In*: Brooks, J. (ed.) *Classic Petroleum Provinces*, Geological Society, London, SpecialPublications, **50**, 379–398, https://doi.org/10.1144/GSL.SP.1990.050.01.24

Van Der Molen, A.S., Dudok, H.W. & Wong, T.E. 2005. The influence of tectonic regime on chalk deposition: examples of the sedimentary development and 3D stratigraphy of the chalk Group in the Netherlands offshore area. *Basin Research*, **17**, 63–81.

van Lochem, H. In press. F17-Chalk: new insights in the tectonic history of the Dutch Central Graben. *In*: Kilhams, B., Kukla, P.A., Mazur, S., McKie, T., Mijnlieff, H.F. & van Ojik, K. (eds) *Mesozoic Resource Potential of the Southern Permian Basin*. Geological Society, London, Special Publications, **469**. https://doi.org/10.1144/SP469.17

Van Wijhe, D.H. 1987. Structural evolution of inverted basins in the Dutch offshore. *Tectonophysics*, **137**, 171–219.

van Winden, M.E. 2015. Salt tectonics in the northern Dutch offshore. MSc thesis, Utrecht University, Faculty of Geosciences, Utrecht, The Netherlands.

Vendeville, B.C. 2002. A new interpretation of Trusheim's classic model of salt diapir growth. *Transactions of the Gulf Coast Association of Geological Societies*, **52**, 943.

Verreussel, R.M.C.H., Bouroullec, R. et al. In press. Stepwise basin evolution of the Middle Jurassic–Early Cretaceous rift phase in the Central Graben area of Denmark, Germany and The Netherlands. *In*: Kilhams, B., Kukla, P.A., Mazur, S., McKie, T., Mijnlieff, H.F. & van Ojik, K. (eds) *Mesozoic Resource Potential of the Southern Permian Basin*. Geological Society, London, Special Publications, **469**, https://doi.org/10.1144/SP469.23

Verweij, J.M., Souto Carneiro Echternach, M. & Witmans, N. 2009. *Terschelling Basin and southern Dutch Central Graben Burial History, Temperature, Source Rock Maturity and Hydrocarbon Generation – Area 2A*. TNO Report TNO-034-UT-2009-02065. TNO (Geological Survey of The Netherlands), Utrecht, The Netherlands.

Voigt, S., Wagreich, M. et al. 2008. Cretaceous. *In*: McCann, T. (ed.) *The Geology of Central Europe. Volume 2: Mesozoic and Cenozoic*. Geological Society, London, 923–997.

Wijker, D. 2014. *Fault Mapping and Reconstruction of the Structural History of the Dutch. Central Graben.* MSc thesis (EBN), Vrije Universiteit Amsterdam, Amsterdam, The Netherlands.

Wilson, M., Neumann, E.-R., Davies, G.R., Timmerman, M.J., Heeremans, M. & Larsen, B.T. 2004. Permo-Carboniferous magmatism and rifting in Europe: introduction. *In*: Wilson, M., Neumann, E.-R., Davies, G.R., Timmerman, M.J., Heeremans, M. & Larsen, B.T. (eds) *Permo-Carboniferous Magmatism and Rifting in Europe*. Geological Society, London, Special Publications, **223**, 1–10, https://doi.org/10.1144/GSL.SP.2004.223.01.01

Worum, G. & Michon, L. 2005. Implications of continuous structural inversion in the West Netherlands Basin for understanding controls on Palaeogene deformation in NW Europe. *Journal of the Geological Society, London*, **162**, 73–85, https://doi.org/10.1144/0016-764904-011

Wride, V.C. 1995. Structural features and structural styles from the Five Countries Area of the North Sea Central Graben. *First Break*, **13**, 395.

Zanella, E. & Coward, M.P. 2003. Structural framework. *The Millennium Atlas: Petroleum Geology of the Central and Northern North Sea*, **357**, 45–59.

Ziegler, P.A. 1982. Triassic rift and facies patterns in western and central Europe. *Geologische Rundschau*, **71**, 747–772.

Ziegler, P.A. 1988. *Evolution of the Arctic-North Atlantic and Western Tethys*. American Association of Petroleum Geologists, Memoirs, **43**.

Ziegler, P.A. 1990. Tectonic and palaeogeogaphic development of the North Sea Rift system. *In*: Blundell, D.J. & Gibbs, A.D. (eds) *Tectonic Evolution of the North Sea Rifts*. Oxford Science, Oxford, 1–36.

Ziegler, P.A. 1992. North Sea rift system. *Tectonophysics*, **208**, 55–75.

Ziegler, P.A. & Dèzes, P. 2007. Cenozoic uplift of Variscan Massifs in the Alpine foreland: Timing and controlling mechanisms. *Global and Planetary Change*, **58**, 237–269.

Zwaan, F. 2018. Lower Cretaceous reservoir development in the North Sea Central Graben, and analogue settings in the Southern Permian Basin and South Viking Graben. *In*: Kilhams, B., Kukla, P.A., Mazur, S., McKie, T., Mijnlieff, H.F. & van Ojik, K. (eds) *Mesozoic Resource Potential of the Southern Permian Basin*. Geological Society, London, Special Publications, **469**. First published online January 4, 2018, https://doi.org/10.1144/SP469.3

Deriving relationships between diapir spacing and salt-layer thickness in the Southern North Sea

KARINA HERNANDEZ*, NEIL C. MITCHELL & MADS HUUSE

School of Earth and Environmental Sciences, University of Manchester, Williamson Building, Oxford Road, Manchester M13 9PL, UK

**Correspondence: karina.hernandez@manchester.ac.uk*

Abstract: In analytical models of salt diapirism, the initial salt-layer thickness and the post-deformation spacing of salt structures are key parameters. Here, 3D seismic data from The Netherlands offshore has enabled these parameters to be measured over large areas which can then be compared with model predictions. Estimates of the original salt-layer thickness were obtained by spatially filtering present thickness, using filters with varied spatial scales that remove local effects. Loss of evaporite minerals by dissolution or erosion during exposure, cannot be ruled out and, as such, thicknesses are minima. Spacing estimates were derived in two dimensions by locating the minimum separation of lines representing ridgelines of diapirs/walls. Because the length scale of spatial filtering was chosen based on the dependent variable, diapir spacing, the results are non-unique. Nevertheless, choosing an apparently optimal filter length of 50 km, a ratio between diapir spacing and original thickness from 12 to 20 is defined. This ratio is greater than has been reported for the pillow province of the UK North Sea Quadrant 44, which is as expected if pillows evolve into diapirs with progressive halokinetic deformation. This work is key to understanding the evolution of salt displacement, a necessity for unlocking remaining hydrocarbon resources.

Various hydrocarbon plays in the Central and Southern North Sea rely on precise descriptions of both salt-thickness and salt-movement history. For example, salt-structure evolution affects adjacent salt-basin geometry containing potential reservoirs (Stewart & Clark 1999). In the North Sea, deformation caused by salt displacement has formed potential dip-closed traps for hydrocarbons in overlying reservoirs (Glennie 1997), as well as turtleback anticline structures (Fontaine *et al.* 1993). A correct understanding of salt thickness also impacts structural reconstruction of the pre-salt section via accurate velocity modelling and is a key parameter in successful hydrocarbon exploration in the Southern Permian Basin (Ten Veen *et al.* 2012). Additionally, salt diapirs influence the temperature fields of sedimentary basins, affecting the maturity of organic matter and hydrocarbon generation (Dronkert & Remmelts 1996). Moreover, the Zechstein salt deposits are a commercial resource for gypsum, anhydrite, halite and potash salts, and have been considered for the storage of hydrocarbons, chemicals and waste.

Early models of salt diapirism considered deformation to be a result of a Rayleigh–Taylor (R–T) gravitational instability, arising from density differences between the overburden and the less dense salt layer (Nettleton 1934; Ramberg 1981). It should be noted that clastic sediments are less dense than halite until they are buried. This burial depth varies from basin to basin; in the Gulf of Mexico, the depth at which the bulk density of the overburden equals that of salt is estimated to be 1500 m (Nelson & Fairchild 1989). However, this is not necessarily valid for the North Sea; a review of Hamilton (1976), describing sediment burial compaction, shows densities approaching halite densities within only a few hundred metres of burial. In R–T models, the sedimentary and salt-rock layers were treated as fluids of differing density (Biot & Odé 1965; Ramberg 1972, 1981; Jackson *et al.* 1990), with the layers assumed to be initially at equilibrium. Perturbations between these layers break the equilibrium, causing the less dense material to rise up through the denser layer. Disturbances between the layers may be induced by faulting, folding, differential erosion or episodic sedimentation, for example. R–T models predict that the spacing of salt diapirs or other salt structures should be proportional to the original thickness of the salt layer; hence, both spacing and original thickness are important parameters. Turcotte & Schubert (1982) suggested that the tectonically or depositionally derived disturbance that produces a faster rise of less dense material dominates the subsequent instability. The growth time is determined by the wavelength, λ, which in the model described by Turcotte & Schubert (1982) is expected

From: Kilhams, B., Kukla, P. A., Mazur, S., McKie, T., Mijnlieff, H. F. & van Ojik, K. (eds) 2018. *Mesozoic Resource Potential in the Southern Permian Basin*. Geological Society, London, Special Publications, **469**, 119–137.
First published online February 26, 2018, https://doi.org/10.1144/SP469.16

to be 2.568 times the original salt thickness for rapid rise.

More recent studies have considered the overburden to be rigid or brittle (Vendeville 1989; Jackson & Vendeville 1994; Talbot 1995) and suggested that buoyancy by itself may not be sufficient to drive salt diapirism through that overburden. However, buoyancy is still an important influence during the development of a salt structure (i.e. during active diapirism, the rise of the diapir is driven by salt buoyancy relative to its host rocks: Hudec & Jackson 2007). Furthermore, field evidence, such as folds in sedimentary rocks and salt structures with periodic spacings, suggest that viscous-like behaviour of the overburden can occur. As such, it is possible that the overburden can deform with both brittle and viscous characteristics, depending on spatial and temporal scales and the rates of deformation (Poliakov *et al.* 1996).

A more recent study by Ismail-Zadeh *et al.* (2002) modified the R–T model with a perfectly plastic overburden layer. It was proposed that the characteristic wavenumber (inverse of the wavelength) of salt structures decreases with increasing thickness ratio (given by the thickness of the overburden in relation to the initial salt-layer thickness) but then increases by a series of abrupt jumps when the thickness ratio reaches a limit given by the viscosity ratio between the salt and sedimentary layers. Ismail-Zadeh *et al.* (2002) also analysed the possible effects of extension and compression within a developing salt structure. This is of particular interest as the Southern North Sea is an area that has undergone several phases of tectonic extension and compression, which are likely to have affected salt tectonics.

Previous studies of salt displacement of the Late Permian Zechstein in Germany (Jaritz 1973, cited in Rönniund 1989), on the UK side of the Southern North Sea (Hughes & Davison 1993) and in the Hormuz salt of Abu Dhabi (O'Brien 1957; Colman-Sadd 1978; Hudec & Jackson 2007) have reported relationships between initial salt-layer thickness and the distance between adjacent diapirs. However, little systematic investigation has been published regarding the relationship between salt thickness and the distance between salt structures within NW Europe.

In the current study, a relationship between wavelength and initial salt thickness was developed from data of the Dutch sector of the Southern North Sea, which is one of the best-characterized areas globally thanks to several decades of hydrocarbon exploration (cf. Doornenbal & Stevenson 2010). Whilst it is recognized that there is a large degree of uncertainty in these estimates, it is hoped that this study furthers previous work (e.g. Turcotte & Schubert 1982; Remmelts 1996). Finally, the values derived here are interpreted in the light of known geological processes and stratigraphy.

Regional geology

The study area is located in the Dutch sector of the Southern North Sea, over an area of approximately 38 000 km^2 including a variety of structural settings: Dutch Central Graben (DCG), Broad Fourteens Basin (BFB), Terschelling Basin (TB), Cleaverbank Platform (CP) and Central Offshore Platform (COP) (Remmelts 1996; Kombrink *et al.* 2012) (Fig. 1). During the Permian period, this area was part of an east–west-trending complex of sedimentary basins from the UK in the west to Poland in the east, referred to as the Southern Permian Basin (Ziegler 1990; Geluk 2007; Doornenbal & Stevenson 2010). Across this large basin system, evaporites were deposited throughout the Late Permian forming the Zechstein evaporitic group (Taylor 1990; Ziegler 1990; Geluk 2000, 2007; Doornenbal & Stevenson 2010). In most salt-tectonic literature, the term 'salt' is given to rocks composed mostly of halite, a convention adopted here; although, in the Zechstein Group, each of the five cycles typically comprises claystone (density 2–2.9 g m^{-3}), carbonate (density 2.3–2.7 g m^{-3}), anhydrite (density *c.* 2.98 g m^{-3}) and halite (density *c.* 2.17 g m^{-3}) (Van Adrichem Boogaert & Kouwe 1993–1997; Geluk 2005; Ten Veen *et al.* 2012). Therefore, salt structures may contain additional evaporitic minerals. The Zechstein evaporites played an important role during the later Mesozoic structural evolution of the basin, during which the ductile evaporite layer became mobilized, and reshaped into pillows, diapirs and walls (Ziegler 1990; de Jager 2007; Stewart 2007; Doornenbal & Stevenson 2010).

During the Early Triassic, regional extension in the North Sea formed the BFB and the Central Graben (Van Wijhe 1987). The principal axes of extension were then orientated north–south and NNE–SSW. Major subsidence of these graben subsequently occurred during the Late Jurassic–Early Cretaceous (Van Wijhe 1987). Major extension across the Central Graben occurred during the mid-Kimmeridgian (Heybroek 1975; Ziegler 1990; de Jager 2007; Doornenbal & Stevenson 2010; Kombrink *et al.* 2012). From then until the Early Cretaceous, subsidence of the West and Central Netherlands basins and the BFB, and uplift of the Texel–IJsselmeer High, the Cleaver Bank Platform and other platforms, occurred with movement on faults trending NW–SE to NNW–SSE. Erosion occurred on these exposed platforms, as indicated by the absence of Jurassic sediments defining the Mid-Cimmerian Unconformity (*c.* 140 Ma) (Brown 1984; Ziegler 1990; Duin *et al.* 2006; Doornenbal

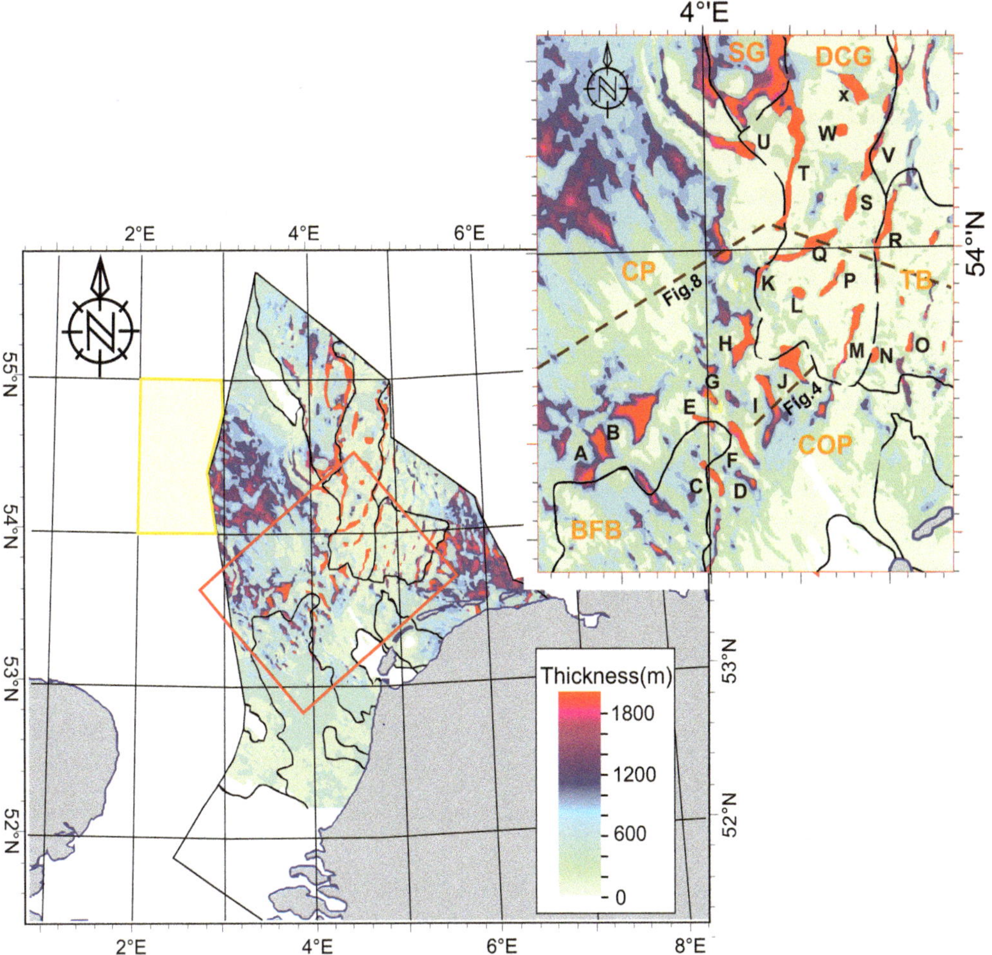

Fig. 1. Main structural elements and salt structures of the Dutch North Sea Basin, including the Broad Fourteens Basin (BFB), the Cleaverbank Platform (CP), the Central Offshore Platform (COP), the Terschelling Basin (TB) and the Dutch Central Graben (DCG). In the inset, the main salt structures studied here are labelled (A–W). Colour shading represents the total Zechstein thickness, varying from 0 to 1800 m. Black lines present the limits between the main structural elements. Blue lines represent coastlines. The orange-outlined rectangle locates the study area and the extent of the seismic coverage. The yellow rectangle locates the area studied by Hughes & Davison (1993) in the UK North Sea Quadrant 44. Dashed brown lines indicate the locations of profiles in Figures 4 and 8.

& Stevenson 2010; Kombrink *et al.* 2012). In basins such as the Central Graben, subsidence continued during the Jurassic, allowing deposition and preservation of a full stratigraphic sequence (Kooi *et al.* 1989; Ziegler 1990; de Jager 2007; Doornenbal & Stevenson 2010). During the Early–Late Cretaceous, regional compression caused basin inversion and reverse movement on the graben-bounding faults. In the Dutch Central Graben, uplift occurred in the centre of the basin (de Jager 2003, 2007) and overthrusting occurred in the BFB (Van Wijhe 1987). Subsequently, a phase of subsidence occurred from the Neogene onwards (Kooi *et al.* 1989; Ziegler 1990; Remmelts 1996; Doornenbal & Stevenson 2010).

Halokinetic processes in the Dutch Southern North Sea

The intensity and timing of salt displacement, salt-structure distribution and strike orientation all vary spatially between different structural domains in the Dutch Southern North Sea, and divergence is even observable between adjacent structures (Taylor

1984). In the Dutch Central Graben, intense displacement of salt has occurred, creating large-amplitude structures (Remmelts 1996; Doornenbal & Stevenson 2010; Ten Veen *et al.* 2012). This structural area is composed of mostly north–south-trending elongated salt walls, as well as isolated salt diapirs with no preferential orientation (Wride 1995; van Winden 2015). Meanwhile, the salt structures located near the southern end of the Dutch Central Graben and in the TB have NW–SE trends (Remmelts 1996; Harding & Huuse 2015). In previous research, the commencement of these salt structures has been assigned to the regional extension phase of the Triassic (228–203 Ma) (Remmelts 1995; Fattah *et al.* 2012; Kombrink *et al.* 2012). Further inversion phases, originating from compressive movements during the Late Cretaceous and Cenozoic, affected the Dutch Central Graben (Heybroek 1975; Kombrink *et al.* 2012), reactivating salt displacement (Ten Veen *et al.* 2012).

In platform areas, such as the CP, the salt has moved, forming salt pillows and other diapiric structures at the edge of the platform (Remmelts 1996; Ten Veen *et al.* 2012).The orientation of salt structures associated with the CP trend mainly NW–SE (Remmelts 1996; Chen 2016). The occurrence of salt displacement here has been suggested to initially commence between the Middle Triassic and Early Cretaceous (Remmelts 1996; Ten Veen *et al.* 2012). Further phases of displacement then occurred during the Late Cretaceous and Eocene–Oligocene periods in which major fold structures, cored by salt pillows, were initiated or amplified (Hughes & Davison 1993; Stewart & Coward 1995; Remmelts 1996; Ten Veen *et al.* 2012).

According to Stewart & Coward (1995), Remmelts (1996) and Chen (2016), most salt structures in the Southern Permian Basin, in particular those with diapiric walls, occur in three different spatial groupings, each with differing structural trends. The first group comprises north–south-trending salt structures, distributed between the North Sea Central Graben and the onshore area of northern Germany. The second group, consisting of pillows and walls and trending northwesterly, is located in the west of the basin, extending from UK waters to onshore Netherlands. The third group, consisting of low salt walls and pillows, has an eastwards trend and is located onshore in Germany.

In the Dutch North Sea, faults located beneath the salt layer have alignment directions varying between east–west and NNE–SSW, although the dominant orientation is NW–SE (Remmelts 1996; Fattah *et al.* 2012; Kombrink *et al.* 2012; Ten Veen *et al.* 2012). The coincidence of episodes of intense salt displacement and acute tectonic phases, as well as a close relationship between salt-structure orientation and the structural grain, has been described by Koyi & Petersen (1993), Remmelts (1995, 1996), Scheck *et al.* (2003), Mohr *et al.* (2005), Maystrenko *et al.* (2005) and Ten Veen *et al.* (2012). Meanwhile, in the TB, salt structures do not coincide with subsalt faults. It has been suggested, therefore, that basement extension has only indirectly affected diapirism by generating space in the overburden (Harding & Huuse 2015). Another mechanism of salt-diapir formation has been proposed by Coward & Stewart (1995) and Buchanan *et al.* (1996), who described salt pillows as buckle folds due to compression and down-building, respectively. In summary, the mechanisms initiating salt diapirism alongside buoyancy in the Dutch North Sea include crustal movement during the main tectonic phases and differential loading, often localized by basement faults.

After salt displacement begins, salt structures typically develop in three stages: (1) the layered stage; (2) the pillow stage; and, (3) the piercing stage (Trusheim 1960; Vendeville 2002). During these stages, salt movement interacts with ongoing sediment deposition as the salt thinning from around the salt structure creates space for sediment accumulation and the rising salt structure thins the overburden. The distinct phases of salt piercement are divided into those referred to as: 'reactive', when the diapir reacts in response to the extension of the overburden; 'active', when the diapir actively intrudes over a relatively thin overburden, a process which is mainly driven by differential loading; and 'passive', where the crest of the diapir reaches the surface (Hudec & Jackson 2007).

Davison *et al.* (1996) compiled data from Trusheim (1960), Hospers *et al.* (1988) and Hughes & Davison (1993), which demonstrated that the spacing between salt structures in the UK area of the North Sea averages 25 km for structures orientated NE–SW and 58 km for structures orientated in a NE–SW direction. Meanwhile, in the Norwegian–Danish Basin, these values are 23 and 39 km for structures orientated in NE–SW and NW–SE directions, respectively. Furthermore, in northern Germany, the average spacing is *c.* 31 and 26 km in east–west and north–south directions, respectively.

Dataset

Data provided by the TNO (Geological Survey of The Netherlands) includes digital depth and thickness maps (resolution 250 × 250 m), calibrated to well markers, of the Zechstein (Late Permian), Lower and Upper Germanic (Triassic), Altena, Schieland, Scruff, Niedersachsen (Jurassic), Rijnland, Chalk (Cretaceous) and North Sea groups, developed from a series of research projects (e.g. Duin *et al.* 2006; Kombrink *et al.* 2012). The conversion of seismic two-way travel time (TWT) to depth and thickness maps was carried out by the TNO using a velocity model developed from logs and checkshot data from 1500 wells (Van Dalfsen *et al.* 2006). For the more complex Zechstein velocity field, an assumption was made that velocities increase at locations where salt thickness is small, as the percentage of halite against carbonates is comparatively lower in areas where salt thickness is small than in areas where it is thicker. The value of interval velocity was set at 4500 m s^{-1} where the salt layer was thicker than 280 ms. In areas where salt thickness was smaller than 280 ms, the interval velocity was calculated as a function of the thickness in TWT. These interval velocities were further corrected from possible deviations of the interval velocities at borehole locations. For further information on the interval velocities and depth conversion methodologies, see Van Dalfsen *et al.* (2006). 3D seismic data were made available by the TNO and PGS (SNS MegaSurvey) with typical bin spacings of between 25 and 50 m, and resolution of the Zechstein interval of the order of 30–40 m.

Methodology

Some published models of salt-structure development associate the distance between salt structures to the initial thickness of the salt layer (at the moment of deposition) (Biot & Odé 1965; Ramberg 1972, 1981; Turcotte & Schubert 1982; Jackson *et al.* 1990). As the depositional thickness of the salt layer has been modified by salt displacement, it is necessary first to calculate the original salt thickness before salt mobilization, with the caveat that dissolution or erosion may have occurred.

Estimating salt thickness before displacement

Evaporite thickness data (Zechstein thickness maps) were analysed with a smoothing filter, which performs a spatial average within a moving window, convolving the input values with a function or kernel. Several algorithms exist to perform this task. The simplest is a moving average filter, which is a simple linear filter with a boxcar kernel utilizing uniform weights. This method replaces each point in the input, with the average of the points within the width of the kernel. Other algorithms apply a weighted filter width: for example, cosine and Gaussian kernels. The moving average with a boxcar kernel is the quickest algorithm, but it lacks the smoothness of Gaussian or cosine-tapered algorithms (Smith 2013).

Ten Veen *et al.* (2012) estimated the original depositional salt thickness of the Dutch North Sea by applying a boxcar filter with a moving window of 5 km iteratively 25 times. The iterations improve the smoothness of a boxcar kernel compared with a single convolution with a larger boxcar filter. Consequently, as the number of iterations increases, the results approach an outcome that would be obtained with a cosine or Gaussian kernel. The algorithm selected for this study was an average moving window cosine-tapered filter. This filter produces a smooth result in one step and the computational time required in this case was not prohibitive as the studied area is smaller (*c.* 37 000 km^2) in comparison to the one considered by Ten Veen *et al.* (2012) (*c.* 57 000 km^2).

The spatial averaging applied to the present salt thickness reverses, in a theoretical sense, some of the effects of salt movement, although not the effects of salt dissolution and erosion. As described above and by various previous studies (Brown 1984; Ziegler 1990; Duin *et al.* 2006; Doornenbal & Stevenson 2010; Kombrink *et al.* 2012), salt dissolution and erosion are known to have occurred in the study area associated with events such as the Cimmerian Unconformity. Unfortunately, salt loss by erosion/dissolution is difficult to quantify. Cartwright *et al.* (2001) speculated that it may have removed as much as 30–40% of the original Zechstein in the Forth Approaches area of the UK North Sea, although how that loss varies spatially is still unknown.

The maximum amount of salt that is available to flow into a salt structure is regulated by the initial wavelength (Davison *et al.* 1996). Therefore, the choice of filter length scale is commonly made assuming that flow between two salt structures occurs predominately within one structural wavelength. The cosine-tapered filter used here has an effective width (width at half the maximum weight) equal to half the full width of the filter (Fig. 2). Experiments were carried out with different filter widths. The results of some these experiments are seen in Figure 3. with a width of 50 km finally

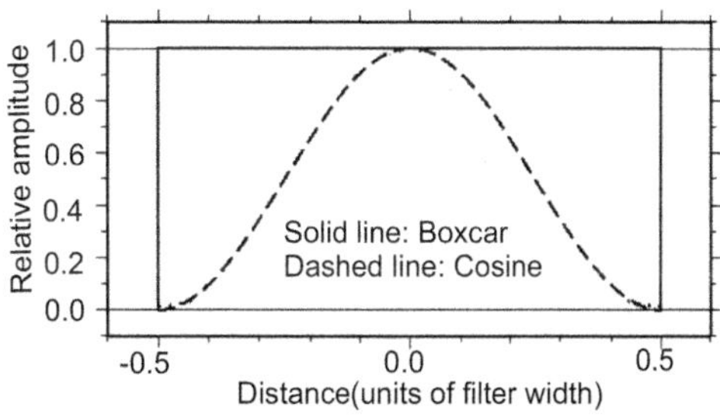

Fig. 2. Shape of a cosine arch filter compared with a boxcar-type filter. Both filters have the same width, but the area affected by the cosine-type filter is half the area affected by the boxcar-type filter, with the weights in the cosine filter concentrated in the inner half of the filter width (Wessel & Smith 1998).

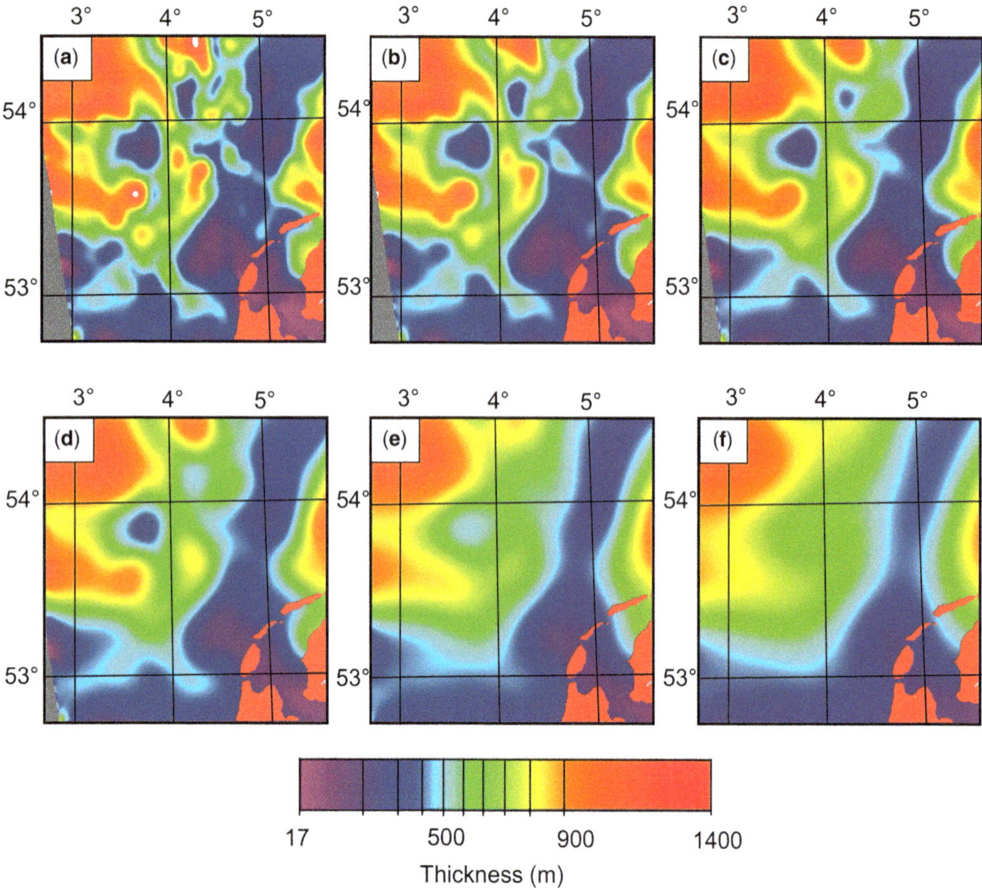

Fig. 3. Thickness of the Zechstein salt group after applying a cosine-tapered filter, results for different filter widths: (**a**) 25 000 m width, (**b**) 30 000 m width, (**c**) 45 000 m width, (**d**) 50 000 m width, (**e**) 75 000 m width and (**f**) 100 000 m width.

chosen. The filter width was chosen to encompass the typical spacing between diapirs, ensuring that the effect of individual diapirs on the filtered grid is muted. The maximum wavelength measured between structures was approximately 22 km. Therefore, to ensure that the data are filtered over at least 22 km, the full width of the filter should be at least 44 km. The results of the filtered thickness are shown in Figure 4.

Spacing between salt structures

The salt-structure wavelength is the distance between the crests of two consecutive structures, where the crests are here defined as the topographical maxima of the salt surface. Before the advent of 3D seismic data, wavelength measurements were made along 2D seismic lines and, in some cases, incorporated gravity data (Hospers *et al.* 1988). Hughes & Davison (1993), for example, made such measurements in the Silver Pit Basin, where salt structures are mainly pillows with approximately symmetrical limbs with the crests of structures well defined. Points representing the crests in each salt structure were located in the seismic line and the wavelength was calculated as the distance between pairs of crestal points. This requires that the seismic lines are perpendicular to the principal elongation axis of the salt structure but, in practice, not all lines have optimal orientations. Moreover, two neighbouring salt structures can have different orientations.

3D seismic data demonstrate that evaporites can flow in three dimensions producing complex geometries (Rowan & Ratliff 2012). Whereas in 2D cross-sections only one salt structure can be related to another, in 3D one salt structure may be surrounded by more than two salt structures, so the choice of the pair over which to make distance measurements can

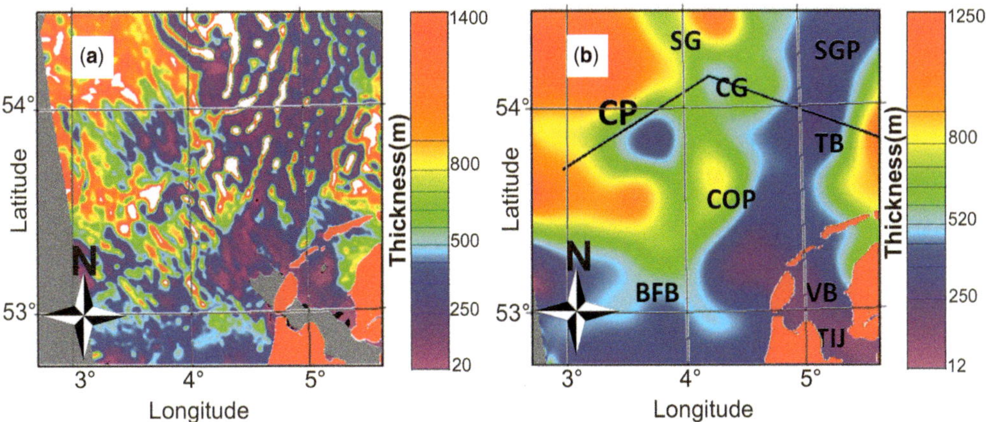

Fig. 4. Thickness map of the Zechstein salt group (**a**) before and (**b**) after application of a cosine-tapered filter of 50 km full width. The location of the cross-section in Figure 8 is represented by a black line. Structural elements are the Central Graben (CG), the Broad Fourteens Basin (BFB), the Terschelling Basin (TB), and the Cleaverbank and Central Offshore platforms (CP and COP, respectively).

be unclear; this can also be observed, although with considerably less certainty, in gravity data. In the present study, the work was carried out in a series of steps. The first step involves locating the crests of the salt structures, before locating the nearest neighbouring structures and measuring the distance between their crests.

Surface curvature was found to be a useful property for identifying crests (Ramsay 1967; Chopra & Marfurt 2007). The point of maximum negative curvature commonly coincides with the highest point of the salt surface. Whereas in 2D seismic data a salt structure may have a well-defined point of maximum negative curvature, in 3D seismic data a salt diapir of irregular shape can have several points of maxima, as demonstrated in Figure 5.

In the studied area, predominantly elongate structures occur, with axes orientated in two major

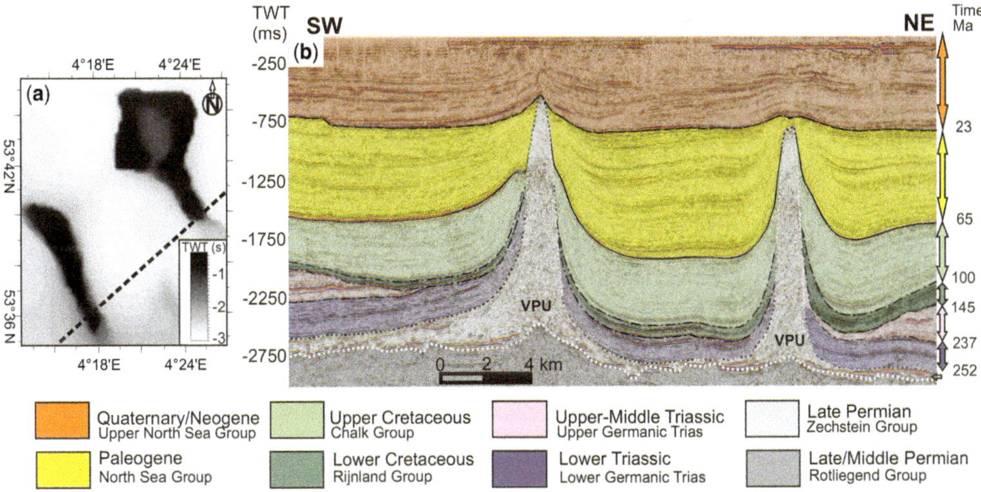

Fig. 5. Salt structures in (**a**) plan view and (**b**) an interpreted seismic cross-section NW–SE through salt diapirs I and J. The 3D seismic coverage allows us to observe in plan view (a) variations in the salt-layer thickness, such as salt structures with irregular shapes: for example, diapir J displays a hooked-shaped crest where several points of salt-thickness maxima can be identified. These variations are not possible to appreciate in the 2D seismic section (b), where diapirs I and J appear to be regular-shaped salt diapirs with a quasi-symmetrical well-defined pointed crest.

directions, NE–SW and north–south (Scheck-Wenderoth *et al.* 2008). Smaller isolated domes also occur in the southern part of the Central Graben, as well as pillows in the Cleaver Bank Platform and diapiric structures at the edge of the platform (Remmelts 1996). Salt flow along a salt wall can be considered to occur by 2D plane flow (Talbot & Jackson 1987) and, as vertical stress from a diapir lifts its roof, internal deformation maintains stress perpendicular to the bounding limits (Schultz-Ela *et al.* 1993). Elongate salt structures in plan view resemble ellipsoids with major axes indicating the direction of preferential salt flow (perpendicular to them). Distance measurements should be made in the direction of flow, as the analytical models, from which conclusions are made, only represent 2D flow. It is recognized that in some areas of this study (e.g. salt structures L and W in Fig. 1), the crestal elongation direction of the salt structures is not well defined. These factors add complexity to the selection of crestal points along the salt structures. In this study, to avoid having more than one point representative of the crest along any cross-section, a series of points along the major axis in the middle of the structure was recorded to represent the crest line of the salt structure (Fig. 6). The crests of other structures, such as diapirs with no elongated axis and a more irregular shape, were considered to be located parallel to the major structural trend. Points were sampled every 100 m along the crests of 24 salt structures.

Limitations of method

Earlier studies, such as that of Hughes & Davison (1993), estimated the original salt thickness by

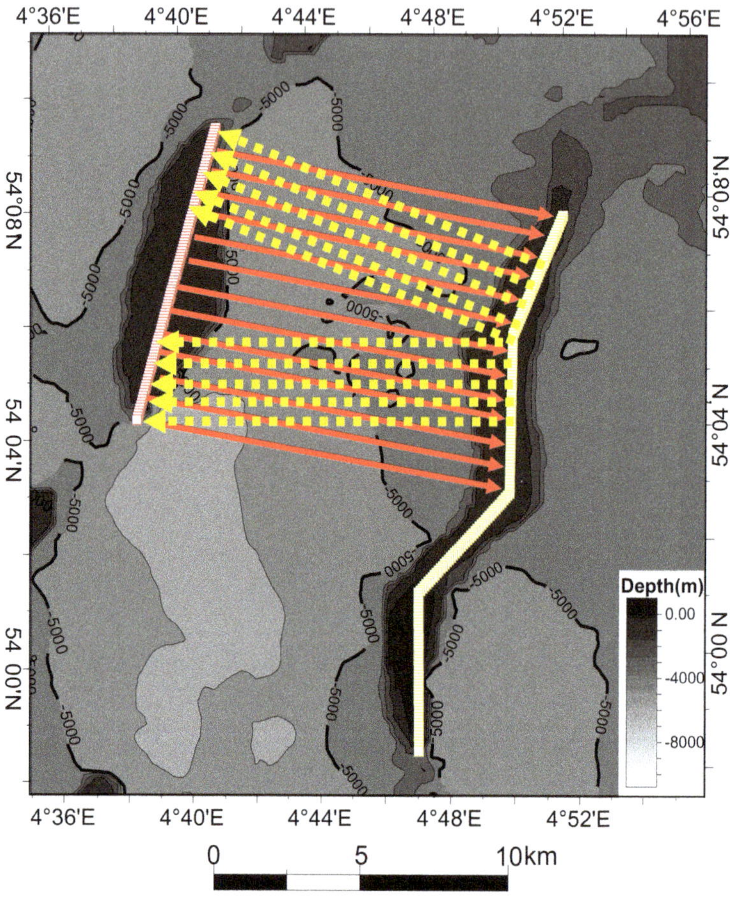

Fig. 6. Depth map of the top Zechstein salt layer. The algorithm employed to determine wavelengths involves calculating the orthogonal distance to the structural trend at each point sampled along the crest of each salt structure. Thus, the arrowed lines (red for the salt structure in the west and blue for the salt structure in the east) represent the shortest orthogonal distances between the crests at each sampled point (squares).

averaging the present salt thickness over the lengths of their 2D sections. This method only considers salt flow along the section. One improvement would be to allow for salt flow in all directions, which is achieved here by averaging thickness in two directions, rather than one. However, there are a number of recognized limitations to the methods utilized in this study. Firstly, the estimated original salt-layer thickness obtained by filtering has edge effects at the borders because the data area is finite. Hence, the moving average filter centred on an edge has less data than for a node within the centre of the dataset. This results in the filter effectively smoothing with a smaller and varied length scale at the edges. The effect can only be avoided by obtaining data to extend the area of thickness data or by leaving out measurements within the affected edge areas. Additionally, the theoretical models outlined in our introduction represent 2D flow. Where salt structures are 3D, interstructure wavelengths may only roughly correspond with these models and it is presently unclear how the measurements should be modified to allow them to be compared with 2D model predictions. Considering the above comments, it is unclear whether the characteristic wavelength of a particular salt structure should be calculated as the average, medium or mode of the individual values calculated along each crest. It is also unclear how all the structures relate to each other in a 3D sense and, therefore, when a particular salt structure is surrounded by more than one salt diapir; the most representative characteristic wavelength is ambiguous.

Results

If dissolution and erosion are assumed to be spatially invariable, the thickness map in Figure 4b represents an estimate of the original depositional salt thickness. It shows that the lowest values of salt thickness (0, i.e. non deposition, to *c.* 20 m) are localized towards the south near the Texel-IJsselmeer High. Thinning also occurs along a NE–SW trend, near to the west margin of the Central Graben, over the Schill Grund Platform (SGP) and at the eastern limit of the TB (thicknesses of 300–400 m). In the Dutch Central Graben, thicknesses are intermediate (*c.* 600 m) but reach maximum values in the Step Graben (SG) in the north (*c.* 1000 m). Along the CP, an area of minimum thickness (*c.* 350 m) occurs in the south over the COP; this is surrounded by an increasing thickness of the salt layer with maxima (*c.* 1000 m) towards the north over the SG and towards the south over the BFB.

In Figure 7, the average estimated salt thickness against wavelength of the present-day salt structures is shown divided into two different groups, with different orientations. The Central Graben Group is mainly orientated north–south and NNE–SSW; whereas the Cleaverbank Platform Group is mainly orientated NW–SE. When structure averages of these data are taken, distinct relationships between

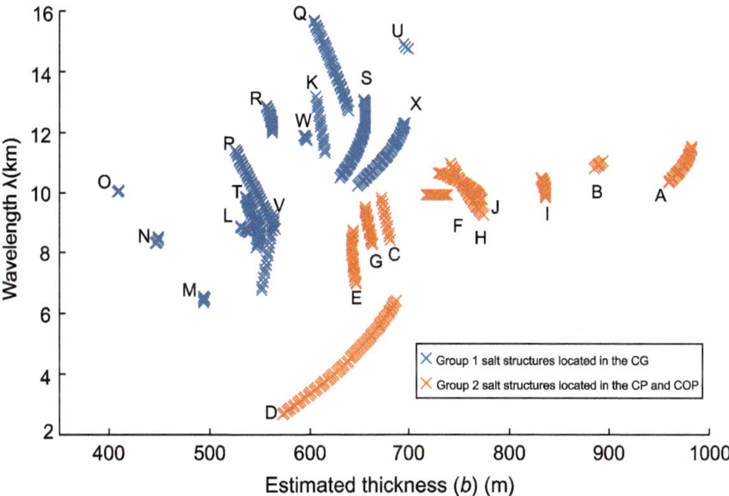

Fig. 7. Plot of spatially filtered salt thickness (b) against measured wavelength (λ). It can be observed that the relationship of wavelength–thickness varies for each salt structure. However, in general, data are shown to be divided into two different groups: blue crosses correspond to salt structures located in the Central Graben (CG); and red crosses represent structures located in the Cleaverbank and Central Offshore platforms (CP and COP, respectively). The original thickness of structures in the CG is minor compared to the thicknesses of structures in the CP and COP. Wavelengths measured for the structures in the CG are greater than those on the CP and COP.

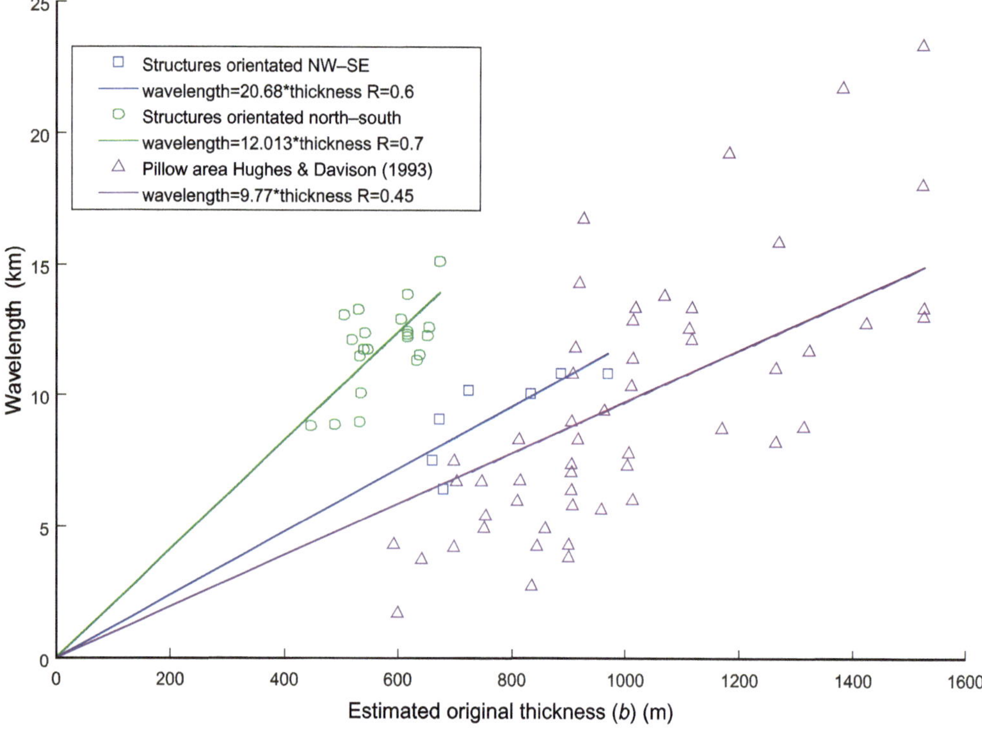

Fig. 8. Depicted with triangles is the Hughes & Davison (1993) plot of estimated salt thickness v. measured wavelength, which shows a ratio between salt structure wavelength (λ) and the thickness of the salt layer (b) of around $\lambda = 9.77b$ (the straight lines in the figure are regressions constrained to pass through (0,0)). Represented with circles and squares are the average variations in the estimated salt thickness (b) against measured wavelength (λ) of the salt structures with north–south and NW–SE trends, respectively, of the present study. Thicknesses were sampled from the 50 km filtered sediment thickness grid at the pillow locations.

wavelength and thickness are observable (Fig. 8). The salt structures trending north–south and NNE–SSW have wavelength to thickness ratios of *c.* 20 (m/m), whilst the salt structures trending NW–SE have ratios of *c.* 12 (m/m).

For all the sampled points, the values of original thickness at the salt structures orientated north–south varies from 400 to 695 m. The wavelength of such structures varies from 6740 to 15 660 m. For the structures orientated NW–SE, the thickness varies from 570 to 980 m, and wavelength varies from 2626 to 11 515 m. The ranges of the original thickness and wavelength values measured at the sampled points of all the salt structures, as well as the range of values of the average per salt structure, are listed in Table 1. The values of the original thicknesses of structures that are orientated north–south are minor compared to the thicknesses of structures orientated NW–SW, which also have a broader range of values. Furthermore, the wavelengths measured for the structures that trend north–south are greater than those measured on structures orientated NW–SE.

Discussion

Varied estimates of the original salt-layer thickness

Ten Veen *et al.* (2012) calculated the depositional salt thickness for the whole Dutch offshore using a similar methodology: application of an average moving filter. However, the filter used had a boxcar-shaped window and was applied iteratively. The iterative boxcar-shaped window filter is faster in terms of computational time than the cosine-shaped window used for the present study, and therefore is more efficient for big datasets, such as the one analysed by Ten Veen *et al.* (2012). The cosine-shaped window filter used for the present study, on the other hand, provides smoother edges without the need to apply it iteratively and is most suitable for smaller datasets.

Ten Veen *et al.* (2012) presented their results on a thickness distribution map, as well as providing one profile section (Fig. 9). Within their map, most of the

Table 1. *Minimum and maximum values of original thickness and wavelength for the salt structures located at the Central Graben, whose principal orientation is north–south, and at the Cleaverbank Platform and Central offshore Platform, whose structures are instead principally orientated NW–SE*

		Central graben Principal orientation north–south		Cleaverbank platform and central offshore platform Principal orientation NW–SE	
		Minimum	Maximum	Minimum	Maximum
Original thickness (m)	General	400	695	570	980
	Average	450	670	660	967
Wavelength (m)	General	6740	15660	2626	11515
	Average	8820	15080	6400	10900

General refers to the limits within all the measurements in each salt structure; average refers to the limits within the average values per salt structure.

present study area is represented by one single colour (blue), corresponding to values of around 600 m; therefore, no discernible variations in thickness are observable. The profile, however, demonstrated depositional thickness variations across the CP, Central Graben, TB and Ameland Platform. In order to compare the results obtained by Ten Veen *et al.* (2012) with the results obtained in this research, the profile presented in Ten Veen *et al.* (2012) has been digitized (Fig. 9) and compared with extracted values for original thickness obtained during this study along a similar profile (Fig. 9). It can be observed in Figure 9 that both profiles follow broadly similar patterns. Some differences between values were demonstrated. At the CP in close proximity to the Dutch Central Graben, Ten Veen *et al.* (2012) estimated the original thickness to be *c.* 875 m, whilst this study estimates it to be *c.* 725 m. Within the Dutch Central Graben, the maximum and minimum values calculated by Ten Veen *et al.* (2012) are *c.* 720 and *c.* 375 m, respectively, whereas maximum and minimum values for this study are *c.* 655 and *c.* 495 m, respectively, providing a variation in difference in thickness of between *c.* 65 and 150 m. Such a difference could be explained by considering that the profile along-thickness values extracted may be positioned at a slightly different location to those that were reported on by Ten Veen *et al.* (2012). Alternatively, such a variation may be related to the type of filter applied and their grade of smoothness, as already outlined in the 'Methodology' section of this paper.

To assess the filtered thicknesses (Fig. 4b), it should be considered that the evaporites deposited during the late Permian would have been subject to the configuration of the basin at the time of deposition. According to Geluk (2007), the topography of the carbonate platform, developed during the

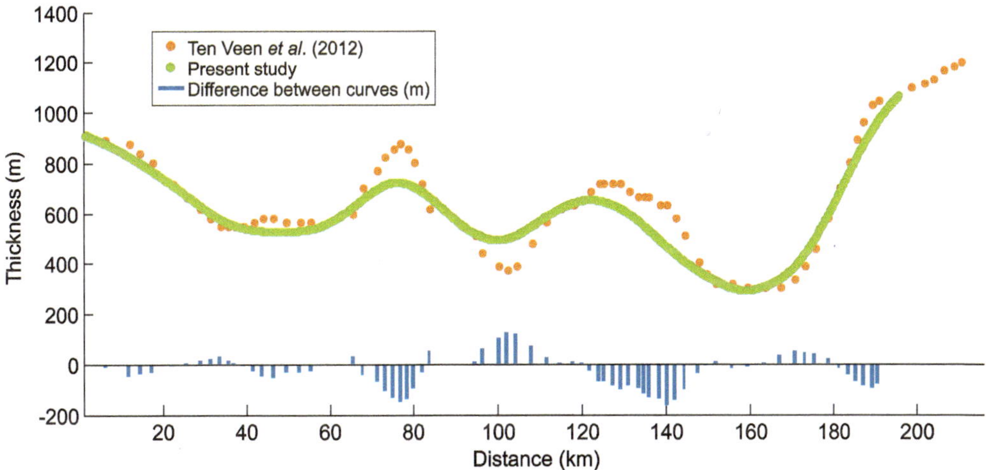

Fig. 9. Cross-section (locality shown in Fig. 3) displaying the salt-thickness values after applying a moving average filter. Results from Ten Veen *et al.* (2012) are displayed in orange; results from the present study are displayed in green.

first two cycles of the Zechstein Group, controlled the distribution and thickness of the overlying salt units. The salt deposits would have probably thickened from the platform areas towards the carbonate basin. The present-day facies distribution of Z2 carbonates (Doornenbal & Stevenson 2010) shows the platform limit of the restricted basin to be located landwards, close to the southern boundary of the BFB. Therefore, most of the northern part of the study area is within the basin depocentre. According to Doornenbal & Stevenson (2010), the structural elements that currently comprise the North Sea Basin were not strongly differentiated in mid-Permian times. Consequently, in the study area, the salt layer can be expected to be thinner landwards and relatively homogeneous offshore towards the north. The observed thinning of the salt layer (from *c.* 10 to *c.* 1200 m) towards the south of the study area (Fig. 4b) corresponds to the expected thin salt around the platform areas. The variations in thickness in other parts of the study area may be due to deep erosion of the Permo-Triassic cover during the Late–Mid-Cimmerian tectonic phase (Remmelts 1996; Geluk 2007), particularly on the CP, the Schill Ground Platform (SGP) and areas of the TB. Understanding the original salt thickness within the Dutch Central Graben is more difficult due to intense halokinesis. For a point of reference, the mean salt thicknesses in each structural area, calculated on map-based average volume, varies from 428 to 870 m (TB, *c.* 428 m; SGP, *c.* 460 m; COP, *c.* 596 m, SG, 750 m; Dutch Central Graben, *c.* 600 m; CP, *c.* 870 m).

In the present study area, the observed filtered thickness in the Dutch Central Graben (*c.* 700–1400 m) is somewhat smaller than over the CP (*c.* 800–1400 m). The original thickness was expected to be similar in these two areas, as both areas were fully developed structurally by the Late Jurassic. In the present distribution of salt thickness, the average thickness value in the CP (*c.* 870 m) is higher by up to *c.* 270 m than the thickness in the Dutch Central Graben (*c.* 600 m). Therefore, even if the filter were to average values within entire structural areas applying a larger window, the thickness in the Dutch Central Graben would still be thinner than in the CP. Consequently, this variation in thickness cannot be explained by salt flow beyond the limits of the filter window. It is proposed that this difference can be explained by extreme salt loss due to erosion and/or dissolution or by salt flow out of the Dutch Central Graben towards the platforms.

Relating results to mechanical models

Methods previously used to obtain wavelength data by Trusheim (1960), Hospers *et al.* (1988) and Remmelts (1996) are not described in detail within the associated publications. The salt R–T model of Turcotte & Schubert (1982) suggests that the characteristic wavelength should be 2.568 times the original salt thickness. However, our results found larger ratios (12–20) that also vary between different structural groups. If the R–T model is appropriate here, there may be several explanations for this discrepancy:

- faster growing salt structures tend to merge with surrounding structures (Ramberg 1981);
- the effect of initial irregularities in the interface between layers can potentially result in salt diapirs growing with wavelengths four times longer than otherwise (Schmeling (1987));
- the effect of viscosity ratio upon the dominant wavelength (the larger the viscosity ratios, the greater the dominant wavelength/thickness ratio: Ramberg 1968, 1972).

The R–T model is considered simplistic as it ignores overburden rigidity (Poliakov *et al.* 1996). Ismail-Zadeh *et al.* (2002) extended the R–T model instability with a plastic layer representing the overburden overlying a viscous salt layer. In the present study area, salt displacement has been suggested by Coward & Stewart (1995) to be related to periods of regional extension that, according to Jackson & Vendeville (1994), weaken the overburden by thinning and fracturing it, initiating diapirism. Therefore, a best approach to analysing diapirism may consider a plastic overburden or a viscoelastic combination where deformation will be influenced by environmental conditions and strain rate (Davison *et al.* 1996).

In the Ismail-Zadeh *et al.* (2002) model, the characteristic wavelength of the salt structures increases with increasing thickness ratio but then decreases in a series of abrupt jumps. The thickness ratio is given by:

$$\frac{H_1}{H_2} \quad \text{where} \quad H_i = \frac{h_i}{h_1 + h_2} \tag{1}$$

where h_1 and h_2 are the plastic and viscous layer thicknesses, respectively.

An increasing wavelength and, therefore, an increasing thickness ratio implies an increasing overburden thickness, so wavelength increases with deposition until the sedimentary layer thickness reaches a critical value given by the ratio between the sedimentary cover thickness and the salt layer thickness. Subsequently, failure leads to the wavelength decreasing. According to the Ismail-Zadeh *et al.* (2002) model, a thin salt layer combined with thick overburden should be associated with a large initial wavelength. In their analysis, the thickness ratio (equation 1) and viscosity ratio (equation 2) between layers are the main influences on the characteristic wavelength of salt structures. The maximum characteristic wavelength occurs when the sedimentary cover reaches a thickness that depends on viscosity ratio. With a smaller viscosity ratio, the maximum growth rate takes place at a larger thickness ratio (equation 1), which means that the growth rate reaches the maximum at later stages of sedimentation.

The effective viscosity ratio, v, is defined as:

$$v = \frac{\eta_2}{\eta_1} \qquad (2)$$

where η_1 and η_2 are the viscosities of the overburden and salt layer, respectively. The thickness ratio increases with the thickness of the overburden, and the viscosity ratio decreases with smaller salt viscosities over sedimentary layers with high viscosities. Therefore, the viscosity ratio decreases as the thickness of the overburden grows because its effective viscosity increases (e.g. the overburden viscosity increases from 1.25×10^{18} to 2×10^{20} Pa s for a thickening by 3 km in the Great Kavir: Ismail-Zadeh *et al.* 2002). Alternatively, Orlic & Buijze (2014) analysed wellbore closure by creep of rock salt, and concluded that creep of the Zechstein salt is both linear and non-linear. Creep strain rates increase with increasing temperature and differential stress, so that creep of the salt layer is expected to increase with depth. Consequently, in a basin with constant sedimentation and rates of sedimentation higher than rates of erosion, the viscosity ratio decreases with time and the limit of the thickness ratio at which the salt structures reach their maximum wavelength increases. This means that the maximum thickness of the sedimentary layer at which the maximum wavelength develops is twice the thickness of the salt layer for viscosity ratios of 10^{-2}, and five times the thickness of the salt layer for viscosity ratios of 10^{-3}.

In the study area, according to Ten Veen *et al.* (2012), salt flow in the Central Graben was more intense, allowing salt to form domes and walls; however, in the platform areas (the Ameland and Cleaverbank platforms), salt movement was minor, and only formed salt pillows and a few diapirs.

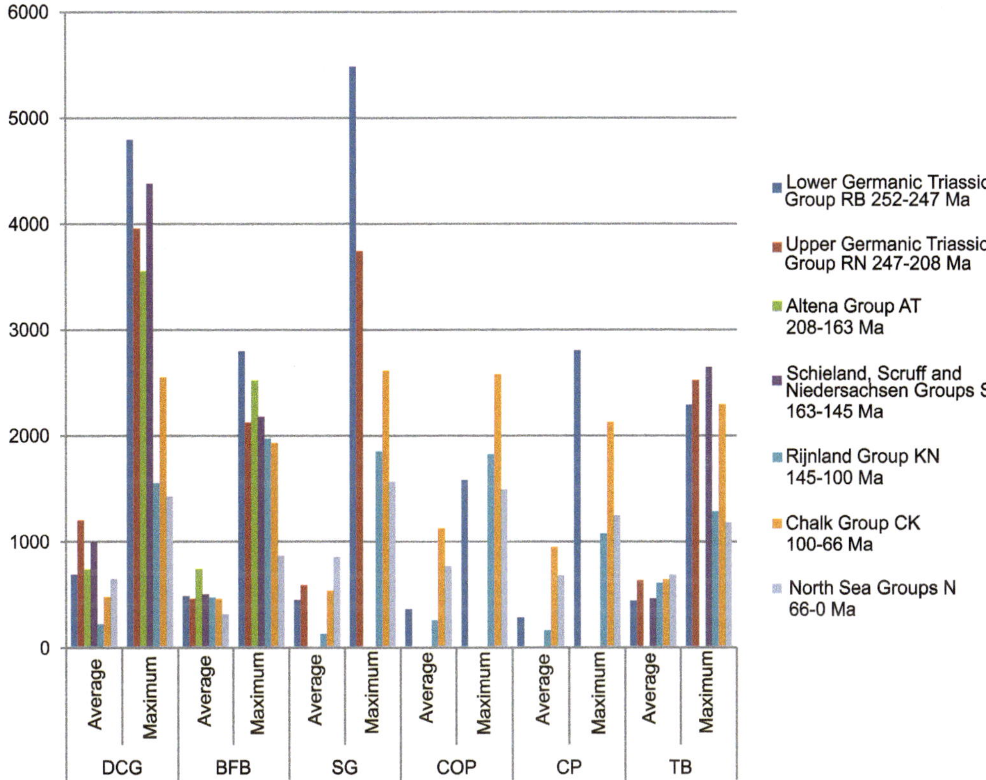

Fig. 10. The average and maximum thicknesses of sediments deposited in the Dutch Central Graben (DCG), the Broad Fourteens Basin (BFB), the Step Graben (SG), the Central Offshore Platform (COP), the Cleaverbank Platform (CP) and the Terschelling Basin (TB) for the post-Zechstein stratigraphy.

Considering that salt-structure evolution occurs in a series of stages, pillows representing an early stage and diapirs along with walls a more evolved stage (Trusheim 1960; Vendeville 2002), it is possible to compare the wavelength–thickness ratios obtained from this study and those in the pillow zone study of Hughes & Davison (1993) as if these represent a sequence in time. It is observed that the wavelength–thickness ratio is lower for the structures located in the pillow zone (*c.* 9.7), increase on the structures located at the platform areas (*c.* 11.5) and reaches the highest value for the more mature structures located The thickness of sediments deposited in the study area varies between the different structural zones. Figure 10 summarizes the maximum and average thickness deposited in each structural region for each sedimentary group of the stratigraphic column above the Zechstein Group. The average thickness was calculated by dividing the total volume of sediments deposited by the area of deposition. In general, sediments deposited in the Dutch Central Graben are thicker than those deposited in the COP and the CP. The results appear to confirm the idea that thicker sedimentary covers produce bigger initial characteristic wavelengths, as salt structures in the Dutch Central Graben have longer wavelengths than the salt structures in the CP and COP.

The sedimentary cover to salt-layer thickness ratios (equation 1) also vary between each structural area (Fig. 11). This was calculated with the average deposited thickness of sediments and salt in each structural region. If 40% of salt has been removed by dissolution or 40% of overburden sediments have been eroded, as suggested by Ten Veen *et al.* (2012), the values of the thickness ratios would be reduced by up to 40% and the maximum values of thickness ratio would be 1.5, 2.9 and 5.8 during the Late Cretaceous period in the CP, COP and Dutch Central Graben, respectively.

The greater wavelengths present in the Dutch Central Graben, compared to the CP, imply smaller viscosity ratios between the cover and the salt layer after the Jurassic in order to increase the wavelengths at such thickness ratios. However, according to TNO (2011) diapirism in the southern part of the Dutch Central Graben (towards the limit with the COP and TB) stopped after the Jurassic period when the thickness ratio reached 3.5–4.0. Effective viscosity ratios of 10^{-3}–10^{-2} correspond with the maximum value of characteristic wavelength at such a thickness ratio. In the SG, diapirism continued until the Cretaceous period; in this case, the wavelength may have continued to grow after the Jurassic period without a decrease in the viscosity ratio due to erosion of sediments in that area related to Early Cretaceous tilting along the Elbow Spit High–Step Graben boundary fault (Remmelts 1996) which decreased

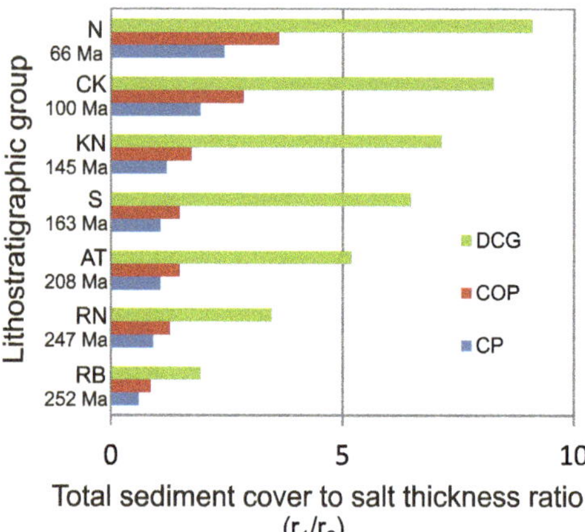

Fig. 11. Total sediment cover to salt thickness ratio (r_1/r_2) in the Dutch Central Graben (DCG), the Central Offshore Platform (COP) and the Cleaverbank Platform (CP). The average thickness of the sedimentary cover was estimated after deposition of the Lower Germanic (RB), Upper Germanic (RN), Altena (AT), Schieland, Scruff and Niedersachsen (S), Rijnland (KN), Chalk (CK) and North Sea (N) groups. The ratio increases with the deposition of each new layer of sediments. The ratio was calculated using the average thickness deposited in each area, ignoring any salt dissolution and erosion of the overlying sediments.

the thickness ratio. Diapirism in the COP and CP continued until the Paleogene, with thickness ratio values from 2 to 4. These smaller thickness ratios are considered to be due to lower sedimentation rates in the area of the CP and COP compared to the thicker overburden in the Dutch Central Graben. These differences in thickness ratios may have resulted in smaller wavelengths in the CP and COP, and higher wavelengths in the Dutch Central Graben.

In the gravitational instability model for two viscous fluid layers, there is a well-established mode of maximum instability (a maximum growth rate at a specific wavelength value). However, in the Ismail-Zadeh *et al.* (2002) plastic–viscous model, the growth-rate curve oscillates. Consequently, the growth-rate curve has similar values for different wavelengths and, rather than a characteristic wavelength, a variety of wavelengths is expected. This is in agreement with the results observed in the present study. The range of values of average wavelength is broader in the structures orientated north–south than in the structures trending NW–SE. The wavelength–thickness relationship has been subject to previous studies in the North Sea. Hughes & Davison (1993) (Fig. 8) reported wavelength values of between 1 and 24 km, and a relationship between pillow wavelength and salt-layer thickness (*b*) of $b = 0.643 + 0.036\lambda$. This relationship implies that wavelength varies at approximately 9.7 times the salt thickness. The range of wavelengths values measured by Hughes & Davison (1993) is wider than the ones measured in this study. This variability may be related to smaller viscosity ratios, a wide range of salt-layer original thickness values and differences in strain histories in the Hughes & Davison (1993) study area.

Schmeling (1987) studied the effects of initial conditions of R–T instabilities, including initial sinusoidal-type perturbations within the interface between the salt and overburden layers. These disturbances could be produced by faulting, folding, differential erosion or episodic sedimentation. It was concluded that the magnitudes of the amplitude and wavelength of these perturbations influence the final salt-structure wavelength, increasing it by a factor of up to 4. Schmeling (1987) also found that disturbances produced by steep reverse or normal faults, modelled as step-like functions, would need to be spaced by 2.66 times the expected wavelength without initial disturbances, if they were to grow. The actual wavelength of the salt structures in the study area is 2600–15 600 m. If these wavelengths have been increased by a factor of 4 due to faults, for example, the equivalent wavelength values that would exist in the absence of the faults would be 650–3900 m. In order to have affected the salt-structure wavelength, therefore, the fault spacing is at least *c*. 1700–*c*. 10 000 m (areas with shorter and longer wavelength, respectively).

Variations in salt thickness have various implications in the styles of deformation: for example, Duffy *et al.* (2013) related the effect of evaporites producing complex modes of structural evolution in a rift under salt influence compared with a rift free from evaporite influence. Studies carried out by Ten Veen *et al.* (2012) extensively described the link between the original thickness of the salt layer with the stages of faulting affecting the pre- and post-salt sedimentary layers. In the present study, the relationship of wavelength to original salt thickness was analysed, determining that this varies within different structural areas of the Dutch Southern North Sea. The results in the present study area agree with the information on the summary of structures in the North Sea by Davison *et al.* (1996), whereby salt structures have larger wavelengths along structures orientated NW to WNW, including structures from the Norwegian–Danish Basin, the UK area of the North Sea and north Germany. Structures orientated NNE to north are related to the Triassic–Jurassic extensional phase where basins such as the BFB and Central Graben formed. Structures orientated NW to WNW are associated with Late Cretaceous and Cenozoic compressional phases. Therefore, subsidence controlled by these structural elements may be an underlying cause of the wavelength variation, as it allowed thicker deposition of sediments in basinal areas; whereas in platforms the sedimentary cover is thinner.

The relationship between original salt thickness and wavelength as described above is important as it contributes to the knowledge of the parameters involved and the history of salt movement in the studied area. Such understanding is necessary to evaluate hydrocarbon prospects and fields related to salt structures. Mesozoic reservoirs of interest within the Dutch North Sea include the Triassic Main Buntsandstein Formation, and the Late Jurassic Detfurth and Scruff Greensand formations in Block L (Dronkert & Remmelts 1996), which are found in four-way dip closures, rim synclines and truncation traps: all related to, and depending on, a good understanding of salt tectonics.

Conclusions

In this study, the relationship between salt-diapir spacing and salt-layer thickness across the Southern North Sea is analysed. It is hoped that these data will provide a useful framework for further hydrocarbon exploration, specifically as a tool for understanding structural evolution plus related trap formation timing and reservoir development.

Using an average moving-window cosine-tapered filter technique, it has been observed that estimated values of original salt thicknesses across the Southern North Sea vary between 20 and 1400 m. Areas with lower thickness correspond to the Texel-IJsselmeer High, where thicknesses range from 0 (non-deposition) to *c*. 20 m, the Schill Grund Platform (SGP) and at the eastern limit of the Terschelling Basin (TB) (thicknesses of 300–400 m). Thickening of the salt layer occurs towards the north in the Central Graben and Step Graben (*c*. 1000 m) and in the Cleaverbank Platform (CP) vicinity. Thinning of the salt layer towards the Texel-IJsselmeer High was expected as it corresponds to a Permian Zechstein carbonate platform area. Thinning in other areas, however, may be related to high salt erosion/dissolution, as the structural elements that currently comprise the North Sea Basin were not strongly differentiated in mid-Permian times and, therefore, the original thickness in those areas is expected to be more uniform.

Previous calculations of original salt thickness, made by Ten Veen *et al.* (2012) using a similar methodology, are in good agreement with the estimations made in the present study. Variations between both results can be explained by differences in the type of filter applied and the degree of smoothness that such filters provide. The original salt thickness estimated using filtering, although efficient, does not remove the effects of salt dissolution or erosion, so the estimates are minima. Moreover, the edge effect effectively leads to less filtering at the dataset edges. Notwithstanding the limitations of the calculations, the depositional salt thickness is useful for analysing the evolution of basins where salt displacement exerts significant control.

Wavelength values vary along each salt structure and between structural domains. This is in good agreement with the Ismail-Zadeh *et al.* (2002) plastic–viscous model of a density instability, where the growth-rate curve oscillates and produces a variety of wavelengths. Larger wavelengths, as well as a broader range of wavelength values observed in the Central Graben, may indicate a thickness ratio that is relatively large in comparison with those of the Central Offshore and Cleaverbank platforms. The analysis of wavelength–salt thickness in the study area found two different wavelength–thickness ratios, 12 and 20. The ratios obtained in this study are larger than the 2.568 proposed by Turcotte & Schubert (1982). This discrepancy can be explained by adding complexity to their simple model, such as the effect of large viscosity ratios between the salt layer and the overburden, irregularities in the interface such as faults, the effect of sedimentation, and the presence of a brittle overburden instead of a fluid one.

The wavelength–thickness ratios observed in this study are associated with two different salt-structure orientations. These orientations not only reflect the different fault patterns in the two areas but also different structural processes and subsequent stages of evolution of the salt structures. Besides being part of the same basin, the CP and the Central Graben are two structural areas with different evolutionary histories and characteristics, such as different erosion and sedimentation rates. In the Central Graben, extension resulted in normal faulting that regionally produced accelerated basement subsidence and allowed deposition of a thick sedimentary cover and localized salt withdrawal by the subsidence of post-salt sediments. The larger amount of sediments deposited in the Central Graben and the earlier development of salt structures led to bigger contrasts in thickness and viscosity between the salt layer and the sedimentary layer that conform to Rayleigh–Taylor (R–T) instability; this is most likely to be reflected in the larger wavelength–thickness ratio of the salt structures orientated north–south located in this area compared to the wavelength–thickness ratio of the salt structures orientated NW–SE located in the CP and COP. The different ratios wavelength–thickness seem to be related not only to the thickness of the salt layer but also to the thickness of the overburden; variations that are most likely to be related to the structural control on deposition.

Further work

Interbedded non-halite mineral units within the salt layer are an important factor when considering salt-structure evolution. According to Davison *et al.* (1996), interlayered evaporitic units, such as anhydrite, produce a relative decrement in the lateral flow of the salt layer and, according to Burliga (1996), such an effect is expected to be greater during the pillow stage where the strength and anisotropy of the evaporitic layer is higher. In this study, the evaporitic layer was considered as a uniform salt layer constituted mostly of halite. Further work should focus on the study of the effect of the composition of the evaporitic layer on the style of deformation.

Regarding changes in strain rates and their effect on the evolution of salt diapirs, the Southern Permian Basin has undergone several phases of tectonism involving strike-slip, extensional and compressional processes (Littke *et al.* 2008), suggesting that the background strain rate is likely to have varied over time. Koyi & Petersen (1993), Remmelts (1995, 1996), Scheck *et al.* (2003), Maystrenko *et al.* (2005), Mohr *et al.* (2005) and Ten Veen *et al.* (2012) have pointed out the contemporaneous occurrence of episodes of intense salt displacement and acute tectonic phases, as well as a close relationship between salt-structure orientation and the structural grain. By adding a background strain rate to their analysis to simulate compression or tension, Ismail-Zadeh *et al.* (2002) encountered oscillations in the growth-rate curve, which grew in amplitude with increasing background strain. Increasing strain rates also resulted in changes in the maximum growth rate. With rapid strain rates ($>10^{-4}$), R–T instability is replaced by a buckling instability. In this study, a broader range of wavelength values in the Central Graben compared to the CP and COP is observed. This may be associated with oscillations in the growth-rate curve produced by larger changes in strain within the Central Graben. It is also feasible that high background strain rates relate to the broader variations in wavelength in the Silver Pit Basin observed by Hughes & Davison (1993) compared to the variations examined in the present study. Shorter wavelengths in the COP and CP compared with the Central Graben may be associated with greater effects of shortening in that area. However, quantitative estimates of strain rate in the study area are difficult to obtain due to a complex history of inversion and salt movement in three dimensions. Therefore, in order to assess accurately the effects of variations in strain rates in the development of salt structures, the magnitude of the strain-rate changes, as well as salt-structure growth rates and their effect on wavelength values, need to be further constrained. Further research should also focus on the study of variations in the wavelength and thickness parameters within specific structural domains, as well as along particular salt structures.

The Mexican Petroleum Institute (Instituto Mexicano del Petroleo (IMP)) is acknowledged for funding Karina Hernandez in this research. The seismic and well data were made available by the Geological Survey of the Netherlands (TNO). The authors would also like to thank Ben Kilhams, Matthijs Van Winden and an anonymous reviewer for their helpful and detailed comments which helped to improve this article, as well as Schlumberger for provision of the interpretation software Petrel™ 2013.

References

Biot, M. & Odé, H. 1965. Theory of gravity instability with variable overburden and compaction. *Geophysics*, **30**, 213–227.

Brown, S. 1984. Jurassic. *In*: Glennie, K. (ed.) *Introduction to the Petroleum Geology of the North Sea*. Blackwell Scientific, Oxford, 103–131.

Buchanan, P.G., Bishop, D.J. & Hood, D.N. 1996. Development of salt-related structures in the Central North Sea: results from section balancing. *In*: Alsop, G.I., Blundell, D.J. & Davison, I. (eds) *Salt Tectonics*. Geological Society, London, Special Publications, **100**, 111–128, https://doi.org/10.1144/GSL.SP.1996.100.01.09

Burliga, S. 1996. Kinematics within the Kłodawa salt diapir, central Poland. *In*: Alsop, G.I., Blundell, D.J. & Davison, I. (eds) *Salt Tectonics*. Geological Society, London, Special Publications, **100**, 11–21, https://doi.org/10.1144/GSL.SP.1996.100.01.02

Cartwright, J., Stewart, S. & Clark, J. 2001. Salt dissolution and salt-related deformation of the Forth Approaches Basin, UK North Sea. *Marine and Petroleum Geology*, **18**, 757–778.

Chen, P. 2016. *Fault reactivation analysis of the Cleaver Bank High based on 3D seismic data*. Master's thesis, Utrecht University, Utrecht, The Netherlands.

Chopra, S. & Marfurt, K. 2007. Curvature attribute applications to 3D surface seismic data. *The Leading Edge*, **26**, 404–414.

Colman-Sadd, S. 1978. Fold development in Zagros simply folded belt, Southwest Iran. *AAPG Bulletin*, **62**, 984–1003.

Coward, M. & Stewart, S. 1995. Salt-influenced structures in the Mesozoic–Tertiary cover of the southern North Sea, UK. *In*: Jackson, M.P.A., Roberts, D.G. & Snelson, S. (eds) *Salt Tectonics: A Global Perspective*. American Association of Petroleum Geologists, Memoirs, **65**, 229–250.

Davison, I., Alsop, I. & Blundell, D. 1996. Salt tectonics: some aspects of deformation mechanics. *In*: Alsop, G.I., Blundell, D.J. & Davison, I. (eds) *Salt Tectonics*. Geological Society, London, Special Publications, **100**, 1–10, https://doi.org/10.1144/GSL.SP.1996.100.01.01

de Jager, J. 2003. Inverted basins in the Netherlands, similarities and differences. *Netherlands Journal of Geosciences*, **82**, 339–349.

de Jager, J. 2007. Geological development. *In*: Wong, T.E., Batjes, D.A. & de Jager, J. *Geology of the Netherlands*. Royal Netherlands Academy of Arts and Sciences, Amsterdam, 5–26.

Doornenbal, H. & Stevenson, A. 2010. *Petroleum Geological Atlas of the Southern Permian Basin Area*. European Association of Geoscientists & Engineers (EAGE), Houten, The Netherlands.

Dronkert, H. & Remmelts, G. 1996. Influence of salt structures on reservoir rocks in Block L2, Dutch continental shelf. *In*: Rondeel, H.E., Batjes, D.A.J. & Nieuwenhuijs, W.H. (eds) *Geology of Gas and Oil under The Netherlands*. Springer, Dordrecht, The Netherlands, 159–166.

Duffy, O.B., Gawthorpe, R.L., Docherty, M. & Brocklehurst, S.H. 2013. Mobile evaporite controls on the structural style and evolution of rift basins: Danish Central Graben, North Sea. *Basin Research*, **25**, 310–330.

Duin, E., Doornenbal, J., Rijkers, R., Verbeek, J. & Wong, T.E. 2006. Subsurface structure of the Netherlands – results of recent onshore and offshore mapping. *Netherlands Journal of Geosciences*, **85**, 245–276.

Fattah, R.A., Verweij, J., Witmans, N. & Ten Veen, J. 2012. Reconstruction of burial history, temperature, source rock maturity and hydrocarbon generation in the northwestern Dutch offshore. *Netherlands Journal of Geosciences*, **91**, 535–554.

Fontaine, J., Guastella, G., Jouault, P. & de la Vega, P. 1993. F15-A: a Triassic gas field on the eastern limit of the Dutch Central Graben. *In*: Parker, J.R. (ed.) *Petroleum Geology of Northwest Europe: Proceedings of the 4th Conference*. Geological Society, London, 583–593, https://doi.org/10.1144/0040583

Geluk, M. 2000. Late Permian (Zechstein) carbonate-facies maps, the Netherlands. *Netherlands Journal of Geosciences*, **79**, 17–27.

Geluk, M. 2007. Permian. *In*: Wong, T.E., Batjes, D.A. & de Jager, J. *Geology of the Netherlands*. Royal Netherlands Academy of Arts and Sciences, Amsterdam, 63–83.

Geluk, M.C. 2005. *Stratigraphy and tectonics of Permo-Triassic basins in the Netherlands and surrounding areas*. PhD thesis, Utrecht University, Utrecht, The Netherlands.

Glennie, K. 1997. History of exploration in the southern North Sea. *In*: Ziegler, K., Turner, P. & Daines, S. (eds) *Petroleum Geology of the Southern North Sea: Future Potential*. Geological Society, London, Special Publications, **123**, 5–16, https://doi.org/10.1144/GSL.SP.1997.123.01.02

Hamilton, E.L. 1976. Variations of density and porosity with depth in deep-sea sediments. *Journal of Sedimentary Research*, **46**, 280–300.

Harding, R. & Huuse, M. 2015. Salt on the move: multi stage evolution of salt diapirs in the Netherlands North Sea. *Marine and Petroleum Geology*, **61**, 39–55.

Heybroek, P. 1975. On the structure of the Dutch part of the Central North Sea Graben. *In*: Woodland, A.W. (ed.) *Petroleum and the Continental Shelf of North-west Europe*. Applied Science, Barking, Essex, UK, 339–351.

Hospers, J., Rathore, J., Jianhua, F., Finnstrøm, E. & Holthe, J. 1988. Salt tectonics in the Norwegian–Danish Basin. *Tectonophysics*, **149**, 35–60.

Hudec, M.R. & Jackson, M.P. 2007. Terra infirma: understanding salt tectonics. *Earth-Science Reviews*, **82**, 1–28.

Hughes, M. & Davison, I. 1993. Geometry and growth kinematics of salt pillows in the southern North Sea. *Tectonophysics*, **228**, 239–254.

Ismail-Zadeh, A.T., Huppert, H.E. & Lister, J.R. 2002. Gravitational and buckling instabilities of a rheologically layered structure: implications for salt diapirism. *Geophysical Journal International*, **148**, 288–302.

Jackson, M. & Vendeville, B. 1994. Regional extension as a geologic trigger for diapirism. *Geological Society of America Bulletin*, **106**, 57–73.

Jackson, M., Cornelius, R., Craig, C., Gansser, A., Stöcklin, J. & Talbot, C. 1990. Salt diapirs of the Great Kavir, central Iran. *In*: Jackson, M.P.A. (ed.) *Salt Diapirs of the Great Kavir, Central Iran*. Geological Society of America, Memoirs, **177**, 1–150.

Jaritz, W. 1973. Zur entstehung der salzstrukturen nordwestdeutschlands. *Geologisches Jahrbuch*, **A10**, 3–77.

Kombrink, H., Doornenbal, J., Duin, E., Den Dulk, M., Ten Veen, J. & Witmans, N. 2012. New insights into the geological structure of the Netherlands; results of a detailed mapping project. *Netherlands Journal of Geosciences*, **91**, 419–446.

Kooi, H., Cloetingh, S. & Remmelts, G. 1989. Intraplate stresses and the stratigraphic evolution of the North Sea Central Graben. *In*: van der Linden, W.J.M., Cloetingh, S.A.P.L., Kaasschieter, J.P.K., van de Graaff, W.J.E., Vandenberghe, J. & van der Gun, J.A.M. (eds) *Coastal Lowlands*. Springer, Dordrecht, The Netherlands, 49–72.

Koyi, H. & Petersen, K. 1993. Influence of basement faults on the development of salt structures in the Danish Basin. *Marine and Petroleum Geology*, **10**, 82–94.

Littke, R., Bayer, U., Gajewski, D. & Nelskamp, S. (eds) 2008. *Dynamics of Complex Intracontinental Basins: the Central European Basin System*. Springer, Berlin.

Maystrenko, Y., Bayer, U. & Scheck-Wenderoth, M. 2005. Structure and evolution of the Glueckstadt Graben due to salt movements. *International Journal of Earth Sciences*, **94**, 799–814.

Mohr, M., Kukla, P., Urai, J. & Bresser, G. 2005. Multiphase salt tectonic evolution in NW Germany: seismic interpretation and retro-deformation. *International Journal of Earth Sciences*, **94**, 917–940.

Nelson, T.H. & Fairchild, L. 1989. Emplacement and evolution of salt sills in the northern Gulf of Mexico. *Houston Geological Society Bulletin*, **32**, 6–7.

Nettleton, L.L. 1934. Fluid mechanics of salt domes. *AAPG Bulletin*, **18**, 1175–1204.

O'Brien, C. 1957. Salt diapirism in south Persia. *Geologie en Mijnbouw*, **19**, 357–376.

Orlic, B. & Buijze, L. 2014. Numerical modeling of wellbore closure by the creep of rock salt caprocks. *Paper presented at the 48th US Rock Mechanics/Geomechanics Symposium*, 1–4 June 2014, Minneapolis, Minnesota, USA.

Poliakov, A.N., Podladchikov, Y.Y., Dawson, E.Ch. & Talbot, C.J. 1996. Salt diapirism with simultaneous brittle faulting and viscous flow. *In*: Alsop, G.I., Blundell, D.J. & Davison, I. (eds) *Salt Tectonics*. Geological Society, London, Special Publications, **100**, 291–302, https://doi.org/10.1144/GSL.SP.1996.100.01.09

Ramberg, H. 1968. Fluid dynamics of layered systems in the field of gravity, a theoretical basis for certain global structures and isostatic adjustment. *Physics of the Earth and Planetary Interiors*, **1**, 63–87.

Ramberg, H. 1972. Theoretical models of density stratification and diapirism in the Earth. *Journal of Geophysical Research*, **77**, 877–889.

Ramberg, H. 1981. *Gravity, Deformation, and the Earth's Crust: In Theory, Experiments, and Geological Application*. 2nd edn. Academic Press, London.

Ramsay, J. 1967. *Folding and Fracturing of Rocks*. McGraw-Hil, New York.

Remmelts, G. 1995. Fault-related salt tectonics in the southern North Sea, the Netherlands. *In*: Jackson, M.P.A., Roberts, D.G. & Snelson, S. (eds) *Salt Tectonics: A Global Perspective*. American Association of Petroleum Geologists, Memoirs, **65**, 261–272.

Remmelts, G. 1996. Salt tectonics in the southern North Sea, the Netherlands. *In*: Rondeel, H.E., Batjes, D.A.J. & Nieuwenhuijs, W.H. (eds) *Geology of Gas and Oil under The Netherlands*. Springer, Dordrecht, The Netherlands, 143–158.

Rönniund, P. 1989. Viscosity ratio estimates from natural Rayleigh-Taylor instabilities. *Terra Nova*, **1**, 344–348.

Rowan, M.G. & Ratliff, R.A. 2012. Cross-section restoration of salt-related deformation: best practices and potential pitfalls. *Journal of Structural Geology*, **41**, 24–37.

Scheck, M., Bayer, U. & Lewerenz, B. 2003. Salt redistribution during extension and inversion inferred from 3D backstripping. *Tectonophysics*, **373**, 55–73.

Scheck-Wenderoth, M., Maystrenko, Y., Hübscher, C., Hansen, M. & Mazur, S. 2008. Dynamics of salt basins. *In*: Littke, R., Bayer, U., Gajewski, D. & Nelskamp, S. (eds) *Dynamics of Complex Intracratonic Basins. The Central European Basin System*. Springer, Berlin, 307–322.

Schmeling, H. 1987. On the relation between initial conditions and late stages of Rayleigh–Taylor instabilities. *Tectonophysics*, **133**, 65–80.

Schultz-Ela, D.D., Jackson, M.P. & Vendeville, B.C. 1993. Mechanics of active salt diapirism. *Tectonophysics*, **228**, 275–312.

Smith, S. 2013. *Digital Signal Processing: A Practical Guide for Engineers and Scientists*. Newnes, Boston, MA.

Stewart, S. 2007. Salt tectonics in the North Sea Basin: a structural style template for seismic interpreters. *Special Publication-Geological Society of London*, **272**, 361–369.

Stewart, S. & Clark, J. 1999. Impact of salt on the structure of the Central North Sea hydrocarbon fairways. *In*: Fleet, A.J. & Boldy, S.A.R. (eds) *Petroleum Geology of Northwest Europe – Proceedings of the 5th Conference*. Geological Society, London, 179–200, https://doi.org/10.1144/0050179

Stewart, S.A. & Coward, M.P. 1995. Synthesis of salt tectonics in the southern North Sea, UK. *Marine and Petroleum Geology*, **12**, 457–475.

Talbot, C. 1995. Molding of salt diapirs by stiff overburden. *In*: Jackson, M.P.A., Roberts, D.G. & Snelson, S. (eds) *Salt Tectonics: A Global Perspective*.

American Association of Petroleum Geologists, Memoirs, **65**, 61–75.

Talbot, C. & Jackson, M. 1987. Internal kinematics of salt diapirs. *AAPG Bulletin*, **71**, 1068–1093.

Taylor, J.C.M. 1984. Late Permian–Zechstein. *In*: Glennie, K.W. (ed.) *Introduction to the Petroleum Geology of the North Sea*. Blackwell Scientific, Oxford, 137–173.

Taylor, J. 1990. Upper Permian–Zechstein. *In*: Glennie, K.W. (ed.) *Petroleum Geology of the North Sea: Basic Concepts and Recent Advances*. 4th edn. Blackwell Scientific, Oxford, 174–211.

Ten Veen, J., Van Gessel, S. & Den Dulk, M. 2012. Thin- and thick-skinned salt tectonics in the Netherlands; a quantitative approach. *Netherlands Journal of Geosciences*, **91**, 447–464.

TNO. 2011. *Tectono-stratigraphic Charts of the Netherlands Continental Shelf*. TNO Built Environment and Geosciences, **9**.

Trusheim, F. 1960. Mechanism of salt migration in northern Germany. *AAPG Bulletin*, **44**, 1519–1540.

Turcotte, D. & Schubert, G. 1982. *Geodynamics: Applications of Continuum Physics to Geological Problems*. Wiley, New York.

Van Adrichem Boogaert, H.A. & Kouwe, W.F.P. (compilers). 1993–1997. *Stratigraphic Nomenclature of the Netherlands; Revision and Update by RGD and NOGEPA*. Mededelingen Rijks Geologische Dienst, **50**.

Van Dalfsen, W., Doornenbal, J., Dortland, S. & Gunnink, J. 2006. A comprehensive seismic velocity model for the Netherlands based on lithostratigraphic layers. *Netherlands Journal of Geosciences*, **85**, 277–292.

Van Wijhe, D.V. 1987. Structural evolution of inverted basins in the Dutch offshore. *Tectonophysics*, **137**, 171–219.

van Winden, M. 2015. *Salt tectonics in the northern Dutch offshore*. Master thesis, Utrecht, The Netherlands.

Vendeville, B.C. 1989. Scaled experiments on the interaction between salt flow and overburden faulting during syndepositional extension. *In*: Ventress, W.P.S., Bebout, D.G., Perkins, B.F. & Moore, C.H. (eds) *Gulf of Mexico Salt Tectonics, Associated Processes and Exploration Potential*. SEPM Volume 10.

Vendeville, B.C. 2002. A new interpretation of Trusheim's classic model of salt-diapir growth. *Gulf Coast Association of Geological Societies Transactions*, **52**, 943–952.

Wessel, P. & Smith, W.H.F. 1998. New, improved version of generic mapping tools released. *Eos, Transactions of the American Geophysical Union*, **79**, 579.

Wride, V. 1995. Structural features and structural styles from the Five Countries Area of the North Sea Central Graben. *First Break*, **13**, 395–407.

Ziegler, P. 1990. *Geological Atlas of Western and Central Europe*. 2nd edn. Shell Internationale Petroleum Maatschappij, The Hague. Geological Society, London.

An introduction to the Triassic: current insights into the regional setting and energy resource potential of NW Europe

MARK GELUK[1]*, TOM MCKIE[2] & BEN KILHAMS[3]

[1]*Gerbrandylaan 18, 2314 EZ Leiden, The Netherlands*

[2]*Shell UK Exploration & Production, 1 Altens Farm Road, Nigg, Aberdeen AB12 3FY, UK*

[3]*Nederlandse Aardolie Maatschappij (NAM), PO Box 28000, 9400 HH Assen, The Netherlands*

**Correspondence: mark.c.geluk@gmail.com*

Abstract: A review of recent Triassic research across the Southern Permian Basin area demonstrates the role that high-resolution stratigraphic correlation has in identifying the main controls on sedimentary facies and, subsequently, the distribution of hydrocarbon reservoirs. The depositional and structural evolution of these sedimentary successions was the product of polyphase rifting controlled by antecedent structuration and halokinesis, fluctuating climate, and repeated marine flooding, leading to a wide range of reservoir types in a variety of structural configurations. Triassic hydrocarbon accumulations form an important energy resource across the basin, not only in the established Buntsandstein fairway but also in Rogenstein oolites and Muschelkalk carbonates. In addition, sand-prone sections in the Late Triassic, such as the Schilfsandstein, have the potential to be hydrocarbon reservoirs. Several Triassic intervals are now the focus for developing geothermal projects. A detailed understanding of Triassic reservoir quality and distribution is one of the main keys to efficiently unlocking the geothermal and remaining hydrocarbon potential across the basin.

A concise overview of the stratigraphic and palaeogeographical development of the Triassic is presented here, including recent insights into the understanding of these complex systems. The chapters in this Special Publication deal mainly with the Triassic as a hydrocarbon reservoir, including exploration aspects in the Dutch and German offshore (Kilhams *et al.* 2018; Kortekaas *et al.*, this volume, in press), the impact of overpressure models in the Terschelling Basin (Peeters *et al.*, this volume, in press), the use of open access well data (van Kempen *et al.*, this volume, in press) and enhanced gas recovery from the mature De Wijk gas field (Goswami *et al.* 2018). Franz *et al.* (this volume, in press) also consider the deltaic systems of the Rhaetian and their use as geothermal reservoirs in northern Germany. An overview of historical Triassic hydrocarbon discoveries and possible future geothermal aspects is introduced as a framework for these chapters.

Triassic basin evolution

Over the last decade, various studies in NW Europe have resulted in new insights into the sedimentary and tectonic evolution and resource potential of the Triassic in the Southern Permian Basin (SPB). High-resolution stratigraphic correlations have been carried out in order to consolidate the different stratigraphic frameworks in NW Europe, and to identify basin-wide reservoir and seal units from the UK in the west to Poland in the east (Dadlez *et al.* 1998; Lokhorst 1998; Evans *et al.* 2003; Geluk 2005; Feist-Burkhardt *et al.* 2008; McKie & Williams 2009; Bachmann *et al.* 2010; Barnasch 2010). These correlations can also be tied, supported by magnetostratigraphy (Szurlies *et al.* 2012) and limited biostratigraphy (Kürschner & Herngreen 2010; Backhaus *et al.* 2013; Scholze *et al.* 2017), to equivalent successions further afield in the North Sea area, thus assisting in developing an understanding of the regional palaeogeographical evolution in response to tectonics and climate. It has also become apparent that, for many regions, much of the Permian and Triassic time periods are missing in the sedimentary record (Fig. 1).

Antecedent Caledonian, Variscan and Permian structures had a great bearing on the structural and depositional processes that operated in these basins. The principal controlling lineaments are the Iapetus Suture and the Trans-European Suture Zone, marking the northern margin of the Caledonian collision (Smit *et al.* 2016). North of this line, focused rift tectonics prevailed; whereas, to the south, a wide extension zone existed in areas of thickened Variscan crust (Praeg 2004). As a result, the SPB is a complex amalgamation of linked depocentres (e.g. the Polish Trough, North German and Sole Pit basins). Fault systems inherited from earlier Permian rift systems formed the backbone of the later Triassic graben systems (Coward *et al.* 2003). During the Triassic,

From: Kilhams, B., Kukla, P. A., Mazur, S., McKie, T., Mijnlieff, H. F. & van Ojik, K. (eds) 2018. *Mesozoic Resource Potential in the Southern Permian Basin*. Geological Society, London, Special Publications, **469**, 139–147.
First published online January 26, 2018, https://doi.org/10.1144/SP469.1

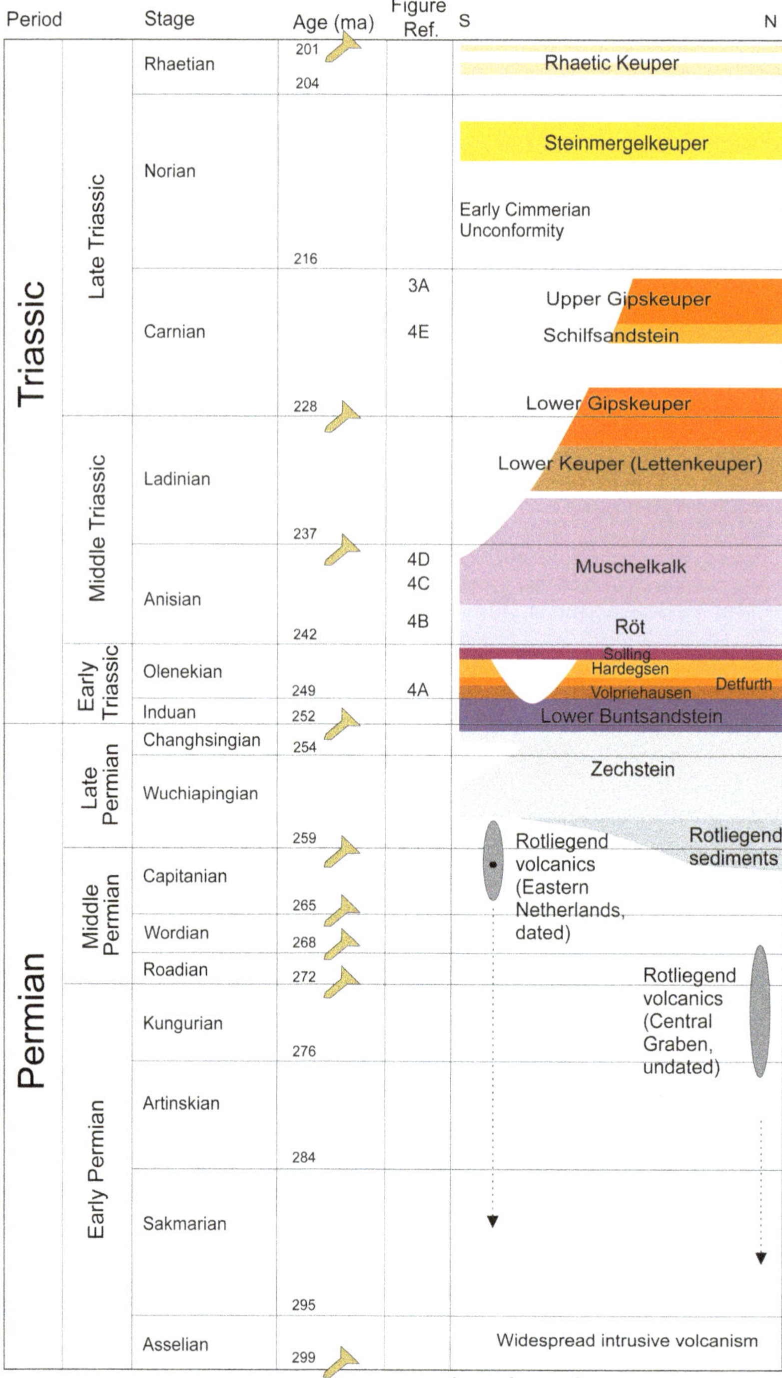

Fig. 1. Chronostratigraphic diagram of the Permian and Triassic deposits in The Netherlands and adjacent areas (after Geluk 2005). Stage names and numerical ages are after Cohen *et al.* (2013). 'Golden spikes' are indicated. The Induan and Olenekian replace the former Scythian. Two intra-Triassic unconformities are present – the Hardegsen Unconformity (Olenekian) and the Early Cimmerian Unconformity (Norian). Over large areas of the Southern Permian Basin, these unconformities result in large intervals of missing section.

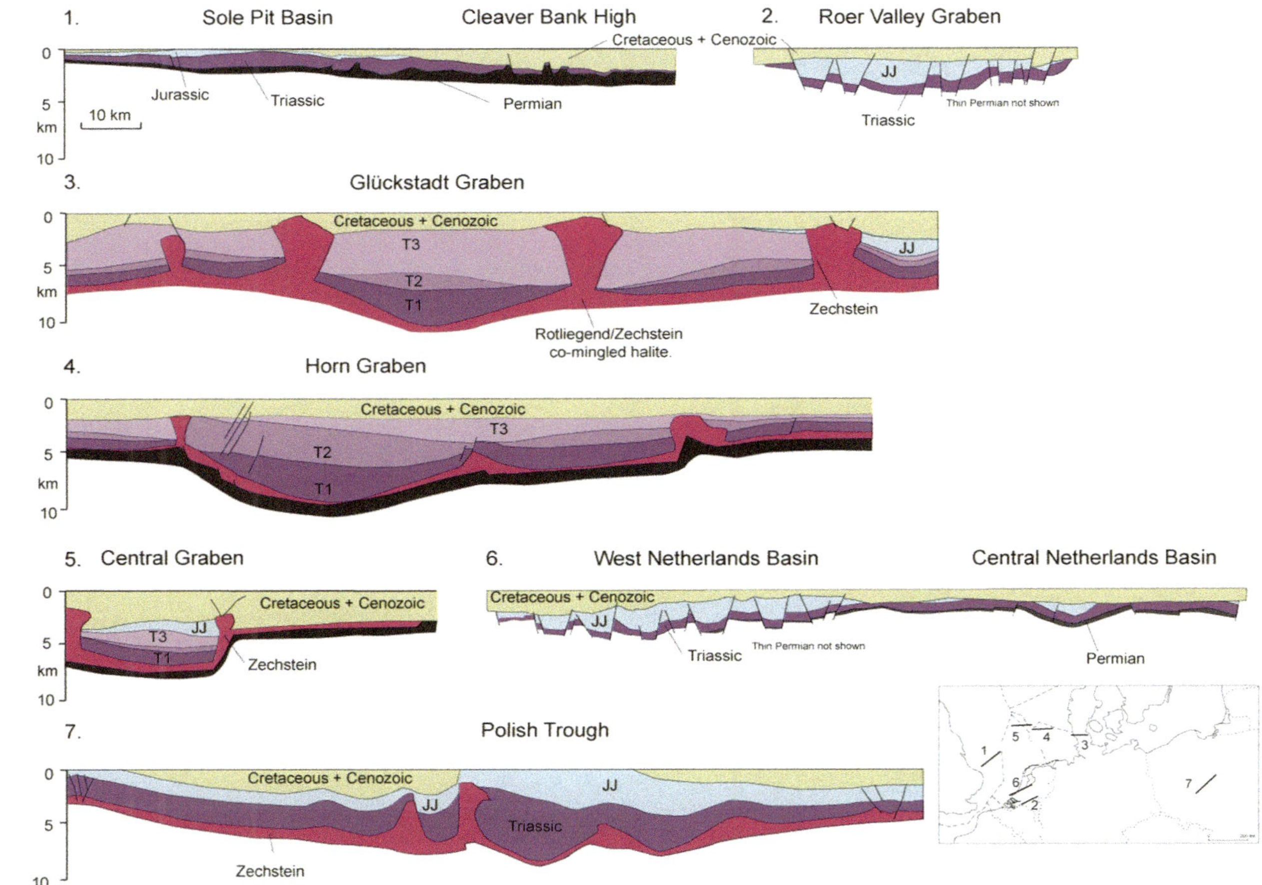

Fig. 2. Geological cross-sections of various Triassic basins in NW Europe: T1, Lower and Main Buntsandstein; T2, Solling, Röt and Muschelkalk; T3, Keuper; JJ, Jurassic. Note the difference in scale between these basins, with the thickest Triassic successions in the Glückstadt and Horn graben. The depth scale is in kilometres. Two-fold vertical exaggeration. Modified after Geluk (2005) and references therein.

progressive basin modification by extensional tectonics occurred as a result of fault propagation from both the Boreal and Tethyan regions (Bachmann *et al.* 2010). From the Middle Triassic onwards, rifting concentrated mainly in the Central and Northern North Sea, and the Horn and Glückstadt grabens (Fig. 2), together with onshore graben in the UK and eastern France (Bourquin *et al.* 1997, 2002; Ruffell & Shelton 1999). The degree of extension across the basins varied and resulted, in combination with the pre-existing structural grain and the presence of underlying mobile Zechstein salt, in a wide variety of tectonic structures (Fig. 2).

The palaeolatitude and location of the SPB within Pangea resulted in generally arid to semi-arid conditions throughout the Triassic (Fig. 3), although the degree of aridity fluctuated across the region, possibly in response to warming induced by volcanism from Large Igneous Provinces and, more locally, by episodic marine flooding increasing humidity. Warming may have influenced monsoon strength, evaporation rates, temperature gradients and the ability of monsoons to penetrate the continental interior (McKie 2014). In response to these changes, fluvial systems cyclically expanded and contracted across the basin, alternating with widespread playa, aeolian dune fields and, more rarely, perennial lakes. Larger-scale eustatic sea-level changes caused episodic flooding of the SPB and provided a fluctuating base level that also controlled

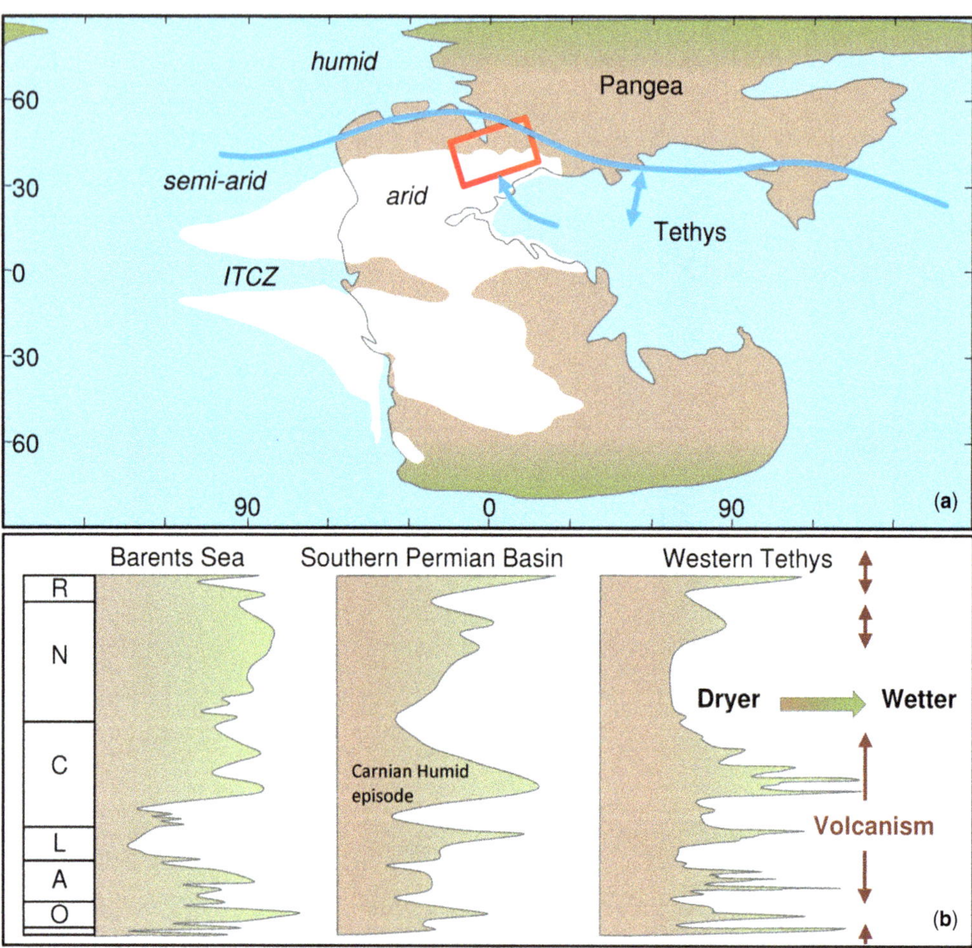

Fig. 3. Synopsis of Triassic climatic settings. (**a**) Late Triassic (Carnian–Norian) Pangean climate zonation (after Sellwood & Valdes 2007), placing the SPB region in the arid to semi-arid zone (red rectangle). (**b**) North to south climate variations through the Triassic (modified after McKie 2017). The plots are qualitative, with relatively drier conditions to the left and wetter to the right. O, Olenekian; A, Anisian; L, Ladinian; C, Carnian; N, Norian; R, Rhaetian; ITCZ, Intertropical Convergence Zone.

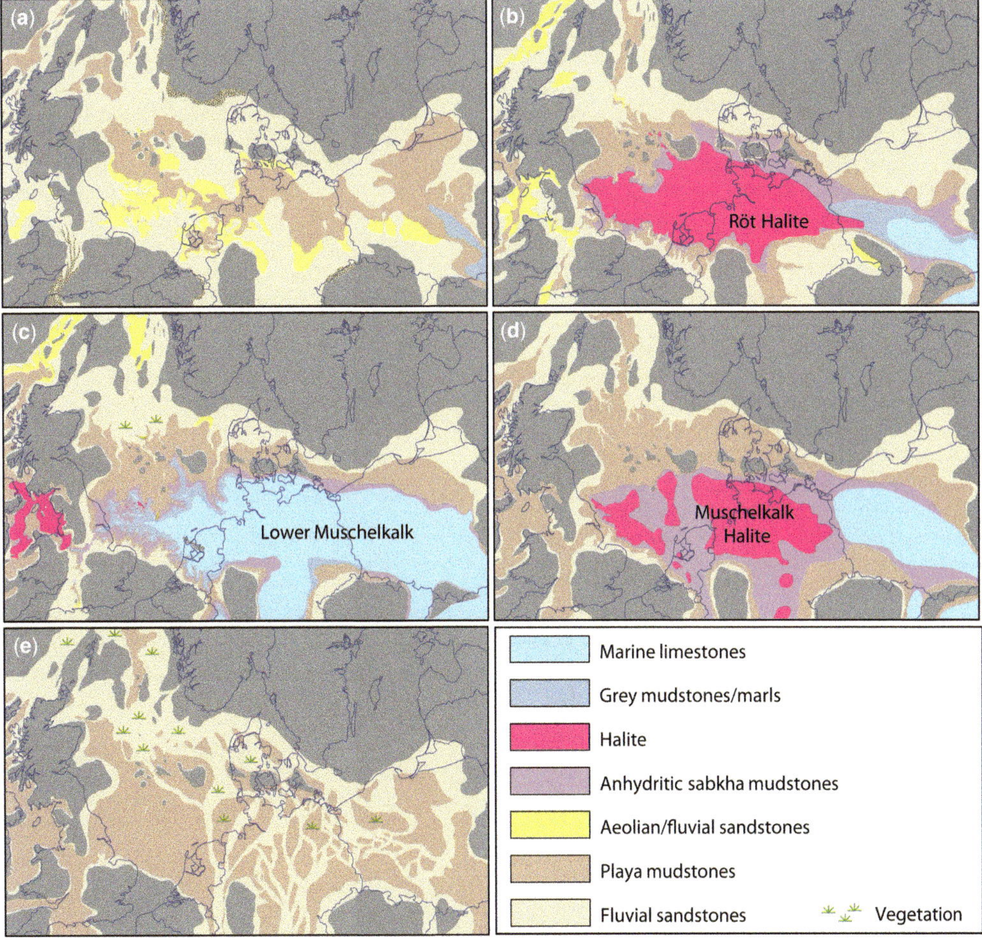

Fig. 4. Illustrative Triassic palaeogeographical reconstructions (after McKie 2017): (**a**) Main Buntsandstein (Olenekian) fluvial–aeolian deposition within continental basins; (**b**) Röt (Lower Anisian) restricted marine connection and evaporites; (**c**) Lower Muschelkalk marine flooding (Anisian); (**d**) Middle Muschelkalk basin restriction and evaporate precipitation; and (**e**) Schilfsandstein (Middle Carnian) expansion of Fennoscandian drainage in response to regional climate wettening.

downstream fluvial architecture at these times (Franz *et al.* 2014). During episodes of marine restriction, short-lived extensional pulses within the main rift zones provided conduits for marine brines to enter accommodation space, resulting in evaporite precipitation and preservation (Barnasch 2010; McKie 2017).

The basal part of the Triassic succession, the Buntsandstein, is dominated by continental clastic deposits. Ephemeral fluvial systems drained northwards off the Variscan fold belt and southwards off Fennoscandia, and flowed towards basinal playa in the central SPB (Geluk & Röhling 1997). In the prevailing arid climate, these fluvial deposits were subject to aeolian reworking in their distal reaches (Fig. 4a). In Late Olenekian–Early Anisian times, the SPB area was episodically inundated by marine water that entered the basin from the Tethys via the Silesian gate in SE Poland (Bachmann *et al.* 2010). Under dry climatic conditions, the evaporites of the Röt Formation were deposited in partially restricted basins (Fig. 4b) (cf. Kovalevych *et al.* 2002). A major transgression occurred during the latter part of the Anisian (Middle Triassic) via the Silesian and Burgundy gates. Ongoing arid conditions and low clastic supply led to the deposition of the shallow-marine carbonates of the Lower Muschelkalk (Fig. 4c). An interruption to the marine connection with the Tethys resulted in basin desiccation and the precipitation of anhydrite and halite of the Late

Anisian Middle Muschelkalk (Fig. 4d) (Bachmann *et al.* 2010) before fully marine conditions were re-established during the Early Ladinian Upper Muschelkalk. During the Late Ladinian, marine-influenced conditions receded from the SPB, although episodic flooding is recorded by thin marine intervals and the penetration of marine brines to form substantial Carnian halite accumulations.

With ongoing rifting and thermal subsidence, the depositional systems gradually expanded and onlapped over the highs around the basin margins in Middle–Late Triassic times (e.g. Paris Basin, onshore UK). However, a major change in basin palaeogeography occurred from the Ladinian/Carnian onwards, when Fennoscandia became the dominant sediment source area and the Variscides ceased to source large fluvial systems (van der Zwan & Spaak 1992). In general, sediments of these Fennoscandian systems were contained within the Central North Sea (represented by the Skagerrak Formation), leaving the SPB dominated by mud-prone and evaporitic playa conditions (Keuper); however, during wetter climate conditions, higher discharge rivers were able to expand across the SPB. The Carnian Schilfsandstein represents the largest of these systems (Fig. 4e) and was possibly driven by a major climate excursion (Fig. 3b) resulting from volcanism associated with the Wrangellia Large Igneous Province (cf. Ruffell *et al.* 2015).

In the latter stages of the Triassic, the climate gradually became more humid (Kozur & Bachmann 2010) as a result of the northwards drift of Pangea and the increasing maritime influence from both the Boreal and Tethyan seaways. Deltaic and shallow-marine conditions dominated during the Rhaetian over the northern half of the area across Denmark and NW Germany (Bachmann *et al.* 2010; Franz *et al.*, this volume, in press).

Triassic energy resources

Hydrocarbons associated with Triassic reservoirs occur in various sub-basins across the Southern

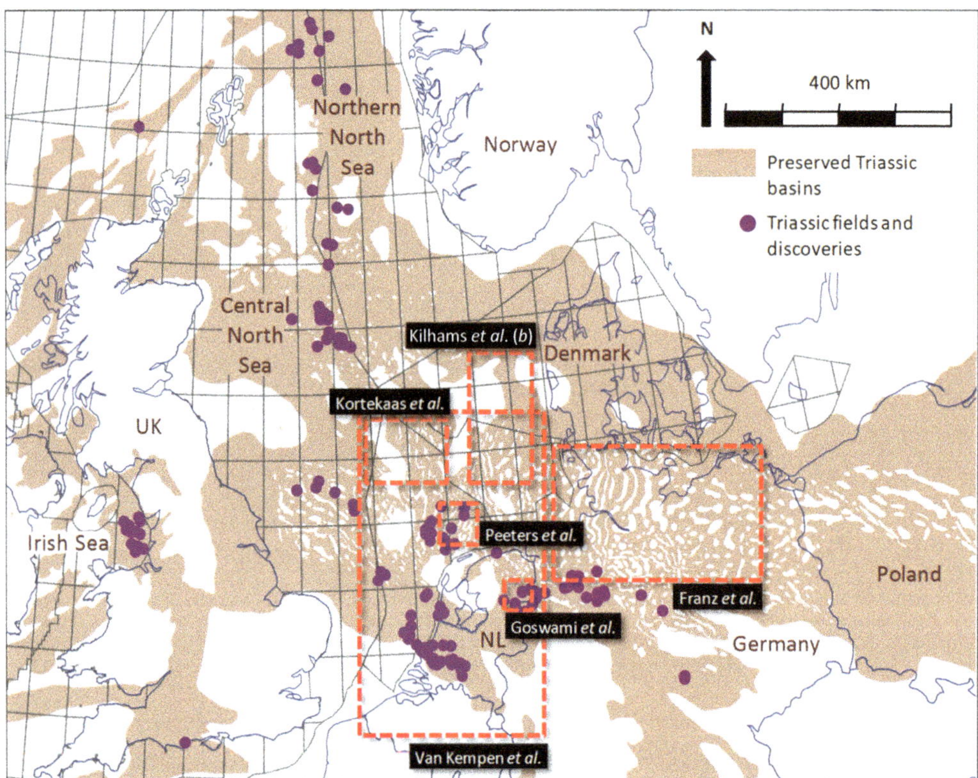

Fig. 5. Distribution of Triassic-hosted hydrocarbon fields in the NW European region. Map after Glennie (1998), Lokhorst (1998), Goldsmith *et al.* (2003) and Bachmann *et al.* (2010). The preserved Triassic basin area represents the extent of the Basal Bunter Shale interval. Red dashed boxes represent the study areas from this Special Publication.

Permian Basin, the East Irish Sea, the Central and Northern North Sea, and the Wessex Basin (Fig. 5), largely associated with sand-prone fluvial systems such as the Sherwood Sandstone Group, and the Skagerrak and Lunde formations. In the Southern North Sea and adjacent onshore areas, hydrocarbons mainly occur in the sandstones of the Lower Triassic Buntsandstein. Other reservoirs of secondary importance are the Rogenstein lacustrine oolites and the marine dolomites of the Lower Muschelkalk (Palermo *et al.* 2008).

Various publications provide more details of the many Triassic fields across the region (e.g. Goldsmith *et al.* 2003; De Jager & Geluk 2007; Bachmann *et al.* 2010). In the Southern North Sea and adjacent onshore areas, Triassic gas volumes are modest when compared to the Rotliegend (see Kilhams *et al.* 2018) but form a number of important fields (e.g. Caister, G14-A and L09). The offshore extension of the West Netherlands Basin also exhibits a number of Triassic-hosted gas fields, such as Q16-A. The onshore equivalent area has been important historically, with a number of Triassic fault blocks charged with hydrocarbons (e.g. Gaag-Monster, Botlek and Pernis West). Another trend of the Triassic fields is associated with the margin of the Lower Saxony Basin in The Netherlands and Germany (e.g. Sleen, Roswinkel and De Wijk (De Jager & Geluk 2007; Goswami *et al.* 2018) in The Netherlands, and Hengstlage, Rehden and Apeldorn in Germany). The underground gas storage capacity in Triassic reservoirs is utilized in Germany, both in depleted gas fields (Allmenhausen, Döttlingen, Kalle and Uelsen) and in aquifers (Berlin and Bucholz).

In the Central and Northern North Sea, widespread fluvial reservoirs extend through the Middle and Late Triassic succession (Anisian–Norian), forming the basis for economic hydrocarbon accumulations in the Alwyn, Beryl, Judy, Marnock and Snorre fields, amongst others (cf. Evans *et al.* 2003). In the East Irish Sea, the fluvial–aeolian Anisian Ormskirk Formation forms the main reservoir (e.g. Morecambe Field; Stuart 1993). In the Wessex Basin, the predominantly fluvial Sherwood Sandstone of Olenekian–Anisian age forms the main reservoir in the Wytch Farm Field (McKie *et al.* 1998).

In general, the best quality reservoirs are either contained within high net-to-gross, well-connected fluvial channel successions, or within aeolian reworked sections associated with dry climatic episodes. Proximal fluvial sections tend to be immature and poorly sorted, and constitute less effective reservoirs; whilst, at distal locations, in the absence of aeolian processes, sections tend to be mud-prone and poorly connected. Middle–Late Triassic reservoirs were associated with generally less arid conditions and can show extensive degradation of reservoir quality by groundwater and palaeosol carbonate cementation.

In the Polish part of the SPB, no hydrocarbon accumulations have been encountered in Triassic rocks. Here, reservoir rocks are present in the Lower Triassic but charge and seal are an issue (Bachmann *et al.* 2010).

Exploration for further Triassic-hosted oil and gas continues across the basin, especially in relatively underexplored areas where data can be reinterpreted in the context of new research (e.g. Kilhams *et al.*, this volume, in press; Kortekaas *et al.*, this volume, in press). One element which could aid further exploration is the investigation of possible Triassic source rock intervals. These are generally considered absent in the Triassic red beds. However, the identification of thin coaly intervals in the Rhaetian of offshore Denmark, NW Germany and The Netherlands could provide a local hydrocarbon source (Nielsen 2003).

The recent demand for clean energy and the development of geothermal energy is an exciting new element for Triassic resources with increasing economic importance. Several Triassic intervals are subject to investigation for future geothermal exploitation. The results of such projects have emphasized the importance of a sound geological understanding of the subsurface to achieve economically attractive projects (e.g. Wolfgramm *et al.* 2009; Pluymaekers *et al.* 2012; Franz *et al.*, this volume, in press).

Conclusions

Studies published during the last decade have supported an increased understanding of Triassic stratigraphy and palaeogeographical evolution, leading to an improved definition of reservoir and seal distribution, trends in reservoir properties, and the potential for hydrocarbon plays in reservoir intervals in addition to the recognized Buntsandstein. Geothermal aspects of similar subsurface targets are now being explored, and require similar understanding. It is hoped that the following chapters will provide further inspiration to unlock the remaining geological questions and the resource potential of the Triassic interval in NW Europe.

Geoff Warrington (University of Leicester) is thanked for suggestions on the text and for information on reservoir ages. Two anonymous reviewers are also thanked for their ideas which greatly improved the manuscript.

References

Bachmann, G.H., Geluk, M.C. *et al.* 2010. Chapter 9. Triassic. *In*: Doornenbal, H. & Stevenson, A. (eds) *Petroleum Geological Atlas of the Southern Permian Basin*

Area. European Association of Geoscientists and Engineers (EAGE), Houten, The Netherlands, 148–173.

Backhaus, E., Hagdorn, H., Heunisch, C. & Schulz, E. 2013. Biostratigraphische Gliederungsmöglichkeiten des Buntsandstein [Biostratigraphic subdivision of the Buntsandstein]. *In*: Lepper, J. & Röhling, H.G. (eds) *Stratigraphie von Deutschland XI*. Schriftenreihe der Deutschen Gesellschaft für Geowissenschaften, **69**, 151–164.

Barnasch, J. 2010. *Der Keuper im Westteil des Zentraleuropäischen Beckens (Deutschland, Niederlande, England, Dänemark): diskontinuierliche Sedimentation, Litho-, Zyklo- und Sequenzstratigraphie [The Keuper in the west part of the Central European Basin (Germany, The Netherlands, Denmark): discontinuous sedimentation, litho, cyclo and sequence stratigraphy]*. Schriftenreihe der Deutschen Gesellschaft für Geowissenschaften, **71**.

Bourquin, S., Vairon, J. & Le Strat, P. 1997. Three-dimensional evolution of the Keuper of the Paris basin based on detailed isopach maps of the stratigraphic cycles: tectonic influences. *Geologische Rundschau*, **86**, 670–685.

Bourquin, S., Robin, C., Guillocheau, F. & Gaulier, J.M. 2002. Three dimensional accommodation analysis of the Keuper of the Paris Basin: discrimination between tectonics, eustacy and sediment supply in the stratigraphic record. *Journal of Marine and Petroleum Geology*, **19**, 469–498.

Cohen, K.M., Finney, S.C., Gibbard, P.L. & Fan, J.-X. 2013. The ICS International Chronostratigraphic Chart. *Episodes*, **36**, 199–204.

Coward, M.P., Dewey, J.F., Hempton, M. & Holroyd, J. 2003. Tectonic evolution. *In*: Evans, D., Graham, C., Armour, A. & Bathurst, P. (eds) *The Millenium Atlas: Petroleum Geology of the Central and Northern North Sea*. Geological Society, London, 17–33.

Dadlez, R., Marek, S. & Pokorski, J. (eds). 1998. *Atlas Paleogeograficzny Epikontynentalnego Permu i Mesozoiku w Polsce [Paleogeographical Atlas of the Epicontinental Permian and Mesozoic in Poland] (1:2 500 000)*. Panstwowy Instytut Geologiczny, Warszawa, Poland.

De Jager, J. & Geluk, M.C. 2007. Petroleum geology. *In*: Wong, Th.E., Batjes, D.A.J. & De Jager, J. (eds) *Geology of the Netherlands*. Royal Dutch Academy of Arts and Sciences, Amsterdam, 237–260.

Evans, D., Graham, C., Armour, A. & Bathurst, P. (eds). 2003. *The Millennium Atlas, Petroleum Geology of the Central and Northern North Sea*. Geological Society, London.

Feist-Burkhardt, S., Götz, A.E. *et al.* 2008. Triassic. *In*: McCann, T. (ed.) *The Geology of Central Europe. Volume 2: Mesozoic and Cenozoic*. Geological Society, London, 749–821.

Franz, M., Nowak, K., Berner, U., Heunisch, C., Bandel, K., Röhling, H.-G. & Wolfgramm, M. 2014. Eustatic control on epicontinental basins: The example of the Stuttgart Formation in the Central European Basin (Middle Keuper, Late Triassic). *Global and Planetary Change*, **122**, 305–329.

Franz, M., Barth, G., Zimmermann, J., Budach, I., Nowak, K. & Wolfgramm, M. In press. Deep geothermal resources of the North German Basin: exploration strategy, development examples and remaining opportunities in Mesozoic hydrothermal reservoirs. *In*: Kilhams, B., Kukla, P.A., Mazur, S., McKie, T., Mijnlieff, H.F. & van Ojik, K. (eds) *Mesozoic Resource Potential of the Southern Permian Basin*. Geological Society, London, Special Publications, **469**, https://doi.org/10.1144/SP469.11

Geluk, M.C. 2005. *Stratigraphy and tectonics of Permo-Triassic basins in the Netherlands and surrounding areas*. PhD thesis, Utrecht University.

Geluk, M.C. & Röhling, H.-G. 1997. High-resolution sequence stratigraphy of the Lower Triassic 'Buntsandstein' in the Netherlands and Northwest Germany. *Geologie en Mijnbouw*, **76**, 227–246.

Glennie, K.W. 1998. *Petroleum Geology of the North Sea, Basic Concepts and Recent Advances*, 4th edn. Blackwell Scientific, Oxford.

Goldsmith, P.J., Hudson, G. & Van Veen, P. 2003. Triassic. *In*: Evans, D., Graham, C., Armour, A. & Bathurst, P. (eds) *The Millenium Atlas: Petroleum Geology of the Central and Northern North Sea*. Geological Society, London, 105–127.

Goswami, R., Seeberger, F. & Bosman, G. 2018. Enhanced gas recovery of an ageing field utilizing N_2 displacement: De Wijk Field, The Netherlands. *In*: Kilhams, B., Kukla, P.A., Mazur, S., McKie, T., Mijnlieff, H.F. & van Ojik, K. (eds) *Mesozoic Resource Potential of the Southern Permian Basin*. Geological Society, London, Special Publications, **469**. First published online January 16, 2018, https://doi.org/10.1144/SP469.2

Kilhams, B., Kukla, P., Mazur, S., McKie, T., Mijnlieff, H., van Ojik, K. & Rosendaal, E. In press. Mesozoic resource potential in the Southern Permian Basin. The geological key to exploiting remaining hydrocarbons whilst unlocking geothermal potential. *In*: Kilhams, B., Kukla, P.A., Mazur, S., McKie, T., Mijnlieff, H. F. & van Ojik, K. (eds) *Mesozoic Resource Potential of the Southern Permian Basin*. Geological Society, London, Special Publications, **469**, https://doi.org/10.1144/SP469.26

Kilhams, B., Stevanovic, S. & Nicolai, C. 2018. The 'Buntsandstein' gas play of the Horn Graben (German and Danish offshore): dry well analysis and remaining hydrocarbon potential. *In*: Kilhams, B., Kukla, P.A., Mazur, S., McKie, T., Mijnlieff, H.F. & van Ojik, K. (eds) *Mesozoic Resource Potential of the Southern Permian Basin*. Geological Society, London, Special Publications, **469**. First published online January 11, 2018, https://doi.org/10.1144/SP469.5

Kortekaas, M., Böker, U., van der Kooij, C. & Jaarsma, B. In press. Lower Triassic reservoir development in the Dutch northern offshore. *In*: Kilhams, B., Kukla, P.A., Mazur, S., McKie, T., Mijnlieff, H.F. & van Ojik, K. (eds) *Mesozoic Resource Potential of the Southern Permian Basin*. Geological Society, London, Special Publications, **469**, https://doi.org/10.1144/SP469.19

Kovalevych, V., Peryt, T.M., Beer, W., Geluk, M. & Halas, S. 2002. Geochemistry of early Triassic seawater as indicated by study of the Röt halite in the Netherlands, Germany, and Poland. *Chemical Geology*, **182**, 549–563.

Kozur, H.W. & Bachmann, G. 2010. The Middle Carnian Wet Intermezzo of the Stuttgart Formation

(Schilfsandstein), Germanic Basin. *Palaeogeography, Palaeoclimatology, Palaeoecology*, **290**, 107–119.

KÜRSCHNER, W.M. & HERNGREEN, G.F.W. 2010. Triassic palynology of central and northwestern Europe: a review of palynofloral diversity patterns and biostratigraphic subdivisions. *In*: LUCAS, S.G. (ed.) *The Triassic Timescale*. Geological Society, London, Special Publications, **334**, 263–283, https://doi.org/10.1144/SP334.11

LOKHORST, A. (ed.) 1998. *NW European Gas Atlas*. CD ROM and report NITG-TNO, Haarlem, The Netherlands.

MCKIE, T. 2014. Climatic and tectonic controls on Triassic dryland terminal fluvial system architecture, Central North Sea. *In*: MARTINIUS, A.W., RAVNÅS, R., HOWELL, J.A., STEEL, R.J. & WONHAM, J.P. (eds) *From Depositional Systems to Sedimentary Successions on the Norwegian Continental Shelf*. International Association Of Sedimentologists, Special Publications, **46**, 19–58.

MCKIE, T. 2017. Paleogeographic evolution of latest Permian and Triassic salt basins in Northwest Europe. *In*: SOTO, J.I., FLINCH, J. & TARI, G. (eds) *Permo-Triassic Salt Provinces of Europe, North Africa and the Atlantic Margins. Tectonics and Hydrocarbon Potential*. Elsevier, Amsterdam, 159–173.

MCKIE, T. & WILLIAMS, B. 2009. Triassic palaeogeography and fluvial dispersal across the northwest European Basins. *Geological Journal*, **44**, 711–741.

MCKIE, T., AGGETT, J. & HOGG, A.J.C. 1998. Reservoir architecture of the upper Sherwood Sandstone, Wytch Farm Field, southern England. *In*: UNDERHILL, J.R. (ed.) *The Development and Evolution of the Wessex Basin and Adjacent Areas*. Geological Society, London, Special Publications, **133**, 399–406, https://doi.org/10.1144/GSL.SP.1998.133.01.21

NIELSEN, L.H. 2003. Late Triassic–Jurassic development of the Danish Basin and the Fennoscandian Border Zone, southern Scandinavia. *Geological Survey of Denmark and Greenland Bulletin*, **1**, 459–526.

PALERMO, D., AIGNER, T., GELUK, M., PÖPPELREITER, M. & PIPPING, K. 2008. Reservoir potential of a lacustrine mixed carbonate/siliciclastic gas reservoir: the Lower Triassic Rogenstein in the Netherlands. *Journal of Petroleum Geology*, **31**, 61–96.

PEETERS, S., ASSCHERT, A. & VERWEIJ, H. In press. Towards a better understanding of the highly overpressured Lower Triassic Bunter reservoir rocks in the Terschelling Basin. *In*: KILHAMS, B., KUKLA, P.A., MAZUR, S., MCKIE, T., MIJNLIEFF, H.F. & VAN OJIK, K. (eds) *Mesozoic Resource Potential of the Southern Permian Basin*. Geological Society, London, Special Publications, **469**, https://doi.org/10.1144/SP469.13

PLUYMAEKERS, M.P.D., KRAMERS, L., VAN WEES, J.D., KRONIMUS, A., NELSKAMP, S., BOXEM, T. & BONTÉ, D. 2012. Reservoir characterisation of aquifers for direct heat production: Methodology and screening of the potential reservoirs for the Netherlands. *Netherlands Journal of Geosciences – Geologie en Mijnbouw*, **91**, 621–636.

PRAEG, D. 2004. Diachronous Variscan late-orogenic collapse as a response to multiple detachments: a view from the internides in France to the foreland in the Irish Sea. *In*: WILSON, M., NEUMANN, E.-R., DAVIES, G.R., TIMMERMAN, M.J., HEEREMANS, M. & LARSEN, B. T. (eds) *Permo-Carboniferous Magmatism and Rifting in Europe*. Geological Society, London, Special Publications, **223**, 89–138, https://doi.org/10.1144/GSL.SP.2004.223.01.05

RUFFELL, A. & SHELTON, R. 1999. The control of sedimentary facies by climate during phases of natural extension: examples from the Triassic of onshore and offshore England and Northern Ireland. *Journal of the Geological Society, London*, **156**, 779–789, https://doi.org/10.1144/gsjgs.156.4.0779

RUFFELL, A., SIMMS, M.J. & WIGNALL, P.B. 2015. The Carnian Humid Episode of the late Triassic: a review. *Geological Magazine*, **153**, 271–284.

SCHOLZE, F, WANG, X. *ET AL.* 2017. A multistratigraphic approach to pinpoint the Permian-Triassic boundary in continental deposits: the Zechstein–Lower Buntsandstein transition in Germany. *Global and Planetary Change*, **152**, 129–151.

SELLWOOD, B.W. & VALDES, P.J. 2007. Mesozoic climate. *In*: WILLIAMS, M., HAYWOOD, A.M., GREGORY, F.J. & SCHMIDT, D.N. (eds) *Deep-Time Perspectives on Climate Change: Marrying the Signal from Computer Models and Biological Proxies*. The Micropalaeontological Society, Special Publications. Geological Society, London, 210–224.

SMIT, J., VAN WEES, J.-D. & CLOETINGH, S. 2016. The Thor suture zone: from subduction to intraplate basin setting. *Geology*, **44**, 707–710.

STUART, I.A. 1993. The geology of the North Morecambe Gas Field, East Irish Sea Basin. *In*: PARKER, J.R. (ed.) *Petroleum Geology of Northwest Europe: Proceedings of the 4th Conference*. Geological Society, London, 883–895, https://doi.org/10.1144/0040883

SZURLIES, M., GELUK, M.C., KRIJGSMAN, W. & KÜRSCHNER, W.M. 2012. The continental PermianTriassic boundary in the Netherlands: implications for the geomagnetic polarity time scale. *Earth and Planetary Science Letters*, **317–318**, 165–176.

VAN DER ZWAN, C.J. & SPAAK, P. 1992. Lower to Middle Triassic sequence stratigraphy and climatology of the Netherlands, a model. *Palaeogeography, Palaeoclimatology, Palaeoecology*, **91**, 277–290.

VAN KEMPEN, B.M.M., MIJNLIEFF, H.F. & VAN DER MOLEN, J. In press. Data mining in the Dutch Oil & Gas Portal: a case study on reservoir properties of the Volpriehausen Sandstone Interval. *In*: KILHAMS, B., KUKLA, P.A., MAZUR, S., MCKIE, T., MIJNLIEFF, H.F. & VAN OJIK, K. (eds) *Mesozoic Resource Potential of the Southern Permian Basin*. Geological Society, London, Special Publications, **469**, https://doi.org/10.1144/SP469.15

WOLFGRAMM, M., OBST, K., BEICHEL, K., BRANDES, J., KOCH, R., RAUPPACH, K. & THORWART, K. 2009. *Produktivitätsprognosen geothermischer Aquifere in Deutschland [Productivity prognosis for geothermal aquifers in Germany]*. *In*: Proceedings des Geothermiekongresses, 17–19 November 2009, Bochum, Germany.

Lower Triassic reservoir development in the northern Dutch offshore

M. KORTEKAAS[1]*, U. BÖKER[2], C. VAN DER KOOIJ[3] & B. JAARSMA[1]

[1]*EBN BV Daalsesingel 1, 3511 SV Utrecht, The Netherlands*

[2]*PanTerra Geoconsultants BV, Weversbaan 1-3, 2352 BZ Leiderdorp, The Netherlands*

[3]*Utrecht University, Budapestlaan 4, 3584 CD Utrecht, The Netherlands*

**Correspondence: marloes.kortekaas@ebn.nl*

Abstract: Sandstones of the Main Buntsandstein Subgroup represent a key element of the well-established Lower Triassic hydrocarbon play in the southern North Sea area. Mixed aeolian and fluvial sediments of the Lower Volpriehausen and Detfurth Sandstone members form the main reservoir rock, sealed by the Solling Claystone and/or Röt Salt. It is generally perceived that reservoir presence and quality decrease towards the north and that the prospectivity of the Main Buntsandstein play in the northern Dutch offshore is therefore limited. Lack of access to hydrocarbon charge from the underlying Carboniferous sediments as a result of the thick Zechstein salt is often identified as an additional risk for this play. Consequently, only a few wells have tested Triassic reservoir and therefore this part of the basin remains under-explored. Seismic interpretation of the Lower Volpriehausen Sandstone Member was conducted and several untested Triassic structures are identified. A comprehensive, regional well analysis suggests the presence of reservoir sands north of the main fairway. The lithologic character and stratigraphic extent of these northern Triassic deposits may suggest an alternative reservoir provenance in the marginal Step Graben system. Fluvial sands with (local) northern provenance may have been preserved in the NW area of the Step Graben system, as seismic interpretation indicates the development of a local depocentre during the Early Triassic. These insights help to improve the chance of finding Lower Buntsandstein reservoir rocks in the northern Dutch offshore.

The Lower Triassic Main Buntsandstein Subgroup (RBM) (Van Adrichem Boogaert & Kouwe 1994) forms part of an established hydrocarbon play in the southern North Sea. In the Netherlands it is the second-most prolific hydrocarbon play after the Permian Rotliegend play (de Jager & Geluk 2007). Regionally widespread mixed aeolian and fluvial sediments of the Lower Volpriehausen and Lower Detfurth Sandstones form the primary reservoir. It is generally perceived that the sediments of these Lower Triassic reservoir rocks are sourced from the southern Variscan mountains, and reservoir presence and quality decrease towards the north (Fig. 1) (e.g. Geluk & Röhling 1997; Geluk 2005, 2007; McKie & Williams 2009; Bachmann *et al.* 2010). Consequently, the prospectivity of the RBM play in the northern Dutch offshore is regarded as limited. Access to charge of hydrocarbons from the underlying Carboniferous deposits is often seen as an additional uncertainty for this play (de Jager & Geluk 2007). As a result, only a few wells have been drilled in the study area (17 000 km^2) with RBM as a primary or secondary target (Fig. 1).

The observation that Triassic sands which have derived their material from a local source area (Olivarius *et al.* 2017) are present in the North German Basin, together with the release of additional seismic and well data in the northern Dutch offshore, has justified a new review of the Lower Triassic prospectivity of this area (EBN 2015, 2016). A regional study using publicly available wells and seismic data from the northern Dutch offshore and surrounding territories was set up to support the investigation of the reservoir character and distribution of Lower Triassic sands north of the main fairway. A comprehensive well review, including stratigraphic correlation and spatial analysis of reservoir development, was conducted to establish the character and geographic distribution of Main Buntsandstein Subgroup reservoir sands in the northern Dutch offshore. Basins and isolated fault blocks with preserved Main Buntsandstein were delineated from platforms and halokinetic highs by seismic interpretation of the Lower Volpriehausen Sandstone Member. These structural depth grids were used to identify closures for potential hydrocarbon entrapment. Although

From: Kilhams, B., Kukla, P. A., Mazur, S., McKie, T., Mijnlieff, H. F. & van Ojik, K. (eds) 2018. *Mesozoic Resource Potential in the Southern Permian Basin*. Geological Society, London, Special Publications, **469**, 149–168.
First published online April 13, 2018, https://doi.org/10.1144/SP469.19

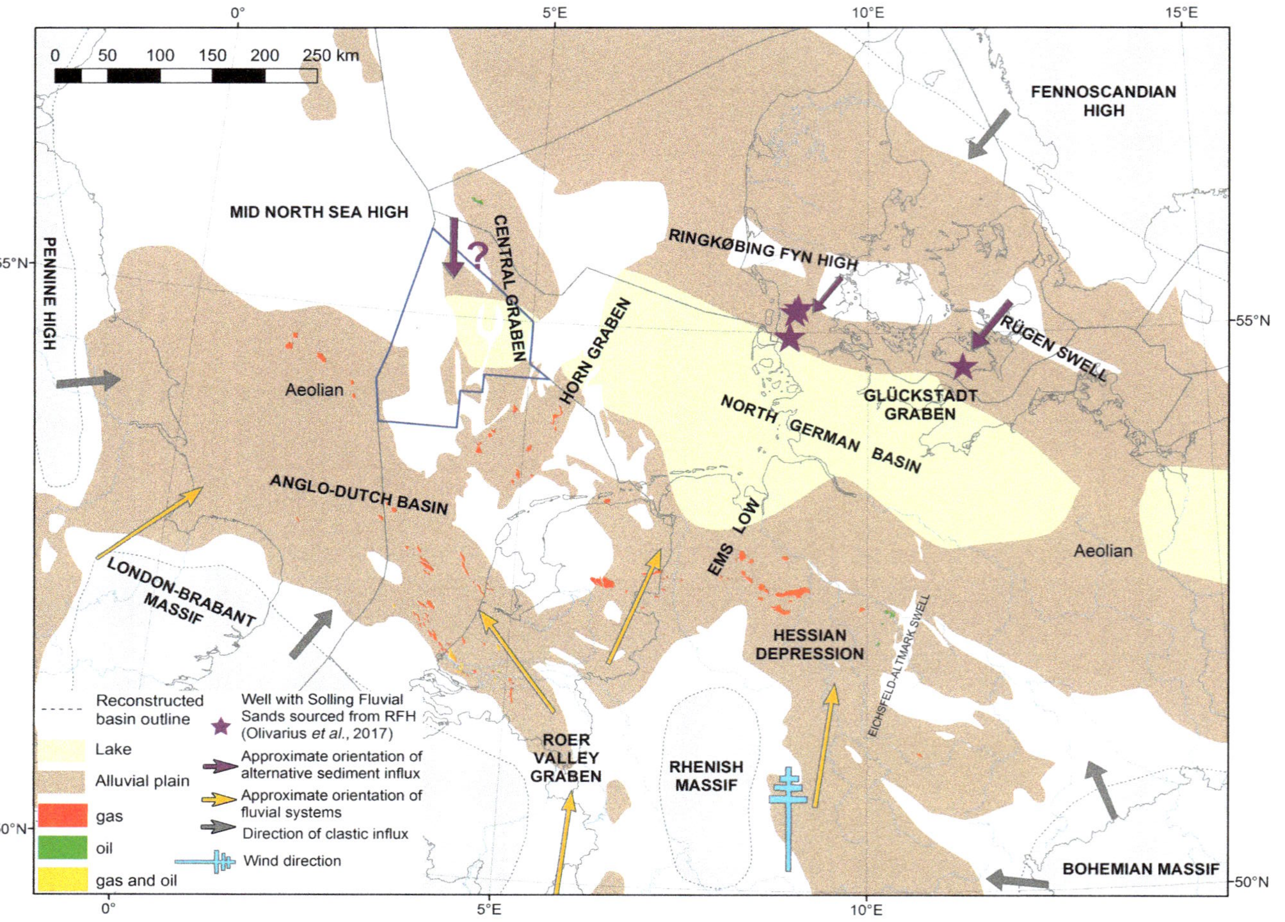

Fig. 1. Palaeogeographical map of the Southern Permian Basin during Early Triassic time, showing the well locations with northern sediment provenance indicated by purple stars (Olivarius *et al.* 2017). The study area is outlined in blue (the palaeogeography is adapted from Bachmann *et al.* 2010, p. 158).

Lower Volpriehausen Sandstone Member thickness cannot be directly mapped on seismic data, intra-Triassic isochron maps do assist in understanding the potential for Early Triassic local depocentre development. A recent study by ter Borgh *et al.* (2018*b*) provides new insights into the source-rock potential in the study area, while the interpretation of the underlying top and base Zechstein horizons in recent three-dimensional (3D) seismic data demonstrates potential access for hydrocarbon migration pathways.

Geological and tectonostratigraphic setting

The study area is located in the northern part of the east–west-trending Southern Permian Basin (Fig. 1) that formed during Late Permian and Early Triassic times (e.g. Ziegler 1990). The northern boundary of the Southern Permian Basin is characterized by a chain of highs, with the Ringkøbing-Fyn High to the east and the Mid North Sea High to the NW of the study area. These highs were relatively stable throughout the Mesozoic (Wride 1995) and are considered non-depositional areas during the Triassic (Geluk 2005, 2007).

The break-up of Pangea during the Early Triassic initiated as a result of rifting between Greenland and Scandinavia that propagated into the North Sea as well as into the North Atlantic Domain (Ziegler 1990). The North Sea rift system transected the Northern and the Southern Permian basins (Ziegler 1990), which ultimately led to the formation of the Dutch Central Graben where differential subsidence began during the Permian (De Jager 2007; Geluk 2007). At the same time the Step Graben was formed to the west where an alternating set of horst and graben structures indicate a lesser degree of subsidence (e.g. ter Borgh *et al.* 2018*a*). The lithosphere of these basins was still considerably thinner and weaker than the lithosphere of Fennoscandia to the north and the Anglo-Brabant and Variscan massifs to the south (Ziegler 1992). The thickness of the RBM therefore reflects a combination of enhanced subsidence in the grabens, and uplift, truncation and erosion elsewhere (Geluk & Röhling 1997).

The Main Buntsandstein Subgroup (RBM)

During the Induan age at the onset of the Triassic, clastic deposition of mud-dominated Lower Buntsandstein Formation (RBSH) took place in the Dutch sector; in the UK, mud-dominated sediments of the contemporaneous Smith Bank and Bunter Shale formations are encountered (e.g. Goldsmith *et al.* 1995; Bachmann *et al.* 2010) (Figs 1, 2). The lithological similarity between these two shale-prone formations that conformably overlie the Permian Zechstein deposits imply that the depositional conditions in the Central and Southern North Sea basins were very similar. Communication between the Central and Southern North Sea basins possibly occurred during the Induan and probably early Olenekian (Fisher & Mudge 1990).

Deposition during the Olenekian in the western part of the Southern Permian Basin is characterized by four widely recognized tectonostratigraphic units bounded by unconformities. The Volpriehausen, Detfurth, Hardegsen and Solling formations each represent large-scale fining-upwards sequences in which mixed fluvial-aeolian sandstones grade upwards into lacustrine clay-siltstone deposits (Geluk & Röhling 1997, 1999). The basal sandstone units of the different formations/sequences have a regional sheet-like character and are commonly referred to as ‘Lower’ (e.g. Volpriehausen) Sandstone Members. The ‘Upper’ Sandstone Members are considered sandy basin-fringe equivalents of the regionally shale-prone (e.g. Volpriehausen Clay-Siltstone) members that are not developed as such in our study area (Van Adrichem Boogaert & Kouwe 1994). The layering cyclicity is believed to record the interplay between the low-frequency build-up and subsequent release of regional tensional stresses that caused broad-scale crustal warping, accompanied by reactivation of NNE-trending swells and troughs along the southern margin of the Southern Permian Basin, and the higher-frequency monsoonal rainfall in the Variscan hinterland (e.g. Geluk 2005; McKie & Williams 2009; Bachmann *et al.* 2010). The strongest of these short-lived rifting pulses (Geluk 2007) was probably the pre-Solling pulse which culminated during the Olenekian and which resulted in the so-called Hardegsen (or H-) Unconformity (Röhling 1991), thereby delineating the Lower Germanic Trias Group from the base of the Upper Germanic Trias Group at the onset of Solling Sandstone deposition (Fig. 2; Van Adrichem Boogaert & Kouwe 1994). Events leading to this Hardegsen Unconformity mainly control preservation of sediments of the Main Buntsandstein Subgroup (Geluk & Röhling 1997; Pharaoh *et al.* 2010).

Figure 1 shows the reconstruction of the Lower Triassic depositional environment. The Variscan mountains in the south and the Fennoscandian High in the north are considered to be the predominant source areas for sediment (e.g. Geluk 2005, 2007; McKie & Williams 2009; Bachmann *et al.* 2010). Based on heavy mineral analysis on core samples, Olivarius *et al.* (2017) identified a southern Variscan source for the Lower Volpriehausen sands, possibly transported through aeolian mechanisms, and a northern (local) source for the Solling Sandstone Formation from the Ringkøbing-Fyn High, possibly transported through fluvial mechanisms.

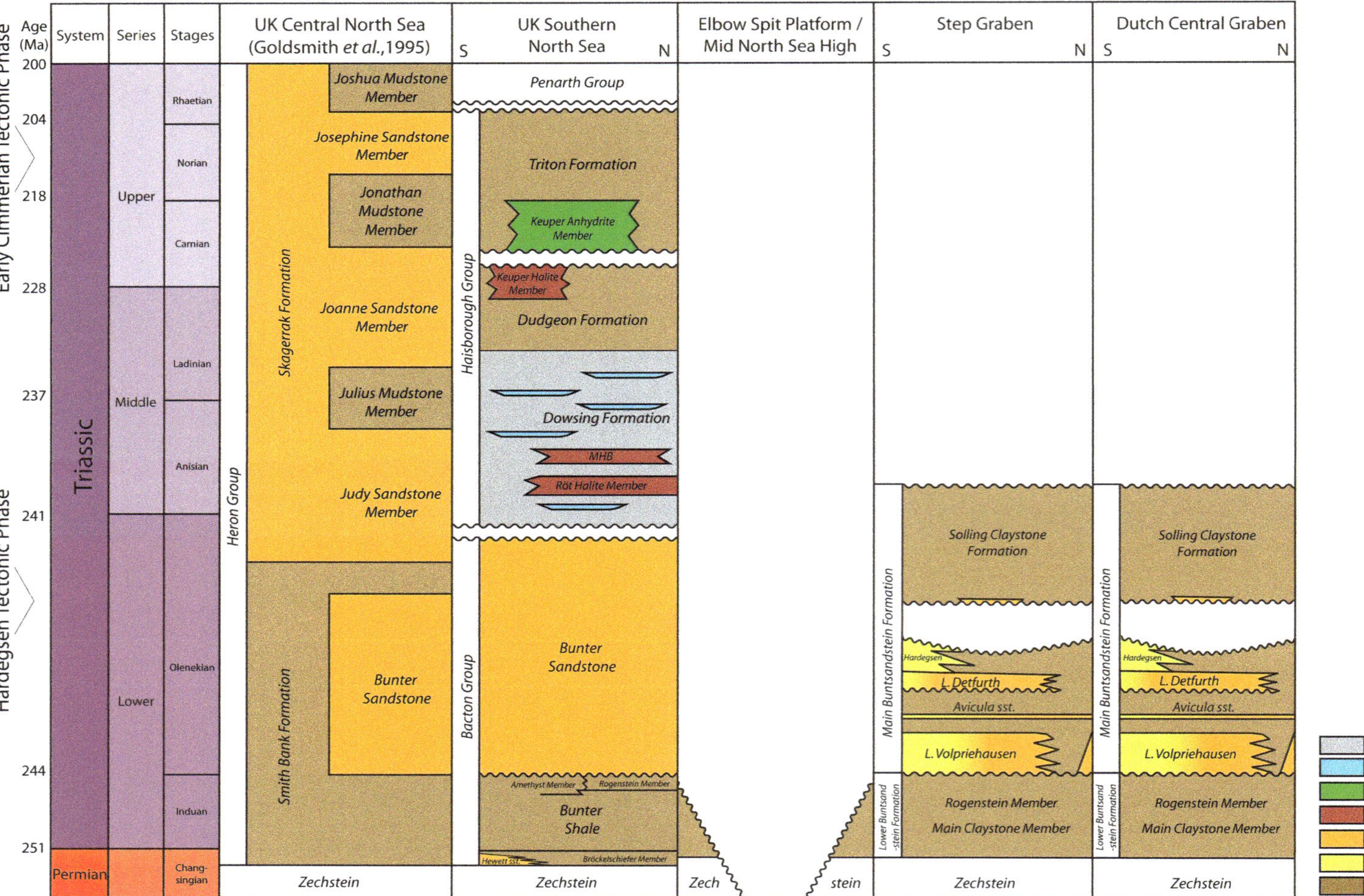

Fig. 2. Lithostratigraphic correlation panel between the UK and the northern Dutch offshore (adapted from van der Kooij 2016).

Post-Hardegsen tectonism and Zechstein salt mobilization

Syndepositional faulting is clearly observable in the Lower Triassic of the NNE–NE-striking Horn Graben, German North Sea (Best *et al.* 1983), Terschelling Basin, Dutch Southern North Sea (Harding & Huuse 2015) and also in the Sole Pit Basin, UK Southern North Sea (Johnson *et al.* 1994). Early generation of salt swells in the Southern Permian Basin area led to synkinematic shortening of the Lower–Middle Triassic cover that detached from Zechstein halite at extensional fault systems (Griffiths *et al.* 1995; Best 1996; Thieme & Rockenbauch 2001). The combination of fault and salt movements resulted in both symmetrical and asymmetrical depocentres that have a strong control on Triassic preservation.

A number of post-Hardegsen rifting pulses took place during the Anisian–Norian. The north–south-oriented fault pattern of this Early Cimmerian tectonic phase suggests east–west- to NE–SW-aligned extension (Zanella *et al.* 2003; de Jager 2007), with accelerated subsidence in the Dutch Central Graben and minor strike-slip movements along a number of fault zones (Geluk 2005, 2007). Basement faulting triggered widespread mobilization of Zechstein salt during the Late Triassic (van Winden 2015).

Middle Cimmerian erosion during Central North Sea doming during the Middle Jurassic led to the separation of Northern and Southern Permian Basins by the Mid North Sea/Ringkøbing-Fyn Highs (Wride 1995). Concurrent erosional truncation of Triassic strata was mostly restricted to the Northern Permian Basin (Underhill & Partington 1993), but also occurred locally on highs such as the eastern shoulder of the German Central Graben (Arfai *et al.* 2014). During the Late Cimmerian phase, which lasted from the Late Jurassic to the Early Cretaceous, the Central Graben subsided and the surrounding platforms were uplifted again (Remmelts 1996), leading to erosion of the Triassic cover (e.g. on the Cleaver Bank High west of the Step Graben system and on the Netherlands Swell to the south of the study area) (Geluk 2007).

The Alpine orogeny led to a number of tectonic inversion events during the Late Cretaceous–Early Paleogene (Spathopoulos *et al.* 2000; de Jager 2007). These were accompanied by NW–SE-oriented dextral strike-slip movements and accelerated salt movements, particularly along lineaments of salt walls that follow regional basement faults (Remmelts 1996; de Jager 2007). Triassic and younger strata were generally displaced and removed during post-Hardegsen erosional events where salt structures pierced the Mesozoic cover.

Data and methods

Seismic and well data

Several 3D and 2D seismic datasets are utilized for structural definition of the Lower Volpriehausen Sandstone Member (Fig. 3). The southern part of the study area (D, E and F blocks) is covered by the 3D Post Stack Time Migrated (PSTM) survey, covering 7950 km^2 (courtesy of Spectrum ASA) and the 3D TerraCube Offshore (2011) seismic dataset. The northern part of the study area is covered by several released 3D surveys and the entire area is covered by the NW–SE- and NE–SW-oriented 2D North Sea Renaissance (NSR) lines (TGS, 2003–07). In the UK sector, public 2D seismic lines from the 2015 Mid North Sea High Oil and Gas Authority (OGA) survey (OGA 2016) were used. Seismic interpretation was performed in the time domain using the non-SEG seismic convention.

Publicly available well data from 61 wells in the northern Dutch offshore (http://www.nlog.nl) are used for reservoir characterization (Figs 3, 4). Public data from wells in the surrounding territory were obtained from the UK OGA and the Danish Geological Survey (GEUS).

Well log interpretation

Wireline logs of all drilled Main Buntsandstein Subgroup sections in the study area were correlated. The vertical profiles were categorized according to log facies display, with the strongest focus on the Lower Volpriehausen Sandstone Member (RBMVL) as the most prominent sand-prone stratigraphic unit. Secondary focus is on the Lower Detfurth Sandstone Member (RBMDL), which is part of the Detfurth tectonostratigraphic cycle that overlies the Volpriehausen cycle (Geluk & Röhling 1997, 1999).

Lithological interpretation and correlation in terms of sand and shale content were carried out for all drilled RBMVL and RBMDL intervals. First, the gross thickness of the RBMVL and RBMDL intervals were determined by isochore extraction of the well tops. True stratigraphic thickness (TST) was calculated using well path information in combination with the structural dip obtained from the depth-converted seismic marker horizon (see the section ‘Seismic interpretation and time–depth conversion’). Wells with clear fault cut-outs at Main Buntsandstein Subgroup (RBM) level (e.g. F17-01, J03-01) and the few truncated sections (E13-01, E13-02) were removed prior to TST gridding (Fig. 5).

In order to obtain an estimate of net sand thickness, the gamma ray (GR) log of each RBM section was analysed using histograms; all show a bimodal distribution of high-GR values marking the shale intervals, and low-GR values marking sand-prone

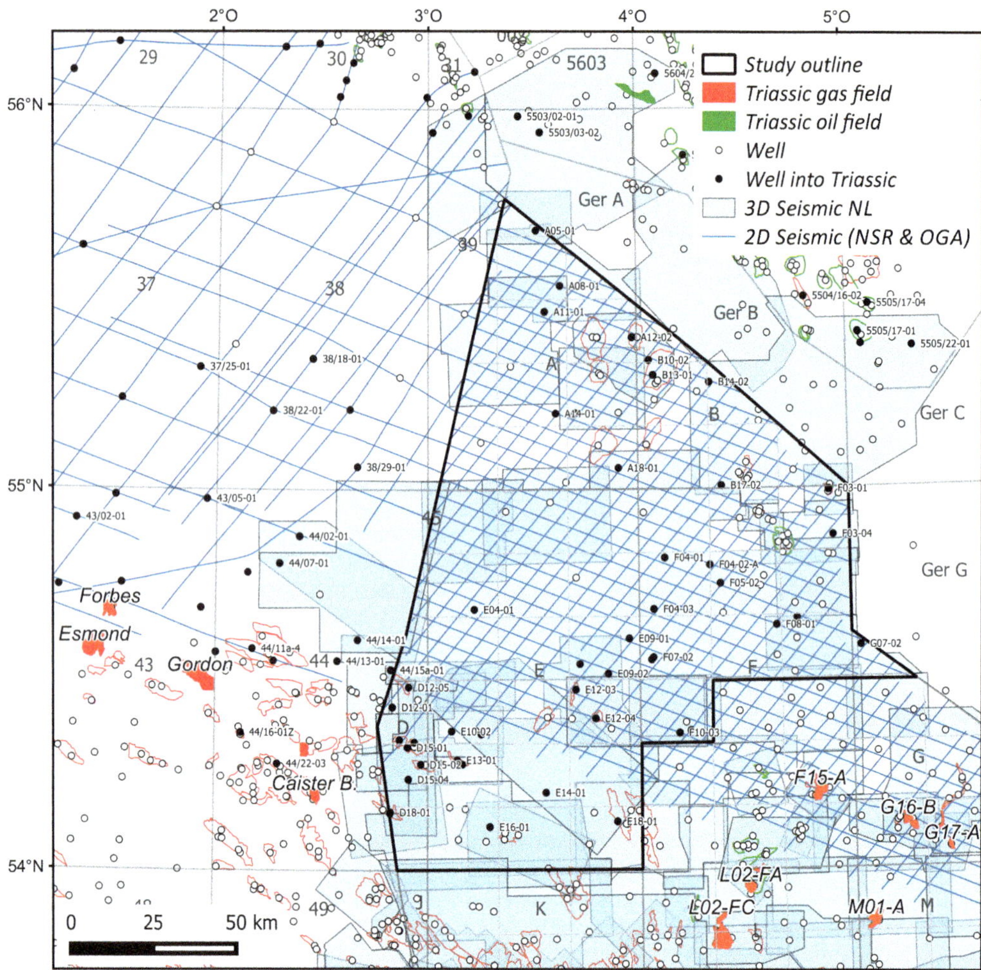

Fig. 3. Overview of study area, available seismic data in the Netherlands, wells and fields (red and green open polygons indicate gas and oil fields, respectively).

intervals. The cut-off for discriminating potential reservoir sandstones from the shales is placed at an intermediate GR value selected for each individual borehole section, and checked against cuttings descriptions. The bimodal GR distribution renders the definition of sand and shale relatively insensitive to the precision of the cut-off value (e.g. Fig. 4), so the cumulative net thicknesses for each section is considered a solid estimate of clean sand content. Regional gridding of the well data is discussed in the section 'Lower Volpriehausen and Detfurth Sandstone lithology mapping'.

Seismic interpretation and time–depth conversion

Seismic interpretation of a Lower Volpriehausen Sandstone marker was performed using the available 3D and 2D seismic surveys. Regional interpretations of the overlying stratigraphic horizons were used for time–depth conversion, including the base Paleocene (base Lower North Sea Group), base Late Cretaceous (base Chalk Group), base Early Cretaceous (base Rijnland Group) and top Triassic. A layer-cake V_0-k model was used with constant k and varying V_0-maps based on regional interpretations of EBN. From base Chalk onwards, the VELMOD-2 V_0-k model was used (Van Dalfsen *et al.* 2007).

Depocentre development

Isochron maps for an early (A) and middle (B) Triassic interval were created for the focus area as indicated by the dashed black outline in Figure 4. The outline of this area has been based on the availability of 3D seismic data (Fig. 3) and on the location

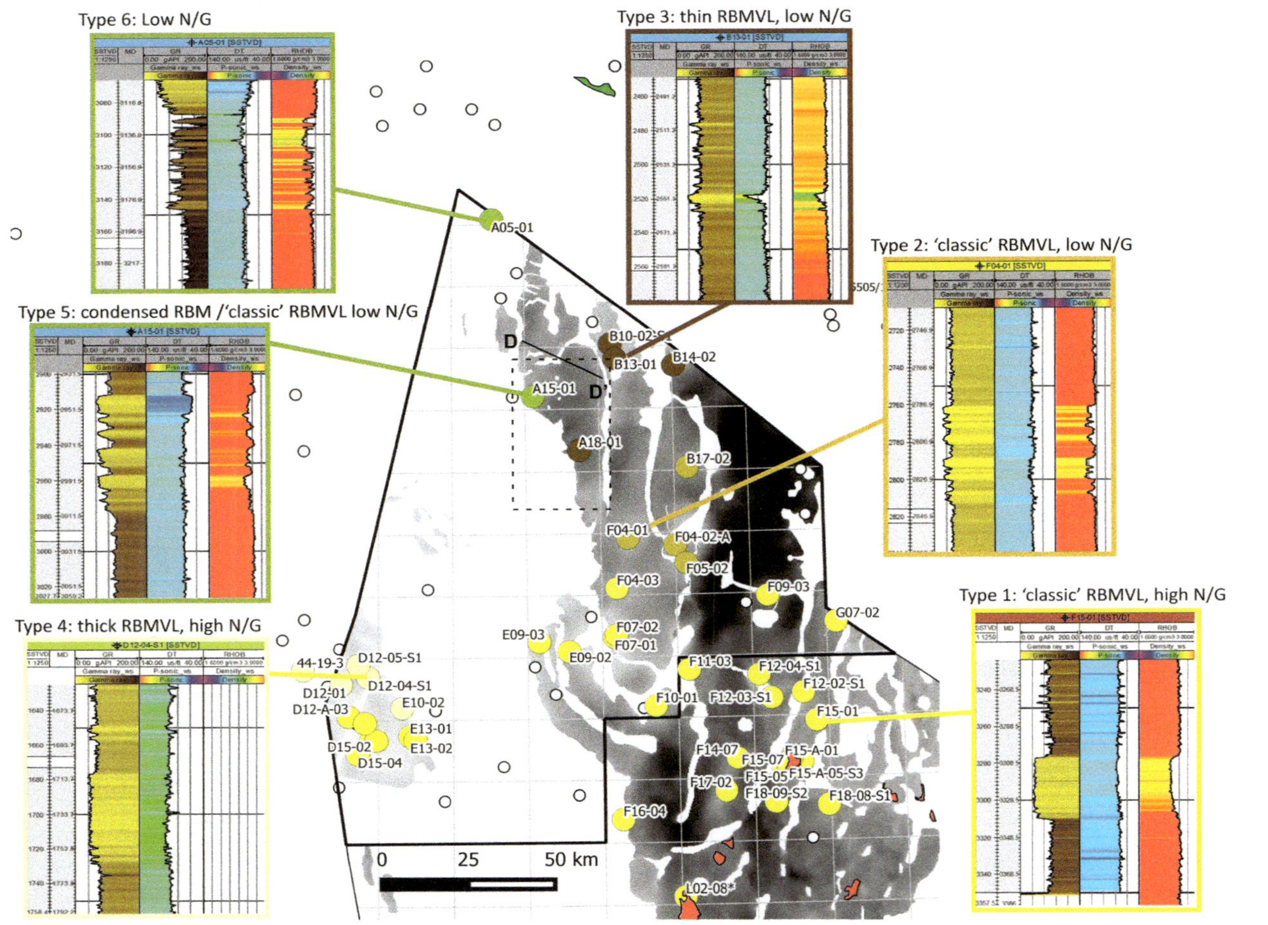

Fig. 4. Distribution of RBM profiles in the northern Dutch offshore. Well locations colour-coded by facies type (numbered according to description in text). Study area of the intra-Triassic isochron mapping (in Fig. 7) is outlined in dashed black. Seismic section from Figure 10 is indicated by black line D–D′. RBMVL reservoir rocks are present in most of the study area. The abundance and thickness of RBMVL decrease from south to north, while sands with local/northern provenance may have developed in the NW area.

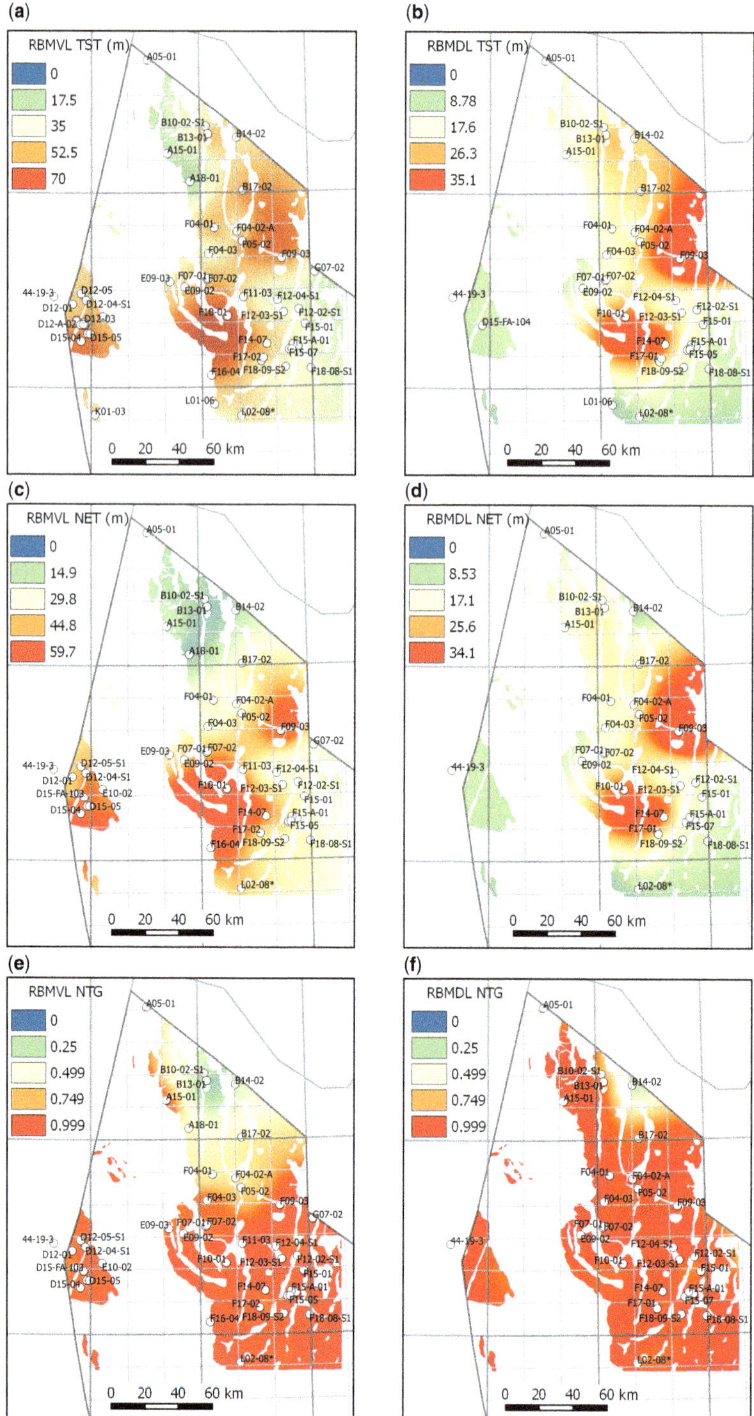

Fig. 5. Series of grids that show (**a**, **b**) true stratigraphic thickness (TST), (**c**, **d**) net sand thickness (NST) and (**e**, **f**) net-to-gross ratios (NTG) for Lower Volpriehausen and Detfurth Sandstone members, respectively. TST and NST are based on well data and both were gridded with the convergent interpolation method. NTG is calculated by dividing the NST grids by TST for each stratigraphic unit. See text for discussion.

of wells A18-01 and A15-01 that both encountered Lower Triassic strata. The isochron maps are used to study accommodation space generation for the two Triassic intervals. The horizons were chosen based on their continuity throughout the TerraCube 2011 seismic dataset. Top Lower Volpriehausen Sandstone was interpreted to represent the top of interval A. The top Röt Evaporite represents the top of interval B.

Lower Triassic reservoir development

Reservoir/sand development

Triassic sandstones are present north of the main fairway, in the Netherlands, UK and Danish territories (Figs 1, 6). Triassic sandstone intervals have been encountered in several wells in the Danish sector (Fig. 6). The Triassic sandstone interval in well 5504/16-02 (U-1X) is approximately 15 m thick and corresponds to deposits of the Main Buntsandstein Subgroup (Michelsen & Andersen 1983; Michelsen & Clausen 2002). Well Bertel-1/1A that drilled the Bertel oilfield possibly drilled Triassic reservoir sandstones. Age dating in this well is inconclusive and the reservoir sandstones might be of Jurassic or Late Triassic (Carnian) age (Schilfsandstein deposits) (GEUS 1993). As the log character of the northernmost well (A05-01) in the Netherlands looks very similar to Bertel-1/1A, it was decided to perform age dating also on A05-01 (Kerstholt & Munsterman 2016). Palynomorphs of both Jurassic and Late Triassic (Carnian) age are found in A05-01. As only a few palynomorphs could be extracted from the cutting samples of A05-01, the age dating remains inconclusive.

Regional mapping of the Lower Volpriehausen Sandstone Member shows that Main Buntsandstein Subgroup reservoir rocks are present in most of the study area (Fig. 4). The Lower Buntsandstein Formation (RBSH) is recognized on seismic cross-section as a uniform low-reflectivity layer overlying the Zechstein evaporites in the entire Dutch offshore sector, unless eroded on the platforms. The reflector representing the Lower Volpriehausen Sandstone Member was clearly traceable in the DEF PSTM survey, but of variable manifestation in the other surveys. As the tectono-cyclic occurrence of sandstones in the Main Buntsandstein Subgroup package cannot be resolved from seismic data in detail, our investigation of the Main Buntsandstein reservoir distribution is largely based on log profile typing and net sand mapping of well data (GR cut-offs). Seismic mapping is mainly used to outline potential traps and areas of post-depositional preservation.

Log profile typing as a means of regional facies analysis. Wireline log profiles of all drilled Main Buntsandstein Subgroup (RBM) sections in the study area were compared and wells were classified, with focus on the Lower Volpriehausen Sandstone Member (RBMVL) as the most prominent sand-prone stratigraphic unit (Fig. 4):

- Type 1: This category represents a regular set of RBM sandstones with typical 'classic Germanic Triassic' expression (type section F15-01). The high net sand content in the RBMVL may contain 1–2 shale stringers (F07-01, F07-02, F04-03). The core description of F16-04 classifies the RBMVL as an amalgamation of aeolian dune, dry aeolian sandflat, damp sandflat, fluvial channel and fluvial sheetflood deposits that represent the uppermost part of a large-scale coarsening-upwards sequence, overlain by distal sheetflood and lake margin sediments of the Volpriehausen Clay-Siltstone Member (RBMVC).
- Type 2: This group is characterized by a regular RBMVL gross thickness with frequent shale intercalations and thus intermediate net-to-gross ratios (type section F04-01). It represents the transition between type 1 and type 3.
- Type 3: This shaly RBMVL section is characterized by a thin sandstone layer at the top of the succession (type section B13-01). This type section is associated with an ephemeral lake environment (Bachmann *et al.* 2010), where only the topmost layers of sand of the coarsening-upwards sequence described in F16-04 (type 1, see above) are preserved. The log character shows a strong resemblance to the Danish U-1 well (15 m sand; Michelsen & Andersen 1983).
- Type 4: This group is characterized by an amalgamated sandstone section with limited sand/shale contrast (type section D12-04-S1). The type section is typical for the Cleaver Bank High and resembles contemporaneous Middle Bunter Sandstone in the adjacent UK sector (e.g. Cameron *et al.* 1992). The absence of intercalated shale formations suggests non-deposition and possibly sediment bypass during tectonic quiescence (e.g. due to a lack of accommodation space near the basin margin) and/or fringe sandstone equivalents that amalgamate stratigraphically with the classic (type 1) sand-shale cyclicity of Volpriehausen, Detfurth and Hardegsen formations.
- Type 5: This type section is characterized by a condensed RBM sandstone package with very thin intercalated shale formations. It has a single occurrence in A15-01. This profile could either represent a transition between the Bunter (type 4) and regular RBM (type 1), or it could be a local expression of the Volpriehausen sandstone similar to type 2 (discussed further in sections

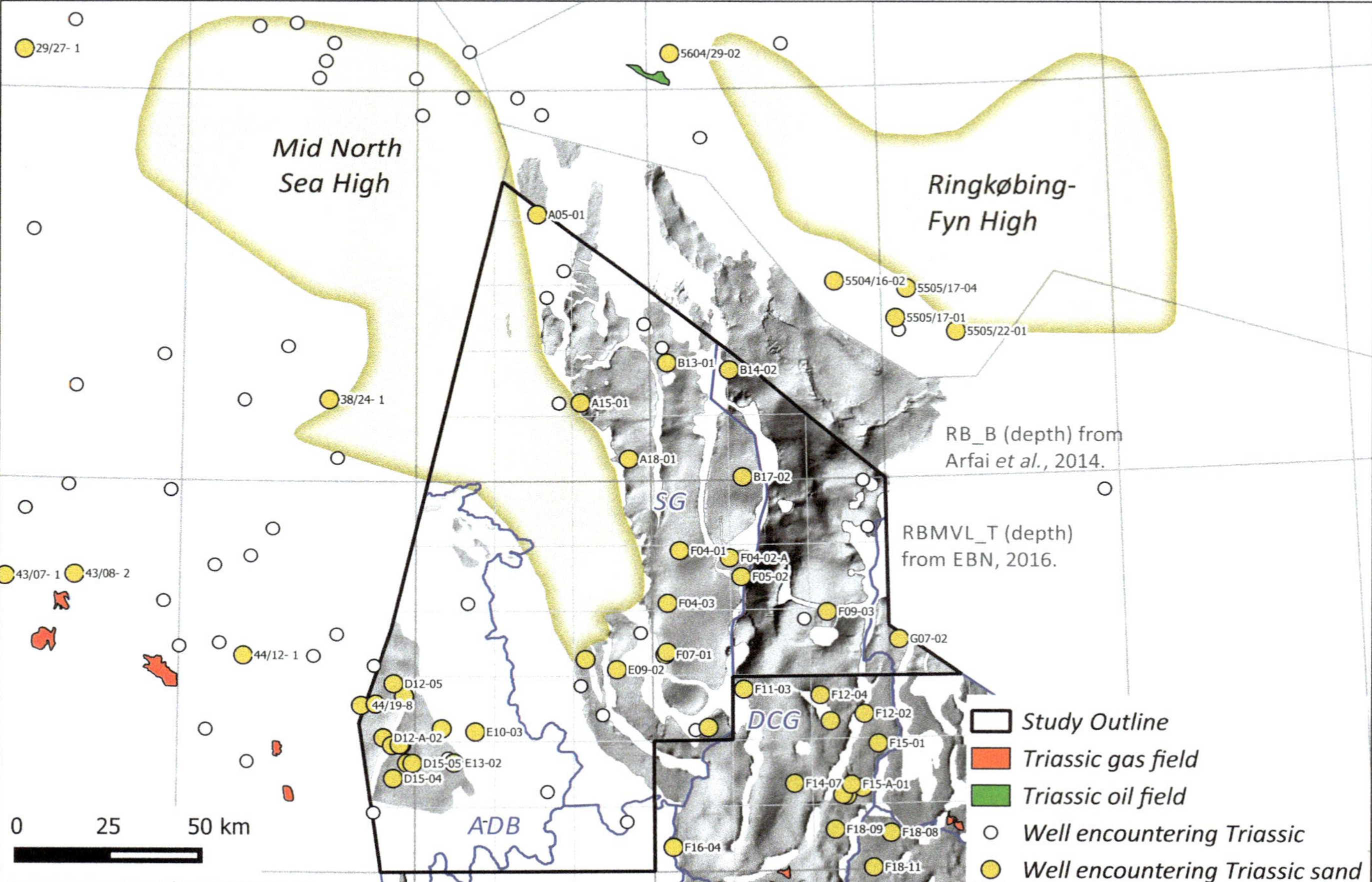

Fig. 6. Five Countries area showing the top Lower Volpriehausen Sandstone Member depth map in the study area outlined in black. The open circles indicate the wells that drilled Triassic rocks and yellow circles indicate the wells that drilled Lower Triassic sands. Structural elements are indicated in blue (SG, Step Graben; DCG, Dutch Central Graben; ADB, Anglo-Dutch Basin) (outlines of structural elements from Kombrink *et al.* 2012).

'Deposition and preservation in the northern sector' and 'Reservoir development').

- Type 6: This category is characterized by a series of relatively thin sandstones with a low GR-contrast to the intercalated and underlying shales. Biostratigraphic analysis is inconclusive, with the exception that Triassic strata is capped by a thin succession of Jurassic sandstone that itself is truncated by the Upper Cimmerian unconformity. This type section has a single occurrence in A05-01 (Kerstholt & Munsterman 2016), and has a strong resemblance to type 5 and to the Bertel-01 oil sandstones in the Danish sector (GEUS 1993).

Lower Volpriehausen and Detfurth Sandstone lithology mapping. Discrimination between Volpriehausen and Detfurth sandstones is equivocal for profiles of type 4 and especially for types 5 and 6. Direct correlation to the regular tectonostratigraphic cyclicity of RBM sandstones in the SW of the study area is hampered by subseismic vertical resolution and commonly barren palynology (e.g. Kerstholt & Munsterman 2016). For gridding purposes we assume that sandstones equivalent to Volpriehausen and Detfurth are preserved in the anomalous RBM profiles, thereby representing overfilled basin conditions where sediment influx outbalances local subsidence and intermittent shale deposits such as the Volpriehausen Clay-Siltstone Member (RBMVC) are preferentially eroded during periods of high-energy sediment bypass (Banham & Mountney 2013). Subdivision of profile types 4–6 into RBMVL and RBMDL equivalents is entirely based on visual correlation of log profiles and should therefore be considered of notional character only.

True stratigraphic thickness (TST) and net sand thickness (NST) are gridded separately using a convergent interpolation algorithm. The interpolation is restricted to the range of the input thicknesses to avoid gridding artefacts. In a second processing step, NST grids were divided by their TST counterparts to obtain a net-to-gross (NTG) map for the RBMVL and RBMDL intervals. TST represents available accommodation space, whereas NTG is considered a proxy for the ratio between net sand influx and relative basin subsidence.

In a general sense the results of the lithology mapping suggest thickness variations within the classic RBM facies expression (type 1), and may also reflect facies change to the north. The interplay between accommodation space and sand deposition in the NW area of the Step Graben is elucidated in the following section; the regional development is discussed in the section on 'Reservoir development'.

Deposition and preservation in the northern sector. Isochron maps of interval A and B in Figure 7 show Lower Triassic thickening on the western margin of the Step Graben in A15 (area of interest, indicated by dashed boundary in Fig. 4); an extra set of reflectors appears at the thickened section (Fig. 8), while no unconformities or truncations are visible at the boundary between interval A and B. This thickening could indicate the presence of a local depocentre and accommodation space for sediments in A15 during Early Triassic times, with a relatively high influx of sand leading to overfilled conditions *sensu* Banham & Mountney (2013); this would explain the high NTG of the RBMVL in A15-01. No Lower Triassic thickening is seen in block A18 (Fig. 7), where well A18-01 indicates low NTG values of the RBMVL.

Hydrocarbon traps at Lower Volpriehausen Sandstone Member

Regional mapping of the Lower Volpriehausen Sandstone Member facilitates the identification of 42 untested structures (leads) in the study area. These leads roughly cluster in three types and areas (Fig. 9): (1) 'classic' leads with proven types of trap, source, seal and reservoir; (2) leads which may be sourced with hydrocarbons via Tertiary volcanic dykes (discussed in the section on 'Northern limits of Lower Triassic sandstones'); and (3) leads with alternative reservoir provenance in the NW Step Graben (as described in the section on 'Reservoir development').

The Zechstein salt deposits play an important role in the formation of Triassic traps, and the four main salt-related traps can be found in the study area: 4-way dip closures above salt domes (e.g. Caister-B field) (Underhill 2009); 3-way dip closures against salt walls (e.g. M1-A and G16-B fields); turtle-back anticlines (e.g. F15-A and L5-FA fields); and a number of truncation traps (e.g. M07-FA field) (e.g. Bachmann *et al.* 2010; van Eijk 2014).

Discussion

Reservoir development

Regional mapping of the Lower Volpriehausen Sandstone Member shows that the member is present in large parts of the study area (Fig. 4). The presence and thickness of the Lower Volpriehausen Sandstone Member decreases from the south to the north. The well locations, together with their type sections as displayed in Figure 4, illustrate the onset of shaling out of the Lower Triassic sandstone deposits (type 1) into an ephemeral lake environment (type 3) in the northern part of the Dutch F Quadrant, as also described by Geluk (2005, 2007).

Figure 5a shows that RBMVL is thickest on the western flank of the Dutch Central Graben (F10 area) with local maxima at the graben axis (F09) and bordering the UK sector in the west (E10).

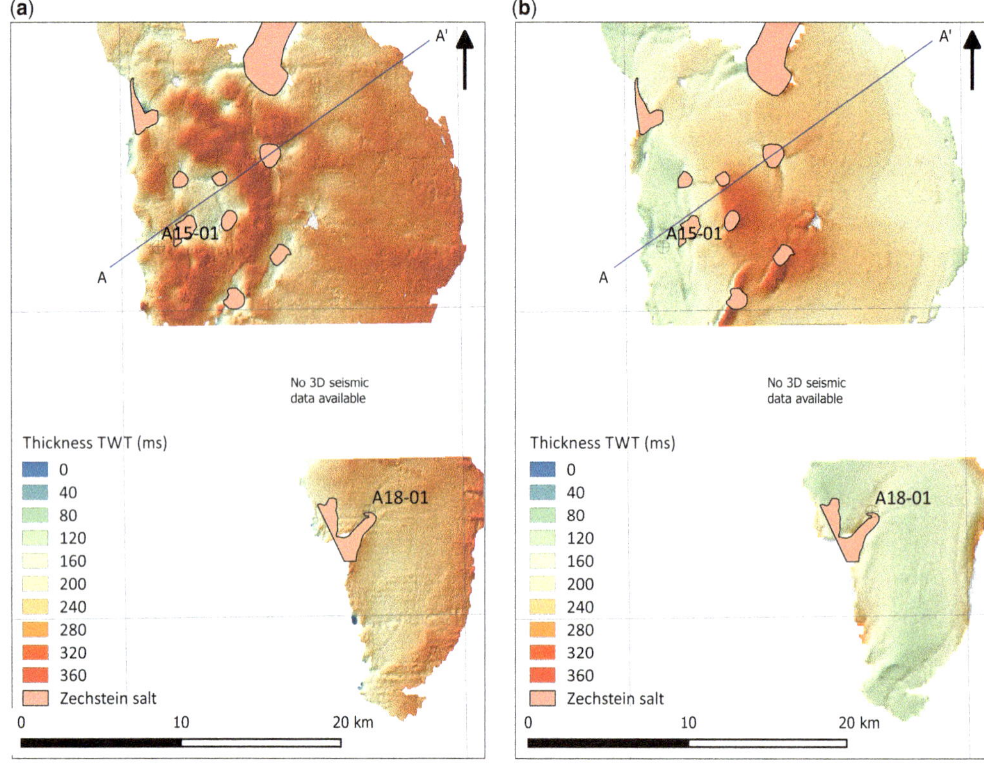

Fig. 7. Time isochore maps of (**a**) interval A (top Lower Volpriehausen Sandstone to top Zechstein) and (**b**) interval B (top Röt Evaporite to top Lower Volpriehausen Sandstone) showing thickening of Lower Triassic strata, suggesting Early Triassic local depocentre development in the A15 area.

Thinning in the NW Step Graben adjacent to the Elbow Spit High is prominent, suggesting that differential subsidence in this area was limited at that time. In Figure 5b, thinning in RBMDL appears to be less localized; this indicates that subsidence in the Step Graben system was more pervasive during deposition of the Detfurth Sandstone Member. In contrast, the SE of the study area appears to

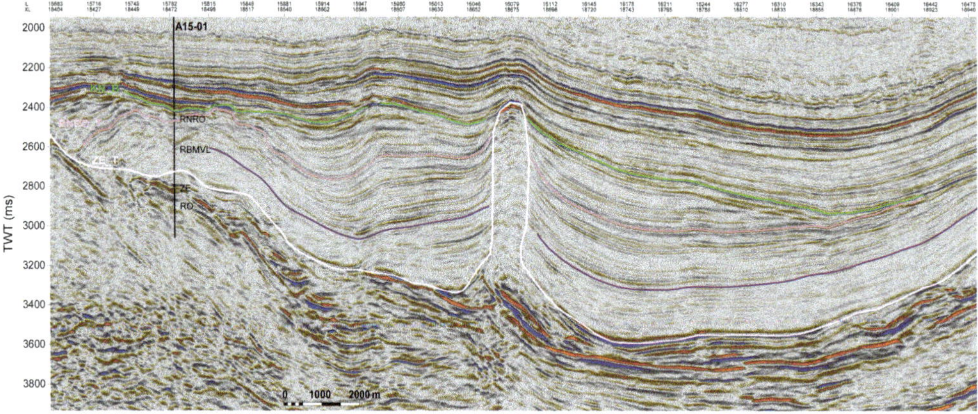

Fig. 8. Seismic section (TWT) along SW–NE transect showing thickening of strata in NW Step Graben area. Location of seismic section indicated in Figure 7.

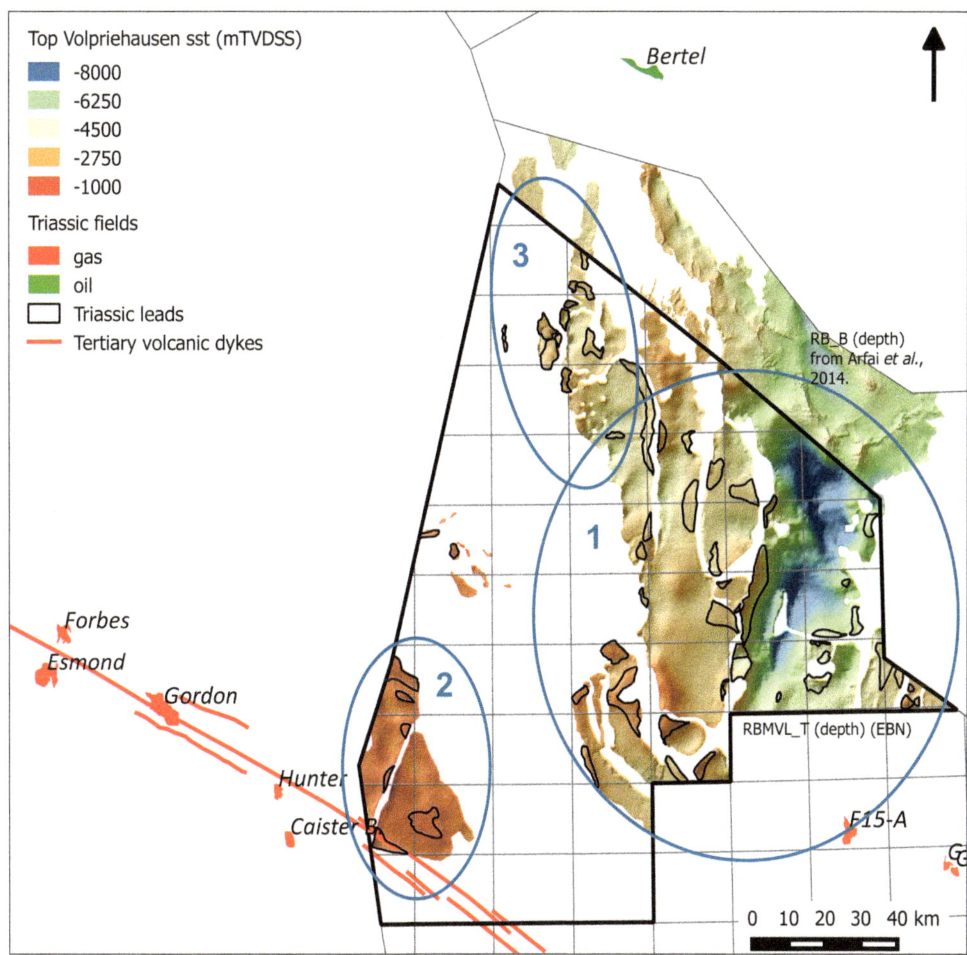

Fig. 9. Top Lower Volpriehausen Sandstone Member depth map in the study area. The identified leads are classified as one of three types: (1) 'classic' with proven types of trap, source, seal and reservoir; (2) leads which may be sourced via Tertiary volcanic dykes; and (3) leads with local/northern reservoir provenance. Triassic fields are shown in red (gas) and green (oil). The UK fields indicated are sourced from the Carboniferous via Tertiary volcanic dykes (Kirton & Donato 1985; Underhill 2009). The volcanic dykes extend into the Dutch sector with leads lining up along their strike.

provide less accommodation space during deposition of RBMDL as compared to RBMVL, suggesting a general shift of RBMVL-concurrent depocentres towards the north. This is supported by the relocation of the RBMVL thickness maxima in F10 towards the graben (F09) axis during Detfurth deposition. NST grids in Figure 5c, d largely follow the TST distribution except for the northern and, in particular, northeastern Step Graben. The NST map of RBMVL (Fig. 5c) clearly shows the limited sand deposition in the north.

The relation between TST and NST is shown in the NTG grids in Figure 5e, f. NTG typically relates to the balance between rate of differential subsidence and the rate of sand supply. Relatively high NTG ratios in the basin fill are preserved if sand influx outpaces the rate at which accommodation space is supplied through subsidence, leading to overfilled conditions as described by Banham & Mountney (2013). NTG grids of both stratigraphic intervals generally confirm the lacustrine conditions in the centre of the Southern Permian Basin (e.g. Geluk 2007). However, shaling out during Volpriehausen deposition was more gradual and widespread with lowest NTG in the central Step Graben. During Detfurth deposition the basin centre facies appears to have been geographically more confined, with focus on the eastern Step Graben shoulder adjacent to the Dutch Central Graben (Fig. 5f). No wells reached the Triassic in the Central Graben to support underfilled conditions as a result of accelerated graben subsidence in the period between RBMVL and RBMDL deposition. However, maximum TST and NTG in F09 during deposition of the Detfurth supports the idea that the Central Graben provided increasing accommodation space and acted as a preferred sand discharge route during the course of the Lower Triassic. The high shale content in B13 and also in the Danish U-1 (Michelsen & Andersen 1983) supports the theory that these northeastern areas were distal to the sediment source and the rate of sand supply was clearly outpaced by differential subsidence, probably even more so in the untested Lower Triassic of the German Central Graben. In contrast, NTG at the northwestern Step Graben boundary fault is relatively high at thin TST, indicating a balance between sediment supply and subsidence that favours the preservation of sand in this area.

Assuming a steady supply of sand, it is evident that the westernmost fault block in the Step Graben was subsiding at low rates during deposition of RBMVL (Fig. 5e) compared to the basins that formed to the east, where more lacustrine deposits were preserved. The northern RBMVL facies may therefore be a distal equivalent of the Bunter Sandstone Formation in the UK sector (Johnson *et al.* 1994), where sediment supply and subsidence were in favour of sand preservation. Sand-prone equivalents to the RBMVC and RBMDC are therefore likely to represent basin fringe facies similar to the Upper Volpriehausen Sandstone Member (RBMVU) and Upper Detfurth Sandstone Member (RBMDU) at the southern margin of the Southern Permian Basin (Van Adrichem Boogaert & Kouwe 1994). The transition between Bunter- and RBM-type Lower Triassic is seen in log profiles of the E Quadrant (type 4; section on 'Log profile typing as a means of regional facies analysis') and can be correlated westwards into the Bunter of the UK Southern North Sea (e.g. well 44/19-3). The provenance of the Bunter Sandstone Formation is the Variscan massif of the London-Brabant Platform in the south, with significant contribution from the UK onshore to the west (Cameron *et al.* 1992; Johnson *et al.* 1994). We suggest that the Bunter bypassed the Elbow Spit High and contributed to the transitional log profile described for A15-01 (type 5; section on 'Log profile typing as a means of regional facies analysis'). The western Step Graben blocks subsided rather uniformly during deposition of the Detfurth (Fig. 5f), whereas new accommodation space developed on the eastern Step Graben and possibly Central Graben areas.

Northern limits of Lower Triassic sandstones

The presence and distribution of reservoir sand in the NW sector of the study area, in particular in view of the condensed sand anomalies in the NW Step Graben (A15-01), and south of the Mid North Sea High (A05-01) requires further investigation.

The NW sector of the Step Graben is marked by an amalgamation of Mesozoic unconformities that truncate Jurassic and Triassic strata increasingly to the north (Mid North Sea High) and the east (Elbow Spit High). The elongated fault blocks of the Step Graben system are increasingly downstepping from the Elbow Spit High in the West towards the Central Graben in the East. Uplift of the western platforms concurrent to subsidence in the Central Graben resulted in erosion of the Mesozoic cover of these platforms (Geluk & Röhling 1999), thereby also affecting the least subsiding Step Graben fault blocks adjacent to the Elbow Spit area.

Evidence of early salt mobilization associated with dextral-transtensional movements at WSW–ENE-striking basement fault lines may be found in the seismic section of Figure 10. Zechstein salt accumulation in the footwalls concurrent with block-rotation and syntectonic Lower Triassic infill of the half-graben-style depocentres developed in the subsiding hanging walls (Fig. 10). Limited halite precipitation at the margin of the Zechstein basin and subsequent withdrawal by salt walls that developed towards the basin axis in the east and SE lead to early grounding of these salt pods (Smith *et al.* 1993; Banham & Mountney 2013). Such a depositional environment is described by Smith *et al.* (1993) as a linked pod synform system that facilitates conduits of clastic transport, whereas ephemeral lakes form in temporarily isolated synforms. Switching of the main conduits represents a major obstacle for predicting reservoir development in this environment.

Clearly post-depositional reactivation of the basement faults took place during the Early Cimmerian phase, prior to deposition of the Main Muschelkalk Formation (RNMU), and is in some instances associated with crestal collapse of the Zechstein and gravity infill of the sub-Muschelkalk strata. Similar incisions on intra-pod highs are seen in the NW part of the Step Graben where Lower Triassic strata have been exposed and eroded at the Middle Cimmerian unconformity. Jurassic sediments are sporadically preserved as infill to incised salt swells (e.g. Kimmeridge Clay Formation drilled in A08-01). The entire Triassic–Jurassic sediment package was truncated at the Late Cimmerian Unconformity (also known as the Base Cretaceous Unconformity or BCU) before the entire northern sector was covered by marine clastics (often subseismic and as thin as 10 m; e.g. wells A08-01, A12-02 and A14-01) and chalk. The collapse of early salt swells during the Jurassic is evidence of salt removal by circulating meteoric waters and possibly sub-aerial exposure south of the Five Countries area

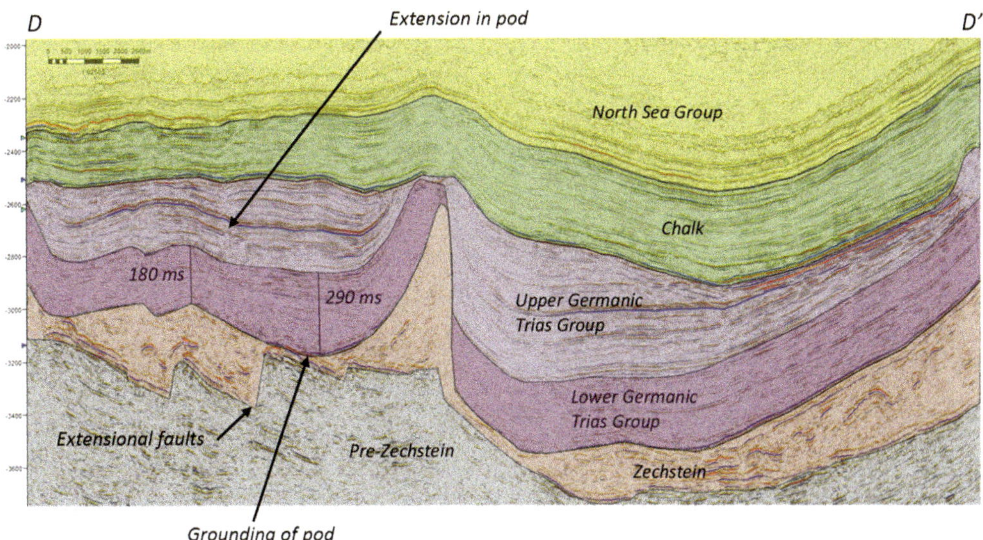

Fig. 10. Seismic section showing extensional faults in the pre-Zechstein deposits with a thin Zechstein layer on top. A thickened Lower Triassic is seen on top of the thin Zechstein layer. See Figure 4 for profile location.

that was repeatedly subject to uplift, non-deposition and erosion, in particular: near the basin margins and at intra-basin swells such the Mid North Sea high during the Hardegsen phase (Röhling 1991; Bachmann *et al.* 2010; Pharaoh *et al.* 2010); and by the rifting pulses of the Early Cimmerian phase (Ziegler 1990), thermal doming during the Middle Cimmerian phase (Geluk & Röhling 1999) and during Late Cimmerian accelerated subsidence of the Central Graben (Remmelts 1995, 1996).

Contrary to the general perception of southern provenance of RBM reservoir sands, we propose an alternative provenance of sediments encountered in the NW part of the Step Graben, deposited in a local depositional system similar to the observations in the L09 block where Solling sandstones present a unique gas-bearing reservoir drilled by L09-08 (de Jager 2012). Here, well and seismic data show that the reservoir package is wedge-shaped, thickening into a listric normal fault that detaches onto the top of the Zechstein salt, adjacent to a salt wall. The aeolian sandstones were clearly preserved in a syn-depositional half-graben, the development of which appears to be related to Early Triassic extension. These locally deposited aeolian wedges were also drilled in L06, but were however salt-plugged (de Jager & Geluk 2007; de Jager 2012). Similar to such localized Solling 'Fat' sand deposits in the L Quadrant, local depocentres could have existed in the NW Step Graben area, preserving reservoir sands as drilled by A15-01.

The seismic data shows potential for local depocentres in rim-synclines in the A15-01 area near the Step Graben western boundary fault (Fig. 8; van der Kooij 2016). Here, a 'pockmark-like field of mini salt domes' records Early Triassic halokinesis, possibly halted by the interplay between (1) restricted halite precipitation at the margin of the Zechstein basin, (2) repeated uplift and removal of cover sediments during Early Triassic rift pulses and subsequent Cimmerian tectonic phases, and (3) withdrawal to salt walls that built up towards the basin centre (Remmelts 1996). The generation of accommodation space by the proposed pod synform system, analogous to the observation of Smith *et al.* (1993) in the UK Central North Sea, and by concurrent rim syncline formation in the A15-01 area, outlines the potential for local sand deposition during the Early Triassic.

The presence of sands in the unique vertical RBM profile of A15-01 (type 5; Fig. 4) confirms the exposure of this northernmost area of the Dutch offshore sector to sand erosion and subsequent transport. It is not clear whether the sands at A15-01 represent (1) a condensed section (described above as type 5) of the full tectono-cyclic sandstone package of RBM as a transition between the regular type 1 and the Bunter-like type 4 or (2) shale intercalated RBMVL (similar to type 2), whereas the younger RBM sandstones are absent. Option 1 would suggest the widespread deposition of sands across the NW area, possibly facilitated by relay ramp systems in the transfer zones of the major north-south–striking faults, whereas option 2 would be in favour of RBMVL facies association with the transition to ephemeral lakes at the basin centre to the east and SE.

The different log character of A05-01 and A15-01 (Fig. 4), the indication of a local depocentre in A15 (Fig. 8) and the bimodal grain-size distributions found within samples of these two wells (Bezemer 2016) could indicate an alternative reservoir provenance in the NW Step Graben area. Bimodal grain-size distributions in A05-01 and A15-01 suggest a different transport mechanism or sediment source, in contrast to the unimodal grain-size distributions of all samples from A18-01, F16-04 and L09-08 (Bezemer 2016) that suggest aeolian transport. It should be noted that the samples from A05-01, A15-01 and A18-01 were taken from cutting material, while the other samples were taken from cores.

Fluvial sediments from the local highs could have been deposited in local depocentres. Olivarius *et al.* (2017) show that heavy mineral analyses may help to determine the provenance of Lower Volpriehausen and Solling Sandstone deposits. A feasibility study by Vonk (2016) shows that heavy minerals can also be obtained in the northern Dutch offshore from cutting material from wells A05-01, A15-01 and A18-01. However, only a few samples were taken; it is not possible to conclude the provenance of RBM sands in these three wells from the current dataset. Heavy mineral analyses of both these three northern wells and a number of wells that drilled typical Variscan-sourced Lower Volpriehausen Sandstones could assist in resolving the provenance of RBM sands in the NW Step Graben, and are suggested for future research.

The current evaluation provides new insights into the northern Dutch offshore and surrounding areas, suggesting that local depocentre development is possible at the northern margin of the Southern Permian Basin where locally, westerly or northwesterly sourced reservoir sands may have been preserved. In addition, the log-typing and facies description described in this paper show that good-quality (likely Variscan sourced) RBM reservoir sands also exist up into the northernmost F-blocks.

Lower Triassic prospectivity in the northern Dutch offshore: source rocks, charge and seal mechanisms

Access to mature source rocks and hydrocarbon charge from the underlying Carboniferous is often identified as an additional risk for the Main Buntsandstein play in the northern Dutch offshore, as thick Zechstein salt would prevent hydrocarbon migration into the Main Buntsandstein reservoir layers. The primary source rock of the hydrocarbons retained in producing Triassic gas fields are Upper Carboniferous (Westphalian) and Namurian coals and organic-rich shales (e.g. Fontaine *et al.* 1993; de Jager & Geluk 2007). Upper Carboniferous coals are present in the southern part of the study area (Kombrink 2008), but the organic-rich shales of the Epen Formation (Namurian) are also expected to be present and mature (Gerling *et al.* 1999; ter Borgh *et al.* 2018*b*). Recent studies provide new insights into the source-rock potential in the northern part of the study area, where both Namurian shales and coals and Lower Carboniferous coals from the Scremerston (Viséan) and Yoredale (Dinantian) formations are interpreted to be present and mature in the northern Dutch offshore (Schroot *et al.* 2006; Arfai & Lutz 2017; ter Borgh *et al.* 2018*b*). Lateral oil migration may happen when RBM reservoir sandstones are juxtaposed against Jurassic Posidonia shale without Zechstein salt entering the fault zone (e.g. Geluk 2007).

The migration of hydrocarbons from the Carboniferous into the overburden may occur along faults, through salt windows, along carrier beds and via volcanic dykes. Figure 11 shows the Zechstein salt isochron which illustrates the presence of a large variability in thickness and the presence of potential salt windows near many of the identified structural closures. In the western area (D, E blocks) where Zechstein salt is generally thick, Lower Triassic structures may be charged from Carboniferous coals via volcanic dykes, analogous to UK Triassic gas fields (Kirton & Donato 1985; Brown *et al.* 1994; Underhill 2009). These dykes have been mapped in the UK (Kirton & Donato 1985; Brown *et al.* 1994; Underhill 2009) and can be followed into the Dutch sector (Figs 9, 11) on seismic data (EBN 2015). These dykes may appear as linear features on an amplitude extraction at base Tertiary (Brown *et al.* 1994; EBN 2015). Expansion and contraction caused by heating and cooling of these Paleogene igneous intrusions has created fractures along the sides but also within these intrusions, allowing vertical migration (Underhill 2009). The UK Bunter fields Forbes, Esmond, Gordon, Hunter and Caister B are considered to have been charged via Tertiary dykes, as all fields are located in an area with thick Zechstein salt deposits. Forbes, Esmond and Caister B in particular, as it is positioned >10 km away from a mapped Tertiary dyke, may suggest that significant lateral hydrocarbon migration took place. Cenozoic movement of salt is considered as a major risk for maintaining charge in underfilled turtleback anticlines such as Caister B (Bachmann *et al.* 2010).

In most Triassic fields in the Netherlands the top seal is formed by Triassic Röt salt and Solling shale and, where the Triassic is truncated at the Base Cretaceous Unconformity, by Lower Cretaceous shales (e.g. de Jager & Geluk 2007). The four-way and three-way dip closures are generally dependent on Upper Triassic sealing shales and/or Zechstein side/top seal. Halokinesis can create traps in the

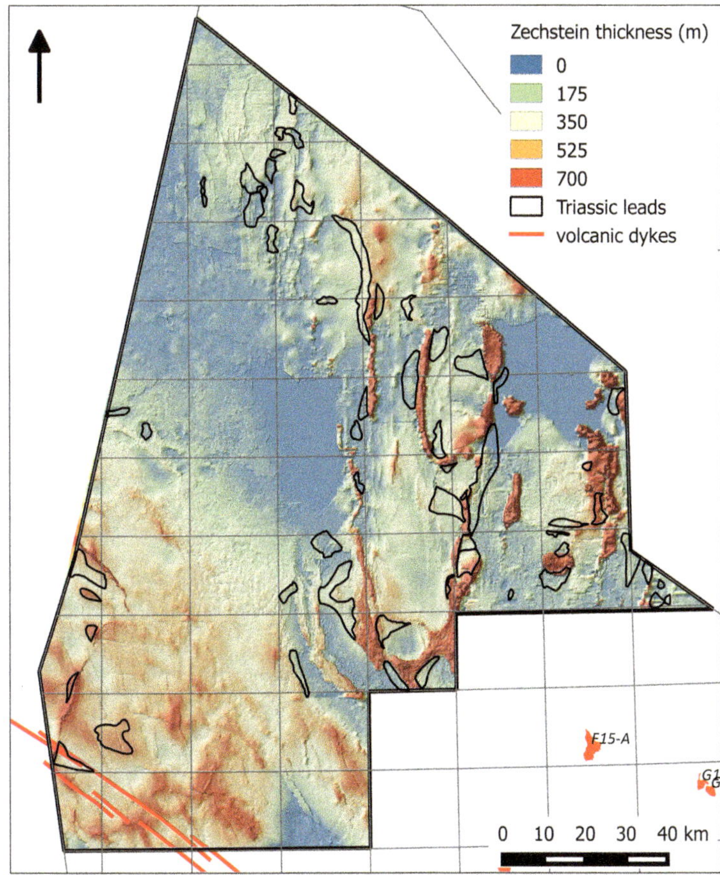

Fig. 11. Zechstein isochron indicating large variability in Zechstein salt thickness (time), which could enable vertical hydrocarbon migration through Zechstein salt windows. Red lines indicate location of Tertiary volcanic dykes. Black polygons indicate the structural closures (leads) at RBMVL level.

Triassic overburden, but the near presence of salt can also cause (partial) salt plugging of reservoir rocks, resulting in reservoir deterioration. Salt-plugged reservoir rocks may even act as a competent side seal (M1-A field) (Bachmann *et al.* 2010). Salt plugging may be recognized on seismic data, but it remains difficult to recognize a salt-plugged reservoir without calibration to nearby well data.

Truncation traps depend on the sealing capacity of overlying Jurassic, Cretaceous or Paleogene strata. The Jurassic part of the Flora field in the UK North Sea sector demonstrates that only a thin veneer (3 m) of Lower Cretaceous marl is sufficient to form the top seal of this truncation trap field (Hayward *et al.* 2003).

Detailed analysis of the sealing potential of the Triassic overburden has not been the focus of this study, but should be performed when evaluating the sealing capacity of the individual leads.

Conclusions

A regional study using wells and seismic data from the northern Dutch offshore and surrounding territories indicates the presence of Triassic reservoir sandstones north of the main, well-understood, fairway. A comprehensive borehole review and log typing in the study area shows the presence of sands in the NW Step Graben area. Seismic interpretation also indicates the presence of Early Triassic local depocentre development in the Step Graben. Syntectonic strata in local depocentres may have been formed in this area due to early halokinesis in the Triassic, in analogy to the Central North Sea described in Smith *et al.* (1993). Grain-size analyses, together with heavy mineral analyses, are proposed to further investigate whether fluvial sands with a (local) northern provenance may have been preserved in the northwestern area of the Step Graben system.

Reservoir gridding allows the investigation of regional trends but does not resolve local depocentres, for example, analogues to the Solling Fat Sand play in the L09 and L06 areas (south of our study area). Depocentres appear in the F09 and F10 areas and may be related to differential subsidence and preferred sediment discharge in the Dutch Central Graben. The northern RBMVL is characterized by high NTG at relatively low stratigraphic thickness, suggesting that differential subsidence in the westernmost Step Graben, adjacent to the Elbow Spit High, controlled sand preservation at rates that were possibly sustained by an influx of fluvial sands from the same source that supplied the Bunter Sandstone province in the west. In contrast, RBMDL thickness is uniform in the Step Graben area and depocentres appear to have shifted eastwards into the Central Graben.

Regional mapping of the Lower Volpriehausen Sandstone in the A, B, D, E and F quadrants in the northern Dutch offshore enabled us to identify 42 untested structures in the study area. Recent studies on the source-rock potential in the northern Dutch territory suggest that source rocks of different ages are present in the study area, while the presence of salt windows (indicated by Zechstein isochron mapping) and dykes (mapped on Base North Sea seismic amplitude extraction maps) allow vertical hydrocarbon migration. We conclude that the remaining prospectivity of the Lower Triassic in this part of the Southern Permian Basin may be rather large.

We thank Kees van Ojik, the reviewers Mark Geluk and Carlo Nicolai, and editor Tom McKie for their constructive comments.

References

Arfai, J. & Lutz, R. 2017. 3D basin and petroleum system modelling of the NW German North Sea (Entenschnabel). *In*: Bowman, M. & Level, B. (eds) *Petroleum Geology of NW Europe: 50 Years of Learning – Proceedings of the 8th Petroleum Geology Conference*. Geological Society, London, 8, 67–86, https://doi.org/10.1144/PGC8.35

Arfai, J., Jähne, F., Lutz, R., Franke, D., Gaedicke, C. & Kley, J. 2014. Late Palaeozoic to Early Cenozoic geological evolution of the northwestern German North Sea (Entenschnabel): new results and insights. *Netherlands Journal of Geosciences*, **93**, 147–174.

Bachmann, G.H., Geluk, M.C. *et al.* 2010. Chapter 9, Triassic. *In*: Doornenbal, H. & Stevenson, A. (eds) *Petroleum Geological Atlas of the Southern Permian Basin Area*. EAGE Publications B.V., Houten, 148–173.

Banham, S.G. & Mountney, N.P. 2013. Evolution of fluvial systems in salt-walled mini-basins: a review and new insights. *Sedimentary Geology*, **296**, 142–166.

Best, G. 1996. Floßtektonik un Norddeutschland: erste Eergebnisse reflexionsseismischer Untersuchungen an der Salzstruktur 'Oberes Allertal'. *Zeitschrift der deutschen geologischen Gesellschaft*, **147**, 455–464.

Best, G., Kockel, F. & Schöneich, H. 1983. Geological history of the southern Horn Graben. *Geologie en Mijnbouw*, **62**, 25–34.

Bezemer, G. 2016. *Grain size analysis of the Triassic Main Buntsandstein Subgroup in the northern Dutch offshore*. BSc thesis, Utrecht University, Utrecht.

Brown, G., Platt, N.H. & McGrandle, A. 1994. The geophysical expression of Tertiary dykes in the southern North sea. *First Break*, **12**, 137–146.

Cameron, T.D.J., Crosby, A., Balson, P.S., Jefferey, D.H., Lott, G.K., Bulat, J. & Harrison, D.J. 1992. *The Geology of the Southern North Sea*. UK offshore regional report Keyworth, Nottingham.

De Jager, J. 2007. Geological development. *In*: Wong, T., Batjes, D.A.J. & de Jager, J. (eds) *Geology of the Netherlands*. Royal Netherlands Academy of Arts and Sciences, Amsterdam, 5–26.

De Jager, J. 2012. The discovery of the Fat Sand Play (Solling Formation, Triassic), Northern Dutch offshore–a case of serendipity. *Netherlands Journal of Geosciences*, **91**, 609–619.

De Jager, J. & Geluk, M.C. 2007. Petroleum geology. *In*: Wong, T., Batjes, D.A.J. & de Jager, J. (eds) *Geology of the Netherlands*. Royal Netherlands Academy of Arts and Sciences, Amsterdam, 241–264.

EBN 2015. The Triassic Main Buntsandstein play: New prospectivity away from the main fairway. *Prospex 2015*, https://www.ebn.nl/wp-content/uploads/2014/11/Poster6_EBN_Prospex2015.pdf

EBN 2016. Focus on Dutch Oil & Gas 2016, https://www.ebn.nl/wp-content/uploads/2016/12/Focus-Dutch-Oil-and-Gas-2016.pdf

Fisher, M.J. & Mudge, D.C. 1990. Triassic. *In*: Glennie, K.W. (ed.) *Introduction to the Petroleum Geology of the North Sea*. Blackwell Scientific Publications, Oxford, 191–218.

Fontaine, J.M., Guastella, G., Jouault, P. & de la Vega, P. 1993. F15-A: a Triassic gas field on the eastern limit of the Dutch Central Graben. *In*: Perker, J.R. (ed.) *Petroleum Geology of Northwest Europe. Proceedings of the 4th Conference*. Geological Society, London, 583–593.

Geluk, M.C. 2005. *Stratigraphy and Tectonics of Permo-Triassic Basins in the Netherlands and Surrounding Areas*. PhD Thesis Utrecht University.

Geluk, M.C. 2007. Triassic. *In*: Wong, T.E., Batjes, D.A.J. & de Jager, J. (eds) *Geology of the Netherlands*. Royal Netherlands Academy of Arts and Sciences, Amsterdam, 85–106.

Geluk, M.C. & Röhling, H.G. 1997. High-resolution sequence stratigraphy of the Lower Triassic 'Buntsandstein' in the Netherlands and northwestern Germany. *Geologie en Mijnbouw*, **76**, 227–246.

Geluk, M.C. & Röhling, H.G. 1999. High-resolution sequence stratigraphy of the Lower Triassic Buntsandstein: a new tool for basin analysis. *Epicontinental Triassic. Zentralblatt fur Geologie und Palaontologie*, **1**, 7–8.

Gerling, P., Geluk, M.C., Kockel, F., Lokhorst, A., Lott, G.K. & Nicholson, R.A. 1999. NW European Gas Atlas – new implications for the Carboniferous

Gasplays in the Western Part of the Southern Permian Basin. *In*: Boldly, S.A.R. & Fleet, A.J. (eds) *Petroleum Geology of the Northwest Europe. Proceedings of the 5th Conference*. Geological Society, London, 799–809, https://doi.org/10.1144/0050799.

GEUS 1993. *Bertel-1, Bertel-1A*. Final Well Report, Volume 1 (Tekst). Report File no 7882, 01/01-1993.

Goldsmith, P.J., Rich, B. & Standring, J. 1995. Triassic correlation and stratigraphy in the South Central Graben, UK North Sea. *In*: Boldy, S.A.R. (ed.) *Permian and Triassic Rifting in Northwest Europe*. Geological Society, London, Special Publications, **91**, 123–143, https://doi.org/10.1144/GSL.SP.1995.091.01.07

Griffiths, P.A., Allen, M.R., Craig, J., Fitches, W.R. & Whittington, R.J. 1995. Distinction between fault and salt control of Mesozoic sedimentation on the southern margin of the Mid-North Sea High. *In*: Boldy, S.A.R. (ed.) *Permian and Triassic Rifting in Northwest Europe*. Geological Society, London, Special Publications, **91**, 145–159, https://doi.org/10.1144/GSL.SP.1995.091.01.08

Harding, R. & Huuse, M. 2015. Salt on the move: multi stage evolution of salt diapirs in the Netherlands North Sea. *Marine and Petroleum Geology*, **61**, 39–55.

Hayward, R.D., Martin, C.A.L., Harrison, D., Van Dort, G., Guthrie, S. & Padget, N. 2003. The Flora field, blocks 31/26a, 31/26c, UK North Sea. *In*: Gluyas, J.G. & Hichens, H.M. (eds) *United Kingdom Oil and Gas Fields Commemorative Millenium Volume*. Geological Society, London, Memoirs, **20**, 549–555, https://doi.org/10.1144/GSL.MEM.2003.021.01.44

Johnson, H., Warrington, G. & Stoker, S.J. 1994. 6. Permian and Triassic of the Southern North Sea. *In*: Knox, R.W.O'B. & Corday, W.G. (eds) *Lithostratigraphic Nomenclature of the UK North Sea*. British Geological Survey, Nottingham.

Kerstholt, S.J. & Munsterman, D.K. 2016. *Jurassic or Triassic?* Palynological dating of well A05-01, interval 10250-10720 Ft. TNO Report 2016 R10459. TNO, Utrecht (http://www.nlog.nl).

Kirton, S.R. & Donato, J.A. 1985. Some buried Tertiary dykes of Britain and surrounding waters deduced by magnetic modelling and seismic reflection methods. *Journal of the Geological Society*, **142**, 1047–1057.

Kombrink, H. 2008. *The Carboniferous of the Netherlands and surrounding areas; a basin analysis*. PhD thesis, Utrecht University.

Kombrink, H., Doornenbal, J.C., Duin, E.J.T., Den Dulk, M., Ten Veen, J.H. & Witmans, N. 2012. New insights into the geological structure of the Netherlands; results of a detailed mapping project. *Netherlands Journal of Geosciences*, **91**, 419–446.

McKie, T. & Williams, B. 2009. Triassic palaeogeography and fluvial dispersal across the northwest European Basins. *Geological Journal*, **44**, 711–741.

Michelsen, O. & Andersen, C. 1983. Mesozoic structural and sedimentary development of the Danish Central Graben. *Geologie en Mijnbouw*, **62**, 93–102.

Michelsen, O. & Clausen, O.R. 2002. Detailed stratigraphic subdivision and regional correlation of the southern Danish Triassic succession. *Marine and Petroleum Geology*, **19**, 563–587.

OGA 2016, https://www.gov.uk/government/organisations/oil-and-gas-authority

Olivarius, M., Weibel, R., Friis, H., Boldreel, L.O., Keulen, N. & Thomsen, T.B. 2017. Provenance of the Lower Triassic Bunter Sandstone formation: implications for distribution and architecture of aeolian v. fluvial reservoirs in the North German Basin. *Basin Research*, **29**, 113–130.

Pharaoh, T.C., Dusar, M. *et al.* 2010. Chapter 3, Tectonic evolution. *In*: Doornenbal, H. & Stevenson, A. (eds) *Petroleum Geological Atlas of the Southern Permian Basin Area*. EAGE Publications B.V., Houten, 25–57.

Remmelts, G. 1995. Fault-related salt tectonics in the southern North Sea, The Netherlands. *In*: Jackson, M.P.A., Roberts, D.G. & Snelson, S. (eds) *Salt Tectonics: a Global Perspective*. AAPG Memoir, **65**, 261–272.

Remmelts, G. 1996. Salt tectonics in the southern North Sea, the Netherlands. *In*: Rondeel, H.E., Batjes, D.A.J. & Nieuwenhuijs, W.H. (eds) *Geology of Gas and Oil Under the Netherlands*. Springer, Netherlands, 143–158

Röhling, H.G. 1991. A Lithostratigraphic subdivision of the Early Triassic in the Northwest German Lowlands and the German Sector of the North Sea, based on Gamma Ray and Sonic Logs. *Geologisches Jahrbuch, A*, **119**, 3–23.

Schroot, B.M., v.Bergen, F., Abbink, O.A., David, P., v.Eijs, R. & Veld, H. 2006. *Hydrocarbon potential of the Pre-Westphalian in the Netherlands on- and offshore*. Report of the PETROPLAY project. TNO report NITG 05-155-C.

Smith, R.I., Hodgson, N. & Fulton, M. 1993. Salt control on Triassic reservoir distribution, UKCS central North Sea. *In*: Parker, J.R. (ed.) *Petroleum Geology of Northwest Europe: Proceedings of the 4th Conference*. Geological Society, London, Petroleum Geology Conference Series, **4**, 547–557.

Spathopoulos, F., Doubleday, P.A. & Hallsworth, C.R. 2000. Structural and depositional controls on the distribution of the Upper Jurassic shallow marine sandstones in the Fife and Angus fields area, Quadrants 31 & 39, UK Central North Sea. *Marine and Petroleum Geology*, **17**, 1053–1082.

ter Borgh, M.M., Jaarsma, B. & Rosendaal, E.A. 2018*a*. Structural development of the northern Dutch offshore: Paleozoic to present. *In*: Monaghan, A.A., Underhill, J.R., Hewett, A.J. & Marshall, J.E.A. (eds) *Paleozoic Plays of NW Europe*. Geological Society, London, Special Publications, **471**. First published online March 20, 2018, https://doi.org/10.1144/SP471.4

ter Borgh, M.M., Eikelenboom, W. & Jaarsma, B. 2018*b*. Hydrocarbon potential of the Visean and Namurian in the northern Dutch offshore. *In*: Monaghan, A.A., Underhill, J.R., Hewett, A.J. & Marshall, J.E.A. (eds) *Paleozoic Plays of NW Europe*. Geological Society, London, Special Publications, **471**. First published online March 20, 2018, https://doi.org/10.1144/SP471.5

Thieme, B. & Rockenbauch, K. 2001. Floßtektonik in der Trias der Deutschen Südlichen Nordsee. *Erdöl-Erdgas-Kohle*, **117**, 568–573.

Underhill, J.R. 2009. Role of intrusion-induced salt mobility in controlling the formation of the enigmatic 'Silverpit Crater', UK Southern North Sea. *Petroleum*

Geoscience, **15**, 197–216, https://doi.org/10.1144/1354-079309-843

Underhill, J.R. & Partington, M.A. 1993. Jurassic thermal doming and deflation in the North Sea: implication of the sequence stratigraphic evidence. *In*: Parker, J.R. (ed.) *Petroleum Geology of Northwest Europe. Proceedings of the 4th Conference*. Geological Society, London, Petroleum Geology Conference Series, 4, 337–346.

Van Adrichem Boogaert, H.A. & Kouwe, W.F.P. 1994. Stratigraphic nomenclature of the Netherlands; revision and update by RGD and NOGEPA, Section E Triassic. Mededelingen Rijks Geologische Dienst 50.

Van Dalfsen, W., van Gessel, S.F. & Doornenbal, J.C. 2007. *Velmod-2, Joint Industry Project*. TNO-report 2007-U-R1272C. TNO, Utrecht.

Van der Kooij, C. 2016. *Triassic reservoir development in the northern Dutch offshore*. MSc thesis, Utrecht University, Utrecht.

Van Eijk, M.M. 2014. *Volpriehausen prospectivity review*. MSc thesis, Vrije Universiteit Amsterdam.

Van Winden, M.E. 2015. *Salt tectonics in the northern Dutch offshore*. MSc thesis, Utrecht University.

Vonk, A.N. 2016. *Provenance of Lower Triassic sands in the northern Dutch offshore*. BSc thesis, Utrecht University, Utrecht.

Wride, V.C. 1995. Structural features and structural styles from the Five Countries Area of the North Sea Central Graben. *First Break*, **13**, 395–407.

Zanella, E., Coward, M.P. & McGrandle, A. 2003. Crustal structure. *In*: Evans, D., Graham, C., Armour, A. & Bathurst, P. (eds) *The Millennium Atlas: Petroleum Geology of the Central and Northern North Sea*. Geological Society, London, 35–43.

Ziegler, P.A. 1990. *Geological Atlas of Western and Central Europe*. 2nd revised edn. Shell Internationale Petroleum Maatschappij B.V., The Hague.

Ziegler, P.A. 1992. North Sea rift system. *Tectonophysics*, **208**, 55–75.

The 'Buntsandstein' gas play of the Horn Graben (German and Danish offshore): dry well analysis and remaining hydrocarbon potential

BEN KILHAMS[1]*, SNEZANA STEVANOVIC[2] & CARLO NICOLAI[1]

[1]*Shell Exploration and Produktion Deutschland GmbH, Suhrenkamp 71–77, Hamburg, Germany*

[2]*Shell Projects and Technology, Kessler Park, 2288 GS, Rijswijk, The Netherlands*

**Correspondence: b.kilhams@shell.com*

Abstract: An analysis of the Lower Triassic 'Buntsandstein' gas play in the underexplored Mesozoic rift system of the German–Danish Horn Graben is presented. Dry hole information from four well penetrations is analysed alongside the development of a 3D basin model. It is demonstrated that the dry holes do not preclude the existence of a working hydrocarbon system. Reservoir and seal elements are present, although details of quality and distribution are uncertain. Carboniferous coal preservation is likely, in a limited area, within the graben and can be constrained through seismic mapping. Vertical gas migration through the Zechstein interval is considered likely due to a large thickness variability (driven by halokinesis and facies changes). The overlap of peak gas-expulsion timing and halokinetic movements make rollover/turtle-back traps risky in terms of breaching or underfill. Dry wells in Denmark are explained by a combination of this relative timing and uncertainty over longer distance migration. This play analysis demonstrates a general agreement with previously published 1D basin models with respect to gas-expulsion timing. However, in contrast to published examples, it is shown that the Zechstein interval can allow for vertical gas migration. Considerable uncertainty in parameters, such as depth conversion, amount of erosion and migration paths, are recognized. Exploration opportunities remain, albeit relatively high risk, in the German area of the graben both in the 'Buntsandstein' play and at other stratigraphic levels.

Tectonics, stratigraphy and hydrocarbon systems

Tectonic evolution

The Horn Graben is a NNE–SSW-trending Late Carboniferous–Late Triassic extensional system formed at the meld between the Avalonia and Baltica tectonic plates, cutting perpendicular to the NNW–SSE-trending Ringkøbing and Schill Grund highs in the present-day offshore areas of Germany and Denmark (Best *et al.* 1983; Clausen & Korstgård 1994; Thybo 1997; Abramovitz & Thybo 1999; Guterch *et al.* 2010; Pharaoh *et al.* 2010) (Fig. 1). The suture between these domains has a complex history, with multiple tectonic phases, but is likely to have formed in the Middle–Late Silurian (Berthelsen 1998; Samuelsson *et al.* 2002) and resulted in the long-lived (dextral strike-slip-dominated) Tornquist fault splay zone (Thybo 2001; Lassen & Thybo 2012).

Late Carboniferous dextral shear, which is likely to be related to the distant Variscan Orogeny (Thybo 2001), and its interaction with existing basement blocks generated a series of rotated basement structures (Ringkøbing-Fyn High/Schill Grund High) and dissecting lows (Central Graben, Horn Graben and Glückstadt Graben) (Clausen & Korstgård 1994; Pharaoh *et al.* 2010) (Fig. 1) with associated volcanism (Best *et al.* 1983; Thybo 2001). The precise timing of these early intra-block lows is largely unknown, with estimates for the opening of a proto-Horn Graben ranging from Late Precambrian (Lassen & Thybo 2012), Devono-Carboniferous (Best *et al.* 1983; Vejbaek 1990; Thybo 2001) to Permian times (southern area: Vejbaek 1990). Figure 2 demonstrates a comparison to the opening of the genetically related Glückstadt Graben (e.g. Maystrenko *et al.* 2005*a*, *b*, 2006; Littke *et al.* 2008; Scheck-Wenderoth *et al.* 2008; Sachse & Littke, this volume, in press). The early evolution of the Horn Graben and Schill Grund High is of interest because it impacts on the development and preservation of Carboniferous source rock intervals during the Permian Variscan uplift phase.

From Late Permian to Late Triassic times, the Horn Graben underwent significant rifting associated with an increased heat flow related to the reorganization of Pangaea (Clausen & Korstgård 1994, 1996; Mogensen 1995). This led to the development of a thick Triassic clastic section (3.5–6 km: Fig. 3a), including a sequence of Lower Triassic sandstones equivalent to the Buntsandstein seen across the

From: Kilhams, B., Kukla, P. A., Mazur, S., McKie, T., Mijnlieff, H. F. & van Ojik, K. (eds) 2018. *Mesozoic Resource Potential in the Southern Permian Basin*. Geological Society, London, Special Publications, **469**, 169–192.
First published online January 11, 2018, https://doi.org/10.1144/SP469.5

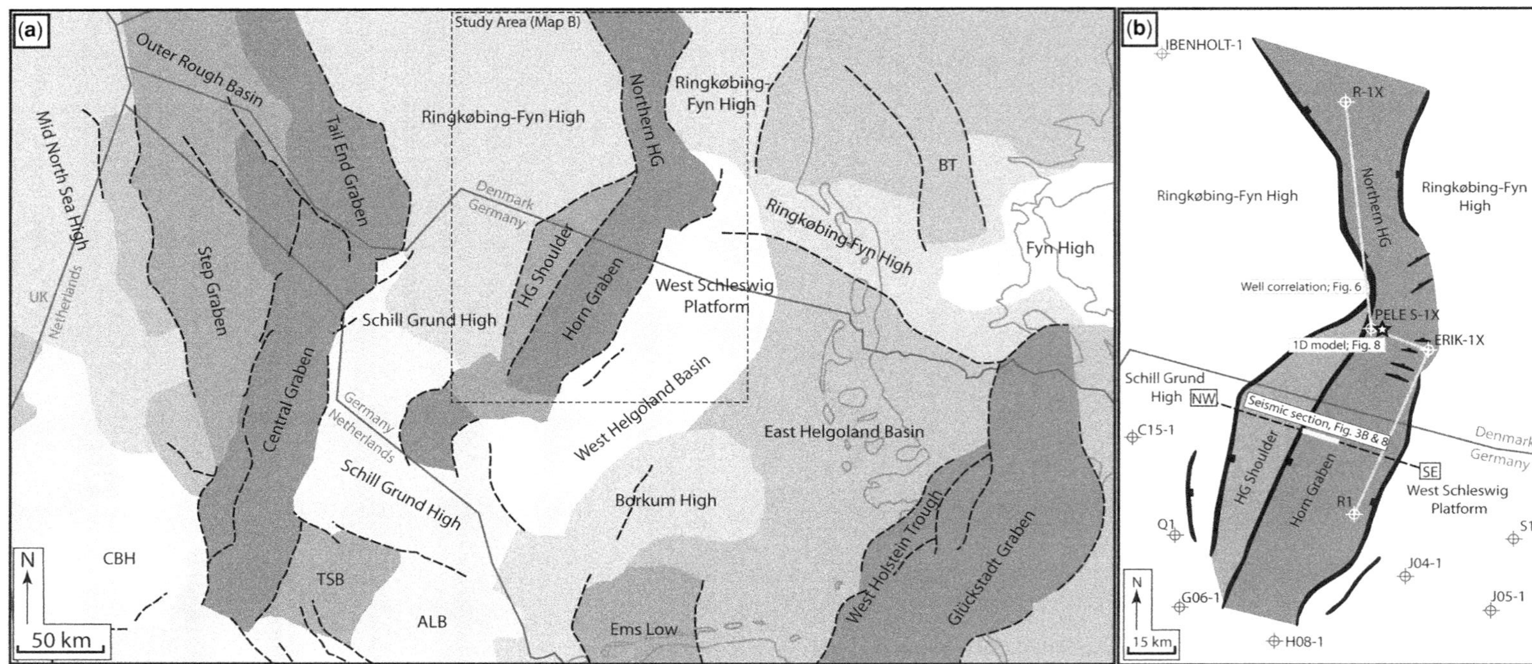

Fig. 1. Tectonic framework and position of the Horn Graben. (**a**) Upper Triassic tectonic elements, simplified after Pharaoh *et al.* (2010). Darker colours represent Upper Triassic major basins grading to platforms, and highs are in lighter colours. Dark dashed lines represent major structural lineaments. Note the three transecting graben systems of the Glückstadt Graben, Horn Graben and Central Graben. ALB, Ameland Block; BT, Brande Trough; CBH, Cleaver Bank High; HG, Horn Graben; TSB, Terschelling Basin. The light dashed box shows the approximate study area illustrated in the detailed map. (**b**) Map of the study area illustrating tectonic elements (simplified fault framework for illustration), well penetrations, line of seismic section (Fig. 3b (dashed black) & Fig. 8 (solid grey)), correlation between wells (Fig. 6) and 1D modelling output (Fig. 10).

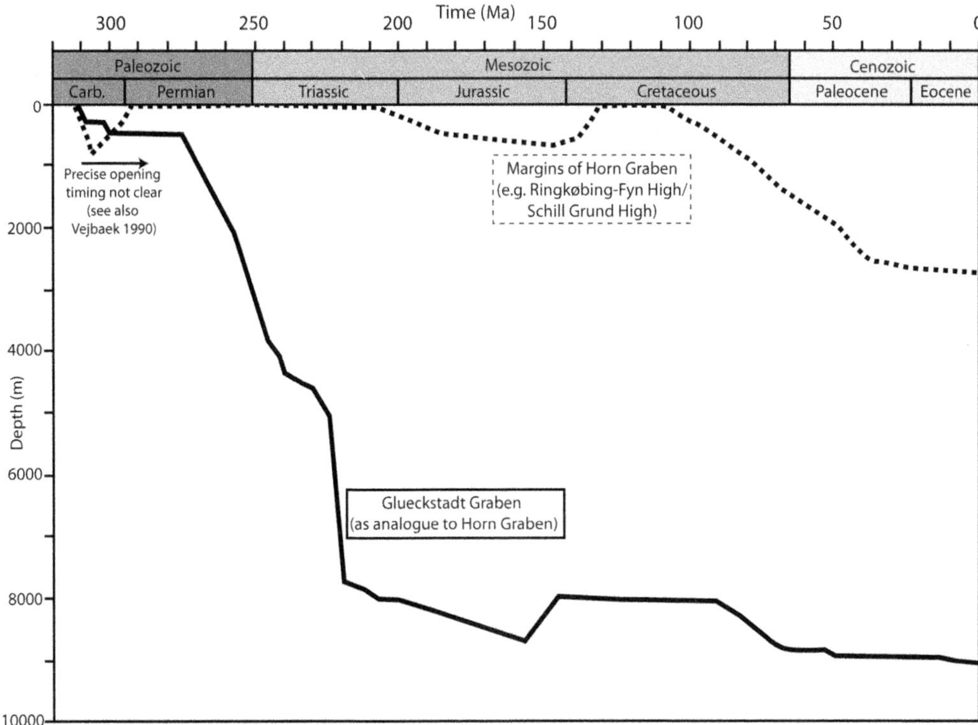

Fig. 2. Basin evolution through time, after Littke *et al.* (2008) and Scheck-Wenderoth *et al.* (2008), illustrating the phases of inversion and graben opening with respect to the graben centre (here using the Glückstadt Graben as an analogue: solid line) and the margins (e.g. Ringkøbing-Fyn High/Schill Grund High: dashed line). Note the uncertainty on the development of the proto-Horn Graben (e.g. Vejbaek 1990).

Southern Permian Basin area (e.g. McKie & Williams 2009). From the Middle Triassic onwards, the Upper Permian Zechstein salt sequences (and possibly Rotliegend evaporites: Jähne-Klingberg 2012) underwent halokinetic movements producing large diapirs and walls (e.g. Fig. 3b), as seen across the centre of the Southern Permian Basin (e.g. Maystrenko *et al.* 2005*b*, 2006; Mohr *et al.* 2005; McCann *et al.* 2008; Kukla *et al.* 2008; Scheck-Wenderoth *et al.* 2008; Warren 2008; Strozyk *et al.* 2017; Warsitzka *et al.* 2017). As in much of northern Germany and The Netherlands, the Jurassic section is largely missing due to a period of shortening known as the Cimmerian Unconformity (latest Middle–Late Jurassic; continuing in places into the Early Cretaceous: Littke *et al.* 2008). This event is variable in extent and magnitude, being driven by a mixture of local tectonics, North Sea doming and sea-level changes (Kyrkjebø *et al.* 2004; Pharaoh *et al.* 2010). The precise amount of uplift and missing stratigraphy is difficult to estimate but is likely to be less than 500 m (Mogensen 1995; Beha *et al.* 2008; Maystrenko *et al.* 2008; Pienkowski *et al.* 2008). The post-Jurassic sequence is relatively tectonically quiescent with occasional through-going faults and continued localized salt diapirism caused by a combination of large sediment loading (Harding & Huuse 2015) and far-field (Alpine and/or Betic) compressional phases (Rasmussen 2013). A generalized structural cross-section of the Horn Graben is shown in Figure 3.

Generalized stratigraphy

Figure 4 illustrates the general stratigraphy of the Horn Graben, including missing sections due to erosional unconformities. This consists of a Devonian or older 'basement', which is exhibited on the Schill Grund High by sections of sedimentary and volcanic strata (Espinosa 2016: see also the Siri-1X well, which tested a short section of 'basement'). The Carboniferous sequence consists of marine shales grading to more continental (coal-rich deltaics) deposits to the south (McCann 1999; Kombrink 2008; Kombrink *et al.* 2010). Due to an Early Permian period of uplift, the Carboniferous is generally missing on the Schill Grund and Ringkøbing-Fyn highs (Hedemann 1980; Kombrink *et al.* 2010; Uffmann & Littke 2011).

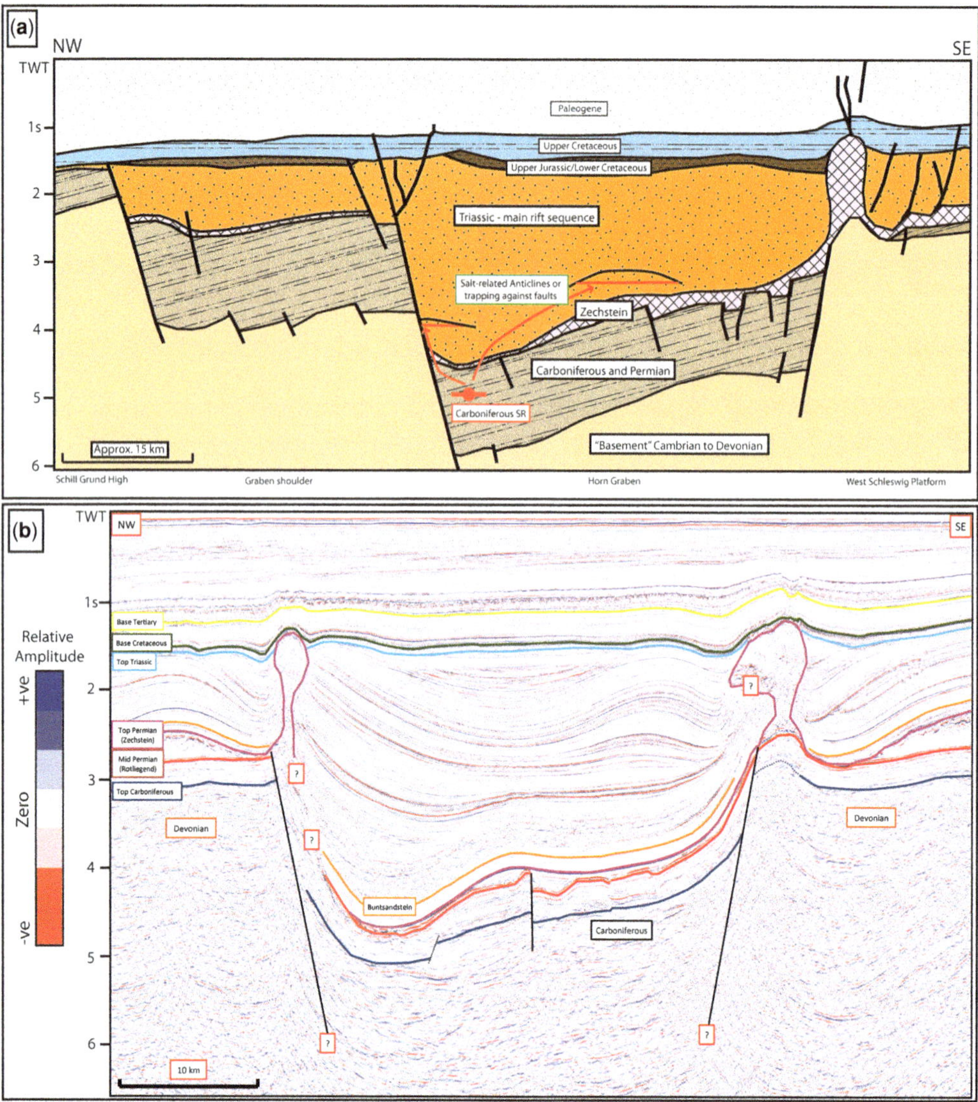

Fig. 3. Hydrocarbon play elements and example seismic section. (**a**) Conceptual section (after Vejbaek 1990) showing the gas generation (red symbol) from Carboniferous source rocks (coals), and migration (red arrows) through thin salt windows into the Lower Triassic Buntsandstein section where salt- and fault-related structures allow trapping of hydrocarbons with shaley sections providing top and lateral seals. (**b**) Example seismic section (see Fig. 1b for the position) including the main seismic markers. Note the large graben-bounding faults, salt structures and thick central Triassic section (between the pink Top Zechstein and the blue Top Triassic).

In this paper, some evidence is provided for the development and subsequent preservations of coals in the proto-Horn Graben but, as yet, no wells constrain these deltaics in this area. The Permian consists of the 'Rotliegend sequence' of sandstones, shales and evaporites capped by carbonates, anhydrites and halites of the Zechstein (e.g. Stemmerik *et al.* 2000; Geluk 2005; Stollhofen *et al.* 2008; Gast *et al.* 2010; Peryt *et al.* 2010).

The Triassic interval includes major sandstone units and associated interbedded mudstones (Volpriehausen, Detfurth, Hardegsen and Solling packages) in the informalized 'Buntsandstein' (Van Adrichem Boogaert & Kouwe 1994; Geluk 2005; Beha *et al.* 2008; Feist-Burkhardt *et al.* 2008; Stollhofen *et al.* 2008; Kortekaas *et al.*, this volume, in press), and thick sequences of Muschelkalk, Keuper shales and evaporites in the Middle and

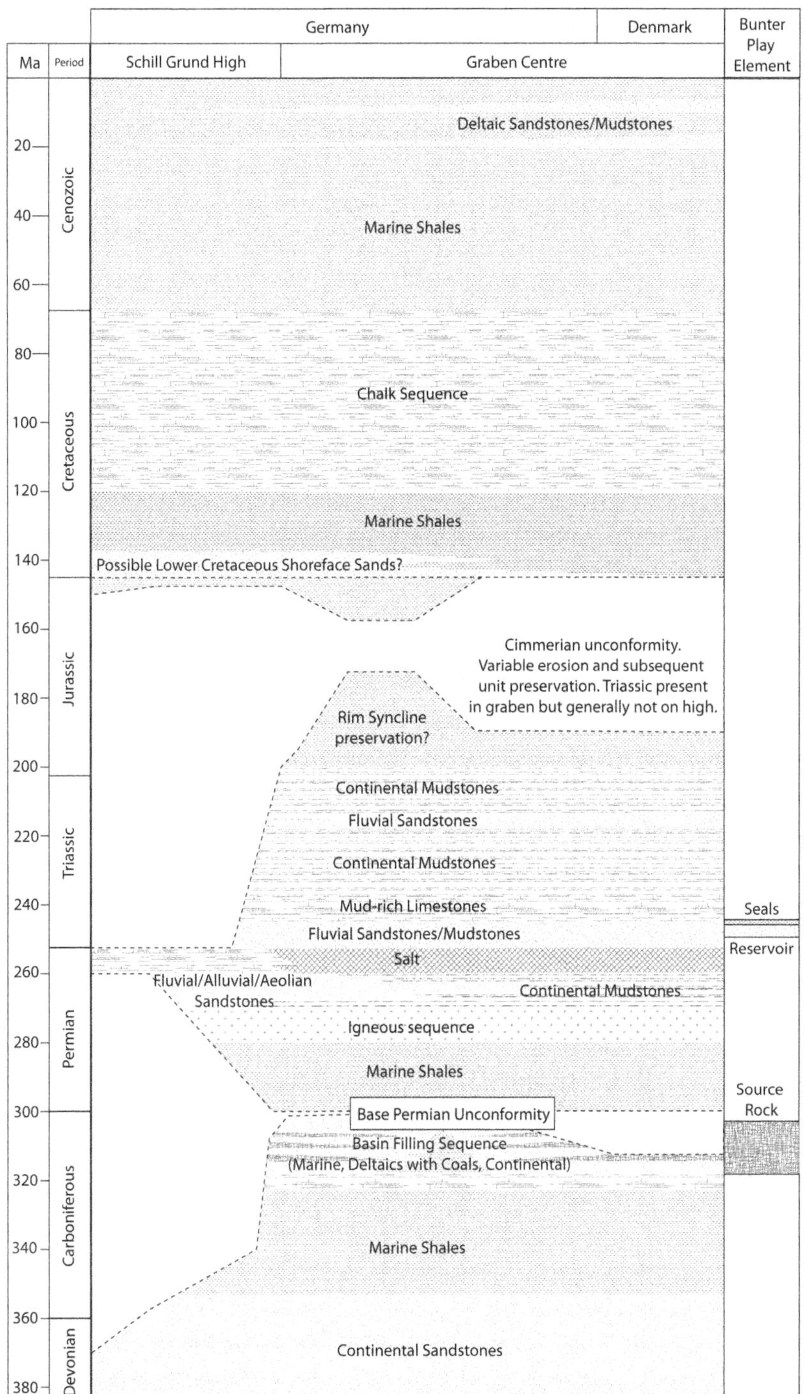

Fig. 4. Simplified stratigraphy and missing sections illustrating, from old to young, Devonian red-bed development, Carboniferous marine and delta systems (including source rock deposition), erosion related to Early Permian inversion, development of Zechstein salt, deposition of a mainly continental Triassic sequence including Buntsandstein reservoir/seal units, further erosion due to Jurassic inversion, and subsequent deposition of the post-Cretaceous predominantly marine section.

Upper Triassic (Beha *et al.* 2008). The thin preserved Jurassic stratigraphy, generally in deeper parts of the Danish Horn Graben and salt-related rim synclines, consists of Upper Jurassic 'Kimmeridge' and equivalent shales (Beha *et al.* 2008). The post-Jurassic consists of Lower Cretaceous sandstones and mudstones, a relatively thin (mostly <1 km) Upper Cretaceous chalk sequence, and a Tertiary sequence of mudstones and deltaic sandstones related to the progradation of the Eridanos delta and equivalents (e.g. Huuse 2002; Rasmussen 2009; Knutz 2010).

Buntsandstein play, previous exploration and remaining questions

The European Mesozoic rifts are well known by the hydrocarbon industry and can be, for example in the case of the Central Graben, prolific (e.g. Brzozowska *et al.* 2003). In contrast, the Horn Graben is largely underexplored and poorly understood. Table 1 and Figure 1b illustrate the well penetrations in the graben and the surrounding highs. The structural cross-section in Figure 3a shows the major play elements relating to the Buntsandstein play, including the reservoir sandstones, structural elements, Carboniferous gas-prone source rocks and possible migration paths (see also Di Primio *et al.* 2008 for a basin-scale hydrocarbon play overview).

Note that, within the Horn Graben, only four wells penetrate the Triassic and only two of these are valid tests of the Buntsandstein interval. All of these wells were dry. Various regional and localized basin models plus seismic interpretations have been published for northern Germany and the German offshore which give context to the potential hydrocarbon systems of the Horn Graben (e.g. Uffmann & Littke 2011; Heim *et al.* 2013; Arfai *et al.* 2014). Beha *et al.* (2008) specifically considered the Horn Graben, using 1D and 2D basin modelling, and suggested that the dry holes can be explained in two ways:

- Any Carboniferous coals were eroded during Variscan uplift phases and, therefore, no gas-generating source rock is present (no wells penetrate within this interval).
- Charge has occurred but has not reached the Triassic because of the sealing Zechstein salt interval.

Here the Buntsandstein play is reviewed, using the available wells and 2D seismic data, including the generation of a 3D basin model with the motivation to consider the remaining hydrocarbon potential. In particular, assumptions regarding source rock presence, charge timing and potential migration paths (Beha *et al.* 2008) are challenged with the aim of considering the validity of well penetrations and to further understand why they were dry.

Table 1. *Well penetrations across the study area, including details of their position relative to the graben, deepest penetrated interval, well result and notes on the Triassic section*

Well name	Country	Year	Position	Deepest interval	Result	Triassic notes
R-1X	**DK**	**1973**	**Graben centre**	**Permian: Rotliegend volcanics**	**Dry**	**500 m of Triassic section**
Pele-S1X	**DK**	**1975**	**Graben centre**	**Triassic: Buntsandstein**	**Dry**	**2500 m of Triassic section**
Erik-1X	**DK**	**2001**	**Graben centre**	**Triassic: Lower Bunter claystones**	**Dry**	**2000 m of Triassic section**
R1	**DE**	**1966**	**Graben centre**	**Triassic: Keuper**	**Dry**	**Upper Triassic only**
H08-1	DE	1975	Graben margin	Triassic: Lower Bunter claystones?	Dry	No released data, believed to be syncline penetration
G06-1	DE	1976	Graben margin	Triassic: Lower Bunter claystones	Dry	1000 m of Triassic section
Ibenholt-1	DK	1987	RKFH	Cambrian	Dry	250 m of Triassic section
C15-1	DE	1977	SGH	Devonian	Dry	No Triassic section.
Q1	DE	1966	SGH	Devonian	Dry	No Triassic section.
J04-1	DE	1977	WSP	Triassic: Lower Bunter claystones?	Dry	No released data, believed to be Bunter test
J05-1	DE	1982	WSP	Permian: Rotliegend volcanics	Dry	No released data, believed to be Rotliegend test.
S1	DE	1976	WSP	Permian: Rotliegend	Dry	1358 m Triassic section (Best *et al.* 1983)

Wells in bold are those within the centre of the Horn Graben, as seen in Figure 1b.
RKFH, Ringkøbing-Fyn High; SGH, Schill Grund High; WSP, West Schleswig Platform.

Datasets and methods

Well penetrations

Four wells have been drilled within the Horn Graben bounding faults, with various other wells on the surrounding highs (Fig. 1b). Table 1 gives an overview of the spud year, deepest penetrated formation, results, available data and other observations.

In summary, the Noordsee R-1 well (German offshore, 1966) did not reach the Bunter section (total depth (TD) in Muschelkalk (Best *et al.* 1983) or in the overlying Keuper (Wolf pers. comm. 2016)). The R-1X well (Danish offshore, 1973) penetrated the Buntsandstein in the far north of the graben. Two wells close to the Danish–German border specifically tested salt-related rollovers at the Buntsandstein level (Pele-S1X in 1975 and Erik-1X in 2001: Danish offshore). All four wells were dry (no hydrocarbon shows are noted in any available well reports). Further wells have been drilled onto the surrounding Rynkøbing-Fyn/Schill Grund highs as illustrated in Figure 1b.

Seismic data

The primary seismic dataset (example in Fig. 3b) is a 2D survey shot in 1997 by Maersk. This is a relatively good-quality migrated dataset (approximate line spacing 4 km, typical vertical resolution of *c.* 20–40 m at the Lower Triassic level based on an internal velocity of approximately 4000 m s^{-1} and a dominant frequency of 40–50 Hz) and includes both the German and Danish areas of the graben. Various noise is recognized in the dataset, especially in the deeper sections, with sideswipe from faults and salt structures likely (e.g. Fig. 3b). However, the main seismo-stratigraphic markers (e.g. the Buntsandstein and Top Rotliegend) are visible and continuous (Fig. 3b).

Comments on seismic interpretation and depth conversion

Various stratigraphic horizons were interpreted across the graben for the purposes of structural mapping, and as input for depth conversion and basin modelling. These included Base Paleocene (Base North Sea Group), Base Chalk Group, Base Cretaceous, Top Triassic, Middle Triassic marker equivalent to the Base Keuper, Top Buntsandstein, Top Zechstein, Top Rotliegend and Top Carboniferous (e.g. Fig. 3b).

One of the key uncertainties is depth conversion, especially in the lower Triassic and deeper strata. This is of particular significance because it acts as an important input to basin modelling and governs reservoir effectiveness.

Well-derived sonic velocity data were generally not available to this study, although marker tops and time picks were utilized, and overburden velocities were available in the Schill Grund High area. The greatest uncertainty is driven by the very thick (3–5 km) Triassic sediment pile within the graben, and what velocity model best represents this interval. For the purposes of this evaluation, velocities were derived from the following sources:

- From mudline to Top Triassic, internal analysis is based on well-tied velocities across the German offshore.
- For the Triassic interval, three end-member models were utilized that were designed to test the likely depth range in the pre-Middle Triassic interval; using a V_0K model (V_0; initial velocity = 2365; K; gradient = 0.49), and constant velocities of 3500 and 4000 m s^{-1}. These models were based on published velocity models (Van Dalfsen *et al.* 2007) and other internal Triassic velocity models from nearby Dutch and Danish offshore blocks. Figure 5a shows an example of the 4000 m s^{-1} constant velocity derived depth map for the top of the Buntsandstein.
- Below the Top Zechstein, standard velocities were used (e.g. a constant velocity of 4470 m s^{-1} for the Zechstein) (e.g. Van Dalfsen *et al.* 2007). This is an oversimplified model because it does not take into account changes in Zechstein facies.

In this basin-modelling study, various sensitivities were tested to understand the impact of different Triassic velocities. A considerable error margin remains (*c.* ±250 m at the Top Bunter level; ±500 m at the Top Carboniferous level). This uncertainty is consistent with other published depth-conversion attempts: for example, GPDN (2016) present a deepest point for the Top Rotliegend approximately 300 m greater than estimated by this study. Such uncertainties can impact both charge timing and reservoir quality.

A 3D basin model was generated using Shell's in-house CAULDRON software (e.g. Burley & Scotchman 1999; Giles *et al.* 1999; Leischner *et al.* 2011). This involves inputting a series of depth horizons, generalized assumptions on lithologies, heat flow/tectonic histories and source rock quality. CAULDRON then uses this information to make an upwards/downwards projection between the associated surfaces. Table 2 gives an overview of the inputs and likely range of uncertainties. Well information, and further calibration data, were gathered from various internal well reports plus the published material listed in Table 2.

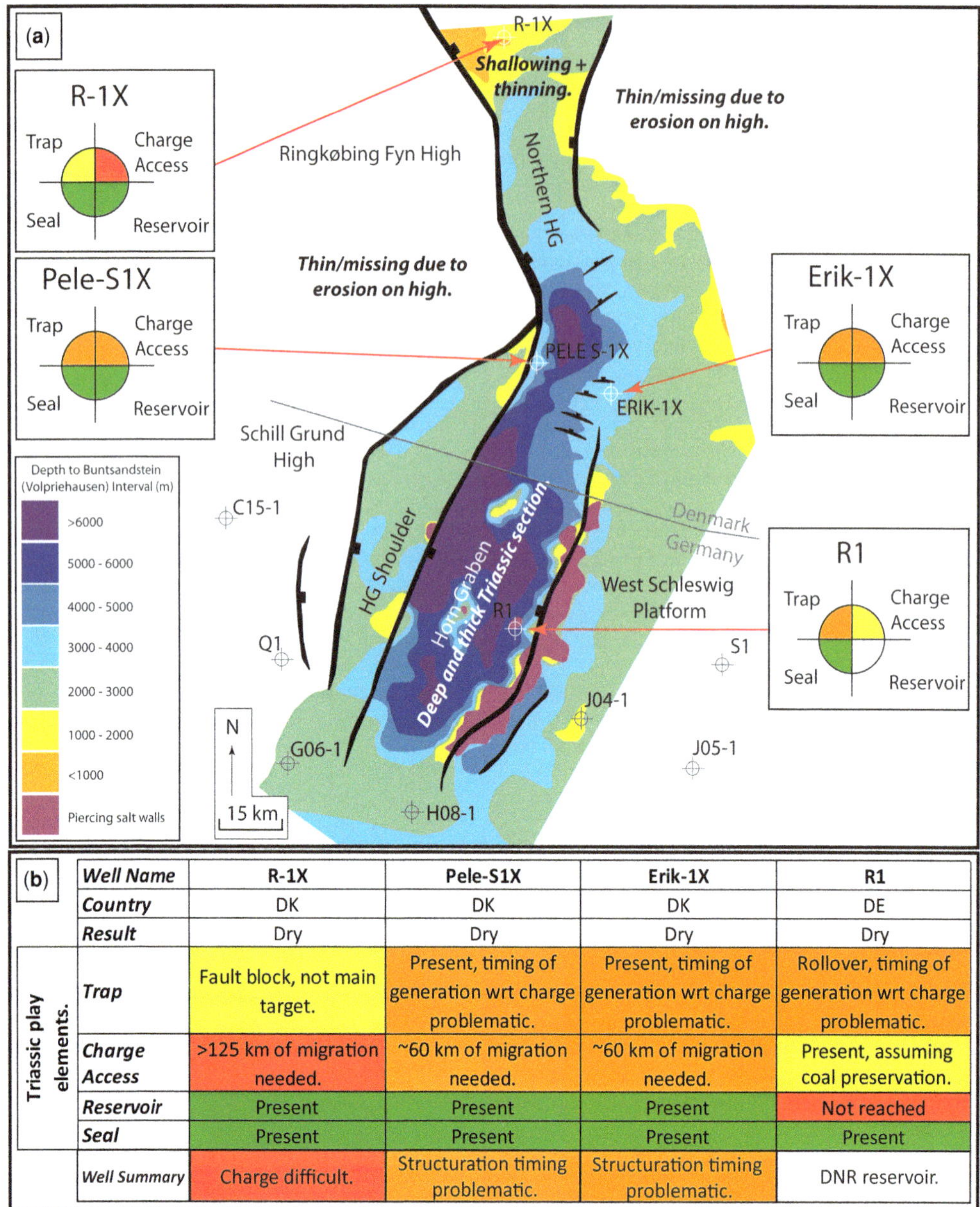

	Well Name	R-1X	Pele-S1X	Erik-1X	R1
	Country	DK	DK	DK	DE
	Result	Dry	Dry	Dry	Dry
Triassic play elements.	Trap	Fault block, not main target.	Present, timing of generation wrt charge problematic.	Present, timing of generation wrt charge problematic.	Rollover, timing of generation wrt charge problematic.
	Charge Access	>125 km of migration needed.	~60 km of migration needed.	~60 km of migration needed.	Present, assuming coal preservation.
	Reservoir	Present	Present	Present	Not reached
	Seal	Present	Present	Present	Present
	Well Summary	Charge difficult.	Structuration timing problematic.	Structuration timing problematic.	DNR reservoir.

Fig. 5. Depth of reservoir section and dry wells. (**a**) Depth to the Buntsandstein reservoir interval (Volpriehausen seismic pick) in metres; note the very deep central area of the graben, interrupted by salt structures. The Triassic section thins and shallows to the north, and is thin or entirely missing on the Schill Grund and Ringköbing-Fyn Highs due to the Jurassic inversion but is present on the West Schleswig Platform. Well segments represent the different Buntsandstein play elements seen in the wells and detailed in the table in (b). (**b**) Table of dry hole analysis results for the four wells in the main part of the Horn Graben. It is proposed that wells R-1X, Pele-S1X and Erik-1X were dry due to a combination of charge access issues (migration distance) and the timing of that charge relative to structuration timing. Colours represent a range of outcomes from present (green) to likely present (yellow), problematic (orange) and not present (red). White represents a stratigraphic section missing from that well. DNR, did not reach (Buntsandstein).

Table 2. *Basin-modelling parameters*

Input	Typical range	Reference comments
Base boundary		
Crustal thickness	28–40 km	Varying across model area (Abramovitz & Thybo 1999; Gregersen & Voss 2002; Grad *et al.* 2009; Bonté *et al.* 2012; Maystrenko & Scheck-Wenderoth 2013) and through time due to rifting (Beha *et al.* 2008).
Crustal density	2790–2900 kg m^{-3}	Varying across model area, transitioning between Avalonian and Baltican crust, with latter typically being denser and older (Bonté *et al.* 2012; Maystrenko & Scheck-Wenderoth 2013)
Crustal strength	1×10^{13}–1.2×10^{13} Pa m	Tesauro *et al.* (2007)
Heat regime		
Crustal thermal conductivity	2.3–2.8 W m^{-1} K^{-1}	Tesauro *et al.* (2007)
Radioactive heat production	0.5–2 μW m^{-3}	Tesauro *et al.* (2007) and Bonté *et al.* (2012)
Crustal heat capacity	100–1100 J kg^{-1} K^{-1}	Tesauro *et al.* (2007)
Heat flow through time	40–95 mW m^{-2}; present day 52 mW m^{-2}	Variable due to rifting and inversion phases. Calibrated using internal data and the heat-flow phases of Beha *et al.* (2008) plus Uffmann & Littke (2011)
Geothermal gradient	Present day 31–35°C km^{-1}	Modelled fit to well data, internal reports and heat flow through time. Present day from Agemar *et al.* (2011) and Bonté *et al.* (2012)
Top boundary		
Water depth through time	Present day 15–50 m	Highly variable through time; framework from Doornenbal & Stevenson (2010)
Surface temperature through time	Present day 5–12°C	Dependent on water depth and climatic variation (e.g. Doornenbal & Stevenson 2010; Bonté *et al.* 2012)
Burial history		
Depth conversion	Deepest point 6–10 km, highly uncertain	Discussed in the text, various models and lack of control data. Geluk (2005) and Maystrenko & Scheck-Wenderoth (2013) as guides
Timing of erosional events	(1) Major Late Carboniferous–Early Permian (310–290 Ma); (2) moderate Late Jurassic (175–150 Ma); (3) minor Early Cretaceous (130–140 Ma)	Variable across basin; estimates from Beha *et al.* (2008) and Uffmann & Littke (2011)
Source rock quality	Westphalian 5–60% total organic carbon (TOC); 150–350 mg g^{-1} of TOC hydrogen index	Various internal reports, highly variable and uncertain over the study area. Typical Westphalian coal values used (e.g. Gautier 2003; Pletsch *et al.* 2010). Overburden calibrated using Beha *et al.* (2008)
VR, temperature + thermal conductivity calibration	Various	Calibrated using internal well data and reports plus Beha *et al.* (2008), Sachse & Littke (2016) and Uffmann & Littke (2011)

Various example input parameters and references were used to build the 3D basin model. Uncertainty remains in many of these categories, especially the degree of erosion in inversion phases and the depth conversion.

Play element analysis

Reservoir

The primary Triassic reservoir interval across the Southern Permian Basin area is the Middle 'Buntsandstein' (specifically the Volpriehausen, Detfurth, Hardegsen and Solling intervals, which include associated mudstones) (Geluk & Röhling 1997; Szurlies 2004; Geluk 2005; Feist-Burkhardt *et al.* 2008). The Horn Graben is likely to represent the northern margin of a playa lake system with fluvial, alluvial and aeolian systems focused into the evolving structural low. Precise sediment sources and their relative contributions are debated but include Fennoscandia to the north (*sensu* Clausen & Pedersen 1999; Geluk 2005; McKie & Williams 2009; Bachmann *et al.* 2010; Wolf *et al.* 2014, 2015), the Variscan front to the south and locally sourced material from the Ringkøbing-Fyn High (Olivarius *et al.* 2015; Kortekaas *et al.*, this volume, in press).

Figure 6 demonstrates a well correlation through the four Horn Graben wells. These show that the Buntsandstein interval gross thicknesses range from 50 to 800 m (net-to-gross 20–50%). Petrophysical analysis (see also Beha *et al.* 2008) confirms good reservoir properties in these wells (15–20% porosity). However, reservoir effectiveness remains a concern when considering the deep graben areas (over 6 km deep) due to the porosity loss via compaction.

Seismically, the Buntsandstein interval (often the top of the lowermost Volpriehausen sandstone marker) can be picked over much of the study area as a relatively higher-amplitude soft loop, suggesting that the sandstones seen in the well penetrations are widespread (e.g. Goldsmith *et al.* 1995). However, it is worth noting that the Lower Triassic section demonstrates complex sedimentation patterns related to bounding fault movements and, subsequently, halokinesis. This has the potential to produce variable depositional facies within the Buntsandstein section and the overlying Lower and Middle Triassic intervals. For example, in places, the western bounding fault of the graben appears to have moved immediately after Buntsandstein deposition, with subsequent Middle–Late Triassic halokinetic phases. These complexities are in contrast to the tramline basin-scale trends generally presented for these sections, especially the pre-Keuper stratigraphy (e.g. Geluk 2005). More work is needed to document these tectonic and halokinetic phases and their relationship to seismic facies variability, especially in relation to depth (porosity variability and noise in the data). Wolf *et al.* (2015) considered some of the possible variabilities in the Triassic interval across the German offshore. Figure 7 shows a simplified conceptual model of the Bunter interval clastic sedimentation as a first attempt to visualize possible depositional facies variabilities, including the influence of tectonism on the Buntsandstein interval. This is based on a generalized understanding of the Buntsandstein interval from local well penetrations and regional reconstructions.

In summary, well penetrations demonstrate that thick Buntsandstein reservoir intervals are present in the Horn Graben. However, the spatial and temporal variability in reservoir quality is unknown.

Seal

The correlation in Figure 6 demonstrates various interbedded and capping shale packages associated with the Volpriehausen, Detfurth, Hardegsen and Solling claystones (e.g. Geluk 2005). In addition, the overlying Röt sequence contains further shales and halites. It has been demonstrated by working gas fields in the Dutch offshore (e.g. F15A: Bachmann *et al.* 2010) and experimental studies (Heinemann *et al.* 2012) that these shales form a competent top seal with a high capillary entry pressure. However, any late salt movements could produce fracture networks that would threaten seal competency. As noted previously, it should be recognized that there are various atypical internal unconformities within the Lower and Middle Triassic section. These have the potential to impact both local depositional patterns and subsequent erosion. In addition, differential overpressure cells are known to impact trap/seal integrity in the Dutch Central Graben (e.g. Gaarenstroom *et al.* 1993). Detailed Horn Graben pressure data were not available to this study; hence, this aspect is not further investigated.

In summary, the well penetrations demonstrate typical Bunter shales that have the capacity to act as good seals for gas. It is expected that these would occur across the area of interest but may be impacted by localized tectonism, causing possible thickness variations or breaching by faults and/or fractures.

Carboniferous source rocks: development and preservation

The primary source rocks within the Southern Permian Basin area are Carboniferous (Namurian–Westphalian) gas-prone Type III coals which are known to develop in progressively prograding delta systems around the basin margins (e.g. Kombrink 2008; Kombrink *et al.* 2010). An Early Permian uplift and erosion event (Saalian and related events: Geluk 2005) then planed off much of the later sequences, especially around highs such as the Ringkøbing-Fyn and Schill Grund. The potential to preserve coal sequences in the Horn Graben area is a major uncertainty because of a lack of well control. It is proposed

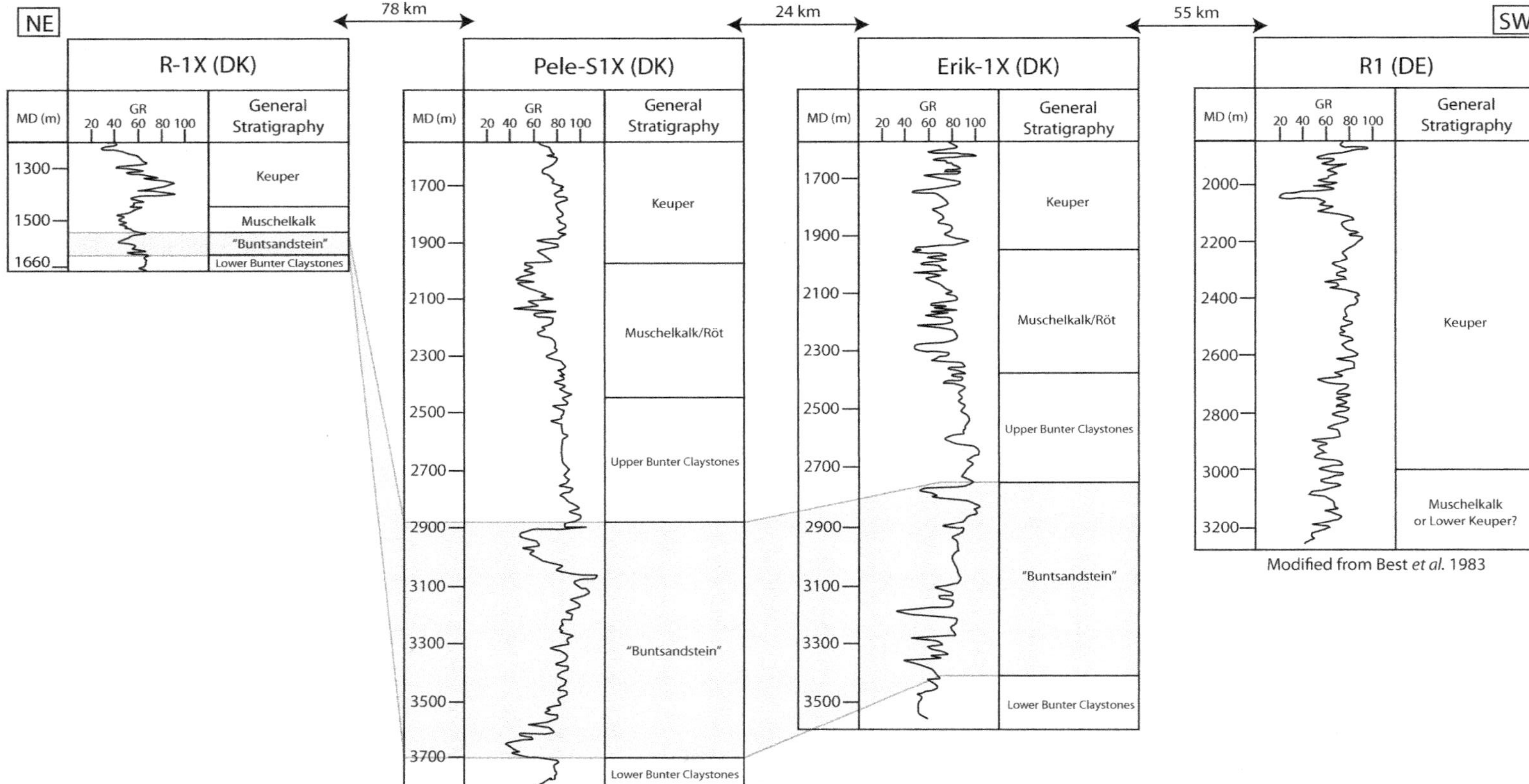

Fig. 6. Large-scale correlation panel of wells in the central part of Horn Graben. See Figure 1 for the line of section. Well data are from internal reports and Best *et al.* (1983). Note the varying thickness of the Triassic sections and the development of a thick Buntsandstein interval in wells Pele-S1X and Erik-1X. Also note that the Buntsandstein, as discussed in the text, consists of various cycles of sands and interbedded claystones (Volpriehausen, Detfurth, Hardegsen and Solling: Geluk 2005).

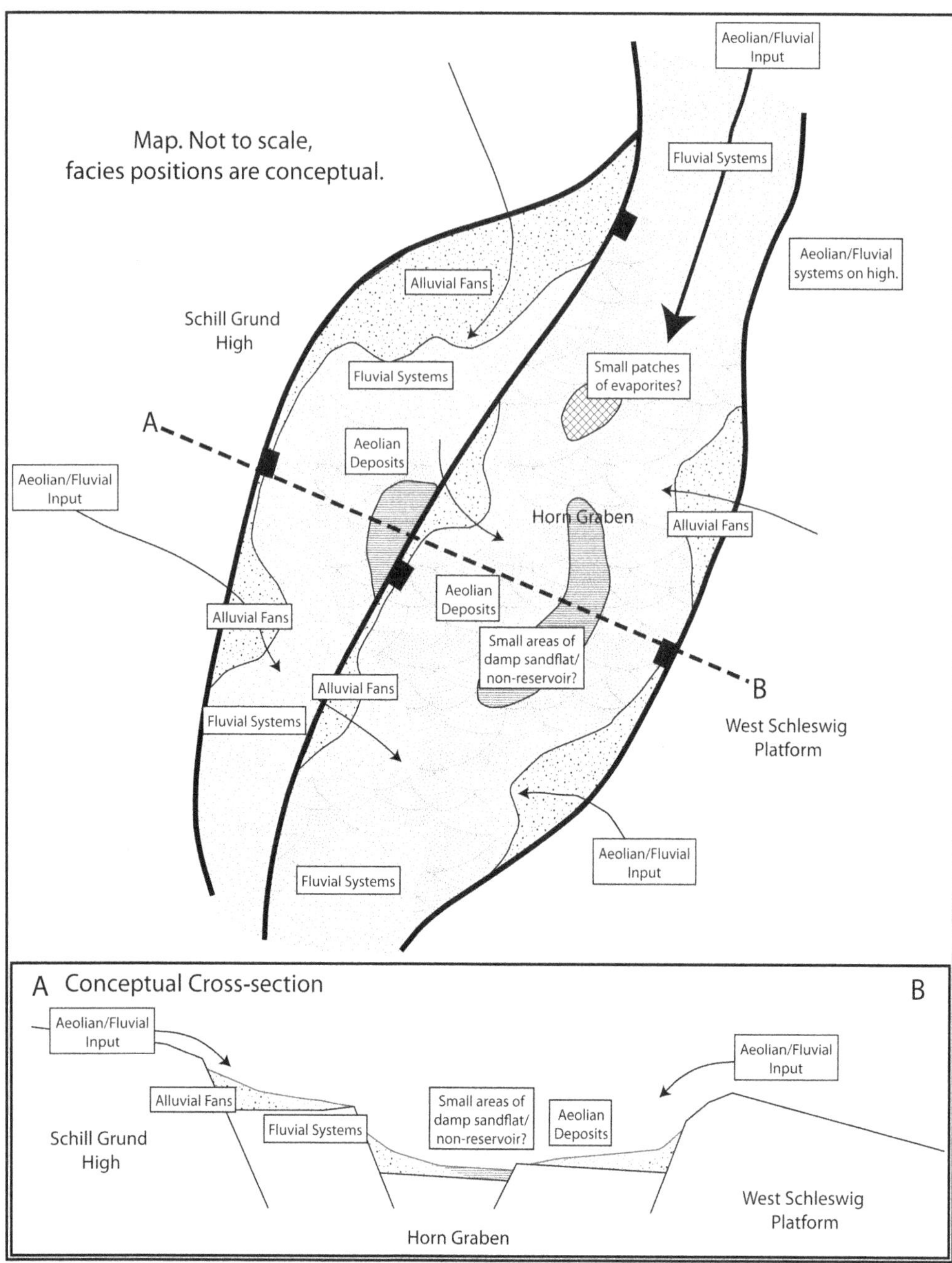

Fig. 7. Conceptual facies map and cross-section of the Buntsandstein interval (e.g. the Volpriehausen sandstone). Not to scale and facies positions are entirely conceptual, illustrating the possible impact of fluvial systems from the north, alluvial fans from bounding faults and areas of damp sandflat/evaporites in-between sand-rich systems. The aim is to illustrate the possible variation in Buntsandstein facies across the graben system.

that a pre-uplift proto-Horn Graben structure would have allowed for the preservation of coal sequences within the lows. Here, seismic observation and conceptual models (e.g. Hedemann 1980) are utilized to suggest that coals could have been preserved up to the German–Danish border.

Figure 8 illustrates details of a seismic line across the deep graben close to the German–Danish border where, according to Kombrink *et al.* (2010), no Westphalian coals should be preserved. It is observed that various bright reflectors occur below the Top Carboniferous surface which can be assigned to three seismic facies groups based on variable character, from shallow to deep (Fig. 8a):

- An upper group of moderately continuous to continuous reflectors with bright amplitudes.
- A middle group of discontinuous reflectors with variable amplitudes.
- A lower laterally discontinuous bright, higher-angle reflector.

In a nearby onshore northern Netherlands production well (Tjuchem-2A), a sequence of Namurian–Westphalian B age has been tested and can be accurately tied to seismic data using sonic velocities and checkshots. Three distinct seismic characters are noted, from shallow to deep (Fig. 8b):

- In the Westphalian B section, a group of moderately continuous to continuous reflectors with moderately bright to bright amplitudes. Well logs and cuttings show this to be a coal-prone interval.
- In the Westphalian A/B section, mostly discontinuous reflectors with occasional more continuous intervals with variable amplitudes. Well logs and cuttings show this interval to represent fluvial channel sands and claystones.
- In the Namurian section, a mixture of continuous, high-amplitude and discontinuous, low–moderate amplitude reflectors.

Although the frequency content of the seismic datasets is different, it is possible to see similarities in the sequences, especially in the upper two units where bright amplitude, continuous markers give way to discontinuous reflectors of variable amplitudes. In addition, if the upper unit contains Westphalian B coals, it may be possible to see acoustically soft intervals. Figure 8c demonstrates a pseudo-acoustic inversion (90° phase rotation with amplitude and frequency balancing) with a colour palette designed to enhance soft markers. The upper bright amplitudes (Fig. 8a, unit 1) appear as a moderately continuous soft to very soft marker, possibly consistent with coals. The third (onshore Namurian) unit is more difficult to see in Figure 8a, with cross-cutting high-angle markers that are likely to be igneous (intrusive) in origin. By comparison, in Figure 8c this interval appears as a hard event with a soft post-cursor.

Mapping of these units suggests that coal-prone intervals could be preserved in the deep graben close to the Danish–German border. Figure 9b demonstrates the northern limit of source rock preservation used in the basin modelling (note that, for simplicity, the model was first run assuming coal development across the whole area and then cut to a realistic extent: Fig. 9).

It is recognized that a number of uncalibrated observations are presented. Alternative source rock scenarios are possible. For example, the Namurian could be the shallower package (unit 1: Fig. 8a) rather than the Westphalian B. That Namurian interval could also consist of coal-prone intervals. In addition, the soft, high-amplitude intervals could be associated with shales or high-porosity sandstones rather than coals. However, it is proposed that, when compared to a calibrated analogue (Fig. 8b), preservation of Westphalian B coals is a realistic scenario. As such, the basin model results presented in the following subsection assume variable early burial from south to north and minimal Early Permian uplift in the southern Horn Graben.

Charge timing and migration; 3D basin modelling insights

A 1D extraction of the 3D basin model for a point close to the Pele-S1X well is illustrated in Figure 10. This example is included, assuming coal preservation, to allow comparison to the work of Beha *et al.* (2008). However, it is reasonable to expect the Carboniferous to be deeply eroded at this position and, as discussed above, coals may not be present. A possible burial history curve is illustrated, although the large uncertainty in depth-conversion and basin-modelling inputs must be acknowledged. A Carboniferous structural low evolves into rapid rifting in the Early–Middle Triassic leading to, based on modelled vitrinite reflectance values, a relatively short oil mature phase between 250 and 225 Ma, and a peak gas-expulsion phase at this particular 1D location between *c.* 225 and 175 Ma immediately followed by burial to an overmature phase. Middle–Late Jurassic uplift (here modelled as *c.* 400 m: *sensu* Beha *et al.* 2008) seems to have had little impact on maturity.

Modelled vitrinite reflectance evolution for various time snapshots and gas-expulsion maps for various time intervals assuming coal preservation across the whole study area (Fig. 10). This demonstrates that the oil window was reached at *c.* 235 Ma, the gas window was reached across much of the graben between 235 and 150 Ma, trending to overmaturity post-100 Ma. As a result, any gas explusion would predominantly have occurred between 235 and 175 Ma. An additional complexity is the role of mobile salt in determining heat flow and conductivity. It is possible that localized patches of coals had a colder Triassic–Jurassic history (e.g. Yu *et al.* 1992; Magri *et al.* 2008), allowing for slightly later

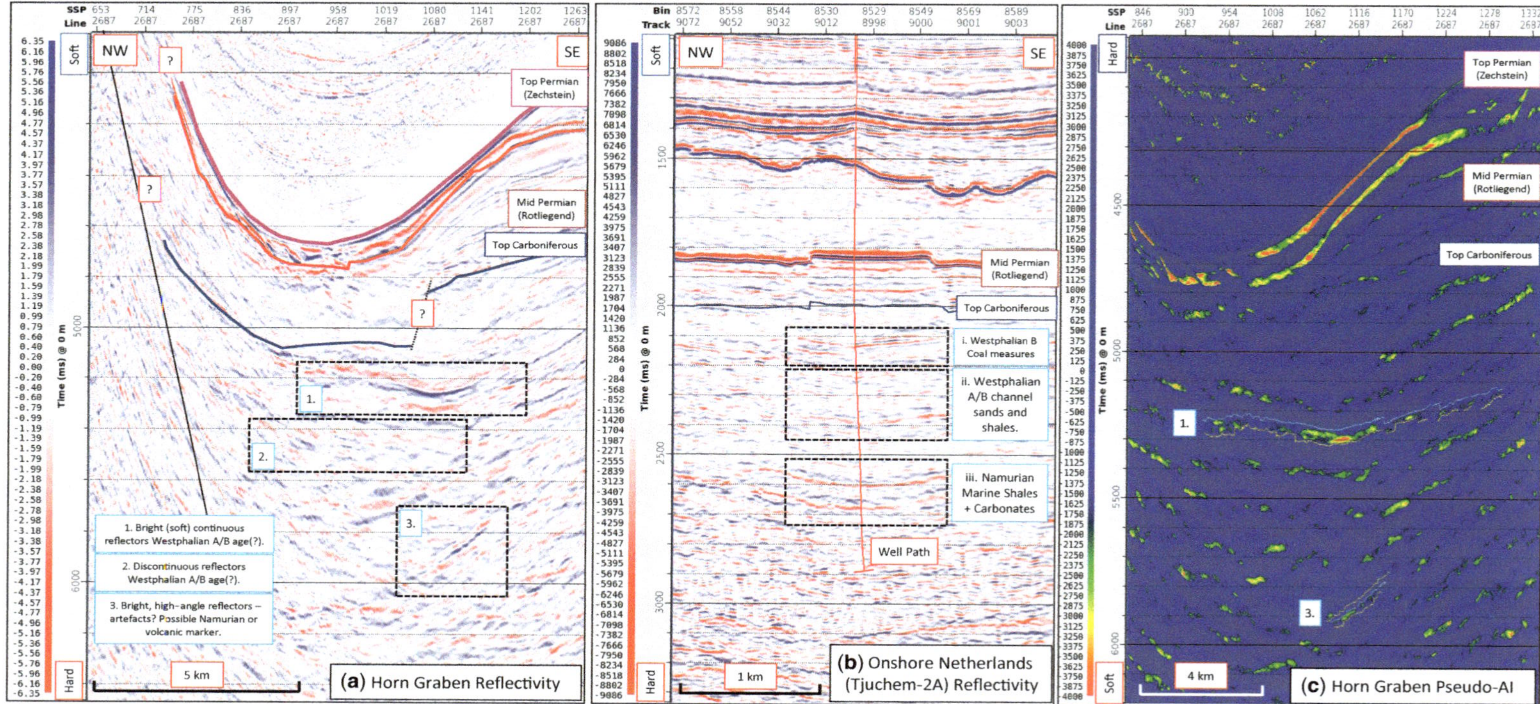

Fig. 8. Seismic observations of the pre-Permian interval pertaining to the NW part of the deep graben seen in Figure 3 and located close to the German–Danish border (Fig. 1 map). (**a**) Horn Graben reflectivity data (two-way time (TWT)) highlighting three intervals of varying seismic character (1–3); details in the text. (**b**) Onshore northern Netherlands reflectivity data (TWT) as a calibrated analogue to the Horn Graben illustrating three intervals of varying seismic character (i–iii); the red line represents the velocity-tied well path; details in the text. (**c**) Horn Graben pseudo-acoustic inversion designed to highlight soft markers with details of 1 and 2 of (a); details in the text.

maturity. This is expected to be a relatively minor effect. The distribution of relative peak gas expulsion, assuming a limited area of preserved coals, is illustrated in Figure 9b. It is noted that relatively high, but early, gas expulsion occurs in the centre of the graben.

It should be acknowledged that overmature coals are associated with a risk of nitrogen generation (Gerling *et al.* 1997; Krooss *et al.* 2008), and the German offshore has a history of high (20–80% mainly in the southern part of the H and J blocks plus the L and M blocks) nitrogen Rotliegend discoveries (Lokhorst 1997). It cannot be ruled out that a later nitrogen-rich charge phase flushed through the petroleum plays in the study area, although no direct data allow confirmation.

Limited basin-modelling sensitivities were run, varying one parameter at a time. For example, it was noted that, although there is a significant depth-conversion uncertainty, there is little impact of the different models on the timing of overmaturity. For example, for a point near well Pele-S1X, the onset of overmaturity varies by 10 myr in the different depth scenarios. In a second example, the amount of Jurassic uplift was also varied but the timing of peak expulsion seems relatively robust when inputting a range of uplift models (e.g. up to 1 km of Jurassic uplift can occur before significant post-inversion charge is generated).

Vertical migration of gas in the Southern Permian Basin from the Carboniferous units to the overburden is thought to occur via various mechanisms including along faults, carrier beds and through salt windows/welds (e.g. Butler & Pullan 1990; Crepieux *et al.* 1998; Pletsch *et al.* 2010). Figure 11a illustrates the large variability in Zechstein salt thickness across the study area which could allow for vertical migration through salt windows. This is partly driven by Mid-Triassic pillowing and subsequent Late Triassic–Mid-Jurassic diapirism (with a later Tertiary reactivation phase in some areas), which can be observed in seismic sections by thickening and thinning relationships. However, this area also marks the transition from evaporitic salt lake to platformal carbonate Zechstein facies (e.g. Vejbaek 1990; Ziegler 1990; Peryt *et al.* 2010). The precise original distribution and thickness of carbonate and halite facies depends on the palaeotopography but a conceptual example is shown in Figure 11b. Lateral migration distances along carrier beds are not well documented but have the potential to be tens of kilometres if the right combination of geometries and reservoir quality occurs. The time needed for short- and long-distance migration is also not clear (modelling within the CAULDRON software suggests a time period of 10–50 myr to migrate from source rock to the Lower Triassic interval). A summary of the critical play elements is shown in Figure 12 illustrating the main structural phases, source rock/reservoir/seal deposition and halokinetic phases. An example 'critical moment' is illustrated where the peak-expulsion/migration phase overlaps with halokinesis.

Buntsandstein traps and trap timing

The majority of structural closures within the Lower Triassic are related to salt rollovers, turtle backs and salt flanks (e.g. Fig. 3) (see Beha *et al.* 2008). In addition, there are a number of cross-cutting faults that could allow for discrete fault blocks, some of these may have been generated by halokinesis, others by relatively late movements of the graben-bounding faults. In addition, variable depositional facies may allow for discrete stratigraphic traps. As such, the timing of Zechstein salt movement is important because it defines the structural development of the majority of Buntsandstein interval traps. Baldschuhn *et al.* (1996) considered salt movement timing across NW Germany and demonstrated a complex history with halokinesis varying in age from the Early Triassic to the Tertiary. In this study, it is noted through variable thickness mapping of overburden horizons that halokinetic movements, although also complex, mostly demonstrate movement histories between the Late Triassic and Middle Jurassic (Figs 3 & 12). It appears that salt movement into pillows started in the Middle Triassic, with further salt withdrawal and diapirism starting from the latest Triassic into the Middle Jurassic, with smaller, localized pulses into the Upper Jurassic and reactivation in the Tertiary (*sensu* Harding & Huuse 2015). Note that the main stage of diapirism coincides with the main gas charge phase (Fig. 12).

Dry hole analysis

A well by well analysis of factors that may have led to a dry hole outcome in the study area penetrations is shown in Figure 5. In summary, two major issues are noted:

- Access to charge, not due to vertical migration but to uncertainty on lateral migration paths and distances. This can be seen in combination with uncertainty on Carboniferous coal distribution.
- The timing of charge relative to halokinesis-related structuration is suboptimal across the area. The existing wells have deliberately targeted large rollovers and other salt-related structures (e.g. salt highs with related faulting). These structures were likely still evolving during, and after, peak expulsion. The time for migration to occur remains an uncertainty but late (Tertiary) reactivation of some salt structures could have caused further loss of hydrocarbons.

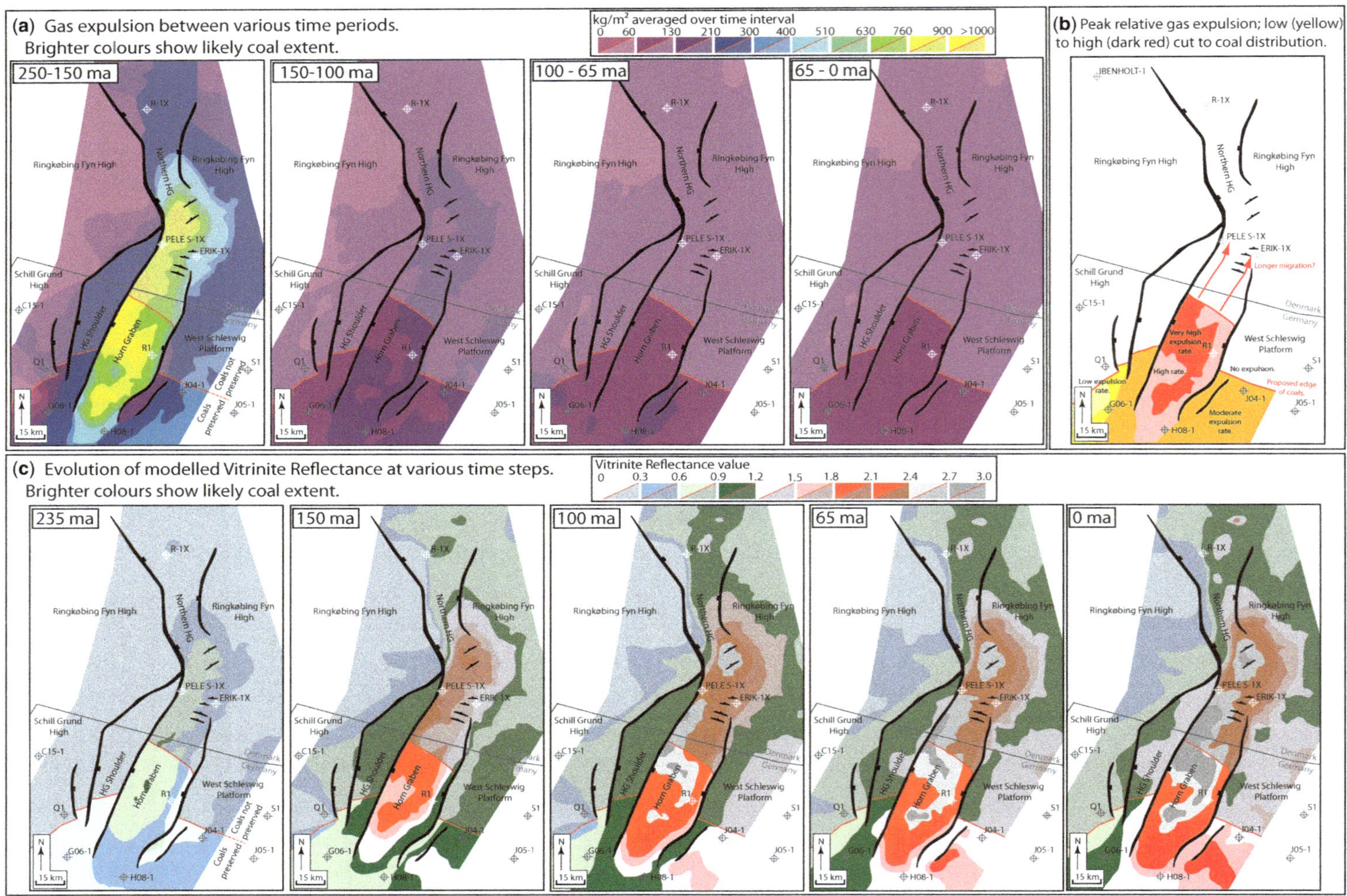
(a) Gas expulsion between various time periods.
Brighter colours show likely coal extent.
kg/m² averaged over time interval
0 60 130 210 300 400 510 630 760 900 >1000
250-150 ma
150-100 ma
100 - 65 ma
65 - 0 ma
Coals not preserved
Coals preserved
(b) Peak relative gas expulsion; low (yellow) to high (dark red) cut to coal distribution.
IBENHOLT-1
Longer migration?
Very high expulsion rate.
High rate.
No expulsion.
Proposed edge of coals.
Low expulsion rate.
Moderate expulsion rate.
(c) Evolution of modelled Vitrinite Reflectance at various time steps.
Brighter colours show likely coal extent.
Vitrinite Reflectance value
0 0.3 0.6 0.9 1.2 1.5 1.8 2.1 2.4 2.7 3.0
235 ma
150 ma
100 ma
65 ma
0 ma
Ringkøbing Fyn High
Northern HG
Horn Graben
HG Shoulder
Schill Grund High
West Schleswig Platform
Denmark
Germany
R-1X
PELE S-1X
ERIK-1X
R1
C15-1
Q1
G06-1
H08-1
J04-1
J05-1
S1
N
15 km

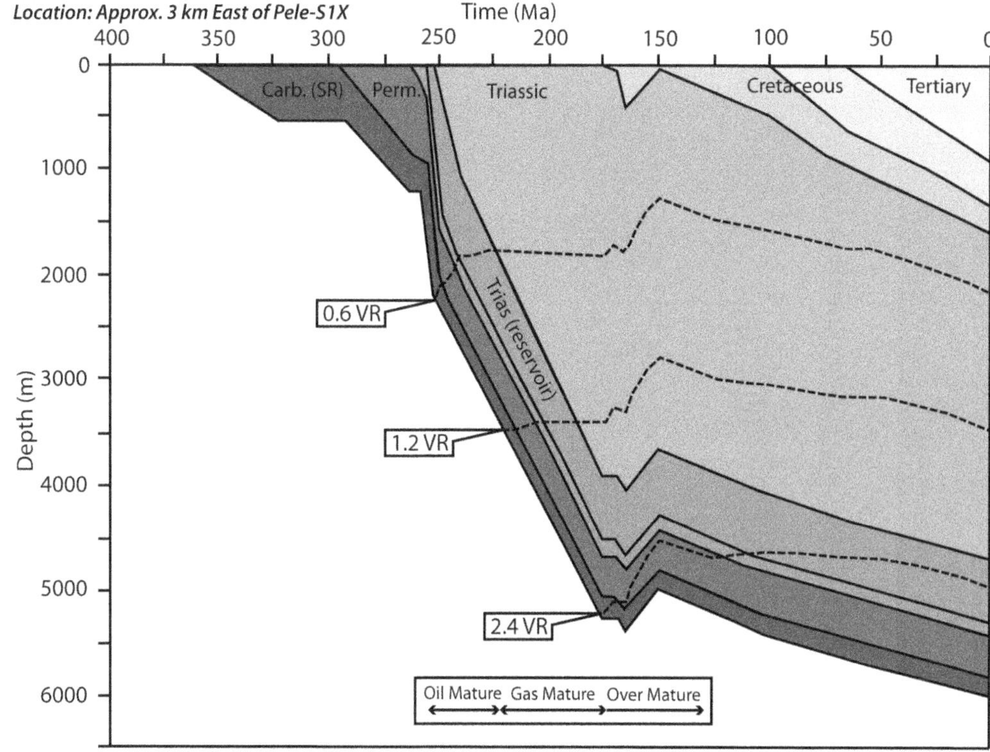

Fig. 10. 1D model from close to Pele-S1X (3 km east; position shown in Fig. 1) illustrating the basin model output for this location. Note the relatively short oil mature phase between 250 and 225 Ma; the peak gas-expulsion phase at this particular 1D location is between *c.* 225 and 175 Ma, immediately followed by burial to an overmature phase. SR, source rock (Carboniferous); VR, vitrinite reflectance.

Discussion of play elements and remaining potential

Play analysis

Comparison of this Buntsandstein play analysis and 3D basin modelling with the Denmark-focused 1D and 2D basin modelling of Beha *et al.* (2008) shows that general charge timing is consistent between the two models if strong early rifting is assumed. Beha *et al.* (2008) demonstrated a large hydrocarbon generation event in the deep graben between the Middle Triassic and Early Jurassic. Uffmann & Littke (2011) produced a regional 3D model that also suggests high transformation ratios in the Horn Graben at 200 Ma, although the extent of present-day overmature areas appears smaller.

Beha *et al.* (2008) noted that migration into the Buntsandstein reservoirs of wells Pele-S1X and Erik-1X from a perceived Carboniferous coaly source rock is tricky. If this observation is valid, this would involve longer-distance migration when it may be easier to migrate gas from the deep basin immediately up adjacent faults. The presence of Carboniferous coals remains a key uncertainty but, in the absence of well penetrations, preservation up to the

Fig. 9. Maturity steps, maps of basin-modelling outputs. (**a**) Gas expulsion between various time periods: the area likely to have preserved coals is south of the red line (brighter colours); the area likely to have preserved coals is north of the red line (washed-out colours). Note that the overwhelming majority of expulsion is early (250–150 Ma), with limited charge post-150 Ma. (**b**) Peak relative gas expulsion cut to the likely coal distribution (e.g. Hedemann 1980: mapping for this study; red line): high in dark red and low in yellow. Note the red arrows which represent the possible migration route that would be needed to charge the PeleS-1X and Erik-1X structures. (**c**) Evolution of modelled vitrinite reflectance at various time steps illustrating the main gas maturity phase occurring between 235 and 150 Ma with overmature coals developing after 150 Ma. The area likely to have preserved coals is south of the red line (brighter colours); the area not likely to have preserved coals is north of the red line (washed-out colours).

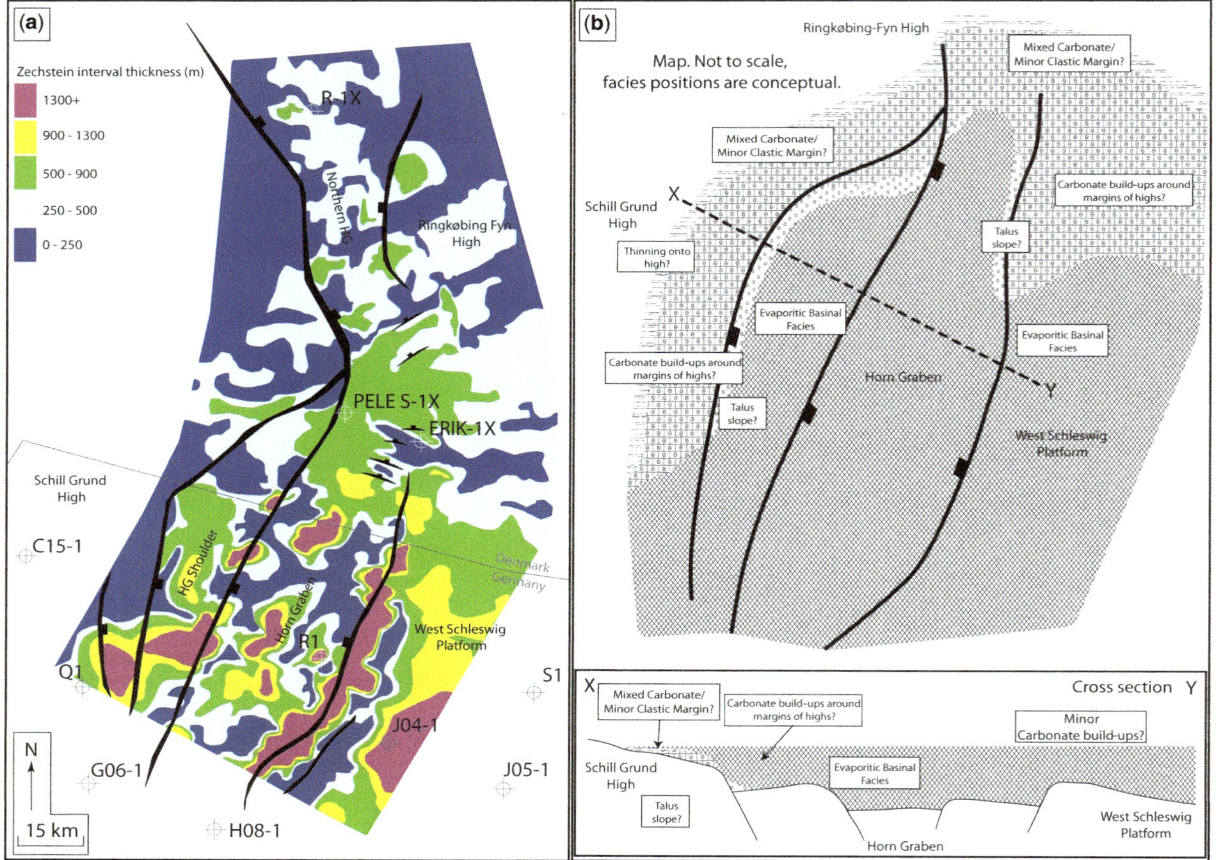

Fig. 11. Zechstein interval migration windows. (**a**) Variability in Zechstein interval thickness across the study area from over 1300 m to less than 250 m. This is partly driven by halokinesis (especially the very thick pink sections) but is originally also a function of depositional facies. (**b**) Conceptual facies map and cross-section of the Zechstein interval (after Vejbaek 1990; Geluk 2005; Peryt *et al.* 2010) illustrating the transition from predominantly evaporitic basinal facies to slope and platform/mixed clastic margin. Not to scale and facies positions are entirely conceptual.

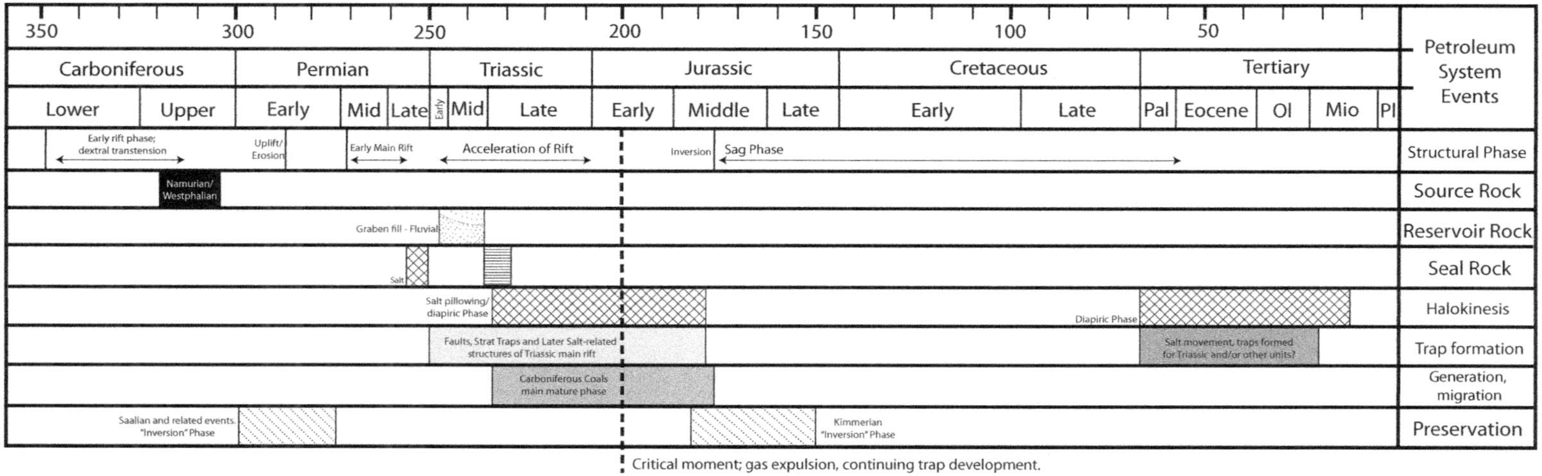

Fig. 12. Petroleum systems event chart illustrating the main phases of structural development, deposition of source/reservoir/seal rocks, and intervals of halokinesis and gas generation. The dashed line shows a possible critical moment for the Horn Graben Buntsandstein play where gas expulsion occurs whilst salt-related traps are still forming and moving. It is suggested that this makes these traps liable to severe underfill or breaching, even if they were once charged.

Danish border is considered to be a realistic scenario based on observations of seismic character (Fig. 8). This reverts to an earlier model of Hedemann (1980) and challenges Kombrink *et al.* (2010). In a similar manner, the Danish well R-1X is unlikely to have had access to any gas charge unless a closer source rock to the north can be proven. Wells on the surrounding highs see various issues of source rock presence, migration routing and structural closure validity. Beha *et al.* (2008) suggested that migration from the Carboniferous to the overburden could be impeded by thick Zechstein halites with localized sweet spots occurring where the salt may be thinner. It is demonstrated (Fig. 11) that the Zechstein varies hugely in thickness (from 100 to 1500 m over 1–2 km), with large areas of very thin section due to pillowing, diapirism and facies changes. The positions of the Pele-S1X and Erik-1X wells mark a change from basinal evaporities to platformal carbonates (Vejbaek 1990), and, therefore, it seems unlikely that the Zechstein could effectively impede vertical migration.

It is noted that the dry holes Pele-S1X (rollover with various reactivated faults) and Erik-1X (rollover/turtle back) both targeted halokinesis-related structures. It is noted that these structures would have been adjusting (due to salt pillowing and diapirism) during peak expulsion and migration times. This temporal overlap is considered risky, and inherently underfilled or breached salt-related structures would be expected, especially if these are reactivated in the Tertiary or salt movements produced faulting and/or fracturing through sealing units.

In summary, the two key valid play tests, wells Pele-S1X and Erik-1X, are likely to be dry due to the charge timing in relation to the structuration timing, with further uncertainty about vertical and horizontal migration from a Carboniferous source rock that is probably not developed in the Danish sector.

Remaining uncertainties and implications for hydrocarbon exploration

In this study, it has been demonstrated that the existing dry holes within the Horn Graben do not preclude a working hydrocarbon system and further exploration potential. Seen in these terms, this Mesozoic rift system would appear to be an exciting frontier, especially in the German licence blocks where no well penetrations test the Buntsandstein play in an area associated with considerable structuration and peak gas expulsion. However, the authors recognize the inherent uncertainty involved in basin modelling and play analysis, especially in an underexplored area. With no pre-Triassic well penetrations there are significant uncertainties, particularly in terms of depth conversion, crustal thickness, heat flow, source rock facies (including preservation) and the amount of Permian/Jurassic erosion. Further work on the model sensitivities would be useful: for example, varying two or more linked parameters (e.g. peak expulsion may be pushed later in a shallow depth-conversion model plus low heat-flow scenario). Migration mechanisms and distances remain a major uncertainty. On a basinal scale, it is often assumed that gas can migrate long distances: for example, from the centre of the Southern Permian Basin to the Groningen structure (e.g. Lokhorst 1997). Here, the Horn Graben was considered as a closed system, with relatively short migration distances favouring major faults the most likely scenario. Any follow-up basin modelling may wish to consider these migration routes and distances in more detail. A review of likely pressure regimes within the Triassic would also help to understand the role of overpressures and trap integrity. For example, in the Dutch Central Graben, various structures are considered dry due to seal breach linked to critical fracture propagation (e.g. Bartholomew *et al.* 1993; Gaarenstroom *et al.* 1993; Peeters *et al.*, this volume, in press). These failures are developed by differential overpressure cells caused by a combination of enclosing salt walls, undercompacted Triassic shales and rapid Tertiary burial. The authors consider this to be a further critical risk to prospectivity in the Horn Graben. However, a lack of detailed pressure data hampered further investigation.

Conclusions

The Triassic Buntsandstein gas play of the Horn Graben Mesozoic rift system was considered in terms of dry hole information and 3D basin modelling. It was noted that:

- Reservoir and seal elements, with some uncertainties, appear robust.
- Traps are present and mainly salt movement related, likely forming and subsequently tilting from the Middle Triassic to the Middle Jurassic.
- Source rock (Carboniferous gas-prone coals) development and subsequent preservation is likely, but uncertain and unconstrained, within the German part of the graben. Although not demonstrated by well penetrations, the geological history and seismic geometries of the Danish area suggest that no Carboniferous source rocks are preserved.
- 3D basin modelling reveals that charge timing is not optimal for the Buntsandstein salt-related traps in that peak expulsion occurs whilst traps are still forming and tilting. In these structures, breaching and underfill is considered likely. The generation of nitrogen-rich gas from overmature coals is possible.

- Four wells have been drilled within the Horn Graben, all dry holes. The Danish well R-1X is likely to be too far from the Carboniferous source rocks to see gas charge. The German Noordsee R1 well does not reach the Buntsandstein interval. The Danish wells Pele-S1X and Erik-1X are likely to be dry for a combination of reasons, including structures that were still forming during gas expulsion/migration and uncertain access to charge from the SW.

Undertaking a dry hole analysis allows the remaining exploration potential to be defined. It is noted that this graben is highly underexplored and the German area, where gas charge is most likely to occur, is largely untested. The Buntsandstein play, due to gas-charge timing, is tricky but has remaining potential. Other plays, both deeper and shallower, are yet to be studied in detail. The authors welcome further insights with regard to model inputs and implications. It is hoped that this paper will regenerate interest in this underexplored section of the prolific North Sea Basin.

The authors would like to thank Maersk Oil for releasing the seismic data. Shell Exploration and Production Deutschland are thanked for allowing this work to be published. We would like to thank Marco Wolf, Fabian Jähne-Klingberg and colleagues at the BGR for their input, ideas and general discussions; we recognize that there is a large amount of ongoing work at the BGR which is eagerly awaited. Marloes Kortekaas, Yuriy Maystrenko, Kees van Ojik and Frank Strozyk all provided excellent feedback during the review process that greatly improved the manuscript. Special thanks go to Andreas Beha for various discussions and feedback.

References

Abramovitz, T. & Thybo, H. 1999. Pre-Zechstein structures around the MONA LISA deep seismic lines in the southern Horn Graben area. *Bulletin of the Geological Society of Denmark*, **45**, 99–116.

Agemar, T., Schellschmidt, R. & Schulz, R. 2011. *Subsurface Temperature Distribution of Germany*, Leibniz Institute for Applied Geophysics, Hannover Germany, https://www.geotis.de/homepage/sitecontent/info/publication_data/local/local_data/Poster_Agemar_FB2011.pdf

Arfai, J., Jähne, F., Lutz, R., Franke, D., Gaedicke, C. & Kley, J. 2014. Late Palaeozoic to Early Cenozoic geological evolution of the northwestern German North Sea (Entenschnabel): new results and insights. *Geologie en Mijnbouw*, **93**, 147–174.

Bachmann, G.H., Geluk, M.C. *et al.* 2010. Triassic. *In*: Doornenbal, J.C. & Stevenson, A.G. (eds) *Petroleum Geological Atlas of the Southern Permian Basin Area*. European Association of Geoscientists and Engineers (EAGE), Houten, The Netherlands, 149–173.

Baldschuhn, R., Binot, F. *et al.* 1996. *Geotektonischer Atlas von NW-Deutschland 1:300 000 [Geotectonic Atlas on NW Germany]*. Bundesanstalt für geowissenschaften und Rohstoffe, Hannover, Germany.

Bartholomew, I., Peters, J. & Powell, C. 1993. Regional structural evolution of the North Sea: oblique slip and the reactivation of basement lineaments. *In*: Parker, J. (ed.) *Petroleum Geology of Northwest Europe: Proceedings of the 4th Conference*. Geological Society, London, 1109–1122, https://doi.org/10.1144/004 1109

Beha, A., Thomsen, R.O. & Littke, R. 2008. Thermal history, hydrocarbon generation and migration in the Horn Graben in the Danish North Sea: a 2D basin modelling study. *International Journal of Earth Sciences*, **97**, 1087–1100.

Berthelsen, A. 1998. The Tornquist Zone northwest of the Carpathians: an intraplate pseudosuture. *Geologiska Foreningens i Stockholm Forhandlingar*, **120**, 223–230.

Best, G., Kockel, F. & Schöneich, H. 1983. Geological history of the Southern Horn Graben. *Geologie en Mijnbouw*, **62**, 25–33.

Bonté, D., van Wees, J.D. & Verweij, J.M. 2012. Subsurface temperature of the onshore Netherlands: new temperature dataset and modelling. *Geologie en Mijnbouw*, **91**, 491–515.

Brzozowska, J., Eriksen, S., Holm, L. & Olsen, S. 2003. Exploration history. *In*: Evans, D., Graham, C., Armour, A. & Bathurst, P. (eds) *Millenium Atlas: Petroleum Geology of the Central and Northern North Sea*. Geological Society, London, 331–343.

Burley, S.D. & Scotchman, I.C. 1999. Basin modelling applications in reducing risk and maximizing reserves. Introduction and review. *In*: Fleet, A.J. & Boldy, S.A.R. (eds) *Petroleum Geology of Northwest Europe: Proceedings of the 5th Conference*. Geological Society, London, 1309–1311, https://doi.org/10.1144/0051309

Butler, M. & Pullan, C.P. 1990. Tertiary structures and hydrocarbon entrapment in the Weald basin of southern England. *In*: Hardman, R.F.P. & Brooks, J. (eds) *Tectonic Events Responsible for Britain's Oil and Gas Reserves*. Geological Society, London, Special Publications, **55**, 371–391, https://doi.org/10.1144/GSL.SP.1990.055.01.19

Clausen, O.R. & Korstgård, J.A. 1994. Displacement geometries along graben bounding faults in the Horn Graben, Offshore Denmark. *First Break*, **12**, 305–315.

Clausen, O.R. & Korstgård, J.A. 1996. Planar detaching faults in the southern Horn Graben, Danish North Sea. *Marine and Petroleum Geology*, **13**, 537–548.

Clausen, O.R. & Pedersen, P.K. 1999. Late Triassic structural evolution of the southern margin of the Ringkøbing-Fyn High, Denmark. *Marine and Petroleum Geology*, **16**, 653–665.

Crepieux, N., Sacleux, M. & Mathis, B. 1998. Influence of pressure on the petroleum system. Example from Triassic in the Netherlands Central Graben. *Bulletin Elf Exploration*, **22**, 123–131.

Di Primio, R., Cramer, B. *et al.* 2008. Petroleum systems. *In*: Littke, R., Bayer, U. *et al.* (eds) *Dynamics of Complex Intracontinental Basins: The Central European Basin System*. Springer, Berlin, 411–432.

Doornenbal, J.C. & Stevenson, A.G. 2010. *Petroleum Geological Atlas of the Southern Permian Basin Area*. European Association of Geoscientists and Engineers (EAGE), Houten, The Netherlands.

Espinosa, G.A. 2016. *Devonian reservoir characterisation on the Schill Grund High and the margins of the Horn Graben*. Masters thesis, University of Manchester, Manchester, UK.

Feist-Burkhardt, S., Götz, A.E. *et al.* 2008. Triassic. *In*: McCann, T. (ed.) *The Geology of Central Europe. Volume 2: Mesozoic and Cenozoic*. Geological Society, London, 749–822.

Gaarenstroom, L., Tromp, R., de Jong, M. & Brandenburg, A. 1993. Overpressures in the Central North Sea: implications for trap integrity and drilling safety. *In*: Parker, J. (ed.) *Petroleum Geology of Northwest Europe: Proceedings of the 4th Conference*. Geological Society, London, 1305–1313, https://doi.org/10.1144/0041305

Gast, R., Dusar, M. *et al.* 2010. Rotliegend. *In*: Doornenbal, J.C. & Stevenson, A.G. (eds) *Petroleum Geological Atlas of the Southern Permian Basin Area*. European Association of Geoscientists and Engineers (EAGE), Houten, The Netherlands, 101–121.

Gautier, D.L. 2003. *Carboniferous–Rotliegend Total Petroleum System Description and Assessment Results Summary*. United States Geological Survey Bulletin, **2211**.

Geluk, M.C. 2005. *Stratigraphy and tectonics of Permo-Triassic basins in the Netherlands and surrounding areas*. PhD thesis, Utrecht University, Utrecht, The Netherlands.

Geluk, M.C. & Röhling, H.G. 1997. High-resolution sequence stratigraphy of the Lower Triassic 'Buntsandstein' in the Netherlands and Northwestern Germany. *Geologie en Mijnbouw*, **76**, 227–246.

Gerling, P., Idiz, E., Everlien, G. & Sohns, E. 1997. New aspects on the origin of nitrogen in natural gas in Northern Germany. *Geologisches Jahrbuch*, **103**, 65–84.

Giles, M.R., Indrelid, S.L. *et al.* 1999. Charge and overpressure modelling in the North Sea: multi-dimensional modelling and uncertainty analysis. *In*: Fleet, A.J. & Boldy, S.A.R. (eds) *Petroleum Geology of Northwest Europe: Proceedings of the 5th Conference*. Geological Society, London, 1313–1324, https://doi.org/10.1144/0051313

Goldsmith, P.J., Rich, B. & Standring, J. 1995. Triassic correlation and stratigraphy in the South Central Graben, UK North Sea. *In*: Boldy, S.A. (ed.) *Permian and Triassic Rifting in Northwest Europe*. Geological Society, London, Special Publications, **91**, 123–143, https://doi.org/10.1144/GSL.SP.1995.091.01.07

GPDN 2016. *Geopotenzial Deutsche Nordsee, Tiefenkarten des Perm und Prä-Perm [Geopotential of the German North Sea, deep maps of the Permian and Pre-Permian]*. Geopotenzial Deutsche Nordsee (GPDN), http://www.gpdn.de

Grad, M., Tiira, T. & ESC Working Group 2009. The Moho depth map of the European Plate. *Geophysical Journal International*, **176**, 279–292.

Gregersen, S. & Voss, P. 2002. Summary of project TOR: delineation of a stepwise, sharp, deep lithosphere transition across Germany–Denmark–Sweden. *Tectonophysics*, **360**, 61–73.

Guterch, A., Wybraniec, S. *et al.* 2010. Crustal structure and structural framework. *In*: Doornenbal, J.C. & Stevenson, A.G. (eds) *Petroleum Geological Atlas of the Southern Permian Basin Area*. European Association of Geoscientists and Engineers (EAGE), Houten, The Netherlands, 11–23.

Harding, R. & Huuse, M. 2015. Salt on the move: multi stage evolution of salt diapirs in the Netherlands North Sea. *Marine and Petroleum Geology*, **61**, 39–55.

Hedemann, H.A. 1980. Die Bedeutung des Oberkarbons fur due Kohlenwasserstoffvorkommen im Nordseebecken [The role of the Upper Carboniferous for future hydrocarbons in the North Sea basin]. *Erdol und Kohle*, **33**, 255–266.

Heim, S., Lutz, R. *et al.* 2013. Geological evolution of the North Sea: cross-border basin modellign study on the Schill Grund High. *Energy Procedia*, **40**, 222–231.

Heinemann, N., Wilkinson, M. *et al.* 2012. CO_2 storage in the offshore UK Bunter Sandstone Formation. *International Journal of Greenhouse Gas Control*, **6**, 210–219.

Huuse, M. 2002. Late Cenozoic paleogeography of the eastern North Sea Basin: climatic vs tectonic forcing of basin margin uplift and deltaic progradation. *Bulletin of the Geological Society of Denmark*, **49**, 145–170.

Jähne-Klingberg, F. 2012. Deformation of Rotliegend salt formations and their influence on the structural pattern in the German North Sea. *Schriftenreihe der Deutschen Gesellschaft für Geowissenschaften*, **80**, 521.

Knutz, P.C. 2010. Channel structures formed by contour currents and fluid expulsion: significance for Late Neogene development of the central North Sea basin. *In*: Vining, B. & Pickering, K. (eds) *Petroleum Geology: From Mature Basins to New Frontiers – Proceedings of the 7th Petroleum Geology Conference*. Geological Society, London, 289–315, https://doi.org/10.1144/0070077

Kombrink, H. 2008. *The Carboniferous of the Netherlands and surrounding areas; a basin analysis*. PhD thesis, Utrecht University, Utrecht, The Netherlands.

Kombrink, H., Besly, B. *et al.* 2010. Carboniferous. *In*: Doornenbal, J.C. & Stevenson, A.G. (eds) *Petroleum Geological Atlas of the Southern Permian Basin Area*. European Association of Geoscientists and Engineers (EAGE), Houten, The Netherlands, 81–99.

Kortekaas, M., Böker, U., van der Kooij, C. & Jaarsma, B. In press. Lower Triassic reservoir development in the Dutch northern offshore. *In*: Kilhams, B., Kukla, P.A., Mazur, S., McKie, T., Mijnlieff, H.F. & van Ojik, K. (eds) *Mesozoic Resource Potential of the Southern Permian Basin*. Geological Society, London, Special Publications, **469**. https://doi.org/10.1144/SP469.19

Krooss, B.M., Plessen, B. *et al.* 2008. Origin and distribution of non-hydrocarbon gases. *In*: Littke, R., Bayer, U., Gajewski, D. & Nelskamp, S. (eds) *Dynamics of Complex Intracontinental Basins: The Central European Basin System*. Springer, Berlin, 433–458.

Kukla, P.A., Urai, J.L. & Mohr, M. 2008. Dynamics of salt structures. *In*: Littke, R., Bayer, U., Gajewski, D. & Nelskamp, S. (eds) *Dynamics of Complex Intracontinental Basins: The Central European Basin System*. Springer, Berlin, 291–306.

Kyrkjebø, R., Gabrielsen, R.H. & Faleide, J.I. 2004. Unconformities related to the Jurassic-Cretaceous synrift – post-rift transition of the northern North Sea. *Journal of the Geological Society, London*, **161**, 1–17, https://doi.org/10.1144/0016-764903-051

Lassen, A. & Thybo, H. 2012. Neoproterozoic and Palaeozoic evolution of SW Scandinavia based on integrated seismic interpretation. *Precambrian Research*, **204**, 75–104.

LEISCHNER, K.R., CHUPAROVA, E. *ET AL*. 2011. Petroleum systems analysis in the Gulf of Mexico – from regional framework to predicting fluid and rock properties. *AAPG Search and Discovery Article #90124, AAPG Annual Convention and Exhibition*, April 10–13, 2011, Houston, Texas, USA.

LITTKE, R., SCHECK-WENDEROTH, M. *ET AL*. 2008. Subsidence, inversion and evolution of the thermal field. *In*: LITTKE, R., BAYER, U., GAJEWSKI, D. & NELSKAMP, S. (eds) *Dynamics of Complex Intracontinental Basins: The Central European Basin System*. Springer, Berlin, 125–154.

LOKHORST, A. (ed.). 1997. *NW European Gas Atlas*. TNO, Haarlem, The Netherlands.

MAGRI, F., LITTKE, R. *ET AL*. 2008. Temperature fields, petroleum maturation and fluid flow in the vicinity of salt domes. *In*: LITTKE, R., BAYER, U., GAJEWSKI, D. & NELSKAMP, S. (eds) *Dynamics of Complex Intracontinental Basins: The Central European Basin System*. Springer, Berlin, 323–344.

MAYSTRENKO, Y. & SCHECK-WENDEROTH, M. 2013. 3D lithosphere-scale density model of the Central European Basin System and adjacent areas. *Tectonophysics*, **601**, 53–77.

MAYSTRENKO, Y., BAYER, U. & SCHECK-WENDEROTH, M. 2005*a*. The Glueckstadt Graben, a sedimentary record between the North and Baltic Sea in north Central Europe. *Tectonophysics*, **397**, 113–126.

MAYSTRENKO, Y., BAYER, U. & SCHECK-WENDEROTH, M. 2005*b*. Structure and evolution of the Glueckstadt Graben due to salt movements. *International Journal of Earth Sciences*, **94**, 799–814.

MAYSTRENKO, Y., BAYER, U. & SCHECK-WENDEROTH, M. 2006. 3D reconstruction of salt movements within the deepest post-Permian structure of the Central European Basin System – the Glueckstadt Graben. *Geologie en Mijnbouw*, **85**, 181.

MAYSTRENKO, Y., BAYER, U. *ET AL*. 2008. The Central European Basin System – an overview. *In*: LITTKE, R., BAYER, U., GAJEWSKI, D. & NELSKAMP, S. (eds) *Dynamics of Complex Intracontinental Basins: The Central European Basin System*. Springer, Berlin, 15–34.

MCCANN, T. 1999. The tectonosedimentary evolution of the northern margin of the Carboniferous foreland basin of NE Germany. *Tectonophysics*, **313**, 119–144.

MCCANN, T., KIERNOWSKI, H. *ET AL*. 2008. Permian. *In*: MCCANN, T. (ed.) *The Geology of Central Europe. Volume 2: Mesozoic and Cenozoic*. Geological Society, London, 531–598.

MCKIE, T. & WILLIAMS, B. 2009. Triassic palaeogeography and fluvial dispersal across the northwest European Basins. *Geological Journal*, **44**, 711–741.

MOHR, M., KUKLA, P.A. *ET AL*. 2005. Multiphase salt tectonic evolution in NW Germany: seismic interpretation and retro-deformation. *International Journal of Earth Sciences*, **94**, 917–940.

MOGENSEN, T.E. 1995. Triassic and Jurassic structural development along the Tornquist Zone, Denmark. *Tectonophysics*, **252**, 197–220.

OLIVARIUS, M., WEIBEL, R. *ET AL*. 2015. Provenance of the Lower Triassic Bunter Sandstone Formation: implications for distribution and architecture of aeolian vs. fluvial reservoirs in the North German Basin. *Basin Research*, **29**, 113–130.

PEETERS, S., ASSCHERT, A. & VERWEIJ, H. In press. Towards a better understanding of the highly overpressured Lower Triassic Bunter reservoir rocks in the Terschelling Basin. *In*: KILHAMS, B., KUKLA, P.A., MAZUR, S., MCKIE, T., MIJNLIEFF, H.F. & VAN OJIK, K. (eds) *Mesozoic Resource Potential of the Southern Permian Basin*. Geological Society, London, Special Publications, **469**, https://doi.org/10.1144/SP469.13

PERYT, T.D., GELUK, M.C. *ET AL*. 2010. Zechstein. *In*: DOORNENBAL, J.C. & STEVENSON, A.G. (eds) *Petroleum Geological Atlas of the Southern Permian Basin Area*. European Association of Geoscientists and Engineers (EAGE), Houten, The Netherlands, 123–147.

PHARAOH, T.C., DUSAR, M. *ET AL*. 2010. Tectonic evolution. *In*: DOORNENBAL, J.C. & STEVENSON, A.G. (eds) *Petroleum Geological Atlas of the Southern Permian Basin Area*. European Association of Geoscientists and Engineers (EAGE), Houten, The Netherlands, 25–57.

PIENKOWSKI, G., SCHUDACK, M.E. *ET AL*. 2008. Jurassic. *In*: MCCANN, T. (ed.) *The Geology of Central Europe. Volume 2: Mesozoic and Cenozoic*. Geological Society, London, 823–922.

PLETSCH, T., APPEL, J. *ET AL*. 2010. Petroleum generation and migration. *In*: DOORNENBAL, J.C. & STEVENSON, A.G. (eds) *Petroleum Geological Atlas of the Southern Permian Basin Area*. European Association of Geoscientists and Engineers (EAGE), Houten, The Netherlands, 225–253.

RASMUSSEN, E.S. 2009. Neogene inversion of the Central Graben and Rynkobing-Fyn High, Denmark. *Tectonophysics*, **465**, 84–97.

RASMUSSEN, E.S. 2013. Cenozoic structures in the eastern North Sea Basin – a case for salt tectonics: discussion. *Tectonophysics*, **601**, 226–233.

SACHSE, V.F. & LITTKE, R. 2016. Burial, temperature and maturation history of the Ringkøbing-Fyn High, Denmark. *Schriftenreihe der Deutschen Gesellschaft für Geowissenschaften*, **167**, 249–268.

SACHSE, V.F. & LITTKE, R. In press. The impact of Quaternary glaciation on temperature and pore pressure in Jurassic troughs in the Southern Permian Basin, northern Germany. *In*: KILHAMS, B., KUKLA, P.A., MAZUR, S., MCKIE, T., MIJNLIEFF, H.F. & VAN OJIK, K. (eds) *Mesozoic Resource Potential of the Southern Permian Basin*. Geological Society, London, Special Publications, **469**, https://doi.org/10.1144/SP469.7

SAMUELSSON, J., VECOLI, M., BEDNARCZYK, W.S. & VERNIERS, J. 2002. Timing of the Avalonia–Baltica plate convergence as inferred from palaeogeographic and stratigraphic data of chitinozoan assemblages in west Pomerania, northern Poland. *In*: WINCHESTER, J.A., PHARAOH, T.C. & VERNIERS, J. (eds) *Palaeozoic Amalgamation of Central Europe*. Geological Society, London, Special Publications, **201**, 95–113, https://doi.org/10.1144/GSL.SP.2002.201.01.05

SCHECK-WENDEROTH, M., MAYSTRENKO, Y. *ET AL*. 2008. Dynamics of salt basins. *In*: LITTKE, R., BAYER, U., GAJEWSKI, D. & NELSKAMP, S. (eds) *Dynamics of Complex Intracontinental Basins: The Central European Basin System*. Springer, Berlin, 307–322.

STEMMERIK, L., INESON, J.R. & MITCHELL, J.G. 2000. Stratigraphy of the Rotliegend Group in the Danish part of the

Northern Permian Basin, North Sea. *Journal of the Geological Society, London*, **157**, 1127–1136, https://doi.org/10.1144/jgs.157.6.1127

Stollhofen, H., Bachmann, G.H. *et al.* 2008. Upper Rotliegend to Early Cretaceous basin development. *In*: Littke, R., Bayer, U., Gajewski, D. & Nelskamp, S. (eds) *Dynamics of Complex Intracontinental Basins: The Central European Basin System*. Springer, Berlin, 181–210.

Strozyk, F., Reuning, L., Scheck-Wenderoth, M. & Tanner, D.C. 2017. The tectonic history of the Zechstein Basin in the Netherlands and Germany. *In*: Soto, J., Flinch, J. & Tari, G. (eds) *Permian–Triassic Salt Provinces in Europe, North Africa and the Atlantic Margins: Tectonics and Hydrocarbon Potential*. Elsevier, Amsterdam, 221–241.

Szurlies, M. 2004. Magnetostratigraphy: the key to a global correlation of the classic Germanic Trias-case study Volpriehausen Formation (Middle Buntsandstein), Central Germany. *Earth and Planetary Science Letters*, **227**, 395–410.

Tesauro, M., Kaban, M.K. *et al.* 2007. 3D strength and gravity anomalies of the European lithosphere. *Earth and Planetary Science Letters*, **263**, 56–73.

Thybo, H. 1997. Geophysical characteristics of the Tornquist Fan area, northwest Trans-European Suture Zone: indication of late Carboniferous to early Permian dextral transtension. *Geological Magazine*, **134**, 597–606.

Thybo, H. 2001. Crustal structure along the EGT profile across the Tornquist Fan interpreted from seismic, gravity and magnetic data. *Tectonophysics*, **334**, 155–190.

Uffmann, A.K. & Littke, R. 2011. 3D petroleum systems modelling of the North German Basin. *First Break*, **29**, 49–63.

Van Adrichem Boogaert, H.A. & Kouwe, W.F.P. (compilers). 1994. *Stratigraphic Nomenclature of the Netherlands; Revision and Update by RGD and NOGEPA, Section E Triassic*. Mededelingen Rijks Geologische Dienst, **50**.

Van Dalfsen, W., van Gessel, S.F. & Doornenbal, J.C. 2007. *Velmod-2 Joint Industry Project*. TNO, Utrecht, The Netherlands.

Vejbaek, O.V. 1990. The Horn Graben and its relationship to the Oslo Graben and the Danish Basin. *Tectonophysics*, **178**, 29–49.

Warren, J.K. 2008. Salt as sediment in the Central European Basin System as seen from a deep time perspective. *In*: Littke, R., Bayer, U., Gajewski, D. & Nelskamp, S. (eds) *Dynamics of Complex Intracontinental Basins: The Central European Basin System*. Springer, Berlin, 247–276.

Warsitzka, M., Kley, J., Jähne-Klingberg, F. & Kukowski, N. 2017. Dynamics of prolonged salt movement in the Glückstadt Graben (NW Germany) driven by tectonic and sedimentary processes. *International Journal of Earth Sciences*, **106**, 131–155, https://doi.org/10.1007/s00531-016-1306-3

Wolf, M., Steuer, S. *et al.* 2014. *3D-Lithofaziesmodell des Buntsandstein in der zentralen deutschen Nordsee [3D lithofacies model for the Buntsandstein in the central german North Sea]*. Geopotenzial Deutsche Nordsee Technical Report Geopotenzial Deutsche Nordsee (GPDN), http://www.gpdn.de/?pgId=323&WilmaLogonActionBehavior= Default

Wolf, M., Steuer, S. *et al.* 2015. Lithofacies distribution in the Central European Basin: a 3D model of the Buntsandstein facies in the central German North Sea. *Zeitschrift der Deutschen Gesellschaft für Geowissenschaften*, **166**, 341–359.

Yu, Z., Lerche, I. & Lowrie, A. 1992. Thermal impact of salt: simulation of thermal anomalies in the Gulf of Mexico. *Pure and Applied Geophysics*, **138**, 181–192.

Ziegler, P.A. 1990. *Geological Atlas of Western and Central Europe*. Shell Internationale Petroleum Maatschappij, The Hague. Geological Society, London.

Geothermal resources of the North German Basin: exploration strategy, development examples and remaining opportunities in Mesozoic hydrothermal reservoirs

MATTHIAS FRANZ[1]*, GREGOR BARTH[2], JENS ZIMMERMANN[3], INGMAR BUDACH[3], KERSTIN NOWAK[3] & MARKUS WOLFGRAMM[3]

[1]*Geowissenschaftliches Zentrum der Universität Göttingen, Abteilung Angewandte Geologie, Goldschmidtstraße 3, 37077 Göttingen, Germany*

[2]*Landesamt für Umwelt, Naturschutz und Geologie (LUNG), Goldberger Straße 12, 18273 Güstrow, Germany*

[3]*Geothermie Neubrandenburg GmbH, Seestraße 7a, 17033 Neubrandenburg, Germany*

**Correspondence: mfranz1@gwdg.de*

Abstract: The North German Basin yields enormous geothermal resources of more than 13 000 EJ (exajoule: 1 EJ = 1 × 10^{18} J) heat in place bound to Paleozoic petrothermal and Mesozoic hydrothermal reservoirs. So far, these resources are only exploited at a few localities. Thus, geothermal energy is considered an underutilized energy resource. Despite long-term experience in exploiting Rhaetian hydrothermal reservoirs, the exploration risk remains high, which is mainly related to high expectations on reservoir thickness and quality. Previous exploration campaigns have identified potential hydrothermal reservoirs in six Mesozoic reservoir complexes. But, as high-resolution subsurface maps are not available, the reliable prediction and targeting of reservoirs remains an unsolved problem.

As such, an exploration strategy integrating methods of sedimentology, palaeontology, petrography and reservoir characterization was applied to a large database of cores and wireline logs. This contribution details the key results of the exploration of Upper Keuper and Middle Jurassic reservoir complexes, including high-resolution subsurface facies, sandstone thickness and reservoir quality maps. Sets of these maps enable the reliable prediction and targeting of individual hydrothermal reservoirs, and, thus, make a significant contribution to a lowered exploration risk.

Considering its thick basin fill and geothermal gradients of 32–90 K per km of depth, the North German Basin yields enormous geothermal resources (Fig. 1) (Wolfgramm *et al.* 2014). Following an intense phase of hydrocarbon exploration, the exploration of geothermal resources started in the early 1970s. Based on these campaigns, significant resources were identified in (1) Paleozoic sedimentary and volcanic rocks, and (2) Mesozoic sedimentary rocks but, mainly for economic reasons, the exploration was focused on Mesozoic hydrothermal reservoirs (see Feldrappe *et al.* 2008). The exploitation of geothermal energy started with operating Rhaetian reservoirs at the sites of Waren (in 1986) and Neustadt-Glewe (in 1994). Based on this experience, an economically successful operation of hydrothermal reservoirs requires the following key parameters: >65°C reservoir temperature, >20 m reservoir thickness and >500 mD permeability (Rockel *et al.* 1997).

Soon after, the development of hydrothermal reservoirs at other localities in North Germany failed. For example, at the planned site of Belzig, the thickness of an identified Rhaetian reservoir was too limited; while at the planned site of Hamburg-Allermöhe, the limited quality (permeability) inhibited the development of another Rhaetian reservoir. These, and other failures, demonstrate the high exploration risk of Mesozoic hydrothermal reservoirs, which is mainly related to the lateral variability of reservoir thickness and quality.

Jung *et al.* (2002) estimated the total geothermal resources of Germany for electricity production to be about 2100 EJ (exajoule: 1 EJ = 1 × 10^{18} J). Considering the total geothermal energy production (heat, electricity) of 8962 TJ (terajoule: 1 TJ = 1 × 10^{12} J) in 2015, geothermal energy is an underutilized energy resource of Germany (AGEB 2015). About 96% of the total geothermal resources are bound to so-called petrothermal reservoirs (2016 EJ) which require stimulation techniques to be developed (Enhanced or Engineered Geothermal Systems: EGS), 4% are bound to faults (84 EJ) and only 1%

From: Kilhams, B., Kukla, P. A., Mazur, S., McKie, T., Mijnlieff, H. F. & van Ojik, K. (eds) 2018. *Mesozoic Resource Potential in the Southern Permian Basin*. Geological Society, London, Special Publications, **469**, 193–222.
First published online April 13, 2018, https://doi.org/10.1144/SP469.11

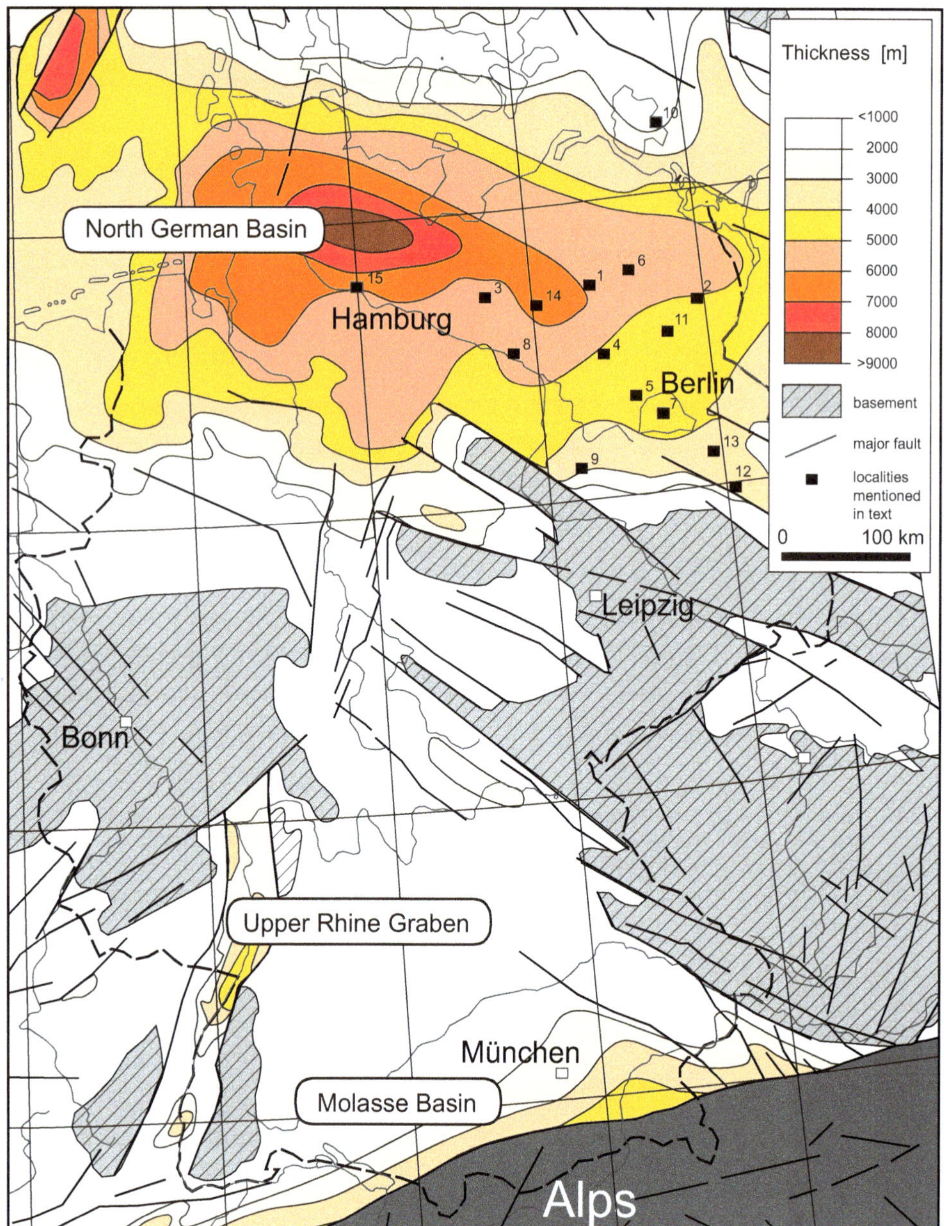

Fig. 1. Overview map of the main German basins: the North German Basin, Molasse Basin and the Upper Rhine Graben. These basins may yield considerable geothermal resources. The isopachs refer to the total sedimentary thickness of the Upper Rotliegend–base Quaternary succession; compiled after Ziegler (1982, 1990). The black squares are geothermal facilities or localities referred to as in the text: 1, Waren; 2, Prenzlau; 3, Neustadt-Glewe; 4, Neuruppin; 5, Velten; 6, Neubrandenburg; 7, Berlin; 8, Bad Wilsnack; 9, Belzig; 10, Binz; 11, Templin; 12, Burg; 13, Bad Saarow (for details see Table 2); 14, Parchim 1; 15, Hamburg-Allermöhe.

(21 EJ) is bound to hydrothermal reservoirs which can be exploited by, for example, conventional Doublet systems (Jung *et al.* 2002). But it is important to note that hydrothermal operations in North Germany have generally been more successful than petrothermal reservoir projects.

Numerous previous studies have evaluated Mesozoic hydrothermal reservoirs of the North

German Basin (e.g. Katzung 1984; Diener *et al.* 1988–1992; Wormbs *et al.* 1988; Katzung *et al.* 1992; Scholle 1992; Beutler *et al.* 1994; Schulz & Röhling 2000; Kabus *et al.* 2003; Feldrappe *et al.* 2008). These studies employed existing databases of well data in order to evaluate either individual localities or reservoir complexes. However, reliable reservoir prediction is hampered by a lack of high-resolution subsurface facies mapping (see Feldrappe *et al.* 2008, fig. 5c). Following an overview of geothermal resources, including development examples in Mesozoic hydrothermal reservoirs, this contribution details the key results of the exploration of Upper Triassic Rhaetian and Middle Jurassic reservoir complexes. The employed exploration strategy combines classical methods of sedimentology, palaeontology, petrography and reservoir characterization, and was applied to a large database of cored and logged wells. The resulting high-resolution GIS-based subsurface facies, sandstone thickness and reservoir quality maps make significant contributions to improving reservoir prediction, and, thus, lower the exploration risk for Mesozoic hydrothermal reservoirs. The set of maps is currently implemented into the Geothermal Information System (GeotIS: Agemar *et al.* 2014).

Geological evolution

The Central European Basin (CEB) is a large epicontinental sag basin (e.g. Bally & Snelson 1980; Bachmann & Grosse 1989; Ziegler 1990). In terms of structure, the CEB is composed of several subbasins, with the North German Basin as the largest one (Gast *et al.* 1998). Following the Variscan Orogeny, the CEB initiated in the Late Carboniferous and has been active up to recent times. During this long time span, a thick volcano-sedimentary basin fill accumulated, with the greatest thickness in the North German Basin (NGB). For example, the sedimentary succession of the Upper Rotliegend I–Tertiary exceeds a thickness of 8000 m in NW Germany (Bachmann & Grosse 1989).

The complex basin history comprises phases of high subsidence alternating with phases of basin reorganization and tectonic pulses (Nöldeke & Schwab 1977; Schwab 1985; Kley *et al.* 2008; Scheck-Wenderoth *et al.* 2008; Bachmann *et al.* 2010). Accordingly, the Upper Carboniferous–Upper Cretaceous basin fill is herein subdivided into three stages: (1) basin initiation; (2) basin fill; and (3) basin differentiation/inversion (Fig. 2). This subdivision partly resembles the subdivision of Nöldeke & Schwab (1977) and Schwab (1985).

Basin initiation in the Late Carboniferous started with the formation of smaller intra-Variscan basins and a larger foreland basin to the north of the Variscan Mountains. The initiation stage was characterized by strong volcanism resulting in accumulation of a 3000 m thick volcano-sedimentary succession (Lower Rotliegend) between the Altmark area and the island of Rügen (Marx *et al.* 1995; Plein 1995; Benek *et al.* 1996; Breitkreuz & Kennedy 1999; Breitkreuz *et al.* 2008) (Fig. 2). The oldest post-volcanic sediments of the Upper Rotliegend I Müritz Subgroup (Artinskian–Kungurian) are restricted to fault-bounded graben-like structures, where up to 600 m thick successions of conglomerates, fluvial sandstones and playa deposits were accumulated (Plein 1993; Hoffmann *et al.* 1997). In the Middle Permian (Guadalupian), continued crustal stretching and intrusions resulted in thermal updoming and partial erosion (Saalian Unconformity: Stille 1924; Ziegler 1990; Bachmann & Hoffmann 1995) of the stratigraphic succession accumulated during the initiation stage of the basin evolution.

Above this unconformity, the onset of the Upper Rotliegend II (Capitanium–Wuchiapingium) marks the change to the second phase of the basin evolution, the basin fill stage. This approximately 100 myr long stage of basin evolution was characterized by high subsidence initially (265–240 Ma) and lower subsidence rates subsequently (240–140 Ma) (Fig. 2). Basin expansion and reorganization formed the Northern and Southern Permian Basins, separated by the Mid North Sea and Ringköbing-Fyn High (Seemann 1982). Following separate development during the depositional periods of the Rotliegend and Zechstein groups, both basins merged into the Central European Basin in the Early Triassic. The succession accumulated during the basin fill stage spans from the Upper Rotliegend II to the Upper Jurassic, and comprises both continental and marine strata. The deposits of the early part of this stage, including the Upper Rotliegend II (Permian)–Buntsandstein (Germanic Triassic) succession, are considerable thicker compared to the deposits of the later part of this stage, including the Muschelkalk (Germanic Triassic)–Upper Jurassic succession (Fig. 2).

During the third basin evolution stage, the basin differentiation and inversion stage, Atlantic rifting and the collision of the African and European plates triggered the inversion of parts of the basin (e.g. Ziegler 1990; Kley & Voigt 2008). Accordingly, salt tectonics attained a maximum intensity during this stage (Jaritz 1974, 1987). This resulted in repeated uplift and erosion of both, the pre- and synkinematic basin fill. As a consequence, stratigraphic units of the differentiation and inversion stage are not always preserved in the North German Basin. In particular in NE Germany, Upper Jurassic and Lower Cretaceous strata are only present in the form of isolated erosional remnants.

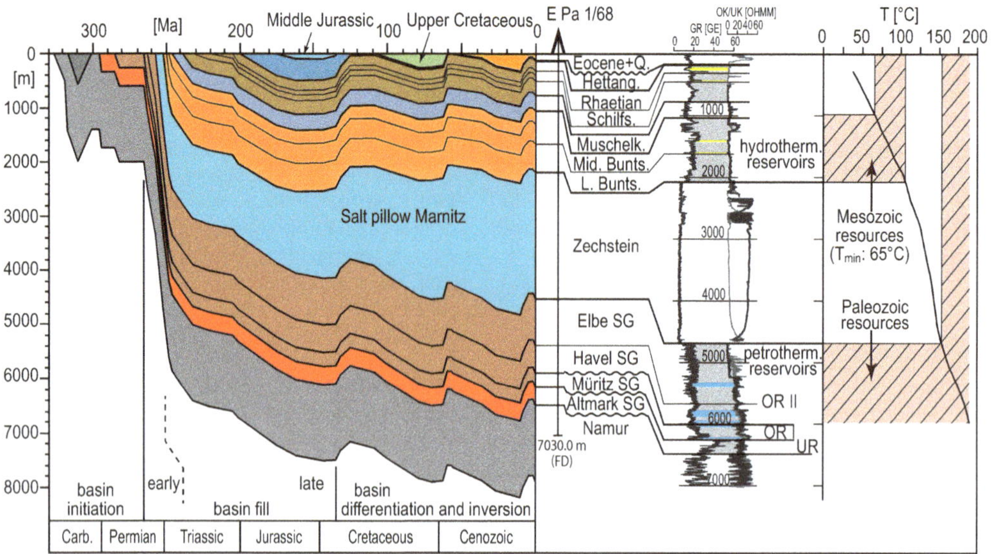

Fig. 2. Basin fill of the NGB. The example of well Parchim1/68 (for the location see Fig. 1) shows Mesozoic and Paleozoic reservoirs separated by the Zechstein. High-permeable Mesozoic reservoirs are operated using conventional hydrothermal methods, whereas low-permeable to tight Paleozoic reservoirs are operated using unconventional petrothermal methods considered under the term Enhanced Geothermal Systems (EGS). SG, Sub-Group; UR, Unterrotliegend; OR, Oberrotliegend; the temperatures are from Hoth (1997) the subsidence curve is modified from Hoth (1997), Friberg (2001) and Wolfgramm (2005).

Geothermal resources of the North German Basin

In the North German Basin, temperatures increase with increasing depth resulting from underlying heat flow values of 50–90 mW m^{-2} (average 70 mW m^{-2}: Lotz 2004; Wolfgramm *et al.* 2014). Due to variations of basin fill thickness and structure, geothermal gradients are subject to regional variations. Accordingly, reported values of average geothermal gradients vary from 33 K per km depth (Katzung 2004) to 45–50 K per km depth (Hoth 1997). Locally, for example, at the top of salt structures, elevated geothermal gradients of up to 90 K per km depth were observed (Ziegenhardt *et al.* 1980; Hoth 1997; Katzung 2004).

The geothermal resources of the North German Basin are subject to several subdivisions. In terms of depth, the geothermal resources are subdivided into shallow resources from 0 to 500 m depth, and deep resources below 500 m depth. The shallow geothermal resources, mainly hydrothermal reservoirs, are increasingly used for the heating/cooling of buildings. The deep geothermal reservoirs may be further subdivided into Mesozoic and Paleozoic geothermal resources, with the Zechstein as a natural barrier in-between (Fig. 2). The Mesozoic geothermal resources are mainly bound to hydrothermal resources, whereas the Paleozoic geothermal reservoirs are mainly bound to petrothermal reservoirs. In terms of temperature, hydrothermal reservoirs may be subdivided into hot (>100°C), warm (60–100°C) and thermal (>20°C) reservoirs, as suggested by Agemar *et al.* (2014).

Paleozoic geothermal resources

Paleozoic geothermal resources are mainly bound to Rotliegend Group sediments and volcanic rocks. Jung *et al.* (2002) estimated a geothermal potential of 13 000 EJ heat in place for Rotliegend-age volcanic rocks, and 340 EJ heat in place for Rotliegend siliciclastic sediments. At least parts of these enormous potentials are available at almost any location in North Germany since the Upper Rotliegend II is regionally distributed. Resulting from a deep burial depth of 2500 m to more than 5500 m (base Zechstein Group) and a complex diagenetic history, potential reservoirs of the basin-initiation stage and early basin fill stage are low permeable to tight, and therefore need EGS techniques to be developed. However, so far, only preliminary experiences have been made with EGS in the North German Basin.

Hydrothermal resources in Mesozoic reservoir complexes

Mesozoic hydrothermal resources are bound to sandstone reservoirs, which are grouped into complexes of the: (1) Middle Buntsandstein; (2) Lower–Middle Keuper; (3) Upper Keuper; (4) Lower Jurassic; (5) Middle Jurassic; and (6) Lower Cretaceous (Fig. 3). With increasing depth, Mesozoic reservoir complexes are characterized by decreasing pore volumes and increasing salinity of the pore fluids. For burial depths up to 1000 m, the decrease in pore volumes is mainly related to increasing sediment compaction. At depths over 1000 m, diagenetic cement precipitation results in further reduction of pore volumes. As an estimate, a burial depth of 1000 m results in the lowering of pore volumes by about 20% (Fig. 4a; Table 1). The salinity of pore fluids is mainly related to: (1) primary fluids salinity (e.g. freshwater or seawater); (2) secondary processes, such as diagenetic mineral alteration; and/or (3) mixing with other saline fluids. On average, the salinity of pore fluids increases by about 10 g l^{-1} per 100 m depth (Fig. 5) (Wolfgramm *et al.* 2014). Within reservoirs that are influenced by highly saline brines related to either *in situ* salt layers (e.g. Lower Triassic Röt Formation) or to Zechstein salt structures, the pore fluid salinity increases by up to 20–25 g l^{-1} per 100 m depth.

Middle Buntsandstein reservoir complex. In the North German Basin, the base Triassic is located at a depth between 1000 and 5500 m (Bachmann *et al.* 2010). In contrast to south and central Germany, where the Buntsandstein is predominantly composed of sandstones, the basinal facies of the North German Basin is mainly composed of shales and siltstones (Röhling 1991, 2013). Only strata from the Middle Buntsandstein Subgroup yield potential sandstone reservoirs, which are located at the southern and northern margins of the North German Basin. Fed

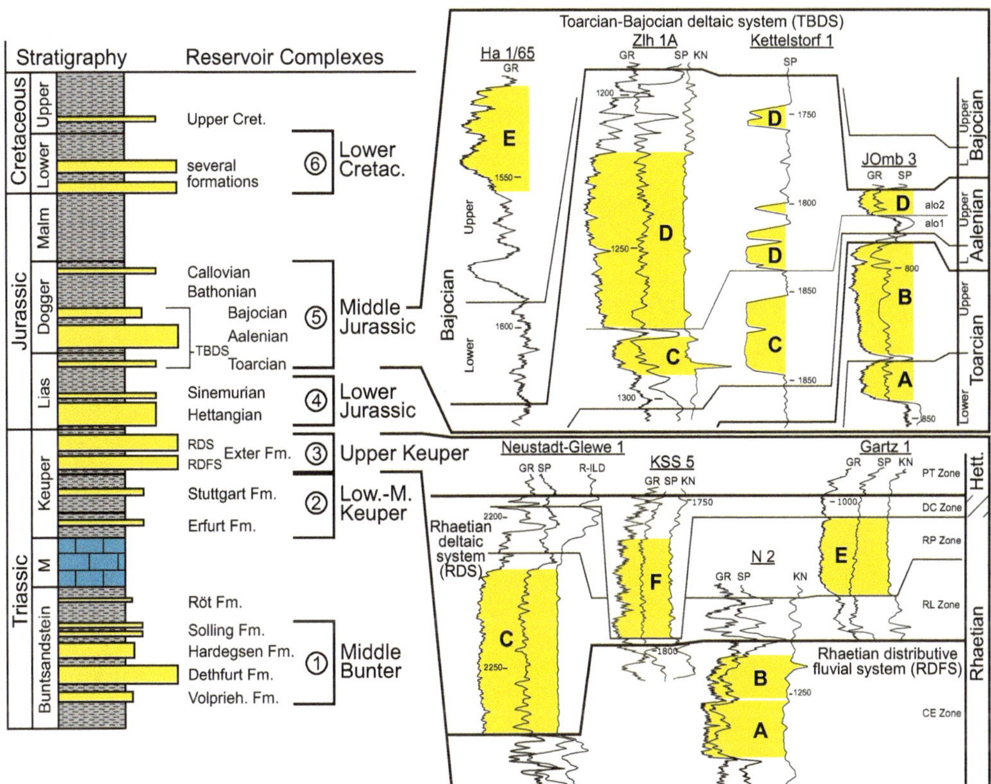

Fig. 3. Mesozoic reservoir complexes with examples of the Rhaetian Deltaic System (RDS) and Toarcian–Bajocian Deltaic System (TBDS). RDS with examples of Rhaetian reservoirs A–F (but D is not shown) and palynomorph zonation: CE Zone, *Cerebropollenites–Enzonalasporites* Zone; RL Zone, *Rhaetipollis–Limbosporites* Zone; RP Zone, *Ricciisporites–Polypodiisporites* Zone; DC Zone, *Deltoidospora–Concavisporites* Zone; PT Zone, *Pinuspollenites–Trachysporites* Zone (Lund 1977; Barth *et al.* 2018*a*) – for the locations see Figure 11. TBDS with examples of reservoirs: A, Lower Toarcian; B, Upper Toarcian; C, Upper Aalenian 1 (alo 1); D, Upper Aalenian 2 (alo 2); E, Upper Bajocian; Ha 1, Hagenow 1; Zlh 1, Zechliner Hütte 1 – for the locations see Figure 14. Hagenow 1 and Kettelsdorf 1 are modified from Brand & Mönnig (2009).

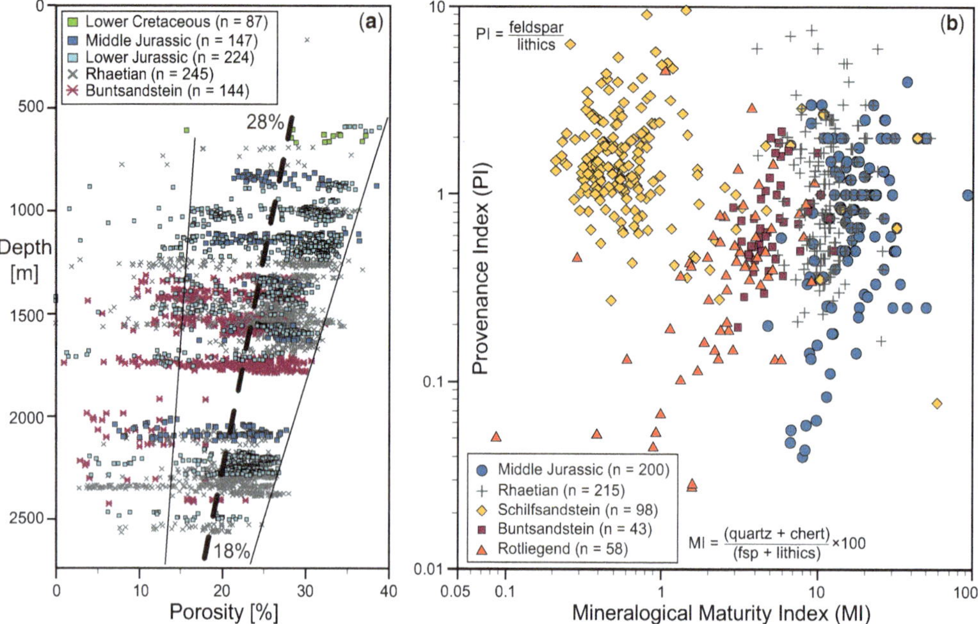

Fig. 4. (**a**) Porosity v. depth and (**b**) mineralogical maturity v. provenance. Increasing depth results in decreasing porosities of Mesozoic reservoirs. Note the average porosities of 28 and 18% at depths of about 600 and 2500 m, respectively. Porosities refer to gas porosities of core samples. Mineralogical maturity and provenance indexes (Pettijohn 1957) of Palaeozoic and Mesozoic reservoirs obtained from point counting of thin sections (modified after Franz *et al.* 2015).

by Variscan sources to the south, sandstone reservoirs of the Volpriehausen to Solling formations form a 50–100 km wide sandy depositional belt stretching from the Weser-Ems area, via Hannover, up to northern Brandenburg (southern Buntsandstein). A corresponding sandy depositional belt stretches from the Darß via Stralsund up to the island of Usedom, which was fed by Scandinavian sources from the north (northern Buntsandstein). Within these sandy depositional belts, individual reservoirs are locally more than 60 m thick, such as: (1) the Doethlingen Sandstone between the towns of Bremen and Celle (Schulz & Röhling 2000); or (2) the Detfurth Sandstone in the vicinity of the town of Karlshagen (Wolfgramm *et al.* 2008).

Table 1. *Decreasing porosity with increasing depth, exemplified by initial porosities of 10–40%*

Depth	Porosity (%)			
0	40	30	20	10
1000 m	32	24	16	8
2000 m	26	19	13	6
3000 m	20	15	10	5

Data were obtained from more than 800 measured porosity values of the North German Basin; see also Figure 4a.

The geothermal potential of the Middle Buntsandstein reservoir complex has been subject to previous investigation. Beutler *et al.* (1994) and Schulz & Röhling (2000) evaluated the reservoirs of the southern Buntsandstein, whereas Katzung (1984) and Wormbs *et al.* (1988) evaluated reservoirs of the northern Buntsandstein. Reservoirs of the Middle Buntsandstein are composed of fine- to medium-grained sandstones of low to intermediate maturities (Fig. 4b). Detrital composition of quartzes, feldspars and lithics result in lithic arkoses, feldspatic litharenites and lithic subarkoses compositions. In the area between the Darß and Usedom, reservoirs of the Detfurth to Solling formations are characterized by an average thickness of more than 20 m, an average porosity of more than 20% and an average permeability of more than 500 mD (Fig. 6a) (Wolfgramm *et al.* 2008). Locally, diagenetic processes contributed to a significant lowering of the primary porosity. The development of Middle Buntsandstein reservoirs is hampered by the limited prediction of reservoir thickness and quality of individual localities. The northern Buntsandstein is currently under evaluation by the 'GeoPoNDD' project (http://www.sandsteinfazies.de), which is being

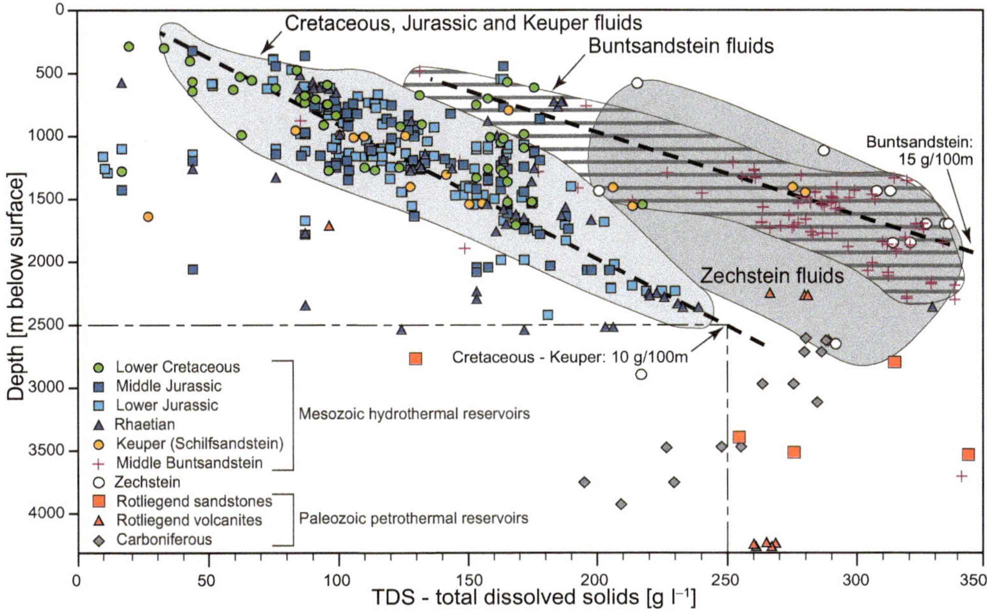

Fig. 5. Salinity v. depth. Increasing depth results in increasing salinities of fluids; *c.* 550 fluid analyses of Mesozoic and Paleozoic reservoirs (modified after Wolfgramm *et al.* 2014).

carried out by the authors with the goal of producing detailed subsurface maps in the near future.

Lower–Middle Keuper reservoir complex. Fluvio-deltaic sandstones of the Ladinian Lower Keuper (Erfurt Formation) and the mid-Carnian Schilfsandstein (Stuttgart Formation) are intercalated within the mainly shaly succession of the Keuper (Fig. 3). The Schilfsandstein and its geothermal potential were recently evaluated (Franz *et al.* 2014, 2015). Sandy channel fills and levee–crevasse splay complexes of the Lower and Upper Schilfsandstein were identified as main targets for further geothermal exploration. For the Lower Schilfsandstein, these reservoirs are 8–31 m thick (average 22.6 m), and are characterized by porosities of 0.9–37% (average 27.3%) and permeabilities of 0.3–3058 mD (average 1060 mD). For the Upper Schilfsandstein, these reservoirs are 8–35 m thick (average 17.7 m), and are characterized by porosities of 23–32% (average 29.2%) and permeabilities of 488–2972 mD (average 1799 mD: Franz *et al.* 2015). The quality of potential reservoirs is limited by the generally low maturity of the sandstones being lithic arkoses and feldspatic litharenites, the high content of shaly matrix within pores, and substantial diagenetic cementation (Fig. 4b) (Förster *et al.* 2010). A further limitation is given by rapid lateral facies shifts within the Stuttgart Formation. Due to the limited reservoir quality and the uncertainties in reservoir prediction, the potential of the Stuttgart Formation was considered to be rather low (Franz *et al.* 2015).

Upper Keuper reservoir complex. The Upper Keuper or Rhaetian reservoir complex comprises distributive fluvial and deltaic depositional systems (Fig. 3) (Franz & Wolfgramm 2008; Bachmann *et al.* 2010; Barth *et al.* 2014; Gaupp 1991). Both systems are from Scandinavian provenance to the north, and stretch across the eastern part of the North German Basin. In general, Rhaetian successions are sand-dominated in NE Germany, successively becoming more shaly towards NW Germany. To the west of Hamburg, the Rhaetian is dominantly composed of shales. The fine- to medium-grained clean sandstones of both depositional systems are characterized by high maturity, high porosities and permeabilities, and form excellent reservoirs (Figs 4b & 6c). Therefore, these have been the main targets for previous geothermal exploration (e.g. Scholle 1992; Beutler *et al.* 1994; Feldrappe *et al.* 2008). The Rhaetian distributive fluvial system forms the Rhaetian reservoirs A and B (previously the Postera Sandstone). At the town of Neubrandenburg, the municipal supplier operates these reservoirs for seasonal heat storage (Table 2) (Kabus *et al.* 2009; Obst & Wolfgramm 2010). The Rhaetian deltaic system forms the Rhaetian reservoirs C–F (previously the Contorta Sandstone). In the towns of Neustadt-Glewe and Waren, municipal suppliers operate these reservoirs for

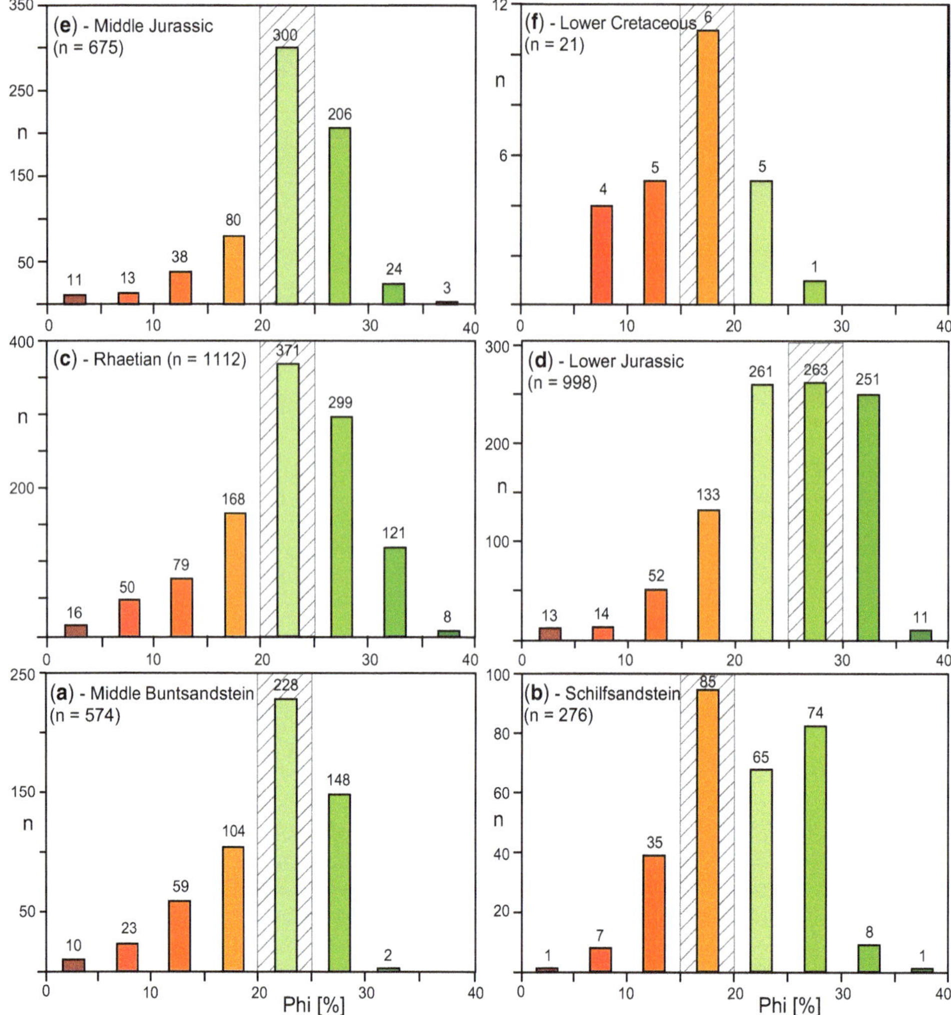

Fig. 6. Porosity (phi) distribution of Mesozoic reservoir complexes. Hatched bars refer to the mean value (modified after Wolfgramm *et al.* 2014).

district heating (Table 2). So far, uncertainties in the prediction of reservoir thickness and quality have contributed to significant exploration risks hampering the development of Rhaetian reservoirs at other localities. More recently, Franz *et al.* (2015) contributed to a lowering of the exploration risk underlining the high potential of Rhaetian hydrothermal reservoirs. Key results of this Rhaetian deltaic system reservoir are detailed in a following subsection.

Lower Jurassic reservoir complex. The base Jurassic is located at depths of between 800 and 4000 m (Lott *et al.* 2010). In terms of depositional environments, the Lower Jurassic resembles much of the Rhaetian deltaic system. Due to a general transgressive trend throughout the Liassic, the coastal and deltaic environments were successively shifted toward the basin margins in the NE (Barth *et al.* 2018*a*, *b*).

Accordingly, the western limit of potential geothermal reservoirs shifted from the area around the town of Schwerin in the Hettangian to the area around the town of Neubrandenburg in the Pliensbachian. Performance wise the Lower Jurassic reservoirs are comparable to Rhaetian reservoirs because of their high reservoir qualities (Figs 4b & 6d) (Rauppach *et al.* 2008; Wolfgramm *et al.* 2008). The Lower Jurassic reservoir complex was previously evaluated by Katzung (1984), Wormbs *et al.* (1988) and Scholle (1992), and is currently under

Table 2. *Geothermal heating stations, plants and facilities in northern Germany operating hydrothermal reservoirs with selected key parameters*

Location	Status	First well	Wells	Reservoir	Depth (m)	T (°C)	TDS ($g\ l^{-1}$)	E_{th} ($GWh\ a^{-1}$)
1 – Waren (D)	Operating	1986	Wa 3, Wa 4	Rhaetian	1540	61	158	2.4
2 – Prenzlau (D – VHE)	Operating	1986	Pr 1	Rotliegend	2800	108	n/a	0.36
3 – Neustadt-Glewe (D, E)	Operating	1987	NG 1, NG 2	Rhaetian	2250	99	221	18.33
4 – Neuruppin (D, S, B)	Operating	2007	NnS 1, NnS 2	Aalenian	1620	63	187	0.52
5 – Velten (D)	Inoperative	1990	Vt 2	Hettangian	1550	63	187	None
6 – Neubrandenburg (H)	Operating	1985	N 1, N 4	Rhaetian	1250	53.4	134	None
7 – Berlin (H)	Operating	1996	AR 1, AR 2	Hettangian	320	19.4	32	None
8 – Bad Wilsnack (S, B)	Operating	1996	Wlk 1	Rhaetian	1000	39.1	162	None
9 – Belzig (S, B)	Operating	1996	Bz 1	Rhaetian	730	27	188	None
10 – Binz (S, B)	Operating	1994	Binz 1	TBm	1020	28.5	110	None
11 – Templin (S, B)	Operating	1995	Tp 1	Hettangian	1620	67.3	163	None
12 – Burg/Spreew. (S, B)	Operating	1998	BuC 1	TBm	1295	54.5	230	None
13 – Bad Saarow (S)	Operating	1996	SrwPk 1, SrwPk 2	Hettangian	428	22	24.6	None

B, balneology; D, district heating (doublet, open system); D – VHE, deep borehole vertical earth heat exchanger (closed system); E, electricity generation; H, heat storage; TBm, Middle Buntsandstein; TDS, total dissolved solids; E_{th}, annual heat production in 2015 (source: http://www.geotis.de).

evaluation by the authors in the GeoPoNDD project (http://www.sandsteinfazies.de). In the past, uncertainties in the prediction of reservoir thickness and quality have hampered the development of Lower Jurassic reservoirs.

Middle Jurassic reservoir complex. Following the second-order maximum flooding in the Early Toarcian Tenuicostatum Ammonite Zone, the Toarcian–Bajocian Deltaic System successively prograded from Pomerania (late Early Toarcian) up to the Weser-Ems area (Bajocian). As individual deltas of the Toarcian, Aalenian and Bajocian are genetically related, forming the Toarcian–Bajocian deltaic system, the reservoirs of Toarcian deltas are included in the Middle Jurassic reservoir complex (Zimmermann *et al.* 2018). The Middle Jurassic was largely affected by post-sedimentary erosion due to a phase of basin inversion (Ziegler 1990). In particular in the western part of the North German Basin, only the Middle Jurassic strata present in rim synclines along salt structures were protected from erosion (Jaritz 1974). Recent evaluations of the Toarcian–Bajocian deltaic system have contributed to improved predictions in the reservoir thickness and quality (Franz *et al.* 2015; Zimmermann 2015); key results are detailed in a following subsection.

Lower Cretaceous reservoir complex. The base of the Cretaceous is located at depths of between 300 m in NE Germany to 4000 m in NW Germany (Vejbæk *et al.* 2010). The Lower Cretaceous, in particular the Berriasian (Wealden) and Valanginian, was affected by intense basin reorganization during inversion (Fig. 2) (Ziegler 1990). This resulted in a patchy distribution of Lower Cretaceous strata within large and small sub basins (Kemper 1979; Voigt *et al.* 2008). Synsedimentary tectonics and salt diapirism resulted in large variations in stratigraphic thickness. Whereas in larger parts of the North German Basin the thickness of the Lower Cretaceous does not exceed 200 m, thicknesses locally increase to more than 1800 m in southern Lower Saxony (Vejbæk *et al.* 2010).

In the North German Basin, the number and quality of potential Lower Cretaceous reservoirs varies strongly from west to east. At the SW margins of the Lower Saxony Sub basin, fluvio-deltaic reservoirs are present in the Berriasian (Bückeberg Formation), and coastal-marine reservoirs occur throughout the Valanginian–Albian (Mutterlose & Bornemann 2000). Towards the north, coastal siliciclastics interfinger with shales and marls but have no reservoir potential. Towards the east and NE, potential reservoirs are limited to the Berriasian–Hauterivian, such as the Wealden Sandstone in Mecklenburg or the Dichotomiten Sandstone in Lower Saxony (Valanginian).

Lower Cretaceous sandstone reservoirs are locally of high quality: in particular, the coastal-marine sandstones of the Valanginian–Albian (Beutler *et al.* 1994; Schulz & Röhling 2000). For example, Kuder *et al.* (2014) reported an average permeability of 1837 mD for the Bentheim Sandstone (Valanginian) in the western Lower Saxony Sub basin. In contrast, Rauppach *et al.* (2008) reported an average permeability of only 226 mD for reservoirs of Wealden-type facies in the eastern part of the North German Basin (Fig. 6f). So far, the uncertainties in distribution of Lower Cretaceous reservoirs and their quality have resulted in significant exploration risks. Lower Cretaceous reservoirs of the eastern part of the North German Basin are currently under evaluation in the GeoPoNDD project (http://www.sandsteinfazies.de).

Development examples in the North German Basin

The exploitation of geothermal energy in northern Germany started in 1986 with the exploitation of a Rhaetian hydrothermal reservoir at 1540 m depth below the city of Waren. The municipal supplier of Waren installed a geothermal doublet and provided thermal energy to customers via district heating (Fig. 7). Since then, many different geothermal facilities have been developed (Table 2). In general, a distinction is made between closed and open systems. The deep borehole heat exchanger operated in the town of Prenzlau is an example of a closed system. Open systems are the 'conventional' geothermal doublets operated in the towns of Waren, Neustadt-Glewe, Neubrandenburg (from 1989 to 1998) and Neuruppin, the geothermal heat storage site operated in Neubrandenburg (since 2003) and Berlin (parliamentary buildings), as well as balneological applications.

Geothermal doublets consist of an injection well and a production well (Fig. 7). Hot or warm saline waters are produced from the hydrothermal reservoir by a submersible pump, filtered, passed through a heat exchanger and reinjected into the reservoir. In order to avoid corrosion and scaling, the thermal loops are pressurized with nitrogen. At Waren, 60 m^3 h^{-1} of warm reservoir water with a temperature of *c.* 61°C is produced from the 1540 m deep Rhaetian reservoir (Fig. 8). The reservoir water is passed through a heat exchanger where the thermal energy is transferred to the secondary or supply

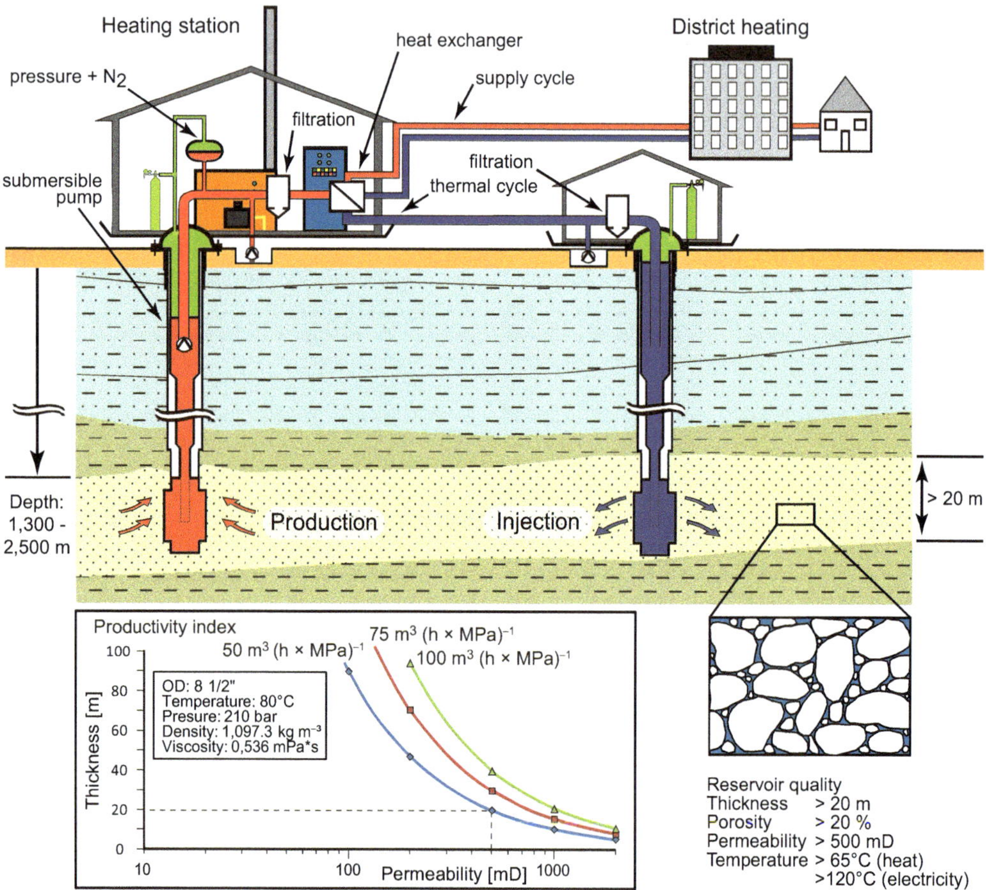

Fig. 7. Conventional operation of hydrothermal reservoirs in the North German Basin using a doublet system (modfied after Rockel *et al.* 1997). The minimum productivity index of 50 m^3 $(h \times MPa)^{-1}$ requires high reservoir qualities (for an explanation see the text).

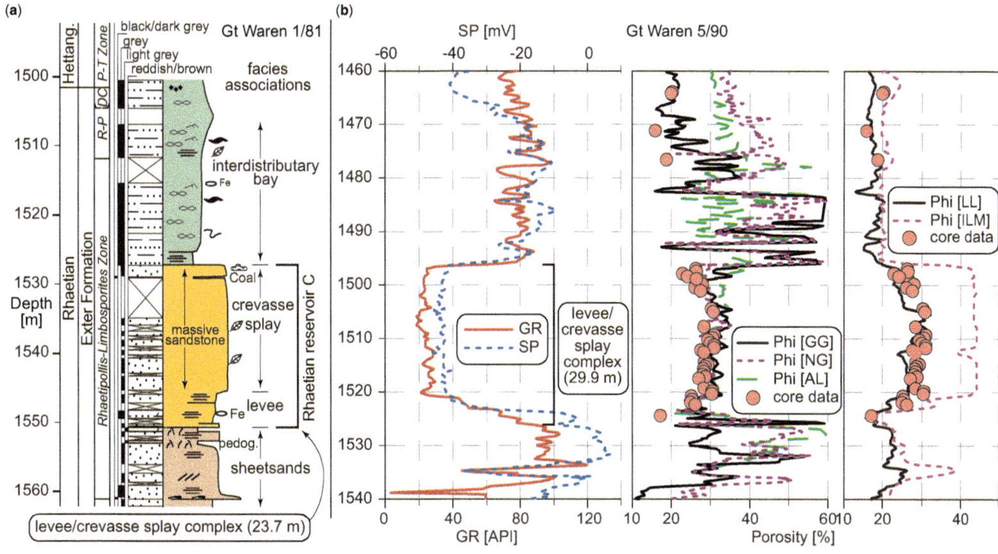

Fig. 8. Example of Rhaetian reservoir. (**a**) Litholog of cored well Waren 1/81 showing the 23.7 m thick levee/crevasse splay reservoir type operated at the Waren site (Table 2). (**b**) GR and SP logs (left), and calculated and measured porosity data of well Waren 5/90 (middle and right). The calibration of the different porosity (phi) calculation models with measured core porosities suggests that the Phi [LL] and Phi [GG] are probably the most realistic calculation models.

cycle. The annual heat production of 2.4 GWh a^{-1} is used to heat 1700 residential units. The reinjection temperature of the thermal cycle is about 40°C. The heating station at Neustadt-Glewe operates with a production rate of 100 m^3 h^{-1} of reservoir water at a 98°C temperature produced from the Rhaetian reservoir F at 2200 m depth (Contorta sandstone). The Neustadt-Glewe heating station had an annual heat production of 18.33 GWh a^{-1} in 2015 (Table 2).

The heat storage concept, currently in operation at Neubrandenburg or at the Reichstag in Berlin, is, however, slightly different. For example, the heat storage of the Reichstag is developed via two wells at a distance of about 300 m (warm side, RT Gt1; cold side, RT Gt 2). Both wells are used for production and injection based on the operational mode. During the summer, excess heat generated by the Reichstag heating power station is stored in a Lower Jurassic reservoir (Hettangian). By means of this, warm water at a temperature of 70°C is injected, stored and produced during the winter (Seibt & Kabus 1997). Since the beginning of the heat storage process, the warm well RT Gt 1 has heated up from around 20°C to about 65°C. The storage is then regenerated again during the summer of the next year by producing water from the cold well RT Gt 2, heating it up and reinjecting it into the warm well RT Gt1. Since 1996, a total of more than 1.5×10^6 m^3 of fluid has been stored (approximately 870 000 m^3 for regeneration and 550 000 m^3 for heating).

The geothermal heating station at Neubrandenburg was reconstructed from a conventional station running a doublet system to a heat storage facility. The excess heat of the power station is stored during the summer and used for heating in the winter. For each storage cycle, approximately 3.0 MW are injected and 2.5–3.0 MW are recovered.

Exploration strategy for Mesozoic reservoir complexes

Mesozoic sandstone reservoirs are preferred for conventional operation: in particular, those of the Jurassic and the Rhaetian (Fig. 3). The following parameters should be fulfilled in order to use these sandstones as geothermal reservoirs. For the long-term operation of a geothermal heating station using a doublet system, temperatures of >65°C (direct heat generation) and circulation rates of 50–100 m^3 h^{-1} are required (Fig. 7). By using heat pumps, operation at even lower temperatures is possible: however, much higher circulation rates are needed for compensation. Warm reservoir water with a temperature of >65°C is usually found in reservoirs at a minimum depth of 1300 m. In order to achieve the required circulation rates, with a reasonable drawdown of the water level, a productivity index of at least 50 $m^3 (h \times MPa)^{-1}$ is necessary. According to the Darcy–Dupuit well equation, the productivity index is primarily dependent on the thickness and permeability of the sandstone (Wolfgramm *et al.* 2014). Therefore, permeability has to be higher than 500 mD and the effective thickness of the aquifer should exceed 20 m, corresponding to an effective permeability thickness of 10 Dm. A lower reservoir thickness has to be compensated by higher permeability, and vice versa (Fig. 7).

So far, the evaluation of hydrothermal reservoirs has been focused on the depth of reservoirs and thus their temperatures (e.g. Katzung 1984; Wormbs *et al.* 1988; Beutler *et al.* 1994; Feldrappe *et al.* 2008). The depth of reservoirs and their temperature were collected for the Mesozoic reservoir complexes and are available in the form of classical maps, and online at http://www.geotis.de (Agemar *et al.* 2014). In addition, numerous contributions focused on the quality of reservoir complexes: for example, the hydraulic parameters (e.g. Beutler *et al.* 1994; Rauppach *et al.* 2008; Wolfgramm *et al.* 2008; Kuder *et al.* 2014). Apart from these rather generalized studies on reservoir complexes, detailed studies on individual reservoirs are missing. This refers, in particular, to high-resolution subsurface facies and sandstone thickness maps. As such maps provide the basis for improved reservoir predictions and reduction of the exploration risk, the exploration strategy designed for Mesozoic hydrothermal reservoirs has a strong focus on the reconstruction of depositional environments. The approach explained below was applied to a large database of cored and logged wells in order to construct large-scale subsurface maps. In the years 2011–15, this approach was employed to investigate the Keuper and Middle Jurassic reservoir complexes (Franz *et al.* 2015). It is currently applied to the Middle Buntsandstein, Lower Jurassic and Lower Cretaceous reservoir complexes

First, cored wells and outcrops are subject to detailed litho- and biofacies analyses. Based on grain size and physical bedding structures, primary and secondary lithofacies types are defined, and together with biofacies analyses, macro- and micropalaeontology, palynology and ichnology, grouped into facies associations. These facies associations are employed to reconstruct the depositional environments of reservoir complexes. Palaeontological data are employed for biostratigraphic dating of sections: in particular, dating of individual reservoirs (Fig. 9). Following this, biostratigraphically dated sections are used to reconstruct stratal pattern architectures along basin axis–basin margin transects, and to define and date genetic surfaces, such as maximum flooding and regression surfaces. Biostratigraphy supplemented by sequence stratigraphy is then employed to define time lines for construction of subsurface maps: for example, the maximum regression surface of a delta progradation. In order to evaluate the quality of dated reservoirs, their petrography, diagenesis and hydraulic parameters are investigated by means of microscopic methods, geochemistry and laboratory measurements. These results are then added to a database.

To allow the construction of subsurface maps from investigated cored wells and outcrops, the motifs of wireline logs are used to abstract from core facies to electrofacies. This knowledge is then applied to a database of several hundred logged wells. The analyses of net-sandstone thickness and facies for each time line resulted in a set of GIS-based subsurface sandstone thickness and facies maps (see Franz *et al.* 2015; Zimmermann *et al.* 2018).

For the herein detailed Rhaetian and Toarcian–Bajocian deltaic systems, channel belts of distributaries and associated levee/crevasse splays, as well as distributary/mouth bar complexes, were identified as key targets for geothermal exploration. In order to control the mapped courses of channel belts and to characterize their palaeohydraulics, the width of distributary channel belts (w_{dcb}) was calculated using empirical equations (see later) (Franz *et al.* 2015; Zimmermann *et al.* 2018).

The integration of reservoir quality and subsurface maps resulted in the identification of which reservoir thickness and quality most probably exceeds the required 20 m and 500 mD, respectively. These maps are herein referred to as reservoir quality maps. However, it is important to note that these maps do not illustrate a possibility of commercial success.

So far, the identification of an individual hydrothermal reservoir body, such as a channel belt, could not be demonstrated from conventional 2D seismic data (Fig. 10). In the future, the acquisition of 3D seismic data should contribute to an improved reservoir exploration if reservoir geobodies can be identified and imaged.

Exploration example 1: the Rhaetian deltaic system

A late Triassic transgression resulted in the brackish/marine Rhaetian Sea, a large inland sea covering the western half of the Central European Basin (e.g. Fischer *et al.* 2012; Barth *et al.* 2018*a*). This inland sea received siliciclastic detritus from surrounding source areas. In particular, Scandinavian source areas supplied large volumes of sediment via southern Sweden to the North German Basin, feeding a large fluvial-dominated deltaic system (Franz & Wolfgramm 2008). This deltaic system, herein referred to as the Rhaetian deltaic system, stretched from Schonen southwards up to

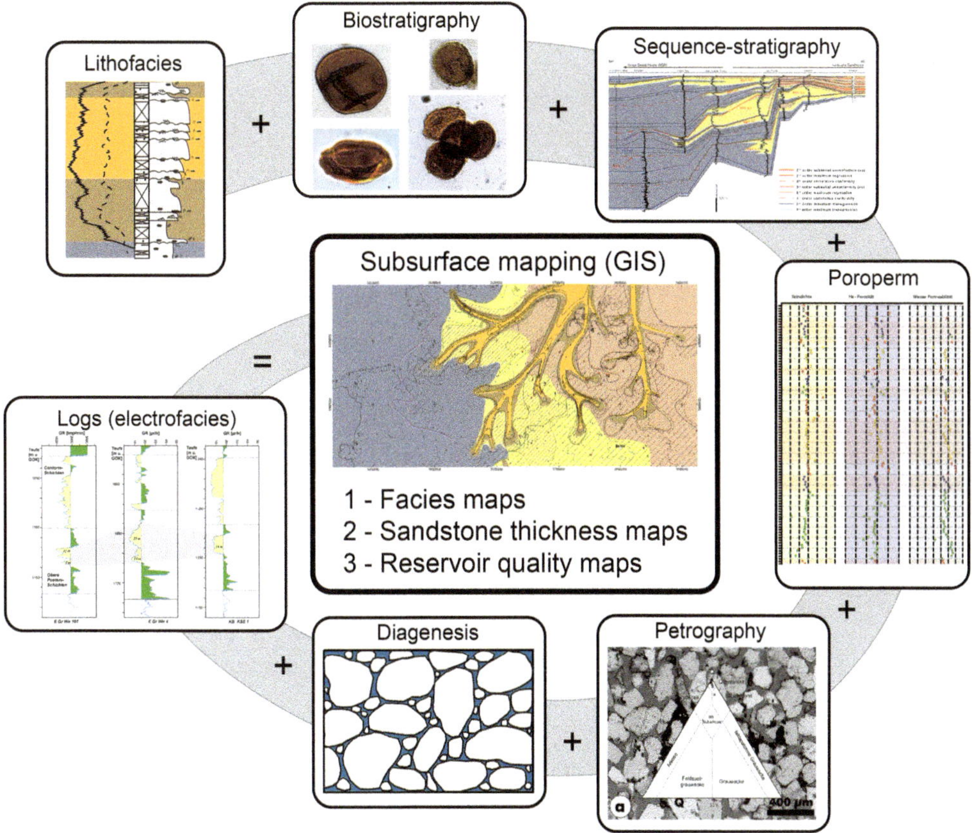

Fig. 9. Simplified exploration workflow applied to Mesozoic hydrothermal reservoirs of the North German Basin, partly modified from Zimmermann *et al.* (2015).

North Brandenburg, and westwards up to the Weser-Ems area (Fig. 11).

Database

The exploration of the Rhaetian deltaic system was based on: (1) facies analysis of more than 2000 m cored intervals from 23 wells of the North German Basin; and (2) five outcrop analogue studies from the southern margins of the basin. High-resolution time control on the Rhaetian deltaic system is provided by palynomorph biostratigraphy (Lund 1977, 2003; Heunisch 1999). Numerous palynostratigraphic datasets were generated during previous subsurface mapping campaigns and exploration for hydrocarbons, geothermal resources or underground storage sites. The majority of these data are present in the form of unpublished core reports, only a few data have been published (e.g. Lund 1977, 2003; Heunisch 1996*a*, *b*; Heunisch *et al.* 2010). The unpublished data are owned by geological state surveys (subsurface mapping) or private companies (exploration) and are available upon request or permission, respectively. In addition to these data, a larger dataset of 158 rock samples from nine wells and several outcrops was recently collected and studied in terms of palynomorph biostratigraphy (Barth *et al.* 2018*a*).

The results of palynomorph biostratigraphy and facies analysis were calibrated to a large database of 515 logged wells (Fig. 11). The petrography and granulometry of Rhaetian reservoirs were studied by means of 110 grain size and 215 thin sections analyses (Fig. 4). The quality of Rhaetian reservoirs of the Neustadt-Glewe, Gartz and KSS 5 deltaic complexes were evaluated using a database of 740 porosity and 262 permeability measurements.

Palynostratigraphic control on the Rhaetian deltaic system

The Rhaetian deltaic system is composed of four individual deltaic complexes that can be assigned to palynomorph zones (Lund 1977; Barth *et al.*

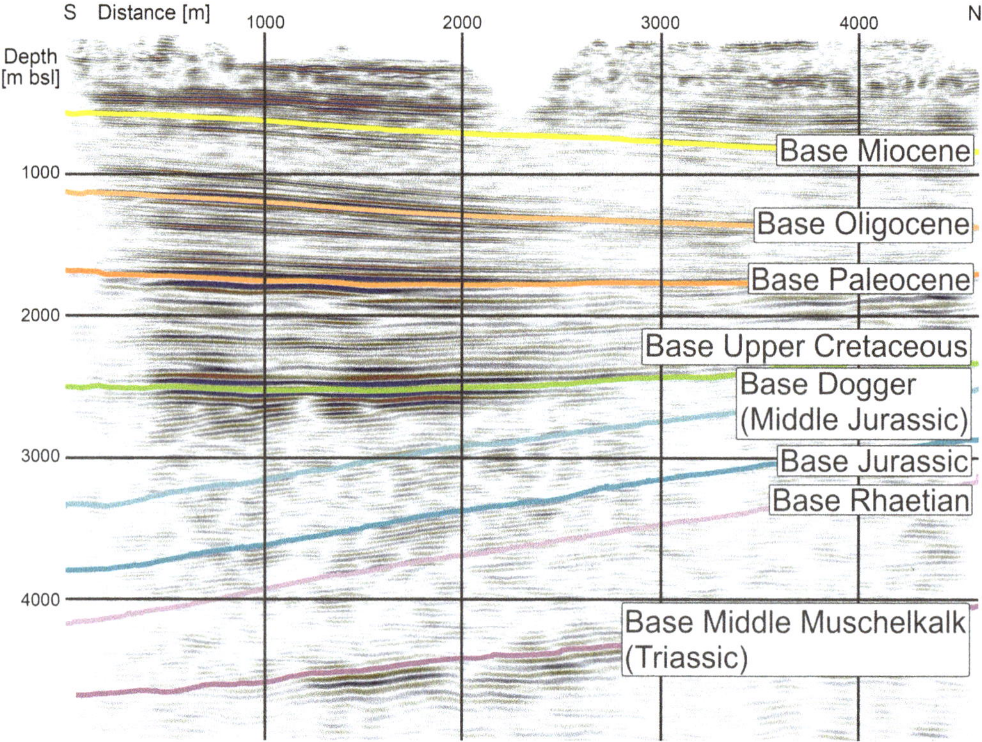

Fig. 10. Interpreted 2D seismic exploration profile. Seismic line HH1001 (Hamburg–Wilhelmsburg), modified from Wolfgramm *et al.* (2010); for the location see Figure 1. m bsl, metres below sea level.

2018*a*). The evolution of the Rhaetian deltaic system started with the deposition of the Neustadt-Glewe and KSS 5 deltaic complexes assigned to the middle *Rhaetipollis–Limbosporites* (RL) Zone (Barth *et al.* 2018*a*). Abandonment of these complexes and the northwards shift of deltaic sedimentation contributed to the formation of the Rødby deltaic complex assigned to the upper RL Zone (Lund 1977). Due to a transgression resulting in the maximum extension of the Rhaetian Sea in the uppermost RL Zone (Barth *et al.* 2018*a*), deltaic sedimentation shifted to the east. The Gartz complex is assigned to the *Ricciisporites–Polypodiisporites* Zone, and parts of the KSS 5 deltaic complex are assigned to the *Deltoidospora–Concavisporites* Zone (Fig. 11) (Barth *et al.* 2018*a*). Thus, the Rhaetian deltaic system study is a particularly good example of high-resolution palynostratigraphy contributing to detailed reconstructions of a deltaic system, such as deposition and abandonment of deltaic complexes.

Based on the correlation of the Rhaetian palynomorph zonation with the chronostratigraphic standard (Ogg 2012), the evolution of the Rhaetian deltaic system is believed to have formed over a 2.5 myr long time interval of the latest Triassic (Barth *et al.* 2018*a*). In terms of sequence stratigraphy, the Rødby, Neustadt-Glewe and parts of the KSS 5 deltaic complexes were formed around the maximum regression of Rhaetian 3rd-order sequence Rh 1 (mrs Rh1) and the Gartz and parts of the KSS 5 deltaic complexes were formed around the maximum regression of Rhaetian sequence Rh2 (mrs Rh2: Fig. 11) (Barth *et al.* 2018*a*).

Morphology of the Rhaetian deltaic system

The progradation of the Rhaetian deltaic system resulted in the deposition of up to 80 m-thick delta plain deposits (Fig. 11). The overall progradational architecture and the elongate shape of individual deltaic complexes correspond to the elongate high-constructive delta type according to Fisher *et al.* (1969).

The consecutive formation of deltaic complexes resulted in a compound delta plain estimated to be 47 000 km^2 in size. The Neustadt-Glewe, Rødby and KSS 5 deltaic complexes were mainly fed by drainage systems located to the west of the island of Rügen; only the Gartz deltaic complex was fed by a drainage system to the east of this island

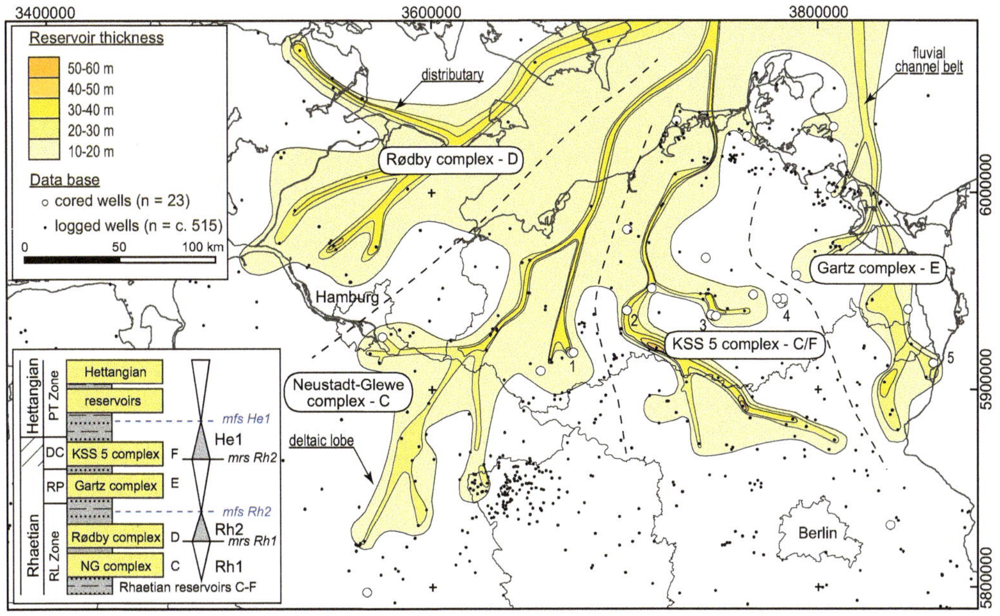

Fig. 11. Subsurface sandstone thickness map of the Rhaetian deltaic system. The high constructive fluvial-dominated deltaic system is composed of four subsequently formed deltaic complexes. The reservoir thickness refers to the total net-thickness of sandstone reservoirs assigned to the interval of the RL to DC Zones; post-sedimentary erosion not shown (see Fig. 12b); cored wells: 1, Neustadt-Glewe 1 and 2; 2, KSS 5; 3, Waren 1–Waren 5; 4, Neubrandenburg 1–Neubrandenburg 6; 5, Gartz 1; RL, *Rhaetipollis-Limbosporites* Zone; RP, *Ricciisporites-Polypodiisporites* Zone; DC, *Deltoidospora-Concavisporites* Zone; PT, *Pinuspollenites-Trachysporites* Zone, for details see Barth *et al.* (2018*a*).

(Fig. 12a). Based on the occurrence of marine phytoplankton, such as acritarchs and dinoflagellate cysts (Barth *et al.* 2018*a*), the Rhaetian delta plain is subdivided into a lower delta plain of about 21 000 km^2 and an upper delta plain of about 26 000 km^2 in size.

Each of the four deltaic complexes was fed by a single distributary channel. Increasing bifurcation in a downstream direction indicated repeated lobe switch and abandonment during the formation of a deltaic complex. The angles between bifurcated channels vary from 25° to 55°.

The upper delta plain is composed of distributary channels, levee/crevasse splay complexes, sheetsands and wetland facies associations (Table 3). The sandy fills of lateral-migrating distributary channels are recognized by up to 30 m thick successions of single to multiple stacked point-bar deposits. As dated by palynomorphs, the distributary channels of the Gartz, KSS 5 and Neustadt-Glewe deltaic complexes demonstrate up to 30 m of incision into older Rhaetian strata. The incision rates decrease downstream and distributaries are only slightly incised at the lower delta plain (Barth *et al.* 2018*a*). The lower delta plain is composed of distributary channels, levee/crevasse splay complexes, sheetsands and interdistributary bay facies associations (Table 3). At the distal end of the lower delta plain, the delta front is composed of terminal distributary channel and mouth bar facies associations forming distributary/mouth bar complexes. Distributary/mouth bar complexes are up to 30 m thick and have a lobe-shaped bar-finger architecture. Lobe switching and increased bifurcation at the distal end of distributaries resulted in lateral associations of sandy distributary/mouth bar complexes and shaly fills of interdistributary bays.

In terms of hydrothermal exploration opportunities, distributary channels and associated lateral levee/crevasse splay complexes are the most relevant morphological units. Based on subsurface facies and sand thickness mapping, these facies associations form up to 20 km broad sandy belts. Accordingly, the calculated medians of meander belt widths are 0.8–9.8 km, roughly corresponding to these values if lateral associated stacked sheetsands are taken into account (Fig. 13). Downdip, towards the prodeltaic environment, sand content deceases and marine conditions prevail, as indicated by the presence of marine phytoplankton and macrofauna with *Rhaetavicula contorta*. Marine processes, responsible for resedimentation of delta front silts and sands, contributed to thin intercalations within laminated shales

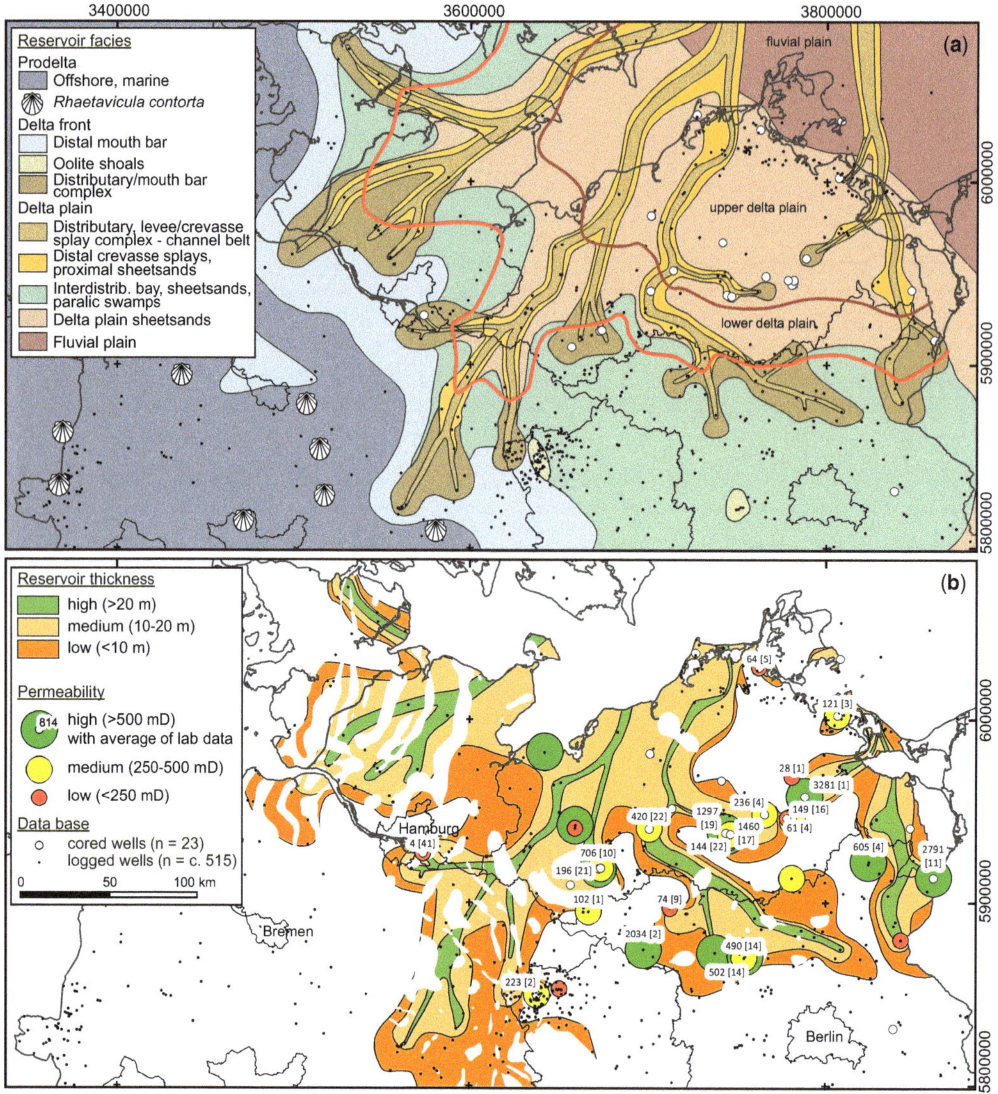

Fig. 12. The Rhaetian deltaic system. (**a**) Subsurface facies map of the high-constructive fluvial-dominated Rhaetian deltaic system; red lines represent the margins of the upper and lower deltaic plains; post-sedimentary erosion not shown. (**b**) Reservoir quality map in terms of thickness and permeability, average of lab data are only shown for mapping and geothermal wells; post-sedimentary erosion according to Baldschuhn *et al.* (2001) and Müller & Reinhold (2011).

in the proximal prodelta. The distal prodelta is predominantly formed of successions of laminated shales that are up to decametres thick.

Reservoir types and reservoir quality

Individual high-quality reservoirs up to a few metres thickness may be present in any of the sand-prone facies associations. However, taking the minimum thickness of 20 m required for long-term operation of hydrothermal reservoirs into account, only compound reservoir types formed of stacked individual reservoirs are considered viable exploration targets. Such compound reservoir types are represented by up to 95 m thick channel belts, resulting from vertical stacking of a distributary channel fill and lateral associated levee/crevasse splay complexes, and by up to 30 m thick progradational distributary/mouth bar complexes (Table 3). Total net-thicknesses of sandstone reservoirs range from 7 m to 56 m for

Table 3. *Facies associations of the Rhaetian deltaic system and their reservoir quality*

	Facies association	Individual reservoirs	Compound reservoirs				
			Type	Net-thickness	Porosity (φ) (M)	Permeability (k) (M)	Target
Delta plain	Wetland	No reservoirs	–	–	–	–	–
	Sheetsand	0.2–1 m	Stacked sheetsands	2–15 m	23.2%	239 mD	No
	Distributary channel	**5–25 m**	**Channel belt**	**7–56 m**	**19.7%**	**602 mD**	**Yes**
	Levee/crevasse splay complex	**5–15 m**					
	Interdistributary bay	No reservoirs	–	–	–	–	–
Delta front	**Terminal distributary channel**	**5–10 m**	**Distributary-mouth bar**	**5–25 m**	**22.3%**	**688 mD**	**Yes**
	Mouth bar	**2–10 m**					
Prodelta	Prodelta, proximal	No reservoirs	–	–	–	–	–
	Prodelta, distal	No reservoirs	–	–	–	–	–

M, mean.

channel belts and from 5 m to 25 m for distributary/mouth bar complexes.

In terms of petrography, granulometry and reservoir quality, the compound reservoir types of the Neustadt-Glewe, Gartz and KSS 5 deltaic complexes are comparable and do not show significant variations. The reservoir types are characterized by fine- to medium-grained, moderately to well sorted and mainly clean quartzarenites (Fig. 4b). The high median permeabilities of 602 mD for channel belts and of 688 mD for levee/crevasse splay complexes are mainly related to high median porosities of 19.7% for channel belts and of 22.3% for levee/crevasse splay complexes (Table 3). Wolfgramm

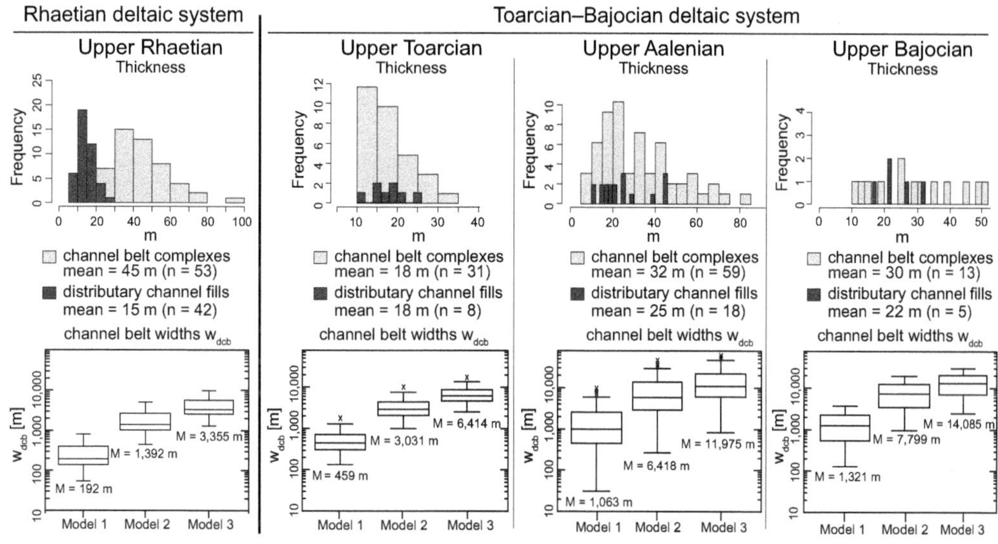

Fig. 13. Thickness distribution plots of distributary channel fills (T_d) and channel belt complexes, as well as calculations of channel belt widths (w_{dcb}) for the Rhaetian and Toarcian–Bajocian deltaic systems. Model 1 ($w_{dcb} = 0.95d_c^{2.07}$) according to Fielding & Crane (1987); Model 2 ($w_{dcb} = 12.1d_c^{1.85}$) according to Fielding & Crane (1987) and Miall (1996); Model 3 ($w_{dcb} = 64.6d_c^{1.54}$) according to Collinson (1978).

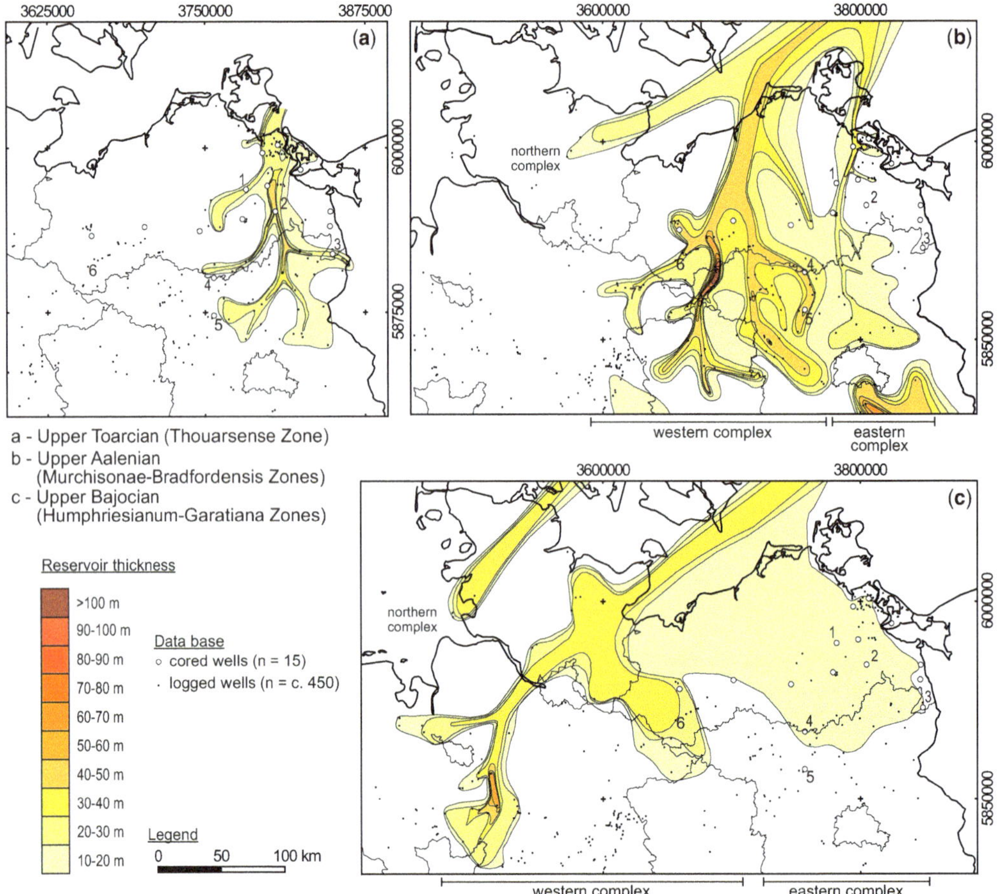

Fig. 14. Subsurface sandstone thickness maps of the Toarcian–Bajocian deltaic system, modified from Zimmermann (2015) and Zimmermann *et al.* (2018): The reservoir thickness refers to the total net-thickness of sandstone reservoirs assigned to the Upper Toarcian (Reservoir B), Upper Aalenian (Reservoirs C and D) and Upper Bajocian (Reservoir E) intervals; post-sedimentary erosion not shown (see Fig. 16); 1, JOmb 1; 2, JOmb 3; 3, JOmb 10; 4, Zechliner Hütte 1A; 5, Neuruppin 2; 6, Hagenow 1; 7, Kettelstorf 1.

et al. (2008) emphasized the dominance of pores larger >5 µm (70%), and reported 13 µm as the median radius of all intergranular pores.

In many cases, the porosity was not significantly affected by authigenic cementation (Franz & Wolfgramm 2008). However, in a few cases, the porosity of Rhaetian reservoirs was partly to completely destroyed by authigenic cementation close to salt structures (Baermann *et al.* 2000).

Reservoir prediction

In the eastern and northern part of the North German Basin, channel belt reservoirs are present and form up to 20 km broad belts trending N–S to NE–SW across the delta plain (Fig. 12b). Based on the evaluated well data, these belts are recorded in 31 wells of the Neustadt-Glewe deltaic complex (Rhaetian C), 11 wells of the Rødby deltaic complex (Rhaetian D), eight wells of the Gartz deltaic complex (Rhaetian E) and 19 wells of the KSS 5 deltaic complex (Rhaetian C–F). According to their broad distribution, the channel belt reservoirs represent the key exploration target. The subsurface facies and sand thickness maps enable an improved reservoir prediction for underexplored localities along the mapped courses of channel belts (Fig. 12b). Lateral to these courses, the reservoir thickness decreases rapidly and, thus, the exploration risk increases, although permeabilities may still be high.

The distributary–mouth bar reservoir type is present in the form of falciform reservoir bodies fringing the distal ends of delta lobes. This reservoir type has been recorded in 25 wells of the Neustadt-Glewe deltaic complex (Rhaetian C), five wells of the Rødby deltaic complex (Rhaetian D), two wells of the Gartz deltaic complex (Rhaetian E) and eight wells of the KSS 5 deltaic complex (Rhaetian C/F). As the occurrence of distributary–mouth bars is limited to the delta front, this reservoir type is restricted to central parts of the North German Basin (Fig. 12b).

Exploration example 2: the Toarcian–Bajocian deltaic system

Following basin-wide deposition of the Lower Toarcian Posidonia shale and equivalents during a second-order sea-level highstand in the Tenuicostatum ammonite Zone, the successive retreat of the Jurassic sea enabled the progradation of coastal-deltaic environments. Fed by Scandinavian sources to the north, subsequent progradation of genetically related deltas from the Lower Toarcian to the Upper Bajocian resulted in the formation of the Toarcian–Bajocian deltaic system (Fig. 14) (Zimmermann 2015; Zimmermann *et al.* 2018).

Database

The assessment of the Toarcian–Bajocian deltaic system was based on facies analysis of *c.* 2000 m of cored interval in 15 wells of the North German Basin and an outcrop analogue study in SE Lower Saxony (Zimmermann *et al.* 2014). High-resolution biostratigraphic time control is available in the form of numerous unpublished core reports from the 1960–80s (see the database of the RDS). All data were added to a large database of biostratigraphically calibrated wells (Zimmermann *et al.* 2015). The petrography, granulometry and quality of Toarcian–Bajocian reservoirs were evaluated based on 108 grain size analyses, about 200 thin sections, and a database of 1704 porosity and permeability measurements (Zimmermann 2015).

Biostratigraphic on the Toarcian–Bajocian deltaic system

Biostratigraphic control on the Lower and Middle Jurassic is based on a high-resolution framework of ammonite, ostracod, foraminifer and palynomorph zonations (Klingler & Neuweiler 1959; Triebel & Klingler 1959; Plumhoff 1963; Störmer & Wienholz 1965; Malz 1966, 1971; Michelsen 1975; Sivhed 1980; Brand & Mönnig 2009; Luppold 2012). In a previous study, Zimmermann *et al.* (2015) calibrated a large set of more than 450 wells to zones (and subzones) of the NW European ammonite zonation, and introduced a sequence stratigraphic framework of third- and fourth-order sequences (Fig. 14). According to this high-resolution chronostratigraphic control of Jurassic ammonite zones, delta formation in the Lower Toarcian (Bifrons), Upper Toarcian (Thouarsense), Upper Aalenian (Murchisonae–Bradfordensis) and Upper Bajocian (Garantiana) could be calibrated to 1.0–2.1 myr long intervals (Zimmermann 2015; Zimmermann *et al.* 2018).

Morphology of the Toarcian–Bajocian deltaic system

During the 14.4 myr long Toarcian–Bajocian interval, third-order sea-level fluctuations triggered the subsequent formation of four individual delta systems. As these deltas are genetically related, they have been treated as one Toarcian–Bajocian deltaic system. High-resolution basin-scale mapping of sandstone thickness and facies enabled the detailed description of delta morphology, and the interpretation of sedimentary processes and their controls (Zimmermann 2015; Zimmermann *et al.* 2018). The following description is focused on the deltas formed in the Upper Toarcian Thouarsense Zone, and the Upper Aalenian Murchisonae–Bradfordensis and Upper Bajocian Garantiana ammonite zones (Fig. 15).

The Upper Toarcian elongated river-dominated delta comprises up to 40 m thick delta plain successions in the eastern part of the North German Basin, stretching about 200 km from northern basin margins (Fennoscandian Border Zone) up to the east Prignitz and north Brandenburg areas. The delta plain, covering an area of about 20 000 km^2, was fed by a network of distributary channels of meandering to anastomosing channel type. Calculated medians of meander belt widths range between 0.5 and 6.4 km, and mapped widths of meander belts are up to 15 km (Figs 13 & 15). The Upper Toarcian distributive channel network shows angles of 30°–70° between successive distributary channel belts, and increasing frequency of bifurcation and decreasing channel belt fill from about 40 m to about 10 m in a downstream direction. Distally, towards the basin, the progradational delta front is composed of 10–50 m thick distributary/mouth bar complexes characterized by bar-finger architecture (Zimmermann 2015; Zimmermann *et al.* 2018).

The Upper Aalenian lobe-shaped river-dominated delta stretches from the northern basin margin up to the Altmark area (Fig. 14). The delta plain, characterized by up to 100 m thick successions, covers an area of about 40 000 km^2. The substantial enlargement, compared with the older Toarcian deltas, resulted from successive shallowing of the Jurassic Sea from Toarcian to Aalenian times (Zimmermann *et al.* 2015). Based on sandstone

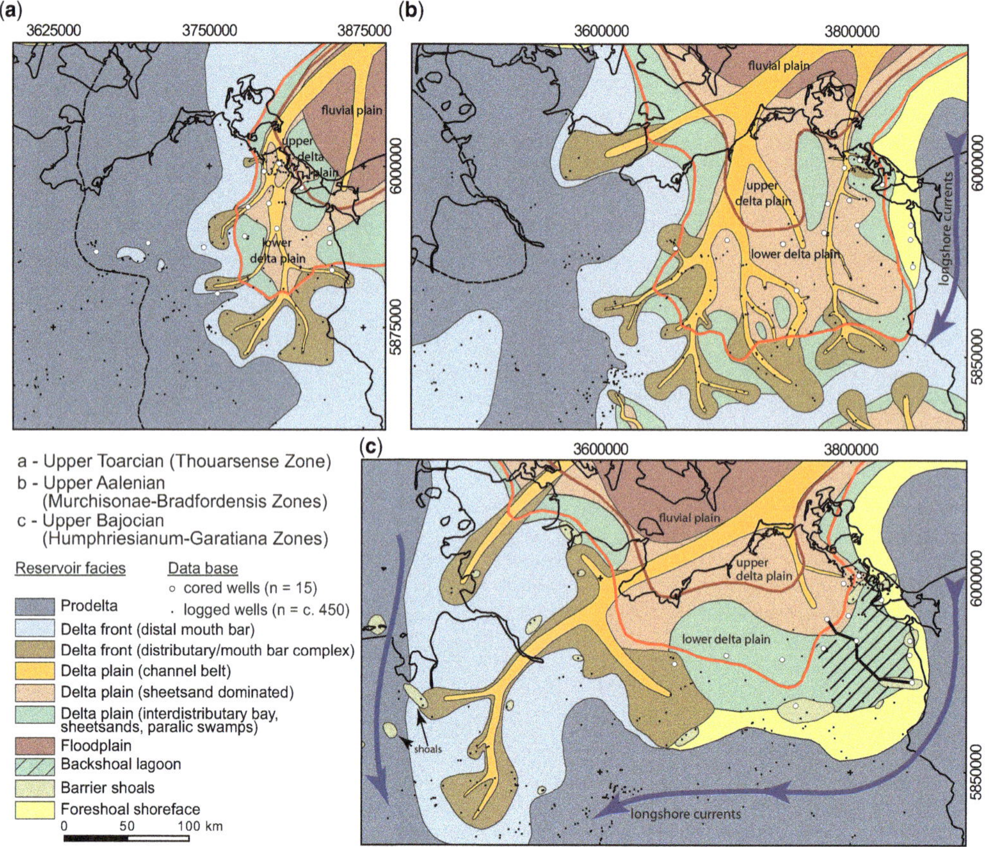

Fig. 15. Subsurface facies maps of the Toarcian–Bajocian deltaic system, modified from Zimmermann *et al.* (2018); post-sedimentary erosion not shown (see Fig. 16).

thickness and facies mapping, at least three deltaic complexes (northern, eastern and western) can be defined (Figs 14 & 15). Here, the focus is on the western and eastern deltaic complexes since much of the northern deltaic complex was subject to later erosion (Fig. 16). The delta plain of the western complex is up to 100 m thick and the delta plain of the eastern complex is up to 30 m thick. The distributive channel networks of both complexes show angles of 50°–100° between bifurcated distributary channels and increasing bifurcation in the downstream direction. An outcrop analogue situated in SE Lower Saxony revealed the lateral shift and stacking architecture of a meandering distributary channel system (Zimmermann *et al.* 2014). For the western deltaic complex, the lateral shift of distributaries resulted in up to 85 m thick successions of distributary channel fills and associated levee/crevasse splay complexes (Fig. 17; Table 4). These channel belts yielded sandstone reservoirs of up to 70 m net-thickness. Locally increased thicknesses of >100 m result from salt tectonics (Fig. 14). Calculated medians of meander belt widths range between 1.1 and 12.0 km, and mapped widths of meander belts range up to 15 km (Figs 13 & 15). Delta fronts comprise up to 70 m thick distributary/mouth bar complexes with up to 50 m thick (net-thickness) sandstone reservoirs of bar-finger sand architecture (Zimmermann 2015; Zimmermann *et al.* 2018). The morphology of the eastern margin of the Upper Aalenian delta reveals an increasing influence of marine processes (Fig. 15).

The Upper Bajocian mixed river-dominated and wave-modified delta stretches from the Fennoscandian Border Zone up to the Ems-Weser area. In accordance to the Aalenian delta, at least three deltaic complexes can be subdivided (Fig. 14). The western deltaic complex shows a cuspate and slightly oblique to elongate overall morphology, differing greatly from the eastern deltaic complex which was reshaped by marine processes (Fig. 15). The western deltaic complex was fed by a

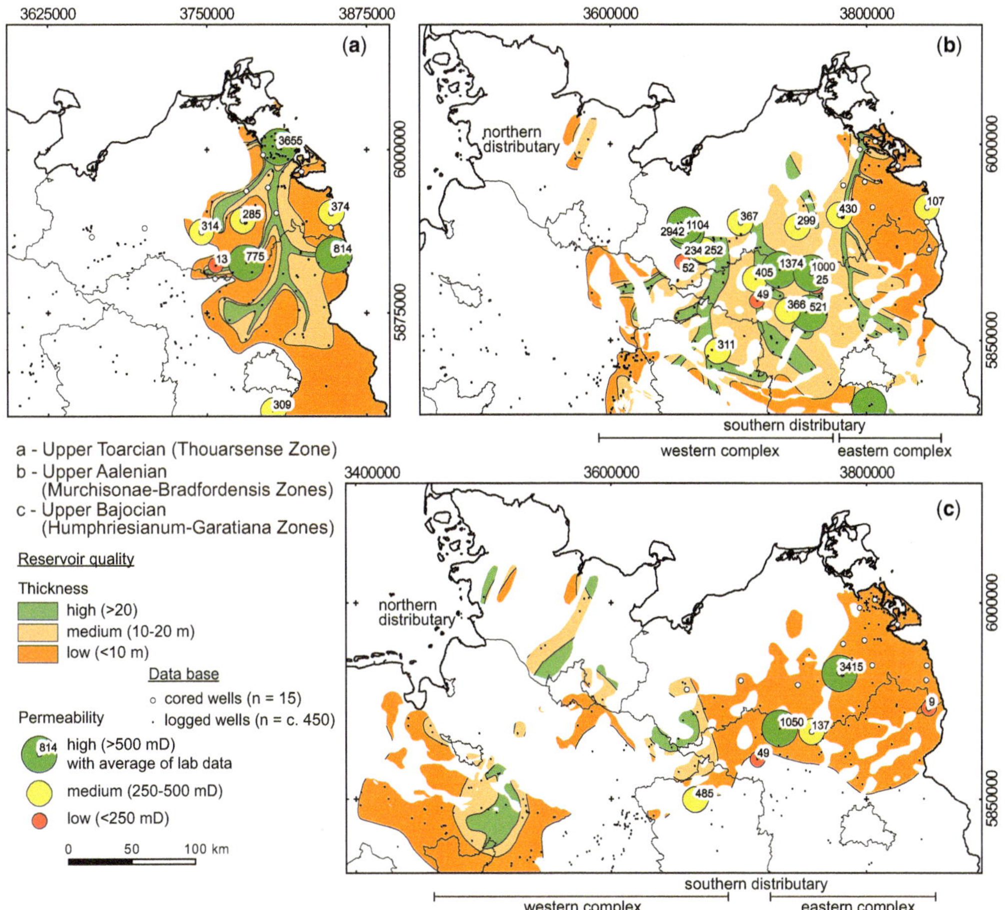

Fig. 16. Reservoir quality map in terms of thickness and permeability, average of lab data are only shown for mapping and geothermal wells, modified from Zimmermann (2015); post-sedimentary erosion according to Baldschuhn *et al.* (2001), Brand & Mönnig (2009).

main distributary channel network characterized by increasing bifurcation in a downstream direction and angles of 50°–100° between bifurcated channels. Up to 30 m thick distributary channel fills form together with lateral associated levee/crevasse splay complexes up to 50 m thick sandy successions of channel belts (Fig. 13). The delta front comprises 30–50 m thick amalgamated distributary/mouth bar complexes of lobate sand bodies with bar-finger sand architecture (Zimmermann 2015; Zimmermann *et al.* 2018).

Reservoir types and reservoir quality

Similar to the Rhaetian deltaic system, compound reservoirs of channel belts and distributary/mouth bar complexes represent the key targets for hydrothermal exploration opportunities (Fig. 17). The channel belt reservoir type is composed of almost univariate medium- to fine-grained sandstone characterized by high maturity (Fig. 4b). A high median porosity of 25% relates to a high median permeability of 1426 mD (Fig. 6; Table 4). The distributary/mouth bar complex reservoir type is composed of fine- to medium-grained sandstone characterized by bivariate to multivariate grain-size distributions due to a greater amount of fine particles resulting in a median permeability of 498 mD (Table 4) (Zimmermann 2015). In general, diagenetic modifications of grains in Aalenian sandstones are rare, and diagenetic cement content is low. Only in local areas (e.g. close to salt structures) has fluid migration contributed to formation of authigenic cements, mainly quartz or patchy dolomite.

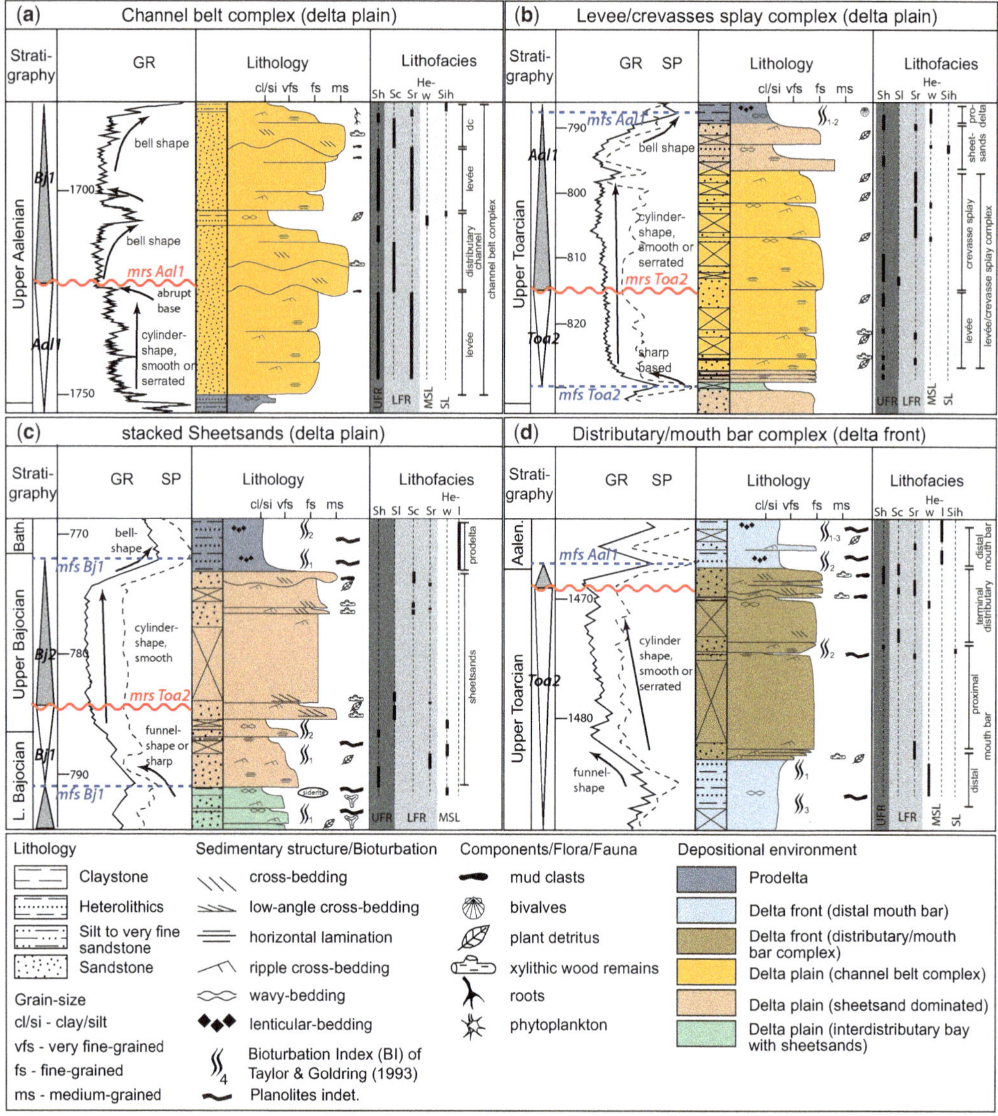

Fig. 17. Reservoir types of the Toarcian–Bajocian deltaic system (core examples). (**a**) Upper Aalenian, Neuruppin 2/86; (**b**) Upper Toarcian, JOmb 3/66; (**c**) Upper Bajocian, JOmb 1/65; and (**d**) Upper Toarcian, JOmb 10/66. For the location see Figure 14. mfs, maximum flooding surface; mrs, maximum regression surface (modified from Zimmermann *et al.* 2018).

Reservoir prediction

For the Toarcian–Bajocian deltaic system, the channel belt reservoir type is present and forms up to 15 km wide depositional belts trending across the delta plains (Fig. 15). The reservoir type was proven by numerous wells for the Lower Toarcian delta (reservoir A, not shown in Figs 15 & 16), the Upper Toarcian delta (reservoir B, Figs 15a & 16a), the Upper Aalenian delta (reservoirs C and D, Figs 15b & 16b) and the Bajocian delta (reservoir E, Figs 15c & 16c). Laterally, in the up to 85 m thick sand-prone channel belts, the thickness of reservoirs decreases rapidly and exploration risk subsequently increases. Up to 40 m thick falciform-shaped reservoirs of distributary/mouth bar complexes mark the distal ends of delta lobes. The reservoir quality maps of the Toarcian–Bajocian deltaic system

Table 4. *Facies associations of the Toarcian–Bajocian deltaic system and their reservoir quality; locally increased thicknesses due to salt tectonics not considered*

	Facies association	Individual reservoirs	Compound reservoirs				
			Type	Net-thickness	Porosity (φ) (M)	Permeability (k) (M)	Target
Delta plain	Wetlands	No reservoirs	–	–	–	–	–
	Sheetsands	0.52–5.0 m	Stacked sheetsands	1.0–15 m	24%	200 mD	No
	Distributary channel	**5–20 m**	**Channel belt**	**5–70 m**	**25%**	**1426 mD**	**Yes**
	Levee–crevasse splay complex	**1–20 m**					
	Interdistributary bay	No reservoirs	–	–	–	–	–
Delta front	**Terminal distributary channel**	**5–10 m**	**Distributary/ Mouth bar**	**10–50 m**	**23%**	**498 mD**	**Yes**
	Mouth bar	**5–30 m**					
Shoreface	Barrier shoal	2–5 m	Shoal sands	5–20 m	24%	82 mD	No
Prodelta	Prodelta, proximal	No reservoirs	–	–	–	–	–
	Prodelta, distal	No reservoirs	–	–	–	–	–

M, mean.

enable an improved reservoir prediction for both reservoir types (Fig. 16a–c).

However, in contrast to the Rhaetian deltaic system, the presence of reservoirs is limited by later erosion. This refers, in particular, to the Upper Aalenian and Upper Bajocian deltas as both were subject to widespread erosion in Schleswig-Holstein, and in larger parts of Mecklenburg-Vorpommern and Lower Saxony (Fig. 16c). A further limitation is given by the shallow burial depths of often less than 1000 m, as the reservoir temperatures are too low.

Discussion

Controls on Mesozoic reservoir complexes

The formation of Mesozoic hydrothermal reservoirs was controlled by tectonic and climatic processes. In the Triassic and Jurassic, the general tectonic regime changed to E–W and NE–SW directed extension (Ziegler 1982; Kockel 2002; Kley *et al.* 2008). Accordingly, the low, but continuous, subsidence rates resulted in generally small thicknesses of hydrothermal reservoirs. As demonstrated herein, Triassic and Jurassic hydrothermal reservoirs exceeding 20 m thickness are not regionally distributed. The present-day distribution of Triassic and Jurassic reservoir complexes may be limited by post-depositional erosion due to erosional unconformities: for example, the Norian Early Cimmerian Unconformity (Beutler & Schüler 1978) or the Aalenian mid-Cimmerian unconformity related to thermal updoming in the North Sea area (Hallam & Sellwood 1976; Underhill & Partington 1993). It is noteworthy that these unconformities are not of basin-wide occurrence, but locally up to some 100 m of strata may be missing (Beutler & Schüler 1978; Barnasch *et al.* 2005; Stollhofen *et al.* 2008; Beutler *et al.* 2012). The change to compressional tectonics in the Cretaceous resulted in complex patterns of uplift and subsidence. Post-depositional erosion during the Cretaceous affected the present-day distribution of the Lower Cretaceous reservoir complex and, in places, also older reservoir complexes (Fig. 2) (Ziegler 1990; Kley & Voigt 2008; Voigt *et al.* 2008; Vejbæk *et al.* 2010).

Besides variabilities in thickness and the presence of strata, the general tectonic regime and basin evolution is reflected by successively increasing maturities of sandstones. Rotliegend sandstones of the early basin fill stage are characterized by low mineralogical maturity (Fig. 4b) (Wolfgramm 2005). Although the evolution of the Permo-Carboniferous climate was complex (Roscher & Schneider 2006), the low maturity of Rotliegend sandstones can be explained simply by an arid to semi-arid climate (e.g. Glennie 1972, 1983). Buntsandstein and Rhaetian to Lower Cretaceous sandstones of the late basin fill stage are characterized by intermediate and high mineralogical maturities, respectively (Fig. 4b). The high maturities

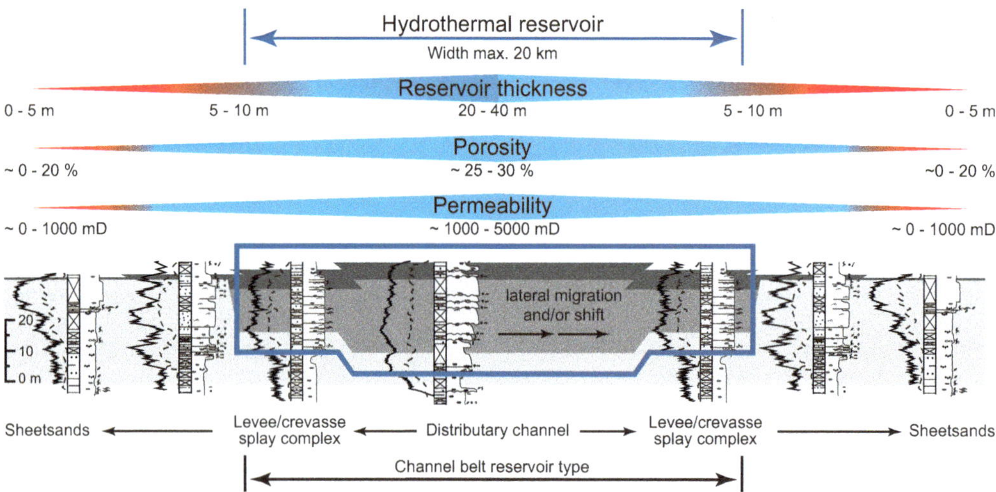

Fig. 18. Potential of the channel belt reservoir type. Synopsis of subsurface facies, sandstone thickness and reservoir quality along a synthetic cross-section.

of Rhaetian and younger sandstones are in agreement with Upper Triassic–Jurassic increased onshore precipitation reconstructed from clay mineral assemblages (Ahlberg *et al.* 2002; Feist-Burkhardt *et al.* 2008; Dera *et al.* 2009). Sandstones of the Lower–Middle Keuper reservoir complex are exceptional as they are characterized by very low mineralogical maturities (Fig. 4b) (Förster *et al.* 2010; Franz *et al.* 2015). Apart from the rather general discussion whether the Schilfsandstein represents a humid climate event (see Simms & Ruffell 1989; Visscher *et al.* 1994; Kozur & Bachmann 2010), the very low maturity – in particular, the high content of lithics – seems not to be controlled by climate fluctuations (Franz *et al.* 2014). Kozur & Bachmann (2010) suggested a provenance association with rift shoulder uplift in the area of the Scandinavian Caledonides but further constraints are required.

The North German Basin comprises numerous salt walls, diapirs and pillows of Rotliegend and/or Zechstein salt (Ziegler 1990; Reinhold *et al.* 2008). Salt diapirism started in the early Triassic and attained maxima in the late Triassic–early Jurassic and early Cretaceous intervals (Trusheim 1957; Kootz & Schumacher 1967; Jaritz 1974). As detailed by Hansen *et al.* (2007) and Scheck-Wenderoth *et al.* (2008), salt diapirism resulted in local differences in subsidence and sedimentation patterns in the NGB, whereas a general impact on depositional environments of Triassic and Jurassic reservoir complexes is not indicated (Franz *et al.* 2014, 2015; Zimmermann *et al.* 2015, 2018; Barth *et al.* 2018*a*). The role of halokinesis in influencing Lower Cretaceous reservoir distribution is subject to an ongoing study.

Reservoir prediction and exploration risk

For the herein detailed Rhaetian and Middle Jurassic reservoir complexes, the compound reservoirs of channel belt and distributary/mouth bar types have been identified as key targets of geothermal exploration (Tables 3 & 4). Only these two reservoir types meet the required key parameters of reservoir quality: >20 m thickness and >500 mD permeability (Rockel *et al.* 1997). The reservoir temperature depends on the local geothermal gradient and, thus, the burial depth. As such, this parameter is not related to the depositional environment of the reservoir.

Subsurface facies mapping of the Rhaetian and Toarcian–Bajocian deltaic systems revealed larger areas covered by the channel belt reservoir type compared to the distributary/mouth bar type for both systems (Figs 12 & 15). Considering the successful operation of the channel belt reservoir type at Neustadt-Glewe and Waren (Fig. 8), further geothermal exploration should focus on this reservoir type. Within the Rhaetian and Toarcian–Bajocian deltaic systems, up to 20 km wide sand-prone strings, herein considered channel belts, have been mapped whose positions were mainly controlled by distributary channels (Fig. 18). Lateral migration and shifting of distributary channels resulted in vertical stacks of sandy channel fills and lateral associated sand-prone levee/crevasse splay complexes. The general high potential of the channel belt reservoir type may be limited by the following aspects. (1) Channel belts are not characterized by a uniform large sandstone thickness. The sandstone thickness often exceeds 20 m but this is not necessarily the case for each point within the channel belt. It is important to note that this aspect cannot be addressed in basin-scale sandstone thickness maps. (2) Laterally, channel belts show facies shifts with delta plain environments often dominated by shaly lithologies. With increasing distance from the distributary channel, the sandstone thickness also decreases and shaly intercalations become more abundant. This is important as substantial shaly intercalations are negatively influencing the productivity and injectivity of the hydrothermal reservoirs: for example, due to the mobilization of fine-grained particles.

Conclusions

Basin-scale subsurface facies, sandstone thickness and reservoir quality maps of the Rhaetian and Toarcian–Bajocian deltaic systems enable the improved exploration of Rhaetian and Middle Jurassic hydrothermal reservoirs for individual areas. These maps have already been used to help plan ongoing and future projects in Lower Saxony and Mecklenburg-Vorpommern.

In the next step, basin-scale subsurface maps should be substantiated by detailed mapping studies. Regional- to local-scale subsurface facies, sandstone thickness and reservoir quality mapping studies will contribute to a further reduction of exploration risks and exploration strategies for individual localities. Such regional- to local-scale studies should routinely implement the acquisition of 3D seismic data. The successful targeting of geobodies by means of 3D seismic campaigns will further contribute to a significant reduction in exploration risks and costs.

The authors acknowledge funding by the Bundesministerium für Wirtschaft und Energie (https://doi.org/10.13039/501100006360: grant numbers 0325285 and 0325920 to Matthias Franz and Markus Wolfgramm) BMWi and Bundesanstalt für Geowissenschaften und Rohstoffe (BGR). Karsten Obst, Juliane Brandes (Landesamt für Umwelt und Naturschutz Mecklenburg-Vorpommern), Michael Göthel (Landesamt für Bergbau, Geologie und Rohstoffe Brandenburg), Thomas Koch, Karl-Heinz Friedel (LAGB Sachsen-Anhalt), Joachim Schubert (†TLUG), Heinz-Gerd Röhling (Landesamt für Bergbau, Energie und Geologie Niedersachsen) and Robert Lippmann (RWE DEA) kindly provided access to well data and core material. The reviews of Renaud Bouroullec (TNO) and an anonymous reviewer, and the editorial handling of Harmen Mijnlieff (TNO) and Ben Kilhams (Shell), are gratefully acknowledged.

References

AGEB 2015. *Satellitenbilanz Erneuerbare Energien 2015 [Annual report on renewable energies 2015]*, http://www.ag-energiebilanzen.de/7-0-Bilanzen-1990-2015.html

Agemar, T., Alten, J., Ganz, B., Kuder, J., Kühne, K., Schumacher, S. & Schulz, R. 2014. The Geothermal Information System for Germany – GeotIS. *Zeitschrift der Deutschen Gesellschaft für Geowissenschaften*, **165**, 129–144.

Ahlberg, A., Arndorff, A. & Guy-Ohlson, D. 2002. Onshore climate change during the Late Triassic marine inundation of the Central European basin. *Terra Nova*, **14**, 241–248.

Bachmann, G.H. & Grosse, S. 1989. Struktur und Entstehung des Norddeutschen Beckens – geologisch-geophysikalische Interpretation einer verbesserten Bouguer-Schwerekarte [Structure and formation of the North German Basin – Interpretation of an improved Bouguer map]. *Niedersächsische Akademie der Geowissenschaften Veröffentlichungen*, **2**, 23–47.

Bachmann, G.H. & Hoffmann, N. 1995. Bildung und Entwicklung des Norddeutschen Rotliegend-Beckens [Formation and evolution of the South Permian Basin]. *In*: Plein, E. (ed.) *Stratigraphie von Deutschland, Norddeutsches Rotliegendbecken.* Rotliegend Monographie Teil II, **183**. Courier Forschungsinstitut Senckenberg, Frankfurt, Germany, 156–169.

Bachmann, G.H., Geluk, M. *et al.* 2010. Triassic. *In*: Doornenbal, H. & Stevenson, A. (eds) *Petroleum Geological Atlas of the Southern Permian Basin Area.* European Association of Geoscientists and Engineers (EAGE), Houten, The Netherlands, 149–173.

Baermann, A., Kröger, J. & Zarth, M. 2000. Anhydritzemente im Rhätsandstein Hamburgs – Röntgen- und kernspintomographische Untersuchungen und Lösungsversuche [Anhydrite cementation of the Rhaetian sandstone of Hamburg – X-Ray and NMR investigations and solution tests]. *Zeitschrift für angewandte Geologie*, **46**, 144–152.

Baldschuhn, R., Binot, F., Fleig, S. & Kockel, F. 2001. Geotektonischer Atlas von NW Deutschland und dem deutschen Nordsee-Sektor [Tectonic atlas of NW Germany and the German North Sea sector]. *Geologisches Jahrbuch Reihe A*, **153**, Hannover.

Bally, A. & Snelson, S. 1980. Realms of subsidence. *In*: Miall, A.D. (ed.) *Facts and Principles of World Petroleum Occurrence.* Canadian Society of Petroleum Geologists, Memoirs, **6**, 9–94.

Barnasch, J., Franz, M. & Beutler, G. 2005. Hochauflösende Gliederung des Keupers der Eichsfeld-Altmark-Schwelle zur Präzisierung der Diskordanzen [High-resolution Keuper stratigraphy and disconformities of the Eichsfeld-Altmark Swell]. *Hallesches Jahrbuch für Geowissenschaften*, **B19**, 153–160.

Barth, G., Franz, M., Heunisch, C., Kustatscher, E., Thies, D., Vespermann, J. & Wolfgramm, M. 2014. Late Triassic (Norian–Rhaetian) brackish to freshwater habitats at a fluvial-dominated delta plain (Seinstedt, Lower Saxony, Germany). *Palaeobiology and Palaeoenvironments*, **94**, 495–528.

Barth, G., Franz, M., Heunisch, C., Ernst, W., Zimmermann, J. & Wolfgramm, M. 2018*a*. Marine and terrestrial sedimentation across the T–J transition in the Central European Basin. *Palaeogeography. Palaeoclimatology. Palaeoecology*, **489**, 74–94.

Barth, G., Pieńkowski, G., Zimmermann, J., Franz, M. & Kuhlmann, G. 2018*b*. Palaeogeographical evolution of the Lower Jurassic: high-resolution biostratigraphy and sequence stratigraphy in the Central European Basin. *In*: Kilhams, B., Kukla, P.A., Mazur, S., McKie, T., Mijnlieff, H.F. & van Ojik, K. (eds) *Mesozoic Resource Potential in the Southern Permian Basin*. Geological Society of London, Special Publications, **469**. First published online January 4, 2018, https://doi.org/10.1144/SP469.8

Benek, R., Kramer, W. *et al.* 1996. Permo-Carboniferous magmatism of the Northeast German Basin. *Tectonophysics*, **266**, 379–404.

Beutler, G. & Schüler, F. 1978. Über altkimmerische Bewegungen im Norden der DDR und ihre regionale Bedeutung (Fortschrittsbericht) [Early Cimerian tectonis in the NE GDR and their importance on regional geology (progress report)]. *Zeitschrift für geologische Wissenschaften*, **6**, 403–420.

Beutler, G., Röhling, H.G., Schulz, R. & Werner, K.H. 1994. *Regionale Untersuchungen von geothermischen Reserven und Ressourcen in Nordwestdeutschland [Geothermal reserves and resources of NW Germany – regional studies]*. Endbericht, Archivnr. 111 758, Hannover (GGA).

Beutler, G., Junker, R., Niediek, S. & Rößler, D. 2012. Tektonische Diskordanzen und tektonische Zyklen im Mesozoikum Nordostdeutschlands [Tectonic unconformities and cycles in the Mesozoic of NE Germany]. *Zeitschrift der Deutschen Gesellschaft für Geowissenschaften*, **163/4**, 447–468.

Brand, E. & Mönnig, E. 2009. Litho- und Biostratigraphie des Mittel-Jura (Dogger) in Bohrungen Norddeutschlands [Litho- and biostratigraphy of the Middle Jurassic (Dogger) in wells of North Germany]. *Schriftenreihe der Deutschen Gesellschaft für Geowissenschaften*, **54**, 5–73.

Breitkreuz, C. & Kennedy, A. 1999. Magmatic flare-up at the Carboniferous/Permian boundary in the NE German Basin revealed by SHRIMP zircon ages. *Tectonophysics*, **302**, 307–326.

Breitkreuz, Ch., Geißler, M., Schneider, J. & Kiersnowski, H. 2008. Basin initiation: Volcanism and sedimentation. *In*: Littke, R., Bayer, U., Gajewski, D. & Nelskamp, S. (eds) *Dynamics of Complex Intracontinental Basins: The Central European Basin System*. Springer, Berlin, 173–180.

Collinson, J. 1978. Vertical sequence and sand body shape in alluvial sequences. *In*: Miall, A.D. (ed.) *Fluvial Sedimentology*. Canadian Society of Petroleum Geologists, Memoirs, **5**, 577–586.

Dera, G., Pellenard, P., Neige, P., Deconinck, J.-F., Pucéat, E. & Dommergues, J.-P. 2009. Distribution of clay minerals in Early Jurassic Peritethyan seas: palaeoclimatic significance inferred from multiproxy comparisons. *Palaeogeography, Palaeoclimatology, Palaeoecology*, **271**, 39–51.

Diener, I., Wormbs, J. *et al.* 1988–1992. *Abschlußberichte Geothermische Ressourcen im Nordteil der DDR 1:200 000 [Report on geothermal resources in the northern GDR 1:200000]*. Zentrales Geologisches Institut, Berlin.

Feist-Burkhardt, S., Götz, A.E. *et al.* 2008. Triassic. *In*: McCann, T. (ed.) *The Geology of Central Europe, Volume 2: Mesozoic and Cenozoic*. Geological Society, London, 749–821.

Feldrappe, H., Obst, K. & Wolfgramm, M. 2008. Die mesozoischen Sandsteinaquifere des Norddeutschen Beckens und ihr Potenzial für die geothermische Nutzung [Mesozoic sandstone aquifers of the North German Basin and their potential]. *Zeitschrift für Geologische Wissenschaften*, **36**, 199–222.

Fielding, C. & Crane, R. 1987. An application of statistical modelling to the prediction of hydrocarbon recovery factors in fluvial reservoir sequences. *In*: Ethridge, F., Flores, R. & Harvey, M. (eds) *Recent Developments in Fluvial Sedimentology*. Society of Economic Paleontologists and Mineralogists, Special Publications, **39**, 321–327.

Fischer, J., Voigt, S. *et al.* 2012. Palaeoenvironments of the late Triassic Rhaetian Sea: implications from oxygen and strontium isotopes of hybodont shark teeth. *Palaeogeography. Palaeoclimatology. Palaeoecology*, **353–355**, 60–72.

Fisher, W.L., Brown, L.F., Scott, A.J. & McGowen, J.H. (eds) 1969. *Delta Systems in the Exploration for Oil and Gas*. Bureau of Economic Geology, University of Texas, Austin, TX, USA.

Förster, A., Schöner, R. *et al.* 2010. Reservoir characterisation of a CO_2 storage aquifer. The Upper Triassic Stuttgart Formation in the Northeast German Basin. *Marine and Petroleum Geology*, **27**, 2156–2172.

Franz, M. & Wolfgramm, M. 2008. Sedimentologie, Petrologie und Fazies geothermischer Reservoire des Norddeutschen Beckens am Beispiel der Exter-Formation (Oberer Keuper, Rhaetium) NE-Deutschlands [Sedimentology, petrology and facies of geothermal reservoirs of the North German Basin on the example of the Exter Formation (Upper Keuper, Rhaetian)]. *Zeitschrift für geologische Wissenschaften*, **36**, 223–248.

Franz, M., Nowak, K., Berner, U., Heunisch, K., Bandel, K., Röhling, H.-G. & Wolfgramm, M. 2014. Eustatic control on epicontinental basins: the example of the Stuttgart Formation in the Central European Basin (Middle Keuper, Late Triassic). *Global and Planetary Change*, **122**, 305–329.

Franz, M., Wolfgramm, M., Barth, G., Nowak, K., Zimmermann, J., Budach, I. & Thorwart, K. 2015. *Verbundprojekt: Identifikation hydraulisch geeigneter Bereiche innerhalb der mesozoischen Sandsteinaquifere in Norddeutschland [R&D project: Hydraulic properties of Mesozoic sandstone aquifers of North Germany]*. Forschungsvorhaben, Dokumentation, Schlussbericht TU Bergakademie Freiberg, Freiberg, Germany.

Friberg, L.J. 2001. *Untersuchungen zur Temperatur- und Absenkungsgeschichte sowie zur Bildung und Migration von Methan und molekularem Stickstoff im Nordostdeutschen Becken [Investigations on temperature and burial history as well as formation and migration of methan and nitrogen in the North German Basin]*. Berichte Forschungszentrum Jülich, **3914**.

Gaupp, R. 1991. Zur Fazies und Diagenese des Mittelrhät-Hauptsandsteins im Gasfeld Thönse [Facies and diagenesis of the Mittelrhät-Hauptsandstein in the Thönse gasfield]. *Niedersächs. Akad. Geowiss. Veröffentl*, **6**, 34–54.

Gast, R., Pasternack, M., Piske, J. & Rasch, H.-J. 1998. Das Rotliegend im Nordostdeutschen Raum: Regionale

Übersicht, Stratigraphie, Fazies und Diagenese [The Rotliegend of NE Germany: regional overview, stratigraphy, facies and diagenesis]. *Geologisches Jahrbuch*, **A149**, 59–79.

Glennie, K.W. 1972. Permian Rotliegendes of Northwest Europe interpreted in light of modern desert sedimentation studies. *AAPG Bulletin*, **56**, 1048–1071.

Glennie, K.W. 1983. Early Permian (Rotliegendes) palaeowinds of the North Sea. *Sedimentary Geology*, **34**, 245–265.

Hallam, A. & Sellwood, B.W. 1976. Middle Mesozoic sedimentation in relation to tectonics in the British area. *Journal of Geology*, **84**, 302–321.

Hansen, M.B., Scheck-Wenderoth, M., Hübscher, C., Lykke-Andersen, H., Dehghani, A., Hell, B. & Gajewski, D. 2007. Basin evolution of the northern part of the Northeast German Basin – insights from a 3D structural model. *Tectonophysics*, **437**, 1–16.

Heunisch, C. 1996*a*. Palynologie des Oberen Muschelkalks bis Unteren Lias [Palynology of the Upper Muschelkalk to Lower Jurassic]. *In*: Beutler, G., Heunisch, C., Luppold, F.W., Rettig, B. & Röhling, H.-G. (eds) *Muschelkalk, Keuper und Lias am Mittellandkanal bei Sehnde (Niedersachsen) und die regionale Stellung des Keupers. Geologisches Jahrbuch*, **A145**, 67–197.

Heunisch, C. 1996*b*. Palynologische Untersuchungen im oberen Keuper Nordwestdeutschlands [Palynological investigations on the upper Keuper of NW Germany]. *Neues Jahrbuch für Geologie und Paläontologie*, **200**, 87–105.

Heunisch, C. 1999. Die Bedeutung der Palynologie für Biostratigraphie und Fazies in der Germanischen Trias. *In*: Hauschke, N. & Wilde, V. (eds) *Trias – Eine ganz andere Welt*. Pfeil, München, Germany, 207–220.

Heunisch, C., Luppold, F.W., Reinhardt, L. & Röhling, H.-G. 2010. Palynofazies, Bio- und Lithostratigrafie im Grenzbereich Trias/Jura in der Bohrung Mariental 1 (Lappwaldmulde, Ostniedersachsen) [Palynofacies, bio- and lithostratigraphy of the T-J boundary interval on the example of the Mariental 1 well (Lappwald syncline, eastern Lower Saxony)]. *Zeitschrift der Deutschen Gesellschaft für Geowissenschaften*, **161**, 51–98.

Hoffmann, N., Pokorski, J., Linders, W. & Bachmann, G.H. 1997. Rotliegend stratigraphy, palaeogeography and facies in the eastern part of the Central European Basin. *In*: Podemski, M. (ed.) *Proceedings of the XIII International Congress on the Carboniferous and Permian*, 28 August–2 September 1995, Kraków, Poland, Part 2, 75–86.

Hoth, P. 1997. *Fazies und Diagenese von Präperm-Sedimenten der Geotraverse Harz – Rügen [Facies and diagenesis of pre-Permian sediments between the Harz mountains and the island of Rügen]*. Schriftenreihe für Geowissenschaften, **4**.

Jaritz, W. 1974. Zur Entstehung der Salzstrukturen Nordwestdeutschlands [On the formation of salt structures in NW Germany]. *Geologisches Jahrbuch*, **10**, 1–77.

Jaritz, W. 1987. The origin and development of salt structures in Northwest Germany. *In*: Lerche, I. & O'Brien, J.J. (eds) *Dynamical Geology of Salt and Related Structures*. Academic Press, Orlando, FL, USA, 479–493.

Jung, R., Röhling, S., Ochmann, N., Rogge, S., Schellschmidt, R., Schulz, R. & Thielmann, T. 2002. *Abschätzung des Technischen Potenzials der geothermischen Stromerzeugung und der geothermischen Kraft-Wärmekopplung (KWK) in Deutschland [Estimation of the geothermal potential for power generation and cogeneration systems in Germany]*. Gutachten im Auftrag des Deutschen Bundestag Bundesanstalt für Geowissenschaften und Rohstoffe, Hannover, Germany.

Kabus, F., Lenz, G., Wolfgramm, M., Hoffmann, F. & Kellner, T. 2003. *Studie zu den Möglichkeiten der Stromerzeugung aus hydrothermaler Geothermie in Mecklenburg-Vorpommern [Study on the potential of power generation from hydrothermal reservoirs in Mecklenburg-Vorpommern]*. GTN-Bericht 4383.

Kabus, F., Wolfgramm, M., Seibt, A., Richlak, U., Beuster, H. 2009. Aquifer thermal energy storage in Neubrandenburg – monitoring throughout three years of regular operation. *In*: Gehlin, S. (ed.) *Effstock 2009, Thermal Energy Storage for Efficiency and Sustainability: 11th International Conference on Thermal Energy Storage*, June 14–17, 2009, Stockholm, Schweden, 1–8.

Katzung, G. (ed.) 1984. *Geothermie-Atlas der Deutschen Demokratischen Republik [Geothermal atlas of the German Democratic Republic]*. Zentrales Geologisches Institut, Berlin.

Katzung, G. 2004. Geothermie [Geothermics]. *In*: Katzung, G. (ed.) *Geologie von Mecklenburg-Vorpommern*. E. Schweizerbart'sche Verlagsbuchhandlung (Nägele u. Obermiller), Stuttgart, Germany, 444–451.

Katzung, G., Diener, I. & Kühn, P. 1992. Temperaturverteilung im Untergrund Ostdeutschlands und für die Nutzung der geothermischen Ressourcen in Betracht kommenden Aquifere [Subsurface temperatures of East Germany and potential geothermal resources]. *Braunkohle*, **1992**, 27–32.

Kemper, E. 1979. Die Unterkreide Nordwestdeutschlands. Ein Uberblick [The Lower Cretaceous of NW Germany. An overview]. *In*: Wiedmann, J. (ed.) *Aspekte der Kreide Europas*. International Union of Geological Sciences Series, **A6**, 1–9.

Kley, J. & Voigt, T. 2008. Late Cretaceous intraplate thrusting in central Europe: effect of Africa–Iberia–Europe convergence, not Alpine collision. *Geology*, **36**, 839–842.

Kley, J., Franzke, H.J. *et al.* 2008. Strain and stress. *In*: Littke, R., Bayer, U., Gajewski, D. & Nelskamp, S. (eds) *Dynamics of Complex Intracontinental Basins: The Central European Basin System*. Springer, Berlin, 97–124.

Klingler, W. & Neuweiler, F. 1959. Leitende Ostracoden aus dem deutschen Lias Beta [Ostracode guide forms of the German Lias beta]. *Geologisches Jahrbuch*, **76**, 373–410.

Kockel, F. 2002. Rifting processes in NW-Germany and the German North Sea sector. *Netherlands Journal of Geosciences*, **81**, 149–158.

Kootz, G. & Schumacher, K.-H. 1967. Der Keuper im Bereich der Altmark-Südwestmecklenburg-Schwelle unter besonderer Berücksichtigung der Bohrung Salzwedel 1 [The Keuper of the Altmark-SW Mecklenburg Swell on the example of the Salzwedel 1 well]. *Jahrbuch für Geologie*, **1**, 89–117.

KOZUR, H. & BACHMANN, G.H. 2010. The Middle Carnian Wet Intermezzo of the Stuttgart Formation (Schilfsandstein), Germanic Basin. *Palaeogeography. Palaeoclimatology. Palaeoecology*, **290**, 107–119.

KUDER, J., BINOT, F., HÜBNER, W., ORILSKI, J., WONIK, T. & SCHULZ, R. 2014. Für die Geothermie wichtige hydraulische Parameter von Gesteinen des Valangin und der Bückeberg-Formation (Wealden) in Nordwestdeutschland. *Zeitschrift der deutschen Gesellschaft für Geowissenschaften*, **165**, 455–467.

LOTT, G.K., WONG, T.E., DUSAR, M., ANDSBJERG, J., MÖNNIG, E., FELDMAN-OLSZEWSKA, A. & VERREUSSEL, R.M. C.H. 2010. Jurassic. *In*: DOORNENBAL, J.C. & STEVENSON, A.G. (eds) *Petroleum Geological Atlas of the Southern Permian Basin Area*. European Association of Geoscientists and Engineers (EAGE), Houten, The Netherlands, 175–193.

LOTZ, B. 2004. *Neubewertung des rezenten Wärmestroms im Nordostdeutschen Becken [Reevaluation of the recent heat flow in the North German Basin]*. Scientific Technical Report STR 04/04.

LUND, J.J. 1977. *Rhaetic to Lower Liassic Palynology of the Onshore South-eastern North Sea Basin*. Geological Survey of Denmark, Series II, **109**.

LUND, J.J. 2003. Rhaetian to Pliensbachian palynostratigraphy of the central part of the NW German Basin exemplified by the Eitzendorf 8 well. *Courier Forschungsinstitut Senckenberg*, **241**, 69–83.

LUPPOLD, F.W. 2012. Ostracod assemblages from the Middle Jurassic of NW Germany with special reference to the Sowerbyi ammonite Zone (Early Bajocian, Jurassic). *In*: LUPPOLD, F.W. & MUTTERLOSE, J. (eds) *Marine Micropalaeontology (Foraminifera, Ostracoda) – Biostratigraphy and Taxonomy – A tribute to Helmut Bartenstein, Erich Brand and Heinz Malz*. Neues Jahrbuch für Geologie und Paläontologie, **266**, 217–238.

MALZ, H. 1966. Zur Kenntnis einiger Ostracoden-Arten der Gattungen Kinkelinella und Praeschuleridea [Contribution to some ostracod species of the genera Kinkelinella and Praeschuleridea]. *Senckenbergiana lethaea*, **47**, 385–404.

MALZ, H. 1971. Zur Taxonomie 'glattschaliger' Lias-Ostracoden [Taxonomy of smooth-shelled ostracods of the Liassic]. *Senckenbergiana lethaea*, **52**, 433–455.

MARX, J., HUEBSCHER, H.D., HOTH, K., KORICH, D. & KRAMER, W. 1995. Vulkanostratigraphie und Geochemie der Eruptivkomplexe [Stratigraphy of volcanites and geochmistry of eruptive successions]. *In*: PLEIN, E. (ed.) *Stratigraphie von Deutschland, Norddeutsches Rotliegendbecken*. Rotliegend Monographie Teil II, **183**. Courier Forschungsinstitut Senckenberg, Frankfurt, Germany, **183**, 54–83.

MIALL, A.D. 1996. *The Geology of Fluvial Deposits: Sedimentary Facies, Basin Analysis, and Petroleum Geology*. Springer, Berlin.

MICHELSEN, O. 1975. Lower Jurassic biostratigraphy and ostracods of the Danish Embayment. *Danmarks Geologiske Undersogelse*, **104**.

MÜLLER, C. & REINHOLD, K. (eds) 2011. Geologische Charakterisierung tiefliegender Speicher- und Barrierehorizonte in Deutschland – Speicher-Kataster Deutschland [Geology of reservoirs and seals in the subsurface of Germany – Reservoir Cadastre Germany]. *Schriftenreihe der deutschen Gesellschaft für Geowissenschaften*, **74**, 243.

MUTTERLOSE, J. & BORNEMANN, A. 2000. Distribution and facies patterns of Lower Cretaceous sediments in northern Germany: a review. *Cretaceous Research*, **21**, 733–759.

NÖLDEKE, W. & SCHWAB, G. 1977. Zur tektonischen Entwicklung des Tafeldeckgebirges der Norddeutsch – Polnischen Senke unter besonderer Berücksichtigung des Nordteils der DDR [On the post-variscan tectonic evolution of the North German-Polish Basin with special focus on the northern GDR]. *Zeitschrift für Angewandte Geologie*, **23**, 369–379.

OBST, K. & WOLFGRAMM, M. 2010. Geothermische, balneologische und speichergeologische Potenziale und Nutzungen des tieferen Untergrundes der Region Neubrandenburg [The deep subsurface of Neubrandenburg – geothermal potential and underground storage]. *Neubrandenburger Geologische Beiträge*, **10**, 145–174.

OGG, J.G. 2012. Triassic. *In*: GRADSTEIN, F.M., OGG, J.G., SCHMITZ, M.D. & OGG, G.M. (eds) *The Geologic Time Scale*. Elsevier, Amsterdam, 681–730.

PETTIJOHN, F.J. 1957. *Sedimentary Rocks*. Harper & Brothers, New York.

PLEIN, E. 1993. Bemerkungen zum Ablauf der palaogeographischen Entwicklung im Stefan und Rotliegend des Norddeutschen Beckens [Notes on the palaeogeographic evolution of the Southern Permian Basin in the Stefanian and Rotliegend]. *Geologisches Jahrbuch*, **A131**, 99–116.

PLEIN, E. (ed.) 1995. *Stratigraphie von Deutschland, Norddeutsches Rotliegendbecken [Stratigraphy of Germany, Southern Permian Basin]*. Rotliegend Monographie Teil II, **183**. Courier Forschungsinstitut Senckenberg, Frankfurt, Germany, **183**.

PLUMHOFF, F. 1963. *Die Ostracoden des Oberaalenium und tieferen Unterbajocium (Jura) des Gifhorner Troges, Nordwestdeutschland [Ostracods of the Upper Aalenian and basal Lower Bajocian (Jurassic) of the Gifhorn Trough, NW Germany]*. Abhandlungen der seckenbergischen naturforschenden Gesellschaft, **503**.

RAUPPACH, K., WOLFGRAMM, W., THORWART, K. & SEIBT, P. 2008. Hydraulic Features of Geothermal Aquifers in the North German Basin. *Zeitschrift für Geologische Wissenschaften*, **36**, 267–280.

REINHOLD, K., KRULL, P. & KOCKEL, F. 2008. *Salzstrukturen Norddeutschlands, Geologische Karte 1:500 000 [Salt structures of North Germany, geological map 1:500 000]*. Bundesanstalt für Geowissenschaften und Rohstoffe, Hannover, Germany.

ROCKEL, W., HOTH, P. & SEIBT, P. 1997. Charakteristik und Aufschluss geothermaler Speicher [Characterisation and development of geothermal reservoirs]. *Geowissenschaften*, **15**, 244–252.

RÖHLING, H.G. 1991. A lithostratigraphic subdivision of the Lower Triassic in the Northwest German Lowlands and the German sector of the North Sea, based on Gamma-Ray and Sonic Logs. *Geologisches Jahrbuch*, **A119**, 3–24.

RÖHLING, H.G. 2013. Der Buntsandstein im Norddeutschen Becken – regionale Besonderheiten [The Buntsandstein in the North German Basin – regional characteristics].

In: Lepper, J. & Röhling, H.G. (eds) *Stratigraphie von Deutschland XI – Buntsandstein. Schriftenreihe der Deutschen Gesellschaft für Geowissenschaften*, **69**, 269–384.

Roscher, M. & Schneider, J.W. 2006. Permocarboniferous climate: Early Pennsylvanian to Late Permian climate development of central Europe in a regional and global context. *In*: Lucas, S.G., Cassinis, G. & Schneider, J.W. (eds) *Non-Marine Permian Biostratigraphy and Biochronology*. Geological Society of London, Special Publications, **265**, 95–136, https://doi.org/10.1144/GSL.SP.2006.265.01.05

Scheck-Wenderoth, M., Krzywiec, P., Zühlke, R., Maystrenko, Y. & Froitzheim, N. 2008. Permian to Cretaceous tectonics. *In*: McCann, T. (ed.) *The Geology of Central Europe, Volume 2: Mesozoic and Cenozoic*. Geological Society, London, 999–1031.

Scholle, T. 1992. *Genese und Diagenese des Rhät + Hettang/Untersinemur im Nordosten Deutschlands – ein Beitrag zur Nutzung geothermischer Energie [Depositional environments and diagenesis of the Rhaetian and Hettangian/Lower Sinemurian of NE Germany – a contribution to the exploitation of geothermal resources]*. PhD thesis, Ernst-Moritz-Arndt-Universität Greifswald.

Schulz, R. & Röhling, H.-G. 2000. Geothermische Ressourcen in Nordwestdeutschland [Geothermal resources of NW Germany]. *Zeitschrift für Angewandte Geologie*, **46**, 122–129.

Schwab, G. 1985. *Palaomobilitat der Norddeutsch-Polnischen Senke [Tectonic evolution of the North German-Polish Basin]*. PhD thesis, Akademie der Wissenschaften der DDR.

Seemann, U. 1982. Depositional facies, diagenetic clay minerals and reservoir quality of Rotliegend sediments in the Southern Permian Basin (North Sea): a review. *Clay Minerals*, **17**, 55–67.

Seibt, P. & Kabus, F. 1997. A large-scale aquifer heat and cold storage system right in the middle of Berlin. *In*: Ochifuki, K. & Nagana, K. (eds) *Proceedings of the 7th International Conference on Thermal Energy Storage MEGASTOCK '97*, Sapporo, Japan, Volume 1, 455–460.

Simms, M.J. & Ruffell, A.H. 1989. Synchroneity of climate change and extinctions in the Late Triassic. *Geology*, **17**, 265–268.

Sivhed, U. 1980. Lower Jurassic ostracods and stratigraphy of western Skane, southern Sweden. *Sveriges geologiska undersökning*, **50**, 1–85.

Stille, H. 1924. *Grundfragen der vergleichenden Tektonik [Principles of comparative studies in tectonics]*. Borntraeger, Berlin.

Stollhofen, H., Bachmann, G. et al. 2008. Upper Rotliegend to Early Cretaceous basin development. *In*: Littke, R., Bayer, U., Gajewski, D. & Nelskamp, S. (eds) *Dynamics of Complex Sedimentary Basins. The Example of the Central European Basin System*. Springer, Berlin, 181–210.

Störmer, N. & Wienholz, E. 1965. Mikrobiostratigraphie an der Lias/Dogger-Grenze in Bohrungen nördlich der Mitteldeutschen Hauptscholle [High-resolution biostratigraphy of the Liassic/Doggerian boundary in wells north of the Cental German Block]. *Jahrbuch für Geologie*, **1**, 533–591.

Triebel, E. & Klingler, W. 1959. Neue Ostracoden-Gattungen aus dem deutschen Lias [New ostracod genera of the German Lias]. *Geologisches Jahrbuch*, **76**, 335–372.

Trusheim, F. 1957. Über Halokinese und ihre Bedeutung für die strukturelle Entwicklung Norddeutschlands [Halotectonics and their importance on the structural evolution of North Germany]. *Zeitschrift der Deutschen Gesellschaft für Geowissenschaften*, **109**, 111–151.

Underhill, J.R. & Partington, M.A. 1993. Jurassic thermal doming and deflation in the North Sea: implication of the sequence stratigraphic evidence. *In*: Parker, J.R. (ed.) *Petroleum geology of Northwest Europe – Proceedings of the 4th Conference*. Geological Society, London, 337–346, https://doi.org/10.1144/0040337

Vejbæk, O.V., Andersen, C. et al. 2010. Cretaceous. *In*: Doornenbal, J.C. & Stevenson, A.G. (eds) *Petroleum Geological Atlas of the Southern Permian Basin Area*. European Association of Geoscientists and Engineers (EAGE), Houten, The Netherlands, 195–209.

Visscher, H., Van Houte, M., Brugman, W.A. & Poort, P.R. 1994. Rejection of a Carnian (Late Triassic) 'pluvial event' in Europe. *Rev. Palaeobot. Palynol.*, **83**, 217–226.

Voigt, S., Wagreich, M. et al. 2008. Cretaceous. *In*: McCann, T. (ed.) *The Geology of Central Europe, Volume 2: Mesozoic and Cenozoic*. Geological Society, London, 923–995.

Wolfgramm, M. 2005. Fluidentwicklung und Diagenese im Norddeutschen Becken – Petrographie, Mikrothermometrie und Geochemie stabiler Isotope [Fluids and diagenesis of the North German Basin – Petrography, microtermometry and geochemistry of stable isotopes]. *Hallesches Jahrbuch, B*, Beiheft, **20**, 1–143.

Wolfgramm, M., Rauppach, K. & Seibt, P. 2008. Reservoir-geological characterization of Mesozoic sandstones in the North German Basin by petrophysical and petrographical data. *Zeitschrift für Geologische Wissenschaften*, **36**, 249–265.

Wolfgramm, M., Karp, T., Lenz, G., Thorwart, K., Taugs, R. & Wessel, K. 2010. Tiefengeothermische Erkundung in Hamburg-Wilhelmsburg [Exploration of deep geothermal reservoirs of Hamburg-Wilhelmsburg]. *Paper presented at the Der Geothermiekongress DGK 2010*, 17–19 November 2010, Karlsruhe, 1–10.

Wolfgramm, M., Franz, M. & Agemar, T. 2014. Explorationsstrategie tiefer geothermischer Ressourcen am Beispiel des Norddeutschen Beckens [Exploration strategy of deep geothermal Resources on the Example of the North German Basin]. *In*: Bauer, M., Freeden, W., Jacobi, H. & Neu, T. (eds) *Handbuch Tiefe Geothermie*. Springer, Berlin, 463–505.

Wormbs, J., Diener, I. & Pasternak, G. 1988. *Geothermische Ressourcen im Nordteil der DDR (I) [Geothermal resources of the northern GDR (I)]*. Zentrales Geologisches Institut, Berlin.

Ziegenhardt, W., Schön, M. & Gilch, M. 1980. Einige Ergebnisse der strukturgeologischen Erkundung von Untergrundspeichern [Tectonic exploration of underground storage sites – some results]. *Zeitschrift für Angewandte Geologie*, **26**, 165–171.

Ziegler, P.A. 1982. Triassic rifts and facies patterns in Western and Central Europe. *Geologische Rundschau*, **71**, 747–772.

Ziegler, P.A. 1990. *Geological Atlas of Western and Central Europe*. Shell Internationale Petroleum Maatschappij, The Hague. Geological Society, London.

Zimmermann, J. 2015. *The Lower and Middle Jurassic System in the North German Basin: sequence stratigraphy, depositional environments and basin evolution*. PhD thesis, TU Bergakademie Freiberg.

Zimmermann, J., Franz, M. & Wolfgramm, M. 2014. The Late Aalenian Polyplocussandstein Formation in SE Lower Saxony, Germany: meandering distributary and crevasse splay sedimentation on a lower deltaic plain. *Neues Jahrbuch für Geologie und Paläontologie, Abhandlungen*, **271**, 69–94.

Zimmermann, J., Franz, M., Heunisch, C., Luppold, F.-W., Mönnig, E. & Wolfgramm, M. 2015. Sequence stratigraphic framework of the Lower and Middle Jurassic in the Central European Basin: epicontinental sequences controlled by Boreal cycles. *Palaeogeography, Palaeoclimatology, Palaeoecology*, **440**, 395–416.

Zimmermann, J., Franz, M., Schaller, A. & Wolfgramm, M. 2018. The Toarcian–Bajocian deltaic system in the North German Basin: Subsurface mapping of ancient deltas – morphology, evolution and controls. *Sedimentology*, **65**, 897–930, https://doi.org/10.1111/sed.12410

Towards a better understanding of the highly overpressured Lower Triassic Bunter reservoir rocks in the Terschelling Basin

STEFAN PEETERS[1,2]*, ANNEMIEK ASSCHERT[3] & HANNEKE VERWEIJ[4]

[1]*Department of Earth Sciences, Utrecht University, Heidelberglaan 8, 3584CS Utrecht, The Netherlands*

[2]*Present address: Sweco Nederland BV, De Holle Bilt 22, 3732HM De Bilt, The Netherlands*

[3]*EBN BV, Daalsesingel 1, 3511SV Utrecht, The Netherlands*

[4]*TNO, Princetonlaan 6, 3584CB Utrecht, The Netherlands*

**Correspondence: stefan.peeters@sweco.nl*

Abstract: Large lateral variations in pore-fluid pressures and associated overpressures occur in the compartmentalized Lower Triassic Bunter reservoir rocks in the Terschelling Basin (northern Dutch offshore). This study describes the main controlling factors on the distribution of these overpressures by combining seismically derived structural interpretations with pressure data. The most important controlling factor is considered to be the presence and type of both lateral and vertical permeability barriers. Lateral permeability barriers are formed by Zechstein salt and faults. The most effective vertical permeability barrier for maintaining high overpressures is the Upper Triassic Röt evaporite. Overpressures of more than 30 MPa are observed in the Bunter compartments where the Röt is present and laterally continuous. Overpressures are lower than 15 MPa in compartments where the Röt was affected by Mid-Cimmerian erosion. Based on structural observations, it is concluded that the highly overpressured compartments became fully hydraulically restricted during the Late Jurassic and Early Cretaceous. Calculations show that subsequent sedimentary loading, during the Cretaceous and Cenozoic, was able to generate the observed overpressures.

Previous work, from the 1980s onwards, has led to a regional geohydraulic characterization of the Dutch subsurface (Verweij 1990, 2003; Verweij & Simmelink 2002; Verweij *et al.* 2012), providing a regional framework in which pore-fluid pressures can be explained and estimated. However, large lateral differences in pore-fluid formation pressures and associated overpressures may occur locally in the Dutch subsurface (where overpressure in this paper is defined as the excess pore-fluid pressure above hydrostatic pressure for a water density of 1020 kg m^{-3}). Such large lateral differences in overpressures occur in the Lower Triassic Bunter reservoir rocks in the Terschelling Basin, where overpressures vary between 10 MPa and 40 MPa (Verweij *et al.* 2012) (Fig. 1). Previous studies have concluded that the presence of permeability barriers plays an important role in the maintenance of such high fluid overpressures in the subsurface (Hubbert 1940; Verweij 2003). Permeability barriers are known to inhibit fluid flow (Hubbert 1940), which means that generated overpressures (e.g. due to an increase in stress or due to a relative increase in fluid volume) are able to persist (Mann & Mackenzie 1990; Neuzil 1995; Osborne & Swarbrick 1997; Verweij *et al.* 2012). In the Terschelling Basin, vertical permeability barriers are formed by Zechstein salt and Triassic evaporites (such as the Röt and Muschelkalk evaporites: Verweij *et al.* 2012). Additionally, lateral permeability barriers formed by Zechstein salt walls and faults have created hydraulically restricted compartments in the Lower Triassic Bunter strata. A 2D basin-modelling study performed by Nelskamp *et al.* (2012) concluded that the formation time of these permeability barriers is of key importance for the generation of overpressures in the Dutch North Sea, including the Bunter compartments in the study area. However, the specific timing of the formation of local permeability barriers in the Terschelling Basin, as well as the exact distribution of overpressures, remains largely unknown.

This study aims to gain better insight into the cause of overpressures in the Lower Triassic Bunter strata in this basin and, in particular, to better understand the factors that control the distribution of these overpressures. This is important for the prediction of gas volumes and for assessing drilling risks for further hydrocarbon exploration. The aims of this study were achieved by:

- The identification of Bunter compartments via seismic mapping;
- The execution of pressure analysis utilizing data from the Southern North Sea (SNS) pressure database;

From: Kilhams, B., Kukla, P. A., Mazur, S., McKie, T., Mijnlieff, H. F. & van Ojik, K. (eds) 2018. *Mesozoic Resource Potential in the Southern Permian Basin*. Geological Society, London, Special Publications, **469**, 223–236.
First published online March 22, 2018, https://doi.org/10.1144/SP469.13

(a)

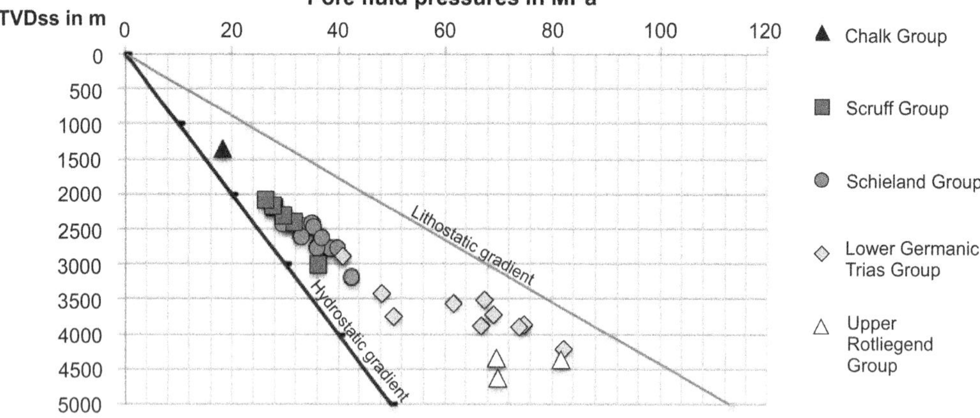

(b)

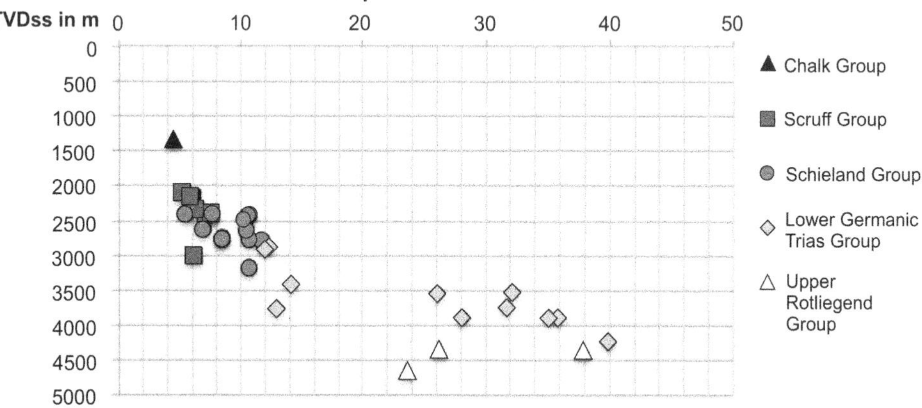

Fig. 1. Illustration of the large variations in pore-fluid formation pressures (**a**) and associated overpressures (**b**) in the Lower Germanic Trias Group in the Terschelling Basin. The hydrostatic seawater gradient represents a seawater density of 1020 kg m^{-3}. The lithostatic gradient represents an average rock density of 2300 kg m^{-3}. TVDss, true vertical depth subsea.

- The interpretation of the main permeability boundaries in Bunter compartments via seismic interpretation, with the purpose of estimating the hydraulic closure time of each structure;
- The calculation of the amount of overpressure that could have been generated by sedimentary loading since the time of hydraulic closure.

Geological history and stratigraphy

During the Permian, the Terschelling Basin and surrounding areas were part of the large east–west trending Southern Permian Basin that stretched from the current British Isles to Poland (Ziegler 1990; Geluk 2005; Geluk 2007; Doornenbal & Stevenson 2010). Seawaters invaded this landlocked basin episodically and led to the cyclic deposition of the evaporites of the Zechstein Group (Ziegler 1990). The end of the Permian and the largest part of the Early Triassic were characterized by tectonic quiescence and generally gentle thermal subsidence (Brückner-Röhling & Röhling 1998; Geluk 2005). The strata of the Lower Germanic Trias Group were deposited under these tectonic conditions. The Lower Germanic Trias Group comprises the Main Buntsandstein Subgroup, which is divided into four formations: the Lower Buntsandstein Formation, the Volpriehausen Formation, the Detfurth Formation and the Hardegsen Formation (Van Adrichem Boogaert & Kouwe 1993; Geluk 2005; Feist-Burkhardt *et al.* 2008) (Fig. 2). The sediments of

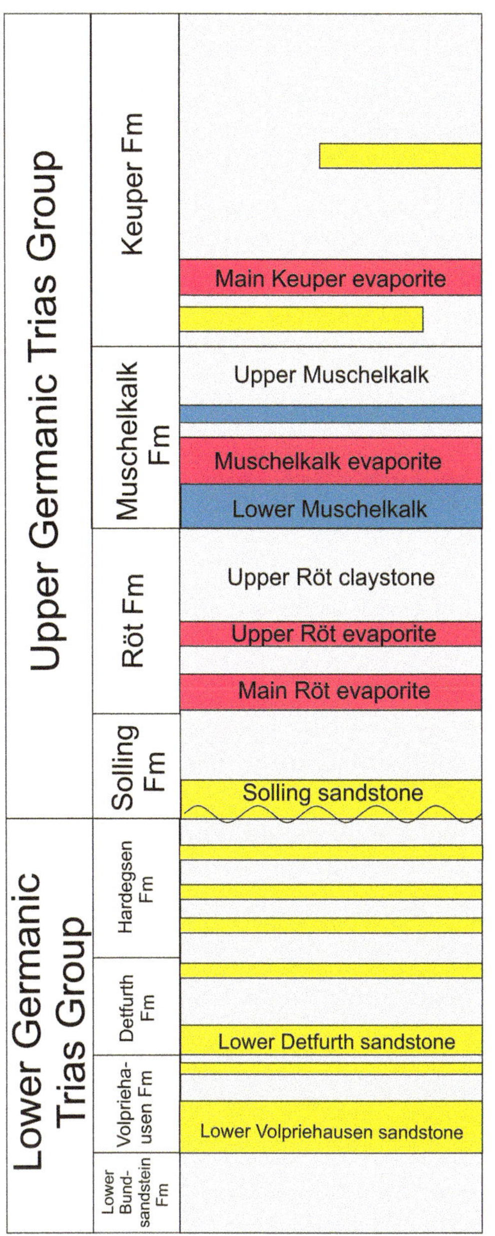

Fig. 2. Schematic diagram indicating the stratigraphy of the Triassic interval in the Terschelling Basin (yellow, sandstone-dominated intervals; grey, silt- and claystone-dominated intervals; pink, evaporite-dominated intervals; blue, limestone-dominated intervals).

these formations were mostly deposited in a lacustrine and fluvial setting with aeolian influences (McKie & Williams 2009; McKie 2014). Each of these formations typically consists of a basal sandstone interval, which fines upwards into siltstones and claystones (Röhling 1991; Geluk 2005; McKie 2014). The thickest and laterally most widespread sandstone unit in the Terschelling Basin is the Volpriehausen Formation (Geluk 2005). However, in this study, pressure data are included for all four formations of the Main Buntsandstein Subgroup.

During the Early Triassic, break-up of the Pangea continent initiated in the North Atlantic (Steel & Ryseth 1990; Ziegler 1990; Geluk 2005). A branch of this rift system extended southwards into the current Southern North Sea area and caused east–west rifting during the Middle and Late Triassic, thereby creating north–south-trending faults (Ziegler 1990; Geluk 2005). Several active rifting pulses occurred in the Southern North Sea area during this time period. Each rifting phase was preceded by a thermal heating event, causing initial uplift, followed by a period of tectonic quiescence due to regional thermal subsidence (Glennie 1986; Brückner-Röhling & Röhling 1998; Geluk 2005). The first hiatus and associated unconformity related to a thermal heating event is the Hardegsen Unconformity, which marks the boundary between the Lower and Upper Germanic Trias Group (Röhling 1991; Van Adrichem Boogaert & Kouwe 1993). The Upper Germanic Trias Group comprises sediments of the Solling, Röt, Muschelkalk and Keuper formations (Van Adrichem Boogaert & Kouwe 1993). During deposition of the Upper Germanic Trias Group, conditions gradually shifted from continental to marine (Kovalevych *et al.* 2002; McKie 2017; Geluk *et al.* 2018). Several times, the area again became land locked, which resulted in the deposition of the Röt, Muschelkalk and Keuper evaporites (Ruffell *et al.* 2015; McKie 2017). Apart from these evaporites, the Upper Germanic Trias Group mainly consists of shales and carbonate sequences (Ziegler 1990; Geluk 2005).

The strongest Mesozoic rifting pulse in the Southern North Sea region occurred during the Middle Jurassic (158–154 Ma), which was preceded by widespread uplift and subsequent erosion, leading to the formation of the Mid-Cimmerian Unconformity (Ziegler 1990; Duin *et al.* 2006). This event was followed by a period of NE–SW extension during the Late Jurassic and Early Cretaceous, forming new NW–SE-trending faults but also reactivating older Variscan fault trends (Rattey & Hayward 1993; Duin *et al.* 2006; de Jager & Geluk 2007). Crustal extension was strongest in the structural lows bounded by older NW–SE alignments, such as the Terschelling Basin and West Netherlands Basin (Duin *et al.* 2006; Kombrink *et al.* 2012). Additionally, significant subsidence occurred locally in the Dutch Central Graben during this time period (Munsterman *et al.* 2012; Bouroullec *et al.* 2018; Verreussel *et al.* 2018). In these subsiding basins, Upper Jurassic sediments of the Schieland and

Scruff groups were deposited, as well as Lower Cretaceous sediments of the Rijnland Group (Munsterman *et al.* 2012). These sediments were deposited during mixed coastal and marine conditions, with clastics being sourced from the rift shoulders (Munsterman *et al.* 2012; Zwaan 2018). Towards the Middle Cretaceous, active rifting waned in the Southern North Sea area and, due to ongoing thermal subsidence, it evolved into a large open shallow-marine basin (Ziegler 1990; Duin *et al.* 2006). The area continued to subside in the Cenozoic but, from the early part of the Late Cretaceous, compressional stresses from the Alpine Orogeny were increasingly transmitted into the Southern North Sea area, thereby inverting older Variscan and Mesozoic faults (Duin *et al.* 2006; de Jager & Geluk 2007; Reicherter *et al.* 2008).

Results

Identification of Bunter compartments

In total, 19 Lower Triassic Bunter compartments were identified in the Terschelling Basin (Fig. 3). The identification of these compartments was based on the top of the overlying Röt evaporite layer, which displays a moderate decrease in acoustic impedance with respect to the overlying Upper Röt Claystone Member. The compartments are separated from each other by Zechstein salt walls and salt diapirs towards the ESE and WNW, and by faults towards the NNE and SSW (Fig. 3). Zechstein evaporites form the base of the compartments. The Röt evaporite layer and the underlying Bunter compartments display a general trend of increasing depth towards the west of the Terschelling Basin, in the direction of the Dutch Central Graben (Fig. 3). The Röt evaporite is present all across the central and western part of the Terschelling Basin and is, therefore, continuously overlying the Bunter reservoir rocks in those areas. The Röt evaporite is partly absent in the eastern part of the Terschelling Basin, where it was eroded during the Mid-Cimmerian uplift. Instead, the Bunter reservoir rocks are overlain by Upper Jurassic shales of the Friese Front Member in the eastern parts of compartments 4, 11 and 15 (Fig. 3).

Pressure analysis

Reliable pressure measurements from drill stem tests (DSTs) and wireline formation tests (WFTs) indicate the pressure regime in six of the 19 Bunter compartments (Fig. 4). The other 13 compartments contain either unreliable test data, mud weight data or no data at all. In order to distinguish between reliable and unreliable pressure measurements, a quality coding developed by Hortle *et al.* (2013) was utilized. Each test type has its own coding system, where the reliability of the performed pressure measurement results from the cumulative answer on a set of questions (Hortle *et al.* 2013). For DSTs, the questions address the following factors: any mechanical problems during the test, the availability of raw data, the stabilization of build-up and the location of gauge relative to the test interval (Hortle *et al.* 2013). For WFTs, the following factors are addressed: any mechanical problems during the test, the difference between the initial and final hydrostatic pressure, the trendline of measured pressures, the permeability of the formation of interest, and the fluctuations in temperature during pressure measurement (Hortle *et al.* 2013). It is proposed that the large number of unreliable pressure measurements in the Bunter reservoir rocks (Fig. 4) is caused by reduced reservoir quality due to widespread secondary salt cementation. This salt plugging inhibits the free flow of reservoir fluids into the well bore, making it difficult to perform reliable pressure tests. The structural characteristics of the six compartments that contain reliable pressure measurements were used to study the main controlling factors on the distribution of overpressures. It was found that these factors are:

- The continuous presence and type of permeability barriers – in compartments where the Röt evaporite is continuously present, overpressures are generally higher than 30 MPa. In compartments where Jurassic shales overlie the Bunter reservoir rocks, overpressures are less than 15 MPa (Fig. 4).
- The depth of the Bunter reservoir rocks – it is observed that the deeper the Bunter reservoir rocks, the higher the overpressure regime in the associated compartments (Figs 1, 3 & 4).
- The location of the Bunter reservoir rocks with respect to the adjacent platforms – generally, compartments that are closer to adjacent platform areas in the north, south and east display lower overpressures (Fig. 4);

To estimate the pressure regime of undrilled compartments, all areas were grouped into five classes based on structural analogies with compartments that contain reliable pressure measurements (Fig. 4). The areas in class 1 are situated on the structural highs towards the south of the Terschelling Basin and are hydraulically the most open, because they are partly overlain by Lower Cretaceous claystones and sandstones, and are lacking the Triassic and Jurassic permeability barriers (Fig. 4). Overpressures in these areas are lower than 10 MPa. The Bunter compartments of class 2 are situated in the eastern part of the Terschelling Basin (Fig. 4). The Bunter reservoir rocks in these compartments are partly overlain by Upper Jurassic shales, and display overpressures between of 10 MPa and 15 MPa. The Lower Triassic Bunter strata of classes 3, 4 and 5 are continuously overlain by the Röt evaporite and

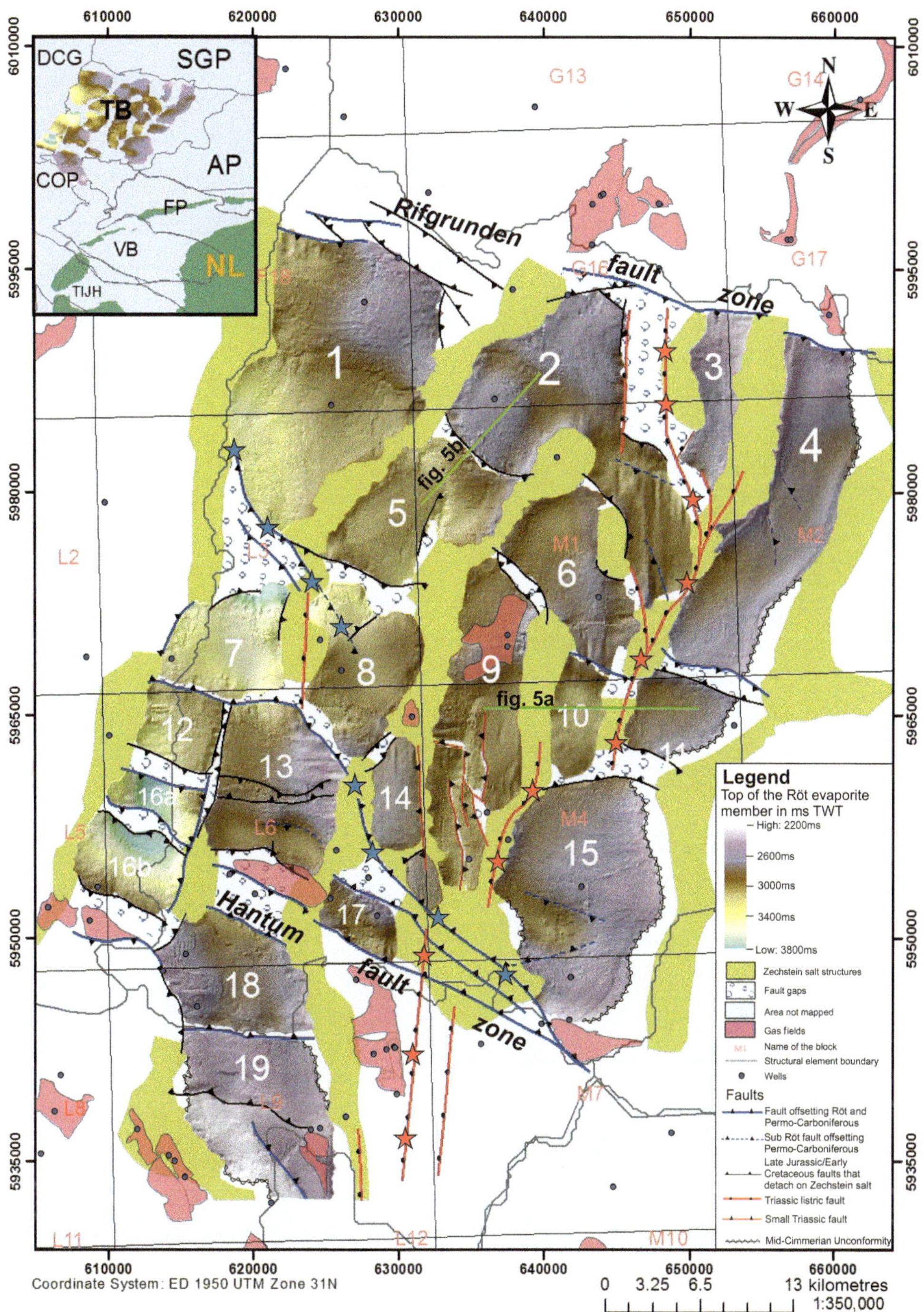

Fig. 3. Two-way time (TWT) map showing the top of the Röt evaporite and thereby indicating the location of the underlying Bunter compartments. Yellow areas represent Zechstein salt walls. The transparent red areas indicate gas fields. The locations of the compartments with respect to the northern Dutch onshore and the regional structural elements are indicated in the inset. Abbreviations: AP, Ameland Platform; COP, Central Offshore Platform; DCG, Dutch Central Graben; FP, Friesland Platform; TB, Terschelling Basin; TIJH, Texel IJsselmeer High; SGP, Schill Grund Platform; VB, Vlieland Basin.

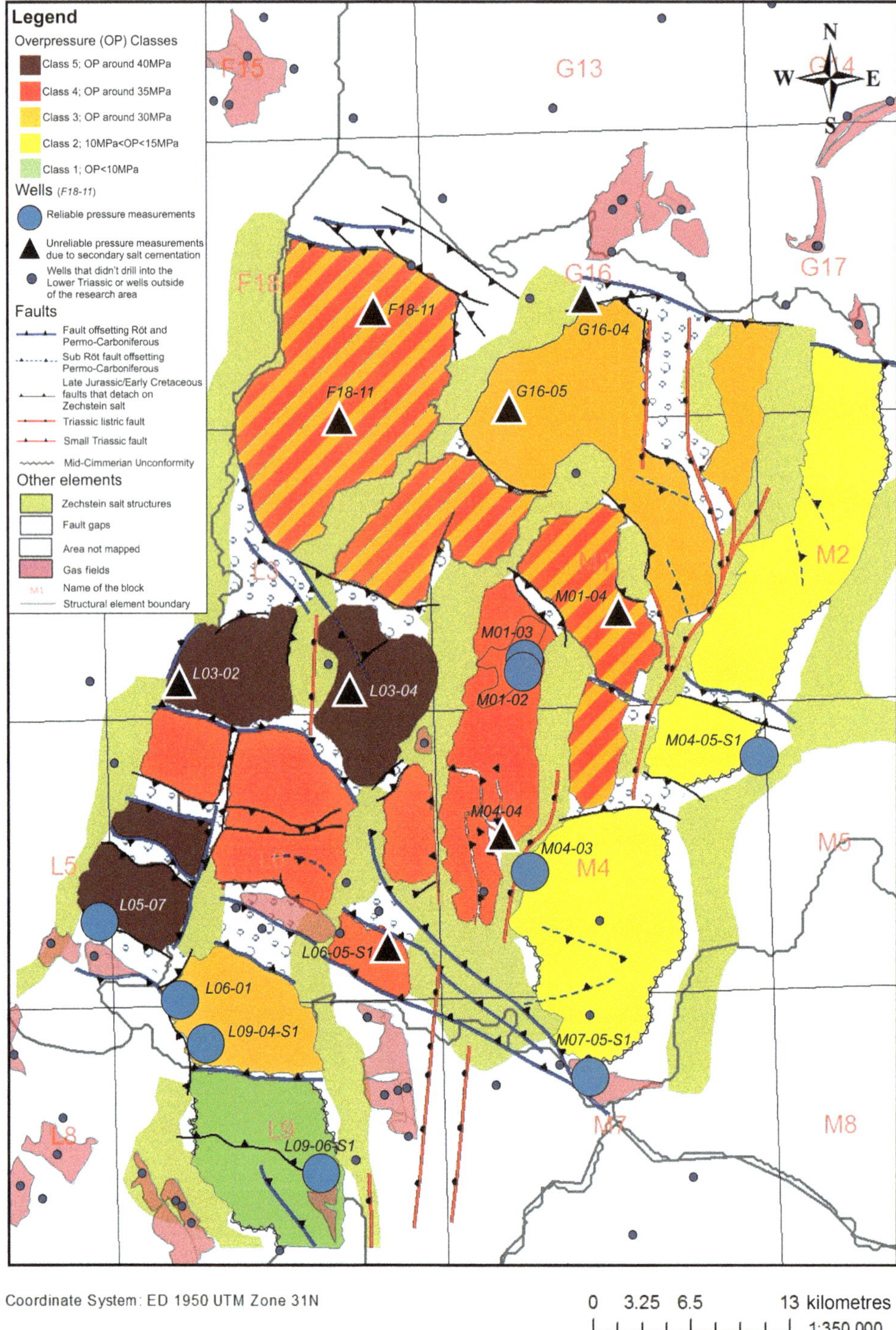

Fig. 4. Map indicating the (estimated) pressure regime of the Bunter reservoir rocks in the Terschelling Basin. The blue circles indicate wells where reliable pressure measurements were obtained. The triangles indicate wells where no reliable pressure measurements were obtained because the reservoir was affected by secondary salt cementation. The hatched orange/red compartments contain characteristics of both class 3 and class 4 compartments.

display overpressures of around 30 MPa in class 3, around 35 MPa in class 4 and up to 40 MPa in class 5 (Fig. 4). The position of the Bunter compartments within the Terschelling Basin and the depth of the Bunter reservoir rocks further distinguishes the compartments of classes 3, 4 and 5. Class 3 compartments are located at the edges of the Terschelling Basin, where they are bounded by (relatively) small-scale faults that separate them from the structural highs. The compartments of class 4 and 5, on the other hand, are either located in the centre of the Terschelling Basin, or they are separated from the structural highs via a major fault system (e.g. compartment 17). The compartments in class 5 are situated deeper in the subsurface than the compartments in class 4 and are hydraulically the most restricted (Figs 3 & 4).

Main permeability boundaries

Lateral permeability barriers of Bunter compartments are observed to be formed by Zechstein salt structures and faults (Figs 3 & 4). Insight into the timing of formation of these lateral permeability barriers was obtained from seismically derived structural observations, leading to the understanding that the Bunter compartments became hydraulically fully restricted in three phases:

- In the centre of the Terschelling Basin, a NNE–SSW-trending zone is present where the Bunter compartments have drifted away from each other (Fig. 3, indicated by small red stars). The overlying Solling-, Röt- and Muschelkalk formations display a significant thickening into the salt structures towards the west of this zone (Fig. 5a, indicated by the curved red arrow). Additionally, the reflectors of these formations display a downlapping onto the underlying Zechstein salt (Fig. 5a, indicated by short red arrows). This zone has been interpreted as a zone of suprasalt, shallow-rooted listric faults that developed in response to east–west regional extension. Stratigraphic thickening related to these listric faults indicates that, in the SE part of compartment 2 and in the eastern part of compartment 17, fault movements occurred during deposition of the Solling Formation. This leads to the local development of the so-called 'Solling Fat Sands' (*sensu* de Jager 2012). From these localities, fault activity has spread northwards and southwards during Röt, Muschelkalk and also, probably, Keuper times. It is proposed that activity along these faults caused some compartments (i.e. compartments 2, 6, 7, 9, 10, 14 and 17) to become hydraulically restricted at their eastern boundary during the Röt and Muschelkalk periods.
- Zechstein salt structures formed in the Terschelling Basin in response to Triassic east–west regional extension and associated faulting. The evolution of these salt structures affected the depositional stratal patterns of the Muschelkalk and Keuper formations. Strata of the Muschelkalk Evaporite Member and Upper Muschelkalk Member generally thin towards the salt structures (Fig. 5a, indicated with black V shapes), demonstrating that the salt structures were in a reactive stage (Vendeville & Jackson 1992). Strata of the Keuper Formation generally thicken towards the salt structures (Fig. 5a, indicated by the curved black arrow), thereby creating a rim syncline. This indicates that the salt structures were in an active piercing stage during deposition of the Keuper Formation (Vendeville & Jackson 1992). This suggests that all compartments were closed towards the east and the west during the Late Triassic Keuper Period.
- The NW–SE-trending faults that bound the compartments towards the NE and SW display a significant thickening of the Scruff, Schieland, and Rijnland groups in the downthrown block (Fig. 5b, indicated by the blue arrow). This indicates that these faults were active during the Late Jurassic and Early Cretaceous. Some of the NW–SE-trending faults are shallow-rooted faults and detach on the Zechstein salt (Fig. 5b), while others penetrate the Zechstein and also offset Permo-Carboniferous strata. The latter is the case at the Rifgrunden and Hantum fault zones that bound the Terschelling Basin in the north and south, respectively (Fig. 3). Additionally, a large NW–SE-trending basement-rooted fault zone is present in the centre of the Terschelling Basin that merges into a north–south trend towards the Dutch Central Graben (Fig. 3, indicated by small blue stars). The basement-rooted fault zones are likely to have formed during Variscan times, and were reactivated during the Late Jurassic and Early Cretaceous in response to regional NE–SW extension. The same regional NE–SW extension also formed new, mainly shallow-rooted, faults. Overall, activity along the NW–SE-trending faults caused the compartments to close in the north and south during the Late Jurassic and the first part of the Early Cretaceous.

Based on the timing of these three phases, the compartments that are bounded on all sides by permeability barriers became fully hydraulically restricted during the Late Jurassic and first part of the Early Cretaceous.

Overpressure calculations and comparisons

The contribution of sedimentary load to observed overpressure can be determined by calculating the part of that load that is carried by the pore fluids. It is recognized that important limiting assumptions

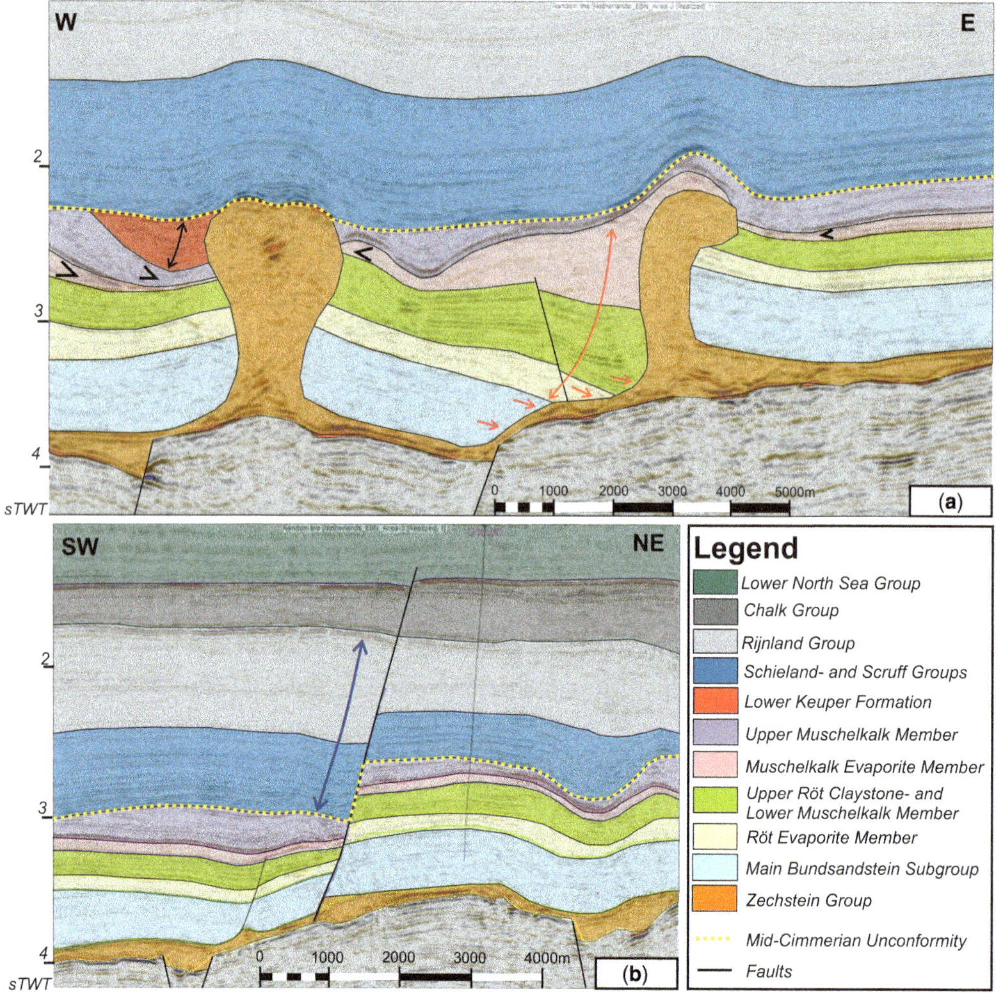

Fig. 5. Seismic sections through the Terschelling Basin showing two Zechstein salt walls and a Triassic listric fault, where the thick synkinematic strata are indicated with the curved red arrow (**a**) and a Late Jurassic–Early Cretaceous NW–SE-trending fault (**b**). The overpressured Bunter reservoir rocks are part of the Main Buntsandstein Subgroup (indicated in light blue). The location of the seismic lines is indicated in Figure 3.

in the calculation are that no fluid flow occurred after the hydraulic closure time of the compartment (i.e. from the Cretaceous onwards), that the fluid-filled Bunter units behave poroelastically and that loading only occurs in a vertical sense. The following formula was used to determine the maximum amount of overpressure that can be generated by sedimentary loading from the time of hydraulic closure (Verweij 2003):

$$P_{\text{excess}} = (P_{\text{lithostatic}}\beta) - P_{\text{hydrostatic}} \quad (1)$$

where P_{excess} is the maximum amount of overpressure (in MPa) that can be generated by sedimentary loading, $P_{\text{lithostatic}}$ is the lithostatic pressure (in MPa), $P_{\text{hydrostatic}}$ is the hydrostatic pressure (in MPa) and β is the pore pressure coefficient or Skempton's coefficient (Skempton 1954). This coefficient describes the part of the incremental load caused by the overburden (sedimentation + water column for offshore) that is carried by the pore fluids for a particular formation of interest. The pore-pressure coefficient depends on the vertical compressibility of the rock matrix (bulk modulus of the rock), the porosity of the rock and the compressibility of the pore fluid (bulk modulus of the fluid) (Skempton 1954; Domenico & Schwartz 1998; Flemings *et al.* 2002; Verweij 2003). A pore-pressure coefficient

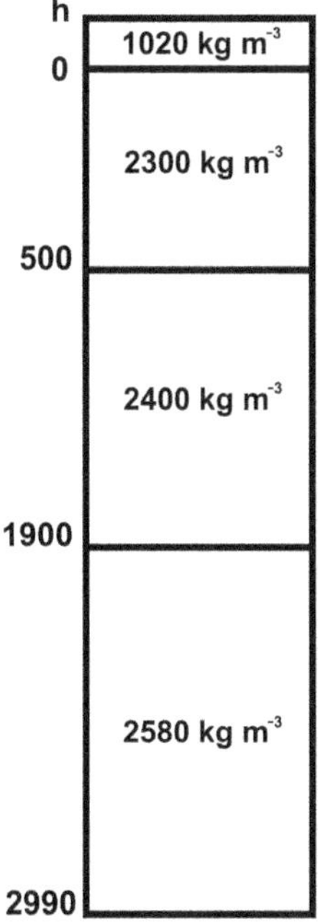

Fig. 6. Schematic diagram showing the different bulk densities that were used to calculate the contribution of Cretaceous and Cenozoic sedimentary loading to the observed overpressures.

of 1 would indicate that all of the incremental load caused by sedimentation is carried by the pore fluids (Verweij 2003). In general, the pore-pressure coefficient tends to decrease with increasing compaction during burial, typically to values below 0.9 (Verweij 2003). For the Triassic reservoir rocks, a pore-pressure coefficient of 0.85 is taken (Verweij 2003). When the expressions for the lithostatic ($P_{\mathrm{lithostatic}}$) and the hydrostatic ($P_{\mathrm{hydrostatic}}$) pressure in equation (1) are rewritten, the following expression is obtained:

$$P_{\mathrm{excess}} = \left(\sum_{4} \int_{Z1}^{z2} (\rho_{\mathrm{bulk}}(z) g \, \mathrm{d}z) + \rho_{\mathrm{w}} g h \right) \beta - \rho_{\mathrm{w}} g z \qquad (2)$$

where ρ_{bulk} is the bulk density (in kg m^{-3}) (i.e. the combined density of both rock and fluids that fill the pore spaces), g is the gravitational constant (in m s^{-2}) (=9.81 m s^{-2}), ρ_{w} is the seawater density (=1020 kg m^{-3}), h is the seawater level above the rock column (in m) (=35 m), β is the pore-pressure coefficient and z is the depth (in m). Because the bulk density differs for different rock types and generally also increases with depth, the total $P_{\mathrm{lithostatic}}$ caused by the overburden is calculated for a few main intervals (characterized by approximately the same bulk density), which are then summated. Using bulk density logs from the wells F18-11, M01-03, L05-07 and L06-1 (see Fig. 4 for the well locations), overpressures were calculated that could be generated by Cretaceous and Cenozoic sedimentary loading for the compartments 1, 9, 16 and 18. The bulk density log of well L06-01 in compartment 18 is shown in Figure 6 as an example. Using equation (2), the bulk density log of well L06-01 and the above stated parameter values, an overpressure of 31.4 MPa is calculated. This value is slightly higher than the observed 28.1 MPa of overpressure for this well, which was obtained from reliable WFTs. Calculations for the other three wells also resulted in overpressures generated by Cretaceous and Cenozoic loading that were slightly higher (*c.* 3 MPa) than the observed overpressures. This suggests that Cretaceous and Cenozoic sedimentary loading was capable of generating the observed overpressures assuming that the limiting assumptions and conditions underlying equation (2) are fulfilled.

Discussion

Secondary salt cementation, conceptual model on its formation

Secondary salt cementation, or salt plugging, is a widespread phenomenon in the Bunter reservoir rocks in the Terschelling Basin and has a significant impact on the hydraulic conditions of the Bunter reservoirs (Dronkert & Remmelts 1996; Mauthe 2002; Wever *et al.* 2010). Given the amount and widespread nature of secondary salt cementation in the area (Fig. 4), it is suggested that such cementation could only have occurred under (brief) hydraulically open conditions: that is, when overpressured fluids were able to escape (Mauthe 2002). It is proposed that a regional outburst of fluids would have decreased the fluid pressures and the solubility of the chemical components in the migrating reservoir fluids, leading to widespread precipitation of minerals and also to salt cementation of the reservoirs. It is possible that such hydraulically open conditions developed due to Mid-Cimmerian uplift and erosion during the Late Jurassic (Fig. 7). During this time, regional uplift caused erosion into Upper Triassic strata on the adjacent platform areas towards the north and south of the Terschelling Basin, providing a fluid-escape pathway towards the north and south of the basin. A prerequisite for this scenario is that the Bunter reservoir rocks were already overpressured before the Mid-Cimmerian. As described here, structural observations indicate that the Bunter reservoir rocks were hydraulically restricted towards the east and west from the Triassic Keuper period

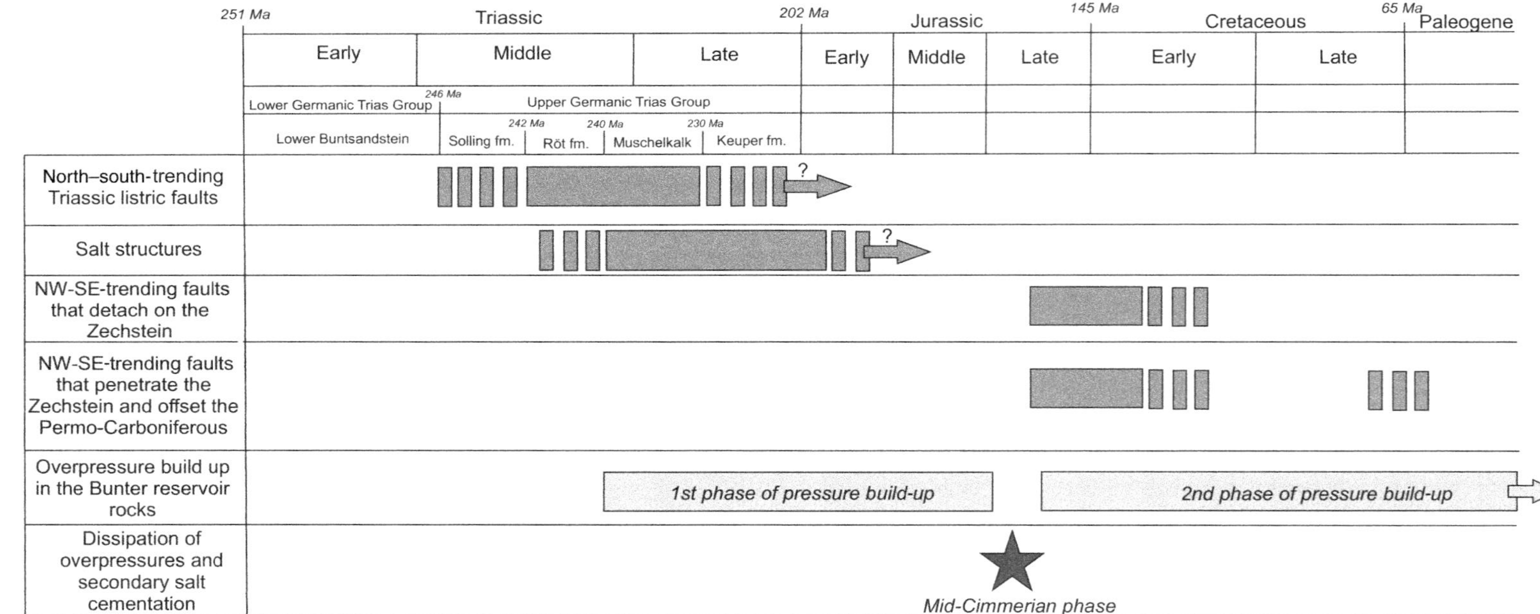

Fig. 7. Schematic diagram indicating the time that the lateral permeability barriers of the Bunter reservoir rocks were active. The diagram also indicates the conceptual model of two phases of overpressure build-up, as well as the Mid-Cimmerian phase of uplift and associated regional secondary salt cementation.

onwards, making it plausible that overpressures had built up during the Late Triassic and Early Jurassic (Fig. 7). This is supported by the basin modelling of overpressure generation along a 2D cross-section in the Terschelling Basin by Nelskamp *et al.* (2012). After the hydraulic opening of the system during the Mid-Cimmerian uplift, the northern and southern permeability boundaries of the compartments developed in response to regional NE–SW extension during the Late Jurassic and Early Cretaceous. From this time onwards, overpressures could build up again. However, available pore space for fluids (including hydrocarbons) was likely to have been considerably lowered due to salt cementation by that time. The observation that Bunter reservoir properties are generally better in the hydraulically more open compartments (Fig. 4) could be the result of later dissolution of secondary salt. This would have been possible in the hydraulically more open compartments due to the fact that the Röt evaporite is partly absent here. Consequently, water would have been able to percolate through the system to some extent during later periods and (partly) wash out the salt from the pores.

Cause of overpressures

It is recognized that the contribution of the sedimentary load to observed overpressures was determined in this study by making various assumptions and simplifications. For example, sedimentary loading started after the hydraulic closure of the compartments that were estimated based on seismically derived structural observations which, by definition, involves subjective interpretation. In addition, the applied equations (1) and (2) are based on limiting assumptions and conditions, and the input parameters, such as bulk density and pore-pressure coefficient, are generalized values subject to uncertainty. The choice of pore-pressure coefficient (β) also has a significant impact on the outcome of the calculation. A pore-pressure coefficient of 0.85 represents a water-saturated sandstone reservoir of relatively good porosity. However, the presence of gas as a pore fluid reduces the pore-pressure coefficient because gas is more compressible than water. Also, a significant part of the Bunter reservoirs are salt-plugged, which influences the mechanical and elastic behaviour, and, subsequently, reduces the pore-pressure coefficient. As such, the calculated overpressure due to sedimentary loading should be considered to be a maximum value.

It is also recognized that sedimentary loading is not the only pressure-generating mechanism that has operated since the beginning of compartment hydraulic closure. The Terschelling Basin underwent a certain amount of shortening due to Late Cretaceous inversion (Duin *et al.* 2006; de Jager & Geluk 2007). Such a transition from a tensile to a compressive tectonic stress regime is accompanied by changes in the mean stress direction that might have led to an increase in pore-fluid pressures if fluid flow was inhibited (Sibson 2000; Verweij 2003). Continued salt movement during Cretaceous and Cenozoic times (e.g. Harding & Huuse 2015) would have also influenced the stress field in the Terschelling Basin. Additionally, the presence of natural gas (i.e. compartment 9, Fig. 3), and variations in heat flow and fluid salinities, could have contributed to the generation of overpressures (Osborne & Swarbrick 1997). Regarding the salinity of the fluids, it is noted that pore-fluid overpressure, as defined in this paper, is the difference between the measured pore-fluid pressure at a certain depth and the hydrostatic pressure for porewater at seawater density (1020 kg m^{-3}) at that same depth. The porewater in the Triassic units in the Terschelling Basin is known to be more saline (>150 000 mg l^{-1}; density of >1100 kg m^{-3}) than seawater (Verweij 2003; Verweij *et al.* 2012). As such, the hydrostatic pressure in the Triassic units may be higher, and the corresponding pore-fluid overpressures somewhat lower, than the values shown in Figure 1b. Because the seawater density was also used in the calculations of overpressure due to sedimentary loading, the comparison values of observed and calculated overpressures are directly comparable.

Distribution of overpressures

In this study, the pressure regime of the Lower Triassic Bunter reservoir compartments that lack (reliable) pressure data was estimated based on structural characteristics and similarities of these compartments with areas of reliable pressure data. However, given the variation in the amount of secondary salt cementation that is present in the pores of the Bunter reservoir rocks, one could discuss the exact validity of the estimated pressure regimes for the Bunter compartments. Secondary salt cementation has greatly reduced the porosity and permeability of the Bunter reservoirs, and, as a consequence, has reduced its hydraulic continuity, inhibiting lateral fluid flow. As a result, numerous wells that were drilled in the so-called ‘highly overpressured compartments’ have encountered tight reservoirs. On the other hand, some wells did encounter highly overpressured reservoirs and even gas (M01-02 in compartment 9, Figs 3 & 4). It is therefore suggested that within the compartments that are identified in this study, isolated pockets of fluids are present in certain areas where secondary salt cementation is less abundant. There might be multiple isolated pockets of fluids in one compartment but, given their similarities in structural characteristics and geological evolution, it is likely that pockets in the same compartment (as identified in this study) have the same overpressure regime. Hence, overpressure estimates should be seen as an approximation for isolated pockets of fluids within that compartment.

Conclusions

The most important factor that controls the distribution of overpressures in the Lower Triassic Bunter reservoirs rocks in the Terschelling Basin is the presence and type of both lateral and vertical permeability barriers. Lateral permeability barriers are formed by Zechstein salt and faults in the area. The vertical permeability barrier that exerts most control on overpressures is the Röt evaporite. In compartments where the Röt evaporite is continuously present, overpressures of around 30 MPa and higher are observed. A sharp decrease in overpressures is observed in those areas where the Röt evaporite is absent and where Lower Triassic Bunter strata are directly overlain by either Upper Jurassic or Lower Cretaceous strata. Additional controlling factors on the distribution of overpressures are the depth at which the Bunter reservoir rocks are present, as well as their location within the basin. Corresponding to these controlling factors, a regional trend of decreasing overpressures can be observed towards the edges of the Terschelling Basin. Structural observations indicate that the compartments became hydraulically closed in roughly three phases:

- activity on listric faults closed some compartments towards the east of the Terschelling Basin during the Middle Triassic Röt and Muschelkalk periods;
- salt structures caused the compartments to close in the east and west of the Terschelling Basin during the Late Triassic Keuper period;
- activity on NW–SE-trending faults caused the compartments to close in the north and south of the Terschelling Basin during the Late Jurassic and the first part of the Early Cretaceous.

Calculations show that subsequent sedimentary loading during the Cretaceous and Cenozoic was able to generate the observed overpressures and is, therefore, likely to be a main overpressure-generating mechanism for the Bunter reservoir rocks in this area.

We would like to thank EBN and TNO for providing the opportunity to publish this MSc. study, which was co-supervised by Utrecht University. Special thanks to Renaud Bouroullec (for sharing his ideas regarding salt tectonics in the area), Marc Hettema (for his input on geomechanics), Fred Beekman, Eveline Rosendaal and Guido Hoetz (all for their constructive feedback during this project). The authors are also very grateful for the comments and improvements that were suggested by the reviewers (Harmen Mijnlieff and Ben Kilhams) and an anonymous reviewer.

References

BOUROULLEC, R., VERREUSSEL, R. ET AL. 2018. Tectonostratigraphy of rift basins affected by halokinetics: the syn-rift Middle Jurassic–Lower Cretaceous in the Dutch Central Graben, Terschelling Basin and neighbouring platforms, Dutch offshore. *In*: KILHAMS, B., KUKLA, P.A., MAZUR, S., MCKIE, T., MIJNLIEFF, H.F. & VAN OJIK, K. (eds) *Mesozoic Resource Potential of the Southern Permian Basin*. Geological Society, London, Special Publications, **469**. First published online March 15, 2018, https://doi.org/10.1144/SP469.22

BRÜCKNER-RÖHLING, S. & RÖHLING, H.G. 1998. Paleotectonics in the Lower and Middle Triassic (Buntsandstein, Muschelkalk) of the North German Basin. *Hallesches Jahrbuch Geowissenschaften, B, Beiheft*, **5**, 27–28.

DE JAGER, J. 2012. The discovery of the Fat Sand Play (Solling Formation, Triassic), Northern Dutch offshore – a case of serendipity. *Geologie en Mijnbouw – Netherlands Journal of Geosciences*, **91**, 609–619.

DE JAGER, J. & GELUK, M.C. 2007. Geological development. *In*: WONG, T.E., BATJES, D.A.J. & DE JAGER, J. (eds) *Geology of the Netherlands*. Royal Netherlands Academy of Arts and Sciences, Amsterdam, 5–26.

DOMENICO, P.A. & SCHWARTZ, F.W. 1998. *Physical and Chemical Hydrogeology*. 2nd edn. Wiley, New York.

DOORNENBAL, J.C. & STEVENSON, A.G. (eds). 2010. *Petroleum Geological Atlas of the Southern Permian Basin Area*. European Association of Geoscientists & Engineers (EAGE), Houten, The Netherlands.

DRONKERT, H. & REMMELTS, G. 1996. Influence of salt structures on reservoir rocks in Block L2, Dutch continental shelf. *In*: RONDEEL, H.E., BATJES, D.A.J. & NIEUWENHUIJS, W.H. (eds) *Geology of Gas and Oil under the Netherlands*. Kluwer, Dordrecht, The Netherlands, 179–189.

DUIN, E.J.T., DOORNENBAL, J.C., RIJKERS, R.H.B., VERBEEK, J.W. & WONG, TH.E. 2006. Subsurface structure of the Netherlands – Results of recent onshore and offshore mapping. *Netherlands Journal of Geosciences*, **85**, 245–276.

FEIST-BURKHARDT, X., GÖTZ, A.E. ET AL. 2008. Triassic. *In*: MCCANN, T. (ed.) *The Geology of Central Europe. Volume 2: Mesozoic and Cenozoic*. Geological Society, London, 749–821.

FLEMINGS, P.B., STUMP, B.B., FINKBEINER, TH. & ZOBACK, M. 2002. Flow focusing in overpressured sandstones: theory, observations, and applications. *American Journal of Science*, **302**, 827–855.

GELUK, M.C. 2005. *Stratigraphy and tectonics of Permo-Triassic basins in the Netherlands and surrounding areas*. PhD thesis, University of Utrecht, Utrecht, The Netherlands.

GELUK, M.C. 2007. Permian. *In*: WONG, TH.E., BATJES, D.A.J. & DE JAGER, J. (eds) *Geology of the Netherlands*. Royal Netherlands Academy of Arts and Sciences, Amsterdam, 63–83.

GELUK, M., MCKIE, T. & KILHAMS, B. 2018. An introduction to the Triassic: current insights into the regional setting and energy resource potential in NW Europe. *In*: KILHAMS, B., KUKLA, P.A., MAZUR, S., MCKIE, T., MIJNLIEFF, H.F. & VAN OJIK, K. (eds) *Mesozoic Resource Potential of the Southern Permian Basin*. Geological

Society, London, Special Publications, **469**. First published online January 26, 2018, https://doi.org/10.1144/SP469.1

Glennie, K.W. 1986. Development of N.W. Europe's southern Permian gas basin. *In*: Brooks, J., Goff, J.C. & Van Hoorn, B. (eds) *Habitat of Palaeozoic Gas in N.W. Europe*. Geological Society, London, Special Publications, **23**, 3–22, https://doi.org/10.1144/GSL.SP.1986.023.01.01

Harding, R. & Huuse, M. 2015. Salt on the move: multi stage evolution of salt diapirs in the Netherlands North Sea. *Marine and Petroleum Geology*, **61**, 39–55.

Hortle, A., Otto, C. & Underschultz, J. 2013. A quality control system to reduce uncertainty in interpreting formation pressures for reservoir and basin pressure system analysis. *Journal of Petroleum Geology*, **36**, 163–177.

Hubbert, M.K. 1940. The theory of ground-water motion. *The Journal of Geology*, **48**, 785–944.

Kombrink, H., Doornenbal, J.C., Duin, E.J.T., Den Dulk, M., Van Gessel, S.F., Ten Veen, J.H. & Witmans, N. 2012. New insights into the geological structure of the Netherlands; results of a detailed mapping project. *Netherlands Journal of Geosciences*, **91**, 419–444.

Kovalevych, V.M., Peryt, T., Carmona, V., Sydor, D.V., Vovnyuk, S. & Halas, S. 2002. Evolution of Permian seawater: Evidence from fluid inclusions in halite. *Neues Jahrbuch für Mineralogie – Abhandlungen*, **178**, 27–62.

Mann, D.M. & Mackenzie, A.S. 1990. Prediction of pore fluid pressures in sedimentary basins. *Marine and Petroleum Geology*, **7**, 55–65.

Mauthe, G. 2002. Halite cementation in sandstone from the Middle Triassic bunter Formation in north-west Germany Hengstlage and Dötlingen gas fields. *Erdöl Erdgas Kohle*, **118**, 917–940.

McKie, T. 2014. Climatic and tectonic controls on Triassic dryland terminal fluvial system architecture, central North Sea. *In*: Martinius, A.W., Ravnås, R., Howell, J.A., Steel, R.J. & Wonham, J.P. (eds) *From Depositional Systems to Sedimentary Successions on the Norwegian Continental Margin*. John Wiley, London, 19–57.

McKie, T. 2017. Paleogeographic evolution of latest Permian and Triassic salt basins in Northwest Europe. *In*: Soto, J.I., Flinch, J. & Tari, G. (eds) *Permo-Triassic Salt Provinces of Europe, North Africa and the Atlantic Margins: Tectonics and Hydrocarbon Potential*. Elsevier, Amsterdam, 159–173.

McKie, T. & Williams, B. 2009. Triassic palaeogeography and fluvial dispersal across the northwest European Basins. *Geological Journal*, **44**, 711–741.

Munsterman, D.K., Verreussel, R.M.C.H., Mijnlieff, H.F., Witmans, N. & Kerstholt-Boegehold, S. 2012. Revision and update of the Callovian-Ryazanian Stratigraphic Nomenclature in the northern Dutch Offshore, i.e. Central Graben Subgroup and Scruff Group. *Netherlands Journal of Geosciences – Geology en Mijnbouw*, **91**, 555–590.

Nelskamp, S., Verweij, J.M. & Witmans, N. 2012. The role of salt tectonics and overburden in generation of overpressure in the Dutch North Sea area. *Netherlands Journal of Geosciences*, **91**, 517–534.

Neuzil, C.E. 1995. Abnormal pressures as hydrodynamic phenomena. *American Journal of Science*, **295**, 742– 786.

Osborne, M.J. & Swarbrick, R.E. 1997. Mechanisms for generating overpressure in sedimentary basins: a reevaluation. *AAPG Bulletin*, **81**, 1023–1041.

Rattey, R.P. & Hayward, A.B. 1993. Sequence stratigraphy of a failed rift system: the Middle Jurassic to Early Cretaceous basin evolution of the Central and Northern North Sea. *In*: Parker, J.R. (ed.) *Petroleum Geology of Northwest Europe: Proceedings of the 4th Conference*. Geological Society, London, 213–249, https://doi.org/10.1144/0040215

Reicherter, K., Froitzheim, N. *et al.* 2008. Alpine tectonics north of the Alps. *In*: McCann, T. (ed.) *The Geology of Central Europe. Volume 2: Mesozoic and Cenozoic*. Geological Society, London, 1233–1285.

Röhling, H.G. 1991. A lithostratigraphic subdivision of the Lower Triassic in the Northwest German lowlands and the German sector of the North Sea, based on Gamma ray and sonic logs. *Geologisches Jahrbuch Reihe A, Band A*, **119**, 3–24.

Ruffell, A., Simms, M.J. & Wignall, P.B. 2015. The Carnian Humid Episode of the late Triassic: a review. *Geological Magazine*, **153**, 271–284.

Sibson, R.H. 2000. Tectonic controls on maximum sustainable overpressure: fluid redistribution from stress transitions. *Journal of Geochemical Exploration*, **69–70**, 471–475.

Skempton, A.W. 1954. The pore pressure coefficients *A* and *B*. *Gotechnique*, **4**, 143–147.

Steel, R. & Ryseth, A. 1990. The Triassic–Early Jurasssic succession in the northern North Sea: megasequence stratigraphy and intra-Triassic tectonics. *In*: Hardman, R.F.P. & Brookes, J. (eds) *Tectonic Events Responsible for Britain's Oil and Gas Reserves*. Geological Society, London, Special Publications, **55**, 139–168, https://doi.org/10.1144/GSL.SP.1990.055.01.07

Van Adrichem Boogaert, H.A. & Kouwe, W.F.P. 1993. Stratigraphic nomenclature of the Netherlands, revision and update by RGD and NOGEPA, Section A, General. *Mededelingen Rijks Geologische Dienst*, **50**, 1–40.

Vendeville, B.C. & Jackson, M.P.A. 1992. The rise of diapirs during thin-skinned extension. *Marine and Petroleum Geology*, **9-4**, 331–354.

Verreussel, R., Bouroullec, R. *et al.* 2018. Stepwise basin evolution of the Middle Jurassic–Early Cretaceous rift phase in the Central Graben area of Denmark, Germany and The Netherlands. *In*: Kilhams, B., Kukla, P.A., Mazur, S., McKie, T., Mijnlieff, H.F. & van Ojik, K. (eds) *Mesozoic Resource Potential of the Southern Permian Basin*. Geological Society, London, Special Publications, **469**. First published online March 15, 2018, https://doi.org/10.1144/SP469.23

Verweij, J.M. 1990. *Outline of the Cenozoic Hydrogeohistory of the Dutch Section of the North Sea Basin*. TNO Report OS 90-41B. TNO Institute of Applied Geoscience, Delft, The Netherlands.

Verweij, J.M. 2003. *Fluid flow systems analysis on geological timescales in onshore and offshore Netherlands. With special reference to the Broad Fourteens Basin*. PhD thesis, Vrije Universiteit, Amsterdam.

Verweij, J.M. & Simmelink, H.J. 2002. Geodynamic and hydrodynamic evolution of the Broad Fourteens Basin

(The Netherlands) in relation to its petroleum systems. *Marine and Petroleum Geology*, **19**, 339–359.
Verweij, J.M., Simmelink, H.J., Underschultz, J. & Witmans, N. 2012. Pressure and fluid dynamic characterisation of the Dutch subsurface. *Geologie en Mijnbouw – Netherlands Journal of Geosciences*, **91**, 465–490.
Wever, A.K.T., Pacek, I., Arts, R.J. & Kemper, M. 2010. Application of solid and fluid substitution to salt-plugged sandstones using generalized Gassman theory. Extended abstract presented at the 72nd EAGE Conference and Exhibition incorporating SPE EUROPEC 2010, 14–17 June 2010, Barcelona, Spain, https://doi.org/10.3997/2214-4609.201401220
Ziegler, P.A. 1990. *Geological Atlas of Western and Central Europe*. 2nd edn. Shell Internationale Petroleum Maatschappij, The Hague. Geological Society, London.
Zwaan, F. 2018. Lower Cretaceous reservoir development in the North Sea Central Graben, and potential analogue settings in the Southern Permian Basin and South Viking Graben. *In*: Kilhams, B., Kukla, P.A., Mazur, S., McKie, T., Mijnlieff, H.F. & van Ojik, K. (eds) *Mesozoic Resource Potential of the Southern Permian Basin*. Geological Society, London, Special Publications, **469**. First published online January 4, 2018, https://doi.org/10.1144/SP469.3

Enhanced gas recovery of an ageing field utilizing N_2 displacement: De Wijk Field, The Netherlands

RAJASMITA GOSWAMI[1]*, FRITZ C. SEEBERGER[2] & GEERT BOSMAN[3]

[1]*Nederlandse Aardolie Maatschappij (NAM), Schepersmaat 2, 9405 TA Assen, The Netherlands*

[2]*Bijkersweg 6, 9497 PV, Donderen, The Netherlands*

[3]*Veendijk 25, 7852 TE, Wezup, The Netherlands*

**Correspondence: Rajasmita.Goswami@shell.com*

Abstract: Enhancing recovery of the remaining gas in ageing fields presents a commercial opportunity. As such, NAM took the decision to apply on-site generated N_2 injection for the first time. The De Wijk gas field was selected based on size and geological setting, allowing the demonstration of N_2 flooding of depleted gas in diverse reservoir types, as well as testing the efficiency of residual gas sweeping in the watered-out reservoir. De Wijk is one of the oldest onshore natural gas fields in The Netherlands, having produced *c.* 15 Bcm (billion cubic metres) of high calorific gas with a natural N_2 content of 5–11% for more than 60 years, bringing it close to the end of field life. Enhanced gas recovery in this field would extend the field life and that of the Ten Arlo gas-processing system, thereby also lengthening tail-end production of other connected fields. In De Wijk, the existing 35 wells were used as dedicated injectors and producers. The production rates from the latter are used to help better understand reservoir flow behaviour. Integration of geological analysis and available production data helped to implement the N_2 enhanced gas recovery technology in this mature field. Two processes of enhanced gas recovery were tested: a gas–gas displacement in the depleted gas leg (Nitrogen Assisted Depletion Drive: NADD); and gas displacing residual gas in the water leg (Nitrogen Enhanced Residual Gas: NERG). Observations after 1 year of N_2 injection showed that planned v. actual performance of the NADD technology is comparable with an increase in production of 100 000–160 000 Nm^3/day (normal cubic metre per day). NERG application restored a watered-out well; however, further investigation of this application is required. It is demonstrated that a detailed geological understanding of, for example, permeable Rogenstein oolite intervals and the lateral connectivity of internal high-permeability streaks in Volpriehausen sandstones is crucial to the successful application of enhanced gas recovery techniques in such ageing fields.

Gas field recovery by depletion is generally seen as a conventional way to recover gas. Primary recovery is implemented by (free) flowing a gas well, which might be aided by multistage compression and deliquification of the well bore in a late field life. This primary recovery process of ‘sucking’ out gas and associated reservoir fluids can achieve a wide range (40–95%) of recovery factors depending on the reservoir permeability and the degree of aquifer influx. This results in a range of stranded gas volumes that are not recovered by primary gas production. However, as recovery factors are generally higher than in oil, historically less attention has been given to enhanced gas recovery (EGR) methods when compared to highly documented secondary oil recovery cases (e.g. Green & Willhite 1998; Shrimali 2015). In cases where EGR cannot be linked up to an existing infrastructure supplying injection gas (e.g. CO_2 or natural non-HC gas), one of the most efficient enhanced gas recovery methods is generating N_2 from air and injecting it to displace the remaining gas and to maintain pressure above the minimum required for flow in wells. The main advantages of using nitrogen for EGR purposes are its abundance in air (available at any location), high energy efficiency, non-corrosiveness and non-toxicity. However, the application of EGR is still a new technology and the details of its application (especially with regards to dynamic flow behaviour) are poorly understood. Studies have been previously undertaken for potential enhanced gas recovery in two offshore Dutch gas fields, utilizing CO_2 or N_2, but have never been implemented (Leeuwenburgh *et al.* 2014).

De Wijk is one of the oldest onshore natural gas fields in The Netherlands and lies entirely in NAM’s Schoonebeek concession in SW Drenthe. It was discovered in 1949 by the WYK-1 well. The well proved economic gas accumulations in multiple pay intervals (Tertiary, Upper and Lower Cretaceous, and the Lower Triassic). Subsequent appraisal extended the field to the deeper Carboniferous clastics and defined the aerial extent of the hydrocarbon column (Fig. 1: at the Lower Cretaceous and Triassic

From: Kilhams, B., Kukla, P. A., Mazur, S., McKie, T., Mijnlieff, H. F. & van Ojik, K. (eds) 2018. *Mesozoic Resource Potential in the Southern Permian Basin*. Geological Society, London, Special Publications, **469**, 237–251.
First published online January 16, 2018, https://doi.org/10.1144/SP469.2

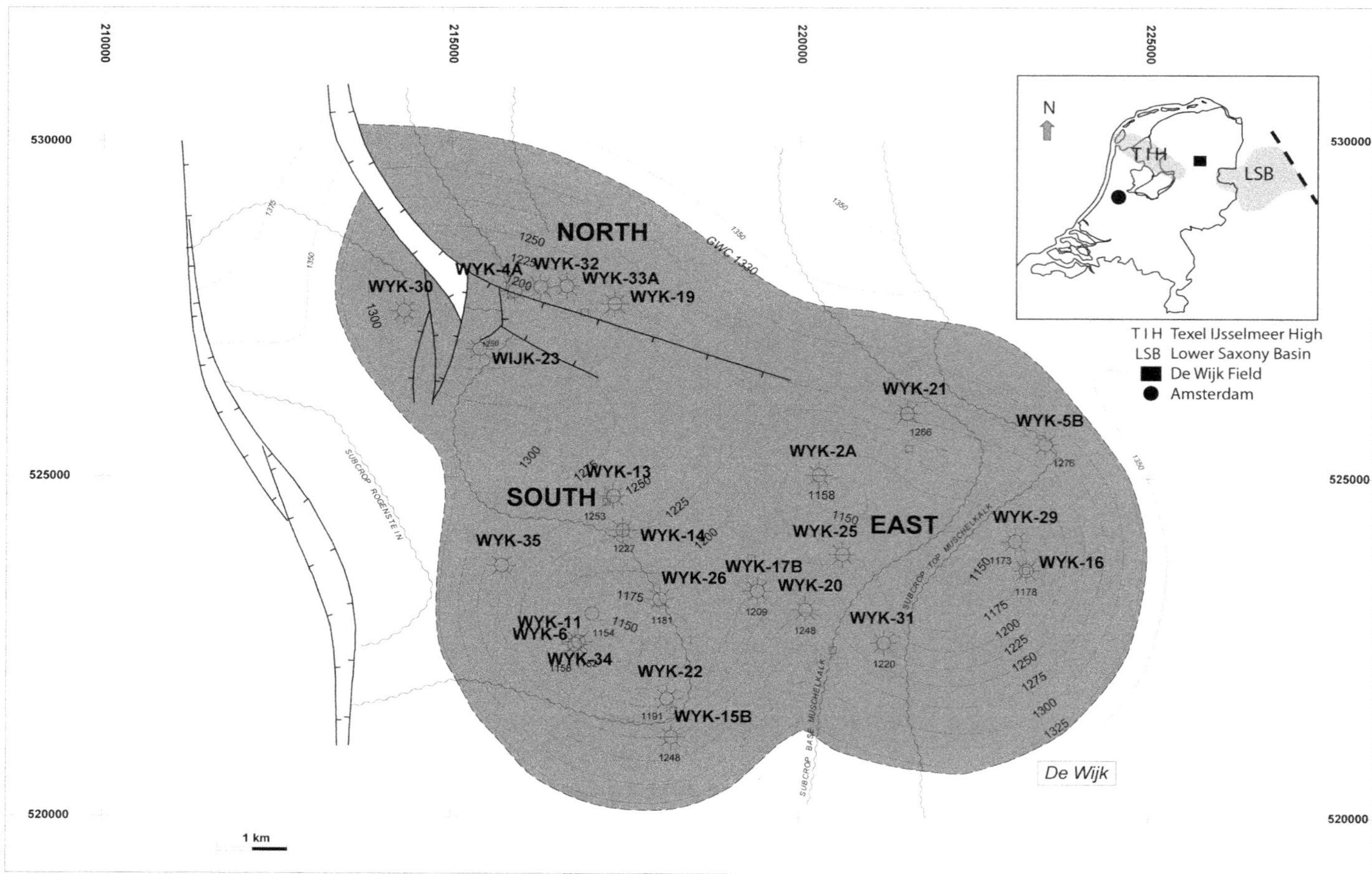

Fig. 1. A top structure map prepared at the top Vlieland Friesland Member. with the outline of the subcrop of Muschelkalk in the east and subcrop of Rogenstein in the south. The gas field is divided into De Wijk North, De Wijk South and De Wijk East. The map also shows the various wells that are present in the area. The location map of the study area is shown in the top right, highlighting the De Wijk Field together with the location of the Texel IJsselmeer High, the Lower Saxony Basin and the city of Amsterdam.

level). The field was developed per horizon with a total production up to 2015 of around 15 Bcm (billion cubic metres) of high calorific gas with a natural N_2 content varying from 5 to 11%. The majority of the wells drilled (from eight surface locations) in the De Wijk Field are completed in the Triassic and Lower Cretaceous reservoirs. Field production capacity has declined from several millions of Nm^3/day (normal cubic metre per day) in the late 1980s to *c.* 0.3 million Nm^3/day in 2015. N_2 enhanced gas recovery technology is expected to extend the project's life by around 15 years, and the field's ultimate gas recovery could rise from 73% to as much as 83%. Between 2014 and 2015, five of the existing wells were converted to be dedicated injectors and the rest to be producers, a new producing well (WYK-35) was drilled and an air separation unit was built next to an identified N_2 injection location. New dedicated pipelines were laid to connect other N_2 injection locations. This provided the surface infrastructure to utilize the different N_2 enhanced recovery techniques. Here we describe the preliminary results of these techniques for the Triassic and Lower Cretaceous intervals, with a focus on geological aspects impacting dynamic gas flow. It is hoped that this project will sustain field production until 2030, and thereby increase the total system gas recovery by 2.8 Bcm.

Geological development of field elements

The geological history of the field is complex, comprising multiple reservoir intervals associated with a variety of depositional and tectonic settings (Bruijn 1996). Figure 2 illustrates a seismic section taken from the study area running west–east, demonstrating source rock and reservoir intervals. The oldest penetrated sediments in the De Wijk area (WYK-1) are of Carboniferous age (Westphalian), including Type III (coaly) gas-prone source rocks that charged the De Wijk Field across all the reservoir intervals. The Permian Rotliegend Play of the Southern Permian Basin is not very extensive in this area. All four cycles of the Zechstein evaporitic sequence that form the top seal of the Permian Play are present (e.g. Ziegler 1990; De Dager & Geluk 2007; Brueren 1959; NAM & RGD 1980). The youngest reservoir intervals in the study area are composed of a series of Upper Cretaceous Chalk (Ruffell 1991) and Tertiary Basal Dongen Tuffite and Landen Clay members (Zagwijn 1989). The top seal for these shallow reservoirs are formed by the Tertiary claystone Ieper Member.

For this study, the main reservoirs focused on are composed of the Lower Triassic sequence (Bunter) (Fig. 3), which marks the change from shallow marine to continental deposition (Geluk & Röhling 1999). In general, Triassic sedimentation took place in an extensive, east–west-running, intracontinental basin that stretched from the eastern coast of the UK across the Southern North Sea and northern Germany into Poland (Geluk 2005; Feist-Burkhardt *et al.* 2008; McKie & Williams 2009).

The evolution of the De Wijk reservoir, and associated sections, is introduced from oldest to youngest (Fig. 3). The lowermost unit is the Main Claystone, which also forms the oldest Triassic interval. As the name suggests, the dominant lithology is claystone with interbedded (but laterally extensive) sand sheets. Palaeogeographically, the Main Claystone was deposited within an extensive playa with episodic sheet floods and wind-blown sands that swept across the basin to form a multitude of shallow channels, terminal splays, aeolian sand sheets, and ephemeral lakes and ponds (McKie 2017). In the latter stages of Main Claystone deposition, the ephemeral lakes were occasionally sufficiently large and persistent enough to form shoreline deposits (the Rogenstein interval). This section is mainly composed of claystones with intercalations of oolite beds and sand-siltstone streaks deposited during the advance and retreat of the lake shoreline (Fig. 3). These oolitic lake-margin deposits were common during deposition of the Rogenstein, and forms an integral part of this reservoir. The oolitic storm deposits, consisting mainly of reworked ooids, are interpreted to indicate the lake-level highstands. The grain size and thickness of the oollitic beds increases towards the top of the Rogenstein Member. These oolitic grainstones show early calcitic cementation preserving large amounts of primary, well-connected pores indicating synsedimentary leaching during lowstand phases over the Rogenstein Member. A secondary leaching of the calcite cement is also observed where cement is partly or wholly dissolved, producing vuggy pores, due to weathering corresponding to the Early Cimmerian Unconformity. Subsequently, Volpriehausen sandstones (Fig. 3) were deposited in an aeolian/fluviatile depositional environment via ephemeral streams, and reworked by wind into small aeolian dunes and adhesion ripples (Van Adrichem Boogaert & Kouwe 1994). A renewed phase of basin-wide claystone deposition was initiated, marking a temporary return to the clay-dominated arid floodplain environment (RGD 1993).

A notable erosional episode in the De Wijk area removed most of the Main Buntsandstein Group down to the Volpriehausen sandstone. Above this Hardegsen erosional surface, playa lake and lake-margin clay-siltstones and sandstone were deposited, directly overlying the sandstone. The Middle Triassic Lower Muschelkalk marks a major marine transgression into the Southern Permian Basin (Pipping *et al.* 2001). The De Wijk area occupied a marginal position with respect to the Muschelkalk Sea, with

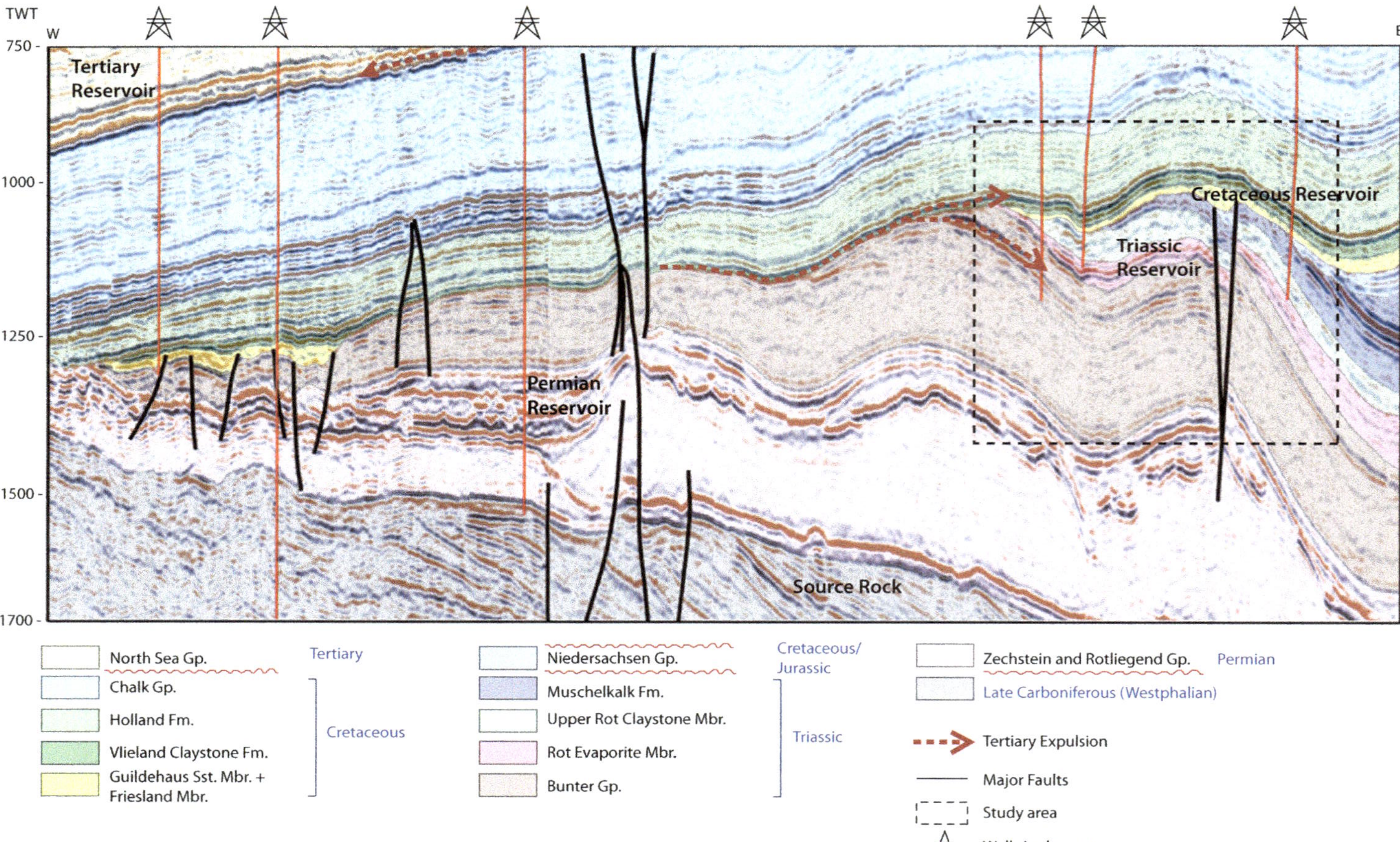

Fig. 2. A two-way time (TWT) regional seismic cross-section highlighting the stratigraphy in the subsurface (as shown in the legend), and the location of the study area (dashed square box). The oldest sediment encountered by the wells (as shown by the wells symbol) is the Carboniferous (including coal-rich source intervals). Vertical (black solid lines; major faults) and lateral (red dashed lines; reservoir intervals) migration paths along are marked.

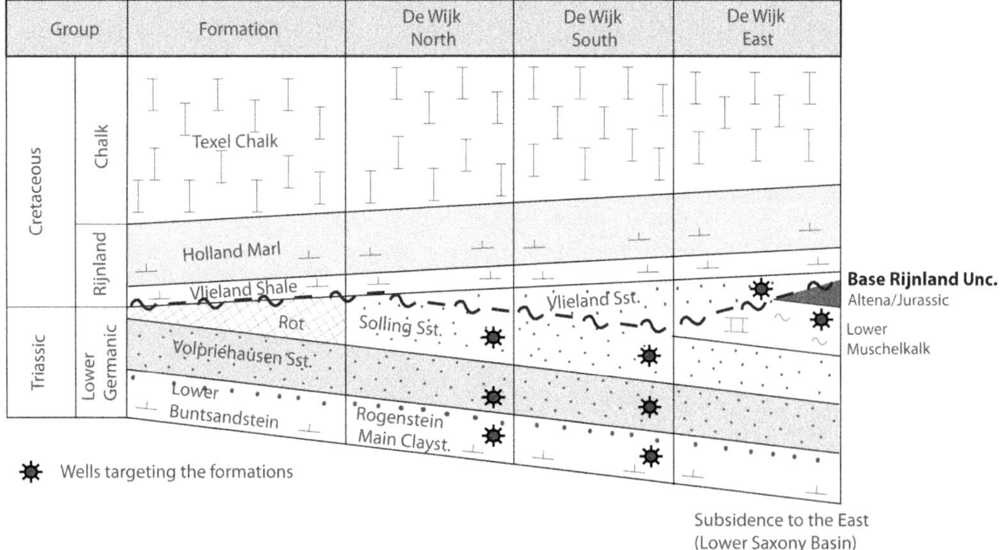

Fig. 3. Chronostratigraphic diagram showing the variable stratigraphy in the De Wijk North, De Wijk South and De Wijk East areas, combined from well data, seismic data and the literature (Bruijn 1996).

deposition occurring in marginal-marine shoreface and lagoonal settings that were fringed by dolomitic mudflats (van der Zwan & Spaar 1992; Palermo *et al.* 2008). Regionally, there is Muschelkalk and Keuper from the Upper Triassic, and Jurassic sediments deposited, but in the De Wijk area this has been removed during Cimmerian events preserving only Muschelkalk in the eastern part of the field (Fig. 3).

The Late Cimmerian event related to a major phase of extension and opening of the Central Graben, with the uplift of highs on the basin margins (Rijkers & Geluk 1996). Most of the uplift occurred west of the present-day De Wijk Field, and induced a general easterly dip forming the Lower Saxony Basin. Due to the higher overburden load in the Lower Saxony Basin, the Zechstein salt migrated in a westwards direction (RGD 1993). As a result, the Triassic formations subcrop out progressively, with the older strata in the west and the younger Triassic strata in the east. On the eastern side of the De Wijk Field, the Niedersachsen and Altena formations (Jurassic sediments) overlie a thick package of Triassic sediments, from Main Claystone to Lower Muschelkallk. Towards the west, the Triassic sediments are gradually eroded and thin out. The youngest Triassic sediments are preserved only in the east, and the Main Claystone subcrops directly the Base Cretaceous Unconformity in the westernmost part of the field.

During the Valanginian (Early Cretaceous), a general marine transgression led to the deposition of the Vlieland sandstones (Fig. 3) in a coastal or shallow-marine environment (Jeremiah *et al.* 2010). This transgressive trend continued through the Early Cretaceous leading to the deposition of basinal Vlieland shales, and the marls plus shales of the Holland Formation. These Lower Cretaceous Vlieland and Holland shales form the top seal to the Triassic and Early Cretaceous reservoirs in the De Wijk Field. Salt movement, once triggered in Late Jurassic times, continued and induced the present-day closures at Lower Cretaceous and Tertiary level, forming the structural trap of the De Wijk Field.

The main reservoirs in the current study, therefore, include the Lower Triassic and sections of the super-cropping Lower Cretaceous reservoirs, sealed by the overlying transgressive marine shales (Vlieland and Holland formations). Laterally, the reservoir complex is controlled by a shallow saddle at base Cretaceous level, separating this complex from a fault-bounded structure to the NE. The spill-point across this saddle is coincident with the gas–water contact (GWC) in the various primary reservoirs, illustrating that they form one reservoir system. Gas migration from Carboniferous source rocks to Mesozoic reservoirs is thought to occur via an area *c.* 30 km to the west where Lower Cretaceous strata directly overly the Rotliegend and Carboniferous due to complete erosion of the Zechstein evaporitic seals (Fig. 3). The timing of migration is considered to be Eocene–Miocene in age. The other reservoirs in the study area are the Upper Cretaceous Ommelanden Chalk and the Tertiary Landen claystone and Basal Dongen Tuffite that follow the same structural configuration and are charged by same Carboniferous source (Fig. 2).

For the purposes of this study, the De Wijk Field is divided into three areas: namely North, South and East (Fig. 4). In the northern and southern area of the field, the gas-bearing reservoirs are the Solling sandstones, the Volpriehausen sandstones, the Rogenstein oolites and the Main Claystone, all unconformably overlain by Vlieland sandstone. The gas in the eastern part of the De Wijk Field is restricted to the sandstone of the Vlieland Formation and the carbonates of the Lower Muschelkalk Formation. For the current EGR study, we consider only the De Wijk East and South areas.

Reservoir intervals

De Wijk East area was discovered by well WYK-5B in 1952, where it produced gas from two separate gas accumulations: the Lower Muschelkalk and the Vlieland reservoirs. The Lower Muschelkalk shows a wide range of reservoir quality within shoreface to supratidal limestone, dolomite and marl beds. The reservoir behaviour of the Lower Muschelkalk is mainly governed by thin high-permeability streaks set in a low-permeability matrix. The overall thickness of the reservoir is *c.* 103 m, with an average net-to-gross of 0.57 and average porosity of 14%. The Muschelkalk is overlain unconformably by Vlieland sandstone. The quality of the sands degrades to the NW. Figure 4a shows the distribution of the reservoir-quality sands in the field. The reservoir is divided into poor-quality lower shoreface sediments (very clay-prone) and good-quality upper shoreface sediments (clean sands). These sands were only penetrated by wells WYK-16, WYK-17, WYK-20, WYK-25 and WYK-29. Core and log data suggest that the shoreface sediments can further be subdivided into a 1–6 m thickness of upper shoreface (average net-to-gross of 0.8 and average porosity of 23%) and a 0.5–10 m thickness of lower shoreface sediments (average net-to-gross of 1 and average porosity of 20%). The shoreface sands are highly permeable (*c.* 300–700 mD).

The De Wijk South area was discovered by well WYK-14 in 1974, where it produced gas from Volpriehausen sandstone (Fig. 4b). WYK-22 is the type log, where three gas-bearing formations are observed: the Solling, Volpriehausen and Upper Rogenstein reservoirs (Fig. 5a). The Volpriehausen sandstone is composed of highly porous and extensive sheet-flood and aeolian sandstone. The sandstone is porous (24% on average), red-brown, and partially cemented by anhydrite, dolomite and calcite. Figure 5b illustrates a thin-section micrograph of cuttings samples as an example of the sandstone grains. High porosity (*c.* 28%) and permeability (1000–3000 mD) probably reflects very little burial diagenesis plus the effect of leaching. The thickness varies from 1.6 to 17 m, probably due to truncation by the Early Cimmerian tectonic event. The average net-to-gross is 0.92. The Solling sandstone is more variable, with a higher shale content and, hence, a relatively poorer reservoir compared to the Volpriehausen sandstone. The thickness varies from 2.5 to 14 m; however, the net gas-bearing thicknesses range from 2 to 7 m, with an average net-to-gross of 0.64 and an average porosity of 10%. The Rogenstein Member is mainly composed of a claystone matrix with intercalations of oolitic beds and sand-siltstone streaks. From the log data, it can typically be divided into upper (better-quality) and lower (poorer-quality) units. The ooid grain size and the thickness of the oolite beds increases towards the upper units. Minor compaction and the growth of calcite cement point to a phreatic, early diagenetic origin for the cementation. The higher reservoir quality, in the top 70–80 m, consists mainly of oolites with minor amounts of siltstone and sandstones. Clean oolitic grainstones and leached ooids (as observed under thin section: Fig. 5c) show early calcitic cementation preserving a large amount of primary, well-connected pores. The sandstones have generally good porosities and, sometimes, good permeabilities; however, as these sandstones occur in very thin layers and contain large amounts of clay drapes, these are of lesser importance to production.

EGR methods

The De Wijk enhanced gas project applied two injection displacement processes (Fig. 6): NADD (Nitrogen Assisted Depletion Drive) and a test of NERG (Nitrogen Enhanced Residual Gas). Modelling, and subsequent dynamic behaviour, was constrained by production data and pressure measurements, with support from well and reservoir data (21 wells, 12 of which are cored across the interval of interest), and 3D seismic data of various vintages:

- NADD uses N_2 to displace any remaining gas in a depleted (non-associated) reservoir. For example, in the Rogenstein (De Wijk South), Vlieland and Lower Muschelkalk reservoirs (De Wijk East), pressure has dropped to almost the economic and technical limit (around 15 bars), leaving a substantial amount of hydrocarbons within the gas phase. By injecting another gas (N_2), pressure is maintained or increased to above the minimum threshold, thus lifting the flow of wells, and the remaining gas is displaced towards producers. N_2 injection started in WYK-22 (De Wijk South), and WYK-20 and WYK-31 (De Wijk East). WYK-35 was drilled in the crestal part of the De Wijk South area as one of the dedicated producers, including

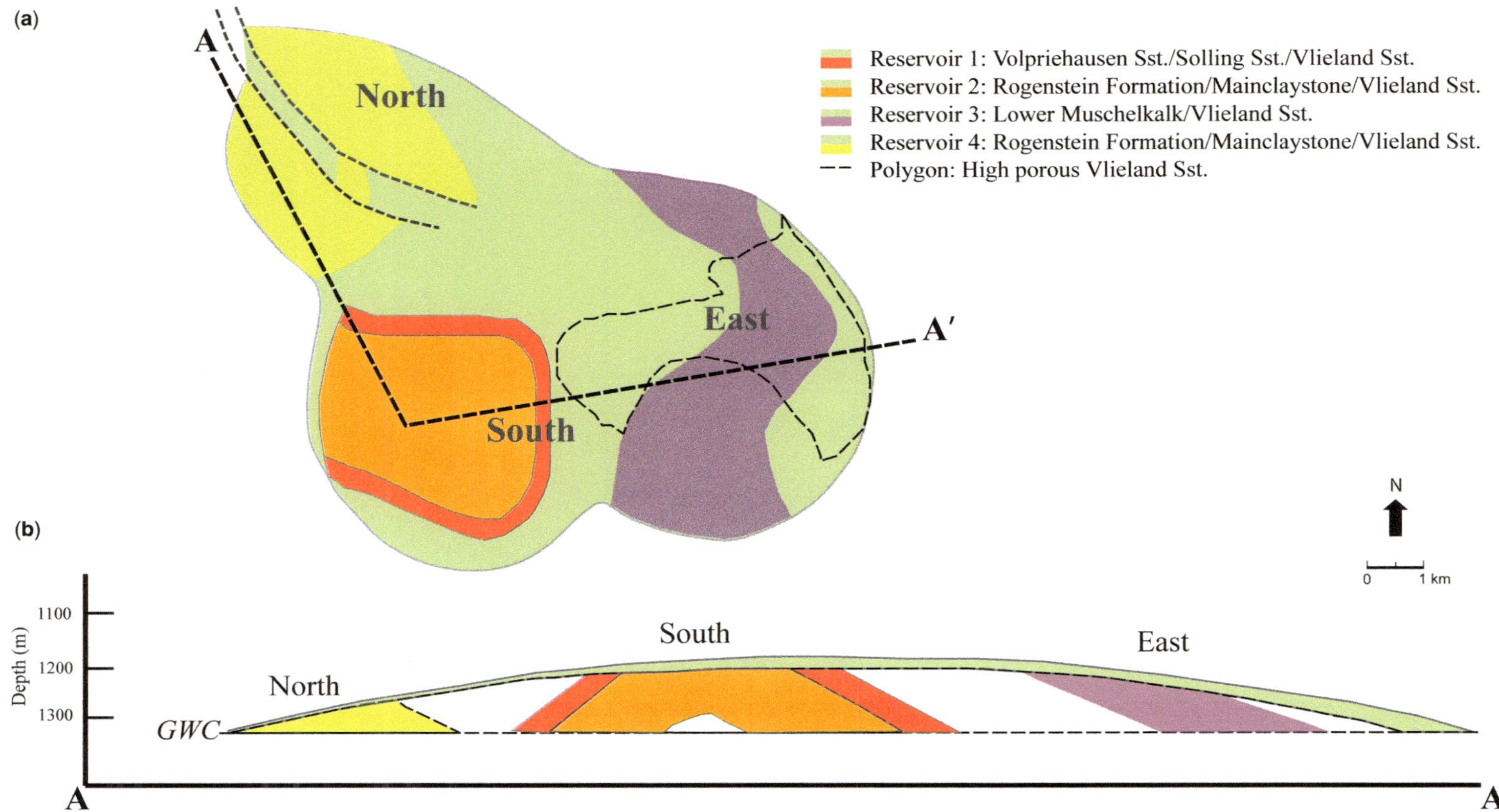

Fig. 4. (**a**) Schematic diagram of a plan view of the De Wijk Field illustrating the outline of the Vlieland Member above the GWC and subcropping reservoir elements. The legend shows the various combination of reservoirs stacked vertically below the Vlieland Member. (**b**) Cross-section illustrating the various stacked reservoirs in different areas of the field.

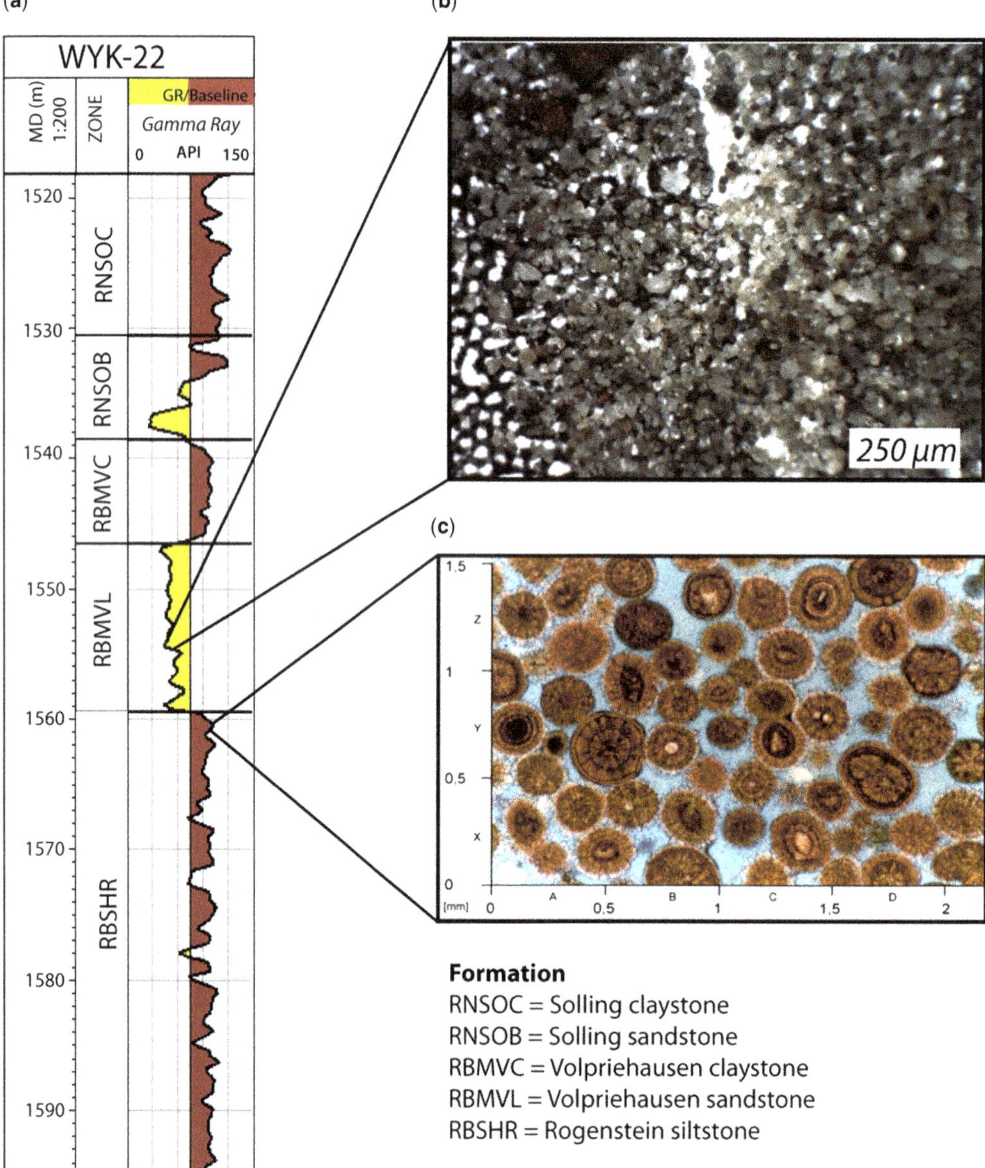

Fig. 5. (**a**) Geological description of the various reservoirs from a type log of the De Wijk South area (location of the well, WYK-22, is shown in Fig. 1). (**b**) Cuttings sample of Volpriehausen sandstone. The sandstone grains are observed as transparent/light pinkish/light greyish quartz, moderate to well-sorted, and rounded to subangular particles of size 150–350 µm. (**c**) Thin-section micrograph under parallel nicols showing the ooids within the Rogenstein, poor compaction and early drusy cementation; the blue stain indicates porosity (plug measurements: 26.3% average porosity and 1317.6 mD average permeability).

WYK-6 and WYK-34. The dedicated producers in De Wijk East area are WYK-29, WYK-17B and WYK-16.

- The NERG process involves N_2 injection that displaces residual gas in watered-out reservoirs (due to aquifer influx during primary production phase, and now trapped by capillary forces). For example, the Volpriehausen interval (De Wijk South) is a watered-out reservoir that experienced strong aquifer support, and all producers watered-out in

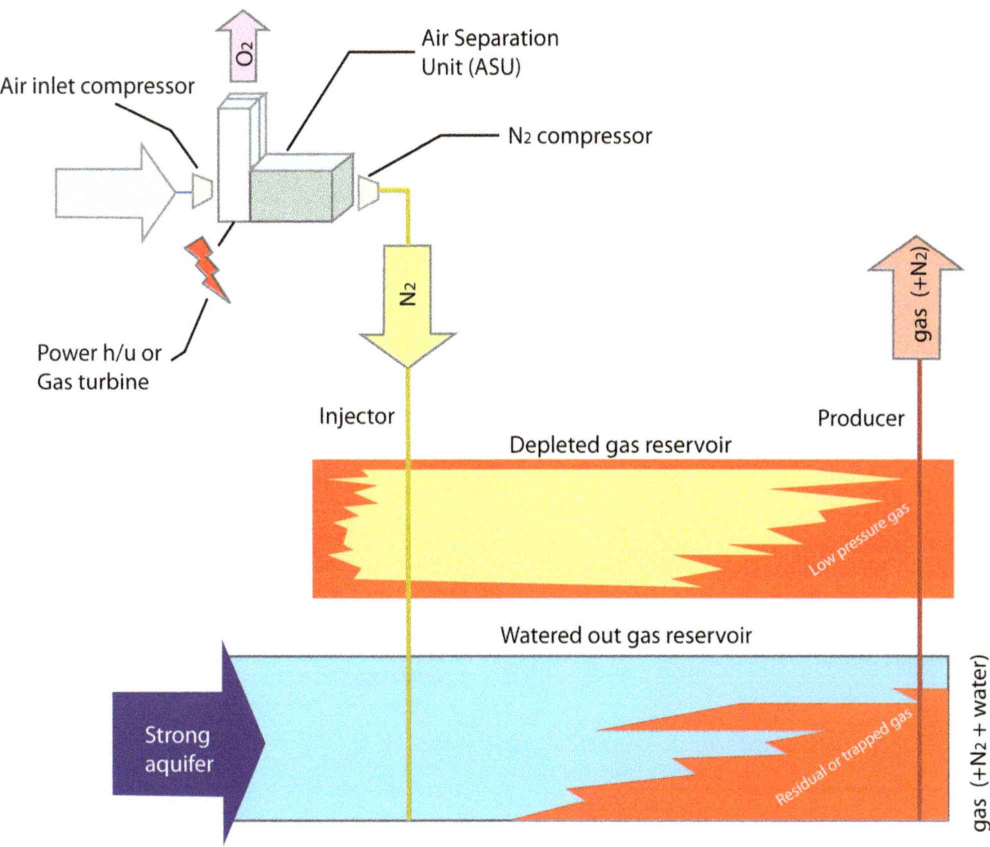

Fig. 6. Schematic diagram showing the processes of enhanced gas recovery using nitrogen: from the generation of the nitrogen gas from the air in the air separation unit (ASU), compressors to increase the gas pressure before injecting it into the target zones of the reservoirs through the injectors (both the depleted-gas leg for NADD and the water-invaded leg for NERG), and, consequently, lifting hydrocarbon gas from the producers. NADD, N_2-assisted depletion drive; NERG, N_2-enhanced residual gas.

the past. A NERG trial was carried out in the watered Volpriehausen sandstone reservoir. The aim was to inject N_2 into the flanks (WYK-15B), and sweep the residual gas in these formations towards a crestal producer (WYK-26), where gas saturation increases and creates a mobile gas phase that is sufficient to enable production.

NADD is a gas–gas displacement process, and sweep efficiency is controlled by the heterogeneity of the conceptual geological model. To forecast performance reservoir modelling, solutions and algorithms that were specifically developed in a numerical reservoir simulator MoReS (Regtien *et al.* 1995) were utilized. Additionally, to correctly capture larger geological features in the geological model, a dispersion function below grid-block scale (based on geological heterogeneity) and a numerical dispersion-free solver were implemented. Modelling of a gas–gas displacement has essentially been proven and successfully applied in NAM's underground gas storage, where indigenous gas has been displaced by dry injected gas with different specifications.

Injection for the EGR in De Wijk is planned from 2013 to 2028. Injections for the NADD and NERG started in August 2013: NADD in the east (WYK-31) and NERG in the south (WYK-15B). WYK-15B NERG stopped mid-2014 and NADD injection in the south (WYK-22) started. An additional NADD injection in WYK-20 started in the autumn of 2014 in the east. NADD simulation of the N_2 cut build-up and gas production showed that after the end of N_2 injection, production is expected to continue for several years by utilizing the lag of injection energy (pressure) in the reservoir. For NERG, injection into the high-permeability Volpriehausen sandstone in the watered-out, downflank well WYK-15B was planned. Pressure responses were continuously monitored in the crestal watered-out well (WYK-26) throughout the implementation, and both N_2 cut build-up and gas production was modelled.

Planned v. actual observation

The actual response of depleted producers (NADD) are comparable with the forecast based on dispersion modelling in MoReS (Shell reservoir simulator). For example, in the De Wijk East area (Fig. 7a), injection into high-permeability Vlieland sandstones in well WYK-20 triggered a good production response in the relatively close-by production well WYK-17B (distance to injector of *c.* 700 m). The injection was increased to maximize early production gains. An overall increase in the daily gas production was observed (of , 100 000–160 000 Nm^3/day). The period of low production in 2016 was caused by low reservoir pressure triggering frequent liquid loading of the well. The modelled and measured sinusoidal build-up of N_2 matches. However, not all wells have seen N_2 breakthrough at this early stage.

The planned response of injected N_2 in NERG (Fig. 7b) v. actual injection and response observed are shown in Figure 7c, d. Injection of N_2 started in August 2013 and lasted until August 2014. Gas production was modelled to produce from February 2014 to February 2015 (considering the delay in the response time of N_2 between injector producer). However, the pressure response in well WYK-26 took about 6 months longer than expected (Fig. 7c). Eventually, a well intervention was performed involving refilling the tubing with gas that contained *c.* 30% N_2. Water completely drained away, and the well was tested and put on production without any formation water production with an average N_2 cut of *c.* 75% (Fig. 7d). The intention was to sweep a high-permeability layer with a rate to minimize gravity override; thus reviving the watered-out well, and producing the residual gas present between injector and producer in a strong aquifer (Fig. 8a). Although the watered-out well (WYK-26) was revived, an unknown amount of N_2 went out of zone (but was observed in well WYK-22), and a relatively less volume of residual gas was produced by creating a secondary gas cap (mixture of residual gas and N_2) (Fig. 8b).

Planned and observed sweep patterns: discussion of geological factors

Figure 9 illustrates a map view of various sweep patterns observed from the two EGR methods (Fig. 9). Pattern A represents NERG in the De Wijk South area comprising a fraction of watered-out Volpriehausen (Fig. 9a). The intended sweep went from downflank well WYK-15B in the south, along the erosional reservoir limit towards the crestal well WYK-26 in a northerly direction (blue outline). Pattern B forms the largest reservoir in this EGR campaign and the intention is to sweep via injection well WYK-22 in the south towards several wells (WYK-6, WYK-11, WYK-34 and WYK-35) in a northwards direction (Fig. 9b). WYK-35 was drilled specifically for the EGR project to provide a new drainage point in the northern part of the Rogenstein. Pattern C is mainly formed by Vlieland sandstone, which has good reservoir properties but a complex pinch-out pattern (the estimated extent above the GWC is outlined by the red line in the De Wijk East accumulation) (Fig. 9c). Injection well WYK-20 is located at the SW fringe of the Vlieland Member, and sweeps the remaining reservoir gas towards the nearby well WYK-17B to the west and to wells WYK-29 and WYK-16 towards the east. WYK-17 shows a fraction of the injection of well WYK-31, which went into the Vlieland only (excluding the injection into the Muschelkalk). Pattern D is the unconformably underlain Muschelkalk reservoir, which has also been partially eroded (Fig. 9d). The estimated fringe of the reservoir is depicted by the green line in the De Wijk East accumulation. Pattern D is swept by injection well WYK-31 in SW part towards producer wells WYK-29 and WYK-16, which are completed in both the Vlieland and Lower Muschelkalk reservoirs.

An efficient recovery from the EGR depends on the reservoir architecture and structural geometry, the chosen injector–producer configuration, and the amount of N_2 that can be accommodated in the production after partial N_2 breakthrough. It has become clear from differences in the planned and observed gas production that a detailed understanding of the reservoir architecture is absolutely key to maximizing sweep efficiency. Here examples are given of ‘good’ predictions and ‘deviations’ from predictions at various reservoir levels to illustrate this point.

Pattern C (Vlieland reservoir) matches the reservoir architecture of the good reservoir Vlieland sandstone. Geological analysis also indicated a similar geometry of the proximal transgressive (good reservoir) sandstone that pinches out and is overlain by an extensive distal transgressive (poor reservoir) sandstone extending to the NW. Although the entire Vlieland sandstone is also part of Pattern C, no specific sweep has been implemented given the poor reservoir properties.

Wells WYK-6 and WYK-11 are completed (in the early 1960s) in relatively poor Rogenstein reservoir intervals, where good reservoir oolitic layers were expected to be absent due to erosion (structurally being at the crestal setting, as seen by contours in Fig. 9). However, production data in response to

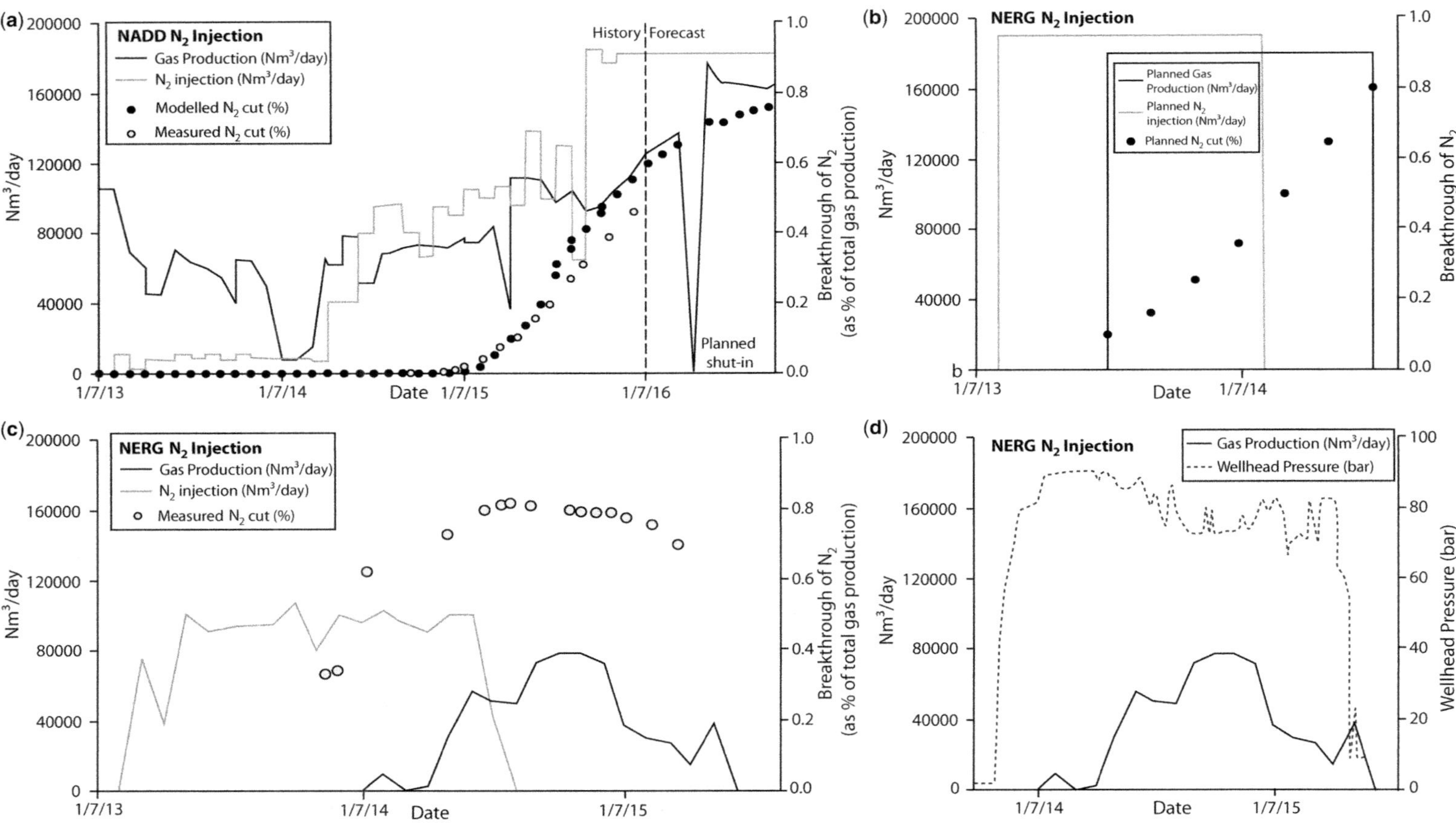

Fig. 7. Illustration of NADD and NERG well test result. (**a**) A producer well test result showing the response to nitrogen from a nearby nitrogen injector. The break in forecasted gas production shows the planned shut-down time of the producer well. (**b**)–(**d**) Comparison of production well response to a nitrogen injection in a nearby well in the water leg (NERG). (a) Shows the planned nitrogen injection and gas production, and the planned nitrogen breakthrough in the producer well. The difference in timing between the nitrogen injection and hydrocarbon production is the response time between the injector and producer. (b) is the measured nitrogen injection and hydrocarbon production, and the subsequently measured nitrogen breakthrough. The comparison between modelled and measured results shows that the model is comparable to the subsurface behaviour despite a slight delay in well response time. (c) Lag time between nitrogen injection and gas production. (d) Pressure response to the nitrogen injection in a watered-out well, allowing for subsequent uplift in gas production.

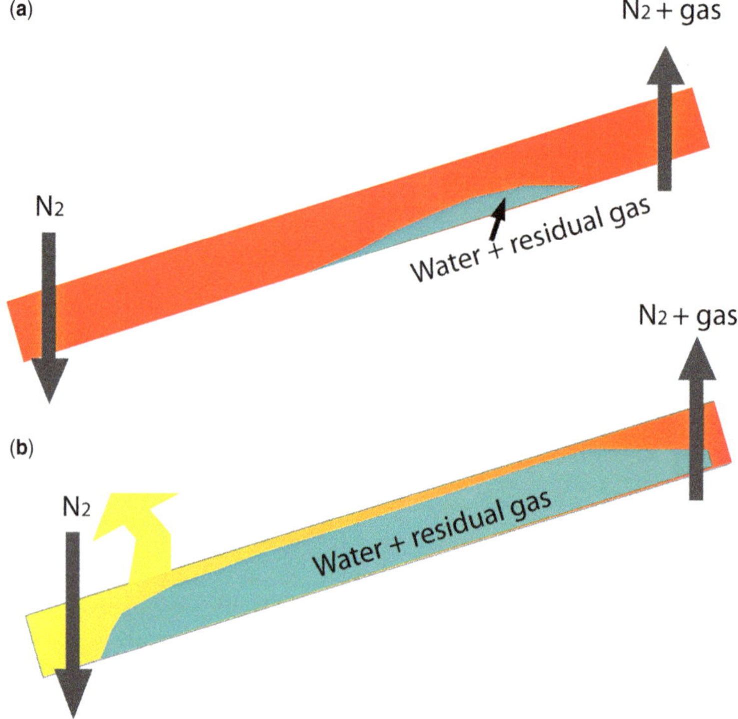

Fig. 8. (**a**) Schematic of what was intended in NERG against (**b**) what was observed. The expectation was to completely stop water production and to produce residual hydrocarbons from the water leg. However, (b) shows an early nitrogen breakthrough from a nearby producer and, thus, sweeps a smaller volume of hydrocarbon from the water leg than predicted.

EGR in WYK-11 showed the likely presence of an unrecognized oolitic layer. Considering that the log data from these old wells can be ambiguous, the reservoir model was updated by extending a high porosity–permeability (poro–perm) oolite layer from well WYK-22 (as observed in the log data in Fig. 5a, b) to the well WYK-11 location. The better production response in Pattern B (Rogenstein reservoir) could immediately be matched by the model. Hence, a better understanding of the reservoir and flow behaviour was gained from the injection and production data.

The reservoir structure map (Fig. 1) shows that not all areas of the reservoirs can be swept efficiently from the existing well patterns. This leads to a wider range of recovery factors of remaining gas, respectively, with varying sweep efficiencies for different patterns. The highest sweep efficiency is estimated for Pattern B (Rogenstein reservoir), with around 70%. This is mainly due to the drilling of the new producer well WYK-35 to extend sweep along the modelled sweep pattern. A lower sweep efficiency in the De Wijk East accumulation results from the presence of a relatively tight reservoir (Vlieland sandstone) that cannot be swept with existing wells. Future infill wells in these areas could improve sweep efficiency.

Currently around 20% of the total planned N_2 has been injected and around 10% of the total planned additional gas recovery has been realized. This is considered 'as planned'. Although analysis is ongoing, from the observed responses over 2015–16, several deductions can be made. For Pattern B (Rogenstein), injection in well WYK-22 was gradually increased to planned injection rates to achieve sweeping of the relatively large remaining gas volume towards wells. An efficient way of sweeping in a relatively tight reservoir matrix with extensive high-permeability layers seems possible, provided that the N_2 injection rate is moderate enough to avoid 'speed flushing' of high-permeability layers, whilst, at the same time, still sufficient N_2 is injected to build up and maintain pressure to keep wells above liquid loading. Production was mainly expected from wells WYK-34 and WYK-35. Currently production rates have stabilized, and decline was arrested in both wells. As expected, no N_2 breakthrough has yet been observed in these two wells.

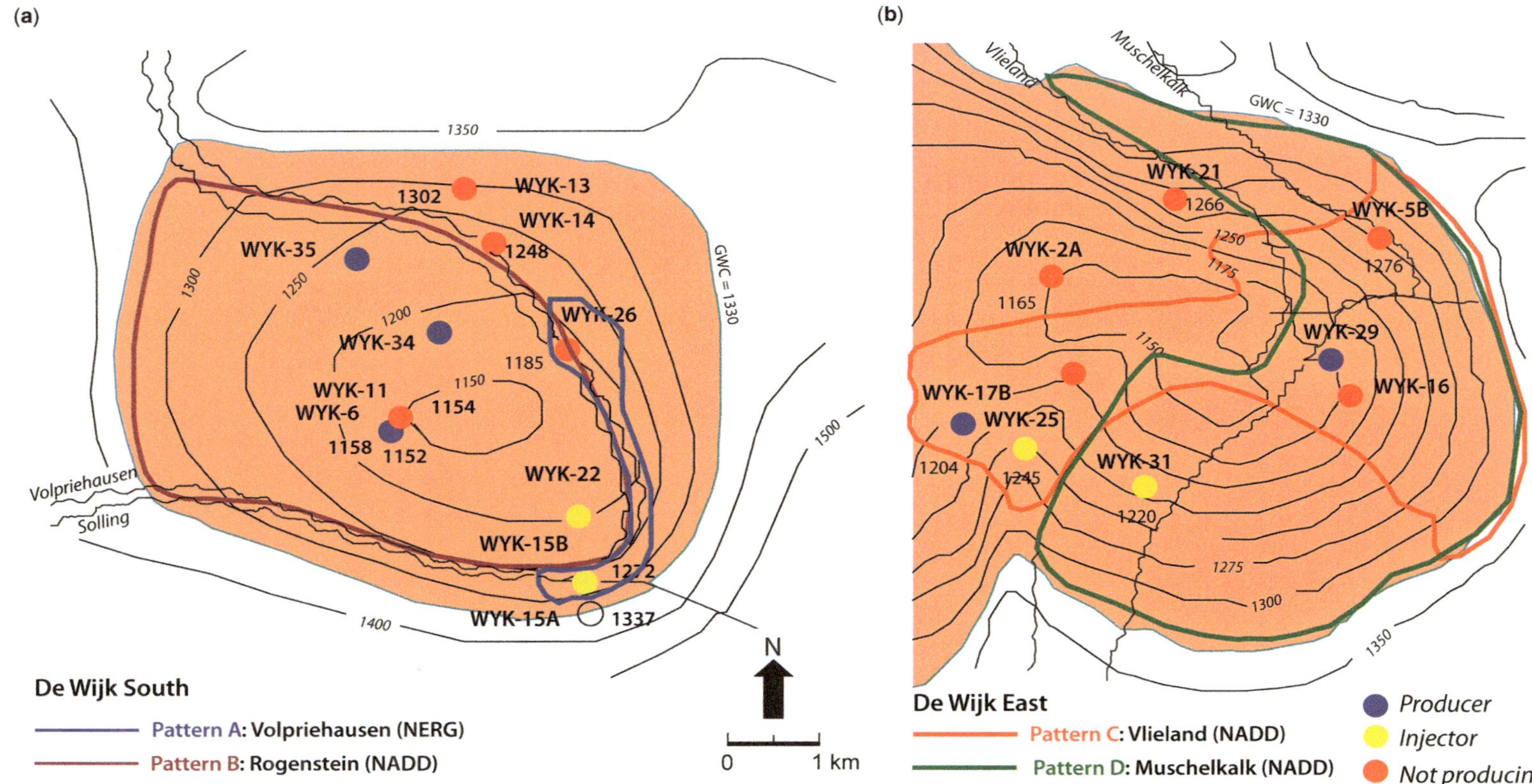

Fig. 9. Sweep patterns of NERG and NADD: (**a**) NERG in watered-out Volpriehausen sandstone (Pattern A) and NADD in depleted lower Rogenstein (Pattern B) in the De Wijk South area; and (**b**) NADD in depleted Vlieland sandstone (Pattern C) and Muschelkalk (Pattern D) in the De Wijk East area.

The first N_2 build-up from Pattern D (Lower Muschelkalk) was observed in well WYK-29. A formation damage problem was encountered in the injection well WYK-31. Prior to the start of injection, a velocity string was removed to reduce friction losses and production logging tests were performed twice (before and after removing the velocity string). By comparing the tests, it was demonstrated that N_2 was injected only into the top part of the reservoir due to unintended reservoir plugging. Injectivity had roughly halved due to this incident. Geological analysis also showed that the main uncertainty was whether Muschelkalk intralayers are vertically in communication due to leaching and karstification or whether they are vertically isolated. The reservoir complexity of the Lower Muschelkalk is mainly governed by the thin high-permeability streaks in a low-permeability matrix, by which small offset (subseismic) faults might result in a large range of juxtaposition scenarios, which either complicate or simplify sweep. This remains difficult to quantify.

Pattern A (Volpriehausen) comprises only a fraction of the watered-out Volpriehausen reservoir as this is designed as a short pilot because the NERG process has more uncertainties The NERG EGR concept has not yet been applied in gas fields, although gas in water injections is a standard industry practice to create underground gas storage in aquifers (e.g. Nigeria: Muonagor & Anyadiegwu 2014). The intended pilot sweep could not be carried out as planned due to an unexpected partial out-of-zone injection. This was indicated by a minor N_2 breakthrough observed unexpectedly in a revived Rogenstein well in combination with the production response in Pattern A, which took much longer than anticipated. Part of the N_2 injected into the Volpriehausen Member is most likely to have flowed via the unconformity into the Vlieland and Rogenstein members. The remaining N_2 most probably experienced gravity override, resulting in residual gas production in well WYK-26 by creating a secondary gas cap (mixture of residual gas and N_2). This reduced the overall recovery of gas, as a large volume of residual gas between the injector and producer was not swept due to gravity override. The upside of creating a secondary gas cap is water-free (and sand-free) production. The downside is production of a gas that contains more N_2 and less residual gas compared to a broader sweep. Production is less complex due to the absence of water production, if a higher N_2 content can be tolerated. However, creating a secondary gas cap within reasonable timescales may only be possible in a higher-permeability reservoir. This implies that the NERG test response needs further work to update reservoir models, and to assess the impact of the observed connectivity with the Volpriehausen interval on the Rogenstein gas recovery forecast.

Conclusions

Integrating geological analysis and production data into a reservoir model was the key factor for the implementation of this project. The positive first reservoir responses demonstrate the feasibility of N_2 EGR in this field; a learning that can be shared for applications in similar mature fields. For example, the second phase of N_2 injection in the shallow basal Dongen Tuffite/Chalk reservoir is expected to commence in late 2017.

NADD (gas–gas displacement using N_2) is considered to be proven (low risk) and forms the basis of this demonstration project. Like De Wijk, the NADD response proved that the depletion gas sweep mechanism works in different reservoir types of a field, as the current study showed that it works for different reservoirs. For example:

- A better understanding of the depositional environment and the reservoir architecture of the better-quality Vlieland sandstone reservoir allowed the preparation of a realistic N_2 breakthrough model that confirmed high sweep efficiency in the single sandstone package.
- Geological data analysis showed that within the low-permeability matrix of the Rogenstein siltstone, high-permeability oolites are present. This understanding helped to optimize the sweeping mechanism by injecting N_2 at a moderate rate. This led to a high EGR recovery efficiency of the remaining gas by moderating the N_2 injection. On the other hand, production data helped in updating the geological model by extending the oolitic layers.
- Uncertainty over whether the Muschelkalk intralayers were vertically interconnected by leaching and via small faults or remained vertically isolated did not significantly affect the modelled EGR recovery efficiency, and both cases were found to be attractive. However, production response data to the injections showed that layers are likely to be vertically isolated and that stimulation of lower layers in an injector is warranted.

The NERG test response is encouraging as a completely watered-out well was restored to formation water-free production. A secondary gas cap was created, resulting in a high N_2 content of the produced gas. The planned alternative of sweeping more volume/residual gas from injection to production point requires more evaluation and testing. The benefits could be a lower N_2 cut and a higher recovery. However, the persisting water production could be a downside. Rebuilding the geological model and further analysis is still a work in progress, with the objectives being to enable evaluation of the NERG test and to assess further development options of the watered-out Volpriehausen Member.

We would like to thank NAM management for agreeing to publication. We are grateful to the Geological Society of London for their invitation to write this paper. Special thanks to Dr B. Kilhams for his critical feedback in an earlier version of the paper. We are grateful to the editor, Dr T. McKie, and reviewers, especially Dr M. Geluk, for their comments that have improved the manuscript tremendously.

References

Brueren, J.W.R. 1959. The stratigraphy of the Upper Permian 'Zechstein' formation in the eastern Netherlands. *In*: *I Giacimenti Gassiferi dell 'Europa Occidentale, 1*. Accadcmia Nazionale dei Lincei, Rome, 243–274.

Bruijn, A. 1996. De Wijk gas field (Netherlands): reservoir mapping with amplitude anomalies. *In*: Rondeel, H.E., Batjes, D.A.J. & Nieuwenhuijs, W.H. (eds) *Geology of Gas and Oil under the Netherlands*. Kluwer, Dordrecht, The Netherlands, 243–253.

De Dager, J. & Geluk, M.C. 2007. Geology of the Netherlands. *In*: Wong, T.E., Batjes, D.A.J. & de Jager, J. (eds) *Petroleum Geology*. Royal Netherlands Academy of Arts and Sciences, Amsterdam, The Netherlands, 241–264.

Feist-Burkhardt, S., Götz, A.E. *et al.* 2008. Triassic. *In*: McCann, T. (ed.) *The Geology of Central Europe, Volume 2: Mesozoic and Cenozoic*. Geological Society, London, 749–822.

Geluk, M.C. 2005. *Stratigraphy and tectonics of Permo-Triassic basins in the Netherlands and surrounding areas*. PhD thesis, Utrecht University.

Geluk, M.C. & Röhling, H.G. 1999. High-resolution sequence stratigraphy of the Lower Triassic 'Buntsandstein' in the Netherlands and Northwestern Germany. *Geologie en Mijnbouw*, **76**, 227–246.

Green, D.W. & Willhite, G.P. 1998. *Enhanced Oil Recovery*. SPE Textbook Series, **6**. Society of Petroleum Engineers, Richardson, TX.

Jeremiah, J.M., Duxbury, S. *et al.* 2010. Lower Cretaceous of the southern North Sea Basins: reservoir distribution within a sequence stratigraphic framework. *Netherlands Journal of Geosciences – Geologie en Mijnbouw*, **89**, 203–237.

Leeuwenburgh, O., Neele, F. *et al.* 2014. Enhanced gas recovery – a potential 'U' for CCUS in The Netherlands. *Energy Procedia*, **63**, 7809–7820.

McKie, T. 2017. Palaegeographic evolution of latest Permian and Triassic salt basins in north west Europe. *In*: Soto, J.I., Flinch, J. & Tari, G. (eds) *Permo-Triassic Salt Provinces of Europe, North Africa and the Atlantic Margins. Tectonics and Hydrocarbon Potential*. Elsevier, Amsterdam, 159–173.

McKie, T. & Williams, B. 2009. Triassic palaeogeography and fluvial dispersal across the northwest European Basins. *Geological Journal*, **44**, 711–741.

Muonagor, C.M. & Anyadiegwu, C.I.C. 2014. Development and conversion of aquifer for underground natural gas storage in Nigeria. *Petroleum & Coal*, **56**, 1–12.

NAM & RGD 1980. *Stratigraphic Nomenclature of The Netherlands*. Verhandelingen van het Koninklijk Nederlands Geologisch Mijnbouwkundig Genootschap, **32**.

Palermo, D., Aigner, T. *et al.* 2008. Reservoir potential of a lacustrine mixed carbonate/siliciclastic gas reservoir: the Lower Triassic Rogenstein in the Netherlands. *Journal of Petroleum Geology*, **31**, 61–96.

Pipping, J.C.P., Carlson, T. *et al.* 2001. Sedimentary cycles are key to improve reservoir performance in carbonates, Triassic Lower Muschelkalk – De Wijk gas field, The Netherlands. *Extended abstract P523 presented at the 63rd Conference of the European Association of Geoscientists & Engineers*, 11 June 2001, Amsterdam, The Netherlands.

Regtien, J.M.M., Por, G.J.A. *et al.* 1995. Interactive reservoir simulation. Paper PE-29146-MS, presented at the SPE Reservoir Simulation Symposium, 2–15 February, San Antonio, Texas, https://doi.org/10.2118/29146-MS

RGD 1993. *Geological Atlas of the Subsurface of The Netherlands, Explanation to Map Sheet V, Sneek – Zwolle*. Geological Survey of The Netherlands, Haarlem, The Netherlands.

Rijkers, R.H.B. & Geluk, M.C. 1996. A tectogenetic mechanism controlling the evolution of the Texel-IJsselmeer High, the Netherlands. *In*: Rondeel, H.E., Batjes, D.A.J. & Nieuwenhuijs, W.H. (eds) *Geology of Gas and Oil Under the Netherlands*. Kluwer Academic, Dordrecht, The Netherlands, 265–284.

Ruffell, A. 1991. Sea-level events during the Early Cretaceous in Western Europe. *Cretaceous Research*, **12**, 527–551.

Shrimali, H.V. 2015. Developments in enhanced oil recovery technologies: a review. *International Journal of Research in Advent Technology*, **3**, 2321–2963.

Van Adrichem Boogaert, H.A. & Kouwe, W.F.P. (compilers). 1994. *Stratigraphic Nomenclature of the Netherlands; Revision and Update by RGD and NOGEPA, Section E Triassic*. Mededelingen Rijks Geologische Dienst, **50**.

van der Zwan, C.J. & Spaar, P. 1992. Lower to Middle Triassic sequence stratigraphy and climatology of the Netherlands, a model. *Palaeogeography, Palaeoclimatology, Palaeoecology*, **91**, 277–290.

Zagwijn, W.H. 1989. The Netherlands during the Tertiary and the Quaternary: a case history of coastal lowland evolution. *Geologie en Mijnbouw*, **68**, 107–120.

Ziegler, P.A. 1990. *Geological Atlas of Western and Central Europe*. 2nd edn. Shell Internationale Petroleum Maatschappij, The Hague. Geological Society, London.

Data mining in the Dutch Oil and Gas Portal: a case study on the reservoir properties of the Volpriehausen Sandstone interval

B. M. M. van KEMPEN*, H. F. MIJNLIEFF & J. van der MOLEN

TNO, Princetonlaan 6, 3584 CB, Utrecht, The Netherlands

**Correspondence: bart.vankempen@tno.nl*

Abstract: This paper presents the results of a data mining exercise on core measurements and reports of the Triassic Lower Volpriehausen Sandstone Member (RBMVL) in The Netherlands. The availability of a large accurate online dataset via the Dutch Oil and Gas Portal (NLOG) enables the definition of general reservoir property trends on a regional–subregional scale. Accurately defined trends are capable of restricting the uncertainty range of key parameters in exploration studies (e.g. porosity and permeability), and thus of reducing risk. Plotting core porosity v. core permeability resulted in a curved data cloud on a log-normal plot. Although porosity–permeability relationships are often defined by straight lines, the dataset in this study shows that a curved relationship results in a better fit with permeability measurements over the entire porosity range. Two methods were applied to define a porosity–depth relationship for sandstones of the RBMVL. The results of both methods confirm, and enable quantification of, the well-known trend of decreasing porosity with depth. By assigning grain density attributes to the core measurements, the dependency of the porosity–depth relationship on sedimentary facies and diagenesis type/grade could be determined. This case study demonstrates the power of this free online dataset in understanding reservoir quality.

Insight regarding reservoir property trends is of great importance for regional exploration studies. In particular, in The Netherlands, thanks to the increasing importance of geothermal resource exploration and exploitation. The presence of a high-quality reservoir sequence, in terms of porosity and permeability, is crucial for a successful geothermal project (e.g. Franz *et al.* this volume, in press). With access to reliable and accurate data, these reservoir property trends are capable of reducing the uncertainty range on estimated porosity and permeability, resulting from the interpolation of reservoir property data from wells. Uncertainty reduction can lead to a significant decrease in exploration risk.

Many studies have been carried out regarding global reservoir property trends, including the porosity–depth and porosity–permeability relationships. The reduction of porosity with burial depth is extensively described in, for example, Ramm & Bjørlykke (1994), Paxton *et al.* (2002), Ehrenberg & Nadeau (2005) and Ajdukiewicz & Lander (2010). Because the trends presented in these studies are often case-specific and not universally applicable, they might not be representative of specific rock formations (and/or locations) in the Dutch subsurface.

A data mining exercise was carried out on the Dutch subsurface databases (NLOG 2017) to determine reservoir property trends on a regional–subregional scale across The Netherlands, predominantly using core measurements. The aim is to create a reliable regional–subregional reference framework to which local reservoir-quality analyses can refer.

The Dutch Lower Volpriehausen Sandstone Member (RBMVL), which is one of the important reservoir units in The Netherlands, is used as a case study. This stratigraphic interval is of Early Triassic age (252–247 Ma) and is part of the Volpriehausen Formation, which is the oldest formation of the Main Buntsandstein Subgroup (Feist-Burkhardt *et al.* 2008). Across The Netherlands, this lithostratigraphically defined unit mainly comprises fine- to medium-grained, light-coloured arkosic sandstones. Although this sandstone interval has experienced a high lateral variation in diagenetic processes, the base is often highly cemented by calcite and dolomite (Geluk 2005). Sediments are predominantly of fluvial origin in the south of The Netherlands and aeolian in the north (see also Geluk *et al.* 2018; Kortekaas *et al.* this volume, in press). Due to the spatial variation in sedimentary facies, the RBMVL is characterized by a variable clay content. As a result, variable reservoir quality is recognized within the datasets.

To be able to define reservoir property trends, the availability of a large dataset is essential. The data used for this study were acquired from the Dutch

From: Kilhams, B., Kukla, P. A., Mazur, S., McKie, T., Mijnlieff, H. F. & van Ojik, K. (eds) 2018. *Mesozoic Resource Potential in the Southern Permian Basin*. Geological Society, London, Special Publications, **469**, 253–267.
First published online March 15, 2018, https://doi.org/10.1144/SP469.15

Oil and Gas Portal (NLOG 2017), from which all publicly available data and information on the Dutch subsurface can be obtained. All core-plug measurements pertaining to the RBMVL were selected. Additional information on petrography and sedimentology of the RBMVL was also downloaded.

When visualizing core data in porosity–depth and porosity–permeability graphs, trends can be recognized and relationships can be defined for the RBMVL specifically. The dataset represents a highly variable reservoir quality that enables the investigation of the porosity–depth relationship, not only for reservoir-quality sandstones but also for reservoirs with limited quality (Ramm & Bjørlykke 1994). Grain density, usually measured on core samples, can be used as a basis for this analysis as it is related to the amount and/or type of porosity-reducing minerals (clay, dolomite, anhydrite, etc.) and is an indirect indication of porosity. Based on these results, not only a relationship between porosity and burial depth is defined, but also a dependence on facies and diagenesis is incorporated (Fontaine *et al.* 1993; Purvis & Okkerman 1996). Because of the growing interest in hydrocarbon reservoirs with limited reservoir quality (or 'tight' reservoirs) in The Netherlands, further insight into the relationship between porosity and depth in these intervals would have additional value.

Data selection and analysis methods

A set of RBMVL core measurements was analysed to investigate the porosity–permeability and porosity–depth relationships of the RBMVL. This publicly available dataset contains ±7000 Volpriehausen interval porosity measurements from 80 different wells (Fig. 1). In this section, the data-selection criteria, applied corrections and methods that are used to analyse porosity–depth relationships are described.

Data selection

Three selection criteria were applied to obtain a representative dataset (from the initial collection of ±7000 measurements) for porosity–depth analysis. First, approximately 900 duplicates were removed because they would influence calculated averages. These are present because some core-analysis reports were received twice, from different sources, and were both added to the database. Secondly, data from wells containing less than 10 core samples were excluded from the analysis, as such a limited number of measurements might not represent the average reservoir properties of the stratigraphic interval. Lastly, data points have been excluded in cases where the cored interval covers less than 25% of the total thickness of the stratigraphic interval. Again, there is a possibility that the measurements do not represent the average reservoir properties, especially when the core is taken at the top or base of the RBMVL interval. These sections are documented as being often highly cemented or clay rich (Geluk 2005).

The remaining dataset for the porosity–depth evaluation exercise contains approximately 5500 porosity measurements from 59 different wells. The coverage of these wells with respect to the distribution of the RBMVL is presented in Figure 1.

The dataset used for the analysis of the relationship between porosity and permeability is a subset of the initial collection of ±7000 measurements, and only contains samples on which both porosity and permeability are measured. It comprises approximately 5100 core plugs from 75 Dutch wells.

Corrections

In order to evaluate porosity–depth trends and to use the results for a direct comparison with reservoir porosities derived from logs, corrections to the core data were applied. An *in situ* correction for porosity and a correction for the maximum burial depth were applied to the individual core measurements.

In situ porosity. When cores are brought to surface, the removal of overburden pressure causes relaxation of the rock, which leads to an increase in porosity (Cosentino 2001). The porosity measurements in the dataset have been collected using conventional core-analysis techniques, which are conducted at atmospheric conditions. Porosity measurements at overburden pressure are relatively scarce and not incorporated into the dataset. Multiple studies have been published in which formulas are presented to correct ambient porosity measurements to *in situ* conditions (Juhasz 1986; Evans *et al.* 1994; Holt *et al.* 2003). The input parameters (total and effective porosity) required for these corrections were not available for this study. Therefore, a correction factor of 0.95, commonly used as a default in the oil and gas industry and based on the uniaxial correction factor by Teeuw (1971), has been applied to all available porosity measurements.

No corrections were applied to the dataset used for the definition of the porosity–permeability relationship because the goal is to present the nature of the porosity–permeability relationship, not to determine exactly the specific porosity–permeability relationship of the RBMVL, corrected for downhole conditions. *In situ* corrections reduce both porosity and permeability, which will not alter the nature of the relationship.

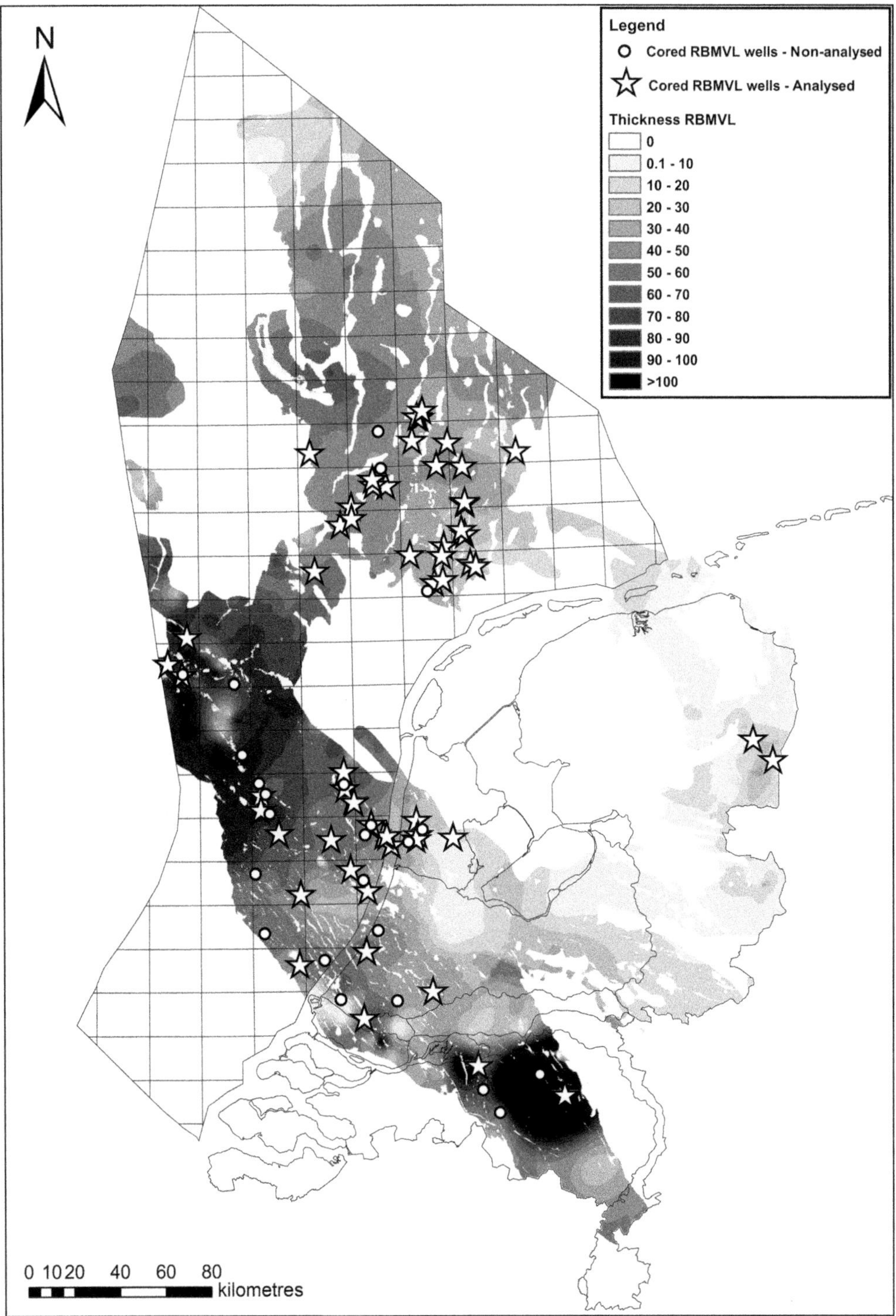

Fig. 1. Thickness map of the RBMVL (inhouse map) showing the locations of the analysed wells (stars) and the non-analysed wells (circles), used for the analysis of the porosity–depth trend. The two sets combined represent all the wells with cored intervals of the RBMVL.

Maximum burial depth. As some of the selected wells are located in inverted basins, such as the West Netherlands Basin (de Jager *et al.* 1997; Duin *et al.* 2006), the difference between the maximum burial depth and present-day depth can be large (>1 km). Due to mechanical and chemical compaction, porosity decreases with increasing burial depth (Waples 2002). Mechanical compaction is generally accepted to be an irreversible process. The use of present-day depth in a porosity–depth plot would result in a large amount of scatter at shallow depths, because low-porosity rocks were possibly buried much deeper than at present day (Fig. 2a). Therefore, present-day depth needs to be corrected to the maximum burial depth (Fig. 2b).

Several studies have been conducted on the burial history of The Netherlands (Van Dalfsen *et al.* 2005; Luijendijk *et al.* 2011; Nelskamp & Verweij 2012). For this study, the burial maps resulting from the basin-modelling study of Nelskamp & Verweij (2012) were utilized to determine the maximum burial depth of the RBMVL at all well locations and this value was assigned to every measurement as an attribute. The average depth correction was 265 m, with a maximum of 1300 m. As a result, a subset of the data points in Figure 2a was shifted downwards, resulting in the porosity v. maximum burial depth plot of Figure 2b, which shows less scatter in the shallow depth domain than Figure 2a.

Data analysis

Compaction (mechanical and chemical) is the driving process for porosity reduction with burial. Compaction is influenced by primary mineralogy and clay volume (e.g. Scherer 1987; Ramm & Bjørlykke 1994). Figure 2b shows a high degree of scatter and a wide spread in porosity measurements, which is attributed to facies differences (varying clay volume and grain sorting) and a varying degree of diagenesis (pore cement). The main evidence of overall decreasing porosity with depth is shown by the 'high-porosity envelope' that represents 'clean' sandstone samples. However, the porosity–depth relationship defined along this envelope may not be applicable to other sandstone facies because porosity can deviate from the 'clean' sandstone porosity–depth trend as a consequence of other processes influencing the degree of compaction (e.g. in case of early gas fill or overpressured reservoirs). In order to define a porosity–depth relationship for reservoir-quality sandstones of the RBMVL, each sample requires an indication of clay content and diagenesis. Two methods were used to assign a reservoir-quality attribute to all samples within the dataset.

Method 1: clay volume and diagenesis. The first method is based on all relevant publicly available information within each selected well, regarding the RBMVL; thereby, focusing on factors that influence porosity other than burial (e.g. clay volume and diagenesis). A gamma-ray log analysis was performed to calculate a clay volume curve over the RBMVL. After applying the required core shift, which was determined per well, the clay volume at the depth of each individual core sample was determined. A 'high clay volume' attribute was assigned to samples with a clay volume higher than 35%.

Furthermore, a broad selection of documents, such as core reports and descriptions, petrography/sedimentology/petrophysics reports, and composite well logs, was used to collect information about the type and degree of diagenesis. By qualitative judgement, based on the available information, 'diagenesis' or 'reservoir-quality' attributes have been assigned to the samples. This resulted in a total of three attributes, with a distinction between reservoir-quality (1) and non-reservoir-quality ('high clay volume' (2) and 'diagenesis' (3)) samples.

Method 2: grain density. The second method was based on grain density, which is measured on each individual core sample, together with the porosity and permeability. Grain density can be regarded as an indirect measure of porosity. The presence of a significant amount of detrital clay minerals generally causes a low porosity (Fig. 3). Although the grain density of clay minerals is variable, clay-rich core samples generally show a higher grain density than quartz (2.65 g cm^{-3}) (Fig. 3). In this study, grain density can be regarded as specific gravity. The majority of the core samples have been hot-oven dried and, therefore, did not contain clay-bound water when the measurements were performed. Also, pore-filling (authigenic) minerals in sandstone (e.g. clay, dolomite and anhydrite) have a high grain density. Hence, an average grain density higher than quartz is usually caused by an increased amount of porosity-reducing minerals in the sample, and therefore indicates a lower porosity than a 'clean' sandstone would have. On the other hand, samples with an average grain density lower than quartz can also have a reduced porosity by the presence of pore cement. In this case, pore-filling minerals are probably halite or light clays, which have a grain density lower than quartz.

In order to make a distinction between reservoir-quality and non-reservoir-quality samples, eight grain-density classes were defined in the range of <2.60 to >2.75 g cm^{-3}, with intervals of 0.025 g cm^{-3}. The grain-density bin attribute was added to each sample. Additionally, a simplification of the bins in two classes was made: 'reservoir quality' and 'non-reservoir quality'. Samples with a grain density in the range of 2.625–2.675 g cm^{-3} were classified as 'reservoir quality' and the remaining samples as 'non-reservoir

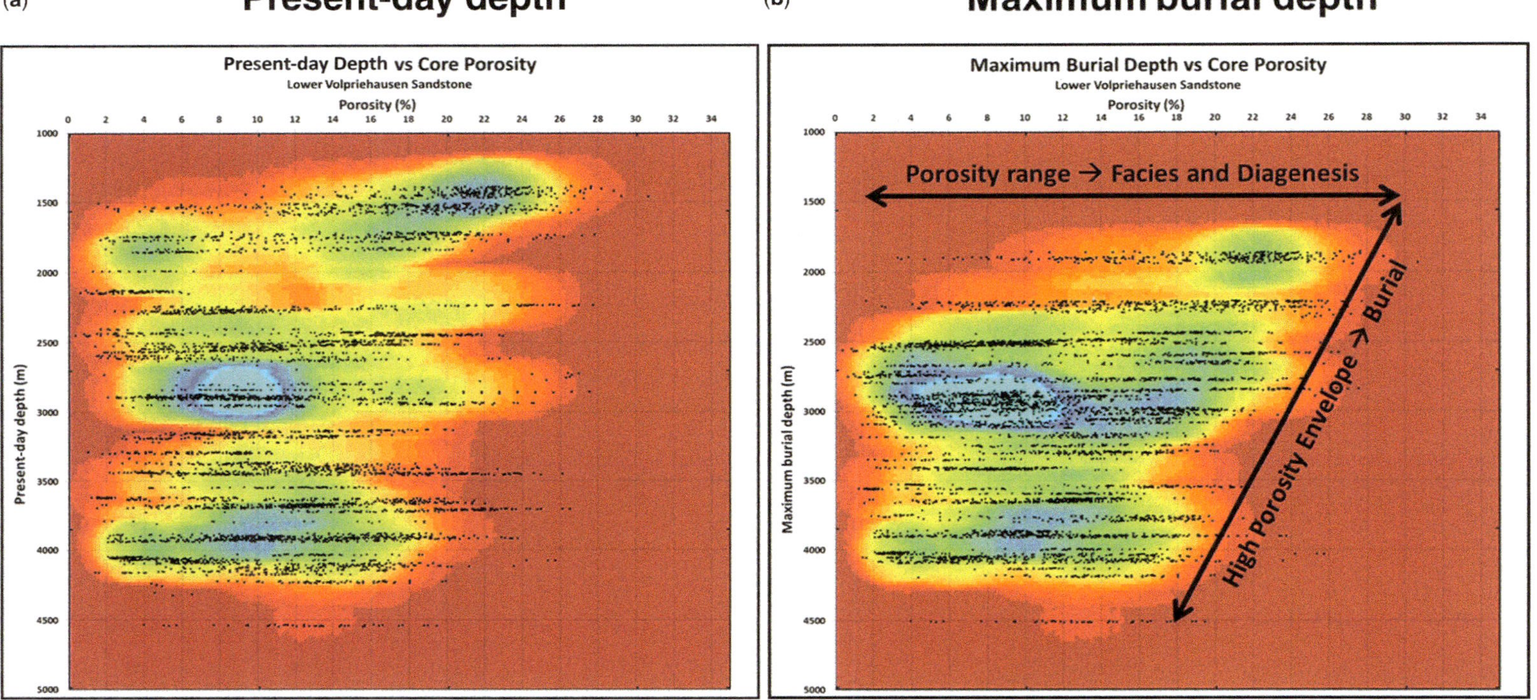

Fig. 2. Porosity v. depth plots of the RBMVL core-plug measurements ($n = 7003$). The coloured background represents the point density (red, low; light blue, high). (**a**) Porosity v. present-day depth (metres true vertical depth subsea (m TVDss)). (**b**) Porosity v. maximum burial depth.

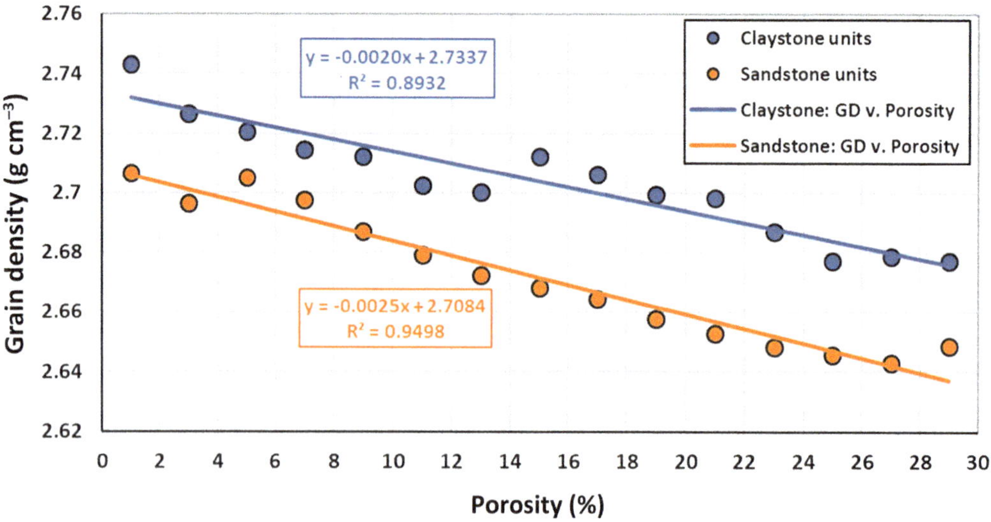

Fig. 3. Grain density v. porosity based on core-plug measurements of Triassic sandstone (orange) and claystone (blue) units. The average grain density is plotted for each porosity bin. The total number of samples underlying the porosity bins are 19729 for the sandstones and 8006 for the claystones.

quality'. Colour attributes based on the detailed and simplified classification have been assigned to all data points in the porosity–depth plot.

The more detailed classification of the porosity measurements was made to investigate whether a porosity–depth relationship is also present within each individual class, and if the relationship changes with increasing grain density. Because the data points per grain-density class show a relatively high scatter in the porosity–depth plot, the porosity measurements in each class were sorted by maximum burial depth and averaged per 50 m depth interval. These averages were plotted v. the mid-depth of the interval, and the linear porosity–depth relationship has been defined per grain-density class.

Results and discussion

Porosity–permeability relationship

A generalized porosity–permeability cross-plot of the RBMVL in The Netherlands (Fig. 4) was created using the selected porosity–permeability dataset (n = 5100). The coloured background plot, created with the 2016 statistical software Isatis (Geovariances 2016), represents the data-point density and was added for an increased visual recognition of trends.

Although the dataset contains samples from the entire depth range, from various sedimentary facies and for various grades of diagenesis, the porosity–permeability relationship is well defined (Fig. 4). The point-density plot shows a relatively narrow cloud of permeability measurements over the entire porosity range. Only at porosities lower than 18% does the spread in permeability broaden to include lower values. This is probably facies-related, as a result of an increased clay volume. Common types of clay (e.g. illite) usually have a larger negative impact on permeability than on porosity. Clays have intragranular microporosity, and only a few volume per cent of these minerals, especially pore-bridging clays, can have a high negative impact on permeability (Rossel 1982; Purvis & Okkerman 1996; Gaupp & Okkerman 2011).

The colour pattern in Figure 4 demonstrates that the increasing trend of the data cloud is not linear on the log-normal plot, but concaves downwards. Porosity–permeability relationships are often defined by an exponential function, which appears as straight lines on the log-normal plot. Small datasets, especially when they do not cover the high-permeability range well, may show an exponential relationship (a straight line on a log-normal plot). However, when expanding the dataset, in most cases the porosity–permeability data show a concaving-downwards trend. Therefore, an exponential relationship will not honour the entire dataset adequately, but overestimates permeability in the low- and high-porosity range, and underestimates permeability

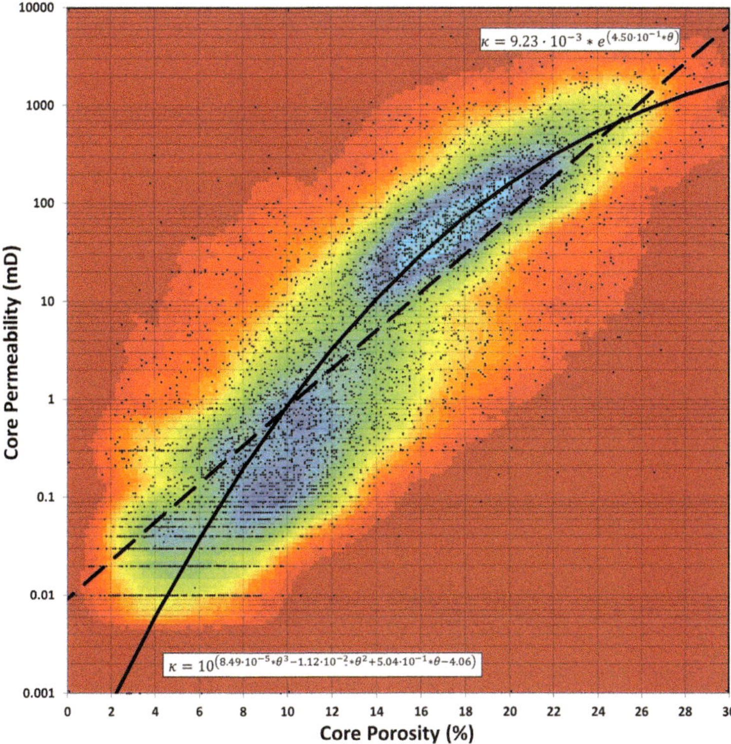

Fig. 4. Porosity v. permeability plot, based on 5131 RBMVL core-plug measurements. The coloured background represents the point density (red, low; light blue, high).

in the intermediate range (Fig. 4). Based on these observations, it can be concluded that the most optimal way to define a porosity–permeability relationship is by a curved (or double exponential) porosity–permeability relationship (e.g. polynomial or power relationship). This results in a smaller permeability error over the entire porosity range.

Porosity–depth relationship

The decrease in porosity with depth is a generally accepted trend predominantly related to compaction (e.g. Ramm & Bjørlykke 1994; Paxton *et al.* 2002; Ehrenberg & Nadeau 2005; Ajdukiewicz & Lander 2010). A large dataset makes it possible to define this relationship on a regional–subregional scale and to make it applicable for specific reservoir intervals, in this case for the RBMVL. However, the large degree of scatter in the porosity–depth plot of Figure 2b impedes the definition of a representative porosity–depth relationship. The reservoir-quality attributes, defined via the two methods described, are used to evaluate the possible existence of different porosity–depth trends per reservoir facies.

Method 1: clay volume and diagenesis. All relevant publicly available information for the RBMVL in each well was used to assign 'high clay volume', 'diagenesis' and 'reservoir-quality' attributes to the data points in the porosity–depth plot (Fig. 5a). The attributes are based on a qualitative judgement, as described in the Methods section.

Figure 5a shows that, after assigning attributes to the data points, a trend of decreasing porosity with increasing burial depth emerges from the scattered data cloud. As porosity is on the x-axis, and carrying the main uncertainty of the average, the relationship itself is defined by a linear x over y trend line. This line is defined through the average porosity of the reservoir-quality core samples, calculated for each individual well (Fig. 5b). The range of the bars in Figure 5b is defined by the P10 and P90 values, and is a measure of the spread in porosity.

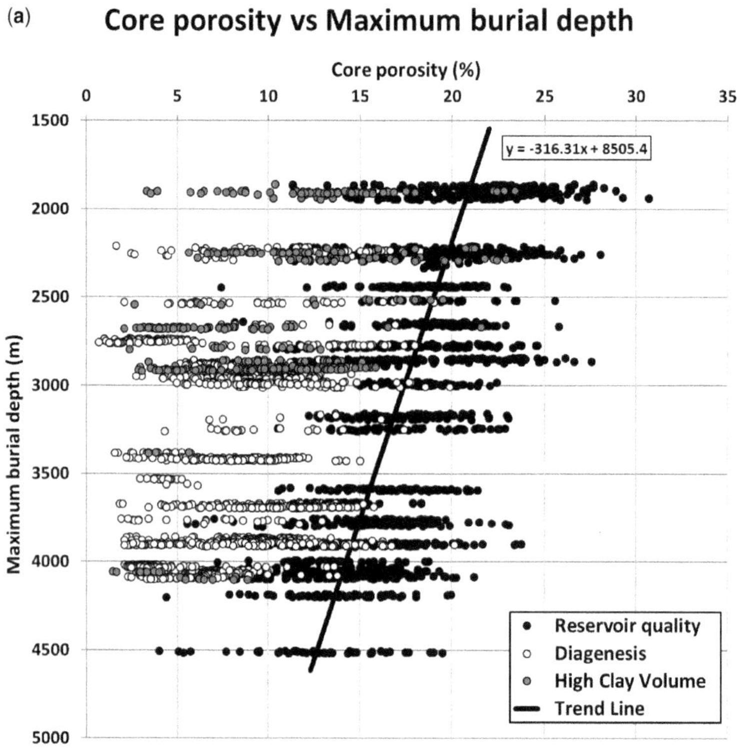

Fig. 5. Porosity v. depth plots (n = 5532). (**a**) All individual measurements, with attributes based on Method 1.

Porosity–depth relationships are often defined by exponential functions (e.g. Ramm & Bjørlykke 1994) because the impact of mechanical compaction is high on very shallow, unconsolidated sediments, but decreases with increasing burial depth. The depth range of the core samples used in this study is between 1500 and 4500 m, and does not cover the very shallow and very deep subsurface. In this range, the decrease in porosity with depth approaches a linear relationship, as shown by the data points in Figure 5a. Therefore, linear functions have been applied in this study to define porosity–depth relationships. The uncertainty of the defined relationships increases significantly outside the depth range of the core samples. Although the linear relationships show a good fit to the data, compared to an exponential relationship they will be likely to underestimate porosity outside the depth range of the dataset, so in the very shallow and deep subsurface.

Some of the average porosities in Figure 5b are higher than the trend line. This might be related to an early migration of gas into the reservoir or an overpressured reservoir, which partly counteracts the porosity decrease due to compaction (Ramm & Bjørlykke 1994). Other averages are lower than the average porosity from the trend line. This can be explained by the presence of clay and/or pore cements. The 'reservoir-quality' samples represent relatively clean sandstone samples, but within this category there is still variation in the (minor) amount of clay and pore cement present.

The effect of grain size or other textural parameters is not taken into account in this study because information regarding these properties is not readily available for most of the core-plug samples. Theoretically, grain size does not affect primary porosity, but grain sorting does (Beard & Weyl 1973). The intensity of compactional and diagenetic processes is related to the textural style of the rock. This is not evaluated separately but is incorporated in Method 1.

Method 1 only results in a porosity–depth relationship for reservoir-quality sandstone. It would also be useful to know how this trend changes with an increasing amount of clay or pore cement. This would require a more quantitative approach: for example, by using the volume percentages of clay and pore cement of each data point, instead of three discrete attributes. Also, more detailed information, such as the timing of diagenesis with respect to burial depth and clay mineral type or origin (authigenic or detrital), would be a valuable addition to this

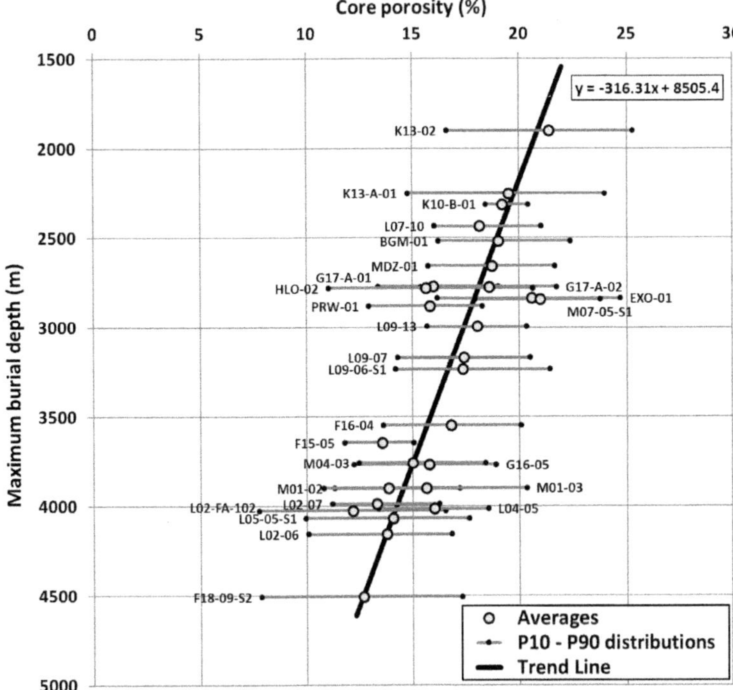

Fig. 5. (*Continued*) (**b**) Average porosity of core samples per well. Horizontal bars represent the P10–P90 range for each well. The trend line through the averages represents the porosity–depth relationship for reservoir-quality sandstones of the RBMVL. The beginning and end points of the bars in the graph depict the P10 and P90 distributions; the circle in the middle represents the average of the core porosity measurements per well.

study. As this kind of information is only available for a very limited number of wells, it falls outside the scope of this study.

Method 2: grain density. Grain density is considered to be a good indirect porosity indicator (Fig. 3). In a sandstone, a grain density higher than quartz (2.65 g cm^{-3}) can be related to the presence of clay or pore cement. These minerals have a detrimental effect on porosity. Figure 6 presents a porosity–depth plot with 'reservoir-quality' and 'non-reservoir-quality' attributes, based on grain density. The reservoir-quality samples demonstrate a decreasing porosity trend with increasing burial depth. Compared to the entire dataset, the reservoir-quality samples show a relatively narrow range of porosity values. Horizontal scatter in the non-reservoir-quality data in Figure 6 is mostly ascribed to samples with a grain density higher than quartz and, therefore, a lower porosity.

Figure 7 shows the resemblance of the 'reservoir-quality' data clouds and porosity–depth trends resulting from Method 1 (left) and 2 (right). Also, the linear relationship, as defined in Figure 5b, matches well with the high point density areas in both plots. The close resemblance of the grain-density results with the results of Method 1, which is based on direct information about clay volume and diagenesis from well data, proves the validity of the grain-density method for defining porosity–depth trends. This is favourable because the grain-density method requires less additional analysis and data than Method 1.

Some data points within the 2.625–2.675 g cm^{-3} range have a lower porosity than that expected from the general trend (marked by ovals in Fig. 7). This can be explained by quartz cementation, which is a process that lowers the porosity but does not result in an increased grain density because the pore cement and matrix both consist of quartz. Figure 7 shows that quartz-cemented samples are only present at depths greater than 2.5 km. This is in correspondence with the results of Bjørlykke & Egeberg (1993), who suggest that, in subsiding basins like the North Sea, sandstones buried to less than 2.5–3.0 km contain little quartz cement because most

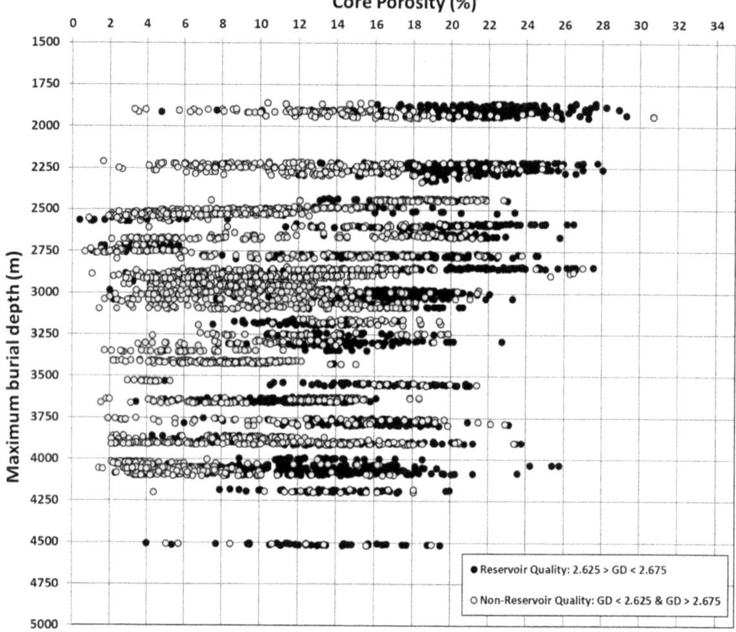

Fig. 6. Porosity v. depth plot, with attributes based on Method 2 ($n = 5532$). The filled data points represent the 2.625–2.675 g cm^{-3} grain-density range, which is defined as reservoir quality. The open data points represent all the remaining grain-density values, defined as non-reservoir quality.

quartz cementation does not form until above 90–100°C, although this process begins at 70–80°C. In theory, the effect of quartz-cemented samples could mask the greater porosity–depth trend. However, the occurrence of quartz cementation in the RBMVL of The Netherlands is considered limited. For example, Purvis & Okkerman (1996) reported that the reduction in porosity in the Triassic Main Buntsandstein Group is caused primarily by dolomite, halite and anhydrite. Quartz cementation occurs, but is less abundant.

Porosity–depth trend per grain-density class. Besides a porosity–depth relationship for high-quality reservoirs, knowledge about the development of this relationship in sandstones with relatively poor reservoir quality, representing more clay-rich sedimentary facies or a varying primary mineralogy, is also useful. As such, the entire RBMVL dataset has been subdivided into eight grain-density classes, and a porosity–depth relationship has been defined for each of these (Fig. 8).

As expected, by its basic definition, Figure 8 shows a decrease in porosity with increasing grain density. The two lowest grain-density classes deviate from this trend, especially the lowermost grain-density class. This can be explained by the presence of halite cement. Halite precipitates from salt-saturated formation water, and significantly reduces porosity (Fontaine *et al.* 1993; Dronkert & Remmelts 1996). It has a very low grain density of 2.17 g cm^{-3} and therefore also reduces the average grain density of a sample. For example, only 10% by volume of halite is required to lower the grain density of a sandstone sample from 2.65 to 2.60 g cm^{-3}.

The decrease in porosity with increasing grain density results in a decrease in the intercept with the x-axis of the individual porosity–depth relationships. This is shown in Figure 9a, where the porosity intercept (in p.u.) is plotted v. the mid-value of each grain-density class. It turns out that a linear relationship can be defined between the grain density and the porosity intercept of the porosity–depth relationship, with a high correlation coefficient ($r^2 > 0.96$). The lowest grain-density class was not taken into account because it is affected by halite cement, as described above.

Figure 8 also shows that the slope of the individual porosity–depth relationships (in p.u. per km) decreases with increasing grain density. This means that the reduction in porosity with depth is less for sandstones with a higher grain density. This may

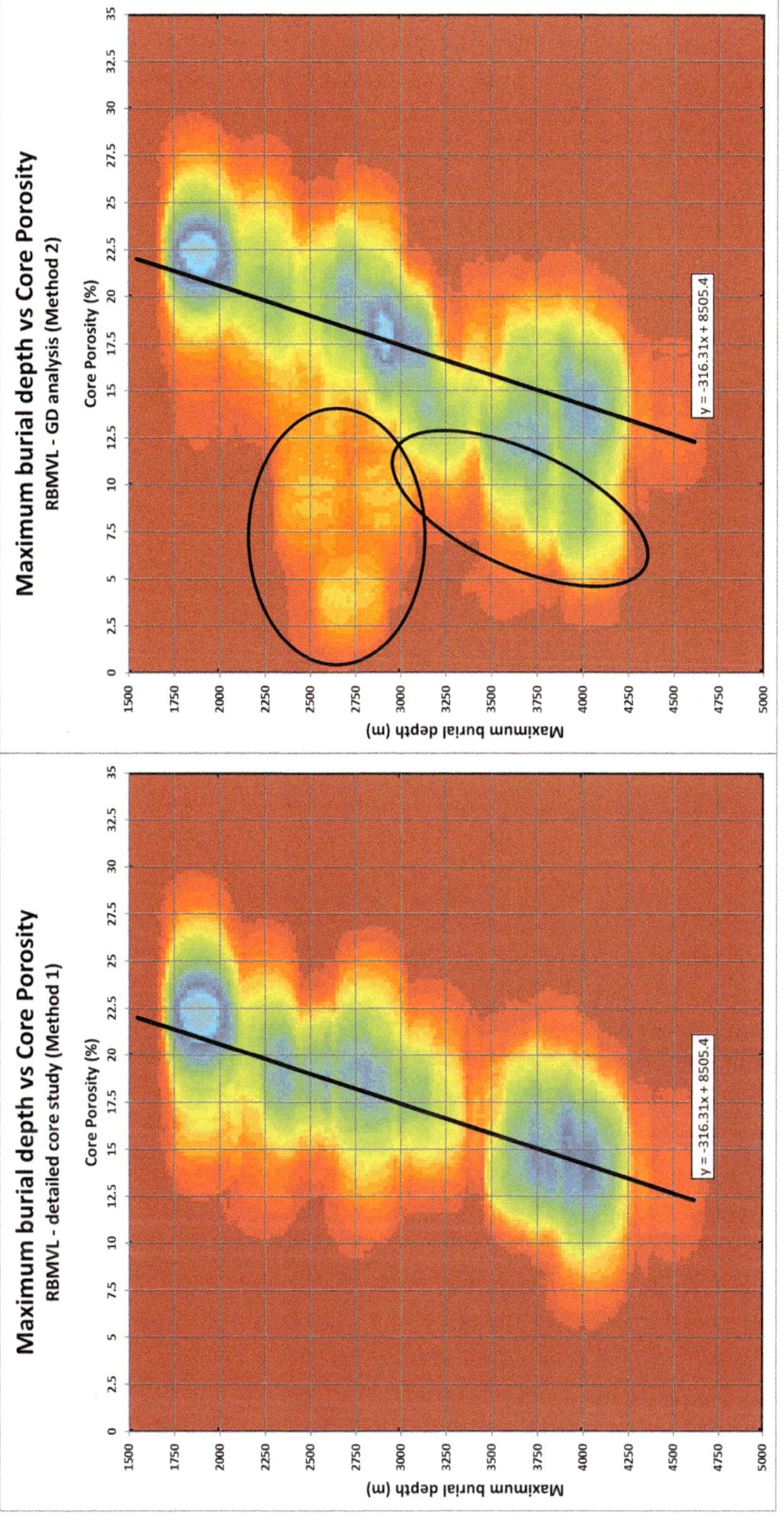

Fig. 7. Point-density plots of porosity v. depth (n = 5532), representing the reservoir-quality samples as defined by Method 1 (left) and Method 2 (right). The relationship defined by Method 1 is presented in both plots. The ovals represent porosity measurements that are likely to be affected by quartz cementation.

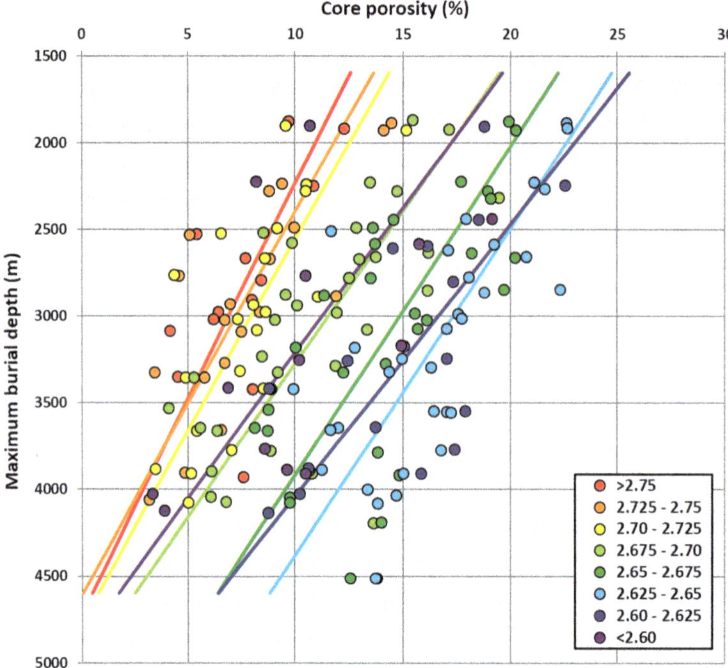

Fig. 8. Average porosity per 50 m depth interval, plotted v. depth ($n = 5532$). Measurements are divided into eight grain-density classes, represented by the colour range red (high grain density), green (moderate grain density) and purple (low grain density). A porosity–depth relationship has been defined for each class (Fig. 9).

be expected because part of the high grain density is caused by authigenic cements, such as dolomite. These cement types tend to strengthen the framework and hamper further compaction. Additionally, if the high grain density is caused by a high clay content, then the associated minerals may have already been compacted in the shallow realm (<1500 m). Figure 9b presents the slope of the porosity–depth relationships v. grain density. As for the porosity intercept, a linear relationship can be defined between grain density and the slope of the porosity–depth curve, with a high confidence coefficient (0.81).

The linear relationships defined in Figure 9a, b provide a continuous spectrum of porosity–depth relationships for varying degrees of reservoir quality. These relationships can be combined to determine the slope and intercept of the linear porosity–depth relationship of the RBMVL reservoir, based on the grain density of the local reservoir. The resulting relationship (with y as porosity in p.u. and x as the depth in km) is not only representative of reservoir-quality RBMVL sandstone, but also of lesser-quality reservoirs consisting of more clay-rich sedimentary facies or an increased grade of diagenesis.

The grain-density method was also applied to unpublished TNO internal studies on core measurements of other clastic formations, such as the other sandstone formations of the Main Buntsandstein Subgroup and the Permian Slochteren Formation. These analyses demonstrate similar outcomes concerning the development of the porosity–depth relationship with varying grain density. However, it should be noted that this technique cannot be applied to all clastic intervals in The Netherlands. A prerequisite is a large collection of core-analysis data (of the order of 1000 or more samples), which most formations are lacking. In which case, or if the measurements only cover a relatively small depth interval, the definition of the porosity–depth relationships per grain-density class may be invalid.

Further work

The porosity measurements used for this study, all within the 1500–4500 m depth interval, approach a linear porosity–depth relationship. However, outside this depth interval, the uncertainty of these

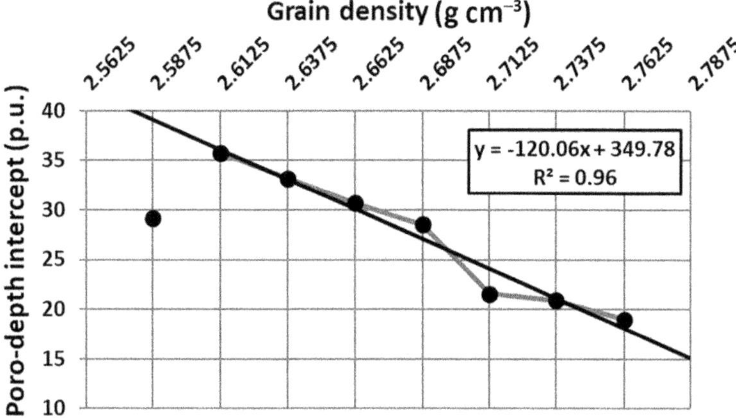

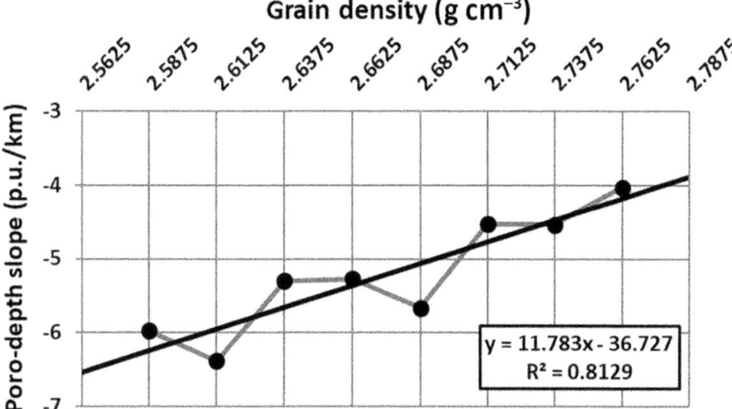

Fig. 9. The linear development of the two coefficients of the porosity–depth relationship with grain density presented in these graphs (n = 5532). The coefficient value is plotted v. the mid-grain-density value of each grain-density class. (**a**) Intercept of porosity–depth v. grain density. (**b**) Slope of porosity–depth v. grain density.

relationships increases significantly because mechanical compaction does not increase linearly with depth. Therefore, further work could include the definition of exponential porosity–depth relationships that fit the data, and thus provide a better porosity estimate outside the depth domain represented in this study.

Currently, the assignment of attributes by Method 1 is based on a qualitative judgement of the available data. Further investigation of, for example, petrographical/point-count measurements could be the basis of a quantitative definition of attributes. However, this is largely dependent on the amount of data available.

The effect of diagenesis is incorporated into this study. However, it was not possible to make a distinction between early and late diagenesis due to a lack of data. The timing of diagenesis is likely to have an effect on mechanical compaction with depth. Therefore, further research could include the determination of the timing of diagenesis with respect to burial, and analysis of the effect on the porosity–depth plot.

Conclusions

In this study, a large set of publicly available Dutch well data (NLOG 2017) was analysed and used to define specific reservoir property relationships for the Early Triassic Lower Volpriehausen Sandstone

Member, on a regional–subregional scale across The Netherlands. It has been demonstrated that:

- Porosity–permeability relationships are generally better described by curved rather than exponential functions, which appear as straight lines on a log-normal plot. The large core data collection of the Triassic sandstones, which covers the entire porosity range, shows a concaving-downwards permeability in the high-porosity range. The dataset is best described using a curved (or double exponential) porosity–permeability relationship (e.g. polynomial or power relationship), which can be formulated for the Triassic sandstones by $k = 10^{(8.49\times10^{-5}\theta^3 - 1.12\times10^{-2}\theta^2 + 5.04\times10^{-1}\theta - 4.06)}$, where k is permeability in millidarcies (mD) and θ is porosity expressed as a percentage (%). Straight lines will most probably overestimate permeability in the high- and low-porosity range, and underestimate permeability in the intermediate-porosity range.
- A porosity–depth plot from core data usually shows a high degree of scatter and no clear relationship, other than, perhaps, a high-porosity envelope. This scatter is caused by a variety in sedimentary facies and degree of diagenesis. A porosity–depth relationship for reservoir-quality sandstone can be revealed from this data cloud after excluding non-reservoir-quality samples (method 1) and correcting for the maximum burial depth. The average trend line for the depth interval analysed is described by: $D = -316.31\,\theta + 8505.4$, where D is the maximum burial depth in metres (m) and θ is porosity expressed as a percentage (%).
- Grain density is a good indirect porosity indicator and can, therefore, be used to distinguish reservoir-quality from non-reservoir-quality samples (method 2). The porosity–depth trend for reservoir-quality sandstone resulting from the grain-density method corresponds very well with the relationship from method 1, which is based on direct well information.
- The definition of multiple grain-density classes allows a more detailed analysis of the porosity–depth relationship. It turns out that the slope and porosity intercept of the different relationships are linearly decreasing with increasing grain density. This allows the definition of a continuous porosity–depth relationship, based on grain density. However, due to the fact that these are linear instead of exponential relationships, the applicability is restricted to the depth range represented in the used dataset. Outside of this range, the uncertainty of the relationships significantly increases.

It is demonstrated in this study that NLOG (2017) represents a free source of data that can be utilized in many projects, including hydrocarbon and geothermal exploration, which is also relatively simple to interrogate. Analysing datasets from this vast amount of subsurface data provides a valuable reference framework for understanding reservoir quality.

The authors acknowledge the reviewers, Pepijn Kole, an anonymous reviewer and the editor, Ben Kilhams, for their constructive remarks. Additionally, we like to express our gratitude to colleagues within TNO for their support and lively discussions.

References

Ajdukiewicz, J.M. & Lander, R.H. 2010. Sandstone reservoir quality prediction: the state of the art. *AAPG Bulletin*, **94**, 1083–1091, https://doi.org/10.1306/intro060110

Beard, D.C. & Weyl, P.K. 1973. Influence of texture on porosity and permeability of unconsolidated sand. *AAPG Bulletin*, **57**, 349–369.

Bjørlykke, K. & Egeberg, P.K. 1993. Quartz cementation in sedimentary basins. *AAPG Bulletin*, **77**, 1538–1548.

Cosentino, L. 2001. *Integrated Reservoir Studies*. Editions Technip, Paris.

de Jager, J., Doyle, M.A., Grantham, P.J. & Mabillard, J.E. 1997. Hydrocarbon habitat of the West Netherlands Basin. *In*: Rondeel, H.E., Batjes, D.A.J. & Nieuwenhuijs, W.H. (eds) *Geology of Gas and Oil under the Netherlands*. Springer, Dordrecht, The Netherlands, 191–209.

Dronkert, H. & Remmelts, G. 1996. Influence of salt structures on reservoir rocks in Block L2, Dutch continental shelf. *In*: Rondeel, H.E., Batjes, D.A.J. & Nieuwenhuijs, W.H. (eds) *Geology of Oil and Gas under the Netherlands*. Springer, Dordrecht, The Netherlands, 159–166.

Duin, E.J.T., Doornenbal, J.C., Rijkers, R.H.B., Verbeek, J.W. & Wong, Th.E. 2006. Subsurface structure of the Netherlands – results of recent onshore and offshore mapping. *Netherlands Journal of Geosciences*, **85**, 245–276.

Ehrenberg, S.N. & Nadeau, P.H. 2005. Sandstone vs. carbonate petroleum reservoirs: a global perspective on porosity–depth and porosity–permeability relationships. *AAPG Bulletin*, **89**, 435–445, https://doi.org/10.1306/11230404071

Evans, R.J., Nieto, J.A. & Yale, D.P. 1994. Improved methods for correcting core porosity to reservoir conditions. *The Log Analyst*, **35**, 21–30.

Feist-Burkhardt, S., Götz, A.E. et al. 2008. Triassic. *In*: McCann, T. (ed.) *The Geology of Central Europe. Volume 2: Mesozoic and Cenozoic*. Geological Society, London, 749–821.

Fontaine, J.M., Guastella, G., Jouault, P. & de la Vega, P. 1993. F15A: a Triassic gas field on the eastern limit of the Dutch Central Graben. *In*: Parker, J.R. (ed.) *Petroleum Geology of Northwest Europe: Proceedings of the 4th Conference*. Geological Society, London, 583–593, https://doi.org/10.1144/0040583

Franz,M., Barth, G., Zimmermann, J., Budach, I., Nowak, K. & Wolfgramm, M. In press. Geothermal resources of the North German Basin: exploration strategy, development examples and remaining opportunities in

Mesozoic hydrothermal reservoirs. *In*: KILHAMS, B., KUKLA, P.A., MAZUR, S., MCKIE, T., MIJNLIEFF, H.F. & VAN OJIK, K. (eds) *Mesozoic Resource Potential of the Southern Permian Basin*. Geological Society, London, Special Publications, **469**, https://doi.org/10.1144/SP469.11

GAUPP, R. & OKKERMAN, J. 2011. Diagenesis and reservoir quality of Rotliegend sandstones in the northern Netherlands: a review. *In*: GRÖTSCH, J. & GAUPP, R. (eds) *The Permian Rotliegend of the Netherlands*. SEPM, Special Publications, **98**, 193–228.

GELUK, M. 2005. *Stratigraphy and tectonics of Permo-Triassic basins in the Netherlands and surrounding areas*. PhD thesis, Utrecht University, Utrecht, The Netherlands.

GELUK, M., MCKIE, T. & KILHAMS, B. 2018. An introduction to the Triassic: current insights into the regional setting and energy resource potential of NW Europe. *In*: KILHAMS, B., KUKLA, P.A., MAZUR, S., MCKIE, T., MIJNLIEFF, H.F. & VAN OJIK, K. (eds) *Mesozoic Resource Potential of the Southern Permian Basin*. Geological Society, London, Special Publications, **469**. First published online January 26, 2018, https://doi.org/10.1144/SP469.1

GEOVARIANCES 2016. *Isatis Software*, http://www.geovariances.com/en/software/isatis439geostatistics-software/

HOLT, R.M., LEHR, C., KENTER, C.J. & SPITS, P. 2003. In-situ porosity from cores: poroelastic correction for stress relief during coring. *Petrophysics*, **44**, 253–261.

JUHASZ, I. 1986. Porosity systems and petrophysical models used in formation evaluation. *Society of Petrophysicists and Well Log Analysts (SPWLA), Transitions of the London Chapter Porosity Seminar, 10 September 1986*.

KORTEKAAS, M., BÖKER, U., VAN DER KOOIJ, C. & JAARSMA, B. In press. The Main Buntsandstein play in the Dutch northern offshore. *In*: KILHAMS, B., KUKLA, P.A., MAZUR, S., MCKIE, T., MIJNLIEFF, H.F. & VAN OJIK, K. (eds) *Mesozoic Resource Potential of the Southern Permian Basin*. Geological Society, London, Special Publications, **469**, https://doi.org/10.1144/SP469.19

LUIJENDIJK, E., VAN BALEN, R.T., TER VOORDE, M. & ANDRIESSEN, P.A.M. 2011. Reconstructing the Late Cretaceous inversion of the Roer Valley Graben (southern Netherlands) using a new model that integrates burial and provenance history with fission track thermochronology. *Journal of Geophysical Research*, **116**, 1–19, https://doi.org/10.1029/2010JB008071

NELSKAMP, S. & VERWEIJ, J.M. 2012. *Using basin modeling for geothermal energy exploration in the Netherlands – an example from the West Netherlands Basin and Roer Valley Graben*. TNO Report TNO-060-UT-2012-00245. TNO, Utrecht, The Netherlands.

NLOG 2017. *Dutch Oil and Gas Portal*, http://www.nlog.nl

PAXTON, S.T., SZABO, J.O., AJDUKIEWICZ, J.M. & KLIMENTIDIS, R.E. 2002. Construction of an intergranular volume compaction curve for evaluating and predicting compaction and porosity loss in rigid-grain sandstone reservoirs. *AAPG Bulletin*, **86**, 2047–2067.

PURVIS, K. & OKKERMAN, J.A. 1996. Inversion of reservoir quality by early diagenesis: an example from the Triassic Buntsandstein, offshore the Netherlands. *In*: RONDEEL, H.E., BATJES, D.A.J. & NIEUWENHUIJS, W.H. (eds) *Geology of Gas and Oil under the Netherlands*. Springer, Dordrecht, The Netherlands, 179–189.

RAMM, M. & BJØRLYKKE, K. 1994. Porosity/depth trends in reservoir sandstones: assessing the quantitative effects of varying pore-pressure, temperature history and mineralogy, Norwegian Shelf data. *Clay Minerals*, **29**, 475–490.

ROSSEL, N.C. 1982. Clay mineral diagenesis in Rotliegend aeolian sandstones of the southern North Sea. *Clay Minerals*, **17**, 69–77.

SCHERER, M. 1987. Parameters influencing porosity in sandstones: a model for sandstone porosity prediction. *AAPG Bulletin*, **71**, 485–491.

TEEUW, D. 1971. Prediction of formation compaction from laboratory compressibility data. *Society of Petroleum Engineers*, **11**, 263–271.

VAN DALFSEN, W., MIJNLIEFF, H.F. & SIMMELINK, H.J. 2005. Interval velocities of a Triassic claystone: Key to burial history and velocity modelling. *Poster presented at the 67th EAGE Conference & Exhibition*, 13–16 June 2005, Madrid, Spain.

WAPLES, D.W. 2002. Evolution of sandstone porosity through time. The modified Scherer model: a calculation method applicable to 1-D maturity modeling and perhaps to reservoir prediction. *Natural Resources Research*, **11**, 257–272.

Tectonostratigraphy of a rift basin affected by salt tectonics: synrift Middle Jurassic–Lower Cretaceous Dutch Central Graben, Terschelling Basin and neighbouring platforms, Dutch offshore

R. BOUROULLEC[1]*, R. M. C. H. VERREUSSEL[2], C. R. GEEL[1], G. DE BRUIN[1], M. H. A. A. ZIJP[2], D. KŐRÖSI[1], D. K. MUNSTERMAN[2], N. M. M. JANSSEN[2] & S. J. KERSTHOLT-BOEGEHOLD[1]

[1]*TNO Netherlands Organisation for Applied Scientific Research, Applied Geosciences, Princetonlaan 6, 3584 CB Utrecht, The Netherlands*

[2]*TNO Netherlands Organisation for Applied Scientific Research, Geological Survey of the Netherlands, Princetonlaan 6, 3584 CB Utrecht, The Netherlands*

**Correspondence: renaud.bouroullec@tno.nl*

Abstract: The Middle Jurassic–Lower Cretaceous in the eastern Dutch offshore provides excellent examples of sand-rich sediments that locally accumulated in the vicinity of rift basin margins affected by salt tectonics. These types of deposits are often geographically restricted and difficult to identify, but can be valuable targets for hydrocarbon exploration. The distribution, thickness and preservation potential of fluvio-lacustrine, shallow- and deep-marine sediments is discussed to provide new insights into the regional and local tectonostratigraphy of the Dutch Central Graben, the Terschelling Basin and their neighbouring platforms. New sedimentological, geochemical, biostratigraphic, stratigraphic and structural information have been analysed and integrated into a new tectonostratigraphic model for the Callovian Lower Graben Formation, Oxfordian Middle and Upper Graben formations, Early–Middle Volgian Terschelling Sandstone and Noordvaarder members, and the Late Volgian–Early Ryazanian Scruff Greensand Formation. It is demonstrated that salt withdrawal at the basin axis was the primary control on the generation of high accommodation during the Callovian–Early Kimmeridgian. Incised valleys developed on the platforms providing lateral sediment input. During the Late Kimmeridgian–Ryazanian salt migration shifted laterally towards the basin margins, providing accommodation adjacent to active salt bodies and deposition of overthickened sandy strata.

During the Middle Jurassic–Early Cretaceous the North Sea area was subject to active rifting that initiated during the Triassic but reached a climax during the Late Jurassic. During this period, several Mesozoic Dutch rift basins were also heavily affected by Zechstein salt withdrawal and migration into autochthonous and shallow salt bodies, especially near rift-bounding faults or other pre-existing faults. Several hydrocarbon fields with reservoirs of Middle Jurassic–Lower Cretaceous age, such as F03-FB and F15A, have been discovered in the Dutch offshore and neighbouring regions. In addition, the Middle Volgian–Early Ryazanian M07-B gas field, discovered in 2007, is located along the southern margin of the Terschelling Basin in a complex setting that involves syntectonic deposition and local sediment erosion and re-deposition. To understand such a complex play type and its economic potential elsewhere in the North Sea basin, a good understanding of the relationship between palaeotopography and sediment distribution and preservation is required.

Three Middle Jurassic–Lower Cretaceous tectonostratigraphic mega-sequences (TMS-1 to -3) are recognized in the Dutch, German and Danish offshore area; these reflect three main stages of basin evolution regarding the Central Graben and adjacent areas (Munsterman *et al.* 2012; Verreussel *et al.* 2018). In the Dutch sector, these sequences contain sand-rich reservoir intervals such as the Lower, Middle and Upper Graben and the Friese Front formations (TMS-1); the Terschelling Sandstone and the Noordvaarder members (TMS-2); and the Scruff Greensand Formation (TMS-3) (Fig. 1). The results of this work validate and support this subdivision by providing new detailed stratigraphic correlation anchored on chronostratigraphic, stratigraphic and seismic information across several structural elements. These elements include two main basins, namely the Dutch Central Graben and the Terschelling Basin, and six flanking platforms, namely the Cleaver Bank Platform, the Step Graben, the Central Offshore Platform, the Friesland Platform, the

From: Kilhams, B., Kukla, P. A., Mazur, S., McKie, T., Mijnlieff, H. F. & van Ojik, K. (eds) 2018. *Mesozoic Resource Potential in the Southern Permian Basin*. Geological Society, London, Special Publications, **469**, 269–303.
First published online March 15, 2018, https://doi.org/10.1144/SP469.22

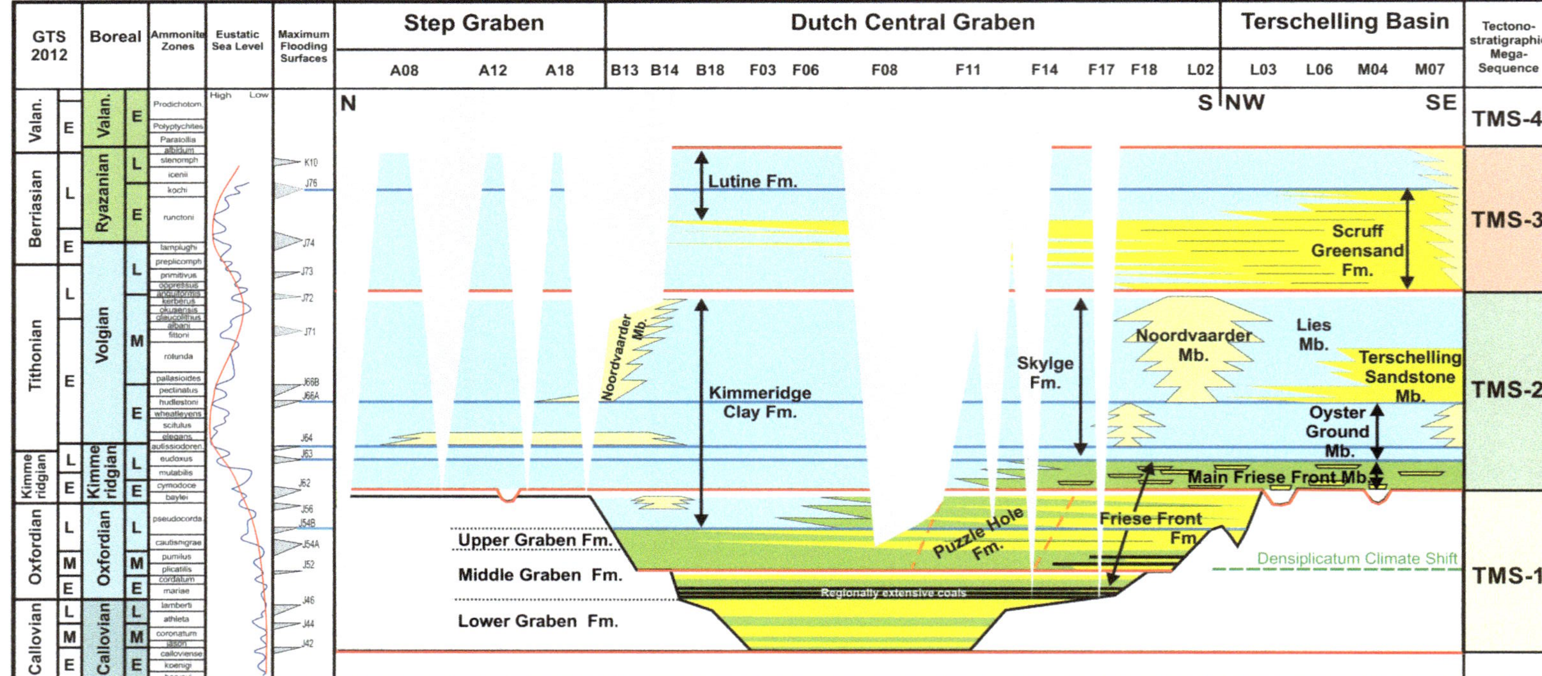

Fig. 1. Lithostratigraphic framework of the Middle Jurassic–Lower Cretaceous in the Terschelling Basin, Dutch Central Graben and Step Graben. The sand-rich intervals are shown as yellow polygons, while pale yellow polygons indicate locally derived depositional systems such as the Noordvaarder Member. Based on core data interpretation and published results (Munsterman *et al.* 2012), fluvial overbanks, delta plain and estuarine/lagoonal claystones are shown in green, while marine claystones are shown in light blue. Unconformities are shown as red lines and maximum flooding surfaces as blue lines. The three regionally extensive Middle Graben Formation coals are shown as thick black lines. Oblique dashed orange lines show the approximate lithostratigraphic limits of time-equivalent formations of the upper part of TMS-1. Proven reservoirs have been found in each TMS but with a predominance of TMS-1. The main source rocks are either Carboniferous coals and/or Middle Jurassic Posidonia Shales. The main seals are either the Valanginian to Albian Rijnland Group or internal Upper Jurassic intervals such as the Late Oxfordian–Middle Volgian (159–147 Ma) Kimmeridge Clay Formation or the Late Kimmeridgian–Middle Volgian (152.5–147 Ma) Lies Member. Ammonite zones after Gradstein *et al.* (2012). Eustatic sea-level curve after Carruthers *et al.* (1996). Maximum flooding surfaces after Partington *et al.* (1993).

Ameland Platform and the Schill Grund Platform (Fig. 2; Heybroek 1975; Schroot 1991; Duin *et al.* 2006; Kombrink *et al.* 2012).

The study area is 270 km long and on average 120 km wide, covering the eastern part of the Dutch A, B and E blocks (east of 3° 20′), the entire B and F blocks, and most of the L, M and G blocks. The area of interest comprises the Dutch Central Graben, the Terschelling Basin and the areas of the flanking platforms located close to the rift basins (Fig. 2). The data used in this research include 102 wells, 11 cores, 5 two-dimensional (2D) seismic surveys, 43 3D seismic surveys, as well as 5500 palynological and 800 stable isotope analyses carried out on core and cuttings samples (see also Verreussel *et al.* 2018). The integration of palynological, stable isotope, core, well and seismic data into a single tectonostratigraphic model for the Dutch offshore will aid further hydrocarbon exploration by supplying an integrated framework for Middle Jurassic–Lower Cretaceous stratigraphy and tectonic events that can be followed at cuttings to seismic scale. This paper is closely linked to that of Verreussel *et al.* (2018); the latter discusses the relevant palynological information and controls used in both papers and presents a stratigraphic overview of the Middle Jurassic–Lower Cretaceous from the Dutch to the Danish sectors. In order to provide further context, an overview of the geological evolution of the area is first presented.

Geological evolution

Tectonic evolution

The Meso-Cenozoic structural evolution of the Dutch offshore region is complex as it involves several extensional, compressional and strike-slip phases, as well as a multiphase salt tectonic history (Van Hoorn 1987; Ziegler 1990*a*, *b*). Unravelling this structural evolution, as well as the stratigraphic response to the changing palaeo-landscapes, is difficult due to the large amount of the stratigraphy being removed during multiple significant erosional events, especially on low-accommodation areas such as the rift shoulders (e.g. Schill Grund Platform, Step Graben, Fig. 2) (Wong 2007). A short description of the key rift-related structural events that impacted the study area is given in the following.

The failed rift system of the Dutch offshore (*sensu* de Jager 2007) is composed of the NNE–SSW-trending Dutch Central Graben and the NW–SE-trending Terschelling Basin, which are bounded by several flanking platforms (Fig. 2). During the Triassic, Jurassic and Early Cretaceous, the structural setting of The Netherlands changed from a single extensional basin configuration (the Southern Permian Basin) to a series of smaller, fault-bounded basins and highs (De Jager 2007). Two main tectonic events shaped the North Sea Basin during the Mesozoic: (1) the break-up of Pangea and associated rifting during most of the Mesozoic (Ziegler 1990*a*); and (2) the closure of the Tethys Ocean/Alpine collision and associated inversion tectonics during the late Mesozoic, which subsequently culminated during the Cenozoic (Ziegler 1990*b*; Kley & Voigt 2008). The initiation of rifting during the Early Triassic was related to the inception of the proto-Atlantic between Greenland and the Fennoscandian High (Lott *et al.* 2010). The southwards propagation of the rift system towards the North Sea can be traced into the northern part of the study area (Ziegler 1990*b*; Roberts *et al.* 1995; Coward *et al.* 2003). As such, Middle and Late Triassic sequences thicken into the newly-formed Dutch Central Graben and Broad Fourteens Basin, where the Zechstein salt was actively withdrawing and was remobilized to form diapirs and rim synclines (De Jager 2007). The Late Triassic extension direction, recorded in faults locally affected by dextral transtension, was E–W aligned (Van Hoorn 1987).

The Early Jurassic was a period of relative tectonic quiescence, with little rifting occurring in the North Sea area (Pharaoh *et al.* 2010). Stagnant-water stratification led to the deposition of the Posidonia Shale Formation during the Toarcian, the principal source rock for the oil provinces of the southern North Sea and northern Germany (De Jager 2007). A key event during the Middle Jurassic was the uplift of the central North Sea domain. This started towards the end of the Aalenian, in response to the emplacement of a mantle plume at the base of the lithosphere (Underhill & Partington 1993) which persisted during the Bajocian and Bathonian. Development of this large thermal dome (700 × 1000 km) caused deep truncation of Lower Jurassic and even Triassic sediments, leading to the development of the regional-scale Mid Cimmerian Unconformity (Ziegler 1990*a*; Underhill & Partington 1993; Surlyk & Ineson 2003). By late Middle Jurassic times, the Central North Sea Dome had subsided sufficiently for open-marine conditions to be restored across the study area. Sedimentation resumed during the Callovian–Late Jurassic in the previously uplifted areas (Ziegler 1990*a*; Callomon 2003).

During the Late Jurassic–Early Cretaceous, crustal extension accelerated across the North Sea rift system driven by the reorganization of the stress field due to the opening of the Arctic–North Atlantic rift system and the related clockwise, westwards rotation of Laurentia–Greenland relative to Eurasia (Ziegler 1988, 1990*a*; Torsvik *et al.* 2002). This reorganization resulted in NW–SE-trending transtensional basins that formed along the southern margin of the Southern Permian Basin. This phase caused large areas outside of the rift basins to be uplifted, exposed and subsequently eroded (Duin

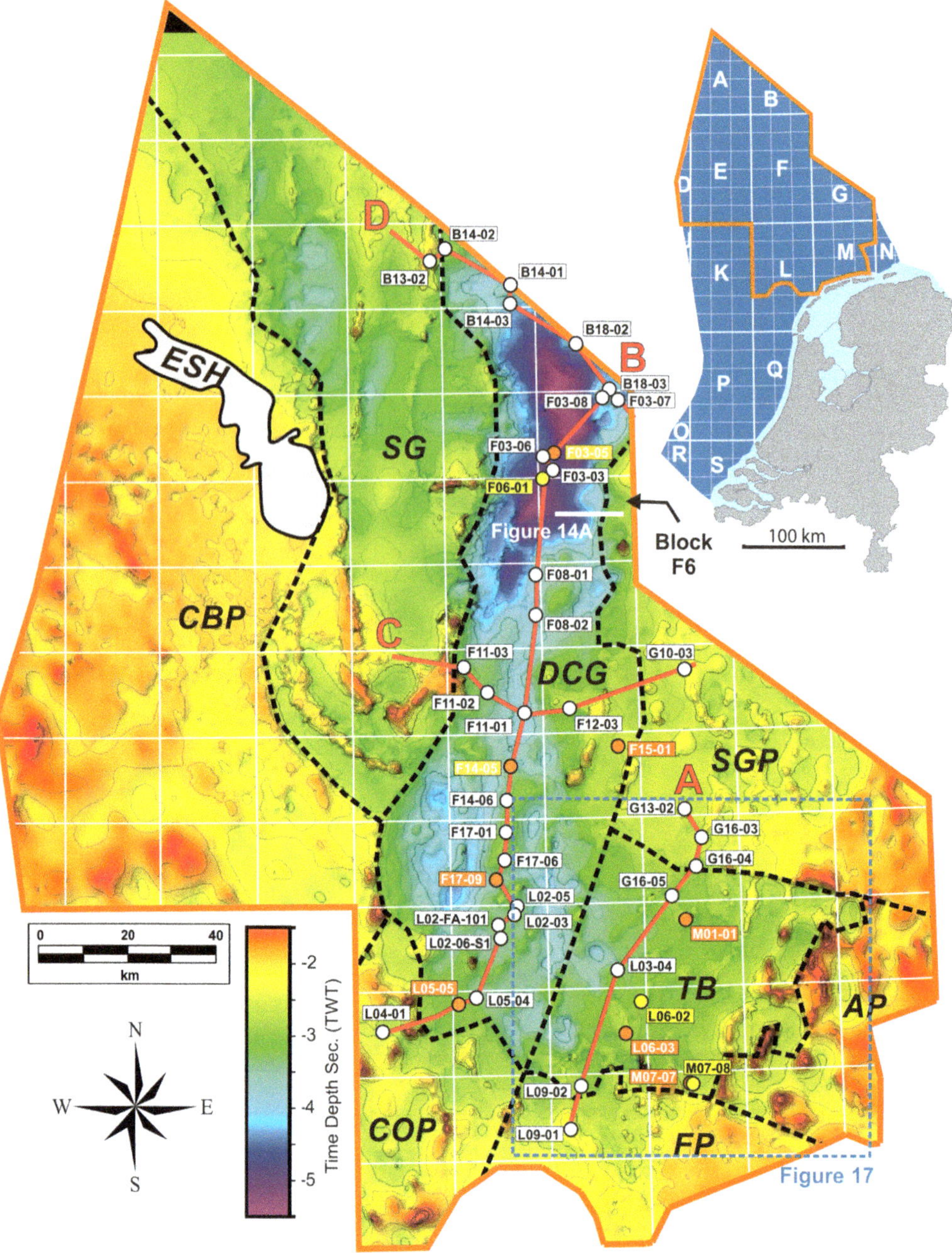

Fig. 2. Base map of the central and northern part of the Dutch offshore. Top Zechstein Group is shown as background surface. The main structural elements are delimited by dashed black lines: CBP, Cleaver Bank Platform; ESH, Elbow Spit High; SG, Step Graben; DCG, Dutch Central Graben; SGP, Schill Grund Platform; TB, Terschelling Basin; AP, Ameland Platform; FP, Friesland Platform; and COP, Central Offshore Platform. The position of the four transects shown in this paper (A, B, C and D; Figs 6–13) are shown as red lines. The regional seismic and correlation panels are shown in Figures 6–13. Key wells displayed in the regional panels are shown as white circles. Wells with studied cores are show as orange circles, while wells F06-01, L06-02 and M07-08 (yellow circles) have cores, the interpretation of which is shown in Figures 3–5.

et al. 2006). In the Dutch sector, up to 1500 m of fluvio-lacustrine and shallow-marine sequences (TMS-1; Fig. 1) accumulated in the Dutch Central Graben during this period; adjacent highs, such as the Step Graben, Friesland and Schill Grund platforms, were uplifted and eroded (see Fig. 2 for location; Munsterman *et al.* 2012). During this period, the Terschelling Basin did not exhibit accommodation since it was not subsiding (Duin *et al.* 2006; Munsterman *et al.* 2012), but subsequently became a sediment catchment area during the following periods (depositional time of TMS-2 and 3; Fig. 1). Sediments of TMS-2 and 3 were also deposited in the Dutch Central Graben and accumulated only as thin intervals on the surrounding platforms (Munsterman *et al.* 2012). Salt pillows, stocks and walls developed along the main bounding faults of the Dutch Central Graben during this period, and created a complex basin configuration and stratigraphy (e.g. van Winden *et al.* 2018).

Contemporaneously to an end-Ryazanian transgression, rifting is documented to have ceased with the region subsequently undergoing post-rift sag. As such, the area of sediment accumulation broadens to encompass the entire North Sea (Cromer Knoll Formation in UK and Norwegian sectors; see Verreussel *et al.* 2018). This interval (TMS-4, Fig. 1) is generally mud-prone, but close inspection often reveals thin sandy deposits at the base of the Cretaceous transgression (e.g. Vlieland Sandstone Formation on the Schill Grund Platform and Vlieland Basin; Munsterman *et al.* 2012).

Salt tectonics

The presence of Permian Zechstein salt, and the multi-stage deformation that affected this mobile substrate, had a pronounced impact on the Mesozoic evolution of the North Sea (Trusheim 1960). Locally, salt tectonics started as early as the latest Permian (Stewart 2007) but became more widespread during the Early Triassic, leading to the formation of minibasins and rafts (Stewart & Clark 1999; Penge *et al.* 1999). In the Dutch sector, such Early Triassic salt-controlled minibasins are rare and only observed in the A Block (insert map, Fig. 2). These minibasins are either due to Early Triassic progradational fluvial systems or to the relative proportion of marginal Zechstein facies compared to the amount of salt (as proposed for the UK sector minibasins by Stewart & Clark 1999). The main period of salt movement in the study area was the Middle and Late Triassic (Geluk 2007), and continued during the Jurassic and Early Cretaceous (van Winden *et al.* 2018). The original depositional thickness of the salt layer was likely variable across the study area, with thicknesses ranging from 400 m in the south to a maximum of 900 m in the north (Ten Veen *et al.* 2012; Hernandez *et al.* 2018). van Winden *et al.* (2018) places the original thickness of the salt as high as 1500 m within the central part of the Dutch Central Graben. Thin-skinned normal faults formed during the Middle Triassic, detaching on the Zechstein salt. Diapirs and rim synclines developed during the Middle Jurassic–Early Cretaceous (De Jager 2007, 2012; Ten Veen *et al.* 2012). The partitioning of the Southern Permian Basin into several basins and platforms during the Triassic and Jurassic was accentuated by intense salt tectonics, primarily along fault-bounded basin margins (Wong 2007). For example, basin compartmentalization and minibasin formation are observed to be associated with salt withdrawal in much of the Dutch Central Graben and the Terschelling Basin (Fig. 2).

Methods

This research involved a series of core descriptions, palynological and geochemical analyses, and seismic and well log interpretation. This new information was combined with previous legacy material to build a robust depositional, stratigraphic, chronostratigraphic and structural framework for Middle Jurassic–Early Cretaceous sequences preserved in the Dutch offshore (Munsterman *et al.* 2012; Verreussel *et al.* 2018).

Sedimentological analysis included the description of 620 m of core data from 11 wells (F03-05, F06-01, F14-05, F15-01, F17-09, L05-05, L06-02, L06-03, M01-01, M07-07 and M07-08; Fig. 2). Based on the lithologies, facies, ichnofacies and resultant facies associations, several depositional settings were identified including fluvial, deltaic, lagoonal, lacustrine, tidal, shoreface and offshore. The cores described include 323 m of Lower Graben/Friese Front formations (TMS-1) (e.g. Fig. 3), 77 m of Terschelling Sandstone Member/Oyster Ground Members (TMS-2) (e.g. Fig. 4), and 211 m of Scruff Greensand Formation (TMS-3) (e.g. Fig. 5).

A new series of palynological analyses was conducted, utilizing core and cuttings material from 24 wells, combined with legacy data from more than 100 exploration wells in the Danish and Dutch offshore (Abbink *et al.* 2006; Munsterman *et al.* 2012; Verreussel *et al.* 2018). This enabled the definition of a regional dinocyst zonation that includes 4 zones and 18 subzones for the Middle Callovian–Barremian (Verreussel *et al.* 2018, fig. 2). The new palynological results gave additional age constraints as well as new information regarding the depositional environments and the climatic variations. Readers are referred to Verreussel *et al.* (2018) for detailed information regarding the construction of this scheme and resulting palaeoclimate observations for the Middle Jurassic–Early Cretaceous. In

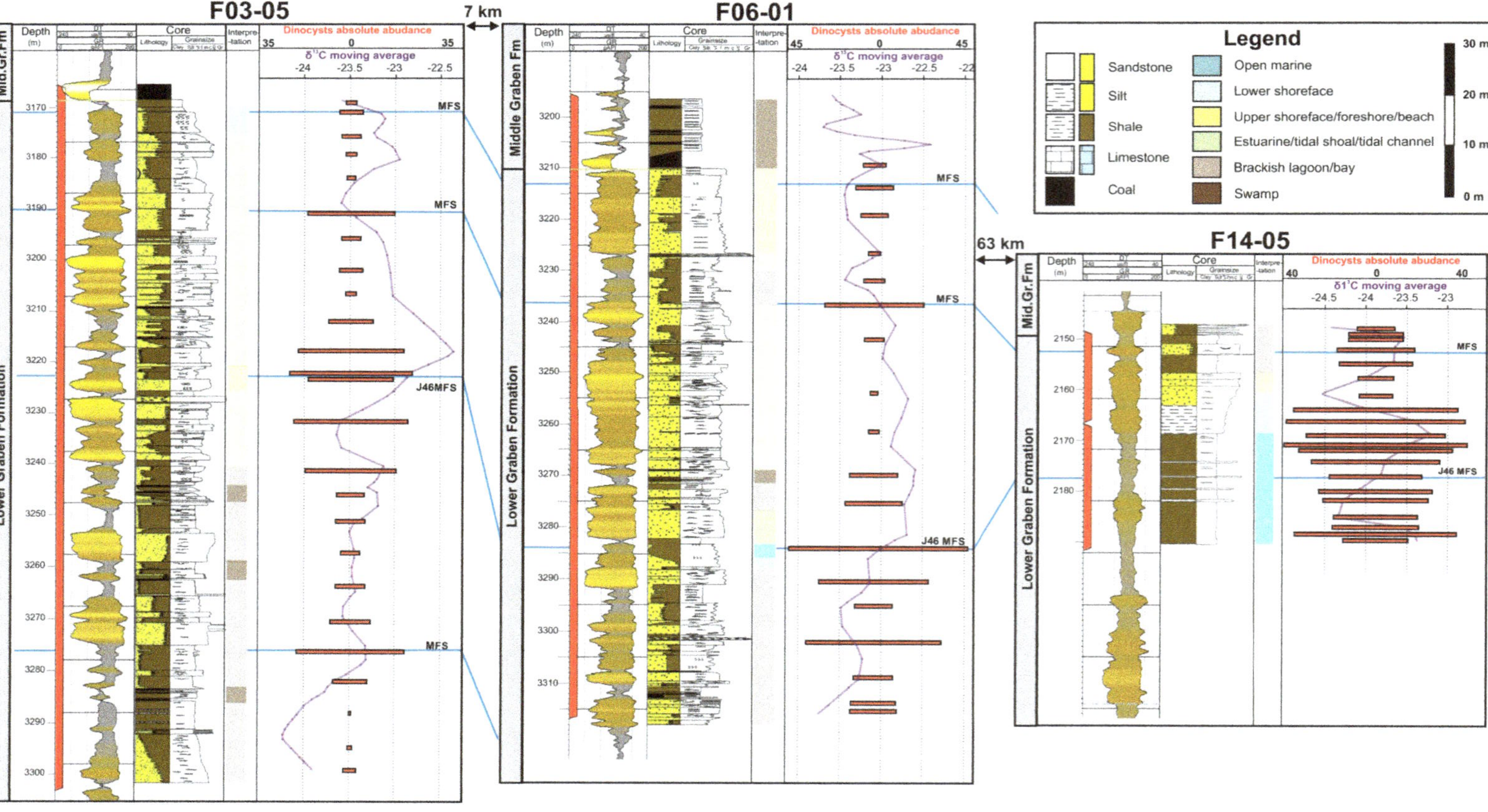

Fig. 3. Correlation panel for the Lower Graben Formation (lower part of TMS-1) between three wells located in the axis of the Dutch Central Graben. Wells F03-05, F06-01 and F14-05 have cores in this stratigraphic interval, which have been interpreted in terms of depositional environment using lithological, facies, ichnofacies and facies association classification. Palynological and stable isotope data were also collected and the results are shown for each well in terms of absolute abundance of dinocysts compared to pollen and spores (zero centred red bars) and $\delta^{13}C$ moving average curves (purple curves). Maximum flooding surfaces (MSF), including the J46 MFS from Partington *et al.* (1993), were identified using core and palynological information, and correlated using additional constraints from stable isotope data. Gamma ray curve is to the left and sonic curve to the right on each display. Depth values are measured depth. See Figure 2 for location.

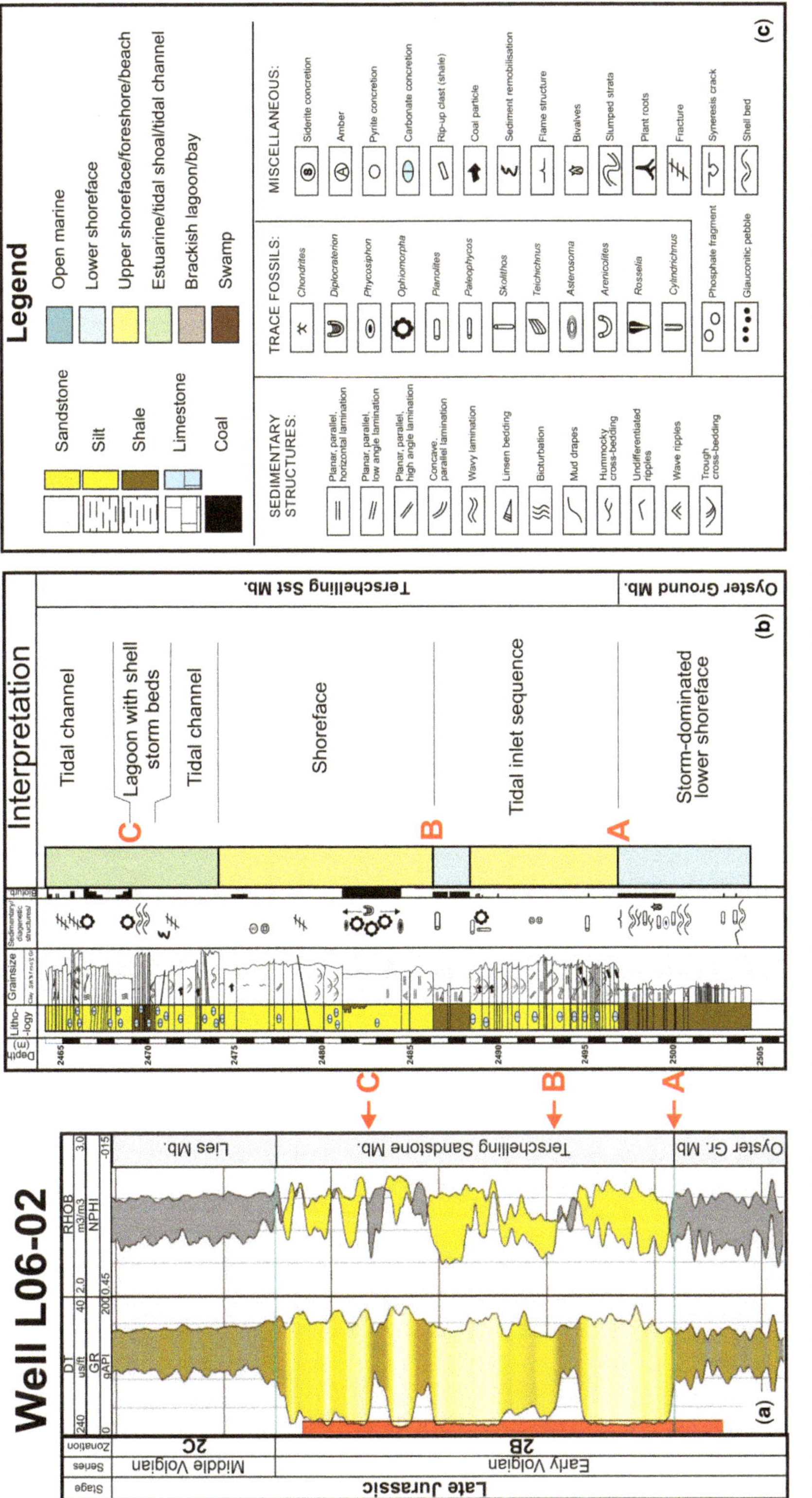

Fig. 4. L06-02 core description and interpretation of the upper part of the Oyster Ground and Terschelling Sandstone members (middle part of TMS-2) in the southern part of the Terschelling Basin. **(a)** Gamma ray/sonic and neutron/density curves of well L06-02 at the core depth. Gamma ray curve is to the left and sonic curve to the right on each display. The red bar shows the position of the core and the red markers (A–C) are placed to help the reader compare wireline log and core information. **(b)** Core description and interpretation. This 40 m long TMS-2 core shows the transition between the Oyster Ground and the Terschelling Sandstone members. **(c)** Legend of Figures 3–5. See text for detailed description and Figure 2 for location. See Verreussel *et al.* (2018, fig. 2) for chronostratigraphic overview, including palynological zonations and events.

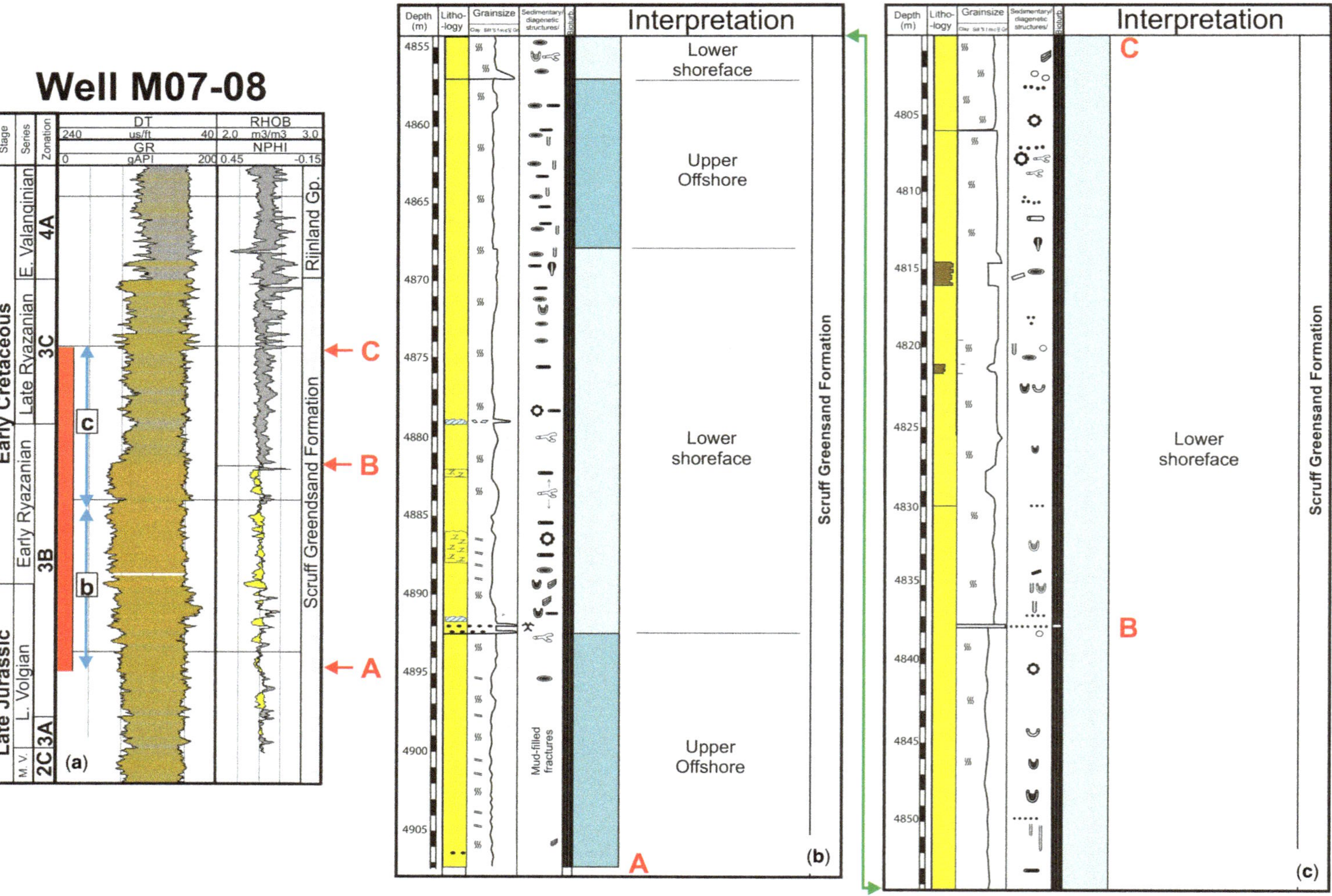

Well M07-08
Stage
Series
Zonation
DT
240 us/ft 40
GR
0 gAPI 200
RHOB
2.0 m3/m3 3.0
NPHI
0.45 -0.15
Late Jurassic
Early Cretaceous
M.V.
L. Volgian
Early Ryazanian
Late Ryazanian
E. Valanginian
2C
3A
3B
3C
4A
Rijnland Gp.
Scruff Greensand Formation
c
b
(a)
C
B
A
Depth (m)
Litho-logy
Grainsize
Sedimentary/ diagenetic structures
Bioturb.
Interpretation
Lower shoreface
Upper Offshore
Lower shoreface
Upper Offshore
Mud-filled fractures
Scruff Greensand Formation
(b)
Interpretation
Lower shoreface
Scruff Greensand Formation
(c)

addition to the palynological analysis, material from nine wells was also used for stable isotope analysis (e.g. Fig. 3). This provided information to allow better stratigraphic correlation constraints.

Tectonostratigraphic analysis was based on an integrated approach, which combined detailed stratigraphic correlation between 102 wells (44 of which are shown in this paper; Fig. 2) and seismic interpretation based on 5 2D surveys and 43 3D surveys (with vintages ranging from 1982 to 2010, with typical sampling rates of 4 ms and zero-phase European polarity; see Table 1 for list of seismic surveys). Well-to-seismic ties were performed using synthetic seismograms, and the seismic interpretation was performed in two-way travel time (TWT). The initial 3D mapping of several key horizons was carried out at TNO (The Netherlands Organisation for Applied Scientific Research) and presented in Duin *et al.* (2006). For the present paper, 11 regional 2D transects were built to analyse the tectonostratigraphic evolution of the study area during the Middle Jurassic–Early Cretaceous using new and refined palynologically constrained age controls. Four of those transects are presented in this paper (Figs 6–13). New 3D seismic mapping was also performed for this research, including the entire Terschelling Basin plus one area (F06 block, Figs 14–16) in the northern part of the Dutch Central Graben. A revised detailed 3D mapping of the rest of the study area is ongoing. The analysis workflow included the identification and mapping of faults, salt features, stratigraphic markers (tied to the wells using new stratigraphic constraints), and regional and local unconformities. Seismically identified stratigraphic terminations (erosion, onlap, downlap) and time–thickness maps of key intervals (e.g. Fig. 17) are also used to analyse the interplay between active structures (faults and salt features) and sediment deposition. In addition, seismic amplitude analysis and horizon slicing techniques were used in a 3D seismic survey in the F06 block (Figs 14, 16).

Middle Jurassic–Lower Cretaceous stratigraphic overview

The Middle Jurassic–Lower Cretaceous interval of The Netherlands is composed of the Schieland, Scruff and Rijnland groups, the former two being the focus of this paper. Abbink *et al.* (2006), Munsterman *et al.* (2012) and Verreussel *et al.* (2018; figs 2–4) divide this interval into three major tectonostratigraphic phases (referred to here as tectonostratigraphic mega-sequences or TMS; Fig. 1).

Here the definitions provided by Verreussel *et al.* (2018) are summarized and contextualized in terms of the main stratigraphic divisions. The Early Callovian–Early Kimmeridgian TMS-1 comprises the Lower, Middle and Upper Graben formations, the Puzzle Hole Formation and the lower part of the Friese Front Formation (Fig. 1). A short description of the formations comprising TMS-1 is given in Table 2. The Early Kimmeridgian–Late Volgian TMS-2 was deposited during a period of increased accommodation in the Dutch Central Graben and Terschelling Basin as well as, to a limited extent, on their neighbouring platforms (e.g. Step Graben, Schill Grund Platforms; Fig. 2). This sequence is mainly composed of the time-equivalent Skylge and Kimmeridge Clay formations. The Main Friese Front Member of the Friese Front Formation is also of TMS-2 age, but is only present in the southern part of the Dutch Central Graben and in the Terschelling Basin (Fig. 1). The Skylge Formation is present in the southern part of the study area, mainly in the Terschelling Basin and the southern part of the Dutch Central Graben. This formation is composed of four members (Table 2), some of which are time-equivalent to each other. The Kimmeridge Clay Formation is present in the central and northern part of the study area, principally in the Dutch Central Graben and as erosional remnants in the Step Graben area (Figs 7b, 15, 16). Note that the lower part of the Kimmeridge Clay Formation is of TMS-1 age (Fig. 1). During the Late Volgian, the adjacent platforms such as the Schill Grund Platform and the Cleaver Bank High were flooded. TMS-3 was subsequently deposited, which comprises the high net-to-gross Scruff Greensand Formation and low net-to-gross Lutine Formation (Fig. 1).

Tectonostratigraphic evolution of the Middle Jurassic–Lower Cretaceous

The interplay between sediment deposition and active structuration creates complex tectonostratigraphic geometries within the study area. For example, the

Fig. 5. M07-08 core description and interpretation of the Scruff Greensand Formation (middle part of TMS-3) in the southern part of the Terschelling Basin. (**a**) Gamma ray/sonic and neutron/density curves of well M07-08 at the core depth. Gamma ray curve is to the left and sonic curve to the right on each display. The red bar shows the position of the core and the red markers (A–C) are placed to help the reader compare wireline log and core information. M.V., Middle Volgian. (**b**) Lower half of the core. (**c**) Upper half of the core. Note that the sediments are becoming coarser above marker B, while the gamma ray curve indicates an apparent fining-upwards trend. This discrepancy is due to the higher glauconite content of the upper part of the Scruff Greensand Formation at that location, resulting in higher gamma ray values. See Figure 4 for legend and Figure 2 for location of the well. See Verreussel *et al.* (2018, fig. 2) for chronostratigraphic overview, including palynological zonations and events.

Table 1. *List of seismic surveys used in this study*

Survey name on nlog.nl	Dutch blocks	Vintage	Type
Z3NAM1993A	A08-A09	1993	3D
Z3NAM1998C	A10-A11-A13-A14	1998	3D
Z3WIN-2000A	A15	2000	3D
Z3FUG-2002A	All B blocks	2002	3D
Z3WIN2001B-1	B10	2001	3D
Z3PET1993A	E09-E12	1993	3D
Z3WIN1997A	E18-F16	1997	3D
Z3NAM1989E	F02-F03-F05-F06	1989	3D
Z3RWE1994A	F02-F03	1994	3D
Z3RWE1994E	F02-F03	1994	3D
Z3NAM1982A	F03-B18	1982	3D
Z3PET1992F	F06	1992	3D
Z3OXY1994A	F08-F09	1994	3D
Z3PET1994B	F10	1994	3D
Z3PET1994A	F15-F16	1994	3D
Z3WIN2003B	F16-F17-L01	2003	3D
Z3NAM1992A	F17-F18	1992	3D
Z3NAM1993C	F18-G16	1993	3D
Z3PET1991A	G10-F12	1991	3D
Z3NAM1997A	G13_G14_G16_G17	1997	3D
Z3GDF2005A-1	G16	2005	3D
Z3CLY1998A	G16-G17	1998	3D
Z3GDF2002A-3	G17-G18	2002	3D
Z3NAM1991E	L01-L02	1991	3D
Z3PET1992A	L01-L04	1992	3D
Z3NAM1994B	L02-L03	1994	3D
Z3NAM-1990B_Zero_European	L04-L05	1990	3D
Z3WIN2003A	L04-L05-L07-L08	2003	3D
Z3NAM1990F	L05-L06-L09-M04	1990	3D
Z3WIN1995A	L07-L08	1995	3D
Z3PEN1985A	L08	1985	3D
Z3NAM1988D	L09-L12	1988	3D
Z3OXY1996A	L11-L12	1996	3D
Z3NAM1988A	L09-L11-L12-M07	1988	3D
Z3WIN2005B	L06 - L02-L03- L05- L08-L09- M01- M04-M07	2005	3D
Z3NAM1990E	M01	1990	3D
Z3NAM1991A	M02	1991	3D
Z3NAM1991D	M04	1991	3D
Z3PET1991B	M04-M05	1991	3D
Z3NAM1996A	M04-M05-M07-M08 M10-M11	1996	3D
Z3NAM1994A	L09-L12-M07-M10	1994	3D
Z3NAM1989B	M09	1989	3D
Z3NAM1989C	M09	1989	3D
SNST-83 (no access on Nlog)	All	1983	2D
SNST-87; Z2NOP1987A	All	1987	2D
NSR10; Z2TGS2010A	All	2010	2D
Z2RWE1996A	F	1996	2D
Z2GTM1998A	F	1998	2D

Note that most survey data are accessible from the nlog website (www.nlog.nl).

stratigraphic record locally shows growth/expanded geometries in the form of stratigraphic wedges and troughs, as well as unconformities within and outside of the Dutch Central Graben and Terschelling Basin (Figs 6–13). The stratigraphic thicknesses and types of depositional systems, as well as their stratigraphic preservation potential, stratigraphic architectures and sediment pathways, also vary greatly depending on location and proximity to active structures, such as salt bodies and syndepositional faults.

Each of the three Middle Jurassic–Lower Cretaceous tectonostratigraphic mega-sequences described below are based on the integration of core descriptions, palynological and geochemical

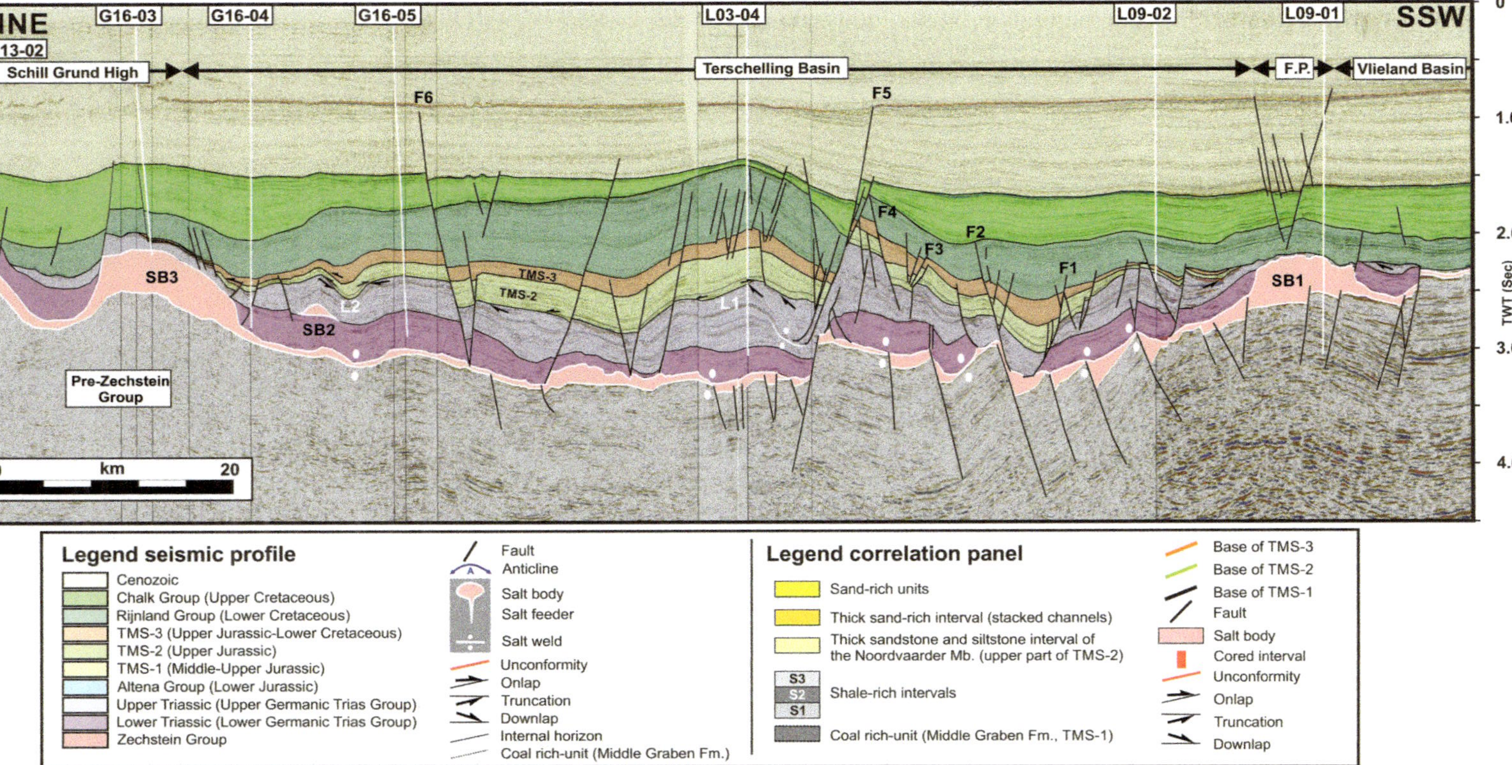

Fig. 6. Interpreted two-way travel time (TWT) for seismic panel A. This section is located in the southeastern part of the study area, oriented NNE–SSW and intercepts four structural elements, namely (from north to south): the Schill Grund Platform, the Terschelling Basin, the Friesland Platform (F.P.) and the Vlieland Basin. This seismic section is 91.5 km long and intercepts seven wells. Note that, along the southern part of the section, the Zechstein Group is offset by several normal faults forming within a large monocline (below salt body SB1). These faults are part of the Hantum Fault Zone (RGD 1991). Several syndepositional faults were active during the Middle Jurassic–Early Cretaceous, such as faults F1–F6. Large growth faults (F4–F7) were also active during the Late Cretaceous and Cenozoic. Most of the faults sole onto the autochthonous Zechstein salt with the exception of F4 that detached on a shallow level, which can be an *in situ* Upper Triassic salt layer (e.g. evaporites of the Röt Formation) or a remobilized upwards allochthonous Zechstein salt sheet that was later welded out. L1 and L2 refer to topographic lows shown in Figure 10. See Figure 2 for location and Figure 7 for equivalent stratigraphic correlation panel.

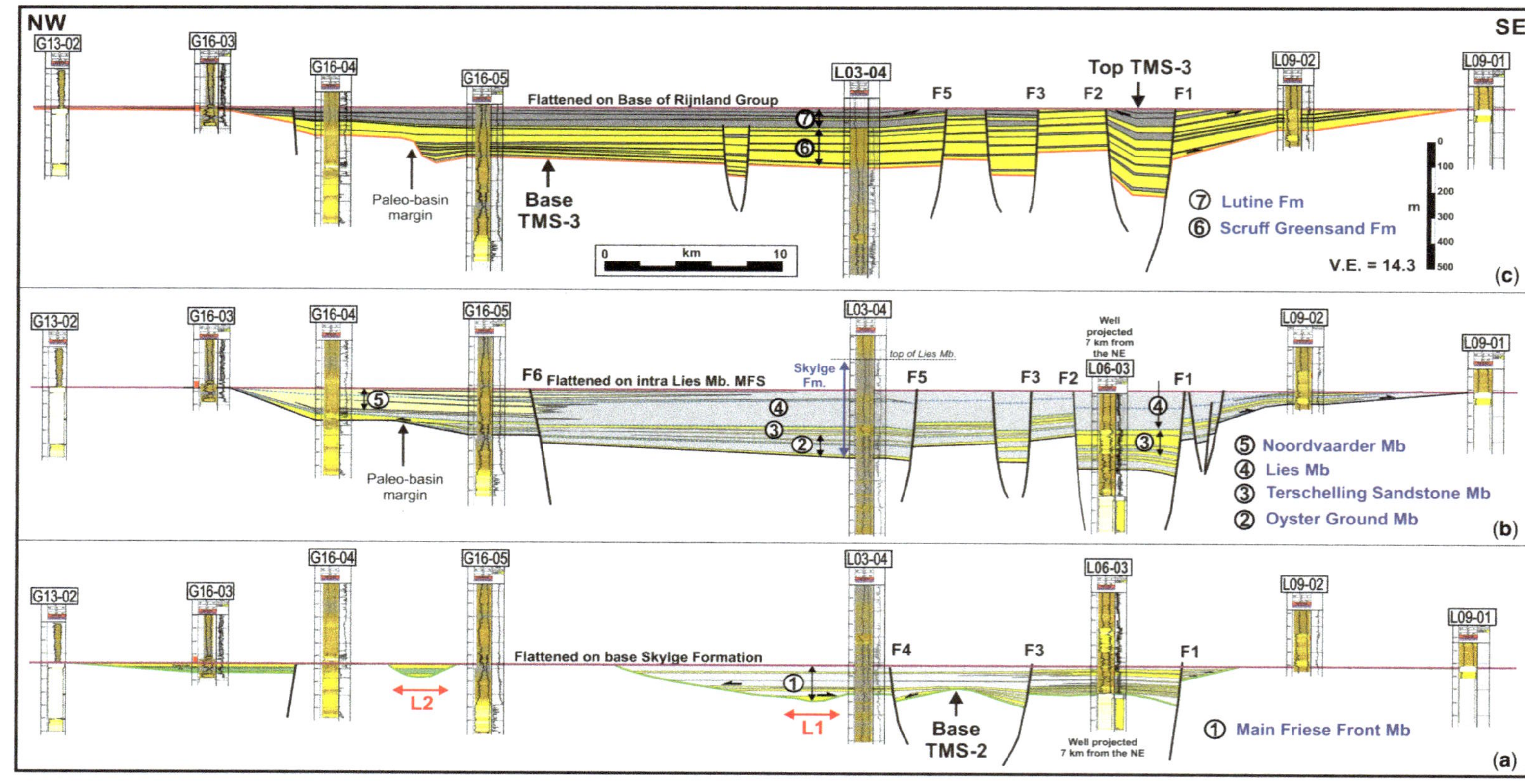
NW
SE
G13-02
G16-03
G16-04
G16-05
L03-04
L09-02
L09-01
Flattened on Base of Rijnland Group
F5
F3
F2
Top TMS-3
F1
Paleo-basin margin
Base TMS-3
km
0
10
⑦ Lutine Fm
⑥ Scruff Greensand Fm
m
0
100
200
300
400
500
V.E. = 14.3
(c)
F6 Flattened on intra Lies Mb. MFS
Skylge Fm.
top of Lies Mb.
Well projected 7 km from the NE
L06-03
⑤ Noordvaarder Mb
④ Lies Mb
③ Terschelling Sandstone Mb
② Oyster Ground Mb
(b)
Flattened on base Skylge Formation
F4
L2
L1
Base TMS-2
Well projected 7 km from the NE
① Main Friese Front Mb
(a)

analyses (Verreussel *et al.* 2018), and seismic plus well data analyses. The following discussion relies on four regional seismic and stratigraphic correlation panels that are used to illustrate the evolution of the basins and their platforms. The stratigraphic correlation and seismic analysis were also performed on seven additional regional panels that are not presented in the present paper, but from which information was extracted to map the distribution of key stratigraphic units (Fig. 18). Panel A (Figs 6, 7) illustrates the evolution of the Terschelling Basin and its southern and northern platforms. Panel B (Figs 8, 9) approximately follows the central part of the Dutch Central Graben (panel linking wells located close to the rift axis, rather than the exact position of the axis), while Panel C (Figs 10, 11) illustrates a strike profile across the Dutch Central Graben and its western and eastern platforms. Finally, Panel D (Figs 12, 13) shows the configuration of the northern part of the Dutch Central Graben, transitioning to the NW to the Step Graben. The distribution of key stratigraphic intervals within the study area is presented in Figure 18.

TMS-1: Callovian–Early Kimmeridgian (165–154.7 Ma)

In general, TMS-1 is limited to the axis of the Dutch Central Graben (Figs 1, 8–13, 18a). The development of the basin at this stage reflects renewed rifting following thermal doming in the early Middle Jurassic (*sensu* Underhill & Partington 1993). Note that the base of TMS-1 in the Danish sector is 3 Ma older than in the Dutch sector, with deposits accumulating as early as during the Bathonian (Verreussel *et al.* 2018, figs 5, 8).

The first deposits recorded above the Mid Cimmerian Unconformity are the fluvio-deltaic, lacustrine and marginal-marine sediments of the lower part of the Middle–Late Callovian (165–164 Ma) Lower Graben Formation (Figs 1, 3, 9, 11, 13, 14, 18a; Verreussel *et al.* 2018, fig. 5). Up to 150 m thick, these deposits accumulated in the rift axis within the narrow (less than 10 km wide) zones in the central/northern part of the basin, such as seen in Figures 9 and 11, onlapping underlying strata. This basin configuration is likely due to active salt withdrawal beneath the basin axis, possibly enhanced by thick-skinned normal faulting occurring on bounding faults along the rift margins (Ten Veen *et al.* 2012). Fluvial channels have been identified in the F06 block using stratal slicing and seismic amplitude extraction techniques, such as ‘root mean square’ and ‘sweetness and frequency’ (Figs 13–16). However, the provenance areas for TMS-1 deposits are unclear since few deposits of that age have been identified outside of the Dutch Central Graben. In the Terschelling Basin for example, only well M01-01 exhibits thin deposits of TMS-1 age (see Fig. 17 for location). This well is located above a salt diapir in the central part of the Terschelling Basin, which could explain the presence of TMS-1 at that location. This is postulated to be a remnant accumulation left in the vicinity of an exposed (or near free surface) salt body. The bulk of the sediments bypassed the area, and was deposited further west into the Dutch Central Graben which was actively subsiding during this period. Several erosional features are identified on seismic in the western part of the Terschelling Basin (e.g. L1, L2, Fig. 6) and are associated with elongated depocentres (I.V., for incised valleys, in Fig. 17). These features are interpreted as incised valleys (or

Fig. 7. Upper Jurassic–Lower Cretaceous stratigraphic correlation for the NW–SE-trending panel A located in the Terschelling Basin and its neighbouring platforms. (**a**) Flattened stratigraphic correlation panel for the Main Friese Front Member (Lower part of TMS-2). Note that no TMS-1 is present along this section. The datum is the base of the Skylge Formation (base of the Oyster Ground Member). The base of TMS-2 is shown as a green line. L1 and L2 correspond to seismically observed topographic lows that are interpreted as incised valleys that formed during TMS-1 and were filled by TMS-2 fluvial deposits of the Main Friese Front Member. Basal onlaps are observed on seismic and indicate that TMS-2 was highly controlled by local palaeotopography. (**b**) Flattened stratigraphic correlation panel for the upper part of TMS-2 that corresponds to the lower part of the Skylge Formation, which is composed of the Oyster Ground, Terschelling Sandstone, Noordvaarder and lower part of the Lies members. Note that the upper part of the Lies Member is not displayed in this figure. The datum is a well-recognized maximum flooding surface (MFS) intra Lies Member. The lower part of the Terschelling Sandstone Member is correlated between wells L06-03 and L03-04, while the upper part of that member in well L06-03 is time-equivalent to the Lies Member in well L03-04. The Noordvaarder Member is only observed in the northern part of the basin and is time-equivalent to both the Terschelling and Lies members further to the south. The Oyster Ground Member (unit 2) does not extend to the northern limit of the basin but pinches out between wells G16-04 and G16-05 where the palaeo-basin margin was contemporaneously located. This can also be seen in Figure 17. (**c**) Flattened stratigraphic correlation panel for TMS-3 composed of the Scruff Greensand and Lutine formations. The datum is the base of the Rijnland Group. The base of TMS-3 is shown as an orange line. As for the Oyster Ground Member, the lower part of the Scruff Greensand Formation was also controlled towards the north by the same palaeo-basin margin. At a later stage during the Scruff Greensand depositional period the basin extended further to the south, reaching the area located around well G16-04. See Figure 2 for location and Figure 6 for legend and corresponding seismic section.

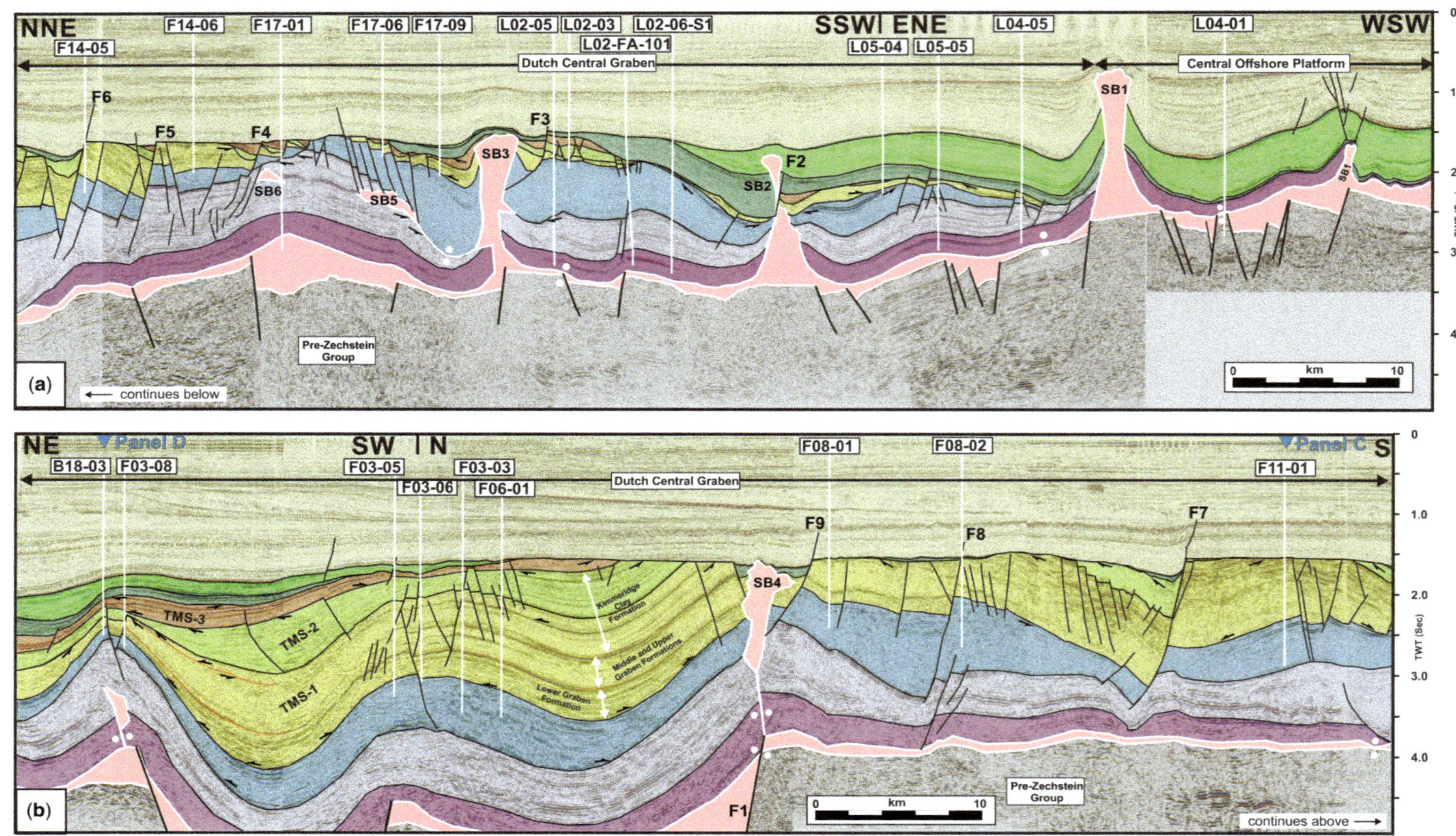
NNE
F14-05
F14-06
F17-01
F17-06
F17-09
L02-05
L02-03
L02-FA-101
L02-06-S1
SSW| ENE
L05-04
L05-05
L04-05
L04-01
WSW
Dutch Central Graben
Central Offshore Platform
F6
F5
F4
F3
F2
SB6
SB5
SB3
SB2
SB1
Pre-Zechstein Group
(a)
continues below
0
km
10
TWT (Sec)
0
1.0
2.0
3.0
4.0
NE
Panel D
B18-03
F03-08
SW | N
F03-05
F03-06
F03-03
F06-01
F08-01
F08-02
Panel C
S
F11-01
Dutch Central Graben
F9
F8
F7
SB4
TMS-3
TMS-2
TMS-1
Kimmeridge Clay Formation
Middle and Upper Graben Formations
Lower Graben Formation
F1
Pre-Zechstein Group
(b)
continues above

canyons) infilled by TMS-2-aged sediments. Similar features have also been identified on the Step Graben and in the NE Dutch Central Graben (Block F06, Figs 2, 16). It is proposed that these palaeo-conduits present in the Terschelling Basin were connected to the Dutch Central Graben area, possibly feeding the basin through a complex network of rivers controlled by structural fabric such as near-surface salt diapirs and fault escarpments. The rate of sediment accumulation is high (up to 330 m Ma^{-1}) during this period as indicated by the thick stratigraphic succession, especially in the northern part of the Dutch Central Graben where the Lower Graben Formation is locally more than 500 m thick (wells F03-03, F03-05-S1 and F03-06; Fig. 9). The youngest sediments of the Lower Graben Formation (164–163.5 Ma) were also confined to the Dutch Central Graben but within a larger zone (15–20 km wide, 110 km long; Fig. 18b), as observed by progressive onlaps continuously further away from the basin axis (Figs 8, 10–12, 15). These deposits are up to 400 m thick, sandy (Fig. 3) and display sedimentary features and ichnofacies consistent with tidally influenced shallow-marine environments. Features alluding to alternating freshwater and saltwater environments, such as synaeresis cracks, are abundant in wells F03-05-S1, F06-01 and F14-05 (Fig. 2). In the F14 block further south within the rift basin the presence of fine-grained offshore mudstones was identified in Upper Callovian strata (well F14-05, Lower Graben Formation core, Fig. 3), indicating drowning and pointing to a relatively lower sediment supply than in the southern/central part of the rift basin (see Verreussel *et al.* 2018 for further discussion of this topic).

The Early–Middle Oxfordian (163.5–160.5 Ma) deposits of the Middle Graben Formation, up to 450 m thickness, accumulated in the Dutch Central Graben within a 40 km (in the south) to 22 km (in the north) wide and 150 km long zone (Fig. 18c). This area extends further south than older units to the L05/F14 blocks. Three regionally extensive coal beds, individually up to 3 m thick, are identified in the lower part of the Middle Graben Formation (Figs 3, 9, 11, 14, 16). They have been identified in numerous wells separated by up to 120 km within the rift system (from well B18-03 to the north to well F02-FA-101 to the south, Fig. 8). These coal beds reflect a humid climate, dominated by wet, lowland palynological indicators (see Verreussel *et al.* 2018). The coal beds also likely reflect sediment starvation in the basin axis. The sudden switch off of sand supply can be ascribed to a regional transgression that flooded the graben shoulders and forced the deposition of marginal-marine sands onto the adjacent plateau areas (Verreussel *et al.* 2018). These sands were likely eroded later, as no Early–Late Oxfordian sandstones are present on these bordering plateaus. A single 200–300 m wide meandering channel is seismically observed in the F06 block where it cuts down into the regionally extensive coal layers (Figs 14–16). The trajectory of this large river is sinuous in the basin axis but shows more angular bends updip towards the eastern basin margin. At this location, the stratigraphic architecture observed on seismic data suggests updip confinement, interpreted as an incised valley (Fig. 16). This river is postulated to have developed on the Schill Grund Platform before reaching the rift axis in the Dutch Central Graben. This type of depositional model is uncommon in rift settings where lateral sediment input usually occurs along fault ramps (e.g. Gawthorpe & Leeder 2000). In the present case, it is proposed that differential subsidence between the rift axis and its shoulders is largely being controlled by smooth and regular salt withdrawal rather than fault propagation. This allows the lateral sediment input to reach the basin through fluvio-deltaic depositional systems that show variable levels of physiographic confinement, from broad unconfined depositional patterns to sediment input through incised valleys or canyons.

The upper part of TMS-1 is composed of several formations that are time-equivalent to each other, from the proximal lower part of the 150–300 m thick Friese Front Formation (blocks L05 to F17), the up to 500 m thick Puzzle Hole Formation (blocks F08 to F14), to the distal up to 1200 m

Fig. 8. Interpreted two-way travel time (TWT) for seismic panel B. This 170 km long section which intercepts 22 wells, is located along the axis of the Dutch Central Graben and extends southwards to the southern part of the Central Offshore Platform. For convenience, the section is split into two parts (**a** and **b**). The base Zechstein dips overall to the north from a highly faulted southern zone to the location of fault F1, which has a 1.5 s (TWT) vertical throw. This large fault bounds the deepest part of the Dutch Central Graben. Four shallow salt diapirs (SB1–4) are located within this section and are emplaced within the Cenozoic (SB1), the Cretaceous (SB2) and the Jurassic (SB3 and 4). Note two smaller salt features (SB5 and SB6) that are either remobilized Zechstein salt bodies emplaced intra-Triassic, or over-thickened *in situ* Triassic salt layers (e.g. Röt Evaporites). Note the over-thickened Lower Jurassic Altena Group north of SB3 that is related to the asymmetric salt withdrawal beneath the F17 turtle structure. Numerous post-Zechstein faults are observed in this section. Seven large listric growth faults (F2–9) were active during the Jurassic and Cretaceous. Only F2 detached on the Zechstein, while other faults detached at the lower part of the Lower Jurassic (F3, F4 and F9) and intra-Triassic (F5-F8). See Figure 2 for location, Figure 6 for legend and Figure 9 for stratigraphic correlation panel of the northern part of this panel.

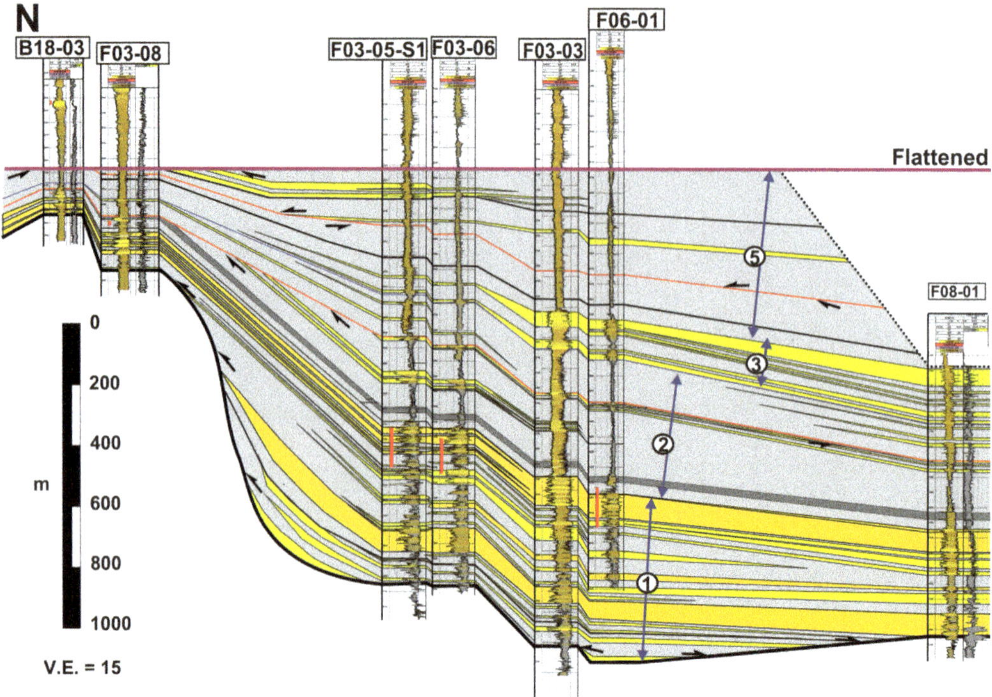

Fig. 9. Stratigraphic correlation panel of TMS-1 in the northern part of the Dutch Central Graben, along the rift axis. This section corresponds to the northern part of panel B shown in Figure 8. The stratigraphic correlation for TMS- 2 and 3 are not shown in the present paper. TMS-1 is thick in this part of the basin, locally reaching up to 1500 m at the location of well F03-03. Basal stratigraphic onlaps are observed on seismic showing that the Lower Graben Formation was filling up palaeotopographic lows inherited from the Middle Jurassic erosional events. Note the presence of the regionally extensive coal measures in the lower part of the Middle Graben Formation (shown as dashed outlined, grey filled unit). The sandy Upper Graben Formation is restricted to the central part of the section, correlates southwards to the time-equivalent Puzzle Hole Formation and pinches northwards toward the northern basin margin where it becomes mud-prone. The Kimmeridge Clay Formation is overall very fine-grained except in the northern part of the section (wells F03-05-S1 and F03-06) where a few 2–10 m thick sandy units are identified. These deposits are interpreted as deep-water sediment gravity flows generated by episodic structural movement along the basin margin. Red lines correspond to seismically constrained regional erosional surfaces. See Figure 2 for location, Figure 6 for legend and Figure 8 for corresponding seismic section. See Figure 3 for correlation of the F03-05, F06-01 and F14-05 Lower Graben Formation cores (red vertical bars along wells on Fig. 9).

thick upper part of the Middle and Upper Graben Formation and lowermost part of the Kimmeridge Clay Formation (blocks F08 to B13) (Fig. 18d–f). In the south the Friese Front Formation is mainly composed of fluvial deposits, while the Puzzle Hole Formation consists of fluvio-deltaic deposits. To the north the Upper Graben Formation is composed of higher net-to-gross strata deposited in a more marginal setting, likely in prodeltaic to shelfal environments (Munsterman *et al.* 2012). Figure 15 demonstrates three such deltas that are 3–5 km wide and are proposed to connect updip to a distributary and fluvial system that originated to the east, likely on the Schill Grund Platform. These fluvio-deltaic systems correlate northwards to the lower part of the marine Kimmeridge Clay Formation.

TMS-2: Late Kimmeridgian–Late Volgian (154.7–146.6 Ma)

TMS-2 is characterized by increased tectonic activity and the development of new secondary basins along the main rift. The Terschelling Basin developed during this period in the Dutch offshore, while the Heno Plateau/Feda Graben/Gertrud Plateau developed into active basins in the Danish western offshore (Andsbjerg & Dybkjær 2003; Verreussel *et al.* 2018, fig. 8). The Terschelling Basin was affected by vertical movements along newly active or reactivated NW–SE-trending faults along its northern (Rifgronden Fault Zone) and southern (Hantum Fault Zone) basin margins (Duin *et al.* 2006). Deposition persisted in the Dutch Central

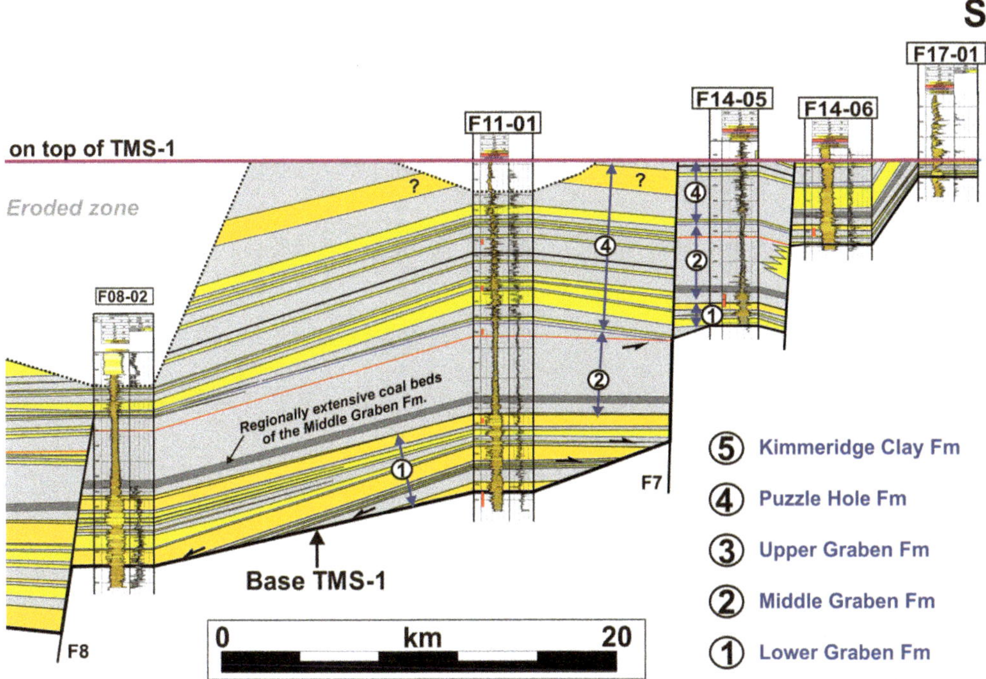

Fig. 9. *Continued.*

Graben during this period, but was primarily composed of fine-grained sediments of the marine Kimmeridge Clay Formation apart from local occurrences of the generally fine- to medium-grained Noordvaarder Member (Figs 1, 13). The Dutch Kimmeridge Clay Formation (159–147 Ma), which is up to 1100 m thick in the northern part of the Dutch Central Graben, represents the deepest water depositional setting of the Middle Jurassic–Lower Cretaceous succession (interpreted as deep-water mudstones with sandy sediment gravity flows of unknown reservoir qualities, since no core is available). Precise water depths are unknown. Note that the older part of the Kimmeridge Clay Formation in the northern part of the Dutch Central Graben is part of TMS-1 (Fig. 1). In a similar fashion, also note that the upper part of the Friese Front Formation (referred as the Main Friese Front Member) belongs to TMS-2 (see Fig. 1).

A significant flooding event occurred during the Late Kimmeridgian (J63 flooding of Partington *et al.* 1993; Fig. 1) that permanently set up marine conditions for the Late Kimmeridgian–Ryazanian in the entire Dutch offshore. In the Terschelling Basin, this flooding is expressed by the occurrence of organic-rich, fossiliferous, laminated shales belonging to the Late Kimmeridgian–Early Volgian Oyster Ground Member (152.4–150.8 Ma), which is up to 320 m thick in the western part of the Terschelling Basin (Figs 1, 4). Following a further significant flooding event, two geographically independent sandy systems became active in the Terschelling Basin. In the southern and central parts of the basin the tidally influenced sand-prone shoreface deposits of the Early–Middle Volgian Terschelling Sandstone Member (150.8–148 Ma), up to 160 m thick, were deposited. Detailed seismic mapping suggest that the palaeo-coastlines were oriented SSW–NNE (from block L09 to block M01; see Fig. 17 for location) during this period in the southern part of the basin. Meanwhile, in the northern part of the Terschelling Basin (blocks G16, L03 and F18) and along the central eastern part of the Dutch Central Graben (blocks F17), the Early Volgian–Middle Volgian finer-grained sandy Noordvaarder Member was deposited in lower shoreface/upper offshore settings (Figs 1, 2, 7b, 18c). The Noordvaarder Member is interpreted as being the result of local erosion off the Schill Grund Platform. A similar, locally sourced, sandy system of the same age is present in the northern part of the Dutch Central Graben in blocks B13 and B14 (Figs 1, 13, 18c), which is also referred to as the Noordvaarder Member. In this area, it is observed to have been deposited on the downthrown side of a large growth fault that seemingly detached on a salt pillow located

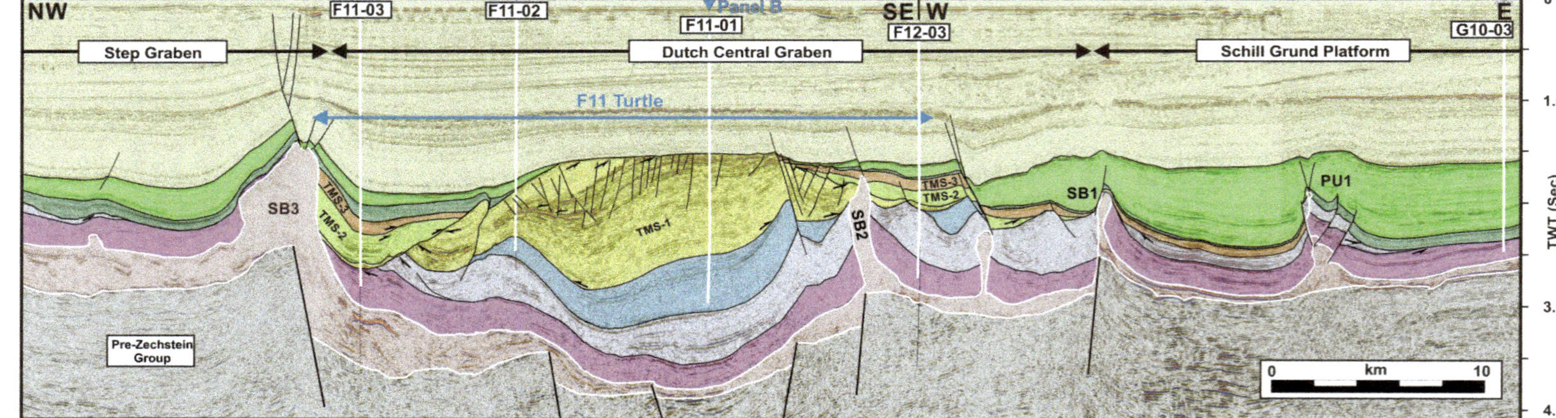

Fig. 10. Interpreted two-way travel time (TWT) for seismic panel C. This 70 km long section is located across the Dutch Central Graben and extends westwards to the southern Step Graben and eastwards to the Schill Grund Platform. Three salt diapirs (SB1-3) are located above large normal faults that were active during and possibly prior to the Triassic–Jurassic rifting. The Altena Group is only present within the Dutch Central Graben and was eroded on the neighbouring platforms and even locally within the Graben (e.g. location of well F12-03 and around SB2). Note the asymmetric geometry of the Middle Jurassic–Lower Cretaceous infill, with a western over-thickened stratigraphy around well F11-03 compared to the eastern side of the graben around well F12-03. The central part of the graben was also uplifted and eroded during the Cretaceous and Cenozoic, likely due to Alpine shortening (e.g. no TMS-2 or -3 are preserved in well F11-01). The Alpine shortening is also illustrated by the presence of a pop-up structure (PU1) and the squeezing of salt body 3 (SB3). See Figure 2 for location, Figure 6 for legend and Figure 11 for equivalent stratigraphic correlation panel.

along the rift basin margin (Fault F1, Figs 12, 13). Figure 17 shows a time–thickness map of TMS-2 in the Terschelling Basin. This sequence is thickest in the western part of the basin (block L03), and thins towards the basin margins. A noticeable ramp geometry is seen in this map (dashed white line, Fig. 17) that represents the basin palaeo-margin at the end of the Oyster Ground Member deposition. The basin later widened, with the Terschelling Sandstone Member being deposited further to the east and north (area between the dashed white and solid white lines in Fig. 17). This widening of the Terschelling Basin continued during the deposition of the following mega-sequence (TMS-3). This is demonstrated in Figure 17 by the orange zone that represents the area located between the maximal extent of TMS-2 (white line) and TMS-3 (blue line).

TMS-3: Late Volgian–Ryazanian (146.6–139 Ma)

The base of TMS-3 is a regional unconformity that erodes into TMS-2 in the Dutch Central Graben (Figs 11, 13) and down to Permo-Triassic strata on the platform areas. TMS-3 is eroded in large parts of the Dutch Central Graben and is only preserved in a few zones, specifically in the south (blocks L02, F18, F17, F16 and E18), along the lateral margins of the graben (blocks F12 and western part of F11) and in the north (blocks F05, F02, F03, B18 and B17). TMS-3 is also locally preserved in the Step Graben (blocks F01, E03, A18, A15 and A12) (Fig. 18h).

TMS-3 is well preserved in the Terschelling Basin with little post-depositional erosion observed. In this basin, the transition from TMS-2 to -3 is mainly conformable in the basin axis and is usually expressed by a coarsening-upwards grain size trend seen in wireline logs (e.g. wells G16-05, L03-04, L09-02 in Fig. 7c; well in F12-03, Fig. 11a; well B18-03 in Fig. 13). At the basin margins, however, this surface is typically observed to be erosional (Figs 11, 13). The vast volumes of sand observed in the Middle Volgian–Early Ryazanian Scruff Greensand Formation (146.6–141.5 Ma) (up to 190 m thick at the axis of the Terschelling Basin, Fig. 7c) indicates substantial erosion in the hinterland and along the basins margins, likely related to denudation and re-deposition.

A geographically restricted sand-prone depositional system is observed along the southern margin of the Terschelling Basin (M07 block) and, based on palynological information, is time-equivalent (147–139 Ma) to the Scruff Greensand and Lutine formations (Figs 1, 5 between markers B and C). This sandy unit has been the subject of recent successful hydrocarbon exploration by Oranje Nassau Energie in (M07-B Field) and has the potential for being a new promising play type in the Dutch offshore and beyond. These deposits are postulated to result from the erosion of the Ameland Platform and Friesland Platform to the south (Zechstein and Triassic-age strata), as suggested by the extra-formational angular clasts of sandstone and limestone observed in the lower shoreface of M07-07 and M07-08 cores (Fig. 5, at depth of 4838 m). In this core, pebble-rich beds are interpreted as debris flows generated by the structurally active basin margin where salt diapirs and faults were active (see block M07, Fig. 17). The gamma ray trend of the well M07-08 (Fig. 5) is heavily biased by glauconite content, which explains the apparent discrepancy between the wireline log response and the lithologies seen in core.

Salt tectonics and synrift stratigraphy

The autochthonous Zechstein salt has been actively remobilized since the Late Triassic. Salt tectonics is often complex and several aspects are discussed below, from the migration and withdrawal of autochthonous (Zechstein) salt to the relationship between deep structures and shallow salt bodies, as well as the specific relationship between shallow salt bodies and Middle Jurassic–Lower Cretaceous stratigraphy. For example, TMS1-3 was heavily affected by contemporaneous salt tectonics, directly around salt bodies and indirectly by normal growth faults that detached into the ductile Zechstein salt.

Autochthonous salt withdrawal

Zechstein salt was actively migrating and withdrawing from its autochthonous level during the Late Jurassic and Early Cretaceous, creating locally strongly subsiding zones notably in the Dutch Central Graben and, in a more limited way, in the Terschelling Basin. In the central part of the Dutch Central Graben (see also Duin *et al.* 2006; van Winden *et al.* 2018) the salt-related strong subsidence occurred mainly during the Middle Triassic–Early Jurassic (245–200 Ma) at the basin axis and shifted to the basin lateral margins during the Middle Jurassic–Early Cretaceous (165–139 Ma), forming large turtle structures (e.g. in blocks F17, F18 and F11 turtles; Figs 8, 15).

Relationship between shallow salt bodies and Middle Jurassic–Lower Cretaceous stratigraphy

Numerous salt bodies were emplaced within the Middle Jurassic–Lower Cretaceous. They often greatly impacted the depositional style, stratal thickness and preservation of sediments due to differential subsidence. This is known to have allowed growth of

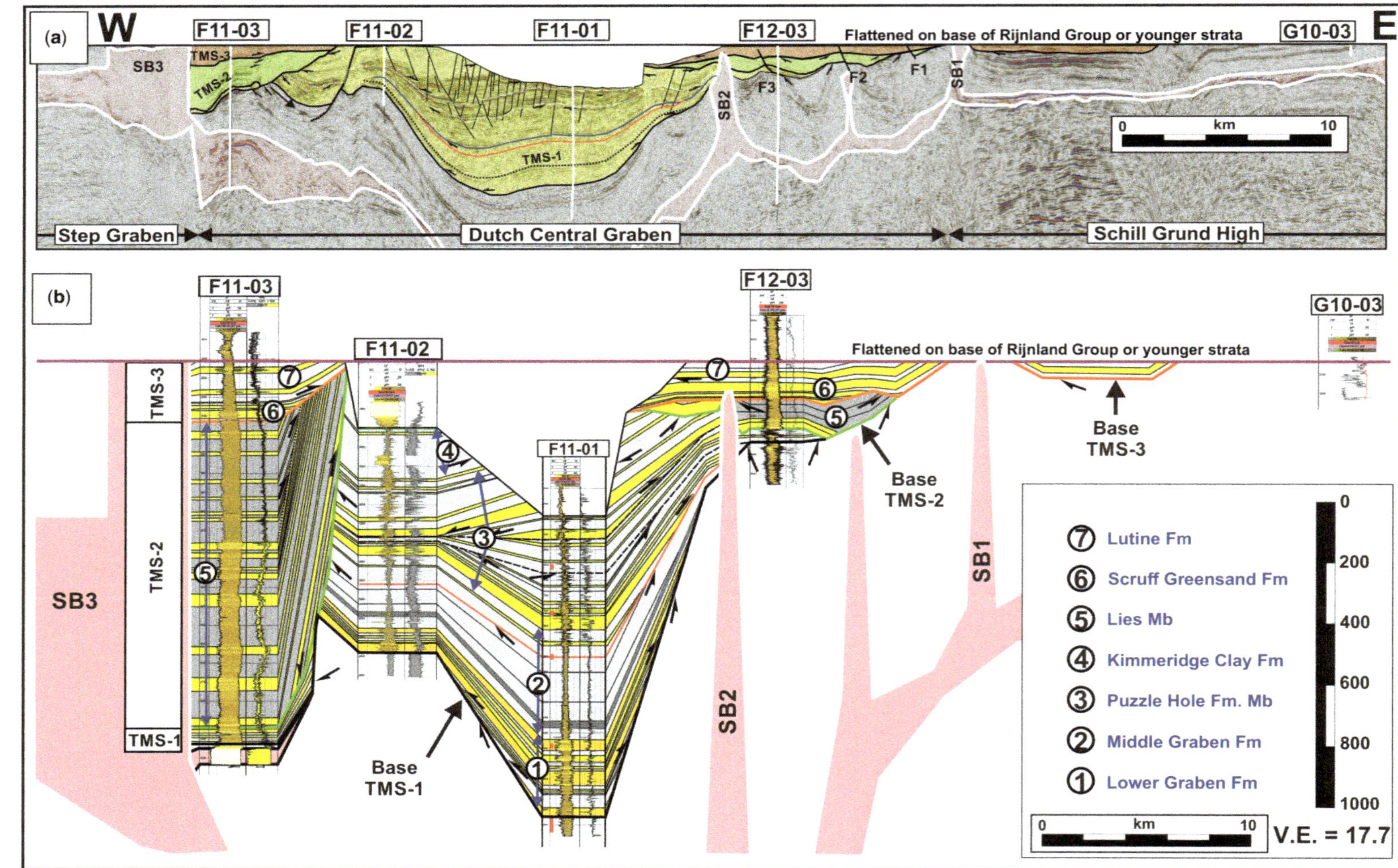

(a)
W
F11-03
F11-02
F11-01
F12-03
Flattened on base of Rijnland Group or younger strata
G10-03
E
SB3
TMS-3
TMS-2
TMS-1
SB2
F3
F2
F1
SB1
0
km
10
Step Graben
Dutch Central Graben
Schill Grund High
(b)
F11-03
F11-02
F11-01
F12-03
G10-03
Flattened on base of Rijnland Group or younger strata
TMS-3
TMS-2
TMS-1
SB3
SB2
SB1
Base TMS-1
Base TMS-2
Base TMS-3
⑦ Lutine Fm
⑥ Scruff Greensand Fm
⑤ Lies Mb
④ Kimmeridge Clay Fm
③ Puzzle Hole Fm. Mb
② Middle Graben Fm
① Lower Graben Fm
0
200
400
600
800
1000
0
km
10
V.E. = 17.7

stratigraphic wedges and troughs (e.g. salt bodies SB2 and SB3, Fig. 7; SB2 and SB3, Figs 10, 11) (see De Jager 2012; van Winden *et al.* 2018).

The evolution of these salt bodies often affected the geometry of surrounding stratigraphic units by forcing the rotation and tilting of neighbouring stratal blocks that were later often eroded, creating complex progressive unconformities and rim synclines. One such example is in the western margin of the Dutch Central Graben in the F11 block (adjacent to salt body SB3, Figs 10, 11). In this example, TMS-1 is only 12 m thick in well F11-03 compared to more than 1000 m thick in well 11-01 at the basin axis (Fig. 11b). In contrast, in Figure 11, TMS-2 and TMS-3 are only present along the basin margins where salt withdrawal created differential subsidence and where rim synclines and troughs formed, likely due to tilted and eroded Triassic stratigraphy. The deformation and withdrawal of allochthonous salt sheets also played a role in the stratigraphic patterns of the Middle Jurassic to Lower Cretaceous. For example, in the B13/B14 area (Figs 2, 12) syndepositional extensional faulting occurred above a deflated allochthonous salt sheet (SB1, Fig. 12). Similar geometries are observed in other salt basins with extensive allochthonous salt systems such as the Deep-water Gulf of Mexico (Bouroullec & Weimer 2017).

Relationship between synsedimentary faults and Middle Jurassic–Lower Cretaceous stratigraphy

Numerous syndepositional faults were active during the deposition of the Middle Jurassic and Early Cretaceous stratigraphy. These faults vary in size and geometry; they detached in various intervals, including on the autochthonous Zechstein halites, along the flanks of shallow salt bodies, at the top or within the Upper Triassic and within the Lower Jurassic. These faults are locally thin-skinned listric growth faults with associated hanging-wall stratigraphic wedges (e.g. fault F1, Fig. 6; fault F7, Fig. 8; fault F1, Fig. 12). It is difficult to estimate the possible local lithological impact of such synsedimentary faults on the surrounding sediment, but previous research (Edwards 1976; Bouroullec 2001) indicates that thicker but non-amalgamated sandstone units are often prevalent in fluvio-deltaic and marginal-marine depositional systems affected by growth faults. The largest growth faults are observed in the Dutch Central Graben (faults F3-9, Fig. 8) and were active during the Middle Jurassic–Cretaceous, with up to 40% thickening during TMS-1 in the case of fault F8 located between well F08-02 and F08-01. Fault F1 (Figs 12, 13) detached on the flank of a salt pillow and was active during the deposition of TMS-2. The hanging wall of that fault is seen to be sandier (relative to the footwall) with up to 160 m of Noordvaarder Member sandstones observed in well B13-02. Such basin margin sand-rich deposits can be attractive stratigraphic plays if the trap geometry can be understood and an efficient seal is present. Small syndepositional faults are observed on the flanks of the salt turtle structure located in block F17 (Fig. 8a), detaching in the Upper Triassic and at the base of the Lower Jurassic. These faults are likely related to the oversteepening of the turtle flank and the sliding of the stratal block northwards on the northern turtle flank and southwards on the southern turtle flank during the deposition of TMS-2 and -3.

Effect of palaeotopography on sediment distribution

Salt migration and fault movement can create palaeotopographic features which can impact reservoir distribution (e.g. Booth *et al.* 2003). One reservoir-scale example is shown in Figures 14–16 where kilometre-scale, laterally offset, marginal deltas and large rivers developed along the basin margin in the F06 block. The geometry of these stratigraphic elements indicates that sediment input into the rift system was mainly lateral instead of longitudinal to the rift axis during those periods. Figure 16 illustrates the decreasing depositional confinement from the updip platform/marginal areas to the basin axis where distributary/unconfined architecture are prevalent. These rapid stratigraphic architecture changes

Fig. 11. Seismic and stratigraphic correlation of panel C for the Middle Jurassic–Lower Cretaceous intervals. (**a**) Interpreted seismic section flattened on top of TMS-3. The unflattened version of this section is shown in Figure 10. TMS-1 shown in yellow, TMS-2 in green and TMS-3 in orange. TMS-1 is present in the axis of the Dutch Central Graben, while TMS-2 and -3 are only present on the lateral margins, within a rim-syncline to the west and a thin succession to the east. Note that post-depositional erosion was compensated for by the flattening procedure between wells F11-03 and F12-03. (**b**) Stratigraphic well correlation of TMS-1, -2 and -3. TMS-1 is thick in the axis of the Dutch Central Graben (980 m). During the deposition of TMS-1 the basin was relatively narrow (10 km wide originally, to 20 km by the end of TMS-1) and was rapidly subsiding, as shown by the trough-shaped geometry of the basin and the complex truncation and onlap configuration along the palaeo-margins. Red lines correspond to seismically constrained regional erosional surfaces. The eastern and western part of the TMS-1 were eroded prior to the deposition of TMS-2 and -3 (see high-angle truncation on the west side and lower angle on the east side). These zones became the loci of deposition for TMS-2 and -3 (wells F11-03 and F12-03). See Figure 2 for location and Figure 6 for legend.

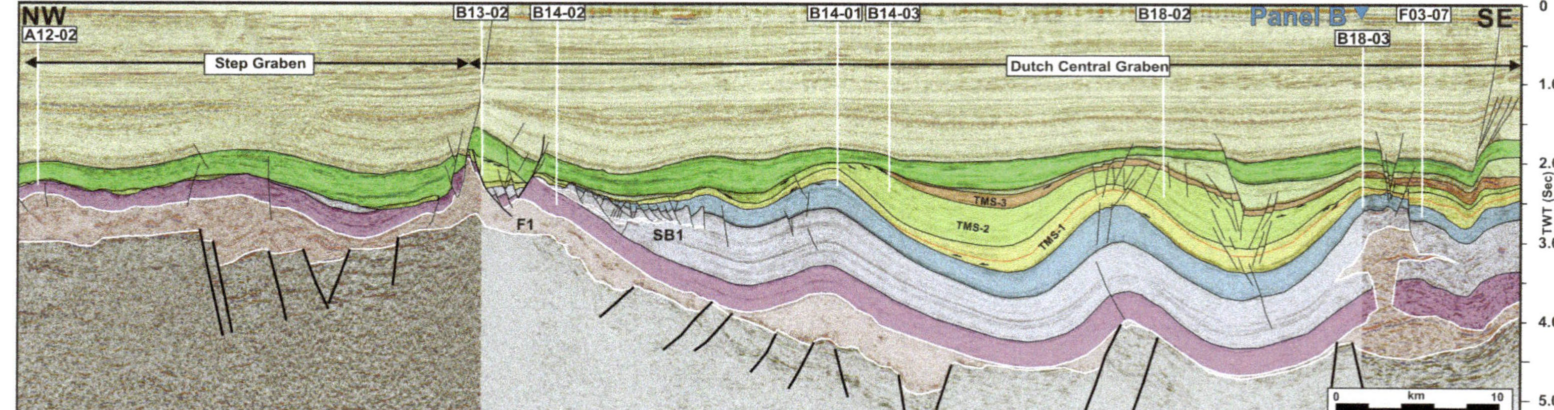

Fig. 12. Interpreted two-way travel time (TWT) for seismic panel D. This 92 km long section is located in the northern part of the Dutch Central Graben and extends northwestwards to the Step Graben. The Mesozoic fill of the graben is folded, mainly due to differential salt withdrawal during the Middle Jurassic–Early Cretaceous, and partially due to Cenozoic Alpine contraction that extended the duration of the fold's growth. TMS-1, -2 and -3 illustrate a thickness change between the over-thickened synclines (between wells B14-03 and B18-02, and between B18-02 and B18-03) and the over-thinned anticlines (at the location of wells B14-01, B18-02 and B18-03). Note that all three mega-sequences are increasingly eroded further north towards the basin margin (e.g. well B14-02). However, an exception is observed at the basin margin proper where TMS-2 is over-thickened on the downthrown side of a large SE-dipping growth fault F1. Note the intra-Triassic allochthonous salt sheet (SB1) located south of well B14-02, interpreted as an extruded Zechstein salt sheet that was nearly fully welded out during the Jurassic. Red line corresponds to an intra-TMS-1 erosional surface. See Figure 2 for location, Figure 6 for legend and Figure 13 for equivalent stratigraphic correlation panel.

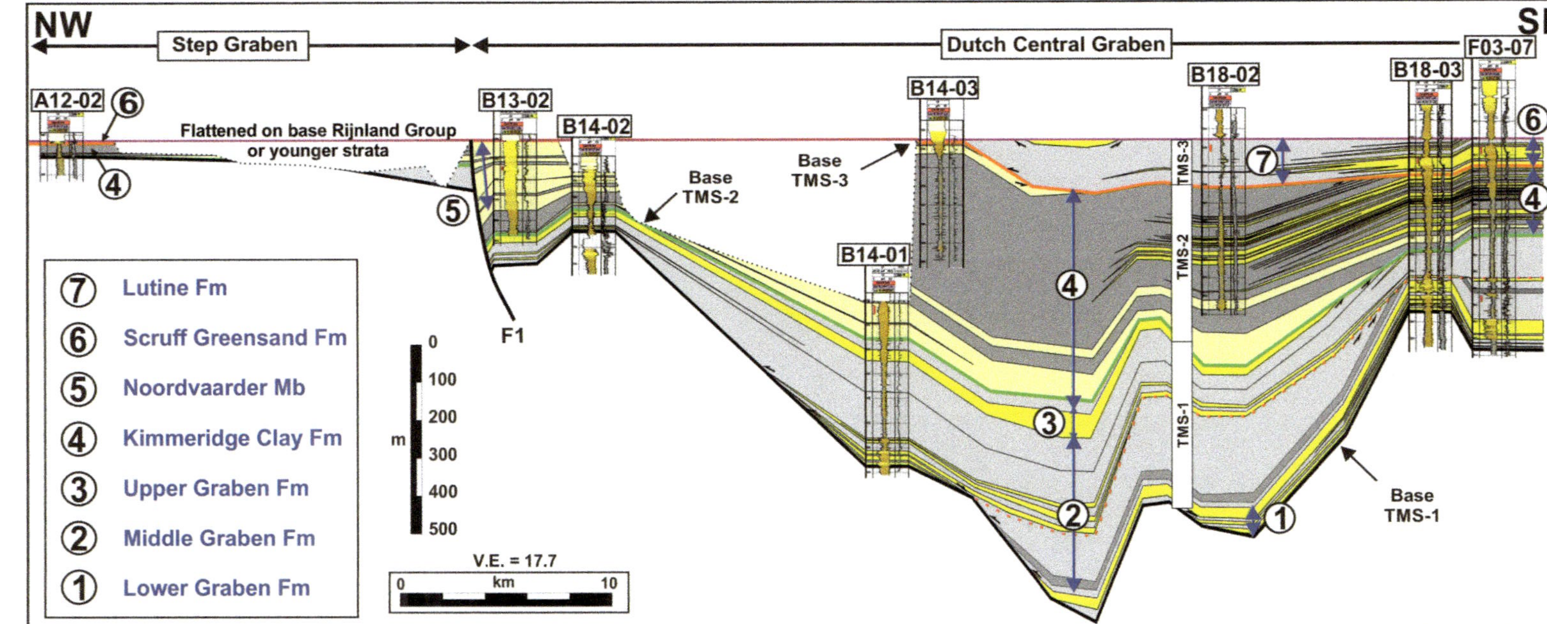

Fig. 13. Stratigraphic correlation of panel D for the Middle Jurassic–Lower Cretaceous. All three mega-sequences are present within the Dutch Central Graben. The lower part of TMS-1 is confined to the axial part of the Dutch Central Graben, with the Lower and Middle Graben formations lapping out onto the palaeo-basin margin towards the NW and filling up palaeo-topography in the basin axis. TMS-2 shows a northwestwards lateral fining trend with well B14-03 showing little sand deposition at the basin axis. The exception is the over-thickened Noordvaarder Member on the downthrown side of a growth fault (F1) that is interpreted as locally derived redeposited sediments from the neighbouring Step Graben area. TMS-3 shows lateral facies change between well B18-03 and B18-02. Dashed red line corresponds to an intra-TMS-1 erosional surface. See Figure 2 for location, Figure 6 for legend and Figure 12 for corresponding seismic section.

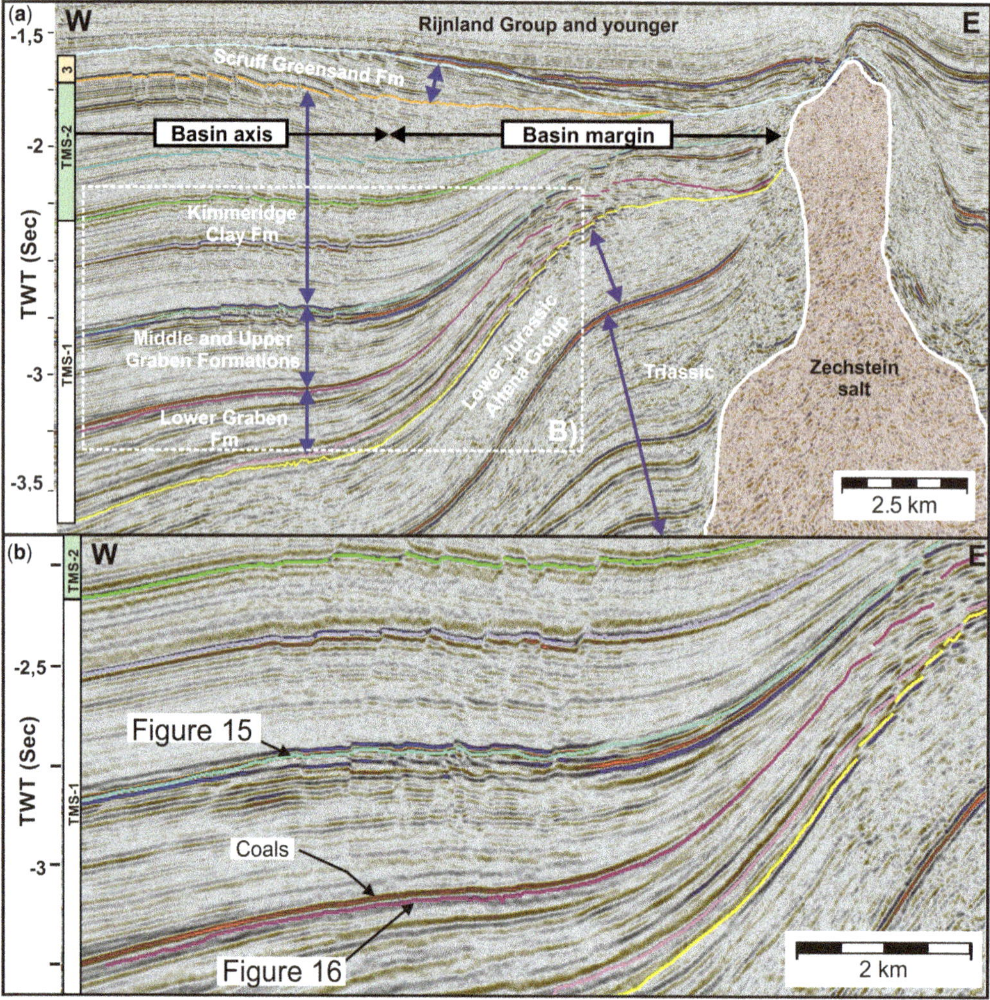

Fig. 14. Depositional systems in block F06 along the eastern basin margin of the Dutch Central Graben. (**a**) Two-way travel time (TWT) E–W-oriented interpreted seismic section. See Figure 2 for location. (**b**) Zoomed-in section showing the stratigraphic position of the maps shown in Figures 15 and 16c. The red amplitude above the purple horizon (top of the Lower Graben Formation) corresponds to the regionally extensive thick coal beds in the lower part of the Middle Graben Formation. Note that this coal beds pinch out onto the basin margin to the east (see green line in Fig. 16c).

illustrate the impact that differential salt withdrawal has on the local accommodation pattern by providing variable subsidence between the basin axis and its margins.

Multiple examples of palaeotopographic control on Middle Jurassic–Lower Cretaceous (165–139 Ma) depositional patterns can be observed in the regional seismic panels (Figs 6–9) where numerous basal and internal stratigraphic onlaps are observed, reflecting the role of local slope and changes in accommodation space and sediment accumulation. The onlap configuration at the base of the Middle–Upper Jurassic at the rift axis (e.g. between wells F03-08 and F03-05, between wells F06-01 and F08-01 in Fig. 8; between wells F11-02 and F12-03 in Figs 10, 11; and between wells B14-03 and B18-02 in Fig. 12) indicates that axial salt withdrawal already played an important role immediately after Middle Cimmerian erosional events. A progressive unconformity at the base of TMS-1 indicates that erosion, and likely re-deposition, of the Lower Jurassic or older strata along the palaeo-basin margins occurred simultaneously as the deposition of the Lower Graben Formation. This is the case between

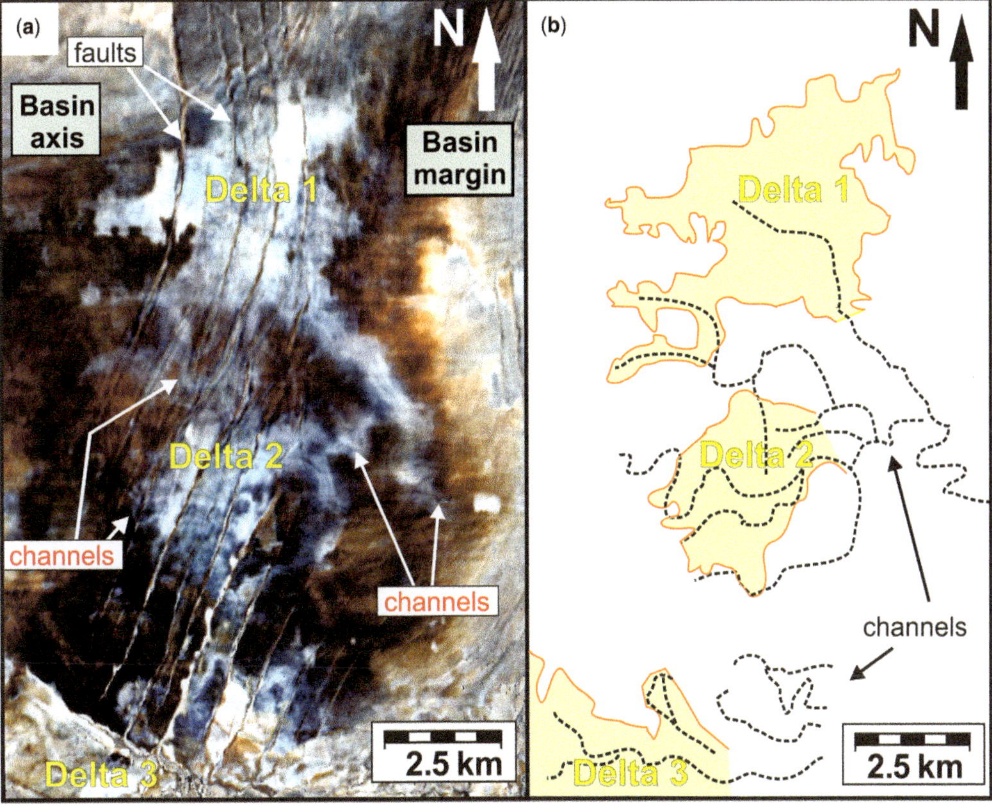

Fig. 15. Fluvio-deltaic depositional systems in the block F06, upper part of the Upper Graben Formation (TMS-1). (**a**) Example of a colour-blended spectral decomposition map and (**b**) line drawing of the main channels and delta edges recognized (see Fig. 14b for stratigraphic position). Note the numerous distributary channels and the three laterally offset bird-foot deltas located in the central part of the map. Deltas are 5–8 km in diameter and show well-developed lobes that extend in multiple direction into the rift basin. The main sediment pathways indicate an eastern origin for these fluvio-deltaic depositional systems, entering the basin from the present-day basin margin area (likely coming from the Schill Grund Platform) and forming a sandy marginal fringe along the basin margin.

wells F11-02 and F11-01 (Fig. 11), where the base TMS-1 is erosional and Lower Graben Formation strata lap onto the western basin margin. During the Early Kimmeridgian the depositional area broadened, reaching the rift shoulders. At the same time, the zones of maximum salt withdrawal shifted from the rift basin axis to the lateral basin margins, triggering rotation and sometimes erosion of the previously deposited Triassic and Jurassic (252–155 Ma, up to Oxfordian age) sediments. Lateral salt migration therefore resulted in the formation of internal and locally extensive unconformities and onlap surfaces, especially during the deposition of TMS-1 (Figs 10, 11). During this period, a type of stratigraphic cannibalism occurred since the depositional area within the rift basin episodically shrunk to narrower zones at the basin axis, possibly resulting in local erosion and re-deposition of Triassic–Upper Jurassic sediments that were originally deposited within the rift basin. Figure 11 shows internal erosional and onlap configuration between wells F11-02 and F11-01, as well as between wells F11-01 and F12-03 (especially within the Puzzle Hole Formation).

Detailed mapping of TMS-2 and -3 in the Terschelling Basin (Fig. 17) also revealed a variable basin margin configuration during the Middle Jurassic–Early Cretaceous. During TMS-2, the basin margin length – including the northern, eastern and southern margins (knowing that to the west the Terschelling Basin transitioned into the southern part of the Dutch central Graben) – was 170 km long and displayed four different basin margin types. Each of these margin types were identified from seismic data and individually mapped around the Terschelling Basin. The most common type of basin margin

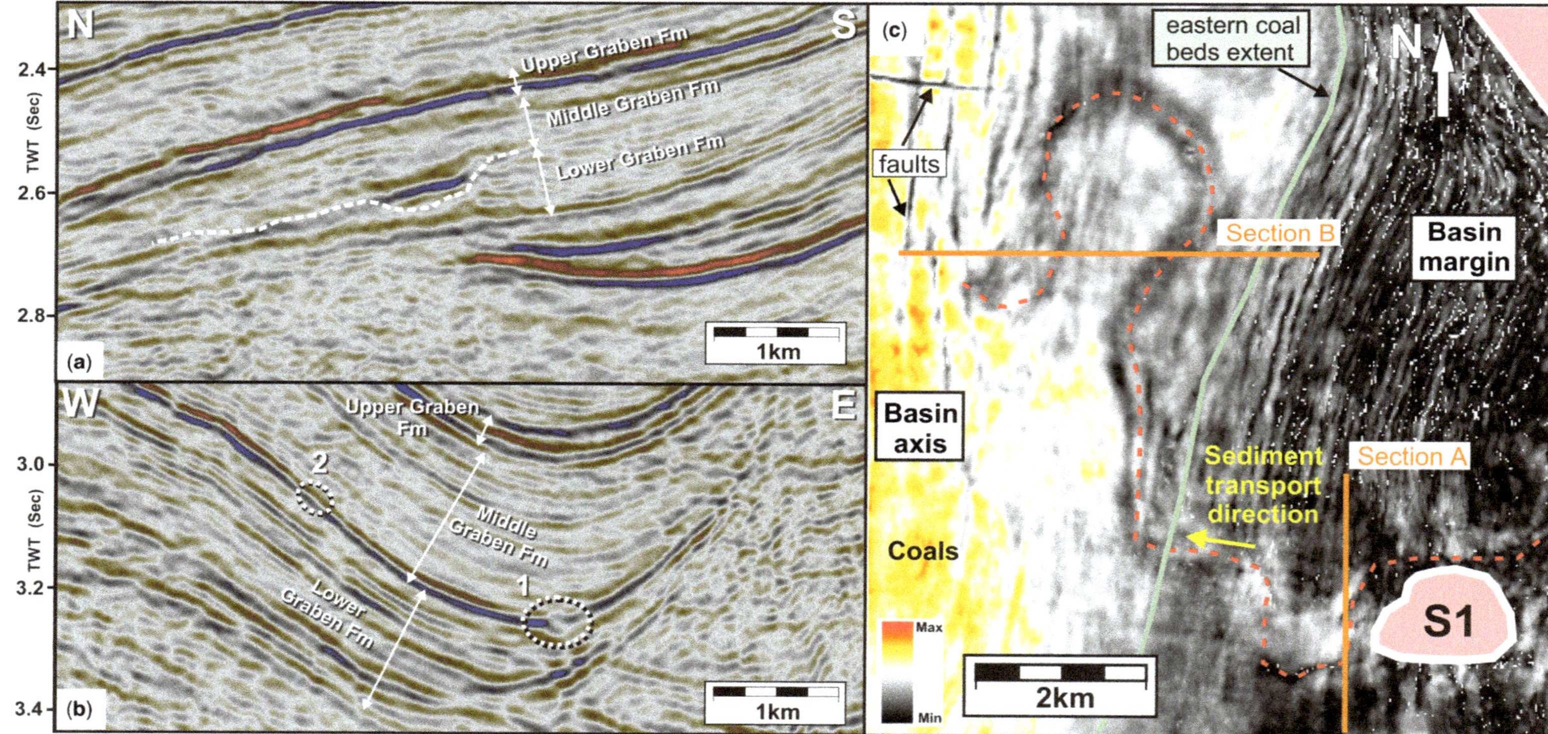

Fig. 16. (**a**) N–S and (**b**) E–W two-way travel time (TWT) seismic sections in the F06 block showing the variable seismic architecture of the fluvial channel shown on map (**c**). At its updip location (a) the channel is interpreted as a wide (1–2 km) confined incised valley (base shown as a dashed white line) that cuts down into the Lower Graben Formation. Downdip (b), the channel is narrower (200–300 m wide), less confined (location 1) and is not recognized further west (location) on this 2D seismic section.
(c) Example of a RMS (root mean square) seismic attribute map of a stratal slice in the lower part of the Middle Graben Formation (see Fig. 14b for stratigraphic position). This stratal slice is located at the level of the regionally extensive coal layers (see Figs 1, 3, 9, 11, 13). The high amplitudes in the basin axis are related to these thick (2–3 m) coal layers sitting below the claystone of the Middle Graben Formation. A 200–300 m wide sinuous fluvial channel is observed (dashed red line) coming from the eastern basin margin area north of the buried salt body 1 (SB1), where it takes a few sharp turns before entering the basin axial zone, where it shows smoother sinuosity such as in a meandering river system (Parker *et al.* 2011). Note the amplitude changes in the channel itself from bright updip to dark downdip, which is attributed to a lithology change of the channel fill from sandy on the basin margin where the channel is confined (a) to muddier at the basin axis where the channel is more unconfined (b) (Brown 2011).

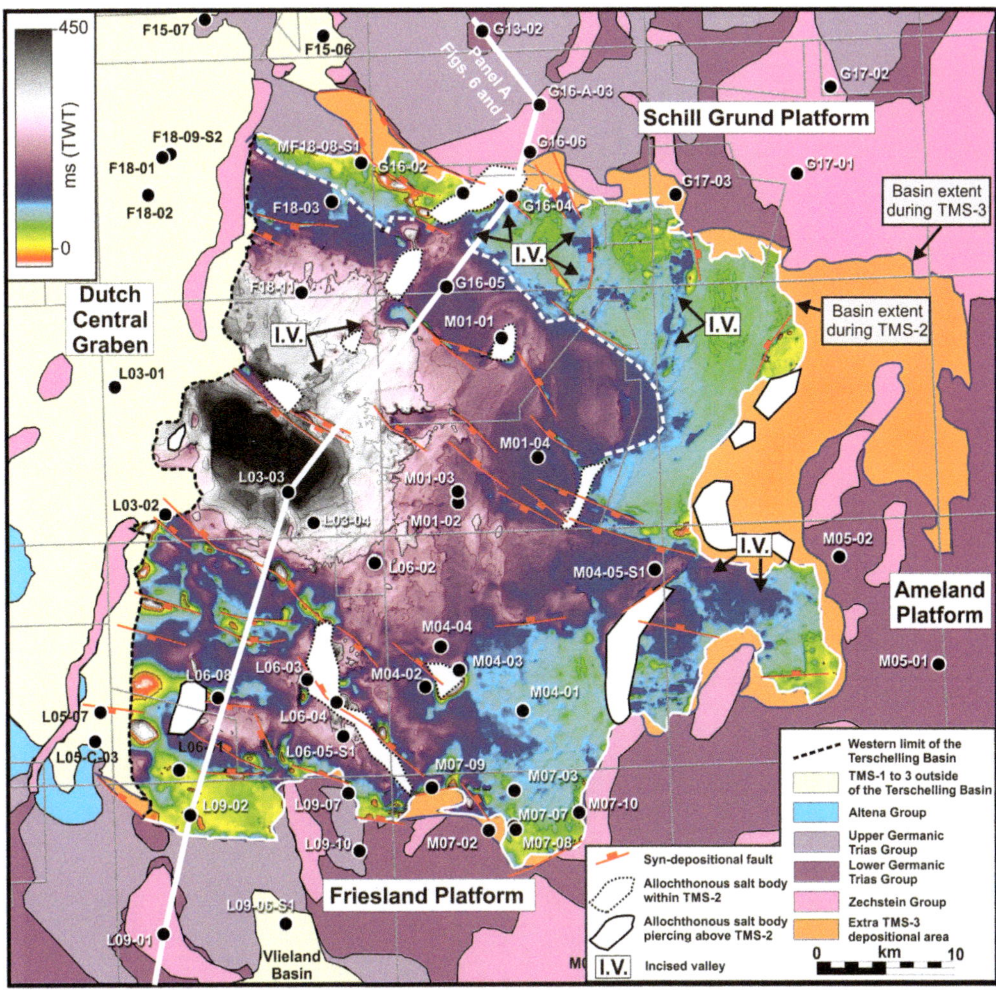

Fig. 17. Composite map of the Terschelling Basin and surrounding areas. Within the Terschelling Basin proper, the time–thickness map of Early Kimmeridgian–Late Volgian TMS-2 is displayed (shown as spectrum colours) as well as the salt bodies (white polygons) and the syndepositional faults (red lines) active during the deposition of this sequence. The base Rijnland Group (base Valanginian) subcrop map is displayed outside of the Terschelling Basin (see legend for details) as a background map. Note that the position of the maximum extent of the basin margins is shown for both TMS-2 (white line) and TMS-3 (blue line). Note that TMS-3 extends locally further than TMS-2, especially in the north and east part of the basin. This area is shown as an orange-coloured zone. The white dashed line shows the position of a sharp transition in time–thickness pattern of TMS-2 that corresponds to the position of the northern basin margin during the deposition of the Oyster Ground Member. See Figure 6 for a cross-sectional view of the basin.

is a thinning geometry (76% of the total length of the palaeo-basin margin) with TMS-2 strata thinning onto gently sloping margins. Rim syncline geometry is the second most common basin margin type (11%), with marginal strata accumulating in local topographic lows developing along the flanks of autochthonous salt pillows. The least common basin margin types are the conduits type (7%), consisting of major entry points for sediment entering the basin, and fault wedge type (6%), consisting of the hanging walls of large growth faults that detached on marginal salt pillows. The basin widened during TMS-3 (see Fig. 17, from the white line to the blue line) and the total length of its margin increased from 170 to 238 km, with the following five types of configurations: thinning (60%); fault wedge (17%); high-angle onlap against salt (14%); rim syncline (7%); and conduit (2%). The

Table 2. *Lithologies, sedimentary structures, ichnofacies and depositional environments of the formations and members present in TMS-1, -2 and -3*

Formation and member	Illustration	Dominant lithology	Sedimentary structures	Burrows	Environment of deposition
TMS-1					
Lower Graben Formation	Cores: Figure 3; others: Figures 1, 9, 11, 13, 14, 16, 18a, b	Claystones, siltstones and fine-grained sandstones; coal layers; siderite bands and nodules; sandstone bodies 1–4 m thick	Cross-bedding, mud drapes (sometimes double), syn-aeresis cracks	In parts only rootlets and *Planolites*; in other parts *Ophiomorpha*, *Asterosoma*, *Rosselia* and *Zoophycos*	Fluvio-deltaic and marginal marine (tidal, lagoonal, storm-influenced shoreface)
Middle Graben Formation	Cores: Figure 3; others: Figures 1, 9, 11, 13, 14, 16, 18c	Grey, locally very silty, carbonaceous claystones with three thick, very extensive coal layers	No core studied	No core studied	Fluvial and lacustrine to marginal marine
Upper Graben Formation	Figures 1, 9, 13, 14, 15, 18d	Greyish-brown, fine-grained, carbonaceous sandstones, separated by silty clay successions	No core studied	No core studied	Marginal marine barrier-island
Puzzle Hole Formation	Figures 1, 9, 11, 18e	Light brownish-grey carbonaceous claystones with intercalations of siltstones and thin sandstones; coal seams are frequent	No core studied	No core studied	Lower delta plain and lagoonal settings with occurrence of bay head deltas, mouthbars, tidal flats and tidal channels
TMS-1 and -2					
Friese Front Formation	Figures 1, 18e, f	Alternating claystones, siltstones, sandstones and some minor coals	No core studied	No core studied	Non-marine delta plain to lagoonal deposits
Kimmeridge Clay Formation	Figures 1, 9, 11, 13, 14, 18g	Olive-grey silty claystones with thin dolomitic streaks	No core studied	No core studied	Lower offshore setting
TMS-2					
Main Friese Front Member (Friese Front Formation)	Figures 1, 7, 18f	Alternating claystones, sandstones and some coals	No core studied	No core studied	Delta plain and lagoonal

Oyster Ground Member (Skylge Formation)	Core: Figure 4; others: Figures 1, 7, 18g	Non- to slightly silty claystones, with a few fossils and lignite fragments	Shell beds, hummocky cross-stratification	Planolites	Lower shoreface
Terschelling Sandstone Member (Skylge Formation)	Core: Figure 4 Figures 1, 7, 18g	Well to poorly sorted, fine- to medium-grained sandstone (occasionally up to coarse sand and even gravel)	Trough cross-bedding, wave ripples, cross-lamination, low-angle lamination	Asterosoma, Diplocraterion, Teichichnus, Ophiomorpha, Skolithos.	Upper shoreface, back barrier and tidal lagoon with channels
Noordvaarder Member (Skylge Formation)	Figures 1, 7, 13, 18g	Well-sorted, greenish-grey, slightly argillaceous, occasionally calcite cemented, glauconitic sandstones	No core studied	No core studied	Lower shoreface to upper offshore
Lies Member (Skylge Formation)	Core: Figure 4. Figures 1, 7, 11, 18g	Bioturbated silty to sandy claystones	No core studied	No core studied	Upper and lower offshore
TMS-3					
Scruff Greensand Formation	Core: Figure 5; others: Figures 1, 7, 11, 13, 14, 18h	Glauconitic, fine- to medium-grained sand	Completely bioturbated	Asterosoma, Diplocraterion, Rosselia, Thalassinoides, Phycosiphon	Lower shoreface to open marine
Lutine Formation	Figures 1, 7, 11, 13, 18h	Olive-grey, grey-brown to black claystones and fine- to very fine-grained argillaceous sandstones	No core studied	No core studied	Upper offshore

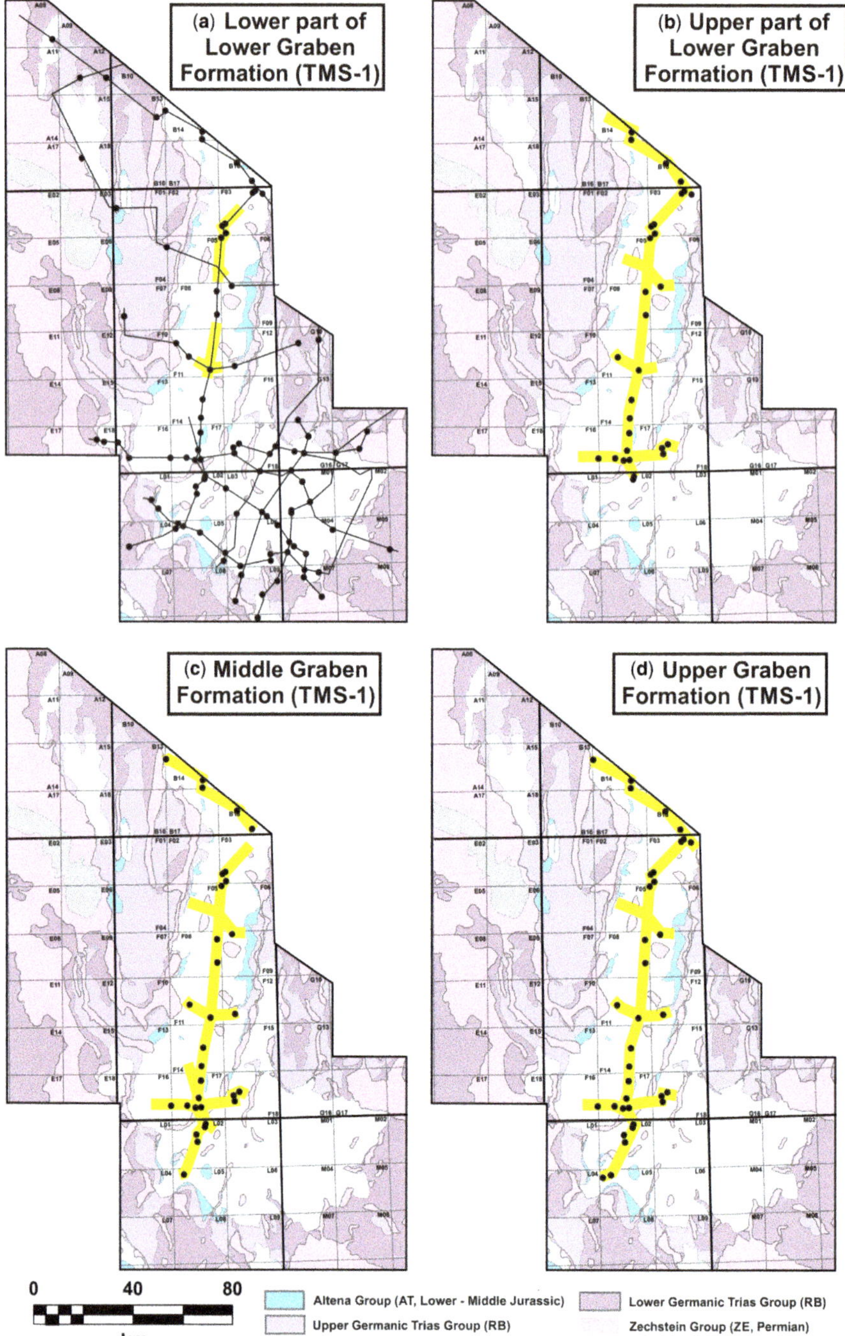

Fig. 18. (**a–h**) Distribution maps for the main lithostratigraphic units of the Middle Jurassic–Lower Cretaceous. Background maps for each time step is the base Rijnland Group subcrop map, in which areas mapped as unspecified Middle Jurassic–Lower Cretaceous strata are shown as white polygons. Each map shows the stratigraphic distribution as coloured thick lines along regional seismic panels (including those not shown in the present paper). Position of all the wells and seismic panels used to build these distribution maps shown in (a). Only the wells that have penetrated the given stratigraphic unit are shown. TMS-1 shown as yellow lines, TMS-2 as green and TMS-3 as orange. Note that the marginal sandy-rich Noordvaarder Member is also shown as dark blue lines in (g).

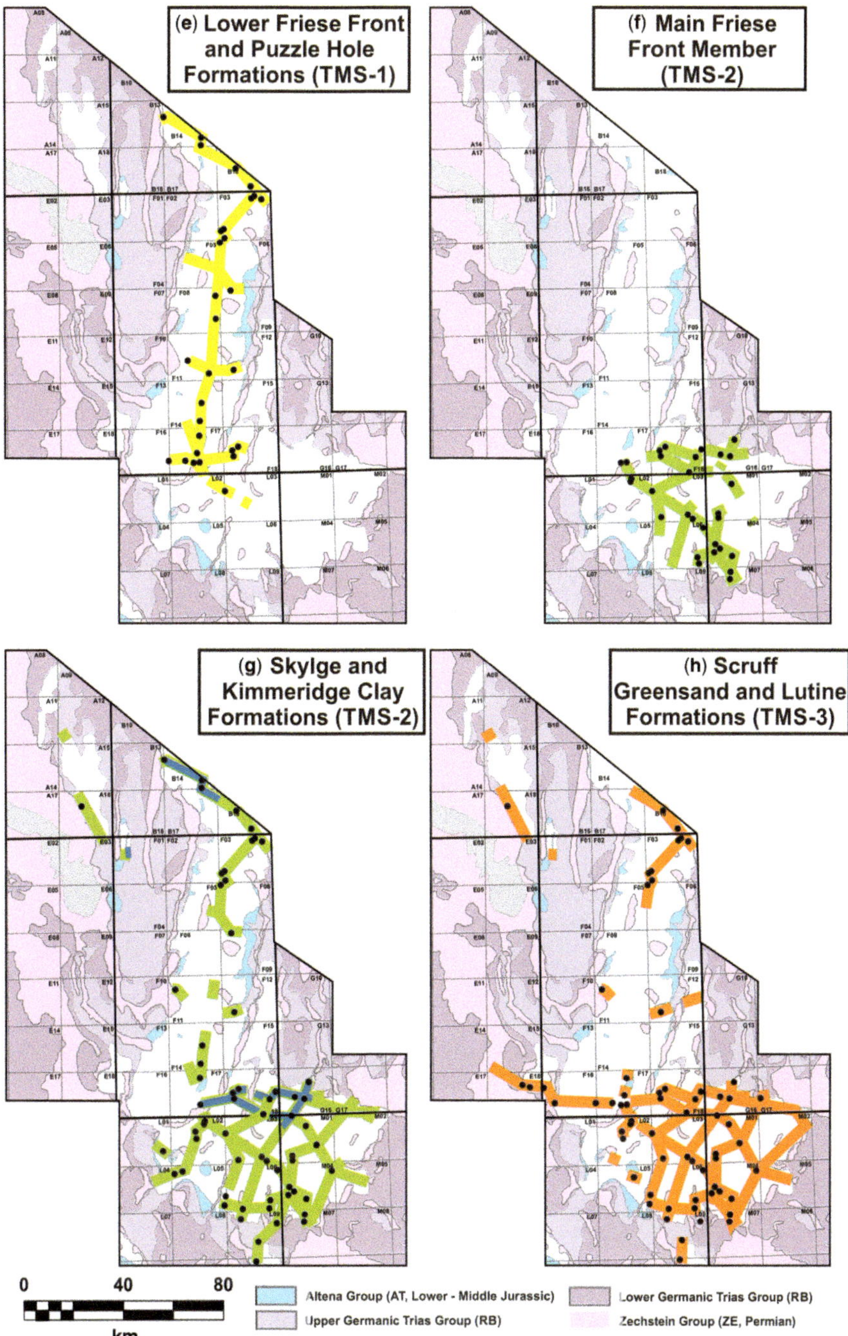

Fig. 18. *Continued.*

M07-B gas discovery is in a fault wedge configuration with reservoir sands of TMS-3.

The western part of the Terschelling Basin, the Step Graben and other low-accommodation zones such as the Schill Grund Platform and the Cleaver Bank Platform were subject to little or no deposition of TMS-1 and -2. Only small remaining areas on the platforms, often located within topographic lows, exhibit strata of that age (e.g. L1 and L2, Fig. 6; I.V. in Fig. 17). Based on the elongated geometry of these lows in map view, their erosional bases and the presence of basal onlaps of their fills, these features are interpreted as incised valleys (or possibly continental canyons) that developed during the deposition of TMS-1, providing sediment conduits into the rift basin within the Dutch Central Graben. The development of these incised valleys (or canyons) started during the deposition of TMS-1 in a few areas, such as the western part of the Terschelling Basin; in most cases however, based on 3D seismic mapping of TMS-1-3 in the Terschelling Basin, the fills of these palaeo-conduits were composed exclusively of TMS-2 sediments during a period of significant sea-level rise. This indicates that these conduits were long-lasting features that provided pathways for sediment transport into the rift basin during TMS-1 and were later drowned and sediment-filled during TMS-2. It is also important to notice that the platform areas suffered high amounts of erosion during multiple phases related to Middle Jurassic doming (Mid Cimmerian Unconformity, Bajocian–Bathonian, 170–166 Ma). The multiple Middle Jurassic–Early Cretaceous rifting phases may have triggered hundreds of metres of rift shoulder uplift and generated relative sea-level drops in the Early Volgian (150.3 Ma) (base of Terschelling Sandstone and Noordvaarder members) and Late Volgian (146.6 Ma) (Scruff Greensand Formation) (Verreussel *et al.* 2018). These erosional events likely frequently reshaped the physiography of the rift shoulders and, subsequently, the hydrographic characteristics of these areas. Most of these palaeosurfaces were also later eroded during the latest Upper Jurassic (base of TMS-3; 146.6 Ma), Early Cretaceous (base of TMS-4, 139 Ma; Verreussel *et al.* 2018) and/or during the Paleocene (Laramide tectonic phase, 63 Ma).

The results of this study, demonstrating the tectonostratigraphic evolution of a failed rift system affected by salt tectonics in the Dutch offshore, are important for local geological knowledge and for hydrocarbon exploration. These results are also relevant for other areas of the North Sea where salt tectonics was contemporaneous to thick-skin extensional tectonics. In this paper, the effects of combined brittle/ductile deformation on continental and shallow-marine depositional systems have been discussed. The study also constitutes an important addition to the topic of syntectonic effects on depositional systems. This topic has been considered extensively in clastic deep-water settings affected by salt tectonics (e.g. Booth *et al.* 2003; Prather 2003), carbonate settings affected by salt tectonics (e.g. Giles & Rowan 2012), continental settings (e.g. Banham & Mountney 2013) and for clastic continental settings affected by brittle deformation (e.g. Gawthorpe & Leeder 2000; Cowie *et al.* 2005). However, relatively little documentation on the impact of rifting and salt tectonics on shallow-marine settings has previously been published (Mannie *et al.* 2016). It is acknowledged that additional tectonostratigraphic analyses are required in other parts of the basin, via detailed 3D mapping and seismic amplitude analysis, to better understand the temporal and spatial evolution of the depositional systems affected by both salt tectonics and brittle deformation. Such topics are the subject of new research by the authors.

The results presented here regarding the Middle Jurassic–Early Cretaceous tectonostratigraphy of the Dutch offshore complement the already extensive knowledge of the interval in the North Sea (Andsbjerg & Dybkjær 2003; Fraser *et al.* 2003; Wonham *et al.* 2014). The article by Verreussel *et al.* (2018) applies knowledge of the Dutch Sector northwards to the offshore German and Danish sectors and, despite being primarily focused on the stratigraphic framework of this larger area, should be considered alongside this work to fully understand the tectonostratigraphic evolution of the entire failed rift system.

Conclusions and perspectives

Based on the integrated results of the tectonostratigraphic analysis, core description and palynological analyses carried out in the Dutch offshore, as well as regional stratigraphic analysis of Verreussel *et al.* (2018), this study acknowledges and strengthens the recently developed tectonostratigraphic threefold subdivision for the Middle Jurassic–Lower Cretaceous interval in the Dutch offshore (Abbink *et al.* 2006). The new chronostratigraphic framework proposed in the present paper (Fig. 1) clarifies the lateral equivalencies of known lithostratigraphic units. The Middle Jurassic–Lower Cretaceous interval was deposited during a period of intense rifting and salt movements. Deep-skinned and thin-skinned tectonic structures were active during this period and resulted in a complex basinal configuration. The withdrawal of Zechstein salt beneath the axis of the Dutch Central Graben and, to a lesser extent the Terschelling Basin, was the primary control affecting high accommodation and subsequent thick accumulation of Middle–Upper Jurassic strata (TMS-1) at the basin axis (e.g. 1.6 km in the F03 block). During TMS-2 and -3, the salt migration shifted laterally from the basin axis towards marginal-shallow salt bodies and locally formed large turtle structures. These salt bodies played a critical role in generating heterogeneous stratigraphy by creating zones of differential subsidence that experienced variable stratigraphic accumulation and preservation. With this improved understanding of the tectonostratigraphy of the Middle Jurassic–Lower Cretaceous in the Dutch offshore, new insights are offered on the complex interplay between active structures and depositional systems. The potential for new play types along the basin margins of rift affected by salt tectonics, as well as possible combined stratigraphic/structural traps in association with salt features in this part of the North Sea Basin, has also been highlighted.

The authors are grateful to O. A. Abbink, B. Vendeville and C. Jackson for their constructive reviews and extremely valuable suggestions, and to B. Kilhams and K. van Ojik for their editorial guidance. We extend our appreciation to Alain Trentesaux, Freek Busschers, Jean-Yves Reynaud, Nicholas Tribovillard, Robin Du Mée, Dario Ventra, Matthijs van Winden, Stefan Peeters, Lamine Diop, Alexander Houben, Jan de Jager, Jeroen Smit and Dimitrios Sokoutis. The authors would like to thank Johan Ten Veen and Harmen Mijnlieff for their valuable comments. Special thanks are due to Thijs Boxem for managing the FOCUS and COMMA projects from which this research originated. Thanks to Friso Veenstra for his business expertise and support. We thank our industry sponsors, EBN B.V., ENGIE E&P Nederland B.V, ONE B.V., Sterling Resources Ltd., Wintershall Noordzee B.V. as well as the Dutch Government for their financial support. The authors are also grateful for valuable discussions with our industry partners during the various meetings, core and poster workshops and the field trip on the Upper Jurassic of the Boulonnais. We would especially like to thank Annemiek Asschert, Leo van Borren, Léna Dauphin, Rutger Gras, Ruud Lambert, Rob Lengkeek, Phil Mollicone, Jeroen Mullink, Hugh Riches, Eveline Rosendaal, Anneliek Vis and Berend Vrouwe for their active roles in the projects.

References

Abbink, O.A., Mijnlieff, H.F., Munsterman, D.K. & Verreussel, R.M.C.H. 2006. New stratigraphic insights in the 'Late Jurassic' of the southern central North Sea Graben and Terschelling Basin (Dutch offshore) and related exploration potential. *Netherlands Journal of Geosciences – Geologie en Mijnbouw*, **85**, 221–238.

Andsbjerg, J. & Dybkjær, K. 2003. Sequence stratigraphy of the Jurassic of the Danish Central Graben. *In*: Ineson, J.R. & Surlyk, F. (eds) *The Jurassic of Denmark and Greenland*. Geological Survey of Denmark and Greenland Bulletin, Copenhagen, **1**, 265–300.

Banham, S.T. & Mountney, N.P. 2013. Controls on fluvial sedimentary architecture and sediment-fill state in salt-walled mini-basins: Triassic Moenkopi Formation, Salt Anticline region, SE Utah, USA. *Basin Research*, **25**, 709–737.

Booth, J.R., Dean, M.C., DuVernay, A.E. III. & Styzen, M.J. 2003. Paleo-bathymetric controls on the stratigraphic architecture and reservoir development of confined fans in the Auger Basin: Central Gulf of Mexico slope. *Marine and Petroleum Geology*, **20**, 563–586.

Bouroullec, R. 2001. *Growth fault kinematics and stratigraphic implications*. PhD thesis, Imperial College, London.

Bouroullec, R. & Weimer, P. 2017. Geometry and kinematics of Neogene allochthonous salt systems of the Mississippi Canyon, northern Atwater Valley, western Lloyd Ridge, and western Desoto Canyon protraction areas, northern deep-water Gulf of Mexico. *AAPG Bulletin*, **101**, 1003–1034.

Brown, A. 2011. *Interpretation of Three-Dimensional Seismic Data*. AAPG, Memoir 42. SEG Investigations no. 9, Tulsa, Oklahoma.

Callomon, J.H. 2003. The Middle Jurassic of western and northern Europe: its subdivisions, geochronology and correlations. *Geological Survey of Denmark and Greenland Bulletin*, **1**, 61–73.

Carruthers, A., Mckie, T., Price, J., Dyer, R., Williams, G. & Watson, P. 1996. The application of sequence stratigraphy to the understanding of Late Jurassic turbidite plays in the Central North Sea, UKCS. *In*: Hurst, A., Johnson, H.D., Burley, S.D., Canham, A.C. & Macertich, D.S. (eds) *Geology of the Humber Group: Central Graben and Moray Firth, UKCS*. Geological Society, London, Special Publications, **114**, 29–45, https://doi.org/10.1144/GSL.SP.1996.114.01.02

Coward, M.P., Dewey, J., Hempton, M. & Holroyd, J. 2003. Tectonic evolution. *In*: Evans, D.J., Graham, C., Armour, A. & Bathurst, P. (eds) *The Millennium Atlas: Petroleum Geology of the Central and Northern North Sea*. The Geological Society, London, 17–33.

Cowie, P.A., Underhill, J.R., Behn, M.D., Lin, J. & Gill, C.E. 2005. Spatio-temporal evolution of strain accumulation derived from multi-scale observations of Late Jurassic rifting in the northern North Sea: a critical test of models for lithospheric extension. *Earth Planetary Sciences Letter*, **234**, 401–419.

De Jager, J. 2007. Geological development. *In*: Wong, T.E., Batjes, D.A.J. & De Jager, J. (eds) *Geology of the Netherlands*. Royal Netherlands Academy of Arts and Sciences, Amsterdam, 5–26.

De Jager, J. 2012. The discovery of the Fat Sand Play (Solling Formation, Triassic), Northern Dutch offshore – a case of serendipity. *Netherlands Journal of Geosciences*, **91**, 609–619.

Duin, E.J.T., Doornenbal, J.C., Rijkers, R.H.B., Verbeek, J.W. & Wong, Th.E. 2006. Subsurface structure of the Netherlands – results of recent onshore and offshore mapping. *Netherlands Journal of Geosciences – Geologie en Mijnbouw*, **85**, 245–276.

Edwards, C.W. 1976. Growth faults in the Upper Triassic deltaic sediments, Svalbard. *AAPG Bulletin*, **60**, 341–355.

Fraser, S.I., Robinson, A.M. *et al.* 2003. Upper Jurassic. *In*: Evans, D. (ed.) *The Millennium Atlas: Petroleum Geology of the Central and Northern North Sea*. The Geological Society, London, 158–189.

Gawthorpe, R.L. & Leeder, M.R. 2000. Tectono-sedimentary evolution of active extensional basins. *Basin Research*, **12**, 195–218.

Geluk, M.C. 2007. Triassic. *In*: Wong, T.E., Batjes, D.A.J. & de Jager, J. (eds) *Geology of the Netherlands*. Royal Netherlands Academy of Arts and Sciences, Amsterdam, 85–106.

Giles, K.A. & Rowan, M.G. 2012. Concepts in halokinetic sequence deformation and stratigraphy. *In*: Alsop,

G.I., Archer, S.G., Hartley, A.J., Grant, N.T. & Hodgkinson, R. (eds) *Salt Tectonics, Sediments and Prospectivity*. Geological Society, London, Special Publications, **363**, 7–31, https://doi.org/10.1144/SP363.2

Gradstein, F.M., Ogg, J.G., Schmitz, M. & Ogg, G. (eds) 2012. *The Geologic Time Scale 2012*. Elsevier.

Hernandez, K., Mitchell, N. & Huuse, M. 2018. Deriving relationships between diapir spacing and salt later thickness in the southern North Sea. *In*: Kilhams, B., Kukla, P., Mazur, S., McKie, T., Mijnlieff, H. & v. Ojik, K. (eds) *Mesozoic Resource Potential in the Southern Permian Basin*. Geological Society, London, Special Publications, **469**. First published online February 26, 2018, https://doi.org/10.1144/SP469.16

Heybroek, P. 1975. On the structure of the Dutch part of the Central North Sea Graben. *In*: Woodland, A.W. (ed.) *Petroleum and the Continental Shelf of North-West Europe*. Applied Science Publishers Ltd, London, 339–349.

Kley, J. & Voigt, T. 2008. Late Cretaceous intraplate thrusting in Central Europe: effect of Africa-Europe-Iberia convergence, not Alpine collision. *Geology*, **36**, 839–842.

Kombrink, H., Doornenbal, J.C., Duin, E.J.T., Den Dulk, M., Van Gessel, S.F., Ten Veen, J.H. & Witmans, N. 2012. New insights into the geological structure of the Netherlands; results of a detailed mapping project. *Netherlands Journal of Geosciences – Geologie Mijnbouw*, **91**, 419–446.

Lott, G.K., Wong, T.E., Dusar, M., Andsbjerg, J., Mönnig, E., Feldman-Olszewska, A. & Verreussel, R.M.C.H. 2010. Jurassic. *In*: Doornenbal, J.C. & Steveson, A.G. (eds) *Petroleum Geological Atlas of the Southern Permian Basin Area*. EAGE Publications bv, Houten, **1**, 175–193.

Mannie, A.S., Jackson, C.A.L., Hampson, G.J. & Fraser, A.J. 2016. Tectonic controls on the spatial distribution and stratigraphic architecture of a net-transgressive shallow-marine synrift succession in a salt-influenced rift basin: Middle to Upper Jurassic, Norwegian Central North Sea. *Journal of the Geological Society*, **173**, 901–915.

Munsterman, D.K., Verreussel, R.M.C.H., Mijnlieff, H.F., Witmans, N., Kerstholt-Boegehold, S. & Abbink, O.A. 2012. Revision and update of the Callovian-Ryazanian Stratigraphic Nomenclature in the northern Dutch Offshore, i.e. Central Graben Subgroup and Scruff Group. *Netherlands Journal of Geosciences-Geologie en Mijnbouw*, **91**, 555–590.

Parker, G., Shimizu, Y. *et al.* 2011. A new framework for modeling the migration of meandering rivers. *Earth Surface Processes and Landforms*, **36**, 70–86.

Partington, M.A., Mitchener, B.C., Milton, N.J. & Fraser, A.J. 1993. Genetic sequence stratigraphy for the North Sea Late Jurassic and Early Cretaceous: distribution and prediction of Kimmeridgian-Late Ryazanian reservoirs in the North Sea and adjacent areas. *In*: Parker, J.R. (ed.) *Petroleum Geology of Northwest Europe. Proceedings of the 4th Conference*. Geological Society, London, Petroleum Geology Conference Series, **4**, 347–370.

Penge, J., Munns, J.W., Taylor, B. & Windle, T.M.F. 1999. Rift-raft tectonics: examples of gravitational tectonics from the Zechstein basins of northwester Europe. *In*: Fleet, A.J. & Boldy, S.A.R. (eds) *Petroleum Geology of Northwest Europe. Proceedings of the 5th Conference*. Geological Society, London, Petroleum Geology Conference Series, **5**, 201–213.

Pharaoh, T.C., Dusar, M. *et al.* 2010. Tectonic evolution. *In*: Doornenbal, H. & Stevenson, A.G. (eds) *Petroleum Geological Atlas of the Southern Permian Basin Area*. EAGE Publications b.v., Houten, 25–57.

Prather, B.E. 2003. Controls on reservoir distribution, architecture and stratigraphic trapping in slope settings. *Marine and Petroleum Geology*, **20**, 529–545.

Rijks Geologische Dienst (RGD). 1991. Geological atlas of the subsurface of the Netherlands, Explanation to Map Sheet II Ameland–Leeuwarden, Haarlem, the Netherlands, 87.

Roberts, A.M., Yielding, G., Kuznier, N.J., Walker, I.M. & Dorn-Lopez, D. 1995. Quantitative analysis of Triassic extension in the northern Viking Graben. *Journal of the Geological Society*, **152**, 15–27.

Schroot, B.M. 1991. Structural development of the Dutch Central Graben. *In*: Michelsen, O. & Frandsen, F. (eds) *The Jurassic in the Southern Central Trough*. Danmarks Geologiske Undersøgelse Series B, **16**, 32–35.

Stewart, S.A. 2007. Salt tectonics in the North Sea Basin: a structural style template for seismic interpreters. *In*: Ries, A.C., Butler, R.W.H. & Graham, R.H. (eds) *Deformation of the Continental Crust: The Legacy of Mike Coward*. Geological Society, London, Special Publications, **272**, 361–396, https://doi.org/10.1144/GSL.SP.2007.272.01.19

Stewart, S.A. & Clark, J.A. 1999. Impact of salt on the structure of the Central North Sea hydrocarbon fairways. *In*: Fleet, A.J. & Boldy, S.A.R. (eds) *Petroleum Geology of Northwest Europe. Proceedings of the 5th Conference*. Geological Society, London, Petroleum Geology Conference Series, **5**, 179–200.

Surlyk, F. & Ineson, J.R. 2003. The Jurassic of Denmark and Greenland: key elements in the reconstruction of the North Atlantic Jurassic rift system. *In*: Ineson, J.R. & Surlyk, F. (eds) *The Jurassic of Denmark and Greenland*. Geological Survey of Denmark and Greenland Bulletin, Copenhagen, **1**, 9–20.

Ten Veen, J.H., van Gessel, S.F. & de Dulk, M. 2012. Thin- and thick salt tectonics in the Netherlands; a quantitative approach. *Netherlands Journal of Geosciences – Geologie en Mijnbouw*, **91**, 447–464.

Torsvik, T.H., Carlos, D., Mosar, J., Cocks, L.R.M. & Malme, T.N.M. 2002. Global reconstructions and North Atlantic paleogeography 440 Ma to recent. *In*: Eide, E.A. (ed.) *Batlas – Mid Norway Plate Reconstruction Atlas with Global and Atlantic Perspectives*. Geological Survey of Norway, Trondheim, 18–39.

Trusheim, F. 1960. Mechanism of salt migration in northern Germany. *AAPG Bulletin*, **44**, 1519–1540.

Underhill, J.R. & Partington, M.A. 1993. Jurassic thermal doming and deflation in the North Sea: implications of the sequence stratigraphic evidence. *In*: Parker, J.R. (ed.) *Petroleum Geology of Northwest Europe. Proceedings of the 4th Conference*. Geological Society, London, Petroleum Geology Conference Series, **4**, 37–345.

Van Hoorn, B. 1987. Structural evolution, timing and tectonic style of the Sole Pit inversion. *Tectonophysics*, **137**, 239–284.

VAN WINDEN, M., DE JAGER, J., JAARSMA, B. & BOUROULLEC, R. 2018. New insights in salt tectonics in the northern Dutch offshore: a framework for exploration. *In*: KILHAMS, B., KUKLA, P., MAZUR, S., MCKIE, T., MIJNLIEFF, H. & V. OJIK, K. (eds) *Mesozoic Resource Potential in the Southern Permian Basin*. Geological Society, London, Special Publications, **469**. First published online January 29, 2018, https://doi.org/10.1144/SP469.9

VERREUSSEL, R.M.C.H., MUNSTERMAN, D.K., BOUROULLEC, R., DYBKJÆR, K., GEEL, C.R., HOUBEN, A.J.P. & KERSTHOLT-BOEGEHOLD, S.J. 2018. Stepwise basin evolution of the Middle Jurassic–Early Cretaceous rift phase in the Central Graben area of Denmark, Germany and the Netherlands. *In*: KILHAMS, B., KUKLA, P., MAZUR, S., MCKIE, T., MIJNLIEFF, H. & V. OJIK, K. (eds) *Mesozoic Resource Potential in the Southern Permian Basin*. Geological Society, London, Special Publications, **469**. First published online March 15, 2018, https://doi.org/10.1144/SP469.23

WONG, T.E. 2007. Jurassic. *In*: WONG, T.E., BATJES, D.A.J. & DE JAGER, J. (eds) *Geology of the Netherlands*. Royal Netherlands Academy of Arts and Sciences, Amsterdam, 107–126.

WONHAM, J.P., RODWEEL, I., LEIN-MATHISEN, T.O.R.E. & THOMAS, M. 2014. Tectonic control on sedimentation, erosion and redeposition of Upper Jurassic sandstones, Central Graben, North Sea. *In:* MARTINIUS, A.W., RAVNÅS, R., HOWELL, J.A., STEEL, R.J. & WONHAM, J.P. (eds) *From Depositional Systems to Sedimentary Successions on the Norwegian Continental Margin*. John Wiley & Sons, Ltd, Chichester, UK, **1**, 473–512.

ZIEGLER, P.A. 1988. *Evolution of the Arctic-North Atlantic and the Western Tethys*. AAPG Memoir **43**.

ZIEGLER, P.A. 1990*a*. *Geological Atlas of Western and Central Europe*. 2nd edn. Shell Internationale Petroleum Maatschappij b.v.; Geological Society Publishing House, Bath.

ZIEGLER, P.A. 1990*b*. Tectonic and palaeogeographic development of the North Sea rift system. *In*: BLUNDELL, D.J. & GIBBS, A.D. (eds) *Tectonic Evolution of the North Sea Rifts*. Oxford Science Publications, Oxford, 1–36.

Stepwise basin evolution of the Middle Jurassic–Early Cretaceous rift phase in the Central Graben area of Denmark, Germany and The Netherlands

R. M. C. H. VERREUSSEL[1]*, R. BOUROULLEC[2], D. K. MUNSTERMAN[1], K. DYBKJÆR[3], C. R. GEEL[2], A. J. P. HOUBEN[1], P. N. JOHANNESSEN[3] & S. J. KERSTHOLT-BOEGEHOLD[2]

[1]*TNO Netherlands Organisation for Applied Scientific Research, Geological Survey of the Netherlands, Princetonlaan 6, 3584 CB Utrecht, The Netherlands*

[2]*TNO Netherlands Organisation for Applied Scientific Research, Applied Geosciences, Princetonlaan 6, 3584 CB Utrecht, The Netherlands*

[3]*GEUS Geological Survey of Denmark and Greenland, Øster Voldgade 10, DK-1350, Copenhagen, Denmark*

**Correspondence: roel.verreussel@tno.nl*

Abstract: This paper presents the results of a cross-border study of the Middle Jurassic–Early Cretaceous rift phase in the Danish–German–Dutch Central Graben area. Based on long-distance correlations of palynologically interpreted wells, a stepwise basin evolution pattern was determined. Four phases are defined and described as tectonostratigraphic mega-sequences (TMS). The TMS are governed by changes in the tectonic regime. TMS-1 reflects the onset of rifting, triggered by regional east–west extension. Rift climax was reached during TMS-1, reflected by thick mudstone accumulations. TMS-2 reflects a change in the tectonic regime from east–west to NE–SW extension. NW–SE-trending normal faults became active during this phase, switching the depocentres from the graben axis into adjacent basins. TMS-3 displays divergent basin development. In the Dutch Central Graben area, it is characterized by a basal unconformity and widespread sandstone deposition, indicating continued salt and fault activity. Organic-rich mudstone deposition prevails in the Danish and German Central Graben area, indicating sediment starvation and water-mass stratification. With TMS-4 the rift phase ended, reflected by regionally uniform mudstone deposition. The basin evolution model presented here coherently places the lithostratigraphic units occurring in a stratigraphic framework and provides a valuable basis for hydrocarbon exploration activities in the region.

In this paper, a basin evolution model for the Middle Jurassic–Early Cretaceous rift phase of the cross-border Central Graben area in Denmark, Germany and The Netherlands is presented (Fig. 1). This study builds on earlier work from TNO (Netherlands Organisation for Applied Scientific Research) on the Dutch Central Graben (Herngreen & Wong 1989; Van Adrichem Boogaert & Kouwe 1993; Abbink *et al.* 2006; De Jager 2007; Lott *et al.* 2010; Munsterman *et al.* 2012) and from the Danish Geological Survey (GEUS) on the Danish Central Graben (Johannessen *et al.* 1996, 2010*a*; Andsbjerg & Dybkjær 2003; Johannessen & Andsbjerg 1993; Møller & Rasmussen 2003). Ideas and concepts from these papers have been refined and integrated into a single cross-border basin evolution model. This was achieved by detailed palynological analysis on approximately 230 exploration wells in the area. The palynological results allowed for the correlation of lithostratigraphic units across the various sub-basins and for the establishment of tectonic phase timing, such as the onset of fault or salt movement. It is apparent that the Middle Jurassic–Early Cretaceous basin evolution of the Central Graben area is mainly controlled by broad-scale changes in its tectonic history, in particular by a change in the extensional regime occurring during the Late Kimmeridgian (Møller & Rasmussen 2003; Zanella & Coward 2003). Absolute sea-level changes also played an important role, especially with respect to sediment and facies distribution (Andsbjerg & Dybkjær 2003; Abbink *et al.* 2006; Bouroullec *et al.* 2018).

This study confirms that this change in extensional direction led to a region-wide shift in the main depocentre position from basin axis to basin margin and later to the adjacent basins. In addition, it is demonstrated for the first time that the basin evolution of the Danish and Dutch Central Graben is

From: Kilhams, B., Kukla, P. A., Mazur, S., McKie, T., Mijnlieff, H. F. & van Ojik, K. (eds) 2018. *Mesozoic Resource Potential in the Southern Permian Basin*. Geological Society, London, Special Publications, **469**, 305–340.
First published online March 15, 2018, https://doi.org/10.1144/SP469.23

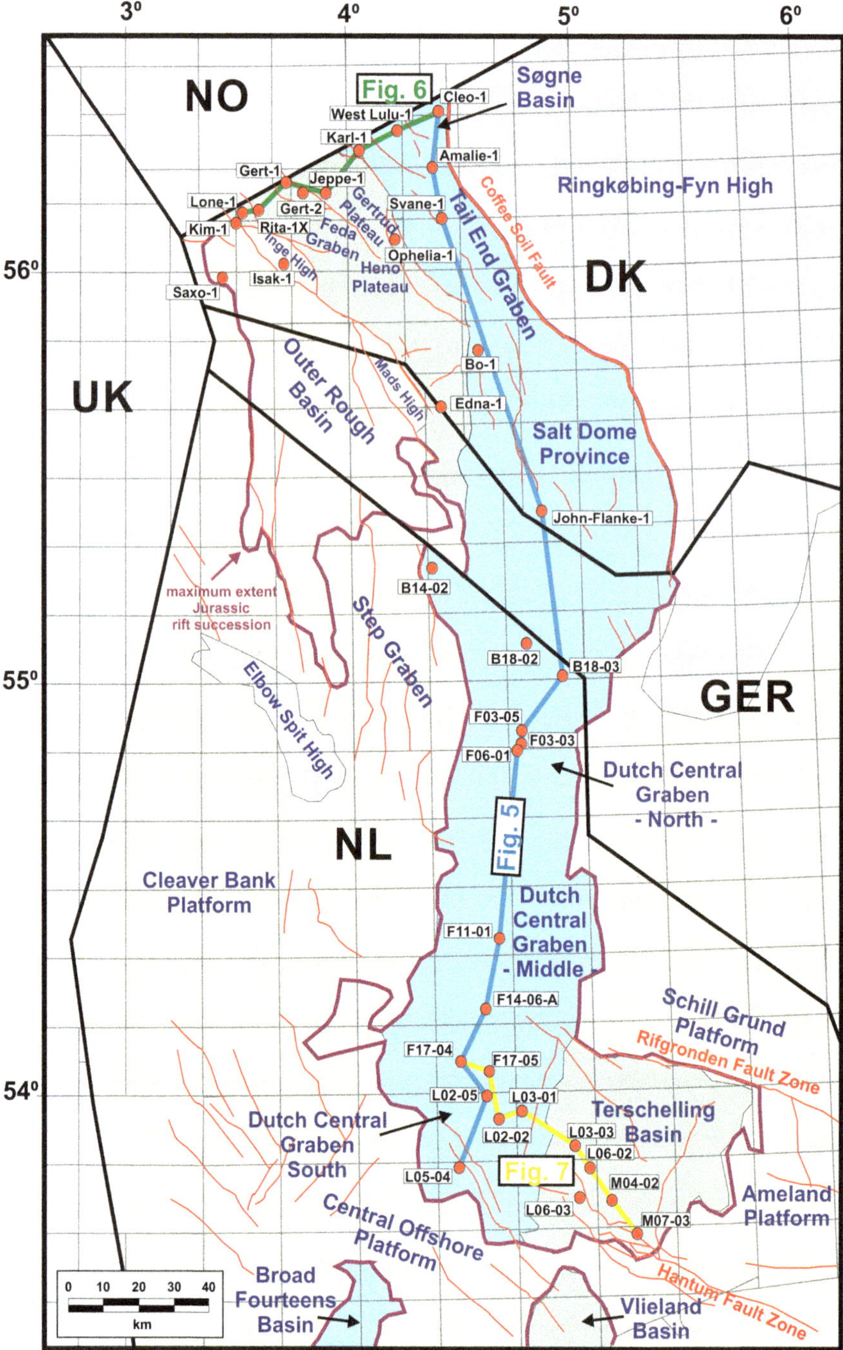

Fig. 1. Location map showing structural elements, well locations and correlation panels. The graben axis is indicated in the blue polygon, the basins adjacent to the graben axis are indicated in olive green polygons, the plateau areas in pale yellow fill and the structural highs in grey polygons. Principal faults are drawn in red. Parts of the offshore sectors of five countries are represented on the map: Norway (NO), Denmark (DK), Germany (GER), United Kingdom (UK) and The Netherlands (NL). The purple line indicates the maximum extent of the Middle Jurassic–Lower Cretaceous sediments. The blue line shows the position of the panel in Figures 5 and 16; the green line corresponds to Figures 6 and 17; the yellow line corresponds to Figures 7 and 18.

very similar in the initial phase of rifting, but diverges considerably in the last phase (latest Volgian and Ryazanian). Four tectonostratigraphic megasequences (TMS) and nine subordinate sequences (TS) are defined and described. Combined, these sequences capture the step-wise basin evolution of the Middle Jurassic–Early Cretaceous rift phase of the Central Graben area (Fig. 2). The numerous lithostratigraphic units from Danish and Dutch stratigraphic nomenclature that occur in the studied time interval are coherently placed and are displayed on Wheeler diagrams.

Geological setting

Tectonic setting

During the Mesozoic, a large rift system developed in the present-day North Sea area as a result of the break-up of Pangea (Zanella & Coward 2003). The first rift phase took place during the Triassic and resulted in large and relatively wide graben structures filled with predominantly non-marine sediments. Zechstein salt was mobilized and, subsequently, influenced Triassic depositional patterns in the Northern and Southern Permian Basin (De Jager 2007; De Jager & Geluk 2007). After a period of relative quiescence and blanketing of the rift structures by fine-grained sediments during the Early Jurassic (Wong 2007), the area was affected by a regional uplift phase during the Middle Jurassic, also known as the North Sea Thermal Doming event (Partington *et al.* 1993*a*; Husmo *et al.* 2003). This uplift and associated volcanism was caused by an active mantle plume in the triple junction area (Underhill & Partington 1993). During this uplift phase, deposition continued in small, fault-controlled sub-basins in the Danish Central Graben (Andsbjerg 2003; Mellere *et al.* 2016). The second and most intense rift phase took place during the Middle Jurassic–Early Cretaceous (Surlyk & Ineson 2003; Zanella & Coward 2003), which is the subject of this study. Rifting activity ceased during the Early Cretaceous (Vejbæk *et al.* 2010); the resultant failed rift consists of three branches that meet in a triple junction, roughly situated 200 km east of Aberdeen (Coward *et al.* 2003). The Central Graben is the southern branch, which runs in a southeasterly direction from the triple junction to the Salt Dome Province in the Danish offshore and due south into the German and Dutch sectors. The Central Graben terminates against the Central Offshore Platform in Block L05 in the Dutch offshore (Fig. 1).

Middle Jurassic–Early Cretaceous eustasy

The Middle Jurassic–Early Cretaceous sedimentary succession in the Central Graben area is deposited in an active rift setting (Fraser *et al.* 2003; Zanella & Coward 2003). As a consequence, the effect of eustatic sea level on the distribution of sediments plays a subordinate role compared to the tectonic component. On a third-order scale, however, facies changes related to eustasy do occur and provide a basis for correlation.

After the publications of Partington *et al.* (1993*a*, *b*), this sequence stratigraphic approach was widely adopted in North Sea exploration (Andsbjerg & Dybkjær 2003). Partington *et al.* (1993*a*, *b*) introduced a sequence stratigraphic framework in which the J-sequences of Rattey & Hayward (1993) were further subdivided and in which the maximum flooding surfaces (MFS) were presented as correlatable horizons, named after the ammonite chronozones these MFS were associated with. In later publications (Duxbury *et al.* 1999; Fraser *et al.* 2003), a nomenclature based on maximum abundances of dinoflagellate cysts was preferred. Regionally important MFS are the J46 in the latest Callovian, the J63 and J64 in the Late Kimmeridgian and the J76 in the Late Ryazanian (Partington *et al.* 1993*a*, *b*; Bouroullec *et al.* 2018, fig. 1).

Middle Jurassic–Early Cretaceous climate

The Jurassic Period is generally considered as a greenhouse world, devoid of major glacial episodes (Donnadieu *et al.* 2011). The start of the Jurassic might even be considered a super-hothouse, with extreme high atmospheric CO_2 levels as a result of volcanic outgassing associated with the Central Atlantic Magmatic Province (Korte & Hesselbo 2011). This generally warm phase, with periodic hyper-thermals, persisted until the end of the Early Jurassic Epoch (Hettangian, Sinemurian, Pliensbachian, Toarcian), with a possible minor cooling during the late Pliensbachian (Korte & Hesselbo 2011). Evidence for cooler climates and possibly even polar ice in the Middle Jurassic Epoch (Aalenian, Bajocian, Bathonian and Callovian) is widespread (Nunn & Price 2010; Dera *et al.* 2011; Dromart *et al.* 2003). An example of such a cooler phase is found in the palynological record of the latest Callovian to earliest Oxfordian from the Dutch offshore. Cored sections from the Lower Graben and Friese Front Formation of wells F03-05-S1, F06-01, F14-05, F17-04 and L05-04 (among others) show relatively cool and humid climates, reflected in the near-absence of the pollen type *Classopollis* (an indicator species for warm-arid conditions; e.g. Abbink 1998; Bonis & Kürschner 2012) and the overall dominance of *Perinopollenites* (an indicator species for wet lowland habitats; Abbink *et al.* 2006). Interestingly, the thick coal occurrences at the base of the succeeding Middle Graben Formation also occur within this relatively cool and wet phase. An important climate

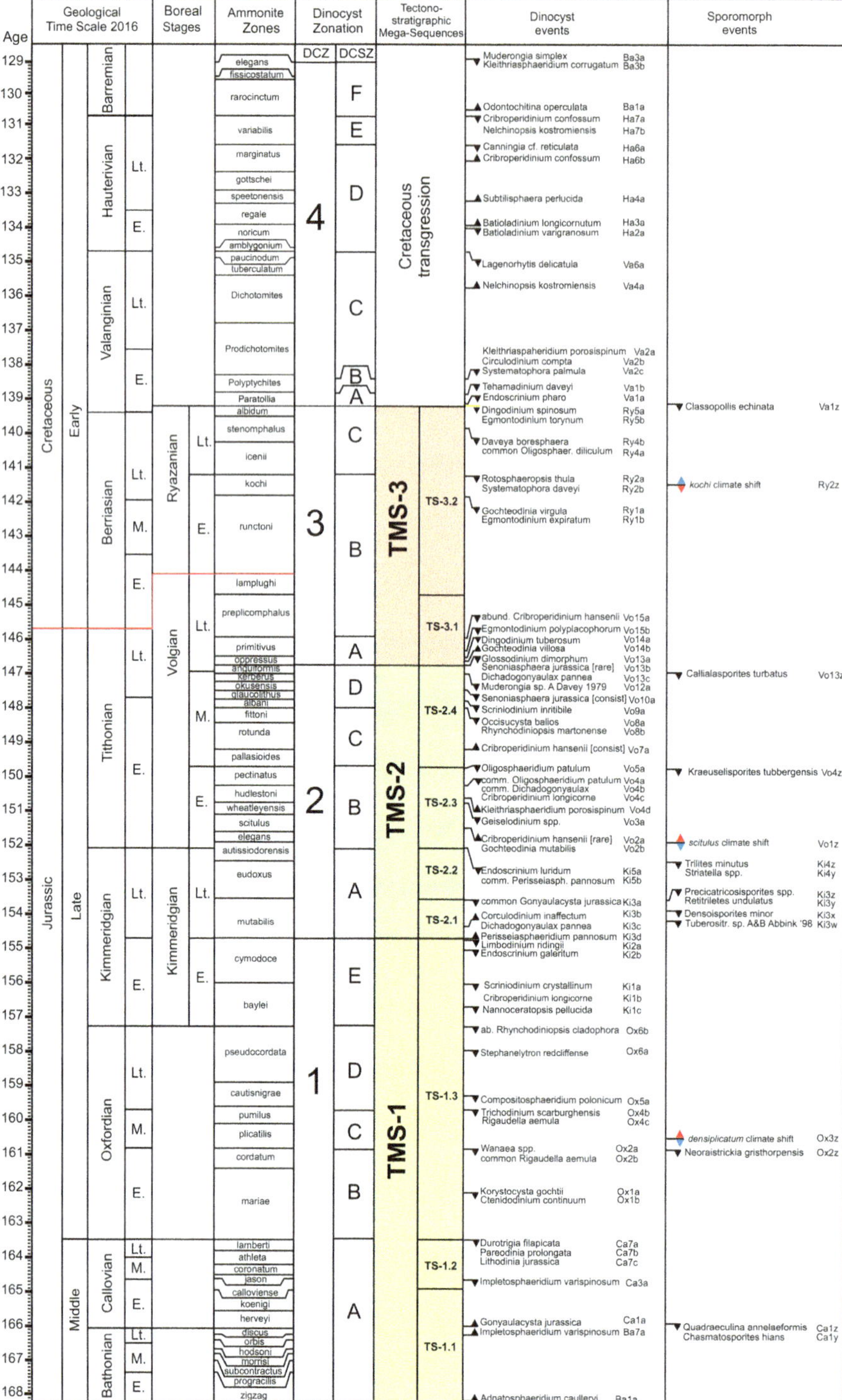

Fig. 2. Age calibration for the stratigraphic framework. The tectonostratigraphic mega-sequences (TMS-1–3) are aligned with the same numbering as the TNO dinoflagellate cyst zonation. The same colour coding to distinguish the three TMSs is used in all figures. The Geological Time Scale 2016 is used for the chronostratigraphic calibration (Ogg *et al.* 2016).

shift is observed during the Middle Oxfordian when relatively warmer and drier conditions were established (Price & Rogov 2009), reflected by a sudden decrease in the occurrence of *Perinopollenites* (Abbink 1998; Abbink *et al.* 2001). This event is known as the *densiplicatum* climate event (Abbink 1998), named after the Densiplicatum Chronozone it is calibrated to (Fig. 2). The climate changed again during the Early Volgian; this time increasingly warmer and arid conditions were established, reflected by the dominance of *Classopollis* (Abbink 1998). This change is referred to as the *scitulus* climate event by Abbink (1998), after the Scitulus Chronozone (Fig. 2). From the Scitulus Chronozone onwards, warm and arid conditions persisted throughout the rest of the Volgian and across the Jurassic–Cretaceous boundary. The Skylge Formation, sections of the Farsund Formation (including the Bo Member) as well as the Kimmeridge Clay Formation were deposited under these warm and arid climatic conditions. Note that in the Danish and Dutch subsurface, the entire period is characterized by siliciclastic sedimentation; in the UK, carbonates and evaporites of the Purbeck Formation characterize this arid phase. The warm and arid phase came to an end at the beginning of the Early Cretaceous Ryazanian Stage, reflected by the decrease of *Classopollis* concomitant with an increase of ornamented trilete spores such as *Cicatricosiporites* (Abbink 1998), indicating still warm but much more humid climate conditions. This climate event is referred to as the *kochi* climate event (Abbink 1998). Tropical and wet (humid) conditions were established during the Late Ryazanian and Valanginian, across a large area from the UK to Germany. This wet phase is reflected in the geological record of NW Europe by widespread fluvial and coastal sediments, rich in plant fossils ('Wealden Facies', Allen *et al.* 1998).

Methodology

The basin evolution model presented in this paper is primarily based on palynological analyses from exploration wells. Over a timespan of 20 years, approximately 230 wells were analysed and interpreted in terms of age and palaeoenvironment. From these 230 wells 32 wells are reported on in this paper, but palynological data are not provided at the detailed level of individual samples. A palynological zonation has been erected to facilitate correlations. The zonation is described in the section 'Palynology' (see also Fig. 2 and Table 1). Seismic interpretations, utilizing both 2D and 3D seismic data (Møller & Rasmussen 2003; De Jager 2007; Bouroullec *et al.* 2018), were also used as supporting information.

Study area

The area of interest is the southern part of the Middle Jurassic–Lower Cretaceous rift system and is loosely referred to as the Central Graben area (Fig. 1). The study area stretches 250 km from north to south, is 50 km wide on average, and includes the offshore territories of Denmark, Germany and The Netherlands. Three depositional domains are distinguished:

1. the primary graben axis, which runs from the boundary of Norway and Denmark all the way to its southernmost tip in the L05 Block in The Netherlands and includes the Søgne Basin, the Tail End Graben, the Salt Dome Province and the Northern Dutch Central Graben;
2. the basins adjacent to the graben, which include the Heno and Gertrud plateaus, the Feda Graben in Denmark and the Terschelling Basin in The Netherlands; and
3. the plateau areas which received little or no sediments during the Middle Jurassic–Early Cretaceous rift phase, including the Inge and Mads highs, the Outer Rough Basin, the Step Graben and the Schill Grund, Central Offshore and Cleaverbank platforms in The Netherlands.

Thicknesses of the Middle Jurassic–Lower Cretaceous sediment succession range from a couple of metres in the Step Graben (Bouroullec *et al.* 2018) to more than 2000 m in the Danish Tail End Graben (Andsbjerg & Dybkjær 2003). The graben axis can be characterized as an en echelon series of sub-basins (Japsen *et al.* 2003) with asymmetric wedge-shaped sediment fills (Møller & Rasmussen 2003). The Danish part of the graben axis roughly comprises the southern tips of the Norwegian/Danish Søgne Basin, the Tail End Graben and the Salt Dome Province. The Danish Graben axis is bound to the east by the Ringkøbing-Fyn High. In the Salt Dome Province, the Central Graben bends sharply to the SSW across the German Entenschnabel into the Dutch offshore. In the Dutch offshore the graben axis comprises the Dutch Central Graben, which is subdivided into northern, middle and southern areas. The Dutch Central Graben is bounded by the Schill Grund Platform to the east and the Step Graben and Cleaverbank Platform to the west. Note that the Step Graben belongs to the plateau domain, with only thin Middle Jurassic–Lower Cretaceous deposits present. The Dutch Central Graben axis pinches out against the Central Offshore Platform in the south. Across the Central Offshore Platform, the tip of two other Mesozoic basins are visible on the location map (Fig. 1): the Broad Fourteens and Vlieland basins. These basins have a NW–SE strike and have a shared tectonic history, but do not belong to the Central Graben area as such.

Table 1. *TNO zonation based on highest and lowest occurrences (HO and LO) of dinoflagellate cysts. Chronostratigraphy and absolute ages are in accordance with Ogg* et al. *(2016)*

Zone	Subzone	Age	Definition	Remarks
DCZ 1		Bathonian–Early Kimmeridgian 168–154.7 Ma	From LO *Adnatosphaeridium caulleryi* to HO *Limbodinium ridingii*	
	DCSZ 1A	Bathonian–Callovian 168–163.5 Ma	From FO *Adnatosphaeridium caulleryi* to HO *Durotrigia filapilicata*, *Pareodinia prolangata* and *Lithodinia jurassica*	FO *Adnatosphaeridium caulleryi* is calibrated to the Zigzag Chronozone (1); HO *Durotrigia filapicata*, *Pareodinia prolongata* and *Lithodinia jurassica* are calibrated to the Lamberti Chronozone (2, 3, 4, 5, 6, 7, 8)
	DCSZ 1B	Early Oxfordian 163.5–161 Ma	From HO *Durotrigia filapicata*, *Pareodininia prolongata* and *Lithodinia jurassica* to HO *Wanaea* spp. and HO common *Rigaudella aemula*	HO *Wanaea* spp. is calibrated to the Cordatum Chronozone (8); HO common of *Rigaudella aemula* is calibrated to the Cordatum Chronozone (7)
	DCSZ 1C	Middle Oxfordian 161–159.8 Ma	From to HO *Wanaea* spp. and HO common *Rigaudella aemula* to HO *Trichodinium scarburghensis* and *Rigaudella aemula*	HO *Trichodinium scarburghensis* is calibrated to the Plicatilis Chronozone; HO *Rigaudella aemula* is calibrated to the Pumilus Chronozone (1, 9, 10). Note that the *densiplicatum* Climate Shift occurs in DCSZ 1C (11, 29, 30).
	DCSZ 1D	Late Oxfordian 159.8–157.5 Ma	From HO *Trichodinium scarburghensis* and *Rigaudella aemula* to HO abundant *Rhynchodiniopsis cladophora*	HO abundant *Rhynchodiniopsis cladophora* and HO *Stephanelytron redcliffense* are in the Pseudocordata Chronozone (7, 9, 12). HO *Compositosphaeridium polonicum* is in the Cautisnigrae Chronozone (13).
	DCSZ 1E	Early Kimmeridgian 157.5–154.7 Ma	From HO abundant *Rhynchodiniopsis cladophore* to HO *Limbodinium ridingii* or HO *Scriniodinium crystallinum*	HO *Limbodinium ridingii* is calibrated to the Cymodoce Chronozone (8). HO *Scriniodinium crystallinum* is calibrated to the Baylei Chronozone (1, 2, 5, 6, 9, 10, 14, 15).
DCZ 2		Late Kimmeridigian–Middle Volgian 154.7–146.6 Ma	From HO *Limbodinium ridingii* or HO *Scriniodinium crystallinum* to HO *Glossodinium dimorphum*, *Senoniasphaera jurassica* and *Dichadogonyaulax pannaea*	
	DCSZ 2A	Late Kimmeridgian 154.5–152.2 Ma	From HO *Limbodinium ridingii* or HO *Scriniodinium crystallinum* to HO *Endoscrinium luridum*	HO *Endoscrinium luridum* is calibrated to the Autissiodorensis Chronozone (1, 9, 15, 16). FO *Dichadogonyaulax pannea* and *Corculodinium inaffectum* is calibrated to the Mutabilis Chronozone (23, 15, 16, 17, 19).

	DCSZ 2B	Early Volgian 152.2–149.8 Ma	From HO *Endoscrinium luridum* to HO *Oligosphaeridium patulum*	HO *Oligosphaeridium patulum* is calibrated to the Pectinatus Chronozone (15, 20, 21). HO abundant *Oligosphaeridium patulum* is calibrated to the Hudlestoni Chronozone (9, 19, 21). LO *Gochteodinia mutabilis* and *Cribroperidinium hansenii* are calibrated to the Scitulus Chronzone (1, 2, 9, 23). HO of *Corculodinium inaffectum* and *Corculodinium paeminosum* are calibrated to the Wheatleyensis Chronozone (19). Note that the *scitulus* Climate Shift occurs in DCSZ 2B (11, 29, 30).
	DCSZ 2C	Middle Volgian (p.p.) 149.8–148 Ma	From HO *Oligosphaeridium patulum* to HO *Occisucysta balios*	HO *Occisucysta balios* is calibrated to the Fittoni Chronozone (2, 23 24)
	DCSZ 2D	Middle Volgian (p.p.) 148–146.6 Ma	From HO *Occisucysta balios* to the HO *Glossodinium dimorphum*, *Senoniasphaera jurassica* and *Dichadogonyaulax pannaea*	HO *Glossodinium dimorphum*, *Senoniasphaera jurassica* and *Dichadogonyaulax pannaea* are calibrated to the Anguiformis Chronozone (2, 9, 15, 19, 21, 22). HO *Scriniodinium inritibile* is calibrated to the Albani Chronozone (2, 15, 16).
DCZ 3		Late Volgian–Late Ryazanian 146.6–139 Ma	From HO *Glossodinium dimorphum*, *Senoniasphaera jurassica* and *Dichadogonyaulax pannaea* to HO *Dingodinium spinosum*	
	DCSZ 3A	Late Volgian (p.p.) 147–146 Ma	From HO *Glossodinium dimorphum*, *Senoniasphaera jurassica* and *Dichadogonyaulax pannaea* to HO abundant *Cribroperidinium hansenii*	HO abundant *Cribroperidinium hansenii* is calibrated to the Primitivus Chronozone (8, 25). HO *Egmontodinium polyplacophorum* is calibrated to the Opressus or Primitivus Chronozone (2, 8, 9).
	DCSZ 3B	Late Volgian (p.p.)–Early Ryazanian 146–141 Ma	From HO abundant *Cribroperidinium hansenii* to HO *Rotosphaeropsis thula* and *Systematophora daveyi*	HO *Rotosphaeropsis thula* and *Systematophora daveyi* are calibrated to the Kochi Chronozone (8, 15, 25, 26, 27). HO *Gochteodinia virgula* and *Egmontodinium expiratum* are calibrated to the Runctoni Chronozone (2, 8, 9, 25). Note that the *kochi* Climate Shift occurs in DCSZ 3B (11, 29, 30).
	DCSZ 3C	Late Ryazanian 141–139.5 Ma	From HO *Rotosphaeropsis thula* and *Systematophora daveyi* to HO *Dingodinium spinosum* and *Egmontodinium torynum*	HO *Dingodinium spinosum* and *Egmontodinium torynum* are calibrated to the Albidum Chronozone (8, 9, 22, 28, 27). LO *Oligosphaeridium diluculum* is calibrated to the Icenii Chronozone Ammonite Zone (2, 9). HO common *Oligosphaeridium diluculum* and HO of *Daveya boresphaera* are calibrated to the Stenomphalus Chronozone (8, 22, 25, 32).
DCZ 4		Valanginian–Early Barremian 139.5–129 Ma	From HO *Dingodinium spinosum* and *Egmontodinium torynum* to HO of *Kleithriasphaeridium corrugatum*	

(*Continued*)

Table 1. *TNO zonation based on highest and lowest occurrences (HO and LO) of dinoflagellate cysts. Chronostratigraphy and absolute ages are in accordance with Ogg* et al. *(2016) (Continued)*

Zone	Subzone	Age	Definition	Remarks
	DCSZ 4A	Early Valanginian (p.p.) 139.5–139 Ma	From HO *Dingodinium spinosum* and *Egmontodinium torynum* to HO of *Tehamadium daveyi*	HO *Tehamadium daveyi* and HO *Endoscrinium pharo* are calibrated to the Paratollia Chronozone (22, 15, 25, 27)
	DCSZ 4B	Early Valanginian (p.p.) 139–138.5 Ma	From HO of *Tehamadium daveyi* to HO *Systematophora palmula*	HO *Systematophora palmula* is calibrated to the Poyptychites Chronozone (22, 27)
	DCSZ 4C	Early (pars.) Valanginian–Late Valanginian 138.5–134.5 Ma	From to HO *Systematophora palmula* to HO *Lagenorthytis delicatula*	HO *Lagenorthytis delicatula* is calibrated to the Amblygonium Chronozone (27, 31). LO *Nelchinopsis kostromiensis* is calibrated to the Dichotomites Chronozone (27).
	DCSZ 4D	Early Hauterivian–Late Hauterivian (p.p.) 134.5–131.5 Ma	From HO *Lagenorthytis delicatula* to HO *Canningia* cf. *reticulata*	HO *Canningia* cf. *reticulata* is calibrated to the base Variabilis Chronozone (22, 27, 28,31). HO *Batioladinium varigranosum* is calibrated to the Noricum Chronozone (27). LO *Subtilisphaera perlucida* is calibrated to the Speetonense Chronozone (27). LO *Cribroperidinium confossum* is calibrated to the Gottschei Chronozone (28, 33, 34).
	DCSZ 4E	Late Hauterivian (p.p.) 131.5–131 Ma	From HO *Canningia* cf. *reticulata* to HO *Nelchinopsis kostromiensis*	HO *Nelchinopsis kostromiensis* and *Cribroperidinium confossum* are calibrated to the Variabilis Chronozone (28, 33, 34, 35).
	DCSZ 4F	Early Barremian 131–129 Ma	From HO *Nelchinopsis kostromiensis* to the HO *Kleithriasphaeridium corrugatum*	HO *Kleithriasphaeridium corrugatum* and *Muderongia simplex* are calibrated to the Elegans Chronozone (28, 35, 37). LO *Odontochitina operculate* is calibrated to the Rarocinctum Chronozone (27).

References: (1) Riding & Thomas (1992); (2) Riding (1987); (3) Prauss (1989); (4) Riding & Bailey (1991); (5) Feist-Burkhardt & Wille (1992); (6) Woollam (1980); (7) Riley & Fenton (1982); (8) Herngreen *et al.* (2000); (9) Partington *et al.* (1993b); (10) Fauconnier (1995); (11) Abbink (1998); (12) Kunz (1990); (13) Poulsen (1998); (14) Århus *et al.* (1989); (15) Poulsen (1996); (16) Poulsen (1994); (17) Nøhr-Hansen (1986); (18) Ioannides *et al.* (1988); (19) Riding & Thomas (1988); (20) Riley (1979); (21) Bailey *et al.* (1997); (22) Davey (1979*a*; *b*); (23) Barron (1989); (24) Munsterman *et al.* (2012); (25) Davey (1982*a*; *b*); (26) Riding & Davey (1989); (27) Costa & Davey (1992); (28) Duxbury (1977); (29) Abbink *et al.* (2001); (30) Abbink *et al.* (2006); (31) Heilmann-Clausen & Birkelund (1987); (32) Birkelund *et al.* (1983); (33) Harding (1990); (34) Kirsch & Below (1995); (35) Duxbury (2001); (36) Mutterlose & Harding (1987); (37) Duxbury (1980); and (38) Heilmann-Clausen & Thomsen (1995).

Attached to the graben axis are several other basins with relatively thick (100–1000 m) Middle Jurassic–Lower Cretaceous deposits. These basins are generally connected to the graben axis on one side and are typically bound by straight and up to 50 km long faults (Fig. 1). In the Danish offshore, west of the Tail End Graben these basins consist of a series of NW–SE-trending plateaus and half-grabens. The basins are connected to the graben axis to the east and bounded by the Mads and Inge highs and by the Outer Rough Basin to the SW. The Outer Rough Basin crosses the German Entenschnabel onto the A and B blocks in the Dutch offshore and contains relatively thin (less than 100 m) Middle Jurassic–Lower Cretaceous successions. In the Dutch offshore, east of the southern Dutch Central Graben, the Terschelling Basin is separated from the graben axis by a salt-filled fault. The Terschelling Basin thickens towards the graben axis and is confined by NW–SE-trending fault zones (Fig. 1).

Palynology

Palynological processing

The large number of samples (estimated to exceed 6000) was processed in different laboratories, but always according to standard palynological processing procedures. The standard processing routine includes treatment with hydrochloric acid to digest the carbonate and hydrofluoric acid, in order to destroy the silicate mineral bonds and release the acid-resistant organic matter. The organic matter is isolated by heavy liquid separation and concentrated by sieving. The remaining organic residue is mounted on glass slides for microscopic analysis.

Palaeoenvironmental interpretation based on palynology

Apart from dolomite stringers in the Farsund and Kimmeridge Clay Formation, the Middle Jurassic–Lower Cretaceous sedimentary succession of the Central Graben area is dominated by siliciclastic sediments (Van Adrichem Boogaert & Kouwe 1993; Andsbjerg & Dybkjær 2003; Lott *et al.* 2010). The depositional environments are known to exhibit strong variation over relatively short distances, due to the active tectonic setting and the relatively complex basin configuration (Johannessen *et al.* 2010*a*; Munsterman *et al.* 2012). The Dutch part of the study area is characterized by marine to non-marine deposits (Abbink *et al.* 2006; Bouroullec *et al.* 2018), while the Danish part, in particular the graben axis, is dominated by deep-marine deposits (Andsbjerg & Dybkjær 2003). In such a setting, palynology is a useful tool for age assessment and palaeoenvironmental interpretation; it provides information from both the marine realm, via the occurrences of dinoflagellate cysts, and the terrestrial realm, via the occurrences of pollen and spores (Abbink 1998). However, palaeoenvironmental interpretations based on palynology reflect generally supra-local to regional conditions and are not directly comparable to sedimentology-based interpretations, as these mainly reflect local conditions (Fig. 3).

Dinoflagellates exhibit a planktonic lifestyle and only the organic-walled cysts end up in the sedimentary record. Dinoflagellate cyst occurrences are not specifically related to palaeo-water depth, but may provide information on sea-surface temperature, productivity and salinity (Sluijs *et al.* 2005). As a rule of thumb, high diversity and low dominance indicates open-marine shelf environments, while low diversity and high dominance indicates restricted, usually marginal, marine conditions (Jansonius & McGregor 1996). Pollen and spore assemblages reflect the regional vegetation from land and, as such, provide information on the palaeo-climate (Abbink *et al.* 2001). As a rule of thumb, the ratio between the marine dinoflagellate cysts and terrestrial pollen and spores indicates the relative distance from shore (Donders *et al.* 2009). Hydrodynamic energy may be reflected in the amount of physical degradation of the organic constituents of the palynological assemblages; very small fragments point to high-energy conditions and intact specimens indicate low-energy conditions (Traverse 2007).

Age calibration

The Late Jurassic and the succeeding Early Cretaceous periods are relatively poorly constrained in a global chronostratigraphic sense. For example, there are no global boundary stratotype section and points (GSSPs) for the base of the Callovian, Oxfordian, Kimmeridgian, Berriasian and Valanginian (Wimbledon *et al.* 2011; Ogg *et al.* 2016). Furthermore, the Late Jurassic is characterized by strong provincialism, in particular with respect to the distribution of ammonites (Gradstein *et al.* 2012). This hampers the application of international standards to the sedimentary successions from the North Sea. Boreal chronostratigraphic schemes are generally applied in North Sea geology, in particular the Russian stages: Kimmeridgian, Volgian and Ryazanian (e.g. Fraser *et al.* 2003). For that reason, reference to Russian chronostratigraphy and to the Sub-Boreal Ammonite Zones is made in this paper (Fig. 2). For practical reasons, the target interval of this paper is loosely referred to as Middle Jurassic–Early Cretaceous although, strictly speaking, this should be late Middle Jurassic–early Early Cretaceous.

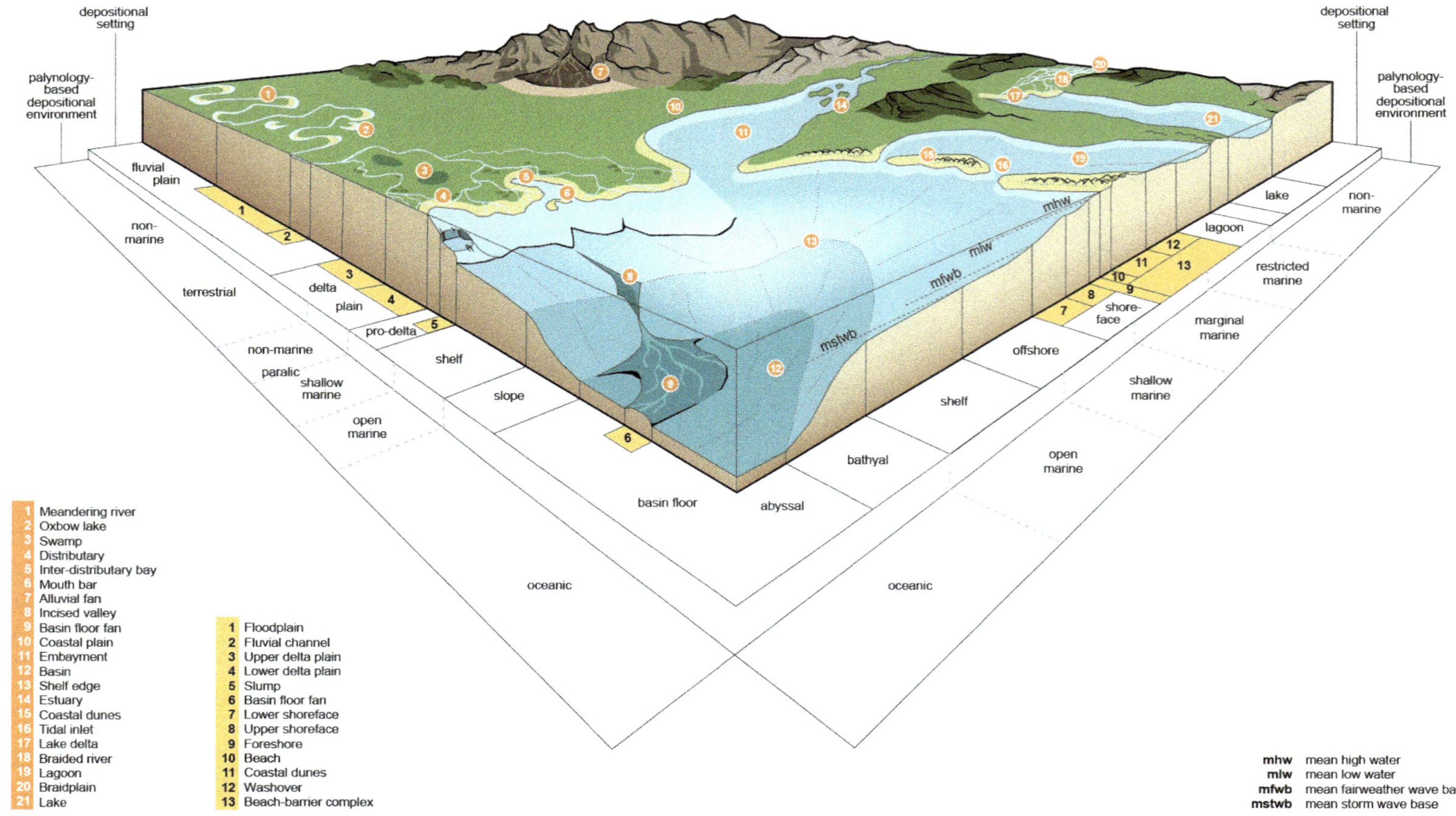

Fig. 3. Schematic overview of depositional environments occurring in the terrestrial and marine realm which are mentioned in the text. These environments form the bulk of the environments that occur in the Middle Jurassic–Lower Cretaceous in the Central Graben area. Note that palaeoenvironmental interpretations based on palynology generally reflect sub-regional to regional conditions, whereas palaeoenvironmental interpretations derived from sedimentology reflect local conditions.

Note that in Ogg *et al.* (2016) the absolute age spans of chronozones vary. For example, the Early Ryazanian Runctoni Chronozone plus the Late Volgian Lamplughi and Preplicomphalus chronozones together represent 4 Ma, whereas the seven chronozones below the Preplicomphalus represent less than 2 Ma. Based on our own observations, the marked difference in absolute age span between the Runctoni–Preplicomphalus and Primitivus–Albani chronozones is questioned.

Palynological zonation

Based on palynological analyses from exploration wells in the Danish and Dutch offshore, a regional dinoflagellate cyst zonation for the Middle Callovian–Barremian is established (Table 1, see also Fig. 2). The zonation includes 4 zones and 18 subzones.

The dinoflagellate cyst zones (DCZ) and subzones (DCSZ) are predominantly based on top occurrences of taxa, to accommodate for use in industry. For practical reasons, the numbering of the independently calibrated dinoflagellate zones has been synchronized with the numbering of the tectonostratigraphic mega-sequences (Fig. 2). The international geological timescale of Ogg *et al.* (2016) is used for chronostratigraphic calibration.

Results

Based on palynological analyses and observations from seismic and sedimentological analyses, the basin development of the Middle Jurassic–Lower Cretaceous from the Central Graben areas of Denmark, Germany and The Netherlands is demonstrated in detail. The basin evolution is seen to follow discrete phases with active depocentres and fault patterns changing through time (Fig. 4). These discrete phases in basin evolution are reflected in the sedimentary record as genetically related accumulations of sediments (Figs 5–7). Four main phases in the basin evolution are distinguished, which are named tectonostratigraphic mega-sequences (TMS) and subordinate sequences (Fig. 4). The TMS are numbered 1–4, from old to young:

- TMS-1: Bathonian–Early Kimmeridgian (168–154.7 Ma) subsidence in the graben axis related to east–west extension;
- TMS-2: Late Kimmeridgian–Late Volgian (154.7–146.6 Ma) subsidence, primarily affecting the basins adjacent to the graben axis, related to NW–SE extension;
- TMS-3: latest Volgian–Ryazanian (146.6–139 Ma) cessation of fault activity and spreading of subsidence to the adjacent plateaus; and
- TMS-4: Valanginian–Barremian (139–126 Ma) regional subsidence.

TMS-1 to TMS-3 combined comprise the Middle Jurassic–Early Cretaceous rift phase. Within these three mega-sequences, nine lower order units of genetically related sedimentary accumulations are recognized (Figs 8–15). These are referred to as tectonostratigraphic sequences (TS). TMS-4 represents a post-rift phase when subsidence was no longer limited to the graben area, reflected by a large-scale regional transgression (Copestake *et al.* 2003). Note that the basin evolution of the narrow Entenschnabel sector of the German offshore (Fig. 1) is integrated into the model but treated less exhaustively, since published subsurface data from the German offshore are scarce.

TMS-1 (168–154.7 Ma)

During the entire timespan of TMS-1 (Bathonian–Early Kimmeridgian) only the graben axis is subjected to major subsidence. Sediments of this age are rarely encountered away from the graben axis in the study area (Fig. 4).

The first synrift deposits are fluvio-deltaic sediments of the Bryne Formation (Aalenian–Bajocian age), which are found in the Danish Søgne Basin and Tail End Graben (Andsbjerg 2003; Andsbjerg & Dybkjær 2003; Mellere *et al.* 2016). Towards the end of the Callovian, the entire graben axis was affected by subsidence. In the succeeding Oxfordian, subsidence related to rifting reached its maximum. Deposition within the graben axis was then dominated by mudstones of the Lola and Middle Graben formations (Fig. 5). In the Danish Graben axis (Søgne Basin, Tail End Graben, Salt Dome Province), the depocentres are situated close to the Coffee Soil Fault, emphasizing the asymmetric style of deformation within this area (Fig. 1). In the Dutch Central Graben axis, the structural style is different since the bounding faults are often overlain by remobilized Zechstein salt (e.g. Pharaoh *et al.* 2010; Bouroullec *et al.* 2018). Because of salt withdrawal, onlaps or truncations along the graben margins are predominant in the Dutch Central Graben (Bouroullec *et al.* 2018, figs 6–13). In the northern part of the Dutch Central Graben a total depositional thickness of 1500 m is observed for TMS-1 (F03 block), while in the southern part of the Dutch Central Graben thicknesses are limited to approximately 200 m (L05 block). This thinner succession is partially due to later erosion, but also related to lower initial accommodation. The thickness variation is accompanied by noticeable sedimentary facies changes. The southern Dutch Central Graben (F17, L02 and L05 Blocks) is characterized by predominantly non-marine, fluvial deposits of the Friese Front Formation (Figs 5–7), while the northern Dutch Central Graben (B18, F03, F05 and F06) displays a succession from

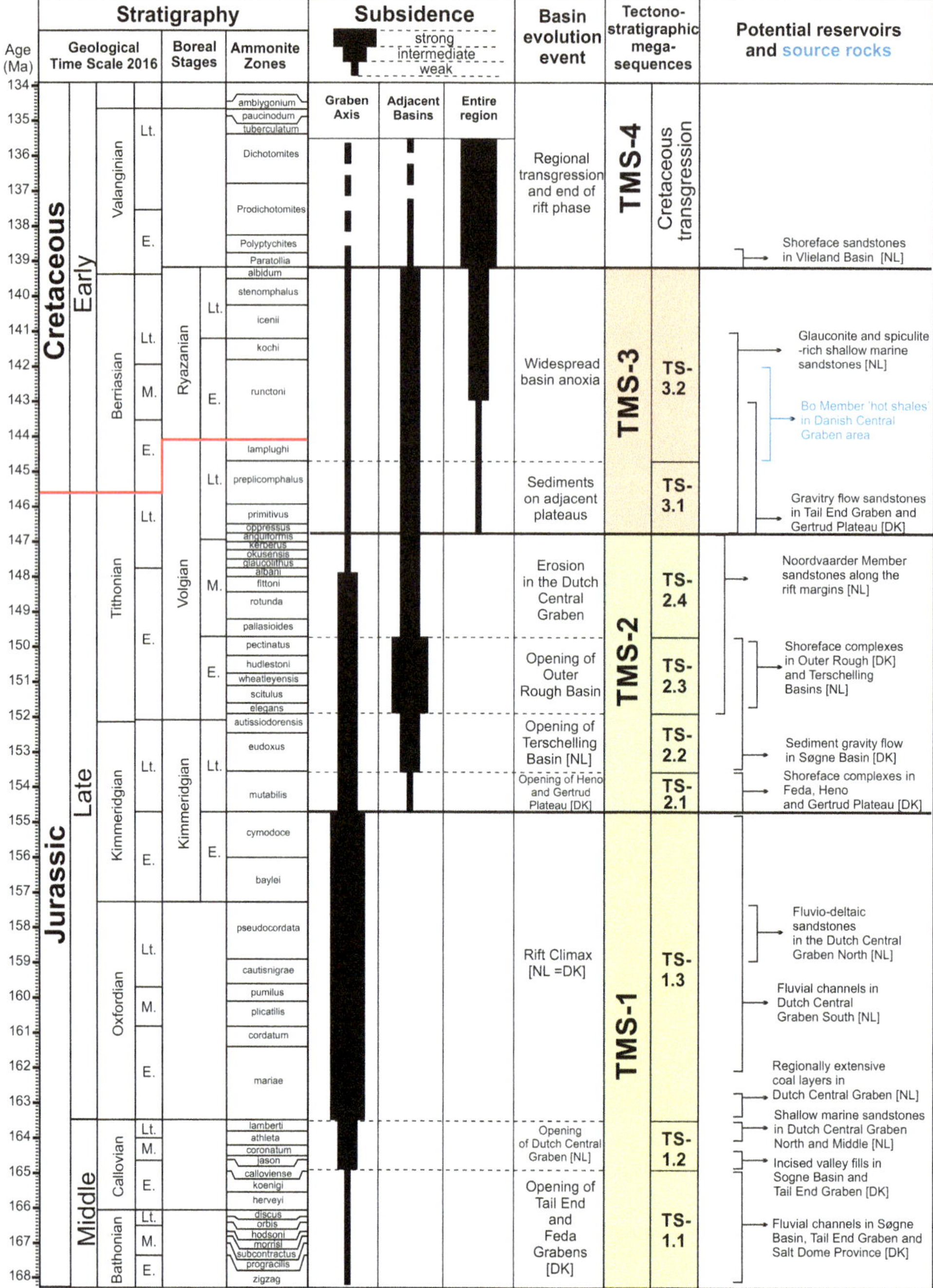

Fig. 4. Schematic representation of the stepwise basin evolution discussed in the text. The vertical bars represent the relative amount of basin subsidence (black) for the different structural domains. The relative amounts of subsidence are estimated from the thicknesses of the sedimentary succession. Timescale after (Ogg *et al.* 2016).

non-marine to marginal-marine sediments (Lower Graben, Middle Graben and Upper Graben formations; Bouroullec *et al.* 2018) to open-marine sediments at the top of TMS-1 (Kimmeridge Clay Formation). Based on these observations a stepwise, or gradual topographic gradient related to

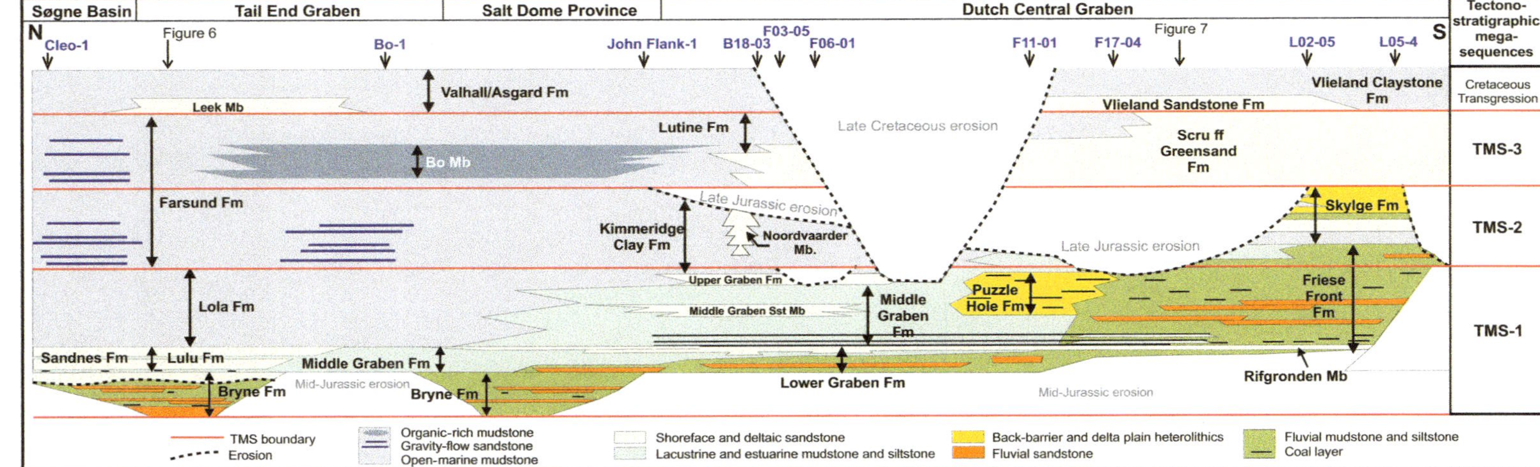

Fig. 5. Wheeler diagram along a N–S transect through the axis of the Central Graben. The diagram displays the generalized time and facies relationships of the Middle Jurassic–Lower Cretaceous lithostratigraphic units. See Figure 1 for location of the panel.

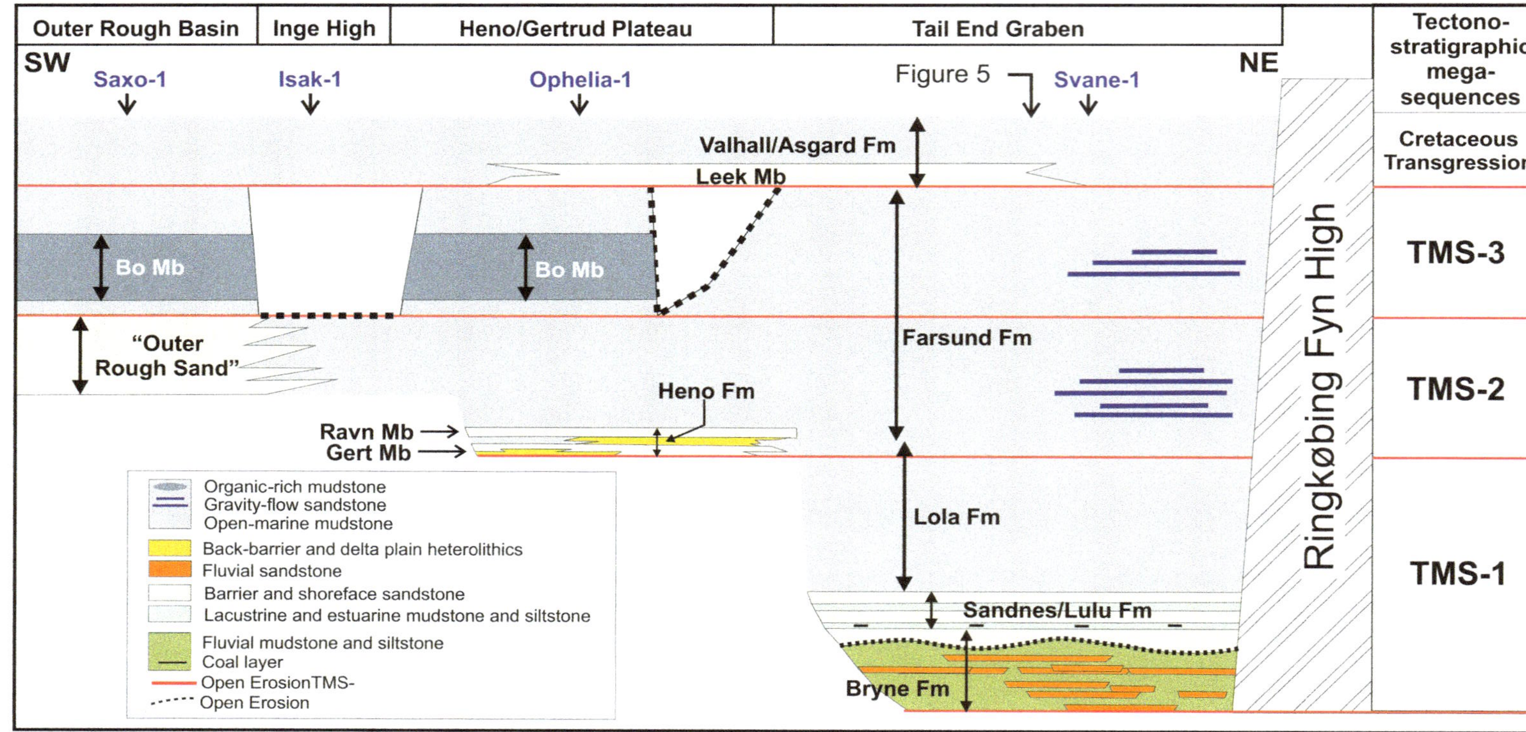

Fig. 6. Wheeler diagram along a W–E transect through the Danish basins. The diagram displays the generalized time and facies relationships of the Middle Jurassic–Lower Cretaceous lithostratigraphic units. See Figure 1 for location of the panel.

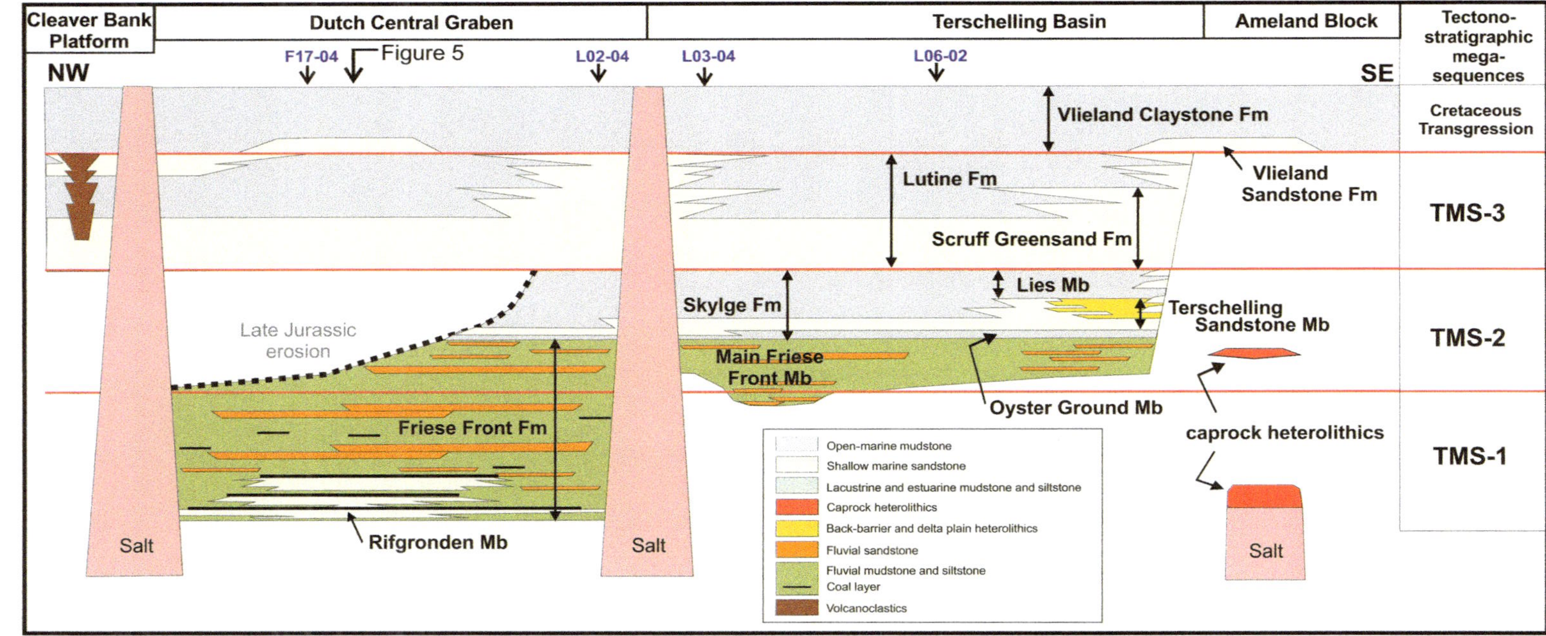

Fig. 7. Wheeler diagram along a W–E transect through the Dutch Terschelling Basin. The diagram displays the generalized time and facies relationships of the Middle Jurassic–Lower Cretaceous lithostratigraphic units. See Figure 1 for location of the panel.

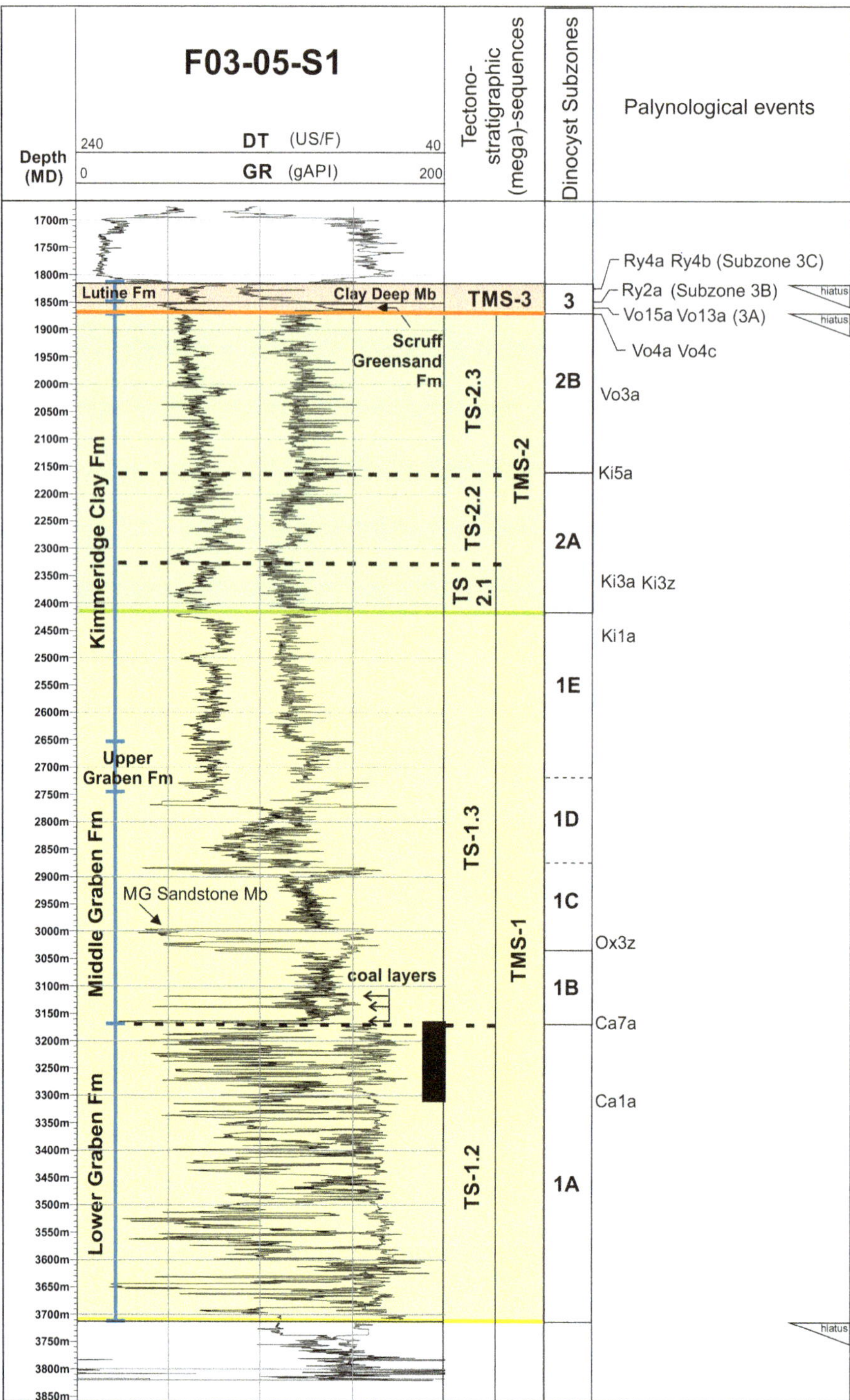

Fig. 8. Age reference for the Dutch Lower Graben, Middle Graben, Upper Graben, Kimmeridge Clay, Scruff Greensand and Lutine formations. Well F03-05-S1 is located in the northern part of the Dutch Central Graben. The wireline log on the left is the gamma ray (GR) and the log on the right is the sonic (DT). See Figure 1 for location and Figure 2 for abbreviations of palynological events.

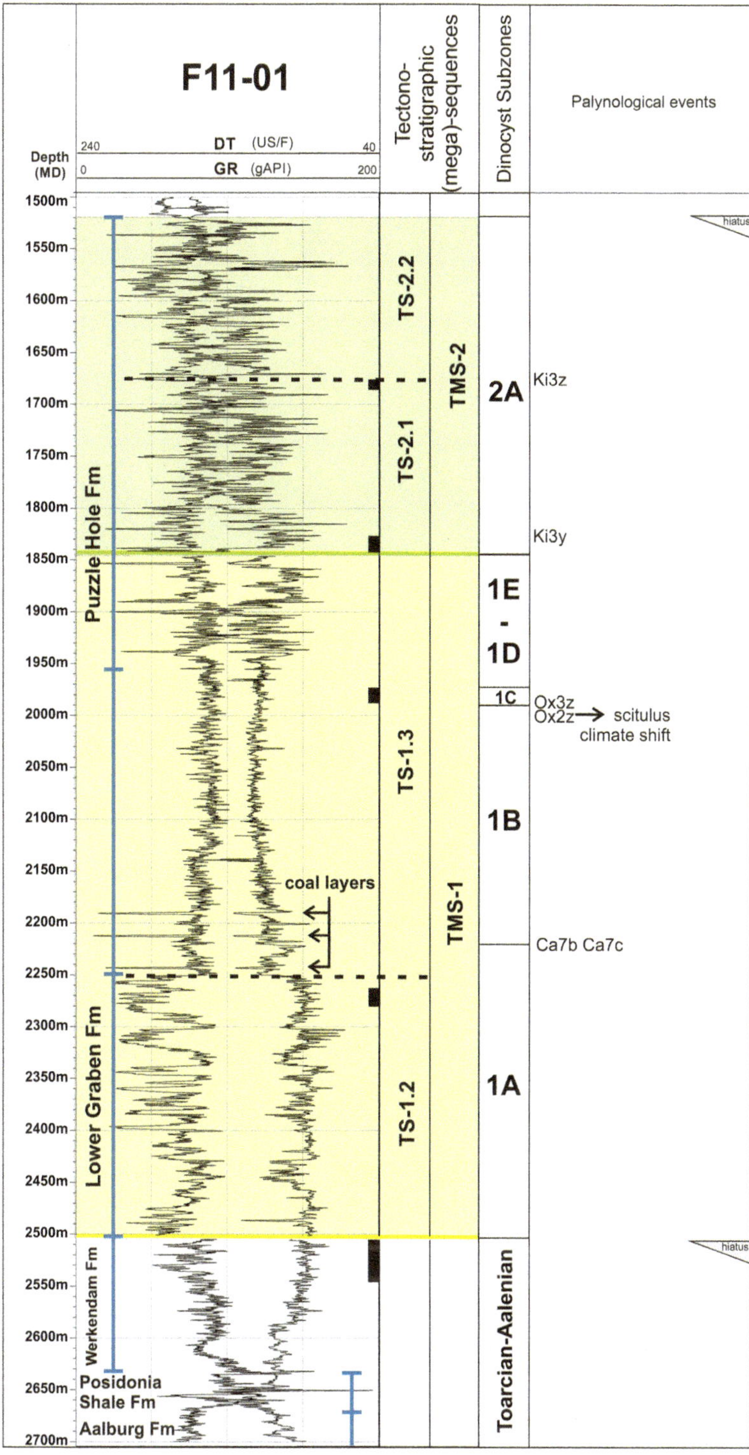

Fig. 9. Age reference for the Dutch Lower Graben, Middle Graben and Puzzle Hole formations. Well F11-01 is located in the middle part of the Dutch Central Graben. Wireline logs as for Figure 8. See Figure 1 for location and Figure 2 for abbreviations of palynological events.

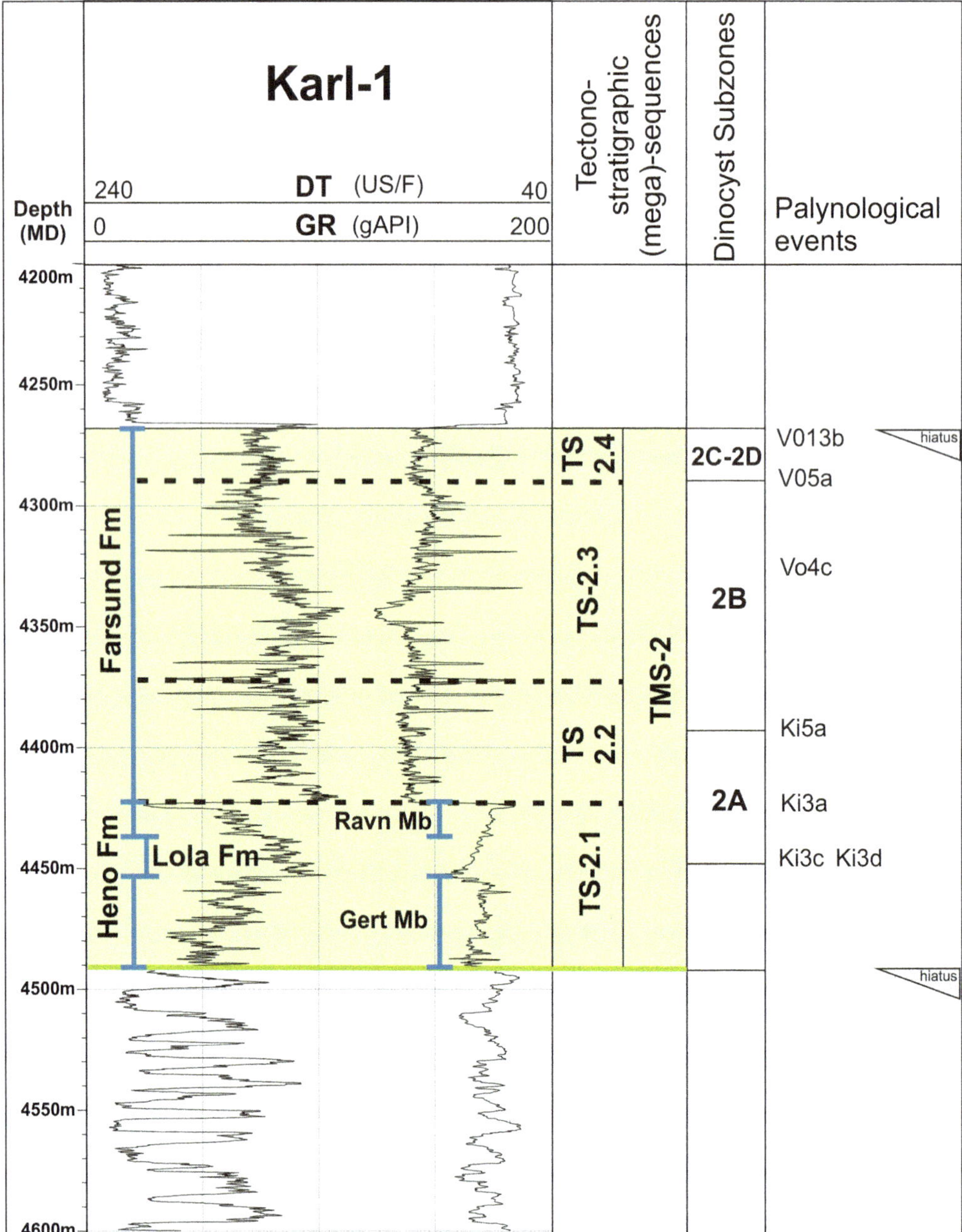

Fig. 10. Age reference for the Danish Heno, Lola and Farsund formations. Well Karl-1 is located in the Danish basins adjacent to the Central Graben axis. Wireline logs as for Figure 8. See Figure 1 for location and Figure 2 for abbreviations of palynological events.

differences in subsidence, is proposed to have existed from south to north.

During TMS-1, basin infill occurred from multiple directions. There is no single, large-scale prograding delta system such as the Brent Delta (*sensu* Husmo *et al.* 2003) filling up the graben along its axis. For example, during the Aalenian–Bathonian, the Danish–Norwegian Søgne Basin was occupied by a tidally influenced delta system from the NE, but the drainage pattern was reversed in the

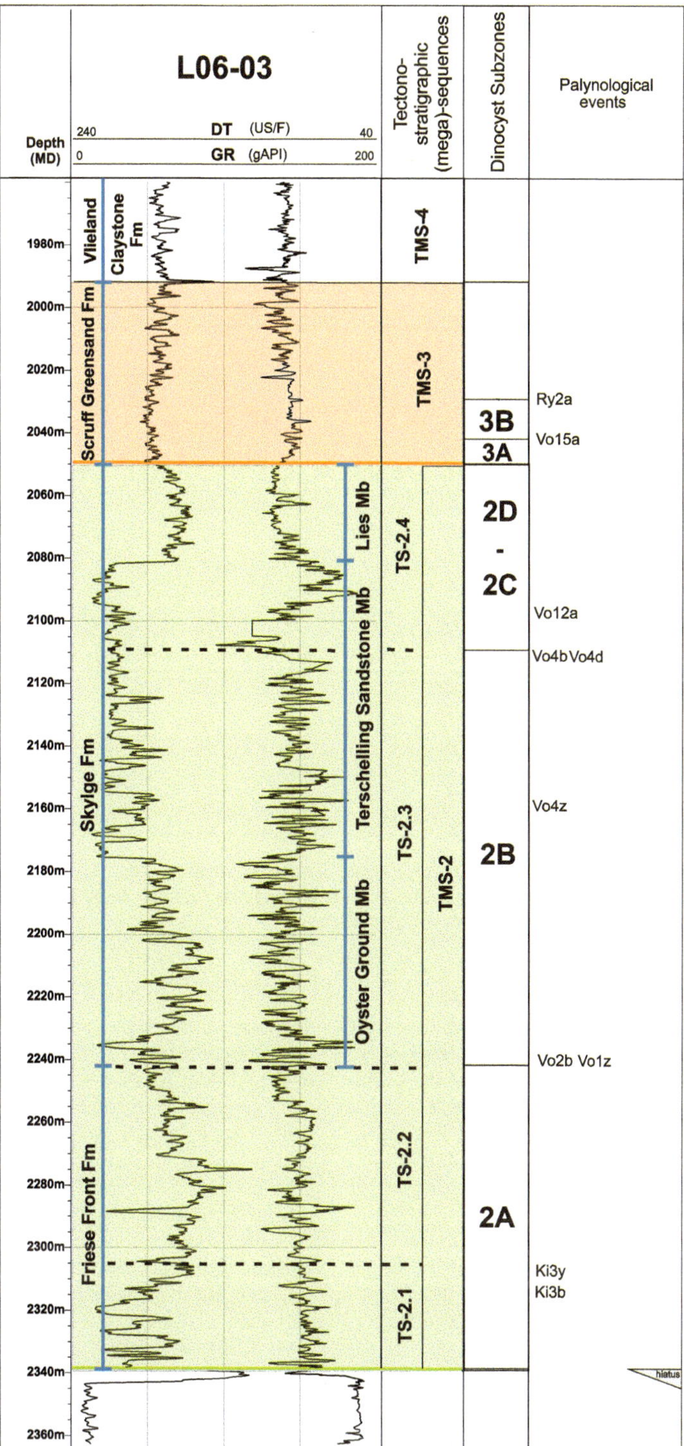

Fig. 11. Age reference for the Dutch Friese Front, Skylge and Scruff Greensand formations. Well L06-03 is located in the Dutch Terschelling Basin, adjacent to the Central Graben axis. Wireline logs as for Figure 8. See Figure 1 for location and Figure 2 for abbreviations of palynological events.

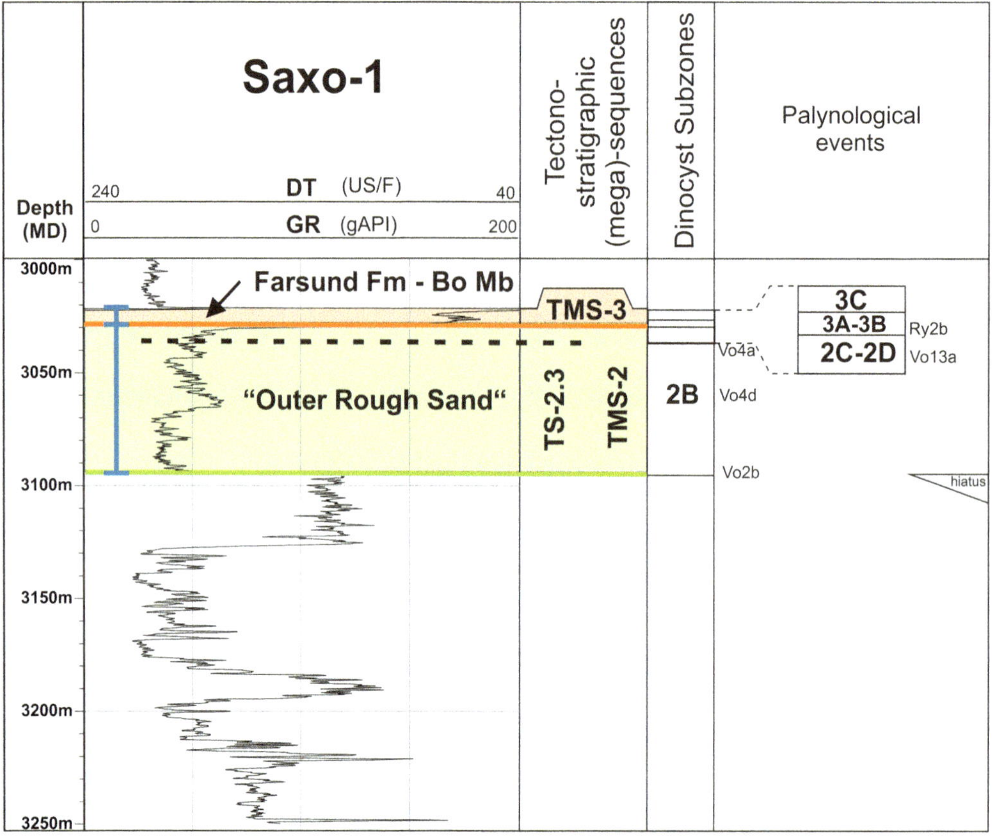

Fig. 12. Age reference for the Danish 'Outer Rough Sand' and the Bo Member of the Farsund Formation. Well Saxo-1 is located in the Danish Outer Rough Basin. The wireline log is the gamma ray (GR). See Figure 1 for location and Figure 2 for abbreviations of palynological events.

Callovian (Bryne–Sandnes Formation, Mellere *et al.* 2016). In the Dutch Central Graben, evidence of lateral sediment input is demonstrated by the occurrence of distributary channels and bird foot deltas on the margin of the Dutch Central Graben, identified on 3D seismic data from the Oxfordian Middle and Upper Graben formations (Bouroullec *et al.* 2018, figs 15, 16).

TS-1.1. Initiation of rifting. Opening of the Søgne Basin, Tail End Graben and Salt Dome Province.

Lithostratigraphy: Denmark: Bryne Formation. The Netherlands: No deposition.

Age: Bathonian–Early Callovian (168–165 Ma; Fig. 2).

The oldest sediments that can be attributed to the TMS-1 are Aalenian or earliest Bajocian sediments from the Bryne Formation (Andsbjerg 2003; Andsbjerg & Dybkjær 2003; Mellere *et al.* 2016; see also Figs 4–6), although Husmo *et al.* (2003) suggest that most of the Bryne Formation can be attributed to the Bathonian. TS-1.1 represents the oldest part of the Middle Jurassic–Early Cretaceous rift phase and expresses the rift initiation. TS-1.1 is limited in occurrence to the Danish Graben area where it is present in the Søgne Basin, the Tail End Graben and in the Salt Dome Province. TS-1.1 may possibly extend into the German sector, but no published information is available to corroborate this. In the Dutch Central Graben, sediments older than Middle Callovian and younger than Aalenian have not been observed (Van Adrichem Boogaert & Kouwe 1993; Munsterman *et al.* 2012; Bouroullec *et al.* 2018). This implies that, during the whole of TS-1.1, subsidence was limited to the Danish Graben.

The Bryne Formation consists of fluvio-deltaic–estuarine deposits (Andsbjerg 2003) and is associated with a clay-rich, tidally influenced delta system that prograded SSE into the Søgne Basin (Mellere *et al.* 2016). The Bryne Formation is in part

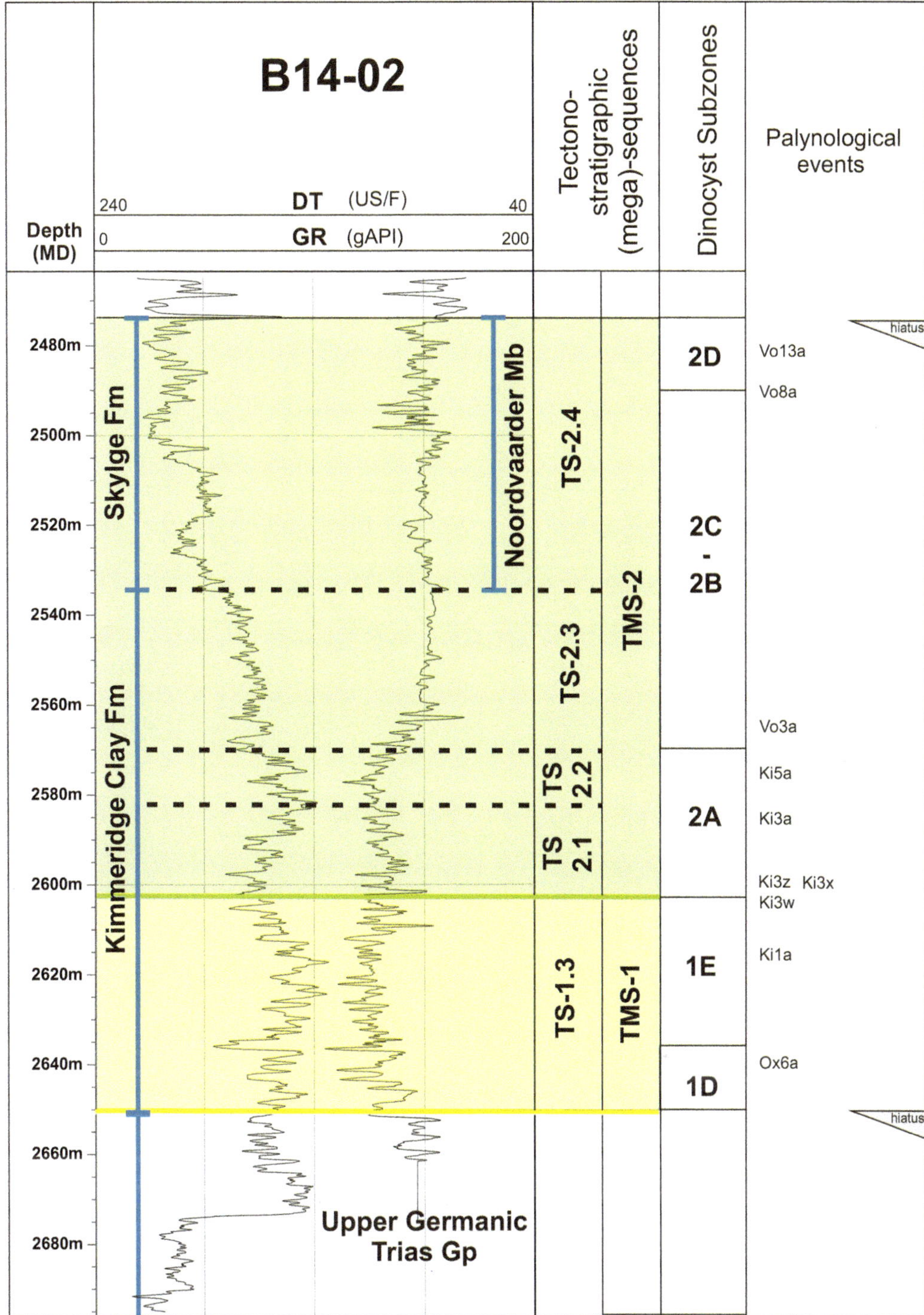

Fig. 13. Age reference for the Dutch Kimmeridge Clay Formation and for the Noordvaarder Member of the Skylge Formation. Well B14-02 is located near the margin of the northern part of the Dutch Central Graben axis. Wireline logs as for Figure 8. See Figure 1 for location and Figure 2 for abbreviations of palynological events.

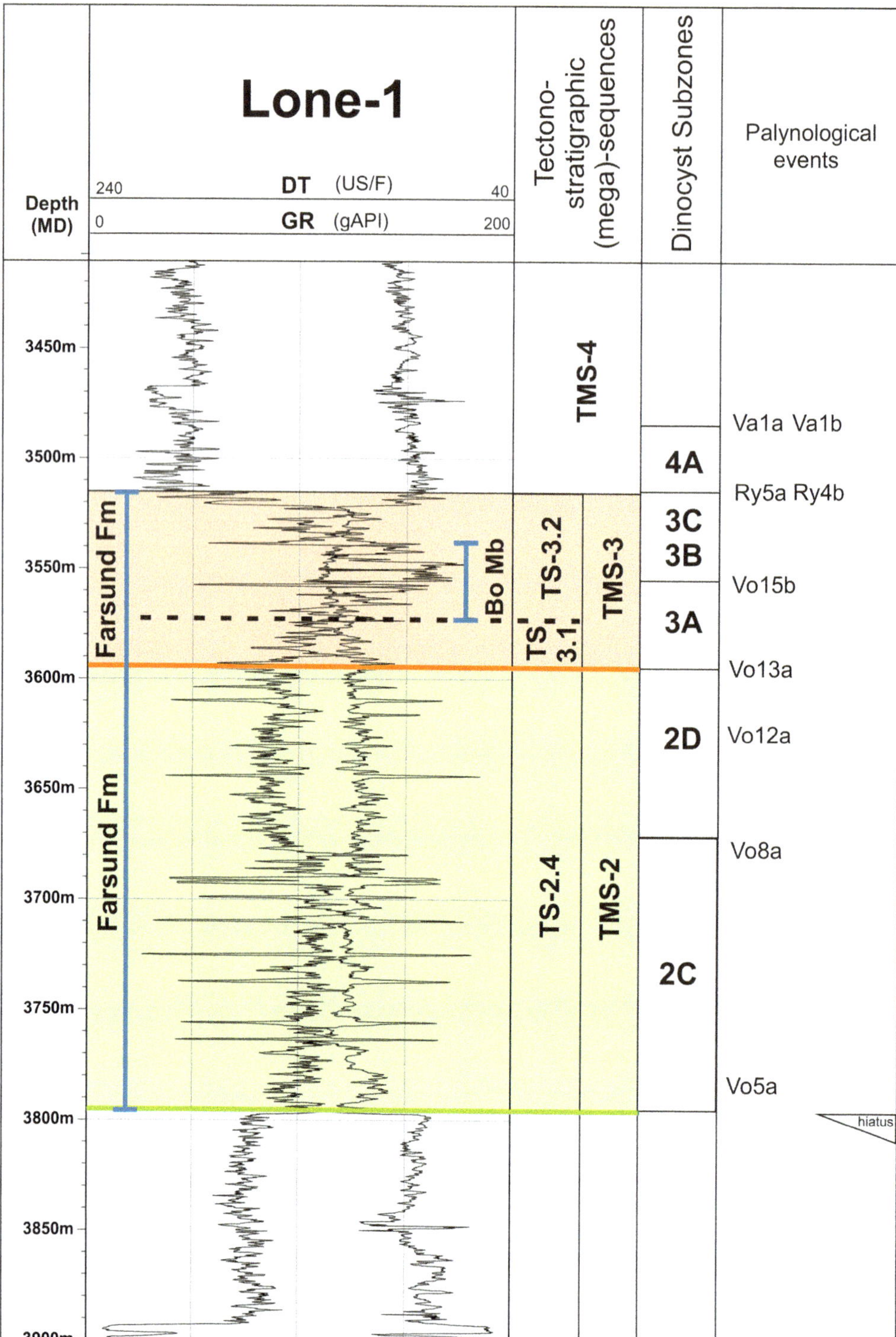

Fig. 14. Age reference for the Danish Farsund Formation and its Bo Member. Well Lone-1 is located in the Danish basins adjacent to the Central Graben axis. Wireline logs as for Figure 8. See Figure 1 for location and Figure 2 for the abbreviations of the palynological events.

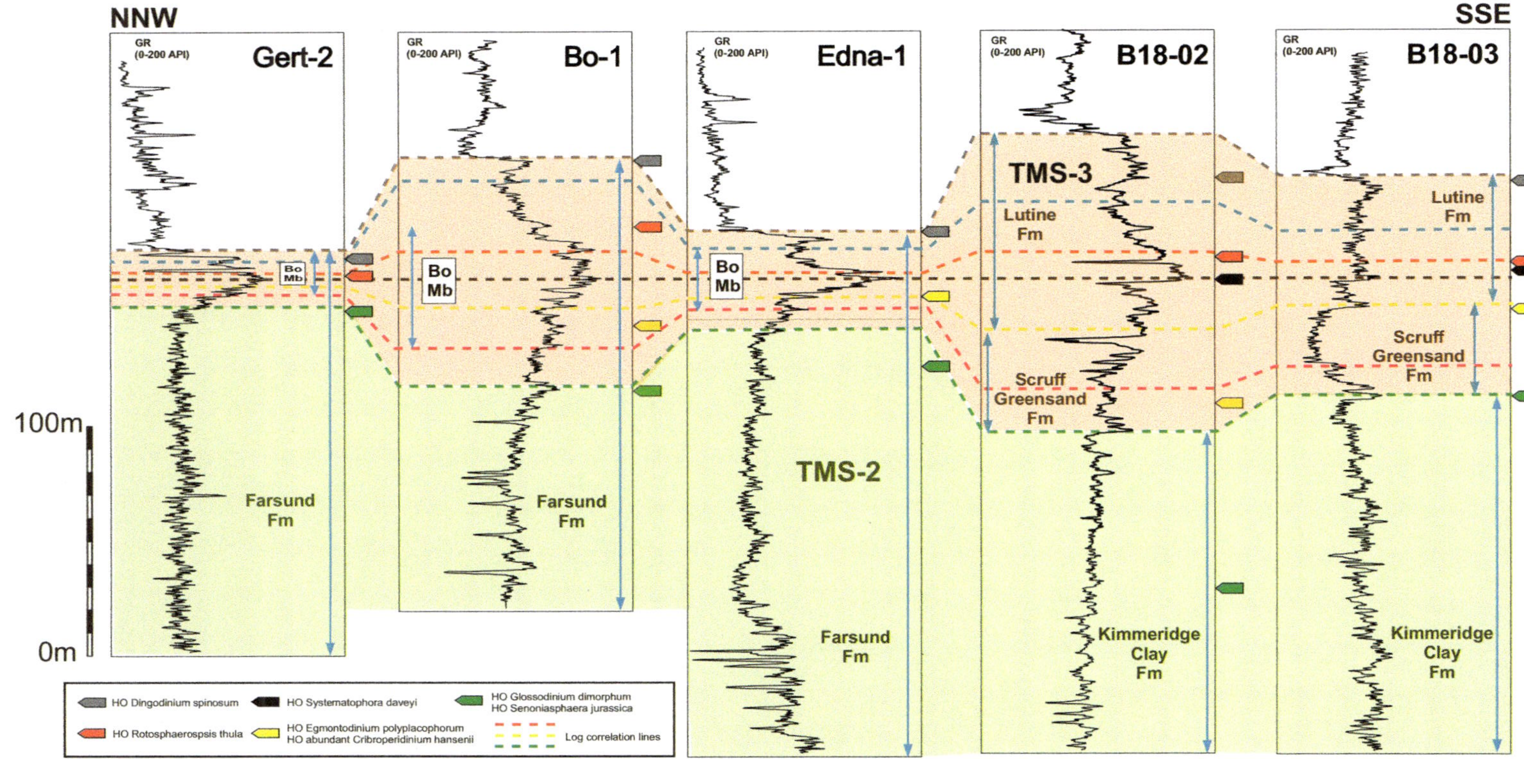

Fig. 15. Panel displaying the facies change of TMS-3 occurring from the Danish offshore to the Dutch offshore. Wireline logs as for Figure 8. The coloured dashed lines correlate the trends of the GR curves from well to well. The log correlations are underpinned by palynological events, indicated on the panel by coloured stars. The tight correlation clearly shows the gradual transition from the glauconitic sandstones of the Scruff Greensand Formation (wells B18-02 and B18-03) in the Dutch offshore to the high gamma-ray, organic-rich mudstones of the Bo Member (Gert-2, Bo-1, Edna-1) in the Danish offshore. See Figure 1 for location of the wells.

equivalent to the coal-bearing non-marine Pentland Formation in the UK sector (Fraser *et al.* 2003). The Bryne Formation is difficult to date since marine microfossils, such as dinoflagellate cysts, are lacking or very rarely found. In addition, the fluvial and estuarine deposits of the succeeding depositional sequence TS-1.2 are similar to the fluvial deposits of TS-1.1. As such, it is challenging to determine the exact timing of the rift initiation and to correlate TS-1.1 across large distances and determine its exact geographic distribution. Unlike the succeeding TS-1.2 sequence, active faulting appears to be lacking or is minimal at during this period. The average thickness observed for TS-1.1 is 100 m; the maximum thickness attained in well penetration is 220 m in West Lulu-1 (Andsbjerg 2003; Andsbjerg & Dybkjær 2003).

TS-1.2. Increasing rift activity. Shoreface complex development in adjacent Danish basins. Opening of the Dutch Central Graben.

Lithostratigraphy: Denmark: Lulu Formation and the top of the Bryne Formation. The Netherlands: Lower Graben Formation and part of the Rifgronden Member of the Friese Front Formation.

Age: Early or Middle Callovian to Late Callovian (165–163.5 Ma; Figs 2, 8, 9).

In contrast to TS-1.1, TS-1.2 is present in the entire graben axis from Denmark to The Netherlands and, as such, reflects the southwards propagation of the Middle Jurassic–Lower Cretaceous rift. At the time of deposition, active faulting occurred along the graben margins. The Central Graben rift system reached its southernmost tip at the border of the Central Offshore Platform, offshore The Netherlands Blocks L05 and L06 (Fig. 1). Seismic sections from the Tail End Graben in Denmark indicate the presence of a small angular unconformity between TS-1.1 and TS-1.2 (Middle Jurassic unconformity in Figs 5, 6) (Møller & Rasmussen 2003). This unconformity likely correlates to the erosive base of fluvial and estuarine sandstones near the top of the Bryne Formation (Andsbjerg & Dybkjær 2003; Mellere *et al.* 2016). In the Dutch Central Graben, TS-1.2 consists mostly of fluvial sandstones, overbank claystones and thin coal layers indicating swamp environments, but towards the top the unit becomes sandier and marine and tidal influence becomes prominent (Bouroullec *et al.* 2018, fig. 3). In well F03-05-S1 (Fig. 8) and F06-01 (Bouroullec *et al.* 2018), TS-1.2 reaches an overall thickness of 500 m with a distinct marine shoreface sandstone at the top (Fig. 16; Bouroullec *et al.* 2018, fig. 3). In the Tail End Graben and Salt Dome Province, TS-1.2 shows more facies variation (Fig. 5; see also Andsbjerg & Dybkjær 2003). The base is usually terrestrial, but marine influence is more prominent than in the Dutch Central Graben. In the Danish/Norwegian Søgne Basin TS-1.2 is largely represented by the Sandnes Formation, a sand-prone, outbuilding series of tidally influenced deltas (Mellere *et al.* 2016).

In the northern part of the Dutch Central Graben, TS-1.2 is represented by the Lower Graben Formation (Figs 8, 9). The basal part of the Lower Graben Formation is non-marine fluvial (Van Adrichem Boogaert & Kouwe 1993; Munsterman *et al.* 2012; Bouroullec *et al.* 2018). The Lower Graben Formation become more marine towards the top, which is sand-rich and interpreted as tidal shoals and tidal channels by Bouroullec *et al.* (2018). In the southernmost tip of the southern part of the Dutch Central Graben, where the Middle Jurassic–Lower Cretaceous sequence pinches out, only the marine flooding surface near the top of TS-1.2 is preserved in the sedimentary record, indicating low accommodation near the hinge of the graben (Fig. 5). This thin, shallow-marine sandstone interval is the Rifgronden Member of the Friese Front Formation.

In the northern part of the Dutch Central Graben, truncation of TS-1.2 occurs locally along the basin margin (Bouroullec *et al.* 2018). Truncation is caused by increased axial subsidence related to the loading on Zechstein salt underneath. In the Danish sector, basin development is controlled by faulting (Møller & Rasmussen 2003).

TS-1.3. Rift climax and east–west-oriented extension phase.

Lithostratigraphy: Denmark: Middle Graben and the Lola Formation. The Netherlands: Middle Graben, Upper Graben, Friese Front formations, Kimmeridge Clay and the Puzzle Hole Formation.

Age: Oxfordian–Early Kimmeridgian (163.5–154.7 Ma; Figs 2, 8, 9).

Characteristic of both the Danish and Dutch Central Graben is the transition from sandstone-dominated successions of TS-1.2 to mudstone-dominated successions of TS-1.3. An important difference between the Danish and the Dutch Central Graben is the amount of marine influence. Marine conditions prevailed in the Søgne Basin, Tail End Graben and Salt Dome Province, while in the Dutch Central Graben marine influence remained weak and was limited to specific horizons, based on the occasional occurrences of dinoflagellate cysts in the otherwise pollen- and spore-dominated Middle Graben Formation. In addition, three distinct and regionally extensive coal layers occur at the base of the TS-1.3 in the Middle Dutch Central Graben (Figs 5, 8, 9, 16; Bouroullec *et al.* 2018, figs 3, 9, 11, 13, 16). The coal layers are up to 3 m thick and can be correlated across large distances. The lateral continuity of the coal layers indicates sediment starvation and a relatively flat basin-floor topography. The sudden lack of sediment supply is notable,

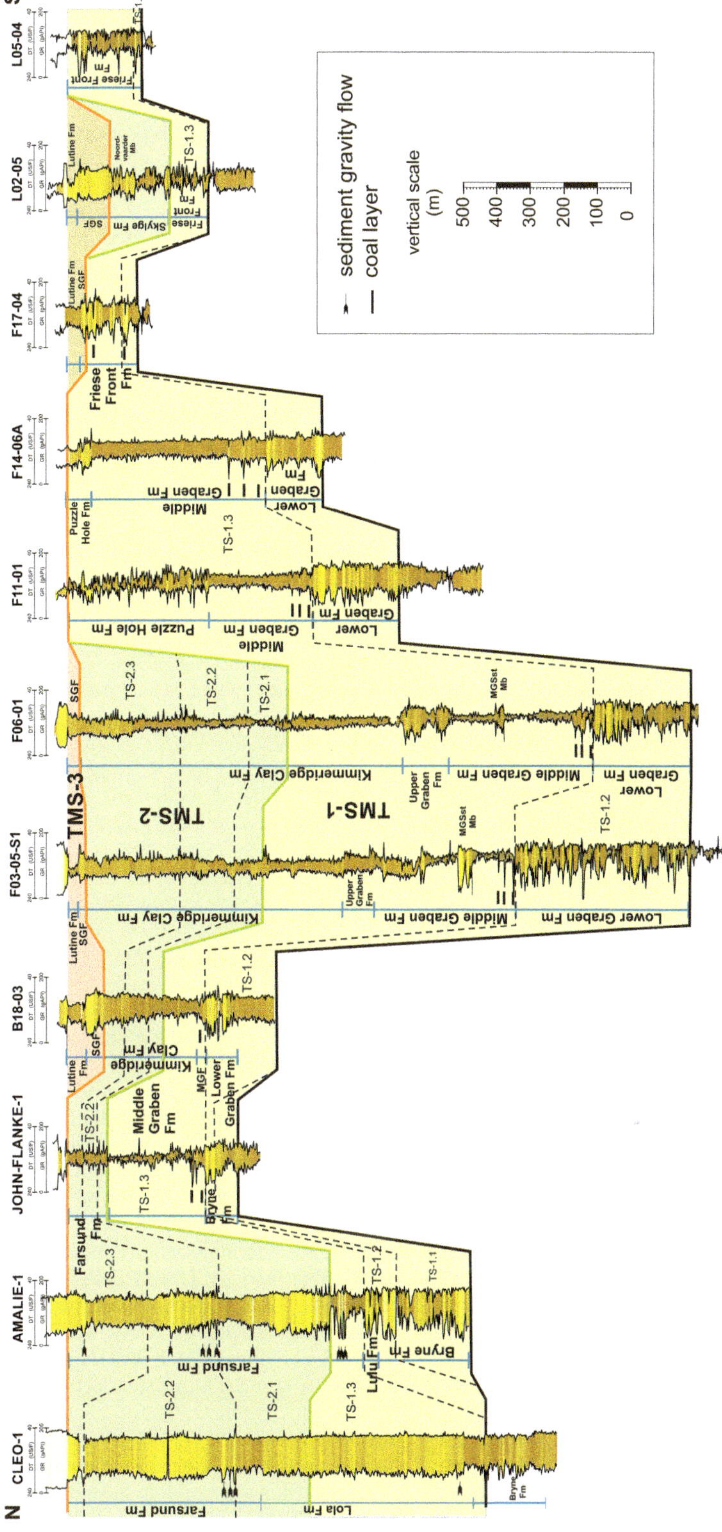

Fig. 16. N–S correlation panel (corresponding to the Wheeler diagram shown in Fig. 5) along the Central Graben axis, from the Danish offshore in the north to the Dutch offshore in the south. Wireline logs as for Figure 8. TMS-1 to TMS-3 are indicated in colour shading. SGF: Scruff Greensand Formation; MGF: Middle Graben Formation. See Figure 1 for location of the wells.

considering the abundant tidal and shoreface sandstones in the underlying TS-1.2 and the narrow (20 km wide) basin configuration during this period. Apparently, the area became starved from any sources of sand for a prolonged period of time, suggesting that the regional transgression at the top of TS-1.2 flooded the graben shoulders and forced the deposition of marginal-marine sands onto the adjacent plateau areas and away from the graben axis. These sands were likely eroded shortly afterwards, as no Early–Late Oxfordian sandstones are present on these bordering plateaus. However, supportive evidence for a marine transgression on the adjacent plateaus is present in so-called 'caprock' sequences. These are heterolithic complexes consisting of anhydrite, breccia, sandstone and shale, together making up the crests of salt diapirs. In the Jurassic, the Schill Grund Platform (Fig. 1) was perforated by a number of salt diapirs. Palynofloras occurring in the basal part of these heterolithics indicate a latest Callovian to earliest Oxfordian age and a marine depositional setting. This suggests that these diapirs have been flooded, partly dissolved and collapsed and filled in with sediments during the latest Callovian–earliest Oxfordian.

The Middle Graben Formation is predominantly non-marine, as indicated by the general lack of marine palynomorphs. Marine palynomorphs are only encountered in specific intervals, associated with maximum flooding surfaces. An estuarine palaeoenvironment with occasional open-marine influence is inferred, probably an embayment with poor connection to the open sea (Fig. 3).

Two sandstone units, the Middle Graben Sandstone Member and the Upper Graben Formation (Fig. 8), were deposited during TS-1.3 in the northern Dutch Central Graben. Both are reservoir levels that form the main producing units in the Dutch offshore F03-FB gas-condensate field (Lott *et al.* 2010). The Upper Graben Sandstone Formation is interpreted as a prograding delta front, based on the coarsening upwards trend of the sediments (Van Adrichem Boogaert & Kouwe 1993). Bouroullec *et al.* (2018) demonstrate that small bird-foot deltas are also present along the eastern margin of the graben within the Upper Graben Formation.

In the middle part of the Dutch Central Graben, a delta-plain environment with sandstones, coals and mudstones developed from the Middle Oxfordian onwards. These deposits are part of the Puzzle Hole Formation (Figs 5, 9).

In the Danish sector, marine mudstones of TS-1.3 are found in the Søgne Basin, Tail End Graben and the Salt Dome Province. The maximum thickness is attained in the Tail End Graben, but many wells do not reach the Lower Kimmeridgian Lola Formation.

TMS-2 (154.7–146.6 Ma)

During this phase, the basins and plateau areas adjacent to the graben axis, such as the Heno Plateau, Gertrud Plateau and the Terschelling Basin, became active and depocentres shifted away from the graben axis to its margins. TMS-2 is characterized by active faulting and salt movement.

From the Early to the Late Kimmeridgian, an important change in the tectonic regime occurred with a change in extension direction from east–west to SW–NE (Zanella & Coward 2003). As a result, Paleozoic NW–SE-trending faults were reactivated and new faults appeared. The adjacent basins started to subside and large amounts of sediments were deposited. In the Danish Graben axis, the depocentres move away from the Coffee Soil Fault hanging wall to the southwestern side of the graben axis and to the adjacent Danish basins (Gertrud Plateau, Gert Ridge and Heno Plateau) which started to subside and eventually became inundated (Fig. 17; 500–1000 m of TMS-2 sediments). These successions comprise the shallow-marine sandstones of the Heno Formation, overlain by open-marine mudstones of the Farsund Formation.

In the Danish Graben axis, in the Søgne Basin and Tail End Graben, gravity-flow sandstones intercalate with offshore mudstones from the Farsund Formation. In the Dutch Graben axis, much of the geological record from this time period is not preserved due to later erosion as a result of a Late Cretaceous inversion phase (Van Adrichem Boogaert & Kouwe 1993; De Jager 2007).

In the northern part of the Dutch Central Graben, mudstone deposition from the Kimmeridge Clay Formation continues without any visible change in lithology from TMS-1 to TMS-2 (Fig. 8). A transitional area is observed in the eastern part of the southern Dutch Central Graben, from non-marine in the western part of the graben to shallow-marine in the Terschelling Basin (Fig. 7). The newly developed Terschelling Basin was subsiding more rapidly than the southern part of the Dutch Central Graben. The average thickness of TMS-2 in the Terschelling Basin is 350 m, with a maximum of 800 m in well L03-01 (Fig. 18). Towards the end of TMS-2, during the Middle Volgian, large parts of the Dutch Central Graben became subjected to erosion. A conformable contact between TMS-2 and TMS-3 is only observed in the Terschelling Basin. In the northern part of the Dutch Central Graben many wells, such as well F03-05-S1, display a hiatus below the base of TMS-3 (Fig. 8). The erosion in the graben axis of the Dutch Central Graben is related to salt movement. Salt migrated from beneath the axial zone towards the lateral margins of the rift basin, which triggered tilt and erosion in the basin axis (Bouroullec *et al.* 2018; Figs 9, 12, 13). In the southern part of

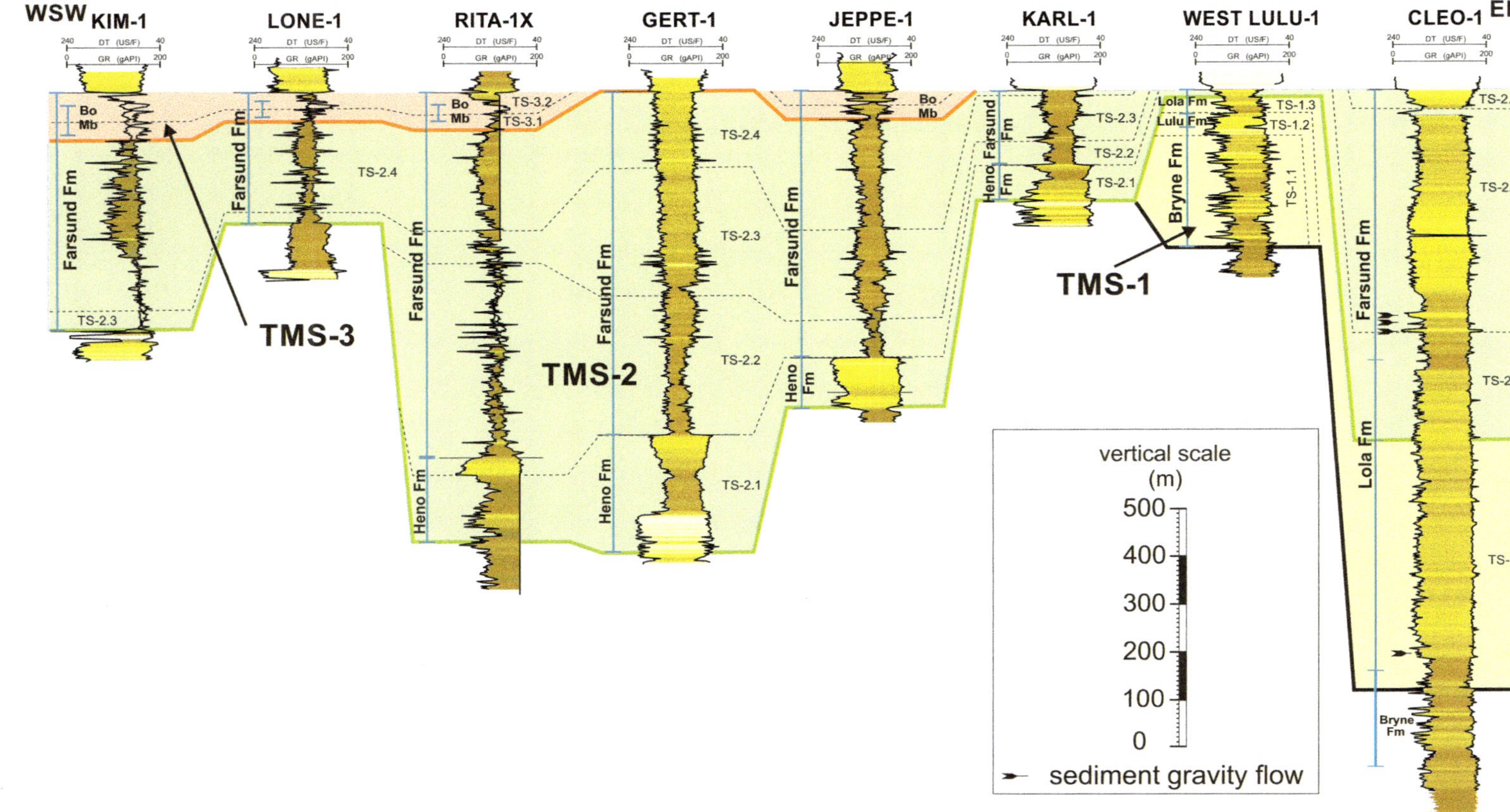

Fig. 17. W–E correlation panel (corresponding to the Wheeler diagram shown in Fig. 6) perpendicular to the Central Graben axis in the Danish offshore. Wireline logs as for Figure 8. TMS-1 to TMS-3 are indicated in colour shading. See Figure 1 for location of the wells.

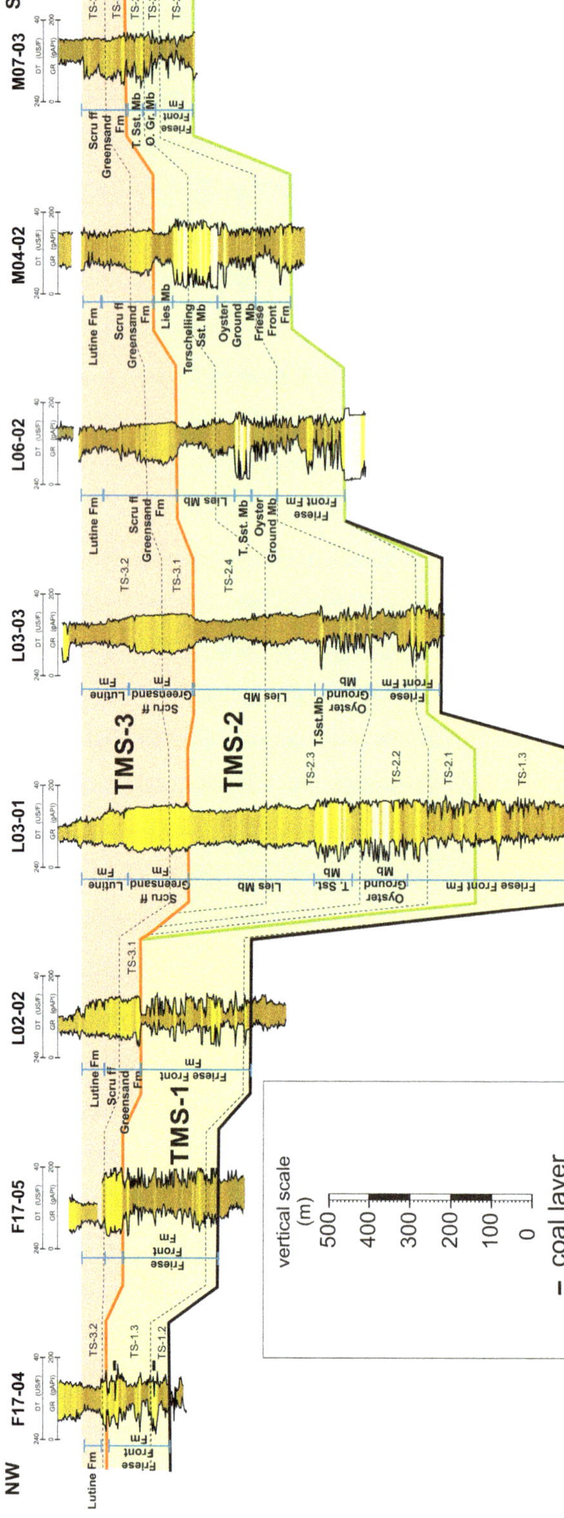

Fig. 18. W–E correlation panel (corresponding to the Wheeler diagram shown in Fig. 6) perpendicular to the Central Graben axis in the Dutch offshore Terschelling Basin. Wireline logs as for Figure 9. TMS-1 to TMS-3 are indicated in colour shading. See Figure 1 for location of the wells.

the Dutch Central Graben, evidence for intra-Middle Jurassic–Lower Cretaceous erosion is found in wells L05-03, L05-04 and F17-04 (Figs 7 and 18), where TMS-1 lies directly underneath TMS-3 or TMS-4. Widespread erosion could also explain the sand-prone Noordvaarder Member in well B13-02 and B14-02 (Fig. 13; see also Bouroullec *et al.* 2018, figs 12, 13).

TS-2.1. Change of extension trend from east–west to NE–SW. Inundation of the Heno and Gertrud Plateaus.

Lithostratigraphy: Denmark: Heno and Farsund formations. The Netherlands: Kimmeridge Clay Formation and lower part of the Friese Front Formation.

Age: Late Kimmeridgian (Mutabilis Chronozone, 154.7–153.5 Ma; Figs 2, 9–11, 13).

On the Gert Ridge and the Heno and Gertrud plateaus, the base of the Middle Jurassic–Lower Cretaceous succession is represented by the base of the Upper Kimmeridgian Heno Formation. The Heno Formation marks the inundation of the Heno and Gertrud plateaus resulting from a change in tectonic regime (Møller & Rasmussen 2003). The Heno Formation consists of two sandstone members, separated by mudstones (Figs 5, 6, 10, 17). The Gert Member is the basal and oldest sandstone member and is composed of non-marine to marginal-marine deposits (Johannessen *et al.* 1996, 2010*a*, *b*; Johannessen 2003). The Gert Member is locally sourced from the Inge and Mads highs (Weibel *et al.* 2010) and by erosion from the edges of the fault-bounded plateaus (Johannessen *et al.* 2010*a*). The youngest Ravn Member is sourced from the Mid North Sea High (Weibel *et al.* 2010). Most of the Danish plateaus were inundated during this period and preserved from erosion.

Mudstone deposition continues in the Dutch Central Graben, but sandier non-marine sediments of the upper part of the Friese Front Formation (sometimes referred to as Main Friese Front Member, Bouroullec *et al.* 2018) are observed in the southern part of the graben and in the western part of the Terschelling Basin. It is assumed that these sediments originated from the eroding plateaus areas and were transported to the graben axis via incised valleys (Fig. 3; see also Bouroullec *et al.* 2018). The base of the rift succession in the Terschelling Basin is dated as Early Kimmeridgian (Mutabilis Chronozone) based on pollen and spores (Fig. 11). In general, TS-2.1 consists of fluvial deposits in the Terschelling Basin.

In the northern part of the Dutch Central Graben, TS-2.1 is represented by mudstones from the Kimmeridge Clay Formation. In that area, no visible change in lithology occurs from TS-1.3 to TS-2.1.

TS-2.2. Opening of the Terschelling Basin. Sandy sediment gravity flows in the Søgne Basin.

Lithostratigraphy: Denmark: Farsund formation. The Netherlands: Kimmeridge Clay Formation and the upper part of the Friese Front Formation.

Age: Late Kimmeridgian (Eudoxus and Autissiodorensis Chronozones, 153.5–152 Ma; Figs 2, 8–11, 13).

In the Danish peripheral basins such as the Heno and Gertrud plateaus the base of TS-2.2 is a regionally traceable flooding event, reflected by the transition from shoreface sandstones of the Heno Formation to the open-marine mudstones of the Farsund Formation (Figs 6, 10, 17). The Farsund Formation consists of distal, open-marine mudstones, often organic-rich. As such, the Farsund Formation can be regarded as a facies equivalent of the Kimmeridge Clay Formation. Open-marine mudstone deposition persisted in the entire Danish Central Graben during TS-2.2. A thickness of 600 m is recorded in the Tail End Graben (Fig. 16). In the deepest parts of the basins, sediment gravity-flow sandstones occur (Johannessen *et al.* 2010*a*), such as seen in well Svane-1 where intercalations of such sandy sediments are frequent.

In the Terschelling Basin, marine influence gradually increases during TS-2.2 (Figs 7, 11, 18). The TS-2.2 sediments in the Terschelling Basin are fluvial deposits and belong lithostratigraphically to the upper part of the Friese Front Formation, referred to as the Main Friese Front Member (Munsterman *et al.* 2012; see also Bouroullec *et al.* 2018).

TS-2.3. Opening of the Outer Rough Basin. Marine incursions and development of sandy shoreface complexes.

Lithostratigraphy: Denmark: 'Outer Rough Sand' and the Farsund Formation. The Netherlands: lower part of the Skylge Formation (Oyster Ground and Terschelling Sandstone Member, lower parts of the Lies and Noordvaarder Member) and the Kimmeridge Clay Formation.

Age: Early Volgian (152–149.6 Ma; Figs 2, 8–14).

In the Terschelling Basin (Fig. 1), TS-2.3 is represented by the Oyster Ground Member and the Terschelling Sandstone Member of the Skylge Formation. The Oyster Ground Member is the oldest unit of the Skylge Formation, which consists of parallel laminated claystones with thin, silt-dominated storm beds (Bouroullec *et al.* 2018, fig. 4) and shelly horizons. The Oyster Ground Member is interpreted as lower shoreface and marks fully marine conditions in the Terschelling Basin during the early stages of TS-2.3. The succeeding Terschelling Sandstone Member represents a shoreface complex with a variety of depositional environments such as tidal inlet sequences, upper shoreface, lagoonal and back-barrier washover fans (Fig. 3; see also Bouroullec *et al.* 2018, fig. 4). The base of the Terschelling Sandstone Member is sharp and reflects a regression. The development of a shoreface complex in the Terschelling Basin is coeval with the development of shoreface sandstones in the Danish and German Outer Rough Basin. This local regressive sandstone unit is known informally as the 'Outer Rough Sand' (Johannessen *et al.* 2010*a*).

During TS-2.3, parts of the structurally elevated area of the Hantum Fault Zone (southern edge of the Terschelling Basin; see Fig. 1) are transgressed where Early Volgian thin, sandy deposits are preserved. Overall, subsidence appears to have decreased in the Dutch Central Graben axis during this period as most of the Zechstein salt withdrew from underneath the basin margin areas (due to welding out axially) rather than from underneath the basin axis itself (as for TMS-1; see Bouroullec *et al.* 2018). Sediment catchment therefore shifted from the Dutch graben axis to the margins.

Two other lithostratigraphic units reflect erosion due to fault activity and salt movement: the 'Outer Rough Sand' in the Outer Rough Basin (Figs 6, 12); and the Noordvaarder Member of the Skylge Formation in the Terschelling Basin and in some parts of the northern Central Graben (Figs 5, 13). The 'Outer Rough Sand' is a shallow-marine sandstone unit sandwiched between the pre-Jurassic and Cretaceous Valhall–Asgard Formation, indicating marine conditions in the Outer Rough Basin (Johannessen *et al.* 2010*a*). The Noordvaarder Member of the Skylge Formation is a sandy unit, directly resulting from fault activity such as displayed in well B14-02 (Fig. 13) but also along the northwestern margin of the Terschelling Basin (Munsterman *et al.* 2012; Bouroullec *et al.* 2018, fig. 7B).

TS-2.4. Local erosion in the graben axis and on the platforms.

Lithostratigraphy: Denmark: Farsund formation. The Netherlands: Skylge (Terschelling Sandstone, Noordvaarder and Lies Members) and the Kimmeridge Clay formations.

Age: Middle Volgian–earliest Late Volgian (149.6–144.6 Ma; Figs 2, 10–14).

The upper part of the Skylge Formation comprises shallow-marine sandstones and shales that are mainly found in the Terschelling Basin and in the southeastern part of the Dutch Central Graben. In the northern part of the Terschelling Basin, the Skylge Formation becomes increasingly sand-rich (Noordvaarder Member). Evidence for erosion at the graben axis can be seen in the northern part of the Dutch Central Graben in well F03-03, where an erosional hiatus separates the Early Volgian shales of the Kimmeridge Clay Formation (TS-2.3) from the Late Volgian sands of the Scruff Greensand Formation (TMS-3) (Fig. 5).

In the Danish area, the overall mud-dominated Farsund Formation is gradually becoming more silty in TS-2.4 than in the underlying TS-2.3, reflected by lower gamma ray values (Figs 14, 17). It is proposed that erosion occurred on a large scale in the area at that time, leading to increased supplies of coarser material even in the deepest parts of the basin.

TMS-3 (146.6–139 Ma)

Sedimentation extends to the adjacent plateaus such as the Schill Grund Plateau and Ameland Block. Sand starvation and organic-rich shale deposition is evident in the Danish sector.

In the Terschelling Basin and Dutch Central Graben, the base of TMS-3 is traceable in well-log seismic reflection patterns due to the lithological contrast between the sandy lower part of TMS-3 and the often mud-prone sediments of TMS-2 (Figs 16, 18; see also Bouroullec *et al.* 2018).

In the Danish Graben axis, it is more difficult to trace the base of the TMS-3. A small excursion towards lower values in the gamma ray trend is observed on wireline logs, pointing to increased silt and sand supply at this time. These intervals are poorly constrained stratigraphically since they are rarely cored and have limited stratigraphic thickness. It is suggested that local uplift affected the entire area, leading to erosion in the shallowest parts of the basin. Indeed, there are indications that the Upper Volgian is missing in the Outer Rough Basin (Johannessen *et al.* 2010*a*).

Large parts of the Danish Central Graben area became subjected to basinal restriction during the Early Ryazanian (Ineson *et al.* 2003), reflected in the organic-rich shale deposition of the Bo Member of the Farsund Formation. Sediment gravity-flow deposits have also been reported from the Gertrud and Tail End grabens around this time interval (Andsbjerg & Dybkjær 2003; Ineson *et al.* 2003).

Sandstone deposition of the Scruff Greensand Formation continued in the Terschelling Basin at around the same time (Fig. 7; see also Bouroullec *et al.* 2018). This sandstone is likely sourced from the Hantum Fault Zone, the Ameland Block and the area to the SE of the Terschelling Basin. Mudstone deposition became dominant throughout most of the Danish and Dutch Graben areas during the Late Ryazanian, indicating a general decrease of tectonic activity and reflecting the final stages of rifting in the area.

TS-3.1. Widespread sand deposition in Dutch sector. Clay deposition and local gravity-flow sands in the Danish sector.

Lithostratigraphy: Denmark: Farsund formation. The Netherlands: Scruff Greensand and Lutine formations.

Age: Late Volgian (146.6–144.6 Ma; Figs 2, 8, 11, 12, 14–16).

The Scruff Greensand Formation is a complex of shallow-marine sandstones (often glauconitic- and sometimes spiculite-bearing) and marine mudstones (Figs 11, 16, 18). This unit occurs over a large area of the Dutch sector, but is absent from the Danish sector. The Danish sector is characterized by clay deposition of the Farsund Formation. Gravity-flow sands are deposited locally along the Coffee Soil Fault. Typically, during TS-3.1 the deposition area widened into plateaus located close to the graben and adjacent basins (e.g. Schill Grund Plateau, Ameland Block, Step Graben and Outer Rough Basin). As a consequence, TS-3.1 reflects the initiation of the Cretaceous transgression.

TS-3.2. Reflects the start of widespread basin anoxia related to the Lower Cretaceous transgression and sand starvation of most of the basins.

Lithostratigraphy: Denmark: Bo Member (Farsund Formation). The Netherlands: Scruff Greensand and Lutine formations.

Age: Latest Volgian–Early Ryazanian (144.6–139 Ma; Figs 8, 12, 14, 15).

TS-3.2 is characterized by widespread shale deposition. In the Danish area, organic-rich shales of the Bo Member (Farsund Formation) are deposited over a large area (Figs 5, 17). The transition from the marine mudstones of the Farsund Formation to the organic-rich facies of the Bo Member is gradual (from the base of TS-3.1 based on gamma ray logs; Ineson *et al.* 2003). The maximum gamma ray values associated with the organic-rich facies are reached in the Early Ryazanian Kochi Chronozone, bounded by the highest occurrence (HO) of the dinoflagellate cyst species *Rotosphaeropsis thula* (Fig. 15). In the northern part of the Dutch sector, the Lutine Formation is also composed of organic-rich shales, but with lower total organic carbon (TOC) values (less than 3%) than the 'hot shale' or Bo Member further north (3–8%, locally exceeding 15%) (see also Fig. 15 for correlations). The absence of bioturbation and well-preserved lamination in the Bo Member suggests anoxic bottom-water conditions (Ineson *et al.* 2003) and therefore water-mass stratification which, in turn, can be linked to changes in basin configuration and/or climate. The first anoxia occurs roughly at the Volgian–Ryazanian transition (Ineson *et al.* 2003), close to the base of TMS-3. Slightly later, near the base of the Early Ryazanian Kochi Chronozone, a climate shift is recorded. This climate change is based on palynology and is reflected in the composition of the pollen and spore assemblages (Abbink 1998). The pollen and spore assemblages change from dominant 'warm and dry' elements to dominant 'warm and moist' elements. The climate change may well have been caused by a change in directional pattern of the proto-North Atlantic seaway connecting the Boreal and Tethys oceans, as suggested by Abbink *et al.* (2001). The same oceanic water-mass reconfiguration could also be responsible for introducing stratification in the Central Graben. Relative sea level was high during most of the Ryazanian, pushing the shoreline southwards and preventing coarse-grained clastics from reaching the central parts of the study area. The transition from the sandstone-dominated Scruff Greensand Formation to the shale-dominated Lutine Formation in the Dutch area indicates that erosion in the Dutch Central Graben decreased, likely due to rift cessation. Subsidence was still high in the Terschelling Basin, however, resulting in a thick succession of TS-3.2.

TMS-4 (139–126 Ma)

Alongside the Cretaceous transgression, the Middle Jurassic–Early Cretaceous rift phase ended. As such, the entire region, including the study area, became subject to subsidence.

The Cretaceous transgression continued and, with the cessation of rifting, the depositional area widened to encompass the entire North Sea (Cromer Knoll Group in the Danish, UK and Norwegian sectors). This interval is generally mud prone, but inspection of these intervals using well data often reveals thin sandy deposits at the base of the Cretaceous transgression (e.g. Vlieland Sandstone Formation on the Schill Grund Platform and Vlieland Basin, or the Leek Member of the Valhall/Asgard Formation in the Tail End Graben; Figs 1, 5–7). This change in lithology can be traced seismically, suggesting that the base of the Cretaceous transgression also reflects a short-lived tectonic event related to a change in intra-plate stress regime. An angular unconformity is occasionally observed at this level (see Abbink *et al.* 2006, fig. 10). The entire area, including the plateau areas, then acted as one structural province subjected to regional subsidence. This caused the sea to transgress much further onto the mainland of the Friesland Platform and the West Netherlands basins, where marginal-marine sandstone deposition prevailed (Jeremiah *et al.* 2010; Zwaan 2018).

Conclusions

Based on detailed palynological analysis and supported by seismic data interpretation (Bouroullec *et al.* 2018), it has been possible to define a series of tectonostratigraphic mega-sequences (TMS) that reflect the stepwise basin evolution of the Late Jurassic rift phase of the Central Graben. Rifting started in the north (Danish sector) and propagated southwards during the initial rift phase. The sediments related to this initial phase are grouped into tectonostratigraphic mega-sequence TMS-1. The oldest sediments of TMS-1 are Bathonian–Middle Callovian in age and are only found in the Danish Søgne Basin, Tail End Graben and Salt Dome Province and belong to the Bryne Formation. At the time of the latest Callovian, marine conditions reached all the way to the southernmost tip of the Central Graben. Evidence from erosional remnants on salt diapirs shows that the latest Callovian marine transgression also extended across the plateau areas east and west of the Dutch Central Graben. During the next phase of basin development, TMS-2, the basins adjacent to the graben were affected. Subsidence in the graben axis continued, but the change in tectonic regime from east–west to NE–SW extension led to the (re-) activation of NW–SE-striking normal faults in certain areas. The Heno and Gertrud plateaus were affected by subsidence first, followed by the Terschelling Basin in the Dutch sector shortly after. Later, during the Early Volgian, the depositional area reached its maximum extent when the Inge and Mads highs, Outer Rough Basin and the Vlieland Basin became active. This transgressive trend is reflected in the shoreface complex occurrences of the Terschelling Sandstone Member in the Terschelling Basin and the 'Outer Rough Sand' in the Outer Rough Basin. At the same time, basin development became increasingly complex as salt welded out in some parts of the Dutch Central Graben axis and subsidence shifted to marginal areas. As a result, thick sandstone occurrences from the Noordvaarder Member are found along the graben margins and erosion is observed at the graben axis. The next phase, TMS-3, demonstrated divergent basin development in the Danish and Dutch sectors. In the Terschelling Basin and parts of the Dutch Central Graben margin, TMS-3 is characterized by widespread sand deposition of the Scruff Greensand Formation. In the Danish area, coeval deposition of organic-rich mudstones of the Bo Member indicates (coarse-grained) sediment starvation and water column stratification. Basin differentiation ended during TMS-4 and the entire area was subjected to regional subsidence due to thermal cooling. Thin sandy units occur locally at the base of TMS-4, but the remainder of sediment accumulation was dominated by mudstones.

Implications and future perspective

The stratigraphic framework presented here coherently places the formations and members from the cross-border area in a stratigraphic framework. The basin evolution model links the spatial and stratigraphical occurrences of reservoirs and seals to regional changes in eustasy, climate and tectonics. These results provide a solid foundation for exploration activities focusing on the Jurassic synrift play. A next step could be the development of detailed cross-border palaeogeographic maps that could serve as a basis for play fairway analysis and for the establishment of common risk segment (CRS) maps. To reach that stage, a few research questions and topics still need to be addressed such as, for example: (1) the establishment of a high-resolution correlation of the Danish/Norwegian Bryne, Lulu and Sandnes formations with the Dutch Lower Graben and Friese Front formations; and (2) a feasible tectonic model to explain the divergent basin development in TMS-3. Additional research on source rocks from the study area, such as the Toarcian Posidonia Shale Formation, the earliest Cretaceous Clay Deep and Bo Members and the coal occurrences from the Middle Graben, Bryne and Sandnes formations would be beneficial for a more detailed understanding of petroleum systems.

The authors would like to thank the two reviewers Bruno Vendeville (University of Lille) and Oscar Abbink (IHS Markit) for their constructive feedback and comments. Oscar is also thanked for the groundbreaking work that paved the way for this study. Ymke van den Berg is thanked for her continuous support. The editor Ben Kilhams is thanked for his patience and scrutiny, the latter of which improved the manuscript significantly. Continued financial support from GEUS and TNO is gratefully acknowledged. Geological insights were also gained via business-to-business and multi-client projects for oil and gas companies. Their contributions are also gratefully acknowledged.

References

ABBINK, O.A. 1998. Palynological investigations in the Jurassic of the North Sea region. *LPP Contributions Series*, **8**, 7–192.

ABBINK, O.A., CALLOMON, J.H., RIDING, J.B., WILLIAMS, P.D.B. & WOLFARD, A. 2001. Biostratigraphy of the Jurassic-Cretaceous boundary strata in the Terschelling Basin, The Netherlands. *Proceedings of the Yorkshire Geological Society*, **53**, 275–302, https://doi.org/10.1144/pygs.53.4.275

ABBINK, O.A., MIJNLIEFF, H.F., MUNSTERMAN, D.K. & VERREUSSEL, R.M.C.H. 2006. New stratigraphic insights in the 'Late Jurassic' of the southern central North Sea Graben and Terschelling basin (Dutch offshore) and related exploration potential. *Netherlands Journal of Geosciences – Geologie en Mijnbouw*, **85**, 221–238.

ALLEN, P., ALVIN, K.L. *ET AL.* 1998. Purbeck–Wealden (early Cretaceous) climates. *Proceedings of the Geologists' Association*, **109**, 197–236.

ANDSBJERG, J. 2003. Sedimentology and sequence stratigraphy of the Bryne and Lulu Formations, Middle Jurassic, northern Danish Central Graben. *In*: INESON, J.R. & SURLYK, F. (eds) *The Jurassic of Denmark and Greenland.*

Geological Survey of Denmark and Greenland Bulletin, **1**, 301–348.

Andsbjerg, J. & Dybkjær, K. 2003. Sequence stratigraphy of the Jurassic of the Danish Central Graben. *In*: Ineson, J.R. & Surlyk, F. (eds) *The Jurassic of Denmark and Greenland*. Geological Survey of Denmark and Greenland Bulletin, **1**, 265–300.

Århus, N., Birkelund, T. & Smelror, M. 1989. Biostratigraphy of some Callovian and Oxfordian cores off Vega, Helgeland, Norway. *Norsk Geologisk Tidsskrift*, **69**, 39–56.

Bailey, D., Milner, P. & Varney, T. 1997. Some dinoflagellate cysts from the Kimmeridge Clay Formation in North Yorkshire and Dorset, UK. *Proceedings of the Yorkshire Geological Society*, **51**, 235–243, https://doi.org/10.1144/pygs.51.3.235

Barron, H.F. 1989. Dinoflagellate cyst biostratigraphy and palaeoenvironments of the Upper Jurassic (Kimmeridgian to basal Portlandian) of the Helmsdale region, east Sutherland, Scotland. *In*: Batten, D.J. & Keen, M.C. (eds) *Northwest European Micropalaeontology and Palynology*. British Micropalaeontological Series, Ellis Horwood Ltd., Chichester, 193–213.

Birkelund, T., Clausen, C.K., Hansen, H.N. & Holm, L. 1983. The Hectoroceras kochi Zone (Ryazanian) in the North Sea Central Graben and remarks on the Late Cimmerian Unconformity. *Danmarks Geologiske Undersøgelse Årbog*, **1982**, 53–72.

Bonis, N.R. & Kürschner, W.M. 2012. Vegetation history, diversity patterns, and climate change across the Triassic/Jurassic boundary. *Paleobiology*, **38**, 240–264.

Bouroullec, R., Verreussel, R. *et al.* 2018. Tectonostratigraphy of a rift basin affected by salt tectonics: synrift Middle Jurassic–Lower Cretaceous Dutch Central Graben, Terschelling Basin and neighbouring platforms, Dutch offshore. *In*: Kilhams, B., Kukla, P.A., Mazur, S., McKie, T., Mijnlieff, H.F. & van Ojik, K. (eds) *Mesozoic Resource Potential in the Southern Permian Basin Basin*. Geological Society of London, Special Publications, **469**. First published online March 15, 2018, https://doi.org/10.1144/SP469.22

Copestake, P., Sims, A.P., Crittenden, S., Hamar, G.P., Ineson, J.R., Rose, P.T. & Tringham, M.E. 2003. Lower Cretaceous. *In*: Evans, D., Graham, C., Armour, A. & Bathurst, P. (eds) *The Millennium Atlas: Petroleum Geology of the Central and Northern North Sea*. Geological Society, London, 191–211.

Costa, L.I. & Davey, R.J. 1992. Dinoflagellate cysts of the Cretaceous System. *In*: Powell, A.J. (ed.) *A Stratigraphic Index of Dinoflagellate Cysts*. British Micropalaeontological Society Publication, Chapman & Hall, London, 99–154.

Coward, M.P., Dewey, J.F., Hempton, M. & Holroyd, J. 2003. Tectonic evolution. *In*: Evans, D., Graham, C., Armour, A. & Bathurst, P. (eds) *The Millennium Atlas: Petroleum Geology of the Central and Northern North Sea*. Geological Society, London, **357**, 17–33.

Davey, R.J. 1979*a*. The stratigraphic distribution of dinocysts in the Portlandian (latest Jurassic) to Barremian (Early Cretaceous) of northwest Europe. *AASP Contributions Series*, **5B**, 48–81.

Davey, R.J. 1979*b*. Two new Early Cretaceous dinocyst species from the northern North Sea. *Palaeontology*, **22**, 427–437.

Davey, R.J. 1982*a*. Die Verbreitung der Palynomorphen im spaten Apt und friihen Alb Nordwestdeutschlands. *Geologisches Jahrbuch, Reihe A*, **65**, 365–403.

Davey, R.J. 1982*b*. Dinocyst stratigraphy of the latest Jurassic to Early Cretaceous of the Haldager No.1 borehole, Denmark. *Danmarks Geologiske Undersogelse, Series B*, **6**, 1–57.

De Jager, J. 2007. Geological development. *In*: Wong, T.E., Batjes, D.A.J. & de Jager, J. (eds) *Geology of the Netherlands*. Royal Netherlands Academy of Arts and Sciences, Amsterdam, 5–26.

De Jager, J. & Geluk, M.C. 2007. Petroleum geology. *In*: Wong, T.E., Batjes, D.A.J. & de Jager, J. (eds) *Geology of the Netherlands*. Royal Netherlands Academy of Arts and Sciences, Amsterdam, 241–264.

Dera, G., Brigaud, B. *et al.* 2011. Climatic ups and downs in a disturbed Jurassic world. *Geology*, **39**, 215–218.

Donders, T.H., Weijers, J.W.H. *et al.* 2009. Strong climate coupling of terrestrial and marine environments in the Miocene of northwest Europe. *Earth and Planetary Science Letters*, **28**, 215–225.

Donnadieu, Y., Dromart, G. *et al.* 2011. A mechanism for brief glacial episodes in the Mesozoic greenhouse. *Paleoceanography*, **26**, 3.

Dromart, G., Garcia, J.P., Picard, S., Atrops, F., Lécuyer, C. & Sheppard, S.M.F. 2003. Ice age at the Middle–Late Jurassic transition? *Earth and Planetary Science Letters*, **213**, 205–220.

Duxbury, S. 1977. A palynostratigraphy of the Berriasian to Barremian of the Speeton Clay of Speeton, England. *Palaeontographica Abteilung B*, **160**, 17–67.

Duxbury, S. 1980. Barremian phytoplankton from Speeton, East Yorkshire. *Palaeontographica Abteilung B*, **173**, 107–146.

Duxbury, S. 2001. A palynological zonation scheme for the Lower Cretaceous United Kingdom Sector, Central North Sea. *Neues Jahrbuch für Geologie und Paläontologie*, **219**, 95–137.

Duxbury, S., Kadolsky, D. & Johansen, S.J. 1999. Sequence stratigraphic subdivision of the Humber Group in the Outer Moray Firth area (UKCS, North Sea). *In*: Jones, R.W. & Simmons, M.D. (eds) *Biostratigraphy in Production and Development Geology*. Geological Society, London, Special Publications, **152**, https://doi.org/10.1144/GSL.SP.1999.152.01.03

Fauconnier, D. 1995. Jurassic palynology from a borehole in the Champagne area, France – correlation of the lower Callovian-middle Oxfordian using sequence stratigraphy. *Review of Palaeobotany and Palynology*, **87**, 15–26.

Feist-Burkhardt, S. & Wille, W. 1992. Jurassic palynology in southwest Germany – state of the art. *Cahiers de Micropaléontologie*, **7**, 2.

Fraser, S.I., Robinson, A.M. *et al.* 2003. Upper Jurassic. *In*: Evans, D., Graham, C., Armour, A. & Bathurst, P. (eds) *The Millennium Atlas: Petroleum Geology of the Central and Northern North Sea*. Geological Society, London, 157–189.

Gradstein, F.M., Ogg, J.G., Schmitz, M. & Ogg, G. 2012. *The Geologic Time Scale 2012*. Elsevier, Oxford, UK.

Harding, I.C. 1990. A dinocyst calibration of the European Boreal Barremian. *Palaeontographica Abteilung B*, 1–76.

Heilmann-Clausen, C. & Birkelund, T.1987. Lower Cretaceous dinoflagellate biostratigraphy in the Danish

Central Trough, *Danmarks geologiske undersøgelse*, **17**. I kommission hos C.A. Reitzels, Copenhagen, Denmark, 1–89.

Heilmann-Clausen, C. & Thomsen, E. 1995. Barremian-Aptian dinoflagellates and calcareous nannofossils in the Ahlum 1 borehole and the Otto Gott clay pit, Sarstedt, Lower Saxony Basin, Germany. *Geologisches Jahrbuch Reihe A*, 257–366.

Herngreen, G.F.W. & Wong, Th.E. 1989. Revision of the 'Late Jurassic' stratigraphy of the Dutch central Graben. *Geologie & Mijnbouw*, **68**, 73–105.

Herngreen, G.F.W., Kerstholt, S.J. & Munsterman, D.K. 2000. Callovian – Ryazanian ('Middle Jurassic to Early Cretaceous') palynostratigraphy of the Central North Sea Graben and Vlieland Basin, The Netherlands. *Mededelingen Nederlands Instituut voor Toegepaste Geowetenschappen TNO*, **63**, 1–99.

Husmo, T., Hamar, G.P., Høiland, O., Johannessen, E.P., Rømuld, A., Spencer, A.M. & Titterton, R. 2003. Lower and middle Jurassic. *In*: Evans, D., Graham, C., Armour, A. & Bathurst, P. (eds) *The Millennium Atlas: Petroleum Geology of the Central and Northern North Sea*. Geological Society, London, 129–156.

Ineson, J.R., Bojesen-Koefoed, J.A., Dybkjær, K. & Nielsen, L.H. 2003. Volgian–Ryazanian 'hot shales' of the Bo Member (Farsund Formation) in the Danish Central Graben, North Sea: stratigraphy, facies and geochemistry. The Jurassic of Denmark and Greenland. *Geological Survey of Denmark and Greenland Bulletin*, **1**, 403–436.

Ioannides, N.S., Colin, J.-P. & Jan du Chene, R. 1988. A preliminary investigation of Kimmeridgian dinoflagellates and ostracods from Quercy, Southwest France. *Bulletin des Centres de recherches exploration-production Elf-Aquitaine*, **12**, 471–491.

Jansonius, J. & McGregor, D.C. 1996. *Palynology: Principles and Applications, Vol. 1 Principles*. American Association of Stratigraphic Palynologists Foundation, Publishers Press, Salt Lake City, Utah, USA, 1–451.

Japsen, P., Britze, P. & Andersen, C. 2003. Middle Jurassic to Early Cretaceous-Lower Cretaceous of the Danish Central Graben: structural framework and nomenclature. *In*: Ineson, J.R. & Surlyk, F. (eds) *The Jurassic of Denmark and Greenland*. Geological Survey of Denmark and Greenland Bulletin, **1**, 233–246.

Jeremiah, J.M., Duxbury, S. & Rawson, P.F. 2010. Lower Cretaceous of the southern North Sea Basins: reservoir distribution within a sequence stratigraphic framework. *Geologie en Mijnbouw*, **89**, 203–237.

Johannessen, P.N. 2003. Sedimentology and sequence stratigraphy of paralic and shallow marine Middle Jurassic to Early Cretaceous sandstones in the northern Danish Central Graben. *In*: Ineson, J.R. & Surlyk, F. (eds) *The Jurassic of Denmark and Greenland*. Geological Survey of Denmark and Greenland Bulletin, **1**, 367–402.

Johannessen, P.N. & Andsbjerg, J. 1993. Middle to Late Jurassic basin evolution and sandstone reservoir distribution in the Danish Central Trough. *In*: Parker, J.R. (ed.) *Petroleum Geology of Northwest Europe: Proceedings of the 4th Conference*. Geological Society, London, 271–283, https://doi.org/10.1144/0040271

Johannessen, P.N., Dybkjær, K. & Rasmussen, E.S. 1996. Sequence stratigraphy of Upper Jurassic reservoir sandstones in the northern part of the Danish Central Trough, North Sea. *Marine and Petroleum Geology*, **13**, 755–770.

Johannessen, P.N., Dybkjær, K., Andersen, C., Kristensen, L., Hovikoski, J. & Vosgerau, H. 2010*a*. Upper Jurassic reservoir sandstones in the Danish Central Graben: new insights on distribution and depositional environments. *In*: Vining, B.A. & Pickering, S.C. (eds) *Petroleum Geology: From Mature Basins to New Frontiers. Proceedings of the 7th Conference*. Geological Society, London, Petroleum Geology Conference Series, **7**, 127–143, https://doi.org/10.1144/0070127

Johannessen, P.N., Nielsen, L.H., Nielsen, L., Møller, I., Pejrup, M. & Andersen, T.J. 2010*b*. Architecture of an Upper Jurassic barrier island sandstone reservoir, Danish Central Graben: implications of a Holocene-Recent analogue from the Wadden Sea. *In*: Vining, B.A. & Pickering, S.C. (eds) *Petroleum Geology: From Mature Basins to New Frontiers. Proceedings of the 7th Petroleum Geology Conference*. Geological Society, London, Petroleum Geology Conference Series, **7**, 145–155, https://doi.org/10.1144/0070145

Kirsch, K.-H. & Below, R. 1995. Quantitative Untersuchung der Dinoflagellatenverteilung in den hell/dunkel-Rhythmiten des Hauterive-Barrême-Grenzbereichs im Niedersächsischen Becken (Norddeutschland) am Beispiel des Profils der Tongrube Otto Gott bei Sarstedt. *Palaeontographica Abteilung B*, **236**, 105–146.

Korte, C. & Hesselbo, S.P. 2011. Shallow marine carbon and oxygen isotope and elemental records indicate icehouse-greenhouse cycles during the Early Jurassic. *Palaeoceonography*, **26**, 1–18.

Kunz, R. 1990. Phytoplankton and Palynofazies im Malm NW-Deutschlands (Hannoversches Bergland). Phytoplankton and palynofacies in the Malm of NW Germany (Hannoversches Bergland). *Palaeontographica Abteilung B*, **216**, 1–105.

Lott, G.K., Wong, T.E., Dusar, M., Andsbjerg, J., Mönnig, E., Feldman-Olszewska, A. & Verreussel, R.M.C.H. 2010. Jurassic. *In*: Doornenbal, J.C. & Stevenson, A.G. (eds) *Petroleum Geological Atlas of the Southern Permian Basin Area*. EAGE Publications b.v., Houten, 175–193.

Mellere, D., Mannie, A., Longhitano, S., Mazur, M., Kulausa, H., Brough, S. & Cotton, J. 2016. Tidally influenced shoal water delta and estuary in the Middle Jurassic of the Søgne Basin, Norwegian North Sea: sedimentary response to rift initiation and salt tectonics. *In*: Hampson, G.J., Reynolds, A.D., Kostic, B. & Wells, M.R. (eds) *Sedimentology of Paralic Reservoirs: Recent Advances*. Geological Society London, Special Publications, **444**, 173.

Møller, J.J. & Rasmussen, E.S. 2003. Middle Jurassic-Early Cretaceous rifting of the Danish Central Graben. *In*: Ineson, J.R. & Surlyk, F. (eds) *The Jurassic of Denmark and Greenland*. Geological Survey of Denmark and Greenland Bulletin, **1**, 247–264.

Munsterman, D.K., Verreussel, R.M.C.H., Mijnlieff, H.F., Witmans, N., Kerstholt-Boegehold, S.J. & Abbink, O.A. 2012. Revision and update of the Callovian-Ryazanian Stratigraphic Nomenclature in the northern Dutch offshore, i.e. Central Graben Subgroup and Scruff Group. *Netherlands Journal of Geosciences – Geologie en Mijnbouw*, **91**, 555–590.

MUTTERLOSE, J. & HARDING, I. 1987. Phytoplankton from the anoxic sediments of the Barremian (Lower Cretaceous) of North-West Germany. *Abhandlungen der Geologischen Bundesanstalt*, **39**, 177–215.

NØHR-HANSEN, H. 1986. Dinocyst Stratigraphy of the Lower Kimmeridge Clay, WesTerschelling Basinury, England. *Bulletin of the Geological Society of Denmark*, **35**, 31–51.

NUNN, E.V. & PRICE, G.D. 2010. Late Jurassic (Kimmeridgian–Tithonian) stable isotopes (δ 18 O, δ 13 C) and Mg/Ca ratios: new palaeoclimate data from Helmsdale, northeast Scotland. *Palaeogeography, Palaeoclimatology, Palaeoecology*, **292**, 325–335.

OGG, J.G., OGG, G.M. & GRADSTEIN, F.M. 2016. *A Concise Geologic Time Scale*. Elsevier, Amsterdam.

PARTINGTON, M.A., MITCHENER, B.C., MILTON, N.J. & FRASER, A.J. 1993*a*. Genetic sequence stratigraphy for the North Sea Late Jurassic and Early Cretaceous: distribution and prediction of Kimmeridgian-Late Ryazanian reservoirs in the North Sea and adjacent areas. *In*: PARKER, J.R. (ed.) *Petroleum Geology of Northwest Europe: Proceedings of the 4th Conference*. Geological Society, London, Petroleum Geology Conference Series, **4**, 347–370, https://doi.org/10.1144/0040347

PARTINGTON, M.A., COPESTAKE, P., MITCHENER, B.C. & UNDERHILL, J.R. 1993*b*. Biostratigraphic calibration of genetic stratigraphic sequences in the Jurassic-lowermost Cretaceous (Hettangian to Ryazanian) of the North Sea and adjacent areas. *In*: PARKER, J.R. (ed.) *Petroleum Geology of Northwest Europe: Proceedings of the 4th Conference*. Geological Society, London, Petroleum Geology Conference Series, **4**, 371–386, https://doi.org/10.1144/0040371

PHARAOH, T.C., DUSAR, M. *ET AL*. 2010. Tectonic evolution. *In*: DOORNENBAL, J.C. & STEVENSON, A.G. (eds) *Petroleum Geological Atlas of the Southern Permian Basin Area*. EAGE Publications b.v., Houten, 25–57.

POULSEN, N.E. 1994. Dinoflagellate cyst biostratigraphy of the Rhaetian-Ryazanian (uppermost Triassic-lowermost Cretaceous) deposits from the Danish Subbasin. *Geobios*, **17**, 409–414.

POULSEN, N.E. 1996. Dinoflagellate cysts from marine Jurassic deposits of Denmark and Poland. *AASP Foundation, Contribution Series*, **31**, 1–227.

POULSEN, N.E. 1998. Upper Bajocian to Callovian (Jurassic) dinoflagellate cysts from central Poland. *Acta Geologica Polonica*, **48**, 237–245.

PRAUSS, M. 1989. Dinozysten-Stratigraphie und Palynofazies im Oberen Lias und DoGertrud Grabener von NW-Deutschland. *Palaeontographica (B)*, **214**, 1–124.

PRICE, G.D. & ROGOV, M.A. 2009. An isotopic appraisal of the Late Jurassic greenhouse phase in the Russian Platform. *Palaeogeography, Palaeoclimatology, Palaeoecology*, **273**, 41–49.

RATTEY, R.P. & HAYWARD, A.P. 1993. Sequence stratigraphy of a failed rift system: the Middle Jurassic to Early Cretaceous basin evolution of the Central and Northern North Sea. *In*: PARKER, J.R. (ed.) *Petroleum Geology of Northwest Europe: Proceedings of the 4th Conference*. Geological Society, London, Petroleum Geology Conference Series, **4**, 215–249, https://doi.org/10.1144/0040215

RIDING, J.B. 1987. Dinoflagellate cyst Stratigraphy of the Nettleton Bottom borehole (Jurassic: Hettangian to Kimmeridgian), Lincolnshire, England. *Proceedings of the Yorkshire Geological Society*, **46**, 231–266, https://doi.org/10.1144/pygs.46.3.231

RIDING, J.B. & BAILEY, D.A. 1991. Durotrigia filapicata, comb. nov. for Gonyaulacysta filapicata (fossil Pyrrhophyta: Dinophyceae). *Taxon*, **40**, 100–102.

RIDING, J.B. & DAVEY, R.J. 1989. *Rotosphaeropsis thula* (Davey 1982) comb. nov. and emend.: a dinoflagellate cyst from the Upper Jurassic-Lower Cretaceous of England. *Journal of Micropalaeontology*, **8**, 109–112, https://doi.org/10.1144/jm.8.1.109

RIDING, J.B. & THOMAS, J.E. 1988. Dinoflagellate cyst Stratigraphy of the Kimmeridge Clay (Upper Jurassic) from the Dorset Coast, southern England. *Palynology*, **12**, 65–88.

RIDING, J.B. & THOMAS, J.E. 1992. Dinoflagellate cysts of the Jurassic System. *In*: POWELL, A. J. (ed.) *A Stratigraphic Index of Dinoflagellate Cysts*. British Micropalaeontological Society Publication, Chapman & Hall, London, 7–98.

RILEY, L.A. 1979. Dinocysts from the Upper Kimmeridgian (*pectinatus* Zone) of Marton, Yorkshire. *Mercian Geologist*, **7**, 219–222.

RILEY, L.A. & FENTON, J.P.G. 1982. A dinocyst zonation for the Callovian to Middle Oxfordian succession (Jurassic) of northwest Europe. *Palynology*, **6**, 193–202.

SLUIJS, A., PROSS, J. & BRINKHUIS, H. 2005. From greenhouse to icehouse; organic-walled dinoflagellate cysts as paleoenvironmental indicators in the Paleogene. *Earth Science Reviews*, **68**, 281–315.

SURLYK, F. & INESON, J.R. 2003. The Jurassic of Denmark and Greenland: key elements in the reconstruction of the North Atlantic Jurassic rift system. *In*: INESON, J.R. & SURLYK, F. (eds) *The Jurassic of Denmark and Greenland*. Geological Survey of Denmark and Greenland Bulletin, **1**, 9–20.

TRAVERSE, A. 2007. Differential sorting of palynomorphs into sediments: palynofacies, palynodebris, discordant palynomorphs. *Paleopalynology*, 543–579.

UNDERHILL, J.R. & PARTINGTON, M.A. 1993. Jurassic thermal doming and deflation in the North Sea: implications of the sequence stratigraphic evidence. *In*: PARKER, J.R. (ed.) *Petroleum Geology of the Northwest Europe: Proceedings of the 4th Conference*. Geological Society, London, Petroleum Geology Conference Series, **4**, 337–345, https://doi.org/10.1144/0040337

VAN ADRICHEM BOOGAERT, H.A. & KOUWE, W.F.P. 1993. Stratigraphic nomenclature of the Netherlands, revision and update by the RGD and NOGEPA. *Mededelingen Rijks Geologische Dienst*, **50**, 1–40.

VEJBÆK, O.V., ANDERSEN, C. *ET AL*. 2010. Cretaceous. *In*: DOORNENBAL, J.C. & STEVENSON, A.G. (eds) *Petroleum Geological Atlas of the Southern Permian Basin Area*. EAGE Publications b.v., Houten, 195–209.

WEIBEL, R., JOHANNESSEN, P.N., DYBKJÆR, K., ROSENBERG, P. & KNUDSEN, C. 2010. Chemostratigraphy of Middle Jurassic to Early Cretaceous reservoir sandstones, Danish Central Graben, North Sea. *Marine and Petroleum Geology*, **27**, 1572–1594.

WIMBLEDON, W.A.P., CASELLATO, C.E. *ET AL*. 2011. Fixing a basal Berriasian and Jurassic/Cretaceous (J/K) boundary – Is there perhaps some light at the end of the tunnel? *Revista Italiana di Paleontologia e Stratigrafia*, **17**, 295–307.

WONG, T.E. 2007. Jurassic. *In*: WONG, T.E., BATJES, D.A.J. & DE JAGER, J. (eds) *Geology of the Netherlands*. Royal Netherlands Academy of Arts and Sciences, Amsterdam, 107–125.

WOOLLAM, R. 1980. Jurassic dinocysts from shallow marine deposits of the East Midlands, England. *Journal of the University of Sheffield Geological Society*, **7**, 243–261.

ZANELLA, E. & COWARD, M.P. 2003. Structural framework. *In*: EVANS, D., GRAHAM, C., ARMOUR, A. & BATHURST, P. (eds) *The Millennium Atlas: Petroleum Geology of the Central and Northern North Sea*. Geological Society, London, **357**, 45–59.

ZWAAN, F. 2018. Lower cretaceous reservoir development in the North Sea Central Graben and potential analogue settings in the Southern Permian Basin and South Viking Graben. *In*: KILHAMS, B., KUKLA, P.A., MAZUR, S., MCKIE, T., MIJNLIEFF, H.F. & VAN OJIK, K. (eds) *Mesozoic Resource Potential in the Southern Permian Basin Basin*. Geological Society, London, Special Publications, **469**. First published online January 4, 2018, https://doi.org/10.1144/SP469.3

Palaeogeographical evolution of the Lower Jurassic: high-resolution biostratigraphy and sequence stratigraphy in the Central European Basin

GREGOR BARTH[1,2]*, GRZEGORZ PIEŃKOWSKI[3], JENS ZIMMERMANN[4], MATTHIAS FRANZ[5] & GESA KUHLMANN[6]

[1]*Federal Institute for Geosciences and Natural Resources (BGR), Wilhelmstraße 25–30, 13593 Berlin, Germany*

[2]*Present address: State Agency for Environment, Nature Conservation and Geology (LUNG), Goldberger Straße 12, 18273 Güstrow, Germany*

[3]*Polish Geological Institute – National Research Institute, Rakowiecka 4, 00-975 Warszawa, Poland*

[4]*Geothermie Neubrandenburg GmbH (GTN), Seestraße 7A, 17033 Neubrandenburg, Germany*

[5]*Geowissenschaftliches Zentrum der Universität Göttingen, Abteilung Angewandte Geologie, Goldschmidtstraße 3, 37077 Göttingen, Germany*

[6]*Federal Institute for Geosciences and Natural Resources (BGR), Wilhelmstraße 25–30, 13593 Berlin, Germany*

**Correspondence: gregor.barth@lung.mv-regierung.de*

Abstract: Basin-scale stratigraphic correlation is the fundamental base for successful reservoir exploration, and especially when dealing with cross-border areas. Differences in lithostratigraphic and chronostratigraphic nomenclature between sub-basins and countries often result in problematic estimations of reservoir geometries and potential. This study combines available biostratigraphic, biofaunal and lithofacies data, together with sequence-stratigraphical correlations of the Lower Jurassic from the Central European Basin (CEB), to propose a genetic-based framework of transgressive and regressive depositional units. The determination of four major biofacies environments, composed of (I) polyhaline open-marine/offshore environments, (II) upper mesohaline marine–brackish environments, (III) lower mesohaline brackish environments and (IV) low oligohaline to freshwater continental environments comprising very rare marine phytoplankton and terrestrial spores and pollens, were translated into 12 biofacies reconstructions of ammonite (sub-) chronozone levels. Variations of biofacies reconstructions in time and space were supplemented by biostratigraphically constrained large-scale progradational and retrogradational sedimentary architecture. Retrogradation is accompanied by increasing polyhaline environments and pinpoint basinwide third-order flooding events, whereas progradation is accompanied by decreasing polyhaline environments pointing to third-order regressions. The outcomes of this study support exploration of Lower Jurassic deep geothermal reservoirs or CO_2 storage sites in the eastern CEB (especially Germany and Poland).

Supplementary material: A list of all documented Liassic ammonites known from the eastern European shelf area (Denmark, The Netherlands, Sweden, Germany, Poland; wells and outcrops) is available at https://doi.org/10.6084/m9.figshare.c.3923467

From the Late Triassic to the Middle Jurassic, the epicontinental Central European Basin (CEB) experienced one of the most fundamental changes in the Earth's history. The transformation of continental environments of the Keuper (Late Triassic) towards fully marine environments of the Lower and Middle Jurassic were accompanied by significant environmental and biofaunal perturbations, such as Large Igneous Provinces, the End-Triassic mass extinction, the Toarcian Oceanic Anoxic Event, carbon and oxygen isotope excursions (Jenkyns 1988; van de Schootbrugge *et al.* 2013), and the invasion of Tethyan and Boreal ammonites (Callomon 1985, 2003; Benton 1986; Hallam 1997; Poulsen & Riding

From: Kilhams, B., Kukla, P. A., Mazur, S., McKie, T., Mijnlieff, H. F. & van Ojik, K. (eds) 2018. *Mesozoic Resource Potential in the Southern Permian Basin*. Geological Society, London, Special Publications, **469**, 341–369.
First published online January 4, 2018, https://doi.org/10.1144/SP469.8

2003). Triggered by global sea-level rise due to the intensified rifting of Pangaea, the continental lowlands of the CEB were successively flooded from the west and, as a consequence, fully marine environments advanced towards the east forming first an enclosed inland sea in the Early Jurassic (e.g. Lott *et al.* 2010) and eventually transforming to an open shelf sea that was well connected to both the Boreal and the Tethyan oceans by Middle Jurassic times (Ziegler 1990; Zimmermann *et al.* 2015; see also the Discussion).

Marine flooding began as early as the Rhaetian (latest Triassic) in the western areas of the offshore UK (Fischer *et al.* 2012; Barth *et al.* 2018) and Denmark (Nielsen 2003), and continued towards the east in the Early Hettangian (Barth *et al.* 2018). However, widespread fully marine environments with ammonites reached the eastern part of the CEB (Poland) about 11.3 myr later in the Early Pliensbachian Jamesoni–Ibex Chronozone (Dadlez 1969; Pieńkowski 2004; Zimmermann *et al.* 2015 amongst others).

Hallam (2001) claimed that the overall pattern appears to be a more or less gradual sea-level rise through the Jurassic interrupted by episodes of comparative stillstands rather than eustatic fall. Temporal sea-level fluctuations resulted in numerous sequence stratigraphic schemes in several sub-basins of the CEB, such as the North Sea Basin (Partington *et al.* 1993; Underhill & Partington 1993; de Graciansky *et al.* 1998; Jacquin *et al.* 1998), the British Jurassic Coast (Callomon & Chandler 1994; Hesselbo & Jenkyns 1998; Jacquin *et al.* 1998), the Danish–Swedish Basin (Pieńkowski 1991*a*, *b*; Koppelhus & Nielsen 1994; Surlyk *et al.* 1995; Andsbjerg & Dybkjær 2003; Michelsen *et al.* 2003; Nielsen 2003; Nielsen *et al.* 2010), the North German Basin (NGB: see Doehler 2005; Bachmann *et al.* 2008; Brand & Mönnig 2009; Zimmermann *et al.* 2015; Barth *et al.* 2018) and the Polish Basin (Pieńkowski 1991*b*, 2004; Feldman-Olszewska 1997*a*, *b*; Pieńkowski *et al.* 2008) (Fig. 1). However, due to different biostratigraphic resolution and applied methods, chronostratigraphic calibration of the sequence stratigraphic bounding surfaces is often varying (Zimmermann *et al.* 2015).

Although extensive research was performed both in the German and Polish part of the CEB, stratigraphical and facies correlations were usually confined to national boundaries or more local areas, and little has been published, as yet, on the trans-border correlation of the Lower Jurassic series (Dadlez 1969; Pieńkowski 1991*b*; Lott *et al.* 2010). Thus, in this study, an attempt is made to combine published and unpublished biostratigraphic and biofaunal information, both in Germany and Poland, supplemented by similar data from southern Denmark and southern Sweden, in order to reconstruct biofaunal provinces in the ammonite subchronozone level for the CEB. In addition, studies on cored and logged wells in the NGB and the Polish Basin were used for the reconstruction of the sedimentary architecture of the Lower Jurassic strata, and then combined with biofaunal reconstructions to forward a synoptic biofaunal and sequence stratigraphic framework of the Lower Jurassic in the CEB.

Geological setting and basin evolution

The early phase of the break-up of Pangaea, which had already initiated in the latest Permian and intensified in the Late Triassic–Early Jurassic, was characterized by west–east-directed extensional stress that resulted in the CEB in the formation of north–south- and NW–SE-trending graben systems (e.g. the Central Graben, Viking Graben, East and West Holstein troughs, Gifhorn Trough) (Wienholz 1967; Ziegler 1982; Dadlez *et al.* 1995; Poprawa 1997; Scheck-Wenderoth *et al.* 2008; Pieńkowski *et al.* 2008), and the activation of the Mid-Polish Trough in the Early Jurassic. The North Sea rift system remained active during the Early Jurassic, as evident in the Viking and Dutch Central graben. Continued regional thermal subsidence of the CEB during the Rhaetian and Hettangian, combined with sea-level rise, controlled the development of a wide, shallow-marine basin that was successively flooded from the west (e.g. Will 1969; Ziegler 1990; Bachmann *et al.* 2008, 2010). This connection coincides with a significant sea-level rise and eastwards transgression that culminated in the Pliensbachian, with fully marine conditions as far east as central Poland (Pieńkowski 2004, 2014). Rising sea-level also facilitated Tethyan waters to enter the CEB (Fig. 2).

The Early Jurassic basin in Poland was generally no deeper than some tens of metres, only occasionally below the storm wave-base (50–80 m in the Early Pliensbachian in western Pomerania), but most frequently it was between 10 and 20 m deep (Pieńkowski 2004). The main centre of Jurassic sedimentation was along the NW–SE-trending Mid-Polish Trough, which extends from western Pomerania to the Holy Cross Mountains and is aligned with the Teisseyre-Tornquist Zone to the north (Dadlez *et al.* 1995; Poprawa 1997; Pieńkowski 2004).

Since the CEB was influenced by recurrent marine environments, the model of transgressive–regressive sequences (T–R sequences) has been applied in former studies and further used in this study (de Graciansky *et al.* 1998; Jacquin *et al.* 1998, and references therein). These T–R sequences are hierarchically ordered, and are composed of first- and second-order sequences (van der Meer *et al.* 2017). The Norian Arnstadt Formation (terrestrial Keuper facies) marks the base sequence boundary of the first-order Ligurian Cycle. The maximum

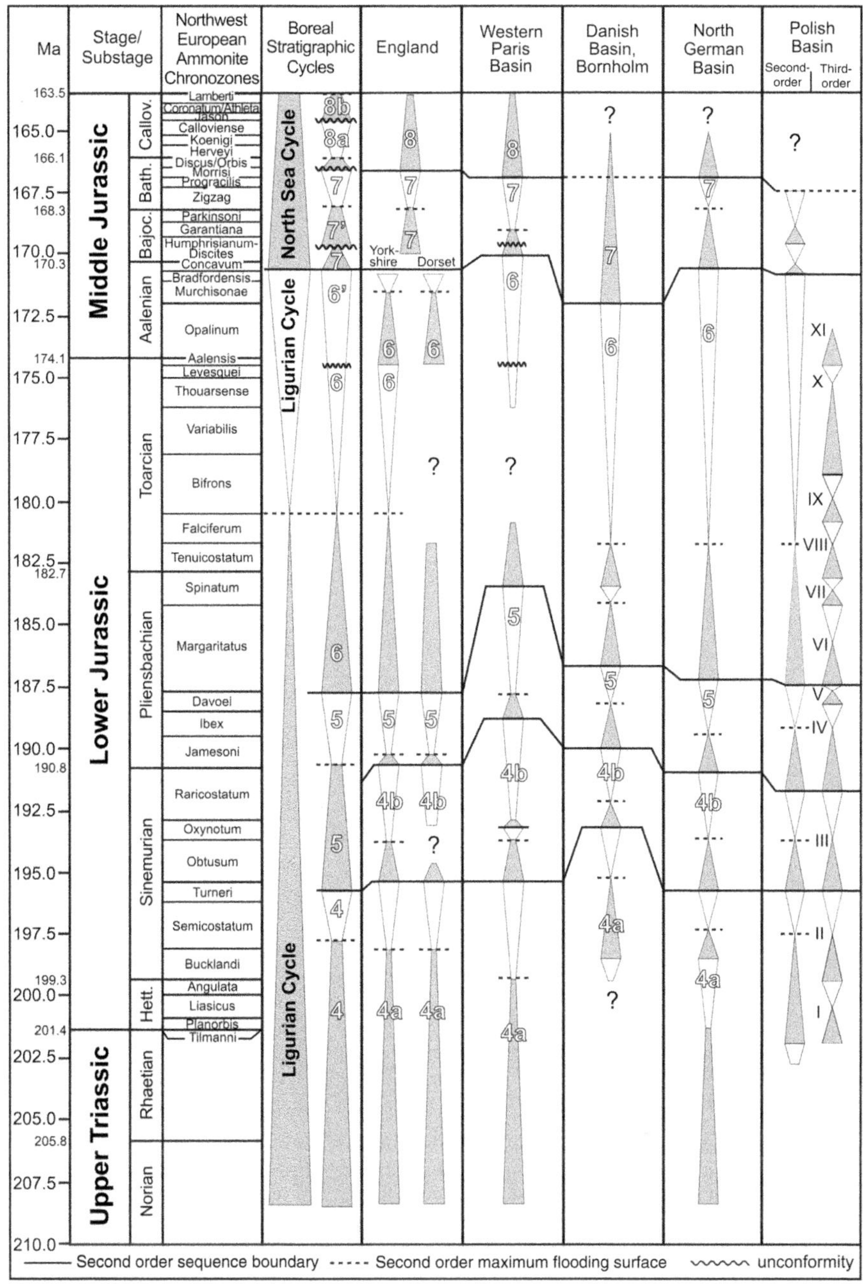

Fig. 1. Compilation of second-order transgressive (T) and regressive (R) sequences in the CEB. First- and second-order Boreal and Tethyan sequences are according to de Graciansky *et al.* (1998) and Jacquin *et al.* (1998). T–R sequences in England are from Dorset and Yorkshire (Hesselbo & Jenkyns 1998; Jacquin *et al.* 1998). Sequences of the western Paris Basin are from de Graciansky *et al.* (1998) and Jacquin *et al.* (1998). Danish sequences are from the Danish Basin (Andsbjerg & Dybkjær 2003; Nielsen 2003) and from Bornholm (Koppelhus & Nielsen 1994; Surlyk *et al.* 1995). T–R sequences from the North German Basin are from Zimmermann *et al.* (2015) and Barth *et al.* (2018). Second- and third-order sequences of the Polish Basin are from Feldman-Olszewska (1997*a*, *b*) – Middle Jurassic; and Pieńkowski (2004) and Pieńkowski *et al.* (2008) – Lower Jurassic. Lower Jurassic sequences are in consecutive order, based on Pieńkowski (2004). Some of these sequences (II, III, IV and VI) coincide with those of second order, while other (I, V, VII, VIII, IX, X and XI) represent the third-order ones.

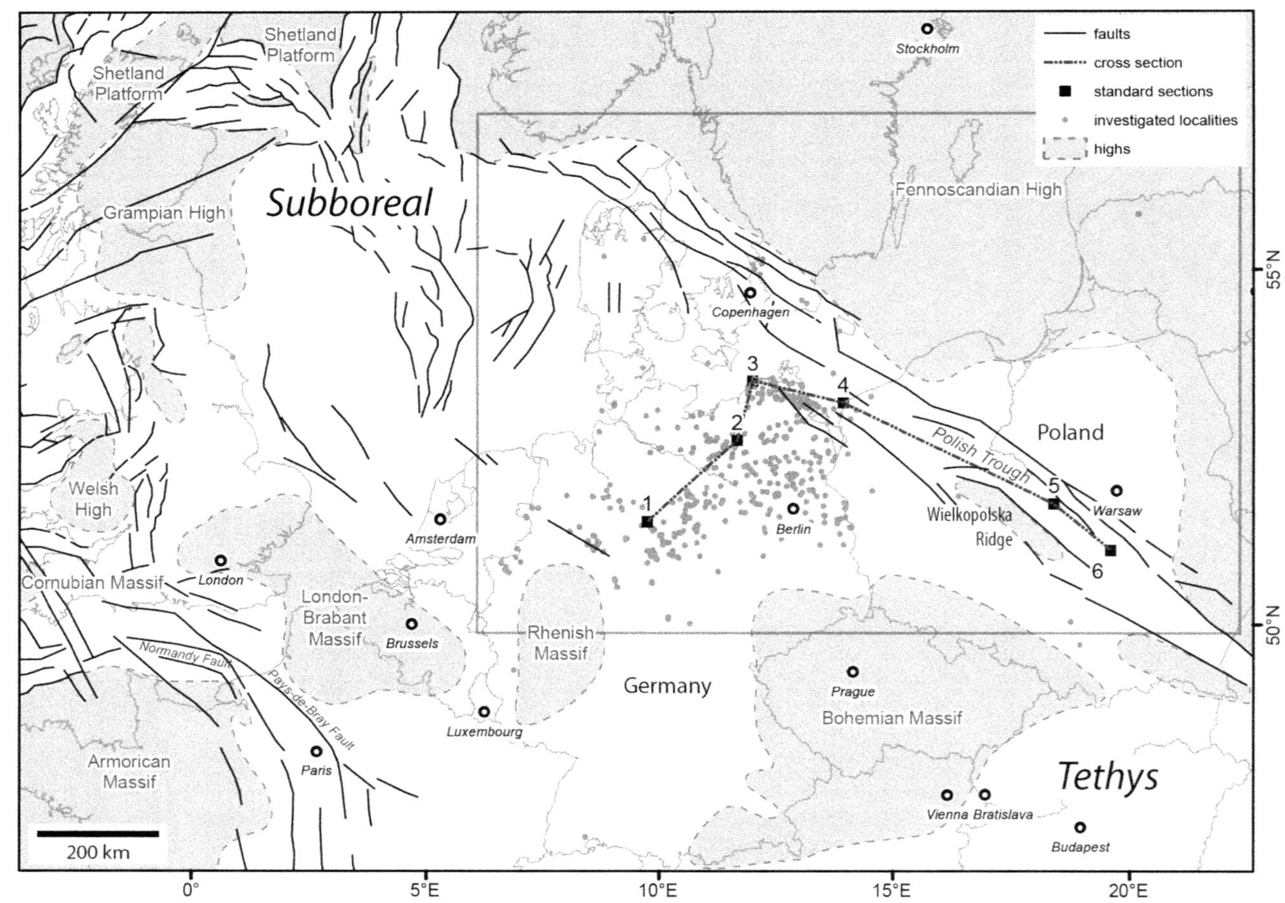

Fig. 2. The Central European Basin with associated sub-basins and surrounding highs during the Early Jurassic (after Ziegler 1990). The inlayed box is the study area. Black lines are the main faults; the dotted black line with numbers is the cross-section (see Fig. 7: 1, Gt Gross-Buchholz 1; 2, Kb KSS 5/66; 3, Kb Barth 10/65; 4, Kamień Pomorski IG1; 5, Kaszewy 1; 6, Przysucha 1); grey dots are wells and outcrops with biofaunal and biostratigraphic information.

transgression of the Ligurian Cycle is placed in the Early Toarcian Tenuicostatum–Serpentinum Chronozone, and regressive systems tract culminates in the Late Aalenian (de Graciansky *et al.* 1998; Jacquin *et al.* 1998 and references therein). The CEB expanded to its largest extent, communicated with Boreal and Tethyan waters through several gates, and experienced many environmental perturbations (Toarcian Oceanic Anoxic Event (T-OAE): Jenkyns 1988; Hesselbo *et al.* 2007; Hesselbo & Pieńkowski 2011; Korte *et al.* 2015; Pieńkowski *et al.* 2016). The following North Sea Cycle is transgressive to its maximum transgression in the Late Jurassic Kimmeridgian, and regressive until the Early Cretaceous (Jacquin *et al.* 1998). For the Early Sinemurian–Early Toarcian, more detailed T–R cycles have been described in the literature and placed into a chronostratigraphic context. A hierarchical division of third-order and, in some sections, fourth-order sea-level cycles was adopted, and proved its applicability in correlation between fully marine and marginal-marine facies (Pieńkowski 2004; Hesselbo & Pieńkowski 2011).

Sequence stratigraphic correlation is made more complex in places where local tectonic movements influenced sedimentation. Commonly, successions above salt structures are reduced and contain more proximal facies, with numerous erosional surfaces associated with hiatuses. Some areas within the CEB area (e.g. the Wielkopolska Ridge) were emerged temporarily and, from time to time, were covered with sediments (Dadlez & Franczyk 1976; Pieńkowski 2004) (Fig. 2). Some older tectonic units, although less marked in Jurassic times, show some distinctiveness in stratigraphical and facies pattern (Pieńkowski 2004). One example is the Gorzów Block (Dadlez 1974; Pieńkowski 2004), which extends from western Poland (Fore-Sudetic Monocline) to eastern Germany (Brandenburg).

Material and methods

Early Jurassic biota recovery resulted in an immense wealth of fossils throughout the Jurassic period. Lower Jurassic sections and their fossil content have a long publication history in Germany, Denmark, Sweden and Poland (e.g. Mestwerdt 1910; Oertel 1922; Lang 1924; Troedsson 1951; Reyment 1959; Kopik 1962, 1964; Schumacher & Sonntag 1964; Dadlez 1969; Karaszewski & Kopik 1970; Hoffmann & Jordan 1982). Biostratigraphy in the Jurassic is commonly achieved by ammonites but, where inadequate facies or preservation potential prohibit ammonite biostratigraphy, ostracods and then lower-resolution palynomorphs and, in parts, foraminifera can be applied. Low-resolution biostratigraphy in marginal-marine deposits makes sequence stratigraphic correlations less certain and precise; however, marginal-marine facies are much more sensitive in registering changes in water depth/shore proximity (Pieńkowski 2004). Stratigraphic subdivision of index fossil-barren deposits has been supported by chemostratigraphy. In some intervals (the Rhaetian–Hettangian transition and the uppermost Pliensbachian–Lower Toarcian section), chemostratigraphy based on carbon isotopes has allowed precise correlation with biostratigraphically and astronomically calibrated western European standards (Hesselbo & Pieńkowski 2011; Pieńkowski *et al.* 2012).

A compilation of published and unpublished (524 wells and 67 outcrops) documentation of Lower Jurassic biofaunal information was summarized in a georeferenced database, which includes the calibration of biofaunal and lithological information to the NW European standard ammonite zones and subzones. Data included 28 ammonite recordings from Denmark, 1903 from Germany, five from The Netherlands, 29 from Poland, 45 from Sweden and 37 from the UK. In total, 2047 ammonite recordings, 2965 foraminifera, 1517 ostracods and 137 echinoderms were collected from the literature and archival materials. A new taxonomic determination has not been made. The dataset was subsequently reviewed following the work of Schlegelmilch (2014) and synonyms were corrected. The updated database was grouped in stratigraphic index fauna (ammonites and ostracods). Abundance, spatial distribution and position of index fauna were displayed for each Lower Jurassic ammonite zone. These index fauna were used to calibrate biofacies data of 220 cored and logged wells to NW European ammonite standard zones. Once all faunal recordings were calibrated to the specific ammonite chronozone or subchronozone, they were subdivided into four groups, defined as 'biofacies': (I) open marine/offshore (polyhaline); (II) marine–brackish (upper mesohaline); (III) brackish (lower mesohaline–upper oligohaline); and (IV) continental (low oligohaline–freshwater). Distinguishing intermediate groups is always approximate: that is, in the sediments deposited in near-coastal lagoons, the recognition of unequivocally freshwater faunas is almost impossible (Hudson *et al.* 1995). Here the group definitions are explained:

(I) Open marine/offshore environments are characterized by stenohaline/polyhaline salinities, and indicated by shaly lithologies yielding cephalopods, echinoderms, marine ostracods, and/or foraminifera and marine phytoplankton, associated with type II marine kerogen. These biofacies occurred mainly in offshore depositional environments.

(II) Marine–brackish (polyhaline–mesohaline) biofacies indicated by sandy to silty lithologies with ostracods, foraminifera, marine molluscs or phytoplankton. Noticeably there are no ammonites, belemnites or echinoderms present. However, the absence of calcium carbonate shells may be also related to their low preservation potential in acidic waters, particularly in the eastern part of CEB which is especially devoid of calcium carbonate (Polish Basin). Marine–brackish biofacies comprise mainly transitional shoreface to offshore depositional environments and prodelta to delta front deposits. Commonly, marine influences are indicated by phytoplankton, such as acritarchs and dinoflagellate cysts (Pieńkowski 2004; Pieńkowski & Waksmundzka 2009), and in places by the dominance of marine kerogen type II, in the Polish Basin often mixed with continental kerogen type III.

(III) Brackish–aquatic (mesohaline–oligohaline) biofacies indicated by sandy to silty lithologies with low-diversity marine phytoplankton and brackish bivalves, and dominantly terrestrial spores and pollen of lowland character, associated with continental kerogen type III. Neither ammonites nor belemnites, echinoderms, ostracods and foraminifera (except for few agglutinated forms) are present. This biofacies characterizes lower deltaic plain, lagoons, sandy barriers, delta front bars and lower distributary channel depositional environments.

(IV) Continental biofacies associated with freshwater low salinities are indicated by sandy to silty lithologies with a dominance of purely terrestrial spores and pollens, xanthophyte alga *Botryococcus*, and conchostracans, and comprise upper deltaic to fluvial plain/lacustrine depositional environments.

The shoreline is inferred to be approximately in the vicinity between brackish to freshwater/continental biofacies (Zimmermann *et al.* 2015). Shifts of shorelines were identified from differences in distances/ranges (intensities) and time (frequencies). Shoreline shifts of high intensities and low frequencies comprise distances of more than 500 km and occurred within periods of more than 10 myr. Less pronounced shifts of shorelines show lower intensities and higher frequencies, and comprise distances of up to 500 km within periods of 1 (sometimes even less) to 6 myr. Consequently, the long-term shifts are considered to be second-order T–R sequences, and the short-term shifts to be third-order T–R sequences. Yet, shorter, fourth-order cycles (within periods of some 10 kyr), were distinguished in the Lower Toarcian section in the Polish Basin (Pieńkowski 2004; Hesselbo & Pieńkowski 2011; Pieńkowski *et al.* 2016). The significance of these cycles (parasequences) has been previously tested and confirmed by chemostratigraphical methods using carbon isotope excursions for correlation (Hesselbo & Pieńkowski 2011). This allowed confident comparison with biostratigraphical standards in the UK, where similar carbon isotope cycles were distinguished, but also on a global astrochronological scale. Current calibration of the orbital Milankovitch cyclicity (Boulila *et al.* 2014) points that the parasequences distinguished in the Polish Basin corresponds approximately with the obliquity cycles (0.03–0.034 myr) although Hesselbo & Pieńkowski (2011) interpreted them as corresponding to short (100 kyr) eccentricity forcing.

Additional to the biofacies-based inferred shoreline shifts, the stratal pattern architectures of more than 250 wells and well logs were analysed. These wells were biostratigraphically constrained to ammonite zones (if possible, to subzones) and interpreted in terms of depositional environments (Fig. 3a). West–east-trending transects connect basinal-marine with marginal-marine and continental sections, and show repeating progradational (regressive) or retrogradational (transgressive) stratal pattern architectures. Biostratigraphic control of the repeated progradational and retrogradational stratal-pattern architecture of sandstones enabled a calibration to well-dated shaly to sandy successions from the basin centre to be made. The combination of biofacies-inferred shoreline shifts (‘time constrained’) and progradational and retrogradational stratal pattern architecture (‘sediment constrained’) result in the identification of sequence stratigraphic bounding surfaces (maximum regressive surface (mrs) and maximum transgressive/flooding surface (mfs)). Maximum retreat of marine biofacies is accompanied by maximum progradation of marginal facies (including the coarsest grain fractions) and marked by the mrs, which corresponds to the sequence boundary (Mitchum *et al.* 1977; Posamentier *et al.* 1988). Accordingly, the maximum advance of marine biofacies is accompanied by the maximum extension of marine shales (often enriched in organic matter) and comprises the mfs (Galloway 1989). The chronostratigraphic position of the mfs determines the nomenclature of the specific sequence. Although sequences are, by definition, unconformity-bound units, the units recognized in this study are essentially T–R cycles (Embry 1993), and unconformities are recorded only to a limited extent, particularly in more distal, basinal settings (Hallam 2001). Both surfaces (mfs and mrs) are also noticeable in petrophysical log responses (gamma-ray values, spontaneous potential); hence, the T–R sequence model of Curray (1964) and Embry (1993) was applied. A hierarchical division of eustatically controlled sea-level cycles proposed by Vail *et al.* (1977), where second-order cycles lasted for 10 ± 80 myr and third-order cycles for 1 ± 10 myr, was adopted. The long-term second-order cycles (in the Boreal realm) were characterized by Zimmermann *et al.* (2015) and, therefore, the main focus of this paper is the correlation of third-order sequences.

Results

Biostratigraphy and biofacies

Transgression in the Planorbis Chronozone (Planorbis Subchronozone). The flooding of the CEB started as early as the Late Triassic Rhaetian (Fischer *et al.* 2012; Barth *et al.* 2018) and continued diachronously eastwards (Barth *et al.* 2018). The maximum flooding is inferred by the maximum eastwards extension and abundance of ammonites of the Early Hettangian Planorbis Subchronozone of the Planorbis Chronozone (74 recordings of Planorbis Chronozone ammonites: Fig. 4a). *Psiloceras psilonotum*, *Psiloceras planorbis*, *Neophyllites antecedens* were documented for the western and central NGB (Hoffmann 1962; Brand & Hoffmann 1963; Sorgenfrei & Buch 1964; Barth *et al.* 2018). The documentation of *Psiloceras psilonotum* and *Psiloceras plicatulum* in the Moseberg section (Barth *et al.* 2018) is an indicator of the open southern marine gate via Thuringia. The transgression reached the northern basin margin as far as northern onshore Denmark (e.g. *Psiloceras planorbis*; well Rodby-1: Sorgenfrei & Buch 1964). Lindström *et al.* (2017) reported also *Psiloceras* cf. *tilmanni* and ?*Nevadaphyllites* sp. from this well, and dated the corresponding horizon to the Tilmanni Chronozone. However, these determinations seem to be disputable considering the poor preservation state of the specimens (Lindström *et al.* 2017, fig. 8). Moreover, only one specimen figured shows a minute part of the suture line, while innermost whorls may well represent other early psiloceratid ammonites. Towards the east, stable marine biofacies terminated in a roughly north–south-trending area around western Mecklenburg-Vorpommern, Brandenburg and Pomerania, as indicated by the occurrence of marine molluscs, foraminifera and marine plankton in the Wernsdorf 2/63 wells (Barth *et al.* 2018). Open-marine conditions terminated westwards of the Lausitz area, where Göthel (1999) described *Pachypteris papillosa*, a mangrove plant, from the clayey/silty deposits of lowermost Hettangian strata from the Seese-Ost 5217/87 well, which is herein interpreted as indicative for brackish–aquatic environments. However, exact biostratigraphic calibration is limited. In central Poland, the occurrence of marine kerogen type II in the Kaszewy 1 borehole (associated with the First Appearance Datum of *Cerebropollenites thiergartii*) and brackish marine bivalves *Cardinia ingelensis* in the nearby Rycerzew borehole (Karaszewski & Kopik 1970) point to the wide extension of a narrow embayment along the Mid-Polish Trough and mark the eastwards extension of the maximum flooding in the Planorbis Subchronozone, most probably in the Psilonotum biohorizon (Barth *et al.* 2018). In a relatively short time, still in the Planorbis Chronozone, a brackish–marine transgression reached the Holy Cross Mountains area (Fig. 4a).

Regression in the Late Planorbis Chronozone (Johnstoni Subchronozone). Marine environments retreated in the Late Planorbis Chronozone, most probably in the Late Johnstoni Subchronozone as evidenced by a decline in the abundance of ammonites (18 ammonite recordings of the Johnstoni Subchronozone: Barth *et al.* 2018). Furthermore, Lower Hettangian sandstones are recorded in several outcrops across the basin (e.g. Sehnde, Kammerbruch and Moseberg in the German part of the CEB; Kvistovta in southern Sweden; and the Zawada ZA-9 and Gliniany Las 2 boreholes in the Holy Cross Mountains region of Poland: e.g. Schumacher & Sonntag 1964; Pieńkowski 1991*a*, 2004; Barth *et al.* 2018). The regression can be recorded in more marginal parts of the basin with the occurrence of prograding fluvio-deltaic facies, usually associated with rootlets and palaeosols (top of the Id or Ie parasequence: Pieńkowski 2004). Biofacies reconstructions were not performed. Assuming the regression took place in the Johnstoni Subchronozone, the exact date would correspond to about 201 Ma (Gradstein *et al.* 2012).

Transgression in the Liasicus Chronozone (Portlocki–Laqueus Subchronozone). A marine transgression then advanced from the west and from the south, as indicated by abundant Liasicus index ammonites in western Lower Saxony (*Alsatites laqueolus* at the Klotzberg outcrop (Jüngst 1928), *Saxoceras* sp. in the Adorf 2 well (Boigk *et al.* 1960), Langelsheim outcrop (Heunisch *et al.* 2010) and in the Schneflingen 4 well (core report)) and Thuringia (*Alsatites laqueolus* at the Moseberg outcrop: Barth *et al.* 2018) (Fig. 4b). The marine (polyhaline) biofacies culminated with a maximum flooding in the Liasicus Chronozone (43 ammonite recordings of the Liasicus Chronozone, and 32 ammonite recordings of the Portlocki and Laqueus subchronozones) and terminated in the area of western Mecklenburg-Vorpommern, as indicated by *Saxoceras* (*Waehneroceras*) *crassicosta* in the abandoned clay-pit Wefensleben (Ermlich 1993) and several occurrences of echinoderm fragments in the Königsberg 7/71 and Zollchow 1/71 wells (core reports). Marine–brackish (mesohaline) environments advanced further eastwards, towards

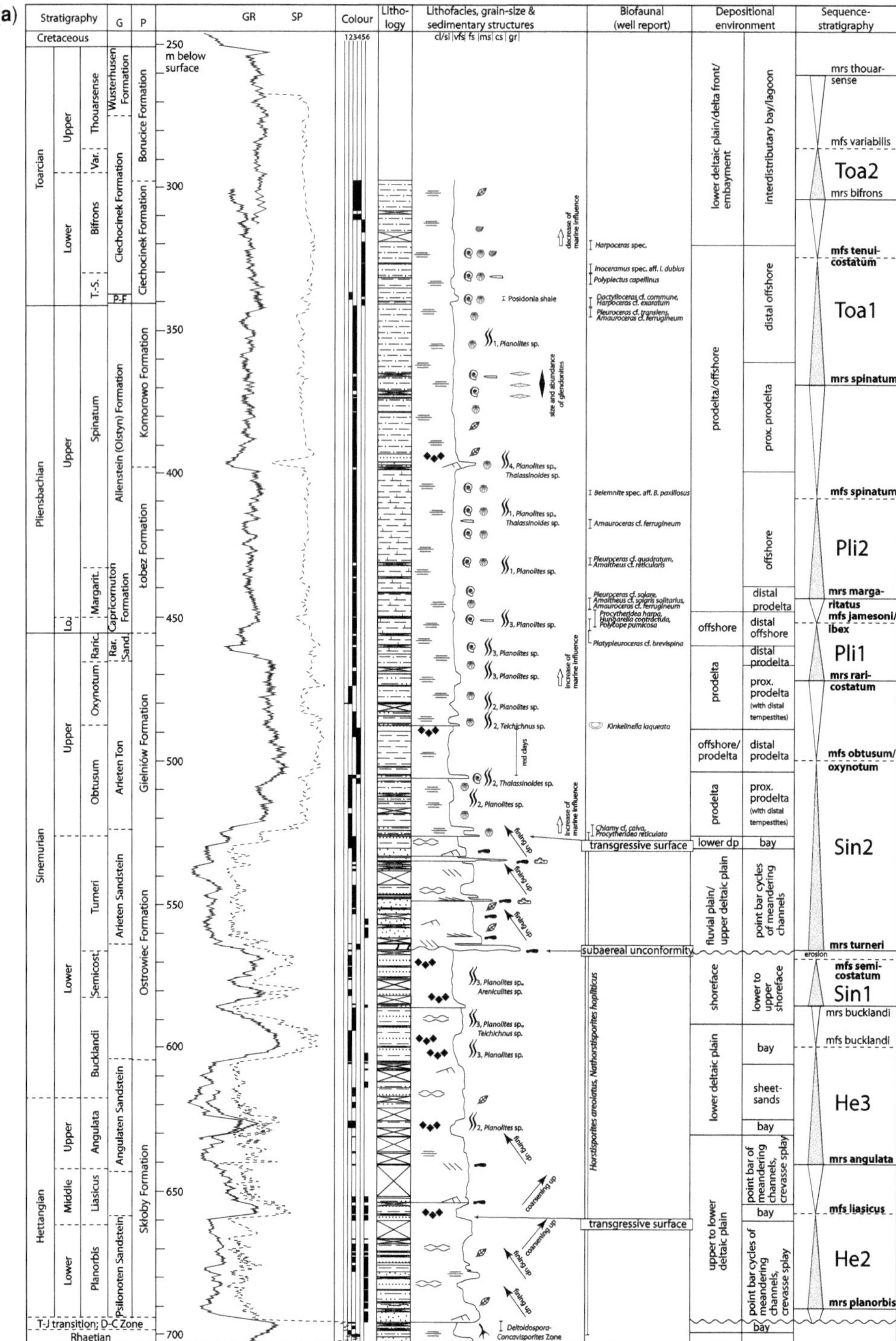

Fig. 3. Detailed log section of the cored Kb Barth 10/65 well in NE Germany. (**a**) Stratigraphic column representing chronostratigraphic stages and substages, NW European ammonite zones plus German (G) and Polish (P) lithostratigraphy. P-F, Posidonienschiefer Formation; GR, gamma ray, SP, spontaneous potential. Modified from Barth *et al.* (2018), bold marked sequence boundaries correspond to spatial biofacies reconstructions (Figs 4–6).

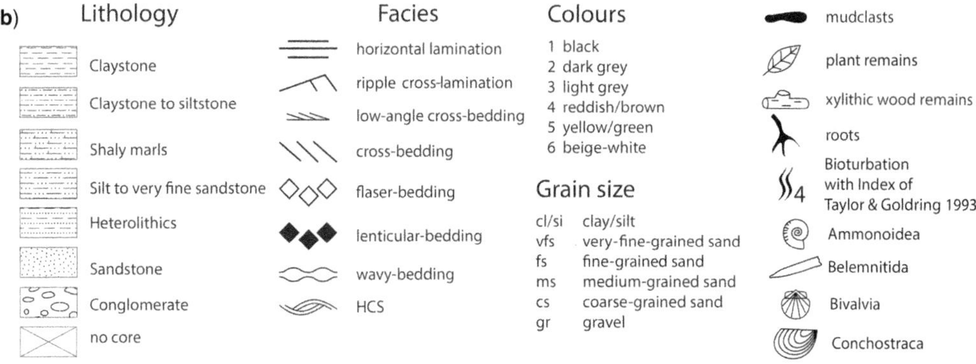

Fig. 3. (*Continued*) (**b**) Legend to (a).

western Pomerania, as indicated by foraminifera fauna and phytoplankton in the Kamień Pomorski IG1 well (Pieńkowski 2004), to the south (Gorzów Wielkopolski IG1) and to the SE as evidenced by the occurrence of marine kerogen type II in central Poland (Kaszewy 1 borehole) and numerous brackish–marine bivalves *Homomya* cf. *venulithus* (well Zawada PA-3), some agglutinated foraminifera fauna and dinoflagellate cysts *Dapcodinium priscum* in the Holy Cross Mountains area (Pieńkowski 2004). These deposits represent a maximum transgressive peak within the Skłoby Formation (Pieńkowski 2004). The Wielkopolska Ridge was covered with siliciclastic deposits probably of continental origin; on the flank of the ridge (Pobiedziska IGH1 borehole) continental or marginal-marine deposits were recorded (Fig. 4b) (Dadlez & Franczyk 1976; Pieńkowski 2004). Assuming the maximum transgression culminated in the Portlocki–Laqueus Subchronozone, the exact date would correspond to about 200.5 Ma (Gradstein *et al.* 2012) and the shoreline advanced for about 500 km (from Kamień Pomorski to the Holy Cross Mountains area).

Regression in the Angulata Chronozone (Complanata–Depressa Subchronozone). Marine biofacies in the Late Hettangian Angulata Chronozone (Complanata–Depressa Subchronozone) show a strong retreat towards the south and SW of associated marine environments, as indicated by the lack of ammonite recordings of the Complanata–Depressa Subchronozone (71 ammonite recordings of the Angulata Chronozone: Fig. 4c) in the central and eastern part of the CEB. *Schlotheimia complanata* was documented by Brand & Hoffmann (1963) in the Weser-Ems area. Although *Schlotheimia angulata* and *S. angulosa* are widespread and abundant in the CEB, index ammonites of the Complanata–Depressa Subchronozone are very rare and limit the marine biofacies to the basin centre; consequently, maximum regression is most likely to have occurred in the Complanata–Depressa Subchronozone. Marine–brackish biofaunal environments shifted contemporaneously towards the south and SW, and only rare occurrence of ostracods (*Haeldia aspinata*) indicate marine–brackish environments. Marine phytoplankton in cored well KSS 5/66 (Barth *et al.* 2018) indicates brackish environments. Further to the SE, in the Holy Cross Mountains area, lagoonal claystones and mudstones (containing siderite bands) developed (the Przysucha Ore-bearing Formation). Lagoons were shallow, which is evidenced by rootlet horizons (Pieńkowski 2004). On the other hand, barrier–lagoon facies still contain brackish fauna, including Cardiniidae (bivalvia) and marine plankton. Assuming the regression took place in the Complanata–Depressa Subchronozone, the exact dating would correspond to about 199.5 Ma (Gradstein *et al.* 2012) and the marine basin retreated for about 300 km (from Kamień Pomorski to KSS 5).

Transgression in the Semicostatum Chronozone (Lyra–Resupinatum Subchronozone). Following the maximum regression in the Late Hettangian Angulata Chronozone, a short-term ingression with the maximum flooding in the early Bucklandi Chronozone was documented in certain sections in the NGB (Fig. 3; Barth *et al.* 2018). This mfs was followed by a regression in the Bucklandi Chronozone with the maximum regression in the Late Bucklandi Chronozone (Fig. 3; Barth *et al.* 2018). However, as this sequence is hardly recognizable in the NGB and absent in the Polish Basin, it is not further described in this paper. Open-marine environments advanced towards the east in the Semicostatum Chronozone from the south, as indicated by *Arnioceras oppeli*, *A. semicostatum*, and *Euagassiceras resupinatum* in Thuringia (Ernst 1986), and from the west, as indicated by *Paracaloceras gmuendense*, *Agassiceras*

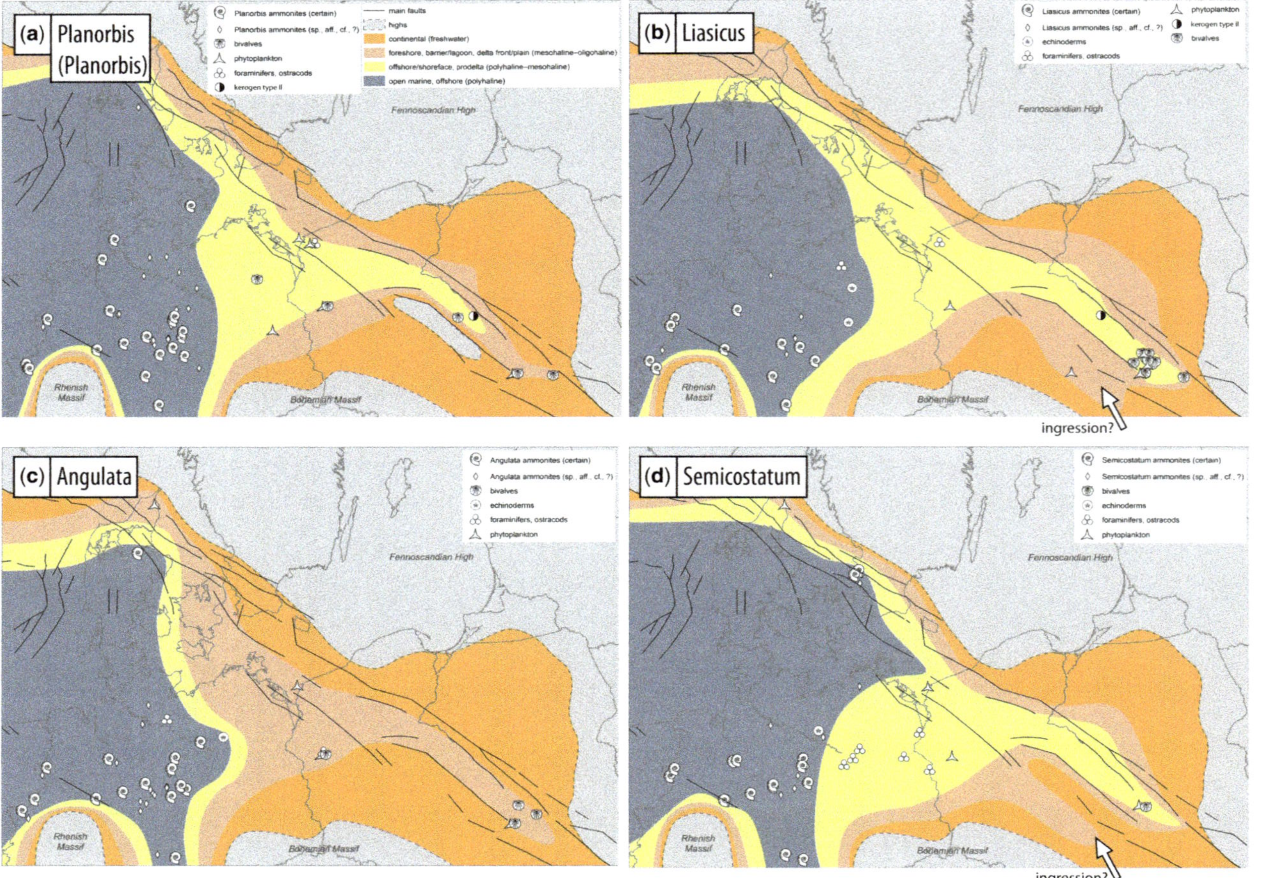

Fig. 4. Biofacies reconstructions for the Early Hettangian–Early Sinemurian. Continental highs are after Ziegler (1990). Biofacies reconstructions for: (**a**) the maximum flooding of the initial transgression in the Early Hettangian Planorbis Chronozone; (**b**) the maximum flooding of the transgression in the Middle Hettangian Liasicus Chronozone; (**c**) the maximum regression of the regression in the Late Hettangian Angulata Chronozone; and (**d**) the maximum flooding of the initial second-order transgression in the Early Sinemurian Semicostatum Chronozone.

scipionanum and *Euagassiceras resupinatum* in the Ems-Weser area (Brand & Hoffmann 1963) and in western Lower Saxony (*A. semicostatum*: Boigk *et al.* 1960) (83 ammonite recordings of the Semicostatum Chronozone: Fig. 4d). Marine environments reached the northern margins of the CEB as far as western Skåne, as indicated by *Agassiceras scipionanum*, *Cymbites striaries* and *Euagassiceras resupinatum* in the Oregården and Döshult quarries (Reyment 1959). The ammonite occurrence of *Arnioceras* sp. in central North Mecklenburg-Vorpommern in the Rostock 17/65 well (Lehmkuhl 1968; Haupt 1967 in Rusbült 1990) indicates the easternmost extension of marine (polyhaline) environments. Consequently, maximum flooding occurred in the Semicostatum Chronozone. Marine–brackish (mesohaline) environments extended further to the cored Barth 10/65 well, where more diversified ichnofauna appears (Fig. 3a), and towards the east as evidenced by marine phytoplankton in the cored Gartz 1/65 well (Barth *et al.* 2018). A brackish–marine transgression advanced far to the SE, along the Mid-Polish Trough, reaching the Holy Cross Mountains area, where brackish–marine bivalves and marine phytoplankton were recorded (Pieńkowski 2004). A brackish–marine embayment also extended to the Gorzów Wielkopolski (the presence of marine phytoplankton). Assuming the transgression took place in the Semicostatum Subchronozone, the exact date would correspond to about 197 Ma (Gradstein *et al.* 2012) and the brackish basin advanced for about 500 km (from Kamień Pomorski to the Holy Cross Mountains area).

Regression in the Turneri Chronozone (Birchi Subchronozone). A regression in the Turneri Chronozone is indicated as marine conditions prevailed only in the western Ems-Weser area, where Brand & Hoffmann (1963) reported *Microderoceras birchi*, *Promicroceras capricornoides* (nine ammonite recordings of the Turneri Chronozone, Fig. 5a) and *Microderoceras birchi.* In addition, *Promicroceras capricornoides* was documented in western Germany (Schubert 2005). In Thuringia, Ernst (1986) reported *Asteroceras brocki* and *Microderoceras birchi*, illustrating that the southern marine gate was still active. Hence, ammonite dispersal in the Turneri Chronozone suggests a significant retreat of marine environments towards the basin centre. However, a report of the ammonite *Microderoceras* sp. (cf. *birchi*) (Bölau 1959) in western Skåne could suggest persistent marine conditions up to the northern basin margins. However, in the light of a basinwide ammonite retreat, a re-evaluation of the assignment of this poorly documented ammonite is needed. Marine–brackish environments retreated from the east towards western Mecklenburg-Vorpommern, as indicated by the occurrence of the ostracods *Procytheridea triebeli*, *P. betzi*, *P. laqueata*, *P. glabellata*, and the foraminifera *Dentalina terquemi*, *D. matutina*, *D. varians and D. primaeva* reported from the Groß Welzin 101/61 well (core report). Brackish environments must have retreated in the same way, although there is little supporting data. Continental environments prevailed throughout the entire Polish Basin. The Wielkopolska Ridge emerged again as evidenced by partial erosion and the removal of older Jurassic deposits to the flanks of the ridge. Assuming the regression took place around the Birchi Subchronozone, the exact date would correspond to about 196 Ma (Gradstein *et al.* 2012) and the shoreline retreated for about 300–400 km (from Gorzów Wielkopolski to KSS).

Transgression in the Obtusum–Oxynotum Chronozone (Denotatus–Simpsoni Subchronozone). Marine environments expanded widely in the CEB during the Obtusum–Oxynotum Chronozone (93 ammonite recordings of the Obtusum and Oxynotum chronozones: Fig. 5b). Flooding occurred from the south via Thuringia as evidenced by the presence of ammonites such as *Asteroceras obtusum* and *Promicroceras planicosta* in the Großer Seeberg area (Ernst 1986), and from the west as evidenced by abundant ammonite records from the western part of the NGB (Schneflingen 2 well core report: Schubert 2005). This event culminated in maximum flooding in the Obtusum–Oxynotum Chronozone. Towards the north, marine environments advanced as far as the northern basin margins as evidenced by ammonite records of *Aegoceras planicosta* in onshore Denmark (Vinding 1 well: see Sorgenfrei & Buch 1964) and *Arnioceras falcaries* in the Pankarp-Strovelstorp well in the area of Skåne (Bölau 1959). Towards the east, marine environments are evidenced by ammonite recordings of *Asteroceras obtusus* from the Waddekath 7 well (Lehmkuhl 1968). In the Pomerania area, *Oxynoticeras* sp. was recorded from Swinhöft/Międzyzdroje (database entry from the Federal Institute for Geosciences and Natural Resources (BGR)), while the endemic Sinemurian bivalve *Tancredia erdmanni* occurs further to the SE, in the Mechowo IG-1 borehole (Kopik 1964; Pieńkowski 2004), marking the easternmost extension of stable stenohaline marine environments. Marine–brackish (mesohaline) environments advanced further to the SE as evidenced by the appearance of marine kerogen type II in central Poland (Kaszewy 1) and the presence of marine phytoplankton and brackish–marine bivalves in the Holy Cross Mountains area (Pieńkowski 2004). The Wielkopolska Ridge was drowned (Pobiedziska borehole) with marginal-marine sediments being recorded on the flanks (Pieńkowski 2004). Assuming the transgression took place around the boundary

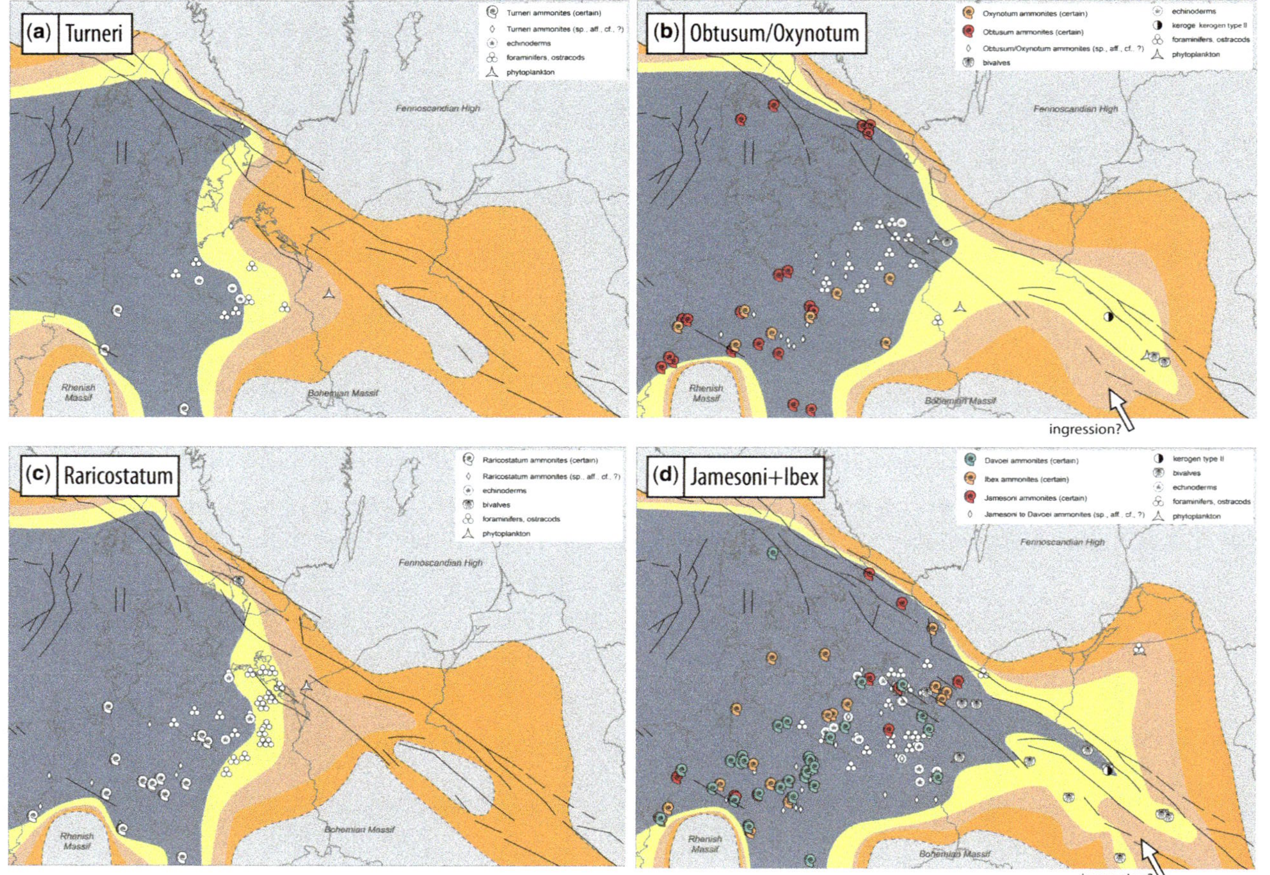

Fig. 5. Biofacies reconstructions for the Early Sinemurian–Early Pliensbachian. Continental highs are after Ziegler (1990). Biofacies reconstructions for: (**a**) the maximum regression of the second-order regression in the late Early Sinemurian Turneri Chronozone; (**b**) the maximum flooding of the transgression in the Late Sinemurian Obtusum–Oxynotum Chronozone; (**c**) the maximum regression of the regression in the Late Sinemurian Raricostatum Chronozone; and (**d**) the maximum flooding of the second-order transgression in the Early Pliensbachian Jamesoni–Ibex Chronozone.

of the Obtusum–Oxynotum Chronozone, the exact date would correspond to about 194 Ma (Gradstein *et al.* 2012) and the fully marine shoreline advanced for about 800 km (from KSS 5 to Kaszewy).

Regression in the Raricostatum Chronozone (Densinodulum–Aplanatum Subchronozone). Marine environments retreated in the Raricostatum Chronozone and prevailed in the basin centre and the southern areas of Germany (61 ammonite recordings of the Raricostatum Chronozone: Fig. 5c). The southern gate via Thuringia remained under stenohaline marine conditions, as indicated by the presence of *Echioceras raricostatum* ammonites in Großer Seeberg (Ernst 1986). The western gate is also considered as still stenohaline marine, as indicated by abundant ammonite recordings of *Echioceras raricostatum* in western Lower Saxony (Brand & Hoffmann 1963; Fischer *et al.* 1986). The retreat of marine environments was governed by a strong retreat of marine–brackish and brackish–aquatic environments from the north towards the south. In the east, after relatively widespread nearshore to brackish–marine sedimentation during most of the Raricostatum age, brackish environments retreated in the latest Raricostatum time from the Polish Basin towards the Pomerania area, and this is indicated by marine phytoplankton only in the Kamień Pomorski IG1 well (Pieńkowski 2004). The Polish Basin was otherwise characterized by continental environments, as numerous levels with rootlets/palaeosols occur and no indications for any marine influence were recorded (Pieńkowski 2004). As such, the Wielkopolska Ridge was probably emerged.

Assuming the regression took place around the Densinodulum–Aplanatum Subchronozone, the exact dating would correspond to about 191.8 Ma (Gradstein *et al.* 2012) and the shoreline retreated for about 600 km (from the Holy Cross Mountains area to Neubrandenburg).

Transgression in the Jamesoni–Ibex Chronozone (Polymorphus, Brevispina, Jamesoni, Masseanum, Valdani and Luridum subchronozones). Marine environments expanded in the Jamesoni–Ibex Chronozone throughout the entire CEB (339 ammonite recordings of the Jamesoni and Ibex chronozones: Fig. 5d). Abundant ammonite recordings in the western CEB, as well as in Thuringia and southern Lower Saxony, prove full marine flooding from the west and the south. Marine environments advanced towards the north up to the northern basin margins as evidenced by *Uptonia jamesoni* ammonite recording from, for example, Kurremölla in the Skåne area (Reyment 1959), and *Beaniceras centaurus* plus *Phricodoceras taylori* from Bornholm (Donovan & Surlyk 2003). Towards the east, stenohaline environments are indicated by *Platypleuroceras aureum*, *Liparoceras* sp., *Tropidoceras* sp., *Peripleuroceras* sp., *Tragophylloceras* sp., *Protogrammoceras* sp. and *Beaniceras* sp. (cf. *senile*) in the Kamień Pomorski IG1 well (Dadlez 1972; Dadlez & Kopik 1972), and *Polymorphites* sp. juv. plus *Acanthopleuroceras maugenesti* in the Mechowo IG-1 well (Dadlez 1972). The most diversified ammonite fauna occurred in the Ibex Zone (Valdani Subzone). It is not clear if stenohaline conditions reached central Poland, where marine kerogen (type II) was found in Lower Pliensbachian shales in the Kaszewy 1 borehole. However, the occurrence of *Cardinia phillea* d'Orbigny in the Parkoszowice 58BN borehole (Kopik 1998; Pieńkowski 2004), along with foraminifera and marine phytoplankton, tends to support that polyhaline–mesohaline conditions occurred across the whole of central Poland. Marginal-marine sedimentation extended over the Wielkopolska Ridge (Dadlez & Franczyk 1976), proven by the occurrence of marine–brackish marine deposits in the Pobiedziska IGH1 borehole (Pieńkowski 2004), and to the NE reaching the Mazury region and Lithuania (Šimkevičius 1998; Pieńkowski 2004). Although ammonites were recorded abundantly in the CEB, the exact timing of the maximum flooding cannot be given precisely; hence, biofacies reconstruction comprises both Jamesoni and Ibex index ammonites. Moreover, in the German and Danish part of the CEB, basinal polyhaline to marine conditions persisted during the Davoei Chronozone (Fig. 5d). Finds of diversified bivalves in the SE of the basin, in the Holy Cross Mountains area (*Pleuromya forchhammeri* Lund, *Nuculana (Dactryomya) zieteni* Brauns and *Pronoella* cf. *elongata* Cox: Kopik 1962, 1964; Pieńkowski 2004) in strata correlated with the Davoei Zone, suggest that marine to brackish–marine conditions also persisted in the Mid-Polish Trough.

Assuming maximum flooding took place around the boundary of the Jamesoni–Ibex Chronozone, the exact dating would correspond to about 189.4 Ma (Gradstein *et al.* 2012) and the shoreline advanced for about 800 km (from Neubrandenburg area to the Holy Cross Mountains area).

Regression in the Margaritatus Chronozone (Stokesi Subchronozone). At the beginning of the Late Pliensbachian, marine environments retreated slightly in the earliest Margaritatus Chronozone (39 ammonite recordings of the Stokesi Subchronozone: Fig. 6a), although only in the northern and the eastern parts of the CEB. In several unpublished well reports in NE Germany, the Margaritatus Chronozone is described as fully marine as evidenced by the recording of all Margaritatus Subchronozone index ammonites: for example, *Amaltheus laevis* in Leversen 1 (core report) and Regina 2 (core report), and *Amaltheus depressus* in Scholen 1 (core report)

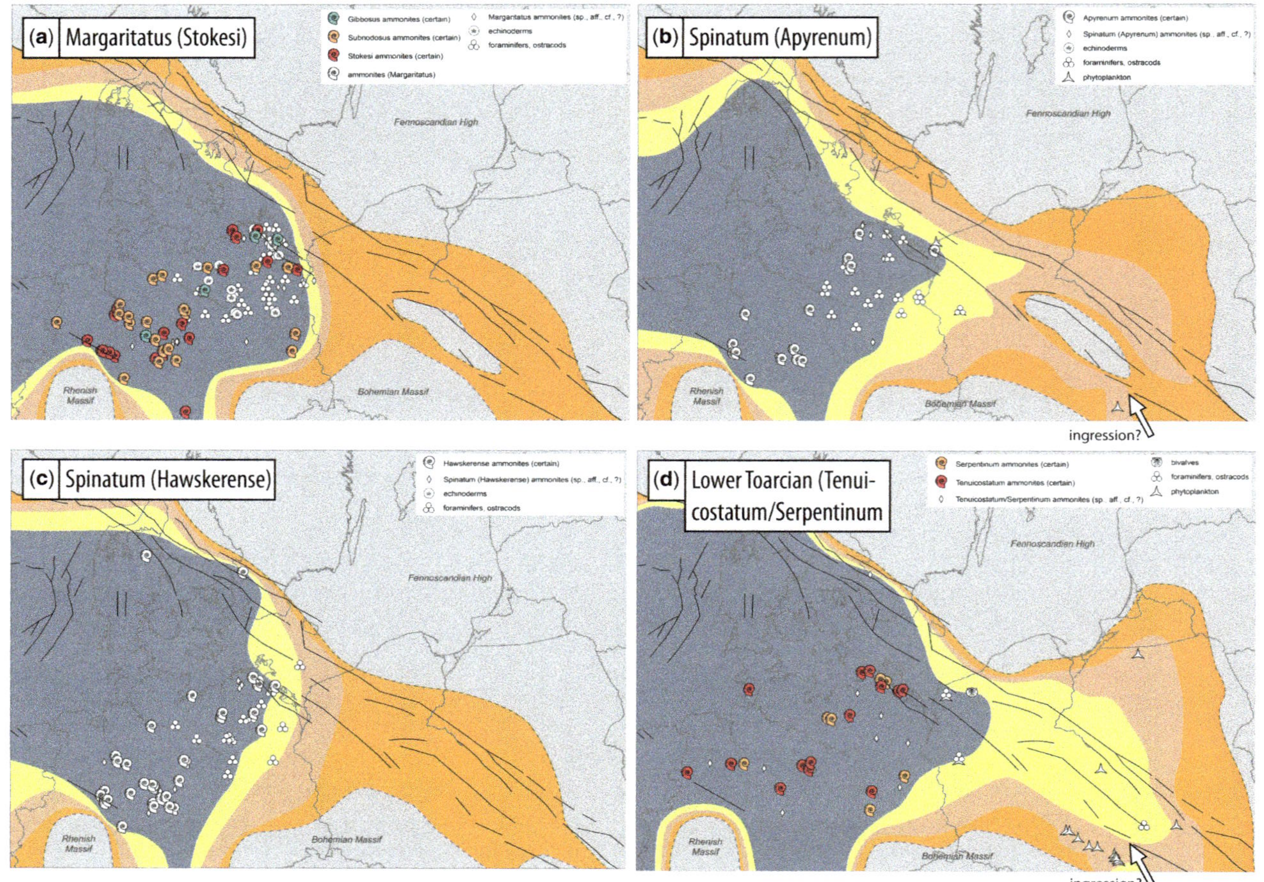

Fig. 6. Biofacies reconstructions for the early Late Pliensbachian–Early Toarcian. Continental highs are after Ziegler (1990). Biofacies reconstructions for: (**a**) the maximum regression of the second-order regression in the early Late Pliensbachian Margaritatus Chronozone (Stokesi Subchronozone); (**b**) the maximum flooding of the transgression in the Late Pliensbachian Spinatum Chronozone (Apyrenum Subchronozone); (**c**) the maximum regression of the regression in the latest Pliensbachian Spinatum Chronozone (latest Hawskerense Subchronozone); and (**d**) the maximum flooding of the first-order transgression in the Early Toarcian Tenuicostatum–Serpentinum Chronozone.

and Wisloh 1 (core report), *Amaltheus margaritatus* in the Dobbertin outcrop (Ernst 1992), *Amaltheus stokesi* in cored well KSS 5/66 (Lehmkuhl 1968), *Amaltheus subnodosus* in cored well Wolgast 1/65 (Lehmkuhl 1968) and *Amaltheus gibbosus* in cored well Wolgast 1/65 (Pschenitzka 1967). The northernmost recording of *Amaltheus margaritatus* in the Danish well Gassum 1 (Sorgenfrei & Buch 1964) indicates marine environments in the northern part during the Margaritatus Chronozone. Nielsen (2003) indicated the erosional surface and sequence boundary (SB-14) at the base of the Margaritatus Chronozone, both in the Danish Basin and in the Fennoscandian Border Zone. Also in Poland, the lower part of the Upper Pliensbachian is associated with a strong regional erosion and ensuing fluvial sedimentation (Pieńkowski 2004). However, the strong regressive period in Poland was followed by an equally strong and widespread marine transgression. In the sediments correlated with the upper parts of the Margaritatus Zone (most probably Subnodosus Subzone), ammonites have been identified in Pomerania *Amaltheus* (?*Proamaltheus*) cf. *wertheri* Lange (Kopik & Marcinkiewicz 1997) and in central Poland *Tragophylloceras* cf. *loscombi* (Sowerby) (Pieńkowski 2014). Marine–brackish environments are inferred from foraminifera and marine phytoplankton recorded in the Kamień Pomorski IG1 well (Pieńkowski 2004), indicating the westwards retreat of marine biofacies. Interestingly, the German part of the CEB was weakly affected by the retreat of marine biofacies in the early Margaritatus Chronozone (Fig. 6a). Assuming the regression took place in the Stokesi Subchronozone, the exact dating would correspond to about 187 Ma (Gradstein *et al.* 2012) and the shoreline retreated in the early Margaritatus Chronozone for about 500 km (from the Holy Cross Mountains area to Kamień Pomorski).

Transgression in the Spinatum Chronozone (Apyrenum Subchronozone). The advance of marine environments was recorded for the Apyrenum Subchronozone (56 ammonite recordings of the Apyrenum Subchronozone: Fig. 6b). Marine environments expanded mainly northwards as indicated by the identification of the ammonite *Pleuroceras spinatum* in central onshore Denmark (Gassum-1: see Sorgenfrei & Buch 1964) and western Schonen (Vilhelmsfält; *Onychoceras laeve* and *Pleuroceras spinatum*: Bölau 1959). In the eastern part of the CEB, the advance of marine environments is documented by the presence of ammonite fauna *Pleuroceras quadratus*, *Pl. solare* and *Amaltheus laevigatus* in northwesternmost Pomerania in well Wolin IG-1 (Kopik 1975). Pomerania and the SE part of Brandenburg were characterized by brackish–marine environments. Marine phytoplankton was recorded in the Kamień Pomorski IG1 and Gorzów Wielkopolski IG1 wells (Pieńkowski 2004). It is possible that the Apyrenum transgression affected southern Poland (Pieńkowski 2004; Gedl 2007) (Fig. 6b). Salt movements occurred in central Poland (the Kaszewy 1 borehole), where strata assigned to that age are strongly syndepositionally reduced and represented by a thin (8 m-thick) interval containing fluvial/deltaic facies (Pieńkowski 2014). The Wielkopolska Ridge was likely to have been elevated at that time due to the same tectonic/salt movements. Assuming the transgression took place around the Subnodosus–Apyrenum Subchronozone, the exact dating would correspond to about 184.2 Ma (Gradstein *et al.* 2012) and the shoreline advanced for about 200 km (from Wolin to Gorzów Wielkopolski).

Regression in the Spinatum Chronozone (latest Hawskerense Subchronozone). Marine environments retreated in the latest Spinatum Chronozone, most probably in the latest Hawskerense Subchronozone, as indicated by a lack of ammonite occurrences in the northwesternmost part of the Polish Basin (61 ammonite recordings of the Hawskerense Subchronozone: Fig. 6c). Marine–brackish environments established in the easternmost parts of Mecklenburg-Vorpommern are indicated by foraminifera genera occurrences in the Baltic well K-5 (*Lenticulina (Astacolus)* and sp., *Marginulina prima* (core report)) and in East Brandenburg (e.g. *Ammodiscus incertus*, *Marginulina prima* and *Eoguttulina liassica* (core report of well Brohm 3/66)). In cored wells KSS 5/66 and Barth 10/65, glendonites (calcite pseudomorphs after ikaite) were recorded (Fig. 3a) and constrained to the upper Spinatum Chronozone by ammonite recordings, indicating cooling in a shallow-marine environment. Brackish–aquatic biofacies were documented in the NW part of the Polish Basin, approximately reaching a limit marked by a line linking the Kamień Pomorski and Gorzów Wielkopolski boreholes; the rest of the Polish Basin was characterized by continental sedimentation with coal-bearing deposits (Pieńkowski 2004; Hesselbo & Pieńkowski 2011; Pieńkowski *et al.* 2016) (Fig. 6c). Assuming the regression took place in the latest Hawskerense Subchronozone, the exact dating would correspond to about 183 Ma (Gradstein *et al.* 2012) and the shoreline retreated for about 100 km to the west (from Gorzów to eastern Mecklenburg-Vorpommern).

Transgression in the Tenuicostatum–Sepentinum Chronozone. Marine environments became widespread in the CEB and advanced far towards the northern basin margins as evidenced by ammonite records of *Dactylioceras* cf. *semicelatum* and *Dactylioceras* cf. *tenuicostatum* in western Skåne

(Reyment 1959; Hoffmann & Martin 1960) (86 ammonite recordings of the Tenuicostatum and Serpentinum chronozones) (Fig. 6d). Brackish–marine environments advanced to the NE as evidenced by marine trace fossils in Bornholm (Koppelhus & Nielsen 1994). In NE Germany, marine environments advanced towards the NE, and abundant Lower Toarcian index ammonites (*Lobolytoceras siemensi*, *Dactylioceras semicelatum* and *Eleganticeras pseudoelegans*) were recorded by Ernst (1992) in the southern Usedom area. At that time, marine and brackish–marine conditions advanced widely across the Polish Basin, making this transgression the most widespread event in the Early Jurassic, although salinity was generally lower than that recorded in the Mid-Polish Trough in Early Pliensbachian times. Lowered salinity during the Early Toarcian in the Polish Basin is attributed to a more pronounced hydraulic cycle and runoff (Pieńkowski 2004; Hesselbo & Pieńkowski 2011; Pieńkowski *et al.* 2016). Diversified foraminifera assemblages, dinoflagellate cysts and marine bivalve *Meleagrinella substriata* (Münster) are reported from Pomerania, pointing to marine (polyhaline) to marine–brackish conditions; while further to the SE (Silesia) and NE (Mazury region), marine phytoplankton points to likely mesohaline conditions over large areas of the Polish Basin (Barski & Leonowicz 2002; Pieńkowski 2004). Numerous finds of marine phytoplankton in the southernmost occurrences of the Lower Toarcian Ciechocinek Formation, in the Silesian-Cracow Upland, suggest an intermittent connection with the Tethys Ocean in the southern part of the Polish Basin (Pieńkowski 2004). Transgression took place around the Pliensbachian–Toarcian boundary at approximately 183 Ma (Gradstein *et al.* 2012; Sell *et al.* 2014). The mfs occurred between the Tenuicostatum and Serpentinum chronozones, at about 182–183 Ma (Gradstein *et al.* 2012; Sell *et al.* 2014). The shoreline advanced for about 600 km in 0.5–1 Ma from westernmost Pomerania to the western part of the Holy Cross Mountains area.

Sequence stratigraphy

The general Mesozoic sea-level curve (first order) shows rather high amplitudes and very low frequencies, was interpreted from independent proxies (e.g. seafloor spreading), and is mainly based on sequence stratigraphic methods (van der Meer *et al.* 2017). Second-order sequences are characterized by a high amplitude and a low frequency of shoreline shift, and are described for the NGB by Zimmermann *et al.* (2015). This paper focuses on the description of the high-amplitude and high-frequency third-order sequences.

Third-order sequence He1 (Upper Rhaetian–Planorbis Chronozone)

Following the prominent end-Triassic regression that culminated, in the NGB, with the formation of a large fluvial-dominated deltaic system (Barth *et al.* 2018), increasing marinity and retrogradational stacking patterns were recognized in the NGB in the Triassic–Jurassic transition (*Deltoidosporites–Concavisporites* Zone: Barth *et al.* 2018), representing the transgressive phase of the sequence He1. The transgression culminated in the widespread flooding of the CEB in the Early Planorbis Subchronozone. Maximum flooding (mfs He1) occurred in the Planorbis Subchronozone (Psilonotum–Plicatulum Biohorizon: Barth *et al.* 2018), and resulted in the deposition of dark clays with ammonites and molluscs in the western part of the NGB. Towards the east, these clays thin out but can be still be found in a relatively narrow embayment along the Mid-Polish Trough, reaching central Poland (Kaszewy 1 borehole) where marine kerogen type II occurs (Fig. 4a). Following the mfs He1, progradational stratal architectures of mostly deltaic sandstones are governed by shrinkage of the marine basin. This regressive phase culminated in the formation of deltaic systems in the central to eastern part of the NGB (Figs 4 & 7). A progradation of coastal–deltaic sandstones in the Johnstoni Subchronozone was recognized in outcrops near Sehnde (Lower Saxony) and near Kammerbruch (Thuringia), and is interpreted as the result of a regressive shoreline (Barth *et al.* 2018). This progradational event can be observed in the Mid-Polish Trough (parasequences Id and Ie: see Pieńkowski 2004).

Third-order sequence He2 (Planorbis–Angulata Chronozone)

Following the mrs He 1, retrogradational stratal patterns and sections with a marked transgressive surfaces (Barth *et al.* 2018) are interpreted as the transgressive phase of the sequence He 2. The transgressive phase is less pronounced in the eastern part of the NGB but marine phytoplankton in cored well KSS 5/55 were recognized in heterolithic deposits and are tentatively constrained to the Middle Hettangian Liasicus Chronozone (Fig. 4b). The mfs He 2 is located within dark clayey to silty deposits, and is also recognizable by very high gamma-ray values in well logs (Fig. 7). In western Mecklenburg-Vorpommern, *Saxoceras* (*Waehneroceras*) *crassicosta* (in the abandoned clay-pit Wefensleben: Ermlich 1993) and several occurrences of echinoderm fragments in dark clayey deposits (in the wells Königsberg 7/71 and Zollchow 1/71 (core reports)) represent deposits of the maximum flooding and are

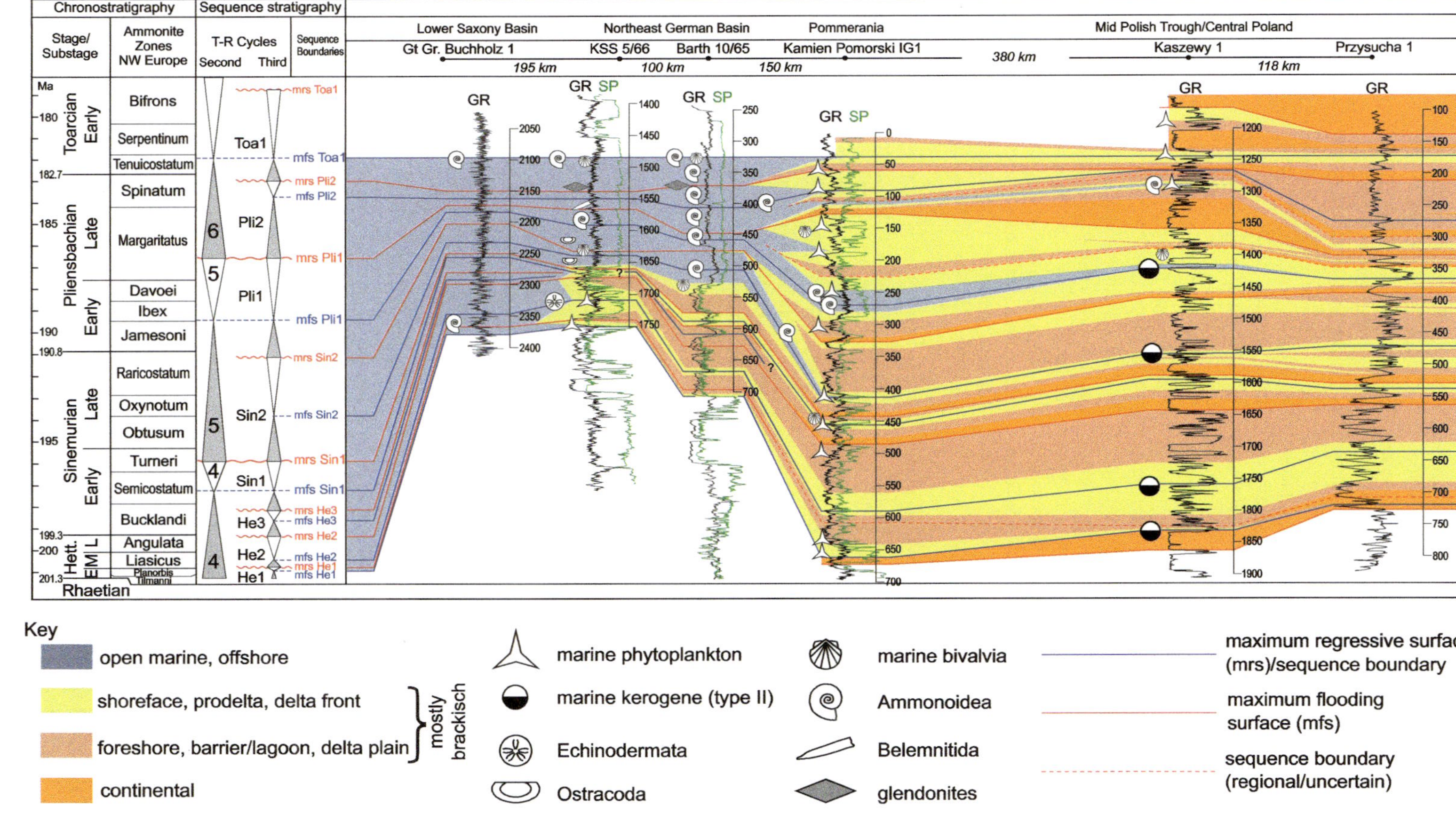

Fig. 7. Sequence stratigraphic cross-section from western Germany across Pomerania to SE Poland showing the facies interfingering between fully marine and terrestrial facies units. Biofauna are adopted from various well documentation. For the location see Figure 2. Each depth scale in metres below surface.

most likely to comprise the maximum flooding surface. Pieńkowski (1991*a*, 2004) documented a widespread flooding event associated with the sedimentation of bioturbated heteroliths that developed in a shallow siliciclastic shelf (offshore shoreface) environment in the Mid-Polish Trough. This flooding event (identified with a maximum flooding surface) is correlated with the Liasicus Chronozone and documented, for example, in the Mechowo IG-1 and the Gorzów Wielkopolski IG-1 boreholes in Pomerania, in the Kaszewy 1 borehole in central Poland where an organic-rich intercalation with marine kerogen type II was found, and, finally, in numerous boreholes in the Holy Cross Mountains area with associated marine phytoplankton and foraminifera. A similar cyclic, transgressive trend culminating in a mfs located between the lower and upper Helsingborg Member was observed in Skåne (Pieńkowski 1991*a*, *b*). Following the maximum flooding, progradational stratal architectures of mostly fluvial deltaic sandstones prograded from the east and NE towards the west in the NGB (Fig. 3a). These sandstones were assigned to the Angulata and Bucklandi chronozones, as overlying claystones in the area of eastern Mecklenburg-Vorpommern (well Lychen 2/59) yield the ammonite *Asaphoceras* cf. *apenninicum* (Lehmkuhl 1968). Maximum progradation of these sandstones culminated in the western Prignitz area, with the mrs He 2 most likely assigned to the Late Hettangian Angulata Chronozone. In Poland, progradation manifested in the development of shallow lagoonal/deltaic facies with siderite bands (Przysucha Ore-bearing Formation), assigned to the Upper Hettangian (Pieńkowski 2004). In Skåne, maximum progradation was correlated with the uppermost Helsingborg Member, which can be assigned to the uppermost Hettangian (Pieńkowski 1991*a*, *b*).

Third-order sequence Sin1 (Angulata–Turneri Chronozone)

Following the mrs He 2, a retrogradational stratal pattern architecture, accompanied with a fining in grain size and increasing marinity, implies a deepening-upwards trend and represents a transgressive phase (Fig. 7). In cored well Barth 10/65, the retrogradational part is further characterized by an increasing marine bioturbation and culminates in dark-grey lenticular heteroliths with abundant *Arenicolites* ichnofauna (Fig. 3a). Heteroliths with *Arenicolites* ichnofauna were interpreted as lower–upper shoreface deposits and comprise the mfs Sin1 in the Semicostatum Zone. Similar heterolithic, bioturbated deposits occur on top of a retrogradational transgressive systems tract in the Mid-Polish Trough (Pieńkowski 2004). Transgressive, marine claystones and mudstones (Döshult Member) in Skåne, with ammonites belonging to the Semicostatum Chronozone (Reyment 1959), represent a coeval counterpart of the transgressive deposits in the Polish and German Basin (Pieńkowski 1991*a*, *b*). The mfs is biostratigraphically constrained in the western part of the NGB and is recognized basinwide utilizing logged wells with high gamma-ray values of broad sections in the western part and spikes in the eastern part. Following the mfs Sin1, progradational stratal architectures of deltaic to fluvial environments are recognized moving progressively from the east to the NE towards the East Prignitz in the NGB (Fig. 7). In the cored Barth 10/65 well, a subaerial unconformity was inferred from cross-bedded conglomeratic sandstones laying on an erosional surface, containing abundant wood debris and showing slight mottling. This section is overlain by a marine transgressive surface containing *Procytheridea reticulata* (Obtusum Chronozone), which suggests that the underlying progradational unit belongs to the Turneri Chronozone. In Poland, progradation occurred in the middle part of the Ostrowiec Formation (upper section of a highstand systems tract in the Lower Sinemurian sequence II), culminating in fluvial sandstones occurring throughout the entire Polish Basin. These fluvial deposits constitute the basal section of the next sequence III (Pieńkowski 2004). The maximum regression of the sequence Sin1 is inferred to be the Turneri Chronozone.

Third-order sequence Sin2 (Turneri–Raricostatum Chronozone)

Following the mrs Sin1, retrogradational stacking patterns with an increasing marine influence characterize the transgressive phase of the sequence Sin2, and culminate in the basinwide deposition of dark laminated clays and silts in the Upper Sinemurian (Figs 4 & 7). The ostracod taxa *Procytheridea reticulata*, together with the mollusc taxa *Chlamys* cf. *calva*, occurred directly above a marine transgressive surface in the cored well Barth 10/65 and enabled the assignment of the transgressive phase to the Obtusum Chronozone. The maximum flooding surface mfs Sin2 is located within laminated clays containing abundant ammonites belonging to the Obtusum and Oxynotum (Fig. 5b). An ostracod assemblage with *Kinkelinella multiforata*, *K. triebeli* and *K. reticulata* occurred in dark-grey, clayey to silty deposits in cored well KSS 5/66 (Zimmermann *et al.* 2015). In the area of Rostock (NE Germany), unfossiliferous laminated red clays are fringed by grey, ammonite- and belemnite-bearing laminated clays (Fig. 3a), and extend towards the Skåne area (Frandsen & Surlyk 2003). Following the maximum flooding, a progradational stratal architecture of mostly fluvial deltaic sandstones results from the regressive phase in the Raricostatum Chronozone (Fig. 7). These sandstones terminated in the southern Usedom to East Prignitz area. The mrs Sin2 is calibrated to the Upper Sinemurian Raricostatum Chronozone (possibly also the Taylori Chronozone of earliest Pliensbachian age), based on the overlying marine clays. These marine clays are widely distributed and are traceable towards the Pomerania area, enabling the basinwide calibration of the underlying progradational unit to the Raricostatum Chronozone.

Third-order sequence Pli1 (Raricostatum–Margaritatus Chronozone)

Following the mrs Sin2, in the latest Raricostatum Chronozone, the marine transgressive surfaces of the transgressive phase of sequence Pli1 with molluscs (cored wells JOmb 1/65 and Gartz 1/65) and ooids (cored wells Saarow-Pieskow 2/60 and Wernsdorf 2/63) is followed by retrogradational stratal-pattern architecture accompanied by an increasing abundance of marine macro- and microfauna. Consequently, marine environments expanded over the basin margins, which led to the deposition of widely distributed marine dark clays with abundant Lower Pliensbachian index fauna (e.g. the 'Capriconu-Ton' in Germany). These clays, mudstones and lenticular heteroliths extend far eastwards, and are traceable towards Pomerania (Łobez Formation) and into central Poland, where dark-grey marine claystones (with marine kerogen type II) were described in the Kaszewy borehole (Fig. 7). The mfs Pli1 is located within this clayey/muddy succession and assigned to the Jamesoni–Ibex Chronozone. Coeval to the open-marine shales of the Łobez Formation, the Gielniów Formation is built of heterolithic/sandy deposits and represents more proximal offshore/shoreface/coastal environments. Following the maximum flooding surface, progradational stratal architectures in western Poland and the eastern NGB illustrate a shoreline retreat, starting from the Davoei Chronozone and culminating at the beginning of the Late Pliensbachian Margaritatus Chronozone. The spatial extent of deltaic and fluvial deposits terminate in the eastern part of the NGB and grade westwards to silty lithologies of distal delta front to lower shoreface environments. The progradation culminated with maximum regression in the Stokesi Subchronozone, and is constrained by overlying clayey deposits containing Subnodosus–Apyrenum Subchronozone index ammonites in cored well JOmb 1/65. The mrs is also recognized in the Danish Basin (SB 14: see Nielsen 2003) and in other European sub-basins (de Graciansky *et al.* 1998; Hesselbo & Jenkyns 1998; Pieńkowski 2004). In Poland, this regression is associated with a sequence boundary marked by widespread fluvial sedimentation preceded by pronounced erosion.

Third-order sequence Pli2 (Margaritatus–Spinatum Chronozone)

Following the mrs Pli1, subordinate retrogradational stratal-pattern architectures indicate the transgressive phase of the sequence Pli2. The transgressive phase culminated in maximum flooding in the Apyrenum Subchronozone of the Spinatum Chronozone with the deposition of 3.5 m of dark-grey to grey laminated clays and silts in NE Germany containing Apyrenum Subchronozone ammonites (cored well JOmb 1/65). These fine-grained deposits, although much thicker (*c.* 85 m), continue to NW Poland (Wolin IG-1 borehole), where ammonites representing the Apyrenum Subchronozone were found (Kopik 1975) (Fig. 7). Such significant regional changes in thicknesses, occurring in the Pliensbachian of Mecklenburg-Vorpommern to Pomerania, are explained by local tectonics. Pliensbachian transgressions were more marked, and deposited thicker (*c.* 265 m in the Wolin IG-1 borehole) sequences of mudstones and sandstones in narrow, syndepositional tectonic graben – such as the Wolin-Recław Zone (Dadlez 1969). The following regressive phase is characterized by progradational stratal architectures of coarsening- and thickening-upwards successions of approximately 60 m-thick fluvio-deltaic sandstones in the Usedom and East Prignitz areas (Zimmermann *et al.* 2015). In the cored well KSS 5/66 (Zimmermann *et al.* 2015) and Barth 10/65, numerous glendonites (calcite pseudomorphs after ikaite) occur in ammonite-bearing laminated clays and silts, and are dated to the Upper Pliensbachian Spinatum Chronozone, most probably the Hawskerense Subchronozone (Fig. 3a). The observation of glendonites lends support to the possibility of a Late Pliensbachian cooling event and polar glaciation at that time (Rogov & Zakharov 2010; Dera *et al.* 2011; Korte & Hesselbo 2011; Bougeault *et al.* 2017; Ruhl *et al.* 2017), although Teichert & Luppold (2013) indicated other possible causes for the origin of the glendonites.

Third-order sequence (Spinatum–Bifrons Chronozone) and fourth-order cycles in Tenuicostatum–Serpentinum Chronozone

Following the maximum regression in the Late Spinatum Chronozone, the most profound environmental change in the Jurassic Period took place during the Early Toarcian Oceanic Anoxic Event (T-OAE)

(Jenkyns 1988). In this event, marine depositional settings across Europe show evidence for widespread anoxia in the form of a coeval black shale. High seawater palaeotemperatures are inferred from isotopic and elemental anomalies (Jenkyns *et al.* 2002; Cohen *et al.* 2007; McArthur *et al.* 2008). A negative carbon-isotope excursion with an average $\delta^{13}C$ amplitude of approximately −7‰ (VPDB – Vienna Pee Dee Belemnite Standard), considered the largest such anomaly in the whole Phanerozoic, has been described from marine and terrestrial materials (Jenkyns 1988; Hesselbo *et al.* 2007). Uppermost Pliensbachian–Lower Toarcian strata are assigned to a single unconformity-bounded depositional sequence (Pieńkowski 2004; Hesselbo & Pieńkowski 2011; Pieńkowski *et al.* 2016). The oldest deposits of the sequence, partly of latest Pliensbachian age resting on a regional erosional surface (sequence boundary), comprise alluvial and deltaic sediments assigned to a number of lithostratigraphic units of local significance. This sequence boundary represents the uppermost of a set of three Upper Pliensbachian sequence boundaries, all marking significant sea-level falls. These are attributed to recurrent Late Pliensbachian glaciations followed by rapid global warming and a coeval transgression at the beginning of Toarcian (Price 1999; Rogov & Zakharov 2010; Suan *et al.* 2010; Korte & Hesselbo 2011; Pieńkowski *et al.* 2016). This warming is associated with a 'precursor' carbon isotope excursion (Hesselbo & Pieńkowski 2011; Pieńkowski *et al.* 2016). The base of the Ciechocinek Formation is a transgressive surface of regional extent, recognized in a number of boreholes across the Polish Basin. In Germany, this transgressive surface is indicated by an increasing marine influence and a significant grain-size decrease to silt and clay (cored wells Zechlinerhütte 1A/65 (Zimmermann *et al.* 2015) and Neubrandenburg 1/82). Following this retrogradational stacking pattern, dark-grey to black laminated shales (with Lower Toarcian index ammonites (Zechlinerhütte 1A/65: western Skåne, Usedom area) and Posidonia shale index molluscs (Barth 10/65 and Neubrandenburg 1/82)) were deposited during the maximum flooding phase. This phase occurred in the upper part of the Tenuicostatum Chronozone, which can be inferred from the reported ammonites *Dactylioceras* cf. *semicelatum* and *Dactylioceras* cf. *tenuicostatum* (Reyment 1959; Hoffmann & Martin 1960), and *Lobolytoceras siemensi*, *Dactylioceras semicelatum* and *Eleganticeras pseudoelegans* (Ernst 1986), plus the upper range of the dinoflagellate cyst *Luehndea spinosa* (Barski & Leonowicz 2002; Pieńkowski 2004). This first-order flooding event had a global impact (Haq *et al.* 1987; Hallam 1988; de Graciansky *et al.* 1998; Jacquin *et al.* 1998; Pieńkowski 2004), and was accompanied by fundamental environmental and climatic perturbations (Rogov & Zakharov 2010; Suan *et al.* 2010; Korte & Hesselbo 2011; van de Schootbrugge *et al.* 2013; Pieńkowski *et al.* 2016). Superimposed on these second- and third-order sea-level cycles are subordinate fourth-order cycles. As such, the Lower Toarcian sequence (Ciechocinek Formation) is subdivided into five or up to seven parasequences: VIIIa, b, c, d, e, f and g, representing 5–20 m-thick shallowing-upwards progradational cycles (Pieńkowski 2004; Pieńkowski *et al.* 2016). These parasequences were correlated with carbon isotope cycles, standard biostratigraphical divisions and published orbital cycles (Hesselbo & Pieńkowski 2011; Pieńkowski *et al.* 2016). In most cases, but not all, the steps in the carbon-isotope curves (correlated with the obliquity orbital cycles [after Boulila *et al.* 2014] or short eccentricity cycles [Huang & Hesselbo 2014]) occur at flooding surfaces at the top of each parasequence. Enhanced sediment delivery, related to increased rainfall and temperature, distort, in higher parasequences, the match between flooding surfaces and carbon isotope steps, particularly at points of sediment delivery, such as river mouths. Nevertheless, stepwise progradation and marked shallowing of the basin is approximately coeval with the onset of the T-OAE and a major stepwise disturbance in the carbon cycle, observed both in oceanic and atmospheric systems (Hesselbo & Pieńkowski 2011; Pieńkowski *et al.* 2016). The basinwide extent of all these parasequences indicates the dominance of regional-scale (Pieńkowski 2004), rather than local-scale, processes. This suggests that autocyclic processes, such as delta lobe switching, are of secondary importance. Exact correlation of this internal depositional architecture, observed in the Polish Basin, could be tentatively compared with the cyclicity observed in the German Posidonia shales within the Tenuicostatum–Serpentinum age (Röhl *et al.* 2001), provided that fourth-order, orbital cycles are distinguished in Germany.

Discussion

Controls and hierarchies of sequences: discrepancies between sea-level records across the CEB

The sequences described above are based on two approaches: first, the time-constrained (ammonite chronozone and subchronozone) advance and retreat of fully marine environments; and, secondly, the material-dependent (e.g. lithology, grain-size, biofacies) stratal-pattern architecture of sandy sedimentary deposits. The advance and retreat of marine environments are accompanied by retrogradational and progradational stratal architectures, and the inferred shift of shorelines vary within distances of 100–500 km in time steps of 1–3 myr. The spatial and temporal fluctuations of the shoreline correspond to third-order sequences (Hallam 2001). This is further constrained by dark clay-rich deposits and marine faunas associated with fully marine environments. The identification of sequence stratigraphic bounding surfaces with ammonite subchronozone biostratigraphic resolution in the CEB was achieved for four sequences (He1, Pli1, Pli2 and Toa1). The Hettangian and Toarcian transgressions (possibly, also, the Early Pliensbachian event) were rapid: for example, the Hettangian transgression, in particular, is associated with a shoreline shift of more than 100 km in approximately 0.1 myr (Hesselbo *et al.* 2004). In addition, the Early Toarcian is associated with rapid flooding, with a shoreline advance of several hundreds of kilometres (towards the east) within less than 1.5 myr.

Special emphasis should be put on the sequence stratigraphic correlation between marine and non-marine (marginal-marine and continental) facies. The sequence stratigraphic correlation should be aimed at determining the correlative significances and ages of key surfaces and derived sea-level curves. Such correlations in the Early Jurassic section were performed in the Polish Basin (Pieńkowski 1991*a*, 2004) and Danish–Swedish Basin (Pieńkowski 1991*a*, *b*; Surlyk *et al.* 1995; Nielsen 2003), and herein extended by correlation with the largely marine German Basin. Proper hierarchy of local cycles, parasequences, systems tracts and sequences can be achieved only by careful regional analysis, supported by chemostratigraphic, palynological and clay mineral analyses. The stratigraphic significance of the transgressive surfaces is enhanced if they are coupled with their continental correlatives. Particularly important are correlatives of transgressive surfaces, which are documented in the Danish Basin and Polish Basin (Surlyk *et al.* 1995; Pieńkowski 2004).

It is important to discuss similarities and discrepancies between this work and other published schemes in adjacent regions. Of note is the similarity between sea-level fluctuations observed in the German–Polish Basin and in England (Dorset and Yorkshire) (Hesselbo & Jenkyns 1998). In addition, similarities occur between inferred sea-level fluctuations in the German–Polish Basin (Pieńkowski 2004) and the sequence order (Ligurian Cycle – Boreal realm) from France (de Graciansky *et al.* 1998) and Denmark (Michelsen *et al.* 2003; Nielsen 2003). However, comparison to published Tethyan cycles illustrates major discrepancies in the Pliensbachian (e.g. Jacquin *et al.* 1998; Gradstein *et al.* 2012). Recently, Bougeault *et al.* (2017) presented climatic and palaeoceanographical changes during the Pliensbachian, inferred from clay mineralogy and stable isotope analysis from the NW Tethyan domain, specifically the Paris Basin. Their conclusions are based on clay mineral characteristics, caused by emersion or flooding of the emerged areas, chiefly reworking sediments from the London–Brabant Massif. These emersion–flooding events are compared to the Tethyan second-order cycles and show (contrary to the Boreal cyclicity and some of the work presented in this paper) that a major sea-level fall occurred between the Jamesoni and Ibex chronozones, while the Davoei and Margaritatus (Stockesi) chronozones are characterized by sea-level highstands. All these sea-level changes are postulated to be linked to glacioeustasy (Bougeault *et al.* 2017). The current authors believe that glacioeustasy is a major forcing mechanism behind frequent and pronounced sea-level fluctuations, particularly in Late Pliensbachian times (see subsection 'Significance of climate controlled 4th order sequences'). However, this study shows reversed sea-level trends, which seems to be unconceivable in the scale of one European Basin. Bougeault *et al.* (2017) based their conclusions on the evolution of one particular provenance area – the London–Brabant Massif. However, the interpretation of eustatic changes from the rock record is a complex and controversial topic. The dilemma as to which changes can be ascribed to global forcing, and which only to regional-scale processes, is not resolved (for a discussion see Cloetingh *et al.* 1985; Miall 1986, 1992; Zimmermann *et al.* 2015; van der Meer *et al.* 2017). Provisionally, a method to reconcile results from England, Denmark and the CEB to the Paris Basin is to invoke local tectonics (differential emersion and submersion) of the London–Brabant Massif. This is similar to the process observed at the Wielkopolska Ridge (Dadlez & Franczyk 1976). The London–Brabant Massif (even assuming its denudation) stands out in its evolution against other elevated 'island' areas – our reconstruction point that the Rhenish and Bohemian Massifs were little affected by the sea-level changes (Figs 5–7), similarly to the Grampian and Armorican massifs. Also, the role of hydraulic cycle, runoff and enhanced delivery of sediments can explain some discrepancies between the 'eustatic' trend and the observed shifts in the proximity of the coastline, such as retrogradational–progradational events (i.e. the Toarcian example described above: see Pieńkowski 2004; Hesselbo & Pieńkowski 2011). Reconciliation of these particular discrepancies between sea-level cycles, particularly in Pliensbachian, will need further studies. It is possible that the resolution of currently available stratigraphic tools is not sufficient in terms of the identification of high-frequency hypothetical glacioeustatic sea-level changes in Pliensbachian times.

Significance of climate-controlled fourth-order sequences

Although marine currents and circulation systems may explain the advance and retreat of marine environments, they would not explain the coeval retrogradational and progradational stratal architectures. Sediment delivery plays a significant role, particularly during periods of enhanced hydraulic cycle, weathering and erosion, commonly associated with warmer and more humid conditions. Concerning second- and third-order cycles, two mechanisms played the most prominent roles: that is, pronounced rifting/volcanic processes (possibly active during the Early Hettangian, which is in the initial phase of Pangaea break-up and activity of the Central Atlantic Magmatic Province: see van der Meer *et al.* 2017); and/or glaciations (indicated as major agents of frequent sea-level changes in the Late Pliensbachian). The frequently occurring shoreline advances of several hundreds of kilometres (mfs He1 and mfs Toa1) within periods of some 100 kyr to about 1.5 myr accounts for strong climatic turnovers (i.e. building and melting of polar ice caps, respectively). A Jurassic polar glaciation, probably of recurrent character (e.g. Price 1999; Suan *et al.* 2010; Korte & Hesselbo 2011; Bougeault *et al.* 2017), was most likely to have been responsible for the sea-level changes in Late Pliensbachian times, and may explain the high-amplitude and high-frequency shoreline shifts. Rapid and pronounced sea-level fluctuations recorded in the Polish part of CEB can also be best explained by recurrent glaciations and deglaciations. Abundant occurrences of Upper Pliensbachian (Spinatum Chronozone) glendonites (Lower Saxony Basin (Teichert & Luppold 2013), KSS 5/66 (Zimmermann *et al.* 2015) and Barth 10/65 (this study)) tend to support the presence of glaciations in the polar regions (Rogov & Zakharov 2010). Teichert & Luppold (2013) raised the question of whether, instead of near-freezing temperatures (their estimation was based on $\delta^{18}O$ pointed to a bottom-water temperature of about 10°C), methanotrophic sulphate reduction was instead the key geochemical process in the near-seafloor sediments leading to ikaite precipitation at the palaeosulphate–methane interface. It should be pointed out that the palaeotemperatures measured solely from $\delta^{18}O$ are less reliable than those based on clumped isotopes (Eiler 2007). Here the authors, based on the current scientific quorum and the wide distribution of glendonites in Upper Pliensbachian strata, tend to support a glacial hypothesis.

It is recognized that more subtle climatic changes related to Milankovitch cyclicity could also influence sedimentation patterns, as has been indicated for Hettangian and Pliensbachian strata in the UK (Ruhl *et al.* 2010, 2017), and for the Lower Toarcian strata in UK, France and Poland (Kemp *et al.* 2005; Suan *et al.* 2010; Hesselbo & Pieńkowski 2011; Boulila *et al.* 2014; Huang & Hesselbo 2014; Pieńkowski *et al.* 2016). However, the distinction of a fourth or even subordinate cycles needs to have continuous profiles, with the application of multidisciplinary, high-resolution bio-, chemo- and astrochronological methods.

Opening of marine gates in SE Poland and the connection to the Tethys

The relative abundance of marine fauna in the Holy Cross Mountains (SE Poland) is governed by sea-level highstands (e.g. during maximum flooding in the Liasicus, Jamesoni–Ibex and Tenuicostatum–Serpentinum chronozones) and may be linked to an opening of a SE connection to the Tethys. Some authors (Karaszewski & Kopik 1970; Pieńkowski 2004) suggested the possibility of an early southern connection with the Tethys in the Hettangian. Additionally, the presence of dinoflagellate cyst *Liasidium variabile* (or its close precursor) in Hettangian deposits correlated with the Liasicus mfs in Gorzów Wielkopolski (P. Gedl pers. comm.; Pieńkowski 2004) might yield some circumstantial evidence for intermittent connections between the Tethys and the Mid-Polish Trough. *Liasidium variabile* has also been identified in slightly younger, Upper Hettangian strata of SW Germany (Brenner 1986). However, this palynomorph is not found in the North Sea Basin earlier than the Upper Sinemurian (Poulsen & Riding 2003). The early appearance of this palynomorph in the eastern part of CEB might suggest that *Liasidium variabile* (or its close precursor) migrated from the Tethys to the west via an intermittent seaway along the Mid-Polish Trough. This would explain the diachronous appearance of this palynomorph between the Mid-Polish Trough and the western CEB and North Sea Basin. However, all direct connections between the Mid-Polish Trough and the Tethys via the Silesian–Moravian or the East Carpathian gate remain hypothetical due to absent, or thin, Lower Jurassic epicontinental deposits in southern and SE Poland (Moryc 2005). This absence can be explained by later erosion, but also bio- and lithofacies become more proximal towards the SE, which suggests the closure of the Polish Basin in that direction (Pieńkowski 2004).

Application to reservoir identification, hydrocarbon exploration and CO_2 sequestration

The results of this study provide essential information for reservoir studies in Germany and Poland as the identified third-order transgressions are traceable basinwide and extend far towards the east in the Polish Basin. The spatial extent of the reconstructed biofacies provides a basic framework for further reservoir exploration. Depending on the location of the target area, specific flooding surfaces can be determined either using marine microfossils (ostracods or foraminifera in the case of marine–brackish biofacies) or palynology (marine phytoplankton in the case of brackish–aquatic biofacies). This information can be directly utilized during drilling and exploration campaigns. Results also demonstrate that third-order regressions are associated with the progradation of sand-bearing units of up to 80 m net sand thickness (Pieńkowski 2004; Zimmermann *et al.* 2015). Sandy successions are associated with shoreface, delta front, delta plain and fluvial plain depositional environments (Pieńkowski 2004; Zimmermann *et al.* 2015; Zimmermann *et al.* 2017; Barth *et al.* 2018); hence, potential target areas are limited to marine–brackish, brackish–aquatic and continental biofacies. Marine–brackish biofacies of regressive phases comprise shoreface sands, delta front bars and terminal distributary channel deposits, which are characterized by fine-grained, well to moderately well-sorted sandstones with porosity and permeability values of 15–25% and 200–700 mD, respectively (Wolfgramm *et al.* 2008), although shaly intercalations may occur and may buffer vertical flow in places (Olariu *et al.* 2012). Brackish–aquatic biofacies comprise deltaic distributary channel, crevasse splay, levee and sheet sands deposits of medium-grained and moderately sorted composition with generally high porosity and permeability values of 20–30% and 500–1500 mD, respectively (Wolfgramm *et al.* 2008). Continental biofacies comprise mostly fluvial-associated environments (Pieńkowski 2004). High-resolution sequence stratigraphy, verified by chemostratigraphical correlation based on $\delta^{13}C$ isotope correlation, proved that one of key sealing units, the Lower Toarcian Ciechocinek Formation composed of clayey-muddy rocks with sandstone intercalation, is consistent in terms of its lithology and spatial extent over the larger part of the Polish Basin and provides an excellent, regional seal (Hesselbo & Pieńkowski 2011). The origin of such favourable properties is attributed to climatic conditions (super greenhouse conditions on land) during the Toarcian Oceanic Anoxic Event (T-OAE) and a sea-level high stand at that time. Moreover, similar conditions occurred four times during the Early–Middle Jurassic times, creating additional regionally correlatable seals in large parts of the Polish Mesozoic epicontinental basin. In terms of potential CO_2 sequestration, this can enable tiered sequestration methods, using several reservoir–seal pairs, and thus allowing much more voluminous and effective storage of CO_2 and methane (for economic purposes) in selected structures (Pieńkowski 2015). Similar properties can be expected in eastern Germany.

Additionally, the results presented herein can help to better understand hydrocarbon resource potential. Besides well-known black shales of Lower Toarcian (Song *et al.* 2015), some results show that source rocks may also occur in other parts of the Lower Jurassic sections in the CEB. Marine flooding, enhanced productivity and anoxia resulting in marine kerogen-rich shales (Suan *et al.* 2010; van de Schootbrugge *et al.* 2013) were not limited to the West European Basin, but for some brief periods extended further to the centre of the Polish Basin. Such events occurred in the Hettangian (twice – in the Planorbis and Liasicus chronozones), the Late Sinemurian (the Obtusum–Oxynotum Chronozone) and the Early Pliensbachian (Jamesoni–Ibex Chronozone). Conditions occurring in narrow marine embayments in central Poland during rapid sea-level rises were conducive for eutrophication and anoxic episodes, resulting in the formation of black shale intervals with elevated total organic carbon (TOC) and the presence of marine kerogen. The intervals with source rocks occur only in thin (up to 4–7 m-thick) intercalations and further investigation would be needed to confirm if such source rocks attain higher thickness elsewhere. However, the upper part of the Lower Toarcian of the Polish Basin (upper Ciechocinek Formation, coeval to the Posidonia shales) is severely depleted in organic matter due to enhanced, fungal-mediated decomposition of terrestrial organic matter on land during extreme greenhouse conditions coeval with the T-OAE (Pieńkowski *et al.* 2016).

Conclusions

Cross-border facies transects within a Lower Jurassic stratigraphic framework allowed spatial correlation of different environments across the epicontinental German and Polish Basin and for the mapping of various palaeogeographical time slices. The high resolution of Jurassic ammonite biostratigraphy allows, in most cases, good calibration of sequence boundaries in the western (German) part of CEB, while data from the shallower eastern (Polish) part of CEB help in the recognition of the magnitude, lateral extent and frequency of sea-level fluctuations.

Results illustrate that third-order flooding events can be constrained to the NW European ammonite Zone to Subzone level and are traceable basinwide towards the eastern Polish Basin. Maximum flooding occurred in the Early Hettangian Johnstoni Subchronozone, the Middle Hettangian Liasicus Chronozone, the Early Sinemurian Semicostatum Chronozone, the Late Sinemurian

Obtusum–Oxynotum Chronozone, the Early Pliensbachian Jamesoni–Ibex Chronozone, the Late Pliensbachian Subnodosus–Apyrenum Subchronozone and the Early Toarcian Tenuicostatum–Serpentinum Chronozone. Corresponding third-order regressions are accompanied by progradational stratal architectures, and culminated in the Early Hettangian Planorbis Subchronozone, the Late Hettangian Angulata Chronozone, the Early Sinemurian Turneri Chronozone, the Late Sinemurian Raricostatum Chronozone, the Late Pliensbachian Stokesi Subchronozone and the Hawskerense Subchronozone. The regressive phases triggered the progradation of fluvial to deltaic depositional environments, and comprise thick sandy potential reservoirs for hydrocarbon, geothermal and sequestration usage.

Changes in ocean and atmosphere dynamics, associated with rapid shifts from icehouse to greenhouse conditions, may have repeatedly contributed to the spread of anoxic bottom-waters and biotic crises (Suan *et al.* 2010; van de Schootbrugge *et al.* 2013). Our interpretation indicated that several maximum flooding surfaces are associated with a wider distribution of organic-rich shales which also have potential as oil-rich source rocks.

This publication results from the GeoPolD II project (BGR project no. A-0602017.A), which is part of the TUNB project: 'Potenziale des unterirdischen Speicher- und Wirtschaftsraumes im Norddeutschen Becken'. Grzegorz Pieńkowski is grateful for financial support from the resources of the Polish National Science Centre, granted on the basis of decision No. DEC-2012/06/M/ST10/00478; this is a contribution to the IGCP project 632 'Continental Crises of the Jurassic'. We also thank Gabriela von Goerne for important corrections and useful comments. Karsten Obst (State Agency for Environment, Nature Conservation and Geology (LUNG)) and Michael Göthel (State Agency for Mining, Geology and Resources (LBGR)) are acknowledged for access to core and well data. Reviewers are acknowledged for helpful reviews that improved the manuscript significantly. The editor is acknowledged for the editing process.

References

ANDSBJERG, J. & DYBKJÆR, K. 2003. Sequence stratigraphy of the Jurassic of the Danish Central Graben. *In*: INESON, J.R. & SURLYK, F. (eds) *The Jurassic of Denmark and Greenland. Geological Survey of Denmark and Greenland Bulletin*, **1**, 265–300.

BACHMANN, G.H., VOIGT, T., BAYER, U., EYNATTEN, V., LEGLER, B. & LITTKE, R. 2008. Depositional history and sedimentary cycles in the Central European Basin System. *In*: LITTKE, R., BAYER, U., GAJEWSKI, D. & NELSKAMP, S. (eds) *Dynamics of Complex Sedimentary Basins. The Example of the Central European Basin System.* Springer, Berlin, 157–172.

BACHMANN, G.H., GELUK, M. *ET AL.* 2010. Triassic. *In*: DOORNENBAL, H. & STEVENSON, A. (eds) *Petroleum Geological Atlas of the Southern Permian Basin Area.* European Association of Geoscientists and Engineers (EAGE), Houten, The Netherlands, 149–173.

BARSKI, M. & LEONOWICZ, P. 2002. Dinoflagellates of Lower Jurassic outcrops at Kozłowice and Boroszów (southern Poland). *Przegląd Geologiczny*, **50**, 411–414.

BARTH, G., FRANZ, M., HEUNISCH, C., ERNST, W., ZIMMERMANN, J. & WOLFGRAMM, M. 2018. Marine and terrestrial sedimentation across the T–J transition in the North German Basin. *Palaeogeography, Palaeoclimatology, Palaeoecology*, **489**, 74–94, https://doi.org/10.1016/j.palaeo.2017.09.029

BENTON, M.J. 1986. More than one event in the last Triassic mass-extinction. *Nature*, **321**, 857–861.

BOIGK, H., DIETZ, C. *ET AL.* 1960. *Zur Geologie des Emslandes [On the geology of the Emsland].* Geologisches Jahrbuch, Beihefte, **37**.

BÖLAU, E. 1959. Der Südwest- und Südostrand des Baltischen Schildes *[The south-western and south-eastern margin of the Baltic Shield]. Geologiska Föreningens i Stockholm Förhandlingar*, **497**, 167–230.

BOUGEAULT, C., PELLENARD, P. *ET AL.* 2017. Climatic and palaeoceanographic changes during the Pliensbachian (Early Jurassic) inferred from clay mineralogy and stable isotope (C–O) geochemistry (NW Europe). *Global and Planetary Change*, **149**, 139–152.

BOULILA, S., GALBRUN, B., HURET, E., HINNOV, L.A., ROUGET, I., GARDIN, S. & BARTOLINI, A. 2014. Astronomical calibration of the Toarcian Stage: implications for sequence stratigraphy and duration of the early Toarcian OAE. *Earth and Planetary Science Letters*, **386**, 98–111.

BRAND, E. & HOFFMANN, K. 1963. Stratigraphie und Fazies des nordwestdeutschen Jura und Bildungsbedingungen seiner Erdöllagerstätten [Stratigraphy and facies of the NW German Jurassic and genesis of its oil deposits]. *Erdöl und Kohle*, **16**, 468–477.

BRAND, E. & MÖNNIG, E. 2009. Litho- und Biostratigraphie des Mittel-Jura (Dogger) in Bohrungen Norddeutschlands. *Schriftenreihe der Deutschen Gesellschaft für Geowissenschaften*, **54**, 5–73.

BRENNER, W. 1986. Bemerkungen zur Palynostratigraphie der Rhät-Lias-Grenze in SW-Deutschland [Remarks on the palynostratigraphy of the Rhaetian to Liassic transition in SW Germany]. *Neues Jahrbuch für Geologie und Paläontologie, Abhandlungen*, **173**, 131–166.

CALLOMON, J.H. 1985. The evolution of the Jurassic ammonite family Cardioceratidae. *Special Papers in Palaeontology*, **33**, 49–90.

CALLOMON, J.H. 2003. The Middle Jurassic of western and northern Europe: its subdivisions, geochronology and correlations. *In*: INESON, J.R. & SURLYK, F. (eds) *The Jurassic of Denmark and Greenland. Geological Survey of Denmark and Greenland Bulletin*, **1**, 61–73.

CALLOMON, J.H. & CHANDLER, R.B. 1994. Some early Middle Jurassic ammonites of Tethyan affinities from the Aalenian of southern England. *Palaeopelagos Special Publication*, **1**, 17–40.

CLOETINGH, S., MCQUEEN, H. & LAMBECK, K. 1985. On a tectonic mechanism for regional sea level variations. *Earth and Planetary Science Letters*, **75**, 157–166.

COHEN, A.S., COE, A.L. & KEMP, D.B. 2007. The late Palaeocene–Early Eocene and Toarcian (Early Jurassic) carbon isotope excursions: a comparison of their time scales, associated environmental changes, causes and consequences. *Journal of the Geological Society, London*, **164**, 1093–1108, https://doi.org/10.1144/0016-76492006-123

Curray, J.R. 1964. Transgressions and regressions. *In*: Miller, R.L. (ed.) *Papers in Marine Geology*. Macmillan, London, 175–203.

Dadlez, R. & Franczyk, M. 1976. Znaczenie paleogeograficzne I paleotektoniczne Garbu Wielkopolskiego w czasie jury dolnej [Palaeogeographical and palaeotectonical significance of the Wielkopolska Ridge in the Early Jurassic]. *Biuletyn Instytutu Geologicznego*, **295**, 27–55 [in Polish].

Dadlez, R. & Kopik, J. 1972. Wybrane problemy stratygrafii i sedymentacji liasu między Świnoujściem a Gryficami [Selected problems of Liassic stratigraphy and sedimentation in the area between Świnoujście and Gryfice (North-West Poland)]. *Kwartalnik Geologiczny*, **16**, 620–637.

Dadlez, R. 1969. Stratigraphy of the Lias in Western Poland. *Prace Instytutu Geologicznego*, **57**, 1–92 [in Polish with an English summary].

Dadlez, R. 1972. *Kamień Pomorski IG 1, Profile Głębokich Otworów Wiertniczych Instytutut Geologicznego, Zeszyt 1 [Kamień Pomorski IG 1. Profiles of Deep Boreholes of the Geological Institute, Volume 1]*. Wydawnictwa Geologiczne, Warszawa, Poland.

Dadlez, R. 1974. Types of local tectonic structures in the Zechstein–Mesozoic complex of northwestern Poland. *Biuletyn Instytutu Geologicznego*, **274**, 149–177.

Dadlez, R., Narkiewicz, M., Stephenson, R.A., Visser, M.T.M. & van Wees, J.-D. 1995. Tectonic evolution of the Mid-Polish Trough: modelling implications and significance for central European geology. *Tectonophysics*, **252**, 179–195.

de Graciansky, P.-C., Jacquin, T. & Hesselbo, S.P. 1998. The Ligurian Cycle: an overview of Lower Jurassic 2nd-order transgressive/regressive facies cycles in western Europe. *In*: de Graciansky, P.-C., Hardenbol, J., Jacquin, T. & Vail, P.R. (eds) *Mesozoic and Cenozoic Sequence Stratigraphy of European Basins*. SEPM, Special Publications, **60**, 467–479.

Dera, G., Brigaud, B. *et al.* 2011. Climatic ups and downs in a disturbed Jurassic world. *Geology*, **39**, 215–218.

Doehler, M. 2005. The Mittelplate oil field. *In*: Dore, A. & Vining, B. (eds) *Petroleum Geology: Northwest Europe and Global Perspectives – Proceedings of the 6th Petroleum Geology Conference*. Geological Society, London, 461–468, https://doi.org/10.1144/0060461

Donovan, D.T. & Surlyk, F. 2003. Lower Jurassic (Pliensbachian) ammonites from Bornholm, Baltic Sea, Denmark. *In*: Ineson, J.R. & Surlyk, F. (eds) *The Jurassic of Denmark and Greenland. Geological Survey of Denmark and Greenland Bulletin*, **1**, 555–584.

Eiler, J.M. 2007. 'Clumped-isotope' geochemistry – the study of naturally-occurring, multiply-substituted isotopologues. *Earth and Planetary Science Letters*, **262**, 309–327.

Embry, A.F. 1993. Transgressive–regressive (TR) sequence analysis of the Jurassic succession of the Sverdrup Basin, Canadian Arctic Archipelago. *Canadian Journal of Earth Sciences*, **30**, 301–320.

Ermlich, C. 1993. *Stratigraphisch-fazielle Untersuchungen im Hettangium von Wefensleben (Sachsen-Anhalt) [Stratigraphy and facies of the Hettangian at Wefensleben (Sachsen-Anhalt)]*. Master thesis, Universität Hannover, Hannover, Germany.

Ernst, W. 1986. Neue Ergebnisse zur Fazies, Ammonitenfauna, Biostratigraphie und Paläogeographie des thüringischen Lias [New results on facies, ammonite fauna, biostratigraphy and paleogeography of the Thuringian Liassic]. *Hallesches Jahrbuch für Geowissenschaften*, **12**, 15–28.

Ernst, W. 1992. Der Lias der Scholle von Dobbertin (Mecklenburg) [The Liassic of the Dobbertin Block (Mecklenburg)]. *Fundgrube*, **28**, 57–70.

Feldman-Olszewska, A. 1997*a*. Depositional systems and cyclicity in the intracratonic Early Jurassic basin in Poland. *Geological Quarterly*, **41**, 475–490.

Feldman-Olszewska, A. 1997*b*. Depositional architecture of the Polish epicontinental Middle Jurassic basin. *Geological Quarterly*, **41**, 491–508.

Fischer, J., Voigt, S. *et al.* 2012. Palaeoenvironments of the late Triassic Rhaetian Sea: implications from oxygen and strontium isotopes of hybodont shark teeth. *Palaeogeography, Palaeoclimatology, Palaeoecology*, **353–355**, 60–72.

Fischer, R., Jäger, M., Konstantinopoulou, A., Kristan-Tollmann, E., Luppold, F.W. & Ohm, H.-H. 1986. Paläontologie einer epikontinentalen Lias-Schichtfolge: Oberes Sinemurium bis Oberes Domerium von Empelde bei Hannover [Paleontology of epicontinental Liassic strata: Upper Sinemurian to Upper Domerian of Empelde near Hannover]. *Facies*, **15**, 53–176.

Frandsen, N. & Surlyk, F. 2003. An offshore transgressive–regressive mudstone-dominated succession from the Sinemurian of Skåne, Sweden. *In*: Ineson, J.R. & Surlyk, F. (eds) *The Jurassic of Denmark and Greenland. Geological Survey of Denmark and Greenland Bulletin*, **1**, 543–554.

Galloway, W.E. 1989. Genetic stratigraphic sequences in basin analysis: architecture and genesis of flooding surface bounded depositional units. *AAPG Bulletin*, **73**, 125–142.

Gedl, P. 2007. Early Jurassic dinoflagellate cysts from the Kraków–Silesia monocline, southern Poland: a record from the Blanowice Formation at Mrzygłód. *Annales Societatis Geologorum Poloniae*, **77**, 147–159.

Göthel, M. 1999. Schlußfolgerungen über Untersuchungen zum Rhät/Lias der Vetschauer Keupermulde (Lausitzer Triasscholle) [Investigations on Rhaetian to Liassic strata of the Vetschauer Keupermulde (Lausitzer Triasscholle)]. *Brandenburgische Geowissenschaftliche Beiträge*, **6**, 65–77.

Gradstein, F.M., Ogg, J.G., Schmitz, M.D. & Ogg, G.M. 2012. *The Geologic Time Scale*. Elsevier, Amsterdam.

Hallam, A. 1988. Reevaluation of Jurassic eustasy in the light of new data and the revised exxon curve. *In*: Wilgus, C.K., Hastings, B.S., Kendall, C.G.S.C., Posamentier, H.W., Ron, C.A. & van Wagner, J.C. (eds) *Sea-Level Changes – An Integrated Approach*. SEPM, Special Publications, **42**, 261–273.

Hallam, A. 1997. Estimates of the amount and rate of sea-level change across the Rhaetian–Hettangian and Pliensbachian–Toarcian boundaries (latest Triassic to early Jurassic). *Journal of the Geological Society,*

London, **154**, 773–779, https://doi.org/10.1144/gsjgs.154.5.0773

Hallam, A. 2001. A review of the broad pattern of Jurassic sea-level changes and their possible causes in the light of current knowledge. *Palaeogeography, Palaeoclimatology, Palaeoecology*, **147**, 23–37.

Haq, B.U., Hardenbol, J. & Vail, P.R. 1987. Chronology of fluctuating sea levels since the Triassic. *Science*, **235**, 1156–1167.

Hesselbo, S.P. & Jenkyns, H.C. 1998. British Lower Jurassic sequence stratigraphy. *In*: de Graciansky, P.-C., Hardenbol, J., Jacquin, T. & Vail, P.R. (eds) *Mesozoic and Cenozoic Sequence Stratigraphy of European Basins*. SEPM, Special Publications, **60**, 561–581.

Hesselbo, S.P. & Pieńkowski, G. 2011. Stepwise atmospheric carbon-isotope excursion during the Toarcian Oceanic Anoxic Event (Early Jurassic, Polish Basin). *Earth and Planetary Science Letters*, **301**, 365–372.

Hesselbo, S.P., Robinson, S.A. & Surlyk, F. 2004. Sea-level change and facies development across potential Triassic–Jurassic boundary horizons, SW Britain. *Journal of the Geological Society, London*, **161**, 365–379, https://doi.org/10.1144/0016-764903-033

Hesselbo, S.P., Jenkyns, H.C., Duarte, L.V. & Oliveira, L.C.V. 2007. Carbon-isotope record of the Early Jurassic (Toarcian) Oceanic Anoxic Event from fossil wood and marine carbonate (Lusitanian Basin, Portugal). *Earth and Planetary Science Letters*, **253**, 455–470.

Heunisch, C., Luppold, F.W., Reinhardt, L. & Röhling, H.-G. 2010. Palynofazies, Bio- und Lithostratigrafie im Grenzbereich Trias/Jura in der Bohrung Mariental 1 (Lappwaldmulde, Ostniedersachsen) [Palynofacies, biostratigraphy and lithostratigraphy of the Triassicâ Jurassic transition in well Mariental 1 (Lappwaldmulde, eastern Lower Saxony)]. *Zeitschrift der Deutschen Gesellschaft für Geowissenschaften*, **161**, 51–98.

Hoffmann, K. 1962. Lias und Dogger im Untergrund der Niederrheinischen Bucht [Lias and Dogger in the subsurface of the Niederrheinische Bucht]. *Fortschritte der Geologie in Rheinland und Westfalen*, **6**, 105–184.

Hoffmann, K. & Jordan, R. 1982. Die Stratigraphie, Paläogeographie und Ammonitenführung des Unter-Pliensbachium (Carixium, Lias Gamma) in Nordwest-Deutschland [Stratigraphy, paleogeography and ammonite records of the Lower Pliensbachian (Carixian, Lias Gamma) in NW Germany]. *Geologisches Jahrbuch*, **A55**, 3–439.

Hoffmann, K. & Martin, G.P.R. 1960. Die Chronozone des *Dactylioceras tenuicostatum* (Toarcien, Lias) in NW- und SW-Deutschland [The chronozone of *Dactylioceras tenuicostatum* (Toarcian, Lias) in NW and SW Germany]. *Paläontologische Zeitschrift*, **34**, 103–149.

Huang, C. & Hesselbo, S.P. 2014. Pacing of the Toarcian Oceanic Anoxic Event (Early Jurassic) from astronomical correlation of marine sections. *Gondwana Research*, **25**, 1348–1356.

Hudson, J.D., Clements, R.G., Riding, J.B., Wakefield, M.I. & Walton, W. 1995. Jurassic paleosalinities and brackish-water communities – a case study. *Palaios*, **10**, 392–407.

Jacquin, T., Dardeau, G., Durlet, C., de Graciansky, P.C. & Hantzpergue, P. 1998. The North Sea Cycle: an overview of 2nd order transgressive/regressive facies cycles in western Europe. *In*: de Graciansky, P.-C., Hardenbol, J., Jacquin, T. & Vail, P.R. (eds) *Mesozoic and Cenozoic Sequence Stratigraphy of European Basins*. SEPM, Special Publications, **60**, 445–466.

Jenkyns, H.C. 1988. The early Toarcian (Jurassic) anoxic event: stratigraphic, sedimentary and geochemical evidence. *American Journal of Science*, **288**, 101–151.

Jenkyns, H.C., Jones, C.E., Gröcke, D.R., Hesselbo, S.P. & Parkinson, D.N. 2002. Chemostratigraphy of the Jurassic System: applications, limitations and implications for palaeoceanography. *Journal of the Geological Society, London*, **159**, 351–378, https://doi.org/10.1144/0016-764901-130

Jüngst, H. 1928. Rhät, Psilonoten- und Schlotheimienschichten im nördlichen Harzvorlande [Rhaetian, Psilonoten- and Schlotheimienschichten (Lower Jurassic) in the northern Harz foreland]. *Geologische und Paläontologische Abhandlungen, Neue Folge*, **16**, 1–194.

Karaszewski, W. & Kopik, J. 1970. Lower Jurassic. The Stratigraphy of the Mesozoic in the margin of the Holy Cross Mountains. *Prace Instytutu Geologicznego*, **56**, 65–89.

Kemp, D.B., Coe, A.L., Cohen, A.S. & Schwark, L. 2005. Astronomical pacing of methane release in the Early Jurassic period. *Nature*, **437**, 396–399.

Kopik, J. 1962. Faunistic criteria of stratigraphical subdivision of the Lias in North-Western and Central Poland. *In*: Passendorfer, E. (ed.) *Księga Pamiątkowa ku czci profesora Jana Samsonowicza*. Polska Akademia Nauk, Warszawa, Poland, 271–312.

Kopik, J. 1964. Stratigraphy of the Lower Jurassic based on the fauna of the Mechowo IG-1 borehole. *Biuletyn Instytutu Geologicznego*, **189**, 43–55 [in Polish with an English summary].

Kopik, J. 1975. Fauna osadów domeru. *In*: Dembowska, J. (ed.) *Wolin IG 1*. Profile Głębokich Otworów Wiertniczych Instytutu Geologicznego, **22**, 53.

Kopik, J. 1998. Jura dolna i środkowa północno-wschodniego obrzeżenia Górnośląskiego Zagłębia Węglowego [Lower and Middle Jurassic of NE fringe of the Upper Silesia Coal Basin]. *Biuletyn Państwowego Instytutu Geologicznego*, **378**, 67–120.

Kopik, J. & Marcinkiewicz, T. 1997. Jura dolna, biostratygrafia [Biostratigraphy of the Lower Jurassic]. *In*: Marek, S. & Pajchlowa, M. (eds) *The Epicontinental Permian and Mesozoic in Poland*. Prace Państwowego Instytutu Geologicznego, **153**, 196–205 [in Polish with an English summary].

Koppelhus, E.B. & Nielsen, L.H. 1994. Palynostratigraphy and palaeoenvironments of the Lower to Middle Jurassic Bagå Formation of Bornholm, Denmark. *Palynology*, **18**, 139–194.

Korte, C. & Hesselbo, S.P. 2011. Shallow marine carbon and oxygen isotope and elemental records indicate icehouse–greenhouse cycles during the Early Jurassic. *Paleoceanography*, **26**, PA4219.

Korte, C., Hesselbo, S.P., Ullmann, C.V., Dietl, G., Ruhl, M., Schweigert, G. & Thibault, N. 2015. Jurassic climate mode governed by ocean gateway. *Nature Communications*, **6**, 10015.

Lang, W.D. 1924. The Blue Lias of the Devon and Dorset coasts. *Proceedings of the Geologists' Association*, **35**, 169–184.

Lehmkuhl, U. 1968. *Bericht über makropaläozoologische Untersuchungsergebnisse im Lias des Nordteils der*

DDR [Report on palaeozoologic investigations on the Liassic (Lower Jurassic) of the NE German Democratic Republic (GDR)]. Forschungsinstitut Erdöl/Erdgas, Außenstelle Schwerin, Abteilung für Geologisch-Geochemische Laboratorien, Archive No. 2019318.

Lindström, S., van de Schootbrugge, B. *et al.* 2017. A new correlation of Triassic–Jurassic boundary successions in NW Europe, Nevada and Peru, and the Central Atlantic Magmatic Province: a time-line for the end-Triassic mass extinction. *Palaeogeography, Palaeoclimatology, Palaeoecology*, **478**, 80–102.

Lott, G.K., Wong, T.E., Dusar, M., Andsbjerg, J., Mönnig, E., Feldman-Olszewska, A. & Verreussel, R.M.C.H. 2010. Jurassic. *In*: Doornenbal, J.C. & Stevenson, A.G. (eds) *Petroleum Geological Atlas of the Southern Permian Basin Area*. European Association of Geoscientists and Engineers (EAGE), Houten, The Netherlands, 175–193.

McArthur, J.M., Algeo, T.J., van de Schootbrugge, B., Li, Q. & Howarth, R.J. 2008. Basinal restriction, black shales, Re–Os dating, and the Early Toarcian (Jurassic) oceanic anoxic event. *Paleoceanography*, **23**, PA4217.

Mestwerdt, A. 1910. Über Faziesverhältnisse im Rät und untersten Lias in Nordwest-Deutschland [On the facies of the Rhaetian and lowermost Liassic in NW Germany]. *Jahrbuch der königlich-preußischen geologischen Landesanstalt*, **31**, 420–429.

Miall, A.D. 1986. Eustatic sea-level changes interpreted from seismic stratigraphy: a critique of the methodology with particular reference to the North Sea Jurassic record. *AAPG Bulletin*, **70**, 131–137.

Miall, A.D. 1992. Alluvial deposits. *In*: Walker, R.G. & James, N.P. (eds) *Facies Models – Response to Sea Level Change*. Geological Association of Canada, St John's, Newfoundland, Canada, 119–142.

Michelsen, O., Nielsen, L.H., Johannessen, P.N., Andsbjerg, J. & Surlyk, F. 2003. Jurassic lithostratigraphy and stratigraphic development onshore and offshore Denmark. *In*: Ineson, J.R. & Surlyk, F. (eds) *The Jurassic of Denmark and Greenland. Geological Survey of Denmark and Greenland Bulletin*, **1**, 147–216.

Mitchum, R.M., Vail, P.R. & Thompson, S. 1977. Seismic stratigraphy and global changes of sea level; Part 2, The depositional sequence as a basic unit for stratigraphic analysis. *In*: Payton, C.E. (ed.) *Seismic Stratigraphy; Applications to Hydrocarbon Exploration*. American Association of Petroleum Geologists, Memoirs, **26**, 53–62.

Moryc, W. 2005. Middle and ?Lower Jurassic deposits in the Księżopol-Lubaczów area (SE Poland). *Biuletyn Państwowego Instytutu Geologicznego*, **408**, 5–72.

Nielsen, L.H. 2003. Late Triassic–Jurassic development of the Danish Basin and the Fennoscandian Border Chronozone, southern Scandinavia. *In*: Ineson, J.R. & Surlyk, F. (eds) *The Jurassic of Denmark and Greenland. Geological Survey of Denmark and Greenland Bulletin*, **1**, 459–526.

Nielsen, L.H., Petersen, H.I., Dybkjær, K. & Surlyk, F. 2010. Lake-mire deposition, earthquakes and wildfires along a basin margin fault; Rønne Graben, Middle Jurassic, Denmark. *Palaeogeography, Palaeoclimatology, Palaeoecology*, **292**, 103–126.

Oertel, W. 1922. Der pommersche Lias [The Liassic of Pomerania]. *Mitteilungen des naturwissenschaftlichen Vereins für Neuvorpommern und Rügen*, **48**, 109–125.

Olariu, M.I., Carvajal, C.R., Olariu, C. & Steel, R.J. 2012. Deltaic process and architectural evolution during cross-shelf transits, Maastrichtian Fox Hills Formation, Washakie Basin, Wyoming. *AAPG Bulletin*, **96**, 1931–1956.

Partington, M.A., Copestake, P., Mitchener, C. & Underhill, J.R. 1993. Biostratigraphic calibration of genetic stratigraphic sequences in the Jurassic–lowermost Cretaceous (Hettangian to Ryazanian) of the North Sea and adjacent areas. *In*: Parker, J.R. (ed.) *Petroleum Geology of Northwest Europe: Proceedings of the 4th Conference*. Geological Society, London, 371–386, https://doi.org/10.1144/0040371

Pieńkowski, G. 1991*a*. Liassic sedimentation in Scania, southern Sweden: Hettangian–Sinemurian of the Helsingborg area. *Facies*, **24**, 39–86.

Pieńkowski, G. 1991*b*. Eustatically-controlled sedimentation in the Hettangian–Sinemurian (Early Jurassic) of Poland and Sweden. *Sedimentology*, **38**, 503–518.

Pieńkowski, G. 2004. The epicontinental Lower Jurassic of Poland. *Polish Geological Institute Special Papers*, **12**, 1–154.

Pieńkowski, G. 2014. The first Early Jurassic ammonite find in central Poland. *Volumina Jurassica*, **12**, 99–104.

Pieńkowski, G. 2015. Geologiczne składowanie ditlenku węgla (CCS) jest metodą bezpieczną – dowody geologiczne [Geological storage of carbon dioxide (CCS) is safe – geological evidence]. *Przegląd Geologiczny*, **63**, 48–54 [in Polish with an English summary].

Pieńkowski, G. & Waksmundzka, M. 2009. Palynofacies in Lower Jurassic epicontinental deposits of Poland – tool to interpret sedimentary environments. *Episodes*, **1**, 21–32.

Pieńkowski, G., Schudack, M.E. *et al.* 2008. Jurassic. *In*: McCann, T. (ed.) *The Geology of Central Europe, Volume 2: Mesozoic and Cenozoic*. Geological Society, London, 823–922.

Pieńkowski, G., Niedźwiedzki, G. & Waksmundzka, M. 2012. Sedimentological, palynological, and geochemical studies of the terrestrial Triassic–Jurassic boundary in northwestern Poland. *Geological Magazine*, **149**, 308–332.

Pieńkowski, G., Hodbod, M. & Ullmann, C.V. 2016. Fungal decomposition of terrestrial organic matter accelerated Early Jurassic climate warming. *Scientific Reports*, **6**, 31930.

Poprawa, P. 1997. Late Permian to Tertiary dynamics of the Polish Trough. *Terra Nostra*, **97**, 104–109.

Posamentier, H.W., Jervey, M.T. & Vail, P.R. 1988. Eustatic controls on clastic deposition II – conceptual framework. *In*: Vilgus, C.K., Hastings, B.S., Kendall, G.C., Posamentier, H.W., Ross, C.A. & Van Wagoner, J.C. (eds) *Sea-Level Changes: An Integrated Approach*. SEPM, Special Publications, **42**, 125–154.

Poulsen, N.E. & Riding, J.B. 2003. The Jurassic dinoflagellate cyst zonation of Subboreal Northwest Europe. *In*: Ineson, J.R. & Surlyk, F. (eds) *The Jurassic of Denmark and Greenland. Geological Survey of Denmark and Greenland Bulletin*, **1**, 115–144.

PRICE, G.D. 1999. The evidence and implications of polar ice during the Mesozoic. *Earth-Science Reviews*, **48**, 183–210.

PSCHENITZKA, U. 1967. *Bericht zur makropaläontologischen Bearbeitung des Lias in der Bohrung Wolgast 1/1A/63 (Objekt Doggersenke Usedom) [Report on macropaleontological investigations of the Liassic in well Wolgast 1/1A/63 (research object Doggersenke Usedom)].* VEB Geologische Erkundung Nord, Schwerin, Archive No. 1948/75.

REYMENT, R. 1959. On Liassic ammonites from Skåne, southern Sweden. *Stockholm Contributions in Geology*, **2**, 103–157.

ROGOV, M.A. & ZAKHAROV, V. 2010. Jurassic and Lower Cretaceous glendonite occurrences and their implication for Arctic paleoclimate reconstructions and stratigraphy. *Earth Science Frontiers*, **17**, 345–346.

RÖHL, H.-J., SCHMID-RÖHL, A., OSCHMANN, W., FRIMMEL, A. & SCHWARK, L. 2001. Erratum to 'The Posidonia Shale (Lower Toarcian) of SW-Germany: an oxygen-depleted ecosystem controlled by sea level and palaeoclimate' [Palaeogeogr., Palaeoclimatol., Palaeocol., 165, 27–52]. *Palaeogeography, Palaeoclimatology, Palaeoecology*, **169**, 273–299.

RUHL, M., DEENEN, M.H.L., ABELS, H.A., BONIS, N.R., KRIJGSMAN, W. & KÜRSCHNER, W.M. 2010. Astronomical constraints on the duration of the early Jurassic Hettangian stage and recovery rates following the end-Triassic mass extinction (St. Audrie's Bay/East Quantoxhead, UK). *Earth and Planetary Science Letters*, **295**, 262–276.

RUHL, M., HESSELBO, S.P. ET AL. 2017. Astronomical constraints on the duration of the Early Jurassic Pliensbachian Stage and global climatic fluctuations. *Earth and Planetary Science Letters*, **455**, 149–165.

RUSBÜLT, J. 1990. *Biostratigraphie des Unteren Lias in Mecklenburg [Biostratigraphy of the Lower Liassic in Mecklenburg].* PhD thesis, University Greifswald, Greifswald, Germany.

SCHECK-WENDEROTH, M., KRZYWIEC, P., ZÜHLKE, R., MAYSTRENKO, Y. & FROITZHEIM, N. 2008. Permian to Cretaceous tectonics. *In*: MCCANN, T. (ed.) *The Geology of Central Europe. Volume 2: Mesozoic and Cenozoic.* Geological Society, London, 999–1031.

SCHLEGELMILCH, R. 2014. *Die Ammoniten des süddeutschen Lias: ein Bestimmungsbuch für Fossiliensammler und Geologen [Ammonites of the South German Liassic: a reference book for fossil collectors and geologists].* Springer, Berlin.

SCHUBERT, S. 2005. Ein Lias-Profil (Hettangium/Sinemurium) vom Bau des Ostwestfalendamm-Tunnels in Bielefeld-Stadtmitte nebst einem Profil von der Finkenstraße in Bielefeld [The Liassic (Hettangian/Sinemurian) of the contruction site of the "Ostwestfalendamm" tunnel (Bielfeld city centre) including the section at "Finkenstraße", Bielefeld]. *Geologie und Paläontologie in Westfalen*, **65**, 5–61.

SCHUMACHER, K.-H. & SONNTAG, H. 1964. Zur Stratigraphie und Ausbildung des Lias im Norden der DDR [Stratigraphy and facies of the Liassic in the northern German Democratic Republic (GDR)]. *Geologie*, **13**, 303–315.

SELL, B., OVTCHAROVA, M. ET AL. 2014. Evaluating the temporal link between the Karoo LIP and climatic–biologic events of the Toarcian Stage with high-precision U–Pb geochronology. *Earth and Planetary Science Letters*, **408**, 48–56.

ŠIMKEVIČIUS, P. 1998. *Jurassic of the South-Eastern Baltic. Lithologu and Clay Minerals.* Geologijos Institutas, Vilnius, Lithuania.

SONG, J., LITTKE, R., WENIGER, P., OSTERTAG-HENNING, C. & NELSKAMP, S. 2015. Shale oil potential and thermal maturity of the Lower Toarcian Posidonia Shale in NW Europe. *International Journal of Coal Geology*, **150–151**, 127–153.

SORGENFREI, T. & BUCH, A. 1964. Deep tests in Denmark 1935–1959. *Danmarks geologiske undersøgelse. III. Række*, **36**, 1–146.

SUAN, G., MATTIOLI, E. ET AL. 2010. Secular environmental precursors to Early Toarcian (Jurassic) extreme climate changes. *Earth and Planetary Science Letters*, **290**, 448–458.

SURLYK, F., ARNDORFF, L. ET AL. 1995. High-resolution sequence stratigraphy of a Hettangian–Sinemurian paralic succession, Bornholm, Denmark. *Sedimentology*, **42**, 323–354.

TAYLOR, A.M. & GOLDRING, R. 1993. Description and analysis of bioturbation and ichnofabric. *Journal of the Geological Society*, **150**, 141–148, https://doi.org/10.1144/gsjgs.150.1.0141

TEICHERT, B.M.A. & LUPPOLD, F.W. 2013. Glendonites from an Early Jurassic methane seep: climate or methane indicators? *Palaeogeography, Palaeoclimatology, Palaeoecology*, **390**, 81–93.

TROEDSSON, G. 1951. On the Höganäs series of Sweden (Rhaeto-Lias). *Lunds Universitets Årsskrift Ny Följd*, **2**, 1–269.

UNDERHILL, J.R. & PARTINGTON, M.A. 1993. Jurassic thermal doming and deflation in the North Sea: implications of the sequence stratigraphic evidence. *In*: PARKER, J.R. (ed.) *Petroleum Geology of Northwest Europe: Proceedings of the 4th Conference.* Geological Society, London, 337–345, https://doi.org/10.1144/0040337

VAIL, P.R., MITCHUM, R.M. & THOMPSON, S. 1977. Seismic stratigraphy and global changes of sea level; Part 4, Global cycles of relative changes of sea level. *In*: PAYTON, C.E. (ed.) *Seismic Stratigraphy: Applications to Hydrocarbon Exploration.* American Association of Petroleum Geologists, Memoirs, **26**, 83–97.

VAN DER MEER, D.G., VAN DEN BERG VAN SAPAROEA, A.P.H., VAN HINSBERGEN, D.J.J., VAN DE WEG, R.M.B., GODDERIS, Y., LE HIR, G. & DONNADIEU, Y. 2017. Reconstructing first-order changes in sea level during the Phanerozoic and Neoproterozoic using strontium isotopes. *Gondwana Research*, **44**, 22–34.

VAN DE SCHOOTBRUGGE, B., BACHAN, A., SUAN, G., RICHOZ, S., JONATHAN, L. & PAYNE, J.L. 2013. Microbes, mud and methane: cause and consequence of recurrent Early Jurassic anoxia following the end-Triassic mass extinction. *Palaeontology*, **56**, 685–709.

WIENHOLZ, R. 1967. Über den geologischen Bau des Untergrundes im Nordostdeutschen Flachland [On the geology of the Northeast German lowlands]. *Zeitschrift für angewandte Geologie*, **1**, 1–87.

WILL, H.-J. 1969. Untersuchungen zur Stratigraphie und Genese des Oberkeupers in Nordwestdeutschland [The Upper Keuper of Nordwest Germany]. *Geologisches Jahrbuch, Beihefte*, **54**, 1–240.

WOLFGRAMM, M., RAUPPACH, K. & SEIBT, P. 2008. Reservoir-geological characterization of Mesozoic sandstones in the North German Basin by petrophysical and petrographical data. *Zeitschrift für geologische Wissenschaften*, **36**, 197–346.

ZIEGLER, P.A. 1982. Triassic rifts and facies patterns in Western and Central Europe. *Geologische Rundschau*, **71**, 747–772.

ZIEGLER, P.A. (ed.). 1990. *Geological Atlas of Western and Central Europe*. Shell Internationale Petroleum Maatschappij, The Hague. Geological Society, London.

ZIMMERMANN, J., FRANZ, M., HEUNISCH, C., LUPPOLD, F.W., MÖNNIG, E. & WOLFGRAMM, M. 2015. Sequence stratigraphic framework of the Lower and Middle Jurassic in the North German Basin: epicontinental sequences controlled by Boreal cycles. *Palaeogeography, Palaeoclimatology, Palaeoecology*, **440**, 395–416.

ZIMMERMANN, J., FRANZ, M., SCHALLER, A. & WOLFGRAMM, M. 2017. The Toarcianâ Bajocian deltaic system in the North German Basin: Subsurface mapping of ancient deltas â morphology, evolution and controls. *Sedimentology*, https://doi.org/10.1111/sed.12410

The impact of Quaternary glaciation on temperature and pore pressure in Jurassic troughs in the Southern Permian Basin, northern Germany

VICTORIA F. SACHSE & RALF LITTKE*

Institute of Geology and Geochemistry of Petroleum and Coal, Energy and Mineral Resources Group (EMR), RWTH Aachen University, Lochnerstrasse 4–20, 52056 Aachen, Germany

**Correspondence: ralf.littke@emr.rwth-aachen.de*

Abstract: A detailed 3D petroleum system model was constructed for the Schleswig-Holstein area in northern Germany. Salt movement and the Quaternary ice episodes were implemented in order to reconstruct their impact on temperature, maturity and pressure. Burial, temperature and maturity histories were calculated for the Jurassic troughs and the Glueckstadt Graben showing both differences and similarities. For example, all locations reached (almost) deepest burial at present day, whilst subsidence and long-term sedimentation rate was highest in Glueckstadt Graben during the Triassic. The Jurassic troughs received their major subsidence and sedimentation pulse later, and were strongly affected by a later salt movement.

The implementation of Quaternary glacial episodes does not have a strong impact on petroleum generation from the major source rock (Lower Toarcian Posidonia Shale). In the case of the Posidonia Shale reaching the stage of petroleum expulsion (outside of the study area), the effect of 'glacial pumping'(i.e. the development of high pore pressures during glaciation followed by expulsion and subsequent pressure release during deglaciation) can be deduced from the model. Petroleum accumulations in the reservoir layers (Dogger sandstones) are also seen to have been affected. This finding is of interest for exploration, as it might control petroleum composition, biodegradation and leakage through cap rocks.

Petroleum systems of the Central European Basin System

Geological studies on the Southern Permian Basin/ Central European Basin System (SPB/CEBS) have a long history, and were strongly driven by oil and gas exploration and production. The first well encountering an oil-bearing layer was drilled as early as 1858–59 in the village of Wietze, north of Hannover, followed by numerous discoveries of gas and oil fields (Maystrenko *et al.* 2008). Major discoveries include the Groningen gas field in the NE Netherlands and the Mittelplate oil field in Schleswig-Holstein, northern Germany. The basin also has a high potential for unconventional shale oil and shale gas (e.g. Bruns *et al.* 2015; BGR 2016; Mohnhoff *et al.* 2016), which remains, however, underexplored due to legal restrictions. Besides exploration activities on natural gas and oil, economic interest in the North German Basin also focused on artificial storage of natural gas and other fluids, geothermic use of permeable layers, and geological repository of nuclear waste in Late Permian Zechstein salt structures.

Aside from detailed studies related to the sedimentary facies, as well as the structural and tectonic evolution of the basin (Littke *et al.* 2008), petroleum source rocks and petroleum systems were studied intensely. Source rocks for thermogenic gas are mainly coal-bearing Pennsylvanian sequences present in the subsurface of much of the area (Lokhorst *et al.* 1998; Maystrenko *et al.* 2008), while it is assumed that marine Carboniferous source rocks also contributed. Gases are methane dominated, except for areas of high thermal maturity, and where the coal-bearing sequence is missing and nitrogen-rich gases can dominate (Littke *et al.* 1995). Source rocks for oil are mainly the Lower Jurassic (Early Toarcian) Posidonia Shale, with additional contributions from the Lower Cretaceous (Berriasian) Wealden shales in the SW (Littke *et al.* 1991; Rippen *et al.* 2013). The Posidonia Shale is a carbonate-rich lithology (often a marlstone rather than a shale) occurring in organic-matter-rich facies over large areas (Song *et al.* 2015). Due to the occurrence of this source rock in quite uniform organic facies at various levels of thermal maturity across northern Germany (immature to gas shale maturity: Bruns *et al.* 2015), various studies on maturity effects have been conducted in the past: for example, on mass balances (Rullkötter *et al.* 1988), maceral transformation (Littke *et al.* 1988), kinetics of petroleum

From: Kilhams, B., Kukla, P. A., Mazur, S., McKie, T., Mijnlieff, H. F. & van Ojik, K. (eds) 2018. *Mesozoic Resource Potential in the Southern Permian Basin*. Geological Society, London, Special Publications, **469**, 371–398.
First published online January 16, 2018, https://doi.org/10.1144/SP469.7

generation (Schaefer *et al.* 1990; Schenk *et al.* 1997) and organic matter porosity evolution (Bernard *et al.* 2012; Klaver *et al.* 2012).

Several large-scale numerical models on the burial history, structural and thermal evolution, as well as on the petroleum systems, have been published during the past 10 years, including the first 3D models (i.e. Maystrenko *et al.* 2005, 2012; Scheck-Wenderoth & Lamarche 2005; Uffmann & Littke 2011, 2013; Bruns *et al.* 2013; Uffmann *et al.* 2014; Bruns *et al.* 2015; Mohnhoff *et al.* 2016; Sachse & Littke 2016). A few published 1D/2D or large-scale 3D petroleum system models also cover the Schleswig-Holstein area, but do not implement its detailed history of salt diapirism. Grassmann *et al.* (2005) published a 2D petroleum system model dealing with the geological history and the petroleum system of the Mittelplate oil field. Another 1D modelling study was presented by Rodon & Littke (2005), which focuses on various deep wells in Schleswig-Holstein. The *Geological Atlas of the Southern Permian Basin* (Lokhorst *et al.* 1998) provides a maturity map, which we consider, however, to be inaccurate for the Schleswig-Holstein area. A much better representation of thermal maturity for this area was published by Uffmann & Littke (2011) based on a large-scale 3D numerical petroleum system model. However, details regarding the geological evolution of this area were not considered there.

Although much of the petroleum systems are well understood in the SPB/CEBS, some parameters received little attention. A reason for this is the resolution of the previous models, which do not consider the complex geological setting of structures, including salt movements or vertical sedimentary heterogeneities. Furthermore, it is widely accepted that ice-sheet load and unload might have had an impact on the petroleum system (e.g. the Barents Sea: Lerche *et al.* 1997; Rodrigues Duran *et al.* 2013; Mittelplate: Grassmann *et al.* 2010). Yet, a 3D model taking this parameter into account does not exist for the north German area. Therefore, we present here 3D basin models comparing the impacts of sequential ice loading and unloading in the Schleswig-Holstein area, considering the Posidonia Shale as the main source rock. Our basic 3D model was conducted based on depth maps from the *Geotectonic Atlas of NW-Germany* (Baldschuhn *et al.* 1996), and was calibrated using thermal information from four wells (Rodon & Littke 2005) (Fig. 1).

Geology of the study area

Tectonic evolution and the sedimentary system

In central Europe, main, large-scale structures developed due to the decomposition of Pangea at the end of the Paleozoic, which initiated the formation of the Central European Basin System (CEBS: Littke *et al.* 2008). The CEBS consists of two main basin systems: the Northern Permian Basin and the Southern Permian Basin, which are separated by a prominent basement swell (Ringkøbing-Fyn High: Sachse & Littke 2016; Middle North Sea High). The Southern Permian Basin consists of several sub-basins, which are separated by swells as well as marginal basins in the south (Fig. 2). Schleswig-Holstein is located in the northern area of the Pompeckj Block in the central Southern Permian Basin (Fig. 2) (Littke *et al.* 2008; Scheck-Wenderoth *et al.* 2008). The study area comprises the East and West Holstein troughs, and is bordered by the West Schleswig Block to the west and the East Holstein Block to the east (Figs 1 & 2). In the following, the East Holstein Trough (EHT) and the West Holstein Trough (WHT) are partly summarized as Jurassic troughs, because Jurassic sedimentary rocks are present there, whereas they are absent in the Glueckstadt Graben. In the entire area, Permo-Carboniferous volcanics and aeolian, fluvial and lacustrine siliciclastics in the North German Basin are covered by Upper Permian Zechstein evaporites (Raith *et al.* 2015; Strozyk *et al.* 2017).

In the Late Permian and Early Triassic, east–west-directed extension led to tectonic movements north and south of the Ringkøbing-Fyn High, resulting in NNW–SSE- to NNE–SSW-orientated rift structures: for example, the Central Graben, the Horn Graben, the Glueckstadt Graben and the Polish Trough (Kockel 2002; Maystrenko *et al.* 2005; Scheck-Wenderoth *et al.* 2008). Its development was mainly controlled by extension during the Upper Triassic Keuper (Kockel 2002), whereas the Glückstadt Graben is the deepest part of the CEBS. Tectonic and halokinetic movements triggered salt diapirism of the Zechstein salt (Maystrenko *et al.* 2005), leading to Zechstein salt walls of Triassic age towards the margin of the graben, becoming younger from the centre to the margin and from north to south. The depocentre of the northern German Zechtein salt province was located in the basin centre, showed a maximum initial thickness of about 2.5 km (Maystrenko *et al.* 2012; Strozyk *et al.* 2014, 2017; Raith *et al.* 2015) and thinned towards the basin margins. Ongoing subsidence and sedimentation led to deposition of mainly siliciclastic layers of the Buntsandstein, marine carbonates of the Muschelkalk, continental siliciclastics of the Keuper, and a mixture of marine mudstones and sandstones, which were deposited under shallow-marine conditions during the Early Jurassic. However, basin-inversion phases in the Middle Jurassic–Early Cretaceous led to significant erosion of Jurassic and Triassic sediments. The depositional thickness for the eroded Jurassic sediments has been estimated (Neunzert *et al.* 1996; Rodon & Littke

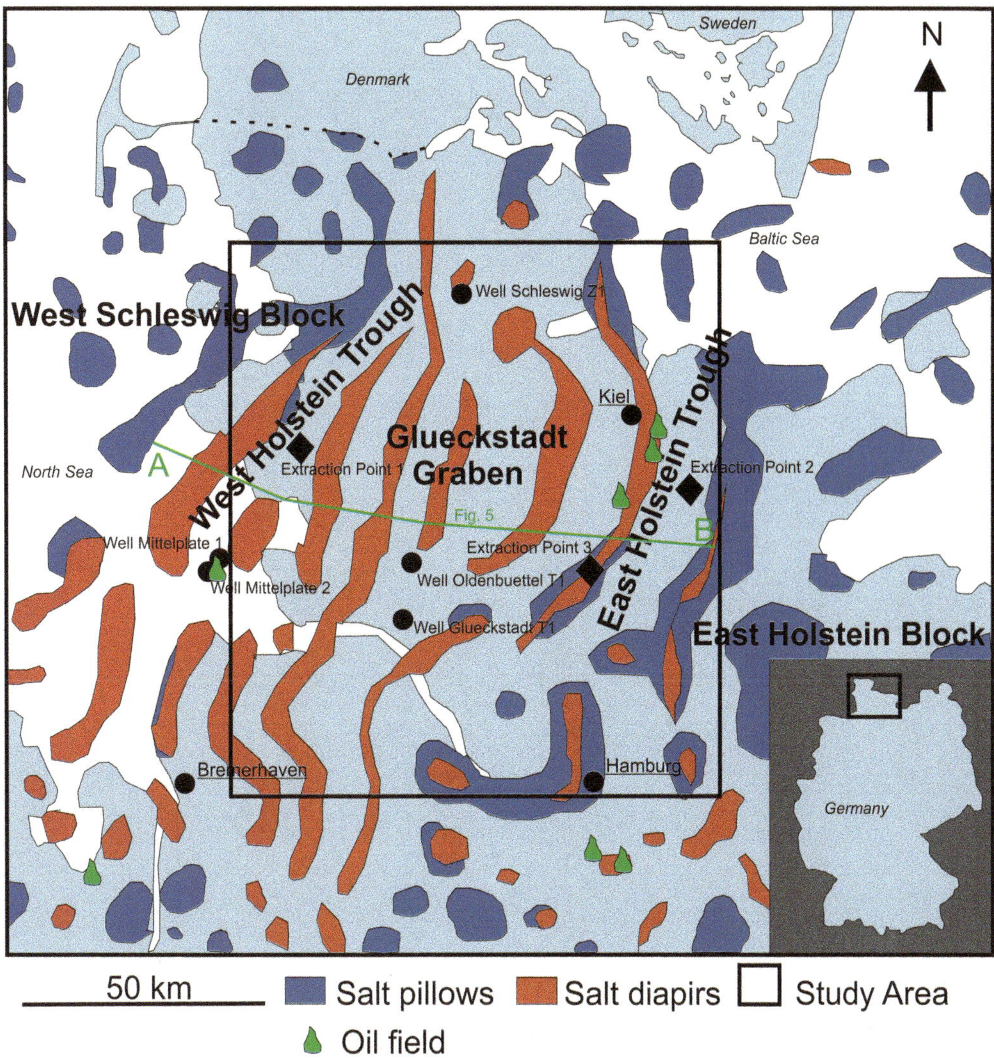

Fig. 1. Location of the study area, including salt structures, wells used for calibration of the model and 1D extraction points in the EHT and WHT.

2005; Japsen *et al.* 2007; Beha *et al.* 2008; Maystrenko *et al.* 2012): ranging from 500 m north of the island of Rügen (Ostholstein Graben), to 700 m for the Glückstadt Graben, to 400 m at the eastern Horn Graben, 900 m for the North German Basin and to 2000 m in the Southern North Sea. These variations of eroded sediment thicknesses have been explained by the North Sea plume (Nielsen 2002). A higher degree of uplift and erosion has been deduced for the area west of the Bay of Kiel towards the centre of the dome (Hansen *et al.* 2005). Layers of Early Cretaceous age consist mainly of marlstones. Subsidence since the early Late Cretaceous led to deposition of marine shales and chalks, discordantly covering Jurassic and Triassic sequences. The Late Cretaceous sequence is dominated by a 400–500 m-thick sequence of chalky sediments (Mutterlose & Bornemann 2000), deposited under shallow-marine conditions, subsequently changing to fairly deep-marine conditions. As a result of the collision of Iberia and Europe, parts of the North German Basin were inverted, but there are no signs for inversion tectonics in the Glueckstadt Graben (Kockel 2002; Kley & Voigt 2008). In the Graben system, additional subsidence occurred, increasing during the Cenozoic and lasting until the present day

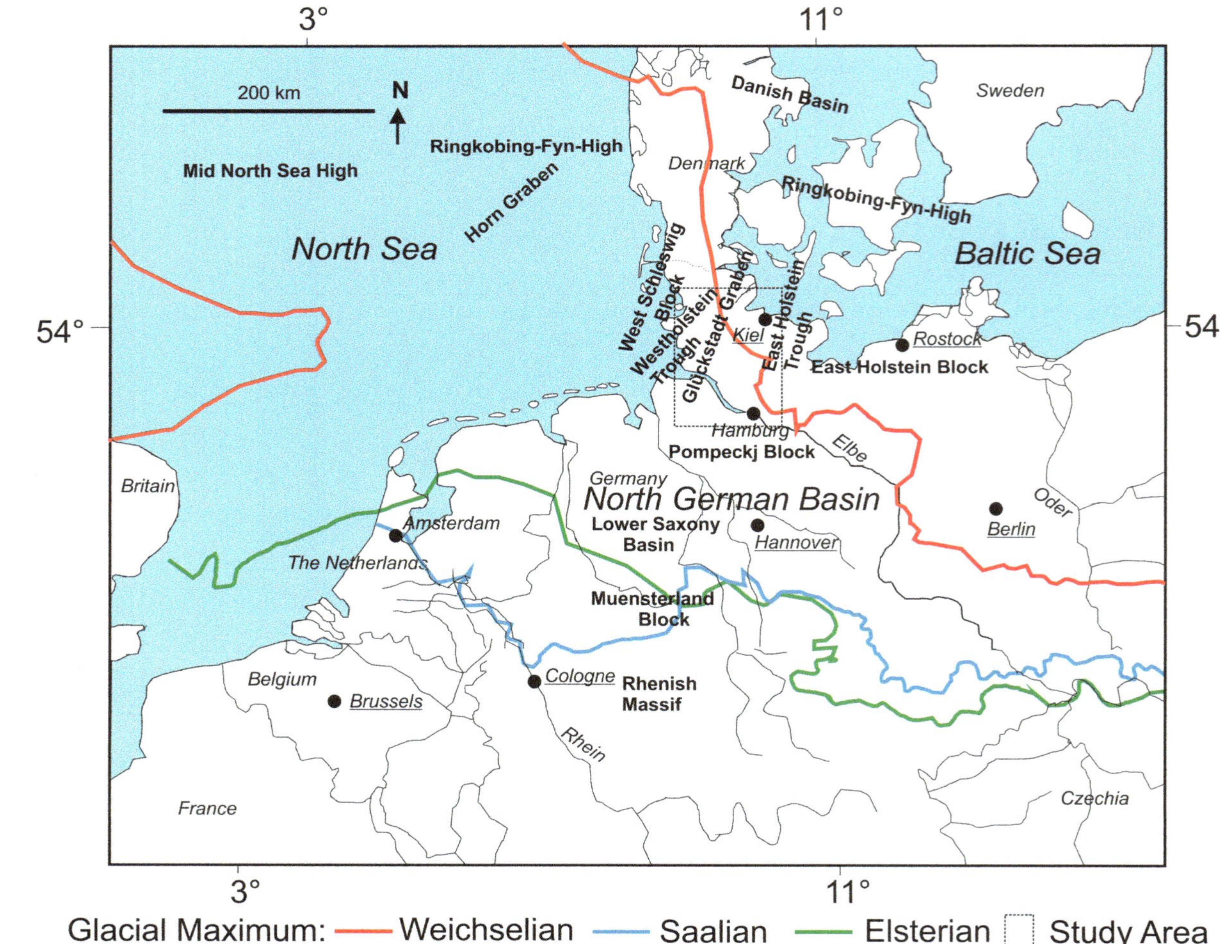

Fig. 2. Overview map of central Europe and the southern maximum extent of the main ice phases (after Ehlers *et al.* 2004).

(Sirocko *et al.* 2008). Simultaneously, salt diapirs rose in the fringe of the graben and existing salt diapirs further advanced (Strozyk *et al.* 2017). Most of the salt diapirs are today covered by Cretaceous and Tertiary sediments, while others are exhumed or appear close to the present-day surface (Segeberg, Elmshorn). Halokinesis influenced the facies distribution during the Paleocene and Eocene, when terrestrial and shallow-marine siliciclastics were deposited. Regressional phases occurred during the lower and middle Paleocene, turning to transgressional phases during the upper Paleocene and Eocene, with a maximum during the Early Oligocene (Rupelian: Reicherter *et al.* 2008). Further minor transgressional phases occurred during the Late Oligocene and the Early Miocene, and led to deposition of sands interfingering with lignites at the basin margins. During the Pliocene, marine transgression covered only the area between Sylt and Hamburg, whereas the larger part of the North German Basin was characterized by fluvial or limnic sediments. The highest sediment thickness of Permian–Quaternary rocks in the North German Basin is 8–10 km and occurs in Schleswig-Holstein and Mecklenburg (Bachmann & Grosse 1989). A generalized stratigraphy is shown in Figure 3.

Petroleum system elements

Two major petroleum systems are known from the NW European Basin: the Paleozoic gas system; and the Mesozoic oil and gas system (di Primio *et al.* 2008). The focus of this study is on the Mesozoic petroleum system, which contains Jurassic and Early Cretaceous petroleum source rocks in northern Germany, of which only the Lower Toarcian Posidonia Shale is relevant for the study area (Fig. 1). Sandstones of Middle Jurassic (Dogger) and Early Cretaceous (Valanginian) age act as reservoirs, and evaporites and Mesozoic shales are the most important seals. The most prominent example for this petroleum system in northern Germany is the Mittelplate Field in the Heide Trough (Grassmann *et al.* 2005), which is part of the West Holstein Trough (WHT). This trough, as well as the East Holstein Trough (EHT), contain thick Early and Middle Jurassic sediments, including the Posidonia Shale. All of the known oil fields of Schleswig-Holstein are located in these Jurassic troughs (Fig. 1).

Glaciation cycles in Schleswig-Holstein

Glaciation and deglaciation may influence sedimentary basins with respect to erosion and deposition, as well as changes in pressure and temperature in the subsurface. Furthermore, ice load and unload led to changes in regional isostasy (Kjemperud & Fjeldskaar 1992). These processes can have an impact on hydrocarbon generation, migration, accumulation and leakage (Lerche *et al.* 1997). Several glaciation phases, advancing from Scandinavia, have covered Schleswig-Holstein since 400 kyr ago (i.e. Litt *et al.* 2008) (Fig. 2): that is, the Elsterian (400–320 ka), the Late Saalian (300–130 ka), the Early Weichselian (100–80 ka), the Middle Weichselian (60–50 ka) and the Late Weichselian (25–15 ka). The spatial and temporal extent of ice coverages are shown in Figures 2 and 3, as well as in Table 1, following Ehlers *et al.* (2004) and Graham *et al.* (2011), with maximum southwards extensions of glaciers occurring during the Saalian and Elsterian periods in north Germany. The Elsterian glaciation advanced in two cycles, first from the north and later from the NE. The total extent of these ice sheets is not well known, while a coverage of the entire Schleswig-Holstein is assumed (e.g. Ehlers 1990; Ehlers & Gibbard 2004) (Fig. 2). The Holstein interglacial separates the Elsterian ice age from the Saalian. A minimum of three great glacial advances (possibly four in Schleswig-Holstein) occurred during the Saalian. Coverage of the entire area of Schleswig-Holstein is likely, possibly covering the area permanently as there is no evidence of interglacial deposits. Ice thicknesses for the Elsterian and Saalian glaciers are not well known, but Mahaney (1995) assumed ice thickness of 800–1000 m for pre-Weichselian times. The Saalian stage was followed by a warm period, the Eemian stage. A subsequent cold period occurred during the Weichselian. It is assumed that glaciers advanced between 70 and 50 ka in the area of the western Baltic Sea/eastern Denmark, although the onset of the main glaciation started at around 25 ka before present in central Europe (Litt *et al.* 2008), and led to at least three ice cycles advancing from the NE. Some authors also subdivide the Weichselian glaciation into five (Stephan *et al.* 1983; Stephan 1997) or six (Piotrowski 1996) ice advances. A conceptual reconstruction of the ice thickness during the Weichselian in NE Germany revealed a maximum thickness of 280 m (Piotrowski & Tulaczyk 1999) and 500 m (Ehlers 1990). However, the Late Weichselian glaciers only covered a small margin of the coastal zone south of the Baltic Sea. The last cooling event prior to the Holocene is the Younger Dryas Stadial between 12.7 and 12.6 ka before present (Litt *et al.* 2008).

Methods and model input

Model input: sequence of events

To assess the influence of ice sheets on the petroleum system in the study area, 3D basin models were constructed using the Schlumberger PetroMod® software v.2014. Details on the software and the workflow are described in Hantschel & Kauerauf

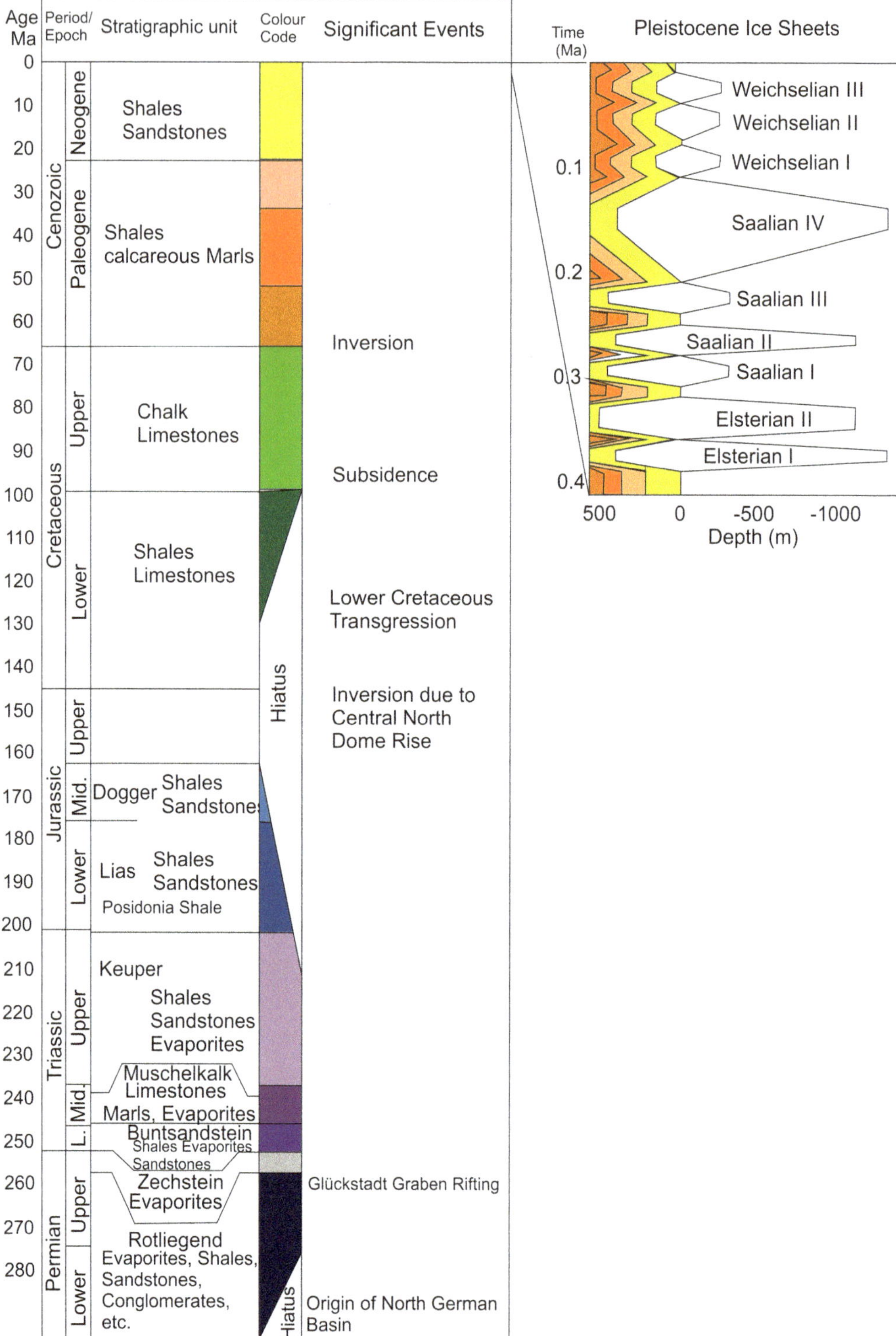

Fig. 3. Stratigraphy, general lithology, major events and assigned ice-sheet history for the example of the Schleswig well area (modified from Rodon & Littke 2005; Warsitzka *et al.* 2017).

Table 1. *Age assignment and lithological properties*

Layer	Main lithology	Depositional age (Ma)	Thermal conductivity at 20°C/100°C ($W m^{-1} K^{-1}$) For ice: at −20°C/0°C	Heat capacity at 20°C/100°C ($kcal kg^{-1} K^{-1}$) For ice: at −20°C/0°C
Quaternary	Sandstone, shale	0–0.01	2.55/2.39	0.20/0.237
Weichselian III	Ice	0.02–0.03	2.39/2.22	0.46/0.49
Weichselian II	Ice	0.05–0.06	2.39/2.22	0.46/0.49
Weichselian I	Ice	0.08–0.1	2.39/2.22	0.46/0.49
Saalian IV	Ice	0.13–0.16	2.39/2.22	0.46/0.49
Saalian III	Ice	0.22–0.23	2.39/2.22	0.46/0.49
Saalian II	Ice	0.26–0.27	2.39/2.22	0.46/0.49
Saalian I	Ice	0.29–0.3	2.39/2.22	0.46/0.49
Elsterian II	Ice	0.3–0.35	2.39/2.22	0.46/0.49
Elsterian I	Ice	0.37–0.38	2.39/2.22	0.46/0.49
Lower Miocene	Sandstone, shale	16–23	2.55/2.39	0.20/0.23
Rupelian–Upper Oligocene	Sandstone, shale	23–28	2.55/2.39	0.20/0.23
Middle Oligocene–Upper Eocene	Sandstone, shale	28–49	2.55/2.39	0.20/0.23
Upper Paleocene–Lower Eocene	Sandstone, shale	49–55	2.55/2.39	0.20/0.23
Upper Cretaceous	Limestone, marl, shale	62–99	2.38/2.25	0.20/0.23
Lower Cretaceous	Shale, siltstone, limestone, sandstone	99–129	2.24/2.13	0.21/0.24
Upper Jurassic	Sandy shale	137–165	2.54/2.36	0.21/0.24
Dogger	Sandstone	165–181	3.95/3.38	0.20/0.24
Posidonia	Shale	181–182	0.9/1.15	0.22/0.26
Liassic	Shale, sandstone	182–200	2.13/2.06	0.2/0.24
Keuper	Sandstone, shale, evaporates	205–235	3.05/2.73	0.2/0.24
Röt/Muschelkalk	Limestone, shale, sandstone	235–244	2.24/2.13	0.20/0.24
Bunter	Sandstone, siltstone, shale	244–251	3.55/3.09	0.20/0.24
Zechstein	Salt, limestone	251–258	6.5/5.25	0.21/0.24
Basement	Basement	258–266	2.72/2.35	0.19/0.22

(2009) and Sachse & Littke (2016). A finite-element forward model was created to simulate the burial history of sediments in the Schleswig-Holstein area taking into account compaction, pressure, temperature and maturation of organic matter, and petroleum generation through time (Hantschel & Kauerauf 2009). This study focuses on the WHT and the EHT, separated by the Glückstadt Graben in-between. The model area is roughly bordered to the west by the present-day coastline of the North Sea, and to the east by the Ostholstein Mecklenburg Block. The modelled area covers approximately 50 600 km^2, 220 km in an east–west direction and 230 km in a north–south extent, with a total of nearly 5.7 million cells and a high horizontal resolution of 150 × 150 m. The model comprises 15 stratigraphic intervals extending between base Zechstein and the present-day topography (Table 1; Fig. 4). Below, a 2000 m-thick basement was implemented representing Permian, Carboniferous and older rocks (Figs 4 & 5). Corresponding horizon depth maps for Mesozoic and Cenozoic strata were adopted from the *Geotectonic Atlas of NW-Germany* (Baldschuhn *et al.* 1996) in their maximum initial resolution of 150 × 150 m. The depositional ages of each stratigraphic unit are defined based on the German Stratigraphic Chart (*Stratigraphische Handtabelle von Deutschland 2012*: Deutsche Stratigraphische Kommission 2012), and their lithologies have been compiled from Ziegler (1990), Rodon & Littke (2005) and Uffmann & Littke (2011) (Table 1). The Posidonia Shale was implemented as a layer of 35 m in thickness using the layer splitting tool. To consider the effects of salt diapirism on thermal conductivity and its effect on the sedimentary rocks, salt movement was implemented into the model using the Salt Piercing Tool of the software (Hantschel & Kauerauf 2009). Based on Warsitzka *et al.* (2017), the main phase of salt diapirism was set for the central study area (Glückstadt Graben) and for the

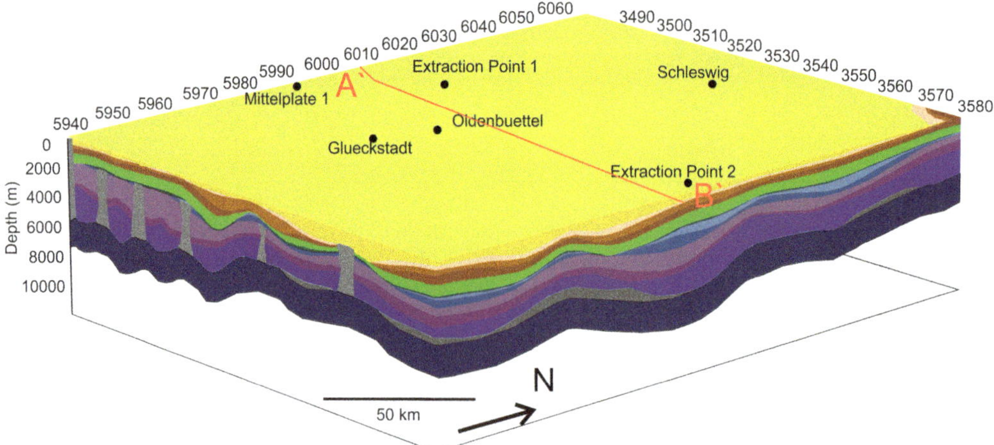

Fig. 4. 3D basin model for the study area including Zechstein salt (grey). The red line shows the location of the 2D cross-section in Figure 5. See Figure 3 for the colour code.

Middle Keuper age, ceasing in the Late Triassic. Early–Middle Triassic-aged salt diapirism was set for the salt structures in the WHT, with a maximum also occurring during the Early Jurassic in the EHT, and continuing during the Cenozoic into the WHT. Based on geological information (Grassmann *et al.* 2005; Scheck-Wenderoth & Lamarche 2005; Uffmann & Littke 2011), three erosional events were assigned as unconformities to the model: Late Cretaceous–Early Paleocene (Fig. 6a); Late Jurassic, which was split for modelling purposes into a Total Jurassic erosion and a Dogger erosion (Fig. 6c) as the erosion cut deeper in some areas; and a Late Triassic (Late Keuper; Fig. 6d) event. In a next step (second model), 11 additional layers were implemented to represent the glaciation history in the Quaternary. These layers represent the various ice sheets of the Elsterian, the Saalian and the Weichselian in northern Germany/Schleswig-Holstein. Ice thicknesses were assumed based on the reconstructed values of Grassmann *et al.* (2010), and in accordance with values published by, for example, Ehlers (1990), Mahaney (1995), Piotrowski & Tulaczyk (1999) and Ehlers *et al.* (2004).

The petrophysical properties assigned to the ice sheets are shown in Table 1. The ice is considered as impermeable, without porosity, incompressible and with the density of frozen water. Ice density was assigned with 0.917 g m^{-3}.

Boundary conditions

The upper thermal boundary conditions of the model were defined using the palaeowater depth and the surface–water interface temperature (SWIT). The SWIT is derived from palaeobathymetric data at the sediment surface–water contact and strongly depends on the palaeogeographical latitude of the

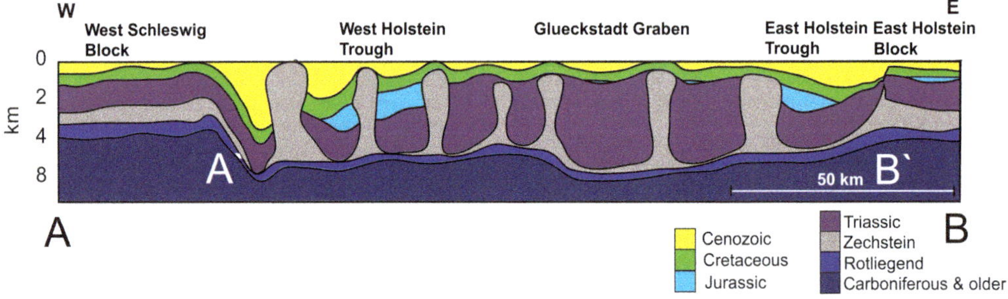

Fig. 5. 2D cross-sections through the West Schleswig Block, the West Holstein Trough (WHT), the Central Glueckstadt Graben, the East Holstein Trough (EHT) and the East Holstein Block (based on Maystrenko *et al.* 2005; Strozyk *et al.* 2017). Note that the location of the cross-section shown in Figure 4 (A′–B′) is shorter than the line presented in this figure (A–B).

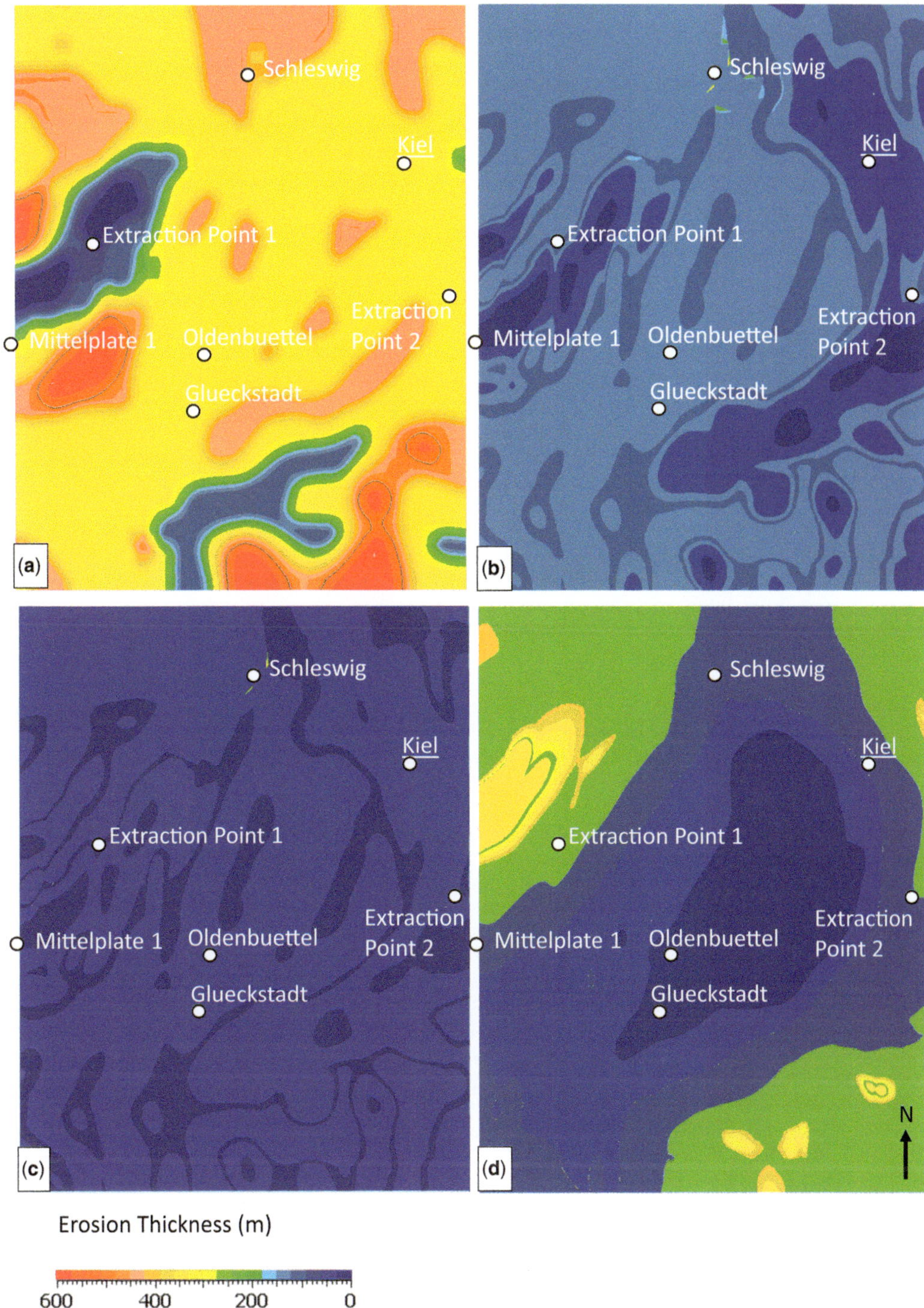

Fig. 6. Erosion maps for the Cretaceous inversion (**a**); the Jurassic erosion event, split into Total Jurassic erosion (**b**) and Dogger erosion (**c**); and Keuper (**d**). For the location see Figure 1.

basin and – in the case of marine conditions – on palaeowater depth (Wygrala 1989). The SWIT was basically calculated for latitude 54°, northern Europe, and adjusted for the glacial and interglacial periods. The interglacial ground temperature was assumed based on literature information (Delisle 1998; Delisle *et al.* 2007; Grassmann *et al.* 2010), and ranges between −5 and 7°C. The southern limit of continuous permafrost in Europe at the Last Glacial Maximum time (18–22 kyr before present) followed approximately the 50° N latitude, with a permafrost temperature of −3°C (Frenzel *et al.* 1992). Permafrost thickness increased toward the north from *c.* 200 m at 50° N (in accordance with Grassmann *et al.* 2010) to more than 800 m on the present-day Arctic Sea coast in ice-free areas (Frenzel *et al.* 1992). While ice-free areas experienced low temperatures during the Weichselian, ground temperatures beneath a glacier may have been substantially higher at the same time. For pure ice, it is −0.9°C for a 1 km-thick glacier and −1.7°C for a 2 km-thick glacier (Paterson 1994). Siegert & Marsiat (2001) proposed mean annual surface temperatures (upper thermal boundary in modelling) from numerical ice-sheet modelling that ranged between −35 and −16°C for ice thicknesses of between 750 and 1000 m, supported by values presented by Cavanagh *et al.* (2006) and Rodrigues Duran *et al.* (2013) (−15°C for 500 m ice thickness and −45°C for 1750 m of ice thickness). Our model followed initially those authors, is in accordance with those values, and results in temperatures of between −2 and 0°C at the sediment–ice boundary. Palaeobathymetric data were implemented from various sources (Ziegler 1990; Kockel 2002; Miller *et al.* 2005; Littke *et al.* 2008). This model assumes a warm base ice sheet. Flowing water below the ice sheets is not considered due to software limitations and the size of the model (see Cavanagh *et al.* 2006).

The lower boundary condition is described by the basal heat flow and followed basically the assumptions made in previous basin and petroleum systems modelling studies in this area: for example, by Neunzert *et al.* (1996), Friberg (2001), Grassmann *et al.* (2005), Rodon & Littke (2005), Senglaub *et al.* (2006), Muñoz *et al.* (2007) and Uffmann & Littke (2011). Rifting phases were considered to be related to high basal heat flows declining over time (McKenzie 1978), while subsequent thermal rebound happens gradually over time until constant heat flow is reached (Waples 2001). The heat flow in our model was set to a constant 50 mW m^{-2} since the Late Permian–Early Triassic and 70 mW m^{-2} before. As no volcanic activity is known for the modelled time frame, extremely high heat flows can be excluded. Volcanic activity, which occurred in pre-Zechstein periods in the study area, is not included in the model.

Calibration

For the calculation of vitrinite reflectance, we utilized the Easy-R_o approach by Sweeney & Burnham (1990), which is applicable for a VRr range between 0.3 and 4.6%. To thermally calibrate the burial and temperature history of the 3D model, 1D extractions of the wells (Glueckstadt, Oldenbuettel, Schleswig and Mittelplate) in the WHT (extraction point 1) and in the EHT (extraction point 2) were created, and calculated vitrinite reflectance and downhole temperature data were compared with measured data from the well locations. The wells (see the data from Rodon & Littke 2005) and extraction points are shown in Figures 1 and 4.

Results and discussion

Burial history of the Glueckstadt Graben and the EHT and WHT

Initially, burial and temperature histories were reconstructed for different wells, for which vitrinite reflectance data were available for calibration (Fig. 7a–c). The burial histories include three phases of burial followed by erosion (Fig. 6); the earliest one occurred in the Late Triassic (Late Keuper) on the margins of the deeper buried structures heralding erosion values of up to 400 m. A second, minor erosional phase was applied to the Late Jurassic, partially truncating the sediments down to Dogger (maximum erosional thickness 150 m). The third phase applies to the Alpine Orogeny and was set for the time of Late Cretaceous–Early Paleocene inversion, causing erosion of Cretaceous sediments up to 450 m, especially above salt diapirs, and forming the Cretaceous–Tertiary Unconformity (Strozyk *et al.* 2017). The erosion maps are shown in Figure 6. However, deepest burial and highest temperatures of sediments were reached at present day in the entire study area (Fig. 7a–c). These results support previous studies by Grassmann *et al.* (2005) and Rodon & Littke (2005), who additionally assumed deep burial and high temperatures at the end of the Jurassic in parts of the area. Such a significant Jurassic deep burial followed by erosion is, however, not necessary for our model calibration. It is still under debate whether Jurassic sediments were eroded (e.g. Rodon & Littke 2005), or only deposited within rim synclines between inflating salt highs (e.g. Duin *et al.* 2006; Wong *et al.* 2007; Strozyk *et al.* 2017, 2018). Missing erosional truncations of Jurassic strata in regional seismic data might support localized deposition in salt rim basins (Maystrenko *et al.* 2005, 2012; Strozyk *et al.* 2017) (see also Fig. 5). On the other hand, such a depositional pattern would imply strong facies differentiation

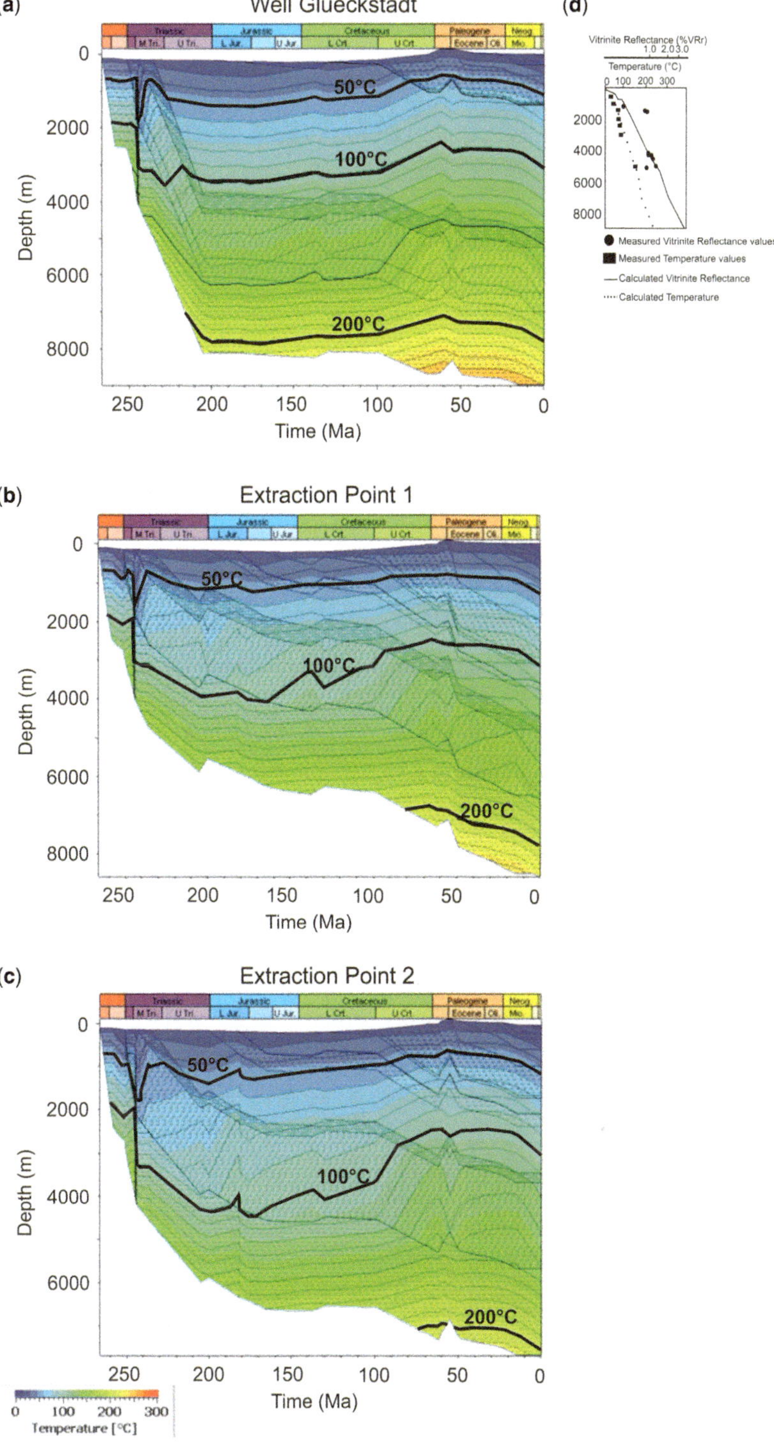

Fig. 7. Burial histories with temperature overlay and thermal isolines for the Glueckstadt well (**a**), extraction point 1 in the WHT (**b**) and extraction point 2 in the EHT (**c**). (**d**) shows the calibration data for the Glückstadt well. For the location see Figure 1.

between terrestial/shallow-marine and deeper-marine environments for the Jurassic which is, at present, not substantiated by any data. In contrast, the marine Toarcian Posidonia Shale is known to represent a rather continuous organic-matter-rich facies throughout the basin indicating widespread, uniform deposition. We here applied the (conservative) idea of very minor Jurassic–Cretaceous erosion (Fig. 6) in order to keep the model as simple as possible. It should be noted, however, that there are important arguments for a stronger early–mid Jurassic sedimentation followed by erosion.

In any case, temperature trends for the Keuper sedimentary sequence are, in general, in accordance with those calculated by Rodon & Littke (2005), but Mesozoic temperatures tend to be lower based on the assumption of lower sedimentation/erosion and slightly lower heat-flow values. A present-day depth of approximately 4400 m is reached in the EHT and WHT for the base of the Keuper stratigraphic unit. Similar present-day depth values for the mid Keuper unit are reached in the central study area (well Glueckstadt: 4600 m). Differences in the burial histories between the central study area (Fig. 7a) and the East and West Holstein troughs (Fig. 7b, c) are observed regarding the sedimentary infill. Whereas the Jurassic troughs show a constant catchment history since the Permian, virtually no sediments were deposited in the central area during the Jurassic and the Cretaceous (unless they were completely eroded; see the discussion above). The dominant sedimentary unit in the central area is the Triassic, thinning out in the Jurassic troughs. In consequence, deposition of the Keuper stratigraphic unit had the strongest impact on the total burial depth in the central area, but not in the Jurassic troughs (Fig. 7). These differences are the result of the initial formation ages of these graben systems, whereof the Glueckstadt Graben is much older than the EHT and WHT. In addition, salt diapirism had an important influence on sediment accommodation space and subsidence, and also on the temperature field (Neunzert *et al.* 1996; Rodon & Littke 2005). As shown by Warsitzka *et al.* (2017), the salt diapirism in the central Glueckstadt Graben was strongest during the Middle Keuper and was mostly finished in the Late Triassic. Early–Middle Triassic initiation of diapirism is evident for the salt structures in the WHT based on Strozyk *et al.* (2017), with a renewal during the Middle Keuper and a maximum during the Early Jurassic. This is in accordance with uplift ages for the salt structures in the EHT (Warsitzka *et al.* 2017). A new phase of diapirism in the WHT occurred in the Cenozoic. Hence, the history of salt diapirism in the Glueckstadt Graben and in the EHT and WHT differs greatly. Furthermore, there are significant thickness differences in Jurassic and Tertiary deposits. Nearly 500 m of additional sediment was deposited in the Jurassic troughs since the Early Paleogene, and much less in the central area (Figs 5 & 7).

Maturity and temperature evolution for the Glueckstadt Graben and the EHT and WHT

Compared to the spatial and temporal evolution of salt diapirism and the related depositional history in the Glueckstadt Graben (Warsitzka *et al.* 2017), similar temperature and maturity trends are observed for the EHT and WHT, but also some differences (Figs 7, 8 & 9). There, temperature and maturity were increasing, especially since the Paleocene, while in the Glueckstadt Graben high temperatures and thermal maturities are dated back to the Late Triassic–Jurassic (Fig. 8a, b). Temperature values of both the mid Keuper unit and the mid Jurassic unit were generally lower since the Jurassic in both Jurassic troughs compared to the Glueckstadt Graben. For the mid Keuper unit in the WHT, highest temperatures of 130°C were calculated for the Miocene, exceeding those for the Glueckstadt Graben, while present-day values are 121°C (i.e. close to maximum values). Since the Late Jurassic, the mid Keuper unit temperature in the EHT was lower than in the WHT, maybe due to slower thermal subsidence related to salt movement. The temperature trend for the mid Keuper unit is similar for all locations with a steady increase during the Triassic and Jurassic. Up to 75°C had been reached in the early Early Cretaceous at well locations Oldenbuettel and Glueckstadt, and 77°C at well location Mittelplate 1. During the Early Cretaceous, a temperature drop occurred due to uplift followed by a period of temperature increase during the Late Cretaceous from renewed burial, reaching a maximum value of 84°C for the Schleswig well. Highest temperatures of 115°C are observed for this layer at the Mittelplate 1 well during the Eocene, while lowest values were calculated for the Schleswig well (see Figs 1 & 8a). Another drop in temperature occurred during the early Paleocene due to the Cretaceous–Tertiary inversion, but in the early Eocene temperatures started to increase slowly again, followed by a drop since the Miocene. This cooling is related to the sediment/atmosphere temperatures, which decreased from approximately 20 to 15°C and to 6.4°C in the Plio-/Pleistocene. Until today, temperatures have slightly decreased, reaching values of between 74°C (Schleswig well) and 110°C (the Mittelplate 1 well). The temperature distribution pattern for the Keuper unit (Fig. 9c) shows highest temperatures in the WHT (extraction point 1), with decreasing temperatures to the fringe areas of the trough. A similar pattern is observed for the EHT, with a temperature peak in the south, NW of the city of

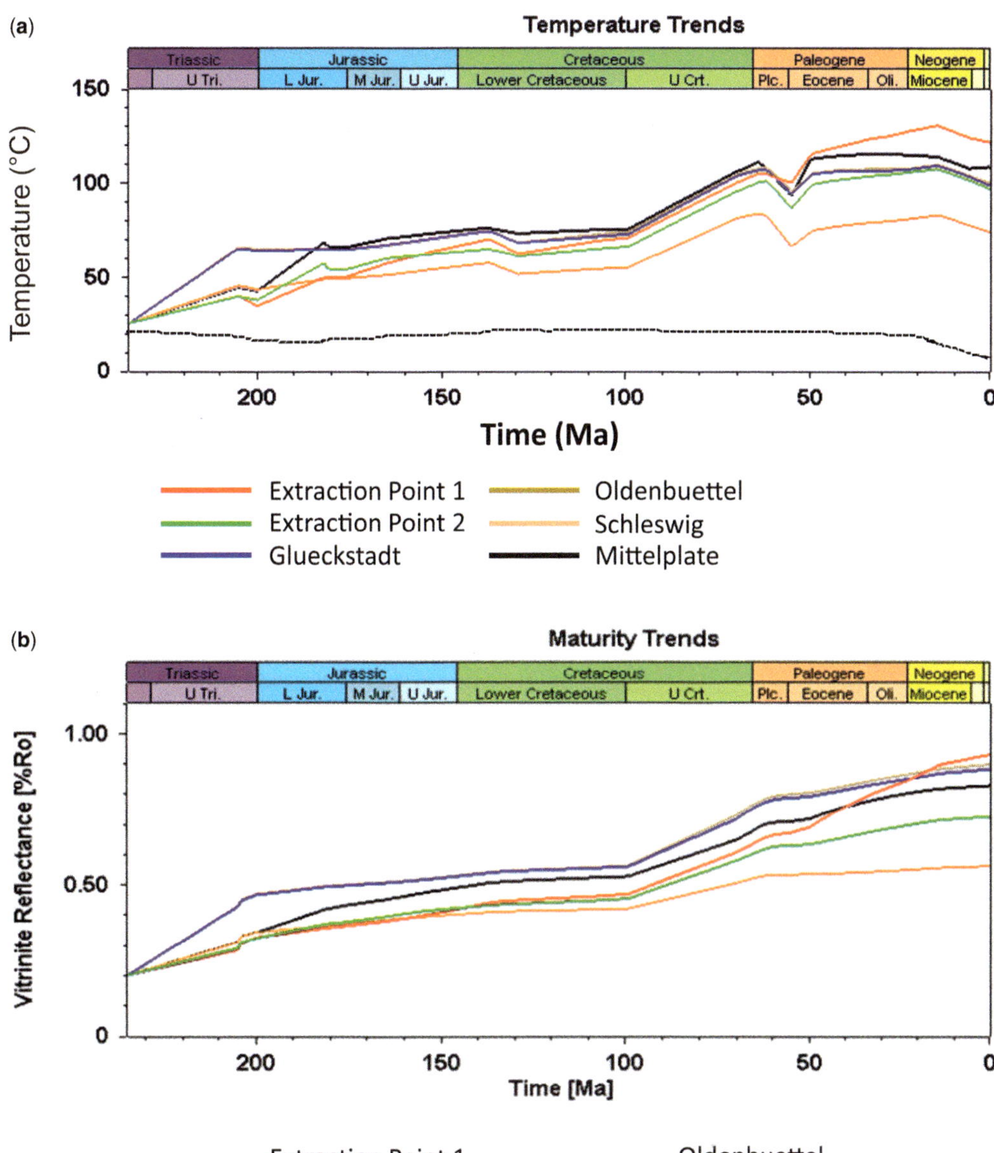

Fig. 8. Temperature, including the sediment–water interface temperature (dotted line) (**a**) and maturity (**b**) development for the mid-Keuper layer for the wells and the extraction points in the EHT and WHT. See Figure 1 for the locations and Figure 6 for the burial histories.

Hamburg. The central Glueckstadt Graben shows relatively cold temperatures, which also occur in the northern study area.

The thermal maturity pattern (Fig. 9a), as expressed by vitrinite reflectance, follows this temperature evolution and strongly relates to the highest temperatures during burial (Figs 7 & 9c). Highest values for the mid Keuper unit are calculated for the area of the Glueckstadt and Oldenbuettel wells (0.9% present-day vitrinite reflectance

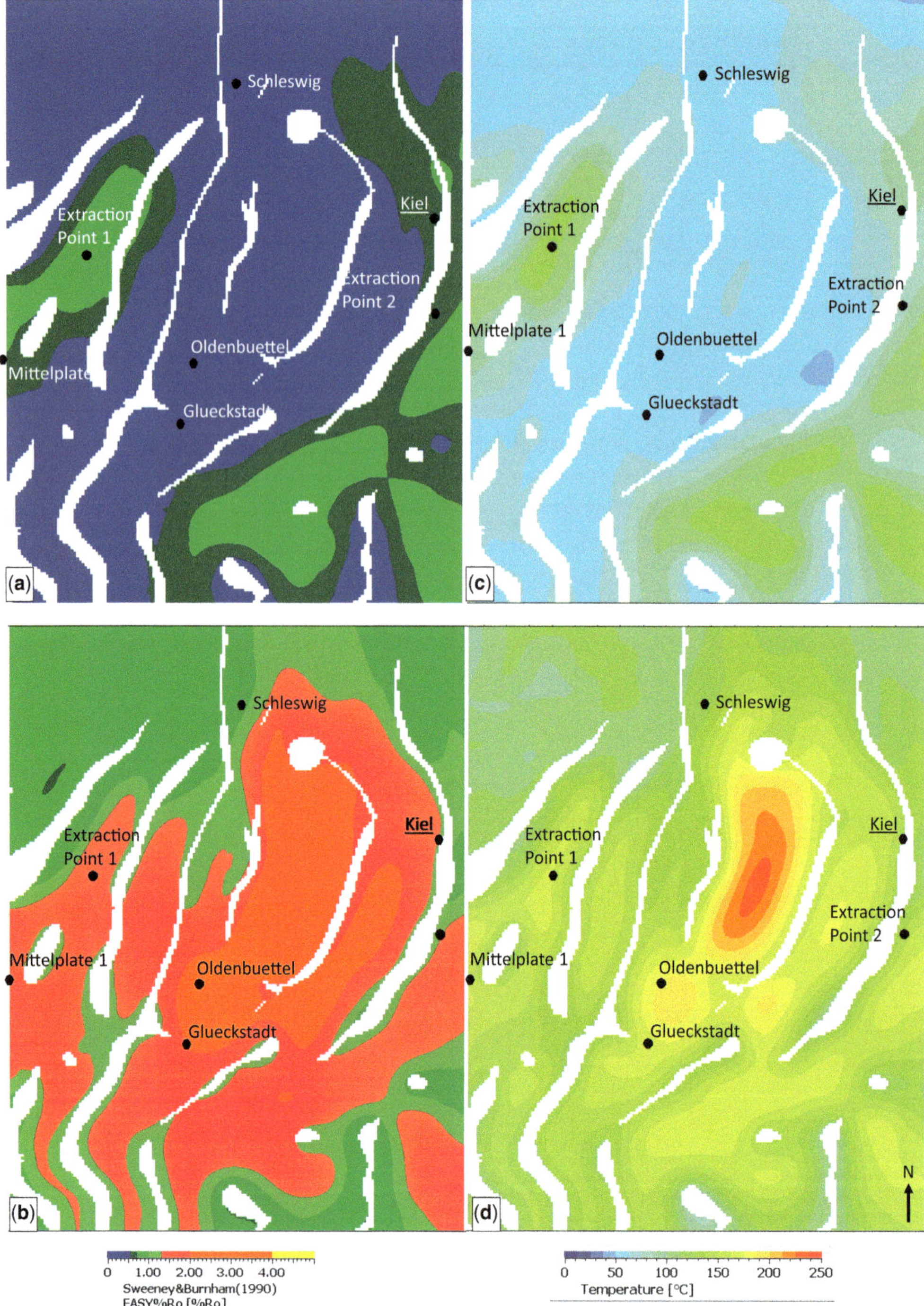

Fig. 9. Calculated present-day maturity maps for the mid-Keuper stratigraphic unit (**a**) and mid-Zechstein stratigraphic unit (**b**); and temperature maps for the mid-Keuper stratigraphic unit (**c**) and mid-Zechstein stratigraphic unit (**d**).

(VRr)), whilst the Schleswig well shows much less burial and thus a lower maturity increase (maximum of 0.56%VRr at the present day: Figs 7, 8 & 9). The maturity map for the study area supports the observation of highest maturity for the Keuper and post-Keuper layers in the Jurassic troughs, whereas the Glueckstadt Graben shows an immature/very early mature Keuper layer (≤0.53%VRr). Values calculated for the WHT tend to be higher than for the EHT (Figs 8 & 9), although there are significant differences between central and marginal parts.

The Zechstein layer shows significantly higher palaeo- and present-day temperatures as well as vitrinite reflectance values: 2.59% VRr at the Oldenbuettel well and 2.15% VRr at the Glueckstadt well. Only 1.3% VRr is calculated for the EHT and 1.6% VRr for the WHT (Fig. 9b): that is, the Early Triassic, Zechstein and older layers are much more mature in the Glueckstadt Graben than in the Jurassic Trough. Comparing the Keuper and Zechstein maturity maps, a change of maturity distribution pattern becomes obvious within the study area. In contrast to the Keuper layer, the Zechstein layer reaches the gas and even dry gas window in the Glueckstadt Graben. This maturity range is present over a wide area expanding from the south (the Glueckstadt well) to the north, close to the Schleswig well (Fig. 9d). Both Jurassic troughs are characterized by a much higher maturity of the Zechstein compared to the Keuper layer, but clearly less than in the Glückstadt Graben (wet gas stage).

This maturity distribution is also depicted in the temperature distribution map for the Zechstein layer, which shows highest temperatures for the central part of the Glueckstadt Graben to the SW of the city of Kiel (Fig. 9d). The differences between the Zechstein and the Keuper layer are based on the different burial histories, as the Glueckstadt Graben experienced a much earlier and deeper burial than the younger West and East Holstein troughs (Fig. 7a–c). However, in contrast to the Glueckstadt Graben, the Jurassic troughs witnessed a much stronger maturity increase since the Jurassic, especially the WHT (Fig. 8). Our results of maturity distribution hence differ slightly from those published by Rodon & Littke (2005) for the Keuper and the Zechstein layers, mainly expressed as a higher maturity for the Zechstein and a lesser maturity for the Keuper in the Jurassic troughs.

Posidonia Shale maturity

The maturity distribution of the Posidonia Shale shows early oil window maturity in the fringe areas and oil window maturity in the deepest parts of the WHT and EHT, especially in the EHT to the NW of Hamburg. The Posidonia Shale (if it were present) in the Glueckstadt Graben and in the northern part of the study area is expected to be immature (Fig. 10a). Based on calculated maturity trends, hydrocarbon generation in the Posidonia Shale occurred since the late Upper Cretaceous, increased during the Paleogene and is reaching highest values at the present day. In the Jurassic troughs, a maximum of 20% transformation ratio would be reached in the southern part of the EHT to the NW of Hamburg, and slightly lower values occur in the central part of the WHT (13%: extraction point 1). All other locations show minor or no transformation (Fig. 10a, b). Therefore, only initial or no hydrocarbon generation can be expected based on the assigned kinetic parameters for the Posidonia Shale in the EHT and WHT (Fig. 10b). However, the Mittelplate oil field was charged by hydrocarbons generated by the Posidonia Shale, buried deeper to approximately 4 km in the western part of the WHT outside the study area (Grassmann *et al.* 2005).

Comparison to previous studies

The model was constructed with less erosion thicknesses for the Jurassic compared to the values given by Rodon & Littke (2005) and is in accordance with the values given by Uffmann & Littke (2011), assuming zero erosion for the Jurassic. However, the timing assumed for erosion of Jurassic strata differs from the latter authors, as they assume an event of erosion down to the Jurassic strata in the Late Cretaceous, which is unlikely for the Schleswig-Holstein area as the Lower Cretaceous seems to be widely distributed (non-eroded) based on seismic interpretation (Fig. 5). However, values presented by Rodon & Littke (2005) assumed between 0 and 1100 m eroded thickness for the Jurassic, with a best fit achieved assuming 700 m of erosion. If such an erosion has occurred and if the seismic interpretation regarding an intact Lower Cretaceous layer is correct/this erosion has to be linked to a Late Jurassic very early Cretaceous (Kimmerian) phase. The heat-flow values assumed in our model are in accordance with Rodon & Littke (2005), with a heat peak of up to 70 mW m^{-2} at well Glueckstadt in the Permian–Triassic, decreasing to 55 mW m^{-2} through time, which is slightly higher than the values assumed in our model (50 mW m^{-2}). As shown by the authors, a slightly higher heat flow of 60 mW m^{-2} would also be acceptable, which is in accordance to Uffmann & Littke (2011). Grassmann *et al.* (2005) postulated a constant heat flow of 50 mW m^{-2} for the area of Mittelplate. These variations in erosion and heat-flow values might result from the models' variances in lateral and vertical heat transport, lithology, radiogenic heat production, and the timing and distribution of salt movements. In any case, only the most recent heat flows are well calibrated by maturity and present-day temperature data.

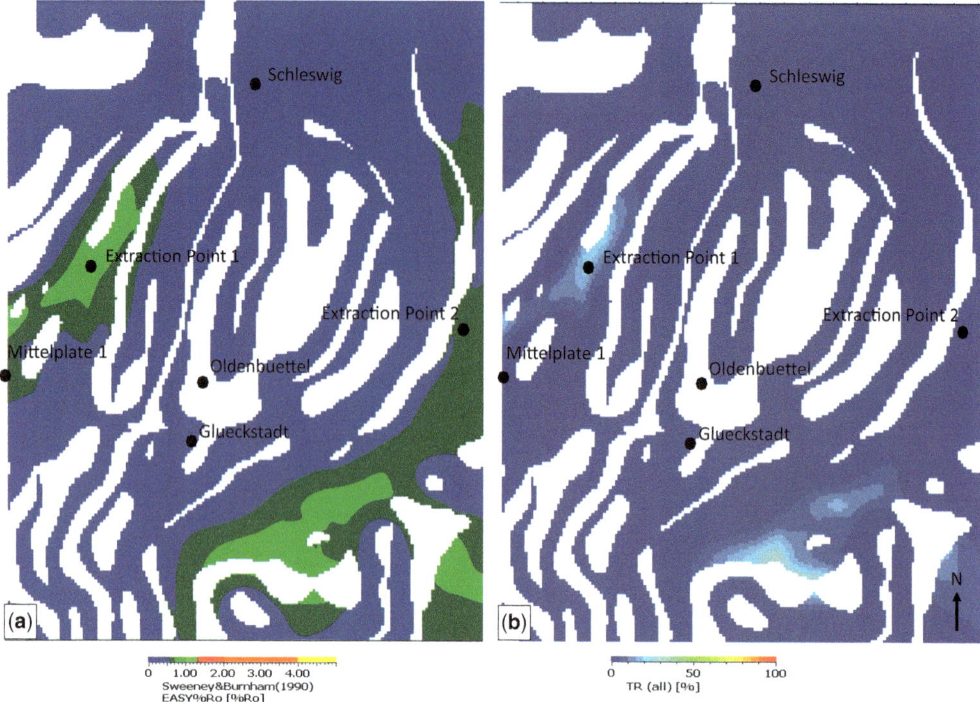

Fig. 10. Calculated present-day maturity (**a**) and transformation ratio (**b**) calculated for a hypothetical Posidonia Shale distribution.

Ice model: influence on temperature and maturity

In addition to the basic model presented and discussed above, a reconstruction of the effects from ice loads was conducted by implementing the various phases of Quaternary glaciation in northern Germany described by several authors (e.g. Ehlers 1990) (Table 1). As inferable from Figures 2 and 3, a spatial and temporal change of the extent and thickness of ice sheets occurred. In contrast to the model without ice-sheet coverage, lower temperatures are observed in the ice-load model since the beginning of the glacial episodes. For source rock maturity development and, thus, petroleum generation, the ice-load history does not have any significant consequence (temperature is reduced during glaciation periods). However, it might have a significant effect on petroleum expulsion and on the petroleum accumulations in the major reservoir rocks (i.e. the Dogger sandstone reservoir layers). These reservoirs in our simulations show temperatures that are 4–7°C lower than in the non-ice model depending on the location and the depth of the Dogger (Figs 11 & 12; shallowest parts east of the Glueckstadt well). This finding matches with observation by Grassmann *et al.* (2010) for the Mittelplate Dogger reservoir and is also in accordance with calculated values of, for example, Fjeldskaar (2014). These authors calculated a temperature drop of up to 10°C in the reservoirs of the Barents Sea. Other studies (Zieba & Grøver 2016) revealed a much higher temperature reduction for the Barents Sea reservoirs, reaching between 20 and 25°C, as each glaciation cycle led to a temperature drop of about 2–3°C. The differences between the temperature calculations thus reflect the uncertainties in the model calculations, being highly dependent on the kind of modelling, locality, lithology, ice thickness, duration of glaciations, etc.

Figure 11a shows the effect of ice load and associated temperature shift for three layers at the location of the Schleswig well (no salt): Liassic, Lower Cretaceous and Upper Cretaceous. The deepest layer, the Liassic, shows only a minor temperature drop during the first ice load (Elsterian I), which indicates that temperature effects from ice sheets can be neglected at great depth, depending on the ice thickness. In contrast, the upper, shallower layers (i.e. Lower/Upper Cretaceous) are much more affected with respect to temperature, in particular during deposition of the thickest ice sheet in the Saalian. Somewhat different impacts on temperature variation

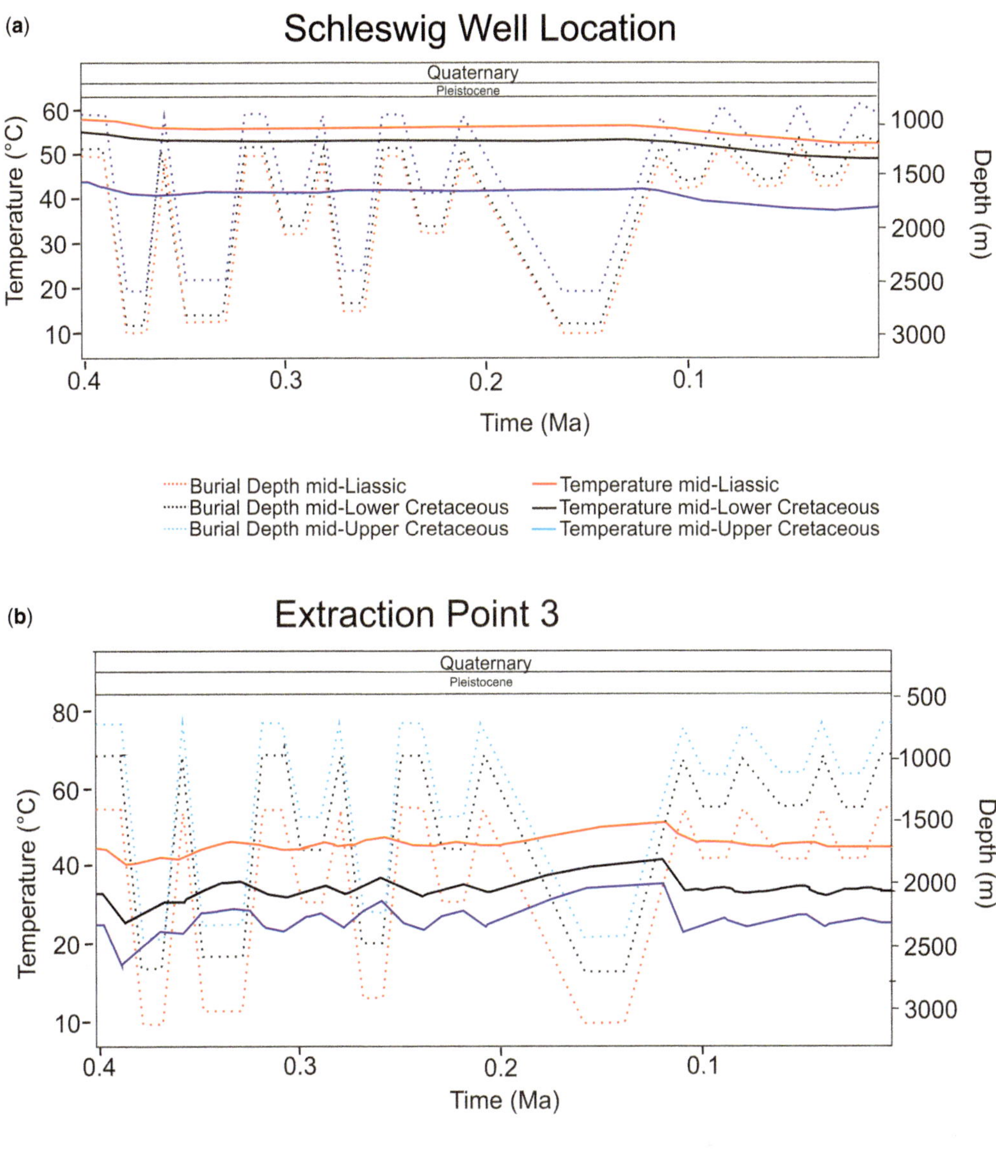

Fig. 11. Temperature and burial history for mid-Upper Cretaceous, mid-Lower Cretaceous and mid-Liassic units showing the effect of ice-load temperatures on depth for the location of the Schleswig well (**a**) and in the area of a salt diapir (extraction point 3, (**b**)) (see Fig. 1 for the location).

during ice load (Fig. 11b) are observed for salt diapirs, as well as in the surrounding sediments, due to the high thermal conductivity of salt (Mello 1995). Basically, the higher salt thermal conductivity leads to lower temperatures below and within the salt compared to the surrounding sediments (Neunzert *et al.* 1996). However, this well-known 'salt chimney' effect also has consequences for temperatures during short-termed glaciation. In the areas of salt diapirs, the temperature effects of glaciation reach deeper into the sedimentary column than in other sedimentary layers and reach the deep sedimentary

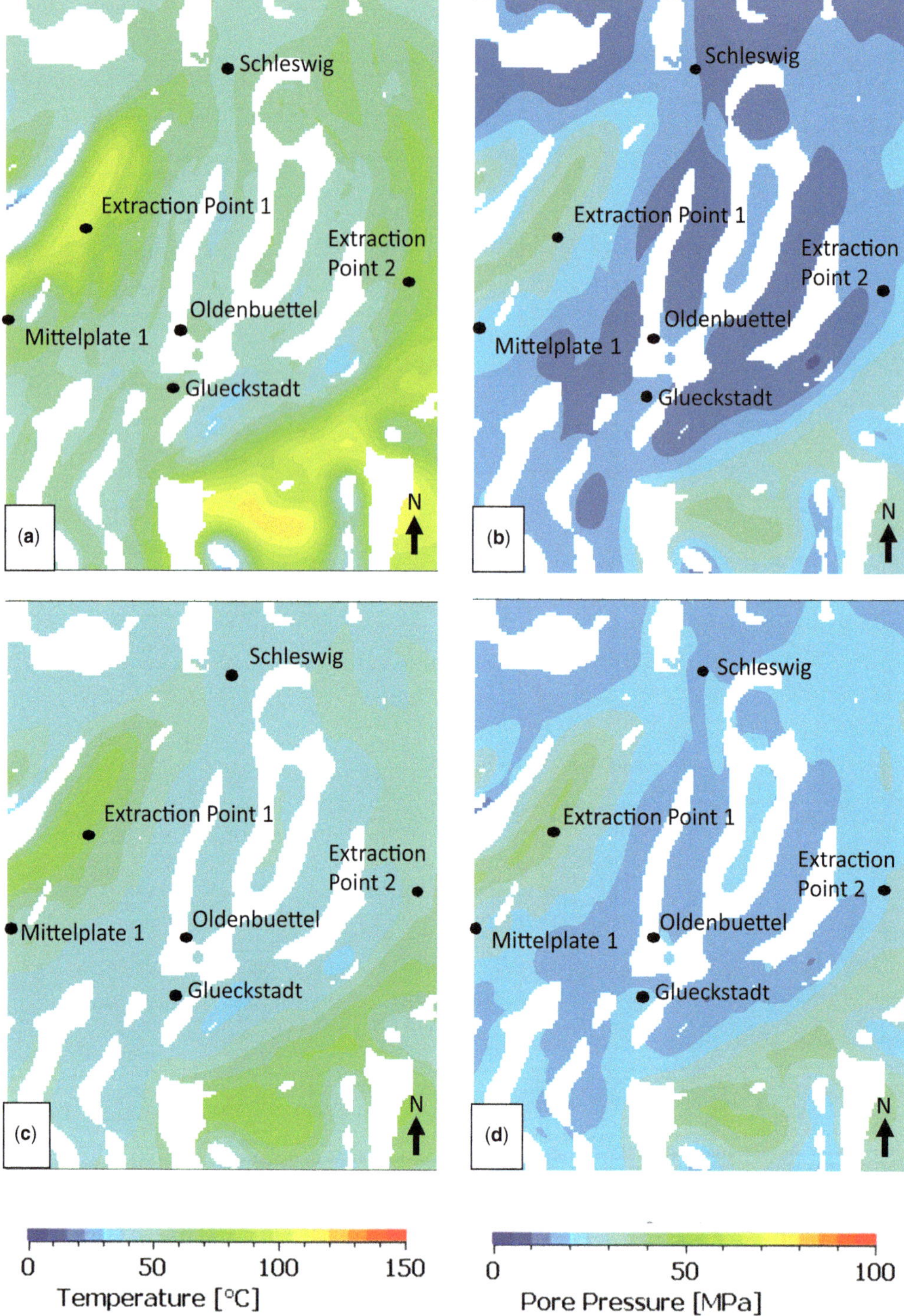

Fig. 12. Temperature (**a**) and pore-pressure (**b**) maps for the Dogger at the present day; the pore-pressure map during the first ice episode (0.38 Ma) (**c**) and immediately after ice retreat (0.36 Ma) (**d**). Compare with Figure 3 for ice episodes.

layers faster (Fig. 11b). A temperature increase immediately after the ice load is observed close to salt areas (Fig. 11b), where the temperature decreases rapidly after ice retreat. The high thermal conductivity of salt leading to a salt chimney effect contrasts with the medium thermal conductivity of ice (much lower than that of salt), leading to heat insulation. Thus, close to the top of salt structures, an increase of temperatures can be observed, in particular between salt and ice. After ice retreat, the original temperature profile is restored faster than in areas without salt (Fig. 13).

As shown, the temperature of the sediments surrounding the salt diapir is influenced both by the salt and by the ice sheet. In consequence, reservoir and even source rock temperatures close to a salt diapir can be lowered or increased significantly. Such temperature changes range between 4°C (depth of the Liassic layer *c.* 2200 m) and 8°C (depth of the Upper Cretaceous layer *c.* 1600 m) in the study area. Effects are expected to be even stronger in areas with thicker ice sheets, such as in Scandinavia (Delisle *et al.* 2007, Sirocko *et al.* 2008; Graham *et al.* 2011).

Furthermore, the lowering of reservoir temperatures can affect oil quality, because reservoir temperature reduction in oil fields below 80°C could have a biodegradation impact. On the other hand, the short duration of the glacial cycles and the many fine-grained, shaly intervals above the Dogger reservoirs might act against such biodegradation (see also Goffey *et al.* 2016; Goffey 2017). The duration and rate of biodegradation are, in general, very poorly constrained. In case the of the Mittelplate Field, there is little evidence of any effects on oil quality.

Ice model: influence on pressure and porosity

The impact of ice sheets on the pressure development is possibly even more important than the temperature effect: for example, on the Liassic Posidonia Shale source rock and the Dogger petroleum reservoirs in the Jurassic troughs. Pressure and stress depend on the maximum ice-sheet thickness, and its spatial and temporal advance and retreat. Highest present-day pore-pressure values for the Dogger layer are revealed for the SW parts of the EHT and WHT, with maximum values of *c.* 50 MPa, while lower values of *c.* 20–30 MPa would be expected in the Glueckstadt Graben for a few locations where the Dogger is present (Fig. 12b). This difference is mainly due to the depth of burial of the Dogger layer, as well as to the presence and thickness of overlying shale.

Figure 12c shows the increase in pore pressure in the entire area during glaciation, indicating a maximum of 20 MPa in the Jurassic troughs and the Glueckstadt Graben. With the retreat of the ice sheets, pore pressures decreased rapidly, almost reaching the initial values (Figs 12d & 14). Nearly initial values (i.e. pressure values without ice loading) have also been reached after retreat of the last glacial ice (Weichselian), when the ice sheets were rather thin compared to the Elsterian and Saalian glaciations (see Fig. 3). The lithostatic pressure in the layer changes from 24 to 40 MPa in the deeper parts of the study area (i.e. in the WHT and, in particular, the EHT) and go almost back to initial values immediately during ice retreat (Fig. 14). Consequently, formation pressures and sediment porosity can be influenced by the ice-sheet cover: that is, the duration of the ice load and the thickness of the sheet. Our model revealed strong porosity changes only in the uppermost layers (Miocene–Upper Cretaceous: Fig. 15a), whereas the Jurassic remained almost unaffected. Interestingly, the Elsterian and early Saalian ice episodes did not affect the porosity in an irreversible way, whereas a strong, constant impact was observed after deposition and retreat of the Saalian IV glacial event (1700 m ice thickness and duration of 0.03 myr; Fig. 15). Porosity decreased by up to 14% for the Miocene and 4% for the Upper Cretaceous in the east and north where glacial episodes (Elsterian I–Weichselian III: Figs 2 & 3) were complete and most intense.

Changes in porosity of layers older than Oligocene are permanent and no poro-elastic recovery is calculated by the software. This is different for the younger layers, as shown in Figure 15 for the Lower Miocene. After the most prominent Saalian IV, permo-elastic recovery of the Lower Miocene layer decreased, resulting in a further decrease in the porosity during the Weichselian ice ages to 38% and leading to a final porosity of 50% in the present intergalacial period (Fig. 15).

Clearly, the modelled porosities and pore pressures bear great uncertainty (e.g. Goffey *et al.* 2016) since this part of the model is not calibrated by real data and since exact knowledge of petrophysical properties is required. With respect to the Miocene, it would also be important to know more about the connections between the surface and the modelled aquifer intervals, and the permeability characteristics of those intervals. The rapid pore-pressure changes in Figure 14 suggest no well-drained aquifers driving the excess pore pressure over hydrostatic. If, for example, subglacial water reached a deep, well-drained aquifer, less overpressure would be expected. A further discussion on groundwater flow below glaciers and related tectonics are found in Thorson (2000) and Moeller *et al.* (2007). In addition, the above results depend, of course, on the real ice-thickness evolution which is another factor with significant uncertainty. As shown in Figure 15b, the effect on porosity depends

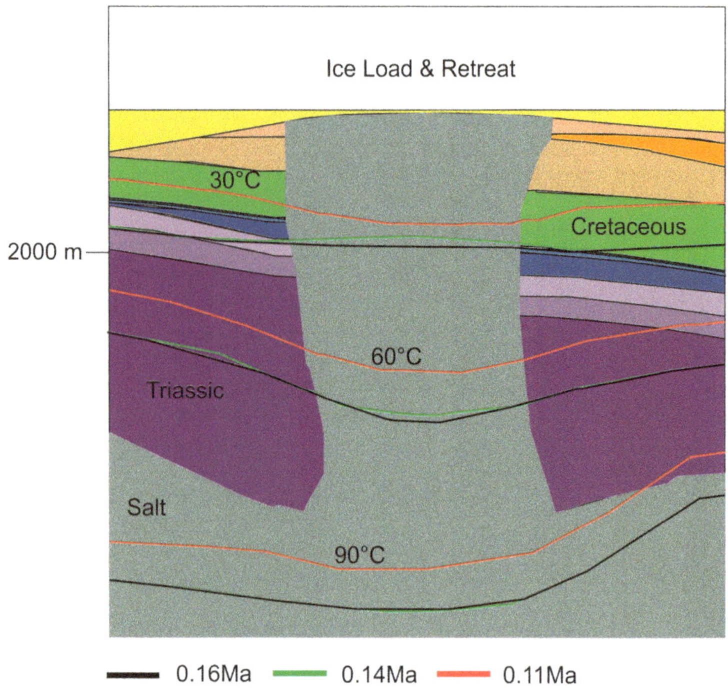

Fig. 13. Thermal feedback of ice retreat calculated for a salt diapir in the EHT.

greatly on the ice-sheet thickness and duration of ice load as Saalian IV is the thickest and longest-lasting event. In consequence, maintenance of pore overpressures is most likely in the case of rather thick and long-lasting ice coverage. Increases and decreases in pore pressures in the source rocks are expected to impact petroleum expulsion. Such 'glacial pumping' might, as an analogy to 'seismic pumping' (Sibson *et al.* 1975; Goffey 2017), trigger primary petroleum migration and expulsion along microfractures. Clearly, more research on geomechanics is necessary to investigate the dynamics of ice-induced poro-elastic decrease/increase with respect to geodynamics – this is not (yet) implemented in the PetroMod software.

Ice-induced consequences on rock mechanics based on the literature

Ice-induced load is known to have consequences on the rock mechanics (fractures, faults, rock failure), as shown, for example, by Atkinson (1987), Johnston *et al.* (1998) and Lund & Näslund (2009). Vertical stress, which is induced by ice load, decreases the differential stress in the crust and increases the stability of faults. After ice retreat, the stress is released and can result in post-glacial faulting. Furthermore, increased pore pressure decreases normal stress on a fault and decreases the shear stress necessary for fault movement. Due to the increased pore pressure, the fault stability lowers, which is largest at the end of glaciations as fault stability is generally decreasing (Lund & Näslund 2009). Johnston *et al.* (1998) tested the dependence of the stress field on the wavelength of the load. They constructed models including different ice-load extents (maximum thickness of 1000 m) and pointed out, based on the amplification of horizontal stress, that stability is dependent on the ice-sheet extent: stability is expected beneath large ice sheets, whilst fault instability is expected beneath small ice sheets. Furthermore, Krautblatter *et al.* (2013) summarized the importance of rock- and ice-induced mechanical processes. Simplified, ice-mechanical processes are important for shallow (low-magnitude: down to 20 m depth) rock-failure processes, while rock-mechanical processes might dominate deep-seated (high-magnitude: higher normal stress) rock failure in warming permafrost rocks. These observations are important, as they might have consequences for earthquakes, groundwater flow and stability of, for example, radioactive waste deposits.

Artifical expulsion model

A further effect was observed for computed petroleum accumulations. In order to generate such

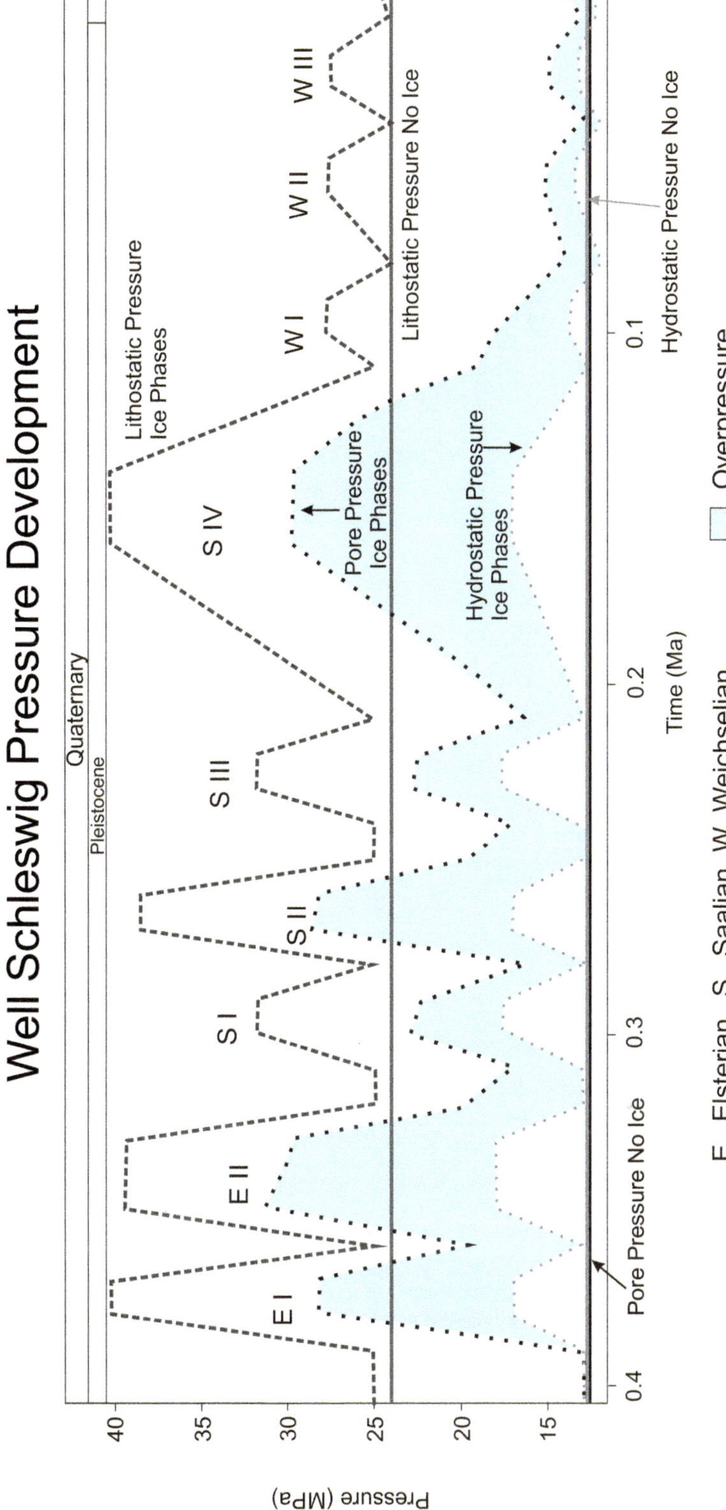

Fig. 14. Computed pore-pressure pattern, hydrostatic pressure and lithostatic pressure for the ice and no ice model at the location of the Schleswig well for the Dogger layer. For the timing of the various ice episodes compare with Figure 3.

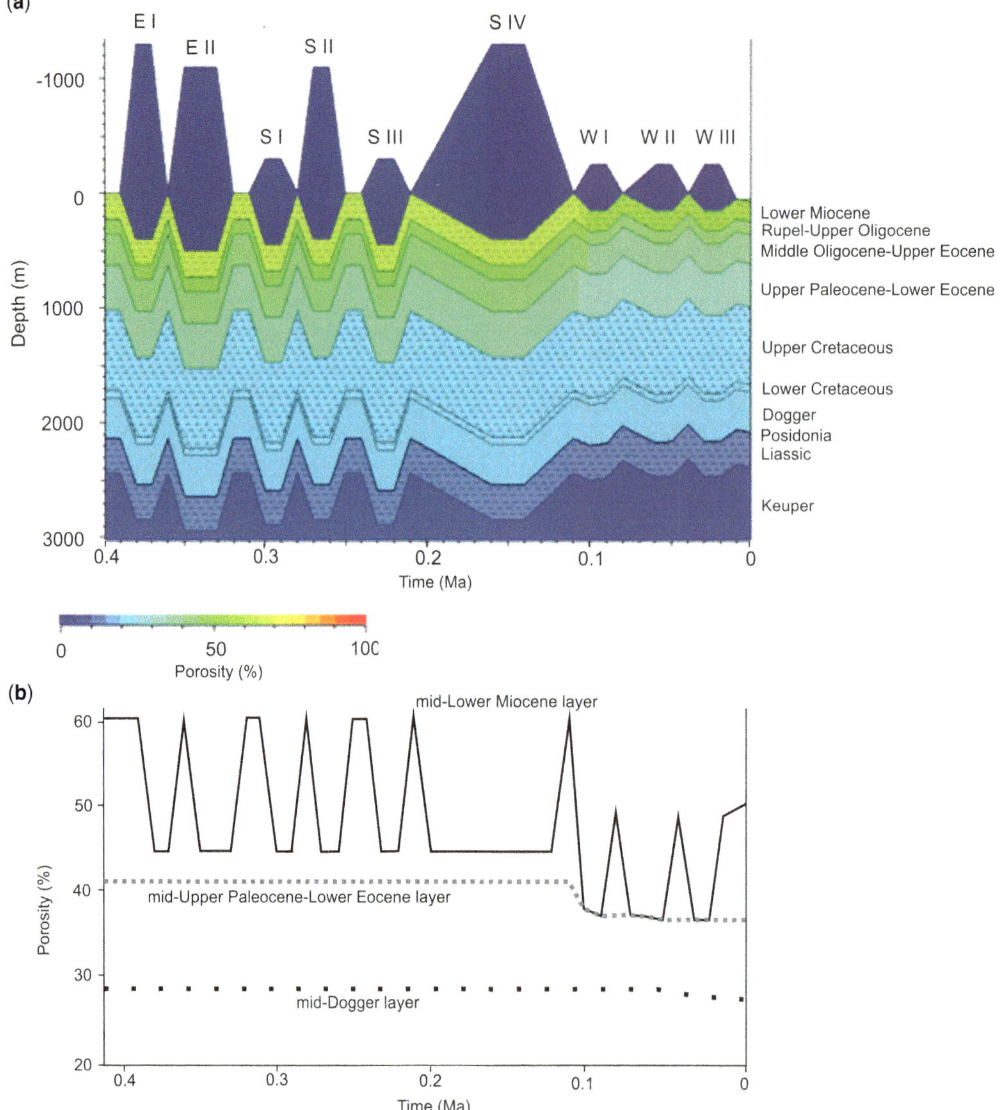

Fig. 15. (**a**) Computed ice-load history with porosity overlay for the Schleswig well and (**b**) 1D extractions for the Schleswig well showing the porosity development over time for the Lower Miocene layer (shale and sandstone), the Upper Paleocene–Lower Eocene layer (shale and sandstone) and the Dogger layer (sandstone).

accumulations in the model, we had to increase the heat flow to 80 mW m^{-2} because the Posidonia Shale is otherwise too immature for significant hydrocarbon generation. (It is situated deeper and within the oil window towards the south (Bruns *et al.* 2015) and just west of our model area; this is the oil kitchen for the large Mittelplate Field (Grassmann *et al.* 2005).) As expected, the state of aggregation and composition of the liquid and gaseous phases of the accumulations containing oil and gas changed immediately after ice load and retreat. The gas-to-oil ratio (GOR) decreases after ice load and recovers to nearly initial values after ice retreat (Fig. 16; Table 2). This is related to the solubility of gas under increased pressure. Temperature also plays an important role for the solubility, as the gas solubility decreases with increasing temperatures. In consequence, the temperature drop due to ice

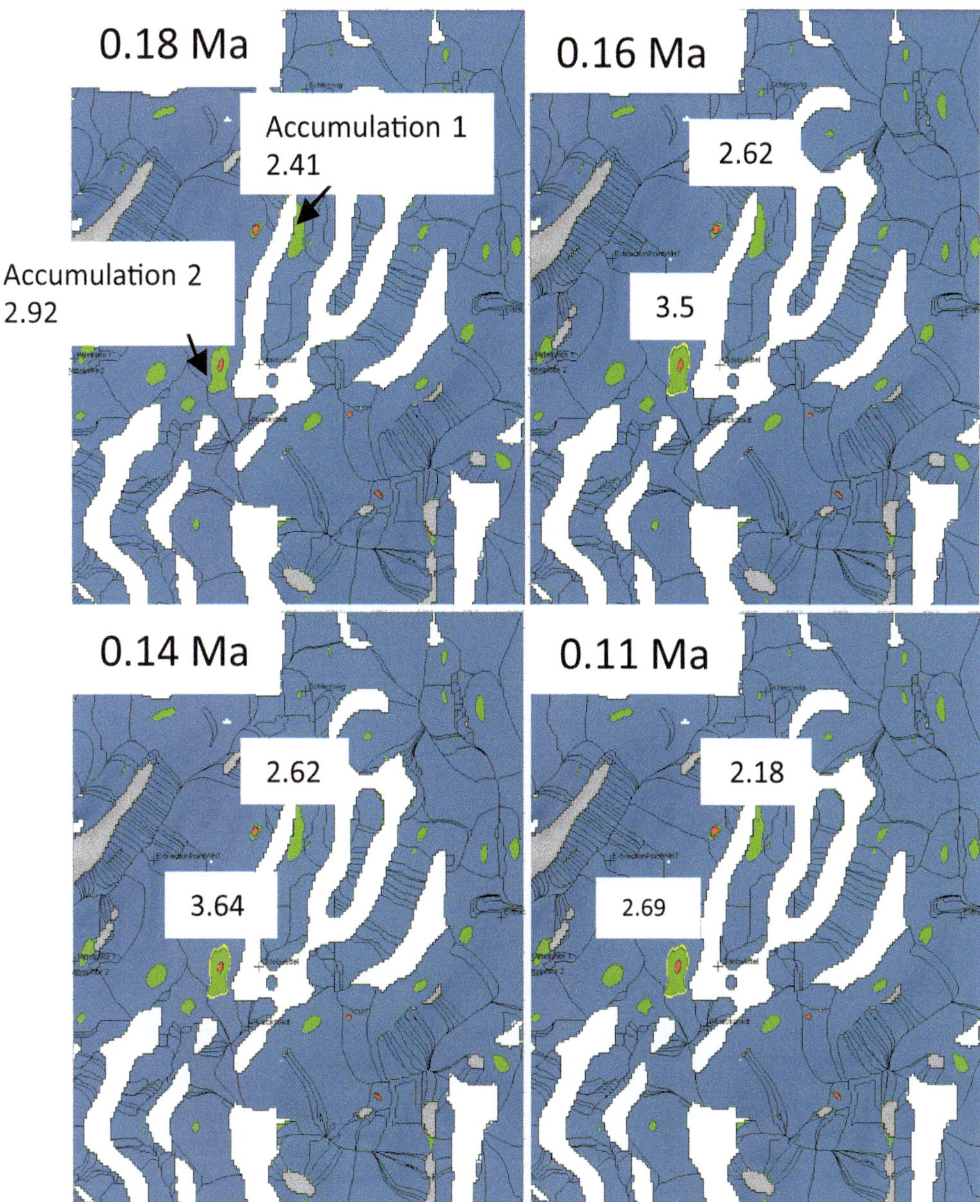

Fig. 16. Dogger reservoir layer, including the theoretically calculated accumulations based on a model with an extremely high, unrealistic heat flow. Values give GOR values for two accumulations through time, showing the increase of gas solubility during ice load. Further values are given in Table 2.

load in the reservoir layers and the pressure increase both lead to an increasing solubility of gas in oil. Taking into account the interaction of ice and salt as described above, the salt further triggers temperature drop down to specific depths and thus the solubility of gases in reservoirs adjacent to the salt, such as the large Mittelplate Field. Within source rocks such as the Posidonia Shale, glaciation increases pore pressure because fluids cannot be expelled easily from low-permeability rocks. Compaction is therefore inhibited, which leads to strong overpressure generation. Such overpressures will ultimately lead to the generation of microfractures and enhanced petroleum primary migration within and

Table 2. *Gas-to-oil ratio (GOR) changes due to glaciation and ice retreat*

Time (Ma)	Accumulation 1	Accumulation 2
0.39	1.59	1.95
0.38	2.62	3.31
0.37	2.62	3.28
0.36	2.06	2.49
0.19	2.41	2.92
0.18	2.62	3.21
0.17	2.62	3.5
0.16	2.62	3.5
0.15	2.62	3.57
0.14	2.62	3.64
0.13	2.62	3.52
0.12	2.62	3.27
0.11	2.18	2.69

Locations of accumulations are shown in Figure 16.

out of the source rock ('glacial pumping'). Microfractures are known to be common conduits for petroleum expulsion (Leythaeuser *et al.* 1988), and are also common in mature and overmature Posidonia Shale (Littke *et al.* 1988). Similarly, fractures might develop in shaly cap rocks above reservoirs. These processes are expected to occur in each petroleum system worldwide that is affected by glaciation and might greatly enhance petroleum expulsion there but possibly also cause cap-rock failure. Thus, Quaternary ice ages might have greatly influenced the presence of petroleum accumulations at high northern latitudes where glaciations have occurred.

Conclusions

A detailed 3D model was constructed for the Schleswig-Holstein area including the East Holstein Trough (EHT) and the West Holstein Trough (WHT) and the central Glueckstadt Graben. Differences in the burial histories between the Jurassic troughs and the Glueckstadt Graben were observed, coupled to variations in the temperature field. Relationships to the timing of salt movements were noted, as well as variations in sediment accumulation rates in time and space. Nevertheless, all locations show deepest or almost deepest burial at the present day. Triassic subsidence and sedimentation rates were higher in the Glueckstadt Graben compared to the Jurassic troughs. Based on the burial histories and the calculated temperature fields, only a little potential for hydrocarbon generation in the Posidonia Shale was calculated: that is, maturity does not reach sufficient values in the study area. Only the deepest parts of the Jurassic troughs might have served as petroleum kitchens: for example, the southern part of the EHT and in the central WHT, slightly west of our study area.

The maturity pattern visualized in several maps (Fig. 9a, b) is dependent on the highly variable burial history of these areas, whereof the main burial phase of the Glueckstadt Graben was in the Triassic. In contrast, the main burial phases of the Jurassic troughs are younger (Fig. 7). In addition, diapirism of the thick Zechstein salt took place earlier in the Glueckstadt Graben (i.e. Triassic), and later in the EHT and WHT (i.e. mainly Jurassic–Cretaceous). Compared to previous work, our model indicates that smaller thicknesses of eroded sediments and lower heat-flow values are required for successful model calibration, which is closely related to the implementation of radiogenic heat and 3D salt diapirism.

Furthermore, the implementation of glacial episodes revealed a significant control of ice loads on the petrophysical properties of rocks. Pore pressure increased during ice load, dependent on ice thickness and duration. This affects the petroleum accumulations: that is, gas solubility increases under increasing pressure and decreasing temperatures. The consequences of the porosity values seem to be related to the spatial and temporal extent of ice-sheet coverage and to the burial depth of the affected layers, as the effect is only irreversible after longer-lasting and thicker ice-sheet coverage in the model. In our model, only a 1700 m-thick ice sheet with a duration of 0.03 myr had an effect. Leakage and migration effects were not observed in the model as the Posidonia Shale is not mature enough for significant hydrocarbon generation.

Glaciation also leads to a significant decrease in the poro-elastic recoverability and a porosity decrease in the sediments, and an increase in the pore pressure in low-permeability lithologies (e.g. within petroleum source rocks). Porosity decreases induced by the various glacial cycles might be reversible to a significant extent, implying that each new glaciation led to renewed pore overpressures that triggered petroleum expulsion ('glacial pumping').

In consequence, there are multiple parameters which are affected by ice loads and thus should be considered in petroleum system models. Besides the impact on the petroleum system, glacial isostatic adjustment and glacial-induced faulting are important processes that should be considered in the context of, for example, earthquakes, groundwater flow or radioactive waste deposits.

The publication of these results was funded by the Excellence Initiative of the German federal and state governments. We thank Reviewers Hanneke Verweij and Graham Goffey for their detailed comments on an earlier draft of this paper.

References

ATKINSON, B.K. 1987. Introduction to fracture machanics and its geophysical applications. *In*: ATKINSON, B.K. (ed.) *Fracture Machanics of Rocks*. Academic Press, London.

BACHMANN, G.H. & GROSSE, S. 1989. Struktur und Entwicklung des Norddeutschen Beckens – geologische und geophysikalische Interpretation einer verbesserten Bouguer-Schwerekarte [Structure and evolution of the North German Basin – geological and geophysical interpretation of a revised Bouguer-gravity map]. *Niedersächsische Akademie der Geowissenschaften*, **2**, 23–47.

BALDSCHUHN, R., KOCKEL, F. *ET AL.* 1996. *Geotektonischer Atlas von NW-Deutschland/Tectonic Atlas of NW-Germany 1:300 000*. Bundesanstalt für Geowissenschaften und Rohstoffe (BGR), Hannover.

BEHA, A., THOMSEN, R.O. & LITTKE, R. 2008. Thermal history, hydrocarbon generation and migration in the Horn Graben in the Danish North Sea: a 2D basin modelling study. *International Journal of Earth Sciences*, **97**, 1087–1100.

BERNARD, S., WIRTH, R., SCHREIBER, A., SCHULZ, H.-M. & HORSFIELD, B. 2012. Formation of nanoporous pyrobitumen residues during maturation of the Barnett Shale (Fort Woth Basin). *International Journal of Coal Geology*, **103**, 3–11.

BGR 2016. *Schieferöl und Schiefergas in Deutschland-Potenziale und Umweltaspekte [Shale oil and shale gas in Germany. Potential and environmental aspects]*. Bundesanstalt für Geowissenschaften und Rohstoffe (BGR), Hannover.

BRUNS, B., DI PRIMIO, R., BERNER, U. & LITTKE, R. 2013. Petroleum system evolution in the inverted Lower Saxony Basin, northwest Germany: a 3D basin modeling study. *Geofluids*, **13**, 246–271.

BRUNS, B., LITTKE, R., GASPARIK, M., WEES, J.D. & NELSKAMP, S. 2015. Thermal evolution and shale gas potential estimation of the Wealden and Posidonia Shale in NW-Germany the Netherlands: a 3D basin modelling study. *Basin Research*, **28**, 2–33, https://doi.org/10.1111/bre.12096

CAVANAGH, A.J., DI PRIMIO, R., SCHECK-WENDEROTH, M. & HORSFIELD, B. 2006. Severity and timing of Cenozoic exhumation in the southwestern Barents Sea. *Journal of the Geological Society, London*, **163**, 761–774, https://doi.org/10.1144/0016-76492005-146

DELISLE, G. 1998. Numerical simulation of permafrost growth and decay. *Journal of Quaternary Science*, **13**, 325–333.

DELISLE, G., GRASSMANN, S., CRAMER, B., MESSNER, J. & WINSEMANN, J. 2007. Estimating episodic permafrost development in northern Germany during Pleistocene. *In*: HAMBREY, M.J., CHRISTOFFERSEN, P., GLASSER, N.F. & HUBBARD, B. (eds) *Glacial Sedimentary Processes and Products*. International Association of Sedimentologists, Special Publications, **39**, 109–120.

DEUTSCHE STRATIGRAPHISCHE KOMMISSION 2012. *Stratigraphische Handtabelle von Deutschland 2012 (STDH 2012) [Stratigraphic Guide for Germany 2012 (STDH 2012)]*. Deutsche Stratigraphische Kommission, Potsdam, Germany.

DI PRIMIO, R., CRAMER, R., ZWACH, C., KROOSS, B.M. & LITTKE, R. 2008. Petroleum systems. *In*: LITTKE, R., BAYER, U., GAJEWSKI, D. & NELSKAMP, S. (eds) *Dynamics of Complex Intracontinental Basins: The Central European Basin System*. Springer, Berlin, 411–432.

DUIN, E.J.T., DOORNENBAL, J.C., RIJKERS, R.H.B., VERBEEK, J.W. & WONG, T.E. 2006. Subsurface structure of the Netherlands – results of recent onshore and offshore mapping. *Netherlands Journal of Geoscience*, **85**, 245–276.

EHLERS, J. 1990. Reconstructing the dynamics of the northwest European Pleistocene ice sheets. *Quaternary Science Reviews*, **9**, 71–83.

EHLERS, J. & GIBBARD, P.L. 2004. *Quaternary Glaciations-Extent and Chronolgy. Part I: Europe*. Elsevier, Amsterdam.

EHLERS, J., EISSMANN, L., LIPPSTREU, L., STEPHA, H.-J. & WANSA, S. 2004. Pleistocene glaciations of North Germany. *In*: EHLERS, J. & GIBBARD, P.L. (eds) *Quaternary Glaciations-Extent and Chronology. Part I: Europe*. Elsevier, Amsterdam.

FJELDSKAAR,W. 2014. The effects of glaciations on the petroleum systems in the Barents Sea. *Hydrocarbon Habitats: Quaternary Processes in NCS Petroleum Systems*. AAPG International Conference and Exhibition, October 24, 2013, Milan. Geological Society of Norway, Trondheim, Norway.

FRENZEL, B., PECSI, M. & VELICHKO, A.A. 1992. *Atlas of Paleoclimates and Paleoenvironments of the Northern Hemisphere, Late-Pleistocene–Holocene*. Gustav Fischer, Stuttgart, Germany.

FRIBERG, L. 2001. *Untersuchungen zur Temperatur-und Absenkungsgeschichte sowie zur Bildung und Migration von Methan und molekularem Stickstoff im Norddeutschen Becken [Studies on the temperature and burial history as well as on the generation and migration of methane and molecular nitrogen in the North German Basin]*. Berichte des Forschungszentrums Jülich, **3914**.

GOFFEY, G. 2017. Under the jackhammer! Glacial impacts on North Sea petroleum systems. London Evening Lecture, 14 March 2017, PESGB, London.

GOFFEY, G., ATTREE, M. *ET AL.* 2016. New exploration discoveries in a mature basin: offshore Denmark. *In*: BOWMAN, M. & LEVELL, B. (eds) *Petroleum Geology of NW Europe: 50 Years of Learning – Proceedings of the 8th Petroleum Geology Conference*. Geological Society, London. First published online September 26, 2016, https://doi.org/10.1144/PGC8.1

GRAHAM, G.C., STOKER, M.S., LONERGAN, L., BRADWELL, T. & STEWART, M.A. 2011. The Pleistocene glaciations of the North Sea basin. *In*: EHLERS, J., GIBBARD, P.L. & HUGHES, P.D. (eds) *Quaternary Glaciations 1 – Extent and Chronology*. 2nd edn. Developments in Quaternary Sciences, **15**. Elsevier, Amsterdam, 261–278.

GRASSMANN, S., CRAMER, B., DELISLE, G., MESSNER, J. & WINSEMANN, J. 2005. Geological history and petroleum system of the Mittelplate oil field, Northern Germany. *International Journal of Earth Sciences*, **94**, 979–989.

GRASSMANN, S., CRAMER, B., DELISLE, G., HANTSCHEL, T., MESSNER, J. & WINSEMANN, J. 2010. pT-effects of Pleistocene glacial periods on permafrost, gas hydrate stability zones and reservoir of the Mittelplate oil

field, northern Germany. *Marine and Petroleum Geology*, **27**, 298–306.
HANSEN, M.B., LYKKE-ANDERSEN, H., GEHGHANI, A., GAJEWSKI, D., HÜBSCHER, C., OLESON, M. & REICHERTER, K. 2005. Mesozoic and Cenozoic geological evolution of the Bay of Kiel, western Baltic Sea. *International Journal of Earth Sciences*, **94**, 1070–1082.
HANTSCHEL, T. & KAUERAUF, A.I. 2009. *Fundamentals of Basin and Petroleum Systems Modeling*. Springer, Dordrecht.
JAPSEN, P., GREEN, P.F., NIELSEN, L.H., RASMUSSEN, E.S. & BIDSTRUP, T. 2007. Mesozoic–Cenozoic exhumation events in the eastern North Sea Basin: a multidisciplinary study based on paleothermal, paleoburial, stratigraphic and sesmic data. *Basin Research*, **19**, 451–490.
JOHNSTON, P., WU, P. & LAMBECK, K. 1998. Dependence of horizontal stress magnitude on load dimension glacial rebound models. *International Geophysical Journal*, **132**, 41–60.
KJEMPERUD, A. & FJELDSKAAR, W. 1992. Pleistocene glacial isostasy-implications for petroleum geology. *In*: LARSEN, R.M., BREKKE, H., LARSEN, B.T. & TALLERAAS, E. (eds) *Structural and Tectonic Modelling and its Applications to Petroleum Geology*. Norwegian Petroleum Society, Special Publications, **1**. Elsevier, Amsterdam, 187–195.
KLAVER, J., DESBOIS, G., URAI, J.L. & LITTKE, R. 2012. BIB-SEM study of the pore space morphology in early mature Posidonia Shale from the Hils area, Germany. *International Journal of Coal Geology*, **103**, 12–25.
KLEY, J. & VOIGT, T. 2008. Late Cretaceous intraplate thrusting in central Europe: effect of Africa-Iberia-Europe convergence, not Alpine collision. *Geology*, **36**, 839–842.
KOCKEL, F. 2002. Rifting processes in NW-Germany and the German North Sea sector. *Geologie en Mijnbouw – Netherlands Journal of Geosciences*, **81**, 149–158.
KRAUTBLATTER, M., FUNK, D. & GÜNZEL, F.K. 2013. Why permafrost rocks become unstable: a rock–ice-mechanical model in time and space. *Earth Surfaces Processes and Landforms*, **38**, 876–887, https://doi.org/10.1002/esp.3374
LERCHE, I., YU, Z., TØRUDBAKKEN, B. & THOMSEN, R.O. 1997. Ice loading effects in sedimentary basins with reference to the Barents Sea. *Marine and Petroleum Geology*, **14**, 227–338.
LEYTHAEUSER, D., LITTKE, R., RADKE, M. & SCHAEFER, R.G. 1988. Geochemical effects of petroleum migration and expulsion from Toarcian source rocks in the Hils syncline area, NW-Germany. *Organic Geochemistry*, **13**, 489–502.
LITT, T., SCHMINCKE, H.-U., FRECHEN, M. & SCHLÜCHTER, C. 2008. Quaternary. *In*: MCCANN, T. (ed.) *The Geology of Central Europe. Volume 2: Mesozoic and Cenozoic*. Geological Society, London, 1287–1340.
LITTKE, R., BAKER, D.R. & LEYTHAEUSER, D. 1988. Microscopic and sedimentologic evidence for the generation and migration of hydrocarbons in Toarcian source rocks of different maturities. *In*: MATTAVELLI, L. & NOVELLI, L. (eds) *Advances in Organic Geochemistry. Organic Geochemistry*, **13**, 549–559.
LITTKE, R., LEYTHAEUSER, D., RULLKÖTTER, J. & BAKER, D.R. 1991. Keys to the depositional history of the Posidonia Shale (Toarcian) in the Hils Syncline, northern Germany. *In*: TYSON, R.V. & PEARSON, T.H. (eds) *Modern and Ancient Continental Shelf Anoxia*. Geological Society, London, Special Publications, **58**, 311–333, https://doi.org/10.1144/GSL.SP.1991.058.01.20
LITTKE, R., KROOS, B.M., IDIZ, E.F. & FRIELINGSDORF, J. 1995. Molecular nitrogen in natural gas accumulations: generation from sedimentary organic matter at high temperatures. *AAPG Bulletin*, **79**, 410–430.
LITTKE, R., BAYER, U., GAJEWSKI, D. & NELSKAMP, S. 2008. *Dynamics of Complex Intracontinental Basins*, Springer, Berlin.
LOKHORST, A., ADLAM, K. *ET AL*. 1998. *NW European Gas Atlas – Composition and Isotope Ratios of Natural Gases*. NITG, Harlem.
LUND, B. & NÄSLUND, J.O. 2009. Glacial isostatic adjustment: implications for glacially induced faulting and nuclear waste repositories. *In*: CONNOR, C.B., CHAPMAN, N.A. & CONNOR, L.J. (eds) *Volcanic and Tectonic Hazard Assessment for Nuclear Facilities*. Cambridge University Press, Cambridge.
MAHANEY, W.C. 1995. Pleistocene and Holocene glacier thicknesses, transport histories and dynamics inferred from SEM microtextures on quartz particles. *Boreas*, **24**, 293–304.
MAYSTRENKO, Y., BAYER, U. & SCHECK-WENDEROTH, M. 2005. The Glueckstadt Graben, a sedimentary record between the North and Baltic Sea in north Central Europe. *Tectonophysics*, **397**, 113–126, https://doi.org/10.1016/j.tecto.2004.10.004
MAYSTRENKO, Y., BAYER, U., BRINK, H.-J. & LITTKE, R. 2008. The Central European Basin system – an overview. *In*: LITTKE, R., BAYER, U., GAJEWSKI, D. & NELSKAMP, S. (eds) *Dynamics of Complex Intracontinental Basins: The Central European Basin System*. Springer, Berlin, 15–34.
MAYSTRENKO, Y., BAYER, U. & SCHECK-WENDEROTH, M. 2012. 3D structural model of the Central European Basin System from the North Sea to Poland. *In*: ALSOP, G.I., ARCHER, S.G., HARTLEY, A.J., GRANT, N.T. & HODGKINSON, R. (eds) *Salt Tectonics, Sediments and Prospectivity*. Geological Society, London, Special Publications, **363**, 409–430, https://doi.org/10.1144/SP363.19
MCKENZIE, D. 1978. Some remarks on the development of sedimentary basins. *Earth and Planetary Science Letters*, **40**, 25–32.
MELLO, U.T. 1995. Role of salt in restraining the maturation of subsalt source rocks. *Marine and Petroleum Geology*, **12**, 697–716.
MILLER, K.G., KOMINZ, M.A. *ET AL*. 2005. The Phanerozoic record of global sea–level change. *Science*, **310**, 1293–1298.
MOELLER, C., MICKELSON, D.M., ANDERSON, M.P. & WINGUTH, C. 2007. Groundwater flow beneath Late Weichselian glacier ice in Njordfjord, Norway. *Journal of Glaciology*, **53**, 180.
MOHNHOFF, D., LITTKE, R. & SACHSE, V.F. 2016. Estimates of shale gas contents in the Posidonia Shale and Wealden of the west-central Lower Saxony Basin from high-resolution 3D numerical basin modelling. *Zeitschrift der Deutschen Gesellschaft für Geowissenschaften*, **167**, 295–314.

MUÑOZ, Y.A., LITTKE, R. & BRIX, M.R. 2007. Fluid systems and basin evolution of the western Lower Saxony Basin, Germany. *Geofluids*, **7**, 335–355.

MUTTERLOSE, J. & BORNEMANN, A. 2000. Distribution and facies patterns of Lower Cretaceous sediments in northern Germany: a review. *Cretaceous Research*, **21**, 733–759.

NEUNZERT, G.H., GAUPP, R. & LITTKE, R. 1996. Absenkungs- und Temperaturgeschichte paläozoischer und mesozoischer Formationen im Nordwestdeutschen Becken [Burial and temperature history of Paleozoic and Mesozoic formations in the northwest German Basin]. *Zeitschrift der Deutschen Geologischen Gesellschaft*, **147**, 183–208.

NIELSEN, S.B. 2002. A post mid-Cretaceous North Sea model. *Bulletin of the Geological Society of Denmark*, **49**, 187–204.

PATERSON, W.S.B. 1994. *The Physics of Glaciers*. Elsevier, Amsterdam.

PIOTROWSKI, J.A. 1996. *Dynamik und subglaziale Paläohydrologie der weichselzeitlichen Eiskappe in Zentral-Schleswig-Holstein [Dynamics and subglacial paleohydrology of the Weichsel ice cap in central Schleswig-Holstein]*. Geologisch-Paläontologisches Institut der Universität Kiel, Berichte, **76**.

PIOTROWSKI, J.A. & TULACZYK, S. 1999. Subglacial conditions under the last ice sheet in northwest Germany: ice-bed separation and enhanced basal sliding. *Quaternary Science Reviews*, **18**, 737–751.

RAITH, A.F., STROZYK, F., VISSER, J. & URAI, J.L. 2015. Evolution of rheologically heterogeneous salt structures: a case study from the northeast of the Netherlands. *Solid Earth Discussions* **7**, 1877–1908, https://doi.org/10.5194/sed-7-1877-2015

REICHERTER, K., FROITZHEIM, N. *ET AL.* 2008. Alpine tectonics north of the Alps. *In*: MCCANN, T. (ed.) *The Geology of Central Europe volume 2: Mesozoic and Cenozoic*. Geological Society, London, 1233–1285.

RIPPEN, D., LITTKE, R., BRUNS, B. & MAHLSTEDT, N. 2013. Organic geochemistry and petrography of Lower Cretaceous Wealden black shales of the Lower Saxony Basin: the transition from lacustrine oil shales to gas shales. *Organic Geochemistry*, **63**, 18–36.

RODON, S. & LITTKE, R. 2005. Thermal maturity in the Central European Basin system (Schleswig-Holstein area): results of 1D basin modeling and new maturity maps. *International Journal of Earth Sciences*, **94**, 815–833.

RODRIGUES DURAN, E., DI PRIMIO, R., ANKA, Z., STODDART, D. & HORSFIELD, B. 2013. 3D-basin modeling of the Hammerfest Basin (southwestern Barents Sea): a quantitative assessment of petroleum generation, migration and leakage. *Marine and Petroleum Geology*, **45**, 281–303.

RULLKÖTTER, J., LEYTHAEUSER, D. *ET AL.* 1988. Organic matter maturation under the influence if a deep intrusive heat source: a natural experiment for quantification of hydrocarbon generation and expulsion from a petroleum source rock (Toarcian shale, Northern Germany). *Organic Geochemistry*, **13**, 847–856.

SACHSE, V.F. & LITTKE, R. 2016. Burial, temperature and maturation history of the Ringkøbing-Fyn High, Denmark. *Zeitschrift der Deutschen Gesellschaft für Geowissenschaften*, **167**, 249–268.

SCHAEFER, R.G., SCHENK, H.J., HARDELAUF, H. & HARMS, R. 1990. Determination of gross kinetic parameters for petroleum formation from Jurassic source rocks of different maturity levels by means of laboratory experiments. *Organic Geochemistry*, **16**, 115–120.

SENGLAUB, Y., LITTKE, R. & BRIX, M.R. 2006. Numerical modelling of burial and temperature history as an approach for an alternative interpretation of the Bramsche anomaly, Lower Saxony Basin. *International Journal of Earth Sciences*, **95**, 204–224.

SCHECK-WENDEROTH, M. & LAMARCHE, J. 2005. Crustal memory and basin evolution in the Central European Basin System-new insights from a 3D structural model. *Tectonophysics*, **397**, 143–165.

SCHECK-WENDEROTH, M., MAYSTRENKO, Y., HÜBSCHER, C., HANSEN, M. & MAZUR, S. 2008. Dynamics of salt basins. *In*: LITTKE, R., BAYER, U., GAJEWSKI, D. & NELSKAMP, S. (eds) *Dynamics of Complex Intracontinental Basins: The Central European Basin System*. Springer, Berlin, 307–322.

SCHENK, H.J., HORSFIELD, B., KROSS, B.M., SCHAEFER, R.G. & SCHWOCHAU, K. 1997. Kinetics of petroleum formation and cracking. *In*: WELTE, D.H., HORSFIELD, B. & BAKER, D.R. (eds) *Petroleum and Basin Evolution*. Springer, Berlin, 233–269.

SIBSON, R.H., MOORE, J.Mc.M. & RANKIN, A.H. 1975. Seismic pumbing-a hydrothermal fluid transport mechanism. *Journal of the Geological Society, London*, **131**, 653–659, https://doi.org/10.1144/gsjgs.131.6.0653

SIEGERT, M.J. & MARSIAT, I. 2001. Numerical reconstruction of LGM by Tertiary uplift and erosion in the Barents Sea. *In*: VORREN, T.O., BERGSAGER, E., DAHL-STAMNES, O.A., HOLTER, E., JOHANSEN, B., LIE, E. & LUND, T.B. (eds) *Arctic Geology and Petroleum Potential*. Norwegian Petroleum Society, Special Publications, **2**, 711–719.

SIROCKO, F., REICHERTER, K., LEHNÉ, R., HÜBSCHER, C., WINSEMANN, J. & STACKEBRANDT, W. 2008. Glaciation, salt and present landscape. *In*: LITTKE, R., BAYER, U., GAJEWSKI, D. & NELSKAMP, S. (eds) *Dynamics of Complex Intracontinental Basins: The Central European Basin System*. Springer, Berlin, 233–245.

SONG, J., LITTKE, R., WENIGER, P., OSTERTAG-HENNING, C. & NELSKAMP, S. 2015. Shale oil potential and thermal maturity of the Lower Toarcian Posidonia Shale in NW Europe. *International Journal of Coal Geology*, **150–151**, 127–153.

STEPHAN, H.-J. 1997. Wie weit reichte die Vergletscherung der letzten Eiszeit in Schleswig-Holstein? [How far did the glaciation reach during the last ice age in Schleswig-Holstein?] *Die Heimat*, **104**, 52–57.

STEPHAN, H.-J., KABEL, CH. & SCHLÜTER, G. 1983. Stratigraphical problems in the glacial deposits of Schleswig-Holstein. *In*: EHLERS, J. (ed.) *Glacial Deposits in North-West Europe*. Balkema, Rotterdam, 305–320.

STROZYK, F., URAI, J.L., VAN GENT, H., DE KEIJZER, M. & KUKLA, P.A. 2014. The internal structure of salt: insights from a regional 3D seismic study of the Permian Zechstein 3 intra-salt stringer in the Northern Netherlands and its implications for salt tectonics. *Interpretation*, **2**, 101–117, https://doi.org/10.1190/INT-2014-0037.1

STROZYK, F., REUNING, L., SCHECK-WENDEROTH, M. & TANNER, D. 2017. *The Tectonic History of the Zechstein*

Basin in the Netherlands and Germany. Elsevier, Amsterdam.

Strozyk, F., Reuning, L., Back, S. & Kukla, P. 2018. Giant pockmark formation from Cretaceous hydrocarbon expulsion in the western Lower Saxony Basin, The Nerherlands. *In*: Kilhams, B., Kukla, P.A., Mazur, S., McKie, T., Mijnlieff, H.F. & van Ojik, K. (eds) *Mesozoic Resource Potential of the Southern Permian Basin*. Geological Society, London, Special Publications, **469**. First published online January 11, 2018, https://doi.org/10.1144/SP469.6

Sweeney, J.J. & Burnham, A.K. 1990. Evaluation of a simple model of vitrinite reflectance based on chemical kinetics. *AAPG Bulletin*, **74**, 1559–1570.

Thorson, R. 2000. Glacial tectonics: a deeper perspective. *Quaternary Science Reviews*, **19**, 1391–1398.

Uffmann, A.K. & Littke, R. 2011. 3D petroleum system modelling of the North German Basin. *First Break*, **29**, 49–63.

Uffmann, A.K. & Littke, R. 2013. Paleozoic petroleum systems of the Münsterland Basin, western Germany: a 3D basin modeling study – Part I: model Input, Calibration and Maturation. *Oil Gas European Magazine*, **39**, 140–146.

Uffmann, A.K., Littke, R. & Gensterblum, Y. 2014. Paleozoic petroleum systems of the Münsterland Basin, western Germany: a 3D basin modeling study – Part II: Petroleum Generation and Storage with Special Emphasis on Shale Gas Resources. *Oil Gas European Magazine*, **40**, 98–103.

Waples, D.W. 2001. A new model for heat flow in extensional Basins: radiogenic heat, asthenospheric heat and the McKenzie Model. *Natural Resources Research*, **10**, 227–238.

Warsitzka, M., Kley, J., Jähne-Klingberg, F. & Kukowski, N. 2017. Dynamics of prolonged salt movement in the Glückstadt Graben (NW Germany) driven by tectonic and sedimentary processes. *International Journal of Earth Sciences*, **106**, 131–155.

Wong, T.E., Batjes, D.A.J. & de Jager, J. (eds). 2007. *Geology of the Netherlands*. Royal Netherlands Academy of Arts and Sciences, Amsterdam.

Wygrala, B. 1989. *Integrated study of an oil field in the southern Po basin, northern Italy*. PhD thesis, University of Cologne, Cologne, Germany (FZJ-2014-03033).

Zieba, K.J. & Grøver, A. 2016. Isostatic response to glacial erosion, deposition and ice loading. Impact on hydrocarbon traps of the southwestern Barents Sea. *Marine and Petroleum Geology*, **78**, 168–183.

Ziegler, P.A. 1990. *Geological Atlas of Western and Central Europe*. Shell Internationale Petroleum Maatschappij, The Hague. Geological Society, London.

The Posidonia Shale of northern Germany: unconventional oil and gas potential from high-resolution 3D numerical basin modelling of the cross-junction between the eastern Lower Saxony Basin, Pompeckj Block and Gifhorn Trough

ALEXANDER T. STOCK & RALF LITTKE*

EMR-Energy and Mineral Resources Group, RWTH Aachen University, Lochnerstraße 4-20, 52064 Aachen, Germany

**Correspondence: ralf.littke@emr.rwth-aachen.de*

Abstract: A high-resolution three-dimensional (3D) numerical basin model, incorporating the eastern part of the Lower Saxony Basin (LSB), the Gifhorn Trough and parts of the southern Pompeckj Block, was built to reconstruct the thermal and structural evolution of this area. The estimation and calculation of the unconventional oil and gas resource density within the Posidonia Shale source-rock unit was the main objective of this study. Incorporating organic–geochemical data for the Posidonia Shale source-rock units, such as compositional petroleum generation kinetics data, allowed a more accurate prediction of hydrocarbon potential compared to large-scale models of the area, as well as a better prediction of bulk adsorption capacity and adsorbed gas content. For the accurate calculation of oil and gas contents within the source-rock lithologies, mineralogy and physical properties of the rocks, such as compressibility, sorption capacity and porosity, are important as well as organic matter quantity, quality and thermal maturity. These properties in turn are strongly dependent on the vastly different burial/uplift histories within the LSB, Gifhorn Trough and the Pompeckj Block. The Gifhorn Trough, large parts of the Pompeckj Block and the flanks of the LSB are interesting concerning the unconventional oil potential, with current source-rock maturities between 0.65% and 1.2% vitrinite reflectance. Central parts of the LSB and small parts of the Pompeckj Block show inherent unconventional gas potential. Methane adsorption capacity is influenced by the burial/uplift history of the basin, which stresses the importance of structural and geochemical interlocking in understanding unconventional hydrocarbon systems.

Due to the ongoing need for fossil fuels and following the 'shale gas boom' in the US, areas around already-exploited German oil fields and potential oil shale plays have been the target of re-evaluation and further investigations. Promising formations for unconventional gas production are Paleozoic strata, including the Namurian Upper and Lower Alum Shale (Uffmann *et al.* 2012; Uffmann & Littke 2013) and the Pennsylvanian sedimentary rocks and coals (Bruns *et al.* 2013; Bruns 2014). Mesozoic strata, including the Lower Jurassic Posidonia Shale (Bruns *et al.* 2015; Mohnhoff 2016), are key horizons and will be the focus of this work. It has to be kept in mind that the Lower Cretaceous Wealden Shales (Berner *et al.* 2010; Rippen *et al.* 2013) are also of interest due to their thickness and organic richness. Although unconventional shale oil and shale gas production are currently legally banned in Germany, investigations into the geological petroleum potential within known German oil-producing provinces can be beneficial; there is still much untapped potential within these areas and the unconventional resources within the basin are as yet poorly constrained. One advantage in this respect is the detailed study of the main Mesozoic source rock, the Posidonia Shale, based on shallow cores taken just south of the petroleum-bearing basin in the Hils Syncline at different maturities ranging from immature to overmature (Littke *et al.* 1988, 1991; Rullkötter *et al.* 1988; Horsfield & Düppenbecker 1991; Frimmel *et al.* 2004; Horsfield *et al.* 2010).

To assess the potential resources on a large scale, three-dimensional (3D) numerical basin modelling can be a powerful tool. Previous studies in the region either focused on a wider area, employing sorption and organic–geochemical data (Bruns *et al.* 2015) using a 1000 × 1000 m cell grid and the source-rock formation as homogenous strata, or focused on only the western LSB considering also the Wealden source rock with a higher resolution of 150 × 150 m (Mohnhoff *et al.* 2015). Other studies focused primarily on the modelling of geochemical and sedimentological aspects of oil reservoirs within the Gifhorn Trough (Schwarzkopf & Leythaeuser 1988; Blumenstein-Weingartz 2012).

From: KILHAMS, B., KUKLA, P. A., MAZUR, S., MCKIE, T., MIJNLIEFF, H. F. & VAN OJIK, K. (eds) 2018. *Mesozoic Resource Potential in the Southern Permian Basin*. Geological Society, London, Special Publications, **469**, 399–421.
First published online April 5, 2018, https://doi.org/10.1144/SP469.21

In this study, the focus is on the Toarcian Posidonia Shale and its conventional oil as well as unconventional shale oil potential. Using high-resolution 3D numerical basin modelling (150 × 150 m horizontal resolution, maximum of 400 m vertical resolution) coupled with compositional petroleum kinetics of the Posidonia Shale, the aim of this study is to assess the quality and quantity of hydrocarbons generated within the study area which is situated at the intersects between the eastern LSB, the Gifhorn Trough and the southern Pompeckj Block (Fig. 1), all being part of the Central European basin system (CEBS) (Littke *et al.* 2008).

Stratigraphic framework

Lower Saxony Basin

The LSB is a sub-basin of the CEBS and is located in northern Germany, bordering the Pompeckj Block to the north and NE, the Central Netherlands Basin to the west, the Cretaceous Münsterland Basin to the south and the Gifhorn Trough and Harz Mountains to the east. Deposition in the LSB, as a part of the CEBS, started in the Paleozoic (Warren 2008). Mostly terrigenous sediments were deposited during the Pennsylvanian and Early Permian, first in a tropical humid climate leading to abundant coal deposits and later in an arid climate (Rotliegend), when strong pyroclastic and ignimbrite volcanism affected the area. This initial rifting stage of the CEBS was followed by alternating shallow-marine to terrigenous deposition from the latest Permian (Zechstein) to the Late Triassic, which lasted until the onset of Jurassic marine sedimentation. During the latest Jurassic–earliest Cretaceous, the paralic to marine Wealden facies was deposited (Rippen *et al.* 2013). For the rest of the Cretaceous, marine deposition prevailed. While the sedimentation pattern is quite uniform (although with different thicknesses) in the CEBS from the Carboniferous until the Late Jurassic, basin differentiation became pronounced thereafter, with the development of several sub-basins along the southern basin margin. In these sub-basins, strong subsidence during the Early Cretaceous was followed by uplift and erosion in the Late Cretaceous. The LSB is one prominent example of this type of basin inversion (Senglaub *et al.* 2005; Voigt *et al.* 2008). Crustal extension across the North Atlantic rift system might have caused this strong basin differentiation (Stollhofen *et al.* 2008; Voigt *et al.* 2008), leading to very high sedimentation rates (Senglaub *et al.* 2005; Littke *et al.* 2011) which were enabled by strong uplift of other basin elements at the same time (Bruns *et al.* 2013). The LSB was subjected to inversion after the onset of the Alpine Orogeny, leading to subsequent erosion (Adriasola-Munoz *et al.* 2007; Sirocko *et al.* 2008).

Pompeckj Block

The Pompeckj Block can be regarded as a central element within the CEBS, bordering the inverted Central Netherlands Basin, the inverted LSB and the Gifhorn Trough to the west and south, and the Ring-Köbing Fyn High and Central Graben to the north. Sharing a uniform sedimentation with the LSB from the Permian until the Middle Jurassic, subsidence for the Pompeckj Block changed drastically during the Late Jurassic. While strong subsidence affected the LSB, the Pompeckj Block experienced uplift and erosion (Littke *et al.* 2008; Fig. 2). This uplift was followed by marine sedimentation during the Late Cretaceous, partly under pelagic depositional conditions. Inversion and erosion of the LSB at the same time contributed to the sediment fill. The Late Cretaceous inversion of the LSB therefore coincided with strong subsidence and sedimentation within the Pompeckj Block and vice versa. The Pompeckj Block also received moderately thick sediments during the Tertiary, when strong sedimentation prevailed in the North Sea area. Salt tectonics affected the Pompeckj Block, in particular in its northern part (Schleswig–Holstein area).

Gifhorn Trough

The Gifhorn Trough, a comparably small Rhenish-striking depression, borders the LSB to the east and shares a similar depositional history as the LSB until the Early Jurassic. Rapid sedimentation during the Jurassic due to strong subsidence was related to salt diapirism leading to the deposition of very thick marine sediments, especially of Liassic and Dogger age. These conditions also favoured the deposition of a very thick source-rock sequence of the Posidonia Shale, which can reach up to 100 m thickness in the area of the Gifhorn Trough (Boigk 1981; Betz *et al.* 1987; Schwarzkopf 1988; Brink *et al.* 1992); it is usually about 35–40 m thick in other parts of the CEBS (e.g. Littke *et al.* 1988). Similarly, thick Dogger β sandstones occur in up to eight layers with decreasing thickness towards the west of the structure, presenting a good reservoir succession. The structural development of the Gifhorn Trough is strongly linked to the Allertal fault line; the area north of this line demonstrated a downwards tilt during the Late Cretaceous inversion of the LSB, allowing for the separation of the Gifhorn Trough from the LSB (Blumenstein-Weingartz 2012). Salt tectonics have strongly influenced the area since the Triassic, continuing until today.

Petroleum systems in the CEBS

The two main petroleum systems in the study area are of Paleozoic and Mesozoic age. The Paleozoic

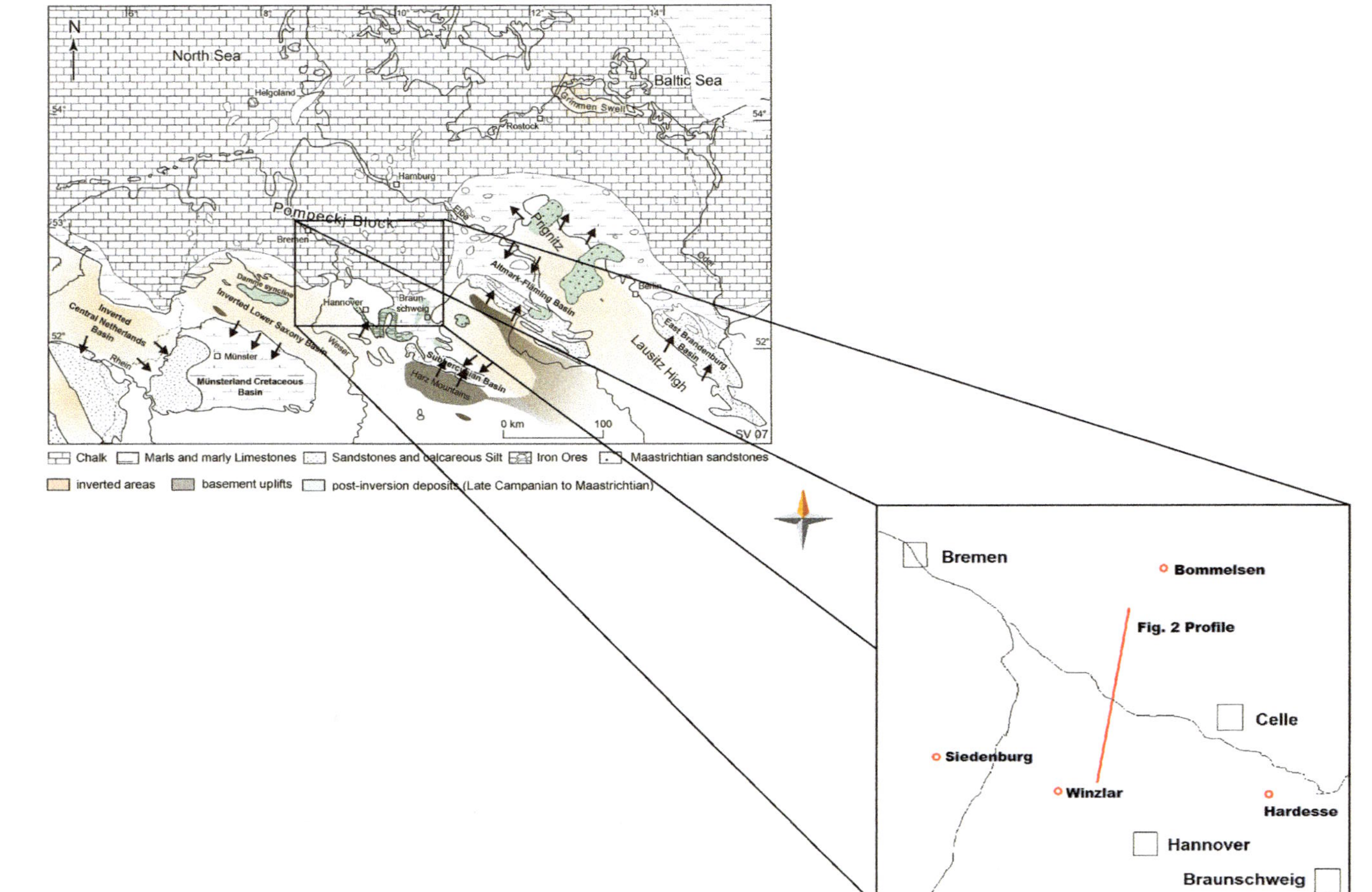

Fig. 1. Palaeogeographic map of the southern margin of the CEBS during the Coniacian–Maastrichtian with locations of inverted areas and post-inversion deposits (after Voigt *et al.* 2008), with the study area marked in black. The red line marks the profile location illustrated in Figure 2 and the red circles mark the well locations of the calibration wells illustrated in Figure 3.

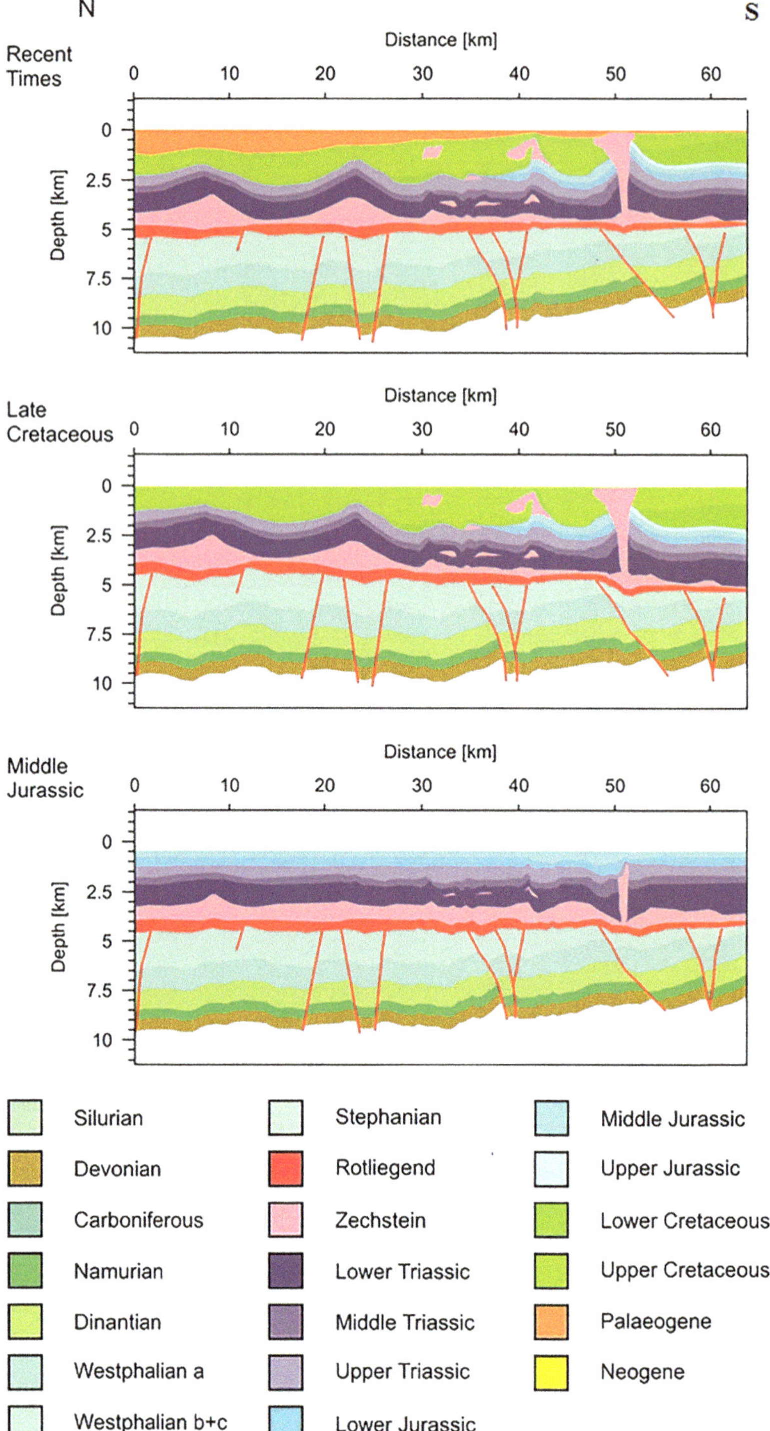

Fig. 2. Profile through the transition zone between the Pompeckj Block in the north and the Lower Saxony Basin in the south. The three time steps shown are the Middle Jurassic, Late Cretaceous and present day (from Littke *et al.* 2008).

Table 1. *Maximum thicknesses and bulk geochemical properties of the Posidonia Shale horizons in the study area*

Stratigraphic unit	Source-rock unit	Thickness (m)	Initial TOC (wt %)	Initial HI (mg HC/g TOC)
Toarcian	Posidonia Shale unit III (Bifrons)	16 (30)	12.0	625
Toarcian	Posidonia Shale unit II (Falciferum)	16 (30)	9.0	600
Toarcian	Posidonia Shale unit I (Tenuicostratum)	8 (20)	10.0	625

Data in parentheses are maximum thicknesses in the area of the Gifhorn Trough.

system consists of mainly Pennsylvanian coal-bearing gas source rocks and mainly Permian sandstone reservoir rocks covered by Permian claystones or salt (Littke *et al.* 1995; Gaupp *et al.* 2008; Bruns 2014). The Mesozoic source-rock units comprise the Toarcian Posidonia Shale and the Berriasian Wealden shales/marlstones. The Toarcian Posidonia Shale, an oil-prone type-II kerogen with an initially high total organic carbon (TOC) of up to 12 wt%, was deposited in an oxygen-depleted environment and is widespread in central and northwestern Europe (Song *et al.* 2015). Reservoir rocks are Dogger sandstones but also younger Jurassic and Cretaceous sandstones and carbonates (Betz *et al.* 1987). The prolific Lias/Dogger group consists of marlstones, carbonates, sandstones and shales, offering the possibility to act as reservoirs (in the case of the sandstones and carbonates) as well as seals (in the case of the marlstones and shales) (Kockel *et al.* 1994).

The Toarcian Posidonia Shale, which is the main horizon under investigation in this study, can be split into three source-rock units (I–III; Table 1) based on their rock mineralogy and kerogen composition (Littke *et al.* 1991; Frimmel *et al.* 2004; Song *et al.* 2016). The uppermost unit III (Bifrons) comprises mostly calcareous shale and exhibits the highest TOC and HI-values (Hydrogen Index), paired with high thicknesses. The middle unit II (Falciferum) has the lowest initial TOC values, but displays similar thicknesses as the overlying unit III. The lowest unit I (Tenuicostratum) has a similar TOC as unit II and is highest in carbonate content, but its initial thickness is the lowest compared to the other two Posidonia Shale units. This relation between the different Posidonia Shale units is of course subject to variable depositional conditions within the study area, especially due to differences in water depth and (salt-diapirism-induced) subsidence rates. Overall, greatest thicknesses are reached in the Gifhorn Trough.

Petroleum generation kinetics

During burial, due to excess compaction pressure and increasing temperature, organic matter is thermally degraded. The rate of this degradation is controlled by several factors, including the exposure time of the kerogen (organic matter) to a productive temperature and kinetic reaction parameters for this thermal degradation (Welte *et al.* 1997; di Primio & Horsfield 2006). Petroleum generation kinetics can either be implemented into petroleum system and basin modelling as bulk petroleum kinetics or as compositional petroleum kinetics. While bulk petroleum kinetics are more frequently used in sedimentary basin modelling for the prediction of petroleum generation, rate and timing (Quigley *et al.* 1987; Tissot *et al.* 1987; Ungerer 1989; Pepper & Corvi 1995), compositional petroleum kinetics offer a more complete assessment of timing and production of petroleum within sedimentary basins (Espitalié *et al.* 1988; Béhar *et al.* 1992, 1997; Sweeney *et al.* 1992, 1995; Dieckmann *et al.* 1998; Vandenbroucke *et al.* 1999). The first publication of petroleum kinetics for the Posidonia Shale was by Schäfer *et al.* (1990), while Schenk & Horsfield (1998) first showed that a kinetic approach for marine type-II source rocks such as the Posidonia Shale is valid. The appraisal of generation rate and timing is of utmost importance for the determination of the conventional/unconventional potential, as well as for the assessment of oil and gas in place in corresponding reservoir systems.

Methods

Bulk geochemistry

Bulk geochemical measurements, including the measurement of TOC and Rock-Eval pyrolysis, were conducted following the methods described in Stock & Littke (2016) for Posidonia Shale samples from three wells from the Hils Syncline (namely Wickensen, Dohnsen and Harderode) and were integrated using mean values listed in Table 1.

Compositional petroleum kinetics

Compositional petroleum kinetics measurements were conducted using a two-stage approach described in

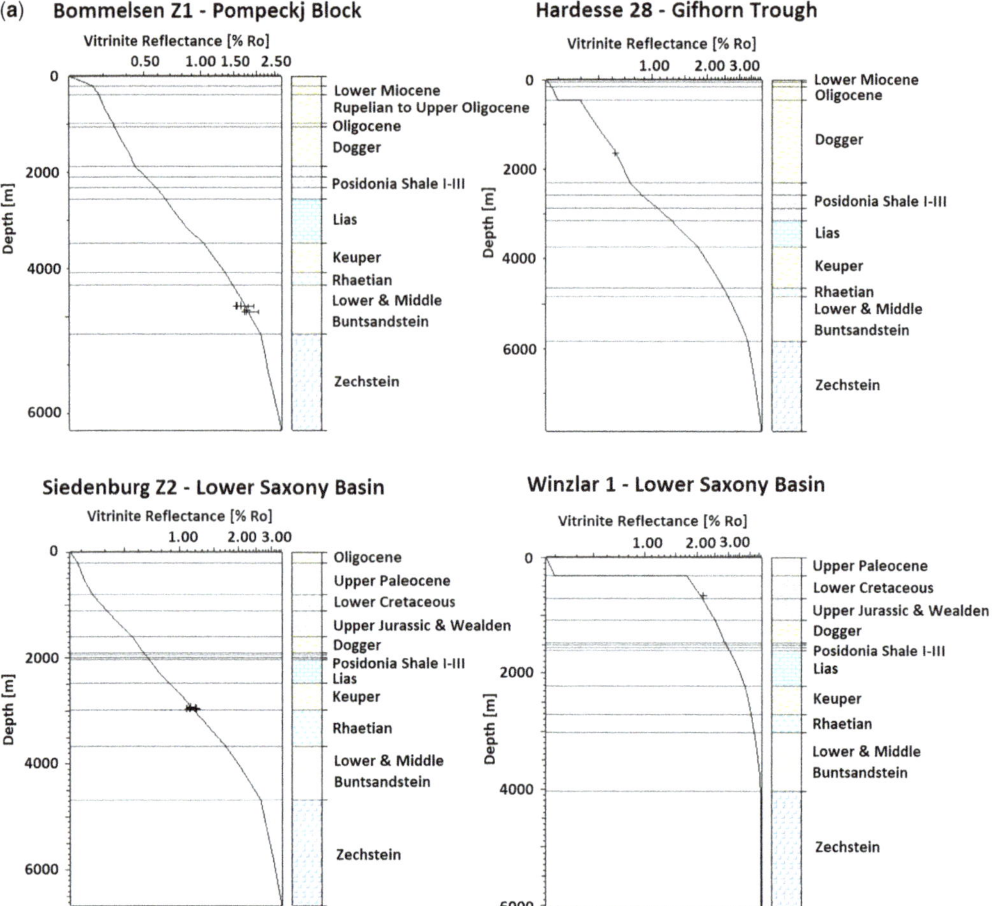

Fig. 3. (**a**) Maturity calibration for wells within the Pompeckj Block, Gifhorn Trough and the Lower Saxony Basin, displaying the comparison of measured and calculated vitrinite reflectance within the study area. The exact well location can be found in Figure 1 and the burial history in (b).

Düppenbecker & Horsfield (1989). For the first stage, kerogen decompositional data were obtained from non-isothermal open-system pyrolysis, described by Burnham *et al.* (1987), heating the samples at 0.1, 0.7 and 5°C min^{-1} and using a Green River Shale sample as a standard, allowing for the determination of the degree of kerogen conversion.

The second stage consists of two procedures: one for the determination of the hydrocarbon composition and one for the degree of conversion, using a closed-system approach. The first step included the microscale sealed vessel (MSSV) pyrolysis technique described in Horsfield *et al.* (1989). Following this method, duplicate aliquots were sealed in glass capillary tubes and heated at three different heating rates in a device described by Schäfer *et al.* (1990) to facilitate heat and mass transfer.

MSSV measurements using one aliquot (cracked open in a helium-flashed furnace at 300°C) allowed for the quantification of C_1-C_{26} n-alkanes, C_{14}-C_{20} isoprenoids, methyl-alkanes, and mono-aromatic and di-aromatic hydrocarbons. For the correct identification, the products were compared with reference internal standards using a single internal gas chromatographic step. A second step, performed with duplicate aliquots using the same set-up but keeping temperatures at 300°C and raising the furnace temperature during the run to 600°C, allowed for the determination of the pyrolysis products from the residual kerogen in a single peak.

The results presented here are summed up for four different ranges – C_1 (methane), C_2-C_5, C_6-C_{14} and C_{15+} – using resolved (valley-to-valley) peaks. The gas/oil ratio was determined by dividing

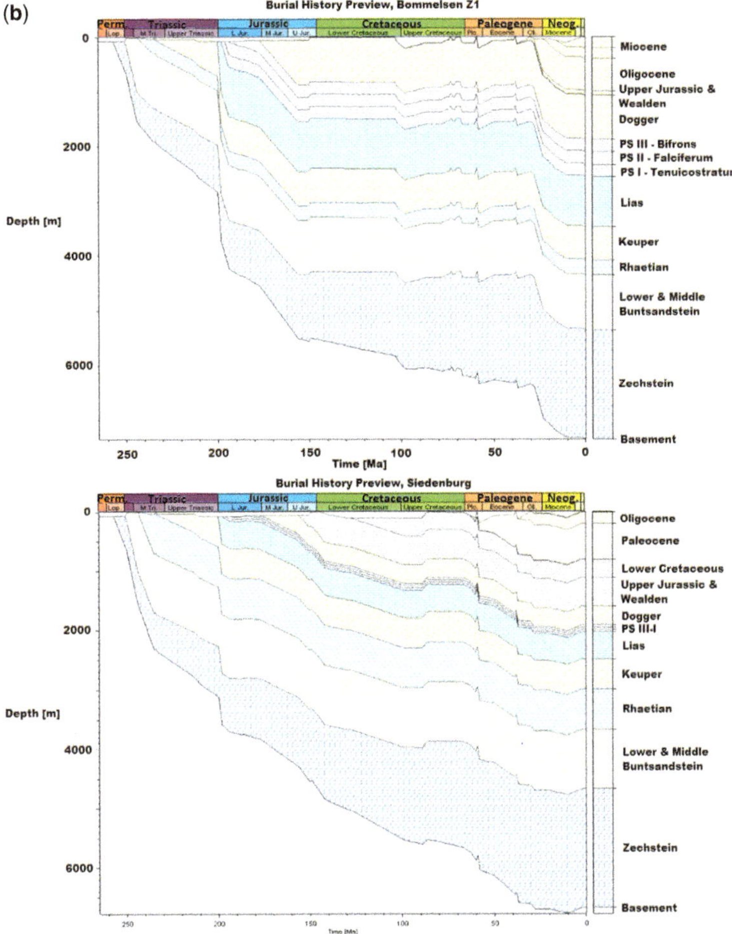

Fig. 3. (*Continued*) (**b**) Vitrinite reflectance calculation according to Sweeney & Burnham (1990). (**I, II**) Burial history plots of four well locations (calibration wells) in (a).

the C_1 and the C_2-C_5 components by all other components (C_{5+}). Although a more exact petroleum phase analysis of up to 14 compound groups is possible (di Primio & Horsfield 2006), simplification into the abovementioned compounds allows for a sufficiently detailed analysis of the hydrocarbons liberated from the Posidonia Shale. The numerical determination regarding the activation energy and frequency factors were conducted using the program Kinetic 2000 at the GFZ Potsdam, allowing for discrete distribution of activation energies and a single value of the pre-exponential factor. The compositional petroleum generation kinetics used for this model were those published by di Primio & Horsfield (2006), and then compared to results obtained on Posidonia Shale samples from the more mature wells Dohnsen and Harderode from the Hils Syncline.

Here, temperature and time are the driving properties in predicting petroleum generation. Within a sedimentary system temperatures influencing sedimentary sequences usually range between 50°C and 300°C, where the reactions are predominantly kinetically controlled.

3D numerical basin modelling and calibration

Hydrocarbon generation and storage potential of the Posidonia Shale was calculated in a 3D numerical basin model using the Schlumberger AaTC PetroMod© software v.2014.1. This software operates on the basis of forward modelling techniques, allowing for the reconstruction of geological events such as sedimentation and erosion in combination with geochemical and petrophysical processes running simultaneously. The study area is defined through

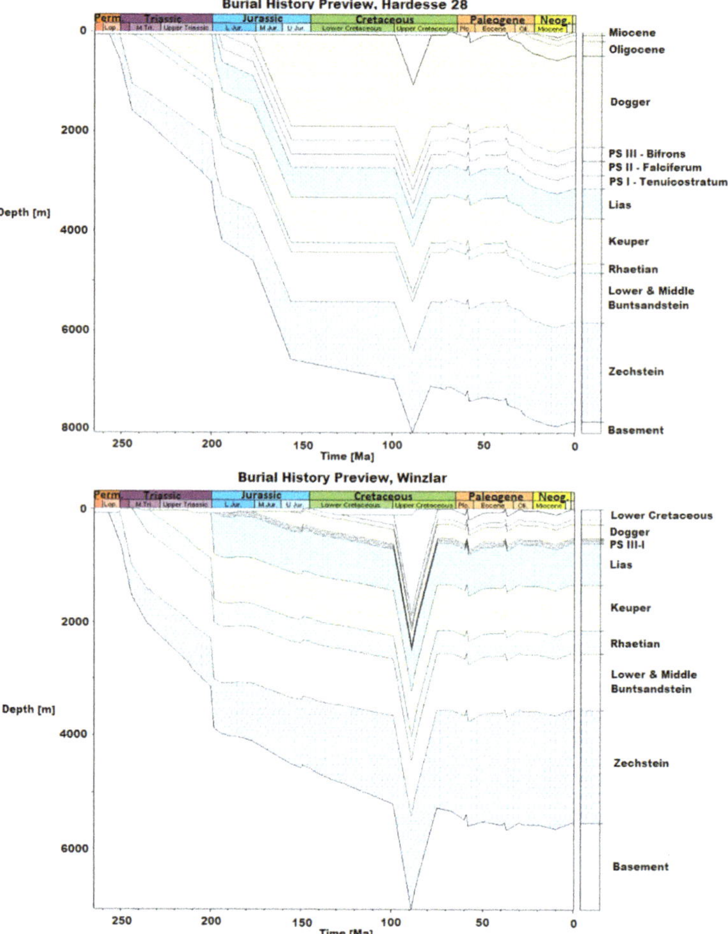

Fig. 3. *(Continued)*

grid cells with a resolution of 150 × 150 m, to which the inherent properties are assigned and for which the calculation results from the progressive forward modelling are given as the model output. This output information includes rock properties such as temperature, compaction, thermal maturity of source rock and other units, total/effective porosity, hydrocarbon generation and expulsion, fluid flow, hydrocarbon composition and adsorption capacity (Bruns *et al.* 2015; Mohnhoff 2016). The model was calibrated using 267 wells within the study area which were assessed based on their quality of data, assigning unreliable temperature data (e.g. fluorescence-based temperature) a lower weight and reliable temperature or maturity data (e.g. vitrinite reflectance) a higher weight. Based on these wells the amount of erosion and heat flow within the study area was determined, and the 3D basin model calibrated (Fig. 3a, b-I, b-II). The calculation of vitrinite reflectance used in this study is based on the Easy-Ro approach by Sweeney & Burnham (1990), utilized by an algorithm in PetroMod that enables calculation of a range of vitrinite reflectance between 0.3% and 4.6% VR_r based on temperature history. Additional properties which are important for the validation and the assessment of the basin model, such as total/effective porosity, thermal maturity, hydrocarbon generation and migration, petroleum composition and adsorption capacity as well as oil and gas potential are determined based on the calibration and the basic simulation results obtained for each cell within the study area.

The thermal state of the basin depends on the basal heat flow, the surface–water interface temperature (SWIT), the thermal conductivity of the different lithologic units and radiogenic heat production inside the sedimentary system, as well as convective processes. The average heat flow of the continental crust at present is close to 60 mW m^{-2} (Allen & Allen 2013). In reconstructions of palaeo-heat flows, often high values are assumed for rifting phases when the mantle–crust boundary is at shallow depth, declining thereafter over periods of 50–100 Ma until a constant heat flow is again achieved (Waples 2001). The sediment–water interface temperature depends on palaeobathymetry and palaeolatitude as well as ocean currents, which are usually neglected in basin models. The PetroMod Software utilizes global mean surface temperatures, based on values from a palaeogeographic reconstruction by Wygrala (1989) with special emphasis on palaeogeographic latitude. Additional information concerning the SWIT was gathered and implemented from Kockel (2002), Miller *et al.* (2005) and Littke *et al.* (2008).

Model outline

Comprising the eastern part of the LSB, parts of the southern Pompeckj Block and the Gifhorn Trough, the study area extends 135 km in an east–west direction and 113 km in a north–south direction, covering a total of 15 255 km^2. The model consists of 18 layers describing Triassic–Quaternary sediments. The depth horizons for these layers were taken from a digitized version of the Geotectonic Atlas of NW Germany by Baldschuhn *et al.* (1996). The resolution of the model is restricted to a maximum cell size of 150 × 150 m with a maximum vertical resolution of 400 m, although it is primarily defined by the individual layer thickness. As a basement structure, a Paleozoic basement with a thickness of roughly 2000 m has been assumed.

Since the Toarcian Posidonia Shale represents the horizon which is of most interest in this study, modifications regarding its resolution within the model were made, allowing for a higher resolution and a more accurate lithology of the Posidonia Shale within the study area. The chosen study area represents an extension of the model created by Mohnhoff (2016) and fits in the scope of work by Bruns *et al.* (2013, 2015). These authors adopted different approaches; the Wealden within the western LSB is favoured in the case of the former, and a broader resolution for the whole study area was used for the latter.

Stratigraphy and lithology

All assigned stratigraphic units and their depositional ages can be found in Table 2 using the German Stratigraphic Chart (STD 2016) as a reference; corresponding lithologies were compiled from Ziegler (1990). To more accurately account for changes within the lithologies, especially throughout such a big area, slight adjustments were made to the facies, especially of the Posidonia Shale. This led to the use of specifically created lithologies and lithology maps of the Posidonia Shale, since pre-defined lithologies from the PetroMod software inadequately described the lithological properties. The individual properties as well as additional relevant lithological characteristics, such as the amount of radiogenic elements (U, Th, ^{40}K), essential for the radiogenic heat production through their radioactive decaying process, were taken from a dataset provided in Bruns (2014). Additional properties which are of importance include the thermal conductivity, the heat capacity and rock-mechanical properties such as density, initial porosity and the Athy's factor.

While all these factors are important concerning the integrity of the basin model, the source-rock characteristics of the Posidonia Shale are the most important regarding the actual determination of the hydrocarbon potential and petroleum generation. Utilizing a subdivision of the Posidonia Shale into three units (I–III), the differences in mineralogical composition as well as slight differences in kerogen composition are accounted for. Mineralogical composition is of importance, having a strong impact on permeability and fraccability, as well as sorption properties for methane (Sone & Zoback 2013; Gasparik *et al.* 2014). The main distinguishing parameters for the Posidonia Shale units are the carbonate and clay content, as well as the TOC content. The highest carbonate contents of up to 60 wt% $CaCO_3$ are achieved within basal unit I of the Posidonia Shale, classifying it as an organic-rich marlstone; on average 10 wt% TOC are present. This unit is usually the thinnest with total thicknesses of up to 8 m and locally up to 20 m within the Gifhorn Trough. The Posidonia Shale unit II possesses a lower carbonate percentage of up to 40 wt% $CaCO_3$ and a slightly lower TOC content of up to 9 wt%. The total thicknesses vary with location for this unit however, reaching a maximum of 16 m in the LSB and Pompeckj Block and up to 30 m in the Gifhorn Trough. Unit III shows similar carbonate contents, classifying it as a carbonaceous shale unit as well as unit II, but the highest TOC contents average 12 wt%. Unit III represents the most productive unit of the Posidonia Shale, averaging similar thicknesses as unit II, but it has a higher productive potential due to the presence of a higher quantity of organic carbon. Organic matter is predominantly composed of finely dispersed alginite (Littke *et al.* 1991). The thicknesses of the Posidonia Shale units were simplified based on values obtained from Schwarzkopf & Leythaeuser (1988) for the Gifhorn Trough and Frimmel *et al.* (2004) for the LSB and Pompeckj Block. In reality, the thickness of the Posidonia Shale unit varies more strongly (Song *et al.* 2015), but small-scale variations as recorded in the

Table 2. *Stratigraphic age assignment and petrophysical properties of the lithologies used in the 3D basin model*

Stratigraphy	Depositional age (Ma)	Lithologies	Thermal conductivity at 20°C/200°C (W (m K)$^{-1}$)	Uranium (ppm)	Thorium (ppm)	Potassium (%)	Heat capacity at 20°C/200°C (kcal (kg K)$^{-1}$)	Density (kg m^{-3})	Initial porosity (%)	Athy's factor k (km^{-1})
Quaternary–Middle Miocene	12–0	Sandstone (typical)	3.95/2.95	1.30	3.50	1.30	0.20/0.27	2720	41.0	0.31
Lower Miocene	23.8–12	Sandstone (typical), shale (typical)	3.31/2.61	1.78	5.20	1.58	0.20/0.27	2716	46.8	0.41
Rupelian–Upper Oligocene	34–23.8	Sandstone (typical), shale (typical)	3.31/2.61	1.78	5.20	1.58	0.20/0.27	2716	46.8	0.41
Middle Oligocene–Upper Eocene	37–34	Sandstone (typical), shale (typical)	1.95/1.90	3.22	10.30	2.42	0.20/0.27	2716	46.8	0.41
Lower Eocene	65–37	Sandstone (typical), shale (typical)	1.95/1.90	2.54	10.45	2.43	0.20/0.27	2704	64.2	0.41
Upper Cretaceous	99–98	Limestone (chalk, typical)	2.90/2.40	1.90	1.40	0.25	0.20/0.27	2680	70.0	0.90
Lower Cretaceous	128–99	Shale (typical), sandstone (typical)	1.79/1.81	3.46	11.15	2.56	0.21/0.27	2702	67.1	0.78
Upper Jurassic and Wealden	156.6–128	Shale (typical), limestone (chalk, typical), shale (organic-rich, organic-lean, typical)	2.18/2.02	2.80	3.34	1.47	0.20/0.27	2690	2690	0.86
Middle Jurassic (Dogger)	178–156.5	Shale (typical), sandstone (typical)	2.54/2.21	2.50	7.75	2.00	0.20/0.27	2710	55.5	0.57

Posidonia Shale III (Bifrons)	180.2–178	Shale (organic-rich) 50%, limestone (chalk, typical) 50%	1.87/1.85	5.95	6.20	1.57	0.21/0.27	2590	70.0	0.86
Posidonia Shale II (Falciferum)	182.4–180.2	Shale (organic-rich) 60%, limestone (chalk, typical) 40%	1.58/1.34	12.76	7.16	1.66	0.22/0.29	2432	70.0	0.86
Posidonia Shale I (Teniucostratum)	183.6–182.4	Shale (organic-rich) 60%, limestone (chalk, typical) 40%	1.91/1.88	3.76	7.76	1.78	0.20/0.27	2638	70.0	0.86
Lower Jurassic (Liassic)	200–183.6	Shale (organic lean, typical), limestone (chalk, typical)	1.79/1.82	3.52	10.94	2.46	0.21/0.27	2698	70.0	08.4
Upper Triassic (Keuper)	235–200	Sandstone (typical), shale (typical)	2.54/2.21	2.50	7.75	2.00	0.20/0.27	2710	55.5	0.57
Middle Triassic (Muschelkalk)	243–235	Limestone (shaly)	2.30/2.08	2.00	4.00	1.00	0.20/0.27	2730	48.0	0.50
Lower Triassic (Buntsandstein)	251–243	Sandstone (clay rich)	3.35/2.66	1.50	5.10	3.60	0.21/0.27	2760	40.0	0.32
Zechstein	258–251	Evaporites (typical)	6.54/4.18	0.02	0.01	0.1	0.2/0.29	2740	1	Uncompactable
Paleozoic Basement	300–258	Shale (typical), sandstone (typical), coal (pure)	2.16/2.01	2.75	8.59	2.12	0.21/0.27	2674	59.5	0.63

outcrop area of the Hils Syncline (Littke *et al.* 1988) are not known in any detail for the deep subsurface of the basin and cannot be properly incorporated into the large-scale model.

Results and discussion

Maturity of the Posidonia Shale units

Figure 3 provides basic information on maturity as related to the structural evolution of the basin, displaying four wells from within the study area: one located in the Pompeckj Block, one in the Gifhorn Trough and two wells from the LSB. Good calibration data were available, but usually from only a few stratigraphic units and commonly from only one stratigraphic unit within the wells (Fig. 3a). It is apparent that during basin evolution two main erosional events, related to stages of rifting and strong basin subsidence in other parts of the basin, have occurred. Only the Pompeckj Block was strongly affected by inversion during the Late Jurassic–Early Cretaceous, leading to erosion within the area. Maturity data there are usually in accordance with present-day burial and temperature, that is, maximum maturation occurred during the Neogene. A value of 0 m erosion was therefore adopted in the model, that is, a hiatus, in order to keep the model as simple as possible (Fig. 4), although it is clear that some erosion has taken place. In contrast, there are restricted areas with high maturities of Jurassic and older rocks, attributed to deep burial during the Late Jurassic followed by strong erosion (Fig. 4). This process also partially affected the Posidonia Shale source rock which, however, is still present in most of the area. An even stronger erosion took place during the Late Cretaceous basin inversion of the LSB and Gifhorn Trough when up to 6500 m (Fig. 5) of sediment was eroded; this is slightly lower than the values calculated by Bruns *et al.* (2013), Mohnhoff (2016) and Senglaub *et al.* (2005) for areas slightly to the west, where maximum erosion reached 6700–7200 m according to these authors. The erosional thicknesses for the Gifhorn Trough are considerably smaller however, having little influence on the reservoir and source-rock thickness within the area. Due to this strong erosional event

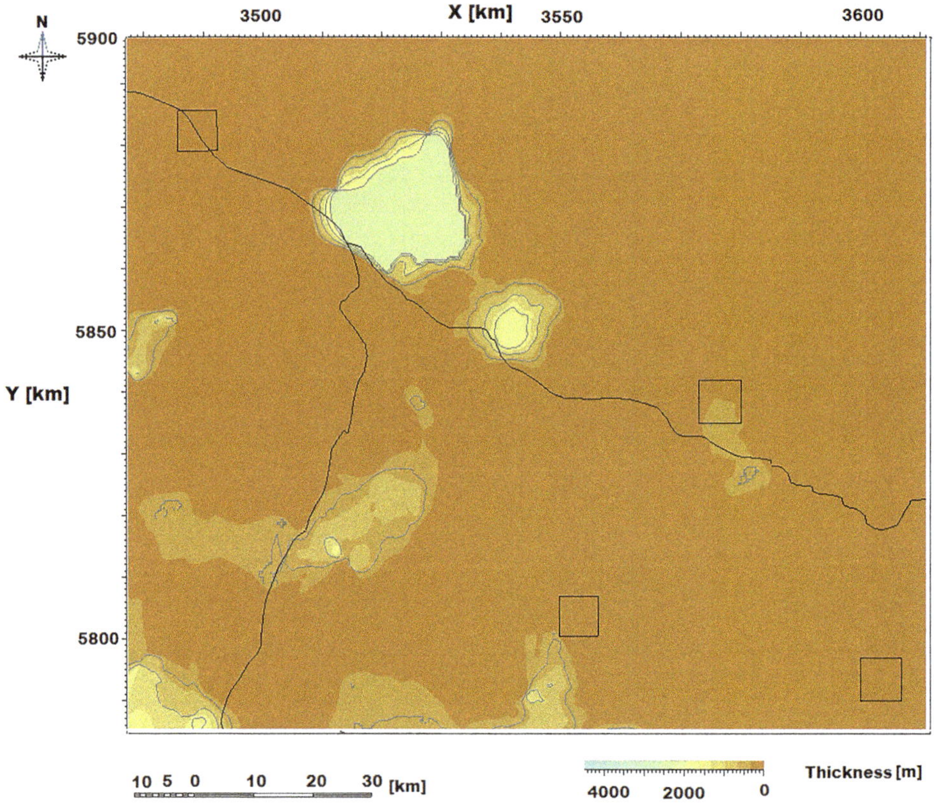

Fig. 4. Maximum erosional thicknesses of the Upper Jurassic erosion event within the study area.

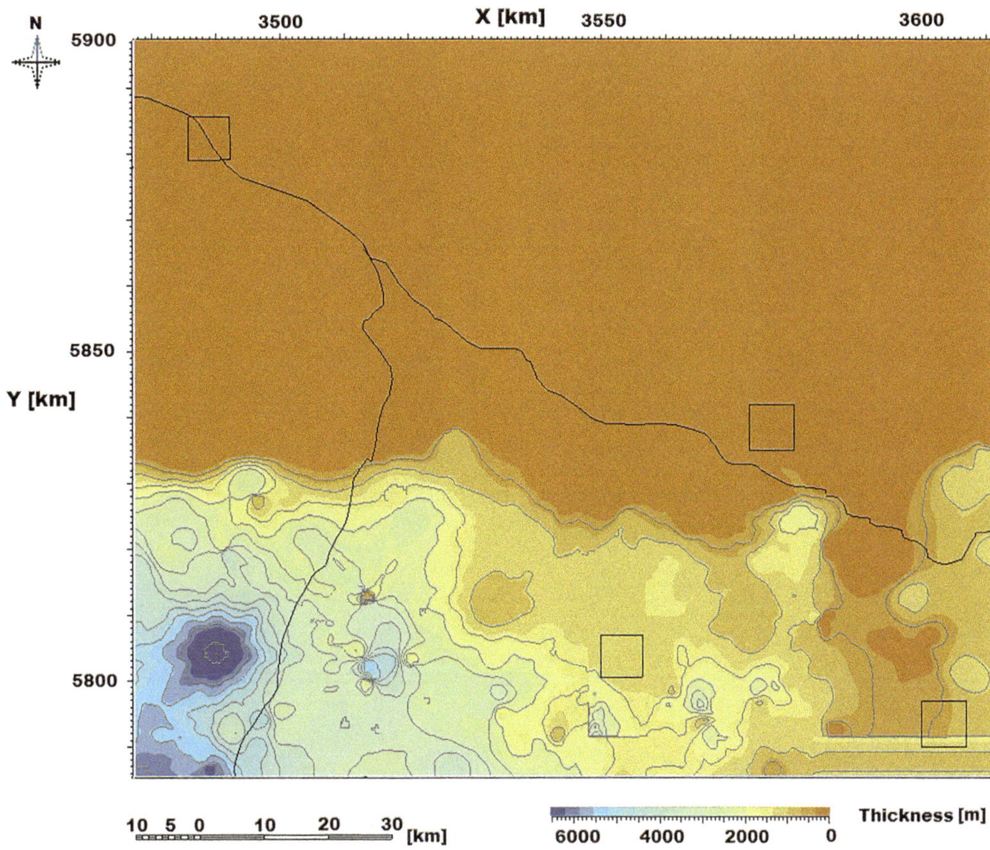

Fig. 5. Maximum erosional thicknesses of the Upper Cretaceous erosion event within the study area.

related to the basin inversion, the highest temperatures and the coinciding point of deepest burial were reached prior to it during the Late Cretaceous at roughly 89 Ma within the area of the Gifhorn Trough and LSB. Most or all parts within the Pompeckj Block, which were only affected by the first Late Jurassic–Early Cretaceous inversion event, reached their point of deepest burial in recent times due to ongoing deposition within the area. Due to these regional erosion events and the presence of Zechstein salt diapirs leading to local erosion, the Posidonia Shale is present in roughly 75% of the area.

The highest thermal maturities achieved for the Posidonia Shale are located within the centre of the LSB (Fig. 6), where vitrinite reflectance values can reach over 4% VRr. There are however small parts within the Pompeckj Block in the NW of the study area where similarly high maturities of up to 2–3% VRr are achieved. This assumption is based on a small quantity of maturity data and should be verified or falsified in the future. The strong Upper Jurassic–Lower Cretaceous erosion (Fig. 4) assumed for this area is also based on these few high-maturity values and needs to be carefully checked in the future. In large parts of the study area however, the thermal maturity of the Posidonia Shale ranges between 0.5% and 1.2% VRr (Fig. 6). This allows for a prediction of hydrocarbon generation as based on transformation ratios, especially on oil generation. While areas of interest for gas production are usually limited to maturity ranges between 1.2% and 3% VRr (except for microbial gas), areas of interest for unconventional oil production have a maturity range of 0.65–1.2% VRr; most conventional oils in the LSB and Gifhorn Trough have maturities corresponding to this range (Stock & Littke 2016).

Unconventional oil and gas potential and alternative scenarios

The present-day transformation ratio of the Posidonia Shale (Fig. 7) allows for an estimation of unconventional hydrocarbon production potential within the study area. While a transformation ratio of 100% would imply that all the reacting kerogen

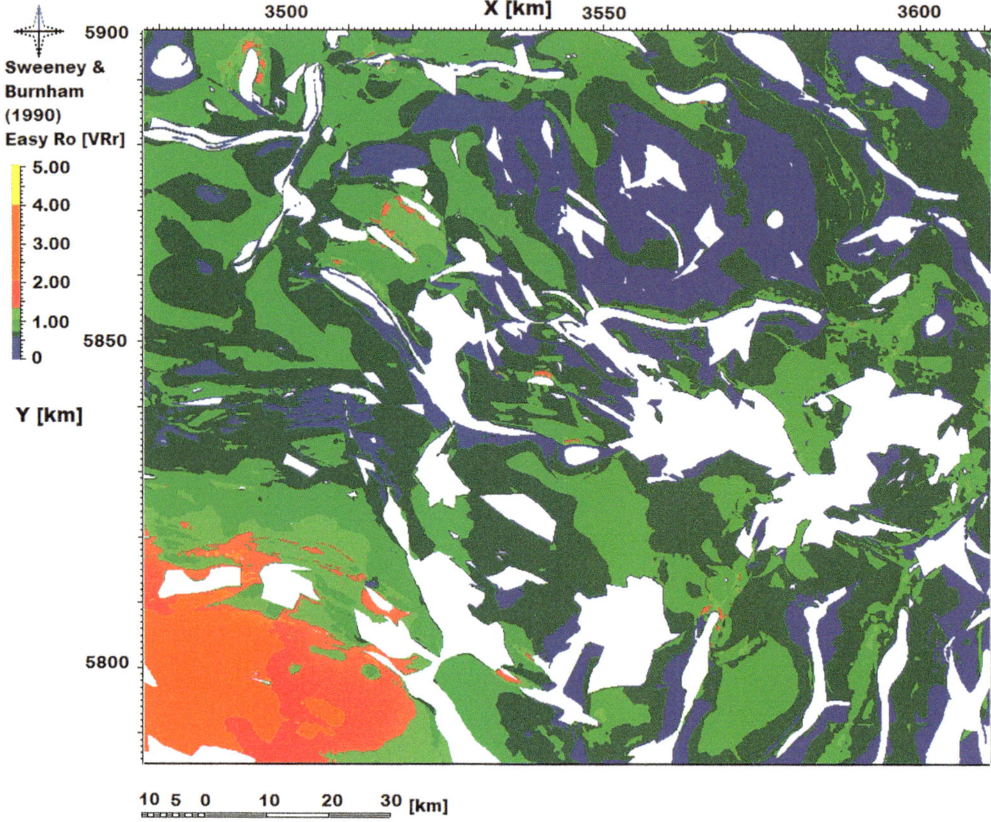

Fig. 6. Thermal maturity of Posidonia Shale within the study area, based on the Easy-Ro approach of Sweeney & Burnham (1990).

within the petroleum system has been converted, it would also imply the presence of much gas or exclusively gas rather than oil. This gas would be stored both as 'free' hydrocarbons within the pore space as well as adsorbed on the remaining kerogen. Such a high transformation rate can be found within areas of the LSB and the above-described local areas of high maturity within the Pompeckj Block, the latter being subject to future verification or falsification. All other areas where the transformation ratio is significantly lower have a remaining potential for further hydrocarbon generation, and are mainly characterized by the presence of oil rather than gas. Clearly, the amount of generated hydrocarbons exceeds the storage capacity of the oil shale at many places, especially where peak oil generation stage has (almost) been reached and where porosity and permeability are much reduced (Gasparik *et al.* 2014; Ghanizadeh *et al.* 2014). Calculations on composition of generated hydrocarbons from the Posidonia Shale (Table 3) show that, throughout the maturation of the Posidonia Shale, hydrocarbons of predominantly higher chain length were liberated (C_{15+}), followed by medium-chain-length hydrocarbons (C_6-C_{14}), volatile short-chain hydrocarbons (C_2-C_5) and lastly methane. These generated products have been divided into expelled and retained hydrocarbons. Based on the PetroMod calculation, higher-chain-length hydrocarbons have been dominantly expelled with a mean total mass of 171 601 Mt, whereas only 919 Mt are retained in the formation itself. These values are somewhat lower for the medium-chain hydrocarbons, where 78 613 Mt of hydrocarbons have been expelled in contrast to the 411 Mt that are retained. Similarly for the volatile short-chain compounds, 16 991 Mt of expelled hydrocarbons compare to 55 Mt retained. This situation is quite different for methane, for which 7837 Mt have been expelled throughout the geological history but 5571 Mt retained within the source rock. This different situation for methane is on the one hand related to the adsorptive gas content, where methane will predominantly adsorb on kerogen but also on clay minerals (Gasparik *et al.* 2014; Ghanizadeh *et al.* 2014), whereas most longer-chain n-alkanes have a lower sorption potential (Maginn *et al.*

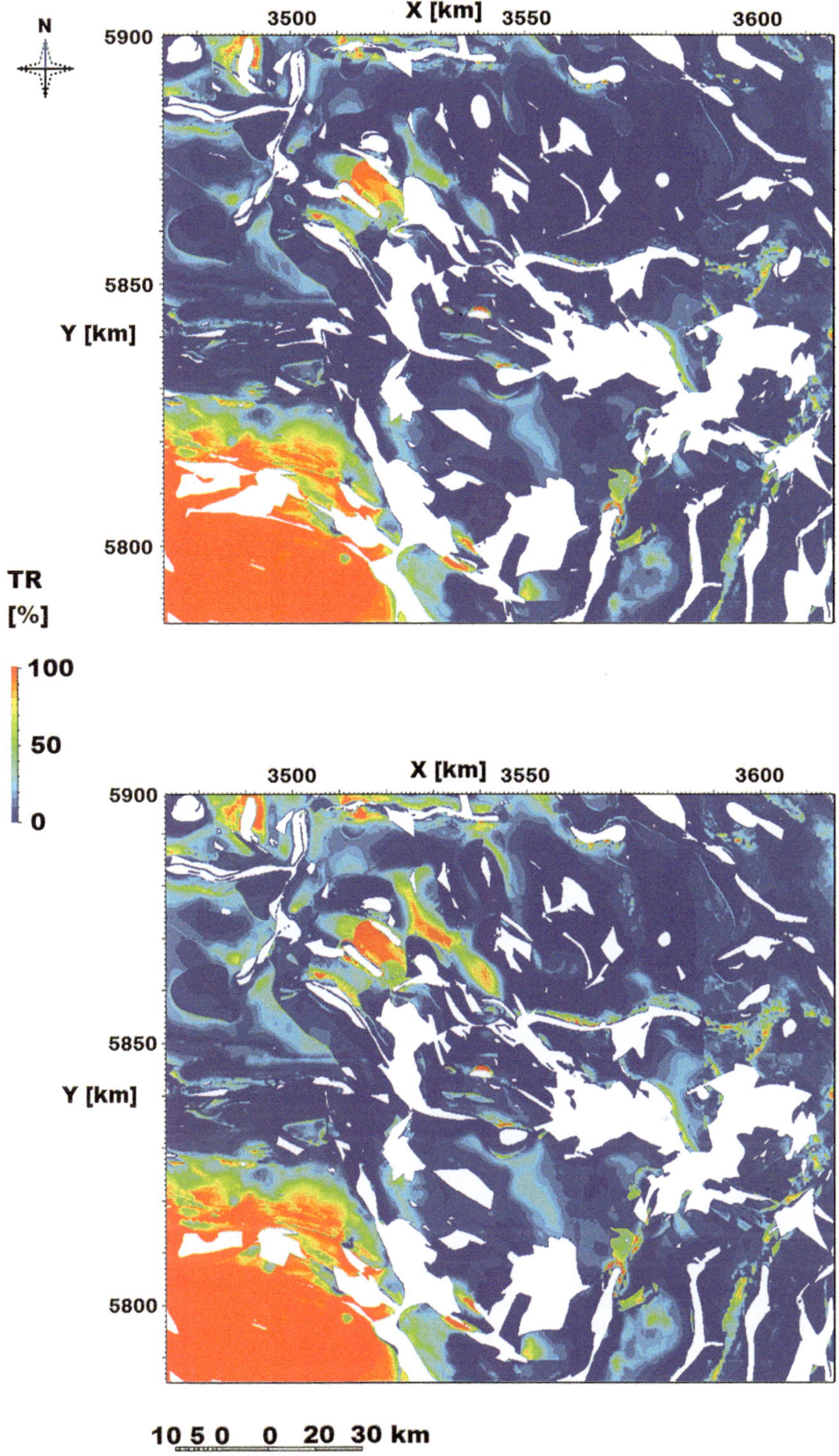

Fig. 7. Transformation ratios (%) of the kerogen for the uppermost and lowermost Posidonia Shale unit. Posidonia Shale units III (top) and I (bottom).

Table 3. *Mean compositional hydrocarbon production balance of the Posidonia Shale within the study area*

	Methane (Mt)		C_2-C_5 (Mt)		C_6-C_{14} (Mt)		C_{15+} (Mt)	
	Retained	Expelled	Retained	Expelled	Retained	Expelled	Retained	Expelled
Total	5571	7837	55	16 991	411	78 613	919	170 601
PS unit III	2340	3291	23	7136	172	33 017	386	71 652
PS unit II	1392	1959	13	4247	102	19 653	229	42 650
PS unit I	1838	2586	18	5607	135	25 942	303	56 298

1995). A second reason is that methane-rich gas has been formed as the last group of hydrocarbons and has had less time for expulsion, whereas long-chain products were generated and expelled first.

Nevertheless, hydrocarbon generation at different maturity levels led to predominantly oil generation from the Posidonia Shale (Figs 8, 9) even for the long-chain hydrocarbons, that is, there is a direct correlation with the transformation ratio. Even the post-peak oil generation mature source rock produced long-chain hydrocarbons (i.e. oil) to a larger degree as compared to short-chain or volatile gas compounds. This feature is related to the character of kerogen within this world-class oil source rock which is dominated by long-chain n-alkanes (Stock *et al.* 2017).

These hydrocarbon generation and transformation ratio maps and data have been compared to results by Mohnhoff (2016). Concerning the amount of free gas within the western LSB, 5571 Mt of free and adsorbed gas are present within the Posidonia Shale in the whole study area based on results from this study. If only the eastern LSB is considered, a similar value of 853 Mt compared to the Mohnhoff (2016) value of 540 Mt results.

Implications for unconventional hydrocarbon production within the eastern LSB, the Gifhorn Trough and the Pompeckj Block

Several types of unconventional hydrocarbon resources are present within the Central European Basin study area (BGR 2016). While the Münsterland Basin situated further to the south provides excellent opportunities for coal-bed methane exploration and production, the central–southwestern LSB offers opportunities for shale gas production, and the Gifhorn Trough and the eastern and northern rim of the LSB contain potential shale oil plays. While large parts of the Posidonia Shale are still immature in between these areas of interest, the resource potential within NW Germany is still enormous (Fig. 10). Large areas where the Posidonia Shale is present and not eroded exhibit excellent oil and gas potential. Our calculations provide evidence of generated, expelled and stored gaseous, mid-chain-length and long-chain hydrocarbons over large areas (Figs 8, 9). Clearly such data based on laboratory experiments need to be tested in future by real data. Nevertheless, we believe that a combination of experimental data obtained under controlled temperature conditions, in open and closed systems on specific source rocks with numerical modelling tools, will be important in the future to understand large-scale geological processes.

The expulsion of hydrocarbons from within a hydrocarbon system and the subsequent adsorption are important parameters, especially for the determination of the oil and gas in place in the source rock and for unconventional production. In particular, longer-chained hydrocarbons are preferentially produced and expelled from a type-II kerogen-bearing source rock such as the Posidonia Shale during the maturation from the immature to the oil-mature stage. The expulsion of these hydrocarbons is a process that is strongly related to pressure and bed thickness; similar to the maturity of the Posidonia Shale, the pressure increased strongly until *c.* 89 Ma when deepest burial followed rapid sedimentation, possibly leading to overpressure in the LSB and Gifhorn Trough.

Pressure also has a crucial influence on the adsorption capacity, leading to its strong increase with increasing pressures; temperature has the opposite effect. Clearly outcrop and shallow core/cutting samples (a few hundred metres depth) can provide excellent information on source-rock properties in general, and on gas storage at depth to a lesser extent, because most gas has been lost there. Another major factor for gas adsorption within a geological system proves to be the TOC content (Gasparik *et al.* 2014; Bruns *et al.* 2015). Highest adsorption capacities are therefore reached in the less mature areas of the Posidonia Shale, where the kerogen has not been thermally degraded to a strong degree due to hydrocarbon generation (Fig. 11). The TOC contents and volume percentages of organic matter are about twice as high as compared to overmature Posidonia Shale (Rullkötter *et al.* 1988). This difference greatly influences adsorption capacity. The reduction in adsorption capacity coincides with an increase in

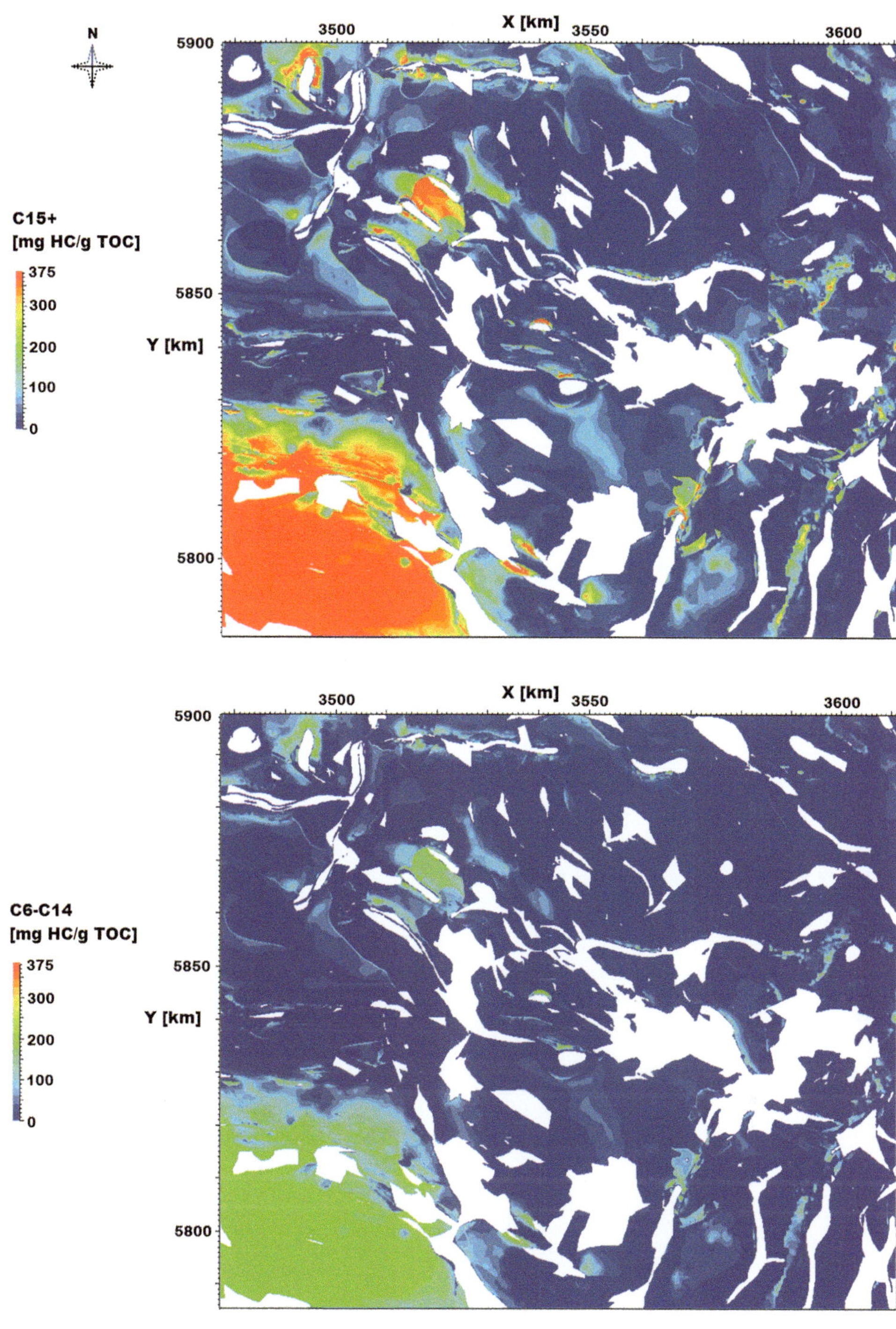

Fig. 8. Hydrocarbon generation for two compound classes of liquid hydrocarbons. Generation of C_{15+} hydrocarbons (top) and generation of C_6-C_{14} hydrocarbons (bottom) combining all three units of the Posidonia Shale. The figure shows the total amount of hydrocarbon generated over time assuming a completely open system.

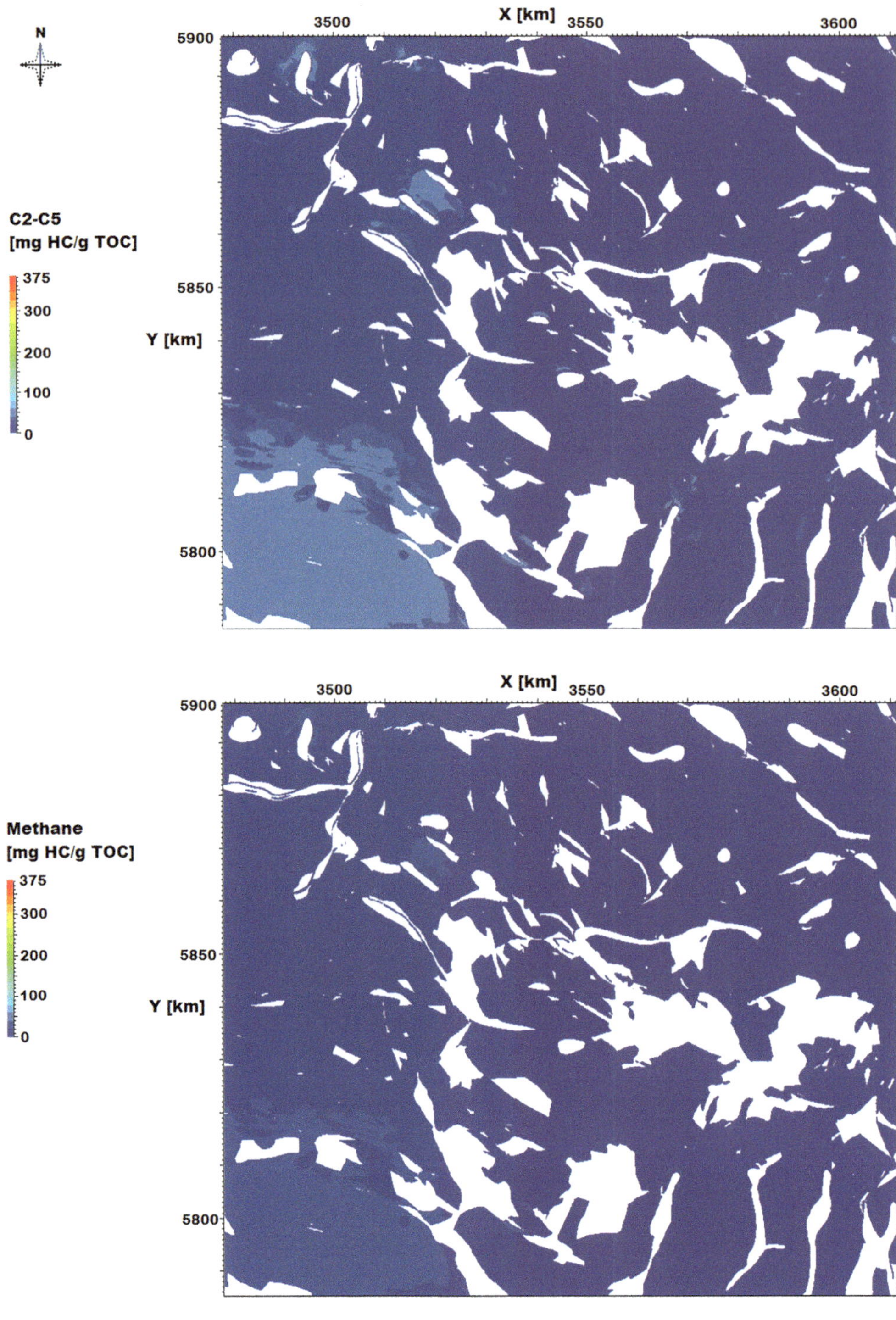

Fig. 9. Hydrocarbon gas generation maps. Generation of C2-C5 hydrocarbons (top) and generation of methane (bottom) combining all three units of the Posidonia Shale.

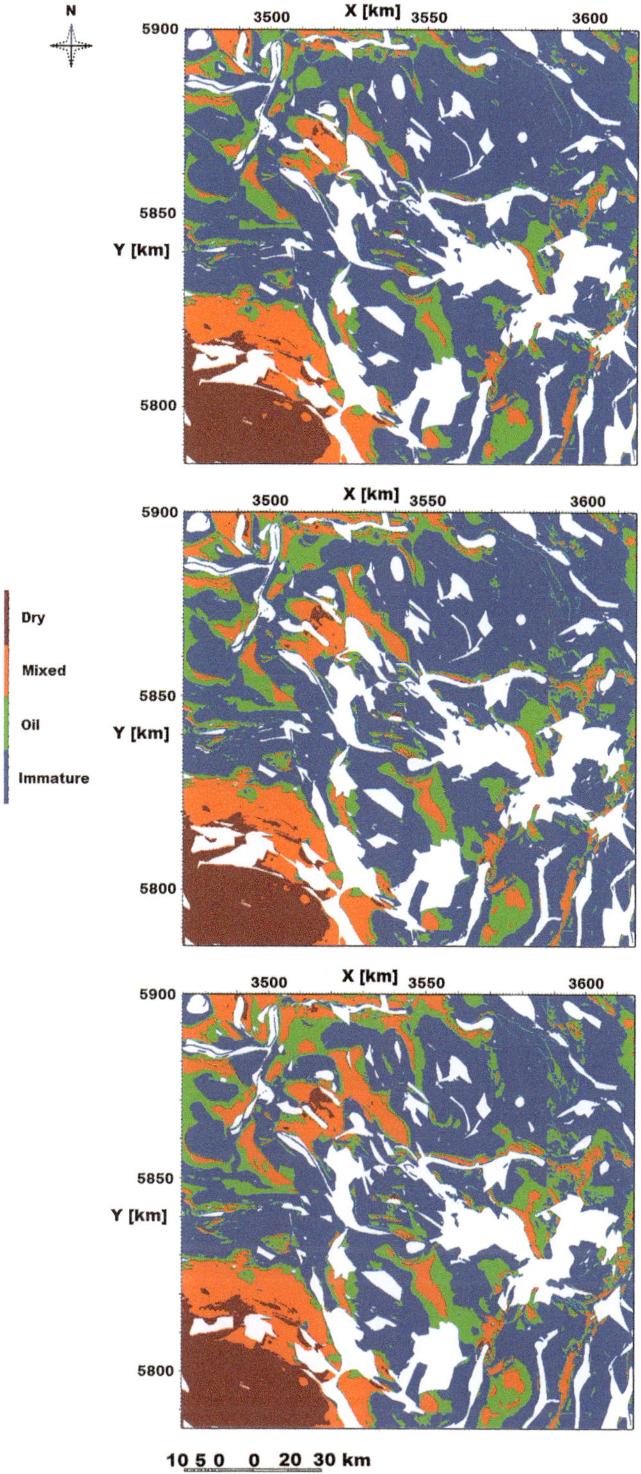

Fig. 10. Hydrocarbon zones of the upper and lower Posidonia Shale units showing the potential for oil, oil/gas (mixed) and dry gas production.

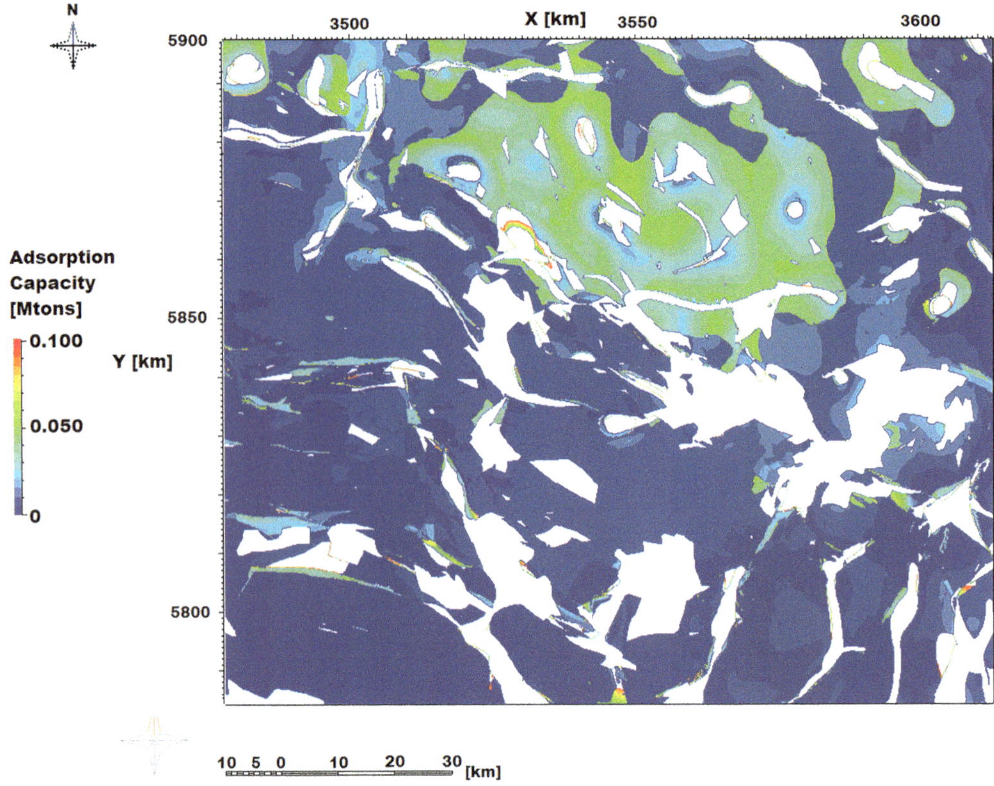

Fig. 11. Adsorption capacity of the Posidonia Shale unit III within the study area in Mt/grid cell. Note the low adsorption capacity along the uplifted/inverted southern margin of the study area.

pore space, which enhances the storage capacity of free gas within the Posidonia Shale units at high maturities. This is due to the fact that new secondary porosity is created at maturities above peak oil stage within the organic matter particles (e.g. Klaver *et al.* 2016).

Nevertheless, we find a strong preference of gas accumulation within the Posidonia Shale lithologies. The adsorption of methane is favoured within the organic matter pore space and also on the clay mineral interfaces. High TOC and clay mineral contents are present in the uppermost unit of the Posidonia Shale, whereas the underlying two units contain more carbonate on the expense of kerogen and clay minerals. The calculations of adsorption capacities within the Posidonia Shale show vastly different values for the different areas (Fig. 11), indicating a strong impact of sorption capacity on total gas in place. Only a fraction of the gas liberated from the kerogen upon maturation due to primary or secondary cracking will actually be adsorbed by the Posidonia Shale, while most methane will be present within the pore space. Porosity is quite high at low levels of maturation when primary pore space exists, reaches a minimum within the oil window and increases again at high maturities due to secondary porosity generation within kerogen when microfractures also open (Littke *et al.* 1988; Klaver *et al.* 2016). This in turn favours unconventional gas production as methane is predominant in the pore space, although the Posidonia Shale is a classical type-II source rock, and as porosity and permeability tend to increase during gas generation stage within the Posidonia Shale (Ghanizadeh *et al.* 2014).

Conclusion and outlook

The Toarcian Posidonia Shale in NW Germany has been proven to be one of the major source rocks and hydrocarbon contributors for conventional oil plays within the CEBS (Boigk 1981; Schwarzkopf & Leythaeuser 1988; Kockel *et al.* 1994; Wehner 1997; Stock & Littke 2016), but not yet as reservoir for unconventional oil and gas. Our study provides a first quantitative petroleum system analysis, indicating a high unconventional petroleum potential for the Posidonia Shale in northern Germany. However, this needs to be confirmed by well and oil flow tests. The socio-ecological and political problems of unconventional oil production in Europe, mainly related to environmental concerns (Ethridge *et al.* 2015; Kaden & Rose 2015), are not part of this study; only the potential resources available from the Posidonia Shale source rock are evaluated here.

The Posidonia Shale units I–III investigated here display some disparities in their productivity. Unit III, with the potentially highest TOC values and a thickness comparable to Unit II, shows the highest theoretical potential for unconventional hydrocarbon production. While the other units also display good inherent qualities for unconventional hydrocarbon production, they might not yield as much as the uppermost unit. From a lithological standpoint, if hydraulic fracturing is a concern then the lowest unit (I) presents a more interesting target due to its higher carbonate content and resulting stronger brittleness, although the other units are still quite high in carbonate at about 25–30% on average. The high content of clay within all Posidonia Shale units lowers the strong production potential however, as clay is mechanically softer compared to quartz and carbonate.

Both high contents of clay and kerogen/TOC lead to a high adsorption potential for methane within the Posidonia Shale, as both factors tend to enhance adsorption capacity values within a geological system. The source rocks reached their maximum thermal maturity in the Upper Cretaceous at their time of deepest burial (89 Ma) in the central LSB and Gifhorn Trough, different from the Pompeckj Block and on the rims of the LSB. Maximum maturities achieved in the Posidonia Shale range up to more than 4% VRr in the centre of the LSB. Other areas, especially in the Gifhorn Trough and parts of the Pompeckj Block, display maturities which are not higher than 1.2% VRr and therefore targets for oil generation.

The authors are grateful to the Geoforschungszentrum Potsdam (GFZ) for the measurements of the source-rock kinetics and to Ferdinand Persson for technical support. Further gratitude is extended to Schlumberger for providing access to the software PetroMod and Kimmeridge Energy for financial support.

References

Adriasola-Munoz, Y., Littke, R. & Brix, M.R. 2007. Fluid systems and basin evolution of the western Lower Saxony Basin, Germany. *Geofluids*, **7**, 335–355.

Allen, P.A. & Allen, J.R. 2013. *Basin Analysis: Principles and Application to Petroleum Play Assessment.* 3rd edn. Wiley-Blackwell, Oxford, UK.

Baldschuhn, R., Kockel, F. *et al.* 1996. *Geotektonischer Atlas von NW-Deutschland/Tectonic Atlas of NW-Germany 1:300 000.* Bundesanstalt für Geowissenschaften und Rohstoffe (BGR), Hannover.

Béhar, F., Kressmann, S., Rudkiewicz, J.L. & Vandenbroucke, M. 1992. Experimental simulation in a confined system and kinetic modelling of kerogen and oil cracking. *In*: Eckhardt, C., Maxwell, J.R., Larter, S.R. & Manning, D.A.C. (eds) *Advances in Organic Geochemistry 1991.* Pergamon Press, Oxford, *Organic Geochemistry*, **19**, 173–189.

Béhar, F., Vandenbroucke, M., Tang, Y., Marquis, F. & Espitalié, J. 1997. Thermal cracking of kerogen in open and closed systems: determination of kinetic parameters and stoichiometric coefficients for oil and gas generation. *Organic Geochemistry*, **26**, 321–339.

Berner, U., Kahl, T. & Scheeder, G. 2010. Hydrocarbon potential of sediments of the German Wealden Basin. *Oil Gas European Magazine*, **2/2010**, 80–84.

Betz, D., Führer, F., Greiner, G. & Plein, E. 1987. Evolution of the Lower Saxony Basin. *Tectonophysics*, **137**, 127–170.

BGR 2016. *Schieferöl und Schiefergas in Deutschland – Potenziale und Umweltaspekte.* Bundesanst. Geowiss. Rohstoffe, Hannover.

Blumenstein-Weingartz, I.O. 2012. *Modelling of in-reservoir petroleum biodegradation: principles and application.* Dissertation, RWTH Aachen University.

Boigk, H. 1981. *Erdöl und Erdgas in der Bundesrepublik Deutschland.* Enke, Stuttgart.

Brink, H.J., Dürschner, H. & Trappe, H. 1992. Some aspects of the late and post-Variscan development of the Northwestern German Basin. *Tectonophysics*, **207**, 65–95.

Bruns, B. 2014. *Unconventional petroleum systems in NW-Germany and the Netherlands: a 3D numerical basin modelling and organic petrography study.* Dissertation, RWTH Aachen University.

Bruns, B., di Primio, R., Berner, U. & Littke, R. 2013. Petroleum system evolution in the inverted Lower Saxony Basin, northwest Germany: a 3D basin modeling study. *Geofluids*, **13**, 246–271.

Bruns, B., Littke, R., Gasparik, M., Wees, J.D. & Nelskamp, S. 2015. Thermal evolution and shale gas potential estimation of the Wealden and Posidonia Shale in NW-Germany and the Netherlands: a 3D basin modelling study. *Basin Research*, **28**, 2–33, https://doi.org/10.1111/bre.12096

Burnham, A.K., Braun, R.L. & Samoun, A.M. 1987. Comparison of methods for measuring kerogen pyrolysis and fitting kinetic parameters. *Energy Fuels*, **1**, 452–458.

di Primio, R. & Horsfield, B. 2006. From petroleum type organofacies to hydrocarbon phase prediction. *American Association of Petroleum Geologists Bulletin*, **90**, 1031–1058.

Dieckmann, V., Schenk, H.J., Horsfield, B. & Welte, D.H. 1998. Kinetics of petroleum generation and cracking by programmed-temperature closed-system pyrolysis of Toarcian shales. *Fuel*, **77**, 23–31.

Düppenbecker, S. & Horsfield, B. 1989. Compositional information for kinetic modelling and petroleum type prediction. *Organic Geochemistry*, **16**, 259–266.

Espitalié, J., Ungerer, P., Irwin, I. & Marquis, F. 1988. Primary cracking of kerogens; experimenting and modeling

C1, C2-C5, C6-C15 and C15+ classes of hydrocarbons formed. *In*: MATTAVELLI, L. & NOVELLI, L. (eds) *Advances in Organic Geochemistry 1987: Part II: Analytical Geochemistry. Proceedings of the 13th International Meeting on Organic Geochemistry*. Pergamon Press, Oxford, *Organic Geochemistry*, **13**, 893–899.

ETHRIDGE, S., BREDFELDT, T., SHEEDY, K., SHIRLEY, S., LOPEZ, G. & HONEYCUTT, M. 2015. The Barnett Shale: from problem formulation to risk management. *Journal of Unconventional Oil and Gas Resources*, **11**, 95–110.

FRIMMEL, A., OSCHMANN, W. & SCHWARK, L. 2004. Chemostratigraphy of the Posidonia Black Shale, SW Germany: I. Influence of sea-level variation on organic facies evolution. *Chemical Geology*, **206**, 199–230.

GASPARIK, M., BERTIER, P., GENSTERBLUM, Y., GHANIZADEH, A., KROOSS, B.M. & LITTKE, R. 2014. Geological controls on the methane storage capacity in organic-rich shales. *International Journal of Coal Geology*, **123**, 34–51.

GAUPP, R., MÖLLER, P., LÜDERS, V., DI PRIMIO, R. & LITTKE, R. 2008. Fluids in sedimentary basins: an overview. *In*: LITTKE, R., BAYER, U., GAJEWSKI, D. & NELSKAMP, S. (eds) *Dynamics of Complex Intracontinental Basins*. Springer-Verlag, Berlin, 347–366.

GHANIZADEH, A., AMANN-HILDENBRAND, A., GASPARIK, M., GENSTERBLUM, Y., KROOSS, B.M. & LITTKE, R. 2014. Experimental study of fluid transport processes in the matrix system of the European organic-rich shales: II. Posidonia Shale (Lower Toarcian, Northern Germany). *International Journal of Coal Geology*, **123**, 20–33.

HORSFIELD, B. & DÜPPENBECKER, S.J. 1991. The decomposition of Posidonia Shale and Green River kerogens using microscale sealed vessel (MSSV) pyrolysis. *Journal of Analytical and Applied Pyrolysis*, **20**, 107–123.

HORSFIELD, B., DISKO, U. & LEISTNER, F. 1989. The microscale simulation of maturation: outline of a new technique and its potential applications. *Geologische Rundschau*, **78**, 361–373.

HORSFIELD, B., LITTKE, R., MANN, U., BERNARD, S., TIEM ANH, T.V., DI PRIMIO, R. & SCHULZ, H.-M. 2010. Shale Gas in the Posidonia Shale, Hils Area, Germany. *AAPG Annual Convention. New Orleans, 11–14 April.*

KADEN, D. & ROSE, T. 2015. *Environmental and Health Issues in Unconventional Oil and Gas Development*. Elsevier Publishing.

KLAVER, J., DESBOIS, G., LITTKE, R. & URAI, J.L. 2016. BIB-SEM pore characterization of mature and post mature Posidonia Shale samples from the Hils area, Germany. *International Journal of Coal Geology*, **158**, 78–89.

KOCKEL, F. 2002. Rifting processes in NW-Germany and the German North Sea sector. *Geologie en Mijnbouw-Netherlands Journal of Geosciences*, **81**, 149–158.

KOCKEL, F., WEHNER, H. & GERLING, P. 1994. Petroleum systems of the Lower Saxony basin, Germany. *In*: MAGOON, L.B. & DOW, W.G. (eds) *The Petroleum System–From Source to Trap, AAPG, Memoir no.* **60**, 573–568.

LITTKE, R., BAKER, D.R. & LEYTHAEUSER, D. 1988. Microscopic and sedimentologic evidence for the generation and migration of hydrocarbons in Toarcian source rocks of different maturities. *Organic Geochemistry*, **13**, 549–559.

LITTKE, R., BAKER, D.R., LEYTHAEUSER, D. & RULLKÖTTER, J. 1991. Keys to the depositional history of the Posidonia Shale (Toarcian) in the Hils Syncline, northern Germany. *In*: TYSON, R.V. & PEARSON, T.H. (eds) *Modern and Ancient Continental Shelf Anoxia*, Geological Society, London, Special Publications, **58**, 311–333, https://doi.org/10.1144/GSL.SP.1991.058.01.20

LITTKE, R., KROOSS, B.M., IDIZ, E. & FRIELINGSDORF, J. 1995. Molecular nitrogen in natural gas accumulations: generation from sedimentary organic matter at high temperature. *American Association of Petroleum Geologists Bulletin*, **79**, 410–430.

LITTKE, R., SCHECK-WENDEROTH, M., BRIX, M.R. & NELSKAMP, S. 2008. Subsidence, inversion and evolution of the thermal field. *In*: LITTKE, R., BAYER, U., GAJEWSKI, D. & NELSKAMP, S. (eds) *Dynamics of Complex Intracontinental Basins – The Central European Basin System*. Springer, Berlin, Heidelberg.

LITTKE, R., KROOSS, B.M., UFFMANN, A.K., SCHULZ, H.-M. & HORSFIELD, B. 2011. Unconventional gas resources in the Paleozoic of Central Europe. *Oil and Gas Science and Technology – Revue d'IPF Energies Nouvelles*, **66**, 953–977.

MAGINN, E.J., BELL, A.T. & THEODOROU, D.N. 1995. Sorption thermodynamics, siting, and conformation of long n-alkanes in silica as predicted by Configurational-Bias Monte Carlo Integration. *Journal of Physical Chemistry*, **99**, 2057–2079.

MILLER, K.G., KOMINZ, M.A. *ET AL.* 2005. The Phanerozoic record of global sea–level change. *Science*, **310**, 1293–1298.

MOHNHOFF, D. 2016. *Oil and gas shales of Northern Germany: implications from organic geochemical analyses, petrophysical measurements and 3D numerical basin modelling*. Dissertation, RWTH Aachen University.

MOHNHOFF, D., LITTKE, R. & SACHSE, V.F. 2015. Estimates of shale gas contents in the Posidonia Shale and Wealden of the west-central Lower Saxony Basin from high-resolution 3D numerical basin modelling. *German Journal of Geology (ZDGG)*, **167**, 295–314.

PEPPER, A & CORVI, P.J. 1995. Simple kinetic models of petroleum formation. Part I: oil and gas generation from kerogen. *Marine and Petroleum Geology*, **12**, 291–319.

QUIGLEY, T.M., MACKENZIE, A.S. & GRAY, J.R. 1987. Kinetic theory of petroleum generation. *In*: TISSOT, B. (ed.) *Migration of Hydrocarbons in Sedimentary Basins*. Editions Technip, Paris, France, *Colloques at Seminares*, **45**, 649–665.

RIPPEN, D., LITTKE, R., BRUNS, B. & MAHLSTEDT, N. 2013. Organic geochemistry and petrography of Lower Cretaceous Wealden black shales of the Lower Saxony Basin: the transition from lacustrine oil shales to gas shales. *Organic Geochemistry*, **63**, 18–36.

RULLKÖTTER, J., LEYTHAEUSER, D. *ET AL.* 1988. Organic matter maturation under the influence of a deep intrusive heat source: a natural experiment for quantification of hydrocarbon generation and expulsion from a petroleum source rock (Toarcian shale, northern Germany). *Organic Geochemistry*, **13**, 847–856.

SCHÄFER, R.G., SCHENK, H.J., HARDELAUF, H. & HARMS, R. 1990. Determination of gross kinetic parameters for petroleum formation from Jurassic source rocks of different maturity levels by means of laboratory experiments. *Organic Geochemistry*, **16**, 115–120.

SCHENK, H.J. & HORSFIELD, B. 1998. Using natural maturation series to evaluate the utility of parallel kinetics models: an investigation of Toarcian shales and Carboniferous coals, Germany. *Organic Geochemistry*, **29**, 137–154.

Schwarzkopf, T. 1988. *Herkunft und Migration des Erdöls in ausgewählten Dogger beta Lagerstätten des Gifhorner Troges: Wechselwirkungen zwischen Kohlenwasserstoffgenese und Sandsteingenese*. Dissertation, RWTH Aachen University.

Schwarzkopf, T. & Leythaeuser, D. 1988. Oil generation and migration in the Gifhorn Trough, NW-Germany. *Organic Geochemistry*, **13**, 245–253.

Senglaub, Y., Brix, M.R., Adriasola, A.C. & Littke, R. 2005. New information on the thermal history of the southwestern Lower Saxony Basin, northern Germany, based on fission track analysis. *International Journal of Earth Sciences*, **94**, 876–896.

Sirocko, F., Reicherter, K., Lehné, R., Hübscher, C.H., Winsemann, J. & Stackebrandt, W. 2008. Glaciation, salt and the present landscape. *In*: Littke, R., Bayer, U., Gajewski, D. & Nelskamp, S. (eds) *Dynamics of Complex Intracontinental Basins*. Springer, Berlin, 211–228.

Sone, H. & Zoback, M.D. 2013. Mechanical properties of shale-gas reservoir rocks – Part 2: ductile creep, brittle strength, and their relation to the elastic modulus. *Geophysics*, **78**, 393–402.

Song, J., Littke, R., Maquil, R. & Weniger, P. 2015. Organic facies variability in the Posidonia Black Shale from Luxembourg: implications for thermal maturation and depositional environment. *Palaeogeography, Palaeoclimatology, Palaeoecology*, **410**, 316–336.

Song, J., Littke, R. & Weniger, P. 2016. Organic geochemistry of the Lower Toarcian Posidonia Shale in NW Europe. *Organic Geochemistry*, **106**, 76–92, https://doi.org/10.1016/j.orggeochem.2016.10.014

STD 2016. *Stratigraphische Tabelle von Deutschland 2016 (Menning, M. and Hendrich, A.)*. Deutsches GeoForschungszentrum, Potsdam.

Stock, A. & Littke, R. 2016. Geochemical composition of oils from the Gifhorn Trough and Lower Saxony Basin in comparison to Posidonia Shale source rocks from the Hils Syncline, Northern Germany. *German Journal of Geology*, **167**, 315–331.

Stock, A., Littke, R., Schwarzbauer, J., Horsfield, B. & Hartkopf-Fröder, C. 2017. Organic geochemistry and petrology of Posidonia Shale (Lower Toarcian, Western Europe) – the evolution from immature oil-prone to overmature dry gas-prone kerogen. *International Journal of Coal Geology*, **176**, 36–48.

Stollhofen, H., Bachmann, G.H. *et al.* 2008. Upper Rotliegernd to Early Cretaceous basin development. *In*: Littke, R., Bayer, U., Gajewski, D. & Nelskamp, S. (eds) *Dynamics of Complex Intracontinental Basins*. Springer, Berlin, 211–228.

Sweeney, J.J. & Burnham, A.K. 1990. Evaluation of a simple model of vitrinite reflectance based on chemical kinetics (1). *AAPG Bulletin*, **74**, 1559–1570.

Sweeney, J.J., Burnham, A.K., Braun, R.L., Talukdar, S. & Vallejos, C. 1992. A test of a kinetic model of petroleum generation, expulsion, and cracking in the Maracaibo Basin, Venezuela. *AAPG Annual Meeting, Calgary, Alberta, Canada, 22–25. June, 1992.* **1**, 128.

Sweeney, J.J., Braun, R.L., Burnham, A.K., Talukdar, S. & Vallejos, C. 1995. Chemical kinetic model of hydrocarbon generation, expulsion, and destruction applied to the Maracaibo Basin, Venezuela. *American Association of Petroleum Geologists Bulletin*, **79**, 1515–1532.

Tissot, B.P., Pelt, R. & Ungerer, P. 1987. Thermal history of sedimentary basins, maturation indices, and kinetics of oil and gas generation. *American Association of Petroleum Geologists Bulletin*, **71**, 1445–1466.

Uffmann, A.K. & Littke, R. 2013. Paleozoic petroleum systems of the Münsterland Basin, western Germany: a 3D basin modeling study: Part I: model input, calibration and maturation. *Oil Gas European Magazine*, **39**, 140–146.

Uffmann, A.K., Littke, R. & Rippen, D. 2012. Mineralogy and geochemistry of Mississippian and Lower Pennsylvanian Black Shales at the Northern Margin of the Variscan Mountain Belt (Germany and Belgium). *International Journal of Coal Geology*, **103**, 92–108.

Ungerer, P. 1989. State of the art of research in kinetic modeling of oil formation and expulsion. *Organic Geochemistry*, **16**, 1–25.

Vandenbroucke, M., Béhar, F. & Rudkiewicz, J.L. 1999. Kinetic modelling of petroleum formation and cracking: implications from the high pressure/high temperature Elgin field (UK, North Sea). *Organic Geochemistry*, **30**, 1105–1125.

Voigt, T., Reicherter, K., Von Eynatten, H., Littke, R., Voigt, S. & Kley, J. 2008. Sedimentation during basin inversion. *In*: Littke, R., Bayer, U., Gajewski, D. & Nelskamp, S. (eds) *Dynamics of Complex Intracontinental Basins*. Springer, Berlin, 211–228.

Waples, D.W. 2001. A new model for heat flow in extensional basins: radiogenic heat, asthenospheric heat and the McKenzie model. *Natural Resources Research*, **10**, 227–238.

Warren, J.K. 2008. Salt as sediment in the Central European Basin System as seen from a deep time perspective. *In*: Littke, R., Bayer, U., Gajewski, D. & Nelskamp, S. (eds) *Dynamics of Complex Intracontinental Basins*. Springer-Verlag, Berlin, 247–276.

Wehner, H. 1997. Source and maturation of crude oils in Northern and Eastern Germany – an organic geochemical approach. *Geologisches Jahrbuch D*, **103**, 85–102.

Welte, D.H., Horsfield, B. & Baker, D.R. 1997. *Petroleum and Basin Evolution*. Springer-Verlag, Berlin.

Wygrala, B. 1989. Integrated study of an oil field in the southern Po basin, northern Italy. *Forschungszentrum Jülich GmbH, Berichte der Kernforschungsanlage Jülich*, **2313**, 328.

Ziegler, P.A. 1990. *Geological Atlas of Western and Central Europe*. Shell Internationale Petroleum Maatschappij BV, Geological Society of London.

Reservoir architecture model of the Nieuwerkerk Formation (Early Cretaceous, West Netherlands Basin): diachronous development of sand-prone fluvial deposits

A. G. VONDRAK[1]*, M. E. DONSELAAR[2] & D. K. MUNSTERMAN[3]

[1]*PanTerra Geoconsultants BV, Weversbaan 1–3, 2352 BZ Leiderdorp, The Netherlands*

[2]*Department of Geoscience and Engineering, Delft University of Technology, PO Box 5048, 2600 GA Delft, The Netherlands*

[3]*TNO – Geological Survey of the Netherlands, PO Box 80015, 3508 TA Utrecht, The Netherlands*

**Correspondence: A.Vondrak@panterra.nl*

Abstract: The sand-prone fluvial Delft Sandstone Member of the Nieuwerkerk Formation is the main target for geothermal exploitation in the West Netherlands Basin. Uncertainty reduction in the modelling of permeability heterogeneity and connectivity of the sandstone reservoir units is of prime importance in the low-profit geothermal energy market. Lithostratigraphic correlation of the Delft Sandstone Member dates back to the 1970s, and considers the fluvial deposits as Valanginian (140–134 Ma) in age and synchronously deposited throughout the basin.

Analysis of newly-acquired palynological data, in combination with well-log correlation and seismic interpretation, enabled the construction of a sequence-stratigraphic framework for the Nieuwerkerk Formation, in which the fluvial sandstones were deposited in a much wider age range than previously believed, from Late Ryazanian to Valanginian (142–134 Ma). The depocentre of fluvial deposition shifted in time from NE to SW across the width of the basin. This diachronous development resulted in a discontinuous spatial arrangement of fluvial sandstones. The new reservoir architecture model will aid in the well placement of geothermal doublets and the assessment of interference risk of adjacent geothermal projects.

The West Netherlands Basin (WNB) is a mature area for hydrocarbon exploitation (Fig. 1) that started production in the 1950s and reached its peak in the 1980s. Production is partly from Lower Cretaceous fluvial and shallow-marine sandstone reservoirs. Since 2007, the WNB has undergone a revival in terms of geothermal energy production, which targets the same reservoirs, with the focus on fluvial reservoirs of the Delft Sandstone Member of the Lower Cretaceous Nieuwerkerk Formation (Den Hartog Jager 1996; Van Adrichem Boogaert & Kouwe 1997; Herngreen & Wong 2007; Donselaar *et al.* 2015; Willems *et al.* 2017). The requirements for economically viable geothermal projects in The Netherlands deviate from those of oil and gas exploitation in that:

- geothermal reservoir targets are in structural lows where the temperature is higher, whereas oil and gas exploration is focused on the highs of structural traps;
- geothermal projects consist of two wells – a production and a re-injection well – that need to be in the same reservoir space for pressure maintenance;
- the profit margins in geothermal projects are much lower than in hydrocarbon production.

The maturity of the WNB as a hydrocarbon production area means that the available datasets include vintage 2D and 3D seismic data plus hydrocarbon well data. These wells are mostly located on structural highs in the subsurface, which contributes to reservoir models with limited spatial resolution and considerable geological uncertainty in terms of shape, size, connectivity and internal permeability heterogeneity of the reservoir units. This is especially so when the subsurface data are extrapolated from the structural highs to the geothermal targets in the structural lows. In addition, the Delft Sandstone Member target is in a low net-to-gross (N/G) lower coastal-plain depositional setting (Donselaar *et al.* 2015; Willems *et al.* 2017). Such poorly constrained datasets illustrate the inherent economic and environmental risk for geothermal energy production projects, and highlight the need for improved reservoir architecture models for the Nieuwerkerk Formation.

The aim of the present study is to construct a reservoir architecture model for the distribution in space

From: Kilhams, B., Kukla, P. A., Mazur, S., McKie, T., Mijnlieff, H. F. & van Ojik, K. (eds) 2018. *Mesozoic Resource Potential in the Southern Permian Basin*. Geological Society, London, Special Publications, **469**, 423–434.
First published online March 15, 2018, https://doi.org/10.1144/SP469.18

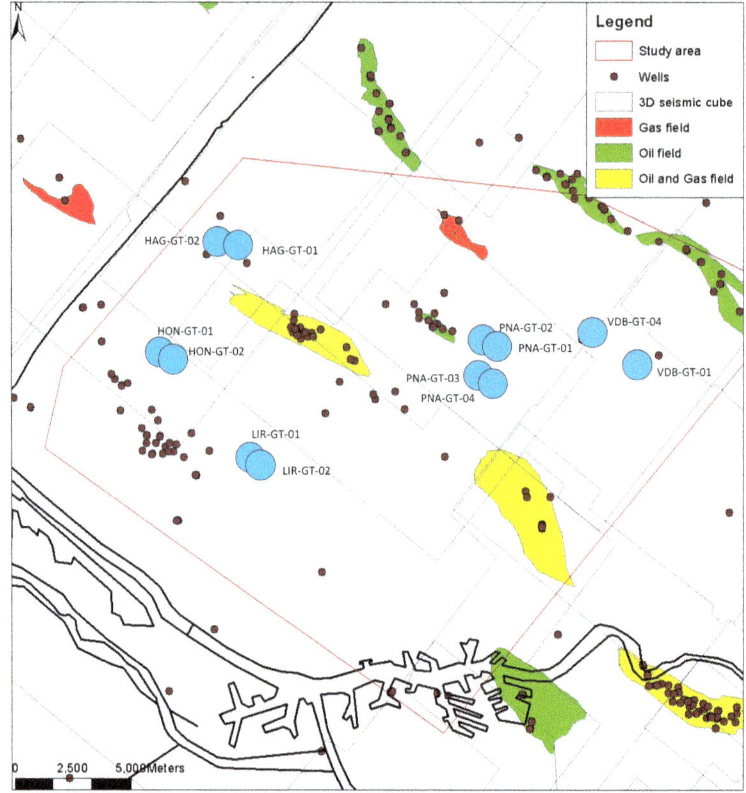

Fig. 1. Overview of the study area, with all drilled wells and fields in the Rijswijk, Rijn, Delft and Alblasserdam reservoirs. Oil (green) and gas (red) fields in the West Netherlands Basin. The black polygons indicate the areas covered by 3D seismic data. Blue circles are the locations of the geothermal wells; HAG-GT-01 and HAG-GT-02 (The Hague: city heating); PNA-GT-01, PNA-GT-02, PNA-GT-03 and PNA-GT-04 (Pijnacker: greenhouse heating); HON-GT-01 and HON-GT-02 (Honselersdijk: greenhouse heating); LIR-GT-01 and LIR-GT-02 (De Lier: greenhouse heating); and VDB-GT-01 and VDB-GT-04 (Bleiswijk: greenhouse heating). Modified from: NLOG (2017).

and time of the sandstone units in the Delft Sandstone Member. For this, a large-scale framework is constructed from the re-evaluation of existing 3D seismic surveys. Newly-acquired palynological data from geothermal wells are used to biostratigraphically constrain the correlation of available well logs from hydrocarbon and geothermal wells. Results from this sequence-stratigraphically constrained reservoir architecture model will aid in reducing the risks of geothermal energy exploitation and in assessing the future aspects of subsurface spatial planning.

Geological setting

The WNB and the adjacent offshore area (Fig. 2) formed during several rift phases from Kimmeridgian (157 Ma) to Barremian (131 Ma) time. Transtensional tectonic movements fragmented the WNB into NW–SE-trending sub-basins. Rift activity ceased in the Hauterivian (131 Ma) and the basin entered a post-rift sag phase. Basin inversion occurred in the Late Cretaceous Laramide compressional phase (Van Wijhe 1987; Den Hartog Jager 1996). Existing basin-bounding faults were reactivated and are visible on seismic as reverse offsets. Many of the oil fields in the WNB are pop-up structures that formed during this compression (Den Hartog Jager 1996; Racero-Baena & Drake 1996).

Syntectonic sedimentation in the WNB is characterized by Upper Jurassic and Lower Cretaceous fluvial deposits in a lower coastal-plain depositional setting, and subsequent lagoonal, plus marine, sediments formed in a large-scale transgressive succession (Den Hartog Jager 1996; Jeremiah *et al.* 2010). The total thickness of the basin fill is up to 3000 m, with large thickness variations as a result of tectonically induced differential subsidence (Van Adrichem Boogaert & Kouwe 1997). The stratigraphic

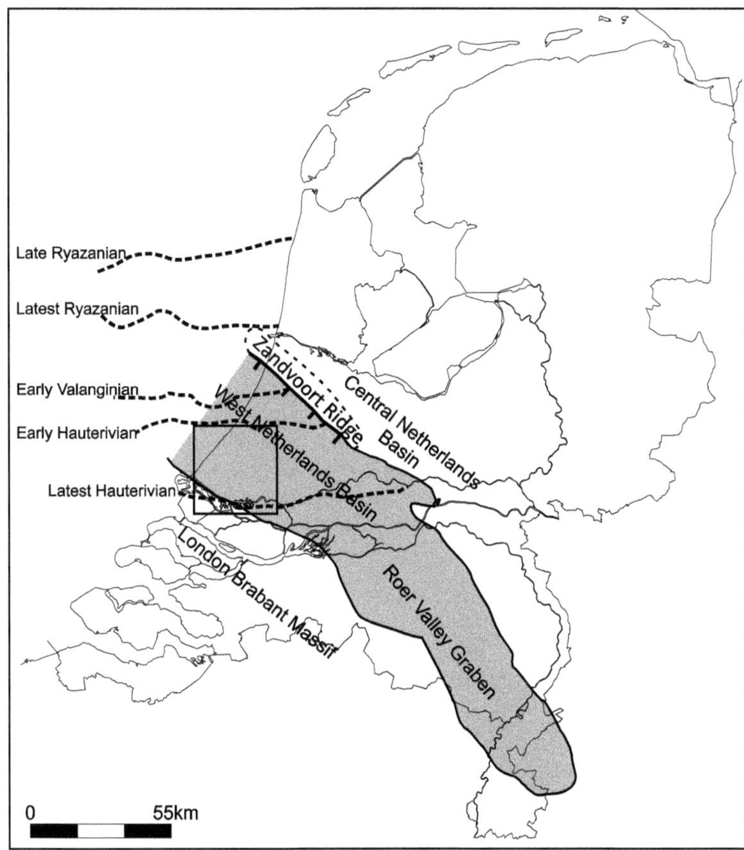

Fig. 2. Geological setting of the West Netherlands Basin (WNB) during the Cretaceous: the London–Brabant Massif is in the south Central Netherlands Basin to the north, and the shoreline is moving into the WNB from the north (indicated by time lines from Jeremiah *et al.* 2010).

subdivision of the Early Cretaceous period (Fig. 3) comprises, from base to top: (1) fluvial to marginal-marine sediments of the Nieuwerkerk Formation deposited during the Cimmerian rift phase; and (2) marine clastics of the Rijnland Group deposited during the post-rift sag phase (Den Hartog Jager 1996; Van Adrichem Boogaert & Kouwe 1997).

The Nieuwerkerk Formation

The Nieuwerkerk Formation consists, from base to top, of the Alblasserdam Member, the Delft Sandstone Member and the Rodenrijs Claystone Member (Van Adrichem Boogaert & Kouwe 1997). The Alblasserdam Member ranges in age from late Oxfordian to Valanginian (*c.* 158–135 Ma: Van Amerom *et al.* 1976). However, the age determination has a high uncertainty because of the low sporomorph content, especially in the lower, oxidized part of the member (Van Adrichem Boogaert & Kouwe 1997).

The Alblasserdam Member is present throughout the WNB, and consists of braided-river deposits with high sandstone/shale ratios (Den Hartog Jager 1996; DeVault & Jeremiah 2002). Sandstone geometries consist of sheets and isolated or stacked channels (Van Adrichem Boogaert & Kouwe 1997). The Alblasserdam Member unconformably overlies Middle Jurassic shallow-marine limestone and shelf mudstone of the Brabant and Werkendam formations, respectively. The total thickness ranges from less than 100 m to over 1300 m, with this large variation believed to be a consequence of syntectonic sedimentation in NW–SE-orientated grabens.

The Delft Sandstone Member conformably overlies the Alblasserdam Member. The Delft Sandstone Member is present in the west and central parts of the WNB. Based on sporomorph assemblages, a Valanginian age (139–134 Ma) has been assigned to the Delft Sandstone Member (Van Adrichem Boogaert & Kouwe 1997). The Delft Sandstone Member consists of meandering-river deposits and associated

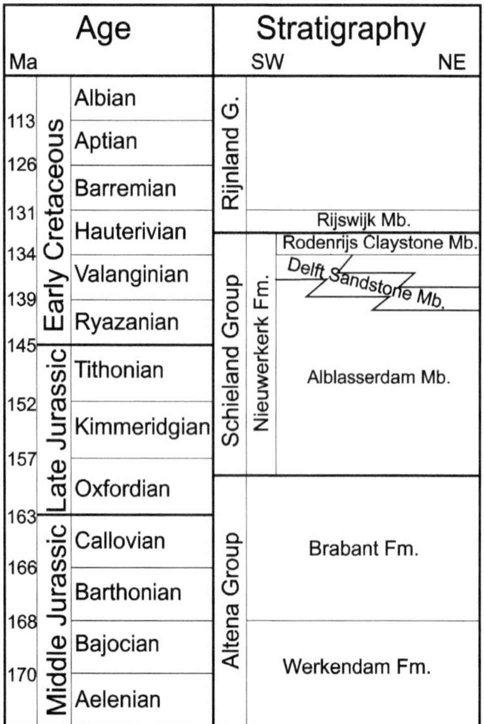

Fig. 3. Stratigraphic column of the Middle Jurassic–Early Cretaceous (based on Van Adrichem Boogaert & Kouwe 1997; Jeremiah *et al.* 2010).

floodplain fines in a lower coastal-plain setting (Van Adrichem Boogaert & Kouwe 1997; Donselaar *et al.* 2015; Willems *et al.* 2017). On the basis of detailed core and cutting analyses from hydrocarbon wells in the area, the Delft Sandstone Member is subdivided into three units with varying N/G and stacking patterns of the fluvial sandstone (Donselaar *et al.* 2015). Unit 1 consists of loosely-stacked single-storey meandering river sandstone bodies with well-preserved fine-grained tops embedded in a floodplain claystone and siltstone matrix. Unit 2 consists of interbedded claystone, siltstone and coal layers formed in a wet floodplain and swamp environment (with very low N/G). Unit 3 is characterized by stacked, multistorey and laterally-amalgamated meandering river sandstone bodies with minor mudstone-dominated floodplain intervals.

Data and methods

Seismic interpretation

Ten 3D seismic surveys were carried out between the 1980s and the 1990s over the WNB. These surveys are in the public domain (available at NLOG 2017).

The 10 3D seismic surveys (all in two-way time (TWT)) were merged into one. The merging was focused on the Nieuwerkerk Formation interval, making sure that the surveys were aligned to this interval. Therefore, bulk vertical shifts were applied during merging to a number of surveys; these shifts were identified in overlapping areas. Based on visual inspection in the areas of overlap, the seismic survey with the best quality was kept in the merged cube. Edge effects were eliminated where possible (i.e. in areas of overlap).

The merged seismic cube was used to visualize the thickness variation of the Nieuwerkerk Formation in the various sub-basins. The horizons of the base Rijnland and base Schieland groups interpreted by RGD (2012) were also used in this study. These two horizons are the top and the base of the Nieuwerkerk Formation and indicate the (seismic) thickness of this formation. To visualize the thickness variation in the Nieuwerkerk Formation, flattening on the Top De Lier horizon was applied. For this purpose, the Top De Lier was interpreted on a number of lines. This is a continuous horizon and was interpreted on a soft kick.

Wells

Four geothermal wells, and their associated gamma-ray (GR) log suites, were utilized for this study (HON-GT-01, HON-GT-02, PNA-GT-02 and VDB-GT-04). The GR logs were loaded into Schlumberger's Petrel 2014 software and normalized for the Nieuwerkerk Formation (as described by Shier 2004). The resulting normalized GR logs were used to make lithostratigraphic correlations from well to well.

No velocity data were present that could be used to tie the wells to the seismic in the structural lows of the basin, where the geothermal wells are located. For simplicity, therefore, the VelMod2 velocity model from TNO (Van Dalfsen *et al.* 2007) was applied. VelMod2 is a layer-cake velocity model based on sonic logs, checkshot data and interpreted horizons across The Netherlands, and uses inputs from hydrocarbon wells. This causes greater uncertainty in the lows, as mostly the existing well penetrations are located on structural highs.

Palynostratigraphy and the Sporomorph EcoGroup method

A total of more than 159 cuttings samples were selected from the geothermal wells VDB-GT-04, PNA-GT-02, HON-GT-01 and HON-GT-02 for palynological analysis. The sample resolution for the Nieuwerkerk Formation section is 10 m. All rock samples were processed at TNO, using the standard sample processing procedures of TNO and the University of Utrecht (Janssen & Dammers 2008). This involves HCl and HF treatment, heavy liquid separation, and sieving over an 18 µm mesh sieve. The microscopy analysis was carried out according to standard procedures. The palynological slides were then quantitatively analysed. The main palynomorph categories (dinoflagellate cysts, spores and pollen) were counted up to 200 specimens. The main miscellaneous categories (e.g. acritarchs and Botryococcus) were calculated separately. Because dinoflagellate cysts are very rare in predominantly coastal- to fluvial-plain settings, the remainder of the slide was scanned for additional taxa. For the dinoflagellate cyst taxonomy, the 'Lentin and Williams index' was utilized (e.g. Williams *et al.* 2017). For pollen and spore taxonomy, reference is made to the TNO sporomorph compilation report 2005-053-C (Herngreen 2005).

The age interpretation is based on the LOD (last occurrence datum) of palynomorphs and refers to a large literature base concerning the Early Cretaceous in NW Europe (e.g. Davey 1979, 1982; Heilmann-Clausen 1987; Costa & Davey 1992; Partington *et al.* 1993; Duxbury *et al.* 1999; Herngreen *et al.* 2000; Duxbury 2001; Munsterman *et al.* 2012; Gradstein *et al.* 2016). The boreal stages of the geological timescale of Gradstein *et al.* (2012) were followed.

As well as providing age assessments, palynological analysis also aided palaeoenvironmental interpretation. The SEG (Sporomorph EcoGroup) method (Abbink 1998) was applied to improve the characterization and differentiation of the non-marine to very-marginal-marine-influenced assemblages of the Nieuwerkerk Formation. The borehole material consisted of cuttings from mud flush drilling; hence, overburden cavings may also be included. Nevertheless, this SEG method provides a continuous development of sequences by calculating the ratio of the sporomorph ecogroups. Abbink (1998) and Abbink *et al.* (2004*a*, *b*) divided the Jurassic–Lower Cretaceous sporomorphs of the Dutch North Sea into six ecogroups: 'Upland', 'Lowland', 'River', 'Pioneering', 'Coastal' and 'Tidally influenced' vegetation. In essence, the method is based on the assumption that, in a non-marine setting (in proximity to the coast), the aerial extent of specific groups will expand or shrink in harmony with sea-level fluctuations. For example, the area of lowland vegetation shrinks during a marine flooding, and, hence, the number of sporomorphs within this group declines and attains only minimal values during a maximum flooding surface (MFS). In this study, a combination of top occurrences of rare marine dinoflagellate cysts and the SEG method was used to provide a well-defined stratigraphic framework.

Results

Seismic interpretation

Results of seismic mapping demonstrate that a number of sub-basins are present within the greater WNB (Fig. 4). Increasing thickness towards faults is caused by syntectonic sedimentation, this wedging indicates at which (seismic) intervals the (then normal) faults were active. These features are best viewed by flattening on the top De Lier horizon, interpreted for this purpose. The thickness differences of the Nieuwerkerk Formation provide an idea of the location of depocentres. The axis of this rift basin is seen to be NW–SE orientated, with the seismic reflectors in the Nieuwerkerk Formation dipping towards this axis and the whole section also thickening towards the axis (Fig. 5). This thickening indicates that the faults forming the axis were active during deposition of the Nieuwerkerk Formation. At the present day, this same axis is aligned with a structural high (the Pijnacker High). During deposition of the Nieuwerkerk Formation and the Rijswijk Member, the Pijnacker High was a depression (pull-apart basin: Racero-Baena & Drake 1996) and thus part of the axial depression in this basin.

Towards the end of the deposition of the fluvial section of the Nieuwerkerk Formation, the layers become more parallel to the top of the formation (and exhibit less wedge-like geometries). Again, this is visualized by flattening the seismic on the top De Lier Member (Fig. 5). This parallel orientation indicates horizontal deposition, which suggests the end of active fault movement in a large part of the basin.

Flattening of the seismic on the top De Lier horizon also shows that not all faults in the WNB were active all the time or simultaneously (see Fig. 5, where the red arrows indicate movement). In addition, fault activity continued throughout the deposition of the Nieuwerkerk Formation during the Lower Cretaceous (Fig. 5).

Well correlation

Gamma-ray (GR) logs have been utilized for lithology identification within the clastic Nieuwerkerk Formation. Donselaar *et al.* (2015) demonstrated that these Nieuwerkerk sandstones are fluviatile and, using that framework, it becomes clear that the distances between individual wells (ranging from 2 to 12 km) are too large to effectively correlate single channel bodies.

The well correlation (Fig. 6) shows that the Rodenrijs Claystone Member covers the fluvial section of the Nieuwerkerk Formation. The Nieuwerkerk Formation is an alternation of sandstone and clay with various thicknesses (Donselaar *et al.*

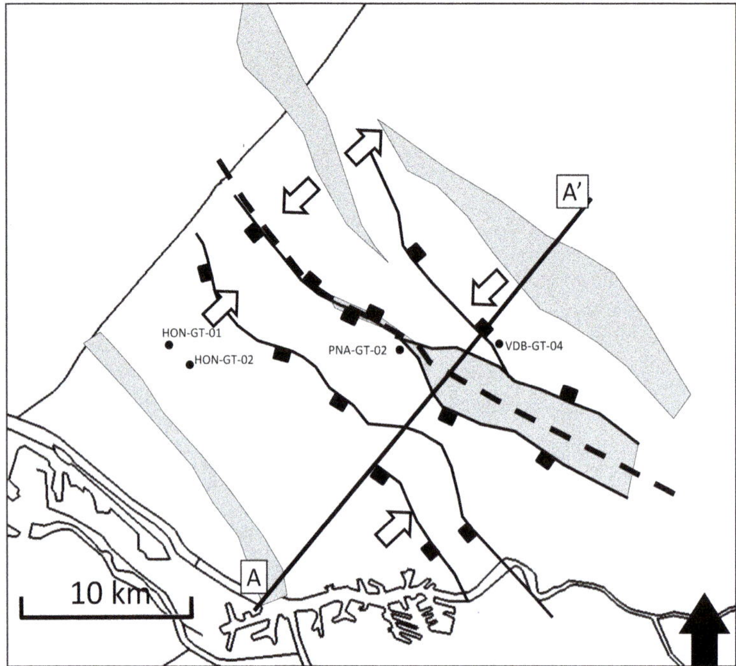

Fig. 4. Structure of the Nieuwerkerk Formation in the West Netherlands Basin (WNB). The map was generated after flattening on the top of the De Lier Member. The (palaeo-) highs are indicated in grey, and the dips in the sub-basins are indicated with arrows. The dotted line shows the axis of this part of the basin. Line A–A′ indicates the location of the seismic section displayed in Figure 5.

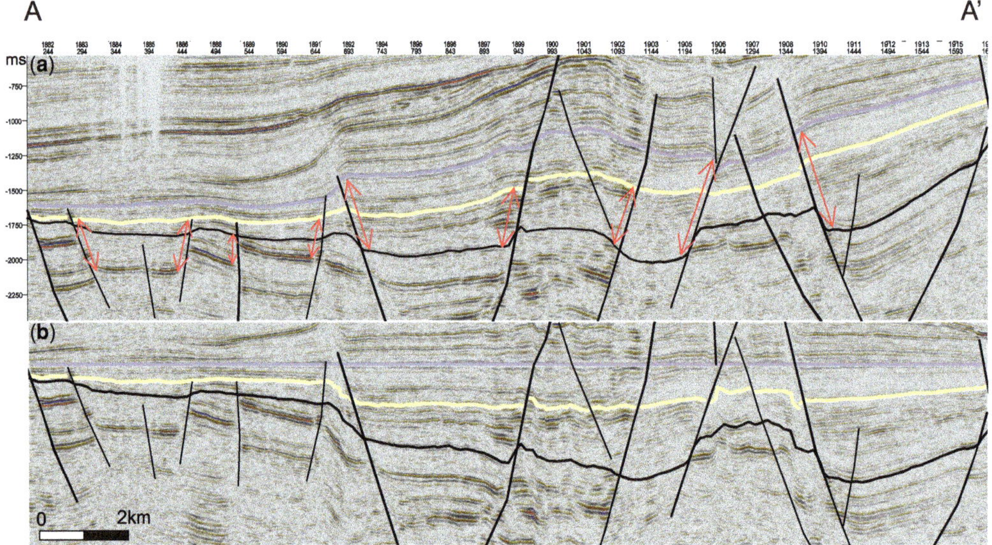

Fig. 5. SE–NW-trending seismic section in time (ms): (**a**) unflattened section; and (**b**) the same section flattened on the top De Lier. The seismic section shows wedging of the Nieuwerkerk Formation towards some of the faults. Purple, Top De Lier Member; yellow, Top Nieuwerkerk Formation; black, Base Nieuwerkerk Formation. The intervals of fault activity are indicated with red arrows. The location of this line is shown in Figure 4.

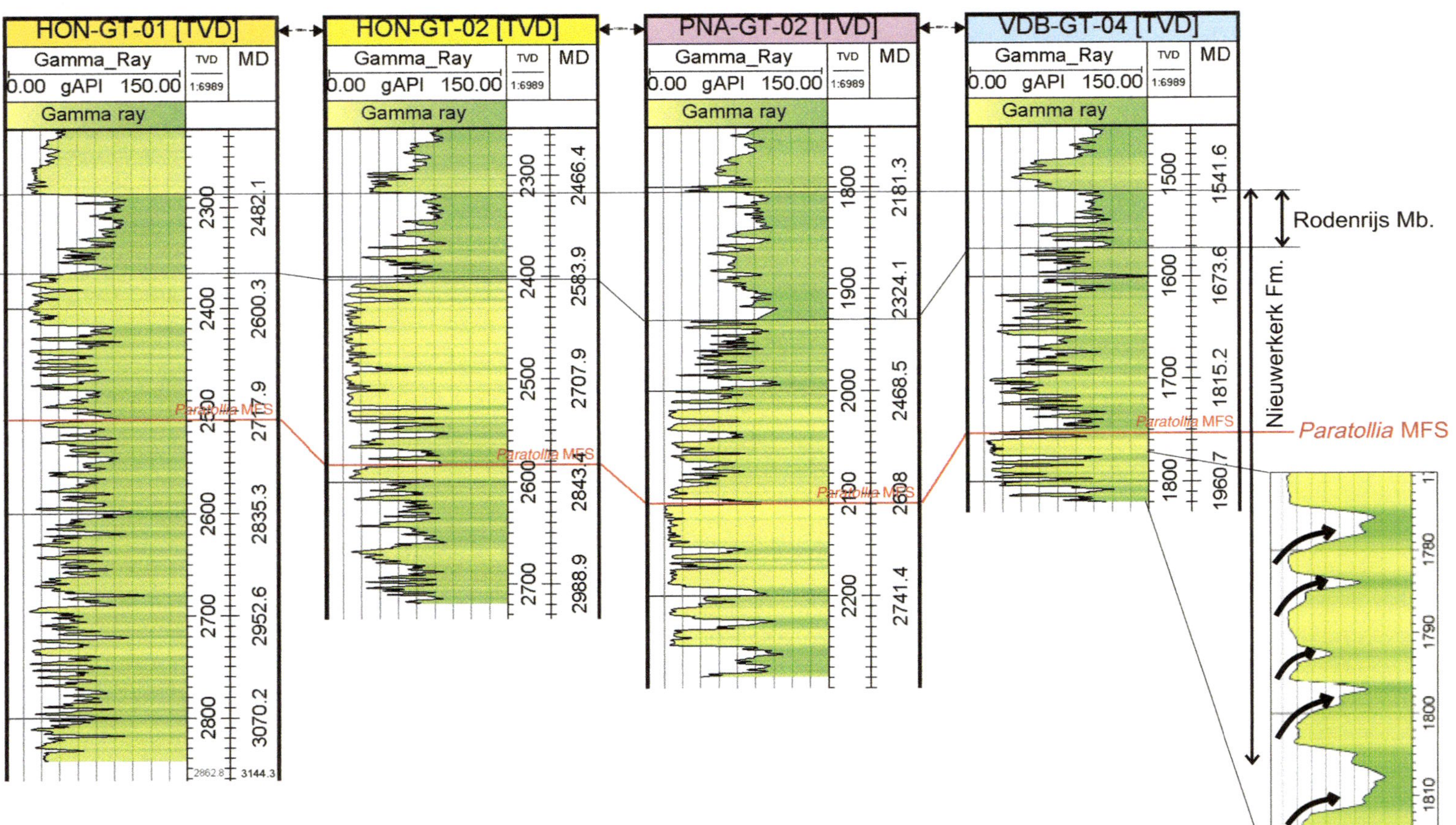

Fig. 6. Well section showing the gamma-ray (GR) logs of four geothermal wells in the WNB; showing a thick sandstone package in each well in the Nieuwerkerk Formation. The well section is flattened on the top of the Nieuwerkerk Formation (top of the Rodenrijs Claystone Member). The red line is the *Paratollia* marker defined by the palynological study. Inset: detail of the GR log, consisting of 5–8 m-thick fluvial sandstone bodies with a fining-upwards succession, characterized by a sharp basal decrease and gradual upwards increase (arrows) of the GR signal. All depths are in metres.

2015; Willems 2017). A thick (high N/G: i.e. higher than 0.65) sandstone package is demonstrated in every studied well (gross thickness ranging from 50 to 150 m: Fig. 6). These sandstone packages were originally interpreted as the Delft Sandstone Member. There are also numerous thinner sand intervals identified in the GR suite that are 5–8 m thick (Fig. 6) and also occur in all the analysed wells. These sand intervals are characterized by an abrupt base of the sequence (a sudden decrease in the GR) and a gradual fining upwards at the top of the sequence (indicated by the black arrows in Fig. 6).

Figure 6 shows the variability of N/G in the studied wells. In the two HON-GT wells, the basal sections have a relatively low N/G (*c.* 0.30), whereas the upper sections have a high N/G. In wells PNA-GT-02 and VDB-GT-04, the trend is reversed, with the basal sections demonstrating a higher N/G than the upper sections.

Palynological results

Palynological analysis demonstrates a marine Rijswijk Member with an underlying lagoonal Rodenrijs Claystone Member. In the Rijswijk Member, a large variety of marine dinoflagellate cysts occurs. The Rodenrijs Claystone Member comprises a relatively monotypical association of the dinoflagellate cyst genus *Subtilisphaera* (very common occurrence), which is associated with restricted marine conditions (Batten & Lister 1988). Below the Rodenrijs Claystone Member, the analysed sections of the Nieuwerkerk Formation are predominantly terrestrial. The preservation of palynomorphs is reasonable and Sporomorphs (pollen and spores) are dominant. *In situ* marine indicators within this section are rare. The last occurrence of the sporomorph taxon *Aequitriradites verrucosum* is generally taken as a marker for the top of the Valanginian (Herngreen 2005). The 'climate shift', seen in the *kochi* Ammonite Zone of the Early Ryazanian, is not reached in the studied interval of the current wells (cf. Abbink *et al.* 2001, 2006). Continuing low numbers of the sporomorph genus *Classopollis* down to the base of the succession confirm an age not older than the Late Ryazanian, *icenii* Ammonite Zone. The Early Valanginian *Paratollia* Ammonite Zone was recorded in all wells, based on (very) rare last occurrences of the dinocyst *Canningia compta* and/or *Stiphrosphaeridium dictyophorum* (Davey 1982; Strauss *et al.* 1993). The LODs of Oligosphaeridium diluculum (Costa & Davey 1992), Batioladinium sp. I *sensu* Davey (1982), Pareodinia sp. I *sensu* Davey (1982) and *Perisseiasphaeridium insolitum* confirm an Early Valanginian age.

The SEG method is applied to the non-marine to very-marginal-marine-influenced assemblages of the Nieuwerkerk Formation using the abundant presence of spores and pollen in all wells (*sensu* Abbink 1998). SEG analysis shows that the 'Pioneer', 'Coastal' and 'Tidally-influenced' vegetation groups are absent or have minimal values. The vegetation is composed of the 'Upland', 'Lowland' and 'River' ecogroups. The River Ecogroup comprises mainly simple psilatrilete spores (various fern families), that also fit within the 'Lowland' EcoGroup (Abbink *et al.* 2004*a*, *b*). The 'Upland' EcoGroup consists of bisaccate pollen formed by gymnosperms (*Gymnospermae*). These higher pine forests are usually considered to flourish in a relatively drier environment (Abbink 1998). However, the assemblages described here show a relatively humid (fluvial) lowland vegetation which is not in direct proximity to the coast. This interpretation is supported by very rare occurrences of marine indicators.

The SEG model of Abbink (1998) assumes that the relative numbers of lowland vegetation decrease during a transgression and show minimal values during a MFS. Accordingly, relative increasing abundances of 'Upland' SEG are interpreted to be correlative to flooding events (Abbink 1998, p. 98). In this way, a well-correlatable (maximum) flooding stands out consistently in each of the four wells during the Early Valanginian *Paratollia* Ammonite Zone (see Table 1). The correlation is not in line with the presence of the thicker sandstones. In the two HON-GT wells, the thick sandstone package is immediately covered by the lagoonal Rodenrijs Claystone Member. In contrast, the thick sandstone package in well VDB-GT-01 is covered by a shale-rich section of the Nieuwerkerk Formation, after which the lagoonal Rodenrijs is seen across the whole basin.

Discussion

To date, the Delft Sandstone Member has been interpreted as the sand-prone part of the Nieuwerkerk Formation (Den Hartog Jager 1996; Van Adrichem Boogaert & Kouwe 1997). A Valanginian age has been assigned to the Delft Sandstone Member and, on the basis of lithostratigraphical correlation, the member has been assumed to be continuous and synchronously deposited throughout the WNB.

The new sequence stratigraphic framework presented in this study deviates from the original lithostratigraphic interpretation. Well-to-well correlation in this framework, substantiated by the results of newly-acquired palynological data, shows a complex reservoir architecture of thick sandstone packages which step up in time and space from Ryazanian age in the NE (wells VDB-GT-04 and PNA-GT-02) to Valanginan age in the SW (wells PNA- GT-02, HON-GT-01 and HON-GT-02). The deposition of the sand-prone sections occurred

Table 1. *Palynological results (depths are in metres)*

Well	LOD *Canningia compta*		LOD *Stiphrosphaeridium dictyophorum*		*Paratollia* MFS	
	MD	TVDss	MD	TVDss	MD	TVDss
PNA-GT-02	2620	2108	2600	2093	2620	2108
HON-GT-01	2740	2510	ND	ND	2730	2502
HON-GT-02	ND	ND	2830	2582	2820	2575
VDB-GT-04	1890	1748	1890	1748	1890	1748

The results are focused on the Early Valanginian *Paratollia* Ammonite Zone, with indications of the last occurrence datum (LOD) of selected taxa and maximum flooding surfaces (MFSs).
MD, measured depth; TVDss, true vertical depth subsea.

in three phases, and the diachronous sequences testifies to a shift in time of the fluvial system from NE to SW.

In Phase 1 (Late Ryazanian: 142–139 Ma), rift fault activity created accommodation along the axis of the basin (Fig. 4), which was filled with closely-stacked fluvial sandstone in wells VDB-GT-04 and PNA-GT-02 (Fig. 6: well PNA-GT-02, 2110–2195 m true vertical depth (TVD); well VDB-GT-04, 1760–1800 m TVD). The high net-to-gross (N/G: 0.70) section in well PNA-GT-02 shows that sediment influx outpaced accommodation increase. This resulted in erosion of the floodplain fines, and the preservation of a stacked sandstone interval (Figs 6 & 7). The slightly lower N/G (0.65) in well VDB-GT-04 is explained by the larger lateral distance (*c.* 6 km) from the depocentre axis, which resulted in a higher preservation potential of the floodplain fines. The bottom section of the gamma-ray (GR) log of well VDB-GT-04 shows 3–8 m-thick sandstone successions, with a sharp decrease in GR at the base and a gradual increase upwards (Fig. 6), which is interpreted as a fining-upwards meandering-river sandstone embedded in floodplain fines (Flood & Hampson 2015). These successions in the GR are encountered in all four wells (within the palynologically defined terrestrial intervals).

Phase 2 (Early Valanginian: 139 Ma) is marked by the *Paratollia* MFS. This phase is a marine incursion that enters the WNB from the north, and is considered to have caused the landwards shift of fluvial facies and the preservation of fine-grained floodplain deposits.

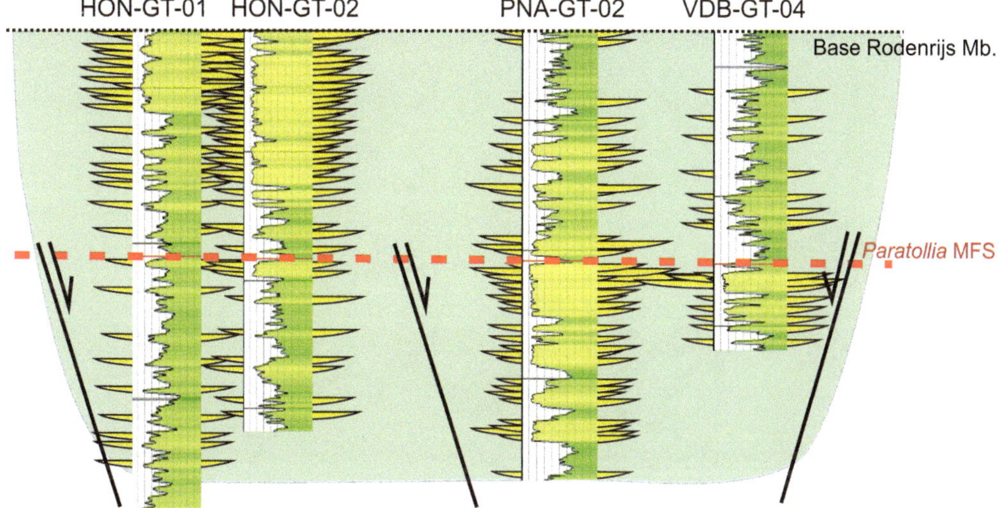

Fig. 7. Schematic drawing of the architectural model of the fluvial part of the Nieuwerkerk Formation including the four analysed wells. Yellow indicates sand-rich intervals; green indicates the shale-rich intervals.

In Phase 3 (Valanginian: 139–134 Ma), based on seismic interpretation observations and according to Den Hartog Jager (1996), major rift fault activity ceased (Fig. 5) and the southern part of the WNB evolved to be an unconfined lower coastal-plain environment (based on palynological analysis) without fault-bounded sub-basins. The study area is considered to have exhibited low-gradient palaeotopography with related meandering fluvial channels (Donselaar *et al.* 2015; Willems *et al.* 2017).

The upwards change in well VDB-GT-04 from closely-stacked sandstone beds to isolated, single sandstone beds in a fine-grained matrix is interpreted as a shift of the depocentre away from this well location. Simultaneously, the accumulation of thick, vertically-connected fluvial channel sandstones continued at the onset of Phase 3 in well PNA-GT-02 and over time gradually changed to isolated sandstone beds in an overall decreasing N/G system. In contrast, in well HON-GT-02, the N/G and stacking pattern changed upwards from isolated sandstone beds in floodplain fines to closely-stacked sandstone beds (Figs 6 & 7). The GR expression in well HON-GT-01 shows a similar trend, but the highest N/G correlates with the upper part of the stacked sandstone interval in well HON-GT-02 (Figs 6 & 7).

The shift of the depocentre of the fluvial system is explained by the initial confinement in a period of rift activity and the corresponding topographical relief. After the marine incursion, the main channel of the fluvial system abruptly shifted to the SW part of the basin (HON-GT and PNA-GT) and abandoned the NE (VDB-GT) area. In the SW part of the basin, the rift activity ceased, causing the fluvial system to freely change its flow path over the unconfined low-gradient lower coastal plain by the process of river avulsion, which resulted in a NE to SW step-up of the sand-prone part of the Nieuwerkerk Formation. Figure 8 shows a schematic top view of the discussed model, with four channel belts that were active during the different terrestrial phases. The avulsions in the SW part of the WNB are the result of the positive topographical relief created by the fluvial branches of the earlier channels.

The reservoir architecture model presented in this paper allows for Delft Sandstone correlation over larger distances (*c.* 10–20 km), and provides a reliable framework for kilometre-scale correlation in geothermal doublet projects. This model shows that sandstone reservoirs are present in the structural lows which are targets for geothermal exploration. The distribution of the sands is bound by the sub-basins trending NW–SE. As such, it is suggested,

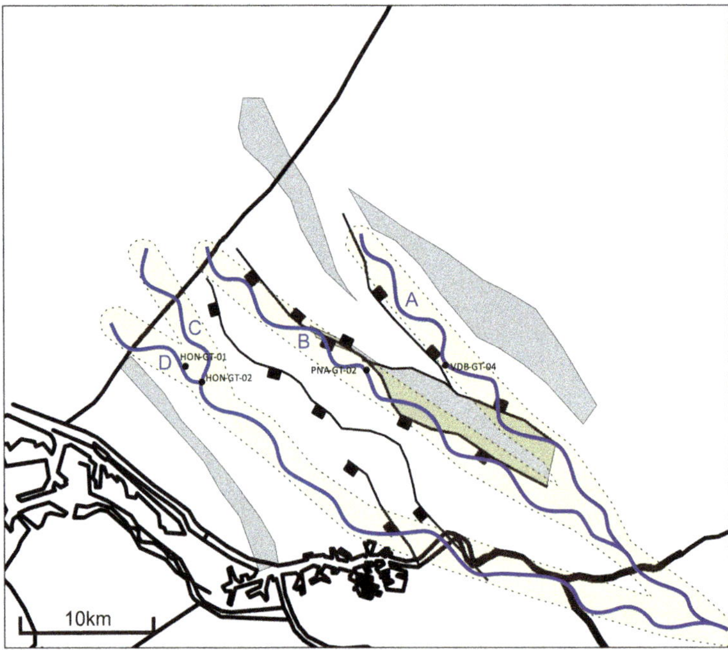

Fig. 8. Map view of the model of the reservoir architecture of the Nieuwerkerk Formation during the fluvial phases, showing the channels in the sub-basins. The channel belts have a width of approximately 2 km, as indicated by the yellow polygon. Channel belts A and B were active during Phase 1, and channel belts B, C and D were active in Phase 3.

for the most efficient realization of a geothermal project, that production and re-injection wells should be orientated in a NW–SE pattern. This would ensure maximum pressure and flow communication.

Conclusions

Palynological and well log analysis on wells HON-GT-01, HON-GT-02, PNA-GT-02 and VDB-GT-04 in the West Netherlands Basin (WNB), in combination with seismic interpretation, yielded a biostratigraphically constrained reservoir architecture model in which the sand-prone sections of the Nieuwerkerk Formation are seen to have been deposited diachronously.

A NE to SW shift of fluviatile belts is demonstrated to have occurred during the Ryazanian–Valanginian (144–134 Ma) period and is proposed to be the result of tectonic activity combined with the cessation of fault movement in a lower coastal-plain domain.

This improved reservoir architecture model will aid in the exploration for new geothermal prospects. This includes a better understanding of reservoir connectivity within geothermal doublet systems, and an increased predictability of sandstones (and reservoir quality) before and during drilling. The model also reduces the uncertainty of pressure and temperature interferences between neighbouring geothermal production projects.

This study was made possible by the Research Agenda Geothermal Energy of the Ministry of Economic Affairs and LTO Glaskracht Nederland in the innovation programme 'Kas als Energiebron'. The authors are grateful to Cees Willems (TU Delft) for inspiring discussions. We kindly thank Green Wells Westland BV in Honselersdijk for access to, and use of, the cuttings of wells HON-GT-01 and HON-GT-02. We thank Daan den Hartog Jager and Leo Van Borren for their constructive comments of the original manuscript.

References

Abbink, O.A. 1998. *Palynological investigations in the Jurassic of the North Sea region*. PhD thesis, Utrecht University, Utrecht, The Netherlands.

Abbink, O.A., Targarona, J., Brinkhuis, H. & Visscher, H. 2001. Late Jurassic to earliest Cretaceous palaeoclimatic evolution of the southern North Sea. *Global and Planetary Change*, **30**, 231–256.

Abbink, O.A., Van Konijnenburg-Van Cittert, C.J. & Visscher, H. 2004*a*. A sporomorph ecogroup model for the NW European Jurassic–Lower Cretaceous: concepts and framework. *Netherlands Journal of Geosciences – Geologie en Mijnbouw*, **83**, 17–38.

Abbink, O.A., Van Konijnenburg-Van Cittert, J.H.A., Van der Zwan, C.J. & Visscher, H. 2004*b*. A sporomorph ecogroup model for the NW European Jurassic–Lower Cretaceous. *Netherlands Journal of Geosciences – Geologie en Mijnbouw*, **83**, 81–92.

Abbink, O.A., Mijnlieff, H.F., Munsterman, D.K. & Verreussel, R.M.C.H. 2006. New stratigraphic insights in the 'Late Jurassic' of the Southern Central North Sea Graben and Terschelling Basin (Dutch Offshore) and related exploration potential. *Netherlands Journal of Geosciences – Geologie en Mijnbouw*, **85**, 221–238.

Batten, D.J. & Lister, J.K. 1988. Early Cretaceous dinoflagellate cysts and chlorococcalean algae from freshwater and low salinity palynofacies in the English Wealden. *Cretaceous Research*, **9**, 337–367, https://doi.org/10.1016/0195-6671(88)90007-9

Costa, L.I. & Davey, R.J. 1992. Dinoflagellate cysts of the Cretaceous system. *In*: Powell, A.J. (ed.) *A Stratigraphic Index of Dinoflagellate Cysts*. Chapman & Hall, London, 99–154.

Davey, R.J. 1979. The stratigraphic distribution of dinocysts in the Portlandian (latest Jurassic) to Barremian (Early Cretaceous) of northwest Europe. *American Association of Stratigraphic Palynologists, Contribution Series*, **5B**, 48–81.

Davey, R.J. 1982. *Dinocyst Stratigraphy of the Latest Jurassic to Early Cretaceous of the Haldager No. 1 Borehole, Denmark*. Geological Survey Denmark, Series B, **6**.

Den Hartog Jager, D.G. 1996. Fluviomarine sequences in the Lower Cretaceous of the West Netherlands Basin: correlation and seismic expression. *In*: Rondeel, H.E., Batjes, D.A.J. & Nieuwenhuijs, W.H. (eds) *Geology of Gas and Oil under the Netherlands*. Springer, Dordrecht, The Netherlands, 229–242.

DeVault, B. & Jeremiah, J. 2002. Tectonostratigraphy of the Nieuwerkerk Formation (Delfland Subgroup), West Netherlands Basin. *AAPG Bulletin*, **86**, 1679–1707.

Donselaar, M.E., Groenenberg, R.M. & Gilding, D.T. 2015. Reservoir geology and geothermal potential of the Delft Sandstone Member in the West Netherlands Basin. *In*: Horne, R. & Boyd, T. (eds) *Proceedings of the World Geothermal Congress 2015, Melbourne, Australia, 19–25, April 2015*, Paper 12054.

Duxbury, S. 2001. A palynological zonation scheme for the Lower Cretaceous United Kingdom Sector, Central North Sea. *Neues Jahrbuch Geologisches Palaontologue Abhandlungen*, **219**, 95–137.

Duxbury, S., Kadolsky, D. & Johansen, S. 1999. Sequence stratigraphic subdivision of the Humber Group in the Outer Moray Firth area (UKCS, North Sea). *In*: Jones, R.W. & Simmons, M.D. (eds) *Biostratigraphy in Production and Development Geology*. Geological Society, London, Special Publications, **152**, 23–54, https://doi.org/10.1144/GSL.SP.1999.152.01.03

Flood, Y.S. & Hampson, G.J. 2015. Quantitative ananlysis of the dimensions and distribution of channelized fluvial sandbodies within a large outcrop dataset: Upper Cretaceous Blackhawk Formation, Wasatch Plateau, central Utah, U.S.A. *Journal of Sedimentary Research*, **85**, 315–336.

Gradstein, F., Ogg, J., Smith, A. & Ogg, G.M. 2012. *The Geologic Time Scale 2012*. Elsevier, Amsterdam.

Gradstein, F., Waters, C. *et al.* 2016. Stratigraphic guide to the Cromer Knoll, Shetland and Chalk groups of the

Norwegian and North seas. *Newsletters on Stratigraphy*, **49**, 71–280.

Heilmann-Clausen, C. 1987. *Lower Cretaceous Dinoflagellate Biostratigraphy in the Danish Central Trough.* Danmarks Geologiske Undersogelse, Serie A, **17**.

Herngreen, G.F.W. 2005. *Jurassic and Cretaceous Sporomorphs of NW Europe: Taxonomy, Morphology, Ranges of Marker Species and Zonation, with Remarks on Botanical Relationship and Ecology.* TNO Report 05-053-C TNO-NITG, Utrecht, The Netherlands.

Herngreen, G.F.W. & Wong, T.E. 2007. Cretaceous. *In*: Wong, T.E., Batjes, D.A.J. & de Jager, J. (eds) *Geology of the Netherlands.* Royal Netherlands Academy of Arts and Sciences, Amsterdam, 127–150.

Herngreen, G.F.W., Kerstholt, S.J. & Munsterman, D.K. 2000. *Callovian–Ryazanian ('Upper Jurassic') Palynostratigraphy of the Central North Sea Graben and Vlieland Basin, The Netherlands.* Mededelingen Nederlands Instituut voor Toegepaste Geowetenschappen TNO, **63**.

Janssen, N.M.M. & Dammers, G. 2008. *Sample Processing for Pre-Quaternary Palynology.* TNO Report 2008-UR1190/A TNO, Utrecht, The Netherlands.

Jeremiah, J.M., Duxbury, S. & Rawson, P. 2010. Lower Cretaceous of the southern North Sea Basins: reservoir distribution within a sequence stratigraphic framework. *Netherlands Journal of Geosciences – Geologie en Mijnbouw*, **89**, 203–237.

Munsterman, D.K., Verreussel, R.M.C.H., Mijnlieff, H.F., Witmans, N., Kerstholt-Boegehold, S. & Abbink, O.A. 2012. Revision and update of the Callovian–Ryazanian Stratigraphic Nomenclature in the northern Dutch offshore, i.e. Central Graben Subgroup and Scruff Group. *Netherlands Journal of Geosciences – Geologie en Mijnbouw*, **91**, 555–590.

NLOG 2017. *Seismic Surveys*, http://www.nlog.nl/en/selection-screen-3d-seismic-surveys

Partington, M.A.P., Copestake, P., Mitchener, B.C. & Underhill, J.R. 1993. Biostratigraphic correlation of genetic stratigraphic sequences in the Jurassic–lowermost Cretaceous (Hettangian to Ryazanian) of the North Sea and adjacent areas. *In*: Parker, J. (ed.) *Petroleum Geology of Northwest Europe: Proceedings of the 4th Conference.* Geological Society, London, 371–386, https://doi.org/10.1144/0040371

Racero-Baena, A. & Drake, S. 1996. Structural style and reservoir development in the West Netherlands oil province. *In*: Rondeel, H., Batjes, D. & Nieuwenhuijs, W.H. (eds) *Geology of Gas and Oil under the Netherlands.* Springer, Dordrecht, The Netherlands, 211–229.

RGD. 2012. *DGM-deep V4 Onshore*, http://www.nlog.nl/en/dgm-deep-v4-onshore

Shier, D.E. 2004. Well log normalization: methods and guidelines. *Petrophysics*, **45**, 268–280.

Strauss, C., Elstner, F., Jan du Chene, R.J., Mutterlose, J., Reiser, H. & Brandt, K.-H. 1993. New micropaleontological and palynological evidence on the stratigraphic position of the 'German Wealden' in NW Germany. *Zitteliana*, **20**, 389–401.

Van Adrichem Boogaert, H.A. & Kouwe, W.F.P. (compilers). 1997. *Stratigraphic Nomenclature of the Netherlands; Revision and Update by RGD and NOGEPA.* Mededelingen Rijks Geologische Dienst, **50**.

Van Amerom, H.W.J., Herngreen, G.F.W. & Romein, B.J. 1976. Palaeobotanical and palynological investigation with notes on the microfauna of some core samples from the Lower Cretaceous of the West Netherlands Basin. *Mededelingen Rijks Geologische Dienst, Nieuwe Serie*, **27**, 41–79.

Van Dalfsen, W., Van Gessel, S.F. & Doornenbal, J.C. 2007. *Velmod-2.* Joint Industry Project TNO Report 2007-U-R1272C TNO, Utrecht, The Netherlands.

Van Wijhe, D.H. 1987. Structural evolution of inverted basins in the Dutch offshore. *Tectonophysics*, **137**, 171–219.

Willems, C.J.L. 2017. *Doublet deployment strategies for geothermal hot sedimentary aquifer exploitation: Application to the Lower Cretaceous Nieuwerkerk Formation, West Netherlands Basin.* PhD thesis, TU Delft, Delft, The Netherlands.

Willems, C.J.L., Nick, H.M., Donselaar, M.E., Weltje, G.J. & Bruhn, D.F. 2017. On the connectivity anisotropy in fluvial Hot Sedimentary Aquifers and its influence on geothermal doublet performance. *Geothermics*, **65**, 222–233.

Williams, G.L., Fensome, R.A. & MacRae, R.A. 2017. Lentin and Williams index of fossil dinoflagellates. *American Association of Stratigraphic Palynologists, Contribution Series*, **48**.

Tectonic control on the Early Cretaceous Bentheim Sandstone sediments in the Schoonebeek oil field, The Netherlands

GEERT-JAN VIS[1]*, WILLEM D. SMOOR[2], KEES W. RUTTEN[3], JAN DE JAGER[2] & HARMEN F. MIJNLIEFF[1]

[1]*TNO, Geological Survey of the Netherlands, PO Box 80015, 3508 TA Utrecht, The Netherlands*

[2]*Faculty of Earth and Life Sciences, VU University Amsterdam, De Boelelaan 1085, 1081 HV Amsterdam, The Netherlands*

[3]*Slokkert Consultancy, Slokkert 8, 9451 TD Rolde, The Netherlands*

**Correspondence: geert-jan.vis@tno.nl*

Abstract: The reservoir rock of the Schoonebeek oil field is formed by the sandstones of the Bentheim Sandstone Member. The sedimentology and depositional environment of this sandstone have been extensively studied, but the relationship between the geometry of the sandstone and tectonic activity in the Schoonebeek area remains poorly understood. 355 boreholes and two three-dimensional (3D) seismic surveys were used to study this relationship. An eroded zone in the west of the field and an area where the original depositional thickness is still intact were identified. Using the ezValidator software package it can be seen that uplift of a local anticline played an important role in the erosion of the sandstone. Deposition of the sands of the Bentheim Sandstone Member and the overlying Vlieland Sandstone and Claystone formations occurred on an unstable changing palaeotopography, whereby the instability was probably driven by halokinetic movement of the underlying Late Permian Zechstein salt. Syndepositional tectonic movements affected local thickness variations in the Bentheim Sandstone Member in the west of the field, leading to westwards thinning.

This study aims to unravel the influence of tectonic activity in the Schoonebeek area on the thickness distribution of the Bentheim Sandstone Member by using novel seismic interpretation techniques in combination with published data, information and models.

The Schoonebeek oil field lies in the NE of The Netherlands along the border with Germany (Fig. 1). Geologically, it is situated on the western flank of the Lower Saxony Basin. The field is contiguous with the German Emlichheim field in the south and the Rühlertwist–Rühlermoor fields in the east. One of the largest onshore oil fields in Europe, it was discovered in 1943 by the Bataafse Petroleum Maatschappij (BPM) with borehole SCH-003. The field came fully on-stream in 1945 and continued producing for nearly 50 years (Knaap & Coenen 1987; Rondeel *et al.* 1996). When about 25% of STOIIP (stock tank oil initially in place) had been produced, production rates declined and the field was shut-in in 1996 (Vejbæk *et al.* 2010). By then 604 boreholes had been drilled (http://www.nlog.nl). The field contained 164 mln m^3 (1 bln bbl) STOIIP of medium to heavy, highly viscous oil (160 cp at 40°C) (Vejbæk *et al.* 2010). The field has been re-developed in recent years by the Nederlandse Aardolie Maatschappij (NAM) with steam injection techniques using 75 new, mainly horizontal, boreholes, and production recommenced in early 2011 (http://www.nlog.nl). At the end of the re-development period in 2040, the field is expected to have produced another 19 million Sm^3 of oil (NAM 2008).

The Bentheim Sandstone Member is the reservoir rock for the Schoonebeek oil field (Figs 1b, 2), which is part of the Lower Cretaceous Vlieland Sandstone Formation. The type-section for The Netherlands is defined in borehole SND-01 (Van Adrichem Boogaert & Kouwe 1993). Possible reservoir analogues for the Bentheim Sandstone Member are found around the town of Bad Bentheim in Germany. During the last decades, numerous studies have been conducted to understand the stratigraphy, sedimentology and depositional environment of the shallow-marine sandstone sequence (Kemper 1968, 1973, 1976; Wittenhagen 1980; Kortmann 1983; Stadtler & Mutterlose 1996; Wonham *et al.* 1996*a*, *b*, 1997; Stadtler 1998; Mutterlose & Bornemann 2000; Malmborg 2002). For an overview of petrophysical studies on the Bentheim Sandstone Member, the reader is referred to Peksa *et al.* (2015).

Various studies have revised and detailed the lithostratigraphic units (or the German counterparts) for the Schoonebeek area in a sequence-stratigraphic framework (Stadtler & Mutterlose 1996; Stadtler

From: KILHAMS, B., KUKLA, P. A., MAZUR, S., MCKIE, T., MIJNLIEFF, H. F. & VAN OJIK, K. (eds) 2018. *Mesozoic Resource Potential in the Southern Permian Basin*. Geological Society, London, Special Publications, **469**, 435–455.
First published online March 15, 2018, https://doi.org/10.1144/SP469.25

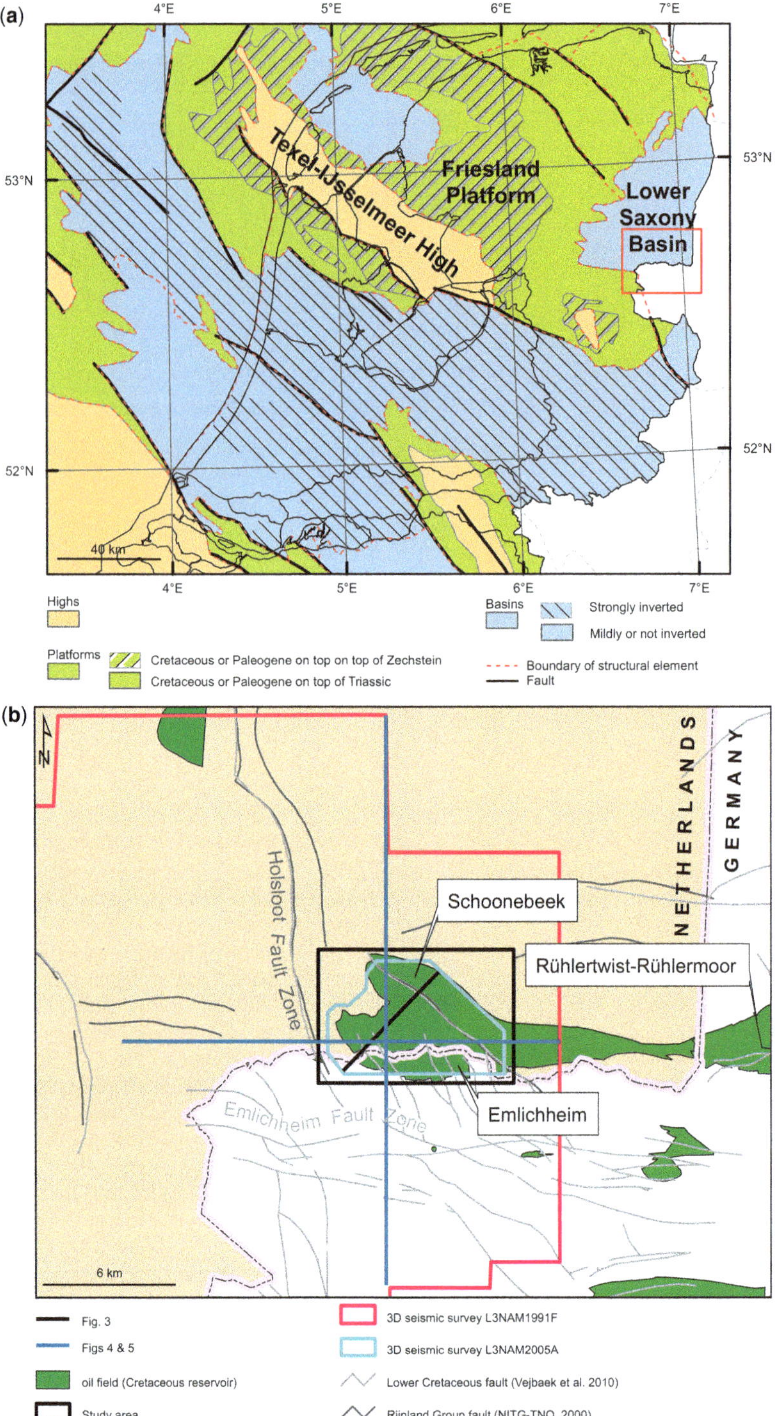

Fig. 1. (**a**) Late Jurassic–Early Cretaceous structural elements of The Netherlands, modified after Kombrink *et al.* (2012). The red box marks the area depicted in (b). For the position within the Southern Permian Basin, see the figure in the introduction by Kilhams *et al.* (this volume, in press). (**b**) Regional overview showing the location of the study area, seismic surveys, Cretaceous oil fields (Pletsch *et al.* 2010), faults and seismic sections (TNO-NITG 2000).

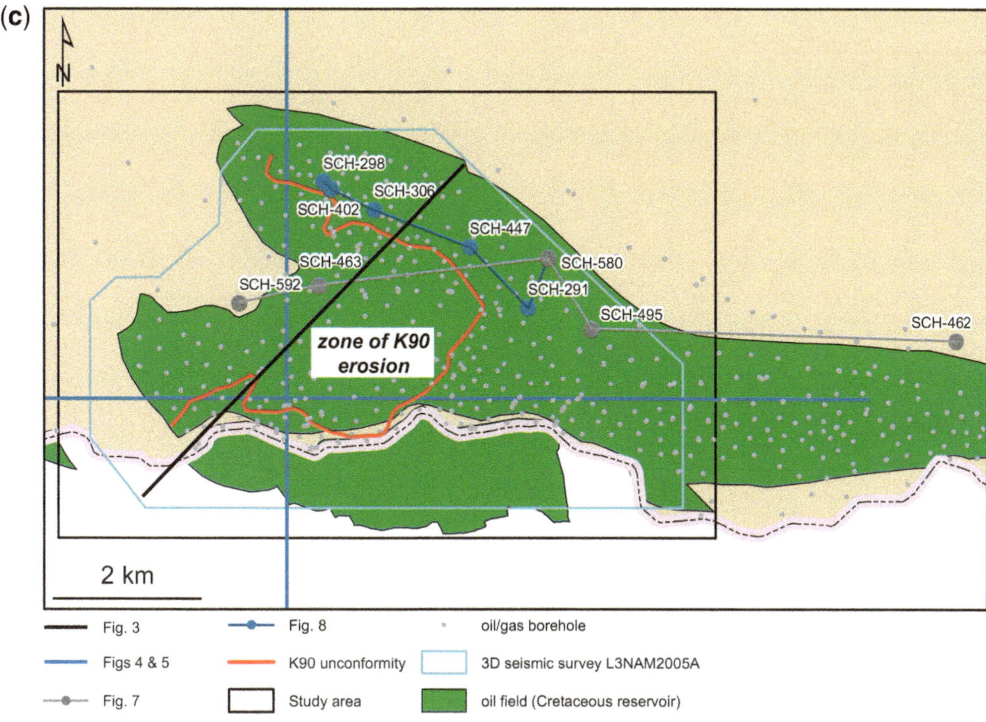

Fig. 1. (*Continued*) (**c**) Detail of the Schoonebeek field showing the study area and the location of seismic and borehole sections.

1998; Mutterlose & Bornemann 2000; Hoedemaeker & Herngreen 2003; Jeremiah *et al.* 2010). In this study the sequence-stratigraphic framework of Jeremiah *et al.* (2010) is used in conjunction with the Dutch lithostratigraphic subdivision (Van Adrichem Boogaert & Kouwe 1993). Therefore the sequence boundaries of Jeremiah *et al.* (2010, figs 2, 3) were integrated with the litho-chronostratigraphic scheme based on the relative position to the lithostratigraphic units defined in Van Adrichem Boogaert & Kouwe (1993, fig. G3). This resulted in a tectonostratigraphic chart of the Upper Jurassic–Lower Cretaceous succession for the study area in the Schoonebeek field (Fig. 2). The succession is characterized by a number of hiatuses which are graphically depicted as white areas. An overview of the lithostratigraphy of the study area is given in Table 1.

In more detailed reservoir studies, the Bentheim Sandstone Member is divided in three units: a (1) Lower, (2) Middle and (3) Upper Bentheim Sandstone unit. These sandstone units are separated by claystone intervals; the Romberg shale is the parting claystone interval between the Middle and Upper Bentheim Sandstone units (Mutterlose & Bornemann 2000). The Romberg shale has been recognized in boreholes from the Schoonebeek area, but has been placed between the Lower and Middle Bentheim Sandstone units (Jeremiah *et al.* 2010). The claystone intervals have a limited lateral extent, either caused by pinchout of the claystone towards the basin fringe or by pinchout of the sandstone in a basinwards direction where the claystone interval becomes part of the undifferentiated Vlieland Claystone Formation.

The Bentheim Sandstone Member is underlain by either the Bentheim Claystone Member or, in the absence of this claystone member, by the strata from the Niedersachsen Group (Coevorden Formation). The Bentheim Claystone Member is generally thin and similar to claystone of the underlying Coevorden Formation, and therefore difficult to interpret on well logs. Biostratigraphical analyses are essential for evaluating the nature and presence or absence of the Bentheim Claystone Member. The base of the Bentheim Sandstone Member is defined by Jeremiah *et al.* (2010) as their 'tectonically accentuated' K20 horizon. However, Hoedemaeker & Herngreen (2003) state that the base of the Bentheim Sandstone Member cannot be a sequence boundary, because they claim there is no hiatus between the Bentheim Sandstone Member and the underlying Bentheim Claystone Member. This is corroborated by Mutterlose & Bornemann (2000), who also suggest that no hiatus exists between these members.

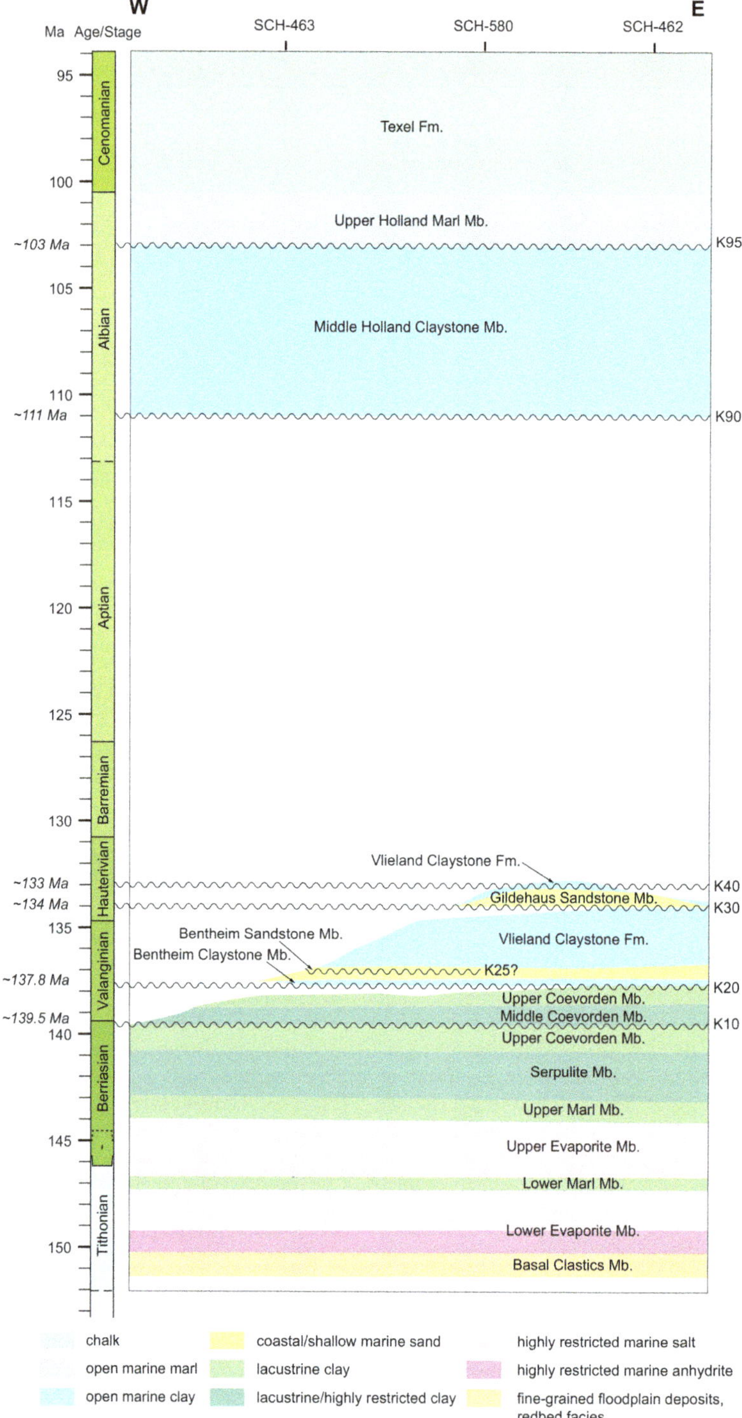

Fig. 2. Tectonostratigraphic chart for the study area. Timescale produced with TSCreator (http://www.tscreator.org), lithostratigraphy based on that of Van Adrichem Boogaert & Kouwe (1993). See http://www.dinoloket.nl for the online version of Dutch lithostratigraphic nomenclature. Sequence boundaries were taken from Jeremiah *et al.* (2010) and plotted in time based on the biostratigraphic zones and their age in Gradstein *et al.* (2012).

Table 1. *Dutch lithostratigraphy and codes as used in this article based on Van Adrichem Boogaert & Kouwe (1993)*

Group	Formation	Member
Upper North Sea Group (NU)*		
Middle North Sea Group (NM)*		
Lower North Sea Group (NL)*		
Chalk Group (CK)		
Rijnland Group (RN)	Holland Formation (KNGL)	Upper Holland Marl Member (KNGLU)*
		Middle Holland Claystone Member (KNGLM)
		Lower Holland Marl Member (KNGLL)
	Vlieland Claystone Formation (KNNC)	Westerbork Member (KNNCW)*
		Ruinen Member (KNNCE)
		Bentheim Claystone Member (KNNCV)
	Vlieland Sandstone Formation (KNNS)	Gildehaus Sandstone Member (KNNSG)
		Bentheim Sandstone Member (KNNSP)
Niedersachsen Group (SK)	Coevorden Formation (SKCF)	Upper Coevorden Member (SKCFU)
		Middle Coevorden Member (SKCFM)
		Lower Coevorden Member (SKCFL)*
	Weiteveen Formation (SKWF)	
Altena Group (AT)*		
Upper Germanic Trias Group (RN)*		
Lower Germanic Trias Group (RB)*		
Zechstein Group (ZE)*		

*Not discussed in the text, but included in figures.

For this study, focusing on the seismic evaluation of the Bentheim Sandstone Member interval, both the detailed subdivision of the member itself and of the contact with the Bentheim Claystone Member, are beyond seismic resolution. The K20 horizon is here named the Base Bentheim Unconformity (BBU), referring to the unconformity below the Bentheim Sandstone Member in the absence of the Bentheim Claystone Member, or referring to the unconformity below the Bentheim Claystone Member.

The Schoonebeek field lies at the western margin of the Lower Saxony Basin (Fig. 1a) which formed as a result of Mesozoic crustal extension and wrench faulting (Guterch *et al.* 2010). The influence of tectonic movements on the distribution of the Bentheim Sandstone Member has been discussed by Wonham *et al.* (1996*a*) to explain facies changes in outcrops in the Bad Bentheim area. In the Bad Bentheim area the lower sandstone layer (Bentheim 1 *sensu* Wonham *et al.* 1996*a*) has been interpreted to be deposited under syndepositional differential tectonic subsidence, to explain the rapid thickness increase of nearly 60 m over 3–4 km distance (Mutterlose & Bornemann 2000). A W–E cross-section across the Schoonebeek field suggests a shingled progradation of the Bentheim Sandstone Member towards the east (Jeremiah *et al.* 2010). A general eastwards thickening of the claystone succession between the Bentheim Sandstone Member and the younger Gildehaus Sandstone Member is clearly illustrated (Jeremiah *et al.* 2010). The interpretation suggests that the westwards thinning is caused by two local intra-Vlieland Formation unconformities (K25 and K30) and finally by the K90/base Holland Formation unconformity. The present study aims to find seismic evidence for the sedimentary-tectonic evolution of the Bentheim Sandstone Member in the western part of the Schoonebeek field. More specifically, the central research question is: did syndepositional tectonic movements affect thickness variations in the Bentheim Sandstone Member of the Schoonebeek area? A new interpretation of processes and mechanisms that led to the present-day distribution of these reservoir sandstones is presented, supported by seismic sections across the Schoonebeek field from which tectonic events have been interpreted. Additionally, a sediment thickness map is presented which, combined with palinspastic reconstructions of the seismic data, supports our findings.

Data and methods

Publicly available 3D seismic data were interpreted to characterize the geometry of the Bentheim Sandstone Member and to identify phases of tectonic activity that possibly affected the present-day thickness variations. Sub-regional seismic sections were interpreted to analyse the structural evolution of the larger Schoonebeek area. Seismic data were also used as input for palinspastic reconstructions of various cross-sections. Borehole data were used to map thickness variations of the Bentheim Sandstone Member.

Seismic data

Six 3D seismic surveys of varying vintages cover the Schoonebeek field. High-resolution survey L3NAM2005A was used for a detailed analysis of the structural setting for the west side the field. This survey has a surface area of 30 km^2 with about 2 s of vertical reach. The sampling interval of the survey was 2 ms, which means that at the reservoir level – varying from 600 to 940 m depth – the resolution is up to 5 m vertically. The bin size is 8 × 8 m, with 56-fold coverage at 120 Hz. The publicly available original processing of the 2005 survey was used. For use in the ezValidator™ software package, this was enhanced (Fig. 3) by a structural discontinuity filter which highlights faults, channels, diapirs and other seismic discontinuities (Landmark software). Survey L3NAM1991F was used to relate local features to the regional structural setting. The regional survey measures roughly 1000 km^2 and has about 3.5 s of seismic reach. The bin size measures 25 × 25 m with 12-fold coverage at 8–72 Hz and a sampling interval of 4 ms. The vertical resolution of the seismic data at Bentheim Sandstone Member level is up to *c.* 10 m.

Survey L3NAM2005A was used to map faults in detail and for palinspastic reconstructions, while survey L3NAM1991F was used for the construction of cross-sections across the Schoonebeek field of which two are presented here: one aligned N–S (Fig. 4) and the other E–W (Fig. 5). The regionally mapped horizons from TNO were plotted on these sections (Kombrink *et al.* 2012). Additionally, the top and base of the Bentheim Sandstone Member and the faults, cutting through the Rijnland Group were interpreted. Schlumberger Petrel 2015 was used for the work on seismic survey L3NAM1991F.

The Bentheim Sandstone Member overlies clays of either the Coevorden Formation or the Bentheim Claystone Member which rests on top of the Coevorden Member. The lithological and density contrast between the sands and the clays generates a seismic reflector at the base of the Bentheim Sandstone Member. To correctly identify the Bentheim Sandstone Member in the seismic surveys a seismic-to-well tie is needed. It was made for borehole SCH-580 (Fig. 6) using the velocity model VelMod 2 for the time–depth relationship of this borehole (Van Dalfsen *et al.* 2007). No VSPs (vertical seismic profiles) or check-shots are available for any of the boreholes in the area in the TNO-DINO database (http://www.nlog.nl). A synthetic seismogram was generated using the available sonic and density logs, as well as a statistical wavelet, extracted from the seismic volume, while taking care that no major faults intersected the borehole trajectory. In borehole SCH-580, sonic velocities have been logged from the Upper Cretaceous succession down to the total depth of the borehole (in the Upper Permian), but some depth intervals are missing. A density log is available from the Upper Cretaceous to the Middle Jurassic succession. Taken together, these logs cover more than 800 m of the succession, with good coverage of the Bentheim Sandstone Member. Despite manual final adjustments to the time–depth relationship after comparison of the synthetic seismogram to the seismic traces surrounding the well bore, a mismatch remains (Fig. 6). Because data from boreholes for depth and thickness maps were used, this is a minor problem for this study.

Borehole data

To avoid time–depth conversion difficulties and possible seismic interpretation problems and errors in this study, and because of the high borehole density, borehole data were used to make a thickness-distribution map of the Bentheim Sandstone Member. Borehole data and stratigraphic interpretations were taken from the TNO-DINO database (http://www.nlog.nl). Only vertical or near-vertical boreholes were used to assess the thickness of the Bentheim Sandstone Member in order to reduce thickness errors introduced by horizontal boreholes. Calculated true-vertical thickness values from the database were used for this study. Two borehole correlations were constructed to visualize the effect of sedimentary and erosive processes (Figs 7, 8).

Palinspastic reconstruction tool

Palinspastic reconstructions for the Schoonebeek area were made based on seismic survey L3NAM2005A. The ezValidator™ software package was used to unravel and clarify the stratigraphy and the local vertical movements around the BBU in the western part of the Schoonebeek field. This software was developed to study detailed stratigraphic configurations in complexly faulted and folded geological settings in seismic data (Rutten 2004).

The ezValidator™ software package offers three algorithms: unfaulting, unfolding and unconformities. The unfaulting algorithm shifts the seismic on either side of a fault to match horizon correlations and/or hand-drawn local correlations. Away from the faults an elastic model is used to handle the deformation (Barnett *et al.* 1987). Unfolding uses a construction method with lines orthogonal to the stratigraphy inspired by Busk (1929). Finally, unconformity gaps may result from unfaulting/unfolding the depositional sequence above the unconformity separately from the one below the unconformity. Such a display shows a Wheeler-diagram-type of representation of the geology, bearing in mind that seismic time is not equal to geological time.

Results

Pre-Valanginian period

Seismic sections across the Schoonebeek field show a sedimentary succession covering the Carboniferous basement (Figs 4, 5). The succession starts with Late Permian carbonates and evaporites (Zechstein Group), including salt of variable thickness. The near-uniform thickness of the overlying Triassic succession and continuous parallel reflectors illustrate the tectonically stable conditions in the area during that period (Figs 4, 5). The basal part of the overlying Lower Jurassic succession (Altena Group) has a similar character to the Triassic succession, indicating the continuation of tectonic quiescence.

The upper part of the Lower Jurassic succession and the overlying Upper Jurassic succession (Niedersachsen Group) has a different seismic character. Seismic reflectors are continuous but the stratigraphic thickness is more variable. The variations in thickness and truncations of seismic reflectors indicate the contemporaneous development of a series of synclines and anticlines with a relatively short wavelength of 5–8 km. The axes of these folds are oriented in an E–W direction, perpendicular to the Holsloot fault zone to the west (Figs 1b, 5). The anticlines of this first folding phase (phase-1) can be identified by the crestal faults which can still be seen (Fig. 4b). The sediments that were deposited during the Late Jurassic (Niedersachsen Group) show sedimentary thinning at the location of the phase-1 anticlines, supporting the interpretation of anticlines and associated reduced accommodation space. At the location of the phase-1 synclines, the Upper Jurassic succession is thickest and the seismic reflectors locally show onlap against the older Jurassic deposits (Fig. 4). Seismic data not shown here demonstrate that the E–W-oriented folds do not occur west of the Holsloot fault zone, which disconnected the tectonic processes taking place to the east from those to the west of the fault.

Valanginian period (Bentheim Sandstone Member)

Local tectonics: seismic reconstruction. A more detailed look at the Schoonebeek field sedimentary and structural features in a reconstructed NE–SW-oriented section shows that, at the member level in the Niedersachsen Group, thickness differences also occur (Fig. 3). First of all, there is remarkable continuity of seismic character and layer thickness across faults, indicating that the faults did not influence sedimentation. Secondly, the angular difference between the base Bentheim Sandstone Member reflector (yellow) and the reflectors below is a clear indication that this reflector represents an unconformity (near-base Bentheim). Thirdly, the reconstructed seismic line documents the subsidence patterns before and after this unconformity.

The seismic section (Fig. 3c) was unfolded/flattened on an interpreted reflector in the Holland Formation (green line) and on the reflector representing the base-Coevorden Formation (blue line). Both formations represent fine-grained lacustrine or marine sediments that were probably deposited horizontally. The (near-) base Bentheim Sandstone Member reflector is set to represent an unconformity (BBU) and a gap is shown because of unfolding the upper succession separately from the lower succession, which unveils the angular unconformity (Fig. 3c). When unfolding only the near-base Bentheim Sandstone Member reflector (Fig. 3d, e) the erosion or peneplenation which caused the unconformity becomes visible, that is, the palaeo-surface is visualized. As such, this display documents the structural development of the area.

On the unconformity displays there is a wedge of Coevorden Formation sediments below the BBU that increases in thickness towards the SW (Fig. 3e). This wedge was protected from erosion by subsidence increasing in magnitude to the SW. Above the BBU there is a wedge of deposits belonging to the Vlieland Sandstone and Claystone formations thickening towards the NE (Fig. 3e). This wedge documents subsidence and increasing accommodation space towards the NE during the Valanginian–Hauterivian period (Fig. 3c).

In summary, there was subsidence on the SW side of the study area, increasing towards the SW before the BBU and subsidence on the NE side increasing towards the NE after the BBU. The two opposite movements probably reflect a change of the local tectonic situation. This observation fits well with the observation of the bowl-shaped erosion of the Bentheim Sandstone Member, which is described at the end of this section.

Seismic reflection terminations at the top of the sediment wedge on the SW side show conformable layering in the Upper Coevorden Member with a sharp erosional boundary at the BBU (Fig. 3e). Seismic reflection terminations in the sediment wedge on the NE side are less clear with signs of onlap onto the Bentheim Sandstone Member and evidence of local channel erosion (Fig. 3e).

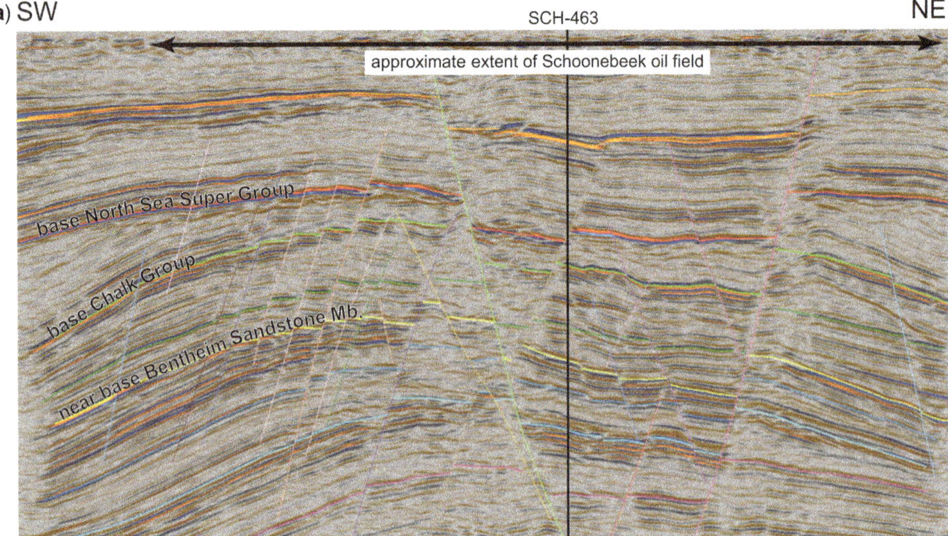

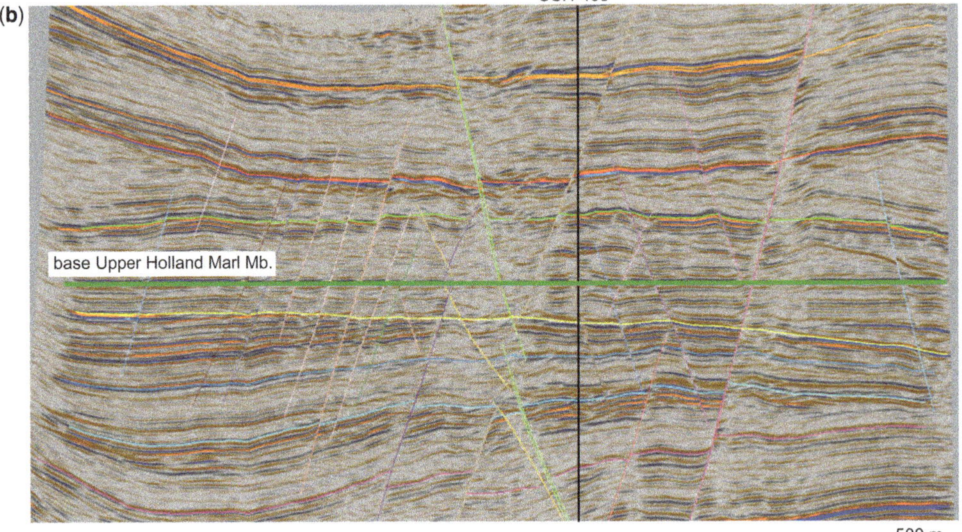

Fig. 3. NE–SW section through the seismic 3D survey L3NAM2005A which was enhanced with a structural discontinuity filter and displayed using ezValidator (see 'Data and methods' section for details). (**a**) Original seismic line with interpretation; (**b**) original seismic line flattened on the base of the Upper Holland Marl Member (green line).

The onlap may reflect a gradual transgression and deepening of the depositional setting.

Seismic data show a graben structure (informally named the Schoonebeek graben) which did not affect the thickness of the Bentheim Sandstone Member despite the major graben boundary faults with a throw of 50–60 m (Figs 4, 5). This indicates that these faults only became active after deposition and partial erosion of the sands much later, during the Cenozoic.

Local tectonics: borehole data. In the study area, the Bentheim Sandstone Member overlies the marine Bentheim Claystone Member (present above the Upper Paratollia MFS (maximum flooding surface); cf. Jeremiah *et al.* 2010). This member has

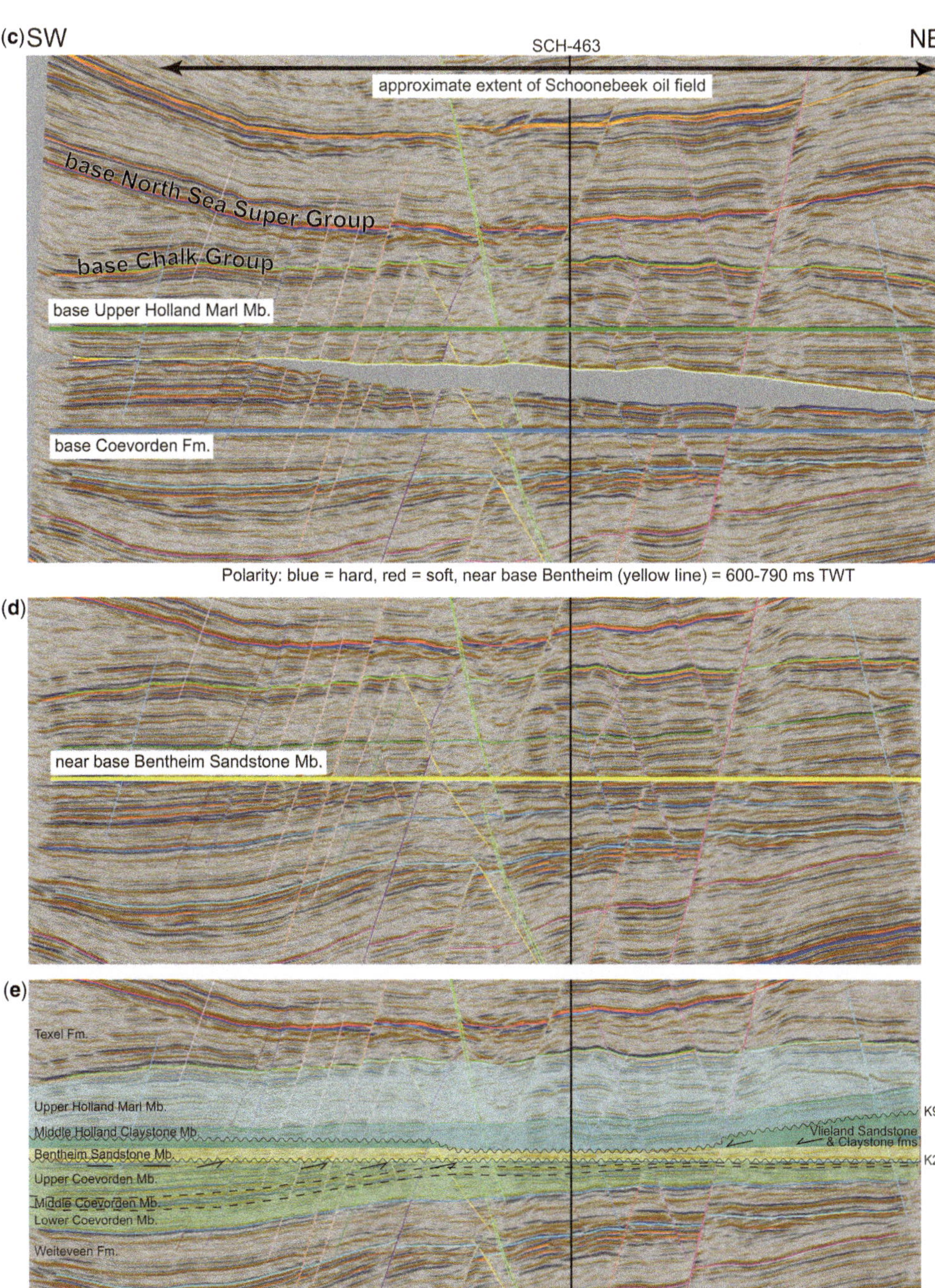

Fig. 3. (*Continued*) (**c**) Wheeler-diagram-type representation generated by unfolding/flattening the base of the Upper Holland Marl Member reflector and the base-Coevorden Formation reflector (blue line); (**d**) palaeo-topography at the time of formation of the Bentheim Sandstone Member, generated by unfolding/flattening the near base Bentheim Sandstone Member reflector; (**e**) interpretation of (d) showing the southwestwards thickening wedge of the Upper Coevorden Member and the northeastwards thickening wedge of the Vlieland Sandstone and Claystone formations and the Middle Holland Claystone Member.

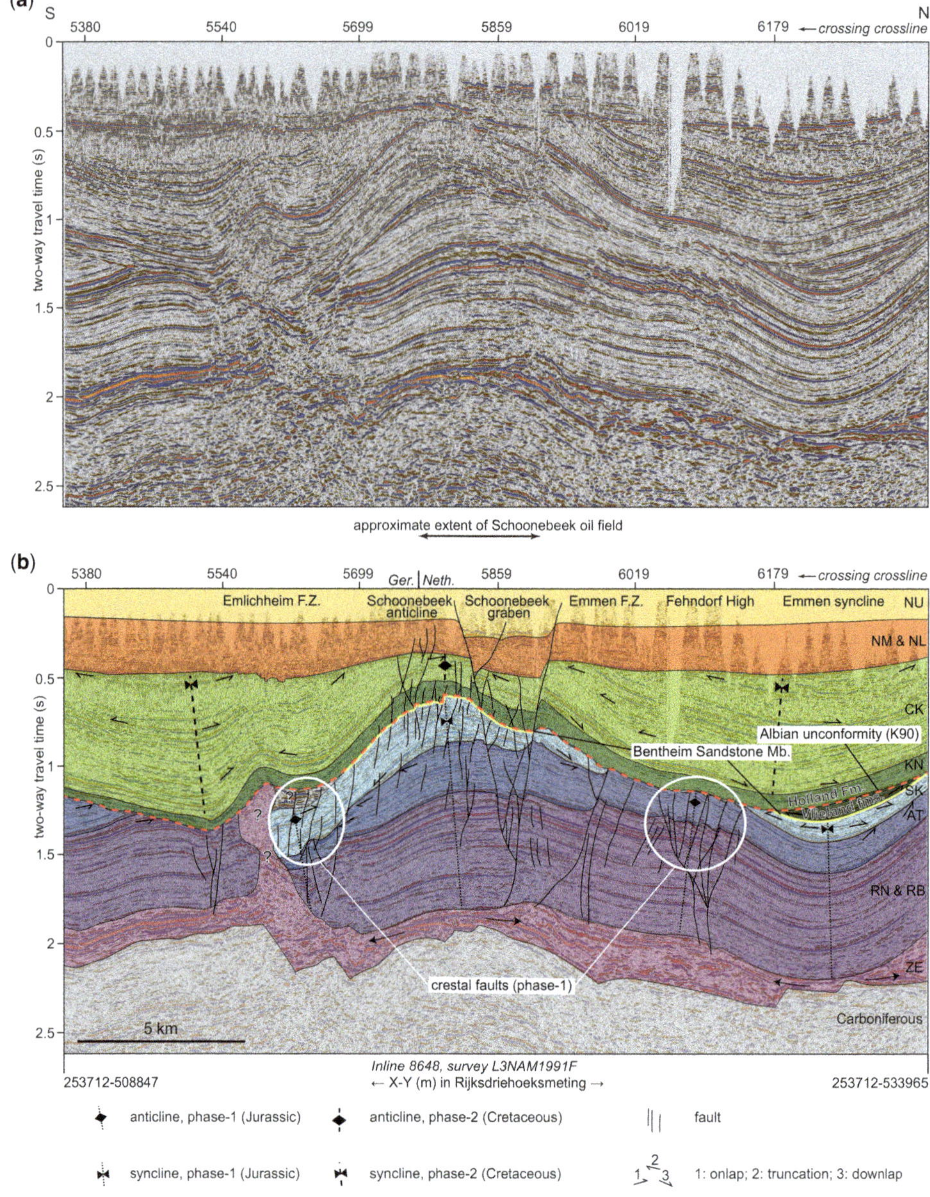

Fig. 4. N–S section through seismic 3D survey L3NAM1991F. (**a**) Original seismic line and (**b**) interpreted seismic line showing Jurassic and Cretaceous folding phases, Late Cretaceous–early Cenozoic peneplanation and the Cenozoic collapse graben (Schoonebeek graben). Also visible is the removal of the Niedersachsen Group and the lower Rijnland Group (i.e. Bentheim Sandstone Member and Vlieland Sandstone and Claystone formations) as a result of Albian erosion (K90). Deposits of the Vlieland Sandstone and Claystone formations were preserved in the Emmen syncline. F.Z., fault zone. For explanation of the stratigraphic codes, see Table 1.

been interpreted based on lithostratigraphy in only 17 boreholes out of the complete set of 837 boreholes drilled in the area (TNO-DINO database, http://www.nlog.nl). Reliable true vertical thickness estimates range up to *c.* 14 m. Biostratigraphic evidence of its presence is only available for well SCH-462 (Fig. 7) (Jeremiah *et al.* 2010). The distribution of boreholes in which this marine clay was

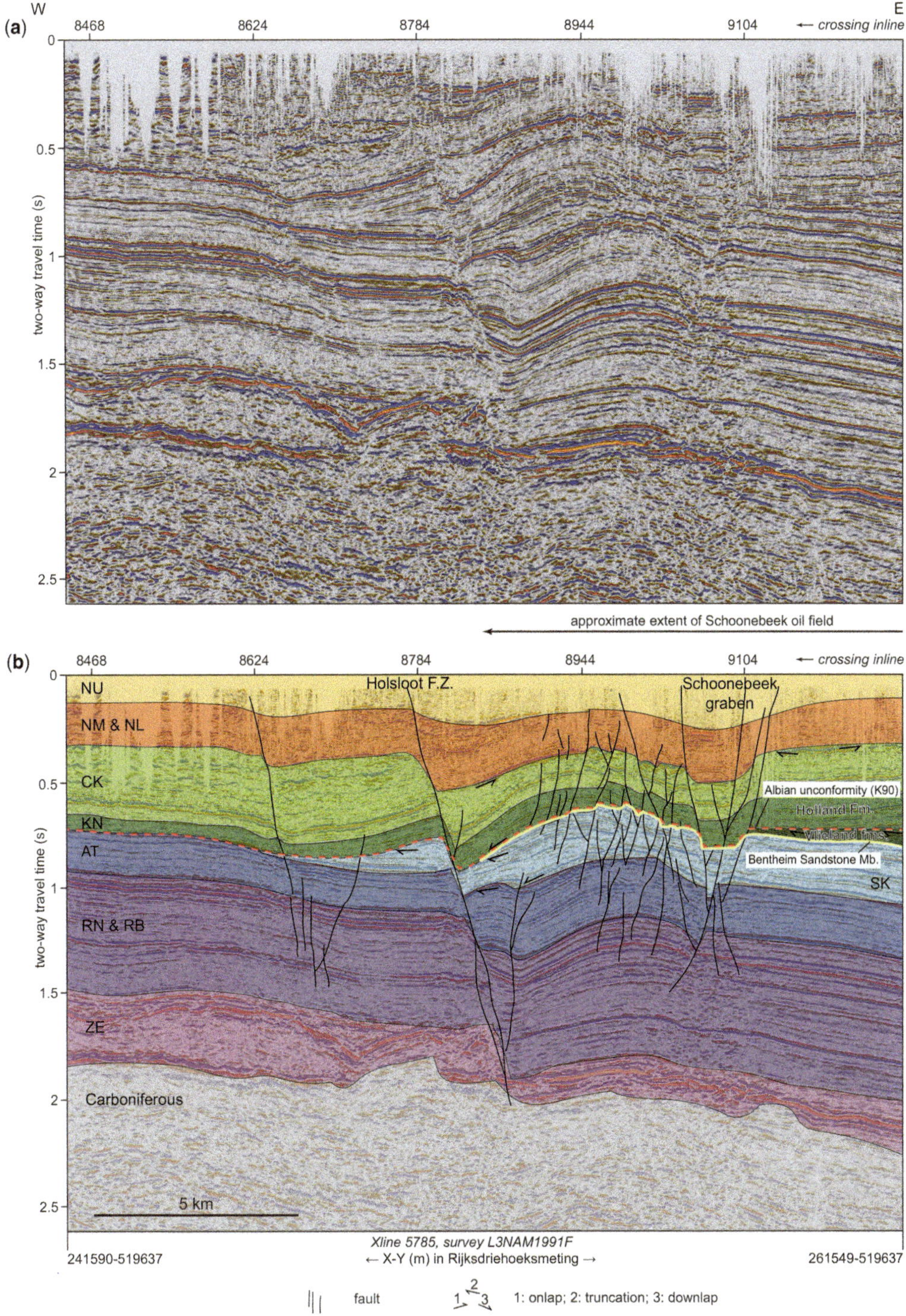

Fig. 5. W–E section through seismic 3D survey L3NAM1991F. (**a**) Original seismic line and (**b**) interpreted seismic line showing the westwards disappearance of the Niedersachsen Group and the lower Rijnland Group (i.e. Bentheim Sandstone Member and Vlieland Sandstone and Claystone formations) as a result of Albian erosion (K90). F.Z., fault zone. For explanation of the stratigraphic codes, see Table 1.

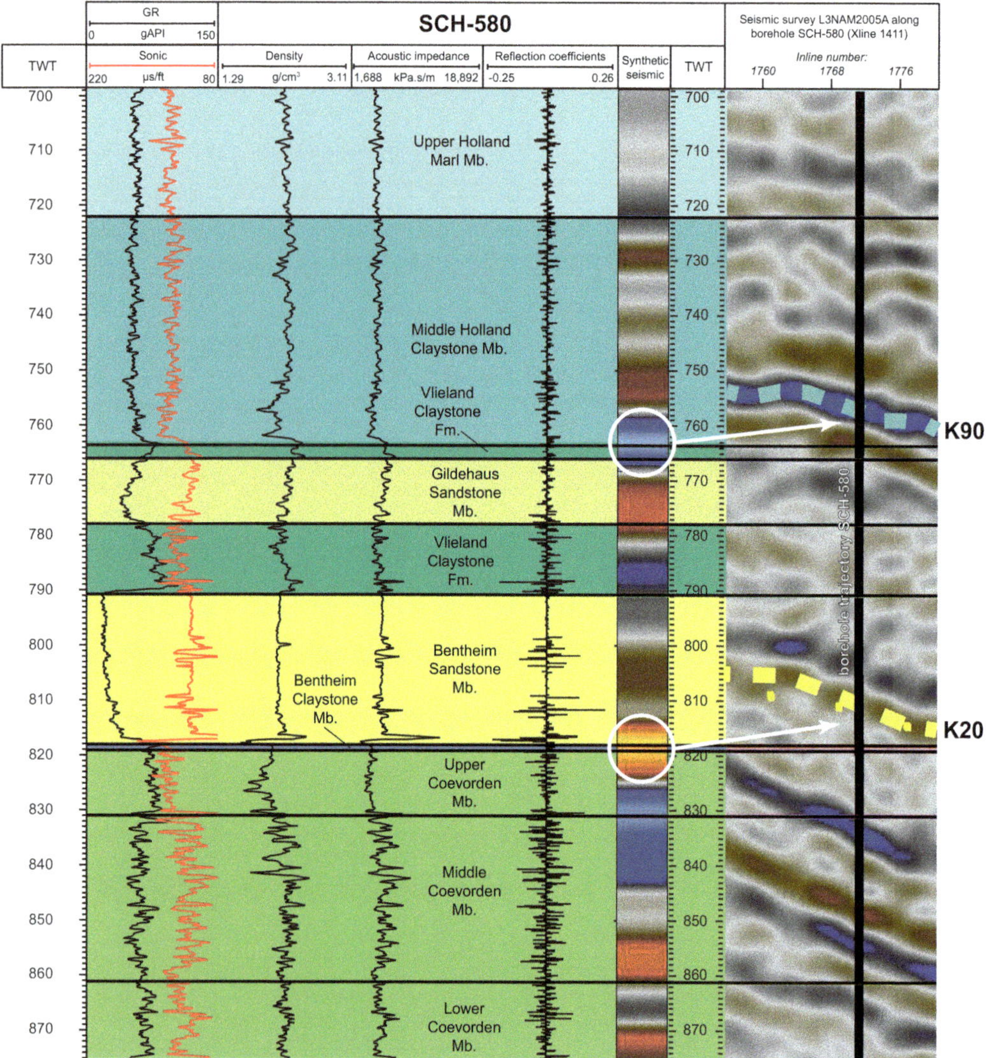

Fig. 6. Synthetic seismogram of borehole SCH-580 including lithostratigraphic interpretation of the well. A vertical mismatch between the depth–time-converted synthetic seismogram of the borehole and the seismic data can be observed. This does not affect the quality of the results of our work, since all depth information was taken from borehole data.

identified shows no pattern; they are randomly distributed. Where it is present, it rests on clays of the Coevorden Formation making it impossible to visualize seismically due to the limited density contrast between the clays. For practical reasons this study works with the base Bentheim Sandstone Member in both seismic and borehole data. Since the majority of boreholes were drilled for production rather than stratigraphic purposes, their lithostratigraphic interpretation may not be very sound. It is therefore likely that the Bentheim Claystone Member is present in more boreholes than currently recognized.

To spatially visualize the thickness distribution of the Bentheim Sandstone Member, true-vertical thickness from 355 (near) vertical boreholes in the area was used as available from stratigraphic interpretations in the TNO-DINO database (http://www.nlog.nl). This approach generates a reliable and accurate thickness map, also for thicknesses close to or below vertical seismic resolution. To interpolate between boreholes inverse distance

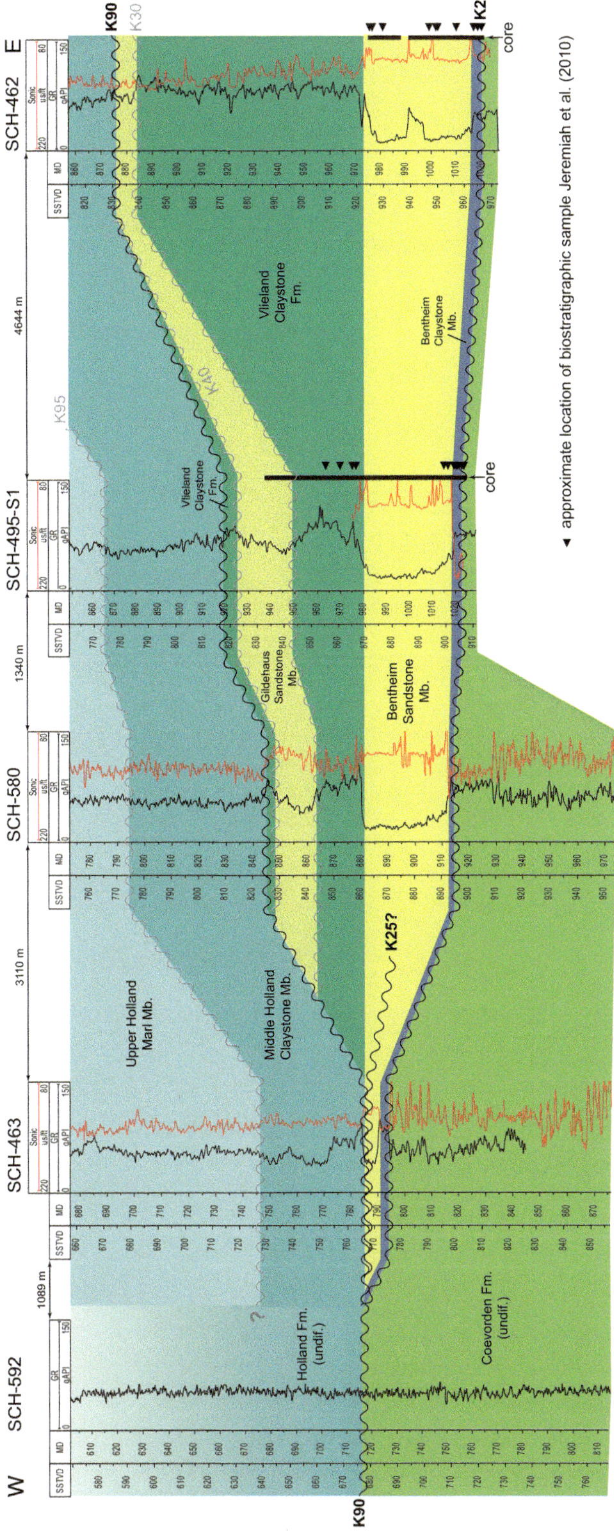

Fig. 7. W–E borehole correlation panel through the northern half of the Schoonebeek field, flattened on the top of the Bentheim Sandstone Member (for location see Fig. 1c). Note the amalgamation of the two major erosional events in the western part of the field: the K20 event at the base Bentheim Claystone Member (BBU) and the main K90 truncation event at the base of the Holland Formation.

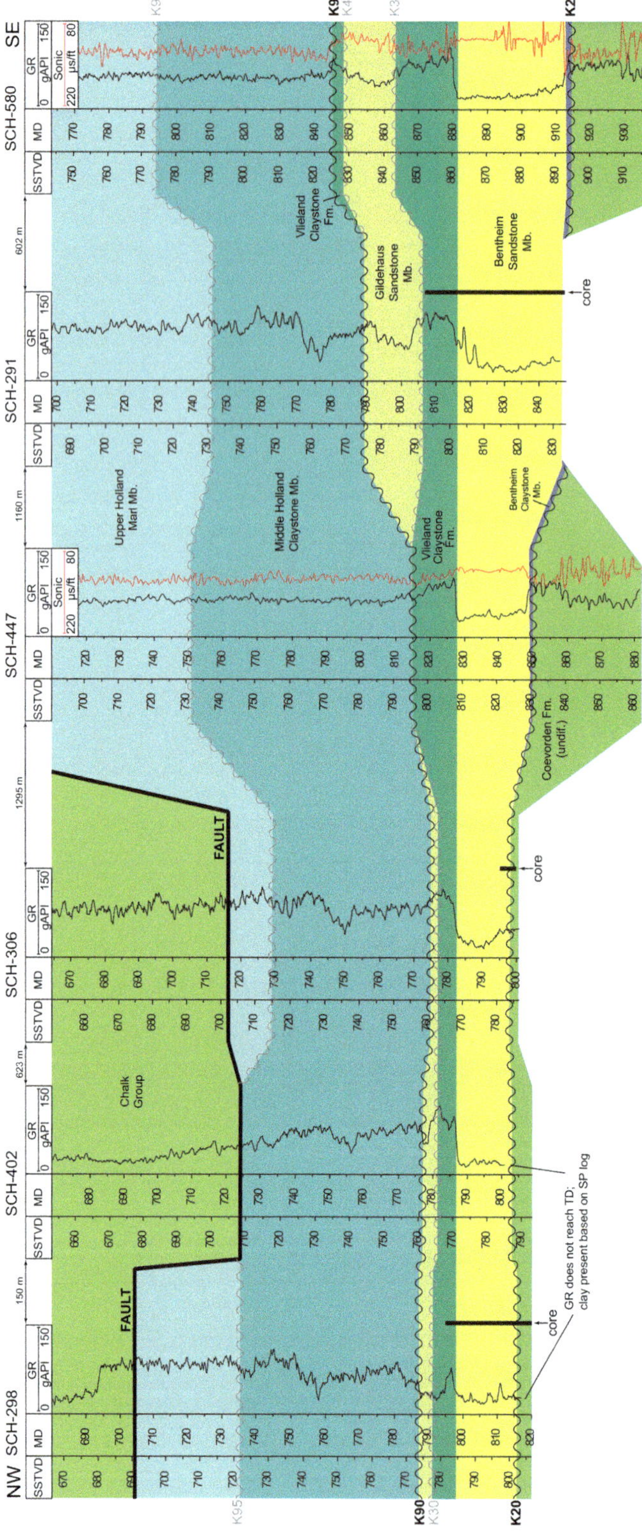

Fig. 8. NW–SE borehole correlation panel through the northern part of the Schoonebeek field, flattened on the top of the Bentheim Sandstone Member (for location see Fig. 1c). Note that at this location the K90 event did not reach the top of the Bentheim Sandstone Member and the sands of this member thin westwards.

weighting (IDW) in ArcGIS was used, generating a grid with cells of 150 × 150 m which honours the thickness in the individual boreholes (Fig. 9a). The mean thickness of the Bentheim Sandstone Member based on the boreholes is 25 m with a standard deviation of ±9 m; the maximum thickness in the used borehole dataset is 56 m. The most extensive zone where the Bentheim Sandstone Member is thickest occurs in the centre of the Schoonebeek field (Fig. 9a). Another small zone with large thickness is present within the field close to the German border in the east, where 20 boreholes were drilled. Generally the member thins westwards and eastwards. In a central western area the Holland Formation rests directly on top of the Bentheim Sandstone Member, which indicates the presence of a hiatus (Fig. 9a). In other areas the succession is stratigraphically (more) complete and the hiatus encompasses less geologic time.

In more detail, boreholes from the northern part of the Schoonebeek field confirm the presence of the Vlieland Claystone Member and the Gildehaus Sandstone Member on top of the Bentheim Sandstone Member (Fig. 8). The more complete succession suggests that no major erosion occurred and that the Bentheim Sandstone Member may have its original depositional thickness. In that case, the westwards thinning trend with *c.* 30 m down to *c.* 16 m TV (true vertical; Figs 8, 9a, B–B′) may represent an original depositional decrease in thickness, possibly due to the presence of palaeo-uplands. The presence of palaeo-uplands is difficult to prove, but may be indicated by the Ruinen Member (Van Adrichem Boogaert & Kouwe 1993). This member is found towards the west of the field and locally contains coaly lagoonal deposits with dispersed iron oolites, indicating a near-terrestrial to coastal environment (Van Adrichem Boogaert & Kouwe 1993). West of the oil field (Fig. 9), the Bentheim Sandstone Member is completely absent and the Holland Formation covers the Coevorden or Weiteveen formations, indicating the presence of a major hiatus (see complete stratigraphy in Table 1). In a seismic section the Holland Formation can be seen to overlie the Coevorden Formation; the reflectors of the latter are truncated by the reflectors of the former, demonstrating the presence of an angular unconformity (Fig. 5, around inline 8784).

In the central western part of the Schoonebeek field, the Vlieland Claystone Member and the Gildehaus Sandstone Member are absent and the Holland Formation directly overlies the Bentheim Sandstone Member. The thickness of the member follows a clear pattern: thicker towards the northern, eastern and southern edges of this area and thinning westwards (Fig. 9a, A–A′ and C–C′). This thickness distribution suggests bowl-shaped erosion, and the presence of palaeo-uplands in the same area suggests that a WNW–ESE-striking anticline plunging to the ESE (Fig. 9b) was growing around the time of deposition of the sands of the Bentheim Sandstone Member and well before later major erosion. The seismic data seem to confirm the presence of a growing anticline, as locally thicker Rijnland Group deposits occur in small accompanying depressions or synclines (Figs 4, 5). Subsequent Albian peneplanation of the area (K90 unconformity cf. Jeremiah *et al.* 2010) was preceded by an episode of folding that locally protected the deposits of the Vlieland Sandstone and Claystone formations in synclines against further erosion (Fig. 3). This folding probably reflects Alpine compressional tectonics (K70-K85 cf. Jeremiah *et al.* 2010).

The varying thickness of the Bentheim Sandstone Member in the Schoonebeek area has a gradual character based on borehole data; no sharp thickness changes can be observed. When looking at seismic lines across the field, the same observation can be made (Figs 4, 5). There seems to be no correlation with the closely spaced fault patterns, or with the major faults of the complex Schoonebeek graben structure.

Hauterivian–Albian period

During the Hauterivian–Albian period, the clastic sediments of the Vlieland Claystone Formation and the Gildehaus Sandstone Member were deposited on top of the Bentheim Sandstone Member as well as the deposits of the overlying clay and marl-dominated Holland Formation (Figs 4, 5, 7, 8). Where the Vlieland Claystone Formation is overlain by the Middle Holland Claystone Member the distinction between the two is difficult to make in seismic data, because of the similar lithology and sub-parallel reflectors. Only where the Vlieland Claystone Formation was folded and tilted prior to deposition of the Middle Holland Claystone Member can an unconformable contact be identified by the truncation of the Vlieland Claystone Formation reflectors (Figs 4, 5; Fig. 3c NE side). Borehole data in the TNO-DINO database show that the Lower Holland Marl Member is virtually absent from the west of the Schoonebeek field; it was only found in three boreholes, which are interpreted to have an incorrect lithostratigraphic interpretation. Because the Lower Holland Marl Member is absent from the study area, and the Middle Holland Claystone Member rests on deposits of Valanginian–Hauterivian age, a hiatus is present. This hiatus is likely the result of erosion that took place around Early Albian time, since the Middle Holland Claystone Member is found in most boreholes (Figs 7, 8).

The Early Albian erosion can be confirmed by seismic data. A syncline that formed north of the Schoonebeek field (Fig. 4, north of crossing line

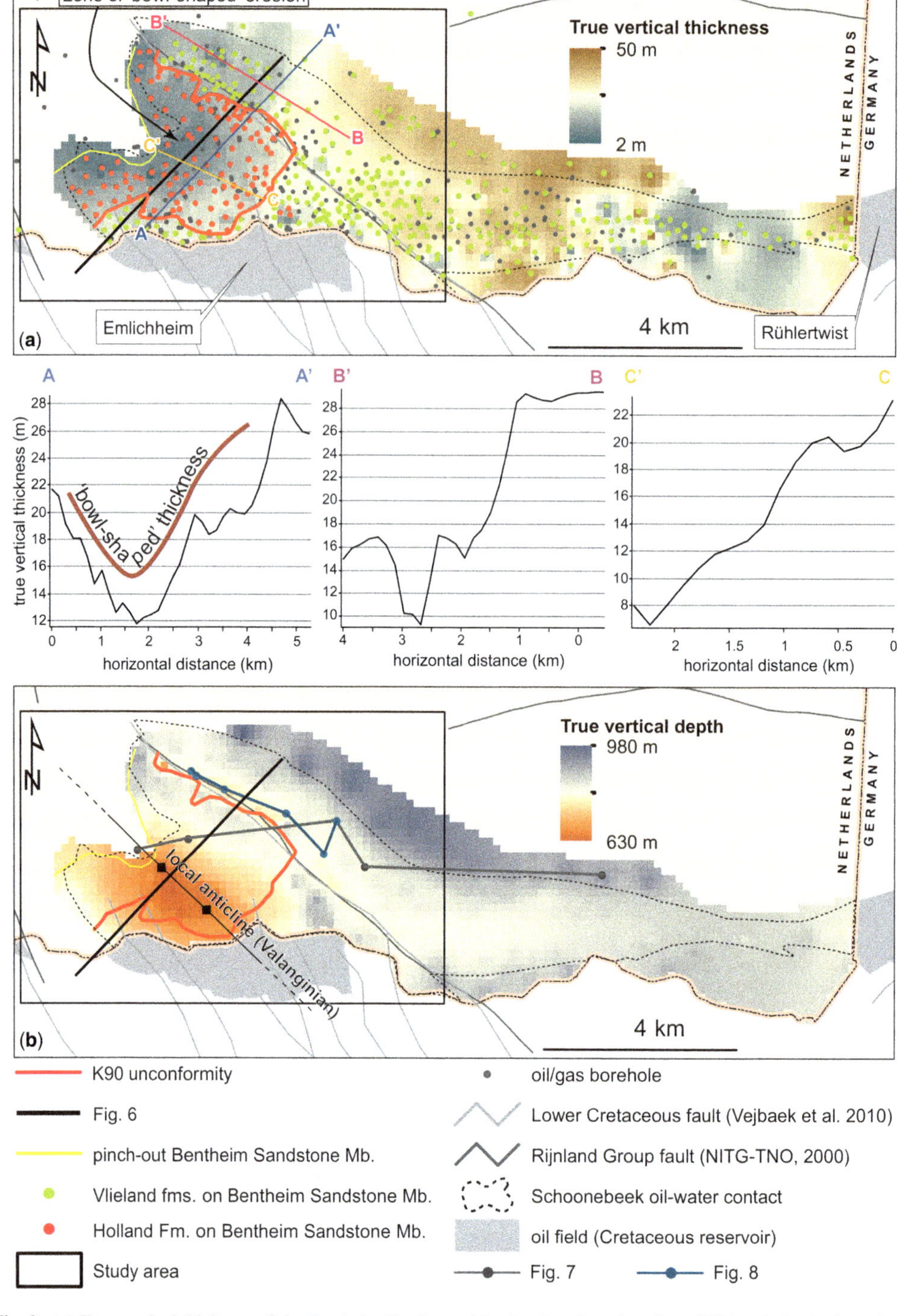

Fig. 9. (**a**) True vertical thickness of the Bentheim Sandstone Member based on data from 355 boreholes (coloured). Three graphs illustrate the changing thickness of the Bentheim Sandstone Member in and next to the eroded zone in the west. Note the bowl-shaped erosion in the west. (**b**) True vertical depth of the top of the Bentheim Sandstone Member (red = shallow). The location of the local anticline responsible for erosion in the west of the field is also indicated.

6179) contains deposits of the Vlieland formations including the Bentheim Sandstone Member. The truncation of reflectors representing these deposits at the top of the syncline-fill demonstrates the presence of an unconformity at the base of the Holland Formation.

Late Cretaceous–Cenozoic period

The carbonates of the Chalk Group have a strongly varying thickness and rest conformably on the deposits of the Rijnland Group. The deposits were taken up in a new series of E–W-striking folds with a wavelength of *c.* 10 km (Fig. 4). This is the second phase (phase-2) of folding to have affected the region. It is notable that the position of the phase-2 anticlines and synclines shifted compared to those of the Jurassic phase-1. The phase-2 Schoonebeek anticline was a syncline during the Late Jurassic, as can be seen by the thickening deposits of that age in the anticline core (Fig. 4). The lower Chalk Group thins above the (presumably) salt-cored asymmetric anticline. That anticline either resulted from Cenomanian reverse fault movement on the Emlichheim Fault Zone (Fig. 4) or from Cenomanian salt escape without reverse fault movement. The large-scale folds most likely result from Sub-Hercynian compression and folding, which started in the late Turonian–Coniacian (Pharaoh *et al.* 2010). The peneplanation of the Chalk Group deposits at the top of the Schoonebeek anticline can be explained by Late Cretaceous–early Cenozoic inversion resulting from the compression (Fig. 4). The large anticlinal structure in the Carboniferous basement below the Late Permian Zechstein salt (Fig. 4) suggests that the Sub-Hercynian compressional stresses also folded Carboniferous deposits.

Cenozoic deposits of the North Sea Super Group subsequently registered a phase of (local) extension, causing faulting in the Schoonebeek area to concentrate on the main graben boundary faults during the Miocene. The resulting Schoonebeek collapse graben caused local thickening of the younger Cenozoic succession. The Holsloot fault zone was reactivated as a normal fault, downthrowing older structures to the east.

Interpretation and discussion

The Schoonebeek field is located at the western margin of the Lower Saxony Basin. This WNW-trending basin developed since the Hettangian and was characterized by transtensional extension and related subsidence which continued until the end of the Aptian (Betz *et al.* 1987; Mazur & Scheck-Wenderoth 2005; Pharaoh *et al.* 2010). The transtensional extension was related to North Sea rifting.

Pre-Bentheim Sandstone Member succession

The western margin of the Lower Saxony Basin forms the stage for deposition of the Bentheim Sandstone Member during the Early Cretaceous. Before that time, sediments of the Niedersachsen Group accumulated in a progressively folded setting (phase-1) as indicated by truncations of seismic reflectors and thickness variations (Fig. 3e). The deposits of the group are thicker at the location of synclines and thinner at the location of anticlines of phase-1. This folding phase is interpreted to reflect initial movement of the Late Permian Zechstein salt as a result of sediment loading. The salt moved laterally to form small cushions and pods (Fig. 4). These salt movements were probably triggered by Middle Cimmerian tectonics which occurred in the Lower Saxony Basin around these times (Pharaoh *et al.* 2010). These tectonic movements caused truncation of the deposits of the Early Jurassic Altena Group (Fig. 4).

As the folding of phase-1 proceeded, more accommodation space was generated in the synclines. Because of that, the phase-1 syncline underneath the Schoonebeek field and the phase-1 Emmen syncline (Fig. 4) were filled with sediments belonging to the Weiteveen Formation and the Coevorden Formation ('Wealden') which onlap the older deposits of the Altena Group (Fig. 4). During the Berriasian, folding affected deposition of sediments of the Middle Coevorden and especially the Upper Coevorden members. Software-aided unfolding and unfaulting of a NE–SW-oriented seismic section unveils clear thickening of the Upper Coevorden Member towards the SW (Fig. 3). The wedge-shaped reflectors in the small depression in this section, which is possibly related to continued development of synclines above salt cushions, indicate that syndepositional differential tectonic movements and/or folding were occurring during the late Berriasian–early Valanginian. The base of the Middle Coevorden Member (Fig. 2) coincides with the base of the K10 sequence (Jeremiah *et al.* 2010). It is also known as the Late Cimmerian Unconformity or the Base Cretaceous Unconformity, but was not identified as an unconformity in this area. As Jeremiah (2000) indicates, in most cases there is no evidence for a major hiatus at this boundary. There are indications that the base of the K10 sequence marks an increase in sea level in the southern North Sea Basin (Jeremiah *et al.* 2010). This sea-level rise hardly influenced our study area; the depositional setting for the sediments of the Middle and Upper Coevorden members which represent the K10 sequence are still dominated by freshwater conditions with rare marine influence (Van Adrichem Boogaert & Kouwe 1993).

The seismic reflectors representing the Upper Coevorden Member are truncated below the BBU

(Fig. 3e). This indicates that, prior to deposition of the overlying deposits, tectonic movement had continued in another direction, causing tilt and uplift. The unfolded seismic section (Fig. 3e) shows that the Bentheim Sandstone Member was deposited on top of the Upper Coevorden Member which has partially tilted and partially horizontal seismic reflectors (Fig. 3d). The hiatus that was created is clearly seen in a Wheeler-diagram-type representation of a seismic line (Fig. 3c). The hiatus is largest in the centre of the section, where uplift as a result of a developing anticline occurred. The area where the hiatus is largest matches well with the zone of greatest erosion of the Bentheim Sandstone Member itself, which took place later. The hiatus probably represents relatively little time (<0.5–1 Ma) as indicated by biostratigraphic data in the east of the field (borehole SCH-495 in Jeremiah *et al.* 2010). The thickening of the Upper Coevorden Member to the SW and erosion prior to deposition of the Bentheim Sandstone Member signals a developing anticline below the west of the study area.

Base Bentheim Unconformity

The hiatus, recognized throughout the North Sea Basin, is of Valanginian age and coincides with the base of the K20 sequence (Jeremiah 2000; Jeremiah *et al.* 2010). These authors state that, just below the base of this sequence, a marine clay is found which contains the Upper Paratollia Maximum Flooding Surface (MFS), characterized by the early Valanginian ammonite *Platylenticeras* sp (Kemper 1961). This MFS is interpreted to correspond to the base of the Bentheim Claystone Member; the tectonically accentuated base of the K20 sequence is placed at the base of the Bentheim Sandstone Member (Jeremiah *et al.* 2010).

The data discussed above allow for another interpretation, in which the base of the K20 sequence (BBU) is placed lower at the base of the Bentheim Claystone Member. In this way, it is combined with the Upper Paratollia MFS. The MFS represents drowning of the area and conversion of the freshwater environment (Middle and Upper Coevorden members) into a shallow-marine environment. In this open-marine environment, progradation of coastal sands belonging to the Bentheim Sandstone Member took place over the fine-grained sediments of the Bentheim Claystone Member. This interpretation is supported by our observation of Schoonebeek cores. In several of them (SCH-590, SCH-495 and SCH-586), the base of the Bentheim Sandstone Member consists of a gradual transition to the underlying claystone. In other wells sharp basal contacts can be observed, which may be caused by erosion or a sudden change in depositional environment due to a rapidly prograding shoreline (Hoedemaeker & Herngreen 2003).

The fact that paralic, lacustrine deposits (Upper Coevorden Member) are directly overlain by marine sands or clays (Bentheim Claystone and Sandstone members), separated by a hiatus, suggests that relative sea-level rise occurred in the area. This was probably caused by the ongoing movement of Late Permian Zechstein salt in the subsurface below the study area against a background of overall Valanginian sea-level rise (Jeremiah *et al.* 2010). The salt movements are confirmed by a depocentre shift to the NE as evidenced by the preservation of a wedge of sediments belonging to the Gildehaus Sandstone Member and the Vlieland Claystone Formation on top of the Bentheim Sandstone Member (Fig. 3e).

Bentheim Sandstone Member

Deposition of the Bentheim Sandstone Member marks a change in the sedimentary system from an almost-sand-starved basin during the past 50 Ma to one with major sand input. The sands were probably generated by erosion from sub-aerial highs to the (north)west such as the Texel-IJsselmeer High and the Friesland Platform. The sediment was probably sourced from Triassic outcrops and uplifted shorefaces belonging to the K10 sequence (Jeremiah *et al.* 2010). Heavy mineral analyses of the outcrops at Bad Bentheim suggest an additional southerly source near the Rhenish Massif (Mutterlose & Bornemann 2000).

Continuous uplift of the local anticline, which is plunging to the ESE (Fig. 9b), caused the top of the Bentheim Sandstone Member to be removed by local erosion in the west of the study area, possibly during the late stage of deposition and shortly after deposition had ended (Fig. 9a). It is hypothesized that the nearness of palaeo-uplands to the west at the time of deposition 'announced' the progressive uplift and approaching erosive phase in the west of the Schoonebeek field. This hypothesis is supported by the preserved succession of synsedimentary thinning towards the palaeo-uplands of deposits of the Bentheim Sandstone Member which was probably caused by the uplift, and by the observation that erosion has caused a bowl-shaped thickness distribution in the Bentheim Sandstone Member (Fig. 9a, A–A′). The bowl shape is the effect of a local anticline being formed after the Late Jurassic folding phase, where most erosion occurred at the top of the anticline. Simultaneously to the erosion of the top of the Bentheim Sandstone Member within the bowl-shaped area, synclines formed along the flanks of this anticline; these shielded sediments belonging to the Bentheim Sandstone Member and later also of the younger Gildehaus Sandstone Member and Vlieland

Claystone Formation from erosion. The fact that these deposits were included in local synclines or depressions also protected them against later erosion during the Albian.

The bowl-shaped erosion of the Bentheim Sandstone Member around the late stages of deposition may correspond with the local tectonically accentuated K25 unconformity, as described by Jeremiah *et al.* (2010). Jeremiah *et al.* state that the unconformity truncates the (informal) Middle and Lower Bentheim Sandstone in a westerly direction and that transgressive sandstones and glauconites of the (informal) Upper Bentheim Sandstone unconformably overlie the Middle Bentheim. However, Jeremiah *et al.* (2010) present no seismic evidence to support the presence of this tectonically accentuated unconformity in the Schoonebeek area.

Biostratigraphic analyses in a core from borehole SCH-495 in the north of the Schoonebeek field (Fig. 7) confirm that there is probably no hiatus between the top of the Bentheim Sandstone Member and the overlying succession of the Vlieland Claystone Formation in that area (Jeremiah *et al.* 2010). Borehole correlations through the area show that the lithostratigraphic succession above the Bentheim Sandstone Member outside the bowl-shaped area is complete and that erosion at the base of the K90 sequence (base of the Middle Holland Claystone Member) has only eroded the Gildehaus Sandstone Member (Fig. 7). The westwards thinning of the Bentheim Sandstone Member of *c.* 30 m (Fig. 7) is therefore interpreted to be a sedimentary pinchout onto the palaeo-uplands, which does not have to be affected by K25 erosion.

Various authors have explained the westwards thinning by erosion related to the K25 and K90 unconformities (Van de Weerd 1996; Jeremiah *et al.* 2010). Our results suggest that the thinning also occurred at the time of sedimentation, unrelated to local K25 erosion and well before the Albian erosion resulting in the K90 unconformity. That insight may affect the distribution of facies and reservoir quality in the area. A trend in reservoir quality related to reservoir thinning has also been found for the adjacent Rühlermoor field, although strong synsedimentary tectonics are postulated to be the main cause of this (Wonham *et al.* 1997).

Post-Bentheim Sandstone Member succession

An important event after deposition of the Bentheim Sandstone Member and the overlying Gildehaus Sandstone Member and Vlieland Claystone Formation was the large-scale erosion which also affected the central western part of the Schoonebeek field in the bowl-shaped eroded zone discussed in the previous section. This event is known as the Early Albian unconformity at the base of the K90 sequence (Jeremiah *et al.* 2010). Biostratigraphic analyses in Schoonebeek boreholes SCH-302 and SCH-309 have identified that Lower Albian clays directly overlie Hauterivian sandstones of the Gildehaus Sandstone Member (Jeremiah *et al.* 2010). According to these authors, the base of the K90 sequence lithostratigraphically corresponds to the top of the Middle Holland Claystone Member (Jeremiah *et al.* 2010). However, based on Dutch lithostratigraphy, sediments of the Middle Holland Claystone Member were deposited during the Early Albian (Van Adrichem Boogaert & Kouwe 1993). The base of the K90 sequence is therefore placed at the base of the Middle Holland Claystone Member. As a result of the K90 unconformity, the Middle Holland Claystone Member rests unconformably on increasingly older deposits in a W–E direction in our study area.

The present-day configuration of the Bentheim Sandstone Member was created during Late Cretaceous–Cenozoic times (Fig. 9b). Sub-Hercynian tectonics caused compression which resulted in phase-2 folding, which also affected the Carboniferous basement. The larger wavelength (*c.* 10 km) of this folding phase was probably the result of a thicker supra-salt (Late Permian Zechstein) succession and less decoupling because salt had already moved into the salt pillows. New salt movements were probably triggered and a small salt dome at the Emlichheim Fault Zone accentuated and deformed the overlying Cretaceous deposits. It is likely that most faults in the Schoonebeek area developed during this time as crestal faults during the development of the Schoonebeek phase-2 anticline. Faults that had been active before were reactivated during the inversion. The Bentheim Sandstone Member at Schoonebeek was pushed into the phase-2 anticline striking WNW–ESE. Ongoing compression finally led to regional inversion and peneplanation of the Chalk Group deposits during the early Cenozoic (Fig. 4). Around Miocene times, extension and oblique rifting resulted in the Schoonebeek graben.

In summary, the sequence of events described above is interpreted as the reflection of continuous uplift in the study area during the Late Jurassic–Late Cretaceous period, with localized differences in expression. In other words, folding was continuing while the sands of the Bentheim Sandstone Member were accumulating. The fact that neither folding nor faulting are clearly registered in the Bentheim Sandstone Member itself suggests that the sands were deposited in a geologically short period of time. The clearest indication for ongoing folding is provided by the westwards thinning of the sands to the north of the bowl-shaped eroded zone (Figs 8, 9a). Tectonic movements were simply not registered in the sands or cannot be observed at present due to the limited detail of the seismic data.

Conclusions

The large amount of publicly available borehole and seismic data from the TNO-DINO database (http://www.nlog.nl) allowed for the study of the nature of thickness variations in the Bentheim Sandstone Member in the area of the western part of the Schoonebeek oil field in unprecedented detail. Using 355 boreholes it was possible to interpolate a map of the true-vertical thickness of the sands of the Bentheim Sandstone Member and to identify eroded zones in the west of the field, as well as an area where the original depositional thickness is still intact. Seismic data assisted in visualizing structural configurations. The ezValidator software package made it possible to reconstruct local tectonics at the time of deposition, clarifying the local subsidence pattern before and after the base-Bentheim unconformity.

It is found that the thickness of the sandstone is variable with a clear thinning trend towards the west of the field caused by syndepositional thinning onto a WNW–ESE-oriented local anticline generating a palaeo-high plunging towards the ESE. Due to continued uplift of the anticline, the palaeo-high was enlarged causing erosion during the late stages of deposition and shortly after that (K25). The major erosion during the Albian (K90) removed an additional portion of the succession in the west. In the rest of the study area this erosion did not remove the Bentheim Sandstone Member. Other thickness changes of the Bentheim Sandstone Member were found to be gradual, without correlation to closely spaced fault patterns.

It is concluded that in the Schoonebeek study area, deposition of the sands of the Bentheim Sandstone Member and the overlying Vlieland Sandstone and Claystone formations occurred on an unstable and changing palaeotopography; such instability was probably driven by halokinetic movement of the underlying Zechstein salt. Syndepositional tectonic movements affected local thickness variations in the Bentheim Sandstone Member in the west of the field.

We thank Nora Witmans, Johan ten Veen, Maryke den Dulk, Kees Geel and various other colleagues at TNO, Geological Survey of The Netherlands for supplying valuable support throughout this research project. Leslie Kramers from NAM (Assen) and John Verbeek (VU University Amsterdam) are thanked for valuable discussions. We thank Jonathan Wonham and Daan den Hartog Jager for their constructive reviews of an earlier draft of this paper. The editors Ben Kilhams and Tom Mckie are thanked for their support and patience.

References

Barnett, J.A.M., Mortimer, J., Rippon, J.H., Walsh, J.J. & Watterson, J. 1987. Displacement geometry in the volume containing a single normal fault. *AAPG Bulletin*, **71**, 925–937.

Betz, D., Fuhrer, F., Greiner, G. & Plein, E. 1987. Evolution of the Lower Saxony Basin. *Tectonophysics*, **137**, 127–170.

Busk, H.G. 1929. *Earth Flexures. Their Geometry and Their Representation and Analysis in Geological Section with Special Reference to the Problem of Oil Finding*. Cambridge University Press, Cambridge, United Kingdom.

Gradstein, F.M., Ogg, J.G., Schmitz, M.D. & Ogg, G. 2012. *The Geologic Time Scale 2012*. Elsevier Science Ltd., Amsterdam.

Guterch, A., Grad, M., Krawczyk, C.M., Wybraniec, S., Ziegler, P.A. & De Vos, W. 2010. Crustal structure and structural framework. *In*: Doornenbal, J.C. & Stevenson, A.G. (eds) *Petroleum Geological Atlas of the Southern Permian Basin Area*. EAGE Publications b.v., Houten, 10–23.

Hoedemaeker, P.J. & Herngreen, G.F.W. 2003. Correlation of Tethyan and Boreal Berriasian-Barremian strata with emphasis on strata in the subsurface of the Netherlands. *Cretaceous Research*, **24**, 253–275.

Jeremiah, J.M. 2000. Lower Cretaceous turbidites of the Moray Firth: sequence stratigraphical framework and reservoir distribution. *Petroleum Geoscience*, **6**, 309–328, https://doi.org/10.1144/petgeo.6.4.309

Jeremiah, J.M., Duxbury, S. & Rawson, P.F. 2010. Lower Cretaceous of the southern North Sea Basins: reservoir distribution within a sequence stratigraphic framework. *Netherlands Journal of Geosciences*, **89**, 203–237.

Kemper, E. 1961. *Die Ammonitengattung Platylenticeras (= Garnieria). Mit einem Beitrag zur Stratigraphie und Bionomie ihrer Schichten (Untere Kreide, mittleres Valendis)*. Bundesanstalt für Bodenforschung und den Geologischen Landesämtern der Bundesrepublik Deutschland, Hannover, **47**.

Kemper, E. 1968. Einige Bemerkungen über die Sedimentationsverhältnisse und die fossilen Lebensspuren des Bentheimer Sandsteins (Valanginium). *Geologisches Jahrbuch*, **86**, 49–106.

Kemper, E. 1973. The Valanginian and Hauterivian stages in northwest Germany. *In*: Casey, R. & Rawson, P.F. (eds) *The Boreal Lower Cretaceous*. Seal House Press, Liverpool, **5**, 327–344.

Kemper, E. 1976. *Geologischer Führer durch die Grafschaft Bentheim und die angrenzenden Gebiete mit einem Abriss der emsländische Unterkreide*. 5th edn. Heimatverein der Grafschaft Bentheim, Bad Bentheim, Germany.

Kilhams, B., Kukla, P., Mazur, S., McKie, T., Mijnlieff, H., van Ojik, K. & Rosendaal, E. In press. Mesozoic Resource Potential in the Southern Permian Basin Area. The geological key to exploiting remaining hydrocarbons whilst unlocking geothermal potential. *In*: Kilhams, B., Kukla, P.A., Mazur, S., McKie, T., Mijnlieff, H.F. & van Ojik, K. (eds) *Mesozoic Resource Potential in the Southern Permian Basin*. Geological Society, London, Special Publications, **469**, https://doi.org/10.1144/SP469.26

Knaap, W.A. & Coenen, M.J. 1987. Exploration for oil and natural gas. *In*: Visser, W.A., Zonneveld, J.I.S. & van Loon, A.J. (eds) *Seventy-Five Years of Geology and*

Mining in the Netherlands (1912–87). Royal Geological and Mining Society of the Netherlands (KNGMG), The Hague, 207–242.

Kombrink, H., Doornenbal, J.C., Duin, E., Den Dulk, M., Van Gessel, S.F., Ten Veen, J.H. & Witmans, N. 2012. New insights into the geological structure of the Netherlands; results of a detailed mapping project. *Netherlands Journal of Geosciences*, **91**, 419–446.

Kortmann, H. 1983. *Sedimentologische Untersuchungen über den Bentheimer Sandstein (Valangin) der Umgebung von Bad Bentheim*. MSc thesis, Universität zu Köln.

Malmborg, P. 2002. *Correlation between diagenesis and sedimentary facies of the Bentheim Sandstone, the Schoonebeek field, The Netherlands – Petrographic and sedimentological re-evaluation of a petroleum reservoir*. MSc thesis, Lunds Universitet.

Mazur, S. & Scheck-Wenderoth, S. 2005. Constraints on the tectonic evolution of the Central European Basin System revealed by seismic reflection profiles from Northern Germany. *Netherlands Journal of Geosciences*, **84**, 389–401.

Mutterlose, J. & Bornemann, A. 2000. Distribution and facies patterns of Lower Cretaceous sediments in northern Germany: a review. *Cretaceous Research*, **21**, 733–759.

NAM 2008. *Winningsplan Schoonebeek Olieveld*. Nederlandse Aardolie Maatschappij, **18**. Available at http://www.nlog.nl.

Peksa, A.E., Wolf, K.-H.A.A. & Zitha, P.L.J. 2015. Bentheimer sandstone revisited for experimental purposes. *Marine and Petroleum Geology*, **67**, 701–719, https://doi.org/10.1016/j.marpetgeo.2015.06.001

Pharaoh, T.C., Dusar, M. *et al.* 2010. Tectonic evolution. *In*: Doornenbal, J.C. & Stevenson, A.G. (eds) *Petroleum Geological Atlas of the Southern Permian Basin Area*. EAGE Publications b.v., Houten, 24–57.

Pletsch, T., Appel, J. *et al.* 2010. Petroleum generation and migration. *In*: Doornenbal, J.C. & Stevenson, A.G. (eds) *Petroleum Geological Atlas of the Southern Permian Basin Area*. EAGE Publications b.v., Houten, 224–253.

Rondeel, H.E., Batjes, D.A.J. & Nieuwenhuijs, W.H. 1996. *Geology of Gas and Oil under the Netherlands*. The Royal Geological and Mining Society of the Netherlands, Kluwer Academic Publishers, Dordrecht.

Rutten, K.W. 2004. Validating seismic correlations by unfaulting and multi-horizon flattening. *66th EAGE Conference & Exhibition*, EAGE, Paris.

Stadtler, A. 1998. *Der Bentheimer Sandstein (Valangin, NW-Deutschland) – Eine palökologische und sequenzstratigraphische Analyse*. Ruhr-Universität Bochum, Bochum, **49**.

Stadtler, A. & Mutterlose, J. 1996. Palaeontological and integrated sequence stratigraphic analysis of the Bentheim Sandstone. *In*: Johnson, H.D., Wonham, J.P. *et al.* (eds) *Final Report European Commission Joule Programme JOU2_CT93_0441*. Imperial College, London, **26**, 1–7.

TNO-NITG 2000. *Explanation to Map Sheet VI: Veendam-Hoogeveen + Map Sheet VI*. Netherlands Institute of Applied Geoscience TNO – National Geological Survey, Utrecht.

Van Adrichem Boogaert, H.A. & Kouwe, W.F.P. 1993. Stratigraphic nomenclature of The Netherlands; revision and update by RGD and NOGEPA, Section G. *Mededelingen Rijks Geologische Dienst*, **50**, 1–80.

Van Dalfsen, W., Van Gessel, S.F. & Doornenbal, J.C. 2007. *Velmod-2 Joint Industry Project*. TNO Built Environment and Geosciences, TNO-report 2007-U-R1272C.

Van de Weerd, A.A. 1996. Petroleum Geology of Lower Cretaceous Fields in the Emsland area (German-Netherlands border region). *In*: Johnson, H.D., Wonham, J.P. *et al.* (eds) *Final Report European Commission Joule Programme JOU2_CT93_0441*. Imperial College, London, **26**, 1–18.

Vejbæk, O.V., Andersen, C. *et al.* 2010. Cretaceous. *In*: Doornenbal, J.C. & Stevenson, A.G. (eds) *Petroleum Geological Atlas of the Southern Permian Basin Area*. EAGE Publications b.v., Houten, 195–209.

Wittenhagen, S. 1980. *Sedimentologisch-petrografische Untersuchung des Bentheimer Sandsteins im Erdölfeld Bramberge/Emsland*. Technischen Universität Clausthal-Zellerfeld.

Wonham, J.P., Mutterlose, J. & Stadtler, A. 1996*a*. Structural controls on Lower Cretaceous tidal sandstone deposits: the Bentheim sandstone (Valanginian) of the Lower Saxony Basin, NW Germany. *In*: Johnson, H.D., Wonham, J.P. *et al.* (eds) *Final Report European Commission Joule Programme JOU2-CT93_0441*. Imperial College, London, **26**, 1–29.

Wonham, J.P., Mutterlose, J. & Stadtler, A. 1996*b*. Reservoir characterisation studies of the Bentheim Sandstone: a synthesis of outcrop and subsurface data. *In*: Johnson, H.D., Wonham, J.P. *et al.* (eds) *Final Report European Commission Joule Programme JOU2_CT93_0441*. Imperial College, London, **26**, 1–23.

Wonham, J.P., Johnson, H.D., Mutterlose, J., Stadtler, A. & Ruffell, A. 1997. Characterization of a shallow marine sandstone reservoir in a syn-rift setting: The Bentheim sandstone formation (Valanginian) of the Ruhlermoor field, Lower Saxony Basin, NW Germany. *GCSSEPM Foundation 18th Annual Research Conference Shallow Marine and Nonmarine Reservoirs*, 427–448.

The impact of heterogeneity on waterflood developments in clastic inner shelf reservoirs: an example from the Holland Greensand Member, Rotterdam Field, The Netherlands

RICHARD J. PORTER*, ALBERTO MUÑOZ ROJAS & MALTE SCHLÜTER

Nederlandse Aardolie Maatschappij B.V., Schepersmaat 2, 9405 TA Assen, The Netherlands

**Correspondence: richard.porter@shell.com*

Abstract: A geological understanding of the way fluids flow through a reservoir is crucial when considering waterflood developments in heterogeneous reservoirs. Recent integrated production geoscience and reservoir engineering studies on the Holland Greensand oil reservoir (Aptian) of the Rotterdam Field in the West Netherlands Basin, are used to illustrate the geological controls on fluid flow through rocks deposited in a clastic inner shelf depositional setting. An integrated understanding of the subsurface has been gained through core, log and production data analysis. Analogous Early Cretaceous Upper Greensand Formation coastal exposures in SW England have been used to gauge the scale of lateral and vertical heterogeneity, and its impact on hydrocarbon and water flow paths. Argillaceous sandstones deposited at the distal margins of a subtidal sand-sheet provide internal reservoir baffles, while fluids are believed to preferentially follow flow paths through associated laterally extensive tempestites. Cemented layers observable in wells are interpreted to be laterally restricted and are not considered as major baffles. These observations have been used to steer static reservoir model construction and subsequent simulation. Results from field testing by production logging tool analysis support the geological concepts developed and the modelling approach used. This work has important implications for defining the approach for future reservoir management and redevelopment in the Rotterdam Field, in particular the well type and configuration and completion strategies post-reservoir flooding. The application of methodologies employed in this study is recommended for similar waterflood developments.

As producing hydrocarbon fields become increasingly mature and production begins to decline, a geological understanding of the way fluids flow through a reservoir becomes crucial. This is particularly true when fields with heterogeneous reservoirs are considered for secondary recovery methods. Such methods include waterflooding, the injection of water to help maintain reservoir pressure. The impact of heterogenous reservoir rock properties on waterflood design and performance has been well documented (Dake 1978). In theory, a reservoir may either be in total vertical equilibrium or display an absence of cross-flow between reservoir zones. Analysis of historical production and forecasts of fields currently undergoing waterflood oil recovery, illustrates that many reservoirs are best described as intermediate between these two end-members (Dake 1978). It is therefore important to recognize the style and degree of heterogeneity contained within the reservoir. This assessment can then be incorporated into the overall field-development strategy. This paper documents integrated geoscience and reservoir engineering studies undertaken on the Early Cretaceous Holland Greensand oil reservoir of the Rotterdam Field, west Netherlands Basin between early 2013 and mid-2015. This is used to illustrate the importance of understanding geological controls on fluid flow in a heterogenous reservoir deposited in a clastic shelf depositional environment.

Rotterdam Field

The Rotterdam Field is located beneath the city of Rotterdam in the hydrocarbon-prone West Netherlands Basin (Fig. 1; de Jager *et al.* 1996; see also Vondrak *et al.* 2018). The field comprises four stacked paralic clastic reservoirs of Early Cretaceous age containing gas and oil (Racero-Baena & Drake 1996). The upper three reservoirs form the main producing intervals and are part of the Rijnland Group. These are namely the IJsselmonde Sandstone Member, the De Lier Member and the Holland Greensand Member (Fig. 2). The field consists of a mildly faulted, NW–SE-trending anticline (Fig. 3), with an aerial extent of approximately 2 km by 4.5 km and a maximum relief of *c.* 190 m. The Rotterdam Field is sited in one of many well-known Late Cretaceous–Early Tertiary inversion structures in the west Netherlands (Winstanley 1993; de Jager 2007; Luijendijk *et al.* 2011). Inversion in Rotterdam is considered to have been relatively

From: Kilhams, B., Kukla, P. A., Mazur, S., McKie, T., Mijnlieff, H. F. & van Ojik, K. (eds) 2018. *Mesozoic Resource Potential in the Southern Permian Basin*. Geological Society, London, Special Publications, **469**, 457–477.
First published online April 16, 2018, https://doi.org/10.1144/SP469.20

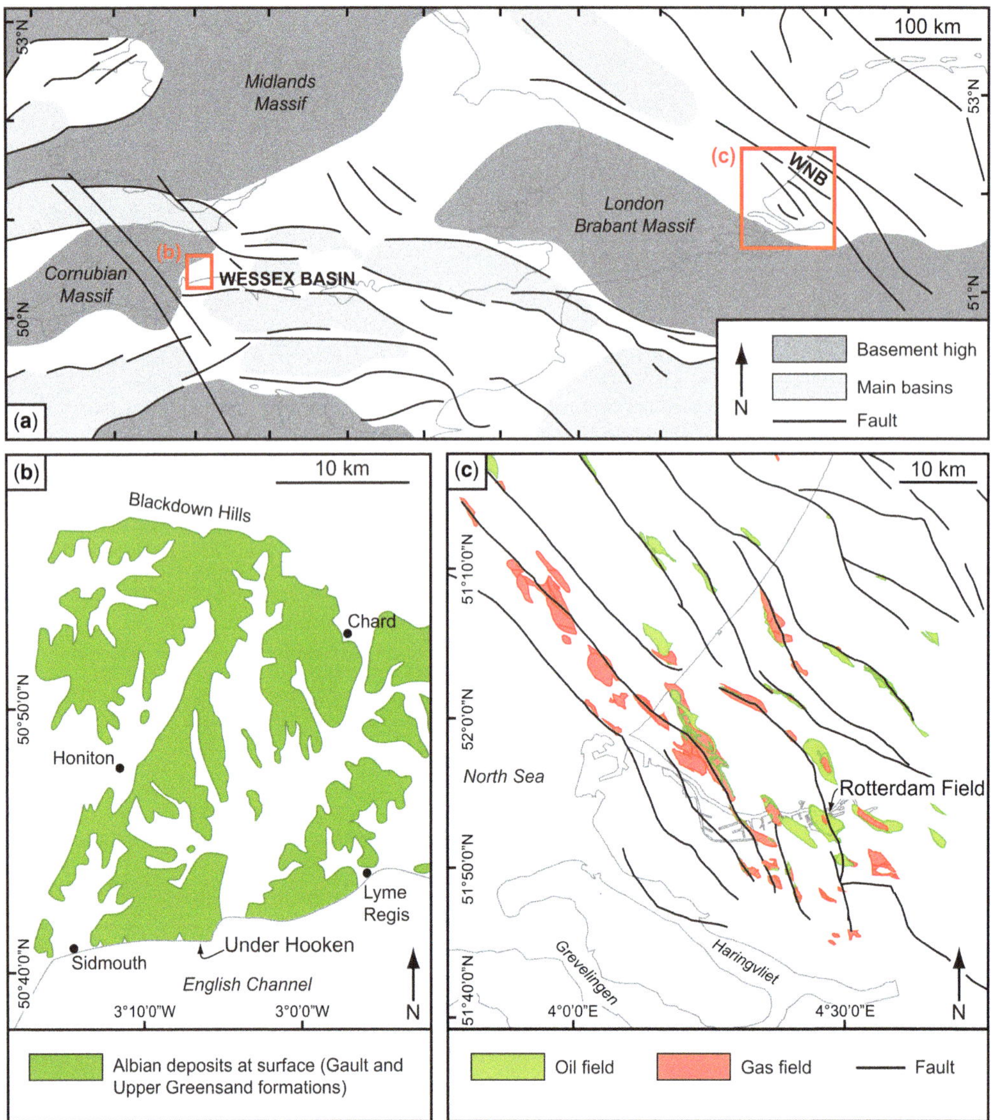

Fig. 1. (**a**) Schematic palaeogeographic map of the southern UK and Netherlands during the Aptian–Albian, including generalized fault patterns (modified from Wells *et al.* 2010). WNB, West Netherlands Basin. (**b**) Geological sketch map showing the location of Under Hooken and the outcrop and subcrop of the Upper Greensand Formation in SW England (modified from Gallois 2004). (**c**) Overview map showing the regional location of the Rotterdam Field and its position relative to other West Netherlands Basin hydrocarbon fields. Faults from Vejbæk *et al.* (2010).

mild (Racero-Baena & Drake 1996), and was controlled by motion along deeper-seated Triassic–Jurassic rift-related faults. One main fault crosses the structure, is located north and east of the crest (Fig. 3), and has a maximum fault throw of 80 m. Each reservoir is sealed by mudrocks of either the Vlieland Claystone Formation or Holland Formation. The main hydrocarbon source is provided by Late Carboniferous Westphalian A-B coals and fissile, organic-rich mudrocks of the Early Jurassic Posidonia Shale Formation (van Balen *et al.* 2000). The sandstones of the Rijnland Group reservoirs, Rotterdam Field, form part of a series of marine transgressive–regressive cycles and include fluvially dominated delta front deposits (IJsselmonde Sandstone Member), lower shoreface deposits (De Lier Member) and shelf sand- and siltstones (Holland Greensand Member) (Racero-Baena & Drake 1996;

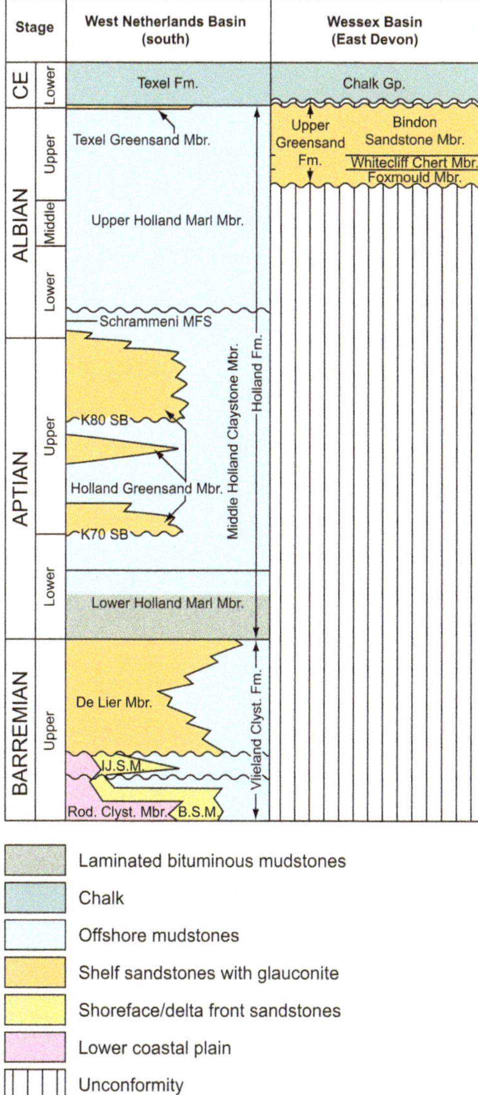

Fig. 2. Aptian–Albian lithostratigraphic framework of the southern West Netherlands Basin and the Wessex Basin in east Devon, UK. West Netherlands stratigraphy from Jeremiah *et al.* (2010), Wessex Basin stratigraphy from Gallois (2004). IJ.S.M, IJsselmonde Sandstone Member; B.S.M, Berkel Sandstone Member; Rod. Clyst. Mbr., Rodenrijs Claystone Member; SB, sequence boundary; MFS, maximum flooding surface.

Jeremiah *et al.* 2010). These arenaceous units were deposited at, or along, a palaeoshoreline which flanked the London Brabant Massif to the south, and pass laterally into the offshore mudrocks of the Vlieland Claystone Formation or Holland Formation (Jeremiah *et al.* 2010).

Development history and study objectives

Discovered in 1984 by well Rotterdam-1 (RTD-1), a total of 21 wells have been drilled in the Rotterdam Field. These range from dual-string multiple reservoir completions in deviated wells, to dedicated reservoir production or injection horizontal and multilateral wells. The field is characterized by the simultaneous presence of several recovery mechanisms, operating concurrently in the different reservoirs. The IJsselmonde Sandstone Member reservoir is supported by a well-established aquifer while the De Lier Member reservoir, having undergone several reservoir management strategies, is currently under waterflood. The Holland Greensand Member reservoir has been developed in two main development phases. A period of primary depletion was followed by the current waterflood drive. This comprises four water-injection wells providing support to up to five hydrocarbon-producing wells (Fig. 3). All wells in the Rotterdam Field require gas lift for oil production. This is due to a near-hydrostatic pressure gradient and the subsequent reservoir depletion and pressure restoration. Typical oil recoveries of greater than 30% have been recorded for reservoirs under a waterflood for favourable rock and fluid properties (Satter *et al.* 2008). To date, the hydrocarbon recovery from the Holland Greensand reservoir has been poor. As of June 2015, the estimated production was approximately 7% of the stock-tank original oil in place.

Given the long production history, a study was initiated which aimed to understand the fundamental reasons for the poor recovery and, to assist the identification of drilling targets, establish the current hydrocarbon saturation distribution. Unfortunately, limited historical reservoir surveillance activities have been undertaken on the Rotterdam Field. An allocated production and injection history was therefore used to describe production performance, alongside basic simulation models to assess reservoir fluid movement. These models suggested a complex fluid displacement process, likely due to the presence of high reservoir heterogeneity, with fluids being displaced in confined layers with a potential for water on top of oil. Comparisons with an outcrop analogue and integrated modelling work, were performed to test the hypothesis of a vertically segregated reservoir displacement process.

Holland Greensand: reservoir geology

The Holland Greensand Member is of Late Aptian–Early Albian age (Fig. 2; Jeremiah *et al.* 2010). In the Rotterdam Field it attains a maximum thickness of 165 m in the northwestern part, thinning to approximately 125 m in the south. The member is underlain and conformably overlain by the Middle

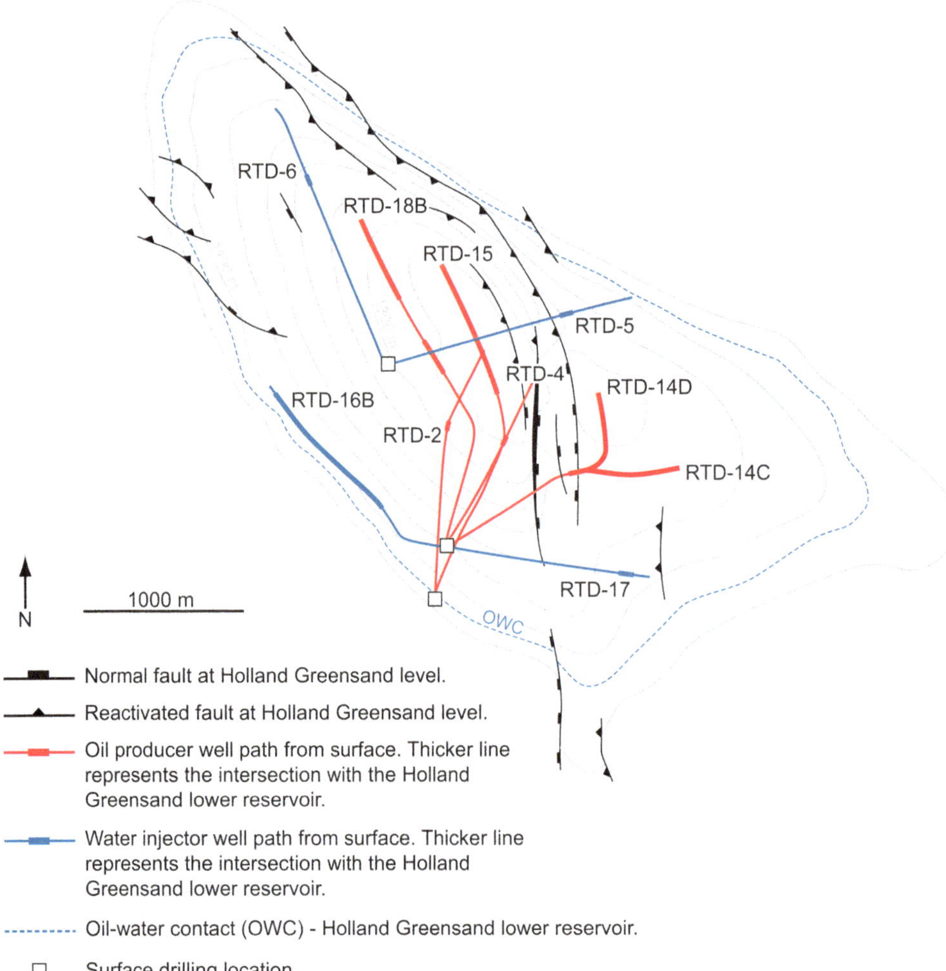

Fig. 3. Depth map of the Rotterdam Field showing the top surface of the Holland Greensand lower reservoir.

Holland Claystone Member (Early Aptian–Early Albian) (Jeremiah *et al.* 2010). Five tuff horizons, which can be correlated across much of the West Netherlands Basin (Jeremiah *et al.* 2010), occur within the upper part of the Holland Greensand Member. In the Rotterdam area, the member has been attributed to the North Sea-wide K70 and K80 tectonic/sequence stratigraphic units (Jeremiah 2000; Jeremiah *et al.* 2010). The hydrocarbon-bearing zone, split into a lower and an upper reservoir unit, is approximately 70 m thick. This is situated between the K80 sequence boundary, interpreted as the result of relative sea-level fall and sediment progradation due to ongoing Austrian tectonism (Jeremiah 2000), and the *Schrammeni* maximum flooding surface, recording renewed sea-level rise at the beginning of the Albian (Jeremiah *et al.* 2010).

The Holland Greensand reservoir has been cored in five wells (RTD-2, -3B, -5, -6 and -13). It consists of glauconitic, argillaceous sandstones, with visual clay contents ranging from 10% to 40%, intercalated with sandy mudstones and carbonate cemented bodies. These sandstones are olive grey, very fine- to fine-grained and are moderately well sorted with predominantly subrounded grains. Bioturbation by a largely horizontal, deposit- or suspension-feeding community occurs throughout the reservoir interval (Fig. 4). This consists principally of *Palaeophycus*, *Planolites*, *Nereites*, *Chondrites*, *Helminthopsis* and *Phycosiphon*. Although infaunal reworking appears to have destroyed almost all primary lamination in the cored intervals (bioturbation index, BI = 4–6; Taylor & Goldring 1993), careful inspection shows the trace fossil-forming fauna followed primary bedding structures. The reservoir interval is therefore dominated by low-angle cross-bedding with lesser proportions of horizontal, ripple and wavy lamination. Porosity values range from 20% to 25% and permeability values from 10 to 50 millidarcies (mD). Intermittent deposition of cleaner, better-quality sandstones with sharp bases and gradational tops exhibiting an upwards-fining grain-size profile punctuate the background sedimentation. These units are also fully bioturbated, with *Helminthopsis*, *Chondrites* and *Nereites* often overprinting initial complete reworking by *Teichichnus zigzag*. The tops of these units are gradational and are often bioturbated by a sub-1 mm scale assemblage of *Helminthopsis*, *Chondrites* and *Nereites* with occasional *Cylindrichnus* and *Rosselia*. These 1–3 m thick packages may occur throughout the vertical extent of the reservoir. The basal 0.5–1 m of these packages may be cemented with siderite and/or calcite and are occasionally associated with shelly fragments. The uncemented sections of these packages have porosity values around 30% and permeability values of 150–400 mD. The contacts between the mudstones and overlying sandstones are often sharp and occasionally slightly undulated. Within the upper reservoir the proportion of mudstone and bioturbated mudstone drapes increases, as does the glauconite content. Five, thin tuffaceous units occur between the base of the upper reservoir and the top of the Holland Greensand Member. These occur either as distinct laminated tuff beds with minor *Chondrites* bioturbation towards their tops, or more subtly reworked into the surrounding argillaceous sandstones.

The sedimentological variations seen in the cored sections described above (Fig. 4) are also reflected in well log responses over these intervals. This calibration allows interpretation of similar small-scale cycles in wells without core. Overall, the lower reservoir has an irregular trending gamma ray log response, reflecting subtle changes in relative grain size (Fig. 5). The upper reservoir displays progressively increasing gamma ray values and corresponds to the overall increasing argillaceous content seen in the cores. The cemented intervals observed in the cores of RTD-3B, -5 and -6 have been calibrated to logs. Thicker cements appear as high-density peaks; these are confirmed with borehole image logs (Formation MicroScanner (FMS) logs). Thinner cement nodules (<15 cm thick) are not seen on conventional logs, but are interpreted on the FMS logs. By using a combination of core and logs, the distribution of cements per reservoir subdivision (zone) has been estimated (Fig. 6). Although observed at most stratigraphic levels, cements are only present in all wells at one level (upper reservoir zone 5).

The Holland Greensand Member is observed to exhibit fine-grained, finely laminated and cross-laminated, silty sandstones and sandy siltstones, authigenic glauconite, scattered whole and abraded shell fragments and sponge spicules. Together with marine bioturbation typical of the archetypal *Cruziana* ichnofacies, this suggests deposition in the Rotterdam region as part of an inner-shelf to subtidal sand-sheet complex (Stride *et al.* 1982; Desjardins *et al.* 2012*a*). The high intensity of the bioturbation, together with the indication of original low-angle cross-stratification with some horizontal, ripple and wavy lamination, suggests deposition under lower-energy conditions in the distal part of the sand-sheet, either the front or towards the margin (Desjardins *et al.* 2012*a*). Here, deposition occurred relatively slowly under low-energy conditions. Cleaner sandstones with slightly erosional bases, upwards-fining grain-size profiles and often appearing reworked by a post-depositional *Teichichnus zigzag* are interpreted as tempestites (Baniak *et al.* 2014). This suggests deposition of the Holland Greensand reservoir was above storm wave base. Although the bioturbation present in the Holland Greensand is comparable to that recorded in the bottom-set deposits of compound dunes (Desjardins *et al.* 2012*b*; Olariu *et al.* 2012), the absence of overlying larger-scale, thickening-upwards sequences of typically non-bioturbated, cross-stratified sandstone (Dalrymple *et al.* 2003; Dalrymple & Choi 2007), suggests that the lower Holland Greensand reservoir in the Rotterdam Field area was unlikely to have been deposited as part of a compound dune field. Similar ichnoassemblages to that seen in the Holland Greensand reservoir have also been recorded from tidal sand ridge flanks and the swales between the ridges (Desjardins *et al.* 2012*b*). A tidal sand ridge model is not favoured here however, since packages of clean, cross-stratified sandstone that often form the cores of the tidal sand ridges (Leva López *et al.* 2016) are not present in the Rotterdam Field area. Further, such sand bodies have not been recorded in Holland Greensand Member cores in the West Netherlands Basin. Core and well log data suggest the lower reservoir of the Holland Greensand to be largely aggregational, whereas the upwards-increasing argillaceous content of the upper reservoir implies a back-stepping of the sand-sheet and an increasing dominance of low-energy inner-shelf deposits.

Outcrop reservoir analogue: Lower Cretaceous Foxmould Member, Upper Greensand Formation, SW England

Analogues were sought to provide guidance to the lateral extent of facies and cements in the Rotterdam Field Holland Greensand reservoir. Particular attention was paid to lateral scales similar to the well spacing of the field, typically 150–600 m and up to 1.5 km. Although fine-grained, highly bioturbated,

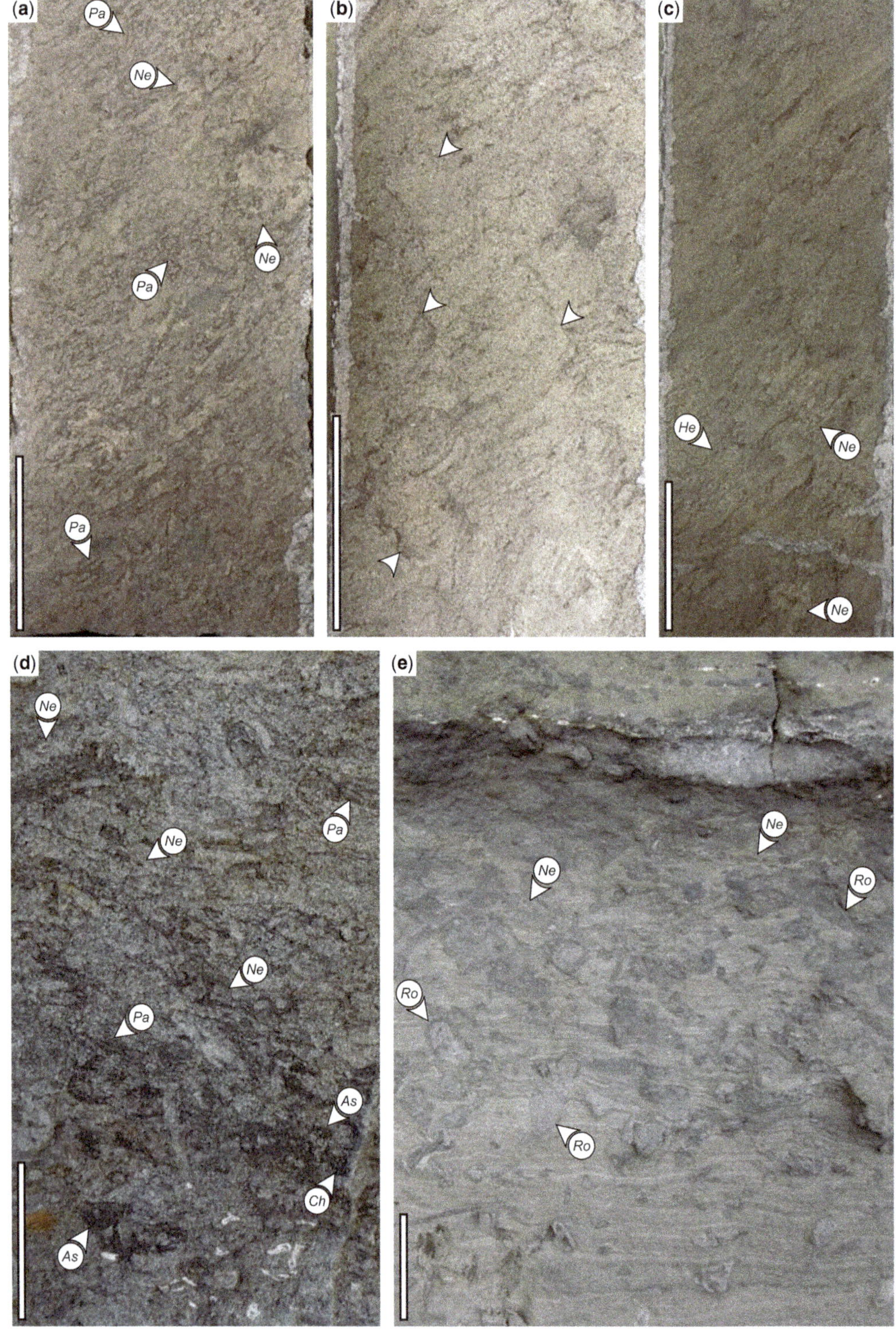
(a)
Pa
Ne
Ne
Pa
Pa
(b)
(c)
He
Ne
Ne
(d)
Ne
Pa
Ne
Ne
Pa
As
Ch
As
(e)
Ne
Ne
Ro
Ro
Ro

glauconitic shelf facies have been described in the literature (e.g. Hocking *et al.* 1988; Burns 2002), and several studies document the dimensions and distribution of individual cement bodies in shallow-marine environments (e.g. McBride *et al.* 1995; Molenaar 1998; Dutton *et al.* 2002; Hampson *et al.* 2015), outcrops were required with depositional characteristics more comparable to the Holland Greensand reservoir. To address this, a pilot study of the analogous Lower Cretaceous, Upper Greensand Formation of the Wessex Basin, UK was undertaken.

The Upper Greensand Formation, which is Late Albian in age (Hart & Williams 1990), is exposed along much of a 35 km section of the east Devon and west Dorset coast in the Wessex Basin, UK (Figs 1, 2). It is predominantly disconformably overlain by the Cenomanian-aged Chalk Group. The base of the formation rests unconformably on bioturbated firmground and hardground surfaces on a series of rocks ranging from Middle Triassic to Early Jurassic in age (Gallois & Goldring 2007). The formation can be divided into three members (Gallois 2004), but only those sediments comprising the lower half of the formation, the Foxmould Member, comprise sedimentary deposits which are facies analogous to those of the Holland Greensand reservoir. One of the best exposed and most accessible sections to allow detailed observation can be found in the cliffs at Under Hooken (UK grid reference SY 2209 8790) (Fig. 1c). Almost the entire 25–30 m thickness of the Foxmould Member can be studied here, with beds that may be traced laterally for 400 m. Sedimentary logging shows the member comprises highly bioturbated, very fine- to fine-grained sandstones and silty sandstones which range in colour from dark, olive green-grey to light greenish-grey and buff. Prominent carbonate concretions and more extensive, tabular cemented horizons occur through the Foxmould Member. The sandstones specifically of the member are however generally weakly cemented. The member is typically thickly to very-thickly bedded, with the clay and silt content varying on a bed-by-bed basis. The bases of the beds are usually sharp, some of which can be observed to down-cut into underlying units. Where present, shallow scours are filled with shell debris, serpulid tubes and small mudstone chips. These are usually found beneath cleaner sandstone beds, 40–75 cm thick, displaying parallel and wavy lamination and hummocky cross-stratification. High levels of bioturbation occur (BI = 5–6) throughout the rest of the Foxmould Member. Like the Holland Greensand Member, this consists of a largely horizontal, deposit- and suspension-feeding community. This is dominated by *Planolites*, *Palaeophycus*, *Asterosoma* and *Nereites*. *Scolicia* and *Chondrites* may also appear commonly at specific horizons throughout the member. Large boxwork burrows such as *Ophiomorpha* have only been identified to date in the basal beds of the member. Hummocky cross-stratified beds are less intensively reworked (BI = 1–4). Bioturbation increases upwards in these beds, with *Rosselia* and *Asterosoma* forming the dominant component of the ichnoassemblage. Individual beds may be traced across the lateral extent of the outcrops (400 m), with little change in thickness or in facies.

The carbonate nodules found within the Foxmould Member are present as individual occurrences, yet can be observed to coalesce to form elongate, irregular nodules or tabular sheets. The nodules and sheets occur along discrete stratigraphic horizons throughout the Foxmould Member (Fig. 7). Individual nodules measured in the lower part of the member range from 5–25 cm thick to 20–60 cm in length. Along layers, nodules are separated at 50–150 cm spacing. Nodules may also coalesce to form larger bodies between 70 and 150 cm (Fig. 8). Carbonate sheets are laterally extensive across the 400 m wide exposure.

Previous analysis of the Upper Greensand coastal exposures in SW England indicated deposition in shallow-marine environments close to or above storm wave base (Williams 1991; Gallois 2004; Gallois & Goldring 2007). More specifically, the deposits at Under Hooken are here believed to form the distal part of a subtidal sand-sheet (Desjardins *et al.* 2012*a*). The fine-grained, highly bioturbated nature of the sandstones is indicative of relatively low-energy deposits with slow, steady sedimentation rates. This has allowed a largely infaunal, deposit- or suspension-feeding community, typical of the *Cruziana* ichnofacies, to become established

Fig. 4. Example ichnofabrics from (a–c) slabbed core of the Holland Greensand reservoir, Rotterdam Field and (d, e) coastal exposures of the Upper Greensand Formation, Under Hooken, east Devon, UK (UK grid reference SY 2209 8790). All scale bars 5 cm. (**a**) Argillaceous sandstones with complete bioturbation dominated by *Palaeophycus* (*Pa*) and *Nereites* (*Ne*). Sand-sheet margin (well RTD-3B, depth 1896.8 m). (**b**) Fine-grained sandstones containing bell-shaped patches of laminated sandstone (margins of patches highlighted), interpreted as *Teichichnus zigzag*. Sand-sheet margin tempestite (well RTD-3B, depth 1907.4 m). (**c**) *Teichichnus zigzag* ichnofabric overprinted by *Nereites* (*Ne*) and *Helminthopsis* (*He*). Sand-sheet margin tempestite (well RTD-3B, depth 1907.0 m). (**d**) Silty-sandstones with scattered serpulid fragments, predominantly reworked by *Palaeophycus* (*Pa*) and *Nereites* (*Ne*) with subordinate *Asterosoma* (*As*) and *Chondrites* (*Ch*). Sand-sheet margin, Under Hooken, UK. (**e**) *Rosselia* (*Ro*) and *Nereites* (*Ne*) occurring at the top of fine-grained, hummocky cross-stratified sandstones. Sand-sheet margin tempestite, Under Hooken, UK.

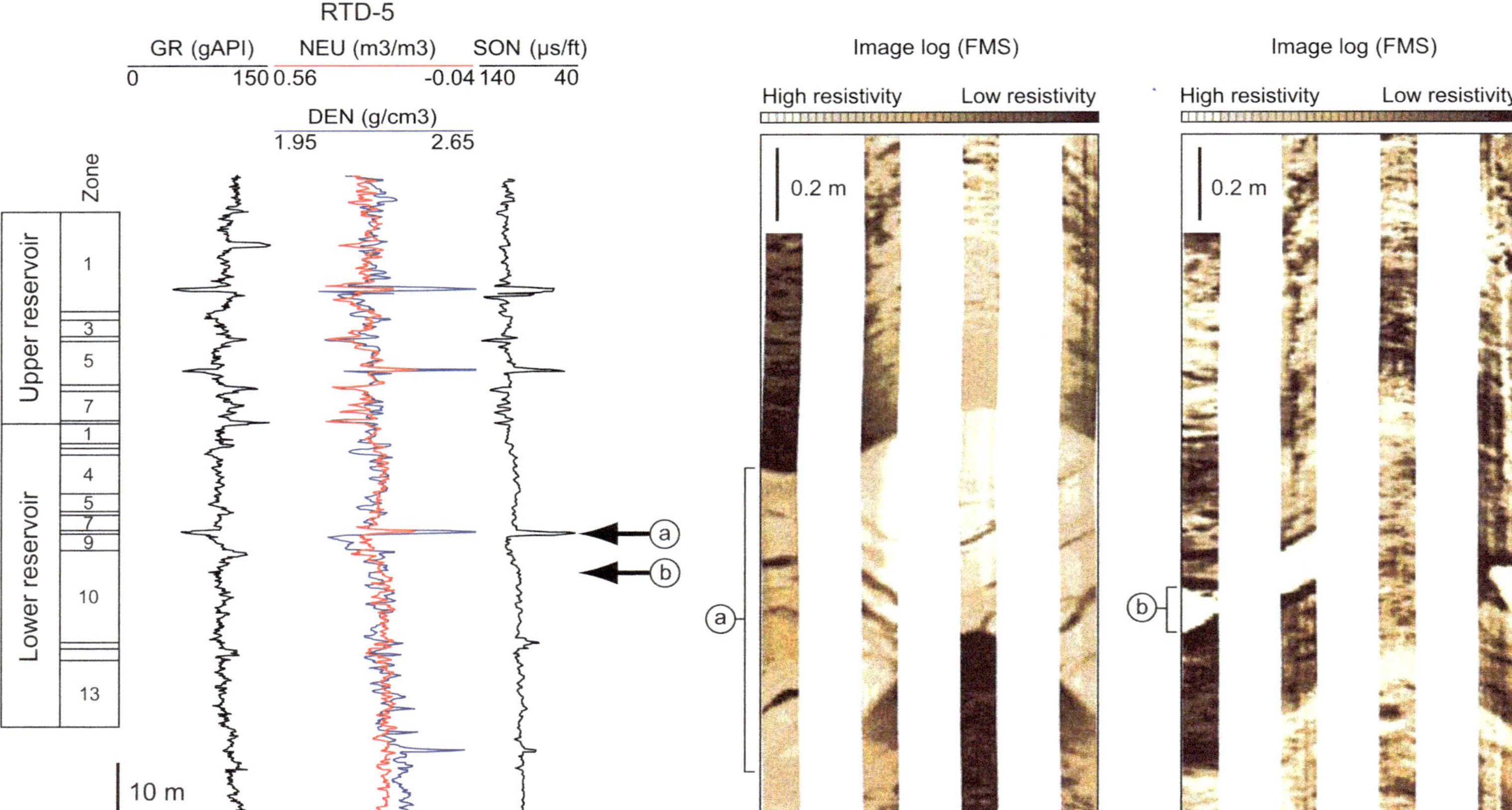

Fig. 5. Cemented intervals occur throughout the Holland Greensand reservoir in the Rotterdam Field. When of sufficient thickness, these may be identified on conventional logs as high-density peaks associated with high sonic velocities and low neutron log values, and as high-resistivity features in image logs ('a'). Smaller nodules of limited thickness may be observed in image logs only ('b'). GR, gamma ray log; NEU, neutron log; DEN, bulk density log; SON, sonic log; FMS, Schlumberger Formation MicroScanner image log.

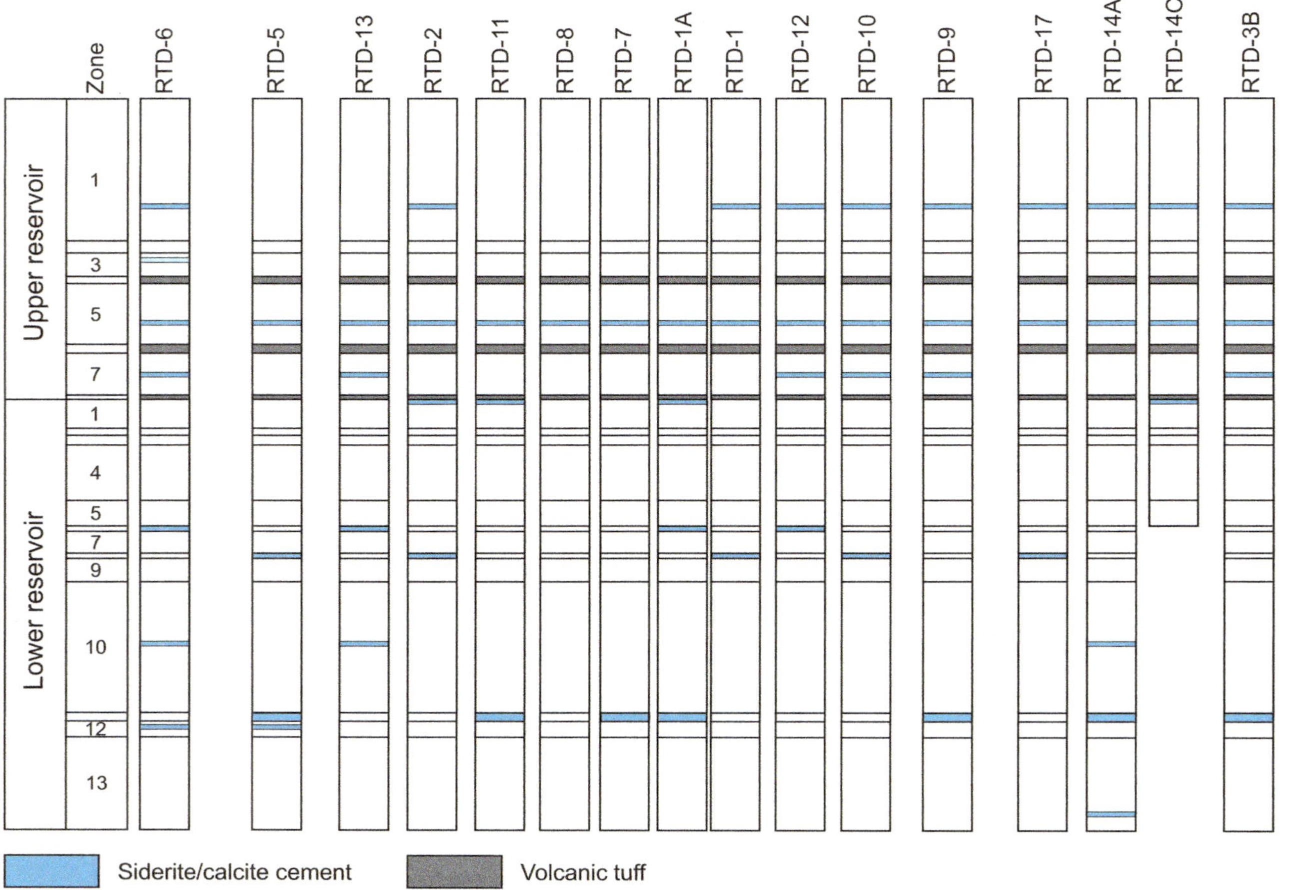

Fig. 6. Schematic summary of the distribution of cemented intervals and volcanic tuff horizons in the Holland Greensand reservoir of the Rotterdam Field. The cemented horizons observed in core and/or interpreted from log responses occur at similar stratigraphic levels. There is however no level at which any particular cemented interval is recorded in every well. Three volcanic tuff beds are recorded in each well, with each located at the same relative stratigraphic position in the reservoir.

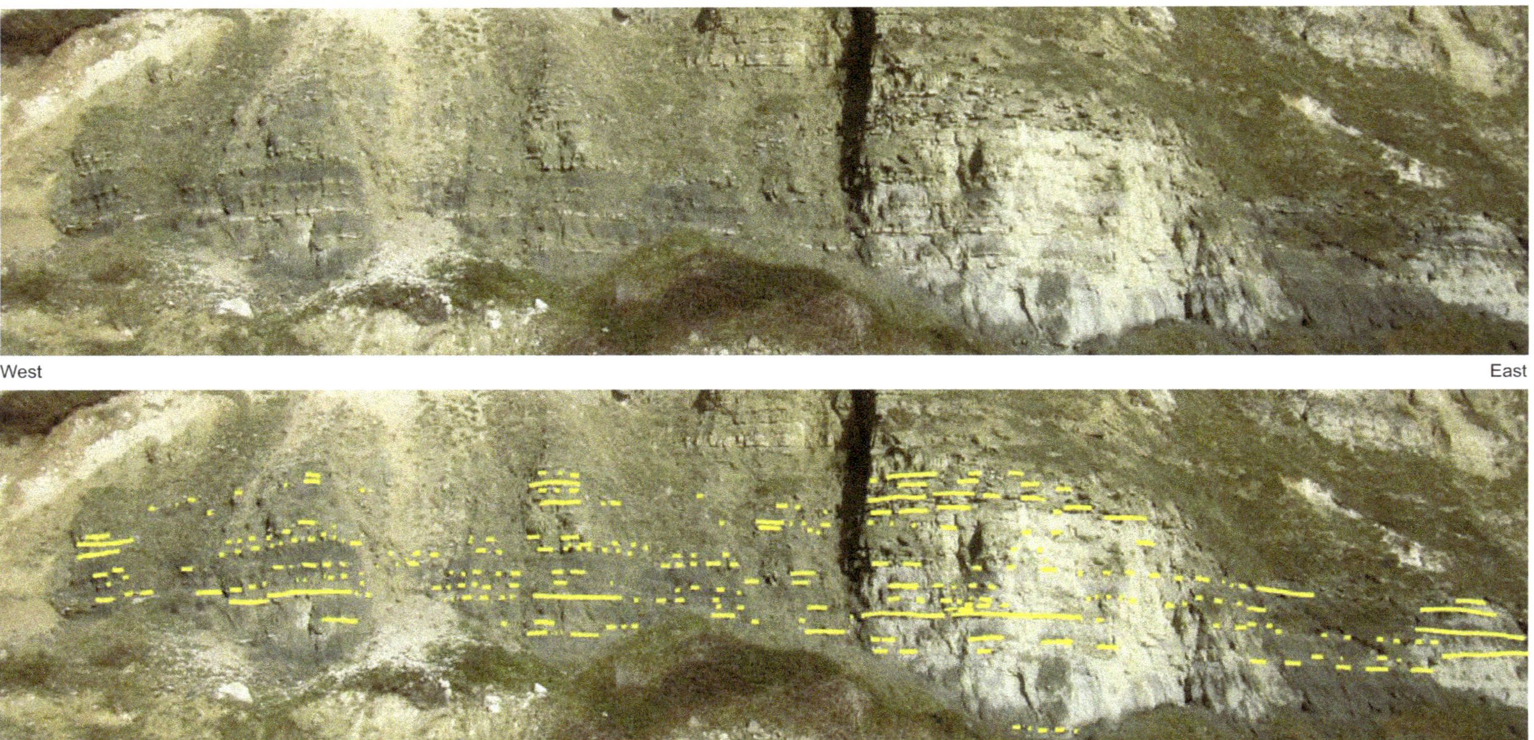

Fig. 7. Coastal exposures of the Early Cretaceous Upper Greensand Formation in the cliffs at Under Hooken, SW England. Cement layers (coloured yellow in tracing) occur at similar stratigraphic levels. Cliff height shown is approximately 20 m and the length of the exposure shown is approximately 100 m. Composite panorama uncorrected for parallax and perspective.

Fig. 8. Thickly interbedded argillaceous sandstones and arenaceous mudstones in the cliffs at Under Hooken, SW England. Basal bedding surfaces are slightly erosive and are often associated with a serpulid lag. Isolated, coalesced and sheet-like nodules occur associated with these surfaces. The cleaner sandstones, deposited under storm conditions, are laterally extensive across the coastal exposure at Under Hooken. These deposits are comparable to those in the Rotterdam Field Holland Greensand oil reservoir, where they are of higher permeability than the surrounding rock and act as the main conduits for fluid movement. Cliff section height shown is 3 m.

(Buatois & Mańgano 2011). The sharp, occasionally erosive bed bases, sometimes associated with lag deposits and higher-energy, hummocky cross-stratified units, colonized in their upper part by a *Rosselia*-dominated opportunistic fauna, suggests deposition during storms (Desjardins *et al.* 2010). Cements, particularly in the lower part of the formation, appear to be associated with the dissolution of shell debris, particularly those from serpulids. The variable shapes of the nodules and sheets, especially those with bases that down-cut into the underlying strata, suggest that possible precursor shell debris was swept by currents into small scours on the sea bottom.

Comparison and implications for correlation and modelling in the Rotterdam Field

Several similarities exist between the sedimentary deposits of the Rotterdam Field Holland Greensand reservoir and those of the Foxmould Member of SE Devon. Both units are formed predominantly of highly bioturbated, glauconitic silty sandstones and sandy siltstones, occasionally punctuated by cleaner sandstones. The ichnofaunal assemblages are also comparable, both being typical of lower-energy fully marine settings associated with the *Cruziana* ichnofacies. Fine, shelly debris is found in varying proportions throughout the two units. Individual and coalesced nodules and tabular sheets are found in the Foxmould Member and are analogous to those observed in core and logs from the Holland Greensand reservoir. Based on the available data however, the frequency of cemented horizons in the Foxmould Member appears higher compared to the Holland Greensand reservoir. Beds deposited under storm conditions also appear to be thicker in the Foxmould Member, the basal surfaces of which truncate underlying beds more prominently than those observed in the Holland Greensand reservoir. This suggests the Foxmould Member was influenced by more intense storm activity during deposition, which could imply a slightly more proximal shelf setting than that of the Holland Greensand Member. Overall, both units represent deposition as part of a shallow, subtidal offshore sandbody. The high bioturbation index seen throughout both units points to a more distal offshore position, such as the margin of a subtidal sand-sheet.

The laterally extensive nature of the facies seen in the Foxmould Member at Under Hooken suggests that correlating sediment packages identified in core and well log data in the Holland Greensand reservoir at Rotterdam can be undertaken with confidence

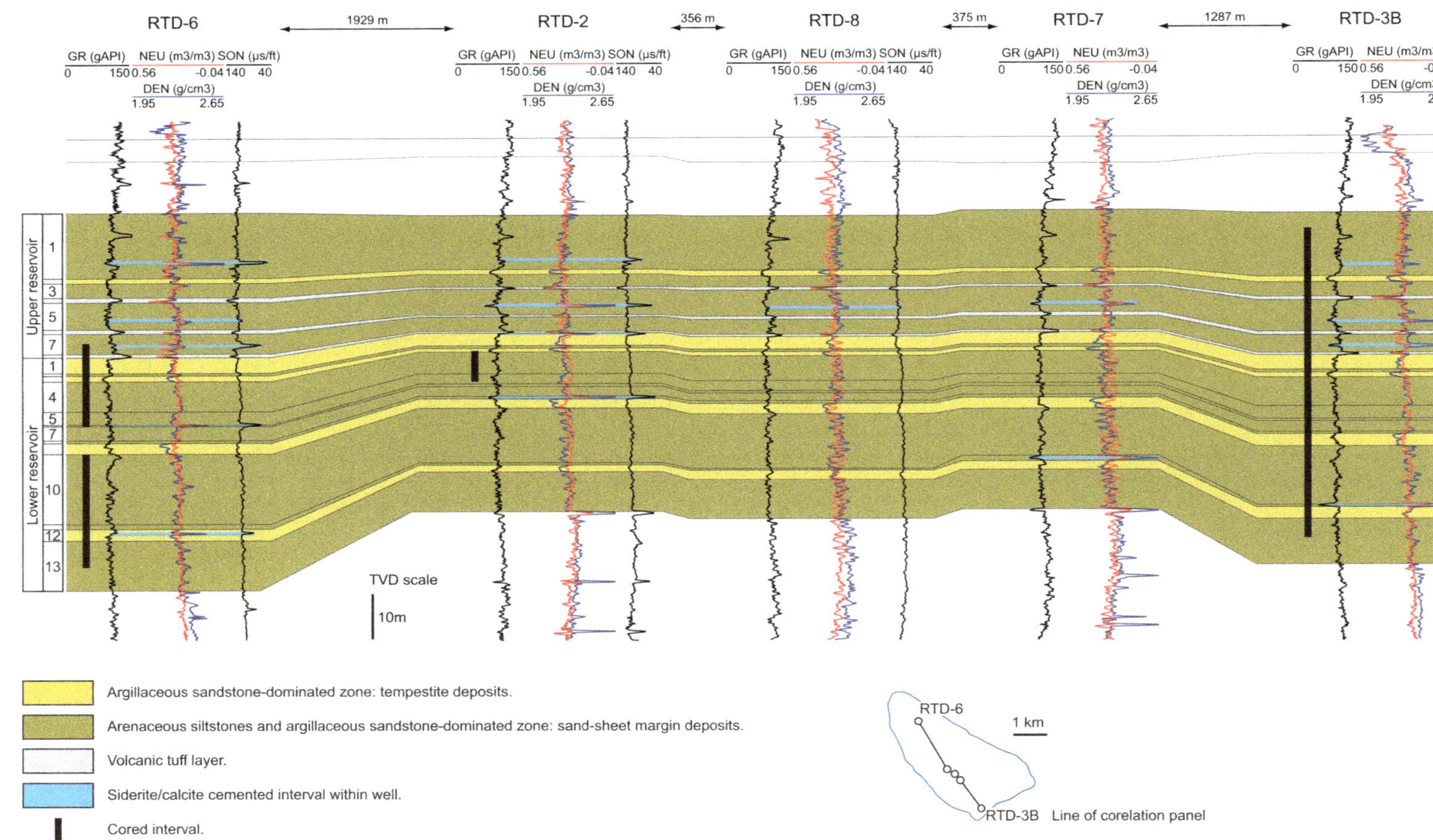

Fig. 9. Correlation panel of the Rotterdam Field Holland Greensand reservoir. The correlation of well log responses after calibration to core shows that facies are laterally continuous across the field. Argillaceous sandstone-dominated reservoir zones with better reservoir properties are layered between zones of poorer-quality rock. Cement bodies occur intermittently along broadly horizontal stratigraphic surfaces. Three tuffaceous units are the only persistent field-wide reservoir marker beds. GR, gamma-ray log; NEU, neutron log; DEN, bulk density log; SON, sonic log.

(Fig. 9). Cement bodies identified in Rotterdam can be expected to occur along broadly horizontal stratigraphic surfaces, and their lateral extent could range from 5–150 cm (individual and coalesced nodules) to over 400 m (discontinuous tabular beds).

Modelling

One static reservoir model was prepared for the Holland Greensand using Schlumberger's Petrel software suite (2011), based on the latest (2011) reprocessed seismic data over the Rotterdam Field. During seismic interpretation, 18 key faults were identified and incorporated into the model. To simplify gridding, the vertical area of interest was limited to 25 m above and below the Holland Greensand reservoir. The grid, which uses a 50 m × 50 m cell size, was orientated NW–SE and SW–NE. This followed the anticipated anisotropy for waterflood testing. Since the fault model was relatively simple, gridding did not require the faults to be oriented and the application of additional trends was not required. A simple zonation and layering scheme was applied during modelling that reflected the correlation (Fig. 9); all zones were proportionally layered. Facies modelling for net and non-net reservoir (cement nodules and tuff horizons) employed a combination of assigning values and stochastic simulation. 'Assign values' was used to ensure capture of field-wide tuff horizons, while 'SIS (GSLIB)' (Sequential Indicator Simulation, Geostatistical Software Library) was used to capture the laterally intermittent carbonate nodules. Vertical variogram analysis was carried out on the data available from 20 well penetrations, which indicated a vertical range of 1.6 m. However, based on the relatively low density of lateral data points, horizontal variogram analysis was inconclusive. Based on analogue comparison with the Foxmould Member a number of lateral cement ranges were deterministically modelled, capturing the geological observations in the Rotterdam Field core and in proportions that honour the well data. These models were as follows: no cements; 50 m (model grid size); 500 m (slightly more than average well spacing); and 5000 m (field scale). Petrophysical property modelling (porosity and permeability) was undertaken using the Sequential Gaussian Simulation (Ringrose & Bentley 2015).

Based on the previously described geological conceptual model and static reservoir model, a dynamic simulation model was built using Shell's Modular Reservoir Simulator (MoReS), a proprietary tool for 3D simulation (Zijlstra *et al.* 2007). Shell's Stochastic Uncertainty Manager tool (SUM), was used to assist in matching the observed production and injection history to the simulation model. The history match showed that the injected water in the lower Holland Greensand reservoir displaces oil through a subset of the multiple sandstone layers in the reservoir (Fig. 10). Given the similar porosity characteristics of the sandstones, the amount of oil and water displaced in those sandstones is controlled by the permeability contrast between layers.

Field surveillance: production logging

To support the simulation results and the integrated reservoir analysis, several surveillance activities were planned from producing wells in the Holland Greensand reservoir. The aim was to identify, if possible, the reservoir zones contributing oil and water production because of the displacement process. Prior to this study, several injection wells had been logged but no oil producer well had been subjected to a Production Logging Test (PLT) evaluation, an industry standard technique for waterflood monitoring (Hill 1990). Although several wells were evaluated, the results obtained from RTD-14 provided the greatest insight into reservoir fluid flow (Fig. 11). This multilateral well intersects the Holland Greensand Member at different stratigraphic heights within the reservoir, achieved by drilling at two different depths from a single, main borehole. The displacement process of this well is supported by two water-injector wells, RTD-5 and RTD-17, with most of the pressure and fluid displacement provided by RTD-17 (Fig. 11). One of the laterals of RTD-14, referred to as RTD-14C, has been completed over an interval that drains a higher section of the reservoir. This lateral is approximately perpendicular to a straight line connecting RTD-5 and RTD-17. The second lateral, RTD-14D, has been completed over a lower reservoir section, and drains a different layer from RTD-14C. RTD-14D runs parallel to a straight line connecting the two injectors.

The main results of the surveillance activity can be summarized as follows. Above the point where the two laterals join (Fig. 12), the total production of the well may be observed. Here the contribution of both laterals gives a comingled mixture of water, oil and small quantities of gas. These were observed flowing at a stable rate towards the surface. At the point of intersection of the two laterals, a 'high' fluid influx was observed during the PLT. In RTD-14C, below the point where the laterals intersect and where only the production from the tempestite-dominated uppermost zones of the lower reservoir can be observed, the surveillance showed close to 100% water; no fluid movement was registered by the components of the logging tool at the well operating conditions during the test.

These observations suggest that RTD-14C, placed at the top of the reservoir, was flooded with water. Most, if not all, of the oil contribution observed

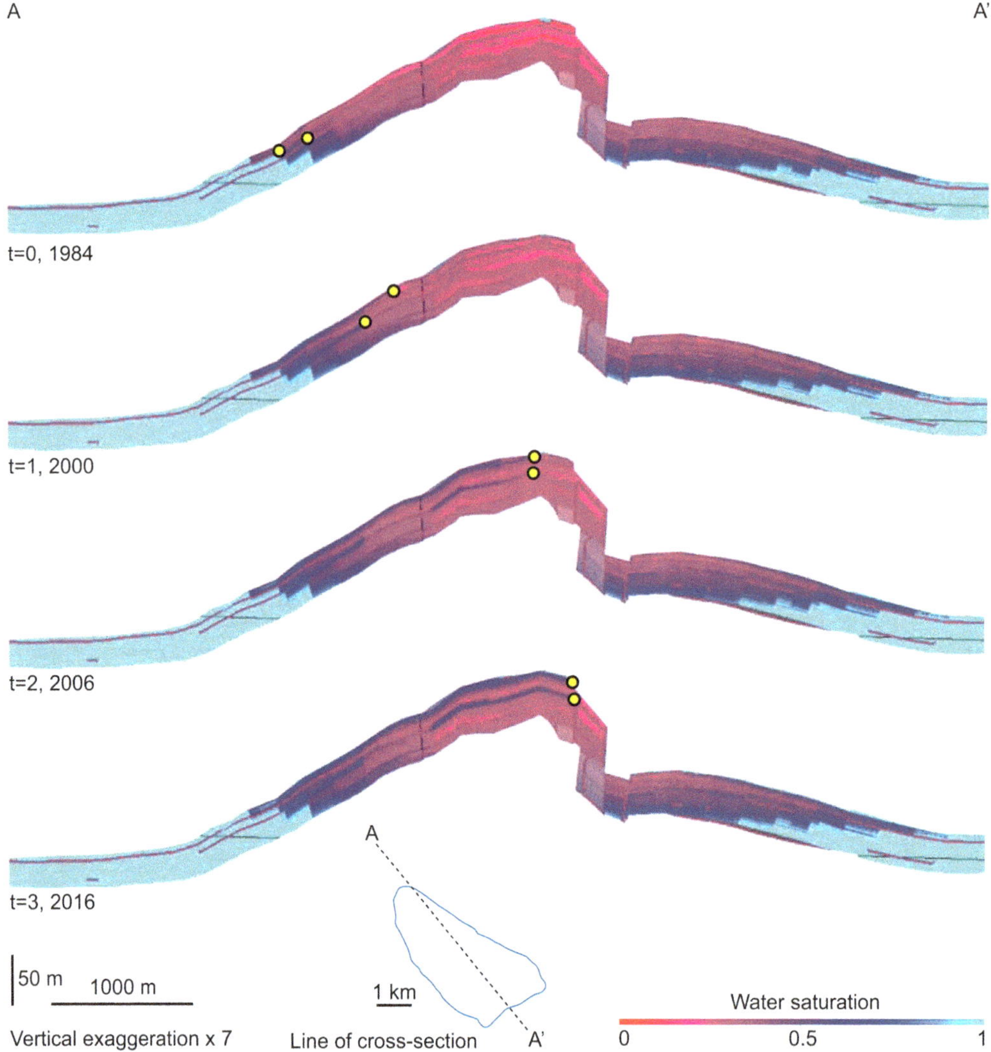

Fig. 10. Time steps through the dynamic simulation saturation model of the lower Holland Greensand reservoir. Blue colours indicate water saturation from the start of production ($t = 0$) to the end of simulation ($t = 3$). Pink colours indicate oil. Preferential water movement occurs along layers with higher permeability. This progressive movement towards the reservoir crest is highlighted by the yellow dots.

comes from RTD-14D, the lateral placed in the lower, distal sand-sheet-dominated portion of the reservoir. This is supported by the outcome of a well stimulation undertaken during the PLT on RTD-14C. This resulted in an increased volume of water entering the lateral after the well was returned to normal operating conditions. Based on the observed oil and water properties, specifically density, it is not expected that water should lie above oil (due to buoyancy effects), unless reservoir properties are such that heterogeneous fluid movement is favoured. This is compounded by the likely direction of the flood front movement provided by the RTD-17 water injection well.

This interpretation is also supported by the known injection profile observed in RTD-17 during a previously acquired PLT (Fig. 13). The profile indicates that most of the injected water enters the lower Holland Greensand reservoir. Even though the well is perforated so that the entire column is opened to injection, water is preferentially injected into reservoir zones 1, 3 and 12, representing storm deposits (Fig. 13). As shown in Figure 12, zones 1 and 3 are also intersected by RTD-14C. This is

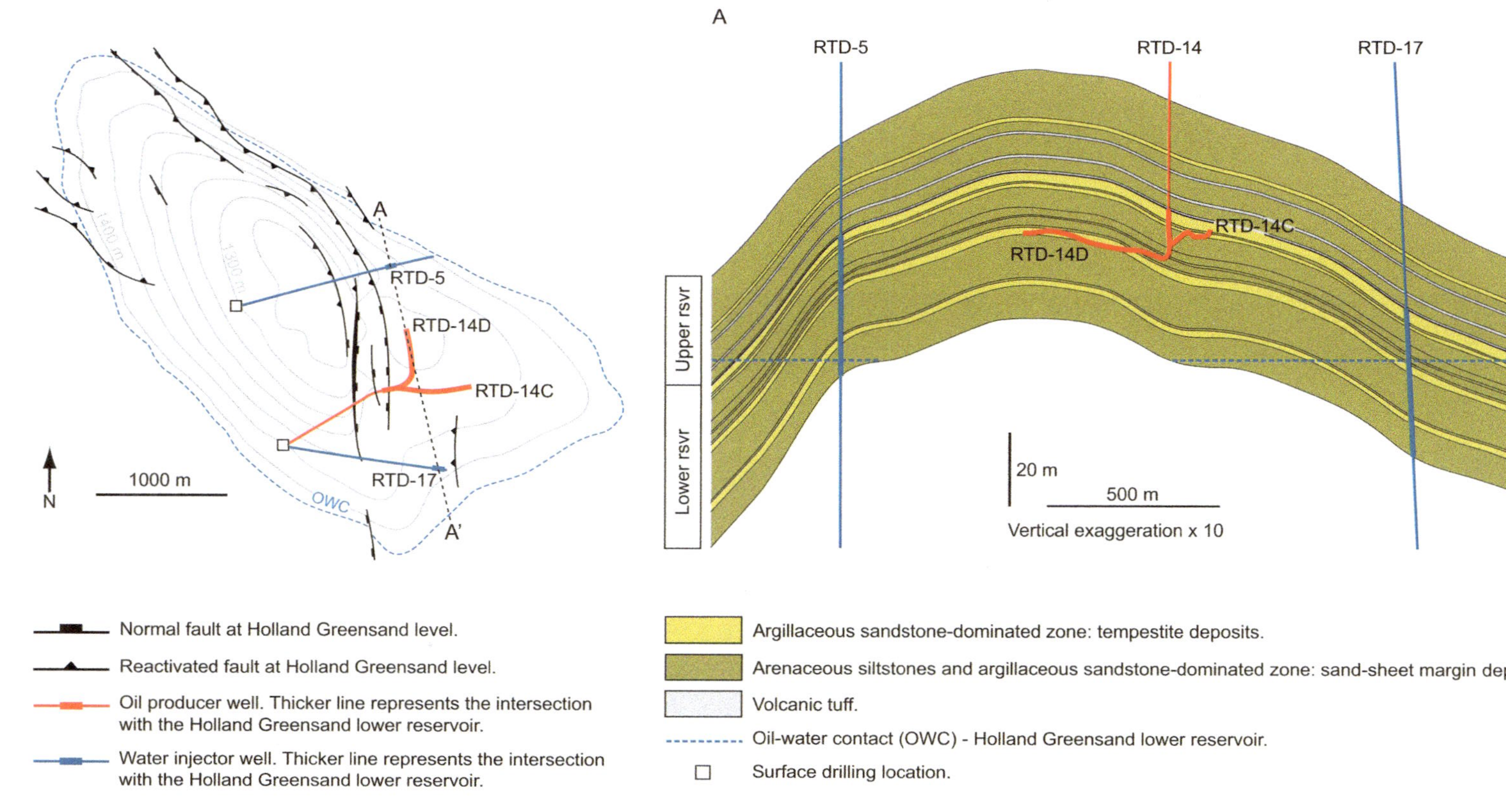

Fig. 11. Holland Greensand lower reservoir top surface map and cross-section with reservoir zones, showing wells RTD-5, RTD-17, RTD-14C and RTD-14D. Note how the two laterals of RTD-14 are placed in different reservoir zones in the upper half of the lower reservoir.

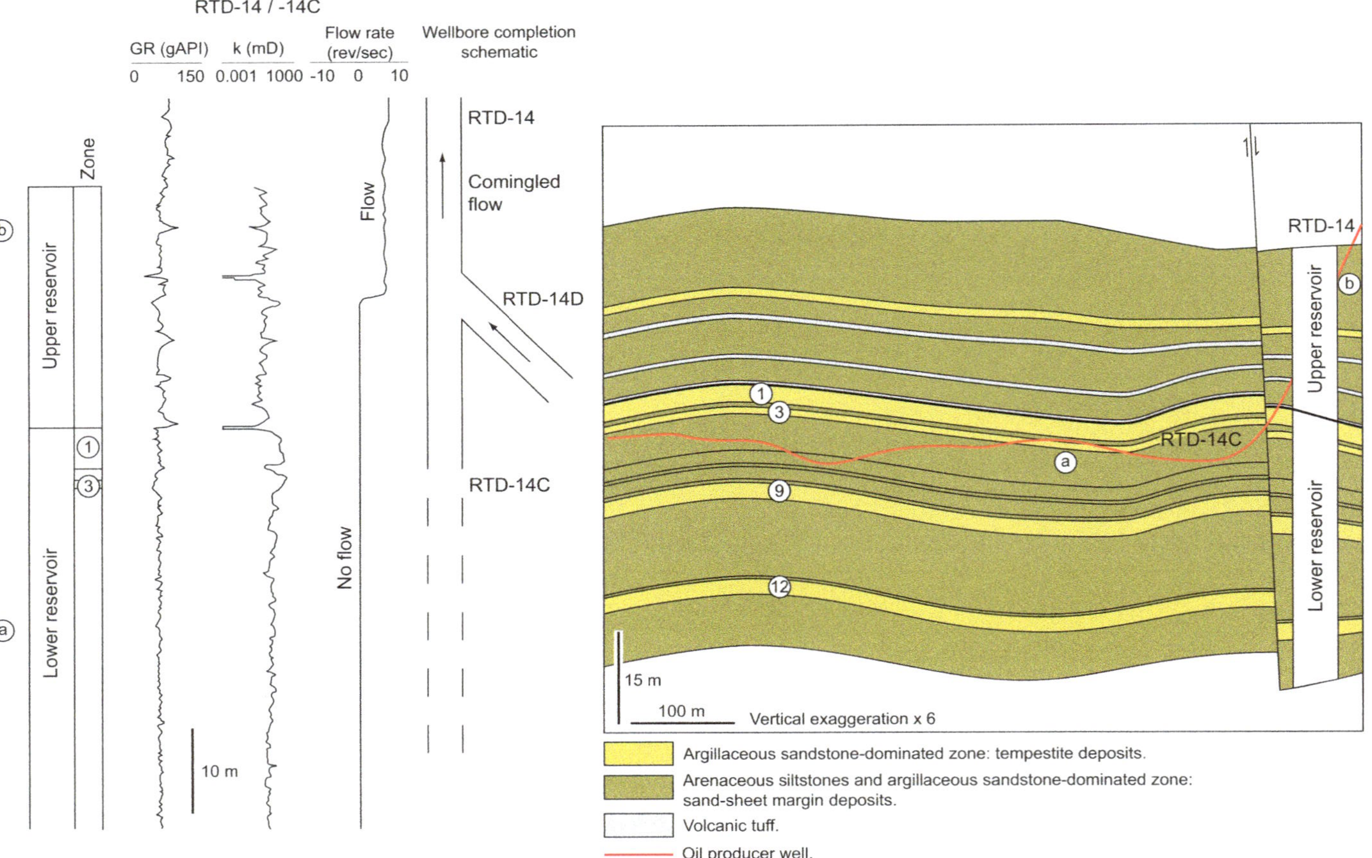

Fig. 12. RTD-14 production logging test results (left) and RTD-14C reservoir well path diagram (right). A steady flow of mixed water, oil and gas was observed at 'b' during the test. This is above the point of intersection of RTD-14C and RTD-14D. A high fluid influx was noted where the two laterals join. Below the point of intersection, no flow and only water was recorded. RTD-14C is placed in the upper part of the lower reservoir and intersects reservoir zones 1 and 3 at 'a'. These units represent tempestite deposits and exhibit better reservoir properties than the surrounding rock.

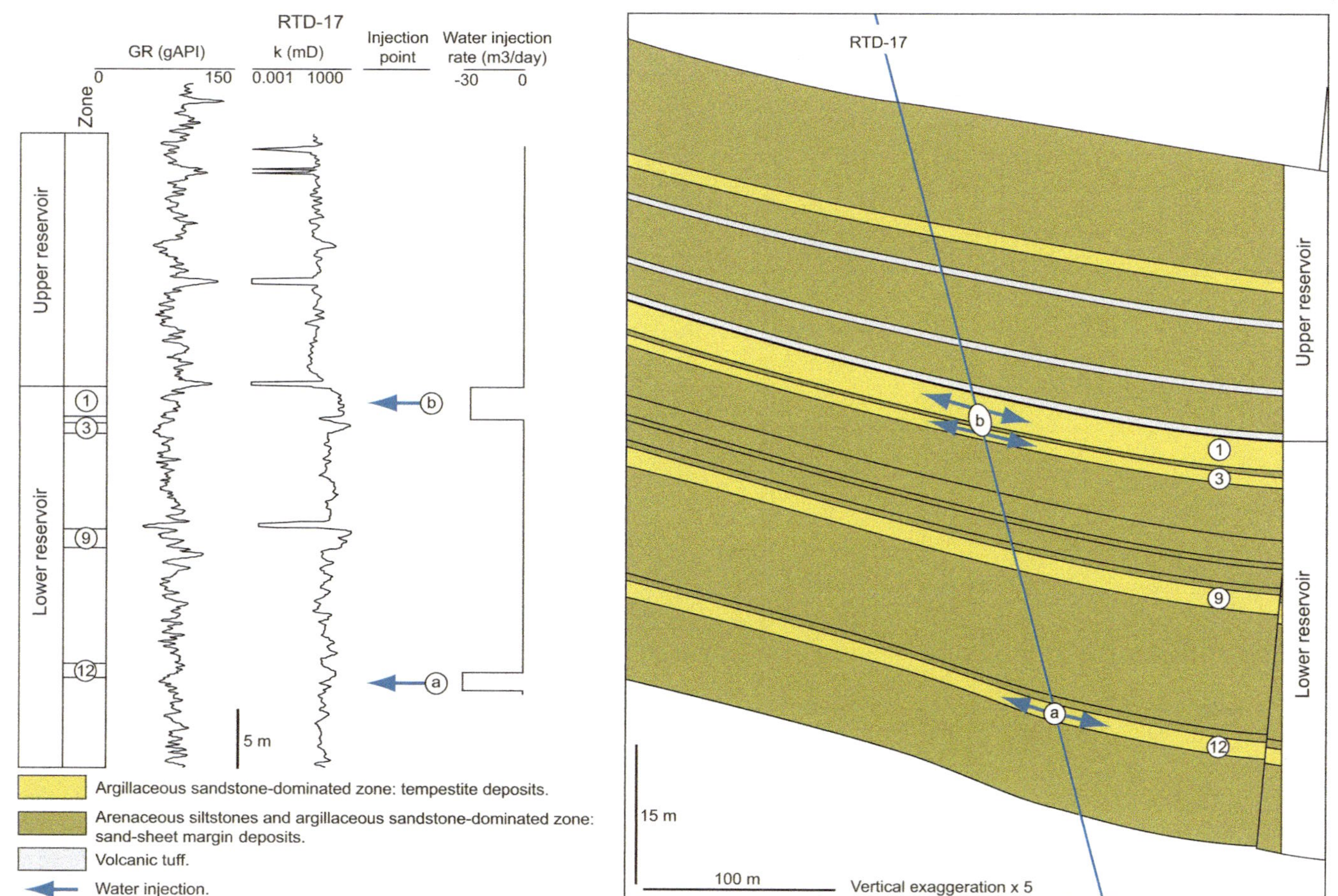

Fig. 13. RTD-17 water injection profile observed during a production logging test (left) and reservoir well path diagram (right). Water is preferentially injected into reservoir zones 1, 3 (at 'b') and 12 (at 'a'). These zones represent laterally continuous storm deposits with associated better rock properties compared to the zones that surround them. Zones 1 and 3 are intersected up-dip by the flooded RTD-14C oil production well (Figs 11, 12). This is consistent with the vertically heterogenous, yet laterally extensive, reservoir prediction.

consistent with the observation that this lateral has been flooded, since most of the water has been injected into those reservoir zones. Since no mechanical obstructions are present, as demonstrated by injection into zone 12, the dominant controlling mechanism for the injected water is the reservoir geology. The presumed connectivity between injection and production wells therefore provides evidence to explain the surveillance observation. The oil within the higher-permeability zones 1 and 3, has been displaced in preference to the reservoir zones dominated by sand-sheet margin deposits. This has occurred to the extent that the flood front has passed the upper lateral, leaving water and residual oil in the pore volume.

In contrast RTD-14D, which is nearly parallel to the direction of the flood (Fig. 11), is still contributing both oil and water from the tempestite-dominated reservoir layers intersected by the nearly-horizontal wellbore of the lateral. This is because very small volumes of water can directly displace oil found at that reservoir level, as shown in the RTD-17 PLT results. Depending on the level of vertical baffling, expressed by the vertical permeability of the sand-sheet margin deposits (or if minor fractures are present), some water will 'drain' by gravity towards the basal zones contributing to the displacement of some of the oil, albeit inefficiently.

The observed movement of fluids is somewhat opposed to that expected of a 'standard' waterflood performance. In such cases, the reservoir is often effectively homogeneous, with good vertical fluid communication existing between reservoir layers. It is therefore anticipated that a bottom-up sweep will occur, first flooding the lower sections of the reservoir as the oil is 'pushed' towards the top of the structure. The remaining oil would consequently be contained in the uppermost intervals of the reservoir. This study shows that underestimating the impact of reservoir heterogeneity can lead to overlooked performance outcomes in a waterflood design. If not properly identified in a timely manner, this can lead to unsatisfactory reservoir recoveries. By using an integrated approach to reservoir analysis, combining reservoir geology, dynamic simulation and surveillance, evidence has been gathered to provide a credible explanation for the reservoir performance of the lower Holland Greensand reservoir in the Rotterdam Field.

Discussion

The exposures of the Upper Greensand at Under Hooken can be used to gain insight into the behaviour of existing hydrocarbon reservoirs, deposited in similar palaeoenvironmental settings. Although varyingly argillaceous on a bed-by-bed basis, the sandstone-dominated nature of the Upper Greensand suggests that reservoirs consisting of similar lithofacies would be fluid-flow connected both vertically and laterally on geological timescales (e.g. more than 10 ka). However, on timescales typical of producing hydrocarbon fields (e.g. 30–40 years), it is envisaged that these subtle vertical changes in clay and silt content are likely to baffle vertical flow. The cleaner sandstones, particularly the laterally extensive tempestites, are likely to provide preferred pathways to horizontal flow. The potential impact of concretions in shallow-marine depositional systems and their vertical flow-inhibiting properties has been well documented (e.g. Ainsworth 2010; Morad *et al.* 2010). The outcrops of the Upper Greensand suggest that where storm-induced erosion occurs, available shell material is swept into the resulting scours. Subsequent diagenetic processes appear to result in concretion formation along these erosion surfaces. This can produce localized concretions which, if present in a subsurface reservoir, would have little impact on vertical fluid flow. This is contrary to the coalesced concretions and tabular cemented sheets observed at other levels in the Upper Greensand, which would result in a vertically compartmentalized reservoir.

In the case of the Holland Greensand reservoir of the Rotterdam Field, it is believed that while the reservoir is likely to be in full lateral and vertical communication, the better-quality, laterally extensive bioturbated tempestite deposits are expected to provide preferred conduits to fluid flow. Correlatable silty sandstone beds of inner-shelf to marginal sand-sheet origin, associated with poorer rock qualities, are envisaged to baffle vertical flow, therefore effectively reinforcing the cleaner sandstones as flow conduits. Cemented intervals seen in the cores and logs, inferred from borehole image logs to be more nodular, are not expected to impact flow despite occurring at similar stratigraphic levels throughout the field.

The reservoir geology has therefore had an important impact on hydrocarbon recovery to date. It is also a central component to future production plans. A potential consequence of the current production-injector well configuration could be that oil, in the reservoir to the east of lateral RTD-14D, has been displaced in both the better-quality tempestite-dominated deposits and poorer-quality distal facies. This oil would effectively be left trapped with a low possibility of recovery. One method of recovering this hydrocarbon volume, and using the existing wells and associated flooding pattern, is to manage production and injection rates into the two main facies types. Another strategy is to consider new producer wells, targeting those areas where oil has been displaced, as predicted by integrated reservoir modelling. Given the understanding of the reservoir geology and the observed field performance, an option would be to drill slanted wells that intersect the entire column. This would ensure the optimal connection of better- and poorer-quality facies. A third approach is to drill horizontal wells placed in tempestite-dominated intervals. Once these are flooded however, due to the poor vertical flow communication in the reservoir, such wells will have a lower chance of draining the surrounding layers with significantly lower rock properties.

To control the levels of possible water production from the storm-dominated units, special well completions could be implemented. This would allow the well drawdown to be distributed, favouring production from oil-bearing zones. For example, Inflow Control Devices (ICDs) (Al-Khelaiwi & Davies 2007) could be spaced along the slant section of the well. Additionally, if the ICDs are equipped with sliding sleeves, this would allow inflow profiles to be 'dynamically' modified over the production history of the well. Already flooded compartments could therefore be shut-off to favour higher oil production. It is also expected, due to the presently observed reservoir permeability contrast, that different layers will be flooded at different times. An active completion policy would allow periodic surveillance runs and provide options to open and close specific compartments.

The experience gained from integrated outcrop and subsurface analysis of the Rotterdam Field can be applied to mature fields and new opportunities with similar heterogenous reservoirs. These include other Holland Greensand Member reservoirs within the West Netherlands Basin (Jeremiah *et al.* 2010) and the Upper Jurassic–Lower Cretaceous reservoirs of the Dutch offshore (Munsterman *et al.* 2012).

Future work

As part of the continuous refinement to understanding and predicting fluid flow in the Holland Greensand reservoir of the Rotterdam Field, several recommendations are made. The inclusion of quantitative seismic interpretation (e.g. forward-modelled reservoir quality estimates from seismic data) in the static geological model could provide enhanced reservoir-wide porosity prediction (Nair *et al.* 2012). Results of additional surveillance performed on this field should be included into the analysis. For instance, the results of tracers added to the water injected into the reservoir, unavailable at the time of study, should be incorporated into future reservoir modelling efforts, together with the results of saturation logs acquired in a subset of the wells. Such logging may reveal additional dynamically important flowing units. In both cases, such surveillance data should provide further understanding of the vertical and areal connectivity of the reservoir, while also providing an indication of the horizontal and vertical sweep efficiency of the waterflood. With such insights, optimization of the horizontal layering scheme in both static and dynamic modelling realms would ensure further improvement to the prediction of fluid pathways, as well as the realistic options of additional development wells for increased recovery.

Conclusions

An integrated reservoir evaluation, including the application of observations derived from analogous exposures of the Upper Greensand Formation in SW England, has provided an improved understanding of the reservoir architecture and geological controls on fluid flow in the Holland Greensand reservoir of the Rotterdam Field.

- Argillaceous sandstones are interpreted as reflecting deposition at the distal margins of a subtidal sand-sheet. Sandstones with lower clay content and therefore better rock properties are related to the clastic input from periodic storms. Cemented intervals, occurring at the base of the better-quality sandstones, are positioned above moderately scoured surfaces.
- Log signatures may be assigned to relatively thin, field-wide, laterally extensive lithofacies packages. These are related to the hydraulic conditions under which they were deposited. Cemented intervals, observable in wells, are seen to occur at similar stratigraphic levels. These are believed to represent laterally restricted nodules rather than extensive sheets.
- The heterogeneous, layered nature of the reservoir, although likely in vertical communication over geological timescales, appears to act as if vertically baffled on hydrocarbon production timescales. Injected fluids are believed to preferentially follow flow paths through tempestites, leaving the surrounding, poorer-quality reservoir less well swept.

The prediction of preferential flow paths and their recognition through reservoir surveillance can have large implications for future reservoir management and redevelopment, including well type, configuration and completion strategies. Integrated geological evaluation, including the use of outcrop analogues, remains crucial in recognizing, understanding and predicting reservoir architecture and the subseismic heterogeneities that impact subsurface fluid flow. The application of methodologies employed in this study is therefore recommended to unlock the remaining energy reserves in similar waterflood developments.

We thank our colleagues for their technical input and contributions to illustrations, especially Nicki Gardner,

Mark Schenkel, Leslie Kramers and Erik Prenger. Ramues Gallois' informed assistance in the field to one of the authors (RJP) is gratefully acknowledged. Ben Kilhams and Tom McKie are thanked for their critical feedback on earlier drafts of this paper. The manuscript was improved by the insightful reviews of Ramues Gallois and Jason Jeremiah. NAM B.V. are thanked for their permission to publish this material.

References

AINSWORTH, R.B. 2010. Prediction of stratigraphic compartmentalization in marginal marine reservoirs. *In*: JOLLEY, S.J., FISHER, Q.J., AINSWORTH, R.B., VROLIJK, P.J. & DELISLE, S. (eds) *Reservoir Compartmentalization*. Geological Society, London, Special Publications, **347**, 199–218, https://doi.org/10.1144/SP347.12

AL-KHELAIWI, F.T. & DAVIES, D.R. 2007. Inflow Control Devices: application and value quantification of a developing technology. Society of Petroleum Engineers. Paper SPE 108700. Presented at the International Oil Conference and Exhibition in Mexico, 27–30 June 2007, Veracruz, Mexico.

BANIAK, G.M., GINGRAS, M.K., BURNS, B.A. & PEMBERTON, S.G. 2014. An example of a highly bioturbated, storm-influenced shoreface deposit: Upper Jurassic Ula Formation, Norwegian North Sea. *Sedimentology*, **61**, 1261–1285.

BUATOIS, L.A. & MAŃGANO, M.G. 2011. *Ichnology: Organism-Substrate Interactions in Space and Time*. Cambridge University Press, Cambridge.

BURNS, F.E. 2002. Trace fossils as tools in glauconitic reservoirs: examples from the Lower Cretaceous of the Carnarvon Basin, North West Shelf. *In*: KEEP, M. & MOSS, S.J. (eds) *The Sedimentary Basins of Western Australia 3. Proceedings of the Petroleum Exploration Society of Australia Symposium*. Petroleum Exploration Society of Australia, Perth, Australia, 203–222.

DAKE, L.P. 1978. *Developments in Petroleum Science, Volume 8: Fundamentals of Reservoir Engineering*. Elsevier, Amsterdam.

DALRYMPLE, R.W. & CHOI, K. 2007. Morphologic facies trends through the fluvial-marine transition in tide-dominated depositional systems: a schematic framework for environmental and sequencestratigraphic interpretation. *Earth-Science Reviews*, **81**, 135–174.

DALRYMPLE, R.W., BAKER, E.K., HARRIS, P.T. & HUGHES, M. 2003. Sedimentology and stratigraphy of a tide dominated foreland-basin delta (Fly River, Papua New Guinea). *In*: SIDI, F.H., NUMMEDAL, D., IMBERT, P., DARMAN, H. & POSAMENTIER, H. (eds) *Tropical Deltas of Southeast Asia – Sedimentology, Stratigraphy and Petroleum Geology*. SEPM, Tulsa, Oklahoma, USA, Special Publications, **76**, 147–173.

DE JAGER, J. 2007. Geological development. *In*: WONG, T.E., BATJES, D.A.J. & DE JAGER, J. (eds) *Geology of the Netherlands*. Royal Netherlands Academy of Arts and Sciences, Amsterdam, the Netherlands, 5–26.

DE JAGER, J., DOYLE, M.A., GRANTHAM, P.J. & MABILLARD, J.E. 1996. Hydrocarbon habitat of the West Netherlands Basin. *In*: RONDEEL, H.E., BATJES, D.A.J. & NIEUWENHUIJS, W.H. (eds) *Geology of Gas and Oil under the Netherlands*. Kluwer Academic Publishers, Dordrecht, the Netherlands, 191–209.

DESJARDINS, P.R., MAŃGANO, M.G., BUATOIS, L.A. & PRATT, B.R. 2010. *Skolithos* pipe rock and associated ichnofabrics from the southern Rocky Mountains, Canada: colonisation trends and environmental controls in an Early Cambrian sand-sheet complex. *Lethaia*, **43**, 507–528.

DESJARDINS, P.R., BUATOIS, L.A. & MÁNGANO, M.G. 2012*a*. Tidal flats and subtidal sandbodies. *In*: KNAUST, D. & BROMLEY, R.G. (eds) *Trace Fossils as Indicators of Sedimentary Environments*. Amsterdam, Elsevier, *Developments in Sedimentology*, **64**, 529–562.

DESJARDINS, P.R., BUATOIS, L.A., PRATT, B.R. & MÁNGANO, M.G. 2012*b*. Sedimentological–ichnological model for tide-dominated shelf sandbodies: Lower Cambrian Gog Group of western Canada. *Sedimentology*, **59**, 1452–1477.

DUTTON, S.P., WHITE, C.D., WILLIS, B.J. & NOVAKOVIC, D. 2002. Calcite cement distribution and its effect on fluid flow in a deltaic sandstone, Frontier Formation, Wyoming. *American Association of Petroleum Geologists Bulletin*, **86**, 2007–2021.

GALLOIS, R.W. 2004. The stratigraphy of the Upper Greensand (Cretaceous) of south-west England. *Geoscience in South-West England*, **11**, 21–29.

GALLOIS, R.W. & GOLDRING, R. 2007. Trace fossils at the basal Upper Greensand (Albian, Cretaceous) unconformity surface in east Devon (south-west England) and the nature of the unconformity surface. *Proceedings of the Geologists' Association*, **118**, 265–275.

HAMPSON, G.J., MORRIS, J.E. & JOHNSON, H.D. 2015. Synthesis of time-stratigraphic relationships and their impact on hydrocarbon reservoir distribution and performance, Bridport Sand Formation, Wessex Basin, UK. *In*: SMITH, D.G., BAILEY, R.J., BURGESS, P.M. & FRASER, A.J. (eds) *Strata and Time: Probing the Gaps in Our Understanding*. Geological Society, London, Special Publications, **404**, 199–222, https://doi.org/10.1144/SP404.2

HART, M.B. & WILLIAMS, C.L. 1990. The Upper Greensand in East Devon: new data but old problems. *Proceedings of the Ussher Society*, **7**, 273–278.

HILL, A.D. 1990. *Production Logging: Theoretical and Interpretive Elements*. SPE, Monograph Series, **14**, Society of Petroleum Engineers, Richardson, Texas, USA.

HOCKING, R.M., VOON, J.W.K. & COLLINS, L.B. 1988. Stratigraphy and sedimentology of the Basal Winning Group, Northern Carnarvon Basin. *In*: PURCELL, P.G. & PURCELL, R.R. (eds) *The North West Shelf, Australia: Proceedings of the North West Shelf Symposium, Perth, Western Australia, 1988*. Petroleum Exploration Society of Australia, Perth, Australia, 203–224.

JEREMIAH, J.M. 2000. Lower Cretaceous turbidites of the Moray Firth: sequence stratigraphical framework and reservoir distribution. *Petroleum Geoscience*, **6**, 309–328, https://doi.org/10.1144/petgeo.6.4.309

JEREMIAH, J.M., DUXBURY, S. & RAWSON, P. 2010. Lower Cretaceous of the southern North Sea Basins: reservoir distribution within a sequence stratigraphic framework. *Netherlands Journal of Geosciences – Geologie en Mijnbouw*, **89**, 203–237.

LEVA LÓPEZ, J., ROSSI, V.M., OLARIU, C. & STEEL, R.J. 2016. Architecture and recognition criteria of ancient shelf ridges; an example from Campanian Almond

Formation in Hanna Basin, USA. *Sedimentology*, **63**, 1651–1676.

LUIJENDIJK, E., VAN BALEN, R.T., TER VOORDE, M. & ANDRIESSEN, P.A.M. 2011. Reconstructing the Late Cretaceous inversion of the Roer Valley Graben (southern Netherlands) using a new model that integrates burial and provenance history with fission track thermochronology. *Journal of Geophysical Research: Solid Earth*, **116**, 2156–2202.

MCBRIDE, E.F., MILLIKEN, K.L., CAVAZZA, W., CIBIN, U., FONTANA, D., PICARD, M.D. & ZUFFA, G.G. 1995. Heterogeneous distribution of calcite cement at outcrop scale in Tertiary sandstones, northern Appenines, Italy. *American Association of Petroleum Geologists Bulletin*, **79**, 1044–1063.

MOLENAAR, N. 1998. Origin of low-permeability calcitecemented lenses in shallow marine sandstones and $CaCO_3$ cementation mechanisms: an example from the Lower Jurassic Luxemburg Sandstone, Luxemburg. *In*: MORAD, S. (ed.) *Carbonate Cementation in Sandstones; Distribution Patterns and Geochemical Evolution*. International Association of Sedimentologists, Blackwell Publishing Ltd, Oxford, Special Publications, **26**, 193–211.

MORAD, S., AL-RAMADAN, K., KETZER, J.M. & DE ROS, L.F. 2010. The impact of diagenesis on the heterogeneity of sandstone reservoirs: a review of the role of depositional facies and sequence stratigraphy. *AAPG* Bulletin, **94**, 1267–1309.

MUNSTERMAN, D.K., VERREUSSEL, R.M.C.H., MIJNLIEFF, H.F., WITMANS, N., KERSTHOLT-BOEGEHOLD, S. & ABBINK, O.A. 2012. Revision and update of the Callovian-Ryazanian Stratigraphic Nomenclature in the northern Dutch offshore, i.e. Central Graben Subgroup and Scruff Group. *Netherlands Journal of Geosciences – Geologie en Mijnbouw*, **91**, 555–590.

NAIR, K.N., KOLBJØRNSEN, O. & SKORSTAD, A. 2012. Seismic inversion and its applications in reservoir characterization. *First Break*, **30**, 83–86.

OLARIU, C., STEEL, R.J., DALRYMPLE, R.W. & GINGRAS, M.K. 2012. Tidal dunes v. tidal bars: the sedimentological and architectural characteristics of compound dunes in a tidal seaway, the lower Baronia Sandstone (Lower Eocene), Ager Basin, Spain. *Sedimentary Geology*, **279**, 134–155.

RACERO-BAENA, A. & DRAKE, S.J. 1996. Structural style and reservoir development of the West Netherlands Basin. *In*: RONDEEL, H.E., BATJES, D.A.J. & NIEUWENHUIJS, W.H. (eds) *Geology of Gas and Oil under the Netherlands*. Kluwer Academic Publishers, Dordrecht, The Netherlands, 211–227.

RINGROSE, P. & BENTLEY, M. 2015. *Reservoir Model Design*. Springer, Houten, The Netherlands.

SATTER, A., IQBAL, G.M. & BUCHWALTER, J.L. 2008. *Practical Enhanced Reservoir Engineering: Assisted with Simulation Software*. PennWell Books, Tulsa.

STRIDE, A.H., BELDERSON, R.H., KENYON, N.H. & JOHNSON, M.A. 1982. Offshore tidal deposits: sand sheet and sand bank facies. *In*: STRIDE, A.H. (ed.) *Offshore Tidal Sands: Processes and Deposits*. Chapman & Hall, New York, 95–125.

TAYLOR, A.M. & GOLDRING, R. 1993. Description and analysis of bioturbation and ichnofabric. *Journal of the Geological Society*, **150**, 141–148.

VAN BALEN, R.T., VAN BERGEN, F., DE LEEUW, C., PAGNIER, H., SIMMELINK, H., VAN WEES, J.D. & VERWEIJ, J.M. 2000. Modelling the hydrocarbon generation and migration in the West Netherlands Basin, the Netherlands. *Geologie en Mijnbouw/Netherlands Journal of Geosciences*, **79**, 29–44.

VEJBÆK, O.V., ANDERSEN, C. *ET AL*. 2010. Cretaceous. *In*: DOORNENBAL, J.C. & STEVENSON, A.G. (eds) *Petroleum Geological Atlas of the Southern Permian Basin Area*. EAGE Publications, Houten, 195–209.

VONDRAK, A., DONSELAAR, M. & MUNSTERMAN, D. 2018. Reservoir architecture model of the Nieuwerkerk Formation (Early Cretaceous, West Netherlands Basin): diachronous development of sand-prone fluvial deposits. *In*: KILHAMS, B., KUKLA, P., MAZUR, S., MCKIE, T., MIJNLIEFF, H. & VAN OJIK, K. (eds) *Mesozoic Resource Potential of the Southern Permian Basin*. Geological Society, London, Special Publications, **469**. First published online March 15, 2018, https://doi.org/10.1144/SP469.18

WELLS, M.R., ALLISON, P.A., PIGGOTT, M.D., HAMPSON, G.J., PAIN, C.C. & GORMAN, G.J. 2010. Tidal modeling of an ancient tide-dominated seaway, Part 2: the Aptian Lower Greensand Seaway of Northwest Europe. *Journal of Sedimentary Research*, **80**, 411–439.

WILLIAMS, C.L. 1991. *The sedimentology of the Upper Greensand of Devon*. Unpublished PhD thesis, University of Plymouth.

WINSTANLEY, A.M. 1993. A review of the Triassic play in the Roer Valley Graben, SE onshore Netherlands. *In*: PARKER, J.R. (ed.) *Petroleum Geology of Northwest Europe: Proceedings of the 4th Conference*. The Geological Society, London, Petroleum Geology Conference Series, **4**, 595–607, https://doi.org/10.1144/0040595

ZIJLSTRA, E.B., REEMST, P.H.M. & FISHER, Q.J. 2007. Incorporation of fault properties into production simulation models of Permian reservoirs from the southern North Sea. *In*: JOLLEY, S.J., BARR, D., WALSH, J.J. & KNIPE, R.J. (eds) *Structurally Complex Reservoirs*. Geological Society, London, Special Publications, **292**, 295–308, https://doi.org/10.1144/SP292.17

Lower Cretaceous reservoir development in the North Sea Central Graben, and potential analogue settings in the Southern Permian Basin and South Viking Graben

FRANK ZWAAN

VU University Amsterdam, Faculty of Earth and Life Sciences, De Boelelaan 1085–1087, 1081 HV Amsterdam, The Netherlands

Present address: Institute of Geological Sciences, University of Bern, Baltzerstrasse 1 + 3, CH-3012 Bern, Switzerland

frank.zwaan@geo.unibe.ch

Abstract: Much of the future hydrocarbon exploration potential in the North Sea lies in locating stratigraphic traps and discrete reservoir intervals. This study assesses the potential for Lower Cretaceous reservoirs, with particular focus on the Norwegian Central Graben and methods to identify future prospects over a wider area. Seismic interpretation and well data reveal the structure and sedimentology of the study area. Although the region was isolated from a large hinterland in the Early Cretaceous, potential local sediment sources, sediment transport routes and areas with possible reservoir development are identified. The greater Mandal High area, where Lower Cretaceous shoreface deposits and submarine fan systems are postulated, is suggested for primary focus. Similar deposits may have developed around the other exposed highs in the region, although several were drowned towards the end of the Early Cretaceous. Detailed seismic and stratigraphic analysis will be necessary to identify individual reservoir units. Since comparable settings may have occurred in the adjacent South Viking Graben and Southern Permian Basin regions during the Early Cretaceous, further reservoir assessment is recommended for the North Sea in general.

The Central Graben of the North Sea represents a prolific and mature hydrocarbon province (e.g. Brooks 1990). Despite 50 years of exploration, further potential continues to be unlocked as demonstrated by the Upper Jurassic shoreface play (e.g. the Edvard Grieg and Johan Sverdrup discoveries at the Norwegian Utsira High: Jørstad 2012; Lundin Petroleum 2017; NPD 2017) and by the Upper Jurassic turbidite play in the UK Moray Firth (e.g. the Buzzard Field: Fraser *et al.* 2003; Ray *et al.* 2010). This paper describes further potential in another stratigraphic horizon, namely the Lower Cretaceous reservoirs of the North Sea. Here shoreface and deep-water reservoir units are developed associated with discrete Mesozoic rifting phases and related localized depocentres.

Lower Cretaceous deep-marine sandstones represent an important play in the UK Moray Firth (e.g. Garrett *et al.* 2000; Johnson *et al.* 2005) (Figs 1a & 2), where large amounts of sand from the exposed East Orkney High and Halibut High were shed into adjacent basins of the Inner and Outer Moray Firth, forming the reservoirs for various hydrocarbon fields (e.g. Scapa and Britannia: McGann *et al.* 1991; Ainsworth *et al.* 2000). Various studies have established sediment transport directions (Hailwood & Ding 2000), sediment provenance areas (Blackbourn & Thomson 2000) and a sequence stratigraphic framework (Jeremiah 2000) in this area.

In contrast to the Moray Firth, the Lower Cretaceous play remains underdeveloped in the Central Graben (Copestake *et al.* 2003; NPD 2017). While the generalized Mesozoic sequence is visible on seismic in the Moray Firth, the thick (2000–4500 m) Cenozoic and Upper Cretaceous overburden in the Central Graben presents an issue for seismically driven reservoir identification within the Lower Cretaceous (up to 30 m net reservoir thickness) (Argent *et al.* 2000; Law *et al.* 2000). Further factors impeding seismic assessment include the general low impedance contrast between sandstones and shales in this interval as well as multiples (induced by the base of the overlying Upper Cretaceous chalk units) (Oakman 2005). Although seismic imaging quality has improved considerably in recent years (e.g. Hampson *et al.* 2010), alternative methods are required to assess the Lower Cretaceous units and to locate potential prospects. Attempts to delineate reservoir development in the UK Central Graben (UKCG) have been made with the use of regional 3D seismic and well data (Milton-Worssell *et al.* 2006) (Fig. 1a), indicating a significant potential

From: Kilhams, B., Kukla, P. A., Mazur, S., McKie, T., Mijnlieff, H. F. & van Ojik, K. (eds) 2018. *Mesozoic Resource Potential in the Southern Permian Basin*. Geological Society, London, Special Publications, **469**, 479–504.
First published online January 4, 2018, https://doi.org/10.1144/SP469.3

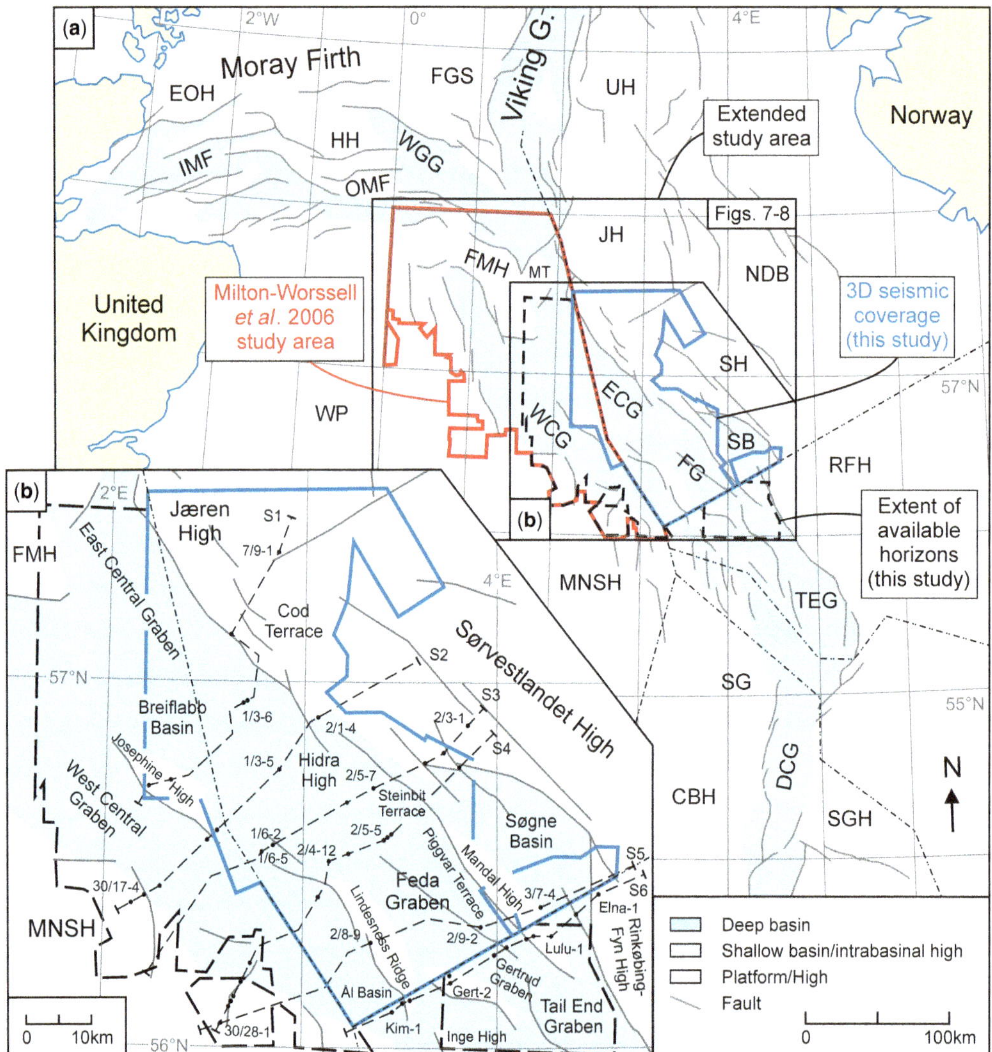

Fig. 1. (**a**) Structural map of the Late Jurassic Central Graben depicting the study area (Norwegian Central Graben area and parts of the UK Central Graben) and the adjacent UK Central Graben study area of Milton-Worssell *et al.* (2006), which in combination define the extended study area. (**b**) Detailed map of the study area, indicating seismic coverage (blue) and the extent of available seismic depth maps (thick dotted outline). Dotted lines indicate interpreted seismic sections S1–S6 (Fig. 5). CBH, Cleaver Bank High; DCG, Dutch Central Graben; EOH, East Orkney High; NDB, Norwegian–Danish Basin; ECG, East Central Graben; FG, Feda Graben; FGS, Fladen Ground Spur; FMH, Forties–Montrose High; HH, Halibut High; IMF, Inner Moray Firth; JH, Jæren High; MNSH, Mid North Sea High; MT, Marnock Terrace; OMF, Outer Moray Firth; RFH, Ringkøbing-Fyn High; SB, Søgne Basin; SH, Sørvestlandet High; SG, Step Graben; SGH, Schill Ground High; TEG, Tail End Graben; UH, Utsira High; WCG, West Central Graben; WGG, With Ground Graben; WP, Western Platform. Modified after Fraser *et al.* (2003), Milton-Worssell *et al.* (2006) and Pharaoh *et al.* (2010).

for Lower Cretaceous sandstone development. Currently, other than local studies (e.g. Rossland *et al.* 2013), no published work has established a similar overview of Lower Cretaceous reservoir potential in the Norwegian sector of the Central Graben.

This study, therefore, aims to assess the potential for Lower Cretaceous sand bodies in the Norwegian Central Graben (NCG), and to link this interpretation to adjacent areas in the UK and to the Danish, German and Dutch parts of the North Sea Rift System

(Southern Permian Basin area), as well as to the South Viking Graben (Fig. 1a).

Geological setting

The geological history of the Central North Sea has generated a diverse stratigraphic record, which is hereby described utilizing a Norwegian stratigraphic nomenclature (Isaksen & Tonstad 1989; NPD 2017) (Fig. 2). The oldest known units in the Norwegian Central Graben are of pre-Permian age; Silurian–Devonian metamorphic basement overlain by Old Red Sandstone (Fossen *et al.* 2008). These units are followed, in stratigraphic order, by varied Carboniferous deposits (NPD 2014), Lower Permian Rotliegend sandstones and thick Upper Permian Zechstein evaporites plus dolomites. The halite sequences of the Zechstein have had a profound influence on the tectonic style; decoupling under- and overlying strata (Hodgson *et al.* 1992; Stewart 2007; Jackson & Lewis 2016; Ge *et al.* 2017; van Winden *et al.* this volume, in press). Lower Triassic Smith Bank shales and Middle–Upper Triassic Skagerrak Sandstone deposition coincided with Late Triassic faulting along the inherited Caledonian structural grain (Bartholomew *et al.* 1993; UKDD 2007). This first rifting phase formed the general Central Graben structure, as these Triassic faults were partially reactivated in the Late Jurassic and Early Cretaceous (Rattey & Hayward 1993), although the main Triassic and Jurassic–Early Cretaceous depocentres do not precisely coincide (Erratt *et al.* 1999). Uplift and erosion due to an Early Jurassic mantle plume formed the Mid-Cimmerian Unconformity in the Central North Sea (Underhill & Partington 1993). Middle Jurassic Bryne Formation coastal plain deposits succeed the hiatus (Bergan *et al.* 1989), followed by a second erosional surface. Eventual dome collapse coincided with the onset of renewed extension in the Northern North Sea (Graversen 2006), propagating into the Central North Sea during Late Jurassic times (Rattey & Hayward 1993). As rifting proceeded, sediment-starved deep-marine basins developed (Copestake *et al.* 2003). This transgression is recorded in the synrift Jurassic Tyne Group by the Ula Formation shoreface sandstones and subsequent deep-marine shales, including the Mandal Formation source rock (Gautier 2005; Nøttvedt & Johannessen 2008).

Major extension ceased towards the end of the Kimmeridgian (Milton 1993), although a secondary phase of rift activity may have continued into the earliest Cretaceous or Ryazanian in the Norwegian area, especially in the NCG (Gowers *et al.* 1993; Sears *et al.* 1993; Zanella *et al.* 2003; Ge *et al.* 2017). Rifting ceased as extensional stresses shifted to the proto-North Atlantic (Coward *et al.* 2003, Oakman 2005). Subsequently, post-rift thermal sag initiated and sediments began to cover the rift topography above the Base Cretaceous Unconformity (BCU) (Rattey & Hayward 1993). The first of these, the Cromer Knoll Group, contains mostly shales but also marly limestones (Tuxen Formation) plus shoreface to deep-marine sandstones (Ran Sandstone units: Isaksen & Tonstad 1989). The sandstones represent potential reservoir bodies but their spatial and temporal extent is poorly constrained, being encountered in only a few wells. In the Aptian, another shift in tectonics and oceanography, the 'Austrian event', occurred. This coincided with the onset of alpine compression and the opening of the North Atlantic, leading to more restricted basins and to the deposition of dark muds of the Sola Formation, overlain by calcareous Rødby Formation sediments (Garrett *et al.* 2000; Copestake *et al.* 2003).

Global sea-level rise and a shift to a tropical climate in the Late Cretaceous saw the development of massive Upper Cretaceous chalk units in the North Sea sag basin (Surlyk *et al.* 2003). Thermal subsidence was, however, interrupted by local inversion pulses and associated Zechstein salt diapirism (Cartwright 1989; Johnson *et al.* 2005; van Winden *et al.* this volume, in press). Renewed sediment input from the eroding North Atlantic rift shoulders gave rise to widespread turbidite systems in the Paleocene and Eocene. From the Oligocene onwards, thermal sag continued, concentrated in the NW of the NCG (Gowers & Sæbøe 1985), while the North Sea Basin gradually filled in with thick clastic sequences.

Established Lower Cretaceous UK reservoirs

Three major sequences of sandstone deposits occur in the Lowermost Cretaceous of the Moray Firth (Jeremiah 2000; Copestake *et al.* 2003) (Fig. 2): Punt sandstones SW of the Halibut Horst, Coracle sandstones south of the East Orkney High and Scapa sandstones east of the Halibut High (locations in Fig. 1). These units were deposited during a phase of low sea level due to tectonic activity related to Austrian compression, ending with a major flooding event in the Barremian (Crittenden *et al.* 1997; Oakman 2005). Rejuvenated tectonic activity associated with the opening of the North Atlantic led to renewed sediment influx in the Aptian (Oakman 2005). During this phase, the Kopervik fairway was established (Law *et al.* 2000) along which large amounts of sand were transported from the East Orkney High to the Outer Moray Firth (Captain and Brittania sandstones, Fig. 2), before an Albian transgression diminished sand influx (Oakman & Partington 1998; Jeremiah 2000).

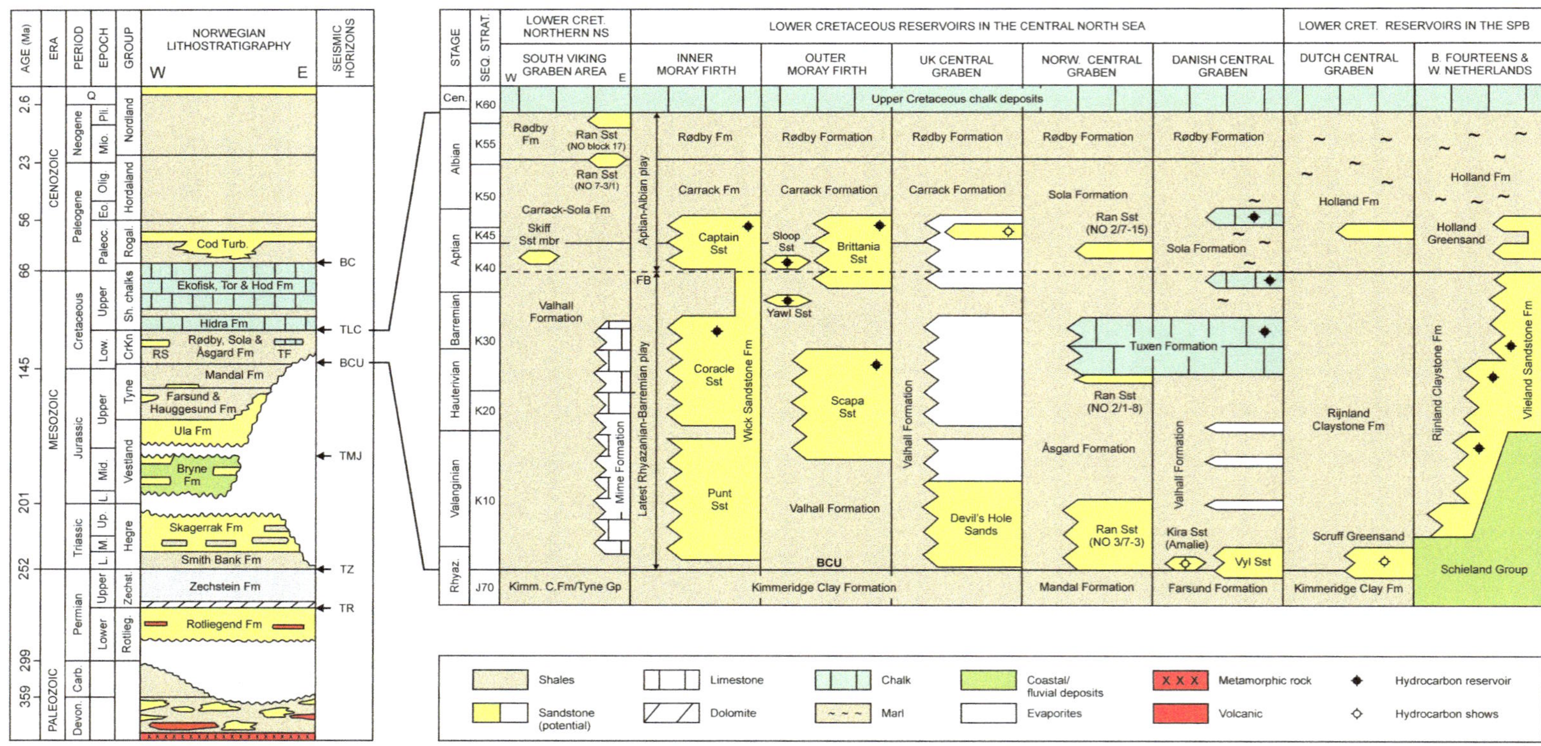

Fig. 2. Norwegian lithostratigraphy for the study area (left) and overview of Lower Cretaceous reservoirs in the Central North Sea (right). CrKn, Cromer Knoll Group; FB, Fischschiefer Bed; NS, North Sea; SPB, Southern Permian Basin. Seismic horizon abbreviations from top to bottom: BC, Base Cenozoic; TLC, Top Lower Cretaceous; BCU, Base Cretaceous Unconformity; TMJ, Top Middle Jurassic; TZ, Top Zechstein; TR, Top Rotliegend. Modified after Vollset & Doré (1984), Van Wijhe (1987), Isaksen & Tonstad (1989), Wong *et al.* (1989), Copestake *et al.* (2003), Jakobsen *et al.* (2005), Milton-Worssell *et al.* (2006), de Jager & Geluk (2007), Herngreen & Wong (2007), UKDD (2007), Wong (2007) and NDD (2012). Geological timescale dates after Walker *et al.* (2012).

These deposits comprise deep-marine sandstones exhibiting a variety of depositional styles including hanging-wall slope-apron fans, linear channel complexes as part of a mini-basin spilling system, or localized mass-flow deposits and mud-dominated slurry-flow deposits (Jones *et al.* 1999; Argent *et al.* 2000). These sedimentary systems demonstrate a high degree of complexity regarding source and transport mechanisms (Eggenhuisen *et al.* 2010). Deposition was strongly influenced by the two Early Cretaceous tectonic phases mentioned above which uplifted and exposed highs and fault scarps, as documented to the north and NW of the main depocentres (e.g. Halibut High and East Orkney High: O'Driscoll *et al.* 1990; Jeremiah 2000; Copestake *et al.* 2003) (Fig. 1a). Tectonic activity furthermore modified the region's bathymetry and redirected sediment transport fairways (Jeremiah 2000; Aas *et al.* 2010). These deep-marine sandstones represent Lower Cretaceous reservoirs in stratigraphic or combination structural–stratigraphic traps in, for instance, the Britannia, Scapa and Captain fields (McGann *et al.* 1991; Jones *et al.* 1999; Pinnock *et al.* 2003).

Although the Forties–Montrose High and Marnock Terrace formed barriers that separated the UKCG depocentres from the Moray Firth during the Early Cretaceous (Fig. 1a), it is possible to extend the Lower Cretaceous Moray Firth reservoir intervals into the UKCG (Milton-Worssell *et al.* 2006) where various wells encounter Lower Cretaceous sands. These well data, in combination with seismically derived maps, allowed Milton-Worssell *et al.* (2006) to postulate the distribution of (mainly deep-marine sandstone) bodies, sourced by the Western Platform, Forties–Montrose High and Jæren High, for both a latest Ryazanian–Barremian play and the Aptian–Albian play in the UKCG. These plays are separated by the Fischschiefer Bed (Fig. 2), an organic-rich mudstone deposited during the Barremian flooding event, that is a regional seismic marker (Ainsworth *et al.* 2000). In this study, a similar division has been made between a 'Latest Ryazanian' interval (near BCU level) and an 'Aptian–Albian' interval (near Top Lower Cretaceous level) to extend the scope into the NCG and to acquire a North Sea-wide overview (Fig. 2).

Data and methods

2D and 3D seismic datasets, provided by Shell Upstream International, combined with data from 474 wells were used to establish a structural framework in the study area (Fig. 1). The 3D data (extent: 11 400 km^2) are a compilation of the Norwegian Carmot dataset, covering the NCG, and part of the UK Megamerge dataset, covering a limited area part of the southern UKCG (Fig. 1). The quality is variable but typically consists of dominant frequencies of *c.* 20–30 Hz, a wavelength of *c.* 60 ms two-way travel time (TWT) with a resulting seismic resolution of *c.* 15 ms TWT. This corresponds to *c.* 30 m vertical resolution, assuming an interval velocity of 3500 m s^{-1}. Water depths range from 40 to 100 m.

Data concerning 319 Norwegian wells in the study area were obtained from the Norwegian Petroleum Directorate (NPD) Factpages (NPD 2017). Additional well data for the UK and Danish sectors (119 and 41 wells, respectively) are from released well log and completion reports, well logs in the Shell archive (e.g. Boirie & Jeannou 1984; Statoil 1991), and published material (e.g. Isaksen & Tonstad 1989) (see Table 1). Additional occurrences of Lower Cretaceous sandstones in UKCG wells are adopted from Milton-Worssell *et al.* (2006).

The following regional seismic horizons were mapped in two-way time (TWT) on 3D seismic and calibrated with time-converted (via well checkshots and calibrated sonic logs) lithostratigraphically defined well tops from the Shell database and the NPD (Fig. 2):

- Base Cenozoic (64 Ma);
- Top Lower Cretaceous (100 Ma);
- Base Cretaceous Unconformity (BCU) (140 Ma);
- Top Rotliegend (270 Ma).

Milton-Worssell *et al.* (2006) mapped the Fischschiefer Bed, which defines the boundary between their two Lower Cretaceous plays (Fig. 2). In this study, this marker could not be traced due to a lack of accurately constrained well picks. The time maps of the four interpreted seismic horizons are combined with existing digital TWT seismic horizon maps provided by Shell Upstream International that allow an extension of the survey further into British and Danish territorial waters (study area: Fig. 1). A time difference assessment between the seismic horizons yields isochron maps, illustrating where the thickest sequences within the Cenozoic, Upper Cretaceous and Lower Cretaceous intervals are situated, revealing the general structural trends in the study area (Figs 3 & 4). Due to the large study area, no time to depth conversion was carried out, which means that these structural trends are somewhat qualitative.

Two more lithostratigraphically constrained horizons have been mapped in TWT on five additional regional 2D seismic transects to provide an additional link to previous studies (S1–S5: Figs 1, 2 & 5):

- Top Middle Jurassic (*c.* 165 Ma);
- Top Zechstein (252 Ma).

Although the available seismic coverage does not include Denmark, an earlier study (Møller &

Table 1. *List of sources for well data*

Type of data	Well	Data source
Lithostratigraphic tops for seismic interpretation	Norway UK Denmark	NPD (2017), Shell database Shell database Shell database
Wells shown in the well panel (Fig. 6)	UK 30/11b-1, UK 29/5a-5 NO 2/1-8 NO 2/7-15 NO 3/7-3 Amalie-1	Milton-Worssell *et al.* (2006) Fjellanger (1986), NPD (2013) Phillips Petroleum (1981), Isaksen & Tonstad (1989) Verolles (1982), Boirie & Jeannou (1984), NPD (2017) Statoil (1991)
Other well data described in the text and in other images	Norway (general) NO 7/3-1 NO 17/10-1 NO 17/11-1 NO 17/11-2 UKCG (general) Denmark (general) Sten-1 Tabita-1 Iris-1 Svane West Lulu-4	NPD (2017) NPD (1979*a*), Strass (1979) NPD (1979*b*), Olsen (1979) A/S Norske Shell (1968) Provan (1976) Milton-Worssell *et al.* (2006) Shell database Kern *et al.* (1983) Bonde *et al.* (1994) Britoil (1985), Bonde *et al.* (1994) Thorsrud *et al.* (2002) Mærsk (1987)
BCU and Top Lower Cretaceous subcrop data	e.g. NO 7/11-8, NO 3/7-1	NPD (2017), Shell database

Rasmussen 2003) provides a useful additional transect across the Danish border (S6: Fig. 5). In combination with the seismic horizon time and isochron maps, these transects offer a detailed insight into the structural framework of the extended study area, revealing the locations of the main basins, highs, diapirs and faults (Figs 3–5). Subsequently, the results of the seismic interpretation are integrated with published data from Copestake *et al.* (2003), Japsen *et al.* (2003), Milton-Worssell *et al.* (2006) and Rossland *et al.* (2013) for an assessment of Lower Cretaceous reservoir potential in the extended study area, of which well data provide a first impression (Figs 4 & 6).

The basic methodology applied by Milton-Worssell *et al.* (2006) has been adopted. Combined isochron maps of the extended study area indicate zones with thin Lower Cretaceous deposits, which were potentially exposed and prone to erosion during the Early Cretaceous (Fig. 7). At these places, well data provide the true thickness of the Lower Cretaceous sequence and the lithology in subcrop below the BCU. Devonian metamorphic rocks and volcanics, and Rotliegend, Triassic Skagerrak, Middle Jurassic Bryne and Upper Jurassic Ula sandstones (Fig. 2) in subcrop indicate whether a specific locality was part of a potential sand source area during the earliest Cretaceous. The presence of sand provenance areas is considered the most important factor controlling sandstone development since the Early Cretaceous was dominated by pelagic mud deposition (Fig. 2). This exercise is repeated for the Aptian–Albian reservoir interval, where the sand-prone lithologies in subcrop below the Top Lower Cretaceous horizon are charted (Fig. 8). The isochron maps subsequently allow the tracing of possible sediment transport fairways by interpreting depocentres as drainage areas and barriers separating them as watersheds. Sediment transport is assumed to have followed the bathymetry given by the isochron maps, leading sediments from the highs to the depocentres. Thus, by combining the isochron map, drainage and sand source areas, potential sand transport routes for the Latest Ryazanian and the Aptian–Albian are mapped (Figs 7 & 8). Well data allow a qualitative check of these interpretations: where sandstones occur in wells, a plausible link with a nearby sand source area can be inferred. If no such well data are available, sediment transport between source and depocentre remains speculative. It is recognized that the sandstones recorded in these wells are not necessarily linked to the postulated source areas, and that these links would need to be proven via further investigation involving advanced

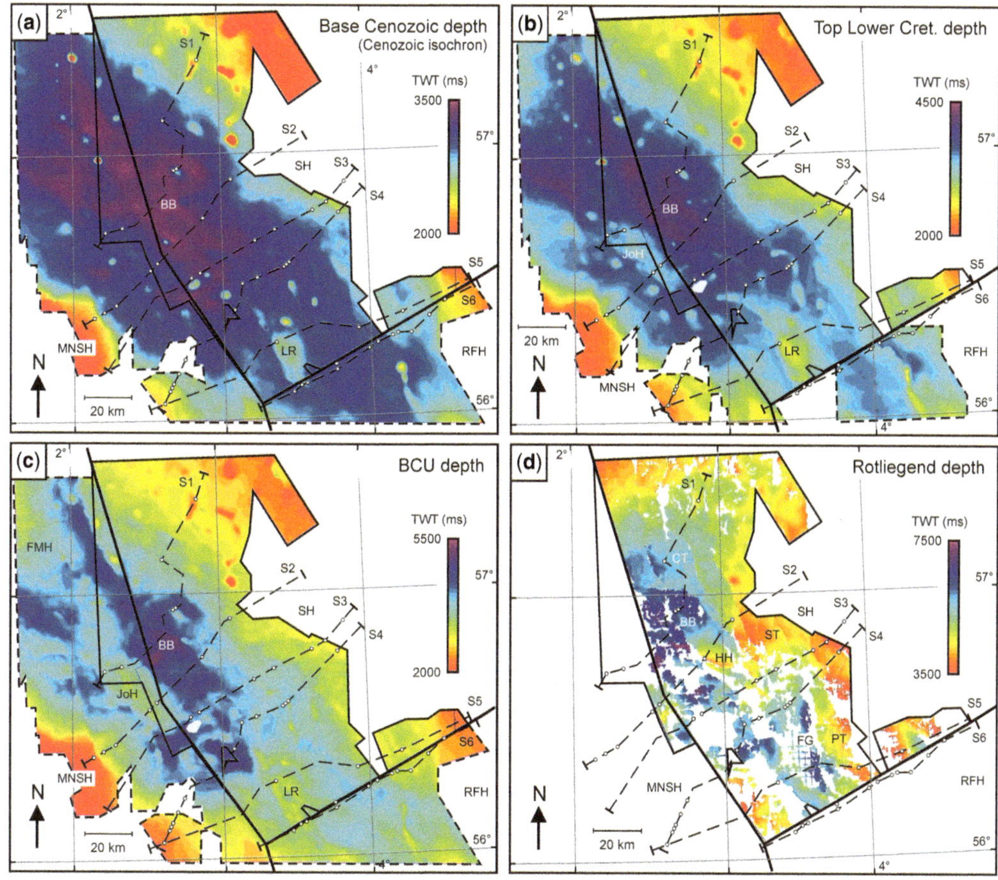

Fig. 3. Time–depth maps of four regional horizons in the study area: (**a**) Base Cenozoic; (**b**) Top Lower Cretaceous; (**c**) Base Cretaceous Unconformity (BCU); (**d**) Top Rotliegend. Note that the Base Cenozoic time–depth map (a) is also the Cenozoic isochron map and that the Top Rotliegend map is incomplete due to locally poor seismic quality. Dotted lines indicate the trace of interpreted transects S1–S6 and white dots indicate well locations along these transects (see Fig. 5). Solid outlines indicate the extent of the 3D seismic survey. Dashed outlines indicate the extent of the available previously interpreted seismic horizons in the UK and Denmark (see Fig. 1). BB, Breiflabb Basin; CT, Cod Terrace; FG, Feda Graben; HH, Hidra High; JoH, Josephine High; FMH, Forties–Montrose High; MNSH, Mid North Sea High; LR, Lindesness Ridge; PT, Piggvar Terrace; RFH, Ringkøbing-Fyn High; SH, Sørvestlandet High; ST, Steinbit Terrace.

seismic and well analysis techniques that are beyond the scope of this study.

Structural framework interpretation

In general, a series of NNW–SSE-orientated en echelon (rift) basins, normal faults and tilted fault blocks follow the larger NW–SE Central Graben trend (Figs 1 & 3–5). The NCG structure is bounded by the Sørvestlandet High and the Ringkøbing-Fyn High to the east, and by the Mid North Sea High to the SW (Figs 3 & 4).

The main depocentres, as identified on the isochron maps, are situated in the Breifflab Basin in the NW (Figs 4a & 5, S1), where up to 8 km of subsidence has occurred (NDD 2012). However, the locations of these depocentres do not coincide with the thickest Upper Jurassic deposits found in the Feda Graben, Søgne Basin and Gertrud Graben to the SE (Erratt *et al.* 1999) (Fig. 5, S5 and S6). This discrepancy is a result of later differential thermal subsidence and sediment infill (Gowers & Sæbøe 1985). Normal faults are omnipresent in the area, but major differences in structural style occur between the pre-Zechstein units, Triassic, Upper

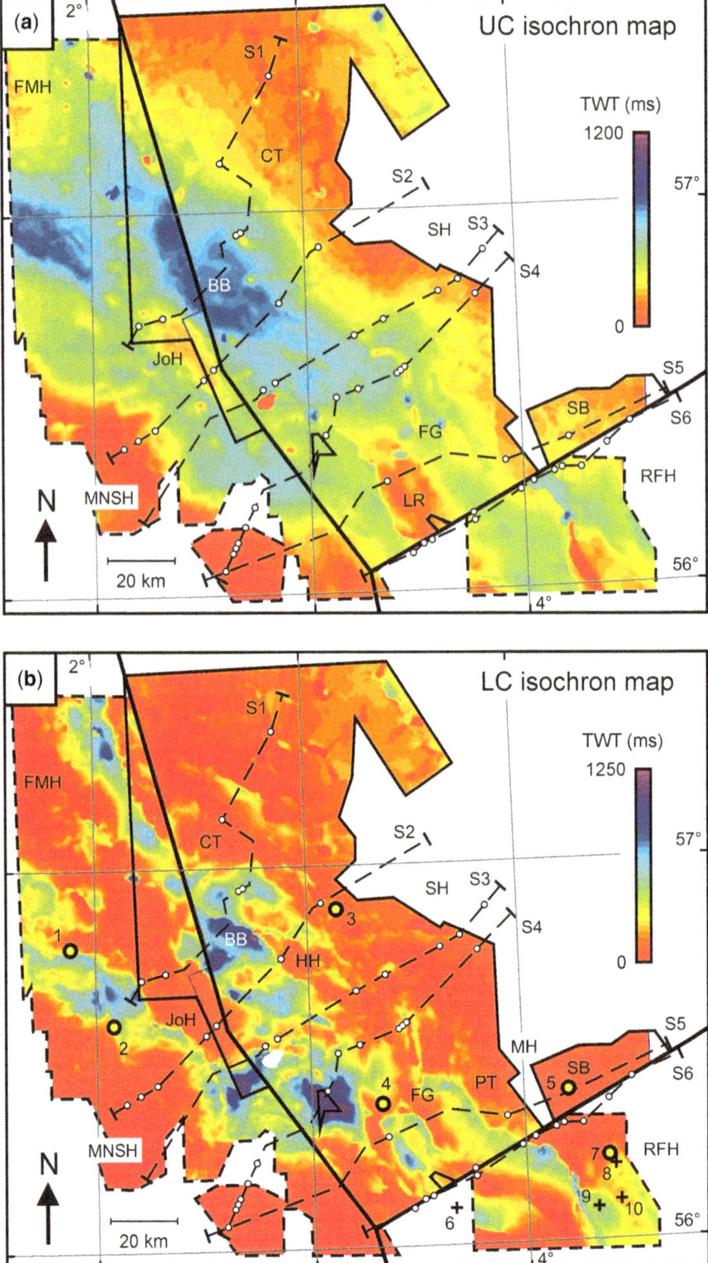

Fig. 4. Isochron maps showing the location and geometry of depocentres in the study area in two-way travel time (TWT). (**a**) UC, Upper Cretaceous isochron map; (**b**) LC, Lower Cretaceous (Cromer Knoll Group). Dotted lines indicate the trace of interpreted transects S1–S6 and wells along these transects (see Fig. 5). Solid outlines indicate the extent of the 3D seismic survey. Dashed outlines indicate the extent of the available previously interpreted seismic horizons in the UK and Denmark (see Fig. 1). The larger yellow dots indicate >3 m sand occurrences in wells within the study area, whereas crosses indicate sandstone traces (<3 m thickness). Wells: (1) UK 29/5a-5; (2) UK 30/11b-1; (3) NO 2/1-8; (4) NO 2/7-15; (5) NO 3/7-3; (6) Sten-1; (7) Amalie-1; (8) Tabita-1; (9) Iris-1; (10) Svane-1. BB, Breiflabb Basin; CT, Cod Terrace; FG, Feda Graben; FMH, Forties–Montrose High; HH, Hidra High; JoH, Josephine High; MH, Mandal High; MNSH, Mid North Sea High; LR, Lindesness Ridge; PT, Piggvar Terrace; RFH, Ringkøbing-Fyn High; SB, Søgne Basin; SH, Sørvestlandet High.

Jurassic, Lower Cretaceous synrift strata and post-rift infill. The Josephine High (Fig. 5, S1), Hidra High (Fig. 5, S2), Border High (Fig. 5, S4), Mandal High (Fig. 5, S5 and S6) Cod Terrace (Fig. 5, S1) and Piggvar Terrace (Fig. 5, S5) represent pre-Zechstein basement blocks forming major structural highs or terraces. Several large salt domes occur within the area as well (e.g. Fig. 5, S1).

Late Jurassic–Early Cretaceous rift structures

Due to Mid-Jurassic thermal doming and associated erosion, few Lower Jurassic units are preserved in the study area. In contrast, significant Upper Jurassic sediments, recording the latest North Sea rift phase, occur locally in extensional basins. These units are best developed in the south of the study area, where the Feda Graben, Gertrud Graben and Søgne Basin half-graben accommodate some 2 km of Upper Jurassic sequences (Fig. 5, S5 and S6) as part of the large-scale left-stepping en echelon Central Graben structure (Erratt *et al.* 1999) (Fig. 1a). Many Triassic faults affect Upper Jurassic strata, indicating fault reactivation: for example, the Skrubbe Fault and the Coffee Soil Fault bounding the Feda Graben and the Søgne Basin, respectively (Fig. 5, S5 and S6). Rifting caused salt movement and diapirism, which impacted Upper Jurassic sedimentation (e.g. in the Søgne Basin).

Subsequently, the major Lower Cretaceous deposits are shifted westwards compared to the Upper Jurassic depocentres (Fig. 5, S1, S3 and S4). A distinct feature is the Early Cretaceous reactivation of the pre-Zechstein half-graben west of the Border High, where Upper Jurassic or Triassic units are absent (Fig. 5, S4). Also striking is the lack of Early Cretaceous tectonic activity in the Søgne Basin; in contrast to significant Triassic and Upper Jurassic syntectonic units, little to no Lower Cretaceous sediments occur (Figs 4b & 5, S5 and S6).

The character of the Early Cretaceous basins varies considerably. The Border High and Breiflabb basins are fault-bounded and show thickening towards the boundary faults, indicating synrift deposition (Fig. 5, S1 and S4). Other rift-bounded basins are found west of the Hidra High (Fig. 5, S2), at well NO 2/4-10 (Fig. 5, S3) and west of the Mandal High (Fig. 5, S6). Yet, the filling of pre-existing deep underfilled Jurassic basins, as well as sediment compaction effects, could partially account for these observations (Rattey & Hayward 1993; Coward *et al.* 2003). At various localities, salt motion affected Early Cretaceous deposition: for example, above the Hidra High (Fig. 5, S2) and at well NO 2/4-3 (Fig. 5, S4). In other parts of the study area, depocentres exhibit sag-type geometries: for example, east of well UK 30/17B-3 and above the Hidra High (both in Fig. 5, S2), west of the NO 1/6-5 diapir (Fig. 5, S3) and in the Ål Basin (Fig. 5, S5 and S6). Faults do not generally continue to the top of the Early Cretaceous, except for those associated with later tectonic inversion.

As such, cessation of rifting is shown to be diachronous. The westwards shift of the Early Cretaceous depocentres with respect to the Jurassic rifts might indicate a change in extensional regime near the start of the Cretaceous, as proposed by previous authors (e.g. Erratt *et al.* 1999), before extension activity ceased altogether due to the opening of the young North Atlantic (Rattey & Hayward 1993).

Post-rift and tectonic inversion structures

The Late Cretaceous and Cenozoic units dominantly show gentle sag geometries along the NW–SE trend of the NCG, indicating further post-rift thermal subsidence. At the Breiflabb Basin on the UK–Norwegian border, thermal subsidence was strongest, creating a major Late Cretaceous/Cenozoic depocentre (Figs 4a, 5, S1, & 6a) (Gowers & Sæbøe 1985). However, signs of inversion are also noted, for instance at the Lindesnes Ridge where Early Cretaceous synrift deposits are uplifted along Skrubbe Fault (Figs 5, S1, & 6e). Inversion-related structures (inverted graben and diapirs/salt domes) disturb not only the Upper Cretaceous deposits, but also Cenozoic strata (Figs 3–5), indicating multiple inversion phases (Gowers *et al.* 1993).

Lower Cretaceous reservoir interpretation

Sandstone occurrences in Norwegian and Danish wells

In contrast to the UKCG, where numerous wells encounter Lower Cretaceous sandstones (Milton-Worssell *et al.* 2006), only three wells in the NCG area (from a total of 160 Lower Cretaceous penetrations) are reported to contain similar deposits (NPD 2017) (Figs 4b & 6). The sandstones in these wells are lithostratigraphically defined as Ran Sandstone units (NPD 2017) and are, in contrast with the deep-marine character of most equivalent Lower Cretaceous sandstones in the UK, interpreted as shallow submarine fans (Isaksen & Tonstad 1989; Milton-Worssell *et al.* 2006).

Well NO 2/1-8 on the Cod Terrace contains a 4 m interval of Ran sandstones, but no further details on lithology or reservoir properties are publicly available (Fjellanger 1986; NPD 2013). These sandstones appear below the Hauterivian–Barremian Tuxen Formation and are, therefore, assigned to the Ryazanian reservoir interval (Fig. 2). Reference well NO 2/7-15 in the Feda Graben (Isaksen & Tonstad 1989) (Fig. 4b) contains a 48 m-thick Ran Sandstone sequence.

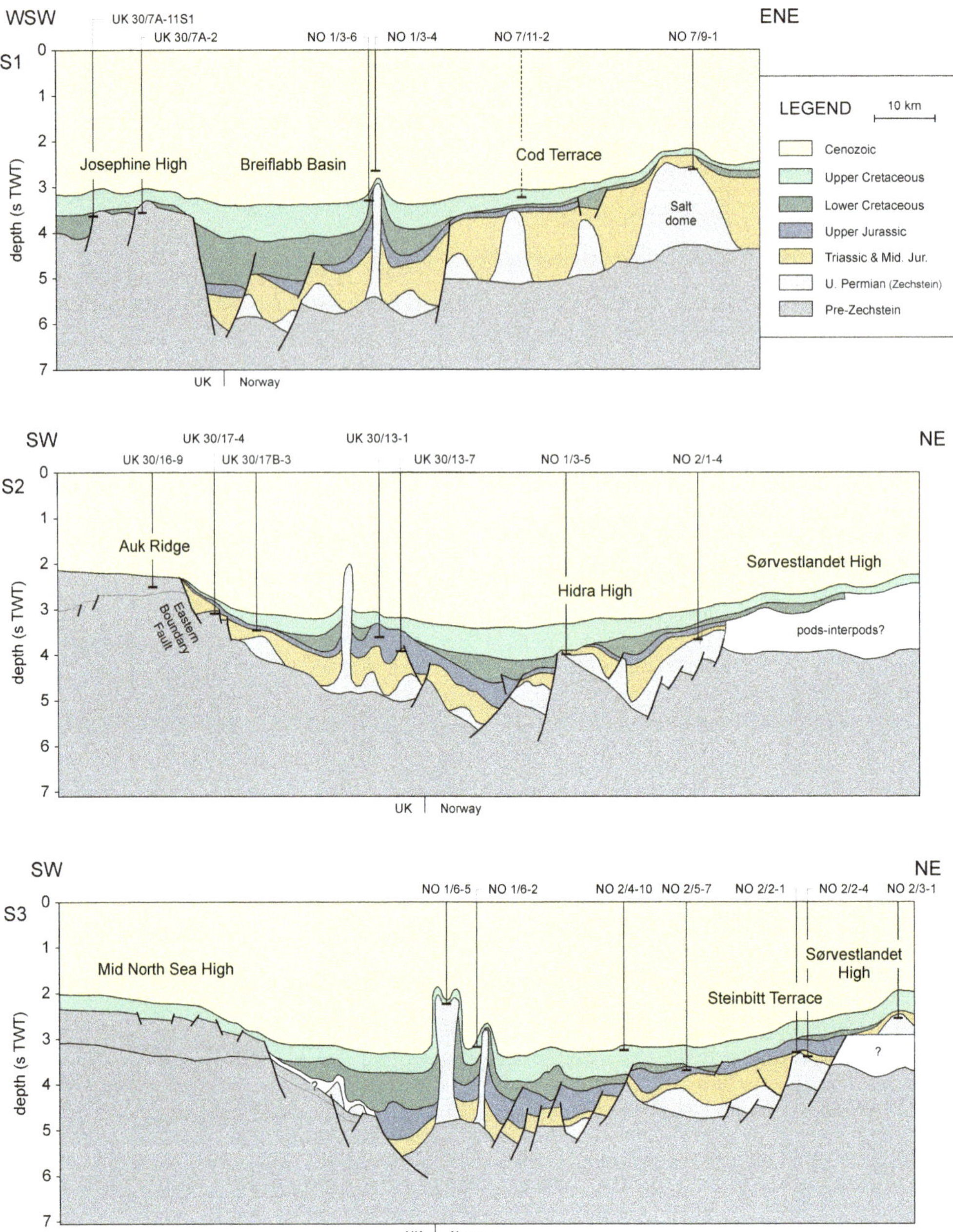

Fig. 5. Interpreted seismic sections S1–S6. Reference datum is mean sea level. For section locations, see Fig. 1b. Section S6 is modified after Møller & Rasmussen (2003).

Cores taken from the lowermost part of this succession are described as dominantly clay-rich siltstones with occasional microporosity and fractures with minor hydrocarbon shows (Phillips Petroleum 1981). However, drill stem tests demonstrated the section to be tight (NPD 2017). The age of these Ran sandstones is poorly constrained, but they are attributed to the Albian–Aptian reservoir interval due to their occurrence directly below the Aptian–Albian Sola Formation (Isaksen & Tonstad 1989) (Fig. 2).

In well NO 3/7-3, east of the Mandal High (Fig. 4b), a 107 m-thick Ran Sandstone sequence

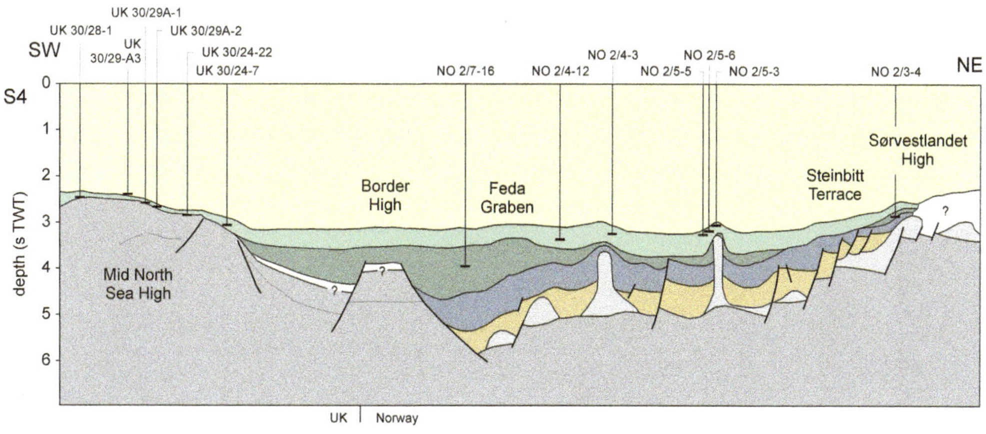

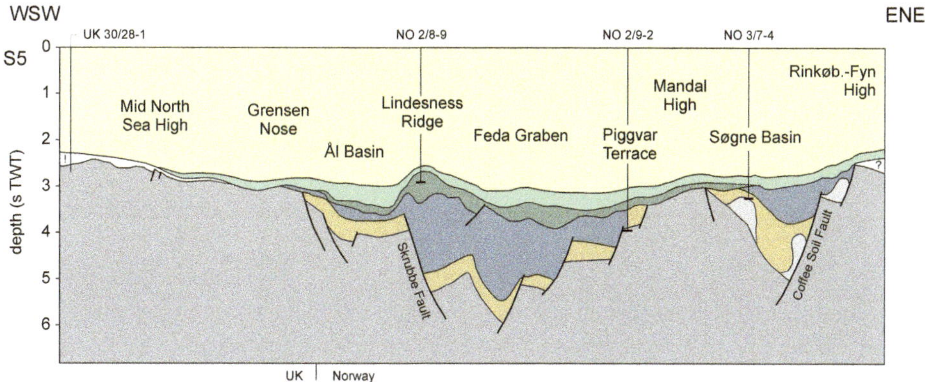

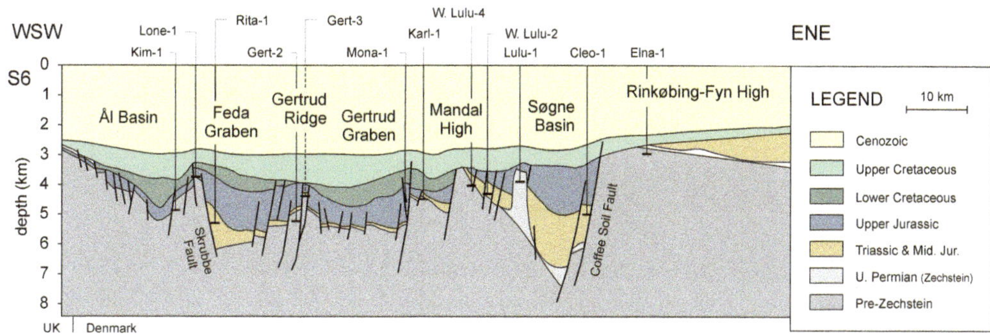

Fig. 5. *Continued.*

occurs on top of the BCU (NPD 2017). These deposits consist of a lower unit of dolomitic and glauconitic sandstones, interbedded with dolomitic and shaley layers, and an upper unit of massive coarse-grained sandstones with occurrences of chalky, sandy limestone, capped by carbonates containing some lignite (Verolles 1982). The massive sandstones (60–70% quartz) are cemented but represent good reservoir potential with porosities and permeabilities of between 20–28% and 0.5–10 mD, respectively (Verolles 1982; Boirie & Jeannou 1984). The NO 3/7-3 Ran sandstones were deposited as lenticular sheets or slope-apron bodies in a restricted and proximal, relatively shallow-marine environment (100–200 m water depth: Verolles 1982), which evolved into an open-marine setting towards the end of the Early Cretaceous (Boirie & Jeannou 1984). Since the Ran sandstones

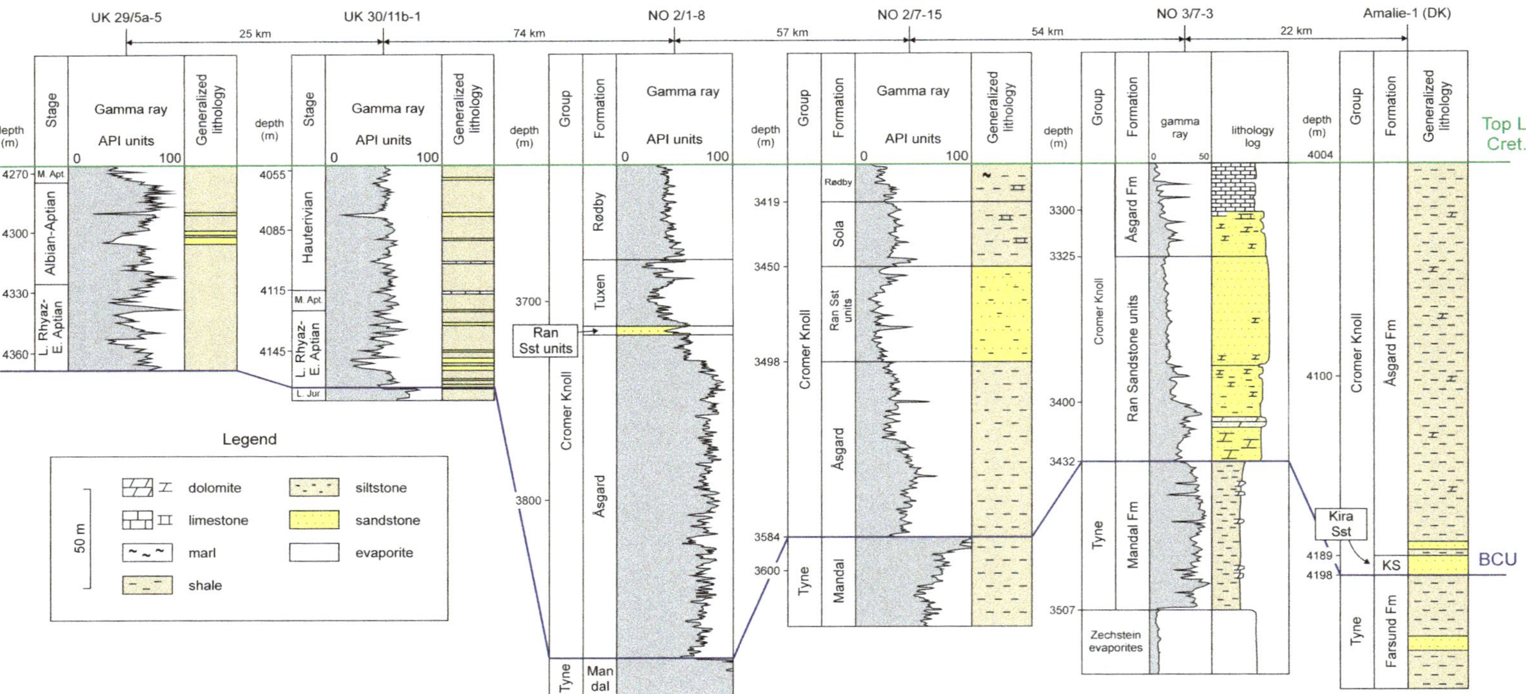

Fig. 6. Well data from wells UK 30/11b-1 and UK 29/5a-5 (modified after Milton-Worssell *et al.* 2006), NO 2/1-8 (modified after NPD 2013), NO 2/7-15 (modified after Isaksen & Tonstad 1989), NO 3/7-3 (modified after Boirie & Jeannou 1984) and Amalie-1 (DK, modified after Statoil 1991), all containing Lower Cretaceous (Ran/Kira) Sandstone units and hung off the Top Lower Cretaceous level (Fig. 2). No lithology data are available for well NO 2/1-8. No gamma-ray data are available for well Amalie-1. Locations are shown in Fig. 4b.

are of Ryazanian age (Boirie & Jeannou 1984), they belong to the Latest Ryazanian reservoir interval.

Four other Norwegian wells encountering Ran Sandstone are situated to the NE, in Block 17, at a considerable distance from the North Sea rift basins and outside the extended study area. The implications of these sandstone occurrences will be addressed in the South Viking Graben regional overview below.

In the Tail End Graben (Denmark), 9 m-thick Lower Cretaceous subangular to subrounded and poor to moderately sorted, fine-grained 'Kira sandstones' are found above the BCU level in the Amalie-1 well, probably deposited as part of a submarine fan system (Statoil 1991) (Fig. 6). These sandstones are oil-bearing and of excellent reservoir quality, with high porosities and permeabilities (0.213 and 319 mD, respectively) and a net-to gross ratio of 0.339 (Statoil 1991). Further Latest Ryazanian sandstones, although thinner, occur in the Tabita-1, Svane-1 and Iris-1 wells south of the Amalie-1 well (Fig. 4b). The Tabita-1 'Kira Sandstone equivalent' at the base of the Lower Cretaceous contains mostly claystone with very-fine-grained siltstone and (quartz) sandstone striae (1–3 cm), as well as cross-bedding with erosional surfaces (Bonde *et al.* 1994). A core from this interval contains conglomeratic intervals of unweathered, angular clasts of metamorphic basement material, as well as folded and disturbed mudstone beds. Both facies are indicative of slope process, whilst the lack of wave-related structures in the core suggests a depositional environment below storm wave base (Bonde *et al.* 1994). In the Svane-1 well, very-fine- to fine-grained, subrounded, poorly sorted calcareous quartz sandstones with an argillaceous matrix and net-to-gross ratios of up to 0.85 are found above the BCU (Thorsrud *et al.* 2002). The Iris-1 well contains various levels of thin sandstone in the Valhall Formation overlying the BCU which are 'a few' metres thick (Britoil 1985). The cored material from this well is predominantly fine grained and similar to that in the Tabita-1 well (Bonde *et al.* 1994). Further to the west, Lower Cretaceous (Latest Ryazanian–Early Hauterivian) fine- to medium-grained, poorly sorted sandstones, belonging to the Latest Ryazanian reservoir interval, are present in the Sten-1 well (Kern *et al.* 1983), making a total of five wells encountering Lower Cretaceous sandstones in the Danish part of the study area (Fig. 4b).

Latest Ryazanian reservoir distribution

An interpretation of the Latest Ryanazian reservoir interval is presented in Figure 7 and depicts the sandstone occurrences in wells, potential source areas with sand-prone lithologies subcropping the BCU and sediment transport fairways to depocentres identified on the Lower Cretaceous isochron map.

Milton-Worssell *et al.* (2006) demonstrated the potential for marine sandstone development in the Ryazanian–Barremian interval of the UKCG, with the Forties–Montrose High and Western Platform interpreted as provenance areas. Closer to the Norwegian–British border, sand-prone lithologies are found in subcrops below the BCU at the Josephine High (Skagerrak Formation), Auk Ridge (Rotliegend) and Argyll Field at the Mid North Sea High (Rotliegend, and the Ula and Skagerrak formations). These represent potential sand source areas for the surrounding depocentres where multiple well penetrations occur (Milton-Worssell *et al.* 2006). The Auk Ridge is also the likely provenance area for the Lower Cretaceous Devil's Hole sandstones to its west (Milton-Worssell *et al.* 2006). These scattered deposits are considered similar to the Norwegian Ran sandstones (Isaksen & Tonstad 1989) and possibly represent a continuation of the Upper Jurassic synrift Fulmar/Ula shoreface or shelf deposits (Bisewski 1990; Johnson & Lott 1993; Copestake *et al.* 2003) (Fig. 2). The UK Flora-Fife Trend area and the Danish Inge High contain Ula Formation and Rotliegend units in subcrop below the BCU. These are potential source areas for the sandstones in the Danish Sten-1 well (Kern *et al.* 1983), which is situated in a Lower Cretaceous depocentre (Fig. 7) and is postulated to be a deep-marine deposit.

The 4 m-thick unspecified sandstone layer in well NO 2/1-8 (NPD 2013) represents an isolated Ran Sandstone occurrence on the Cod Terrace (Figs 2 & 6). The most probable origin would be either the Mandal High or the Cod terrace, where well NO 7/11-8 encounters the Skagerrak Formation in subcrop below the BCU (NPD 2017), indicating a possible small-scale sediment provenance area. Any material originating from the Scandinavian mainland to the NE would most likely be caught in the Norwegian–Danish Basin region, where major Lower Cretaceous depocentres are situated (Copestake *et al.* 2003) (Fig. 1a). Similarly, sediments from the Josephine High would first have had to cross the Breifflab Basin depocentres (Fig. 7). However, the exact nature and provenance of these Ran sandstones cannot be established with the data currently available.

The thickest Ran sandstones in the study area occur in well NO 3/7-3 (107 m: Fig. 6), and these relatively shallow- to open-marine sandstones were deposited at the fringe of the Søgne Basin (Verolles 1982; Boirie & Jeannou 1984), which was tectonically inactive during the Lower Cretaceous (Rossland *et al.* 2013) (Fig. 5, S5 and S6). The adjacent Mandal High and its metamorphic basement units were largely exposed during the Early Cretaceous (Verolles 1982; Copestake *et al.* 2003; Rossland *et al.* 2013) (Figs 7 & 9) and are the probable source

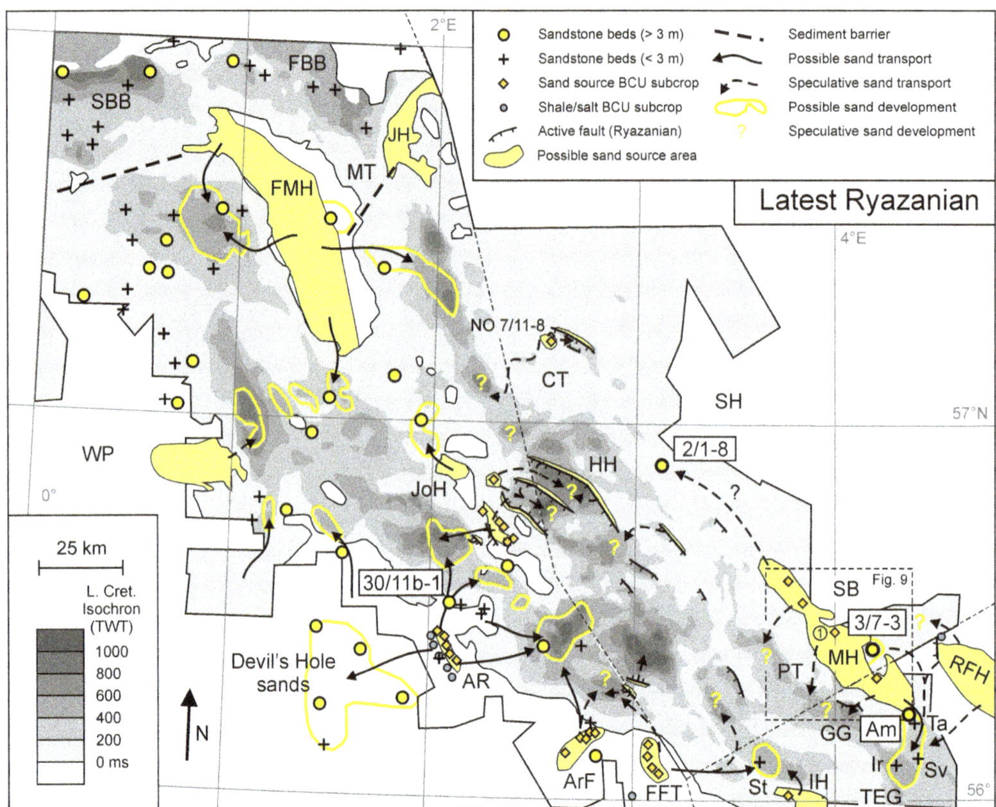

Fig. 7. Interpretation of reservoir potential in the extended study area for the latest Ryazanian (BCU level). Well data, possible sand source areas, Ryazanian fault activity, possible sediment transport fairways and areas of possible sandstone development are projected on top of the Lower Cretaceous isochron map. Well identifiers are: Ir, Iris-1; St, Sten-1; Sv, Svane-1; Ta, Tabita-1; ①: well NO 3/7-1. AR, Auk Ridge; ArF, Argyl Field; CT, Cod Terrace; FBB, Fisher Bank Basin; FFT, Flora-Fife Trend; FMH, Forties–Montrose High; GG, Gertrud Graben; HH, Hidra High; IH, Inge High; JH, Jæren High; JoH, Josephine High; MH, Mandal High; MT, Marnock Terrace; PT, Piggvar Terrace; RFH, Ringkøbing-Fyn High; SB, Søgne Basin; SBB, South Buchan Basin; SH, Sørvestlandet High; TEG, Tail End Graben; WP, Western Platform. Modified after Japsen *et al.* (2003), Milton-Worssell *et al.* (2006) and Rossland *et al.* (2013).

for these proximal Ran sandstones (Verolles 1982). Alternatively, Rossland *et al.* (2013) suggest, on the basis of dip directions, that these sandstones are related to a turbidite system sourced from the Rynkøbing-Fyn High to the east. It should, however, be stressed that their dip-meter data may be affected by post-sedimentary salt movement associated with the large salt dome below the Søgne Basin (Verolles 1982) (Fig. 5, S5 and S6), or could simply represent a deviation in transport direction as frequently observed within local submarine fan systems (e.g. Normark *et al.* 1979).

The presence of thick Ran sandstones in well NO 3/7-3 (Boirie & Jeannou 1984) indicates a promising reservoir development in the area, yet none of the other wells in the vicinity encounter Lower Cretaceous sandstones (NPD 2017) (Fig. 9). This is in accordance with the depositional character of the Ran Sandstone units described by Verolles (1982) and Boirie & Jeannou (1984), who suggested that reservoir bodies in the area, although potentially of significant thickness, may have a restricted lateral extent (Fig. 9). Furthermore, the Mandal High area is little studied, potentially harbouring reservoirs in various other stratigraphic intervals (Rossland *et al.* 2013) (Fig. 10), and a detailed analysis will be required to identify these.

In the east of the study area, the Kira sandstones and their equivalents in the Amalie-1 and Tabita-1 wells (Fig. 6) probably represent submarine fan or slope deposits (Statoil 1991), associated with erosion at the BCU level and the nearby boundary fault

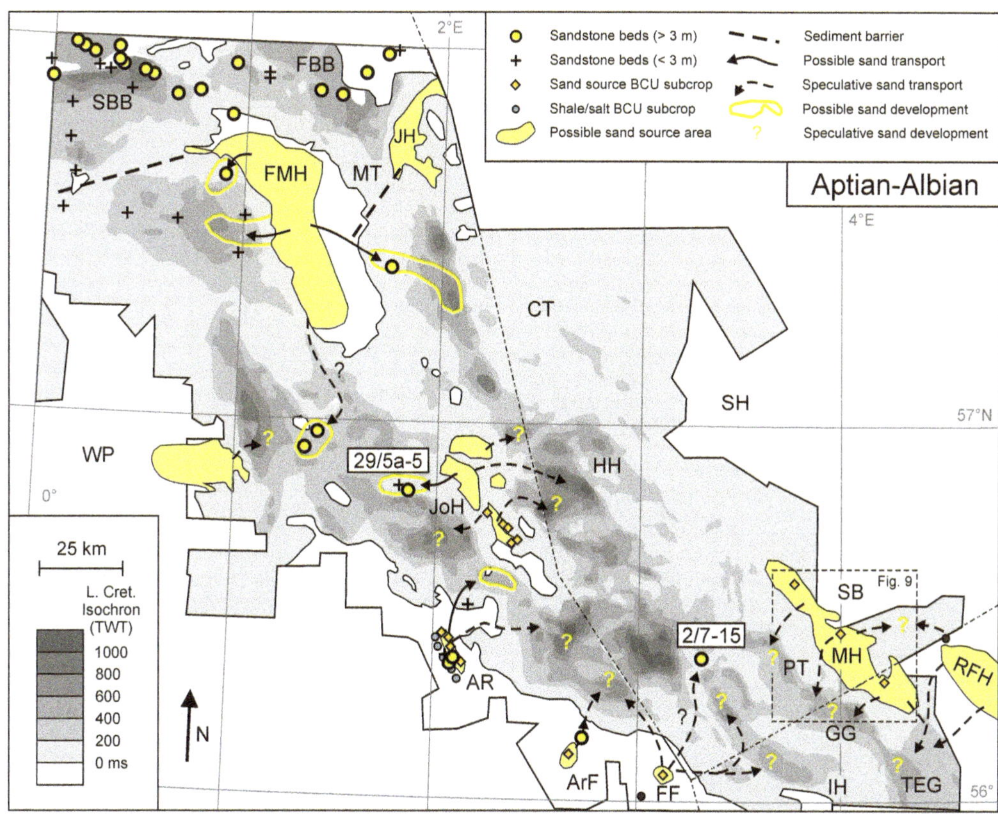

Fig. 8. Interpretation of reservoir potential throughout the extended study area for the Aptian–Albian (near Top Lower Cretaceous level). Well data, possible sand source areas, interpreted possible sediment transport fairways and areas of possible sandstone development are projected on top of the Lower Cretaceous isochron map. AR, Auk Ridge; ArF, Argyl Field; CT, Cod Terrace; FBB, Fisher Bank Basin; FF, Flora Field; FMH, Forties–Montrose High; GG, Gertrud Graben; HH, Hidra High; IH, Inge High; JH, Jæren High; JoH, Josephine High; MH, Mandal High; MT, Marnock Terrace; PT, Piggvar Terrace; RFH, Ringkøbing-Fyn High; SB, Søgne Basin; SBB, South Buchan Basin; SH, Sørvestlandet High; TEG, Tail End Graben; WP, Western Platform. Modified after Japsen *et al.* (2003), Milton-Worssell *et al.* (2006) and Rossland *et al.* (2013).

between the Tail End Graben and the Ringkøbing-Fyn High (Bonde *et al.* 1994) (Fig. 7). Although no rock samples are available from the Amalie-1 well, the metamorphic clasts in cores from the Tabita-1 well are reported to be similar to the basement rocks on the Ringkøbing-Fyn High and on the Mandal High (well NO 3/7-1) (Bonde *et al.* 1994). A possible supply from the Ringkøbing-Fyn High may have involved submarine erosion of the footwall basement, whereas alternative sediment transport from the Mandal High may have bypassed the NO 3/7-3 and Amalie-1 wells before reaching the Tabita-1 well location (Bonde *et al.* 1994) (Fig. 7). The sand-prone intervals in the Svane-1 and Iris-1 wells are possibly correlatable to the Kira sandstones (Bonde *et al.* 1994; Thorsrud *et al.* 2002), which, if correct, may indicate a regional deep-marine fan system (Fig. 7). It should be noted, however, that except for the Amalie-1 well, no Lower Cretaceous reservoir-quality sandstones have been found. Yet, a few localized sandy apron or lobe units may have developed as a continuation of the Jurassic deep-marine sandstones in the area (Bonde *et al.* 1994; Nielsen *et al.* 2015). Similar deposits could also have developed in the Gertrud Graben and Feda Graben to the south and SW of the exposed Mandal High (Rossland *et al.* 2013), but there is currently no evidence to support this interpretation and identifying such reservoirs, if present, will be highly challenging.

In contrast to the UK and Danish Central Graben areas, no Latest Ryazanian sandstones appear in wells within the NCG proper (NPD 2017) and most Lower Cretaceous depocentres are isolated

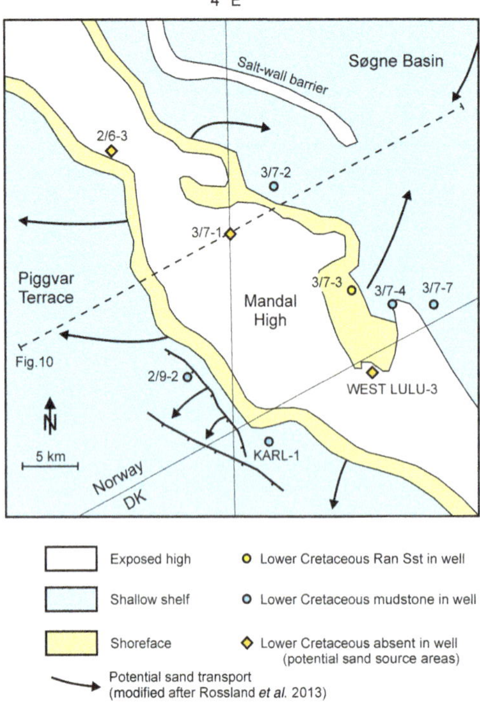

Fig. 9. Proposed Early Cretaceous palaeogeographical situation around the Mandal High area. Image modified after Rossland *et al.* (2013).

from the identified sand source areas (Fig. 7). However, various faults were still active, of which some could have exposed sand-prone lithologies to erosion. Of these, the Hidra High block next to the Breifflab Basin, where Rotliegend units are present in the footwall, is the best example (Figs 5b, S2 & 7). However, it is possible that such smaller sand source areas (e.g. Argyll Field area: 10–100 km^2 and less for exposed fault scarps) might not have produced enough sand-prone material for reservoir-size deposits (McArthur *et al.* 2016*b*). By contrast, the exposed Mandal High amounts to 500–600 km^2, and is associated with the thick Ran sandstones in well NO 3/7-3 and the postulated Amalie fan system; thus representing a significant reservoir potential.

Aptian–Albian reservoir distribution

The interpretation of the Latest Ryazanian reservoir interval is presented in Figure 8, and depicts the sandstone occurrences in wells, potential source areas with sand-prone lithologies subcropping the Top Lower Cretaceous and sediment transport fairways to depocentres identified on the Lower Cretaceous isochron map.

Towards the end of the Early Cretaceous, sandstone occurrences are rarely seen in UKCG wells (Milton-Worssell *et al.* 2006). However, important sediment provenance areas (e.g. the Forties–Montrose High, Auk Ridge and Josephine High) were still in place and exposed, providing sand influx into the adjacent depocentre as recorded in some penetrations (Milton-Worssell *et al.* 2006) (Fig. 8). However, several of the smaller source areas were flooded and covered with Lower Cretaceous deposits (Cod Terrace and Inge High), and potential sourcing from fault scarps was strongly diminished with the cessation of rift activity. Other Ryazanian provenance areas were reduced but remained partially exposed towards the end of the Early Cretaceous as indicated by subcrop data (e.g. the Argyll Field area, Flora-Fife Trend) (cf. Fig. 8 with Fig. 7); yet, no sandstone well occurrences are recorded in the adjacent Aptian–Albian depocentres.

Ran Sandstone units belonging to the Aptian–Albian reservoir interval are found in only one Norwegian well: NO 2/7-15 (Isaksen & Tonstad 1989, NPD 2017) (Figs 6 & 8). These clay-rich silt/sandstones are somewhat isolated from the interpreted sediment provenance areas. The Flora Field area, where the Rotliegend is found in subcrop below the Upper Cretaceous chalk deposits, is proposed as the most likely origin for these units (Fig. 8). However, the character of these NO 2/7-15 Ran sandstones remains poorly constrained and demands further assessment.

It should be noted that the wells in the Søgne Basin area, where thick Ryazanian sandstones were previously deposited (well NO 3/7-3), record only (chalky) mudstone sequences in the Aptian–Albian (Rossland *et al.* 2013; NPD 2017). Also, the potential Amalie fan system in the Danish Tail End Graben to the south is absent in well reports. Yet, the Mandal High was still prone to erosion during the Aptian–Albian, as indicated by metamorphic basement and Bryne Formation subcropping the Upper Cretaceous chalk units (wells NO 2/6-5, NO 3/7-1 and West-Lulu 4: Mærsk 1987; NPD 2017). In addition, large parts of the Ringkøbing-Fyn High have no or thin (only a few metres) Lower Cretaceous cover (Japsen *et al.* 2003). Both highs may, therefore, have produced sand-prone material leading to localized reservoir development (Fig. 8), although there is currently no evidence to support this suggestion.

Overall, the Aptian–Albian reservoir interval provides significantly less potential for Lower Cretaceous sandstone deposits than the Latest Ryazanian, due to the drowning of sand source areas. Still, the Ran Sandstone present in well NO 2/7-15 and the sandstone occurrences in various other wells in the UKCG indicate some reservoir potential.

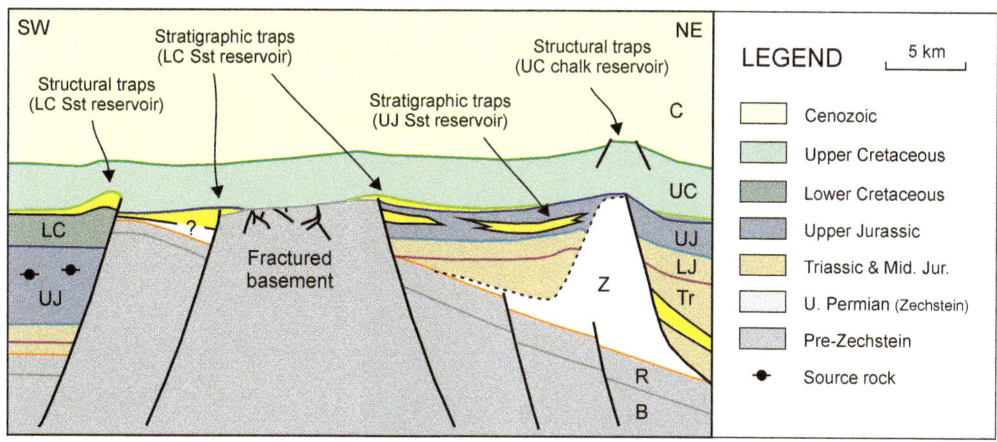

Fig. 10. Idealized cross-section proposing the main potential reservoirs and traps in the Mandal High–Søgne Basin area. C, Cenozoic; UC, Upper Cretaceous; LC, Lower Cretaceous; UJ, Upper Jurassic; LJ, Lower Jurassic; Tr, Triassic; Z, Zechstein (evaporites); R, Rotliegend; B, pre-Permian sediments and/or (metamorphic) basement. Image modified after Rossland *et al.* (2013). Location shown in Figure 9.

Potential analogue settings in the Southern Permian Basin area

This study shows that the UKCG and NCG harbour potential for Lower Cretaceous sandstone reservoir units, suggesting further exploration possibilities. As the Central Graben structure continues south into Danish, German and Dutch territorial waters (Figs 1 & 11), where the geological setting was quite similar during the Early Cretaceous (Voigt *et al.* 2008; Pharaoh *et al.* 2010), it would be worthwhile to extend the scope of a future case study to these areas.

In Denmark, for example, the Ringkøbing-Fyn High along the eastern margin of the Tail End Graben has no or limited Upper Jurassic–Lower Cretaceous sedimentary cover (Japsen *et al.* 2003), and is known to have been the source of various Late Jurassic fan deposits (Johannessen & Andsbjerg 1993; Andsbjerg & Dybkjær 2003). Such conditions are likely to have continued into at least the earliest Cretaceous, as illustrated by the deposition of Vyl sandstones (Figs 2 & 11a). These submarine fan units with moderate reservoir potential are found adjacent to the Coffee Soil Fault and were supplied by the Ringkøbing-Fyn High (Michelsen *et al.* 2003) (Fig. 11a). In addition, the Lower Cretaceous chalks of the Tuxen Formation form the reservoirs in the Danish Valdemar and Adda fields (Copestake *et al.* 2003; Jakobsen *et al.* 2005) (Fig. 2) indicating another attractive target for continued exploration in the area.

Further to the south, the German and Dutch sectors of the Central Graben are flanked by the Schill Grund High to the east, and the Step Graben and Cleaver Bank High to the west (Fig. 11a), areas that were exposed highs during the Late Jurassic and the Early Cretaceous (Pharaoh *et al.* 2010). However, intense Late Cretaceous and Cenozoic basin inversion has caused significant erosion (de Jager 2007) and most of the Lower Cretaceous in the southern sector of the Dutch Central Graben was removed. In the northern sector of the Dutch Central Graben, where inversion and associated erosion was less drastic (Dronkers & Mrozek 1991), Lower Cretaceous sediments are better preserved and hydrocarbon-bearing Scruff sandstones are found (de Jager 2003; de Jager & Geluk 2007) (Fig. 2). Additionally, the adjacent Terschelling Basin, where moderate inversion is recorded (Verweij & Witmans 2009), contains relatively thick Lower Cretaceous deposits (Duin *et al.* 2006; EBN *et al.* 2015).

On the southern fringes of the Southern Permian Basin, the Broad Fourteens Basin and West Netherlands Basin form a continuation of the Lower Cretaceous North Sea basins (Fig. 11a). Although these basins also experienced strong post-rift inversion (Van Wijhe 1987; de Jager 2003), significant parts of the Lower Cretaceous deposits are preserved in the area (over 900 m thick locally: Duin *et al.* 2006) and contain various hydrocarbon fields (de Jager & Geluk 2007). Similar to the situation in the Moray Firth, the associated reservoirs are documented to be visible on seismic due to a relatively thin Upper Cretaceous–Cenozoic overburden (Oakman 2005). The Early Cretaceous depositional environment was, however, rather different from the

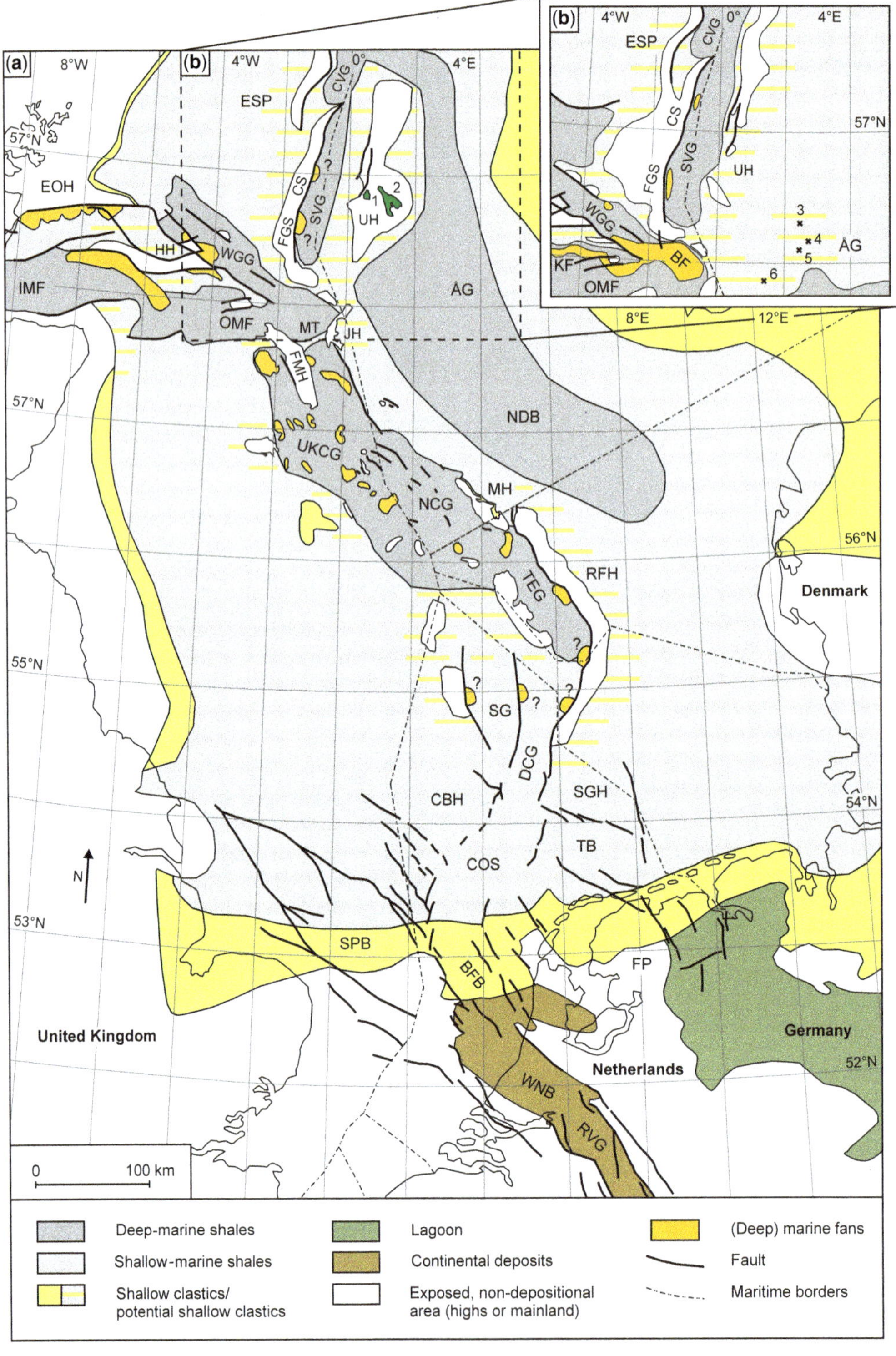

(a)
(b)
8°W
4°W
0°
4°E
8°E
12°E
57°N
56°N
55°N
54°N
53°N
52°N
EOH
HH
IMF
WGG
OMF
MT
JH
FMH
ESP
CVG
CS
SVG
FGS
UH
1
2
ÅG
NDB
UKCG
NCG
MH
TEG
RFH
SG
DCG
SGH
CBH
TB
COS
SPB
BFB
FP
WNB
RVG
KF
BF
3
4
5
6
N
0
100 km
United Kingdom
Netherlands
Germany
Denmark
Deep-marine shales
Shallow-marine shales
Shallow clastics/ potential shallow clastics
Lagoon
Continental deposits
Exposed, non-depositional area (highs or mainland)
(Deep) marine fans
Fault
Maritime borders

situation in the Central Graben and Moray Firth. Instead of isolated shale-dominated basins, receiving limited sand influx from small exposed highs nearby, the area received ample sediment input from the large London–Brabant Massif to the south (Jeremiah *et al.* 2010) (Fig. 11a). Therefore, extensive continental to shallow-marine shelf clastics were deposited in relatively shallow basins, in contrast to the deep-marine basin settings in the Central and Northern North Sea (Figs 2 & 11). The abundance of sand-prone material in the depositional systems on the fringe of the Southern Permian Basin could potentially have fed submarine fans in the rift depocentres further north. The area experienced tectonically induced rejuvenation of clastic input, progradation and the development of a widespread shelf system at the K30 sequence boundary, which is associated with increased Hauterivian deep-marine reservoir development in the Moray Firth (DeVault & Jeremiah 2002). However, except for the Lower Barremian (Wanneperveen) turbidite units found in association with the Friesland Platform near the Dutch–German border (Jeremiah *et al.* 2010) (Fig. 11a), no such deposits are recorded in the Southern Permian Basin. This scarcity of deep-marine sand development may be related to the area's relatively gentle bathymetry during the Early Cretaceous (Fig. 11a), although various other factors are known to affect turbidite systems, such as shelf width, surrounding geomorphology and hinterland lithologies (Martinsen *et al.* 2005; Mudge 2014).

Analogue settings in the South Viking Graben area

Another potential analogue region to the NCG is the South Viking Graben (SVG: Fig. 11). In contrast to the Southern North Sea, the area harboured deep-marine basins (flanked by exposed highs) during the earliest Cretaceous (Fig. 11a). Shallow-marine or terrestrial sandstones were deposited on the Utsira High, forming parts of the reservoirs in the Edvard Grieg and Johan Sverdrup fields (NPD 2017), and may be directly comparable with the Mandal High in the NCG (Rossland *et al.* 2013). The SVG is documented to include Upper Jurassic turbidites (Partington *et al.* 1993; Fraser *et al.* 2003; Jackson *et al.* 2011) and the associated Fladen Ground Spur, Crawford Spur and Utsira High sand provenance areas continued to be exposed in the earliest Cretaceous (Copestake *et al.* 2003) (Fig. 11a). However, no earliest Ryazanian deep-marine sands are reported from the SVG area, potentially providing exploration opportunities.

The situation was different during the Aptian, as Skiff Sandstone units are reported along the fringes of the Fladen Ground Spur and the Crawford Spur (Johnson & Lott 1993; Johnson *et al.* 2005) (Figs 2 & 11b). To the south, the Kopervik fairway supplied the reservoirs of the giant Britannia Field with sands derived from the East Orkney High in the west (Jeremiah 2000). Oakman (2005) suggested that these deep-marine sands represent a fundamentally different depositional system for the Aptian–Albian interval, rather similar to the Cenozoic situation and involving sediment transport over long distances sourced by the exposed North Atlantic rift shoulders, in contrast to the preceding confined Upper Jurassic turbidite fans. The Kopervik system is, however, separated from the SVG by a halokinetically induced high that was in place throughout the Early Cretaceous, so that potential sandstone deposits in the SVG can only be derived from the adjacent highs (Bisewski 1990) (Fig. 11). Further to the north, in the North Viking Graben, deep-marine slumps of Albian age form reservoirs of the Agat Field (Skibeli *et al.* 1995) but these deposits were derived from the main Scandinavian massif (Gulbrandsen & Nyborkken 1991), whereas the SVG remained relatively isolated.

Other wellbore-calibrated sandstone occurrences in the area are reported from the Åsta Graben, SE of the Utsira High (3–6 in Fig. 11b; Table 1). These Ran Sandstone units all occur in the uppermost part of the

Fig. 11. (**a**) Gross depositional environment overview of the Central and Southern North Sea in Ryazanian times (K10). (**b**) Gross depositional environment of the South Viking Graben area in Aptian times (K40–K50), corresponding to the Aptian–Albian reservoir interval. Lower Cretaceous sand presence: (1) Edvard Grieg Field; (2) Johan Sverdrup Field; (3) well NO 7/10-1; (4) well NO 7/11-1; (5) well NO 7/11-2; (6) well NO 7/3-1; ÅG, Åsta Graben; BF, Britannia Field; BFB, Broad Fourteens Basin; CBH, Cleaver Bank High; COS, Central Offshore Saddle; CS, Crawford Spur; CVG, Central Viking Graben; DCG, Dutch Central Graben; EOH, East Orkney High; ESP, East Shetland Platform; FP, Friesland Platform; HH, Halibut High; IMF, Inner Moray Firth; JH, Jæren High; FMH, Forties–Montrose High; KF, Kopervik fairway; MH, Mandal High; MT, Marnock Terrace; NCG, Norwegian Central Graben; NDB, Norwegian–Danish Basin; OMF, Outer Moray Firth; RFH, Ringkøbing-Fyn High; RVG, Roer Valley Graben; SG, Step Graben; SGH, Schill Grund High; SPB, Sole Pit Basin; TB, Terschelling Basin; TEG, Tail End Graben (Danish Central Graben); UH, Utsira High; UKCG, UK Central Graben; WGG, Witch Ground Graben; WNB, West Netherlands Basin. Modified after Copestake *et al.* (2003) and NPD (2017) for the South Viking Graben area; after Copestake *et al.* (2003), Milton-Worssell *et al.* (2006) and Rossland *et al.* (2013) for the UKCG and NCG; after Pharaoh *et al.* (2010) and Vejbæk *et al.* (2010) for the Danish, German and (parts of) the Dutch Central Graben; and after Jeremiah *et al.* (2010) for the Southern North Sea.

Lower Cretaceous, directly underneath the Upper Cretaceous chalk deposits, and are likely to have been deposited in a shallow-marine environment (Olsen 1979; Isaksen & Tonstad 1989).

Potential methods for further detailed reservoir interpretation

As demonstrated by Milton-Worssell *et al.* (2006), detailed seismic analysis is required to distinguish potential reservoir units. However, the presence of thick Upper Cretaceous chalk and the low impedance contrasts between Cromer Knoll shales and sandstones renders seismic imaging of any relatively thin (typically <30 m) Lower Cretaceous sandstone reservoir problematic. To do so requires good-quality 3D seismic data combined with an understanding of the likely depositional systems to be encountered (Crittenden *et al.* 1998; Law *et al.* 2000; McKie *et al.* 2015). With such data available, sedimentary systems such as deep-marine fans may be traceable on time-slice amplitude maps (e.g. Posamentier & Kolla 2003; Martinsen *et al.* 2005; Kilhams *et al.* 2011, 2014*a*). Amplitude v. offset (AVO) techniques could help to distinguish differences in lithology and reservoir fluid content (e.g. Oakman 2005; Veeken & Rauch-Davies 2006; Milton-Worssell *et al.* 2008; Othman *et al.* 2017). Such a study would be recommended for the Tail End Graben area, as there is potential for small-scale reservoir development. Furthermore, the seismic response of Lower Cretaceous sandstone well occurrences in the NCG, as well as in the UKCG where sandstones are more common (Milton-Worssell *et al.* 2006), should be compared to seismic facies in undrilled depocentres. Detailed seismic sequence stratigraphy of Lower Cretaceous depocentres could allow the identification of sea-level-driven erosional unconformities on highs, associated with lowstand fan systems in basinal areas (Posamentier & Vail 1988).

Methods to further assess sand source areas and to localize associated shallow- to deep-marine sandstones might include palynological (or similar biostratigraphic) analysis of cored wells to determine to what degree a high was exposed (e.g. O'Driscoll *et al.* 1990; Mudge & Jones 2004; McArthur *et al.* 2016*b*). Since cores from wells NO 2/7-15 and NO 3/7-3 are available (NPD 2017), magnetic analysis could provide sediment transport directions of these specific Early Cretaceous sandstone occurrences (Hailwood & Ding 2000). Additional petrological and geochemical analysis of heavy minerals (e.g. garnets or zircons) might reveal their provenance area (e.g. Morton *et al.* 2005; Kilhams *et al.* 2014*b*; Nielsen *et al.* 2015), if cuttings/cores of nearby sand source areas are available (e.g. well NO 3/7-1 on the Mandal High, and wells NO 3/7-3 and Tabita-1 in the Søgne Basin and Tail End Graben, respectively: Verolles 1982; Bonde *et al.* 1994) (Fig. 7). Furthermore, it will be important to consider the factors influencing the behaviour and geometries of shoreface systems and deep-marine fans (e.g. sand-to-mud ratio, flow discharge, slope gradient, sea-level changes and fault activity) and where sand deposits occur in these depositional environments (e.g. Posamentier & Kolla 2003; Martinsen *et al.* 2005; McKie *et al.* 2015; McArthur *et al.* 2016*a*). Recently developed software for the simulation of turbidite deposition in combination with palaeorelief reconstructions on 3D seismic could be a powerful tool to predict the distribution of deep-marine fans (Aas *et al.* 2010).

Conclusions

Here, a structural framework of the NCG area has been presented reflecting a diverse geological history including Triassic extension and salt movement, Late Jurassic–Early Cretaceous rifting, and subsequent basin inversion with salt diapirism. Late Jurassic rifting was most intense in the south of the study area, while Early Cretaceous rifting was more important in the north, possibly representing an Early Cretaceous change in tectonic regime before rifting halted altogether. An assessment of the Lower Cretaceous indicates a fair potential for reservoir development. Although the study area is isolated from a large hinterland, local sediment sources and potential sediment transport routes are identified. Most potential is expected around the highs that were exposed in Ryazanian times, while many sand source areas were drowned at the end of the Early Cretaceous (Aptian–Albian). The underexplored Mandal High area, where restricted shallow-marine sandstone deposits around the exposed Mandal High and in the Søgne Basin provide the best potential, is suggested for further focus. Similar depositional environments could have existed around other exposed highs (e.g. Josephine High and Auk Ridge), although they may have been too small to have produced significant reservoir units. Furthermore, the postulated Amalie fan system in Denmark illustrates the possibilities for good-quality deep-marine sandstones, which may have also formed in the depocentres south and SW of the Mandal High. Analogous settings to those in the study area are also recognized in the Southern Permian Basin area to the south and the South Viking Graben to the north. Further analysis of the Lower Cretaceous reservoir intervals of these areas would be an interesting next step. A detailed effort, including the use of advanced seismic techniques and detailed well analysis, will be necessary to accurately define such reservoirs, if present. The discovery of the Edvard Grieg and Johan Sverdrup fields illustrates the importance of continued exploration, especially the reassessment of available well and seismic data, in the context of this mature hydrocarbon province (Jørstad 2012).

This project was originally undertaken as an internship at Shell Upstream International in Assen, The Netherlands as the basis for a master's project (VU University Amsterdam). I would first like to express my gratitude to Carlo Nicolai and Ben Kilhams (Royal Dutch Shell), as well as Jan de Jager (VU University Amsterdam), for their support and guidance during this project, and for helping me to prepare this manuscript for publication. I would also like to thank my colleagues from the Shell UK/Netherlands exploration teams, the NAM IT team and colleagues from Shell Norway for their help and support. I am grateful to Adam McArthur and Jason Jeremiah for their detailed and constructive comments, and to Royal Dutch Shell for funding this project and for allowing me to publish its results.

References

AAS, T.E., HOWELL, J.A., JANOCKO, M. & JACKSON, C.A.-L. 2010. Control of Aptian palaeobathymetry on turbidite distribution in the Buchan Graben, Outer Moray Firth, Central North Sea. *Marine and Petroleum Geology*, **27**, 412–434, https://doi.org/10.1016/j.marpetgeo.2009.10.014

AINSWORTH, N.R., RILEY, L.A. & GALLAGHER, L.T. 2000. An Early Cretaceous lithostratigraphic and biostratigraphic framework for the Britannia Field reservoir (Late Barremian–Late Aptian), UK North Sea. *Petroleum Geoscience*, **6**, 345–367, https://doi.org/10.1144/petgeo.6.4.345

ANDSBJERG, J. & DYBKJÆR, K. 2003. Sequence stratigraphy of the Jurassic of the Danish Central Graben. *In*: INESON, J.R. & SURLYK, F. (eds) *The Jurassic of Denmark and Greenland. Geological Survey of Denmark and Greenland Bulletin*, **1**, 265–300.

ARGENT, J.D., STEWART, S.A. & UNDERHILL, J.R. 2000. Controls on the Lower Cretaceous Punt Sandstone Member, a massive deep-water clastic deposystem, Inner Moray Firth, UK North Sea. *Petroleum Geoscience*, **6**, 275–285, https://doi.org/10.1144/petgeo.6.3.275

A/S NORSKE SHELL 1968. *Exploration Well Resumé. Well 17/11-1*, A/S Norske Shell Exploration & Petroleum Engineering Departments http://www.npd.no/engelsk/cwi/pbl/wellbore_documents/906_01_Exploration_well_resume.pdf

BARTHOLOMEW, I.D., PETERS, J.M. & POWELL, C.M. 1993. Regional structural evolution of the North Sea: oblique slip and the reactivation of basement lineaments. *In*: PARKER, J.R. (ed.) *Petroleum Geology: North-West Europe and Global Perspectives – Proceedings of the 4th Petroleum Geology Conference*. Geological Society, London, 1109–1122, https://doi.org/10.1144/0041109

BERGAN, M., TØRUDBAKKEN, B. & WANDÅS, B. 1989. Lithostratigraphic correlation of Upper Jurassic sandstones within the Norwegian Central Graben: sedimentological and tectonic implications. *In*: COLLINSON, J. (ed.) *Correlation in Hydrocarbon Exploration*. Graham & Trotman, London, 243–251, https://doi.org/10.1007/978-94-009-1149-9_19

BISEWSKI, H. 1990. Occurrence and depositional environment of the Lower Cretaceous sands in the southern Witch Ground Graben. *In*: HARDMAN, R.F.P. & BROOKS, J. (eds) *Tectonic Events Responsible for Britain's Oil and Gas Reserves*. Geological Society, London, Special Publications, **55**, 325–338, https://doi.org/10.1144/GSL.SP.1990.055.01.15

BLACKBOURN, G.A. & THOMSON, M.E. 2000. Britannia Field, UK North Sea: petrographic constraints on Cretaceous provenance, facies and the origin of slurry-flow deposits. *Petroleum Geoscience*, **6**, 329–343, https://doi.org/10.1144/petgeo.6.4.329

BOIRIE, J.M. & JEANNOU, A. 1984. *Well 3/7-3 Lithostratigraphy and reservoir aspect of the pre-chalk sediments (Permian–Lower Cretaceous; Central Graben) Southern District – Offshore Norway*. Boussens Well Report, 062/84 RP, http://www.npd.no/engelsk/cwi/pbl/wellbore_documents/293_03_3_7_3_LITHOSTRATIGRAPHY_AND_RESERVOIR_ASPECT_OF_THE_PRE-CHALK_SEDIMENTS.pdf

BONDE, K., SANDVIK, R., SLOT-PETERSEN, C. & OLSEN, T. 1994. *Tabita-1, Tabita-1A Completion Report*. Statoil Efterforskning OG Produktion A/S, Copenhagen, Denmark.

BRITOIL 1985. *Iris-1 Composite Well Log*. Britoil plc Exploration Drawing Office, Glasgow, UK.

BROOKS, J. 1990. Classic petroleum provinces. *In*: BROOKS, J. (ed.) *Classic Petroleum Provinces*. Geological Society, London, Special Publications, **50**, 1–8, https://doi.org/10.1144/GSL.SP.1990.050.01.01

CARTWRIGHT, J.A. 1989. The kinematics of inversion in the Danish Central Graben. *In*: COOPER, M.A. & WILLIAMS, G.D. (eds) *Inversion Tectonics*. Geological Society, London, Special Publications, **44**, 153–175, https://doi.org/10.1144/GSL.SP.1989.044.01.10

COPESTAKE, P., SIMS, A.P., CRITTENDEN, S., HAMAR, G.P., INESON, J.R., ROSE, P.T. & TRINGHAM, M.E. 2003. Lower Cretaceous. *In*: EVANS, D., GRAHAM, C., ARMOUR, A. & BATHURST, P. (eds) *The Millennium Atlas: Petroleum Geology of the Central and Northern North Sea*. Geological Society, London, 191–211.

COWARD, M.P., DEWEY, J.F., HEMPTON, M. & HOLROYD, J. 2003. Tectonic evolution. *In*: EVANS, D., GRAHAM, C., ARMOUR, A. & BATHURST, P. (eds) *The Millennium Atlas: Petroleum Geology of the Central and Northern North Sea*. Geological Society, London, 17–33.

CRITTENDEN, S., COLE, J.M. & KIRK, M.J. 1997. The distribution of Aptian sandstones in the central and northern north sea (UK sectors): a lowstand systems tract 'play'. Part 1. Stratigraphy, biostratigraphic age determination, and genesis of the sandstones. *Journal of Petroleum Geology*, **20**, 3–25.

CRITTENDEN, S., COLE, J.M. & KIRK, M.J. 1998. The distribution of Aptian sandstones in the central and Northern North Sea (UK sectors) – a lowstand systems tract play. Part 2. distribution and exploration strategy. *Journal of Petroleum Geology*, **21**, 187–211.

DE JAGER, J. 2003. Inverted Basins in the Netherlands, similarities and differences. *Netherlands Journal of Geosciences – Geologie en Mijnbouw*, **82**, 355–366, https://doi.org/10.1017/S0016774600020175

de Jager, J. 2007. Geological development. *In*: Wong, Th.E., Batjes, D.A.J. & de Jager, J. (eds) *Geology of the Netherlands*. Royal Netherlands Academy of Science and Art, Amsterdam, 5–26.

de Jager, J. & Geluk, M.C. 2007. Petroleum geology. *In*: Wong, Th.E., Batjes, D.A.J. & de Jager, J. (eds) *Geology of the Netherlands*. Royal Netherlands Academy of Science and Art, Amsterdam, 241–264.

DeVault, B. & Jeremiah, J. 2002. Tectonostratigraphy of the Nieuwerkerk Formation (Delfland subgroup), West Netherlands Basin. *AAPG Bulletin*, **86**, 1679–1707, https://doi.org/10.1306/61EEDD50-173E-11D7-8645000102C1865D

Dronkers, A.J. & Mrozek, F.J. 1991. Inverted basins of The Netherlands. *First Break*, **9**, 409–425, https://doi.org/10.3997/1365-2397.1991019

Duin, E.J.T., Doornebal, J.C., Rijkers, R.H., Verbeek, J.W. & Wong, Th.E. 2006. Subsurface structure of the Netherlands – results of recent onshore and offshore mapping. *Netherlands Journal of Geosciences – Geologie en Mijnbouw*, **85**, 355–366.

EBN, Ministry of Economic Affairs & TNO 2015. *Prospectivity G and M Blocks – Analysing the Upper Jurassic Play*, http://www.nlog.nl/cmis/browser?id=workspace%3A//SpacesStore/f84ec4ac-b9a6-4771-9212-6f6959233ead

Eggenhuisen, J.T., McCaffrey, W.D., Haughton, P.D.W., Butler, R.W.H., Moore, I., Jarvie, A. & Hakes, W.G. 2010. Reconstructing large-scale remobilisation of deep-water deposits and its impact on sand-body architecture from cored wells: The Lower Cretaceous Britannia Sandstone Formation, UK North Sea. *Marine and Petroleum Geology*, **27**, 1595–1615, https://doi.org/10.1016/j.marpetgeo.2010.04.005

Erratt, D., Thomas, G.M. & Wall, G.R.T. 1999. The evolution of the Central North Sea Rift. Geological Society. *In*: Fleet, A.J. & Boldy, S.A.R. (eds) *Petroleum Geology: North-West Europe and Global Perspectives: Proceedings of the 5th Petroleum Geology Conference*. Geological Society, London, 63–82, https://doi.org/10.1144/0050063

Fjellanger, E. 1986. *Geological Completion Report, Well 2/1-8*. BP Petroleum Development (Norway) Ltd, Stavanger, Norway, http://www.npd.no/engelsk/cwi/pbl/wellbore_documents/459_2_1_8_Completion_report.pdf

Fossen, H., Dallman, W. & Andersen, T.B. 2008. The mountain chain rebounds and founders. *In*: Ramberg, I.B., Bryhni, I., Nøttvedt, A. & Rangnes, K. (eds) *The Making of a Land – Geology of Norway*. Norsk Geologisk Forening, Trondheim, Norway, 232–259.

Fraser, S.I., Robinson, A.M. *et al.* 2003. Upper Jurassic. *In*: Evans, D., Graham, C., Armour, A. & Bathurst, P. (eds) *The Millennium Atlas: Petroleum Geology of the Central and Northern North Sea*. Geological Society, London, 155–189.

Garrett, S.W., Atherton, T. & Hurst, A. 2000. Lower Cretaceous deep-water sandstone reservoirs in the UK Central North Sea. *Petroleum Geoscience*, **6**, 231–240, https://doi.org/10.1144/petgeo.6.3.231

Gautier, D.L. 2005. *Kimmeridge Shales Total Petroleum System of the North Sea Graben Province*. United States Geological Survey Bulletin, **2204-C**, https://pubs.usgs.gov/bul/2204/c/pdf/B2204C.pdf

Ge, Z., Gawthorpe, R.L., Rotevatn, A. & Thomas, M.B. 2017. Impact of normal faulting and pre-rift salt tectonics on the structural style of salt-influenced rifts: the Late Jurassic Norwegian Central Graben, North Sea. *Basin Research*, **29**, 674–698, https://doi.org/10.1111/bre.12219

Gowers, M.B. & Sæbøe, A. 1985. On the structural evolution of the Central Trough in the Norwegian and Danish sectors of the North Sea. *Marine and Petroleum Geology*, **2**, 298–318, https://doi.org/10.1016/0264-8172(85)90026-1

Gowers, M.B., Holtar, E. & Swensson, E. 1993. The structure of the Norwegian Central Trough (Central Graben area). *In*: Parker, J.R. (ed.) *Petroleum Geology: North-West Europe and Global Perspectives: Proceedings of the 4th Petroleum Geology Conference*. Geological Society, London, 1245–1254, https://doi.org/10.1144/0041245

Graversen, O. 2006. The Jurassic–Cretaceous North Sea rift dome and associated basin evolution. *AAPG Search and Discovery Article #3004, AAPG Annual Convention*, June 19–22, 2005, Calgary, Alberta, Canada, http://www.searchanddiscovery.com/documents/2006/06013graversen/images/graversen

Gulbrandsen, A. & Nyborkken, S. 1991. Agat Field – Norwegian North Sea, Subplatform North Viking Graben. *AAPG Search and Discovery Article #91004, AAPG Annual Convention*, April 7–10, 1991, Dallas, Texas, USA, http://www.searchanddiscovery.com/abstracts/html/1991/annual/abstracts/0586b.htm

Hailwood, E. & Ding, F. 2000. Sediment transport and dispersal pathways in the Lower Cretaceous sands of the Britannia Field, derived from magnetic anisotropy. *Petroleum Geology*, **6**, 369–379, https://doi.org/10.1144/petgeo.6.4.369

Hampson, J.M., Walden, S.F. & Bell, C. 2010. Maximizing production and reserves from offshore heavy oil fields using seismic and drilling technology: Alba and Captain Fields, UKNS. *In*: Vining, B.A. & Pickering, S.C. (eds) *Petroleum Geology: From Mature Basins to New Frontiers – Proceedings of the 7th Petroleum Geology Conference*. Geological Society, London, 337–347, https://doi.org/10.1144/0070337

Herngreen, G.F.W. & Wong, Th.E. 2007. Cretaceous. *In*: Wong, Th.E., Batjes, D.A.J. & de Jager, J. (eds) *Geology of the Netherlands*. Royal Netherlands Academy of Science and Art, Amsterdam, 127–150.

Hodgson, N.A., Fansworth, J. & Fraser, A.J. 1992. Salt-related tectonics, sedimentation and hydrocarbon plays in the Central Graben, North Sea, UKCS. *In*: Hardman, R.F.P. (ed.) *Exploration Britain: Geological Insights for the Next Decade*. Geological Society, London, Special Publications, **67**, 31–63, https://doi.org/10.1144/GSL.SP.1992.067.01.03

Isaksen, D. & Tonstad, K. 1989. *A Revised Cretaceous and Tertiary Lithostratigraphic Nomenclature for the Norwegian North Sea*. Norwegian Petroleum Directorate Bulletin, **5**, http://www.npd.no/Global/Norsk/3-Publikasjoner/NPD-Bulletin/Bulletinnr5.pdf

Jackson, C.A.-L. & Lewis, M.M. 2016. Structural style and evolution of a salt-influenced rift basin margin; the impact of variations in salt composition and the role of polyphase extension. *Basin Research*, **28**, 81–102, https://doi.org/10.1111/bre.12099

JACKSON, C.A.-L., LARSEN, E., HANSLIEN, S. & TJEMSLAND, A.-E. 2011. Controls on synrift turbidite deposition on the hanging wall of the South Viking Graben, North Sea rift system, offshore Norway. *AAPG Bulletin*, **95**, 1557–1587, https://doi.org/10.1306/010311 10037

JAKOBSEN, F., INESON, J.R., KIRSTENSEN, L., NYTOFT, H.P. & STEMMERIK, L. 2005. The Valdemar Field, Danish Central Graben: field compartmentalization and regional prospectivity of the Lower Cretaceous chalk play. *In*: DORÉ, A.G. & VINING, B.A. (eds) *Petroleum Geology: North-West Europe and Global Perspectives – Proceedings of the 6th Petroleum Geology Conference*. Geological Society, London, 177–186, https://doi.org/10.1144/0060177

JAPSEN, P., BRITZE, P. & ANDERSEN, C. 2003. Upper Jurassic–Lower Cretaceous of the Danish Central Graben: structural framework and nomenclature. *In*: INESON, J.R. & SURLYK, F. (eds) *The Jurassic of Denmark and Greenland*. Geological Survey of Denmark and Greenland Bulletin, **1**, 233–246, http://www.geus.dk/DK/publications/geol-survey-dk-gl-bull/Documents/nr1_p231-246.pdf

JEREMIAH, J.M. 2000. Lower Cretaceous turbidites of the Moray Firth: sequence stratigraphical framework and reservoir distribution. *Petroleum Geoscience*, **6**, 309–328, https://doi.org/10.1144/petgeo.6.4.309

JEREMIAH, J.M., DUXBURY, S. & RAWSON, P. 2010. Lower Cretaceous of the southern North Sea Basins: reservoir distribution within a sequence stratigraphic framework. *Netherlands Journal of Geosciences – Geologie en Mijnbouw*, **89**, 203–237, https://doi.org/10.1017/S0016774600000706

JOHANNESSEN, P.N. & ANDSBJERG, J. 1993. Middle to Late Jurassic basin evolution and sandstone reservoir distribution in the Danish Central Trough. *In*: PARKER, J.R. (ed.) *Petroleum Geology: North-West Europe and Global Perspectives: Proceedings of the 4th Petroleum Geology Conference*. Geological Society, London, 271–283, https://doi.org/10.1144/0040271

JOHNSON, H., LESLIE, A.B., WILSON, C.K., ANDREWS, I.J. & COOPER, R.M. 2005. *Middle Jurassic, Upper Jurassic and Lower Cretaceous of the UK Central and Northern North Sea*. British Geological Survey Research Report, RR/03/001, http://nora.nerc.ac.uk/3680/1/RR03001.pdf

JOHNSON, J. & LOTT, G.K. 1993. 2. Cretaceous of the Central and Northern North Sea. *In*: KNOX, R.W.O'B. & CORDEY, W.G. (eds) *Lithostratigraphic Nomenclature of the UK North Sea*. British Geological Survey, Nottingham, UK, http://www.bgs.ac.uk/downloads/start.cfm?id=2044

JONES, L.S., GARRETT, S.W., MACLEOD, M., GUY, M., CODON, P.J. & NOTMAN, L. 1999. Britannia Field, UK Central North Sea: modelling heterogeneity in unusual deep-water deposits. *In*: FLEET, A.J. & BOLDY, S.A.R. (eds) *Petroleum Geology: North-West Europe and Global Perspectives – Proceedings of the 5th Petroleum Geology Conference*. Geological Society, London, 1115–1124, https://doi.org/10.1144/0051115

JØRSTAD, A.J. 2012. Johan Sverdrup – offshore Norway: the story behind the Giant Sverdrup Discovery. *AAPG Search and Discovery Article #20177, AAPG International Conference and Exhibition*, September 16–19, 2012, Singapore, https://www.google.ch/url?sa=t&rct=j&q=&esrc=s&source=web&cd=1&cad=rja&uact=8&ved=0ahUKEwiL4v6vwurVAhXGJMAKHXveBpQQFggyMAA&url=http%3A%2F%2Fwww.searchanddiscovery.com%2Fdocuments%2F2012%2F20177jorstad%2Fndx_jorstad.pdf&usg=AFQjCNErkXDHLyXjqcp8dlbV_ZJuqMtKYQ

KERN, J.C., RØRVIG, M., LIBORIUSSEN, J., BLOM, N.S. & MUELDER, J. 1983. *Sten-1 Completion Report*. Chevron Petroleum Company of Denmark, Copenhagen, Denmark.

KILHAMS, B., GODFREY, S., HARLTEY, A. & HUUSE, M. 2011. An integrated 3D seismic, petrophysical and analogue core study of the Mid-Eocene Grid channel complex in the greater Nelson Field area, UK Central North Sea. *Petroleum Geoscience*, **17**, 127–142, https://doi.org/10.1144/1354-079310-022

KILHAMS, B., HARTLEY, A., HUUSE, M. & DAVIS, C. 2014*a*. Characterizing the Paleocene turbidites of the North Sea: Maureen Formation, UK Central Graben. *In*: MCKIE, T., ROSE, P.T.S., HARTLEY, A.J., JONES, D.W. & ARMSTRONG, T.L. (eds) *Tertiary Deep-Marine Reservoirs of the North Sea Region*. Geological Society, London, Special Publications, **403**, 43–62, https://doi.org/10.1144/SP403.1

KILHAMS, B., MORTON, A., BORELLA, R., WILKINS, A. & HURST, A. 2014*b*. Understanding the provenance and reservoir quality of the Sele Formation sandstones of the UK Central Graben utilizing detrital garnet suites. *In*: SCOTT, R.A., SMYTH, H.R., MORTON, A.C. & RICHARDSON, N. (eds) *Sediment Provenance Studies in Hydrocarbon Exploration and Production*. Geological Society, London, Special Publications, **386**, 129–142, https://doi.org/10.1144/SP386.16

LAW, A., RAYMOND, A. *ET AL*. 2000. The Kopervik fairway, Moray Firth, UK. *Petroleum Geoscience*, **6**, 265–274, https://doi.org/10.1144/petgeo.6.3.265

LUNDIN PETROLEUM 2017. *Johan Sverdrup Development*. Lundin Petroleum press releases, https://www.lundin-petroleum.com/download/ot_johansverdrup_present_e/?wpdmdl=6265/

MÆRSK 1987. *Final Well Report West Lulu-4*. Mærsk Olie OG Gas A/S Petroleum Engineering and Operations Department.

MARTINSEN, O.J., LIEN, T. & JACKSON, C.J. 2005. Cretaceous and Palaeogene turbidite systems in the North Sea and Norwegian Sea Basins: source, staging area and basin physiography controls on reservoir development. *In*: DORÉ, A.G. & VINING, B.A. (eds) *Petroleum Geology: North-West Europe and Global Perspectives – Proceedings of the 6th Petroleum Geology Conference*. Geological Society, London, 1147–1164, https://doi.org/10.1144/0061147

MCARTHUR, A.D., HARTLEY, A.J., ARCHER, S.G., JOLLEY, D.W. & LAWRENCE, H.M. 2016*a*. Spatiotemporal relationships of deep-marine, axial, and transverse depositional systems from the synrift Upper Jurassic of the central North Sea. *AAPG Bulletin*, **100**, 1469–1500, https://doi.org/10.1306/04041615125

MCARTHUR, A.D., JOLLEY, D.W., HARTLEY, A.J., ARCHER, S.G. & LAWRENCE, H.M. 2016*b*. Palaeoecology of synrift topography: A Late Jurassic footwall island on the Josephine Ridge, Central Graben, North Sea. *Palaeogeography, Palaeoclimatology, Palaeoecology*, **459**, 63–75, https://doi.org/10.1016/j.palaeo.2016.06.033

McGann, G.J., Green, S.C.H., Harker, S.D. & Romani, R.S. 1991. The Scapa Field, Block 14/19, UK North Sea. *In*: Abbotts, I.L. (ed.) *United Kingdom Oil and Gas Fields, 25 Years Commemorative Volume*. Geological Society, London, Memoirs, **14**, 369–376, https://doi.org/10.1144/GSL.MEM.1991.014.01.46

McKie, T., Rose, P.T.S., Hartley, A.J., Jones, D.W. & Armstrong, T.L. 2015. Tertiary deep-marine reservoirs of the North Sea region: an introduction. *In*: McKie, T., Rose, P.T.S., Hartley, A.J., Jones, D.W. & Armstrong, T.L. (eds) *Tertiary Deep-Marine Reservoirs of the North Sea Region*. Geological Society, London, Special Publications, **403**, 1–16, https://doi.org/10.1144/SP403.12

Michelsen, O., Nielsen, L.H., Johannessen, P.N., Andsbjerg, J. & Surlyk, F. 2003. Jurassic lithostratigraphy and stratigraphic development onshore and offshore Denmark. *In*: Ineson, J.R. & Surlyk, F. (eds) *The Jurassic of Denmark and Greenland*. Geological Survey of Denmark and Greenland Bulletin, **1**, 147–216, http://www.geus.dk/DK/publications/geol-survey-dk-gl-bull/Documents/nr1_p145-216.pdf

Milton, N.J. 1993. Evolving depositional geometries in the North Sea Jurassic rift. *In*: Parker, J.R. (ed.) *Petroleum Geology: North-West Europe and Global Perspectives – Proceedings of the 4th Petroleum Geology Conference*. Geological Society, London, 425–442, https://doi.org/10.1144/0040425

Milton-Worssell, R.J., Stoker, S.J. & Cavill, J.E. 2006. Lower Cretaceous deep-water sandstone plays in the UK Central Graben. *In*: Allen, M.R., Goffey, G.P., Morgan, R.K. & Walker, I.M. (eds) *The Deliberate Search for the Stratigraphic Trap*. Geological Society, London, Special Publications, **254**, 169–186, https://doi.org/10.1144/GSL.SP.2006.254.01.09

Milton-Worssell, R., Fisher, G., Cameron, D., Stoker, S., Avu, A. & Basford, H. 2008. Where are those Lower Cretaceous sands? Geobody detection in a depocentre. *PETEX Conference Proceedings, November 25–27, 2008, London*. Petroleum Exploration Society of Great Britain (PESGB), London, https://www.researchgate.net/publication/237484853_WHERE_ARE_THOSE_LOWER_CRETACEOUS_SANDS_GEOBODY_DETECTION_IN_A_DEPOCENTRE

Møller, J.J. & Rasmussen, E.S. 2003. Middle Jurassic – Early Cretaceous rifting of the Danish Central Graben. *In*: Ineson, J.R. & Surlyk, F. (eds) *The Jurassic of Denmark and Greenland*. Geological Survey of Denmark and Greenland Bulletin, **1**, 247–264, http://www.geus.dk/DK/publications/geol-survey-dk-gl-bull/Documents/nr1_p247-264.pdf

Morton, A.C., Whitman, A.G. & Fanning, C.M. 2005. Provenance of Late Cretaceous to Paleocene submarine fan sandstones in the Norwegian Sea: integration of heavy mineral, mineral chemical and zircon age data. *Sedimentary Geology*, **182**, 3–28, https://doi.org/10.1016/j.sedgeo.2005.08.007

Mudge, D.C. 2014. Regional controls on Lower Tertiary sandstone distribution in the North Sea and NE Atlantic margin basins. *In*: McKie, T., Rose, P.T.S., Hartley, A.J., Jones, D.W. & Armstrong, T.L. (eds) *Tertiary Deep-Marine Reservoirs of the North Sea Region*. Geological Society, London, Special Publications, **403**, 17–42, https://doi.org/10.1144/SP403.5

Mudge, D.C. & Jones, S.M. 2004. Palaeocene uplift and subsidence events in the Scotland–Shetland and North Sea region and their relationship to the Iceland Plume. *Journal of the Geological Society, London*, **161**, 381–386, https://doi.org/10.1144/0016-764 903-038

NDD 2012. The HPHT and UHPHT play in the Norwegian Central Graben. *Norway Discovery Digest*, **13**, 12–33, http://norwaydigest.canadiandiscovery.com/?q=node/33144

Nielsen, M.T., Weibel, R. & Friis, H. 2015. Provenance of gravity-flow sandstones from the Upper Jurassice–Lower Cretaceous Farsund Formation, Danish Central Graben, North Sea. *Marine and Petroleum Geology*, **59**, 371–389, https://doi.org/10.1016/j.marpetgeo.2014.09.016

Normark, W.R., Piper, D.J.W. & Hess, G.R. 1979. Distributary channels, sand lobes, and mesotopography of Navy Submarine Fan, California Borderland, with applications to ancient fan sediments. *Sedimentology*, **26**, 749–774, https://doi.org/10.1111/j.1365-3091.1979.tb00971.x

Nøttvedt, A. & Johannessen, E.P. 2008. The source of Norway's oil wealth. *In*: Ramberg, I.B., Bryhni, I., Nøttvedt, A. & Rangnes, K. (eds) *The Making of a Land – Geology of Norway*. Norsk Geologisk Forening, Trondheim, Norway, 384–417.

NPD 1979*a*. *Interpreted Lithology Log. Well 7/3-1*. NPD Paper, **20**. Norwegian Petroleum Directorate (NPD), Stavanger, Norway, http://www.npd.no/engelsk/cwi/pbl/NPD_papers/164_02_NPD_Paper_No.20_Interpreted_Lithology_log__Well_7_3_1.pdf

NPD 1979*b*. *Interpreted Lithology Log. Well 17/10-1*. NPD Paper, **21**. Norwegian Petroleum Directorate (NPD), Stavanger, Norway, http://www.npd.no/engelsk/cwi/pbl/NPD_papers/161_02_NPD_Paper_No.21_Interpreted_Lithology_log__Well_17_10_1.pdf

NPD 2013. *Wellbore 2/1-8. Composite Log*. Norwegian Petroleum Directorate (NPD), Stavanger, Norway, http://www.npd.no/engelsk/cwi/pbl/wellbore_composite_logs/459.pdf

NPD 2014. *Lithostratigraphic Chart Norwegian North Sea*. OD 1409001. Norwegian Petroleum Directorate (NPD), Stavanger, Norway, http://www.npd.no/Global/Engelsk/2-Topics/Geology/Lithostratigraphy/NS-OD1409001.pdf

NPD 2017. *Norwegian Petroleum Directorate website and FactPages/Maps*. Norwegian Petroleum Directorate (NPD), Stavanger, Norway, http://www.npd.no and http://factpages.npd.no/factpages/Default.aspx?culture=en

Oakman, C.D. 2005. The Lower Cretaceous plays of the Central and Northern North Sea: Atlantean drainage models and enhanced hydrocarbon potential. *In*: Doré, A.G. & Vining, B.A. (eds) *Petroleum Geology: North-West Europe and Global Perspectives – Proceedings of the 6th Petroleum Geology Conference*. Geological Society, London, 187–198, https://doi.org/10.1144/0060187

Oakman, C.D. & Partington, M.A. 1998. Cretaceous. *In*: Glennie, K.W. (ed.) *Petroleum Geology of the North Sea – Basic Concepts and Recent Advances*. 4th edn. Blackwell, Oxford, 294–349.

O'Driscoll, D., Hindle, A.D. & Long, D.C. 1990. The structural controls on Upper Jurassic and Lower

Cretaceous reservoir sandstones in the Witch Ground Graben, UK North Sea. *In*: Hardman, R.F.P. & Brooks, J. (eds) *Tectonic Events Responsible for Britain's Oil and Gas Reserves*. Geological Society, London, Special Publications, **55**, 299–323, https://doi.org/10.1144/GSL.SP.1990.055.01.14

Olsen, R.C. 1979. *Lithology. Well 17/10-1*. NPD Paper, **21**. Norwegian Petroleum Directorate (NPD), Stavanger, Norway, http://www.npd.no/engelsk/cwi/pbl/NPD_papers/161_01_NPD_Paper_No.21_Lithology__Well_17_10_1.pdf

Othman, A.A.A., Bakr, A. & Maher, A. 2017. Integrated seismic tools to delineate Pliocene gas-charged geobody, offshore west Nile delta, Egypt. *NRIAG Journal of Astronomy and Geophysics*, **6**, 81–89, https://doi.org/10.1016/j.nrjag.2017.04.008

Partington, M.A., Mitchener, B.C., Milton, N.J. & Fraser, A.J. 1993. Genetic sequence stratigraphy for the North Sea Late Jurassic and Early Cretaceous: distribution and prediction of Kimmeridgian–Late Ryazanian reservoirs in the North Sea and adjacent areas. *In*: Parker, J.R. (ed.) *Petroleum Geology: North-West Europe and Global Perspectives – Proceedings of the 4th Petroleum Geology Conference*. Geological Society, London, 347–370, https://doi.org/10.1144/0040347

Pharaoh, T.C., Dusar, M. *et al.* 2010. Tectonic evolution. *In*: Doornenbal, J.C. & Stevenson, A.G. (eds) *Petroleum Geological Atlas of the Southern Permian Basin Area*. European Association of Geoscientists and Engineers (EAGE), Houten, The Netherlands, 25–58.

Phillips Petroleum 1981. *Completion Book for the Exploratory Well 2/7–15*. Phillips Petroleum Company Norway, http://www.npd.no/engelsk/cwi/pbl/wellbore_documents/225_01_2_7_15_Completion_report.pdf

Pinnock, S.J., Clitheroe, A.R.J. & Rose, P.T.S. 2003. The Captain Field, Block 13/22a, UK North Sea. *In*: Gluyas, J.G. & Hichens, H.M. (eds) *United Kingdom Oil and Gas Fields, Commemorative Volume*. Geological Society, London, Memoirs, **20**, 431–441, https://doi.org/10.1144/GSL.MEM.2003.020.01.35

Posamentier, H.W. & Kolla, V. 2003. Seismic geomorphology and stratigraphy of depositional elements in deep-water settings. *Journal of Sedimentary Research*, **73**, 367–388, https://doi.org/10.1306/111302730367

Posamentier, H.W. & Vail, P.R. 1988. Eustatic controls on clastic deposition II – sequence and systems tract models. *In*: Wilgus, C.K., Posamentier, H.W., Ross, C.A. & Kendall, C.G.St.C. (eds) *Sea-Level Changes: An Integrated Approach*. SEPM (Society for Sedimentary Geology), Special Publications, **42**, 125–154.

Provan, D.M.J. 1976. *Composite Well Log 17/11–2*. A/S Norske Shell, http://www.npd.no/engelsk/cwi/pbl/wellbore_documents/338_02_Composite_well_log. pdf

Rattey, R.P. & Hayward, A.B. 1993. Sequence stratigraphy of a failed rift system: the Middle Jurassic to Early Cretaceous basin evolution of the Central and Northern North Sea. *In*: Parker, J.R. (ed.) *Petroleum Geology: North-West Europe and Global Perspectives – Proceedings of the 4th Petroleum Geology Conference*. Geological Society, London, 215–249, https://doi.org/10.1144/0040215

Ray, F.M., Pinnock, S.J., Katamish, H. & Trunbull, J.B. 2010. The Buzzard Field: anatomy of the reservoir from appraisal to production. *In*: Vining, B.A. & Pickering, S.C. (eds) *Petroleum Geology: From Mature Basins to New Frontiers – Proceedings of the 7th Petroleum Geology Conference*. Geological Society, London, 369–386, https://doi.org/10.1144/0070369

Rossland, A., Escalona, A. & Rolfsen, R. 2013. Permian–Holocene tectonostratigraphic evolution of the Mandal High, Central Graben, North Sea. *AAPG Bulletin*, **97**, 923–957, https://doi.org/10.1306/12031212093

Sears, R.A., Harbury, A.R., Protoy, A.J.G. & Stewart, D.J. 1993. Structural styles from the Central Graben in the UK and Norway. *In*: Parker, J.R. (ed.) *Petroleum Geology: North-West Europe and Global Perspectives – Proceedings of the 4th Petroleum Geology Conference*. Geological Society, London, 1231–1243, https://doi.org/10.1144/0041231

Skibeli, M., Barnes, K., Straume, T., Syvertsen, S.E. & Shanmugam, G. 1995. A sequence stratigraphic study of Lower Cretaceous deposits in the northernmost North Sea. *In*: Steel, R.J. (ed.) *Sequence Stratigraphy on the Northwest European Margin*. Norwegian Petroleum Society (NPF), Special Publications, **5**, 389–400.

Statoil 1991. *Well Completion Report Amalie-1 and Amalie 1A*. Statoil Efterforskning OG Produktion A/S, Copenhagen, Denmark.

Stewart, S.A. 2007. Salt tectonics in the North Sea Basin: a structural style template for seismic interpreters. *In*: Ries, A.C., Butler, R.W.H. & Graham, R.H. (eds) *Deformation of the Continental Crust: The Legacy of Mike Coward*. Geological Society, London, Special Publications, **272**, 361–396, https://doi.org/10.1144/GSL.SP.2007.272.01.19

Strass, I.F. 1979. *Lithology. Well 7/3-1*. NPD Paper, **20**. Norwegian Petroleum Directorate (NPD), Stavanger, Norway, http://www.npd.no/engelsk/cwi/pbl/NPD_papers/164_01_NPD_Paper_No.20_Lithology__Well_7_3_1.pdf

Surlyk, F., Dons, T., Clausen, C.K. & Higham, J. 2003. Upper Cretaceous. *In*: Evans, D., Graham, C., Armour, A. & Bathurst, P. (eds) *The Millennium Atlas: Petroleum Geology of the Central and Northern North Sea*. Geological Society, London, 213–233.

Thorsrud, A., Jensen, J.R. & Rasmussen, B.L. 2002. *Final Well Report – Well Svane-1 &-1A, 5604/26–4&4A*. Phillips Petroleum International Corporation-Denmark, Copenhagen, Denmark.

UKDD 2007. Triassic reservoirs of the Central Graben. *UK Discovery Digest*, **9**, 181–189, http://ukdigest.canadiandiscovery.com/?q=node/35892

Underhill, J.R. & Partington, M.A. 1993. Jurassic thermal doming and deflation in the North Sea: implications of the sequence stratigraphic evidence. *In*: Parker, J.R. (ed.) *Petroleum Geology: North-West Europe and Global Perspectives – Proceedings of the 4th Petroleum Geology Conference*. Geological Society, London, 337–345, https://doi.org/10.1144/0040337

Van Wijhe, D.H. 1987. Structural evolution of inverted basins in the Dutch offshore. *Tectonophysics*, **137**, 171–219, https://doi.org/10.1016/0040-1951(87)90320-9

van Winden, M., de Jager, J., Jaarsma, B. & Bourollec, R. In press. New insights into salt tectonics in the northern Dutch offshore: a framework for hydrocarbon exploration. *In*: Kilhams, B., Kukla, P.A., Mazur, S., McKie,

T., Mijnlieff, H.F. & van Ojik, K. (eds) *Mesozoic Resource Potential of the Southern Permian Basin*. Geological Society, London, Special Publications, **469**, https://doi.org/10.1144/SP469.9

Veeken, P. & Rauch-Davies, M. 2006. AVO attribute analysis and seismic reservoir characterisation. *First Break*, **2**, 7–18, https://doi.org/10.3997/1365-2397.2006004

Vejbæk, O.V., Andersen, C. *et al.* 2010. Cretaceous. *In*: Doornenbal, J.C. & Stevenson, A.G. (eds) *Petroleum Geological Atlas of the Southern Permian Basin Area*. European Association of Geoscientists and Engineers (EAGE), Houten, The Netherlands, 194–209.

Verolles, F. 1982. *Geological Completion Report. Well 3/7-3*. Elf Aquitaine Norge A/S Exploration Division, 311D/82/79-R, http://www.npd.no/engelsk/cwi/pbl/wellbore_documents/293_03_3_7_3_Completion_Report.pdf

Verweij, J.M. & Witmans, N. 2009. *Terschelling Basin and southern Dutch Central Graben. Mapping and Modelling – Area 2A*. TNO Report, TNO-034-UT-2009-01569. TNO, Utrecht, The Netherlands.

Voigt, S., Wagreich, M. *et al.* 2008. Cretaceous. *In*: McCann, T. (ed.) *Geology of Central Europe, Volume 2: Mesozoic and Cenozoic*. Geological Society, London, 923–998.

Vollset, J. & Doré, A.G. 1984. *A Revised Triassic and Jurassic Nomenclature for the Norwegian North Sea*. Norwegian Petroleum Directorate Bulletin, **3**, http://www.npd.no/Global/Norsk/3-Publikasjoner/NPD-Bulletin/Bulletinenr3.pdf

Walker, J.D., Geissman, J.W., Bowring, S.A. & Babcock, L.E. 2012. *GSA Geologic Time Scale v. 4.0*. Geological Society of America, Boulder, CO, http://www.geosociety.org/documents/gsa/timescale/timescl.pdf

Wong, Th.E. 2007. Jurassic. *In*: Wong, Th.E., Batjes, D.A.J. & de Jager, J. (eds) *Geology of the Netherlands*. Royal Netherlands Academy of Science and Art, Amsterdam, 241–264.

Wong, Th.E., van Doorn, H.M. & Schroot, B.M. 1989. ‘Late Jurassic’ petroleum geology of the Dutch Central North Sea Graben. *Geologische Rundschau*, **78**, 319–336, https://doi.org/10.1007/BF01988367

Zanella, E., Coward, M.P. & McGrandle, A. 2003. Structural framework. *In*: Evans, D., Graham, C., Armour, A. & Bathurst, P. (eds) *The Millennium Atlas: Petroleum Geology of the Central and Northern North Sea*. Geological Society, London, 44–60.

Erosional valleys at a major Late Jurassic–Early Cretaceous unconformity offshore Germany and The Netherlands: potential reservoirs or deteriorated seals?

MARCO WOLF[1]*, ANNELIEKE VIS[2] & ANNEMIEK ASSCHERT[2]

[1]*Federal Institute for Geosciences and Natural Resources (BGR), Stilleweg 2, 30655 Hannover, Germany*

[2]*EBN BV, Daalsesingel 1, 3511 SV, Utrecht, The Netherlands*

**Correspondence: marco.wolf@bgr.de*

Abstract: Joint results of two independent international studies on the erosional impact of the 'Late Cimmerian Unconformity' (LCU), a Late Jurassic–Early Cretaceous erosional event, are presented. The potential impact on the reservoir and sealing potential of adjacent strata are discussed. Results are based on a seismic mapping campaign in the NE part of the Dutch offshore region and the Central German North Sea sector. The geological evolution of potential traps at the level of the LCU was studied to obtain a better understanding of reservoir potential. Erosion or leaching of the evaporitic Triassic intervals at the LCU caused a locally strong topographical relief with prominent elongated depressions along Permian salt structures (interpreted as erosional valleys). The infill of the erosional valleys may consist of remnants of initial strata as breccia or material from nearby eroded strata, possibly Lower Triassic sandstone or clays. These valleys may be overlain conformably by Upper Jurassic–Early Cretaceous sediments or may be filled with a combination of both. The lithological composition of the infill determines the sealing or the reservoir potential of the valleys, which is essential for further evaluation of the resource exploration possibilities in this locality and across the basin.

The authors of this publication, from the German Federal Institute for Geosciences and Natural Resources (BGR) and EBN (Energie Beheer Nederland, the Dutch state participant in the oil and gas sector), present combined results from independent seismic mapping studies of the regional distribution of erosional and subrosional features at the 'Late Cimmerian Unconformity' (LCU) in the Central German North Sea sector and the NE part of the Dutch offshore region. The research is carried out within the scope of the BGR project 'Subsurface Potentials for Storage and Economic Use in the North German Basin' (TUNB: BGR 2017) and a G&M quadrants exploration study in the Dutch offshore region performed by EBN (Nlog 2016). The concept of this joint cross-border publication was developed at the 'Mesozoic Resource Potential of the Southern Permian Basin' conference in London in September 2016. For this publication, it was agreed to utilize German stratigraphic terms (Fig. 1).

Intensive erosion and near-surface subrosion of evaporites are worldwide-known phenomena. A present-day analogue of subrosion features due to halite leaching close to the surface can be studied, for example, in the Dead Sea area (Shalev *et al.* 2006; Oz *et al.* 2016). In Germany, contemporary local subrosion commonly occurs in Triassic evaporitic strata (Klose 2015). Mesozoic subrosion features came to the attention of researchers during the first regional mapping of the deeper subsurface in northern Germany in the project 'Tectonic Atlas of Northwest Germany and the German North Sea Sector' (GTA) (Kockel & Baldschuhn 1995; Baldschuhn *et al.* 2001; Müller *et al.* 2016). In The Netherlands, interest in the subject of Triassic evaporites increased with the development of the 'De Wijk' onshore field (Bruijn 1996; see also Goswami *et al.* 2018 for details), where gas is produced from multiple reservoir intervals including the Vlieland Sandstone Member. This unit was deposited in a depression formed due to erosion of Triassic evaporites, mainly of the Röt Formation (Bruijn 1996). Zwart & Mijnlieff (2006) also considered these depressions as part of a Dutch G-quadrant prospect evaluation.

Valley structures formed by leached evaporitic strata are especially important for the determination of potential reservoirs and seals. The validation of this potential in the deeper subsurface is crucial for many tasks, such as CO_2 storage in saline aquifers, exploitation of geothermal energy and the evaluation of hydrocarbon potential. Potential sealing formations in the SE North Sea consist of claystones and evaporites. The latter occur predominantly in

From: Kilhams, B., Kukla, P. A., Mazur, S., McKie, T., Mijnlieff, H. F. & van Ojik, K. (eds) 2018. *Mesozoic Resource Potential in the Southern Permian Basin*. Geological Society, London, Special Publications, **469**, 505–517.
First published online January 25, 2018, https://doi.org/10.1144/SP469.4

Int. Strat.				GER	NL	Typ. Lithofacies
Upper Triassic	Rhaetian	Keuper	U	Exter Formation	Sleen Formation	Alluvial to deltaic enviroment, clastic rocks with rising marine influence
Upper Triassic	Norian	Keuper	M	Arnstadt Formation	Upper Keuper Claystone Member Dolomitic Keuper Member Red Keuper Claystone Member	Playa enviroment, fine clastic rocks with few evaporitic contents
Upper Triassic	Norian	Keuper	M	Weser Formation	Red Keuper Evaporite Member	Playa enviroment, fine clastic rocks and evaporites
Upper Triassic	Carnian	Keuper	M	Stuttgart Formation	Middle Keuper Claystone Member	Fluvial plain, mud- and mostly sandstones on German side with fewer sandstones on the Dutch side
Upper Triassic	Carnian	Keuper	M	Grabfeld Formation	Main Keuper Evaporite Member	Playa enviroment, fine clastic rocks and evaporites
Upper Triassic / Middle Triassic	Ladinian	Keuper	L	Erfurt Formation	Lower Keuper Claystone Member	Claystone alternating with fine grained sands and coals
Middle Triassic	Ladinian	Muschelkalk	U	Upper Muschelkalk	Upper Muschelkalk Member	Shallow-marine carbonates and clastics
Middle Triassic	Ladinian / Anisian	Muschelkalk	M	Middle Muschelkalk	Middle Muschelkalk Evaporite Member	Evaporites with anhydrite
Middle Triassic	Anisian	Muschelkalk	L	Lower Muschelkalk	Lower Muschelkalk Member	Shallow-marine carbonates (Marl)
Lower Triassic	Olenekian	Buntsandstein	U	Röt-Clay Röt-Salt	Röt Formation	Sabkha enviroment with mainly claystones Sabkha enviroment with mainly halite and some anhyrite
Lower Triassic	Olenekian	Buntsandstein	M	Solling Formation	Solling Formation	Fluvial, aeolian or playa enviroment with prominent fining-upwards cycles from coarse sandstones at the base to claystones at the top
Lower Triassic	Olenekian	Buntsandstein	M	Hardegsen Formation	Hardegsen Formation	
Lower Triassic	Olenekian	Buntsandstein	M	Detfurth Formation	Detfurth Formation	
Lower Triassic	Olenekian	Buntsandstein	M	Volpriehausen Formation	Volpriehausen Formation	
Lower Triassic	Induan	Buntsandstein	M	Quickborn Formation		
Lower Triassic	Induan	Buntsandstein	L	Bernburg Formation Calvörde Formation	Lower Buntsandstein Formation	Mainly fine clastic fluvial and lacustrine clastics with some evaporitic contents on the German side

Permian and Triassic formations (e.g. the Upper Rotliegend, Zechstein, Röt, Middle Muschelkalk and Middle Keuper). Potential reservoir units mostly consist of clastic rocks, like the basal sandstones of the Middle Buntsandstein, the Stuttgart and Exter formations of the Keuper interval, and the Upper Jurassic Scruff Group (in The Netherlands). Additionally, carbonatic reservoir rocks can occur in Jurassic or Cretaceous units. Depending on the lithological infill, such valley structures can manifest as possible seal or reservoirs.

The purpose of this study is to evaluate the distribution of these structural features, which are possibly related to evaporate leaching, and their potential impact on reservoir and seal qualities (of both under- and overlying strata) in the area of investigation. For simplicity, these structural features are referred to as 'erosional valleys' because of the assumed genesis and the valley-like form of the incisions along the salt structures, noticeable in the seismic data. Here the subordinate term 'erosional' is chosen based on the apparent 'vanishing' of evaporitic intervals. However, further differentiation of pure physical erosion and weathering due to dissolution is not possible based on the available well and seismic data.

Regional geological setting

Here the tectonic context is described as a framework to understand how erosional valleys may have formed, what processes caused them and their possible infill stratigraphy. Varying tectonic and climatic conditions during Triassic times often led to cyclic deposition of clastic and evaporitic formations in the SE North Sea, which is part of the approximately east–west-trending Southern Permian Basin (Ziegler 1990; McCann *et al.* 2008; Stollhofen *et al.* 2008; Doornenbal & Stevenson 2010). Triassic basin development comprised a combination of regional subsidence (e.g. during deposition of the Lower and Upper Buntsandstein) (Fig. 1) and additional rift pulses, mainly during Olenekian and Norian times, leading to uplift and several regional unconformities (Geluk & Röhling 1997; Röhling 2013; Röhling & Lepper 2013). Most pronounced, of these surfaces, are the Buntsandstein H-Unconformity and the Keuper K4/K5 Unconformity (the Hardegsen and EKI unconformities, respectively, in The Netherlands: Geluk 2005). The Southern Permian Basin is divided by several graben structures. Prominent are the roughly north–south trending Central Graben, the Horn Graben, and the Glückstadt Graben (Fig. 2) (see Kilhams *et al.* 2018; Sachse & Littke 2018; Verreussel *et al.* this volume, in press). The graben structures are separated by several structural highs, such as the Ringkøbing-Fyn High, the Schillgrund High and platform areas (with constant sedimentation) such as the Westschleswig Block. The regional-scale basin was surrounded by several massifs that acted as a source area for the clastic basin infill. Marine evaporitic sediments were deposited as a result of varying sea-level and tectonic movements that allowed connections to the Boreal Sea and Tethys (Feist-Burkhardt *et al.* 2008). Cyclic alternations of marine facies and thick evaporitic beds reflect multiple, relatively short-lived, marine incursions (Szulc 2000; Franz 2008; Littke *et al.* 2008; Barnasch 2009; Doornenbal & Stevenson 2010; Röhling & Lepper 2013; Bachmann *et al.* 2010). The erosion of these evaporitic beds are key elements in understanding the erosional valleys described in this publication.

The Buntsandstein is subdivided into three distinct parts (Geluk 2005; Geluk & Röhling 2013; Röhling & Lepper 2013; Wolf *et al.* 2015). The Lower Buntsandstein consists of fine clastic sediments and some intervals of carbonatic ooids. The Middle Buntsandstein contains alternating fluvial coarse and fine clastic sediments as regional fining-upwards cycles. Aeolian influences during sedimentation of the Middle Buntsandstein deposits are typical for this interval (Maaß *et al.* 2010; Beyer *et al.* 2014; Olivarius *et al.* 2015, 2017). The Upper Buntsandstein consists of two members: evaporitic halite beds (sabkha) and fine clastic sediments with abundant sulphates (mudflat, sandflat facies). Deposition occurred in a semi-closed evaporitic basin with marine incisions from the Tethys via the Silesian–Moravian Gate (Feist-Burkhardt *et al.* 2008). Various studies demonstrate that local halokinetic movements of the Zechstein deposits started coevally with Buntsandstein deposition (Mohr *et al.* 2005, 2007).

During deposition of the Muschelkalk (Fig. 1) tectonic activity ceased, leading to mostly homogeneous thicknesses across the study area. With the onset of a major transgression, the Muschelkalk switched to a marine-dominated system. The rapid transgression led to starved basin conditions, which are reflected in the deposition of dark fine limestones of the Lower Muschelkalk. With the onset of deposition of the Middle Muschelkalk succession, a minor regression started. This led to increased salinity and,

Fig. 1. The stratigraphic terms used on the German and Dutch side of the area of investigation are illustrated. Additionally, lithofacies information for the border area is given. Units highlighted in yellow indicate formations that could be influenced by leaching, generating the described erosional valleys. The red dotted lines indicate major unconformities.

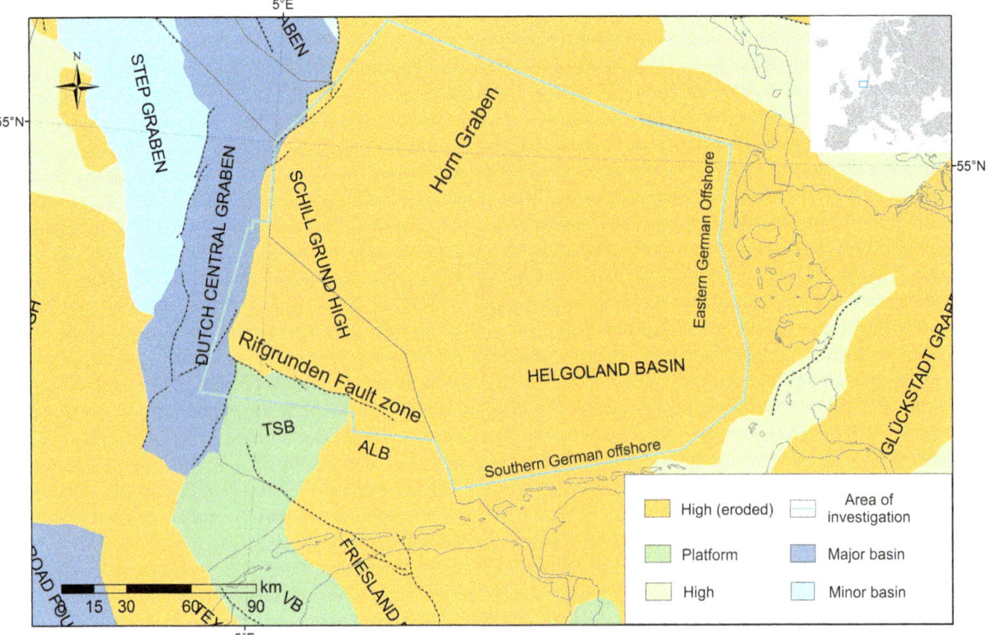

Fig. 2. The main structural units of the Late Jurassic Kimmeridgian in the area of investigation (modified from Pharaoh *et al.* 2010, fig. 3.19). Nearly the whole area was a structural high (orange colour) at this time, with deep erosion of older units. Only small parts of the Dutch area are part of the Central Graben or the Terschelling Basin. The blue line indicates the area of investigation. Abbrevations: ALB, Ameland Block; TSB, Terschelling Basin, VLB, Vlieland Basin.

hence, deposition of halite and gypsum. The marine influence increased with the Ladinian transgression. However, in contrast to the Lower Muschelkalk, the limestones were deposited in a shallow-marine environment (Szulc 2000). Muschelkalk deposition ended with a slow regression, resulting in a small remnant ocean in the southernmost Central European Basin (Franz *et al.* 2015).

Keuper stratigraphy in the basin reflects alternating shallow-marine and continental environments and/or fluctuations between arid and humid climates (Fig. 1). Basinwide unconformities have been identified by various authors (e.g. Bachmann *et al.* 1999; Beutler 2005). The deposition of the Lower Keuper started with an overall regressive trend, resulting in increasing importance of terrestrial facies assemblages and an unconformity at the base of the Keuper (Aigner & Bachmann 1992). Brackish mudstones were deposited in the area of investigation. The Grabfeld Formation shows a sequence of evaporitic and clastic sediments. The base is dominated by evaporites deposited in a very restricted marine or a playa/sabkha environment (Feist-Burkhardt *et al.* 2008). At the beginning of Middle Keuper deposition, strong tectonically driven movements of Zechstein halites started (Best *et al.* 1983; Beutler 2005; Kukla *et al.* 2008; Barnasch 2009). These salt movements, in combination with regional extension, led consequently to the heterogeneous structuration of the Triassic strata (manifested as horsts and graben, synforms and turtle-back structures). Strong tilting of the surrounding blocks (adjacent to salt structures) is responsible for the later exposure of evaporitic Triassic formations at the LCU. In salt-related rim synclines and in major graben structures, salt withdrawal and increased tectonic subsidence led to the deposition of huge thicknesses (over 3000 m) of Triassic and Jurassic sediments (Kockel & Baldschuhn 1995; Doornenbal & Stevenson 2010). The subsequent Stuttgart Formation consists of fluvial/deltaic/estuarine facies belts, mostly consisting of fine-grained sandstones and claystones, related to a basinwide fluvial system with associated channels and floodplains (Ricken *et al.* 1998; Beutler 2005). This system was directed from north to south towards the Tethys Ocean (Rüffer & Bechstädt 1998). Increasing aridity led to the deposition of claystones and evaporites of the overlying Weser Formation. The Weser, Arnstadt and lower parts of the Exter formations are terrestrial with broad alluvial plains. Towards the end of the deposition of the Keuper, the accumulation of evaporites ceased due to increased precipitation. The Exter Formation is separated from the Middle Keuper by an unconformity (Beutler 1998) caused by fluvial erosion. Fluvial and lacustrine organic-rich sandstones were deposited (Feist-Burkhardt *et al.* 2008). Several transgression phases led to Jurassic open-marine conditions.

From the Bajocian–Bathonian onwards, the growth and development of the Mid North Sea triple rift junction led to uplift and erosion (Fig. 2). The area was a local high with a warm humid climate up to the Late Jurassic (Ziegler 1990; Underhill & Partington 1993; Zeiss 2003; Pienkowski *et al.* 2008). From the Late Jurassic to the Early Cretaceous, multiple erosional events led to widespread, often angular, unconformities. These are generally referred to as one supra-regional unconformity, the 'Late Cimmerian Unconformity' (LCU). Along this unconformity, widespread erosion of Mesozoic sediments took place, especially in areas with tilted Mesozoic units (Doornenbal & Stevenson 2010). The unconformity is covered by Upper Jurassic–Lower Cretaceous sediments (Evans *et al.* 2003; Doornenbal & Stevenson 2010; Vejbaek *et al.* 2010; Müller *et al.* 2016). In most parts of the Southern North Sea, from Kimmeridgian to Wealden times, the LCU is covered by condensed, non-marine sediments with many intraformational unconformities (Jeremiah *et al.* 2010). From Hauterivian times, transgressions resulted in marine conditions (Mutterlose & Bornemann 2000; Jeremiah *et al.* 2010). In The Netherlands offshore and in the southern German offshore, the Keuper or Jurassic occur only as remnants in rim synclines.

Methods and data

The area of investigation comprises approximately 23 000 km^2 in the Dutch offshore region and central part of the German North Sea Sector (Fig. 2) (not including the so-called Entenschnabel: e.g. Arfai *et al.* 2014). This is defined by the German offshore border with Denmark in the north, by the main fault zone of the German and Dutch Central Graben in the west, the Rifgrunden Fault Zone in the Dutch south, and the border of the EEZ (Exclusive Economic Zone) in the German southern and eastern offshore (Fig. 2).

For this mapping study, 2D and 3D seismic data were utilized. In the Central German North Sea, 105 heterogeneous (quality and quantity) 2D seismic surveys of different age with several hundred seismic lines and two 3D cubes were interpreted. Most of the Dutch offshore region is covered by public 3D seismic data available at NLog (NLog 2017), leaving only minor gaps in the areas without 3D seismic coverage (Fig. 3). For interpretation of these gaps, EBN

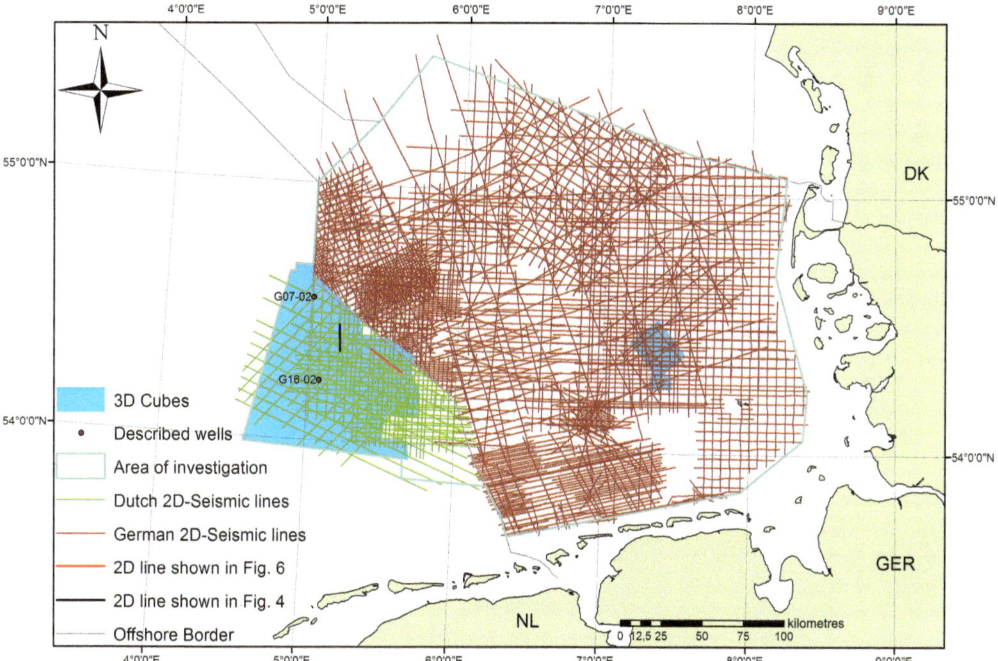

Fig. 3. Seismic 2D lines and 3D survey areas in the German and Dutch North Sea sector used for this publication (based on spatial data from NIBIS 2017; NLog 2017). The 2D seismic line featured in Figure 4 and the wells mentioned in this publication are also highlighted.

used 2D surveys. At the BGR, Paradigm SeisEarth was used for the interpretation of the erosional valleys. For the Dutch part of the area of investigation, EBN interpreted the seismic data using the Schlumberger software Petrel. Both interpretations were converted into grids and co-visualized using Esri ArcGIS.

In both studies, all available geological and geophysical information from 132 wells that reached the Triassic (106 German and 26 Dutch: NIBIS 2017; NLog 2017, respectively) were used to correlate and interpret seismic data and to get an understanding of the over- and underlying strata of the LCU. Spatial information on the erosional valleys is based on seismic interpretation. For seismic interpretation of stratigraphic units, a consistent concept is fundamental. However, in areas like the Central German North Sea with limited well control and sparse 3D seismic data, consistency was only achievable through careful correlation. Erosional valleys delivered important information for this correlation. Few evaporitic stratigraphic formations occur in the Triassic: the Röt-Evaporite, Middle Muschelkalk, Grabfeld and Weser formations (Reinhold *et al.* 2008; Reinhold & Hammer 2016). It was considered, based on previous studies (e.g. Baldschuhn *et al.* 2001), that these formations were the likely cause of erosional valleys due to their solubility and, therefore, provide some control for correlation and seismic interpretation.

Results

By correlating away from well penetrations, it was demonstrated that erosional valleys are only observed when layers prone to dissolution were exposed at or near to the surface during the Late Jurassic–Early Cretaceous (Fig. 4). As the known evaporitic sections are Middle–Late Triassic in age, it is postulated that valleys only occur where the entire Jurassic sequence has been eroded. The erosional valleys were identified as deep and partly steep incisions into Mesozoic strata caused by the LCU. The infill shows onlaps onto the flanks of the incision. The base of the Triassic evaporitic formations are always defined by negative seismic polarity (hard kick) and medium–high amplitudes, constrained by all wells in the area of investigation. The width of the described valleys in the hanging wall of the LCU in the Dutch and Central German North Sea varies from hundreds of metres up to several kilometres. They are 5–45 km long and locally up to 400 m deep (Fig. 5). In the German offshore region, the erosional valleys occur especially in the southern parts of the Horn Graben, SW of the Horn

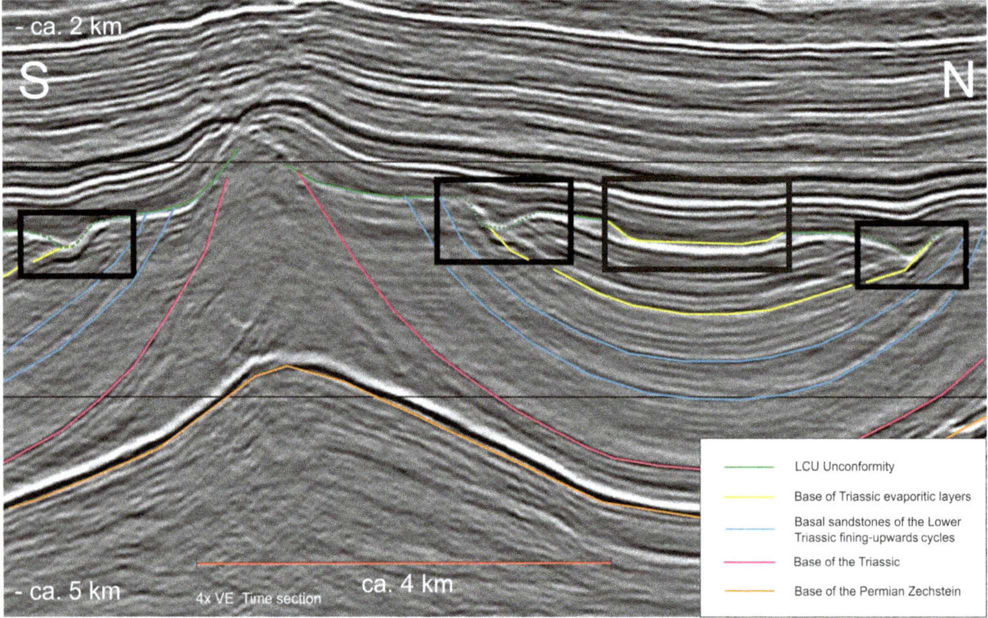

Fig. 4. Typical structural relationship in the study area, with tilted Triassic strata, eroded by the LCU in the top, in the hanging wall of halotectonic-deformed Permian salt (NLog 2017). The strong relief and the changes in the dip of Triassic units below the LCU is related to intensive salt tectonics in the Upper Triassic. The valleys in Triassic sediments are highlighted with black squares, and the reflectors highlighted in yellow illustrate the base of the Triassic evaporate units. VE = vertical exaggeration.

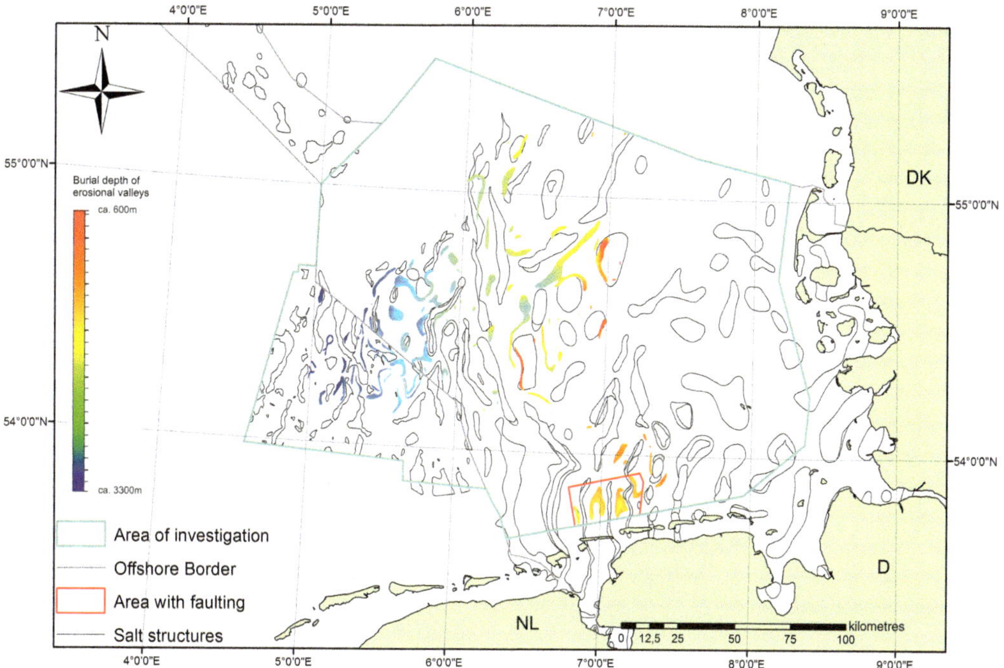

Fig. 5. Distribution of mapped erosional valleys at the LCU in the Southern North Sea. Additionally salt structures are illustrated to show the relationship between erosional features, nearby salt structures and their rim synclines. The range of the depth values were taken from published maps of the base Lower Cretaceous (Baldschuhn *et al.* 2001; Kombrink *et al.* 2012). The area with faulting (red square) represents an area with 2D seismic lines in which faults above erosional valleys are clearly visible. Other erosional valleys may feature subseismic faults above erosional valleys.

Graben and along the eastern plus the southern part of the German North Sea between salt walls (Fig. 5). In the Dutch offshore, the valleys are present in the middle part of the Schillgrund High (where most of the Jurassic is eroded and the Upper Triassic cropped out at the LCU). On the eastern part of the Schillgrund High, near its structural margin (adjacent to the German border), there are no comparable valleys observed (Fig. 5). This can be explained by the absence of the Triassic evaporites due to erosion. In the western part of the Schillgrund High, close to the Dutch Central Graben, the valleys are mostly absent due to the preservation of a thick Jurassic sedimentary succession on top of the Triassic, preventing the evaporites from leaching.

A typical example for the situation is shown in Figure 4. The movements of the Zechstein salts led to tilting of the Triassic sequence. The LCU eroded the tilted Triassic strata, and the relatively less stressed Cretaceous and Tertiary strata were deposited. At the LCU, the described incisions are highlighted by black boxes. In this particular case, two stratigraphic formations (the Röt and Middle Muschelkalk) caused noticeable incisions. More proximal to the diapir, hence higher in the palaeotopography relative to the new erosional valleys, formations that feature known sandstone beds (e.g. the Volpriehausen and Detfurth formations) also crop out and may act as a sediment source (Fig. 4).

Discussion

Genesis of the erosional valleys

The exposure and dissolution of Triassic evaporitic rocks resulted in the formation of erosional valleys in the strike direction along the flanks of salt structures. Increased erosion around the rising salt diapirs led to proximal sediment sources for the infill of the erosional valleys. Dissolution of evaporitic strata along unconformities is observed in numerous places in the North Sea region (Jenyon 1984; Clark *et al.* 1999). The erosional cut down along this unconformity to evaporitic layers facilitated a downwards percolation of solvent water. This led to direct erosion, leaching and, hence, subrosion of evaporites, including halite, which is easily soluble, and possibly anhydrite (Jenyon 1984). The resulting valleys track the outcropping evaporitic strata and form narrow valley-like synforms in the Early Cretaceous palaeotopography that are clearly recognizable in seismic sections (Figs 4 & 6). The distribution of erosional

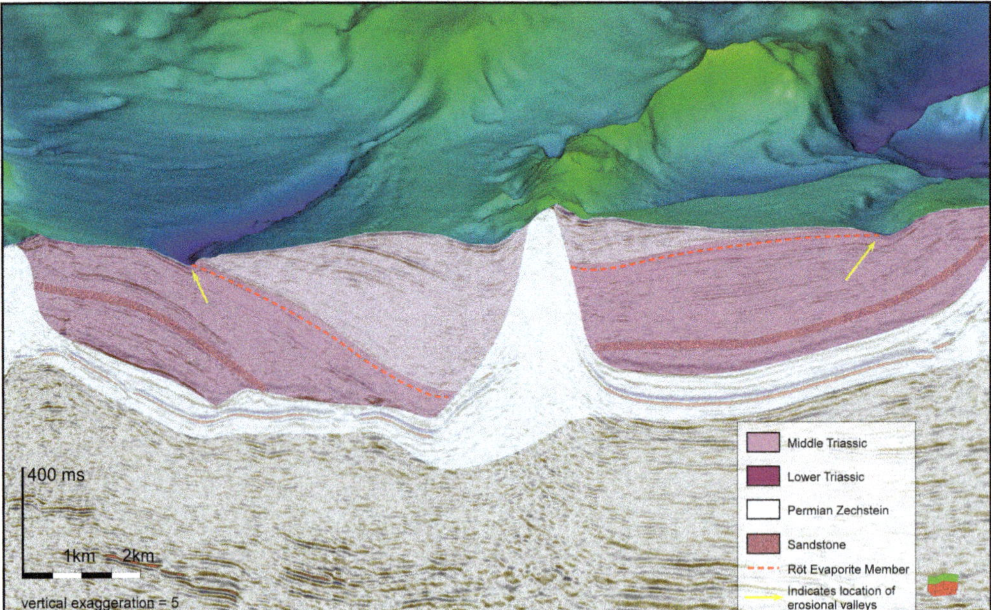

Fig. 6. Spatial 3D view of the autopicked LCU with clearly visible erosional valleys in the Dutch part of the study area. Additionally, the interpretation of the Triassic units is projected onto the seismic section.

valleys is strongly related to salt structures and the subsequent tilting of the Triassic successions in the deeper subsurface (Kockel & Baldschuhn 1995). Most erosional valleys are observed in areas with stronger tilting or uplifting of Triassic sediments. The precise geometries and dimensions of the erosional valleys depend on the magnitude of these structural movements. Local halotectonic movements in the Early Cretaceous are postulated to have had an influence on the local composition of sediments in and around the erosional valleys.

Infill facies of the erosional valleys: reservoirs or seals?

The infill of the erosional valleys governs their status as potential reservoir or sealing intervals but a decisive answer on the infill lithology is difficult since there is no direct well information to calibrate seismic observations. The following points cover the main infill possibilities. Other aspects, like ongoing cementation of the infill (e.g. with evaporitic minerals or burial diagenesis), that would affect their character as reservoir or seal are not considered.

Potential reservoir intervals. The concept of reservoir development is illustrated in Figure 7a, b, and is driven by redeposition of Middle Buntsandstein (Volpriehausen or Detfurth formations) coarse-grained units. This is applicable across most of the area of investigation due to the erosional niveau of the LCU (the sandstone formations are located locally on a higher structural level due to tilting and uplift of Triassic strata along the flanks of salt structures). During dissolution of the evaporites, contemporaneous Lower Triassic sediments, at the uplifted flanks of salt structures, are being eroded and, with a short transportation distance along the structural dip, deposited in the newly created valleys. These processes may be driven by fluvial or aeolian processes. However, the authors postulate that, due to the more humid conditions prevalent in the Jurassic, fluvial transport is more likely. A possible example of this kind of reservoir is found at the De Wijk Field (Bruijn 1996; Goswami *et al.* 2018). Here gas is found in sandstones situated in a depression caused by leaching of the Buntsandstein evaporites. However, whether the infill consists of reworked Lower Triassic sandstones or possible Upper Jurassic Vlieland sandstones is not clear.

Dissolution of other Triassic evaporitic sediments, comprising Middle Muschelkalk or Middle Keuper deposits, could also generate erosional valleys (Fig. 7c). For example, in the southern Central German North Sea, strong subsidence in the Middle Keuper led to large depositional thicknesses (over 2000 m) of the Grabfeld and Weser formations associated with up to 500 m-thick salt deposits. In this area, the LCU only eroded down to the Middle Keuper, exposing these salt deposits and generating

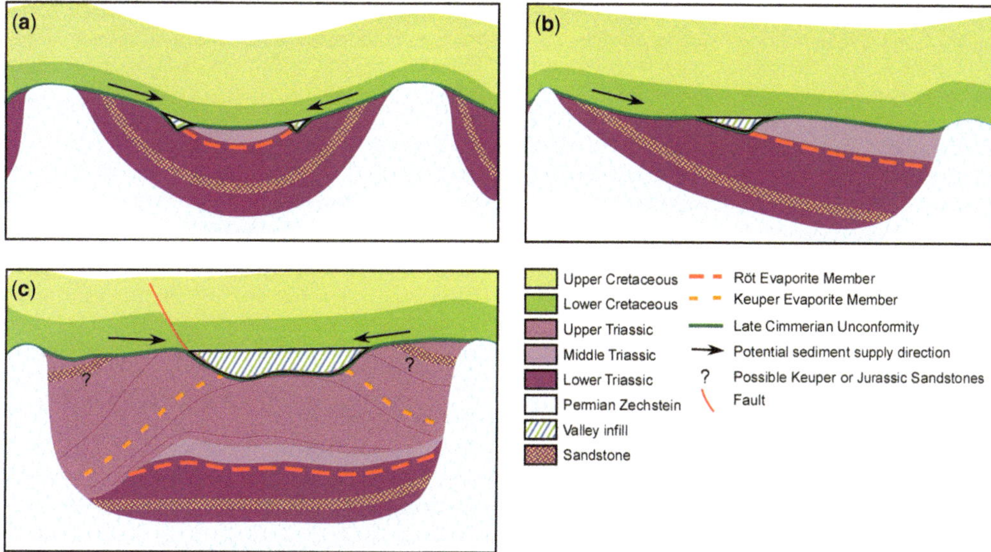

Fig. 7. Principal geological structural situation with erosional valleys due to the solution of evaporites and the refill of these valleys with resedimented older sandstones at the LCU. (**a**) & (**b**) reflect the most common structural style in the area of investigation. Here the Triassic is eroded down to the structural level of the Buntsandstein, leaching Röt Salt and eroding Middle Buntsandstein sandstone layers. The geometry is defined by the distance to the next salt structure and their development. (**c**) shows the situation in the southern German North Sea. Massive subsidence due to salt withdrawal in the Middle Keuper led to increased thicknesses of the Keuper stratigraphy. In this area, the erosion at the LCU reached only until the Keuper, resulting in leaching of thick Weser or Grabfeld Formation salt layers. In (c) possible sources for sandstones may derive from the Exter Formation, and faulting may compromise the sealing capacity of the Cretaceous claystones.

erosional valleys through subsequent dissolution (Fig. 7c). These valleys are stratigraphically above the Middle Buntsandstein and, therefore, must be filled by overlying strata such as the Exter Formation (Fig. 7c), which could also act as a source for possible sandstone infill.

Triassic evaporitic formations consist mainly of halite. However, other evaporites including anhydrite are also present (Reinhold & Hammer 2016). These tend to dissolve in an incomplete manner (relative to halite), resulting in brecciated remnants. This process is comparable to the development of salt diapir cap rocks (e.g. Vendeville 2002; Warren 2006). For example, in The Netherlands, remnants of Zechstein evaporites, together with the breccia of overlying sandstones, form the Zechstein Caprock Formation. The field G16-FA (well G16-2) produces gas from a comparable Zechstein reservoir (Fig. 3) (Kampmeinert 2003). At the base of erosional valleys, the breccia could form comparable reservoir intervals. A favourable reservoir case could involve both reservoir types: underlying remnant breccia with coarse clastic redeposited sandstones on top.

Potential sealing units. The other possibility is that the valleys are solely filled with Lower Cretaceous strata (Jeremiah *et al.* 2010; Vejbaek *et al.* 2010) or, alternatively, reworked Triassic clays. The Lower Cretaceous interval can consist of clays (providing excellent seals to underlying Jurassic or Triassic reservoir units) or locally developed sandstones (poor seals). The potential seals can also be cut by subsequent faulting.

In Early Cretaceous times, claystones were widely deposited in the area of investigation (Doornenbal & Stevenson 2010). However, locally, the Vlieland Sandstone Formation is also present (e.g. in the Dutch onshore region, the Vlieland sandstone is widely deposited and it serves as a proven reservoir – the Kotter Field: DeJong & Laker 1992). The source area of the Vlieland sandstones is suspected to be towards to the south, particularly the Rhenish Massif (Jeremiah *et al.* 2010). Several wells in the offshore region of The Netherlands show Vlieland sands at the base of the Lower Cretaceous strata, directly on top of the LCU. An example being the Dutch well G07-02, where unexpected sands (possibly acting as thief sands) are encountered on top of the Röt accommodation space (Figs 3 & 7b). The anticipated Triassic prospect was found with only minor gas shows, suggesting a gas charge but a deteriorated seal. It is proposed that this was due to the presence of the overlying erosional valley containing reworked sandstones.

Where Lower Cretaceous clays occur directly above erosional valleys, they are generally considered as good sealing units. However, syn- and post-sedimentary subrosion in the subsurface (due to groundwater movements even under marine conditions) led to ongoing subsidence in the valleys. This sometimes caused normal faults on the flanks of an erosional valley that are visible in seismic data (Fig. 7). In the area of investigation, this occurs often in the southern Central German North Sea (Fig. 5). These faults can cut through the overlying claystones, indicating ongoing subsidence and subrosion up to the Upper Cretaceous; hence, compromising their sealing property and causing possible migration pathways (Jenyon 1988).

Conclusion

Erosional valleys are an observed feature of the Dutch and German offshore related to strong unconformities over evaporitic strata and can have different impacts on the reservoir and sealing potential. Four points are important:

- Seismic and well observations demonstrate that valleys at the 'Late Cimmerian Unconformity' (LCU) relate to the exposure and dissolution of Triassic evaporites. Such structures can be seen across the Dutch and German North Sea. Here, these features are termed 'erosional valleys' due to their shape and association with physical and/or chemical erosion processes.
- Thanks to their relationship with evaporitic formations and their exposure (subaerial or submarine), these subrosion features can be used as a seismostratigraphic tool to help constrain the stratigraphy in an area with little well control.
- These erosional valleys would have been exposed and filled by variable strata (from Lower Triassic to Cretaceous intervals) depending on the adjacent hinterland and the precise timing of exposure.
- As such, these structures are important to consider when evaluating the reservoir and sealing potential of eroded successions with interbedded evaporitic strata. Various possibilities are discussed, where both excellent reservoirs and either good seals or deteriorated seals can occur. This can only be differentiated by detailed mapping and careful understanding of the different geological processes at play.

Consequently, the consideration of erosional valleys in regional hydrocarbon and storage potential assessments is strongly recommended. The authors acknowledge that this study is a first description of such erosional valley features in the area of investigation. Further investigations on their potential use could focus on elements such as their internal seismic facies and velocities to help differentiate lithological fill.

We want to thank the conveners of the 'Mesozoic Resource Potential of the Southern Permian Basin' conference, in London 2016, who made this publication possible by bringing two research groups together. Additionally, the authors want to thank the BVEG (Bundesverband Erdgas, Erdöl und Geoenergie e.V.), Maersk, NOPEC, BP and E. On for allowing their seismic and well data to be used for academic purposes, and the BGR and EBN colleagues for their support and scientific input. Furthermore, we want to thank the two reviewers, Dr M. Olivarius and Dr K. Anderskouv, and the editors, Prof. P. A. Kukla and Dr B. Kilhams, for their very useful reviews.

References

Aigner, T. & Bachmann, G.H. 1992. Sequence-stratigraphic framework of the German Triassic. *Sedimentary Geology*, **80**, 115–135.

Arfai, J., Jähne, F., Lutz, R., Franke, D., Gaedicke, C. & Kley, J. 2014. Late Palaeozoic to Early Cenozoic geological evolution of the northwestern German North Sea (Entenschnabel): new results and insights. *Geologie en Mijnbouw – Netherlands Journal of Geosciences*, **93**, 147–174.

Bachmann, G.H., Beutler, G., Hagdorn, A. & Hauschke, N. 1999. Stratigraphie der Germanischen Trias [Stratigraphy of the Germanic Triassic]. *In*: Hauschke, N. & Wilde, V. (eds) *Trias, eine ganz andere Welt: Mitteleuropa im frühen Erdmittelalter.* Frierdrich Pfeil, München, Germany, 81–104.

Bachmann, G.-H., Geluk, M.C. *et al.* 2010. Triassic. *In*: Doornenbal, J.C. & Stevenson, A.G. (eds) *Petroleum Geological Atlas of the Southern Permian Basin Area.* European Association of Geoscientists and Engineers (EAGE), Houten, The Netherlands, 149–173.

Baldschuhn, R., Binot, F., Fleig, S. & Kockel, F. 2001. Geotektonischer Atlas von Nordwest-Deutschland und dem deutschen Nordsee-Sektor [Tectonic Atlas of Northwest Germany and the German North Sea Sector (GTA)]. *Geologisches Jahrbuch, Reihe A*, **153**, 1–88.

Barnasch, J. 2009. *Der Keuper im Westteil des Zentraleuropäischen Beckens (Deutschland, Niederlande, England, Dänemark): diskontinuierliche Sedimentation, Litho-, Zyklo- und Sequenzstratigraphie [The Keuper in the western part of the Central European basin (Germany, Netherlands, England, Denmark): Discontinuous sedimentation, Litho-, Cyclo- and Sequence Stratigraphy].* Dissertation, Martin-Luther-Universität Halle-Wittenberg, Halle, Germany.

Best, G., Kockel, F. & Schoeneich, H. 1983. Geological history of the southern Horn Graben. *Geologie en Mijnbouw – Netherlands Journal of Geosciences*, **62**, 25–33.

Beutler, G. 1998. Keuper. *In*: Bachmann, G.H., Beutler, G. & Lerche, I. (eds) *Excursions of the International Symposium on the Epicontinental Triassic.* Hallesches Jahrbuch für Geowissenschafen, Reihe B, **6**, 45–58.

BEUTLER, G. 2005. *Stratigraphie von Deutschland IV – Keuper [Stratigraphy of Germany IV – Keuper]*. Courier Forschungsinstitut Senckenberg, **253**.

BEYER, D., KUNKEL, C., AEHNELT, M., PUDLO, D., VOIGT, T., NOVER, G. & GAUPP, R. 2014. Influence of depositional environment and diagenesis on petrophysical properties of clastic sediments (Buntsandstein of the Thuringian Syncline, Central Germany). *Zeitschrift der Deutschen Gesellschaft für Geowissenschaften*, **165**, 345–365.

BGR 2017. *Potenziale des unterirdischen Speicher- und Wirtschaftsraumes im Norddeutschen Becken (TUNB) [Storage and barrier potentials of the deeper underground in the North German Basin]*. Bundesanstalt für Geowissenschaften und Rohstoffe (BGR), Hanover, Germany, https://www.bgr.bund.de/DE/Themen/Nutzung_tieferer_Untergru nd_CO2 Speicherung/Projekte/Nutzungspotenziale/Laufend/TUNB.html?nn=7813522

BRUIJN, A.N. 1996. De Wijk Gas field (Netherlands): reservoir mapping with amplitude anomalies. *In*: RONDEEL, H.E., BATJES, D.A.J. & NIEUWENHUIJS, W.H. (eds) *Geology of Gas and Oil under the Netherlands*. Kluwer Academic, Dordrecht, The Netherlands, 243–254.

CLARK, J.A., CARTWRIGHT, J.A. & STEWART, S.A. 1999. Mesozoic dissolution tectonics on the West Central shelf, UK Central North Sea. *Marine and Petroleum Geology*, **16**, 283–300.

DEJONG, M. & LAKER, N. 1992. Reservoir modeling of the Vlieland sandstone of the Kotter field (Block-K18b), offshore, the Netherlands. *Geologie en Mijnbouw*, **71**, 173–188.

DOORNENBAL, J.C. & STEVENSON, A.G. (eds). 2010. *Petroleum Geological Atlas of the Southern Permian Basin Area*. European Association of Geoscientists and Engineers (EAGE), Houten, The Netherlands.

EVANS, D., GRAHAM, C., ARMOUR, A. & BATHURST, P. (eds). 2003. *The Millennium Atlas: Petroleum Geology of the Central and Northern North Sea*. Geological Society, London.

FEIST-BURKHARDT, S., GÖTZ, A.E. *ET AL*. 2008. Triassic. *In*: MCCANN, T. (ed.) *The Geology of Central Europe. Volume 2: Mesozoic and Cenozoic*. Geological Society, London, 749–821.

FRANZ, M. 2008. *Litho- und Leitflächenstratigraphie, Chronostratigraphie, Zyklo- und Sequenzstratigraphie des Keupers im östlichen zentraleuropäischen Becken (Deutschland, Polen) und dänischen Becken (Dänemark, Schweden) [Litho-, Chrono-, Cyclo- and Sequencestratigraphy of the Keuper in the eastern part of the Central European (Germany and Poland) and Danish basin (Denmark and Sweden)]*. Dissertation, Martin-Luther-Universität Halle-Wittenberg, Halle, Germany.

FRANZ, M., KAISER, S.I. *ET AL*. 2015. Eustatic and climatic control on the Upper Muschelkalk Sea (late Anisian/Ladinian) in the Central European Basin. *Global and Planetary Change*, **135**, 1–27, https://doi.org/10.1016/j.gloplacha.2015.09.014

GELUK, M.C. 2005. *Stratigraphy and tectonics of Permo-Triassic basins in the Netherlands and surrounding areas*. Dissertation, University of Utrecht, Utrecht, The Netherlands.

GELUK, M.C. & RÖHLING, H.-G. 1997. High-resolution sequence stratigraphy of the Lower Triassic 'Buntsandstein' in the Netherlands and northwestern Germany. *Geologie en Mijnbouw*, **76**, 227–246.

GELUK, M.C. & RÖHLING, H.-G. 2013. Der Buntsandstein in den Niederlanden und Nordost-Belgien [The Buntsandstein in the Netherlands and north-east Belgium]. *In*: LEPPER, J. & RÖHLING, H.-G. (eds) *Stratigraphie von Deutschland XI. Buntsandstein*. Schriftenreihe der Deutschen Gesellschaft für Geowissenschaften, **69**, 583–597.

GOSWAMI, R., SEEBERGER, F. & BOSMAN, G. 2018. Enhanced gas recovery of an ageing field utilizing N_2 displacement: De Wijk Field, The Netherlands. *In*: KILHAMS, B., KUKLA, P.A., MAZUR, S., MCKIE, T., MIJNLIEFF, H.F. & VAN OJIK, K. (eds) *Mesozoic Resource Potential of the Southern Permian Basin*. Geological Society, London, Special Publications, **469**. First published online January 16, 2018, https://doi.org/10.1144/SP469.2

JENYON, M.K. 1984. Seismic response to collapse structures in the Southern North Sea. *Marine and Petroleum Geology*, **1**, 27–36.

JENYON, M.K. 1988. Overburden deformation related to the pre-piercement development of salt structures in the North Sea. *Journal of the Geological Society, London*, **145**, 445–454, https://doi.org/10.1144/gsjgs.145.3.0445

JEREMIAH, J., DUXBURY, S. & RAWSON, P. 2010. Lower Cretaceous of the southern North Sea Basins: reservoir distribution within a sequence stratigraphic framework. *Netherlands Journal of Geosciences – Geologie en Mijnbouw*, **89**, 203–237.

KAMPMEINERT, D.J. 2003. *Winningsplan G16a-A/public version*. GDF Production Nederland BV, Den Helder, The Netherlands.

KILHAMS, B., STEVANOVIC, S. & NICOLAI, C. 2018. The 'Buntsandstein' gas play of the Horn Graben (German and Danish offshore): dry well analysis and remaining hydrocarbon potential. *In*: KILHAMS, B., KUKLA, P.A., MAZUR, S., MCKIE, T., MIJNLIEFF, H.F. & VAN OJIK, K. (eds) *Mesozoic Resource Potential of the Southern Permian Basin*. Geological Society, London, Special Publications, **469**. First published online January 11, 2018, https://doi.org/10.1144/SP469.5

KLOSE, M. 2015. *Landslide databases as tools for integrated assessment of landslide risk*. Dissertation, University of Vechta, Vechta, Germany.

KOCKEL, F. & BALDSCHUHN, R. 1995. *Structural and Palaeogeographical Development of the German North Sea Sector – Beiträge zur regionalen Geologie der Erde*. Gebrüder Bornträger, Stuttgart.

KOMBRINK, H., DOORNENBAL, J.C., DUIN, E.J.T., DEN DULK, M., VAN GESSEL, S.F., TEN VEEN, J. & WITMANS, N. 2012. New insights into the geological structure of the Netherlands; results of a detailed mapping project. *Netherlands Journal of Geology – Geologie en Mijnbouw*, **91**, 419–446.

KUKLA, P., URAI, J. & MOHR, M. 2008. Dynamics of salt structures. *In*: LITTKE, R., BAYER, U., GAJEWSKI, D. & NELSKAMP, S. (eds) *Dynamics of Complex Intracontinental Basins: The Central European Basin System*. Springer, Berlin, 291–306.

LITTKE, R., BAYER, U., GAJEWSKI, D. & NELSKAMP, S. (eds). 2008. *Dynamics of Complex Intracontinental Basins: The Central European Basin System*. Springer, Berlin.

MAAß, K., VOIGT, T. & GAUPP, R. 2010. Äolische und fluviatile Sedimentation im Unteren Buntsandstein (Calvörde-Formation) Ost-Thüringens [Aeolian and fluvial sedimentation of the Lower Buntsandstein (Calvörde formation) in East-Thuringia]. *Beiträge zur Geologie von Thüringen, Neue Folge*, **17**, 169–199, https://doi.org/10.13140/RG.2.1.4587.2489

MCCANN, T., KIERSNOWSKY, H. *ET AL.* 2008. Permian. *In*: MCCANN, T. (ed.) *The Geology of Central Europe. Volume 1: Precambrian and Palaeozoic*. Geological Society, London, 531–597.

MOHR, M., KUKLA, P., URAI, J. & BRESSER, G. 2005. Multiphase salt tectonic evolution in NW Germany: seismic interpretation and retro-deformation. *International Journal of Earth Sciences*, **94**, 917–940, https://doi.org/10.1007/s00531-005-0039-5

MOHR, M., WARREN, J.K., KUKLA, P.A., URAI, J.L. & IRMEN, A. 2007. Subsurface seismic record of salt glaciers in an extensional intracontinental setting (Late Triassic of northwestern Germany). *Geology*, **35**, 963, https://doi.org/10.1130/g23378a.1

MÜLLER, C., JÄHNE-KLINGBERG, F., VON GOERNE, G., BINOT, F. & RÖHLING, H.-G. 2016. Vom Geotektonischen Atlas ('Kockel-Atlas') zu einem 3D-Gesamtmodell des Norddeutschen Beckens: basisinformationen zum tieferen Untergrund von Norddeutschland [From the Tectonic Atlas ('Kockel Atlas') to a comprehensive 3D model of the North German Basin: basic information on the deep subsurface of northern Germany]. *Zeitschrift der Deutschen Gesellschaft für Geowissenschaften*, **167**, 65–106, https://doi.org/10.1127/zdgg/2016/0072

MUTTERLOSE, J. & BORNEMANN, A. 2000. Distribution and facies patterns of Lower Cretaceous sediments in northern Germany: a review. *Cretaceous Research*, **21**, 733–759.

NIBIS 2017. *Kartenserver*, https://nibis.lbeg.de/cardomap3/

NLOG 2016. *Prospex 2016*, http://www.nlog.nl/en/prospex2016

NLOG 2017. *NLog Seismic Data*, http://www.nlog.nl/en/seismic-data

OLIVARIUS, M., WEIBEL, R., HJULER, M.L., KRISTENSEN, L., MATHIESEN, A., NIELSEN, L.H. & KJØLLER, C. 2015. Diagenetic effects on porosity–permeability relationships in red beds of the Lower Triassic Bunter Sandstone Formation in the North German Basin. *Sedimentary Geology*, **321**, 139–153, https://doi.org/10.1016/j.sedgeo.2015.03.003

OLIVARIUS, M., WEIBEL, R., FRIIS, H., BOLDREEL, L.O., KEULEN, N. & THOMSEN, T.B. 2017. Provenance of the Lower Triassic Bunter Sandstone Formation: implications for distribution and architecture of aeolian v. fluvial reservoirs in the North German Basin. *Basin Research*, **29**, 113–130, https://doi.org/10.1111/bre.12140

OZ, I., EYAL, S., YOSEPH, Y., ITTAI, G., ELAD, L. & HAIM, G. 2016. Salt dissolution and sinkhole formation: results of laboratory experiments. *Journal of Geophysical Research: Earth Surface*, **121**, 1746–1762, https://doi.org/10.1002/2016jf003902

PHARAOH, T.C., DUSAR, M. *ET AL.* 2010. Tectonic evolution. *In*: DOORNENBAL, J.C. & STEVENSON, A.G. (eds) *Petroleum Geological Atlas of the Southern Permian Basin Area*. European Association of Geoscientists and Engineers (EAGE), Houten, The Netherlands, 25–57.

PIENKOWSKI, G., SCHUDACK, M. *ET AL.* 2008. Jurassic. *In*: MCCANN, T. (ed.) *Geology of Central Europe. Volume 2: Mesozoic and Cenozoic*. Geological Society, London, 822–925.

REINHOLD, K. & HAMMER, J. 2016. Steinsalzlager in den salinaren Formationen Deutschlands [Halite beds in the saliniferous formations in Germany]. *Zeitschrift der Deutschen Gesellschaft für Geowissenschaften*, **167**, 167–190, https://doi.org/10.1127/zdgg/2016/0067

REINHOLD, K., KRULL, P. & KOCKEL, F. 2008. *Salzstrukturen Norddeutschlands [Salt structures of Northern Germany]*. BGR Map. Bundesanstalt für Geowissenschaften und Rohstoffe (BGR), Hanover, Germany, https://geoviewer.bgr.de/mapapps/resources/apps/geoviewer/index.html?lang=de

RICKEN, W., AIGNER, T. & JACOBSEN, B. 1998. Levee-crevasse deposits from the German Schilfsandstein. *Neues Jahrbuch fur Geologie und Palaontologie-Monatshefte*, **1998**, 77–94.

RÖHLING, H.-G. 2013. Der Buntsandstein im Norddeutschen Becken – regionale Besonderheiten [The Buntsandstein in the North German Basin – regional features]. *In*: LEPPER, J. & RÖHLING, H.-G. (eds) *Stratigraphie von Deutschland XI. Buntsandstein*. Schriftenreihe der Deutschen Gesellschaft für Geowissenschaften, **69**, 269–384.

RÖHLING, H.-G. & LEPPER, J. 2013. Paläogeographie des Mitteleuropäischen Beckens während der tieferen Trias (Buntsandstein) [Paleogrography of the Central European basin in the Lower Triassic (Buntsandstein)]. *In*: LEPPER, J. & RÖHLING, H.-G. (eds) *Stratigraphie von Deutschland XI. Buntsandstein*. Schriftenreihe der Deutschen Gesellschaft für Geowissenschaften, **69**, 43–67.

RÜFFER, T. & BECHSTÄDT, T. 1998. Triassic sequence stratigraphy in the western parts of the Northern Calcareous Alps. *In*: HARDENBOL, J., DE GRACIANSKY, P.-C., JACQIN, T., FARLEY, M. & VAIL, P.R. (eds) *Mesozoic and Cenocoic Sequence Stratigraphy of European Basins*. Society for Sedimentary Geology (SEPM), Special Publications, **60**, 755–765.

SACHSE, V.F. & LITTKE, R. 2018. The impact of Quaternary glaciation on temperature and pore pressure in Jurassic troughs in the Southern Permian Basin, northern Germany. *In*: KILHAMS, B., KUKLA, P.A., MAZUR, S., MCKIE, T., MIJNLIEFF, H.F. & VAN OJIK, K. (eds) *Mesozoic Resource Potential of the Southern Permian Basin*. Geological Society, London, Special Publications, **469**. First published online January 16, 2018, https://doi.org/10.1144/SP469.7

SHALEV, E., LYAKHOVSKY, V. & YECHIELI, Y. 2006. Salt dissolution and sinkhole formation along the Dead Sea shore. *Journal of Geophysical Research: Solid Earth*, **111**, B03102, https://doi.org/10.1029/2005JB004038

STOLLHOFEN, H., BACHMANN, G. *ET AL.* 2008. Upper Rotliegend to early cretaceous basin development. *In*: LITTKE, R., BAYER, U., GAJEWSKI, D. & NELSKAMP, S. (eds) *Dynamics of Complex Intracontinental Basins: The Central European Basin System*. Springer, Berlin, 181–210.

SZULC, J. 2000. Middle Triassic evolution of the northern Peri-Tethys area as influenced by early opening of the

Tethys Ocean. *Annales Societatis Geologorum Poloniae*, **70**, 1–48.

UNDERHILL, J.R. & PARTINGTON, M.A. 1993. Jurassic thermal doming and deflation in the North Sea: implications of the sequence stratigraphic evidence. *In*: PARKER, J.R. (ed.) *Petroleum Geology of Northwest Europe: Proceedings of the 4th Conference*. Geological Society, London, 337–345, https://doi.org/10.1144/0040337

VEJBAEK, O.V., ANDERSEN, C. ET AL. 2010. Cretaceous. *In*: DOORNENBAL, J.C. & STEVENSON, A.G. (eds) *Petroleum Geological Atlas of the Southern Permian Basin Area*. European Association of Geoscientists and Engineers (EAGE), Houten, The Netherlands, 195–209.

VENDEVILLE, B.C. 2002. A new interpretation of Trusheim's classic model of salt-diapir growth. *Gulf Coast Association of Geological Societies Transactions*, **52**, 943–952.

VERREUSSEL, R.M.C.H., BOUROULLEC, R. ET AL. In press. Stepwise basin evolution of the Middle Jurassic to Early Cretaceous rift phase in the Central Graben area of Denmark, Germany and the Netherlands. *In*: KILHAMS, B., KUKLA, P.A., MAZUR, S., MCKIE, T., MIJNLIEFF, H.F. & VAN OJIK, K. (eds) *Mesozoic Resource Potential of the Southern Permian Basin*. Geological Society, London, Special Publications, **469**, https://doi.org/10.1144/SP469.23

WARREN, J.K. 2006. *Evaporites: Sediments, Resources and Hydrocarbons*. Springer, Berlin.

WOLF, M., STEUER, S., RÖHLING, H.-G., REBSCHER, D. & JÄHNE-KLINGBERG, F. 2015. Lithofacies distribution in the Central European Basin: a 3D model of the Buntsandstein facies in the central German North Sea. *Zeitschrift der Deutschen Gesellschaft für Geowissenschaften*, **166**, 341–359, https://doi.org/10.1127/zdgg/2015/0039

ZEISS, A. 2003. The Upper Jurassic of Europe: its subdivision and correlation. *In*: INESON, J.R. & SURLYK, F. (eds) *The Jurassic of Denmark and Greenland*. Geological Survey of Denmark and Greenland Bulletin, Copenhagen, Denmark, 75–114.

ZIEGLER, P.A. 1990. *Geological Atlas of Western and Central Europe*. 2nd edn. Shell Internationale Petroleum Maatschappij, The Hague. Geological Society, London.

ZWART, L. & MIJNLIEFF, H.F. 2006. *Petroleum Geological and Prospect Evaluation of Three Open Blocks in the G-quad of the Netherlands Offshore*. TNO, Utrecht, The Netherlands.

Giant pockmark formation from Cretaceous hydrocarbon expulsion in the western Lower Saxony Basin, The Netherlands

FRANK STROZYK*, LARS REUNING, STEFAN BACK & PETER KUKLA

Geological Institute, Energy & Mineral Resources (EMR) Group, RWTH Aachen University, Wüllnerstrasse 2, D-52062 Aachen, Germany

**Correspondence: frank.strozyk@emr.rwth-aachen.de*

Abstract: A field of giant pockmarks was discovered at the base of the Upper Cretaceous Chalk unit in the westernmost Lower Saxony Basin in The Netherlands. 3D seismic and well data show that mostly circular, 300–850 m-wide and 10–50 m-deep, pockmarks formed at the top of the Lower Cretaceous Upper Holland Marl Formation, which overlies oil- and gas-filled Lower Cretaceous sandstone reservoirs in the vicinity of the study area. Based on our interpretations, we present a scenario of early gas generation in Carboniferous coals and a localized migration of the gas from its original subsalt reservoirs through a salt weld in the Zechstein evaporites into the shallow Cretaceous sandstone reservoirs and the fine-grained marl above. Diapiring salt walls thereby limited the gas migration and trapping to a 150 km^2-sized basin. A sea-level drawdown during Base Chalk formation possibly led to excess pore pressure in the reservoir and the breaching of the seal close to the seafloor, which caused a short-lived expulsion of the gas and pockmark formation. While hydrocarbon generation, migration and trapping are common processes in this region, gas escaping at the seafloor with pockmark generation appears to be a rather rare and complex phenomenon. In general, the presence of pockmarks associated with salt welds may be used to constrain the timing and migration pathway of hydrocarbons from subsalt into shallower reservoir levels. Both features may imply a general reservoir potential for regions where suitable source rocks are missing in the post-salt succession.

Pockmarks are crater-like, circular or non-circular depressions in the seabed that, in many cases, originate from localized eruption-like seepage of gas or interstitial liquids (e.g. King & MacLean 1970; Hovland 1982, 1984; Hovland & Judd 1988; van Weering *et al.* 1997; Judd & Hovland 2007). Pockmarks of regular size are 2–200 m in diameter and metres to tens of metres deep (e.g. Hovland & Judd 1988; Fader 1991; Haskell *et al.* 1999), while discoveries of giant pockmarks image sizes of 200–1000 m (e.g. Anka *et al.* 2014). Mega-pockmarks of 1000–1500 m diameter and 50–150 m depth – as, for example, described for the Gabon and Equatorial Guinea basins at the West African continental margin (Pilcher & Argent 2007) – are amongst the largest observed types. Pockmarks can occur in a variety of marine environments from estuarine to marine shelf, slope and deep-water settings, and have been recorded at water depths from <2 to 5000 m (Hovland & Judd 1988).

Since buried pockmarks may indicate spontaneous fluid leakage at the seafloor for a very distinct time interval, they can be used to reconstruct the timing and pathways of fluid migration (Heggland 1997; Hovland *et al.* 2002; Andresen *et al.* 2008). Pockmarks also represent indicators for excessive leakage of fluids, like hydrocarbons, into the hydrosphere and atmosphere, or where sequestration within the sediments took place (e.g. Hornbach *et al.* 2007; Judd & Hovland 2007; Anka *et al.* 2014). Such seafloor fluid emissions are mostly sourced from either biogenic gas (Judd & Hovland 2007) or by thermogenic hydrocarbons from buried source rocks (Heggland 1997; MacDonald *et al.* 2000).

In this study, we present the discovery of a field of giant pockmarks at the Upper Cretaceous Base Chalk level at the westernmost edge of the Lower Saxony Basin in the eastern Netherlands (Fig. 1a, b). We commence with a 3D seismic and well-data-based description of the distribution and structure of the pockmarks, as well as the documentation of lithostratigraphic and structural features of the study area. Based on the results, it will be shown that hydrocarbon generation, migration and trapping in this area is a likely scenario, while the formation of pockmarks exclusively in the central study area seems enigmatic. We, hence, compared our results with lithostratigraphic data and petroleum system studies for this region, and discuss a possible scenario of the mechanisms and timing that led from deep-sourced fluid migration to surface expulsion and to the formation of pockmarks in the Cretaceous.

From: KILHAMS, B., KUKLA, P. A., MAZUR, S., MCKIE, T., MIJNLIEFF, H. F. & VAN OJIK, K. (eds) 2018. *Mesozoic Resource Potential in the Southern Permian Basin*. Geological Society, London, Special Publications, **469**, 519–536.
First published online January 11, 2018, https://doi.org/10.1144/SP469.6

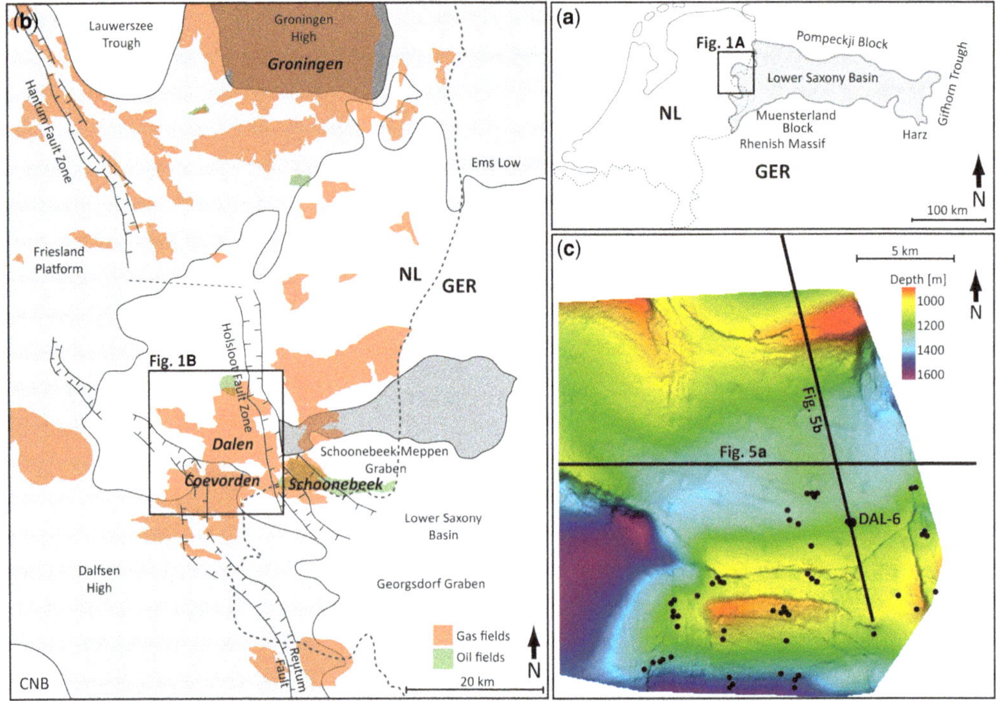

Fig. 1. (**a**) Location of the Lower Saxony Basin in the Southern Permian Basin system (modified after TNO-NITG 2000) and the location of the extended study area at the westernmost edge of the basin close to the eastern border of The Netherlands. (**b**) The extended study area with the main structural elements in the vicinity of the study area (black box: modified after TNO-NITG 2000). Gas (red) and oil fields (green) occur within (i.e. Coevorden, Dalen) and close to the study area (e.g. Schoonebeek: modified after de Jager & Geluk 2007). (**c**) The study area with a map plot of the Base Chalk horizon. Black dots indicate wells used in the study (i.e. COV, DAL); the two black lines represent seismic profiles shown in Figure 6a, b. Well DAL-6 is imaged in Figures 2b & 5b, and part of its logged data are shown in Figure 3.

Geological setting

General tectonostratigraphic evolution

The study area (Fig. 1c) is located at the eastern border of The Netherlands and covers the westernmost extension of the Lower Saxony Basin (Fig. 1b) (TNO-NITG 2000), attributed to the wider Southern Permian Basin (SPB) (Fig. 1a). The main structural elements around the study area are the Dalfsen High in the SW, the Friesland Platform in the NW, the Lauwersee Trough and the Groningen High in the north, the Ems Low in the NE, and the Lower Saxony Basin in the east (Fig. 1b). Subordinated structural elements are NW–SE- to north–south-trending fault zones and graben structures (i.e. the Holsloot Fault Zone and the Reutum Fault Zone) (Fig. 1b), crossing the NW and east of the study area (Fig. 1c).

Within the SPB, up to 1000 m of rock salt and other evaporites of the Late Permian Zechstein Group were deposited on top of Permian Rotliegend siliciclastics (Fig. 2a, b) (e.g. Ziegler 1990; Taylor 1998; Hübscher *et al.* 2007; Doornenbal & Stevenson 2010). In The Netherlands, the Rotliegend commonly contains sandstone reservoirs, which were sourced by gas from Carboniferous coals and are sealed by the Zechstein evaporites (Fig. 1b) (e.g. van Wijhe *et al.* 1980; Glennie 1998; de Jager & Geluk 2007; Kombrink *et al.* 2012). Today, the thick Zechstein evaporites are largely deformed to salt diapirs, pillows and walls (e.g. Duin *et al.* 2006; Geluk 2007; Biehl *et al.* 2014; Strozyk *et al.* 2014, 2017). The incremental inflating of salt highs has been attributed to several extensional and compressional tectonic phases from the Permian to the Paleogene (e.g. Geluk 2007; Kukla *et al.* 2008; Strozyk *et al.* 2017 and references therein). The halokinesis greatly affected the deposition and deformation of an, on average, 2–3 km-thick sedimentary post-salt section, whereof Triassic and Upper Cretaceous sediments represent the thickest units (e.g. Duin *et al.* 2006; de Jager 2007; Kombrink *et al.* 2012).

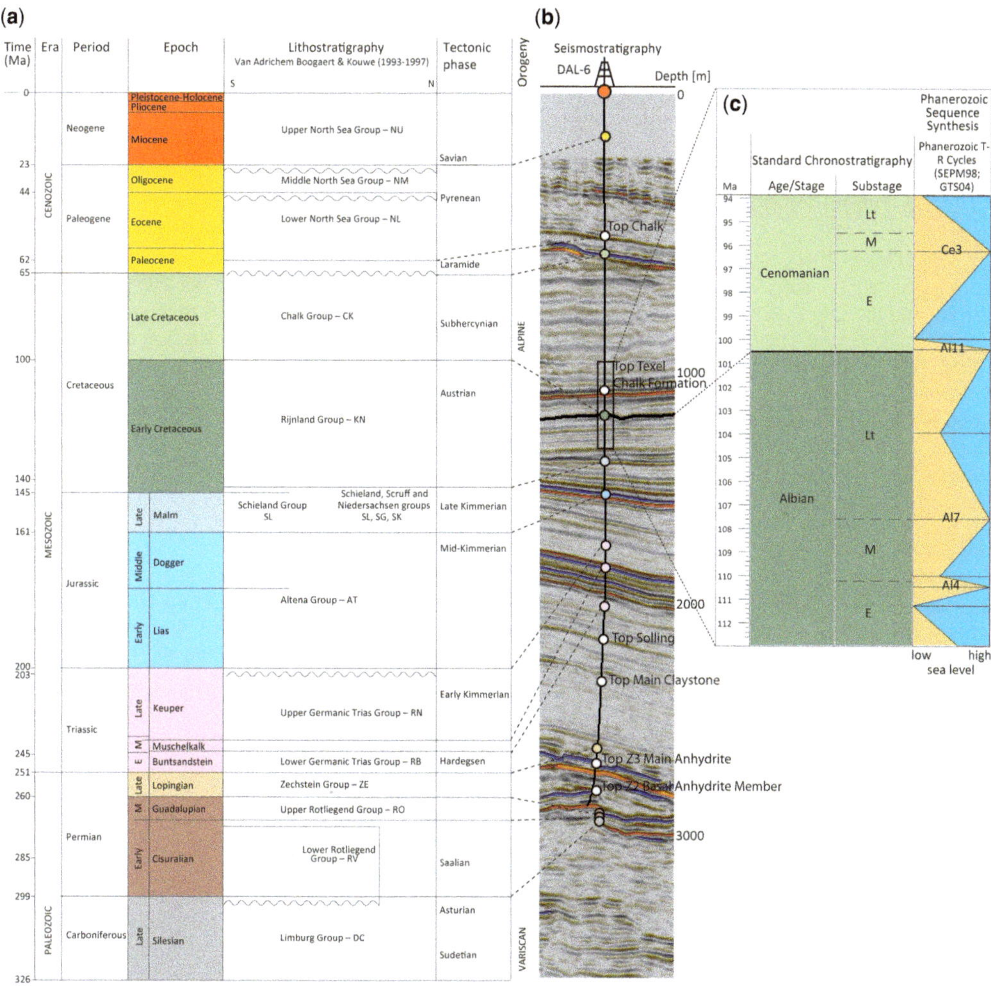

Fig. 2. (**a**) Lithostratigraphic chart covering the Carboniferous to present-day sedimentary succession in the subsurface of The Netherlands (modified after Duin *et al.* 2006). (**b**) Cropped seismic profile at well Dalen-6 (DAL-6: for the location see Fig. 1c), imaging the seismostratigraphic well tie to the lithostratigraphy in the study area; note the depression in the Base Chalk reflection besides the well, which is interpreted as one of the pockmarks. (**c**) General Phanerozoic sequence synthesis cropped to the transition from the Early Cretaceous Albian to the Late Cretaceous Cenomanian (E, Early; M, Middle; Lt, Late).

After the deposition of the Zechstein evaporites and the thick Middle–Upper Triassic siliciclastics (Fig. 2b), extensional tectonics started in the Middle–Late Triassic (i.e. Early Kimmerian) (Fig. 2a) (e.g. Mohr *et al.* 2005; de Jager 2007; Stollhofen *et al.* 2008; Strozyk *et al.* 2017). This triggered thin-skinned raft tectonics and passive salt diapirism (Strozyk *et al.* 2014; also see Vackiner *et al.* 2013). The salt diapirism continued in several phases during the sediment-poor Jurassic and Early Cretaceous (cf. Fig. 2b) (e.g. Geluk 2007), with sediment deposited as thin layers laterally restricted to depressions between the inflating salt highs (e.g. Wong 2007; Strozyk *et al.* 2017). Subsidence during the Late Cretaceous caused the deposition of thick, carbonate-rich Chalk sediment (Fig. 2b), which draped the pre-existing structural framework during a phase of tectonic quiescence (e.g. Herngreen & Wong 2007). A change to compressional tectonics in the Late Cretaceous, which continued until the early Paleogene and has been mainly attributed to the Alpine collision (de Jager 2007), caused inversion tectonics with folding, faulting and erosion, especially of the Upper Cretaceous sediments (Fig. 2b), as well as a reactivation of salt diapirism (Geluk 2007; Geluk *et al.* 2007; Strozyk *et al.* 2014, 2017).

The incrementally planar and even bedding of the overlying Paleogene and Neogene sediments indicate that a cessation of the compressional tectonic regime and a move towards steady basin subsidence started in the Paleocene (Fig. 2b) (Wong 2007; de Gans 2007; Strozyk *et al.* 2017).

Cretaceous lithostratigraphy

Rijnland Group. The Cretaceous in The Netherlands is represented by the Lower Cretaceous Rijnland Group and the Upper Cretaceous–lowermost Paleogene Chalk Group (Figs 2a & 3) (Herngreen & Wong 2007). The Rijnland Group generally comprises marine claystones, marls, siltstones and sandstones (Fig. 3). Its total thickness does not exceed 200 m in the study area (Fig. 2b) (also see Duin *et al.* 2006). The Rjinland Group can be further subdivided into the Vlieland Sandstone Formation, the Vlieland Claystone Formation, which together form the Vlieland Subgroup, and the Holland Formation (Fig. 3) (Herngreen & Wong 2007). Sea-level fluctuations during the deposition of the Rijnland Group formed various transgressive basal sands or prograding coastal-barrier sands (i.e. Vlieland Sandstone Formation (VSF)). The VSF sandstones are argillaceous to clean and commonly contain glauconite (Herngreen & Wong 2007). Depending on the location, the age of the formation ranges from the Early Valanginian for the Bentheim Sandstone and from the Hauterivian for the Gildenhaus Sandstone, and, hence, has been subdivided into several members, some of which are known hydrocarbon reservoirs (see the subsection on 'Hydrocarbon geology' later in this section) (de Jager & Geluk 2007). The Vlieland Claystone Formation comprises brownish grey to grey claystones that are often slightly calcareous (Herngreen & Wong 2007). In places, the formation is silty to sandy, with numerous siltstones and very fine sandstone beds. Its age ranges from Valanginian to early Aptian, and its depositional environment was a shallow- to fairly deep-marine setting, whereby sand beds were deposited mainly during storms and lowstand periods (Herngreen & Wong 2007).

Holland Formation. Deeper marine conditions resulted in the deposition of the Aptian–Albian/earliest Cenomanian Holland Formation (Fig. 3) (Herngreen & Wong 2007), which consists of marls (i.e. Lower and Upper Holland Marl) and marly to calcareous claystones (i.e. Middle Holland Claystone). With only local and thin intercalations of shales, the influx of fine-grained clastics into the marine realm was limited (TNO-NITG 2000; Herngreen & Wong 2007). Sediment geometries, thickness variations and intense reddening suggest the existence of several hiatuses. One considerable hiatus reflects the Austrian tectonic phase and the subsequent Albian transgression (Herngreen & Wong 2007). The unconformable contact of the Holland Formation with the underlying Vlieland Claystone Formation is highlighted by a distinct upwards increase in carbonate content, as reflected by a decrease in gamma-ray values, and an increase in resistivity and sonic velocity (see Fig. 3). The total thickness of the Holland Formation is usually some tens of metres, but increases to more than 300 m in the central parts of the Lower Saxony Basin (TNO-NITG 2000; Herngreen & Wong 2007). It was mainly deposited in a middle to outer neritic environment. Bituminous deposits indicate periods of stagnant basin-floor circulation. The Upper Holland Marl Member (Fig. 3) is the thickest of the three members of the Holland Formation. Its thickness in the vicinity of the study area ranges between *c.* 100 and 200 m, and its age is middle–late Albian, locally ranging into the earliest Cenomanian (TNO-NITG 2000; Herngreen & Wong 2007).

Chalk Group. The Chalk Group (Fig. 3) includes the Upper Cretaceous and lowermost Tertiary rocks, with an age range from the Cenomanian to the Danian. It consists of a succession of marine, predominantly bioclastic and, in parts, marly limestones. Locally, other rock types such as glauconitic sandstones occur. Fossils revealed that sedimentation took place entirely in marine conditions in water depths of 50–300 m (TNO-NITG 2000). Late Cretaceous compression and inversion led to the partial erosion of the sediments (Van Adrichem Boogaert & Kouwe 1993–97), particularly above anticlines associated with salt diapirism (Fig. 4). The Chalk sediments do not exceed more than 1400 m in thickness in the study area (also see Wong 2007).

Hydrocarbon geology

Subsalt. A number of hydrocarbon plays are present in the sedimentary succession, especially of the northern onshore and offshore of The Netherlands. The thick Permian Zechstein evaporites provide a generally effective seal between a Paleozoic gas system and a mixed oil and gas Mesozoic hydrocarbon system (e.g. de Jager & Geluk 2007). Main subsalt gas reservoirs in the Dutch onshore are either in the sandstones of the Permian Rotliegend (e.g. the Slochteren Formation), which is thin (<50 m) or even lacking in the study area (Lokhorst 1998; TNO-NITG 2000; Geluk 2007), or in the Z2 Carbonate Member of the Zechstein Group. The latter is a known producing reservoir in the study area, and was sourced by Upper Carboniferous gas-producing coals of the Westphalian and Stephanian (Fig. 4a), which has generated hydrocarbons since the Triassic

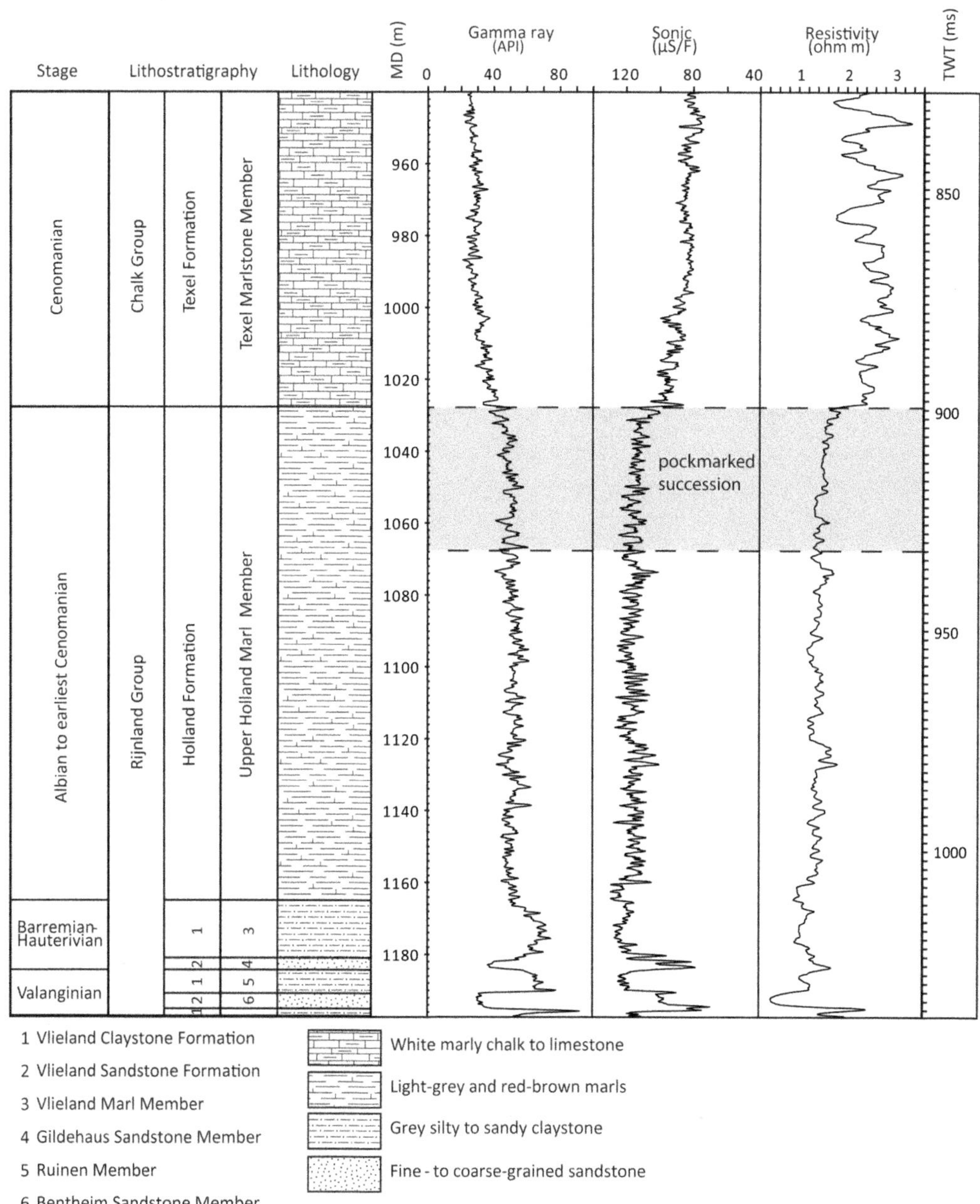

Fig. 3. Lithology and geophysical logging response at well Dalen-6 (DAL-6). Pockmarks have formed in the marly sediments at the top of the Holland Formation (grey interval) during the Albian transgression. During the Cenomanian, the influx of siliciclastic sediments further diminished, leading to the deposition of the marls and limestones of the Chalk Group. The Holland Formation is underlain by the Vlieland Claystone and Sandstone formations. The Bentheim and Gildehaus sandstones of the Vlieland Sandstone Formation form important oil and gas reservoirs less than 10 km WSW of the well location. MD, along-hole depth measured from the rig table; TWT, two-way time.

(TNO-NITG 2000) (Fig. 4c). Westphalian sandstones (i.e. the Tubbergen Formation within the Limburg Group) also provide very productive gas reservoirs; the Coevorden Field is the largest example in the study area (Figs 1b & 4d) (de Jager & Geluk 2007).

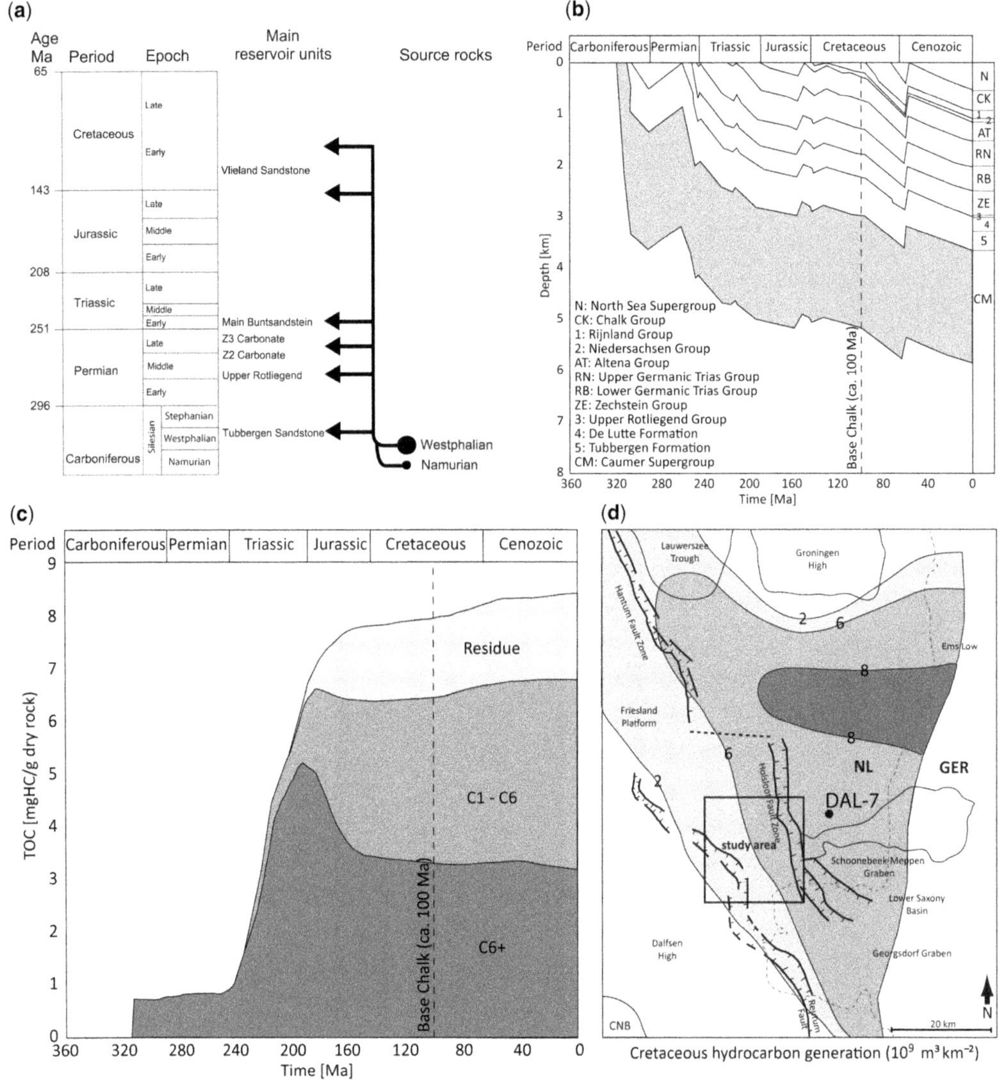

Fig. 4. (**a**) Graph illustrating the charging of Carboniferous coal gas into subsalt and suprasalt reservoirs in the subsurface of the Dutch onshore. Arrows indicate which main reservoirs have been charged with the gas. In the region of the study area, gas was mainly stored in the Tubbergen Sandstone and the Z2 Carbonate in the subsalt, and in the Lower Cretaceous Gildehaus Sandstone Member and the Vlieland Sandstone Formation in the suprasalt. Other reservoirs, like the Upper Rotliegend sandstones, are produced further north of the study area (e.g. at the Groningen High: Fig. 1b). (**b**) Subsidence model for well DAL-7 (for the location see c) east of the study area (modified after TNO-NITG 2000). The Carboniferous unit that contains gas-producing coals is highlighted in a grey colour. (**c**) Hydrocarbon generation model designed for well DAL-7. Gas generation took place here from the Triassic onwards (for the location of the well see d: modified after TNO-NITG 2000). (**d**) Map of Cretaceous hydrocarbon generation in the extended study area, as shown in Figure 1b (modified after TNO-NITG 2000). Contours and numbers represent the total organic carbon (TOC) levels ($\times 10^9$ m^3 km^{-2}) in the Cretaceous.

Suprasalt. Significant oil and gas accumulations were also discovered in Jurassic and Lower Cretaceous sandstone reservoirs, as well as in Triassic siliciclastics and the Chalk Group (e.g. Lokhorst 1998; Doornenbal & Stevenson 2010) (Fig. 4a). While most of these reservoirs show gas accumulations, producible oil in The Netherlands is mostly limited to the Late Jurassic and Early Cretaceous rift basins in a variety of sandstone reservoirs and trap styles (TNO-NITG 2000; de Jager & Geluk 2007)

(Fig. 1b). In the German part of the Lower Saxony Basin, one of the main oil-producing source rocks is the Lower Jurassic Posidonia Shale Formation; while in The Netherlands, the remnants of this Type-II source rock are restricted to Mesozoic rift basins and are missing in the extended study area (see Fig. 1b). In the Dutch part of the Lower Saxony Basin, in the Schoonebeek Field (Fig. 1b), oil is produced from the Bentheim Sandstone at *c.* 800 m depth (TNO-NITG 2000). This accumulation is rather exceptional as it is sourced from lacustrine algal Type-I source rocks of Early Cretaceous age (i.e. the Coevorden Formation). These source rocks have a limited distribution in The Netherlands, but extend into the Lower Saxony Basin in Germany (Binot *et al.* 1991) (Fig. 1a). In the extended study area, a number of gas fields are produced, including reservoirs in oolites in the Lower Buntsandstein, in sandstone in the Volpriehausen, Solling and Vlieland sandstone formations, in limestone in the Muschelkalk Formation, and in tuffite in the Basal Dongen Tuffite (TNO-NITG 2000).

For reservoirs in the vicinity of the study area, Gerling *et al.* (2016) showed that the composition of gaseous hydrocarbons indicates that natural gas from the Upper Carboniferous coals can be mixed with common oil-associated gases from Jurassic source rocks in the Cretaceous reservoirs. Salt tectonics and deep-reaching fault systems have been suggested to provide sufficient migration pathways through the Zechstein section for natural gas originating from deeper strata (Lohr *et al.* 2007; Gerling *et al.* 2016) (Fig. 4a). However, alternative fluid-migration pathways may have established at salt welds, where suprasalt sediments can get very close to the base of the salt section (e.g. Geluk 2007; Gerling *et al.* 2016).

Data and methods

The base for this study is an approximately 600 km^2-sized 3D seismic volume of 2.6 s two-way time (TWT) depth (zero phase following European polarity convention; 25 m line spacing, *c.* 15–20 m vertical resolution) at the eastern border of The Netherlands (Fig. 1b, c), provided by NAM. We incorporated stratigraphic formation tops, lithostratigraphic information and basic well logs (DEN, GR) of more than 70 exploration and production wells in the study area, provided by TNO and NAM (i.e. COV-1–COV-57, DAL-1–DAL-15 and BNT-1). Well logs and formation tops were used to identify the seismic-reflection response of the broad stratigraphic units, as well as of more detailed layers within the Lower–Upper Cretaceous target units (Fig. 2b).

In preparation for seismic interpretation, a general depth conversion of the seismic volume was carried out using check shot data along a number of ‘DAL’ (Dalen) and ‘COV’ (Coevorden) wells, combined with general velocity data of the Dutch subsurface provided by van Dalfsen *et al.* (2006). The resulting depth-domain seismic volume shows a total depth of *c.* 4000 m imaging down to Upper Carboniferous units (cf. Fig. 5).

For seismostratigraphic horizon interpretation, we used combined manual and automated 3D horizon tracking along the Base Chalk reflection (i.e. top Upper Holland Marl Member/Base Chalk) (Figs 1c & 6a, b), a reflection (‘base pockmarks’) that represents the maximum depth of all interpreted pockmarks *c.* 50 m below Base Chalk (see Fig. 7a), and reflections that represent the tops of the Vlieland Claystone Member, the Bentheim Sandstone Formation and the Lower Coevorden Member *c.* 180 m below Base Chalk (cf. Fig. 7). We additionally interpreted reflections that represent tops of the main stratigraphy, including the top of Permian Rotliegend, the base of the Zechstein-2 rock salt layer, top Zechstein, top Triassic, top Lower/Upper Jurassic, top Lower/Upper Cretaceous, an intra-Cretaceous unconformity (i.e. top Ommeland Formation), top Paleogene, and the present-day surface (Neogene) (Figs 2b and 6). Fault interpretation was carried out by high-density fault picking along selected reflector offsets. High-resolution horizon and fault interpretation was conducted in the central study area (Figs 5–7).

Results

Tectonostratigraphic interpretation

The seismic interpretation shows that suprasalt sediments were locally folded, faulted and eroded by movements of the underlying Zechstein salt (Fig. 5). The present-day subsurface is characterized by anticlines and crestal-collapse graben structures above salt highs alternating with sedimentary basins above salt lows, following the top of the deformed Zechstein salt (Figs 5a, b & 6a, b). The main structural trend of salt structures, as well as of the related structures in the suprasalt sediments (i.e. folds and faults), is east–west (Fig. 6b). Another structural trend in the NW–SE to north–south direction reflects mostly very thin salt structures related to fault zones in the suprasalt sediments (Fig. 5a, right). While the anticlinal shape of the suprasalt sediments resembles salt structures trending east–west (Fig. 5b), the north–south-trending salt walls are only indicated by crestal-collapse graben structures (Fig. 5a, right). Here, the inflation of the salt highs caused faulting in the suprasalt sediments, which show localized fault offsets of the order of 50–250 m.

The relatively constant thickness of most of the Triassic sediments indicates their rather planar

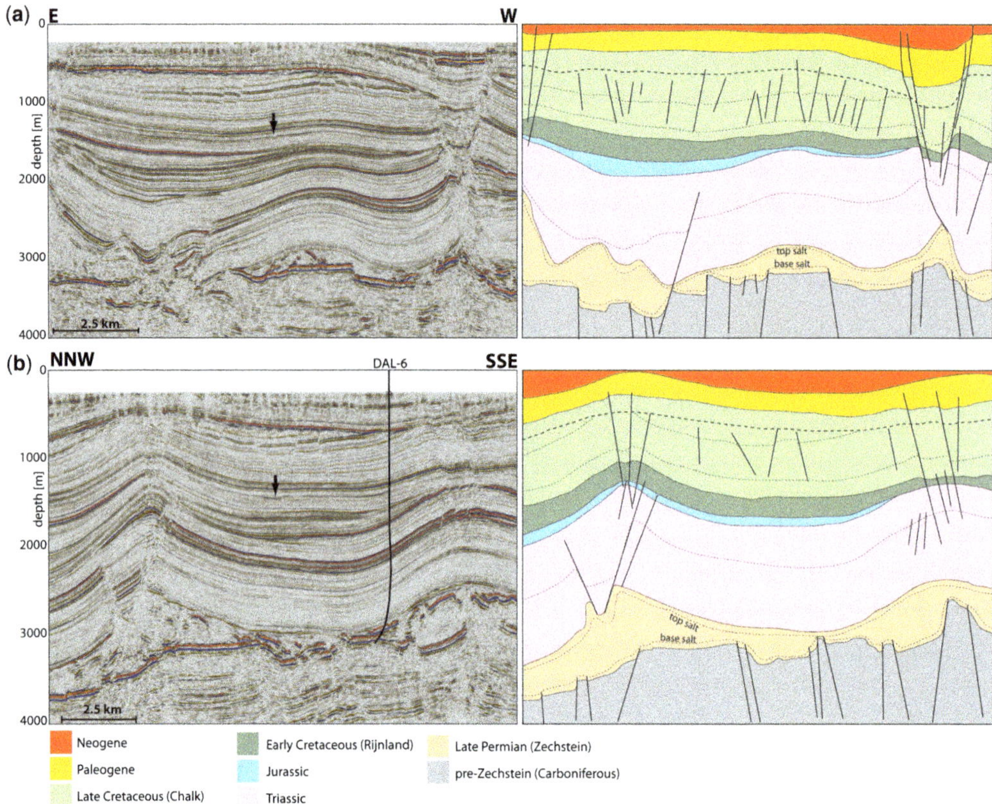

Fig. 5. (**a**) East–west seismic section across the central study area (exaggeration ×5: for the location see Fig. 1c). Note the large crestal-collapse graben in the eastern study area, as well as complex structures in the Triassic sediments in the west and the prominent Base Paleogene unconformity in the upper section. Pockmarks along the Base Chalk reflection are marked by arrows. (**b**) NNW–SSE seismic section (left) across the study area with an interpretation sketch of its main stratigraphic horizons and faults (right; exaggeration ×5: for the location see Fig. 1c). The sedimentary basin in the centre is limited by approximately east–west-trending, pillow-shaped salt walls in the north and south (east–west-trending salt anticlines in Fig. 6). Note the crestal-collapse graben structures within salt-related sediment anticlines above both salt highs.

deposition (cf. Fig. 5b), while the onset of extension during the Middle–Late Triassic caused their disruption, tilting and rafting, which is imaged in a number of faults and the blocky fragmentation of the unit (e.g. Fig. 5a, b). Variations in the thickness and distribution of Upper Triassic sediments indicate the onset of differential sedimentation and salt movement in the Muschelkalk and Keuper (Fig. 5a). The associated rafting of the Triassic strata caused early thin-skinned, passive salt diapirism with salt plugging the faults between the rafts (cf. Mohr *et al.* 2005; Vackiner *et al.* 2013; Strozyk *et al.* 2014, 2017; Strozyk 2017). The continuation of salt diapirism during the Jurassic and Early Cretaceous with a reduced sediment deposition is indicated by lens-shaped bodies limited to bowl-shaped depressions between Triassic anticlines (Fig. 5), which coevally formed due to salt diapirism. Thick, and initially almost planar, sediments of the Chalk Group above (Fig. 5) indicate a temporal quiescence of salt diapirism, which was renewed during the Late Cretaceous under a compressional stress regime. The compression-related structural inversion caused uplift and erosion of mainly Chalk sediments above the reactivated salt highs (see Fig. 5). Another change towards compressional tectonics in Late Cretaceous times (i.e. top Ommeland Formation) formed a major unconformity across the entire study area. All Upper Cretaceous, Paleogene and Neogene sediments above the unconformity only show a minor impact of salt diapirism and regional tectonic stresses. However, some of the salt-related structures – for example, the crestal-collapse graben structures in the northern and eastern study area – indicate recent activity (e.g. Fig. 5a).

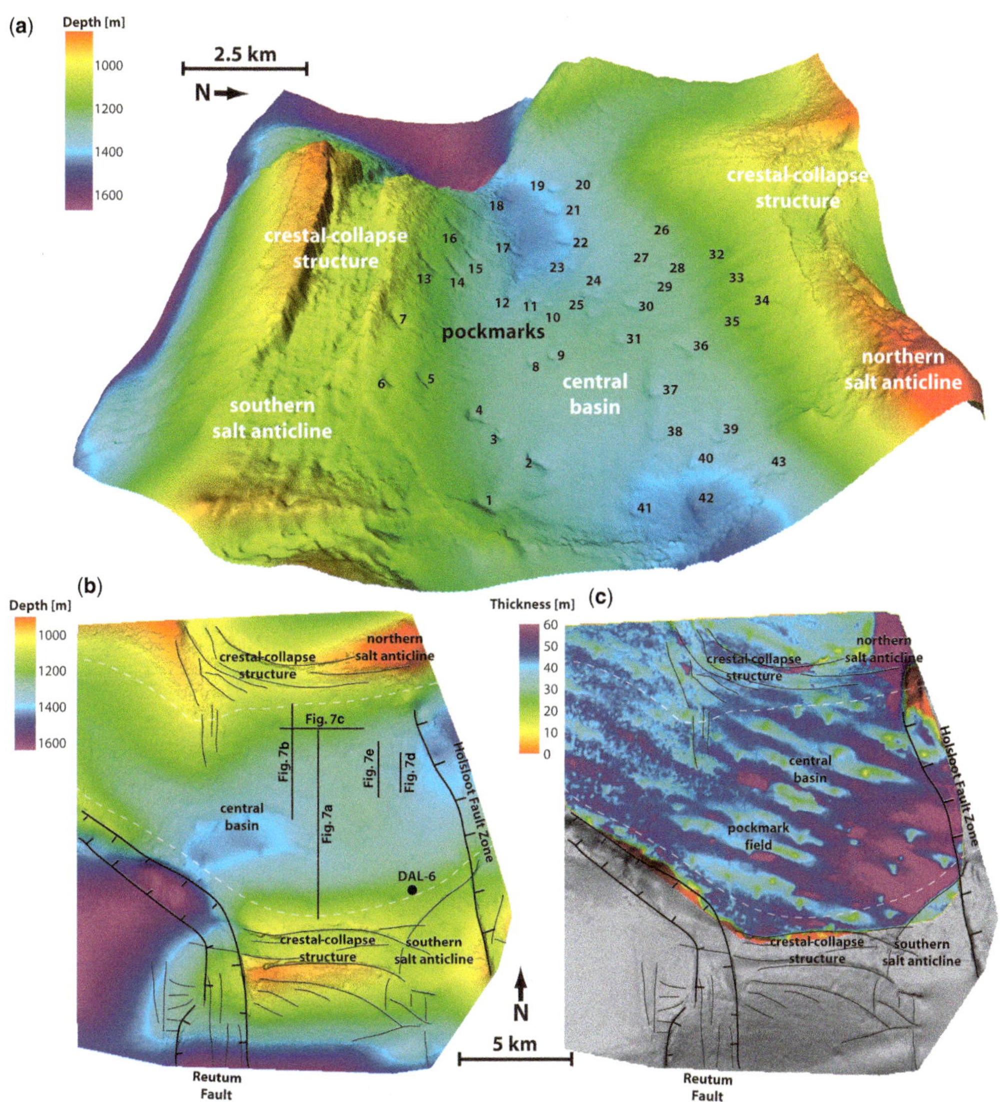

Fig. 6. (**a**) 3D view from the east on the interpretation surface of the base of the Upper Cretaceous Chalk unit. The surface indicates crestal-collapse graben structures above the east–west-trending salt highs in the northern and southern study area. In-between, more than 40 roundish depressions are interpreted as pockmarks. (**b**) Top view of the Base Chalk surface. Thin black lines indicate the major faults, which are mostly associated with approximately east–west-trending crestal-collapse structures. (**c**) Map plot of the stratigraphic thickness between the Base Chalk horizon and the 'base pockmark horizon' (also see Fig. 7a), which represents the maximum depth of pockmarks and was only interpreted across the central to northern study area. Note the prominent NW–SE-trending lineaments, which represent slight changes in the unit thicknesses and developed during faulting of the Chalk unit. Pockmarks are partially indicated by a reduced thickness (i.e. greenish to yellow colours).

Within the Rijnland Group and its transition to the Chalk Group, sub-parallel to parallel, partly wavy reflections of varying continuity, amplitude and frequency occur (Figs 5 & 7a). In the upper part of the Chalk unit, the sediments are commonly faulted. In general, these units are thinner above salt-related anticlines. Hence, their geometry follows the shape of the substratum, which is characterized by two prominent anticlines above the east–west-trending Zechstein salt walls. The fold amplitudes of the Cretaceous anticlines are lower than the fold amplitudes of the Triassic units below (Fig. 5). The

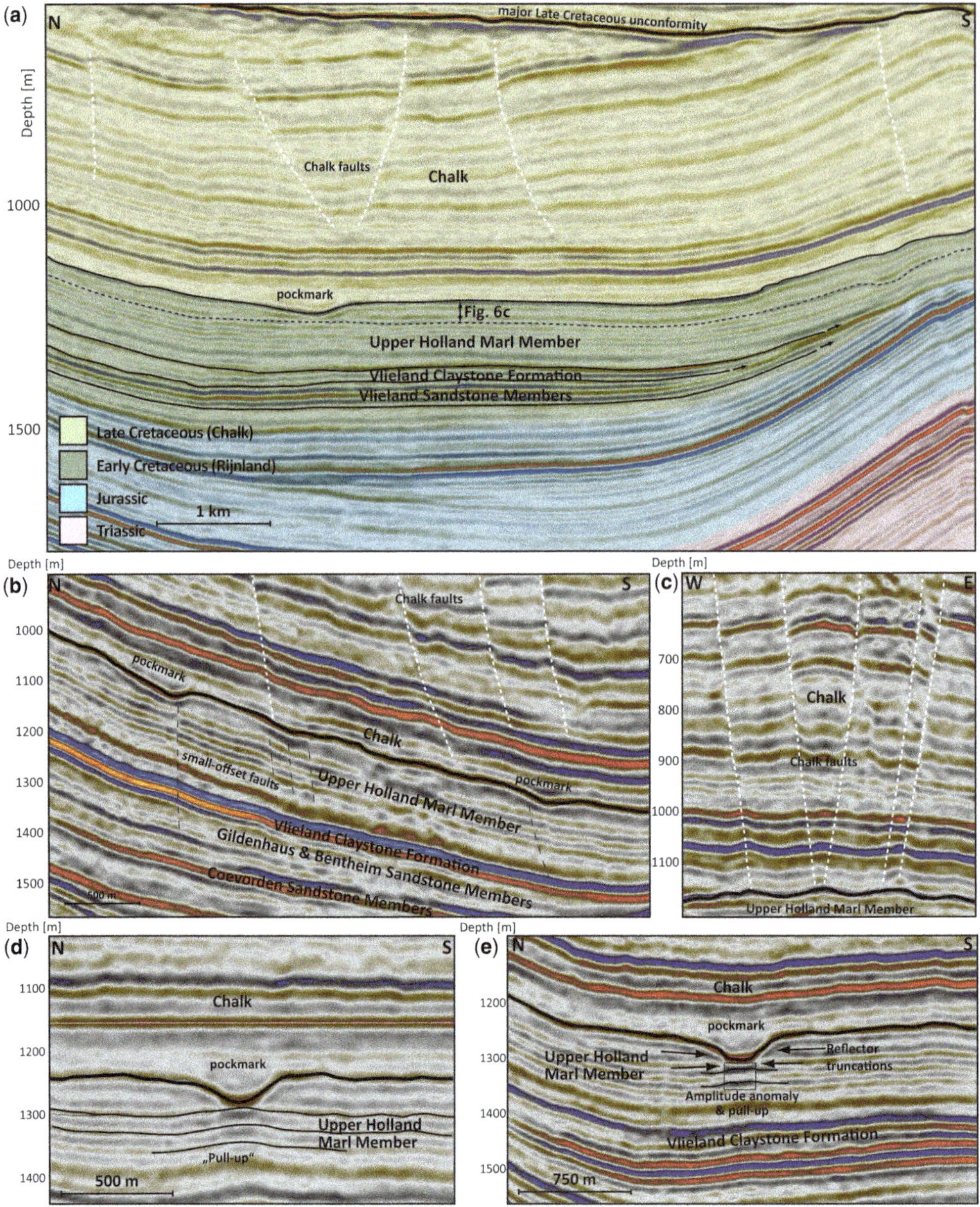

Fig. 7. (**a**) North–south seismic profile highlighting the stratigraphy around the Base Chalk reflection. The relatively thin Jurassic sediments are overlain by Lower Cretaceous sandstones and clays that show clear onlap geometries towards Triassic anticlines formed during Zechstein salt diapirism. Above, the top of the Upper Holland Marl shows the pockmarks, which are covered by the Upper Cretaceous Chalk Group, that experienced inversion, uplift and associated faulting (i.e. NW–SE) during Late Cretaceous compressional tectonics. (**b**) In places, we identify small-offset faults directly below pockmarks. These faults might reach down into the clays and sandstones. (**c**) Cropped section showing faulting of the Chalk unit and the fault correlation to structures observed in the Base Chalk horizon. (**d**) Cropped section flattened to the marked intra-Chalk horizon. It shows a close-up of a pockmark (here: red, soft kick) with amplitude anomalies and pull-up effects of the reflections below the pockmarks. (**e**) Cropped section indicating reflection truncations due to pockmark formation, as well as pull-up effects and amplitude anomalies below the pockmark.

decrease in fold amplitudes through time is due to the Jurassic sediments that fill the salt-rim synclines as lens-shaped deposits (Fig. 5). The units of the Vlieland Formation were deposited similarly, as shown by a lens-shaped basin fill that onlaps onto Triassic anticlines. Distinct onlaps of Lower Cretaceous sediments above Altena Group and Niedersachsen Group strata on the flanks of salt highs (e.g. Figs 5 & 7) indicate that these formed during the Jurassic. In contrast, the Holland Formation above has less thickness variation and can be traced across all anticlines (Fig. 5), indicating low, or the absence of, salt diapirism during this period.

The Base Chalk reflection is generally of high continuity and medium to high amplitudes, except above salt structures where extensional faulting limits the reflector continuity (Figs 5 & 6a, b). Since the tops of the anticlines in the Upper Cretaceous sediments are partially eroded, as indicated by reflection truncations (Fig. 5), a prominent unconformity separates the Chalk unit from the overlying sediments. A reactivation of salt diapirism and associated uplift occurred during the Late Cretaceous compression and renewed folding of sediments above the salt highs (Fig. 5). Erosion of the tops of the anticlines during uplift formed a prominent unconformity (Fig. 5). The upper two-thirds of the Chalk Group show a number of faults (also see data by van Gent *et al.* 2010) that are limited to strata between the Upper Cretaceous unconformity at the top and a high-amplitude reflector *c.* 50 ms above Base Chalk. Average offsets of the mostly NW–SE- to east–west-trending faults are of the order of 20–40 m.

Base Chalk and pockmarks

The Base Chalk horizon in the central study area is characterized by a multitude of roundish, mostly circular to slightly elliptical depressions of 300–850 m width and 10–50 m depth that are interpreted as pockmarks (Fig. 6a) (cf. with data from King & MacLean 1970; Hovland & Judd 1988; Judd & Hovland 2007; Andresen *et al.* 2008; Anka *et al.* 2014; Masoumi *et al.* 2014). In the southern part of the basin, several pockmarks cluster together and form approximately east–west-elongated depressions several kilometres in length (Fig. 6). The larger and deeper pockmarks incise up to four reflectors (*c.* 50 m) into the Upper Holland Marl Formation (Fig. 7c). In places, the pockmark reflector shows higher amplitudes in the deeper, central part of the depressions (i.e. an increased soft kick: Fig. 7c, left). Amplitude anomalies can also be observed in reflections below the centre of the pockmarks (Fig. 7c, right). In total, more than 40 pockmarks occur exclusively in the east–west-trending depression between the two Zechstein salt walls in the central study area (Fig. 6a). While most of them are located within the basin area, some also occur at the flanks of the salt-related, basin-bounding anticlines (Figs 6a & 7b).

Most of the pockmarks are ‘U’-shaped, wavy structures of symmetrical to slightly asymmetrical geometry (Fig. 7), while ‘V’-shaped geometries are rarely observed. Pockmarks with an asymmetrical shape mostly occur as mixed ‘V’- and ‘U’-shaped type structures (Fig. 7b). The depressions are filled with a diffuse to transparent fill or display a local thickening of the reflector above (Fig. 7c, d). The increase in amplitude of the overlying reflector within the depressions (see Fig. 7c, d) is interpreted as passive sediment infill into the topographical depressions created by the pockmarks. In general, the internal geometry of the pockmarks is poorly resolved due to limitations in the resolution of the seismic data. Comparing the observable reflection geometries to the seismic infill facies for pockmarks, as, for example, presented in Andresen *et al.* (2008), these examples would represent a ‘transparent fill’ type (cf. Fig. 7), which could be simply due to a relatively homogeneous composition of the Chalk infill. However, some of the larger and deeper pockmarks show some slightly chaotic reflection geometries (e.g. Fig. 7d, e), which would rather apply to the ‘complex fill’ type (cf. Andresen *et al.* 2008). The reflectors aside, the pockmarks show terminations onto the flanks of the depressions (Fig. 7c). Below the largest and deepest pockmarks, reflection terminations can impact up to four reflectors below. The fifth reflector below Base Chalk (‘base pockmark horizon’) remains in all cases unaffected, defining the maximum depth of the pockmarks (*c.* 50 m: Fig. 7d). Velocity pull-up effects within several reflectors below the pockmarks are also common, as well as anomalously amplified amplitudes within the reflectors directly below the pockmarks (Fig. 7d).

While larger faults are lacking within the Rijnland and Upper Holland Marl formations (Fig. 7), a number of faults are observed in the Chalk Formation above (Fig. 7a), as well as within the anticlines and crestal-collapse structures above salt highs (Figs 5 & 6b). However, close-ups of the almost undeformed Lower Cretaceous sediments show a number of low-offset faults below and between pockmarks, which are close to the limits of seismic resolution and often difficult to detect (Fig. 7b).

A thickness map between Base Chalk and the ‘base pockmark horizon’ reveals a pattern of parallel, elongated structures, which are highlighted as slight vertical variations in layer thickness (Fig. 6c). These features are individually several kilometres long, limited to an area between the basin centre and the northern salt-related anticline, and show a general NW–SE orientation. Close to the top of the western part of the northern salt anticline, these elongated features slightly bend northwards in a direction approximately parallel to the collapse faults above the salt wall (cf. Fig. 6b, c). Similar elongated features have previously been interpreted as sedimentary structures (i.e. scours) formed from bottom-water currents (e.g. Andresen *et al.* 2008). However, the distribution, orientation and positions of these lineaments in the Base Chalk horizon fit the location and orientation of the faults in the Chalk unit above (Fig. 7c). They can be vertically traced towards the Chalk faults above, where they form kink-like structures below converging pairs of the faults (e.g. Fig. 7c). Therefore, these features can be best interpreted as having formed during faulting of the Chalk in the Late Cretaceous compressional phase, and that they are, since they postdate the pockmarks, unrelated to pockmark formation. However, the fold-like structures have a high impact on the shape of the Base Chalk horizon (cf. Fig. 6a, c) and partially overprint the original geometry of a number of pockmarks (e.g. Fig. 7b).

Discussion

Determination and formation of the pockmarks

The size, shape, incision depth and amplitude anomalies of the observed depressions within the Base Chalk horizon in the central study area strongly suggest a pockmark interpretation. An alternative interpretation would be karst structures related to subaerial exposure, which can also generate subcircular depressions visible on 3D seismic data (Stewart 1999; Rosleff-Soerensen *et al.* 2012). However, the depressions in our study area have formed in sediments at the top of the Upper Holland Marl Formation, which was deposited in middle to outer neritic water depth during the Albian sea-level highstand (Fig. 2c). This sea-level rise continued into the Cenomanian, placing the study area in deep water, with no evidence of subaerial exposure associated with the sea-level fall (i.e. AL11) (Fig. 2c) around the Albian–Cenomanian transition (TNO-NITG 2000). Evaporites, which are prone to dissolution and karst formation, occur in the Weiteveen Formation of the Upper Jurassic–Lower Cretaceous Niedersachsen Group (Fig. 2). However, these units are separated from the base of the interpreted depressions by a number of horizontal and continuous reflectors (Fig. 7), and are at least 180 m deeper, precluding any effect on their generation. Subsurface hypogenic karstification, in turn, can create similar morphologies to subaerial epi-karst and can occur independently of the palaeogeographical setting. Hypogenic carbonate karst is characterized by dissolution, where aggressive fluids are continuously delivered to the dissolution front (Palmer 1991; Tóth 1999). However, such fluids, which are undersaturated with respect to carbonate, would rather lead to dissolution in the more carbonate-rich Chalk Group than in the Upper Holland Marl Formation. Consequently, karst processes can be ruled out as the generation mechanism for the observed depressions.

In the studied data, there are no structural features along the edges of the Base Chalk depressions, in particular any fault-related offsets, which exclude an interpretation as fault-bound circular depressions, such as those observed by Clark *et al.* (1999) along the Base Cretaceous horizon at the West Central Shelf in the Central North Sea. The combination of the size and distribution of the depressions, as well as the lithology of the sediments in which they formed, leads to the conclusion that these structures are pockmarks which formed from fluid escape leading to sediment removal at the seafloor.

The pockmarks in the study area show truncations of two–four reflections (Fig. 7c), which indicates a massive loss of sediment within the depression. We calculated volume losses of >0.01 km^3 for the largest pockmarks, and *c.* 0.003 km^3 as an average value for 425 m width and 30 m depth, which results in a total volume of *c.* 0.12 km^3 for the entire pockmark field. The reflections below larger pockmarks commonly show velocity pull-ups below the centre of the depression, and amplifications of the reflection amplitudes below the depressions are common (Fig. 7c). The pull-up effect and the amplitude anomalies, which are observed especially for the larger and deeper pockmarks, can be interpreted to either result from residual fluids that might have caused the generation of the pockmarks (e.g. Rollet *et al.* 2006; Cathles *et al.* 2010) or from changes in the underlying sediment caused by fluid charge (e.g. Cathles *et al.* 2010; Riboulot *et al.* 2014). However, the absence of reflection anomalies in the deeper substratum of the pockmarks (cf. Fig. 7) point at a short-lived expulsion event with no significant amounts of residual fluids or carbonate precipitation in the sediment below the pockmarks (cf. with data from Andresen *et al.* 2008). An alternative and simple explanation for the 'pull-up' anomalies could be seismic velocity of the Chalk sediment infill in the pockmark depressions, which is generally higher than the seismic velocity of the Holland Formation below and besides (van Dalfsen *et al.* 2006). The fact that the pull-up effect is not the same everywhere could be related to the varying depth of the pockmarks and the respective thickness of the Chalk infill, as well as local variations in Base Chalk lithology: for example, the amount of intercalated chert or clay (e.g. TNO-NITG 2000). However, the variability of intensity and geometry of pull-up structures may indicate that an interplay between the mentioned effects may be the best explanation.

As the pockmarks are likely to have formed during a short-term fluid-expulsion event, in total *c.* 0.12 km^3 of fine-grained marl could have been released within a short time into the water column. As the interpreted infill of the pockmarks is probably Chalk, it can be assumed that the dispersed marl was either deposited as a very thin layer across the study area or redistributed by currents. Previous studies have shown that marine currents may affect the geometry of pockmarks after their formation, and that they can be modified to elongated structures (e.g. see Andresen *et al.* 2008; Kilhams *et al.* 2011). Since the Base Chalk pockmarks in this study are mostly circular, a lack of stronger currents is likely. This also means that a very distant deposition of the removed sediment seems unlikely.

The distribution of pockmarks within the study area reveals that most of them occur within a syncline flanked by two east–west-trending Triassic anticlines (Figs 5 & 6), while a number of pockmarks (e.g. Nos 5, 6, 7, 13, 14, 16, 32, 33, 34 and 35 in Fig. 6a) are located on the flanks of the anticlines. Since the Chalk sediments, and this includes the Base Chalk horizon with the embedded pockmarks, were deposited as a planar layer of more or less constant thickness (cf. Fig. 5), salt diapirism can be interpreted as rather low or inactive at the time of pockmark generation (see the earlier subsection on 'Tectonostratigraphic interpretation'). The erosional features of Cretaceous sediments above the salt walls and anticlines in their present-day configuration can be attributed to the Late Cretaceous reactivation of salt diapirism. The pockmarks therefore probably formed in a more or less flat-top setting (also see the conceptual sketch in Fig. 8) and their present-day position can be solely attributed to later deformation processes. Yet, the concentration of pockmarks in a basin flanked by anticlines above salt walls may have focused the vertical migration of fluids (Fig. 8) (see data from Lewis & Holness 1996; Reuning *et al.* 2009; Andresen *et al.* 2011). The Lower Cretaceous Vlieland Sandstone Formation, which was limited to the basin, laterally bounded by anticlines above inflating salt highs and sealed by the marl of the Upper Holland Formation (Figs 5 & 8), had a high potential for fluid trapping and makes it a likelyreservoir for the fluids that eventually caused the generation of pockmarks at the time of Base Chalk formation.

Source of fluids

An analysis of hydrocarbons within the Lower Cretaceous sandstones in well DAL-7 close to the study area (Fig. 4) (TNO-NITG 2000) revealed that part of the gas phase in these reservoirs originated from Carboniferous coals (Krüger *et al.* 2014). However, early biogenic gas production has been described for the Posidonia Shale and the Wealden Formation (de Jager & Geluk 2007), which provides another potential two sources for the fluids. However, the Posidiona Shale is lacking in the study area (de Jager & Geluk 2007) and the transformation ratio for the Wealden Formation has been rated as low during the Cretaceous (Krüger *et al.* 2014). It is therefore unlikely that these formations provided the amount of fluids required for Cretaceous surface expulsion and pockmark formation. Thermogenic and biogenic hydrocarbon generation from Triassic, Jurassic or Cretaceous source rocks during the Early Cretaceous can be excluded for the study area (TNO-NITG 2000). Petroleum system analysis from well DAL-7 (Fig. 4a, b) indicates that, in the study area, Carboniferous coal seams were located at 2–3 km depth in the Early Cretaceous and their thermogenic gas generation has been active since the Triassic, which therefore represents these as the most likely fluid source (Fig. 4a).

Alternative sources for the fluids could be located in various regions around the study area: for example, in the deeper Lower Saxony Basin to the east. If such reservoirs represent the source of the fluids related to the formation of the pockmarks, this would then require an extensive lateral migration of hydrocarbons towards the Lower Cretaceous reservoirs in the study area. Such a fluid migration would have probably resulted in significant remnants of fluids in the various siliciclastics of the study area, which is not the case (TNO-NITG 2000). Furthermore, the analysis of the hydrocarbon composition in the Cretaceous reservoirs in the western Lower Saxony Basin (Krüger *et al.* 2014) revealed that early charging with Carboniferous gas and later charging with oil and gas from the Wealden Formation occurred; while during the Early Cretaceous, the Wealden Formation was also immature. Consequently, an Early Cretaceous charging with hydrocarbons from distant sources seems unlikely.

Migration and shallow trapping of hydrocarbons

While the pockmarks at Base Chalk indicate a single and probably short-lived event of fluid expulsion, a sufficient and long-term active system of gas generation, migration and storage is underlying this region (e.g. de Jager & Geluk 2007; Krüger *et al.* 2014; Gerling *et al.* 2016). In the study area, potential pathways for the gas from the subsalt to the suprasalt can be assumed where a connection between brittle uppermost Zechstein units and the Z2 carbonate gas reservoirs at the base of the salt section formed (Figs 5 & 8). Such a setting is, for example, observed in the central study area where synclines in the thick Triassic layers are overlain by Jurassic and Lower Cretaceous basins, and are close to, or even touching, horst structures in the subsalt strata (Fig. 5a, b, centre). Such a salt weld would be likely to allow large

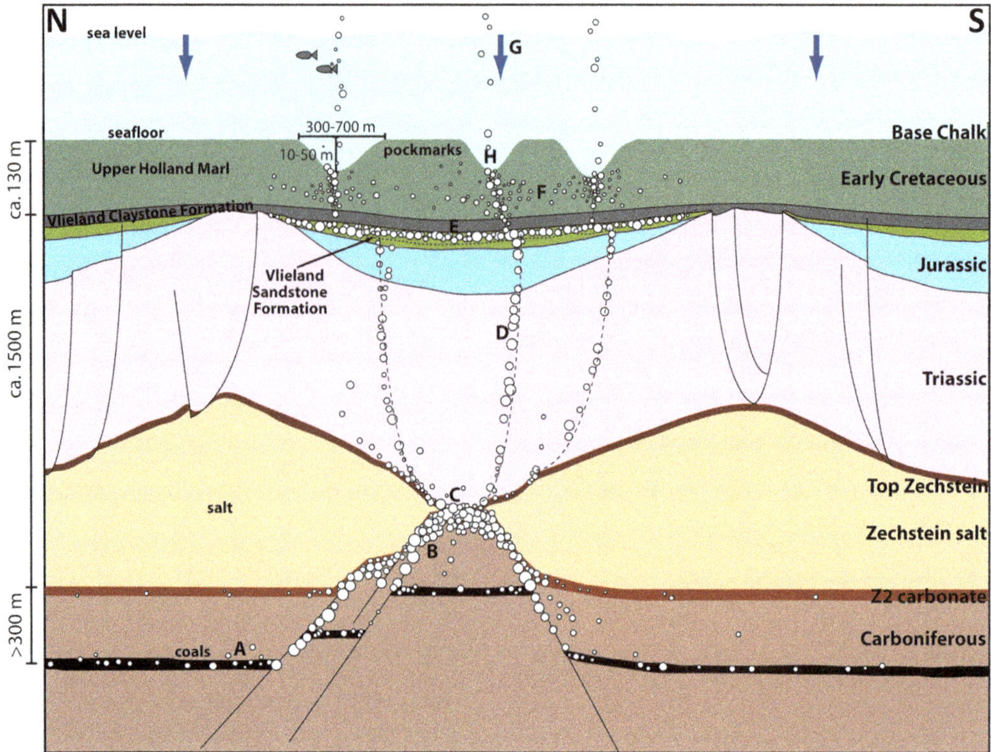

Fig. 8. Conceptual sketch illustrating the formation of the Base Chalk pockmark field: thermogenic gas (white bubbles), which was originally sourced from Carboniferous coals (black: A) after deposition of the thick Triassic sediments (pink), was charged and stored in Z2 carbonate reservoirs (dark red: B). Below a subsiding, sedimentary basin, Zechstein salt (khaki) was withdrawn and a salt weld formed, and the subsalt reservoir was connected to the suprasalt sediments (C). A system of faults and fractures, as well as the permeability of the Triassic and Jurassic (blue) siliciclastic sediments, then allowed the gas to migrate (D) into shallow Lower Cretaceous sandstones (green), which were covered by the Vlieland Clay Formation (grey: E). However, the gas also migrated into the assembling Upper Holland Marl Formation (dark green: F). During the diffuse gas charging of the growing marl section, a sea-level drop (light blue: G) and associated excess pore pressure of the gas was most likely to have exceeded the sealing capacity of the marl and caused a spontaneous, explosive and short-lived gas expulsion, and the formation of pockmarks at the seafloor (H).

amounts of fluids to migrate from the subsalt to the suprasalt sediments (also see the data from Kus *et al.* 2005; Lohr *et al.* 2007); while, in other places, the thickness of the salt layers and, perhaps, salt plugging of the carbonate reservoirs can effectively seal the subsalt reservoirs (Strozyk 2017; also see the data in Schoenherr *et al.* 2009). For the study area, gas production from Carboniferous coals, its accumulation in fractured Z2 carbonate reservoirs and the salt weld provide a likely scenario that allowed contact between the top Zechstein and the subsalt siliciclastics in the Cretaceous, and formed an effective fluid-migration pathway from the subsalt reservoirs into the suprasalt sediments (Fig. 8). Due to the complex deformation history of the Triassic sediments (cf. Fig. 5), a large number of faults and fractures, amongst visible structures in the seismic data (Fig. 5), are expected. In combination with a high matrix permeability for these sediments, suitable fluid-migration pathways through the sedimentary succession into the shallow Bentheim sandstones, which eventually charged the Upper Holland Marl Formation (Fig. 8), are very likely.

During gas charging, the shallow sandstone reservoirs had to be already in place and were partially sealed, so that the timing of gas migration and storage has to be set to a time during the deposition of the Upper Holland Marl Formation. Additional charging after the formation of the pockmarks is likely, while we can exclude charging prior to the deposition of the Vlieland Claystone Member, which appears as the oldest, potentially working, seal for such a shallow reservoir type. Furthermore, the lack of gas in the present-day reservoirs (TNO-NITG

2000) implies that either gas migration stopped or that all the gas in the reservoirs escaped.

Base Chalk fluid-expulsion scenario

Preceding the fluid expulsion and pockmark formation event at Base Chalk, we suggest a scenario in which gas leaked from the sandstone reservoirs into the overlying, continuously deposited marl section (Fig. 8). Small offset faults (Fig. 6b) and the poorly compacted and permeable character of the Vlieland Clay Member (Fig. 3) are likely to have provided fluid-migration pathways into the marl. Slow, but continuous, gas leakage into the marl with diffuse gas dispersion within its fine pore space might have occurred in parallel to the Albian sea-level rise during deposition of the Upper Holland Formation. We suggest that the gas column was rising contemporaneously with the marl deposition, which was a reservoir and seal at the same time. Betzler *et al.* (2011) pointed out that such fine-grained carbonate sediments in basinal settings could act as a capillary seal (*sensu* Cathles *et al.* 2010). A capillary seal would also allow the formation of pressurized gas pockets close to the seafloor. The formation of a hiatal surface at the top of the marl (cf. Fig. 3) could have allowed the gas column to eventually rise close to the seafloor. At the same time, water unloading from a drawdown of the sea level after the Albian–Cenomanian transition (see Fig. 2c) probably decreased the hydrostatic pressure and could have increased the gas volume inside the marl, eventually exceeding its sealing capacity, causing a short-lived gas expulsion at the seafloor (Fig. 8). Once the seal was broken – for example, by volume expansion and bubble formation of the gas – significant amounts of the gas rapidly escaped into various locations at the seafloor. This produced the large number of crater-like pockmarks in the study area (Fig. 8), which were subsequently filled and draped by Chalk sediment.

Conclusions

An approximately 150 km^2-sized field of giant pockmarks was discovered at approximately 1–1.5 km depth at the base of the Upper Cretaceous Chalk Group in the westernmost Lower Saxony Basin, eastern Netherlands. More than 40 of these mostly circular to slightly elliptical depressions of 300–850 m diameter and 10–50 m depth formed at the top of a >100 m-thick marl of the Upper Holland Formation, which is underlain by Lower Cretaceous sandstones that are known as oil and gas reservoirs in the vicinity of the study area. The Mesozoic oil and gas in these reservoirs was mixed with Paleozoic gas sourced from Carboniferous coals. As the coal seams were located at 2–3 km depth in Early Cretaceous times and had generated thermogenic gas since the Triassic, they are the most likely source for the fluids associated with the formation of the pockmarks in the Cretaceous. The coal gas was first trapped in Late Permian Zechstein carbonate reservoirs below the sealing Zechstein salt. Due to early salt tectonic movements, a localized salt weld with a touch-down of suprasalt sediments on base salt formed probably in Jurassic–Early Cretaceous times, and enabled the migration of some of the gas from a subsalt reservoir through the Zechstein and the thick, faulted and fractured Triassic strata into Lower Cretaceous sandstone reservoirs. These shallow sandstone reservoirs were limited to a sedimentary basin between two diapiring Zechstein salt walls and already sealed by the deposition of the fine-grained Upper Holland Marl Member. Subsequently, the gas was charged from the sandstone reservoirs into the Upper Holland Marl Member, which was then both reservoir and seal at the same time. A drawdown of the sea level at Base Chalk eventually led to excess pore pressure, which broke the sealing capacity of the marl, and caused a short-lived and complete expulsion of the gas to the seafloor. This event formed the giant pockmarks.

This study presents an example of how the localized migration of deep-seated hydrocarbons and their storage in shallow reservoirs can relate to salt tectonic processes, while not necessarily associated with the diapiring salt. We suggest that, within salt basins, an excessive salt withdrawal below subsiding suprasalt sediments may produce salt welds, and the suprasalt sediments may connect to the subsalt, which then can lead to a discharge of large volumes of hydrocarbons from subsalt into shallow suprasalt reservoirs. Consequently, the presence of salt welds and fluid-escape structures, like pockmarks, can be used to constrain the timing and potential for fluid migration from deep-seated source rocks into shallower reservoirs. However, fluid-escape structures also imply that hydrocarbon storage in shallower reservoirs requires an effective seal, otherwise significant amounts of hydrocarbons escape at surface.

The authors thank the Geological Survey of the Netherlands (TNO) and Nederlandse Aardolie Maatschappij (NAM) for providing the seismic and well data. Daan den Hartog Jager and Katrin Juul Andresen are thanked for their detailed and very fruitful reviews that greatly improved our manuscript. Frank Strozyk further thanks the German Science Foundation (DFG) for project funding via grant STR-1215/2-1.

References

Andresen, K.J., Huuse, M. & Clausen, O.R. 2008. Morphology and distribution of Oligocene and Miocene pockmarks in the Danish North Sea – implications for

bottom current activity and fluid migration. *Basin Research*, **20**, 445–466.
Andresen, K.J., Huuse, M., Schodt, N.H., Clausen, L.F. & Seidler, L. 2011. Hydrocarbon plumbing systems of salt minibasins offshore Angola revealed by three-dimensional seismic analysis. *AAPG Bulletin*, **95**, 1039–1065.
Anka, Z., Loegering, M.J., di Primio, R., Marchal, D., Rodriguez, J.F. & Vallejo, E. 2014. Distribution and origin of natural gas leakage in the Colorado Basin, offshore Argentina Margin, South America: seismic interpretation and 3D basin modelling. *Geologica Acta*, **12**, 269–285.
Betzler, C., Lindhorst, S., Hübscher, C., Lüdmann, T., Fürstenau, J. & Reijmer, J. 2011. Giant pockmarks in a carbonate platform (Maldives, Indian Ocean). *Marine Geology*, **289**, 1–16.
Biehl, B.C., Reuning, L., Strozyk, F. & Kukla, P.A. 2014. Origin and deformation of intra-salt sulphate layers: an example from the Dutch Zechstein (Late Permian). *International Journal of Earth Sciences*, **103**, 697–712.
Binot, F., Gerling, P., Hiltmann, W., Kockel, F. & Wehner, H. 1991. The petroleum system in the Lower Saxony Basin. *In*: Spencer, A.M. (ed.) *Generation, Accumulation and Production of Europe's Hydrocarbons III*. European Association of Petroleum Geoscientists and Engineers, Special Publications, **3**, 121–139.
Cathles, L.M., Su, Z. & Chen, D. 2010. The physics of gas chimney and pockmark formation, with implications for assessment of seafloor hazards and gas sequestration. *Marine and Petroleum Geology*, **27**, 82–91.
Clark, J.A., Cartwright, J.A. & Stewart, S.A. 1999. Mesozoic dissolution tectonics on the West Central Shelf, UK Central North Sea. *Marine and Petroleum Geology*, **16**, 183–300.
de Gans, W. 2007. Quaternary. *In*: Wong, T.E., Batjes, D.A.J. & de Jager, J. (eds) *Geology of the Netherlands*. Royal Netherlands Academy of Arts and Sciences, Amsterdam, 173–195.
de Jager, J. 2007. Geological Development. *In*: Wong, T. E., Batjes, D.A.J. & de Jager, J. (eds) *Geology of the Netherlands*. Royal Netherlands Academy of Arts and Sciences, Amsterdam, 5–26.
de Jager, J. & Geluk, M.C. 2007. Petroleum geology. *In*: Wong, T.E., Batjes, D.A.J. & de Jager, J. (eds) *Geology of the Netherlands*. Royal Netherlands Academy of Arts and Sciences, Amsterdam, 241–264.
Doornenbal, J.C. & Stevenson, A.G. (eds). 2010. *Petroleum Geological Atlas of the Southern Permian Basin*. European Association of Geoscientists and Engineers (EAGE), Houten, The Netherlands.
Duin, E.J.T., Doornenbal, J.C., Rijkers, R.H.B., Verbeek, J.W. & Wong, Th.E. 2006. Subsurface structure of the Netherlands – results of recent onshore and offshore mapping. *Netherlands Journal of Geosciences – Geologie en Mijnbouw*, **85**, 245–276.
Fader, G.B.J. 1991. Gas-related sedimentary features from the eastern Canadian continental shelf. *Continental Shelf Research*, **11**, 1123–1153.
Geluk, M.C. 2007. Permian. *In*: Wong, T.E., Batjes, D.A.J. & de Jager, J. (eds) *Geology of the Netherlands*. Royal Netherlands Academy of Arts and Sciences, Amsterdam, 63–84.
Geluk, M.C., Paar, W.A. & Fokker, P.A. 2007. Salt. *In*: Wong, T.E., Batjes, D.A.J. & de Jager, J. (eds) *Geology of the Netherlands*. Royal Netherlands Academy of Arts and Sciences, Amsterdam, 283–294.
Gerling, J.P., Blumenberg, M., Eckhard, F. & Stahl, W. 2016. Palaeozoic natural gas in Mesozoic North German oilfields – a remarkable observation. *Z. Dt. Ges. Geowiss. (German J. Geol.)*, **167**, 281–293.
Glennie, K.W. 1998. Lower Permian – Rotliegend. *In*: Glennie, K.W. (ed.) *Petroleum Geology of the North Sea. Basic Concepts and Recent Advances*. 4th edn. Blackwell Science, Oxford, 137–174.
Haskell, N., Grindhaug, J. *et al.* 1999. Delineation of geological drilling hazards using 3-D seismic attributes. *Leading Edge*, **18**, 373–382.
Heggland, R. 1997. Detection of gas migration from a deep source by the use of exploration 3D seismic data. Gas in marine sediments, geology/geochemistry/microbiology. *Marine Geology*, **137**, 41–47.
Herngreen, G.F.W. & Wong, Th.E. 2007. Cretaceous. *In*: Wong, T.E., Batjes, D.A.J. & de Jager, J. (eds) *Geology of the Netherlands*. Royal Netherlands Academy of Arts and Sciences, Amsterdam, 127–150.
Hornbach, M.J., Ruppel, C. & Van Dover, C.L. 2007. Three dimensional structure of fluid conduits sustaining an active deep marine cold seep. *Geophysical Research Letters*, **34**, L05601.
Hovland, M. 1982. Pockmarks and the recent geology of the central section of the Norwegian Trench. *Marine Geology*, **47**, 283–301.
Hovland, M. 1984. Gas-induced erosion features in the North-Sea. *Earth Surface Processes and Landforms*, **9**, 209–228.
Hovland, M. & Judd, A.G. 1988. *Seabed Pockmarks and Seepages. Impact on Geology, Biology and the Marine Environment*. Graham & Trotman, London.
Hovland, M., Gardner, J.V. & Judd, A.G. 2002. The significance of pockmarks to understanding fluid flow processes and geohazards. *Geofluids*, **2**, 127–136.
Hübscher, C., Cartwright, J., Cypionka, H., De Lange, G., Robertson, A., Suc, J.P. & Urai, J.L. 2007. Global look at Salt Giants. *Eos, Transactions of the American Geophysical Union*, **88**, 177–179.
Judd, A. & Hovland, M. 2007. *Seabed Fluid Flow – The Impact on Geology, Biology and the Marine Environment*. Cambridge University Press, New York.
Kilhams, B., McArthur, A., Huuse, M., Ita, E. & Hartley, A. 2011. Enigmatic large-scale furrows of Miocene to liocene age fom the central North Sea: current-scoured pockmarks? *Geo-Marine Letters*, **31**, 437–449.
King, L.H. & MacLean, B. 1970. Pockmarks on the Scotian Shelf. *Geological Society of America Bulletin*, **81**, 3141–3148.
Kombrink, H., ten Veen, J.H. & Geluk, M.C. 2012. Exploration in the Netherlands, 1987–2012. *Netherlands Journal of Geosciences – Geologie en Mijnbouw*, **91**, 403–418.
Krüger, M., van Berk, W., Arning, E.T., Jimenez, N., Schovsbo, N.H., Straaten, N. & Schulz, H.-M. 2014. The biogenic methane potential of European gas shale analogues: results from incubation experiments and thermodynamic modelling. *International Journal of Coal Geology*, **136**, 59–74.

KUKLA, P.A., URAI, J.L. & MOHR, M. 2008. Dynamics of salt structures. *In*: LITTKE, R., BAYER, U., GAJEWSKI, D. & NELSKAMP, S. (eds) *Dynamics of Complex Intracontinental Basins – The Central European Basin System*. Springer, Berlin, 291–306.

KUS, J., CRAMER, B. & KOCKEL, F. 2005. Effects of a Cretaceous structural inversion and a postulated high heat flow event on petroleum system of the western Lower Saxony Basin and the charge history of the Apeldorn gas field. *Netherlands Journal of Geoscience*, **84**, 3–24.

LEWIS, S. & HOLNESS, M. 1996. Equilibrium halite–H_2O dihedral angles: high rock-salt permeability in the shallow crust? *Geology*, **24**, 431–434.

LOHR, T., KRAWCZYK, C.M. *ET AL.* 2007. Strain partitioning due to salt: insights from interpretation of a 3D seismic data set in the NW German Basin. *Basin Research*, **19**, 579–597.

LOKHORST, A. (ed.). 1998. *The Northwest European Gas Atlas*. Netherlands Institute of Applied Geoscience TNO – National Geological Survey, Haarlem, The Netherlands.

MACDONALD, I.R., BUTHMAN, D.B., SAGER, W.W., PECCINI, M.B. & GUINASSO, N.L., JR. 2000. Pulsed oil discharge from a mud volcano. *Geology*, **28**, 907–910.

MASOUMI, S., REUNING, L., BACK, S., SANDRIN, A. & KUKLA, P. 2014. Buried pockmarks on the Top Chalk Surface of the Danish North Sea and their potential significance for interpreting palaeocirculation patterns. *International Journal of Earth Sciences*, **103**, 563–578.

MOHR, M., KUKLA, P.A., URAI, J.L. & BRESSER, G. 2005. Multiphase salt tectonic evolution in NW Germany: seismic interpretation and retro-deformation. *International Journal of Earth Sciences*, **94**, 914–940.

PALMER, A.N. 1991. Origin and morphology of limestone caves. *Geological Society of America Bulletin*, **103**, 1–21.

PILCHER, R. & ARGENT, J. 2007. Mega-pockmarks and linear pockmark trains on the West African continental margin. *Marine Geology*, **244**, 15–32.

REUNING, L., SCHOENHERR, J., HEIMANN, A., URAI, J.L., LITTKE, R., KUKLA, P.A. & RAWAHI, Z. 2009. Constraints on the diagenesis, stratigraphy and internal dynamics of the surface-piercing salt domes in the Ghaba Salt Basin (Oman): a comparison to the Ara Group in the South Oman Salt Basin. *Geoarabia*, **14**, 83–120.

RIBOULOT, V., THOMAS, Y., BERNÉ, S., JOUET, G. & CATTANEO, A. 2014. Control of Quaternary sea-level changes on gas seeps. *Geophysical Research Letters*, **41**, 4970–4977.

ROLLET, N., LOGAN, G.A., KENNARD, J.M., O'BRIEN, P.E., JONES, A.T. & SEXTON, M. 2006. Characterisation and correlation of active hydrocarbon seepage using geophysical data sets: An example from the tropical, carbonate Yampi Shelf, Northwest Australia. *Marine and Petroleum Geology*, **23**, 145–164.

ROSLEFF-SOERENSEN, B., REUNING, L., BACK, S. & KUKLA, P. 2012. Seismic geomorphology and growth architecture of a Miocene barrier reef, Browse Basin, NW-Australia. *Marine and Petroleum Geology*, **29**, 233–254.

SCHOENHERR, J., REUNING, L., KUKLA, P.A., LITTKE, R., URAI, J.L., SIEMANN, M. & RAWAHI, Z. 2009. Halite cementation and carbonate diagenesis of intra-salt reservoirs from the Late Neoproterozoic to Early Cambrian Ara Group (South Oman Salt Basin). *Sedimentology*, **56**, 567–589.

STEWART, S.A. 1999. Seismic interpretation of circular geological structures. *Petroleum Geoscience*, **5**, 273–285, https://doi.org/10.1144/petgeo.5.3.273

STOLLHOFEN, H., BACHMAN, G.H. *ET AL.* 2008. Upper Rotliegend to Early Cretaceous basin development. *In*: LITTKE, R., BAYER, U., GAJEWSKI, D. & NELSKAMP, S. (eds) *Dynamics of Complex Intracontinental Basins – The Central European Basin System*. Springer, Berlin, 181–210.

STROZYK, F. 2017. The internal structure of the Zechstein salt and related drilling risks in the northern Netherlands. *In*: SOTO, J.I., FLINCH, J. & TARI, G. (eds) *Permo-Triassic Salt Provinces of Europe, North Africa and the Atlantic Margins: Tectonics and Hydrocarbon Potential*. Elsevier, Amsterdam, 115–128.

STROZYK, F., URAI, J.L., VAN GENT, H., DE KEIJZER, M. & KUKLA, P.A. 2014. Regional variations in the structure of the Permian Zechstein 3 intra-salt stringer in the Northern Netherlands: 3D seismic interpretation and implications for salt tectonic evolution. *Interpretation*, **2**, 1–17.

STROZYK, F., REUNING, L., SCHECK-WENDEROTH, M. & TANNER, D. 2017. The tectonic history of the Zechstein Basin in the Netherlands and Germany. *In*: SOTO, J.I., FLINCH, J. & TARI, G. (eds) *Permo-Triassic Salt Provinces of Europe, North Africa and the Atlantic Margins: Tectonics and Hydrocarbon Potential*. Elsevier, Amsterdam, 221–241.

TAYLOR, J.C.M. 1998. Upper Permian–Zechstein. *In*: GLENNIE, K.W. (ed.) *Petroleum Geology of the North Sea. Basic Concepts and Recent Advances*. 4th edn. Blackwell Science, Oxford, 174–211.

TNO-NITG 2000. *Geological Atlas of the Subsurface of the Netherlands, Explanation to Map Sheet VI: Veendam-Hoogeveen*. TNO-NITG, Utrecht, The Netherlands.

TÓTH, J. 1999. Groundwater as a geologic agent: an overview of the causes, processes, and manifestations. *Hydrogeology Journal*, **7**, 1–14.

VACKINER, A.A., ANTRETT, P., STROZYK, F., STOLLHOFEN, H., BACK, S. & KUKLA, P.A. 2013. Salt kinematics and regional tectonics across a Permian gas field: a case study from East Frisia, NW Germany. *International Journal of Earth Sciences*, **102**, 1701–1716.

VAN ADRICHEM BOOGAERT, H.A. & KOUWE, W.F.P. (compilers). 1993–97. *Stratigraphic Nomenclature of the Netherlands; Revision and Update*. Mededelingen Rijks Geologische Dienst.

VAN DALFSEN, W., DOORNENBAL, J.C., DORTLAND, S. & GUNNINK, J.L. 2006. A comprehensive seismic velocity model for the Netherlands based on lithostratigraphic layers. *Netherlands Journal of Geosciences*, **1**, 277–292.

VAN GENT, H., BACK, S., URAI, J.L. & KUKLA, P.A. 2010. Small-scale faulting in the Upper Cretaceous of the Groningen block (The Netherlands): 3D seismic interpretation, fault plane analysis and regional paleostress. *Journal of Structural Geology*, **32**, 537–553.

VAN WEERING, T.C.E., KLAVER, G.T. & PRINS, R.A. 1997. Gas in marine sediments. *Marine Geology*, **137**, 1–3.

van Wijhe, D.H., Lutz, M. & Kaasschieter, J.P.H. 1980. The Rotliegend in The Netherlands and its gas accumulations. *Geologie en Mijnbouw*, **59**, 3–24.

Wong, Th.E. 2007. Jurassic. *In*: Wong, T.E., Batjes, D.A.J. & de Jager, J. (eds) *Geology of the Netherlands*. Royal Netherlands Academy of Arts and Sciences, Amsterdam, 107–125.

Ziegler, P.A. 1990. *Geological Atlas of Western and Central Europe*. 2nd edn. Shell Internationale Petroleum Maatschappij, The Hague. Geological Society, London.

F17-Chalk: new insights in the tectonic history of the Dutch Central Graben

HENK VAN LOCHEM
Wintershall Noordzee B.V., Rijswijk, The Netherlands
henk.van-lochem@partners.wintershall.com

Abstract: Wintershall Noordzee BV recently discovered and appraised two Chalk oil fields (Rembrandt and Vermeer) in the Dutch North Sea with wells F17-10, F17-11, F17-12 and F17-13x. Extensive core material is available, and biostratigraphical analysis plus sedimentological evaluation results are presented here. Integration of these data with detailed 3D seismic interpretation results in interesting conclusions on the tectonic inversion of the Dutch Central Graben. It can be determined that the main inversion event is the Sub-Hercynian Phase, after which an island was formed in the Chalk sea during the Campanian and Maastrichtian. Around this island, erosion products can be found in the Chalk intervals. These sediments are time and facies comparable to the Vaals Formation in the south of The Netherlands, adjacent to the inverted Roer Valley Graben. Maastrichtian sediments are seen to onlap onto the island, decreasing it in size and influence as a sediment source. It is proposed that the Laramide inversion phase in the Dutch Central Graben is a period of non-deposition and, possibly, this is also the case for most of the other Dutch inverted basins.

In 2012 Wintershall Noordzee BV, as operator of Joint Venture Group F17a Deep (Netherlands offshore), discovered oil in the Cretaceous Chalk interval with well F17-10 (Fig. 1). This discovery was later renamed the Rembrandt Field, and was subsequently appraised in 2014 by vertical well F17-11 and horizontal well F17-13x. In the same year, well F17-12 discovered oil in a separate Chalk structure, known as the Vermeer Field. The information gained from these wells, and the studies and seismic interpretation associated with these oil discoveries, provide some new insights into the geological history of the Dutch Central Graben and, in particular, the tectonic inversion of this basin.

Block F17 is located at the southern end of the Dutch Central Graben (Fig. 2), centrally on the inversion axis of the basin (Duin *et al.* 2006). The Dutch Central Graben was an important depocentre during the Triassic and Jurassic, where thick packages of continental to shallow-marine sediments have been deposited (e.g. Bouroullec *et al.*, this volume, in press; van Winden *et al.* 2018). During the Late Cretaceous–Paleogene, the basin was inverted; it was uplifted and eroded to such an extent that on the axis of the basin the Cretaceous and Upper Jurassic sediments have largely been eroded (Heybroek 1975; Clark-Lowes *et al.* 1987; Van Wijhe 1987; Kooi *et al.* 1989; Ziegler 1990; Burgers & Mulder 1991; Dronkers & Mrozek 1991; de Jager 2003; Pharaoh *et al.* 2010). The precise timing of inversion is debated, two main events have been defined: the Late Cretaceous Sub-Hercynian Phase (Santonian–Campanian age) and the Early Paleogene Laramide Phase (Danian–Selandian age). de Jager (2007) suggests, in review, that both phases were active in the Dutch Central Graben, and to a similar extent. The main focus of this paper is to consider which inversion events have been active in the southern part of the Dutch Central Graben, and to document what was their tectonic expression and influence on the Upper Cretaceous sedimentary systems.

Chalk stratigraphy

The Upper Cretaceous Chalk interval, informally referred to hereby as 'the Chalk interval', is a large carbonate system extending over most of NW Europe during the Cenomanian–Danian (Ziegler 1990; Surlyk *et al.* 2003). Chalks are generally very homogeneous, mainly composed of nannofossils (mostly coccoliths), with limited admixtures of microfossils and larger bioclasts, and only locally containing siliciclastics. Since the Chalk interval is homogeneous, lithostratigraphic subdivision is often difficult and generally has a low resolution (van Buchem *et al.* 2017). Biostratigraphy based on nannoplankton and foraminifera can be helpful in subdividing and correlating the Chalk interval in more detail. Correlations can be further improved by interpreting seismic sequences on 3D seismic data and then to tie these to well penetrations.

In the Dutch sector, the Chalk interval ranges in thickness from 0 to 1800 m (Duin *et al.* 2006) and

From: Kilhams, B., Kukla, P. A., Mazur, S., McKie, T., Mijnlieff, H. F. & van Ojik, K. (eds) 2018. *Mesozoic Resource Potential in the Southern Permian Basin*. Geological Society, London, Special Publications, **469**, 537–558.
First published online February 19, 2018, https://doi.org/10.1144/SP469.17

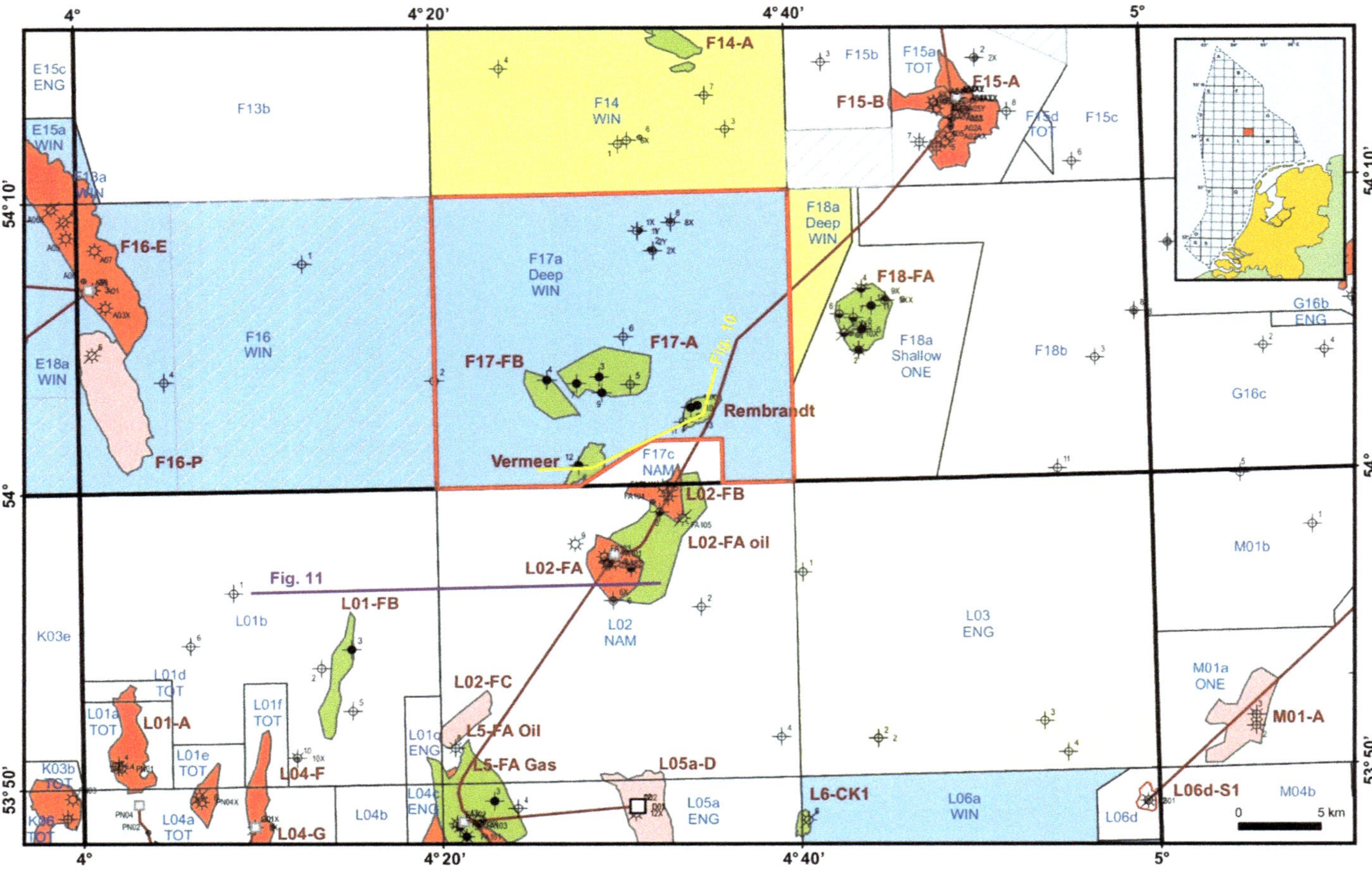

Fig. 1. Location of Block F17a, the Rembrandt and Vermeer fields, and wells F17-10, F17-11, F17-12 and F17-13x. Oil fields are in green; gas fields are in red. Production licences in blue; exploration licences in yellow. Licence operators: WIN, Wintershall Noordzee BV; TOT, Total; ENG, ENGIE; ONE, Oranje Nassau; NAM, Nederlandse Aardolie Maatschappij. Chalk oil fields: Rembrandt and Vermeer. Upper Jurassic oils fields: F14-A, F17-FA, F17-FB, F18-FA, L01-FB, L02-FA oil and L5-FA oil.

Fig. 2. (**a**) Typical cross-section of the Dutch Central Graben, after de Jager (2007); (**b**) shows the location of Block F17 and the main structural provinces.

is essentially subdivided in three units (Van Adrichem Boogaert & Kouwe 1997): the Cenomanian Texel Chalk Formation capped by the argillaceous Plenus Marl Member; the Turonian–Maastrichtian Ommelanden Chalk Formation, which is the thickest unit; and the locally present Ekofisk Formation of Danian (Lower Paleocene) age. This stratigraphic subdivision is not detailed enough to perform stratigraphic or sedimentological studies. To improve stratigraphic detail, Van der Molen & Wong (2007) mapped 11 seismic sequences in the Chalk interval of the Dutch Sector (Fig. 3).

In the Norwegian and Danish sectors, the Turonian–Maastrichtian interval is subdivided into more formations (Fig. 3). In the Norwegian sector, Gennaro *et al.* (2013) performed a detailed seismostratigraphic study mapping seven seismic sequences. In the Danish sector, van Buchem *et al.* (2017) gives an extensive and detailed account of the stratigraphy and depositional history of the Chalk interval. Their seismic stratigraphic study demonstrates that the Chalk interval is subdivided into two basin styles: a thermal subsidence basin from the Cenomanian to Late Santonian with limited deformation and continuous sedimentation; and a basin influenced by inversion tectonics during the Campanian–Danian with active uplift and tectonic processes, which influenced the sedimentary processes. The transition occurred during the Late Santonian–Early Campanian, which resulted in a hiatus that was named the Early Campanian Unconformity (ECU).

Chalk hydrocarbon play

The Upper Cretaceous Chalk hydrocarbon play is well known in the North Sea (Surlyk *et al.* 2003). It was one of the first major plays to be discovered, and it contains well-known oil fields in the Norwegian, Danish and UK sectors: e.g. Ekofisk and Valhall (NO), Dan and Skjold (DK), and Machar and Banff (UK). Nearly all of the Chalk fields are faulted salt-induced four-way dip closures sealed by Paleocene claystones and (oil) charged by Upper Jurassic Kimmeridge Clay source rock. In the Dutch offshore, only one commercial Chalk oil field had been found before the discovery of the Rembrandt and Vermeer fields: the Hanze Field in offshore Block F2 (Price *et al.* 2002). Since source-rock quality Kimmeridge Clay is absent in the Dutch Central Graben, charge has to come from the Lower Jurassic Posidonia Shale Formation (Clark-Lowes *et al.* 1987; Dubey & Mishra 2013). Oil charge from the Posidonia is proven in the southern Dutch Central Graben by the presence of (as yet) uncommercial discoveries in Upper Jurassic sandstone reservoirs (e.g. fields F17-FA, F17-FB and F18-FA: Fig. 1).

Upper Jurassic and Lower Cretaceous claystones often impede oil charge from the Posidonia source rock to the Upper Cretaceous Chalk reservoir, resulting in unsuccessful Chalk play exploration wells.

The presence of two significant Late Cretaceous inversion and erosion phases be could an important downgrading factor for the Chalk play prospectivity of the Dutch Central Graben, since this would be the first reservoir to be eroded, certainly during the Laramide event.

Database

Data from recently (2012–15) drilled wells F17-10, F17-11, F17-12 and F17-13x (Wintershall, F17a Deep Group) are used to reassess the inversion history of the Dutch Central Graben. Since these wells have not, at the point of writing, passed their 5 year confidentiality period, only selected information of these wells is presented.

In 2012, well F17-10 (Fig. 4) was drilled on a salt-induced Chalk closure and found a relatively thin (38 m) oil-bearing Ommelanden Formation (Fig. 3) (Van Adrichem Boogaert & Kouwe 1997). Biostratigraphic analysis proved this Chalk section to be of Maastrichtian age. Unconformably below the Chalk lies a 38 m section of the Vlieland Claystone Formation, which in turn lies above a 50 m-thick Zechstein cap rock associated with an underlying salt dome. A small section of core (5 m) was recovered from the upper part of the Chalk reservoir. A 2 m conglomeratic interval was found at the base of the Maastrichtian.

In 2014, the discovery (now the Rembrandt Field) was appraised by vertical well F17-11 (Fig. 5). One of the objectives of this well was to fully core the Chalk reservoir interval, including part of the Paleocene overburden (Landen Formation). This resulted in a 71 m cored section, although there was a 13 m lost section in the middle of the Chalk interval. A surprise in this well was the presence of an 8 m conglomerate and sandstone interval at the base of the Chalk. The total reservoir interval is 56 m, overlying an interval of 19 m of Early Cretaceous Vlieland Claystone Formation, which rests on Zechstein cap rock.

Also in 2014, well F17-12 (Fig. 6) was drilled on a nearby Chalk closure and made a separate discovery, now called the Vermeer Field. This well completely cored the Chalk reservoir, including part of the Paleocene overburden (Landen Formation) and the underlying Early Cretaceous Holland Formation. The well was drilled at an inclination of 30° through the reservoir. In this well, the Chalk interval was found to have a true thickness of 72 m. Again, a siliciclastic-rich interval (15 m true vertical thickness) was found at the base of the Ommelanden

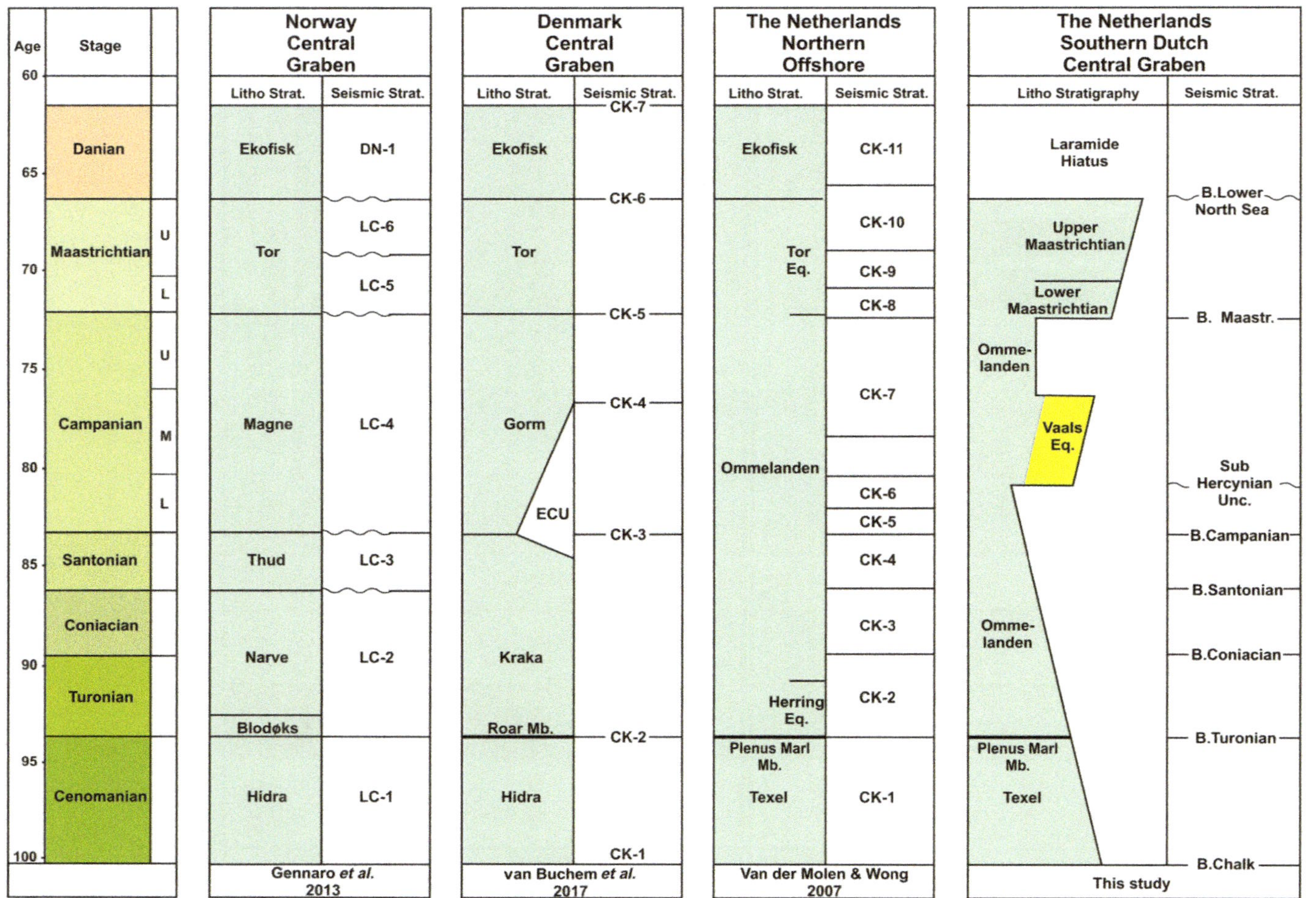

Fig. 3. Stratigraphic scheme of the Chalk in the Dutch (after Van der Molen & Wong 2007), Norwegian (after Gennaro *et al.* 2013) and Danish (van Buchem *et al.* 2017) sectors presenting (informal) lithostratigraphic units and interpreted seismic sequences. The final column represents the subdivision used in this study.

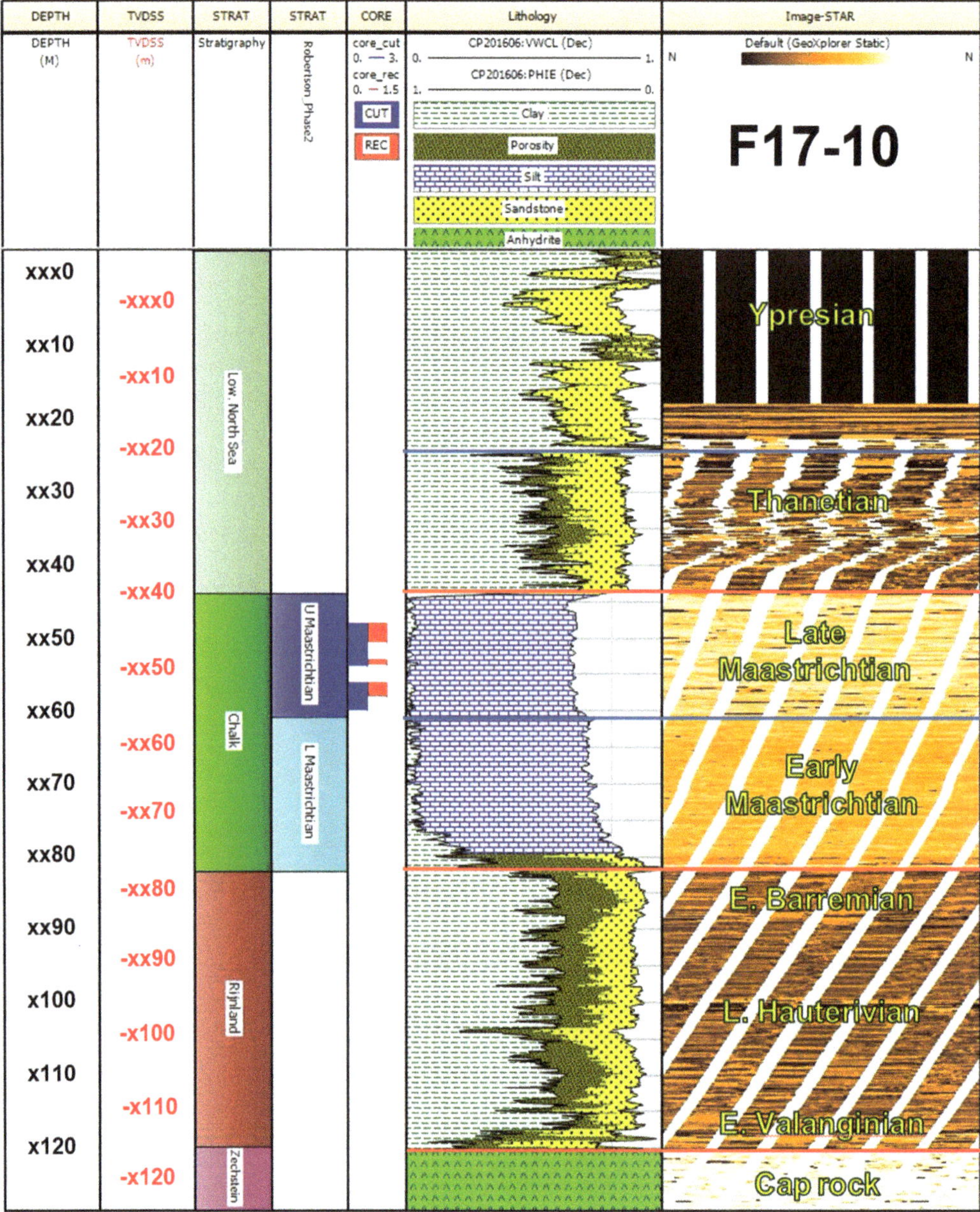

Fig. 4. Interpreted well results for F17-10. Tracks (from left to right): depth (in m MD (metres measured depth)) and depth (in m TVDss (in metres true vertical depth subsea)), lithostratigraphy, biostratigraphy, cores, lithology and image log with biostratigraphical age. Depths are deliberately masked for confidentiality purposes. Note the large time gap of the Sub-Hercynian Unconformity between the Maastrichtian and Barremian. Also note the lower porosity and siliciclastics in the basal part of the Chalk.

Chalk. Unconformably below these clastics, 5 m of chalk of the Texel Formation (Cenomanian age) was found, which was directly underlain by the Holland Formation. In this well, a total length of 109 m core is available.

Well F17-13x is not discussed in this article, as it was drilled horizontally in the Chalk reservoir. Reference is also made in this article to some released well data in the Dutch Central Graben (available via NLOG 2017).

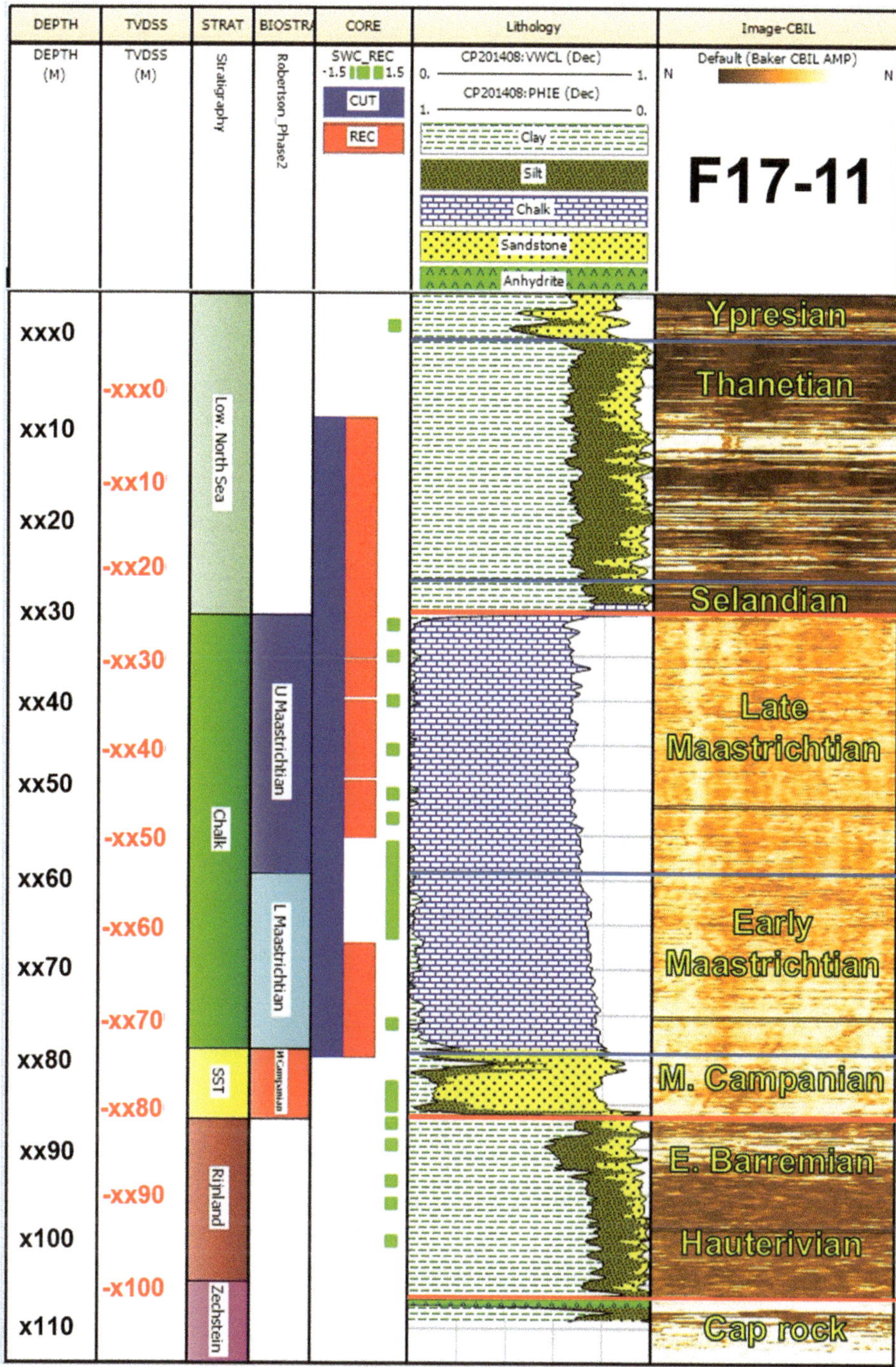

Fig. 5. Interpreted well results for F17-11. Tracks (from left to right): depth (in m MD (metres measured depth)) and depth (in m TVDss (in metres true vertical depth subsea)), lithostratigraphy, biostratigraphy, cores, lithology and image log with biostratigraphical ages. Depths are deliberately masked for confidentiality purposes. Note the presence of a Middle Campanian sandstone between the Maastrichtian Chalk and the Sub-Hercynian Unconformity and above the Barremian claystones.

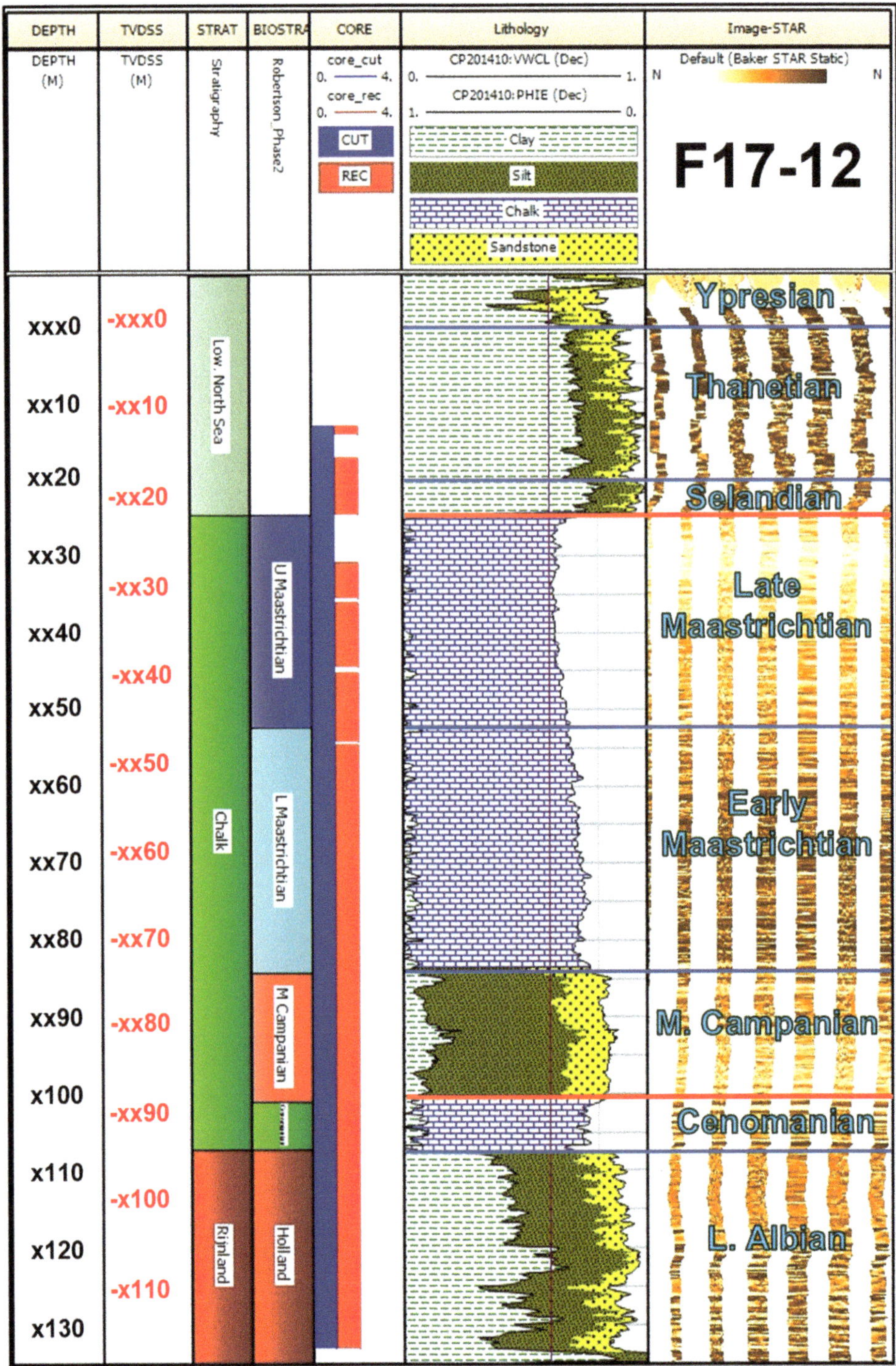

Fig. 6. Interpreted well results for F17-12. Tracks (from left to right): depth (in m MD (metres measured depth)) and depth (in m TVDss (in metres true vertical depth subsea)), lithostratigraphy, biostratigraphy, cores, lithology and image log with biostratigraphical ages. Depths are deliberately masked for confidentiality purposes. Note the Sub-Hercynian Unconformity between the Middle Campanian sandstone and the Cenomanian Texel Chalk.

The Rembrandt and Vermeer fields, and the whole of the southern Dutch Central Graben, have been covered by released 3D seismic data. Since the Chalk is relatively shallow (1250–1700 m) and does not have a complex overburden, the seismic quality and resolution is generally very good, with a high signal to noise ratio and with a vertical resolution of at least 25 m. A total area of nearly 5000 km^2 was available for regional evaluation.

In 2014, a new broadband 3D seismic survey was acquired covering both fields and surroundings, resulting in a high-quality 3D seismic dataset (550 km^2) to map the structural and sedimentological features. The vertical resolution of the survey is 16 m at Top Chalk level. All seismic data are displayed in the European convention: an increase in impedance results in a negative number and is displayed in a red colour.

All biostratigraphic results presented herein (Chalk interval of wells F17-10, F17-11, F17-12 and F17-13x: Figs 4–6) are sourced from internal reports with the analysis performed by CGG Robertson. Age dating in the Chalk interval is mainly based on foraminifera and nannoplankton. In Lower Cretaceous and Paleocene intervals, palynology was also applied. The resolution in these wells is high, a sample interval of 3 m was used, locally even finer in the key cored sections. In conjunction with these studies, the biostratigraphy of the Chalk in wells in surrounding blocks was re-evaluated. This resulted in a consistent and detailed biostratigraphical subdivision of the Chalk and the overlying Paleocene in the Dutch F-quadrant and northern L-quadrant.

The cores cut in the F17-10, F17-11 and F17-12 wells have been analysed by Panterra Geoconsultants BV for their sedimentological facies development. Prior to this work, Wintershall Noordzee BV had commissioned GEUS (the Geological Survey of Denmark and Greenland) to describe the Chalk cores of all available released wells in The Netherlands. Predominantly Danian and Maastrichtian intervals have been cored in around 20 wells.

Biostratigraphic evaluation

The basal part of the Lower North Sea Group consists of the Paleocene Landen Formation (Van Adrichem Boogaert & Kouwe 1997). The formation is overlain by the Basal Dongen Tuffite Member. This interval consists of tuffaceous clays, which provide a clear lithological marker in an otherwise monotonous claystone interval. The member has an Ypresian age and is time-equivalent to the Balder Formation in the Danish, UK and Norwegian sectors. The Landen Formation is found in all wells in the southern F-quadrant and has a thickness of 30–50 m. The formation consists of dark-grey to, sometimes, red upper bathyal claystones. The formation can be correlated to the Sele, Lista and Maureen/Vale formations in the northern part of the North Sea. Detailed biostratigraphical analysis shows that some hiatuses are present in the Landen interval. The upper part of the formation has a Thanetian age, whereas the basal part is placed in the (upper) Selandian.

The basal part of the Selandian and the Danian are absent in the Rembrandt and Vermeer fields and in the surrounding wells. This hiatus coincides with the Laramide tectonic event. No Ekofisk Formation is found on the inverted Central Graben. This formation is, however, present around the Central Graben in the Step Graben, Outer Rough Basin and the Schill Grund High (Van der Molen 2004).

On the axis of the Dutch Central Graben, the Chalk Group is generally very thin (<100 m) or in some areas absent (Duin *et al.* 2006). Biostratigraphic analysis shows that this thin Chalk interval is the upper part of the Ommelanden Formation, Maastrichtian in age. In the Rembrandt and Vermeer fields, the Chalk interval can further be subdivided in an Upper Maastrichtian and a Lower Maastrichtian interval, with a small hiatus between them. In wells with an even thinner interval (e.g. F17-6), the Chalk is often only Upper Maastrichtian in age. A general trend of decreasing water depths, from middle–outer to inner–middle neritic, is observed in foraminiferal assemblages going upwards through the Maastrichtian.

Below the Maastrichtian Chalk, a conglomeratic- and siliciclastic-rich interval is found locally. This interval was (partly) cored in wells F17-11 and F17-12 (Figs 5 & 6), but can also be seen in the logs of e.g. wells F17-4, F18-2 and F14-7. The conglomeratic section has been dated by foraminifera and nannoplankton as Lower Maastrichtian in F17-10 and F17-11 (part), and Middle Campanian in F17-11 (part) and F17-12. The underlying siliciclastic intervals in F17-11 and F17-12 have a Middle Campanian age.

Below the Maastrichtian Chalk, or below the Middle Campanian interval, if present, a significant unconformity is observed. Below this unconformity, rocks of Early Cretaceous (Barremian)–Cenomanian age, based on biostratigraphic dating, have been found in the Rembrandt and Vermeer wells.

Sedimentological evaluation

The Maastrichtian Chalk cored by the three wells is typically a structureless or bioturbated wackestone with occasional chert nodules. Locally, stylolites can be observed, especially at the base of the formation. Rare macrofossils like bivalves and belemnites have been found. Most of the chalk rock consists of a mud matrix of coccolith platelets. Floating in this matrix are skeletal grains, like calcispheres and

Fig. 7. Well F17-11: thin sections of Upper Maastrichtian clean chalk: (**a**) (measured depth (MD) xx42.6 m), Lower Maastrichtian sandy chalk; (**b**) (MD depth xx77.6 m) and Lower Maastrichtian conglomeratic chalk; and (**c**) (MD depth xx79.9 m). Depths are deliberately masked for confidentiality purposes, but compare to Figure 5. Note that the chalk mud matrix is dark in these thin sections; the *c.* 40% microporosity in the chalk mud is not resolved in the thin section.

foraminifera, and fragments of bivalves and bryozoans (Fig. 7a). The basal 10 m of the Chalk contains more siliciclastic grains. This can be observed microscopically (Fig. 7b) and by XRD analysis. The upper part of the Chalk contains typically more than 97% of calcite in the XRD analysis, whereas in the basal part of the formation more than 10% of siliciclastics in the form of admixed quartz and glauconite grains. The increase in siliciclastic and skeletal grains towards the base of the formation decreases the porosity of the Chalk from around 38% at the top to 25% at the base of the formation. The depositional environment is interpreted to be a low-energy, mid- to outer-ramp/shelf setting below the storm wave base, based on the observed trace fossils and the absence of planktonic foraminifera. The higher content of siliciclastic material at the base of the formation indicates a clastic input source and increased hydraulic energy able to transport sand-sized particles.

In well F17-11, the contact between the Maastrichtian Chalk and the Paleocene Landen Formation has been cored (Fig. 8a). It consists of a 25 cm intensely bioturbated interval. Trace fossils contain subvertical or inclined burrows seen in the Chalk, filled with dark claystone material.

In the basal 2 m of the Maastrichtian Chalk interval (cores F17-11 and F17-12), siliciclastic pebbles are present in the chalk matrix. In general, a fining-upwards sequence can be observed with a clast-supported texture at the base and a chalk matrix-supported texture towards the top (Fig. 8b, c). The conglomerate clasts consist of sandstone, glauconite and claystone fragments (Fig. 7c). The depositional environment is interpreted as a shallow-marine setting based on the presence of glauconite, in which coarse material was transported by episodic mass flows or high-energy mass-gravity currents across/into the low-energy chalk environment.

Below the conglomeratic layer, a sandstone interval was found in the cores of wells F17-11 and F17-12. In well F17-11 (Fig. 9a), this sandstone can be described as a poorly to moderately sorted, structureless, medium to coarse sublitharenite. Grains

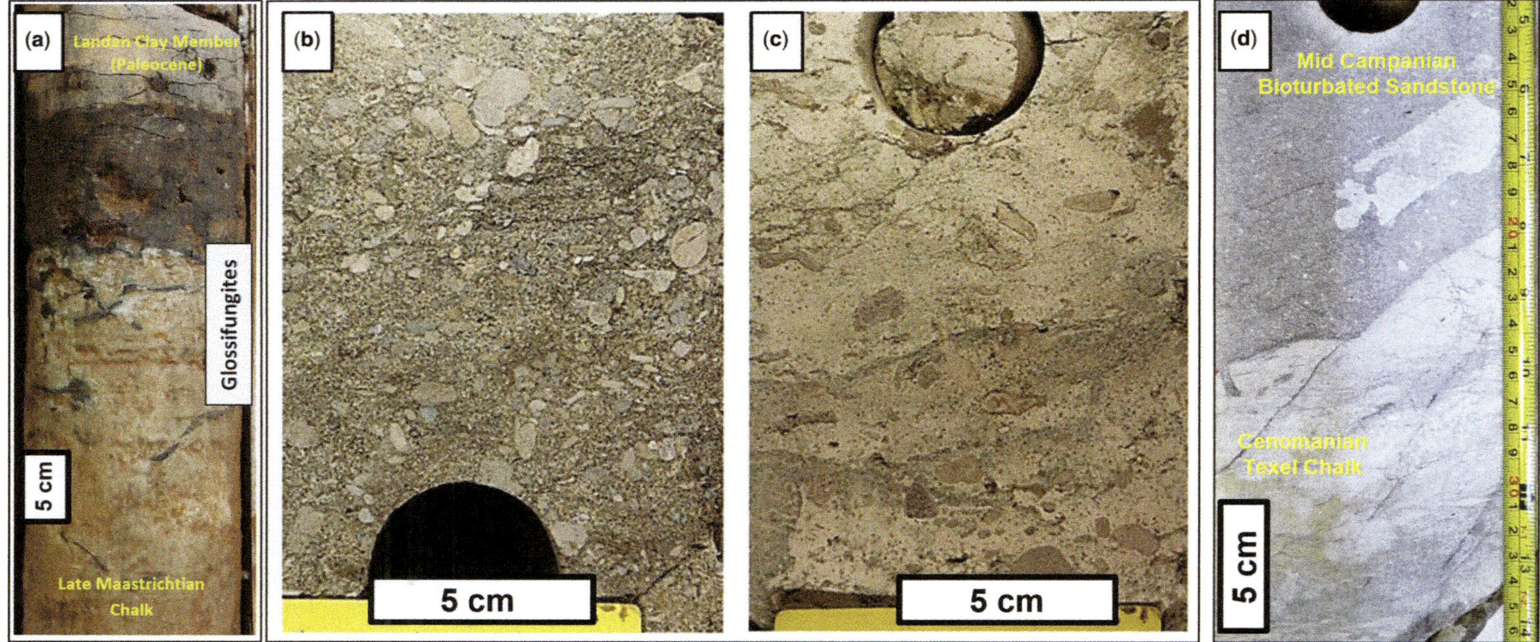

Fig. 8. Core photographs. (**a**) Well F17-11 (measured depth (MD) xx30.0 m): contact between the Landen Formation and Late Maastrichtian Chalk. Note the burrows in the Chalk. (**b**) Well F17-11 (MD depth xx79.8 m): clast-supported conglomerate with a Chalk matrix. (**c**) Well F17-11 (MD depth xx78.7 m): matrix-supported conglomerate. (**d**) Well F17-12 (MD depth x100.9 m): the Sub-Hercynian Unconformity is between the Mid-Campanian bioturbated sandstone and the fractured Cenomanian Texel Chalk. Depths are deliberately masked for confidentiality purposes, but compare to Figures 5 and 6.

consist of abundant subangular to subrounded fine- to coarse-grained quartz and sparse subrounded feldspar, glauconite/green clay pellets, and silt/sandstone clasts. No chalk mud matrix is present. In well F17-12, besides these siliciclastic fragments, soft chalk clasts, thick-walled bivalves and echinoderm fragments are also present (Fig. 9b). At this location, a chalk mud matrix is present in this grain-supported sandstone. The presence of glauconite, marine fossils and prevailing moderately to poorly sorted, fine to medium siliciclastic grains suggest fully marine, downslope transport by storm currents or proximal mass-gravity flows. The grain-supported fabric in well F17-11, without matrix and skeletal content, suggests rapid deposition of clastic sediments in a proximal setting, possibly triggered by mass-gravity flows. However, the admixture of siliciclastic grains with chalk matrix and skeletal fragments in F17-12 suggests concomitant chalk deposition during siliciclastic input in a more distal setting.

In well F17-12, below the sandstone interval, an eroded chalk section of the Texel Formation was found (Fig. 8d). Below the unconformity surface, a fractured marly chalk can be seen, with blocks of chalk floating in the sandstone above the unconformity. The Texel Chalk has been deposited in deeper water (outer neritic), as evidenced by the presence of a greater proportion of planktonic foraminifera relative the Maastrichtian Chalk.

Regional 3D seismic evaluation

A large number of wells were available to tie the geology to seismic reflectors, including vertical seismic profiles (VSPs) from wells F17-11 and F17-12. In general, the Base Lower North Sea Group–Top Chalk Group reflector (Fig. 10) is a conspicuous marker (red/yellow trough, hard kick on seismic sections) that demonstrates increasing impedance from the Landen Formation claystones to the Ommelanden Formation chalks. Meanwhile, the Base Chalk Group–Top Rijnland Group reflector is characterized by a decrease in impedance (blue peak, soft kick). To better evaluate the stratigraphic development of the Chalk Group, and any intraformational unconformities induced by inversion tectonics for a large area around the Rembrandt and Vermeer fields, a detailed 3D seismic interpretation was performed of intra-Chalk markers (Fig. 11). The biostratigraphic evaluation described above, formed the framework for well to seismic ties (Fig. 3) and the boundaries of the interpreted seismic sequences.

Based on seismic interpretation, it can be concluded that most of the Chalk has been eroded in the Dutch Central Graben (Figs 10 & 12). However, a thin Maastrichtian interval (generally <100 m) can be mapped over most of the area. This means that the Maastrichtian is located above the main unconformity surface. Below the Maastrichtian, going from the fringes towards the core of the inverted Central Graben, the subcrop (Fig. 13) varies from Chalk of pre-Maastrichtian stages to Lower Cretaceous marls and claystones (Rijnland Group), Upper Jurassic Scruff Group and Schieland Group sandstones and claystones to Lower Jurassic claystones of the Altena Group. At some locations in the core of the inverted basin, the Maastrichtian Chalk is also absent. This is confirmed by, for example, wells F17-1, F17-2, F14-1, F14-3 and F11-1, where the Paleocene Landen Formation rests directly on Lower Cretaceous or Jurassic sediments. These areas can be mapped on the seismic data by the dimming of the Base Lower North Sea Group reflector, which is caused by the Lower Cretaceous and Jurassic claystone subcrop having a lower impedance than the base of the Chalk interval (Fig. 12).

Detailed 3D seismic evaluation

More detailed mapping in the Rembrandt and Vermeer area (Fig. 14) revealed intra-Chalk thickness variations. The Chalk interval (mostly Maastrichtian) has a thickness of 40–80 m on the Rembrandt and Vermeer salt dome, whereas it is only 20–30 m thick around the structure. The thickness trends on the salt dome show clear channel features (Fig. 14). A relationship between the thickness of the Chalk and the presence of the Campanian sandstone at the base of Chalk can be envisioned. Wells F17-11 and F17-12 seem to have been drilled in a location where the Chalk is thicker and where the sandstone is present, whereas well F17-10 found a thinner Chalk without sandstone. The sandstone was also found in well F17-4 (Fig. 12) to the north of the Vermeer Field. Here, the deposition of the sandstone seems to be determined by a small fault-bounded graben. Near well F18-2, on the east side of the inverted Central Graben (Fig. 12), channels can be mapped at the base of the Sub-Hercynian Unconformity. This well contains the same type of glauconitic sandstone below the Maastrichtian Chalk as found in the F17 wells to the west.

From these observations, it can be proposed that the presence of the sandstone is determined by pre-existing lows. In the case of F17-4, this low was created by synsedimentary tectonism. In the case of the Rembrandt and Vermeer salt dome, it is most likely that subrosion of the salt during the Sub-Hercynian Phase has created a topographical low with subaerial channels cutting down into this feature. It is proposed that the fluvial channels were then later filled with marine sediments, including the sandstones.

Fig. 9. Thin sections of Middle Campanian sandstone: (**a**) well F17-11 (measured depth (MD) xx84.1 m); and (**b**) well F17-12 (MD xx87.0 m) Depths are deliberately masked for confidentiality purposes but compare to Figures 5 and 6. Note the high content of quartz grains in well F17-11 compared to the more carbonate and chalk mud-rich sample from well F17-12.

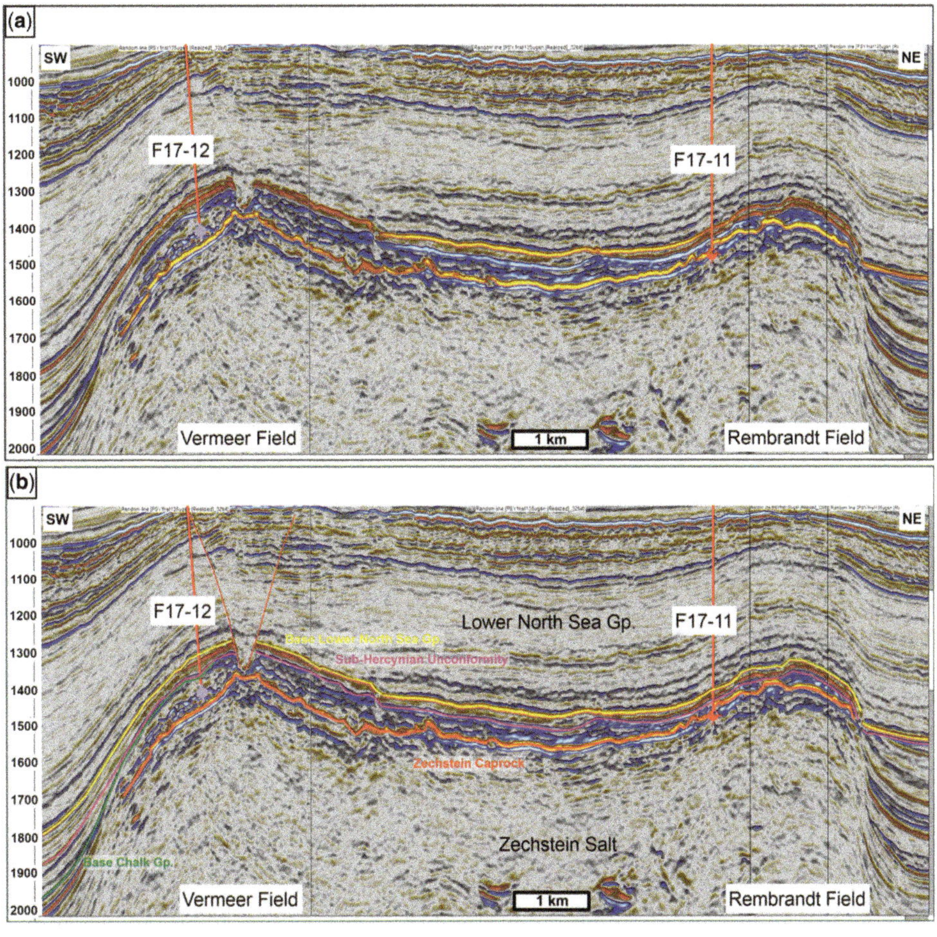

Fig. 10. Uninterpreted (**a**) and interpreted (**b**) two-way time (TWT) seismic section across the Vermeer and Rembrandt fields, showing the thin development of the Maastrichtian Chalk. Yellow marker, Base Lower North Sea Group; pink marker, Sub-Hercynian Unconformity; green marker: Base Chalk Group (only picked where Base Chalk coincides with Base Texel Chalk); red marker, Top Zechstein cap rock. The location is indicated in Figure 1.

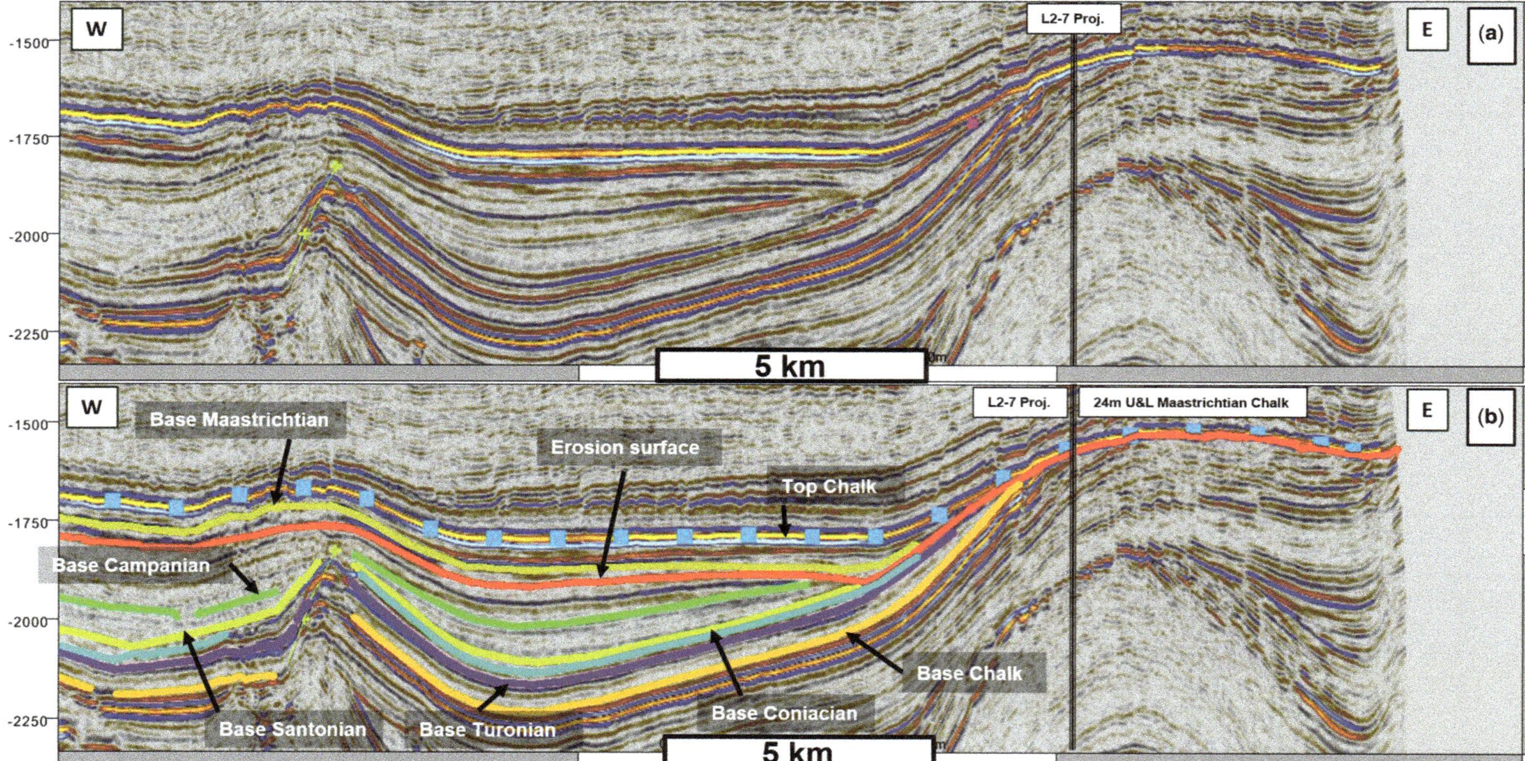

Fig. 11. Uninterpreted (**a**) and interpreted (**b**) two-way time (TWT) seismic section across the western flank of the Dutch Central Graben (blocks L1 and L2), demonstrating erosion by the Sub-Hercynian Unconformity on the flank of the inverted graben and the presence of thin Maastrichtian Chalk (10–25 m) on the crest, as confirmed by the L2 wells. The location is indicated in Figure 1.

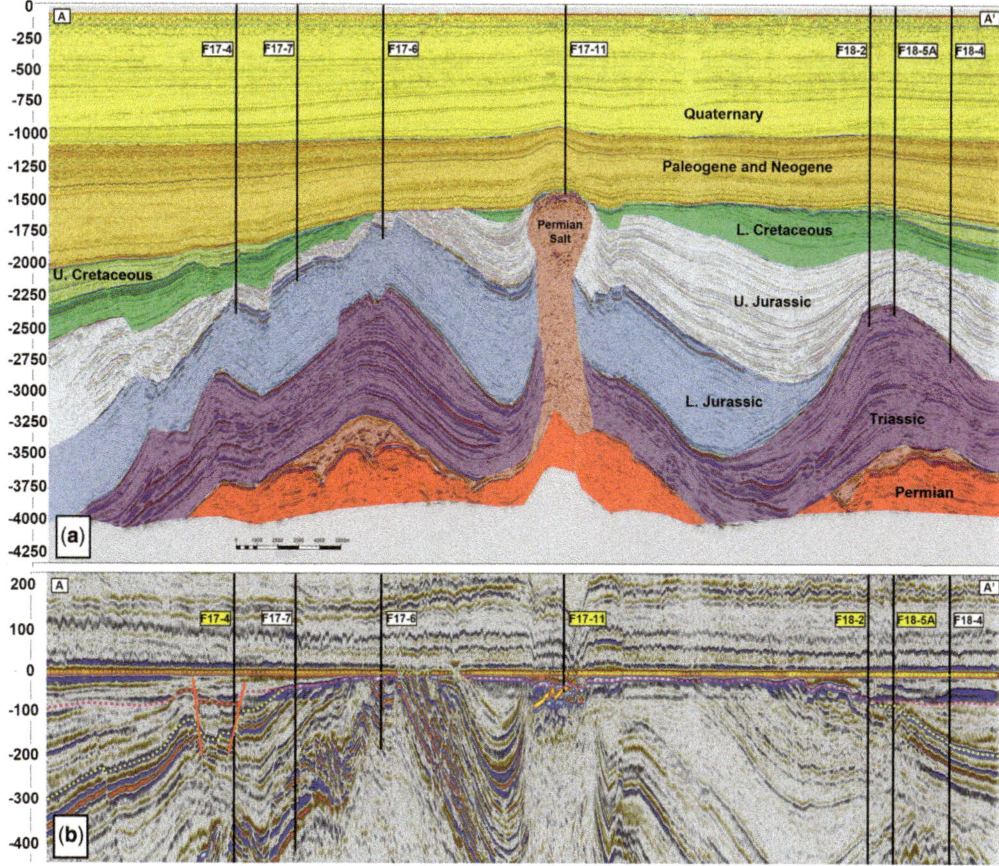

Fig. 12. (**a**) West–east interpreted seismic section (in two-way time (TWT)) across the inverted Dutch Central Graben in blocks F17 and F18. Note that the Upper Cretaceous is present over much of the section, but not visible on this scale. (**b**) Same section as (a) but flattened on Base Lower North Sea and stretched vertically to show the Chalk interval. The Sub-Hercynian Unconformity is in a pink colour; the Base Texel Chalk is in a yellow colour. Wells indicated in yellow have a Campanian Sandstone above the Sub-Hercynian Unconformity. Note the topography of the Sub-Hercynian Unconformity: fault-bounded graben at F17-4, depression above the salt dome near F17-11, channel shaped west of F18-2 and rapidly thickening near F18-2. Maastrichtian Chalk is absent east of well F17-6 where highly reflective Upper Jurassic sandstone, claystone and coal packages subcrop (note the dimming of Base Lower North Sea reflector). The location of this seismic line is given in Figure 13.

Reconstruction of geological history

Based on the observations presented above, the following geological history of the Chalk interval can be proposed (Fig. 15). During the early part of the Late Cretaceous (Cenomanian–Santonian), pelagic chalk was deposited continuously in the Central Graben area (Ziegler 1990), as it was located in part of a larger, fully marine basin (Fig. 15a). The basin edge is thought to have been at least 250 km to the south (Noord Brabant), where Gras & Geluk (1999) indicated shallow-marine sandstones of Santonian age (Fig. 16). The rest of the basin was developed in the typical Chalk facies.

During the Sub-Hercynian tectonic phase (Late Santonian–Early Campanian), the Dutch Central Graben became inverted (de Jager 2007). This resulted in a broad anticline (Figs 2, 12 & 15b), which is proposed to have manifested itself as an island in the Chalk sea extending from Block F3 in the north to Block L3 in the south. It is important to note that in the area around this island, Chalk sedimentation is believed to have continued unimpeded, no large hiatuses are seen in wells outside the Central Graben during this time interval. As soon as the island was formed, erosion of the exposed sediments is proposed to have taken place. The shallowest and youngest chalk intervals were

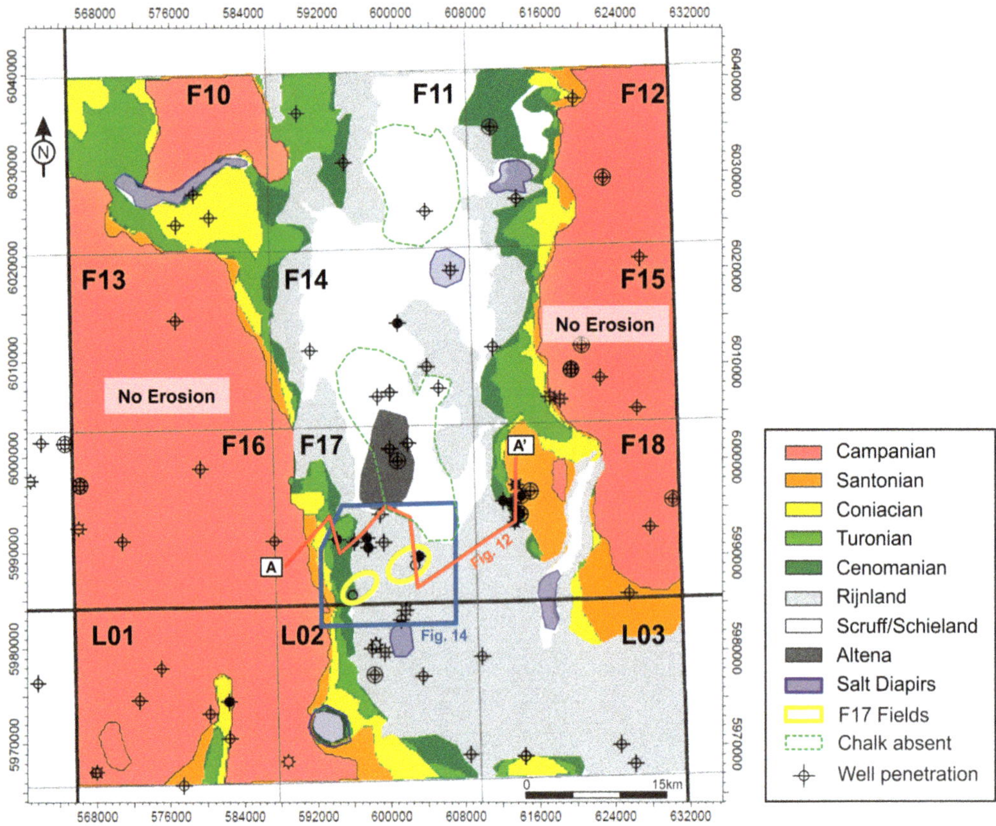

Fig. 13. Subcrop map of the Sub-Hercynian Unconformity southern Dutch Central Graben. Note the increase in age of the subcropping layers towards the centre of the north–south-orientated inversion structure. The subcrop area of the most erosive resistant sandstone-bearing Scruff and Schieland groups is in the centre of the structure (white colour). Where Chalk is absent (green dashed line), the subcrop below the Base Lower North Sea Group is given.

probably poorly lithified, thus easily eroded and taken up in the surrounding Chalk sea. The deeper chalk intervals – for example, the Cenomanian in well F17-12 – were lithified (Fig. 8d). In this well, a fractured unconformity surface can be seen, with blocks of chalk lying above the unconformity. The marls and claystones of the Rijnland Group were probably easy to erode and were deposited as clay clasts in the surrounding Chalk basin. The Scruff and Schieland groups contain sandstones and claystones, and are present in the core of the anticline (Fig. 13). These formations were likely to have been more resistant to erosion and they can be the source of the siliciclastic intervals found above the Sub-Hercynian Unconformity. Continued erosion is proposed to have created a fluvial drainage pattern on the island (Fig. 15c), preferentially using zones of weakness, such as fault-bounded lows (e.g. around well F17-4), or topographical lows created by subaerial salt subrosion as on the Rembrandt and Vermeer salt dome (Figs 12 & 13). Possibly, alluvial sediments were deposited in and around these channels, but these have not been encountered in wells as yet.

During the Middle–Late Campanian (Fig. 15d) part of the island is proposed to have been submerged. During this time, some of the eroded material seems to have been (re-) deposited around the island, probably preferentially in and around the topographical lows of the channels. Well F17-11 is interpreted to be closer to the source area than well F17-12, as it contains more siliciclastic material and hardly any basin-sourced bioclastic material. Note that the sediments in the channels are clearly marine: hence, they are interpreted to be transgressive. In the rest of the surrounding Chalk basin, the Campanian sediments are typically clean pelagic chalk deposits (Van der Molen 2004).

Due to this transgression, the exposed island decreased in size. It is possible, therefore, that storms and waves may have washed off fringing subaerially created clastic deposits (quartz sand grains and

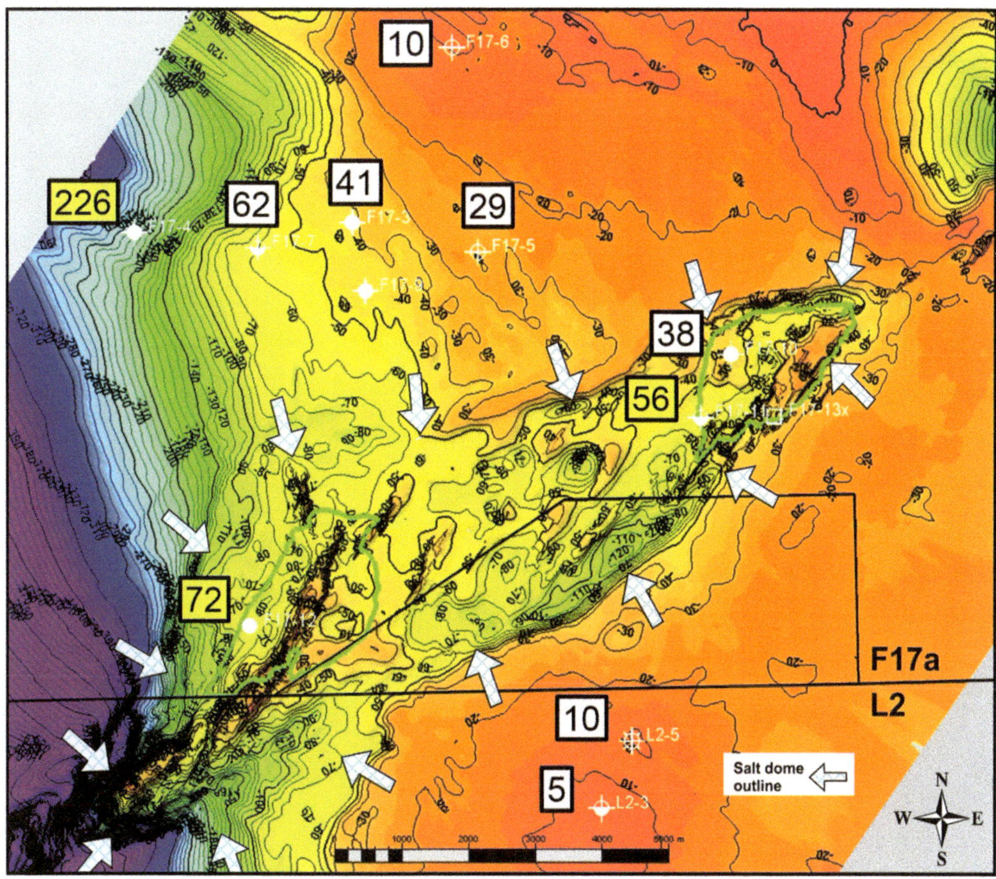

Fig. 14. Isochore map of the Chalk Group Vermeer and Rembrandt areas. Isochores at the well locations are indicated in metres. The location is indicated in Figure 13. Note the greater thickness of the Chalk on the Rembrandt and Vermeer salt dome (indicated by arrows), and the channel-like topography. Wells indicated in yellow have a Campanian Sandstone above the Sub-Hercynian Unconformity.

pebbles) into the shallow-marine area around the basin, giving rise to the conglomerates found within the Middle Campanian–lowermost Maastrichtian Chalk facies (Fig. 15e). During the Early Maastrichtian, the transgressive phase continued, which resulted in an upwards decrease of siliciclastic and bioclastic grains observed in the thin sections from the cores of wells F17-11 and F17-12 (Fig. 7a, b). In the Late Maastrichtian (Fig. 15f), barely any siliciclastic material was deposited in the Chalk of the F17 wells, indicating a limited terrigenous supply and erosion from the island or, at least, a greater distance to the coastline. Only in the core of the anticline, where erosion resistive Upper Jurassic sandstones are present, was no Chalk deposited during the Late Maastrichtian (Fig. 12).

During the Latest Maastrichtian, the Danian and Early Selandian (Fig. 15g), no deposition took place in an area which coincides with the inverted Dutch Central Graben. Detailed biostratigraphic analysis, in wells where Maastrichtian Chalk is present, does not show any variation of the subcrop below this time gap. Also, the variation in strata above the disconformity is very small. In all wells, including those on the crest of the inversion island where no Chalk is present, the strata above the disconformity are Upper Selandian and Thanetian upper bathyal claystones (Fig. 15h). Based on the deep-water setting of the overlying rocks and the limited erosional evidence below the disconformity, the Laramide Phase is not seen here as an important tectonic event. However, most probably, the disconformity took place below sea level and is seen more as period of non-deposition (Fig. 8a), after which fast subsidence and deepening of the North Sea Basin started in the Paleocene.

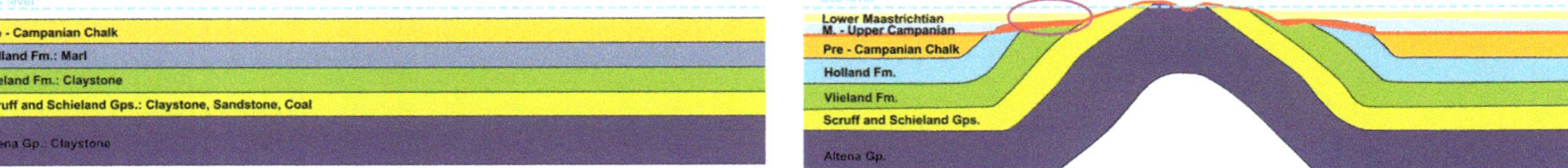

(**a**) Pre-Campanian stratigraphy

(**e**) Continued transgression on island. Storms wash off clastic deposits from island into sea. Deposition Lower Maastrichtian Chalk containing clastics

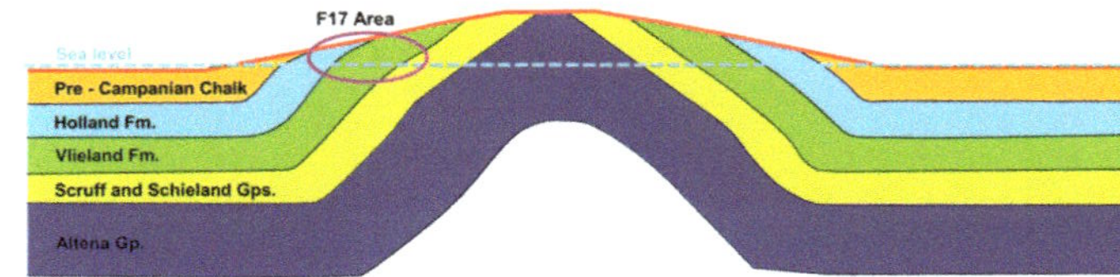

(**b**) Sub-Hercynian tectonic phase (Late Santonian-Early Campanian) creates island in Chalk sea (continued chalk deposition in basins)

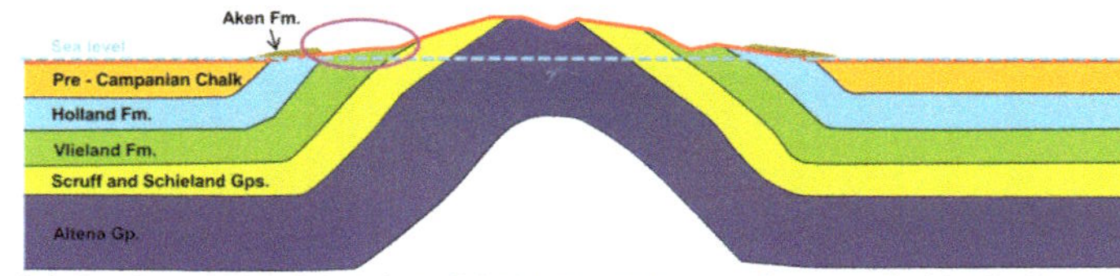

(**c**) Erosion of island, incision of drainage pattern, partly controlled by faults and subroded salt domes, possible deposition fluvial/estuarine Aken Fm. eq.

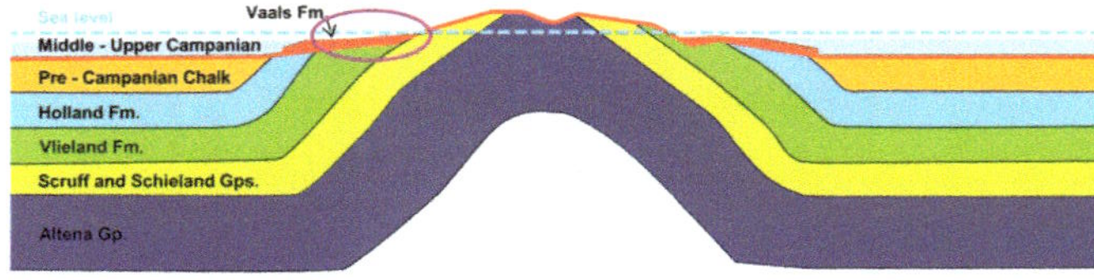

(**d**) Transgression on island. Deposition clastics (Vaals Fm. eq.) around island and in palaeolows. Chalk deposition in basin (Middle-Upper Campanian)

(**f**) Continued transgression on island. Deposition clean Upper Maastrichtian Chalk

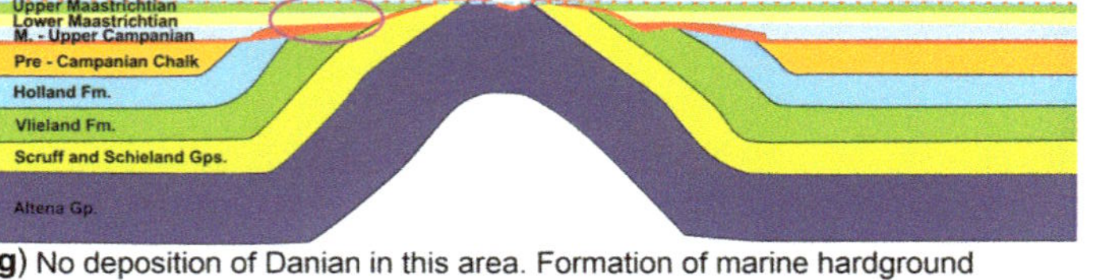

(**g**) No deposition of Danian in this area. Formation of marine hardground

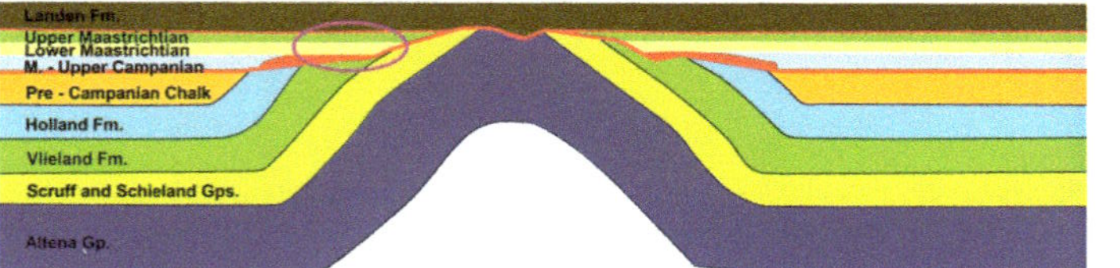

(**h**) Transgression during Selandian. Chalk factory closed down. Continued deposition of thick deep-marine claystone

Fig. 15. Conceptual cartoons (**a**)–(**h**) showing the schematic geological development of the southern Dutch Central Graben from the Campanian to the Paleocene (west–east cross-section, excluding salt domes; no scales implied). The pink circle represents area of the Rembrandt and Vermeer fields.

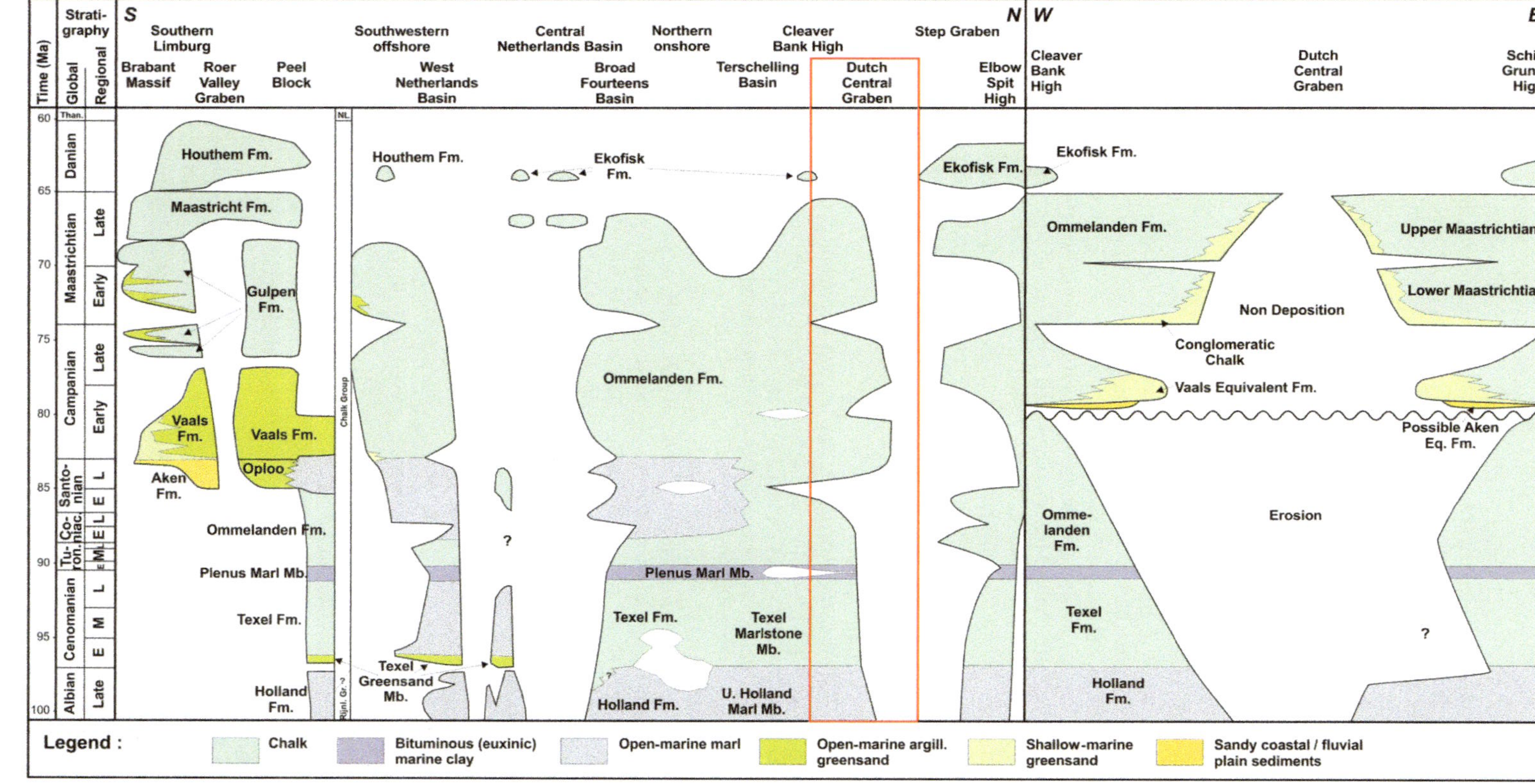

Fig. 16. Lithochronostratigraphic section of the Late Cretaceous of The Netherlands, based on Van Adrichem Boogaert & Kouwe (1997) (left panel), incorporating Gras & Geluk (1999) (Peel Block area) and the results of this study (right panel). The figure shows the stratigraphic position of newly described clastics (Vaals equivalent) in the Dutch Central Graben, and the effects of erosion and non-deposition related to the Sub-Hercynian event.

Discussion and conclusions

The drilling of wells F17-10, F17-11 and F17-12 has contributed to a new understanding of the geology of The Netherlands Chalk interval during tectonic inversion (Fig. 16). For the Dutch Central Graben, it can be demonstrated that this inversion is mainly related to the Sub-Hercynian Phase, which can be dated here as Early Campanian. This dating correlates with the main inversion event (Early Cretaceous Unconformity), as described by van Buchem *et al.* (2017) in the central part of the North Sea (Fig. 3).

The Laramide Phase was here merely a period of non-deposition and not of major inversion, as postulated in other studies (e.g. de Jager 2007). Also, in other inverted basins in The Netherlands (the Broad Fourteens Basin, the Central Netherlands Basin and the Roer Valley Graben: e.g. de Jager 2007; Pharaoh *et al.* 2010), thin Maastrichtian and/or Danian deposits have been found above the major unconformity, suggesting that, also here, the Sub-Hercynian Phase was the main inversion event.

Another interesting observation is the presence of an island in the Chalk sea at the location of the inverted Central Graben. On this island, it is proposed that subaerial drainage patterns can be seismically mapped. The Maastrichtian Chalk is seen onlapping onto the island, diminishing its size and influence over time. For the Dutch Central Graben, the presence of an island has not previously been documented (e.g. Ziegler 1990; Vejbæk *et al.* 2010). It is proposed that more of these islands could have existed in the Chalk sea on other inverted basins related to the Sub-Hercynian tectonic event.

Around the proposed island, siliciclastic sediments have been found which can be linked to erosion after the Sub-Hercynian Phase. These are time-equivalent to the glauconitic shallow-marine sandstones of the Vaals Formation (Fig. 16) in the south of The Netherlands (Zuid Limburg: Kuyl 1983; Bless *et al.* 1986; Felder & Bosch 2000; and the Peel Horst: Gras 1995; Gras & Geluk 1999), which are thought to be deposited next to the inverting Roer Valley Graben partly filling palaeotopographical depressions. In the south of The Netherlands, the Aken Formation is also found locally, which is defined as fluvial and estuarine sandstones on top of the underlying unconformity (Felder & Bosch 2000). The transgressive Vaals Formation may have locally eroded the underlying Aken Formation in South Limburg (Kuyl 1983). The presence of an equivalent to the Aken Formation can also be envisioned in the Dutch Central Graben, possibly preserved in the thicker channels mapped on the Vermeer and Rembrandt salt dome.

In view of the significant angular unconformities related to inversion tectonics seen in seismic sections from the Dutch Central Graben, it has always been enigmatic as to where erosion products related to these unconformities were (re-)deposited. If the main inversion was the Laramide event, then significant Upper Jurassic and Chalk erosion products should be present in the Paleocene. However, in this interval in the surrounding area, no siliciclastic or chalk erosion products have been found; in fact, it is characterized by monotonous deep-marine claystones. A Sub-Hercynian dating of the unconformities and erosion provides a better explanation; the (partly unlithified) chalk erosion products are reworked in the Campanian and Maastrichtian Chalk intervals, whereas siliciclastic erosion products have now been identified in this interval. The total lack of erosion products in or at the base of the Paleocene in the rest of the Dutch offshore and onshore, except near the basin edge near the border with Belgium, suggests that the Laramide Phase did not result in significant erosion in most of The Netherlands, further strengthening the suggestion that the main inversion event in The Netherlands was during the Early Campanian.

The results of this study raise questions for further research. What could not be determined with certainty, due to erosion after the Sub-Hercynian Unconformity in the inversion axis, is the sedimentological development of the Cenomanian–Santonian interval over the area. Was a submarine or subaerial high already present during this time interval in the Dutch Central Graben? In the area around the southern Dutch Central Graben, no indications can be found for this hypothesis; however, thin or absent Texel Chalk in wells north of the study area (e.g. wells F12-1 and F5-1) suggests that this could be the case (Fig. 16). Also, the character and extent of the small hiatus between the Lower and Upper Maastrichtian, and between the Middle Campanian and Lower Maastrichtian deposits, is unclear. In thicker, deeper-water settings in Norway (Gennaro *et al.* 2013) and Denmark (van Buchem *et al.* 2017), stratigraphic heterogeneity has been described in the Maastrichtian interval, which is attributed to active tectonic uplift and increased bottom-current intensity. Since in the Dutch Central Graben the Chalk interval is very thin, detailed seismic intra-Chalk sedimentology is often beyond seismic resolution. This makes it hard to verify the character of the smaller hiatuses.

These new findings on the inversion of the Dutch Central Graben can also have an impact on the further hydrocarbon exploration potential in the inverted basins off NW Europe:

- Large-scale regional mapping may have missed (thin) Campanian–Maastrichtian Chalk in areas where no Chalk was expected due to (Laramide) inversion.

- The clastic sediments related to the erosion after the Sub-Hercynian Phase may locally serve as an additional hydrocarbon exploration target.
- The erosion after the Sub-Hercynian event resulted in a shorter migration path from the Lower Jurassic Posidonia source to the overlying the Maastrichtian Chalk with fewer intermediate seals.
- Basin modelling has to take into account the correct timing and intensity of inversion movements and associated erosion. Overestimating the Laramide event could lead to incorrect results for source-rock maturity and hydrocarbon migration.

Special thanks are due to Catalin Petrea of PanTerra Geoconsultants BV, the biostratigraphical staff of CGG Robertson and colleagues at Wintershall Noordzee BV for their technical contributions, and to the F17a Deep Partner Group (Wintershall Noordzee BV, EBN BV, ENGIE E&P Nederland BV, Rosewood Exploration Ltd & TAQA Offshore BV) for their permission to publish this paper. The reviewers are thanked for their suggestions for improvements to this paper.

References

BLESS, M.J.M., FELDER, P.J. & MEESSEN, J.P.M.T. 1986. Late Cretaceous sea level rise and inversion: their influence on the depositional environment between Aachen and Antwerp. *Annales de Société Géologique de Belgique*, **109**, 333–355.

BOUROULLEC, R., VERREUSSEL, R. *ET AL.* In press. Tectonostratigraphy of rift basins affected by halokinesis: the synrift Middle Jurassic–Lower Cretaceous in the Dutch Central Graben, Terschelling Basin and neighbouring platforms, Dutch offshore. *In*: KILHAMS, B., KUKLA, P.A., MAZUR, S., MCKIE, T., MIJNLIEFF, H.F. & VAN OJIK, K. (eds) *Mesozoic Resource Potential of the Southern Permian Basin*. Geological Society, London, Special Publications, **469**, https://doi.org/10.1144/SP469.22

BURGERS, W.F.J. & MULDER, G.G. 1991. Aspects of the Late Jurassic and Cretaceous history of The Netherlands. *Geologie en Mijnbouw*, **70**, 347–354.

CLARK-LOWES, D.D., KUZEMKO, N.C.J. & SCOTT, D.A. 1987. Structure and petroleum prospectivity of the Dutch Central Graben and neighbouring platform areas. *In*: BROOKS, J. & GLENNIE, K.W. (eds) *Petroleum Geology of North-West Europe: Proceedings of the 3rd conference on Petroleum Geology of North West Europe, Volume 1*. Graham & Trotman, London, 337–356.

DE JAGER, J. 2003. Inverted basins in the Netherlands, similarities and differences. *Geologie en Mijnbouw – Netherlands Journal of Geosciences*, **82**, 339–349.

DE JAGER, J. 2007. Geological development. *In*: WONG, TH. E, BATJES, D.A.J. & DE JAGER, J. (eds) *Geology of The Netherlands*. Royal Netherlands Academy of Arts and Sciences, Amsterdam, 5–26.

DRONKERS, A.J. & MROZEK, F.J. 1991. Inverted basins of The Netherlands. *First Break*, **9**, 409–425.

DUBEY, S. & MISHRA, A. 2013. Petroleum systems modelling: a case study from Dutch Central Graben (Offshore Netherlands). *Extended abstract P35 presented at the PGCE 2013 Kuala Lumpur*, 18–19 March 2013, Malaysia.

DUIN, E.J.T., DOORNENBAL, J.C., RIJKERS, R.H.B., VERBEEK, J.W. & WONG, TH.E. 2006. Subsurface structure of the Netherlands – results of recent onshore and offshore mapping. *Netherlands Journal of Geosciences – Geologie en Mijnbouw*, **85**, 245–276.

FELDER, W.M. & BOSCH, P.W. 2000. *Krijt van Zuid-Limburg. Geologie van Nederland 5*. Nederlands Instituut voor Toegepaste Geowetenschappen (TNO), Delft, The Netherlands.

GENNARO, M., WONHAM, J.P., GAWTHORPE, R. & SÆLEN, G. 2013. Seismic stratigraphy of the Chalk Group in the Norwegian Central Graben, North Sea. *Marine and Petroleum Geology*, **45**, 236–266.

GRAS, R. 1995. Late Cretaceous sedimentation and tectonic inversion, southern Netherlands. *Geologie en Mijnbouw*, **74**, 117–127.

GRAS, R. & GELUK, M.C. 1999. Late Cretaceous–Early Tertiary sedimentation and tectonic inversion in the southern Netherlands. *Geologie en Mijnbouw*, **78**, 1–19.

HEYBROEK, P. 1975. On the structure of the Dutch part of the Central North Sea Graben. *In*: WOODLAND, A.W. (ed.) *Petroleum and the Continental Shelf of North-West Europe, Volume 1*. Applied Science, Barking, Essex, UK, 339–351.

KOOI, H., CLOETINGH, S.A.P.L. & REMMELTS, G. 1989. Intraplate stresses and the stratigraphic evolution of the North Sea Central Graben. *Geologie en Mijnbouw*, **68**, 49–72.

KUYL, O.S. 1983. The inversion of part of the southern border of the Central Graben in South Limburg during the Late Cretaceous. *Geologie en Mijnbouw*, **62**, 401–408.

NLOG 2017. *Released Dutch Borehole Data*, http://www.nlog.nl/en/listing-boreholes

PHARAOH, T.C., DUSAR, M. *ET AL.* 2010. Tectonic evolution. *In*: DOORNENBAL, J.C. & STEVENSON, A.G. (eds) *Petroleum Geological Atlas of the Southern Permian Basin Area*. European Association of Geoscientists & Engineers (EAGE), Houten, The Netherlands, 25–57.

PRICE, A., HOFMANN, A., VAN DALEN, E., MCKELLAR, D. & KAFFENBERGER, G. 2002. Hanze Field in the Dutch North Sea. *Oil Gas European Magazine*, **2**, 15–20.

SURLYK, F., DONS, T., CLAUSEN, C.K. & HIGHAM, J. 2003. Upper Cretaceous. *In*: EVANS, D., GRAHAM, C., ARMOUR, A. & BATHURST, P. (eds) *The Millennium Atlas: Petroleum Geology of the Central and Northern North Sea*. Geological Society, London, 213–233.

VAN ADRICHEM BOOGAERT, H.A. & KOUWE, W.F.P. (compilers). 1997. *Stratigraphic Nomenclature of the Netherlands; Revision and Update by RGD and NOGEPA*. Mededelingen Rijks Geologische Dienst, **50**.

VAN BUCHEM, F.S.P., SMIT, F.W.H., BUIJS, G.J.A., TRUDGILL, B. & LARSEN, P.-H. 2017. Tectonostratigraphic framework and depositional history of the Cretaceous–Danian succession of the Danish Central Graben (North Sea) – new light on a mature area. *In*: BOWMAN, M. & LEVELL, B. (eds) *Petroleum Geology of NW Europe: 50 Years of Learning – Proceedings of the 8th Petroleum Geology Conference*. Geological Society, London. 9–46, https://doi.org/10.1144/PGC8.24

VAN DER MOLEN, A.S. 2004. *Sedimentary development, seismic stratigraphy and burial compaction of the Chalk Group in the Netherlands North Sea area.*

Geologica Ultraiectina, No. 248. PhD thesis, Utrecht University, Utrecht, The Netherlands.

Van der Molen, A.S. & Wong, Th.E. 2007. Towards an improved lithostratigraphic subdivision of the Chalk Group in the Netherlands North Sea area – A seismic stratigraphic approach. *Netherlands Journal of Geosciences – Geologie en Mijnbouw*, **86**, 131–143.

Van Wijhe, D.H. 1987. Structural evolution of inverted basins in the Dutch offshore. *Tectonophysics*, **137**, 171–219.

van Winden, M., de Jager, J., Jaarsma, B. & Bouroullec, R. 2018. New insights into salt tectonics in the northern Dutch offshore: a framework for hydrocarbon exploration. *In*: Kilhams, B., Kukla, P.A., Mazur, S., McKie, T., Mijnlieff, H.F. & van Ojik, K. (eds) *Mesozoic Resource Potential of the Southern Permian Basin*. Geological Society, London, Special Publications, **469**. First published online January 29, 2018, https://doi.org/10.1144/SP469.9

Vejbæk, O.V., Andersen, C. *et al.* 2010. Cretaceous. *In*: Doornenbal, J.C. & Stevenson, A.G. (eds) *Petroleum Geological Atlas of the Southern Permian Basin Area*. European Association of Geoscientists & Engineers (EAGE), Houten, The Netherlands, 195–209.

Ziegler, P.A. 1990. *Geological Atlas of Western and Central Europe*. 2nd edn. Shell Internationale Petroleum Maatschappij, The Hague. Geological Society, London.

Index

Page numbers in *italics* refer to Figures. Page numbers in **bold** refer to Tables.